## The Greek Alphabet

| | | | | | | | | | | |
|---|---|---|---|---|---|---|---|---|---|---|
| Alpha | A | $\alpha$ | Iota | I | $\iota$ | Rho | P | $\rho$ |
| Beta | B | $\beta$ | Kappa | K | $\kappa$ | Sigma | $\Sigma$ | $\sigma$ |
| Gamma | $\Gamma$ | $\gamma$ | Lambda | $\Lambda$ | $\lambda$ | Tau | T | $\tau$ |
| Delta | $\Delta$ | $\delta$ | Mu | M | $\mu$ | Upsilon | $\Upsilon$ | $\upsilon$ |
| Epsilon | E | $\epsilon$ | Nu | N | $\nu$ | Phi | $\Phi$ | $\phi$ |
| Zeta | Z | $\zeta$ | Xi | $\Xi$ | $\xi$ | Chi | X | $\chi$ |
| Eta | H | $\eta$ | Omicron | O | $o$ | Psi | $\Psi$ | $\psi$ |
| Theta | $\Theta$ | $\theta$ | Pi | $\Pi$ | $\pi$ | Omega | $\Omega$ | $\omega$ |

## Abbreviations for Units

| | | | |
|---|---|---|---|
| A | ampere | L | litre |
| Å | angstrom ($10^{-10}$ m) | m | metre |
| atm | atmosphere | MeV | megaelectronvolts |
| C | coulomb | Mm | megametre ($10^6$ m) |
| °C | degree Celsius | mi | mile |
| cal | calorie | min | minute |
| cm | centimetre | mm | millimetre |
| dyn | dyne | ms | millisecond |
| eV | electronvolt | N | newton |
| °F | degree Fahrenheit | nm | nanometre ($10^{-9}$ m) |
| fm | femtometre, fermi ($10^{-15}$ m) | pt | pint |
| ft | foot | qt | quart |
| Gm | gigametre ($10^9$ m) | rev | revolution |
| G | gauss | s | second |
| g | gram | T | tesla |
| H | henry | u | unified mass unit |
| h | hour | V | volt |
| Hz | hertz | W | watt |
| in | inch | Wb | weber |
| J | joule | y | year |
| K | kelvin | yd | yard |
| kg | kilogram | $\mu$m | micrometre ($10^{-6}$ m) |
| km | kilometre | $\mu$s | microsecond |
| keV | kiloelectronvolts | $\mu$C | microcoulomb |
| lb | pound | $\Omega$ | ohm |

# PHYSICS

SECOND EDITION

# Physics

## PAUL A. TIPLER

Oakland University

Rochester, Michigan

WORTH PUBLISHERS, INC.

**Physics,** Second Edition

Paul A. Tipler

Copyright © 1976, 1982 by Worth Publishers, Inc.

All rights reserved

Printed in the United States of America

Library of Congress Catalog Card No. 81-70205

ISBN: 0-87901-135-1

Third Printing, July 1986

Designed by Malcolm Grear Designers

Composition by Progressive Typographers, Inc.

Illustrated by Felix Cooper and John Kramer

Cover by Harold E. Edgerton, MIT, and J. Kim
Vandiver, MIT: Vortex formed at the tip of a fan
blade by heated air above an alcohol lamp
(stroboscopic color schlieren photograph at $\frac{1}{3}$ $\mu$s
exposure)

**Worth Publishers, Inc.**

33 Irving Place

New York, NY 10003

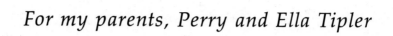

*For my parents, Perry and Ella Tipler*

# Preface

This second edition of *Physics* reflects the many helpful responses and suggestions from reviewers and users of the first edition. Like the first edition, it is a textbook for the standard two-semester or three-semester elementary physics course for engineering and science majors. It is assumed that the students have taken calculus, or are concurrently taking it. In the second edition, the order of topics has been changed to correspond more closely to the structure of most courses in elementary physics. In particular, two chapters on mechanical waves and sound (Chapters 14 and 15) now come right after mechanics so they can be taught in the first semester; and electricity and magnetism is now discussed before optics. The overall organization is now mechanics (Chapters 1–13), mechanical waves and sound (Chapters 14 and 15), thermodynamics (Chapters 16–19), electricity and magnetism (Chapters 20–31), optics (Chapters 32–34), and modern physics (Chapters 35–37). For those who prefer (as I do) to teach all waves together, Chapters 14, 15, 32, 33, and 34 can be taught easily as a unit, either before or after electricity and magnetism.

Many chapters, particularly in the mechanics portion of the book, have been rewritten, reordered, and combined for greater clarity and ease in reading and teaching. Some examples are

1. Chapter 2 ("Motion in One Dimension") has been reorganized and rewritten so that it is easier for the student who is just beginning calculus. Calculus topics such as the initial-value problem and integration are now in optional sections at the end of the chapter.

2. Chapters 4 and 5 ("Newton's Laws" and "Applications of Newton's Laws") have been rearranged so that the techniques of problem solving with examples of constant forces in one dimension appear in Chapter 4, after a statement of Newton's laws and a discussion of force and mass. The material on the basic interactions (formerly in Chapter 6) is now at the beginning of Chapter 5, before the applications of Newton's laws. In that chapter there is also a new section on torque and the static equi-

librium of rigid bodies. This early introduction of the concept of torque allows students to work problems in statics immediately after studying Newton's laws, and it also makes the study of rotational dynamics in Chapters 9 and 10 easier.

3. The chapters on work and energy have been rewritten and rearranged so that the somewhat formal and mathematical material formerly in Chapter 8 has become a more qualitative discussion that appears after energy has been defined in Chapter 6.

4. All the material on the center of mass is now presented together in Chapter 8, "Systems of Particles," which also includes collisions. Collisions are first discussed in a general reference frame, with the treatment of the center-of-mass reference frame relegated to an optional section.

5. The discussion of damped and driven oscillators (formerly in Chapter 15) has been shortened and simplified and placed at the end of the chapter on simple harmonic motion (Chapter 11).

6. The material on Gauss' law is now consolidated in Chapter 21, which also includes a discussion of conductors in electrostatic equilibrium.

There are three new chapters and several new sections. Chapter 13, "Solids and Fluids," contains material on elasticity and on the statics and dynamics of fluids. Chapter 18, "Thermal Properties and Processes," contains new material on the van der Waals gas, phase diagrams, and heat transfer. Chapter 37, "Nuclear Physics," is an elementary introduction to the properties of nuclei, radioactivity (including carbon dating), and nuclear reactions (including both fission and fusion). Among the new sections are "Scientific Notation and Significant Figures," in Chapter 1; "Systems with Varying Mass: Rocket Motion," in Chapter 8; and "General Relativity," in Chapter 35.

There are over 2,000 exercises and problems in the new edition. The number of problems has been increased by about 33%. Most of the new problems are fairly easy; they will provide a smooth transition in level between the exercises and the more challenging problems. However, a few of the new problems should challenge even the best students.

SI units are now used almost exclusively throughout the text, with all examples, exercises, and problems given in SI units. The only exceptions are in Chapter 4, where there are a few exercises on the conversion of force units (pounds to newtons, etc.), and in Chapter 18, where $R$ factors for insulation materials are used with U.S. customary units.

Three new essays have been written for this edition, "Solar Energy" by Laurent Hodges, "Exponential Growth" by Albert Bartlett, and "James Clerk Maxwell" by C. W. F. Everitt.

A feature new to this edition is the short list of learning objectives at the beginning of each chapter, to help students focus on the important information to be presented, to give them an idea of the depth of understanding that will be required, and to serve as an aid for later review. Individual instructors may wish to add or subtract from these lists according to their preferences.

Some features of the first edition that have been retained are:

1. The book is written at a level at which it can be read and enjoyed by the less well-prepared students, and yet be challenging to the better-prepared students. To make the book appeal to a wide range of student abilities, many new concepts are introduced in an intuitive, nonmathe-

matical way, often proceeding from a specific example to a more general relation. Typically, the general or formal derivation follows in an optional paragraph at the end of the section or chapter.

2. Questions are presented at the ends of various sections within each chapter. Some of these questions are routine and can be easily answered from the material in the preceding section; others are open-ended and can serve as a basis for classroom discussion.

3. A review section consisting of a checklist of words and phrases and a set of true-false questions appears at the end of each chapter. Each term in the checklist is accompanied by the page number where it is defined, and answers to all the true-false questions are given at the back of the book.

4. There is an extensive set of relatively easy exercises followed by a set of more difficult problems at the end of each chapter. The exercises are organized by sections within the chapter. Answers to all the odd-numbered exercises and problems are given at the back of the book. (Answers to the even-numbered exercises and problems, and some worked-out solutions, are in the Instructor's Manual.)

5. Fourteen essays from the first edition have been retained. Of the seventeen essays now in the book, four are biographical (on Isaac Newton, Benjamin Franklin, James Maxwell, and Albert Einstein) and the rest are on applications (e.g., xerography, transistors, radar astronomy) or topics of current interest (e.g., thermal pollution, exponential growth, black holes, sonic booms). These brief essays are included for the enjoyment and enlightenment of students and instructors. There are no exercises or problems connected with the essays.

6. The book's readability is enhanced by its attractive physical layout: its large pages, wide margins, two-color line drawings, many photographs, margin comments highlighting important concepts and equations, ruled color lines setting off important equations, and optional material in the right-hand column bounded on each side by colored rules.

## Acknowledgments

Many people have contributed to one or both editions of this book. I would like to thank everyone who used the first edition and sent me comments and corrections, and all those who tried out early drafts of the book. Some of the people who were particularly helpful with the first edition were James Gerhart, Jerry Griggs, John McKinley, Granvil Kyker, and Stanley Williams. I would like to thank Oakland University for granting me a leave during which much of the work on the new edition was done, and the physics department at the University of California at Berkeley for its kind hospitality during my stay there. Peter Matthews of the University of British Columbia contributed most of the interesting new problems, and Mark Scheuern of Oakland University was a great help in the preparation of the answer section. Ron Brown of California Polytechnic State University deserves special thanks for his review of the first edition, the entire revised manuscript, and the galleys of the second edition. He made many important and useful suggestions for improvement.

Others who reviewed various parts of the first edition to offer suggestions for the second edition, or who reviewed the manuscript for the second edition, were: Zvi Bar-Yam, Southeastern Massachusetts University; Henry Bass, University of Mississippi; Ernst Bleuler, Pennsylvania State University, University Park; Sumner Davis, University of California, Berkeley; Paul Doherty, Oakland University; Lowell Eliason, California State University, Long Beach; David A. Glocker, Rochester Institute of Technology; Richard Haracz, Drexel University; Richard T. Kouzes, Princeton University; Andrew C. Kowalik, Nassau Community College; Alan J. Lazarus, Massachusetts Institute of Technology; Sherman Lowell, Washington State University; Robert Luke, Boise State University; W. Anthony Mann, Tufts University; Paul F. Nichols, San Diego State University; Pierre Piroue, Princeton University; L. Arthur Read, Wilfrid Laurier University; Charles Rogers, Southwestern Oklahoma University; John Stevens, Sacramento State University; Jacques Templin, San Diego State University; T. A. Wiggins, Pennsylvania State University, University Park; Robert Williamson, Oakland University; John Van Zytveld, Calvin College.

I was fortunate in having two excellent typists for this edition, Julie Thompson and Marshall Tuttle. As usual, my wife Sue has given me continual support throughout this long project. Finally, I would like to thank everyone at Worth Publishers for their help and encouragement, and particularly June Fox, for her tireless efforts in coordinating this project.

Paul A. Tipler
Rochester, Michigan

February 1982

# Contents

# Contents in Brief

*Physics* is also available in two volumes; Chapters 1-19 in Volume 1 and Chapters 20–37 in Volume 2.

# Contents

# PHYSICS

CHAPTER 1    Introduction

---

**Objectives**    After studying this chapter, you should:

1.  Be able to define the units of length, time, and mass.

2.  Know what is meant by *SI units, mks units, U.S. customary units,* and *cgs units.*

3.  Know what *conversion factors* are and be able to use them to convert from one system of units into another.

4.  Know what is meant by the *dimensions* of a quantity.

5.  Be able to express large or small numbers in scientific notation.

6.  Know the definitions of the trigonometric functions sine, cosine, and tangent and be able to derive the relation $\sin^2 \theta + \cos^2 \theta = 1$.

7.  Know the small-angle approximations $\sin \theta \approx \tan \theta \approx \theta$ and be able to illustrate each with a diagram.

---

We have always been curious about the world around us. Since the beginnings of recorded thought, we have sought ways to impose order on the bewildering diversity of observed events. This search for order takes a variety of forms. One is religion; another is art; a third is science. Although the word "science" has its origins in a Latin verb meaning "to know," science has come to mean not merely knowledge but knowledge specifically of the natural world—most importantly, a body of knowledge organized in a specific and rational way.

Although the roots of science are as deep as those of religion or art, its traditions are much more modern. Only in the last few centuries have there been methods for studying nature systematically. They include techniques of observation, rules for reasoning and prediction, the idea of planned experimentation, and ways for communicating experimental and theoretical results—all loosely referred to as the *scientific method.* An essential part of the advance of our understanding of nature is the open communication of experimental results, theoretical calculations, speculations, and summaries of knowledge. A textbook is one of these forms. An elementary textbook like this has two purposes. It is designed, first, to introduce newcomers in the field of science to material which is already widely known in the scientific and technical community and which will form the basis of their more advanced studies of this knowledge. It may also serve to acquaint students not majoring in science with information and a way of thinking that is having a cumulative effect upon our way of life.

This book concentrates on the subjects of *classical physics*, i.e., mechanics, light, heat, sound, electricity, and magnetism, subjects well understood in the late nineteenth century before the advent of relativity and quantum theory, which were developed in the early years of the twentieth century. The application of special relativity, and particularly quantum theory, to the description of such microscopic systems as atoms, molecules, and nuclei and to a detailed understanding of solids, liquids, and gases is often referred to as *modern physics*. Although modern physics has made many important contributions to technology, the great bulk of technical knowledge and skill is still based squarely on classical physics. While concentrating on classical physics, we shall often discuss modifications based on quantum theory and special relativity and compare the predictions of the classical and modern theories with each other and with the results of experiments.

Except for the interior of the atom and for motion at speeds near the speed of light, classical physics correctly and precisely describes the behavior of the physical world. It is through applications of classical physics that we have been able to exploit natural resources successfully, and it is largely through applications of classical physics that we shall find the technical means necessary to preserve our environment for successful and controlled future use.

## 1-1    Mathematics and Physics

The laws of physics are generalizations from observations and experimental results. For example, Newton's law of universal gravitation was based on a variety of observations: the paths of planets in their motion about the sun, the acceleration of objects near the earth, the acceleration of the moon in its orbit, the daily and seasonal variation in the tides, etc. Physical laws are usually expressed as mathematical equations, which are then used to make predictions about other phenomena and to test the range of validity of the laws. For example, the law of gravitation states that two objects attract each other with a force which is proportional to the mass of each object and inversely proportional to the square of the distance between them. It is written $F = Gm_1m_2/r^2$. Newton's law of gravitation, together with Newton's laws of motion, can be used to predict the orbits of planets, comets, and satellites. Understanding such predictions requires knowledge of elementary calculus and the ability to manipulate and solve simple differential equations. Since the laws of motion are expressed as second-order differential equations and the laws of wave motion are expressed as partial differential equations, understanding physics on any level beyond a qualitative description requires a considerable amount of mathematics. It is usually easiest to learn the physics and the necessary mathematics at about the same time, since the immediate application of mathematics to a physical situation helps you understand both the physics and the mathematics. In this book, it is assumed that you have taken or are now taking a course in calculus. Much of the calculus needed to understand physics is presented in outline form. Since calculus is a new and unfamiliar tool for most students beginning physics, it is used here as simply as possible. As this book is about physics and not mathematics, mathematical details and rigor are not stressed. In the more difficult mathematical problems, plausibility arguments are often substituted for rigorous derivations in order to appeal to your intuition.

## 1-2  How to Learn Physics

A textbook is only one tool for learning physics. A good teacher, lecture demonstrations, films, and experimental work in the laboratory are indispensable. Outside reading is highly recommended. While you concentrate on obtaining a rigorous understanding of classical physics in your introductory course, you should be broadening your familiarity with contemporary physics by reading widely in the many excellent popular and semipopular accounts of modern science, e.g., those in *Scientific American*.

At the end of each chapter are a review, exercises, and problems. The importance of solving problems in learning physics cannot be overemphasized. Only in working problems can you find out whether you have really grasped the text material. Many details can be brought out in problems that cannot be treated in any other way. You should do as many problems as possible, whether assigned or not. One way to gain practice and experience in problem solving is to use the examples in the text as problems. Read the statement or question in the example and then attempt to answer the question or work through the example without looking at the text. When you finish or get stuck, look at the worked example. This approach will demonstrate how well you understand the material and what mistakes you may be making. (Don't be discouraged if you are often stuck. Although many examples are rather direct applications of material discussed in the text, some introduce a new method or solution or approach. A few examples cover famous results and are really an extension of the text. In these cases, you cannot expect to work through the examples on your first try without difficulty.)

Scattered throughout the text are questions. Some are very easy and are answered in the previous discussion. If you have understood the discussion, you should be able to answer them with little trouble. Other questions are meant to extend or apply the discussion in the text. Some have no simple answers but are food for thought. Whether or not you are taking a simultaneous laboratory course, you should be alert to simple experiments or observations relating the concepts in the book to your experience in the real world. The questions should stimulate your ability to relate your study to everyday experience.

It is possible to state the laws of physics in concise statements and equations and use them to deduce the behavior of many systems under various conditions. Although there is a certain aesthetic appeal in this deductive approach, it is not an easy way to learn physics or to develop an understanding of the workings of nature. What we now know of nature results from the efforts of many different people and many years of experimentation, theoretical proposals, and debate. Sometimes discussing the history of an idea helps us understand the presently accepted view. Considering rather special cases can be a helpful preliminary to the general discussion of a physical law. For example, some familiarity with many new concepts can be obtained by treating one-dimensional problems. The extension to three dimensions is then much easier than starting with the general three-dimensional problem would be. We shall consider one-dimensional motion first to introduce some of the concepts of velocity and acceleration before generalizing to two- and three-dimensional motion. Similarly, the ideas of work, potential energy, conservative forces, and many of the properties of waves are first introduced in one dimension before the general three-dimensional situations are treated.

## 1-3   Units

The laws of physics express relationships between physical quantities such as length, time, force, energy, and temperature. Measuring a quantity of this kind involves comparison with some unit value of the quantity. The most elementary measurement is probably that of distance. In order to measure the distance between two points, we need a standard unit, e.g., a metrestick or a ruler. The statement that a certain distance is 25 metres means that it is 25 times the length of the unit metre; i.e., a standard metrestick fits into that distance 25 times. It is important to include the unit metre along with the number 25 in expressing a distance, because there are other units of distance in common use. To say that a distance is 25 is meaningless.

All physical quantities can be expressed in terms of a small number of fundamental units. For example, a unit of speed, e.g., metres per second or miles per hour, is expressed in terms of a unit of length and a unit of time. Similarly, any unit of energy can be expressed in terms of the units of length, time, and mass. In fact, all quantities occurring in the study of mechanics can be expressed in terms of these three fundamental units. The choice of standard units for these fundamental quantities determines a system of units for all mechanical quantities. In the system used universally in the scientific community, which we shall use most often, the standard length is the metre, the standard time is the second, and the standard mass is the kilogram. This system of units is called the mks system (after the *m*etre, *k*ilogram, and *s*econd). The standard of length, the metre (abbreviated m), was originally indicated by two scratches on a platinum-iridium alloy bar kept at the International Bureau of Weights and Measures in Sèvres, France. This length was chosen so that the distance between the equator and the North Pole along the meridian through Paris would be 10 million metres (Figure 1-1). (After construction of the standard metre bar, it was found that this distance differs by a few hundredths of a percent from $10^7$ m.) The standard metre was used to construct secondary standards, which are used to calibrate measuring rods throughout the world. The standard metre is now defined in terms of the wavelength of the red-orange spectral line of the krypton isotope $^{86}$Kr. It is 1,650,763.73 wavelengths of this light. This change made comparison of lengths throughout the world easier and more accurate.

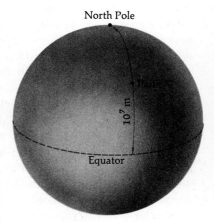

**Figure 1-1**
The standard of length, the metre, was chosen so that the distance from the equator to the North Pole along the meridian through Paris would be $10^7$ m.

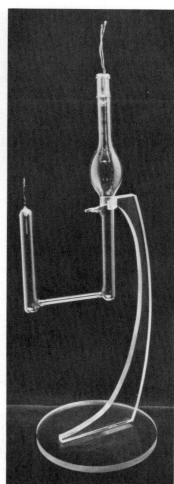

(*Above*) Standard metre bar kept at the National Bureau of Standards, Washington, D.C. This bar was the United States national standard from 1893 to 1960. Made of a platinum-iridium alloy, it is an exact duplicate of the international prototype metre housed at Sèvres, France. (*Right*) Krypton-86 lamp used to define the metre in terms of the wavelength of light. (*Both photos courtesy of the National Bureau of Standards.*)

The unit of time, the second (s), was originally defined in terms of the rotation of the earth to be $\frac{1}{60}(\frac{1}{60})(\frac{1}{24})$ of the mean solar day. The second is now defined in terms of a characteristic frequency associated with the cesium atom.

The unit of mass, the kilogram (kg), equal to 1000 grams (g), is defined to be the mass of a particular standard body, also kept at Sèvres. We shall discuss the concept of mass in detail in Chapter 4.

The units of other quantities in mechanics, such as speed and momentum, are derived from the three fundamental units of length, time, and mass. The choice of these three as the fundamental quantities is somewhat arbitrary. For example, we could replace the choice of a standard length with a standard speed, e.g., the speed of light in a vacuum, and give this unit a name. The unit of length would then be derived from the product of this speed unit and the unit of time.

In our study of thermodynamics and electricity, we shall need three more fundamental physical units, the unit of temperature, the kelvin (K) (formerly the degree Kelvin); the unit for the amount of a substance, the mole (mol); and the unit of current, the ampere (A). There is another fundamental unit, the candela (cd) for luminous intensity, which we shall have no occasion to use in this book. These seven fundamental

*SI units*

units, the metre (m), the second (s), the kilogram (kg), the kelvin (K), the ampere (A), the mole (mol), and the candela (cd), constitute the international system of units (SI).* The unit of every physical quantity can be expressed in terms of these fundamental SI units. Some important combinations are given special names. For example, the SI unit of force, $kg \cdot m/s^2$, is called a newton (N). Similarly, the SI unit of power, $kg \cdot m^2/s^3 = N \cdot m/s$, is called a watt (W). Prefixes for common multiples and submultiples of SI units are listed in Table 1-1. These multiples are all powers of 10. Such a system is called a *decimal system*; the decimal system based on the metre is called the *metric system*. The prefixes can be applied to any SI unit; for example, $10^{-3}$ second is 1 millisecond (ms); $10^6$ watts is 1 megawatt (MW).

In another system of units used in the United States, the *U.S. customary system*, the unit of force, the pound, is chosen to be a fundamental unit instead of mass. (We shall discuss the relation between force and mass in detail in Chapter 4.) The unit of force is defined in terms of the gravitational attraction of the earth at a particular place for a standard body. The other fundamental units in this system are the foot and the second. The second is defined as in the International System. The foot is defined as exactly one-third of a yard, which is now legally defined in terms of the metre:

$$1 \text{ yd} = 0.9144 \text{ m}$$

$$1 \text{ ft} = \tfrac{1}{3} \text{ yd} = 0.3048 \text{ m}$$

making the inch exactly 2.54 cm. This system is not a decimal system. It is less convenient than the SI or other decimal systems because common multiples of the unit are not powers of 10. For example, 1 ft $= \tfrac{1}{3}$ yd, and 1 in $= \tfrac{1}{12}$ ft. We shall see in Chapter 4 that mass is a better choice of a fundamental unit than force because mass can be defined without reference to the gravitational attraction of the earth. Relations between the U.S. customary system and SI units are given in Appendix C.

Another decimal system still in use but gradually being replaced by SI units is the cgs system, based on the centimetre, gram, and second. The centimetre is now defined as $10^{-2}$ m and the gram as $10^{-3}$ kg. Originally the gram was defined as the mass of 1 cm$^3$ of water. (The kilogram is then the mass of $10^3$ cm$^3$ = 1 litre of water). We shall generally work with SI units, giving the names of the corresponding cgs units and the appropriate conversion factors for any quantities you are likely to encounter in your studies.

### Questions

1. What are the advantages and disadvantages of using the length of a person's arm for a standard of length?

2. A certain clock is consistently 10 percent fast compared with the standard cesium clock. A second clock varies in a random way by 1 percent. Which clock would make a more useful secondary standard for a laboratory? Why?

**Table 1-1**
Prefixes for powers of 10

| Multiple | Prefix | Abbreviation |
|---|---|---|
| $10^{18}$ | exa | E |
| $10^{15}$ | peta | P |
| $10^{12}$ | tera | T |
| $10^9$ | giga | G |
| $10^6$ | mega | M |
| $10^3$ | kilo | k |
| $10^2$ | hecto* | h |
| $10^1$ | deka* | da |
| $10^{-1}$ | deci* | d |
| $10^{-2}$ | centi* | c |
| $10^{-3}$ | milli | m |
| $10^{-6}$ | micro | $\mu$ |
| $10^{-9}$ | nano | n |
| $10^{-12}$ | pico | p |
| $10^{-15}$ | femto | f |
| $10^{-18}$ | atto | a |

* The prefixes hecto (h), deka (da), and deci (d) are not powers of $10^3$ or $10^{-3}$ and are rarely used. The other prefix which is not a power of $10^3$ or $10^{-3}$ is centi (c), now used only with the metre, as in 1 cm = $10^{-2}$ m.

---

* SI stands for Système International. Complete definitions of each of these units are given in Appendix A.

# 1-4 Conversion of Units

We have said that the magnitude of a physical quantity must include both a number and a unit. When such quantities are added, subtracted, multiplied, or divided in an algebraic equation, the unit can be treated like any other algebraic quantity. For example, suppose you wish to find the distance traveled in 3 hours (h) by a car moving at a constant rate of 80 kilometres per hour (km/h). The distance is the product of the speed and the time:

$$x = vt = 80 \frac{km}{\not{h}} \times 3 \not{h} = 240 \text{ km}$$

We cancel the unit of time, the hours, just as we would any algebraic quantity to obtain the distance in the proper unit of length, the kilometre. This method makes it easy to convert from one unit of distance into another. Suppose we want to convert our answer of 240 km into miles (mi). We use the fact that

$$1 \text{ mi} = 1.61 \text{ km}$$

If we divide each side of this equation by 1.61 km, we obtain

$$\frac{1 \text{ mi}}{1.61 \text{ km}} = 1$$

We can now change 240 km to miles by multiplying by the factor (1 mi)/(1.61 km):

$$240 \text{ km} = 240 \not{km} \times \frac{1 \text{ mi}}{1.61 \not{km}} = 149 \text{ mi}$$

We have rounded the result to three significant figures, consistent with the fact that the distance 240 km is given to only three significant figures. The factor (1 mi)/(1.61 km) is called a *conversion factor*. All conversion factors have a value of 1 and are used to convert a quantity expressed in one unit of measure into the equivalent in another unit of measure. By writing out the units explicitly, we need not think about whether we multiply by 1.61 or divide by 1.61 to change kilometres to miles because the units tell us whether we have chosen the correct or incorrect factor.

Roman-Greek tablet dating from A.D. 300–500 showing the length of the standard foot. (*Courtesy of the Science Museum, London.*)

*Conversion factor*

---

**Example 1-1** What is the equivalent of 90 km/h in metres per second and in miles per hour?

We use the fact that 60 s = 1 min, and 60 min = 1 h. Then

$$90 \frac{\not{km}}{\not{h}} \times \frac{1000 \text{ m}}{1 \not{km}} \times \frac{1 \not{h}}{60 \not{min}} \times \frac{1 \not{min}}{60 \text{ s}} = 25 \text{ m/s}$$

The quantity 90 km/h is multiplied by a set of conversion factors each having the value 1, so that the value of the speed is not changed. To convert this speed into miles per hour we use the fact that 1 mi = 1.61 km. Then

$$90 \frac{\not{km}}{h} \times \frac{1 \text{ mi}}{1.61 \not{km}} = 55.9 \text{ mi/h}$$

---

## 1-5   Dimensions of Physical Quantities

The area of a plane figure is found by multiplying one length by another. For example, the area of a rectangle of sides 2 and 3 m is $A = (2 \text{ m}) (3 \text{ m}) = 6 \text{ m}^2$. The units of this area are square metres. Because area is the product of two lengths, it is said to have the *dimensions* of length times length, or length squared, often written $L^2$. The idea of dimensions is easily extended to other nongeometric quantities. For example, speed is said to have the dimensions of length divided by time, or $L/T$. The dimensions of any quantity in mechanics are written in terms of the fundamental quantities of length, time, and mass. Adding two physical quantities makes sense only if the quantities have the same dimensions. For example, we cannot add an area to a speed to obtain a meaningful sum. If we have an equation like

$$A = B + C$$

the quantities $A$, $B$, and $C$ must all have the same dimensions. Addition of $B$ and $C$ requires that these quantities be in the same units. For example, if $B$ is an area of 500 $\text{in}^2$ and $C$ is 4 $\text{ft}^2$, we must either convert $B$ into square feet or $C$ into square inches in order to find the sum of the two areas.

We can often find mistakes in a calculation by checking the dimensions or units of the quantities in the calculation. Suppose, for example, we are using a formula for distance $x$ given by

$$x = x_0 + vt + \tfrac{1}{2}at$$

where $t$ is the time, $x_0$ is the initial distance at $t = 0$, $v$ is the speed, and $a$ is the acceleration, which (as we shall see in the next chapter) has dimensions of $L/T^2$ and SI units of metres per second squared $(\text{m/s}^2)$. We can see immediately that this formula cannot be correct: since $x$ has dimensions of length, each term on the right side of the equation must have dimensions of length. Both $x_0$ and $vt$ have dimensions of length, but the dimensions of $\tfrac{1}{2}at$ are $(L/T^2)T = L/T$. Since the last term does not have the correct dimensions, an error has been made somewhere in obtaining the formula. Dimensional consistency is a necessary condition for an equation to be correct. It is, of course, not sufficient. An equation can have the correct dimensions in each term without describing any physical situation.

Our knowledge of the dimensions of quantities often enables us to recall a formula or sometimes even guess at it. Suppose we remember that there is a simple formula relating the velocity of a wave $v$ (dimensions $L/T$), the wavelength $\lambda$* (dimensions $L$), and frequency $f$ (dimensions $T^{-1}$). There is only one dimensionally consistent equation expressing one of these quantities as the product of the other two; it is

$$v = f\lambda$$

This equation happens to be correct. If we were to write down $f = v\lambda$, we could see immediately by checking the dimensions that it is incorrect.

---

* $\lambda$ is the Greek letter lambda. The names of all the letters in the Greek alphabet will be found inside the front cover.

## 1-6   Scientific Notation and Significant Figures

Handling very large or very small numbers is simplified by using powers-of-10, or scientific, notation. In this notation, the number is written as a product of a number between 1 and 10 and a power of 10, such as $10^2 = 100$, or $10^3 = 1000$, etc. The power of 10 is called the *exponent*. For example, the number 12,000,000 is written $1.2 \times 10^7$; the distance from the earth to the sun, about 150,000,000,000 m, is written $1.5 \times 10^{11}$ m. For numbers smaller than 1, the power of 10 or exponent is negative. For example, $0.1 = 10^{-1}$, and $0.0001 = 10^{-4}$. (In this notation, $10^0$ is defined to be 1.) For example, the diameter of a virus is about $0.00000001$ m $= 1 \times 10^{-8}$ m.

In the multiplication of two numbers, the powers of 10, or exponents, are added; in division they are subtracted. These rules can be seen from some simple examples:

$$10^2 \times 10^3 = 100 \times 1000 = 100,000 = 10^5$$

Similarly,

$$\frac{10^2}{10^3} = \frac{100}{1000} = \frac{1}{10} = 10^{2-3} = 10^{-1}$$

---

**Example 1-2** Using scientific notation, compute (*a*) $120 \times 6000$ and (*b*) $3,000,000/0.00015$.

(*a*)   $(1.2 \times 10^2)(6 \times 10^3) = (1.2)(6) \times 10^{2+3} = 7.2 \times 10^5$

(*b*)   $(3 \times 10^6)/(1.5 \times 10^{-4}) = \dfrac{3}{1.5} \times 10^{6-(-4)} = 2 \times 10^{10}$

The two minus signs in (*b*) arise from the subtraction of $-4$ from 6.

---

The addition or subtraction of two numbers written in scientific notation is slightly tricky. Consider, for example,

$$1.2 \times 10^2 + 8 \times 10^{-1} = 120 + 0.8 = 120.8$$

To find the sum without converting both numbers into ordinary decimal form, it is sufficient to rewrite either of the numbers so that its power of 10 is the same as that of the other. In this case, we can find the sum either by writing $1.2 \times 10^2 = 1200 \times 10^{-1}$ and then adding $1200 \times 10^{-1} + 8 \times 10^{-1} = 1208 \times 10^{-1} = 120.8$  or  by  writing  $8 \times 10^{-1}$  as $0.008 \times 10^2$   and   adding   $0.008 \times 10^2 + 1.2 \times 10^2 = 1.208 \times 10^2 = 120.8$. If the exponents are very different, one of the numbers is much smaller than the other and can often be neglected in addition or subtraction. For example,

$$2 \times 10^6 + 9 \times 10^{-3} = 2,000,000 + 0.009$$
$$= 2,000,000.009 \approx 2 \times 10^6$$

Many of the numbers in science are the result of measurement and are therefore known only within some experimental uncertainty. The magnitude of the uncertainty depends on the skill of the experimenter and on the apparatus used and often can only be estimated. It is customary to give a rough indication of the uncertainty in a measurement by the number of digits used. For example, if we say that a table is

2.50 m long, we are implying that its length is probably between 2.495 m and 2.505 m; i.e., we know the length to about ±0.005 m. If we used a metrestick with millimetre markings and measured the table length carefully, we might estimate that we could measure the length to ±0.5 mm rather than ±0.5 cm. We would indicate this precision by using four digits, such as 2.503 m, to give the length. A reliably known digit (other than a zero used to locate the decimal point) is called a *significant figure*. The number 2.503 m has four significant figures. The number 0.00103 has three significant figures. The first three zeros are not significant figures but merely locate the decimal point. In scientific notation, this number is $1.03 \times 10^{-3}$. A common student error, particularly since the advent of hand calculators, is to carry many more digits than are warranted. Suppose, for example, that you measure the area of a circular playing field by pacing off the radius and using the formula area $= \pi r^2$. If you estimate the radius to be 8 m by pacing and use a 10-digit calculator to compute the area, you obtain $\pi(8 \text{ m})^2 = 201.0619298 \text{ m}^2$. The digits after the decimal point are not only bulky to carry around but misleading about the accuracy with which you know the area. You would probably round the result off to 201 m². However, even this is misleading. If you found the radius by pacing, you might expect that your measurement was accurate to only about 0.5 m. That is, the radius could be as great as 8.5 m or as small as 7.5 m. If the radius is 8.5 m, the area is $\pi(8.5 \text{ m})^2 = 226.9800692 \text{ m}^2$, whereas if it is 7.5 m, the area is $\pi(7.5 \text{ m})^2 = 176.7145868 \text{ m}^2$. A general rule when combining several numbers in multiplication or division is that *the number of significant figures in the result is no greater than the least number of significant figures in any of the numbers*. In this case, the radius is known to only one significant figure, so the area is also known only to one significant figure. It should be written as $2 \times 10^2 \text{ m}^2$. If instead of pacing you had used a metrestick to measure the radius and had obtained, say, $r = 8.23$ m, the area would be known to three significant figures: $\pi(8.23 \text{ m})^2 = 2.13 \times 10^2 \text{ m}^2$.

In a textbook and in your homework it may be cumbersome always to write every number with the proper number of significant figures. Most examples and exercises will be done with data to three (or sometimes four) significant figures, but occasionally we can allow ourselves to be sloppy and say, for example, that a table top is 3 ft by 8 ft and has an area of 24 ft², rather than taking the time and space to say it is 3.00 ft by 8.00 ft and has an area of 24.0 ft². Any data you see in an example or exercise can be assumed to be known to three significant figures unless otherwise indicated.

In doing rough calculations or comparisons, we sometimes will round off a number to one or even no significant figures. A number rounded to the nearest power of 10 is called an *order of magnitude*. For example, the height of a small insect, say an ant, might be $8 \times 10^{-4}$ m $\approx$ $10^{-3}$ m. We would say that the order of magnitude of the height of an ant is $10^{-3}$ m. Similarly, though the height of most people is about 2 m, we might round that off and say that the order of magnitude of the height of a person is $10^0$ m. By this we do not mean to imply that a typical height is really 1 m but that it is closer to 1 m than to 10 m or to $10^{-1} = 0.1$ m. We might say that a typical human being is 3 orders of magnitude taller than a typical ant, meaning that the ratio of heights is about $10^3$.

Tables 1-2 to 1-4 give some typical order-of-magnitude values for some masses, sizes, and time intervals encountered in physics.

**Table 1-2**
Order of magnitude of some masses

| Mass | kg |
| --- | --- |
| Electron | $10^{-30}$ |
| Proton | $10^{-27}$ |
| Amino acid | $10^{-25}$ |
| Hemoglobin | $10^{-22}$ |
| Flu virus | $10^{-19}$ |
| Giant amoeba | $10^{-8}$ |
| Raindrop | $10^{-6}$ |
| Ant | $10^{-2}$ |
| Human being | $10^2$ |
| Saturn 5 rocket | $10^6$ |
| Pyramid | $10^{10}$ |
| Earth | $10^{24}$ |
| Sun | $10^{30}$ |
| Milky Way galaxy | $10^{41}$ |
| Universe | $10^{52}$ |

**Table 1-3**
Order of magnitude of some time intervals

| Interval | s |
| --- | --- |
| Time for light to cross nucleus | $10^{-23}$ |
| Period of visible-light radiation | $10^{-15}$ |
| Period of microwaves | $10^{-10}$ |
| Half-life of muon | $10^{-6}$ |
| Period of highest audible sound | $10^{-4}$ |
| Period of human heartbeat | $10^0$ |
| Half-life of free neutron | $10^3$ |
| Period of earth's rotation (day) | $10^5$ |
| Period of revolution of earth (year) | $10^7$ |
| Lifetime of a human | $10^9$ |
| Half-life of plutonium 239 | $10^{12}$ |
| Lifetime of a mountain range | $10^{15}$ |
| Age of earth | $10^{17}$ |
| Age of universe | $10^{18}$ |

# 1-7 Review of Algebra

In the next two sections we review some of the basic results of algebra and trigonometry. You should read through these sections to be sure you are familiar with these results. If you encounter unfamiliar material, study it carefully. These sections will also be a useful reference as you read further or work on problems.

### Equations

The following operations can be performed on mathematical equations to facilitate their solution.

1. The same quantity can be added or subtracted from each side of the equation.

2. Each side of the equation can be multiplied or divided by the same quantity.

3. Each side of the equation can be raised to the same power, e.g., each side can be squared or cubed, and the square (or any other) root of each side can be taken.

4. The reciprocal of each side of the equation can be taken.

It is important to understand that the above rules apply to each *side* of the equation and not to each *term* in the equation.

---

**Example 1-3** Solve the following equation for $x$.

$$(x - 3)^2 + 7 = 23$$

We first subtract 7 from each side to obtain $(x - 3)^2 = 16$. We then take the square root of each side to obtain $\pm(x - 3) = \pm 4$. We have included the plus-or-minus signs because either $(+4)^2 = 16$ or $(-4)^2 = 16$. We do not need to write the $\pm$ on each side of the equation as all the possibilities are included in $x - 3 = \pm 4$. We can now solve for $x$ by adding 3 to each side. There are two solutions: $x = 4 + 3 = 7$ and $x = -4 + 3 = -1$. They are easily checked by substituting these values into the original equation.

---

**Example 1-4** Solve the following equation for $x$:

$$\frac{1}{x} + \frac{1}{4} = \frac{1}{3}$$

This type of equation occurs both in geometric optics and in electric-circuit analysis. Although it is easy to solve, errors are often made. We solve by first subtracting $\frac{1}{4}$ from each side to obtain

$$\frac{1}{x} = \frac{1}{3} - \frac{1}{4} = \frac{4}{12} - \frac{3}{12} = \frac{1}{12}$$

We then take the reciprocal of both sides to obtain $x = 12$. A typical mistake in handling this equation is to take the reciprocal of each term first to obtain $x + 4 = 3$. This operation is not allowed; it changes the relative values of each side of the equation and leads to incorrect results.

---

**Table 1-4**
Order of magnitude of some lengths

| Length | m |
|---|---|
| Radius of proton | $10^{-15}$ |
| Radius of atom | $10^{-10}$ |
| Radius of virus | $10^{-7}$ |
| Radius of giant amoeba | $10^{-4}$ |
| Radius of walnut | $10^{-2}$ |
| Height of human being | $10^{0}$ |
| Height of highest mountains | $10^{4}$ |
| Radius of earth | $10^{7}$ |
| Radius of sun | $10^{9}$ |
| Earth-sun distance | $10^{11}$ |
| Radius of solar system | $10^{13}$ |
| Distance to nearest star | $10^{16}$ |
| Radius of Milky Way galaxy | $10^{21}$ |
| Radius of visible universe | $10^{26}$ |

## Direct and Inverse Proportion

The relationships of direct proportion and inverse proportion are so important in physics that they deserve special consideration. Often much algebraic manipulation can be avoided by a simple knowledge of these relationships. Suppose, for example, that you work for 5 days, 8 hours per day, at a certain pay rate and earn $200. How much would you earn at the same pay rate if you worked 8 days (d)? In this problem the money earned is directly proportional to the time worked. We can write an equation

$$M = Rt$$

where $M$ is the money earned, $t$ is the time, and $R$, the proportionality constant, is the rate. We can express $R$ in dollars per day or dollars per hour. Since $200 was earned in 5 d, the value of $R$ is $R = \$200/(5\text{ d}) = \$40/\text{d}$. In 8 d the amount earned is then $M = (\$40/\text{d})(8\text{ d}) = \$320$. However, we do not have to find the rate explicitly to work the problem. Since the amount earned in 8 d is $\frac{8}{5}$ times that earned in 5 d, this amount is $M = (\frac{8}{5})(\$200) = \$320$.

We can use the same type of example to illustrate inverse proportion. If we get a 25 percent raise, how long would we need to work to earn $200? Here we consider $R$ to be a variable and wish to solve for $t$. We see from the above equation that the time $t$ is inversely proportional to the rate $R$. Then if the new rate is $\frac{5}{4}$ the old rate, the new time will be $\frac{4}{5}$ of the old time, or 4 d.

There are some situations in which one quantity varies as the square or other power of another quantity where the ideas of proportion are still very useful. Suppose, for example, that a 10-in pizza costs $4.50. How much would you expect a 12-in-diameter pizza to cost if it contains the same topping? We expect the cost of a pizza to be approximately proportional to the amount of material, which is proportional to the area of the pizza. Since the area is in turn proportional to the square of the diameter, the cost should be proportional to the square of the diameter. If we increase the diameter by a factor of $\frac{12}{10}$, the area increases by a factor of $(\frac{12}{10})^2 = 1.44$, so we expect the cost to be $(1.44)(\$4.50) = \$6.48$.

---

**Example 1-5** The intensity of light from a point source varies inversely with the square of the distance from the source. If the intensity is 3.20 watts per square metre at 5 m from the source, what is it at 6 m from the same source?

The equation expressing the fact that the intensity varies inversely with the square of the distance can be written

$$I = \frac{C}{r^2}$$

where $C$ is some constant. Then if $I_1 = 3.20$ W/m² at $r_1 = 5$ m, and $I_2$ is the unknown intensity at $r_2 = 6$ m, we have

$$\frac{I_2}{I_1} = \frac{C/r_2^2}{C/r_1^2} = \frac{r_1^2}{r_2^2} = \left(\frac{5}{6}\right)^2 = 0.694$$

The intensity at 6 m from the source is then $I_2 = 0.694(3.20 \text{ W/m}^2) = 2.22$ W/m².

---

# 1-8   Review of Trigonometry

The angle between two intersecting straight lines is measured as follows. A circle is drawn with its center at the intersection of the lines, and the circular arc is divided into 360 parts, called degrees. The number of degrees on the arc between the two lines is the measure of the angle between the lines. For scientific work, a more useful measure of an angle is the radian (rad), defined as the length of the circular arc between the lines divided by the radius of the circle (see Figure 1-2). If $s$ is the arc length and $r$ is the radius of the circle, the angle $\theta$, measured in radians, is defined as

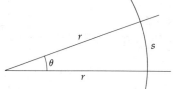

**Figure 1-2**
The angle $\theta$ is defined to be the ratio $s/r$, where $s$ is the arc length intercepted on a circle of radius $r$. The angle so defined is measured in radians.

$$\theta = \frac{s}{r} \qquad\qquad 1\text{-}1$$

*Angle in radians*

Since the angle measured in radians is the ratio of two lengths, it is dimensionless. The radian is therefore a dimensionless unit. In dimensional analysis it can be ignored. We can relate these two measures of angle by noting that a complete circle contains 360°, and since the circumference is $2\pi r$, its radian measure is $2\pi r/r = 2\pi$ rad. The conversion relation is then

$$360° = 2\pi \text{ rad}$$

or

$$1 \text{ rad} = \frac{360°}{2\pi} = 57.3° \qquad\qquad 1\text{-}2$$

Figure 1-3 shows a right triangle formed by drawing the line $BC$ perpendicular to $AC$. The lengths of the sides are labeled $a$, $b$, and $c$. The trigonometric functions $\sin\theta$, $\cos\theta$, and $\tan\theta$ are defined for an acute angle $\theta$ by

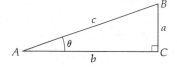

**Figure 1-3**
Right triangle used to define the trigonometric functions $\sin\theta$, $\cos\theta$, and $\tan\theta$.

$$\sin\theta = \frac{a}{c} = \frac{\text{opposite side}}{\text{hypotenuse}} \qquad \cos\theta = \frac{b}{c} = \frac{\text{adjacent side}}{\text{hypotenuse}}$$

$$1\text{-}3$$

$$\tan\theta = \frac{a}{b} = \frac{\sin\theta}{\cos\theta} = \frac{\text{opposite side}}{\text{adjacent side}}$$

Three other trigonometric functions are defined by the inverse of these functions:

$$\sec\theta = \frac{c}{b} = (\cos\theta)^{-1} \qquad\qquad \csc\theta = \frac{c}{a} = (\sin\theta)^{-1}$$

$$\cot\theta = \frac{b}{a} = (\tan\theta)^{-1} = \frac{\cos\theta}{\sin\theta}$$

The pythagorean theorem gives some useful identities:

$$a^2 + b^2 = c^2 \qquad\qquad 1\text{-}4$$

If we divide each term in this equation by $c^2$, we obtain

$$\frac{a^2}{c^2} + \frac{b^2}{c^2} = 1 \qquad\qquad 1\text{-}5$$

or, from the definitions of $\sin\theta$ and $\cos\theta$,

$$\sin^2\theta + \cos^2\theta = 1 \qquad\qquad 1\text{-}6$$

Similarly, we can divide each term in Equation 1-4 by $a^2$ or $b^2$ and obtain

$$1 + \cot^2 \theta = \csc^2 \theta \qquad\qquad\qquad 1\text{-}7$$

and

$$1 + \tan^2 \theta = \sec^2 \theta \qquad\qquad\qquad 1\text{-}8$$

---

**Example 1-6** Use the isosceles right triangle shown in Figure 1-4 to find the sine, cosine, and tangent of 45°.

It is clear from the figure that the two acute angles of this triangle are equal. Since the sum of the three angles in a triangle must equal 180° and the right angle is 90°, each acute angle must be 45°. If we multiply each side of any triangle by a common factor, we obtain a similar triangle with the same angles as before. Since the trigonometric functions involve only ratios of two sides of a right triangle, we can choose any convenient length for one side. Let the equal sides of this triangle have the length 1 unit. The length of the hypotenuse is then found from the pythagorean theorem (Equation 1-4):

$$c = \sqrt{a^2 + b^2} = \sqrt{1^2 + 1^2} = \sqrt{2} \text{ units}$$

The trigonometric functions for the angle 45° are then given by Equations 1-3:

$$\sin 45° = \frac{1}{\sqrt{2}} = 0.707 \qquad \cos 45° = \frac{1}{\sqrt{2}} = 0.707$$

$$\tan 45° = \frac{1}{1} = 1.00$$

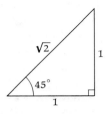

**Figure 1-4**
Isosceles right triangle for Example 1-6. This triangle is also called a 45-45-90 triangle.

---

**Example 1-7** The sine of 30° is exactly $\frac{1}{2}$. Find the ratios of the sides of a 30-60 right triangle.

This common triangle is shown in Figure 1-5. We choose the length 1 unit for the side opposite the 30° angle. The hypotenuse is then obtained from the fact that $\sin 30° = 0.5$:

$$c = \frac{a}{\sin 30°} = \frac{1}{0.5} = 2 \text{ units}$$

The length of the side opposite the 60° angle is found from the pythagorean theorem:

$$b = \sqrt{c^2 - a^2} = \sqrt{2^2 - 1} = \sqrt{3}$$

From these results we can obtain the other trigonometric functions for the angles 30° and 60°:

$$\cos 30° = \frac{b}{c} = \frac{\sqrt{3}}{2} = 0.866$$

$$\tan 30° = \frac{a}{b} = \frac{1}{\sqrt{3}} = 0.577$$

$$\sin 60° = \frac{b}{c} = \cos 30° = \frac{\sqrt{3}}{2} = 0.866$$

$$\cos 60° = \frac{a}{c} = \sin 30° = \frac{1}{2} = 0.500$$

$$\tan 60° = \frac{b}{a} = \frac{\sqrt{3}}{1} = 1.732$$

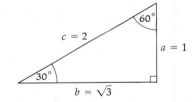

**Figure 1-5**
A 30-60-90 triangle for Example 1-7.

The calculation of the trigonometric functions for another common right triangle, the 3-4-5 right triangle, is left as an exercise.

For small angles, the length $a$ is nearly equal to the arc length $s$, as can be seen from Figure 1-6. The angle $\theta = s/c$ is therefore nearly equal to $\sin \theta = a/c$:

$$\sin \theta \approx \theta \qquad \text{small } \theta \qquad\qquad\qquad\qquad 1\text{-}9$$

*Small-angle approximations*

Similarly, the lengths $c$ and $b$ are nearly equal, so $\tan \theta = a/b$ is nearly equal to both $\theta$ and $\sin \theta$ for small $\theta$:

$$\tan \theta \approx \sin \theta \approx \theta \qquad \text{small } \theta \qquad\qquad 1\text{-}10$$

Since $\cos \theta = b/c$ and these lengths are nearly equal for small $\theta$, we have

$$\cos \theta \approx 1 \qquad \text{small } \theta \qquad\qquad\qquad\qquad 1\text{-}11$$

Equations 1-9 and 1-10 hold only if $\theta$ is measured in radians.

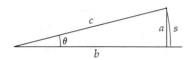

**Figure 1-6**
For small angles, $\sin \theta = a/c$, $\tan \theta = a/b$, and the angle $\theta = s/c$ are all approximately equal.

**Example 1-8** By how much do $\sin \theta$, $\tan \theta$, and $\theta$ differ when $\theta = 15°$? This angle in radians is

$$\theta = 15° \, \frac{2\pi \text{ rad}}{360°} = 0.262 \text{ rad}$$

From a table of trigonometric functions,

$$\sin 15° = 0.259 \qquad \tan 15° = 0.268$$

Thus $\sin \theta$ and $\theta$ (in radians) differ by 0.003, or about 1 percent, and $\tan \theta$ and $\theta$ differ by 0.006, or about 2 percent. For smaller angles, the approximation $\theta \approx \sin \theta \approx \tan \theta$ is even more accurate.

This example shows that if accuracy of a few percent is needed, these approximations can be used only for angles of about 15° or smaller. Figure 1-7 graphs $\theta$, $\sin \theta$, and $\tan \theta$ versus $\theta$ for small $\theta$.

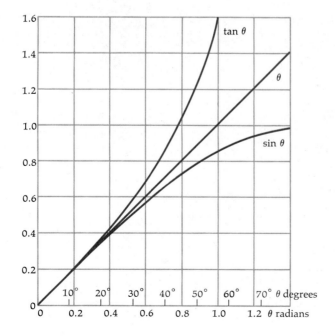

**Figure 1-7**
Plots of $\sin \theta$, $\tan \theta$, and $\theta$ for small angles.

Other trigonometric identities which we shall use occasionally follow directly from the definitions given:

$$\sin(\theta_1 + \theta_2) = \sin\theta_1 \cos\theta_2 + \cos\theta_1 \sin\theta_2$$

$$\cos(\theta_1 + \theta_2) = \cos\theta_1 \cos\theta_2 - \sin\theta_1 \sin\theta_2$$

$$\sin\theta_1 + \sin\theta_2 = 2\sin\tfrac{1}{2}(\theta_1 + \theta_2)\cos\tfrac{1}{2}(\theta_1 - \theta_2)$$

$$\cos\theta_1 + \cos\theta_2 = 2\cos\tfrac{1}{2}(\theta_1 + \theta_2)\cos\tfrac{1}{2}(\theta_1 - \theta_2)$$

1-12

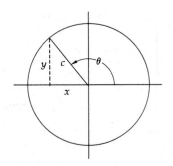

**Figure 1-8**
Diagram defining trigonometric functions for an obtuse angle.

Figure 1-8 shows an obtuse angle with vertex at the origin and one side along the $x$ axis. The trigonometric functions are defined for a general angle by

$$\sin\theta = \frac{y}{c} \qquad \cos\theta = \frac{x}{c} \qquad \tan\theta = \frac{y}{x}$$

1-13

Figure 1-9 plots these functions versus $\theta$. Some useful relations which can be obtained from these figures are

$$\sin(\pi - \theta) = \sin\theta \qquad \cos(\pi - \theta) = -\cos\theta$$

$$\sin(\tfrac{1}{2}\pi - \theta) = \cos\theta \qquad \cos(\tfrac{1}{2}\pi - \theta) = \sin\theta$$

1-14

All trigonometric functions have a period of $2\pi$. That is, when an angle changes by $2\pi$ or $360°$, the original value of the function is obtained again. Thus $\sin(\theta + 2\pi) = \sin\theta$, etc.

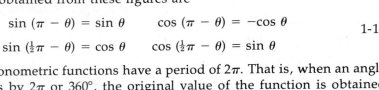

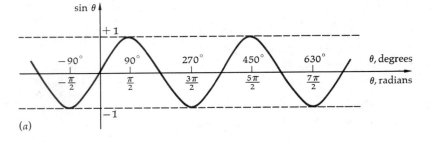

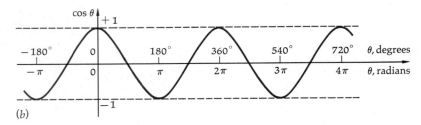

(a)

(b)

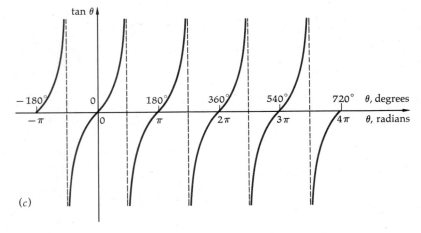

(c)

**Figure 1-9**
The trigonometric functions $\sin\theta$, $\cos\theta$, and $\tan\theta$ versus $\theta$.

## 1-9   The Binomial Expansion

The binomial theorem is very useful for making approximations. One form of this theorem is

$$(1 + x)^n = 1 + nx + \frac{n(n-1)}{2} x^2$$
$$+ \frac{n(n-1)(n-2)}{(3)(2)} x^3$$
$$+ \frac{n(n-1)(n-2)(n-3)}{(4)(3)(2)} x^4 + \cdots \qquad \text{1-15}$$

If $n$ is a positive integer, there are just $n + 1$ terms in this series. If $n$ is a real number other than a positive integer, there are an infinite number of terms. The series converges (is valid) for any value of $n$ if $x^2$ is less than 1. It also converges for $x^2 = 1$ if $n$ is positive. The series is particularly useful if $|x|$ is much less than 1. Then each term in Equation 1-15 is much smaller than the previous term, and we can drop all but the first two or three terms in the equation. If $|x|$ is much less than 1, we have

$$(1 + x)^n \approx 1 + nx \qquad |x| \ll 1 \qquad \text{1-16}$$

*Binomial approximation*

**Example 1-9** Use Equation 1-16 to find an approximate value for the square root of 101.

First state the problem to give an expression of the form $(1 + x)^n$ with $x$ much less than 1:

$$(101)^{1/2} = (100 + 1)^{1/2} = (100)^{1/2}(1 + 0.01)^{1/2} = 10(1 + 0.01)^{1/2}$$

Now we can use Equation 1-16 with $n = \frac{1}{2}$ and $x = 0.01$:

$$(1 + 0.01)^{1/2} \approx 1 + \tfrac{1}{2}(0.01) = 1.005$$
$$(101)^{1/2} = 10(1 + 0.01)^{1/2} \approx 10(1.005) = 10.05$$

We can get an idea of the accuracy of this approximation by looking at the first term that has been neglected in Equation 1-15. This term is

$$\frac{n(n-1)}{2} x^2$$

Since $x$ in this case is 1 percent of 1, the quantity $x^2$ is 0.01 percent of 1. By dropping this term, we are making an error of the order of

$$\frac{n(n-1)}{2} x^2 = -\tfrac{1}{8}(0.01)^2 \approx -0.001\%$$

We thus expect our answer to be correct to within about 0.001 percent. The value of $(101)^{1/2}$ to eight significant figures is 10.049876. The difference between this value and our approximation is 0.000124, which is about 0.001 percent of 10.05.

**Example 1-10** Using $\sin \theta \approx \theta$ and $\cos \theta = (1 - \sin^2 \theta)^{1/2}$, find an approximation for $\cos \theta$ for small $\theta$.

We have

$$\cos \theta = (1 - \sin^2 \theta)^{1/2} \approx (1 - \theta^2)^{1/2} \approx 1 + \tfrac{1}{2}(-\theta^2) = 1 - \tfrac{1}{2}\theta^2$$

This approximation is often useful as a correction to the more drastic approximation $\cos \theta \approx 1$ given by Equation 1-11.

Review

A. Define, explain, or otherwise identify the following (the numbers give the page on which the term is discussed or defined):

| | |
|---|---|
| Units, 4 | Angle, 13 |
| Metric system, 4 | $\sin \theta$, 13 |
| SI units, 5 | $\cos \theta$, 13 |
| U.S. customary system, 6 | $\tan \theta$, 13 |
| Conversion factor, 7 | Small-angle approximations, 15 |
| Dimensions, 8 | Binomial theorem, 17 |
| Significant figure, 10 | |

B. True or false: State whether each statement is true or false and explain your reasoning. If you can, give an example supporting a true statement and a counterexample contradicting a false statement.

1. Two quantities to be added must have the same dimensions.

2. Two quantities to be multiplied must have the same dimensions.

3. All conversion factors have the value 1.

4. If the angle $\theta$ is very small, $\sin \theta$ and $\tan \theta$ are approximately equal.

5. If the angle $\theta$ is very small, $\cos \theta$ and $\theta$ are approximately equal.

Exercises

**Section 1-3, Units; Section 1-4, Conversion of Units; and Section 1-5, Dimensions of Physical Quantities**

1. Write the following using the prefixes listed in Table 1-1 and the abbreviations listed on the inside cover; for example, 10,000 metres = 10 km.
(a) 1,000,000 watts          (b) 0.002 gram
(c) $3 \times 10^{-6}$ metre          (d) 30,000 seconds

2. Write the following without using prefixes:
(a) 40 $\mu$W          (b) 4 ns          (c) 3 MW          (d) 25 km

3. Write out the following (not SI units) without using any abbreviations, for example, $10^3$ metres = 1 kilometre.
(a) $10^{-12}$ boo          (b) $10^9$ low          (c) $10^{-6}$ phone
(d) $10^{-18}$ boy          (e) $10^6$ phone          (f) $10^{-9}$ goat
(g) $10^{12}$ bull

4. In the following equations, the distance $x$ is in metres, the time $t$ in seconds, and the velocity $v$ in metres per second. What are the SI units of the constants $C_1$ and $C_2$?
(a) $x = C_1 + C_2 t$          (b) $x = \frac{1}{2}C_1 t^2$          (c) $v^2 = 2C_1 x$
(d) $x = C_1 \cos C_2 t$          (e) $v = C_1 e^{-C_2 t}$
*Hint:* The arguments of trigonometric functions and exponentials must be dimensionless.

5. What are the dimensions of the constants in Exercise 4?

6. In Exercise 4, if $x$ is in feet, $t$ in seconds, and $v$ in feet per second, what are the units of the constants $C_1$ and $C_2$?

7. From the original definition of the metre in terms of the distance from the equator to the North Pole, find in metres (a) the circumference and (b) the radius of the earth. (c) Convert your answers to (a) and (b) from metres into miles.

8. Complete the following:
(a) 100 km/h = _____mi/h
(b) 60 cm = _____in
(c) 100 yd = _____m

9. In the following, $x$ is in metres, $t$ in seconds, $v$ in metres per second, and the acceleration $a$ in metres per second squared. Find the SI units of each combination:
(a) $v^2/xa$  (b) $\sqrt{x/a}$  (c) $\frac{1}{2}at^2$

10. Find the conversion factor to convert from miles per hour into kilometres per hour.

11. (a) Find the number of seconds in a year. (b) If one could count $1 per second, how many years would it take to count 1 billion dollars (1 billion $= 10^9$)? (c) If one could count one molecule per second, how many years would it take to count the molecules in a mole? (The number of molecules in a mole is Avogadro's number, $N_A = 6 \times 10^{23}$.)

12. The SI unit of force, the kilogram-metre per second squared (kg·m/s²), is called the newton (N). Find the dimensions and the SI units of the constant $G$ in Newton's law of gravitation $F = Gm_1m_2/r^2$.

13. Sometimes a conversion factor can be derived from the knowledge of a constant in two different systems. (a) The speed of light in vacuum is 186,000 mi/s $= 3 \times 10^8$ m/s. Use this fact to find the number of kilometres in a mile. (b) The weight of 1 ft³ of water is 62.4 lb. Use this and the fact that 1 cm³ of water has a mass of 1 g to find the weight in pounds of a 1-kg mass.

**Section 1-6, Scientific Notation and Significant Figures, and Section 1-7, Review of Algebra**

14. Express as a decimal number without using power-of-10 notation:
(a) $3 \times 10^4$  (b) $6.2 \times 10^{-3}$  (c) $4 \times 10^{-6}$  (d) $2.17 \times 10^5$

15. Write in scientific notation:
(a) 100,000  (b) 0.0000000303
(c) 602,000,000,000,000,000,000,000  (d) 0.0014

16. Calculate the following, express your result in scientific notation, and round off to the correct number of significant figures:
(a) $(2.00 \times 10^4)(6.10 \times 10^{-2})$  (b) $(3.141592)(4.00 \times 10^5)$
(c) $(2.32 \times 10^3)/(1.16 \times 10^8)$  (d) $(5.14 \times 10^3) + (2.78 \times 10^2)$
(e) $(1.99 \times 10^2) + (9.99 \times 10^{-5})$

17. Calculate the following, express your result in scientific notation, and round off to the correct number of significant figures:
(a) $(1.14)(9.99 \times 10^4)$  (b) $(2.78 \times 10^{-8}) - (5.31 \times 10^{-9})$
(c) $12\pi/(4.56 \times 10^{-3})$  (d) $27.6 + (5.99 \times 10^2)$

18. Calculate the following, express your result in scientific notation, and round off to the correct number of significant figures:
(a) $(200.9)(569.3)$  (b) $(0.000000513)(62.3 \times 10^7)$
(c) $28,401 + (5.78 \times 10^4)$  (d) $63.25/(4.17 \times 10^{-3})$

19. Solve for $x$:
(a) $3x + 15 = 45$  (b) $\dfrac{2}{x} + 9 = \dfrac{1}{5}$  (c) $\dfrac{1}{x} + \dfrac{2}{3x} = 15$

20. Solve for $x$:
(a) $x^2 + 4 = 7x - 8$  (b) $2x = 1/2x$
(c) $6x^2 + 12x = 0$  (d) $2x^2 + 6x + 1 = 0$

21. The intensity of sound from a point source varies inversely with the square of the distance from the source. If the intensity at 3 m is 4 mW/m², what is it at (a) 4 m, (b) 6 m, (c) 10 m, (d) 2 m?

22. A spherical balloon has a surface area of 1.13 m² and a volume of 0.113 m³. If it is blown up so that its radius doubles, what is its new surface area and volume?

## Section 1-8, Review of Trigonometry

23. Find:
(a) $\sin(\pi/2)$      (b) $\cos(\pi/2)$      (c) $\tan\pi$
(d) $\tan(\pi/4)$      (e) $\sin(\pi/4)$      (f) $\cos(3\pi/4)$

24. Convert from radians to degrees:
(a) $\pi/4$      (b) $3\pi/2$      (c) $\pi/2$
(d) $\pi/6$      (e) $5\pi/6$

25. Convert from degrees to radians:
(a) $60°$      (b) $90°$      (c) $30°$
(d) $45°$      (e) $180°$      (f) $37°$
(g) $720°$

26. Find conversion factors for revolutions per minute to radians per second and degrees per second.

27. A record rotates at a rate of 33.3 revolutions per minute (rev/min). Find the angle in radians and in degrees through which the record rotates in 1 s.

28. The diameter of a record is 30 cm, and the record rotates at 33.3 rev/min. (a) What is the distance traveled by a point on the rim (measured along the circular arc) during 1 rev? (b) At what rate in cm/s does this point travel?

29. A right triangle has sides 3, 4, and 5 m. If $\theta_1$ is the smaller acute angle and $\theta_2$ the larger acute angle of the triangle, find the sine, cosine, and tangent of each of these angles. Find the measures of these angles in degrees using Appendix F.

30. A right triangle has sides $a = 2$, $b = 8$, and $c = ?$ (a) Find the length of the hypotenuse $c$. (b) Find the sine, cosine, and tangent of each angle of the triangle. (c) From Appendix F find the angles.

31. Use the small-angle approximations to find approximate values for (a) $\sin 8°$ and (b) $\tan 5°$. *Hint:* First convert these angles to radian measure.

## Section 1-9, The Binomial Expansion

32. Write out the complete binomial expansion (Equation 1-15) for $(1 + x)^3$ and check by finding $(1 + x)^3$ directly by multiplication.

33. Use the approximation $(1 + x)^n \approx 1 + nx$, $|x| \ll 1$, to find approximate values for:
(a) $\sqrt{99}$      (b) $1/1.01$      (c) $124^{1/3}$

34. The quantity $1/(1 + x)^2$ can be written as an infinite power series

$$1/(1 + x)^2 = 1 + a_1 x + a_2 x^2 + a_3 x^3 + \cdots + a_n x^n + \cdots ,$$

where $a_n$ are constants. Find these constants using the binomial equation 1-15.

35. Use the approximation $(1 + x)^n \approx 1 + nx$ to find approximate values for:
(a) $(9.8)^2$      (b) $(0.999)^5$      (c) $(1.002)^{-2}$

## Problems

1. The angle subtended by the moon's diameter at a point on the earth is about $0.524°$ (Figure 1-10). Use this and the fact that the moon is about 384 Mm away to find the diameter of the moon.

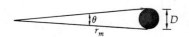

**Figure 1-10**
The angle subtended by the moon $\theta$ is approximately $D/r_m$, where $D$ is the diameter of the moon and $r_m$ is the distance to the moon.

2. A car wheel has an outside diameter of 0.62 m. (a) How many times does the wheel rotate when the car travels 1.0 km? (b) Through how many radians does the wheel rotate in 1 s when the car is traveling at 50 km/h? (This rate of angular rotation in radians per second is called the angular velocity.)

3. Use Equations 1-12 to show that $\sin 2\theta = 2 \sin\theta \cos\theta$.

4. Given that sin 64° = 0.90, use this information to find the following quantities without using any other source of trigonometric data:

(a) cos 64°                          (b) cos 32°
(c) cos 128°                         (d) sin 26°
(e) sin 244°                         (f) sin 96°
(g) tan 64°

5. A rectangular parallelepiped has a base that is a square of side 15.0 cm, and the long edge has length 40.0 cm. Find the angle between a diagonal of the parallelepiped and a diagonal of the base where they meet at one corner.

6. By trial and error using your calculator, find to the nearest degree the largest angle that will satisfy the following approximations to an accuracy of 10 percent or better:

(a) $\cos \theta \approx 1$              (b) $\sin \theta \approx \theta$
(c) $\cos \theta \approx 1 - \theta^2/2$       (d) $\sin \theta \approx \tan \theta$

7. The table below gives experimental results for a measurement of the period of motion $T$ of an object of mass $m$ suspended on a spring versus the mass of the object. These data are consistent with a simple equation expressing $T$ as a function of $m$ of the form $T = Cm^n$, where $C$ and $n$ are constants and $n$ is not necessarily an integer. (a) Find $n$ and $C$. (There are several ways to do this. One is to guess the value of $n$ and check by plotting $T$ versus $m^n$ on ordinary graph paper. If your guess is right, the plot will be a straight line. Another is to plot $T$ versus $m$ on log-log paper. The slope of the straight line on this plot is $n$.) (b) Which data points deviate the most from a straight-line plot of $T$ versus $m^n$?

| Mass $m$, kg | 0.10 | 0.20 | 0.40 | 0.50 | 0.75 | 1.00 | 1.50 |
|---|---|---|---|---|---|---|---|
| Period $T$, s | 0.56 | 0.83 | 1.05 | 1.28 | 1.55 | 1.75 | 2.22 |

8. The table below gives the period $T$ and orbit radius $r$ for the motions of four satellites orbiting a dense, heavy asteroid.

| Period $T$, y | 0.44 | 1.61 | 3.88 | 7.89 |
|---|---|---|---|---|
| Radius $r$, Gm | 0.088 | 0.208 | 0.374 | 0.600 |

(a) These data can be fitted by the formula $T = Cr^n$. Find $C$ and $n$. (b) A fifth satellite is discovered to have a period of 6.20 y. Find the radius for the orbit of this satellite, which fits the same formula.

9. The period $T$ of a simple pendulum depends on the length $L$ of the pendulum and the acceleration of gravity $g$ (dimensions $L/T^2$). (a) Find a simple combination of $L$ and $g$ which has the dimensions of time. (b) Check the dependence of the period $T$ on the length $L$ by measuring the period (time for a complete to-and-fro swing) of a pendulum for two different values of $L$. (c) The correct formula relating $T$ to $L$ and $g$ involves a constant which is a multiple of $\pi$ and cannot be obtained by the dimensional analysis of part (a). It can be found by experiment as in part (b) if $g$ is known. Using the value $g = 9.81$ m/s² and your experimental results from part (b), find the formula relating $T$ to $L$ and $g$.

10. A projectile fired at an angle of 45° travels a total distance $R$, called the range, which depends only on the initial speed $v$ and the acceleration of gravity $g$ (dimensions $L/T^2$). Using dimensional analysis, find how $R$ depends on the speed and on $g$.

11. A ball thrown horizontally from a height $H$ with speed $v$ travels a total horizontal distance $R$. (a) Do you expect $R$ to increase or decrease with increasing $H$? With increasing $v$? (b) From dimensional analysis, find a possible dependence of $R$ on $H$, $v$, and $g$.

12. The volume of a sphere of radius $r$ is $\frac{4}{3}\pi r^3$. (a) Find an expression for the volume of spherical shell contained between radii $r$ and $r + \Delta r$ and use the binomial expansion to find an approximate expression to first order in $\Delta r/r$; that is, drop terms of order $(\Delta r/r)^2$ and higher-order terms. (b) As $\Delta r \to 0$, the volume of the shell tends to $A \, \Delta r$, where $A$ is the surface area of the sphere. Use this to find an expression for the surface area of a sphere.

13. The Great Pyramid of Cheops has a square horizontal base and four triangular faces that join along sloping edges 222 m in length. Each face is inclined at an angle 51° to the horizontal. Find the length of a side of the base.

14. From Equation 1-12 and the identity $\tan \theta = (\sin \theta)/(\cos \theta)$ find an expression for $\tan (\theta_1 + \theta_2)$ in terms of $\tan \theta_1$ and $\tan \theta_2$.

15. A man begins at an intersection of a north-south road with an east-west road and walks 10 km in the direction north of east along a line which makes an angle of 30° with the easterly direction. (a) How far is he from the north-south road? From the east-west road? (b) He then heads along a line perpendicular to his original direction. How far must he walk to reach the north-south road?

16. (a) For each case in Exercise 33 compute the next term $\frac{1}{2}n(n - 1)x^2$ in the binomial expansion used and express this term as a percentage of one. (b) Find the difference between your approximate answer for each case in that exercise and the exact calculation (from tables or from any other method) and express this difference as a percentage of the exact answer. Compare with (a).

17. Repeat Problem 16 using the data of Exercise 35.

18. Often an equation involving trigonometric functions is difficult or impossible to solve analytically but fairly easy to solve graphically or numerically by trial and error using a calculator. (a) Find $\theta$ for which $\theta = 2 \sin \theta$. First sketch $y_1 = \theta$ and $y_2 = 2 \sin \theta$ on the same graph and note that $y_1 = y_2$ at $\theta$ between $\frac{1}{2}\pi$ and $\pi$. Use this rough graphical solution to obtain your first guess; then by trial and error using a calculator, find a solution to three or four significant figures. (b) Find the angle $\theta$ in degrees for which $\tan \theta - \sin \theta = 0.200$. Use Figure 1-7 to obtain your first guess. (c) Find the smallest angle $\theta$ in radians (greater than zero) for which $\tan \theta = \theta$. Obtain your first guess from a plot of $y_1 = \theta$ and $y_2 = \tan \theta$ (see Figure 1-9c).

# CHAPTER 2　　Motion in One Dimension

---

**Objectives**　After studying this chapter, you should:

1.　Know the definitions of displacement, velocity, and acceleration.

2.　Be able to distinguish between velocity and speed.

3.　Be able to calculate the instantaneous velocity from a graph of $x$ versus $t$.

4.　Be able to discuss qualitatively how the acceleration of a particle varies with time from a graph of $x$ versus $t$ and also from a graph of $v$ versus $t$.

5.　Know the important equations which apply when the acceleration is constant and be able to use them to work problems.

6.　Be able to calculate the displacement of a particle from the $v$-versus-$t$ curve and the change in the velocity of a particle from the $a$-versus-$t$ curve by finding the areas under the appropriate curves.

---

To simplify our discussion of motion, we start with objects whose position can be described by locating one point. Such an object is called a *particle*. One tends to think of a particle as a very small object, e.g., a piece of shot, but actually no size limit is implied by the word "particle." If we are not interested in the extension of an object or in its rotational motion, any object can be considered as a particle. For example, it is sometimes convenient to consider the earth as a particle moving around the sun in a nearly circular path. In this case we are interested only in the motion of the center of the earth and are ignoring the size of the earth and its rotation. In some astronomical problems the solar system or even a whole galaxy is treated as a particle.

When we are interested in the internal motion or internal structure of an object, it can no longer be treated as a particle, but our study of particle motion is useful even in these cases because any complex object can be treated as a system of particles. Even so small a thing as an atomic nucleus, with a diameter of only about $10^{-15}$ m (1 fm), turns out to be a rather complicated system of particles when its structure is examined in detail.

To describe the motion of a particle, we need the concepts of *displacement*, *velocity*, and *acceleration*. In the general motion of a particle in three dimensions, these quantities are vectors, which have direction as well as magnitude. We study vectors and the general motion of a particle in Chapter 3; in this chapter, we confine our discussion to motion along a straight line, i.e., motion in one dimension. For such restricted motion, there are only two possible directions, distinguished by designating one positive and the other negative. A simple example of one-dimensional motion is a car moving along a flat, straight, narrow road. We can choose any convenient point on the car for the location of the "particle."

**Question**

1. Give several examples of the motion of large objects where treating the object as a particle is an adequate approximation. Give other examples where it is not.

## 2-1    Speed, Displacement, and Velocity

We are all familiar with the concept of speed. We define the average speed of a particle as the ratio of the total distance traveled and the total time taken:

$$\text{Average speed} = \frac{\text{total distance}}{\text{total time}}$$

The SI units of average speed are metres per second, written m/s, and the U.S. customary units are feet per second (ft/s). A familiar everyday unit is kilometres per hour (km/h). For example, if you drive 200 km in 5 h, your average speed is (200 km)/(5 h) = 40 km/h. Note that the average speed tells you nothing about the details of the trip. You may have driven at a steady rate of 40 km/h for the whole 5 h, or you may have driven faster part of the time and slower the rest of the time; or you may have stopped for an hour to eat and then driven at varying rates during the other 4 h.

The concept of velocity is similar to that of speed but differs because it includes the *direction* of motion. To understand this concept, we first introduce the idea of displacement. Let us set up a coordinate system by choosing some reference point on a line for the origin $O$. To every other point on the line we assign a number $x$ which indicates how far the point is from the origin. The value of $x$ depends on the unit chosen as the measure of distance. The sign of $x$ depends on its position relative to the origin $O$; if it is to the right, it is positive; to the left, negative.

Suppose that our particle is at position $x_1$ at some time $t_1$ and at point $x_2$ at time $t_2$. The change in the position of the particle, $x_2 - x_1$, is called the *displacement* of the particle. It is customary to use the Greek letter $\Delta$ (capital delta) to indicate the change in a quantity. Thus the change in $x$ is written $\Delta x$. The notation $\Delta x$ (read "delta $x$") stands for a single quantity, the change in $x$ (it is not a product of $\Delta$ and $x$ any more than $\cos x$ is a product of cos and $x$):

$$\Delta x = x_2 - x_1 \qquad\qquad 2\text{-}1 \qquad\qquad \textit{Displacement defined}$$

The average velocity of the particle is defined to be the ratio of the displacement $\Delta x$ and the time interval $\Delta t = t_2 - t_1$:

$$v_{av} = \frac{\Delta x}{\Delta t} = \frac{x_2 - x_1}{t_2 - t_1}$$

2-2    *Average velocity defined*

Note that the displacement and the average velocity may be either positive or negative, depending on whether $x_2$ is greater or less than $x_1$ (assuming positive $\Delta t$). A positive value indicates motion to the right; a negative value indicates motion to the left.

**Example 2-1** If $x_1 = 18$ m at $t_1 = 2$ s and $x_2 = 3$ m at $t_2 = 7$ s, find the displacement and the average velocity for this time interval.

By the definition, the displacement is

$$\Delta x = x_2 - x_1 = 3 \text{ m} - 18 \text{ m} = -15 \text{ m}$$

and the average velocity is

$$v_{av} = \frac{\Delta x}{\Delta t} = \frac{x_2 - x_1}{t_2 - t_1} = \frac{3 \text{ m} - 18 \text{ m}}{7 \text{ s} - 2 \text{ s}} = \frac{-15 \text{ m}}{5 \text{ s}} = -3 \text{ m/s}$$

The displacement and average velocity are negative, indicating that the particle moved to the left, toward decreasing values of $x$.

Note that the unit metres per second is included as part of the answer for the average velocity found in Example 2-1. Since there are many other possible choices for units of length and time (feet, inches, miles, light-years, etc. for length, and hours, days, years, etc. for time), it is essential to include the unit with a numerical answer. The statement "the average velocity of a particle is $-3$" is meaningless.

**Example 2-2** How far does a car go in 5 min if its average velocity is 80 km/h?

In this example we are interested in the displacement during a time interval of 5 min. From Equation 2-2 the displacement $\Delta x$ is given by

$$\Delta x = v_{av} \, \Delta t$$

Thus

$$\Delta x = \frac{80 \text{ km}}{\text{h}} \times 5 \text{ min} = 400 \, \frac{\text{km} \cdot \cancel{\text{min}}}{\cancel{\text{h}}} \, \frac{1 \, \cancel{\text{h}}}{60 \, \cancel{\text{min}}} = 6.67 \text{ km}$$

We use the conversion factor 1 h = 60 min to change the unit of displacement from km·min/h to kilometres.

**Example 2-3** A runner runs 100 m in 10 s, then turns around and jogs 50 m back toward the starting point in 30 s. What are her average speed and average velocity for the total trip?

The total distance traveled is 100 m + 50 m = 150 m, and the total time taken is 40 s. The average speed is therefore (150 m)/(40 s) = 3.75 m/s. Note that this is not the average of the running and jogging speeds because she ran for 10 s but jogged for 30 s. To find the average *velocity* we first find the total displacement, which is 50 m (if $x_1$ is taken to be 0, then $x_2 = 50$ m). The average velocity is then

$$v_{av} = \frac{\Delta x}{\Delta t} = \frac{50 \text{ m}}{40 \text{ s}} = 1.25 \text{ m/s}$$

Again, this is *not* the average of the running velocity ($+10$ m/s) and the jogging velocity ($-1.67$ m/s) because the times are different.

It is often useful to interpret physical quantities graphically. Suppose we determine the position $x$ of a particle at various times $t$. Figure 2-1 shows a graph of $x$ versus $t$ for some arbitrary motion along the $x$ axis. The measured points have been connected with a smooth curve. How accurately this curve represents the motion of the particle depends upon the complexity of the motion. On the figure we have drawn a straight line between the initial position labeled point $P_1$ and the final position labeled point $P_2$. The displacement $\Delta x = x_2 - x_1$ and the time interval $\Delta t = t_2 - t_1$ for these points are indicated. The line between $P_1$ and $P_2$ is the hypotenuse of the triangle with sides $\Delta t$ and $\Delta x$. The ratio $\Delta x/\Delta t$ is called the *slope* of this straight line. In geometric terms, it is a measure of the steepness of the straight line in the graph. For a given interval $\Delta t$, the steeper the line, the greater the value of $\Delta x/\Delta t$. Since the slope of this line is just the average velocity for the time interval $\Delta t$, we have a geometric interpretation of average velocity. It is the slope of the straight line connecting the points $(x_1, t_1)$ and $(x_2, t_2)$. In general, the average velocity depends on the time interval chosen. For example, in Figure 2-1, if we chose a smaller time interval by choosing a time $t_2'$ closer to $t_1$, the average velocity would be greater, as indicated by the greater steepness of the line connecting points $P_1$ and $P_2'$:

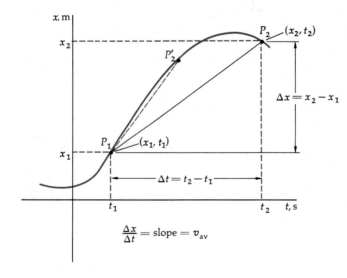

$$\frac{\Delta x}{\Delta t} = \text{slope} = v_{av}$$

**Figure 2-1**
Graph of $x$ versus $t$ for a particle moving in one dimension. The initial and final points $P_1$ and $P_2$ are connected by a straight line. The average velocity is the slope of this line $\Delta x/\Delta t$ and depends on the time interval, as indicated by the fact that the line from $P_1$ to $P_2'$ has a greater slope than that from $P_1$ to $P_2$.

**Questions**

2. What sense, if any, does the following statement make? "The average velocity of the car at 9 A.M. was 60 km/h."

3. Is it possible that the average velocity for some interval may be zero although the average velocity for a shorter interval included in the first interval is not zero? Explain.

## 2-2   Instantaneous Velocity

At first glance, it might seem impossible to define the velocity of a particle at a single instant, i.e., at a specific time. At a time $t_1$, the particle is at a single point $x_1$. If it is at a single point, how can it be moving? On the other hand, if it is not moving, shouldn't it stay at the same point?

Motion depicted by an artist in Duchamp's "Nude Descending a Staircase, No. 2." (*Marcel Duchamp, 1912, Louise and Walter Arensberg Collection, Philadelphia Museum of Art.*) What features in the photograph of a runner imply motion? Can we tell with certainty from a single high-speed photograph that the athlete is moving?

A. J. Wyatt                    Ken Regan/Camera 5

This is an age-old paradox, which can be resolved when we realize that to observe motion and thus define it, we must look at the position of the object at more than one time. It is then possible to define the velocity at an instant by a limiting process.

Figure 2-2 is the same $x$-versus-$t$ curve as Figure 2-1, showing a sequence of time intervals indicated $\Delta t_1$, $\Delta t_2$, $\Delta t_3$, ..., each smaller than the previous one. For each time interval $\Delta t$, the average velocity is the slope of the dashed line appropriate for that interval. This figure shows that as the time interval becomes smaller, the dashed lines get steeper but never incline more than the line tangent to the curve at point $t_1$. We define the slope of this tangent line to be the *instantaneous velocity* at the time $t_1$. The instantaneous velocity is the limit of the ratio $\Delta x/\Delta t$ as $\Delta t$ approaches zero:

$$v(t) = \lim_{\Delta t \to 0} \frac{\Delta x}{\Delta t} = \text{slope of line tangent to } x(t) \text{ curve} \qquad 2\text{-}3$$

*Instantaneous velocity defined*

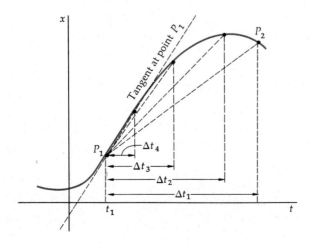

**Figure 2-2**
Plot of $x(t)$ from Figure 2-1. As the time interval beginning at $t_1$ is decreased, the average velocity for that interval approaches the slope of the line tangent to the curve at time $t_1$. The instantaneous velocity at time $t_1$ is defined to be the slope of the line tangent to the curve at $t_1$.

This limit is called the *derivative of x with respect to t* at the time $t_1$. In the usual calculus notation* the derivative is written $dx/dt$:

$$v(t) = \lim_{\Delta t \to 0} \frac{\Delta x}{\Delta t} = \frac{dx}{dt} \qquad\qquad 2\text{-}4$$

It is important to realize that the displacement $\Delta x$ depends on the time interval $\Delta t$. As $\Delta t$ approaches zero, $\Delta x$ does too (as can be seen from Figure 2-2) and the ratio $\Delta x/\Delta t$ approaches the slope of the line tangent to the curve. In computing this limit, we do not set $\Delta t$ equal to zero since then $\Delta x$ would also be zero and the ratio $\Delta x/\Delta t$ would not be defined.

In your study of calculus, you will learn methods for computing derivatives of various functions $x(t)$. For now, the important point to remember is that *the instantaneous velocity is the slope of the tangent line to the x-versus-t curve*. Since this slope may be positive ($x$ increasing) or negative ($x$ decreasing), the instantaneous velocity may be positive or negative in one-dimensional motion. The magnitude of the instantaneous velocity is called the *instantaneous speed*.

**Example 2-4** The position of a particle is given by the function shown in Figure 2-3. Find the instantaneous velocity at the time $t = 2$ s. When is the velocity greatest? When is it zero? Is it ever negative?

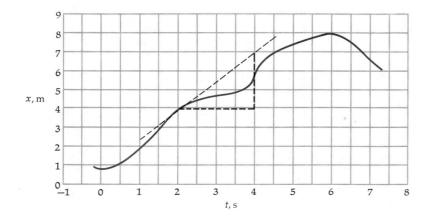

**Figure 2-3**
Plot of $x$ versus $t$ for Example 2-4. The instantaneous velocity at time $t = 2$ s can be found by measuring the slope of the line tangent to the curve at that time.

In the figure we have sketched the line tangent to the curve at time $t = 2$ s. The slope of this line is measured from the figure to be $(3 \text{ m})/(2 \text{ s}) = 1.5$ m/s. Thus $v = 1.5$ m/s at time $t = 2$ s. According to the figure, the slope is greatest at about $t = 4$ s. The velocity is zero at times $t = 0$ and $t = 6$ s, as indicated by the fact that the tangent lines at these times are horizontal, with zero slope. After $t = 6$ s, the curve has a negative slope, indicating that the velocity is negative. (The slope of the line tangent to a curve is often referred to merely as the *slope of the curve*.)

---

* Strictly speaking, the notation $dx/dt$ stands for the single quantity, the derivative of $x$ with respect to $t$, which is the slope of the line tangent to the $x(t)$ curve. Later, we shall define the quantities $dx$ and $dt$ (called *differentials*) so that their ratio is in fact this slope. If $\Delta x$ and $\Delta t$ are chosen small enough, the ratio $\Delta x/\Delta t$ is *nearly* equal to the slope of the tangent line. Therefore it is convenient, though not mathematically rigorous, to think of $dt$ as a very small change in $t$ and $dx$ as the corresponding very small change in $x$.

**Example 2-5** The position of a stone dropped from rest from a cliff is given by $x = 5t^2$, where $x$ is in metres measured downward from the original position at $t = 0$ and $t$ is in seconds. Find the velocity at any time. (We omit explicit indication of the units to simplify the notation.)

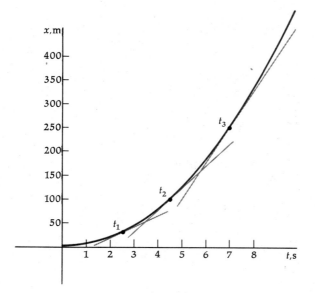

**Figure 2-4**
Plot of the function $x(t) = 5t^2$. Tangent lines are drawn at times $t_1$, $t_2$, and $t_3$. The slope of these tangent lines increases steadily from $t_1$ to $t_2$ to $t_3$, indicating that the instantaneous velocity increases steadily with time.

This function is shown in Figure 2-4, where tangent lines are drawn for three different times, $t_1$, $t_2$, and $t_3$. The slopes of these lines differ. As time passes, the slope of the curve increases, indicating that the instantaneous velocity is increasing with time. We can compute the velocity at some time $t_1$ by computing the derivative $dx/dt$ directly from the definition in Equation 2-3. At time $t_1$ the position is

$$x_1 = 5t_1^2$$

At a later time $t_2 = t_1 + \Delta t$ the position is $x_2$, given by

$$x_2 = 5t_2^2 = 5(t_1 + \Delta t)^2 = 5[t_1^2 + 2t_1 \Delta t + (\Delta t)^2]$$
$$= 5t_1^2 + 10t_1 \Delta t + 5(\Delta t)^2$$

The displacement for this time interval is thus

$$\Delta x = x_2 - x_1 = 10t_1 \Delta t + 5(\Delta t)^2$$

The average velocity for this time interval is

$$v_{av} = \frac{\Delta x}{\Delta t} = 10t_1 + 5 \Delta t$$

As we consider shorter and shorter time intervals, $\Delta t$ approaches zero and the second term $5 \Delta t$ approaches zero, though the first is unaltered. The instantaneous velocity at time $t_1$ is thus

$$v = \lim_{\Delta t \to 0} \frac{\Delta x}{\Delta t} = 10t_1$$

This function, the instantaneous velocity, is proportional to the time. For any general time $t$, the instantaneous velocity is $v(t) = 10t$.

It is instructive to examine the limiting process numerically by computing the average velocity for smaller and smaller time intervals. Table 2-1 gives the average velocity for Example 2-5 computed for $t_1 = 2$ s and various time intervals $\Delta t$, each smaller than the previous one. The table shows that for very small time intervals, the average velocity is very nearly equal to the instantaneous velocity 20 m/s. The difference between $\Delta x/\Delta t$ and $\lim_{\Delta t \to 0} (\Delta x/\Delta t)$ can be made arbitrarily small by choosing $\Delta t$ sufficiently small.

It is important to distinguish carefully between average and instantaneous velocity. By custom, however, the word velocity alone is assumed to mean instantaneous velocity.

**Questions**

4. If the instantaneous velocity does not change from instant to instant, will the average velocities for different intervals differ?

5. If $v_{av} = 0$ for some time interval $\Delta t$, must the instantaneous velocity $v$ be zero at some time in the interval? Support your answer by a sketch of a possible $x$-versus-$t$ curve which has $\Delta x = 0$ for some interval $\Delta t$.

**Table 2-1**
Displacement and average velocity for various time intervals $\Delta t$ beginning at $t = 2$ s for the function $x = 5t^2$

| $\Delta t$, s | $\Delta x$, m | $\Delta x/\Delta t$, m/s |
|---|---|---|
| 1.00 | 25 | 25 |
| 0.50 | 11.25 | 22.5 |
| 0.20 | 4.20 | 21.0 |
| 0.10 | 2.05 | 20.5 |
| 0.05 | 1.0125 | 20.25 |
| 0.01 | 0.2005 | 20.05 |
| 0.005 | 0.100125 | 20.025 |
| 0.001 | 0.020005 | 20.005 |
| 0.0001 | 0.00200005 | 20.0005 |

## 2-3  Acceleration

When the instantaneous velocity of a particle is changing with time, as in Example 2-5, the particle is said to be *accelerating*. The average acceleration for a particular time interval $\Delta t = t_2 - t_1$ is defined as the ratio $\Delta v/\Delta t$, where $\Delta v = v_2 - v_1$ is the change in instantaneous velocity for that time interval:

$$a_{av} = \frac{\Delta v}{\Delta t} \qquad \qquad 2\text{-}5$$

*Average acceleration defined*

The dimensions of acceleration are length divided by (time)². Convenient units are metres per second per second, more compactly written metres per second squared (m/s²), or feet per second squared (ft/s²). *Instantaneous acceleration* is the limit of this ratio as $\Delta t$ approaches zero. If we plot the velocity versus time, the instantaneous acceleration is defined to be the slope of the tangent line to this curve:

$$a = \lim_{\Delta t \to 0} \frac{\Delta v}{\Delta t} = \text{slope of } v\text{-versus-}t \text{ curve} \qquad 2\text{-}6$$

*Instantaneous acceleration defined*

The acceleration is thus the derivative of the velocity with respect to time. The calculus notation for this derivative is $dv/dt$. Since the velocity is also the derivative of the position $x$ with respect to $t$, the acceleration is the second derivative of $x$ with respect to $t$. The common notation for the second derivative of $x$ with respect to $t$ is $d^2x/dt^2$. We can see the origin of this notation by writing the acceleration as $dv/dt$ and replacing $v$ with $dx/dt$:

$$a = \frac{dv}{dt} = \frac{d(dx/dt)}{dt} = \frac{d^2x}{dt^2} \qquad 2\text{-}7$$

The acceleration of a particle is important in physics because it is related to the forces acting on the particle. We shall see in Chapter 4 that the acceleration is directly proportional to the net force acting on the particle. If the velocity is constant, the acceleration is zero since $\Delta v = 0$ for all time intervals. In this case the slope of the $x$-versus-$t$ curve does not change. In Example 2-5 we found that for the position function $x = (5 \text{ m/s}^2)t^2$ the velocity increases linearly with time according to $v = (10 \text{ m/s}^2)t$. In this case the acceleration is constant and equal to $10 \text{ m/s}^2$, the constant slope of the $v$-versus-$t$ curve. We can get at least a qualitative idea of the acceleration from the $x$-versus-$t$ curve. From Figure 2-4 we can see that the velocity is increasing with time, as indicated by the increasing steepness of the tangent lines; i.e., this curve shows that the acceleration is positive.

---

**Example 2-6** From the $x$-versus-$t$ curve in Figure 2-5, discuss qualitatively the velocity and acceleration at the point $t = 5$ s.

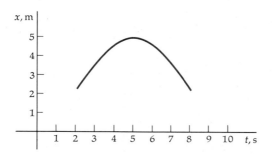

**Figure 2-5**
$x$-versus-$t$ curve for Example 2-6. The slope of the tangent line at $t = 5$ s is zero, indicating that the instantaneous velocity is zero at this time. Just before $t = 5$ s, the slope is positive; just after $t = 5$ s, the slope is negative. Since the slope is decreasing at $t = 5$ s, the acceleration at this time is negative.

At $t = 5$ s the tangent line to the curve is horizontal with zero slope. Thus the instantaneous velocity at this point is zero. However, the fact that the velocity is zero does *not* imply that the acceleration must also be zero. The acceleration is the rate of change of the instantaneous velocity. To find the acceleration at some time we must know how the velocity is changing at that time. For this example, to find the acceleration at $t = 5$ s we must know the velocity at other times in the neighborhood of $t = 5$ s. From the figure we see that the slope of the curve is positive just before $t = 5$ s and negative just after. Thus the velocity changes from a positive value to zero to a negative value, and acceleration at this time is negative. In Figure 2-6 we have sketched the velocity-versus-$t$ curve corresponding to the position function given in Figure 2-5. Note that at $t = 5$ s the velocity is zero; nevertheless the slope of the $v$-versus-$t$ curve is not zero but negative.

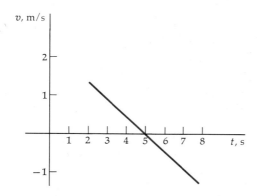

**Figure 2-6**
Velocity-versus-time curve corresponding to the position curve in Figure 2-5. At $t = 5$ s the velocity is zero and decreasing.

**Example 2-7** The position of a particle is given by $x = Ct^3$, where $C$ is a constant with the dimensions of m/s$^3$. Find the velocity and acceleration as functions of time.

The graph of $x$ versus $t$ (not shown) resembles the graph in Example 2-5 except that the increase of $x$ with time is faster. The slope of the curve increases with time, indicating an increasing velocity. As in that example, we can compute the velocity by computing the derivative $dx/dt$ directly from its definition (Equation 2-3). At some time $t_1$ the position is $x_1 = Ct_1^3$. At a later time $t_2 = t_1 + \Delta t$ the position is

$$x_2 = Ct_2^3 = C(t_1 + \Delta t)^3 = Ct_1^3 + 3Ct_1^2 \, \Delta t + 3Ct_1(\Delta t)^2 + C(\Delta t)^3$$

The displacement is thus

$$\Delta x = x_2 - x_1 = 3Ct_1^2 \, \Delta t + 3Ct_1(\Delta t)^2 + C(\Delta t)^3$$

The average velocity for this time interval is

$$v_{av} = \frac{\Delta x}{\Delta t} = 3Ct_1^2 + 3Ct_1 \, \Delta t + C(\Delta t)^2$$

As $\Delta t \to 0$, the two right-hand terms also approach zero and the term $3Ct_1^2$ remains unchanged. The instantaneous velocity at time $t$ is thus

$$v = 3Ct^2$$

(Having computed the velocity at time $t_1$, we can drop the subscript 1 on the time and write the velocity for a general time $t$.) We find the acceleration by repeating the process and thus finding the derivative of $v$ with respect to $t$, which is the second derivative of $x$ with respect to $t$. We omit some of the algebra in this derivation because it is similar to that shown above and in Example 2-5. The change in velocity for the time interval from $t$ to $t + \Delta t$ is

$$\Delta v = 3C(t + \Delta t)^2 - 3Ct^2 = 6Ct \, \Delta t + 3C(\Delta t)^2$$

We divide this expression by $\Delta t$ to obtain the average acceleration for this interval:

$$a_{av} = \frac{\Delta v}{\Delta t} = 6Ct + 3C \, \Delta t$$

The instantaneous acceleration is thus

$$a = \lim_{\Delta t \to 0} \frac{\Delta v}{\Delta t} = 6Ct$$

In this example the acceleration is not constant but increases with time.

In the examples so far we have computed derivatives directly from the definition by taking the appropriate limit explicitly. It is useful to examine various properties of the derivative and to develop rules permitting us to calculate the derivatives of many functions quickly without applying the definition each time. Table E-1 in Appendix E contains a list of such rules, followed by a brief discussion of their origin. We use these rules without comment and leave their detailed study to your calculus course.

Rule 7 in Table E-1 is used so often we repeat it here. If $x$ is a simple power function of $t$, such as

$$x = Ct^n$$

where $C$ and $n$ are any constants, the derivative of $x$ with respect to $t$ is given by

$$\frac{dx}{dt} = Cnt^{n-1}$$

We have already seen applications of this rule for $n = 1, 2$, and $3$ in the examples; e.g., for the position function of Example 2-7, $x = Ct^3$, we found $v = dx/dt = 3Ct^2$, and $a = dv/dt = 6Ct$.

### Questions

6. Give an example of a motion for which the velocity is negative but the acceleration is positive; e.g., sketch a graph of $v$ versus $t$.

7. Give an example of a motion for which both the acceleration and velocity are negative.

8. Is it possible for a body to have zero velocity and nonzero acceleration?

## 2-4  Motion with Constant Acceleration

The motion of a particle which has constant acceleration is important for several reasons. For one, this type of motion is common in nature. For example, near the surface of the earth all objects fall vertically with the constant acceleration of gravity if air resistance can be neglected and if there are no forces acting on the objects other than the pull of gravity. The acceleration of gravity is designated by $g$ and has the approximate value

$$g = 9.81 \text{ m/s}^2 = 32.2 \text{ ft/s}^2 \qquad 2\text{-}8$$

(We often approximate this by $9.8$ m/s$^2$ or $32$ ft/s$^2$, or even $10$ m/s$^2$ if no greater precision is required.) Even when the acceleration of a particle is not constant, we can sometimes learn much about the motion by using the constant-acceleration results developed in this section.

As we have mentioned, if the acceleration is constant, the velocity changes linearly with time. If the velocity is $v_0$ at time $t = 0$, its value $v$ at a later time is given by

$$v = v_0 + at \qquad 2\text{-}9$$

*Constant acceleration: v versus t*

If the particle starts at $x_0$ at time $t = 0$ and its position is $x$ at time $t$, the displacement $\Delta x = x - x_0$ is given by

$$\Delta x = v_{av}t$$

(This is the same as Equation 2-2 with $t$ replacing $\Delta t$ because we have chosen the initial value of $t$ to be zero.)

For constant acceleration, the average velocity is the mean value of the initial and final velocities.* We can obtain this mean value by adding half the increase ($\frac{1}{2}at$) to the initial velocity, as indicated in Figure 2-7:

$$v_{av} = v_0 + \tfrac{1}{2}at$$

---

* This is not true if the acceleration is not constant.

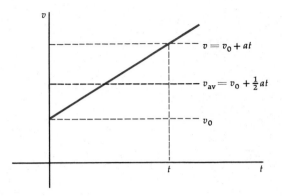

**Figure 2-7**
For constant acceleration the velocity varies linearly with time. The average velocity is then just the mean of the initial and final velocities. If $v_0$ is the initial velocity and $v_0 + at$ is the final velocity, the average velocity is $v_{av} = v_0 + \frac{1}{2}at$.

The displacement is then

$$\Delta x = (v_0 + \tfrac{1}{2}at)t = v_0t + \tfrac{1}{2}at^2$$

and the position function is

$$x = x_0 + v_0t + \tfrac{1}{2}at^2 \qquad \text{2-10}$$

*Constant acceleration:
x versus t*

In terms of the initial and final velocities, the average velocity is

$$v_{av} = v_0 + \tfrac{1}{2}at = v_0 + \tfrac{1}{2}(v - v_0)$$

or

$$v_{av} = \tfrac{1}{2}(v_0 + v) \qquad \text{2-11}$$

*Constant acceleration: $v_{av}$*

In particular, if the initial velocity is zero, the average velocity is half the final velocity.

**Example 2-8**  A ball is thrown upward with an initial velocity of 30 m/s. If its acceleration is 10 m/s² downward, how long does it take to reach its highest point, and what is the distance to the highest point?

If we take the upward direction as positive, we have $v_0 = 30$ m/s and $a = -10$ m/s². The acceleration is negative because it is in the downward direction. As the ball moves upward ($v$ positive), the velocity decreases from its initial value until it is zero. When the velocity is zero, the ball is at its highest point. It then falls, and the velocity becomes negative, indicating that the ball is moving downward. We can find the time $t$ to reach the top of its flight by using Equation 2-8 and setting $v = 0$. We have

$$v = v_0 + at$$

$$0 = 30 \text{ m/s} + (-10 \text{ m/s}^2)t$$

$$t = \frac{30 \text{ m/s}}{10 \text{ m/s}^2} = 3.0 \text{ s}$$

Note that the units work out correctly. Since the initial velocity is +30 m/s and the final velocity is 0, the average velocity for the upward motion is 15 m/s. The distance traveled is then

$$\Delta x = v_{av}t = (15 \text{ m/s})(3.0 \text{ s}) = 45 \text{ m}$$

We could have also found $\Delta x$ from Equation 2-10, but the calculation is somewhat more difficult and there is more chance of making a mistake:

$$\Delta x = v_0t + \tfrac{1}{2}at^2 = (30 \text{ m/s})(3.0 \text{ s}) + \tfrac{1}{2}(-10 \text{ m/s}^2)(3.0 \text{ s})^2$$
$$= +90 \text{ m} - 45 \text{ m} = 45 \text{ m}$$

**Example 2-9** What is the total time the ball in Example 2-8 is in the air?

We could guess the answer to be 6 s since, by symmetry, if it takes 3.0 s to rise 45 m, it should take the same time to fall. This is correct. We can also find the time from Equation 2-10 by setting $\Delta x = 0$. We have

$$\Delta x = v_0 t + \tfrac{1}{2}at^2 = 0$$

Factoring, we obtain

$$t(v_0 + \tfrac{1}{2}at) = 0$$

The two solutions are $t = 0$, which corresponds to our initial conditions, and

$$t = -\frac{2v_0}{a} = -\frac{2(30\text{ m/s})}{-10\text{ m/s}^2} = 6\text{ s}$$

Figures 2-8 and 2-9 show $x$-versus-$t$ and $v$-versus-$t$ curves for the ball in Examples 2-8 and 2-9. Note that at the time 3.0 s the velocity of the ball is zero but the slope of the $v$-versus-$t$ curve is not. The slope of the $v$-versus-$t$ curve has the value $-10$ m/s$^2$ at this time and at all other times because the acceleration is constant.

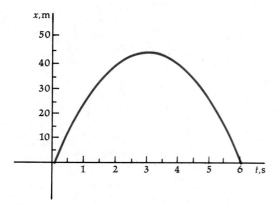

**Figure 2-8**
$x$ versus $t$ for the ball thrown into the air with initial velocity of 30 m/s (Examples 2-8 and 2-9). The curve is a parabola, $x = (30\text{ m/s})t - \tfrac{1}{2}(10\text{ m/s}^2)t^2$.

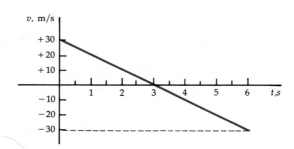

**Figure 2-9**
$v$ versus $t$ for the ball thrown into the air in Examples 2-8 and 2-9. The velocity decreases steadily from its initial value of 30 m/s to its final value $-30$ m/s just before the ball hits the ground. At $t = 3.0$ s, when the ball is at its highest point, the velocity is zero, but its rate of change is $-10$ m/s$^2$, the same as at any other time.

We sometimes want to find the final velocity of a particle after it moves through a given distance with constant acceleration. For example, we may want to know the velocity of a ball dropped from rest from some height $x$. We could find it by first finding the time from $x = \tfrac{1}{2}at^2$ and then substituting the result into $v = at$ (for $v_0 = 0$). But if we are not interested in knowing the time of fall, it is convenient to eliminate the time from these two equations and develop a relationship between $v$, $a$, and $\Delta x$. We can find a general relationship for any initial velocity $v_0$ by solving Equation 2-9 for $t = (v - v_0)/a$ and substituting this result into Equation 2-10. The result after some simplification and rearrangement is

$$v^2 = v_0^2 + 2a\,\Delta x \qquad\qquad 2\text{-}12$$

*Constant acceleration:*
*v versus x*

**Example 2-10** A car traveling at 30 m/s (about 67 mi/h) brakes to a stop. If the acceleration is −5 m/s², how far does the car travel before stopping? This distance is called the *stopping distance*.

In this example we choose the original direction of motion to be positive. The stopping distance will then also be positive, but the acceleration will be negative. (A negative acceleration is sometimes called deceleration.) Setting $v = 0$ in Equation 2-12, we have

$$0 = (30 \text{ m/s})^2 + 2(-5 \text{ m/s}^2)\,\Delta x$$

$$\Delta x = 90 \text{ m}$$

Note that this is a considerable distance. The force that produces this deceleration is that of friction between the car tires and the road. On wet pavement or on gravel, the frictional force is smaller and the magnitude of the acceleration is even less than 5 m/s².

**Example 2-11** What is the stopping distance under the same conditions as Example 2-10 if the car is initially traveling at 15 m/s?

From Equation 2-12 with $v = 0$ we see that the stopping distance is proportional to the square of the initial speed. If we double the speed, the stopping distance is increased by a factor of 4. Similarly, if we halve the initial speed, the stopping distance is reduced by a factor of 4. The stopping distance at 15 m/s is therefore one-fourth that at 30 m/s, or $\frac{1}{4}(90 \text{ m}) = 22.5 \text{ m}$.

Sometimes, even when the acceleration is not constant, valuable insight can be gained about the motion by *assuming* that the constant-acceleration formulas still apply.

**Example 2-12** A car traveling 100 km/h (about 62 mi/h) crashes into a concrete wall, which does not move. How long does it take to bring the car to rest, and what is its acceleration?

In this example, it is not accurate to treat the car as a particle because different parts of it have different accelerations. Moreover, the accelerations are not constant. Nevertheless, let us assume constant acceleration of a point particle. We need more information in order to find either the stopping time or the acceleration. The quantity missing is the stopping distance. We can estimate this from our practical knowledge. The center of the car certainly moves less than half the length of the car. A reasonable estimate for the stopping distance is probably between 0.5 and 1.0 m. Let us use 0.75 m as our estimate. We can then find the time to stop from $\Delta x = v_{av}\,\Delta t$ with $v_{av} = \frac{1}{2}v_0 = 50 \text{ km/h} = 14 \text{ m/s}$ (since we are making estimates, two significant figures is certainly sufficient). Then

$$\Delta t = \frac{\Delta x}{v_{av}} = \frac{0.75 \text{ m}}{14 \text{ m/s}} = 0.054 \text{ s}$$

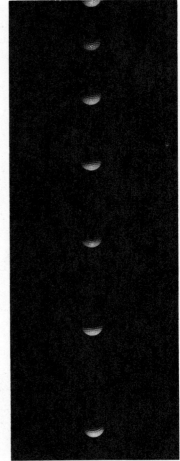

Fundamental Photographs

Multiflash photograph of a golf ball falling with constant acceleration. The position of the ball is recorded at $\frac{1}{25}$-s intervals using a strobe light that flickers 25 times per second. The increase in spacing between successive positions indicates that the speed is increasing.

Since the car is brought from $v_0 = 100$ km/h $= 28$ m/s to rest in this time, the acceleration is

$$a = \frac{\Delta v}{\Delta t} = \frac{0 - 28 \text{ m/s}}{0.054 \text{ s}} = -5.2 \times 10^2 \text{ m/s}^2$$

To get a feeling for the magnitude of this acceleration, we note that it is about 50 times the acceleration of gravity.

---

**Questions**

9. Two boys standing on a bridge throw rocks straight down into the water below. They throw two rocks at the same time, but one hits the water before the other. How can this be if the rocks have the same acceleration $g$?

10. A ball is thrown straight up. What is its velocity at the top of its flight? What is its acceleration at that point?

*Optional*

## 2-5  The Initial-Value Problem and Integration

We have learned how to obtain the velocity and acceleration functions from a given position function by differentiation. The inverse problem is to find the position function $x$ given the velocity $v$ or the acceleration $a$. The inverse procedure is called antidifferentiation or *integration*. The problem of integration is related to finding the area under a given curve; you will study this problem in the latter part of your calculus course. We shall look briefly at the problem from both the algebraic and graphic points of view.

Consider, for example, the problem of finding the velocity and position from a given constant acceleration. We already know the results from the previous section. Here we are interested in the mathematical relations. The velocity function is related to the acceleration by

$$a = \frac{dv}{dt}$$

We wish to find that function $v(t)$ which has a constant derivative. From our experience with differentiation we know that $v(t)$ must vary linearly with $t$. One such function is

$$v = at$$

This is not a general solution to our problem, however, because we can add any constant to the right side without changing the value of the derivative. Calling this constant $v_0$, we have for the general velocity function for constant acceleration

$$v = v_0 + at$$

The additive constant $v_0$ is just the initial velocity.

We find the position function from its relation to the velocity $v = dx/dt$. We therefore wish to find the function $x(t)$ whose derivative is $v_0 + at$. We can treat each term separately (rule 2, Appendix Table E-1).

The function whose derivative is $v_0$ is $v_0 t$ plus any constant. The function whose derivative is $at$ is slightly harder to find. We note from rule 7 in Table E-1 that the derivative of $t^n$ is $nt^{n-1}$. The function whose derivative is $at$ must therefore vary as $t^2$. We can easily verify that $\frac{1}{2}at^2$ (plus any constant) has the derivative $at$. Combining these results and writing $x_0$ for the combined arbitrary constants, we have for the position function

$$x = x_0 + v_0 t + \tfrac{1}{2}at^2$$

This is, of course, the same as Equation 2-10. Whenever we find a function from its derivative, we must include an arbitrary constant in the general function. Since we go through the antidifferentiation process twice to find $x(t)$ from the acceleration, two constants arise. These constants are usually determined from the velocity at some given time and the initial position at some given (and possibly different) time. We usually choose the given time to be $t = 0$. These constants are therefore called the *initial conditions*. The problem "given $a(t)$, find $x(t)$" is called the *initial-value problem*. The solution depends on the form of the function $a(t)$ *and* on the values of $v$ and $x$ at some particular time. This problem is particularly important in physics because the acceleration of a particle is determined by the forces acting on it. Thus if we know the forces acting on a particle and the position and velocity of the particle at some particular time, we can find its position uniquely at all other times.

The problem of integration is related to that of finding the area under a curve. Consider the case of constant velocity $v_0$. The change in position $\Delta x$ during some time interval $\Delta t$ is just the velocity times the time interval:

$$\Delta x = v_0\, \Delta t$$

This is the area under the $v$-versus-$t$ curve, as can be seen in Figure 2-10. This geometric interpretation of the displacement as the area under the $v$-versus-$t$ curve is quite general. Figure 2-11 shows a general $v$-versus-$t$ curve. The area under the curve is approximated by dividing the time into a number of small intervals $\Delta t_1$, $\Delta t_2$, ... and drawing a set of rectangular areas. We can make the approximation as good as we wish by taking enough rectangles and making each $\Delta t$ very small. The

*Initial-value problem*

**Figure 2-10**
The displacement for the interval $\Delta t$ equals the area under the velocity-versus-time curve for that interval. For $v(t) = v_0 = $ constant, the displacement equals the area of the rectangle shown.

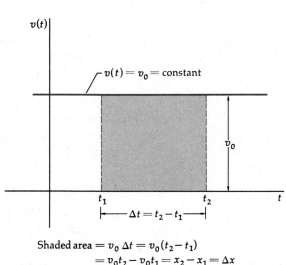

Shaded area $= v_0\, \Delta t = v_0(t_2 - t_1)$
$= v_0 t_2 - v_0 t_1 = x_2 - x_1 = \Delta x$

**Figure 2-11**
Plot of general $v(t)$ versus $t$. The displacement for the interval $\Delta t_i$ is approximately $v(t_i)\,\Delta t_i$, indicated by the shaded rectangular area. The total displacement from $t_1$ to $t_2$ is the area under the curve for this interval, which can be approximated by summing the areas of the rectangles.

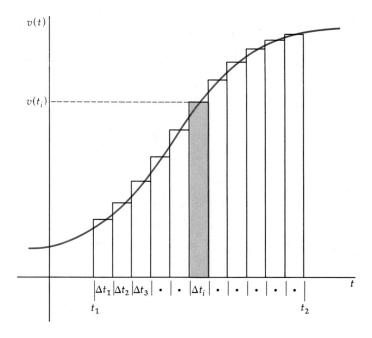

area of one particular rectangle is shaded. It has the value $v(t_i)\,\Delta t_i$, which is equal to the displacement $\Delta x_i$ of the particle during that interval. The sum of the rectangular areas is therefore the sum of the displacements during the time intervals and is equal to the total displacement from time $t_1$ to $t_2$. The area under the curve is therefore equal to the total displacement. Similarly, the area under the $a$-versus-$t$ curve equals the total change in the velocity for the particular time interval concerned.

The average velocity has a simple geometric interpretation in terms of area under a curve. Consider the $v$-versus-$t$ curve in Figure 2-12. The displacement $\Delta x$ during the time interval $\Delta t = t_2 - t_1$ is indicated by the shaded area. By definition of the average velocity for this interval (Equation 2-2), the displacement is also the product of $v_{\mathrm{av}}$ and $\Delta t$:

$$\Delta x = v_{\mathrm{av}}\,\Delta t$$

The average velocity is indicated in Figure 2-12 by the horizontal line, drawn so that the area under it from $t_1$ to $t_2$ equals the area under the actual $v$-versus-$t$ curve.

**Figure 2-12**
Geometric interpretation of average velocity. By definition, $v_{\mathrm{av}} = \Delta x/\Delta t$. Thus the rectangular area $v_{\mathrm{av}}(t_2 - t_1)$ must equal the displacement in the time interval $t_2 - t_1$. It follows that the rectangular area $v_{\mathrm{av}}(t_2 - t_1)$ and the shaded area under the curve must be equal.

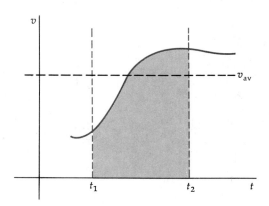

*Optional*

## 2-6   The Differential

An important quantity related to the derivative and useful in making approximations is the differential. Consider the position function we have studied several times:

$$x = 5t^2$$

The distance traveled between times $t_1$ and $t_2 = t_1 + \Delta t$ is

$$\Delta x = x(t_1 + \Delta t) - x(t_1) = 5(t_1 + \Delta t)^2 - 5t_1^2$$
$$= 10t_1 \Delta t + 5(\Delta t)^2 \qquad \text{2-13}$$

The first term in the equation is $\Delta t$ times the derivative of $x$ with respect to $t$. This quantity is called the *differential of x*, written $dx$,

*Differential defined*

$$dx = \left(\lim_{\Delta t \to 0} \frac{\Delta x}{\Delta t}\right) \Delta t \qquad \text{2-14}$$

Equation 2-13 is then

$$\Delta x = dx + 5(\Delta t)^2$$

If the time interval $\Delta t$ is very small, the second term in this equation is small compared with the first because $(\Delta t)^2$ will be small compared with $\Delta t$. Thus, to a good approximation

$$\Delta x \approx dx$$

This approximation is useful because it is often easier to calculate $dx$ than $\Delta x$ when $\Delta x$ is wanted.

Figure 2-13 illustrates the difference between the actual change $\Delta x$ in $x$ and its approximation, the differential $dx$. This approximation is essentially that of replacing the curve $x(t)$ by the straight line tangent to the curve at $t_1$. The validity of the approximation obviously depends on how small $\Delta t$ is and on how much the curve $x(t)$ deviates from a straight line in the range $t_1$ to $t_1 + \Delta t$. For the curve $x(t) = Ct$, the differential $dx = C \Delta t$ equals $\Delta x$ for all $\Delta t$.

It is customary to write $dt$ for the change $\Delta t$ in the independent variable $t$. Then the differential $dx$ is just the derivative of $x$ with respect to $t$ times $dt$. This is the origin of the traditional notation for the derivative we have been using:

$$\frac{dx}{dt} = \lim_{\Delta t \to 0} \frac{\Delta x}{\Delta t}$$

**Figure 2-13**
Comparison of the differential approximation $dx$ with the actual change $\Delta x$ for the interval $t_1$ to $t_2$. The differential $dx$ is the change in $x$ which would result if the slope of $x(t)$ versus $t$ did not change in the interval $t_1$ to $t_2$. The approximation is good if the time interval is small.

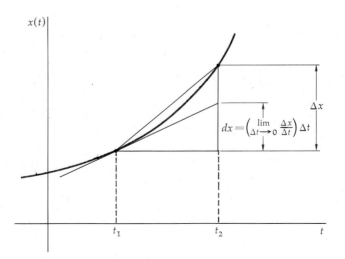

We may regard $dx/dt$ as the single quantity, the derivative of $x$ with respect to $t$, as we have been doing up to now, or we may regard it as the ratio of the differentials $dx$ and $dt$.

---

**Example 2-13** For a falling body whose position is given by $x = (5 \text{ m/s}^2)t^2$, find the distance fallen during the 0.5-s time interval between $t_1 = 10$ s and $t_2 = 10.5$ s.

The derivative of this function is just the velocity

$$v = \frac{dx}{dt} = (10 \text{ m/s}^2)t$$

and the differential $dx$ is

$$dx = (10 \text{ m/s}^2)t \, dt$$

Substituting $t = 10$ s and $dt = 0.5$ s gives

$$dx = (10 \text{ m/s}^2)(10 \text{ s})(0.5 \text{ s}) = 50 \text{ m}$$

This is the distance the particle would fall during 0.5 s if its velocity were equal to 100 m/s, the value of the instantaneous velocity at $t = 10$ s. Thus the differential approximation here neglects the fact that the velocity increases slightly during the 0.5-s interval from $t = 10$ s to $t = 10.5$ s. Since the velocity at $t = 10.5$ is $10(10.5) = 105$ m/s, the average velocity during this interval is 102.5 m/s. The actual distance the body falls during this interval is 51.25 m. This result is greater by 1.25 m than that found using the differential approximation, an error of about $2\frac{1}{2}$ percent.

---

## Review

A. Define, explain, or otherwise identify:

Particle, 23
Displacement, 24
Speed, 24
Average velocity, 24
Instantaneous velocity, 27
Derivative, 28

Slope, 28
Average acceleration, 30
Instantaneous acceleration, 30
Initial-value problem, 38
Differential, 40

B. True or false: If the statement is true, explain; if it is false, give a counterexample, i.e., a known example which contradicts the statement.

1. The equation $x = x_0 + v_0 t + \frac{1}{2}a_0 t^2$ holds for all motion in one dimension.

2. If the acceleration is zero, the particle cannot be moving.

3. The average velocity always equals the mean value of the initial and final velocities.

4. The displacement always equals the product of the average velocity and the time interval.

## Exercises

### Section 2-1, Speed, Displacement, and Velocity

1. A runner runs 2 km in 5 min and then takes 10 min to walk back to the starting point. (*a*) What is the average velocity for the first 5 min? (*b*) What is the average velocity for the time spent walking? (*c*) What is the average velocity for the total trip? (*d*) What is the average speed for the total trip?

2. Work Exercise 1 if the runner walked only halfway back in 10 min and then stopped.

3. A particle is at $x = +5$ m at $t = 0$, $x = -7$ m at $t = 6$ s, and $x = +2$ m at $t = 10$ s. Find the average velocity of the particle during the intervals (a) $t = 0$ to $t = 6$ s; (b) $t = 6$ s to $t = 10$ s; (c) $t = 0$ to $t = 10$ s.

4. A driver begins a 200-km trip at noon. (a) She drives nonstop and arrives at her destination at 5:30 P.M. Calculate her average velocity for the trip. (b) She drives for 3 h, rests for $\frac{1}{2}$ h, and continues driving, arriving at 5:30 P.M. Calculate her average velocity. (c) After resting 2 h, she drives back home, taking 6 h for the return trip. What is her average velocity for the total round trip? (d) What is her displacement?

5. A car travels in a straight line with average velocity of 80 km/h for 2.5 h and then with average velocity of 40 km/h for 1.5 h. (a) What is the total displacement for the 4-h trip? (b) What is the average velocity for the total trip?

6. Figure 2-14 shows the position of a particle versus time. Find the average velocity for the time intervals $a$, $b$, $c$, and $d$ indicated.

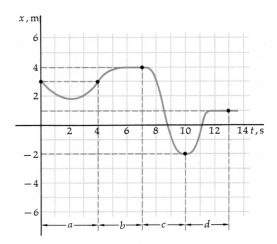

**Figure 2-14**
$x$-versus-$t$ curve for Exercise 6.

### Section 2-2, Instantaneous Velocity

7. For each of the four graphs of $x$ versus $t$ in Figure 2-15 indicate whether the velocity at time $t_2$ is greater than, less than, or equal to the velocity at time $t_1$. Compare the speeds at these times.

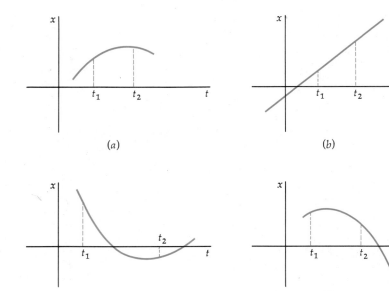

**Figure 2-15**
$x$-versus-$t$ curves for Exercise 7.

8. In the graph of $x$ versus $t$ in Figure 2-16, ($a$) find the average velocity between the times $t = 0$ and $t = 2$ s. ($b$) Find the instantaneous velocity at $t = 2$ s by measuring the slope of the tangent line indicated.

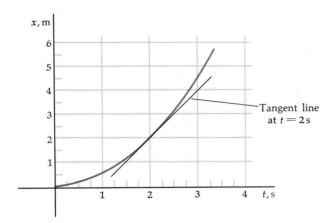

**Figure 2-16**
$x$-versus-$t$ curve for Exercise 8 with tangent line drawn at $t = 2$ s.

9. For the graph of $x$ versus $t$ in Figure 2-17 find the average velocity for the time intervals $\Delta t = t_2 - 0.75$ s when $t_2$ is 1.75, 1.5, 1.25, and 1.0 s. What is the instantaneous velocity at $t = 0.75$ s?

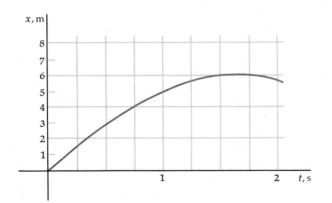

**Figure 2-17**
$x$-versus-$t$ curve for Exercise 9.

10. For the graph of $x$ versus $t$ shown in Figure 2-18, ($a$) find the average velocity for the interval $t = 1$ s to $t = 5$ s. ($b$) Find the instantaneous velocity at $t = 4$ s. ($c$) At what time is the velocity of the particle zero?

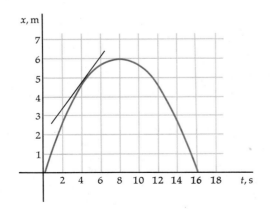

**Figure 2-18**
$x$-versus-$t$ curve for Exercise 10 with tangent line drawn at $t = 4$ s.

11. The position of a particle depends on the time according to $x = (1 \text{ m/s}^2)t^2 - (5 \text{ m/s})t + 1 \text{ m}$. (a) Find the displacement and average velocity for the interval $t = 3$ s to $t = 4$ s. (b) Find a general formula for the displacement for the time interval from $t$ to $t + \Delta t$. (c) Find the instantaneous velocity for any time $t$.

12. The height of a certain projectile is related to time by $y = -5(t - 5)^2 + 125$, where $y$ is in metres and $t$ in seconds. (a) Sketch $y$ versus $t$ for $t = 0$ to $t = 10$ s. (b) Find the average velocity for each of the 1-s time intervals between integral time values from $t = 0$ to $t = 10$ s. Sketch $v_{av}$ versus $t$. (c) Find the instantaneous velocity as a function of time.

### Section 2-3, Acceleration

13. A fast car can accelerate from 0 to 90 km/h in 5 s. What is the average acceleration during this interval? What is the ratio of this acceleration to the free-fall acceleration of gravity?

14. A car is traveling at 45 km/h at time $t = 0$. It accelerates at a constant rate of 10 km/h·s. (a) How fast is the car going at $t = 1$ s? At $t = 2$ s? (b) What is its speed at a general time $t$?

15. At $t = 5$ s an object is traveling at 5 m/s. At $t = 8$ s its velocity is $-1$ m/s. Find the average acceleration for this interval.

16. A particle moves with velocity given by $v = 8t - 7$, where $v$ is in metres per second and $t$ in seconds. (a) Find the average acceleration for the 1-s intervals beginning at $t = 3$ s and $t = 4$ s. (b) Sketch $v$ versus $t$. What is the instantaneous acceleration at any time?

17. State whether the acceleration is positive, negative, or zero for each of the *position* functions $x(t)$ in Figure 2-19.

18. The position of a particle versus time is given by

| $t$, s | 0 | 1 | 2 | 3 | 4 | 5 | 6 | 7 | 8 | 9 | 10 | 11 |
|---|---|---|---|---|---|---|---|---|---|---|---|---|
| $x$, m | 0 | 5 | 15 | 45 | 65 | 70 | 60 | −30 | −50 | −50 | −55 | −55 |

Plot $x$ versus $t$ and draw a smooth curve $x(t)$. Indicate the times or time intervals for which (a) the velocity is greatest; (b) the velocity is least; (c) the velocity is zero; (d) the velocity is constant; (e) the acceleration is positive; (f) the acceleration is negative.

19. The position of an object is related to time by $x = At^2 - Bt + C$, where $A = 8 \text{ m/s}^2$, $B = 6 \text{ m/s}$, and $C = 4 \text{ m}$. Find the instantaneous velocity and acceleration as functions of time.

20. The velocity of a particle under the influence of viscous forces (such as air resistance) is given by $v = Ae^{-bt}$, where $A$ and $b$ are constants. (a) What is the physical significance of the constant $A$? What are the dimensions of $b$? (b) Show that the acceleration is proportional to the velocity. (See rule 10 in Table E-1.)

### Section 2-4, Motion with Constant Acceleration

21. A car accelerates from rest at a constant rate of 8 m/s². (a) How fast is it going after 10 s? (b) How far has it gone after 10 s? (c) What is its average velocity for the interval $t = 0$ to $t = 10$ s?

22. An object with initial velocity of 5 m/s has a constant acceleration of 2 m/s². When its speed is 15 m/s, how far has it traveled?

23. An object with constant acceleration has velocity $v = 10$ m/s when it is at $x = 6$ m and $v = 15$ m/s when it is at $x = 10$ m. What is its acceleration?

24. An object has constant acceleration $a = 4$ m/s². Its velocity is 1 m/s at $t = 0$, when it is at $x = 7$ m. How fast is it moving when it is at $x = 8$ m? At what time is this?

**Figure 2-19**
$x$-versus-$t$ curves for Exercise 17.

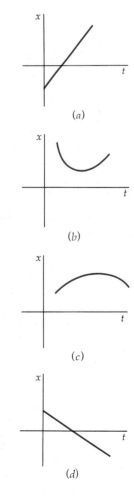

(a)

(b)

(c)

(d)

25. How long does it take for a particle to travel 100 m if it begins from rest and accelerates at 10 m/s²? What is the velocity when it has traveled 100 m? What is the average velocity for this time?

26. A ball is thrown upward with initial velocity of 20 m/s. Its acceleration is constant, equal to 10 m/s² downward. (*a*) How long is the ball in the air? (*b*) What is the greatest height reached by the ball? (*c*) When is the ball 15 m above the ground?

27. A ball is dropped from a height of 3 m and rebounds from the floor to a height of 2 m. (*a*) What is the velocity of the ball just as it reaches the floor? (*b*) What is the velocity just as it leaves the floor? (*c*) If it is in contact with the floor for 0.02 s, what are the magnitude and direction of its average acceleration during this interval? (Use $g \approx 10$ m/s² for the acceleration of gravity.)

28. The minimum distance for a controlled stop with no wheels locked for a certain car is 50 m for level braking from 95 km/h. Find the acceleration, assuming it to be constant, and express your answer as a fraction of the free-fall acceleration of gravity. How long does it take to stop?

### Section 2-5, The Initial-Value Problem and Integration

29. The velocity of a particle is given by $v = 6$ m/s for all time $t$. Find the most general position function $x(t)$.

30. The velocity of a particle in metres per second is given by $v = 7t + 5$, where $t$ is in seconds. Find the most general position function $x(t)$.

31. The acceleration of a certain rocket is given by $a = Ct$, where $C$ is a constant. (*a*) Find the most general position function $x(t)$. (*b*) Find the position and velocity at $t = 5$ s if $x = 0$ and $v = 0$ at $t = 0$, and $C = 3$ m/s³.

32. The velocity of a particle is given by $v = 6t + 3$, where $t$ is in seconds and $v$ in metres per second. (*a*) Sketch $v(t)$ versus $t$ and find the area under the curve for the interval $t = 0$ to $t = 5$ s. (*b*) Find the most general position function $x(t)$. Use it to calculate the displacement during the interval $t = 0$ to $t = 5$ s.

33. The velocity of a particle in metres per second is given by $v = 7 - 4t$, where $t$ is in seconds. (*a*) Sketch $v(t)$ versus $t$ and find the area between the curve and the $t$ axis from $t = 2$ s to $t = 6$ s. (*b*) Find the position function $x(t)$ by integration and use it to find the displacement during the interval $t = 2$ s to $t = 6$ s. (*c*) What is the average velocity for this interval?

34. Figure 2-20 shows the velocity of a particle versus time. (*a*) What is the magnitude in metres of the area of the rectangle indicated? (*b*) Find the displacement of the particle for the 1-s intervals beginning at $t = 1$ s and $t = 2$ s. (*c*) What is the average velocity for the interval from $t = 1$ s to $t = 3$ s?

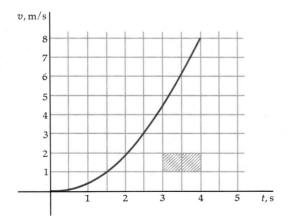

**Figure 2-20**
$v$-versus-$t$ curve for
Exercise 34.

35. Figure 2-21 shows the acceleration of a particle versus time. (*a*) What is the magnitude of the area of the rectangle indicated? (*b*) The particle starts from rest at $t = 0$. Find the velocity at $t = 1, 2,$ and 3 s by counting the squares under the curve. (*c*) Sketch the curve $v(t)$ versus $t$ from your results of part (*b*) and estimate how far the particle traveled in the interval $t = 0$ to $t = 3$ s.

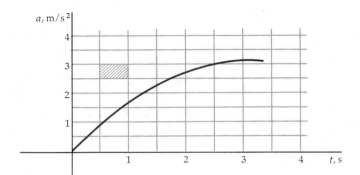

**Figure 2-21**
*a*-versus-*t* curve for Exercise 35.

36. The equation of the curve shown in Figure 2-20 is $v = 0.5t^2$ m/s. Find the displacement of the particle for the interval $t = 1$ s to $t = 3$ s by integration and compare with your answer for Exercise 34. Is the average velocity equal to the mean of the initial and final velocities for this case?

**Section 2-6, The Differential**

37. The distance a body falls from rest in a time $t$ is $x = (5 \text{ m/s}^2)t^2$. (*a*) Find the exact distance the body falls from time $t_1 = 20$ s to time $t_2 = 20.3$ s. (*b*) Find the velocity at time $t = 20$ s and use the differential approximation to find the distance fallen during this interval.

38. The area of a circle of radius $r$ is $A = \pi r^2$. Use the differential approximation to find the change in area $\Delta A$ for a small change in the radius $\Delta r$. Interpret the quantity $dA$ geometrically.

39. Show that the fractional change in area of a circle is approximately related to the fractional change in radius by $\Delta A/A = 2\Delta r/r$. If the radius changes by 1 percent, what is the percentage change in the area?

40. The volume of a sphere of radius $r$ is $V = (4\pi/3)r^3$. (*a*) Find the approximate small change in the volume $dV$ for a small change in radius $dr$ and interpret $dV$ geometrically. (*b*) Show that the fractional changes are related by $dV/V = 3dr/r$. (*c*) The volume of a sphere is determined by measuring its radius. If a 2 percent error is made in this measurement, what is the resulting percentage error in the volume?

**Problems**

1. A car traveling at a constant speed of 20 m/s passes an intersection at time $t = 0$, and 5 s later another car passes the same intersection traveling 30 m/s in the same direction. (*a*) Sketch the position functions $x_1(t)$ and $x_2(t)$ for the two cars. (*b*) Find when the second car overtakes the first. (*c*) How far have the cars traveled when this happens?

2. Hare and Tortoise begin a 10-km race at time $t = 0$. Hare runs at 4 m/s and quickly outdistances Tortoise, who runs at 1 m/s (about 10 times faster than a tortoise can actually run but convenient for this problem). After running 5 min, Hare stops and falls asleep. His nap lasts 135 min. He awakes and again runs at 4 m/s but loses the race. Plot an $x$-versus-$t$ graph for each on the same axes. At what time does Tortoise pass Hare? How far behind is Hare when Tortoise crosses the finish line? How long can Hare nap and still win the race?

3. A particle moves with constant acceleration of 3 m/s². At $t = 4$ s it is at $x = 100$ m. At $t = 6$ s it has a velocity $v = 15$ m/s. Find its position at $t = 6$ s.

4. Figure 2-22 shows the position of a car plotted as a function of time. At which of the times $t_0$ to $t_7$ is (a) the velocity negative, (b) the velocity positive, (c) the velocity zero, (d) the acceleration negative, (e) the acceleration positive, (f) the acceleration zero?

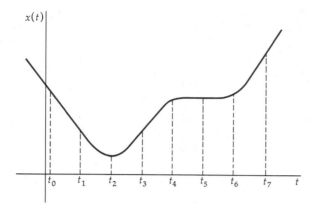

**Figure 2-22**
Position of a car versus time for Problem 4.

5. Sketch a single $v$-versus-$t$ curve in which there are points or segments for which (a) the acceleration is zero and constant while the velocity is not zero; (b) the acceleration is zero but not constant; (c) the velocity and acceleration are both positive; (d) the velocity and acceleration are both negative; (e) the velocity is positive and the acceleration negative; (f) the velocity is negative and the acceleration positive; (g) the velocity is zero, but the acceleration is not.

6. For each of the graphs in Figure 2-23 indicate (a) at what times the acceleration of the object is positive, negative, and zero, (b) at what times the acceleration is constant, (c) at what times the instantaneous velocity is zero.

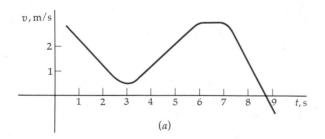

**Figure 2-23**
(a) $v$-versus-$t$ curve and (b) $x$-versus-$t$ curve for Problem 6.

7. A car is traveling 80 km/h in a school zone. A police car starts from rest just as the speeder passes it and accelerates at a constant rate of 8 km/h·s. (a) Make a sketch of $x(t)$ for both cars. (b) When does the police car catch the speeding car? (c) How fast is the police car traveling when it catches the speeder?

8. The police car in Problem 7 chases a speeder traveling 125 km/h. The maximum speed of the police car is 190 km/h. The police car travels from rest with

constant acceleration, 8 km/h·s, until its speed reaches 190 km/h. It then moves at constant speed. (*a*) When does the police car catch the speeder if it starts just as the speeder passes? (*b*) How far has each car traveled? (*c*) Sketch $x(t)$ for each car.

9. When the police car of Problem 8 traveling at 190 km/h is 100 m behind the speeder (traveling 125 km/h), the speeder sees the car and slams on his brakes, locking the wheels. (*a*) Assuming that each car can brake at 6 m/s² and that the driver of the police car brakes as soon as he sees the brake lights of the speeder, i.e., with no reaction time, show that the cars collide. (*b*) What is the speed of the police car relative to the speeder when they collide? At what time do they collide after the brakes are applied? (*c*) The interval between the time the police-man sees the brake lights and puts on his own brakes is called his *reaction time* $T$. Estimate or measure your reaction time and discuss how it affects this problem.

10. When a car traveling at speed $v_1$ rounds a corner, the driver sees another car traveling at a slower speed $v_2$ a distance $d$ ahead. (*a*) If the maximum accelera-tion her brakes can provide is $a$, show that the distance $d$ must be greater than $(v_1 - v_2)^2/2a$ if a collision is to be avoided. (*b*) Evaluate this distance for $v_1 = 90$ km/h, $v_2 = 45$ km/h, and $a = 6$ m/s². (*c*) Estimate or measure your reaction time and calculate its effect on the distance found in part (*b*).

11. A passenger is running at his maximum velocity of 8 m/s to catch a train. When he is a distance $d$ from the nearest entry to the train, the train starts from rest with constant acceleration $a = 1.0$ m/s² away from the passenger. (*a*) If $d = 30$ m and he keeps running, will he be able to jump onto the train? (*b*) Sketch the position function $x(t)$ for the train, choosing $x = 0$ at $t = 0$. On the same graph sketch $x(t)$ for the passenger for various values of initial separation distances $d$, including $d = 30$ m and the critical value $d_c$ such that he just catches the train. (*c*) For the critical separation distance $d_c$, what is the speed of the train when the passenger catches it? What is its average speed for the time interval from $t = 0$ until he catches it? What is the value of $d_c$?

12. A load of bricks is being lifted by a crane at the steady velocity of 5 m/s when one brick falls off 6 m above the ground. Describe the motion of the free brick by sketching $x(t)$. (*a*) What is the greatest height the brick reaches above the ground? (*b*) How long does it take to reach the ground? (*c*) What is its speed just before it hits the ground?

13. A train starts from a station with a constant acceleration of 0.40 m/s². A pas-senger arrives at the station 6.0 s after the end of the train left the very same point. What is the least constant speed at which she can run and catch the train? Sketch curves for the motion of the passenger and train as functions of time.

14. A ball is dropped from the top of a building. At the same instant a ball is thrown vertically upward from the ground level at a speed of 9.0 m/s. The balls collide 1.8 s later. How high is the building?

15. A glider moves along a slightly inclined air track with constant acceleration $a$. It is projected from the bottom ($x = 0$) with initial speed $v_0$. At time $t = 8$ s it is at $x = 100$ cm moving down the track at speed $v = -15$ cm/s. Find the initial speed $v_0$ and the acceleration $a$.

16. The faculty resident of a dormitory sees an illegal water-filled balloon fall past her window. She quickly runs to the window and times a second water balloon which takes 0.10 s to fall the 1.2-m height of the window. If the water balloon was dropped from rest, how high above the bottom of the window was it released?

17. Ball $A$ is dropped from the top of a building at the same instant that ball $B$ is thrown vertically upward from the ground. When the balls collide, they are moving in opposite directions and the speed of $A$ is twice the speed of $B$. At what fraction of the height of the building did the collision occur?

18. Solve Problem 17 if the collision occurs when the balls are moving in the same direction and the speed of $A$ is 4 times that of $B$.

19. Figure 2-24 is a graph of $v$ versus $t$ for a particle moving along a straight line. The position of the particle at time $t = 0$ is $x_0 = 5$ m. (a) Find $x$ for various times $t$ by counting squares and sketch $x$ versus $t$. (b) Sketch the acceleration $a$ versus $t$.

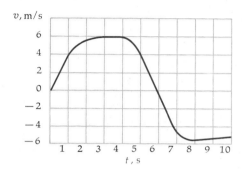

**Figure 2-24**
$v$-versus-$t$ curve for Problem 19.

20. Figure 2-25 shows a plot of $x$ versus $t$ for a body moving along a straight line. Sketch rough graphs of $v$ versus $t$ and $a$ versus $t$ for this motion.

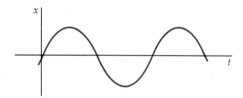

**Figure 2-25**
$x$-versus-$t$ curve for Problem 20.

21. The position of a body oscillating on a spring is given by $x = A \sin \omega t$, where $A$ and $\omega$ are constants and have the values $A = 5$ cm and $\omega = 10°/s = 0.175$ rad/s. Sketch $x$ versus $t$ for times from $t = 0$ to $t = 36$ s. (a) Measure the slope of your graph at $t = 0$ to find the velocity at this time. (b) Using sine tables, calculate the average velocity for a series of intervals beginning at $t = 0$ and ending at $t = 6, 3, 2, 1, 0.5$, and 0.25 s. (c) Compute $dx/dt$ and find the velocity at time $t = 0$.

22. A car has maximum acceleration $a$, which remains constant to high speeds, and it has a maximum deceleration $2a$. The car has to travel a short distance $L$, starting and ending at rest, in the minimum time $T$. (The distance is so short that the car can never reach top speed.) After what fraction of $L$ should the driver move her foot from the gas pedal to the brake, and what fraction of the time for the trip has elapsed at this point?

23. Table 2-2 lists some world track records for short sprints. A simple model of sprinting assumes that a sprinter starts from rest, accelerates with constant acceleration $a$ for a short time $T$, and then runs with constant speed $v_0 = aT$. According to this model, for times $t$ greater than $T$, the distance $x$ varies linearly with time. (a) Make a graph of distance $x$ versus time $t$ from the data in the table. (b) Set up an equation for $x$ versus $t$ according to the simple model described and show that for $t > T$, $x$ can be written $x = v_0(t - \frac{1}{2}T)$. (c) Connect the points on your graph with a straight line and determine the slope and the intercept of the line with the time axis. From the fact that the slope is $v_0$ and the intercept is $\frac{1}{2}T$, compute the acceleration $a$. (d) The record for $x = 200$ m is 19.5 s. Discuss the applicability of this simple model to races of 200 m or more.

**Table 2-2**

| yd | m | $t$, s |
|---|---|---|
| 50 | | 5.1 |
| | 50 | 5.5 |
| 60 | | 5.9 |
| | 60 | 6.5 |
| 100 | | 9.1 |
| | 100 | 9.9 |

24. A Sprint ABM missile can accelerate at $100g$. If an ICBM is detected at an altitude of 100 km moving straight down at a constant speed of $3 \times 10^4$ km/h and the ABM missile is launched to intercept it, at what time and altitude will the interception take place? *Note:* You can neglect the acceleration of gravity in this problem; why?

25. The acceleration of a particle falling under the influence of gravity and a resistive force such as air resistance is given by

$$a = \frac{dv}{dt} = g - Bv$$

where $g$ is the free-fall acceleration of gravity and $B$ is a constant which depends on the mass and shape of the particle and on the medium. Suppose the particle begins with zero velocity at time $t = 0$. (*a*) Discuss qualitatively how the speed $v$ varies with time from your knowledge of the rate of change $dv/dt$ given by this equation. What is the value of the velocity when the acceleration is zero? This is called the *terminal velocity*. (*b*) Sketch the solution $v(t)$ versus $t$ without solving the equation. This can be done as follows. At $t = 0$, $v$ is zero and the slope is $g$. Sketch a straight-line segment, neglecting any change in slope for a short time interval. At the end of the interval the velocity is not zero, and so the slope is less than $g$. Sketch another straight-line segment with a smaller slope. Continue until the slope is zero and the velocity equals the terminal velocity.

26. (*a*) Show that the equation in Problem 25 can be written $df/f = -B\ dt$, where $f = g - Bv$. (*b*) Find the indefinite integral of each side of this equation and solve for $v$, using the fact that $\exp(\ln f) = f$. Find the constant of integration in terms of $g$ and $B$ from the initial condition $v = 0$ at $t = 0$. (*c*) Write your solution in the form $v = v_t(1 - e^{-t/T})$, where $v_t$ is the terminal velocity and $T = 1/B$. (*d*) Plot the points for $t = 0, 1, 3, 6, 10,$ and 20 s for $B = 0.1$ s$^{-1}$.

27. Show that if $y = Cx^n$, where $C$ and $n$ are any constants, the differential $dy$ is related to $dx$ by

$$\frac{dy}{y} = n\frac{dx}{x}$$

# CHAPTER 3

# Motion in Two and Three Dimensions

---

**Objectives**   After studying this chapter you should:

1. Be able to add and subtract vectors using the parallelogram law of addition.

2. Be able to obtain the rectangular components of vectors and use them in addition and subtraction.

3. Know that in projectile motion the horizontal and vertical motions are independent and be able to use this fact in working projectile problems.

4. Know what is meant by the *range* of a projectile and be able to derive an expression for it in terms of the initial speed and angle of projection.

5. Know that for motion along any curve the acceleration of a particle can in general be resolved into a component $dv/dt$ tangential to the curve and a component $v^2/r$ perpendicular to the tangent to the curve.

---

We now extend our description of the motion of a particle to the more general cases of motion in two and three dimensions. For this more general motion, displacement, velocity, and acceleration are vectors, which have direction in space as well as magnitude. In this chapter we investigate the properties of vectors in general and of these three vectors in particular. We also discuss two important special cases of motion in a plane, projectile motion and circular motion.

When the motion of a particle is confined to a plane, its position can be described by two numbers. For example, we might choose the distance $x$ from the $y$ axis and the distance $y$ from the $x$ axis, where the $x$ and $y$ axes are perpendicular axes which intersect at origin $O$, as in Figure 3-1. Alternatively, we can specify the same position by giving the distance $r$ from the origin, and the angle $\theta$ made by the $x$ axis and the line from the origin to the point. There are of course other choices, but in each case two numbers are needed to specify the location of a point on a plane.

If the particle is not confined to a plane but moves in three dimensions, three numbers are needed to specify its position. A convenient

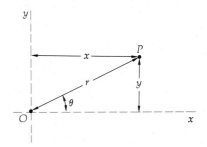

**Figure 3-1**
Coordinates of a point $P$ in a plane. The rectangular coordinates $(x, y)$ and polar coordinates $(r, \theta)$ are related by
$x = r \cos \theta$
$y = r \sin \theta$
$r = \sqrt{x^2 + y^2}$
$\tan \theta = y/x$

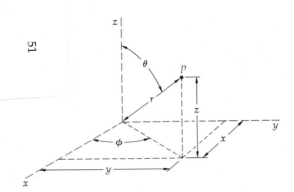

**Figure 3-2**
Coordinates of a point $P$ in three dimensions. The rectangular coordinates $(x, y, z)$ and spherical coordinates $(r, \theta, \phi)$ are related by
$$x = r \sin \theta \cos \phi$$
$$y = r \sin \theta \sin \phi$$
$$z = r \cos \theta$$
$$r = \sqrt{x^2 + y^2 + z^2}$$
$$\tan \theta = \sqrt{x^2 + y^2}/z$$
$$\tan \phi = y/x$$

method is to use the three coordinates $x$, $y$, and $z$. An alternative is to use the spherical coordinates $r$, $\theta$, and $\phi$. The relations between spherical and rectangular coordinates are shown in Figure 3-2. Spherical coordinates are convenient when there is spherical symmetry. For example, if a particle moves on the surface of a sphere, these coordinates are useful because $r$ is constant and only $\theta$ and $\phi$ vary.

Most of the interesting features of motion in more than one dimension can be brought out in two-dimensional motion. Since this motion is easily illustrated on paper or a blackboard, most of our examples will be limited to two dimensions, with an indication of the extension to three-dimensional motion.

## 3-1 The Displacement Vector

Consider a particle which moves from some point $P_1$ to another $P_2$. As in one-dimensional motion, we call the change in position of a particle its *displacement*. In two or three dimensions we indicate the displacement by an arrow from the original position $P_1$ to the new position $P_2$, as in Figure 3-3. We see that the displacement has direction as well as magnitude. We should note also that the displacement as we have defined it does not depend on the path taken by the particle as it moves from $P_1$ to $P_2$ but only on the endpoints $P_1$ and $P_2$. A second displacement is indicated in Figure 3-3 by an arrow from point $P_2$ to $P_3$. The arrow from the original point $P_1$ to $P_3$ indicates the *resultant* displacement from the original position $P_1$ to the last position $P_3$; it is the sum of the two successive displacements. Quantities which have magnitude

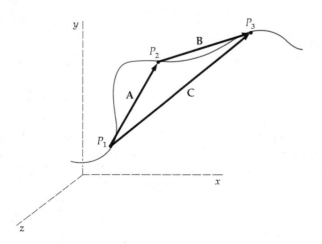

**Figure 3-3**
The addition of vectors. The displacement **C** is equivalent to the successive displacements **A** and **B**; that is, **C** = **A** + **B**.

and direction *and* which obey the addition law illustrated by these displacements are called *vectors*. Many quantities in physics besides displacements are vectors, e.g., velocity, acceleration, force, momentum, and electric field. We denote vector quantities by boldface type, as in **A**, **B**, and **C** for the three displacements in Figure 3-3. (In handwriting, we indicate a vector by drawing an arrow over the symbol.) The addition law for vectors illustrated in this figure is simply

*Vectors*

$$\mathbf{C} = \mathbf{A} + \mathbf{B} \qquad\qquad 3\text{-}1$$

The *magnitude*, or length, of the vector **A** is written $|\mathbf{A}|$ or $A$. The magnitude of a vector ordinarily has physical units, like speed or acceleration, discussed in Chapter 2. A displacement vector, for example, has a magnitude which can be expressed in feet or metres or any other measure of distance. Note that the sum of the magnitudes of **A** and **B** does not equal the magnitude of $\mathbf{C} = \mathbf{A} + \mathbf{B}$ unless **A** and **B** are in the same direction:

$$C \neq A + B \qquad \text{unless } \mathbf{A} \text{ and } \mathbf{B} \text{ are in the same direction}$$

Quantities which are completely specified by only a number (with units), e.g., distance, mass, or temperature, are called *scalars*.

**Questions**

1. Can the displacement of a particle in any time interval have a magnitude which is less than the distance traveled by the particle along its path in that interval? Can its magnitude be more than the distance traveled? Explain.

2. Give an example in which the distance traveled is a significant amount yet the corresponding displacement is zero.

3. Is displacement always in the direction of motion?

4. If $A = 7$ m and $B = 3$ m, what are the greatest and least values possible for $C$ if $\mathbf{C} = \mathbf{A} + \mathbf{B}$?

## 3-2   Components of a Vector

The projection of a vector on a line is called a *component* of the vector in the direction of that line. An important example of vector components is the projections of the vector on the axes of a rectangular coordinate system. These projections, called the *rectangular components* of the vector, are denoted by $A_x$, $A_y$, and $A_z$ for the vector **A**. Figure 3-4 illustrates the rectangular components of the displacements **A**, **B**, and **C** of Figure 3-3. We denote the coordinates of $P_1$ by $(x_1, y_1)$ and use similar designations for $P_2$ and $P_3$; then the components of these vectors are

*Rectangular components*

$$A_x = x_2 - x_1 \qquad A_y = y_2 - y_1$$
$$B_x = x_3 - x_2 \qquad B_y = y_3 - y_2 \qquad\qquad 3\text{-}2$$
$$C_x = x_3 - x_1 \qquad C_y = y_3 - y_1$$

It should be clear from the figure that

$\mathbf{C} = \mathbf{A} + \mathbf{B}$ implies both

$$C_x = A_x + B_x \text{ and } C_y = A_y + B_y \qquad\qquad 3\text{-}3$$

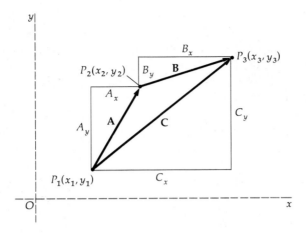

**Figure 3-4**
The $x$ and $y$ components of vectors **A**, **B**, and **C** from Figure 3-3.
$A_x = x_2 - x_1$
$A_y = y_2 - y_1$
$B_x = x_3 - x_2$
$B_y = y_3 - y_2$
$C_x = x_3 - x_1 = A_x + B_x$
$C_y = y_3 - y_1 = A_y + B_y$

The magnitude of a vector can be written in terms of the vector's rectangular components. As shown in Figure 3-4, the rectangular components $A_x$ and $A_y$ form the two legs of a right triangle whose hypotenuse is the magnitude $A$. Hence, from the theorem of Pythagoras,

$$A = \sqrt{A_x^2 + A_y^2} \qquad\qquad 3\text{-}4$$

We can also specify the direction in terms of the rectangular components. If $\theta$ is the angle between the vector and a line parallel to the $x$ axis,

$$\tan \theta = \frac{A_y}{A_x} \qquad\qquad 3\text{-}5$$

We can write a vector conveniently in terms of its rectangular components by making use of *unit vectors*. A unit vector is a dimensionless vector which has the magnitude of 1 and points in any convenient direction. For example, let **i**, **j**, and **k** be unit vectors which point in the $x$, $y$, and $z$ directions, respectively. A general vector **A** can then be written as the sum of three vectors each parallel to a coordinate:

*Unit vectors*

$$\mathbf{A} = A_x\mathbf{i} + A_y\mathbf{j} + A_z\mathbf{k} \qquad\qquad 3\text{-}6$$

The vector $A_x\mathbf{i}$ is the product of the component $A_x$ and the unit vector **i**. It is a vector parallel to the $x$ axis with magnitude $A_x$. The vector sum indicated in Equation 3-6 is illustrated in Figure 3-5. The addition of two vectors **A** and **B** can be written in terms of these unit vectors as

$$\begin{aligned}\mathbf{A} + \mathbf{B} &= (A_x\mathbf{i} + A_y\mathbf{j} + A_z\mathbf{k}) + (B_x\mathbf{i} + B_y\mathbf{j} + B_z\mathbf{k}) \\ &= (A_x + B_x)\mathbf{i} + (A_y + B_y)\mathbf{j} + (A_z + B_z)\mathbf{k} \qquad 3\text{-}7\end{aligned}$$

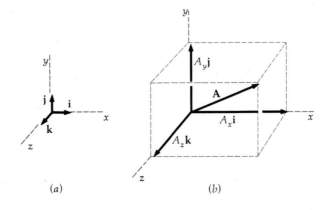

(a)                    (b)

**Figure 3-5**
(*a*) The unit vectors **i**, **j**, and **k** in a rectangular coordinate system. (*b*) The vector **A** can be written in terms of the unit vectors **i**, **j**, **k**; that is, $\mathbf{A} = A_x\mathbf{i} + A_y\mathbf{j} + A_z\mathbf{k}$.

**Example 3-1** A man walks 4.00 km northeast and then 3.00 km south-east. If the origin of coordinates is taken at his starting point and the $x$ axis points straight east, (*a*) what are the rectangular components of his two displacements? (*b*) What are the rectangular components of his overall displacement? (*c*) What are his final distance and direction from the starting point?

(*a*) The successive displacements are shown in Figure 3-6. Displacement **A** has magnitude 4.00 km and is 45° north of east. Displacement **B** has magnitude 3.00 km and is 45° south of east. The components of **A** are

$$A_x = A \cos 45° = (4.00 \text{ km})(0.707) = 2.83 \text{ km}$$

$$A_y = A \sin 45° = (4.00 \text{ km})(0.707) = 2.83 \text{ km}$$

The components of **B** are

$$B_x = B \cos 45° = (3.00 \text{ km})(0.707) = 2.12 \text{ km}$$

$$B_y = -B \sin 45° = -(3.00 \text{ km})(0.707) = -2.12 \text{ km}$$

The component $B_y$ is negative, indicating that it is in the direction of decreasing, or negative, values of $y$ rather than that of increasing, or positive, values of $y$.

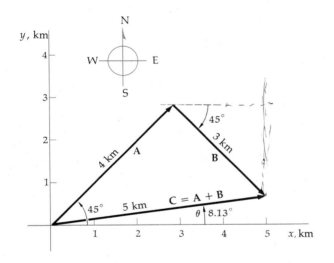

**Figure 3-6**
A plot of the displacements from Example 3-1. North is plotted in the $+y$ direction; east in the $+x$ direction.

(*b*) The components of the resultant displacement **C** = **A** + **B** are calculated as follows:

$$C_x = A_x + B_x = 2.83 \text{ km} + 2.12 \text{ km} = 4.95 \text{ km}$$

$$C_y = A_y + B_y = 2.83 \text{ km} - 2.12 \text{ km} = 0.71 \text{ km}$$

(*c*) The man's final distance from his starting point is the magnitude of **C**:

$$C = \sqrt{C_x^2 + C_y^2} = \sqrt{(4.95)^2 + (0.71)^2} = 5.0 \text{ km}$$

The direction of the man's final position can be specified by giving the angle $\theta$:

$$\tan \theta = \frac{C_y}{C_x} = \frac{0.71}{4.95} = 0.143$$

or

$$\theta = 8.13°$$

That is, the direction of the net displacement is 8.13° north of east. In terms of the unit vectors this resultant displacement can be written

$$\mathbf{C} = (4.95 \text{ km})\mathbf{i} + (0.71 \text{ km})\mathbf{j}$$

---

### Questions

5. Can a component of a vector have a magnitude greater than the magnitude of the vector? Under what circumstances can a component of a vector have a magnitude equal to the magnitude of the vector?

6. Can a vector be equal to zero and still have one or more components not equal to zero?

7. Are the components of $\mathbf{C} = \mathbf{A} + \mathbf{B}$ necessarily larger than the corresponding components of either $\mathbf{A}$ or $\mathbf{B}$?

## 3-3   Properties of Vectors

This section summarizes some important properties of vectors.

### Equality

Two vectors $\mathbf{A}$ and $\mathbf{B}$ are defined to be equal if they have the same magnitude *and* the same direction:

$\mathbf{A} = \mathbf{B}$     if $A = B$ *and* their directions are the same          3-8

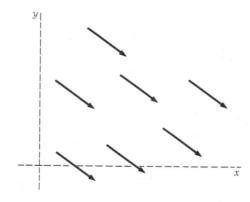

**Figure 3-7**
Vectors are equal if their magnitudes and directions are the same. The vectors in this figure are all equal.

Note that we do not require the endpoints of the vectors to be the same for them to be equal. All the vectors in Figure 3-7 are equal. Thus a vector is completely specified by its magnitude and direction. Since these properties can be written in terms of the rectangular components of the vector, we can also specify the vector completely by giving its rectangular components. Two vectors are equal if their corresponding rectangular components are equal:

$\mathbf{A} = \mathbf{B}$     if $A_x = B_x$, $A_y = B_y$, and $A_z = B_z$          3-9

### Vector Addition

The addition of two vectors $\mathbf{A}$ and $\mathbf{B}$ is defined geometrically as in Figure 3-8. Here, the tail of vector $\mathbf{B}$ is placed at the head of vector $\mathbf{A}$. The sum $\mathbf{C} = \mathbf{A} + \mathbf{B}$ is the vector from the tail of $\mathbf{A}$ to the head of $\mathbf{B}$, as shown.

**Figure 3-8**
Addition of two vectors. The vector $\mathbf{C} = \mathbf{A} + \mathbf{B}$ is drawn from the tail of $\mathbf{A}$ to the head of $\mathbf{B}$ when the tail of $\mathbf{B}$ is placed at the head of $\mathbf{A}$.

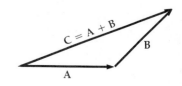

An alternative construction for finding the sum of two vectors is to put the tails of the vectors together and construct a parallelogram, as in Figure 3-9. The resultant vector is along the diagonal of the parallelogram. The vector-addition law is sometimes referred to as the *parallelogram law of addition*. Figure 3-9 shows that this construction gives the same result as Figure 3-8, that is, putting the tail of **B** at the head of **A** and drawing **C = A + B** from the tail of **A** to the head of **B**.

By geometric construction we can see that the sum of the two vectors is independent of the order of the vectors. This is the commutative law of addition:

$$\mathbf{A} + \mathbf{B} = \mathbf{B} + \mathbf{A} \qquad\qquad 3\text{-}10$$

The sum of three or more vectors is independent of the order in which they are added. This *associative law of addition*,

$$(\mathbf{A} + \mathbf{B}) + \mathbf{C} = \mathbf{A} + (\mathbf{B} + \mathbf{C}) \qquad\qquad 3\text{-}11$$

is illustrated for three vectors in Figure 3-10.

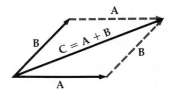

**Figure 3-9**
Parallelogram method of addition of vectors. The vector **C = A + B** is along the diagonal of the parallelogram formed by **A** and **B**. The result is the same as that in Figure 3-8. It can be seen from this figure that **A + B** is the same as **B + A**.

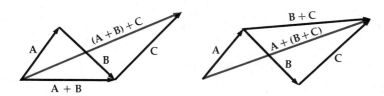

**Figure 3-10**
Vector addition is associative. **(A + B) + C = A + (B + C)**.

### Negative of a Vector

The vector $-\mathbf{A}$ is defined to be the vector which when added to **A** gives zero:

$$\mathbf{A} + (-\mathbf{A}) = 0$$

The vector $-\mathbf{A}$ has the same magnitude as **A** but points in the opposite direction (sometimes stated by saying that $-\mathbf{A}$ is *antiparallel to* **A**). The components of $-\mathbf{A}$ are the negatives of the components of **A**. is, if **A** has components $(A_x, A_y, A_z)$, then $-\mathbf{A}$ has components $(-A_x, -A_y, -A_z)$.

### Subtraction of Vectors

The definition of the negative of a vector allows us to define vector subtraction in a simple way. **B** subtracted from **A** is defined to be $-\mathbf{B}$ added to **A**:

$$\mathbf{A} - \mathbf{B} = \mathbf{A} + (-\mathbf{B}) \qquad\qquad 3\text{-}12$$

Vector subtraction is illustrated in Figure 3-11. It is often useful to think of $\mathbf{A} - \mathbf{B}$ as the vector which is added to vector **B** to obtain vector **A**. This idea is illustrated in Figure 3-11b, where the vectors **A** and **B** are drawn tail to tail. Then the vector $\mathbf{A} - \mathbf{B}$ is drawn from the tip of **B** to the tip of **A**.

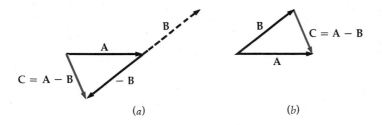

(a) (b)

**Figure 3-11**
Subtraction of vectors. (*a*) **C = A − B** is found by adding $-\mathbf{B}$ to **A**. (*b*) An alternative method of finding **C = A − B**. Here **C** is the vector which when added to **B** gives the vector **A**.

### Multiplication of a Vector by a Scalar

If we multiply a vector $\mathbf{A}$ by a positive scalar $c$, we obtain the vector $c\mathbf{A}$, which has the same direction as $\mathbf{A}$ and magnitude $cA$.* We used this definition when we expressed a vector in terms of its rectangular components and the unit vectors $\mathbf{i}$, $\mathbf{j}$, and $\mathbf{k}$. Note that this definition is consistent with the result that when we add two equal vectors $\mathbf{A} + \mathbf{A}$, we obtain a vector parallel to $\mathbf{A}$ with twice the magnitude, which is $2A$.

### Questions

8. Can two vectors of unequal magnitudes be added to give zero? *No*

*may are 2 just diff direction*

9. Are the magnitudes of $\mathbf{A}$ and $-\mathbf{A}$ equal, or do they have opposite signs? What about the components of $\mathbf{A}$ and $-\mathbf{A}$? — *opp. sign*

10. Does the magnitude of a vector change if the units in which it is expressed change? Does the magnitude of $c\mathbf{A}$ have the same units as $\mathbf{A}$? *no*

## 3-4   The Velocity Vector

We are now ready to extend our concepts of velocity and acceleration to the general motion of a particle. Consider a particle moving along a curve in two dimensions, as in Figure 3-12. At some time $t_1$ it is at point $P_1$, and at a later time $t_2$ it is at point $P_2$. We can describe the position of the particle by a vector $\mathbf{r}$ from the origin to the position of the particle. This *position vector*, or *radius vector* as it is sometimes called, is special in that we specify in its definition the location of its tail. From the figure we see that the displacement vector is the difference of the two position vectors $\mathbf{r}_2 - \mathbf{r}_1$. By analogy with our one-dimensional notation, in which we wrote $\Delta\mathbf{r}$ for a displacement in two (or three) dimensions,

$$\Delta\mathbf{r} = \mathbf{r}_2 - \mathbf{r}_1 \qquad\qquad 3\text{-}13$$

*Position vector*

*Displacement vector*

The new position vector $\mathbf{r}_2$ is thus the sum of the original position vector $\mathbf{r}_1$ and the displacement vector $\Delta\mathbf{r}$.

We define the average velocity vector as the ratio of the displacement vector $\Delta\mathbf{r}$ and the time interval for the displacement $\Delta t = t_2 - t_1$:

$$\mathbf{v}_{av} = \frac{\Delta\mathbf{r}}{\Delta t} \qquad\qquad 3\text{-}14$$

*Average-velocity vector*

* If $c$ is a negative scalar, the vector $c\mathbf{A}$ has the magnitude $|c|A$ and is antiparallel to $\mathbf{A}$.

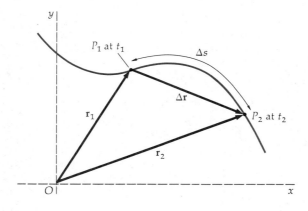

**Figure 3-12**
The position, or radius, vector $\mathbf{r}$ is the vector from the origin to the position of the particle. The displacement vector $\Delta\mathbf{r}$ for the time interval from $t_1$ to $t_2$ is then the difference in the position vectors $\Delta\mathbf{r} = \mathbf{r}_2 - \mathbf{r}_1$.

We note from Figure 3-12 that the magnitude of the displacement vector is *not* equal to the distance of travel $\Delta s$ as measured along the curve. It is, in fact, less than this distance (unless the particle travels in a straight line between points $P_1$ and $P_2$). However, if we consider smaller and smaller time intervals, as indicated in Figure 3-13, the magnitude of the displacement approaches the distance traveled by the particle along the curve and the direction of $\Delta \mathbf{r}$ approaches the direction of the line tangent to the curve at point $P_1$. We define the instantaneous-velocity vector as the limit of the average velocity as the time interval $\Delta t$ approaches zero:

$$\mathbf{v} = \lim_{\Delta t \to 0} \frac{\Delta \mathbf{r}}{\Delta t} = \frac{d\mathbf{r}}{dt} \qquad \qquad 3\text{-}15$$

*Instantaneous-velocity vector*

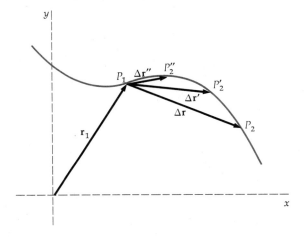

**Figure 3-13**
As smaller and smaller time intervals are considered, the magnitude of the displacement vector approaches the distance traveled along the curve, and the direction of the displacement vector approaches the direction of the line tangent to the curve at point $P_1$.

The instantaneous velocity is thus the derivative of the position vector with respect to time. The direction of the instantaneous velocity is along the line tangent to the curve traveled by the particle in space. The magnitude of the instantaneous velocity is the *speed $ds/dt$*, where $s$ is the distance measured along the curve. The position vector $\mathbf{r}$ can change in magnitude, direction, or both. If the particle moves along a radial line through the origin, as in Figure 3-14, $\mathbf{r}$ changes in magnitude only and the velocity vector is parallel to $\mathbf{r}$. Our discussion of one-dimensional motion in Chapter 2 applies for this special case.

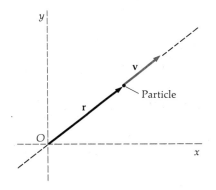

**Figure 3-14**
A particle moving along a radial line has a position vector which does not change in direction but may change in magnitude.

Figure 3-15 shows a particle moving around the origin in a circle of radius $r$. In this case, the position vector has a constant magnitude but is changing in direction. The velocity vector for this motion is tangent to the circle and therefore perpendicular to the radius vector. In the general case illustrated in Figure 3-12 the radius vector is changing in both magnitude and direction.

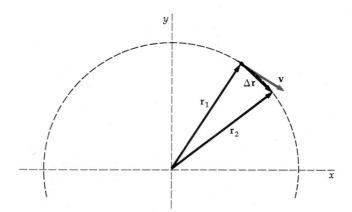

**Figure 3-15**
A particle moving in a circle has a position vector which does not change in magnitude but does change in direction. The velocity vector is tangent to the circle and perpendicular to the radius vector.

---

**Example 3-2** A sailboat has initial coordinates $(x_1, y_1) = (100 \text{ m}, 200 \text{ m})$; 2.00 min later, it has the coordinates $(x_2, y_2) = (120 \text{ m}, 210 \text{ m})$. What are the components, magnitude, and direction of its average velocity for this 2.00-min interval?

$$v_{x,\text{av}} = \frac{x_2 - x_1}{\Delta t} = \frac{120 - 100 \text{ m}}{2.00 \text{ min}} = 10.0 \text{ m/min}$$

$$v_{y,\text{av}} = \frac{y_2 - y_1}{\Delta t} = \frac{210 - 200 \text{ m}}{2.00 \text{ min}} = 5.0 \text{ m/min}$$

$$v_{\text{av}} = \sqrt{(v_{x,\text{av}})^2 + (v_{y,\text{av}})^2}$$
$$= \sqrt{10.0^2 + 5.0^2}$$
$$= \sqrt{125} = 11.2 \text{ m/min}$$

$$\tan \theta = \frac{v_{y,\text{av}}}{v_{x,\text{av}}} = \frac{5.0 \text{ m/min}}{10.0 \text{ m/min}} = 0.500$$

$$\theta = 26.6°$$

---

**Questions**

11. For an arbitrary motion of a given particle does the direction of the velocity vector have any particular relation to the direction of the position vector?

12. Give examples in which the directions of the velocity and position vectors are (*a*) opposite, (*b*) the same, (*c*) mutually perpendicular.

13. Can the velocity vector change direction without changing magnitude? If so, give an example.

## 3-5   The Acceleration Vector

The *average-acceleration vector* is defined as the ratio of the change in the instantaneous-velocity vector $\Delta\mathbf{v}$ and the time interval $\Delta t$:

$$\mathbf{a}_{av} = \frac{\Delta\mathbf{v}}{\Delta t} = \frac{\mathbf{v}_2 - \mathbf{v}_1}{\Delta t} \qquad\qquad 3\text{-}16$$

*Average-acceleration vector*

The *instantaneous-acceleration vector* is defined as the derivative of the velocity vector with respect to time:

$$\mathbf{a} = \lim_{\Delta t \to 0}\frac{\Delta\mathbf{v}}{\Delta t} = \frac{d\mathbf{v}}{dt} \qquad\qquad 3\text{-}17$$

*Instantaneous-acceleration vector*

It is particularly important to note that the velocity vector may be changing in magnitude, direction, or both. If the velocity vector is changing in any way, the particle is accelerating according to our definition. We are perhaps most familiar with acceleration in which the velocity vector changes in magnitude. In this case the speed changes (since the speed is just the magnitude of the velocity vector). However, a particle can be traveling with constant speed and still be accelerating if the direction of the velocity vector is changing. (A particularly important example of this is circular motion, discussed in Section 3-7.) This acceleration is just as real as when the speed is changing. We shall see in the next chapter that a force is necessary to produce an acceleration of a particle. The force required to produce a given acceleration (say $1\ \mathrm{m/s^2}$ downward) on a given particle is the same whether this acceleration is associated with a change in the magnitude of the velocity vector, a change in its direction, or both.

### Questions

14. How is it possible for a particle moving at constant speed to be accelerating? Can a particle with constant velocity be accelerating at the same time?

15. Is it possible for a particle to round a curve without accelerating?

16. Show that the acceleration of a particle always points toward the concave side of its path.

17. Describe a circumstance in which a particle's velocity is horizontal and its acceleration is vertical.

## 3-6   Motion with Constant Acceleration: Projectile Motion

An important special case of motion in two or three dimensions occurs when the acceleration is constant in both magnitude and direction. An example of motion with constant acceleration is that of a projectile near the surface of the earth if air resistance can be neglected.

Let $\mathbf{a}$ be the instantaneous-acceleration vector, which is constant. The equations for the velocity and position vectors are simple generalizations of the one-dimensional constant-acceleration equations 2-9 to 2-11. They are

$$\mathbf{v} = \mathbf{v}_0 + \mathbf{a}t \qquad\qquad 3\text{-}18$$

where $\mathbf{v}_0$ is the initial velocity,

$$\mathbf{r} = \mathbf{r}_0 + \mathbf{v}_0 t + \tfrac{1}{2}\mathbf{a}t^2 \qquad\qquad 3\text{-}19$$

where $\mathbf{r}_0$ is the initial position vector, and

$$\mathbf{v}_{av} = \tfrac{1}{2}(\mathbf{v}_0 + \mathbf{v}) \qquad\qquad 3\text{-}20$$

Figure 3-16 shows the relation between the displacement vector $\mathbf{r} - \mathbf{r}_0$, the initial velocity $\mathbf{v}_0$, and the acceleration $\mathbf{a}$. The displacement at any time $t$ is in the plane formed by the vectors $\mathbf{v}_0$ and $\mathbf{a}$, as can be seen from the figure. The motion is therefore two-dimensional.

Let us choose the $z$ axis to be perpendicular to the plane of $\mathbf{v}_0$ and $\mathbf{a}_0$ and assume that the initial position of the particle is in the $xy$ plane. The motion is then in the $xy$ plane. The $x$ and $y$ components of Equations 3-18 and 3-19 are

$$v_x = v_{0x} + a_x t$$
$$x = x_0 + v_{0x}t + \tfrac{1}{2}a_x t^2 \qquad\qquad 3\text{-}21$$

and

$$v_y = v_{0y} + a_y t$$
$$y = y_0 + v_{0y}t + \tfrac{1}{2}a_y t^2 \qquad\qquad 3\text{-}22$$

When the acceleration is constant, the motion takes place in a plane and the $x$ and $y$ motions can be described separately by equations identical to those for motion in one dimension with constant acceleration.

Let us apply these results to the motion of a projectile, i.e., any body launched into the air and allowed to move freely. The general motion of a projectile is complicated by air resistance, the rotation of the earth, and the variation in the acceleration of gravity. For simplicity, we shall neglect these complications. The projectile then has a constant acceleration directed vertically downward with magnitude $g = 9.81$ m/s² $= 32.2$ ft/s². If we take the $y$ axis as vertical with the positive direction upward and the $x$ axis horizontal in the direction of the original horizontal component of the projectile's velocity, we have for the acceleration

$$a_y = -g \qquad a_x = 0 \qquad\qquad 3\text{-}23$$

Since there is no horizontal acceleration, the horizontal component of the velocity is constant. On the other hand, the vertical motion is simply motion with constant acceleration, identical to that studied in Chapter 2. If we choose the origin at the initial position of the projectile, the components of velocity and position are given by

$$v_x = v_{0x} \qquad v_y = v_{0y} - gt \qquad\qquad 3\text{-}24$$

and

$$x = v_{0x}t \qquad\qquad 3\text{-}25$$

$$y = v_{0y}t - \tfrac{1}{2}gt^2 \qquad\qquad 3\text{-}26$$

**Figure 3-16**
The displacement vector $\mathbf{r} - \mathbf{r}_0$ is the sum of $\mathbf{v}_0 t$ and $\tfrac{1}{2}\mathbf{a}t^2$. Since the displacement is in the plane formed by the vectors $\mathbf{v}_0$ and $\mathbf{a}$, the motion is confined to this plane and is therefore two-dimensional.

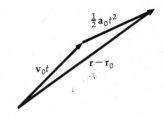

*Equations for projectile motion*

If the initial velocity vector $\mathbf{v}_0$ makes an angle $\theta_0$ with the horizontal axis, the initial velocity components are as shown in Figure 3-17:

$$v_{0x} = v_0 \cos \theta_0 \qquad v_{0y} = v_0 \sin \theta_0$$

**Figure 3-17**
The components of the initial velocity of a projectile are $v_{0x} = v_0 \cos \theta_0$ and $v_{0y} = v_0 \sin \theta_0$, where $\theta_0$ is the angle made by $v_0$ and the horizontal $x$ axis.

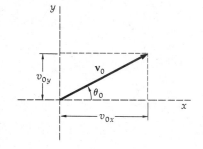

**Example 3-3** A ball is thrown into the air with initial velocity of 50 m/s at 37° to the horizontal. Find the total time the ball is in the air and the total horizontal distance using the approximation $g = 10$ m/s². 

The components of the initial velocity vector are

$$v_{0x} = (50 \text{ m/s}) \cos 37° = 40 \text{ m/s}$$
$$v_{0y} = (50 \text{ m/s}) \sin 37° = 30 \text{ m/s}$$

3-27

Figure 3-18 shows the height $y$ versus $t$ for this example. This curve is identical to Figure 2-8 for Example 2-8 because the vertical acceleration and velocities are the same for these two examples. Since the projectile moves 40 m horizontally during each second, we can interpret this curve as a sketch of $y$ versus $x$ if we change the horizontal axis from a time scale to a distance scale by multiplying the time values by 40 m/s.

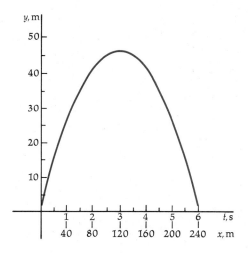

**Figure 3-18**
Plot of $y(t)$ and $y(x)$ for Example 3-3. The time scale can be converted into a horizontal distance scale by multiplying each time by 40 m/s because $x$ is related to $t$ by $x = (40 \text{ m/s})t$.

The curve $y$ versus $x$ is a parabola. The total time the projectile is in the air is twice the time it takes to reach its highest point. We find this time from

$$v_y = v_{0y} - gt = 30 \text{ m/s} - (10 \text{ m/s}^2)t$$

and solving for the time $t$ when $v_y$ is zero. The result is

$$t = \frac{30 \text{ m/s}}{10 \text{ m/s}^2} = 3.0 \text{ s}$$

This is, of course, the same result found in Example 2-8. The total time the projectile is in the air is then 6 s. Since it moves horizontally with the constant velocity of 40 m/s, the total distance traveled is 40 m/s × 6 s = 240 m. This distance is called the *range* of the projectile.

We can apply these methods to find the range $R$ for a general $v_0$ and $\theta_0$. The time to reach the highest point is found by setting $v_y = 0$ in Equation 3-24:

$$t = \frac{v_{0y}}{g}$$

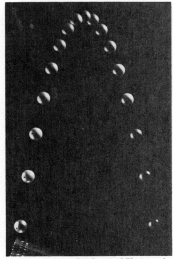

Fundamental Photographs

Multiflash photograph of a ball thrown into the air. The position of the ball is recorded at approximately 0.43-sec intervals.

Illuminated fountains, St. Louis, Missouri, showing parabolic paths of water.

Ralph Krubner/Black Star

The range is then the horizontal distance traveled in twice this time:

$$R = v_{0x} \frac{2v_{0y}}{g} = \frac{2v_{0x}v_{0y}}{g} \qquad \text{3-28}a$$

In terms of the initial speed $v_0$ and angle $\theta_0$ the range is

$$R = \frac{2(v_0 \cos \theta_0)(v_0 \sin \theta_0)}{g} = \frac{v_0^2}{g}(2 \sin \theta_0 \cos \theta_0) \text{ or}$$

$$R = \frac{v_0^2}{g} \sin 2\theta_0 \qquad \text{3-28}b \qquad \textit{Range of a projectile}$$

where we have used the trigonometric identity $2 \cos \theta_0 \sin \theta_0 = \sin 2\theta_0$.

Since the maximum value of $\sin 2\theta_0$ is 1 when $2\theta_0 = 90°$ or $\theta_0 = 45°$, the range is a maximum $v_0^2/g$ when $\theta_0 = 45°$. We note that the horizontal distance traveled is the product of the initial horizontal velocity component $v_{0x}$ and the time the projectile is in the air, which in turn is proportional to $v_{0y}$. The maximum range occurs when these components are equal. In some practical applications, other considerations are important. For example, in the shot put, the ball is projected from an initial height of about 6 or 7 ft rather than from the ground level. This initial height has the effect of increasing the time the ball is in the air. The range is maximum when $v_{0x}$ is somewhat greater than $v_{0y}$, that is, at an angle somewhat smaller than 45°. Studies of the form of the best shot putters show that maximum range occurs at an initial angle of about 42°. With artillery shells, air resistance must be taken into account to predict the range accurately. As expected, air resistance reduces the range for a given angle of projection. It also decreases the optimum angle of projection slightly. An interesting effect due to the rotation of the earth is important for long-range ballistic missiles. The motion does not take place entirely in the plane formed by the vertical and the initial velocity. Instead, there is a slight drift to the right in the northern hemisphere and to the left in the southern hemisphere. It is due to the *Coriolis effect*, which arises because the surface of the earth is accelerating as a consequence of the earth's rotation.

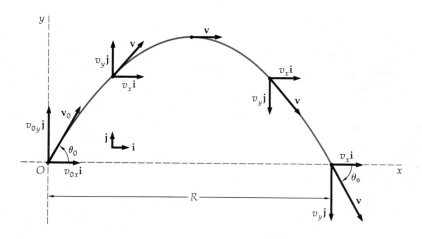

**Figure 3-19**
Path of a projectile with the velocity vector and its rectangular components indicated at several points. The total horizontal distance traveled is the range $R$.

The general equation for the path $y(x)$ can be obtained from Equations 3-25 and 3-26 by eliminating the variable $t$ between these equations. Choosing $x_0 = y_0 = 0$ and using $t = x/v_{0x}$ in Equation 3-26, we obtain

$$y = v_{0y}\frac{x}{v_{0x}} - \frac{1}{2}g\left(\frac{x}{v_{0x}}\right)^2$$

or

$$y = \frac{v_{0y}}{v_{0x}}x - \frac{1}{2}\frac{g}{v_{0x}^2}x^2 \qquad\qquad 3\text{-}29$$

This equation is of the same form as $y(t)$ in Equation 3-26. Figure 3-19 shows the path of a projectile with the velocity vector and its components indicated at several points.

If we view the motion of the projectile in a second coordinate system parallel to the first but moving with speed $v_{0x}$ relative to it, the motion is in one dimension. Let us call the origin of our second coordinate system $O'$ and the axes $x'$ and $y'$. If we let the two coordinate systems be coincident at $t = 0$, the coordinates are related by

$$y' = y \qquad x' = x - v_{0x}t \qquad\qquad 3\text{-}30$$

The equations for the projectile in this coordinate system are found from Equations 3-25 and 3-26. Choosing $x_0$ and $y_0$ to be zero, we have

$$y' = v_{0y}t - \tfrac{1}{2}gt^2 \qquad x' = 0 \qquad\qquad 3\text{-}31$$

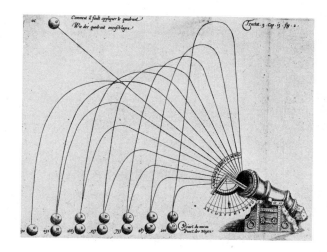

Path of cannonballs for various angles of projection, drawn by Diego Ufano in 1621. Which features of the motion are correct, and which are incorrect? (*Courtesy of the New York Public Library, Rare Book Division, Astor, Lenox, and Tilden Foundations.*)

In this coordinate system, the projectile moves in one dimension along the $y'$ axis.

According to our analysis of projectile motion, an object dropped from a height $h$ above the ground will hit the ground in the same time as one projected horizontally from the same height. In each case, the distance the object *falls* is given by $d = \frac{1}{2}gt^2$ (measuring $d$ downward from the original height). This remarkable fact can easily be demonstrated.

It was first commented upon during the Renaissance. Galileo Galilei (1564–1642), the first person to give the modern, quantitative description of projectile motion we have discussed, used this observation to illustrate the validity of treating the horizontal and vertical components of a projectile's motion as independent motions. Galileo wrote*

Portrait of Galileo.

> Let the ship be motionless and the fall of the stone from the mast take two pulse beats. Then cause the ship to move, and drop the same stone from the same place; from what has been said, it will take two pulse beats to arrive at the deck. In these two pulse beats, the ship will have gone, say, twenty yards so that the natural motion of the stone will have been a diagonal line much longer than the first straight and perpendicular one, which was merely the length of the mast; nevertheless, it will have traversed this distance in the same time. Now, assuming the ship to be speeded up still more, so that the stone in falling must follow a diagonal line very much longer still than the other, eventually the velocity of the ship may be increased by any amount while the falling rock will describe always longer and longer diagonals, and still pass over them in the same two pulse beats. Similarly, if a perfectly level cannon on a tower were fired parallel to the horizon, it would not matter whether a small charge or a great one was put in, so that the ball would fall a thousand yards away, four thousand, or six thousand, or ten thousand or more; all these shots would require equal times and each time would be equal to that which the ball would have taken in going from the mouth of the cannon to the ground if it were allowed to fall straight down without any other impulse.

* Galileo Galilei, *Dialogue Concerning Two Chief World Systems—Ptolemaic and Copernican,* pp. 154–155, trans. Stillman Drake, University of California Press, Berkeley, 1953; reprinted by permission of The Regents of the University of California.

---

**Example 3-4** A hunter with a blowgun wishes to shoot a monkey hanging from a branch. The hunter aims right at the monkey, not realizing that the dart will follow a parabolic path and thus fall below the monkey. The monkey, however, seeing the dart leave the gun, lets go of the branch and drops out of the tree, expecting to avoid the dart. Show that the monkey will be hit regardless of the initial velocity of the dart so long as it is great enough to travel the horizontal distance to the tree before hitting the ground.

This problem is often demonstrated using a target suspended by an electromagnet. When the dart leaves the gun, the circuit is broken and the target falls. Let the horizontal distance to the tree be $x$ and the original height of the monkey be $H$, as shown in Figure 3-20. Then the dart will be projected at an angle given by $\tan \theta = H/x$. If there were no gravity, the dart would reach the height $H$ in the time $t$ taken for it to travel the horizontal distance $x$:

$$y = v_{0y}t = H \qquad \text{in time } t = \frac{x}{v_{0x}} \qquad \text{with no gravity}$$

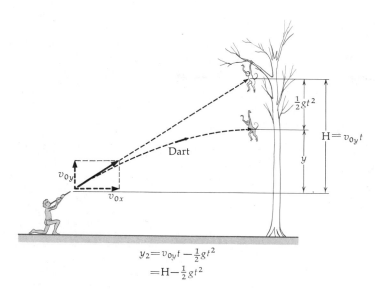

$$\frac{1}{2}gt^2$$

$$H = v_{0_y}t$$

Dart

$v_{0_y}$

$v_{0_x}$

$$y_2 = v_{0_y}t - \tfrac{1}{2}gt^2$$
$$= H - \tfrac{1}{2}gt^2$$

**Figure**
The m
Examp
the da
$v_{0_y}t -$
monke
$H - \frac{1}{2}$
as tha
$H = v_{0_y}t$. The dart will therefore always hit the monkey if
the monkey is released the
instant the dart is fired.

However, because of gravity, the dart has an acceleration vertically
down. In time $t = x/v_{0x}$, the dart reaches a height given by

$$y = v_{0_y}t - \tfrac{1}{2}gt^2 = H - \tfrac{1}{2}gt^2$$

This is lower than $H$ by $\tfrac{1}{2}gt^2$, which is just the amount the monkey falls
in this time. In the usual lecture demonstration, the initial velocity of
the dart is varied so that for large $v_0$ the target is hit very near its original height and for small $v_0$ it is hit just before it reaches the floor.

**Questions**

18. What is the acceleration of a projectile at the top of its flight?

19. Can the velocity of an object change direction while its acceleration
is constant in both magnitude and direction? If so, cite an example.

20. A projectile is launched horizontally at the top of a cliff. Describe
how the angle between its velocity and the acceleration varies as it falls.

## 3-7  Circular Motion

A particle moving in a circular path illustrates many of the important
features of the velocity and acceleration vectors in two dimensions. In
many natural phenomena the motion is circular or nearly circular, e.g.,
satellite motion or motion of the earth around the sun.

We first consider circular motion with constant speed. In everyday
usage, we would say that if the speed is constant, there is no acceleration. However, we have defined acceleration as the rate of change of the
*velocity vector*. Even when the speed is constant, a particle moving in a
circle is accelerating because the direction of the velocity is constantly
changing. As we have said, the reason for defining acceleration in this
way is that it takes a force to produce any acceleration whether it is associated with a change in speed or merely a change in the direction of
motion. A familiar example is rounding a curve in a car at high speed.

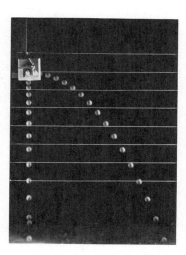

Comparison of a ball dropped
with one projected horizontally. The vertical position is
independent of the horizontal
motion. (*From PSSC Physics,
3d ed., p. 248, D. C. Heath and
Company, Lexington, Mass.,
1971.*)

rce needed to change the direction of the car's velocity is usually
vided by friction. If this force is not great enough to provide the
eded acceleration, the car will not make the curve but will move off in
its original direction. In this section we shall show that for circular motion at constant speed the acceleration is toward the center of the circle
and has the magnitude $v^2/r$, where $v$ is the speed and $r$ the radius of the
circle. Because the acceleration is toward the center of the circle, it is
called *centripetal acceleration*.

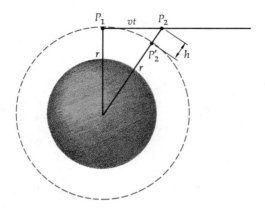

**Figure 3-21**
Satellite moving with speed $v$
in a circular orbit of radius $r$
about the earth. If the satellite
did not accelerate toward the
earth, it would move in a
straight line from point $P_1$ to
$P_2$ in time $t$. Because of its
acceleration, it falls a distance
$h$ in this time. For small $t$,
$h = \frac{1}{2}(v^2/r)t^2 = \frac{1}{2}at^2$.

Figure 3-21 shows a satellite moving in a circular orbit around the
earth. Why doesn't the satellite fall toward the earth? The answer is *not*
that there is no force of gravity on the satellite. At 200 km above the surface of the earth, the gravitational force on the satellite is about 94 percent of that when the satellite is at the earth's surface. The satellite *does*
"fall" toward the earth. But because of its tangential velocity, it continuously misses the earth. Consider Figure 3-21. If the satellite were not
accelerating, it would move from point $P_1$ to point $P_2$ in some time $t$. Instead, it arrives at point $P_2'$ on its circular orbit. Thus in a sense, the satellite "falls" the distance $h$ shown. If we take the time $t$ to be very small,
the points $P_2$ and $P_2'$ are approximately on a radial line, as shown in the
figure, and we can use the approximation that $h$ is much smaller than
orbit radius $r$. We can then calculate $h$ from the right triangle of sides $vt$,
$r$, and $r + h$:

$$(r + h)^2 = (vt)^2 + r^2$$

$$r^2 + 2hr + h^2 = v^2t^2 + r^2$$

or

$$h(2r + h) = v^2t^2$$

If we now neglect the $h$ in the parentheses compared with $2r$, we obtain

$$2rh \approx v^2t^2$$

or

$$h \approx \frac{1}{2}\left(\frac{v^2}{r}\right)t^2$$

Comparing this with the constant-acceleration expression $h = \frac{1}{2}at^2$, we
see that the magnitude of the acceleration of the satellite is

$$a = \frac{v^2}{r}$$

3-32   *Centripetal acceleration*

and the direction is inward toward the center of the circle.

We can show that this result holds in general for circular motion with constant speed by considering the position and velocity vectors (Figure 3-22). Here the initial velocity vector $\mathbf{v}_1$ is perpendicular to the initial position vector $\mathbf{r}_1$. A short time later the velocity is $\mathbf{v}_2$, which is perpendicular to $\mathbf{r}_2$. The angle between the velocity vectors $\Delta\theta$ is the same as that between the position vectors because the position and velocity vectors are mutually perpendicular. If we take the time interval to be very small, the magnitude of the displacement $|\Delta\mathbf{r}|$ is approximately equal to the distance traveled along the arc $\Delta s$. The average acceleration is the ratio of the velocity change $\Delta\mathbf{v} = \mathbf{v}_2 - \mathbf{v}_1$ and the time interval $\Delta t$.

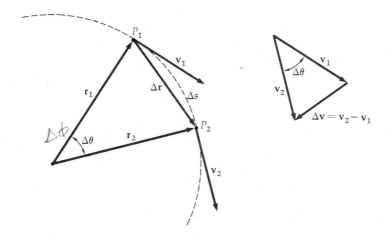

**Figure 3-22**
Calculation of the acceleration of a particle moving with constant speed in a circle of radius $r$. The position and velocity vectors are shown at times $t_1$ and $t_2$. Since $\mathbf{v}_1$ is perpendicular to $\mathbf{r}_1$ and $\mathbf{v}_2$ to $\mathbf{r}_2$, the angle between $\mathbf{v}_1$ and $\mathbf{v}_2$ is $\Delta\theta$, the same as between $\mathbf{r}_2$ and $\mathbf{r}_1$. For very short time intervals, the angle $\Delta\theta$ is very small and the change in velocity $\Delta\mathbf{v}$ is nearly perpendicular to $\mathbf{v}_1$ and points toward the center of the circle.

From the figure we see that for very small $\Delta t$, the velocity change (and therefore the average acceleration) is approximately perpendicular to the velocity vectors and directed toward the center of the circle. We can find the magnitude of the acceleration by expressing the angle $\Delta\theta$ in the similar triangles in Figure 3-22 in radians. We have

$$\Delta\theta = \frac{\Delta s}{r} \approx \frac{|\Delta\mathbf{v}|}{v}$$

If we now use the fact that the distance is given by $\Delta s = v\,\Delta t$, we have

$$\frac{v\,\Delta t}{r} \approx \frac{|\Delta\mathbf{v}|}{v} \qquad \text{or} \qquad \frac{|\Delta\mathbf{v}|}{\Delta t} \approx \frac{v^2}{r}$$

---

**Example 3-5** A satellite moves at constant speed in a circular orbit about the center of the earth and near the surface of the earth. If its acceleration is 9.81 m/s², what is its speed and how long does it take for one complete revolution?

This acceleration is the same as for any body falling freely near the surface of the earth. We take the radius of the earth, about 6370 km, to be the approximate radius of the orbit. (For actual satellites put into orbit a few hundred kilometres above the earth's surface, the orbit radius will be slightly greater, of course, and the acceleration will be slightly less than 9.81 m/s² because of the decrease in the gravitational force with distance from the center of the earth.) The speed of the satellite can be found from Equation 3-32:

$$v^2 = rg = (6370 \text{ km})(9.81 \text{ m/s})$$

$$v = 7.91 \text{ km/s}$$

The time for the satellite to make one complete revolution is called the period $T$. Since it travels a distance $2\pi r$ in this time $T$ at speed $v$, the period is

$$T = \frac{2\pi r}{v} = \frac{2\pi(6370 \text{ km})}{7.91 \text{ km/s}} = 5060 \text{ s} = 84.3 \text{ min}$$

Figure 3-23 is a drawing from Newton's *System of the World* illustrating the connection between projectile motion and satellite motion. Newton's description reads*

> That by means of centripetal forces the planets may be retained in certain . . . orbits, we may easily understand, if we consider the motions of projectiles; for a stone that is projected is by the pressure of its own weight forced out of the rectilinear path, which by the initial projection alone it should have pursued, and made to describe a curved line in the air; and through that crooked way is at last brought down to the ground; and the greater the velocity is with which it is projected, the farther it goes before it falls to the earth. We may therefore suppose the velocity to be so increased, that it would describe an arc of 1, 2, 5, 10, 100, 1000 miles before it arrived at the earth, till at last, exceeding the limits of the earth, it should pass into space without touching it.

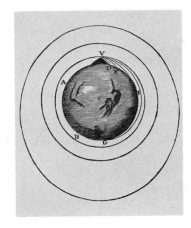

**Figure 3-23**
Drawing from Newton's *System of the World*, published in 1728, illustrating the connection between projectile motion and satellite motion. (*Courtesy of the Niels Bohr Library, American Institute of Physics.*)

**Example 3-6** A car rounds a curve of radius 100 m at a speed of 90 km/h. What is the magnitude of its centripetal acceleration?

To obtain the acceleration in metres per second squared, we need to change the speed to metres per second. Using the result of Example 1-1 that 90 km/h = 25 m/s, we have $v = 25$ m/s. The acceleration is then

$$a = \frac{v^2}{r} = \frac{(25 \text{ m/s})^2}{100 \text{ m}} = 6.25 \text{ m/s}^2$$

*Circular motion with varying speed*

If a particle moves in a circle with speed that is varying, there is a component of acceleration tangent to the circle as well as the centripetal acceleration inward. The tangential component of the acceleration is simply the rate of change of the speed $dv/dt$, whereas the radially inward component has the magnitude $v^2/r$. For any general motion along a curve we can treat a portion of the curve as an arc of a circle of some radius $r$ (Figure 3-24). The particle then has centripetal acceleration $v^2/r$ toward the center of curvature, and if the speed is changing, it has tangential acceleration of magnitude $dv/dt$.

* *System of the World,* trans. Andrew Motte, 1728, University of California Press, Berkeley, 1960; reprinted by permission of The Regents of the University of California.

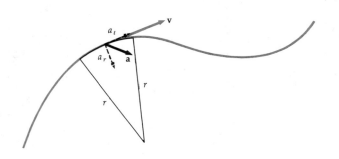

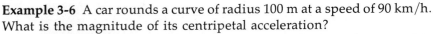

**Figure 3-24**
Particle moving along an arbitrary curve in space. A small segment of any curve in space can be considered to be the arc of a circle of radius $r$. The instantaneous-acceleration vector has a component $a_t$ of magnitude $dv/dt$ tangent to the curve and a component $a_r$ of magnitude $v^2/r$ toward the center of curvature of the arc.

## Review

A. Define, explain, or otherwise identify:

Displacement, 52
Vector, 53
Scalar, 53
Component of a vector, 53
Unit vector, 54
Vector equality, 54

Radius vector, 58
Position vector, 58
Range, 63
Centripetal acceleration, 68
Tangential acceleration, 70

B. True or false:

1. The instantaneous velocity vector is always in the direction of motion.

2. The instantaneous acceleration vector is always in the direction of motion.

3. If the speed is constant, the acceleration must be zero.

4. If the acceleration is zero, the speed must be constant.

5. The component of a vector is a vector.

6. The magnitude of the sum of two vectors must be greater than the magnitude of either vector.

7. If a vector is zero, each of its rectangular components must be zero.

8. It is impossible to go around a curve without acceleration.

9. The time required for a bullet fired horizontally to reach the ground is the same as if it were dropped from rest from the same height.

## Exercises

### Section 3-1, The Displacement Vector

1. A bear walks northeast for 10 m and then east for 10 m. Show each displacement graphically and find the resultant displacement vector.

2. (a) A man walks along a circular arc from the position $x = 5$ m, $y = 0$ to a final position $x = 0$, $y = 5$ m. What is his displacement? (b) A second man walks from the same initial position along the $x$ axis to the origin and then along the $y$ axis to $y = 5$ m and $x = 0$. What is his displacement?

3. A hiker sets off at 8 A.M. in level terrain. At 9 A.M. she is 2 km due east of her starting point. At 10 A.M. she is 1 km northwest of where she was at 9 A.M. At 11 A.M. she is 3 km due north of where she was at 10 A.M. (a) Make a drawing showing these successive displacements as vectors, the tail of each being at the head of the previous one. What are the magnitudes and directions of these displacements? (Specify the direction of vectors by giving their angle with the eastward direction.) (b) What are the north and east components of these displacements? (c) How far is the hiker from her starting point at 11 A.M.? In what direction? (d) Add the three displacement vectors by drawing them to scale. Do these successive straight lines represent the actual path the hiker followed? Is the distance she walked the sum of the lengths of the three displacement vectors?

4. A circular path has a radius of 10 m. An $xy$ coordinate system is established so that the center of the circle is on the $y$ axis and the circle passes through the origin. A man starts at the origin and walks around the path at a steady speed, returning to the origin exactly 1 min after he started. (a) Find the magnitude and direction of his displacement from the origin 15, 30, 45, and 60 s after he started. (b) Find the magnitude and direction of his displacement for each of the four successive 15-s intervals of his walk. (c) How is his displacement for the first 15 s related to that for the second 15 s? (d) How is his displacement for the second 15-s interval related to that for the last 15-s interval?

**Section 3-2, Components of a Vector**

5. What are the rectangular components of the displacement vector for part (*a*) of Exercise 2? Write this vector in terms of the unit vectors **i** and **j**.

6. Find the rectangular components of the vectors which lie in the *xy* plane, have the magnitude *A*, and make an angle $\theta$ with the *x* axis, as shown in Figure 3-25 for the following values of *A* and $\theta$:
(*a*) *A* = 10 m, $\theta$ = 30°
(*b*) *A* = 5 m, $\theta$ = 45°
(*c*) *A* = 7 km, $\theta$ = 60°
(*d*) *A* = 5 km, $\theta$ = 90°
(*e*) *A* = 15 km/s, $\theta$ = 150°
(*f*) *A* = 10 m/s, $\theta$ = 240°
(*g*) *A* = 8 m/s², $\theta$ = 270°

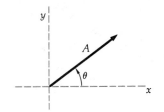

**Figure 3-25**
Exercise 6.

7. Find the magnitude and direction of the following vectors:
(*a*) **A** = 5**i** + 3**j**
(*b*) **B** = 10**i** − 7**j**
(*c*) **C** = −2**i** − 3**j** + 4**k**

8. A cube of side 2 m has its faces parallel to the coordinate planes with one corner at the origin. A fly begins at the origin and walks along the three edges until it is at the far corner. Write the displacement vector of the fly using the unit vectors **i**, **j**, and **k** and find the magnitude of this displacement.

9. A plane is inclined at an angle of 30° with the horizontal. Choose the *x* axis parallel to the plane pointing down the slope and the *y* axis perpendicular to the plane pointing away from the plane. Find the *x* and *y* components of the acceleration of gravity, which has the magnitude 9.8 m/s² and points vertically down.

10. Find the magnitude and direction of **A**, **B**, and **A** + **B** for (*a*) **A** = −4**i** − 7**j**; **B** = 3**i** − 2**j**; (*b*) **A** = 1**i** − 4**j**, **B** = 2**i** + 6**j**.

11. Describe the following vectors by using the unit vectors **i** and **j**: (*a*) a velocity of 10 m/s at an angle of elevation of 60°; (*b*) a vector **A** of magnitude *A* = 5 m and $\theta$ = 225°; (*c*) a displacement from the origin to the point *x* = 14 m, *y* = −6 m.

**Section 3-3, Properties of Vectors**

12. The displacement vectors **A** and **B** shown in Figure 3-26 both have magnitude 2 m. Find their *x* and *y* components. Find the components, magnitude, and direction of the sum **A** + **B**. Find the components, magnitude, and direction of the difference **A** − **B**.

13. For the two vectors **A** and **B** shown in Figure 3-26 find graphically:
(*a*) **A** + **B**
(*b*) **A** − **B**
(*c*) 2**A** + **B**
(*d*) **B** − **A**
(*e*) 2**B** − **A**

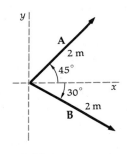

**Figure 3-26**
Exercises 12 and 13.

14. For the vector **A** = 3**i** + 4**j** find any three other vectors **B** which also lie in the *xy* plane and have the property that *A* = *B* but **A** ≠ **B**. Write these vectors in terms of their components and show them graphically.

15. If **A** = 2**i** − 6**j**, find 5**A** and −7**A**.

16. Two vectors **A** and **B** lie in the *xy* plane. Under what conditions does the ratio *A*/*B* equal $A_x/B_x$?

17. If **A** = 5**i** − 4**j** and **B** = −7.5**i** + 6**j**, write an equation relating **A** to **B**.

18. Draw any three nonparallel vectors **A**, **B**, and **C** which lie in a plane and show that the sum (**A** + **B**) + **C** equals (**A** + **C**) + **B**.

**Section 3-4, The Velocity Vector**

19. A vector **A**(*t*) has constant magnitude but is changing direction in a uniform way. Draw the vectors **A**(*t* + $\Delta t$) and **A**(*t*) for a small time interval $\Delta t$ and find the difference $\Delta$**A** = **A**(*t* + $\Delta t$) − **A**(*t*) graphically. How is the direction of $\Delta$**A** related to **A** for small time intervals?

20. A stationary radar operator determines that a ship is 10 km south of him. An hour later the same ship is 20 km southeast of him. If the ship moved at constant speed always in the same direction, what was its velocity during this time?

21. A particle's position coordinates $(x, y)$ are $(2 \text{ m}, 3 \text{ m})$ at $t = 0$; $(6 \text{ m}, 7 \text{ m})$ at $t = 2$ s; and $(13 \text{ m}, 14 \text{ m})$ at $t = 5$ s. ($a$) Find $\mathbf{v}_{av}$ from $t = 0$ to $t = 2$ s. ($b$) Find $\mathbf{v}_{av}$ from $t = 0$ to $t = 5$ s.

22. A particle travels with constant speed in a circular path of radius 5 m around the origin. It begins at $t = 0$ at $x = 5$ m, $y = 0$ and takes 100 s for a complete revolution. ($a$) What is the speed of the particle? ($b$) Give the magnitude and direction of the position vector $\mathbf{r}$ at the times $t = 50$ s, $t = 25$ s, $t = 10$ s, and $t = 0$. ($c$) Find the magnitude and indicate graphically the direction of $\mathbf{v}_{av}$ for each of the following time intervals: $t = 0$ to $t = 50$ s; $t = 0$ to $t = 25$ s; $t = 0$ to $t = 10$ s. ($d$) How does $\mathbf{v}_{av}$ for the interval $t = 0$ to $t = 10$ s compare with the instantaneous velocity at $t = 0$?

23. The position vector of a particle is given by $\mathbf{r}(t) = 5t\mathbf{i} + 10t\mathbf{j}$, where $t$ is in seconds and $r$ in metres. ($a$) Draw the path of the particle in the $xy$ plane. ($b$) Find $\mathbf{v}(t)$ in component form and find its magnitude.

**Section 3-5, The Acceleration Vector**

24. A ball is thrown directly upward. Consider the 2-s time interval $\Delta t = t_2 - t_1$, where $t_1$ is 1 s before the ball reaches its highest point and $t_2$ is 1 s after it reaches its highest point. Find ($a$) the change in speed, ($b$) the change in velocity, and ($c$) the average acceleration for this time interval.

25. Figure 3-27 shows the path of an automobile, made up of segments of straight lines and arcs of circles. The automobile starts from rest at point $A$. After it reaches point $B$, it travels at constant speed until it reaches point $E$. It comes to rest at point $F$. ($a$) At the middle of each segment ($AB$, $BC$, $CD$, $DE$, and $EF$) what is the direction of the velocity vector? ($b$) At which of these points does the automobile have an acceleration? In those cases, what is the direction of the acceleration? ($c$) How do the magnitudes of the acceleration compare for segments $BC$ and $DE$?

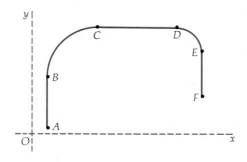

**Figure 3-27**
Exercise 25.

26. Initially a particle is moving due west with a speed of 40 m/s, and 5 s later it is moving north with a speed of 30 m/s. ($a$) What was the change in the magnitude of the particle's velocity during this time? ($b$) What was the change in direction of the velocity? ($c$) What are the magnitude and direction of $\Delta\mathbf{v}$ for this interval? ($d$) What are the magnitude and direction of $\mathbf{a}_{av}$ for this interval?

27. At $t = 0$ a particle located at the origin has a speed of 40 m/s at $\theta = 45°$. At $t = 3$ s the particle is at $x = 100$ m, $y = 80$ m with speed of 30 m/s at $\theta = 50°$. Calculate ($a$) the average velocity and ($b$) the average acceleration of the particle during this interval.

28. A particle has a position vector given by $\mathbf{r} = 30t\mathbf{i} + (40t - 5t^2)\mathbf{j}$, where $\mathbf{r}$ is in metres and $t$ in seconds. Find the instantaneous velocity and acceleration vectors as functions of time $t$.

## Section 3-6, Motion with Constant Acceleration: Projectile Motion

29. A bullet is fired horizontally with an initial velocity of 245 m/s. The gun is 1.5 m above the ground. How long is the bullet in the air?

30. A supersonic transport is flying horizontally at an altitude of 20 km with speed of 2500 km/h when an engine falls off. (a) How long does it take the engine to hit the ground? (b) How far horizontally is the engine from where it fell off when it hits the ground? (c) How far is the engine from the aircraft (assuming it continues to fly as if nothing had happened) when the engine hits the ground? Neglect air resistance.

31. A cannon is elevated at an angle of 45°. It fires a ball with a speed of 300 m/s. (a) What height does the ball reach? (b) How long is the ball in the air? (c) What is the horizontal range?

32. Sketch the trajectory of a projectile and indicate the velocity and acceleration vectors at several points. (a) What is the magnitude of the acceleration at the top of the path? (b) Does the magnitude or direction of the acceleration change from point to point? (c) How does the angle between $\mathbf{a}$ and $\mathbf{v}$ change during the motion?

33. A projectile is launched with speed $v_0$ and at an angle $\theta_0$ with the horizontal. Find an expression for the maximum height it reaches above its starting point in terms of $v_0$, $\theta_0$, and $g$.

34. A projectile is fired with initial velocity of 30 m/s at 60° to the horizontal. At the highest point what is the velocity? The acceleration?

## Section 3-7, Circular Motion

35. A particle travels in a circular path of radius 5 m with a constant speed of 15 m/s. What is the magnitude of its acceleration?

36. An object on the equator has an acceleration toward the center of the earth because of the earth's rotation and an acceleration toward the sun because of the earth's motion along the orbit. Calculate the magnitude of both accelerations and express them as fractions of the free-fall acceleration of gravity $g$.

37. A particle moves in a circle of radius 4 cm. It takes 8 s to make a complete round trip. Draw the path of the particle to scale and indicate the positions at 1-s intervals. Draw the displacement vectors for these 1-s intervals. These vectors also indicate the average velocity vectors for these intervals. Find graphically the change in the average velocity $\Delta\mathbf{v}$ for two consecutive 1-s intervals. Compare $\Delta\mathbf{v}/\Delta t$ measured in this way with the instantaneous acceleration computed from $a_r = v^2/r$.

38. A boy whirls a ball on a string in a horizontal circle of radius 1 m. How many revolutions per minute must the ball make if its acceleration toward the center of the circle is to have the same magnitude as the acceleration of gravity?

39. An airplane pilot pulls out of a dive by following an arc of a circle whose radius is 300 m. At the bottom of the circle, where her speed is 180 km/h, what are the direction and magnitude of her acceleration?

40. An object travels with a constant speed $v$ in a circular path of radius $r$. (a) If $v$ is doubled, how is the acceleration $a$ affected? (b) If $r$ is doubled, how is $a$ affected? (c) Why is it impossible for an object to travel around a perfectly sharp angular turn?

41. In Figure 3-28 the particles are traveling counterclockwise in a circle of radius 5 m with speeds which may be varying. The acceleration vectors are indicated at certain times. Find the values of $v$ and $dv/dt$ for each of these three times.

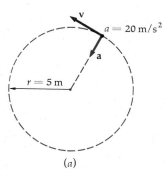

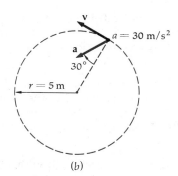

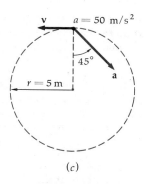

**Figure 3-28**
Exercise 41.

## Problems

1. In parts (a) to (c) of Figure 3-29 particles are traveling in circular paths with varying speed. The velocity vectors are indicated. Find the magnitude of the average-acceleration vector between the two given positions in each case.

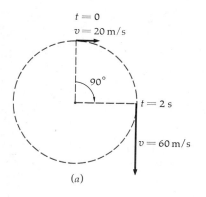

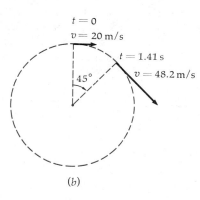

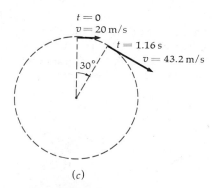

**Figure 3-29**
Problem 1.

2. A particle moves clockwise in a circle of radius 1 m with center at $(x, y) = (1\,m, 0)$. It starts at rest at the origin at time $t = 0$. Its *speed* increases at the constant rate of $(\pi/2)$ m/s². (a) How long does it take to travel halfway around the circle? (b) What is its speed at that time? (c) What is the direction of its velocity at that time? (d) What is its radial acceleration then? Its tangential acceleration? (e) What are the magnitude and direction of the total acceleration halfway around the circle?

3. A baseball is struck by a bat, and 3 s later it is caught 30 m away. (a) If it was 1 m above the ground when struck and caught, what was the greatest height it reached above the ground? (b) What were its horizontal and vertical components of velocity when it was struck? (c) What was its speed when it was caught? (d) At what angle with the horizontal did it leave the bat? (Neglect air resistance.)

4. A gun shoots bullets that leave the muzzle at 250 m/s. If the bullet is to hit a target 100 m away at the level of the muzzle, the gun must be aimed at a point above the target. How far above the target is this point. (Neglect air resistance.)

5. A baseball is thrown toward a player with an initial speed of 20 m/s and 45° with the horizontal. At the moment the ball is thrown, the player is 50 m from the thrower. At what speed and in what direction must he run to catch the ball at the same height at which it was released?

6. Galileo showed that if air resistance is neglected, the ranges are equal for projectiles whose angles of projection exceed or fall short of 45° by the same amount. Prove this.

7. A stone thrown horizontally from the top of a tower hits the ground at a distance 18 m from the base of the tower. (*a*) Find the speed at which the stone was thrown if the tower is 24 m high. (*b*) Find the speed of the stone just before it hits the ground.

8. Two balls are thrown with equal speeds from the top of a cliff of height $H$. One ball is thrown upward at an angle $\alpha$ above the horizontal. The other ball is thrown downward at an angle $\beta$ below the horizontal. Show that each ball strikes the ground with the same speed and find that speed in terms of $H$ and the initial speed $v_0$.

9. A trail bike comes to a ditch. A ramp of angle 10° has been built so that the bike can jump the ditch. If the bike needs to jump a horizontal distance of 7 m to clear the ditch, how fast must it travel off the ramp?

10. A projectile is launched at an angle $\theta$ with the horizontal up a hill of constant slope $\alpha$ ($\alpha < \theta$). Show that the range measured along the slope of the hill is

$$\frac{2v_0^2 \cos \theta \sin (\theta - \alpha)}{g \cos^2 \alpha}$$

11. In Figure 3-30 a trail bike takes off from a ramp at angle $\theta$ to clear a ditch of width $x$ and land on the other side, which is elevated at a height $H$. (*a*) For a given angle $\theta$ and distance $x$, what is the upper limit on $H$ such that the bike has any chance of making the jump? (*b*) For $H$ less than this upper limit, what is the minimum takeoff speed $v_0$ for a successful jump? (Neglect the size of the bike and assume that covering a horizontal distance $x$ and a vertical distance $H$ is sufficient to clear the ditch.)

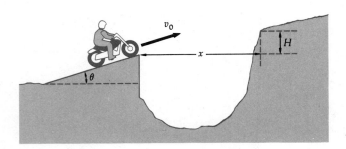

**Figure 3-30**
Problem 11.

12. A hockey puck struck at ice level just clears the top of the glass wall 2.80 m high. The flight time to this point was 0.650 s and the horizontal distance 12.0 m. Find (*a*) the initial speed of the puck and (*b*) the maximum height it will reach.

13. Find the angle of projection such that the maximum height of a projectile is equal to the horizontal range.

14. A freight train is moving at a constant speed of 10 m/s. A man standing on a flatcar throws a ball into the air and catches it as it falls. Relative to the flatcar the initial velocity of the ball is 15 m/s straight up. (*a*) What are the magnitude and direction of the initial velocity of the ball as seen by a second man standing next to the track? (*b*) How long is the ball in the air according to the man on the train? According to the man on the ground? (*c*) What horizontal distance has the ball traveled by the time it is caught according to the man on the train? According to the man on the ground? (*d*) What is the minimum speed of the ball during its flight according to the man on the train? According to the man on the ground? (*e*) What is the acceleration of the ball according to the man on the train? According to the man on the ground?

15. A car is traveling down a highway at 25 m/s. Just as the car crosses a per-pendicularly intersecting crossroad, the passenger throws out a beer can at a 45° angle of elevation in a plane perpendicular to the motion of the car. The initial speed of the can relative to the car is 10 m/s. It is released at a height of 1.2 m above the road. (*a*) Write the initial velocity of the beer can (relative to the road) in terms of the unit vectors **i**, **j**, and **k**. (*b*) Where does the can land?

16. For short time intervals any path can be considered an arc of a circle. How can the radius of curvature of a path segment be determined from the instanta-neous velocity and acceleration? Consider a projectile at the top of its path. In-dicate the velocity vector just before and just after this point. Is the speed changing? What is the radius of curvature of the path segment at this point?

17. The position of a particle as a function of time is

$$\mathbf{R} = 4 \sin 2\pi t \, \mathbf{i} + 4 \cos 2\pi t \, \mathbf{j}$$

where $R$ is in metres and $t$ is in seconds. (*a*) Show that the path of this particle is a circle of radius 4 m with its center at the origin. (*b*) Compute the velocity vector. Show that $v_x/v_y = -y/x$. (*c*) Compute the acceleration vector and show that it is in the radial direction and has the magnitude $v^2/R$.

18. A particle has constant acceleration $\mathbf{a} = 6\mathbf{i} + 4\mathbf{j}$ m/s². At time $t = 0$ the velocity is zero and the position vector is $\mathbf{r}_0 = (10 \text{ m})\mathbf{i}$. (*a*) Find the velocity and position vectors at any time $t$. (*b*) Find the equation of the path in the $xy$ plane and sketch the path.

19. The position of a particle is given by

$$\mathbf{R} = 3 \sin 2\pi t \, \mathbf{i} + 2 \cos 2\pi t \, \mathbf{j}$$

where $R$ is in metres and $t$ is in seconds. (*a*) Plot the path of the particle in the $xy$ plane. (*b*) Find the velocity vector. (*c*) Find the acceleration vector and show that its direction is along $\mathbf{R}$; that is, it is radial. (*d*) Find the times for which the speed is a maximum or minimum.

20. A large boulder rests on a cliff 400 m above a small village in such a position that if it should roll off, it would leave with a speed of 50 m/s (Figure 3-31). There is a pond, diameter 200 m, with its edge 100 m from the base of the cliff, as shown. The village houses are at the edge of the pond. (*a*) A physics student says that the boulder will land in the pond. Is she right? (*b*) How fast will it be going when it hits? What will the horizontal component of its velocity be when it hits? (*c*) How long will the boulder be in the air?

**Figure 3-31**
Problem 20.

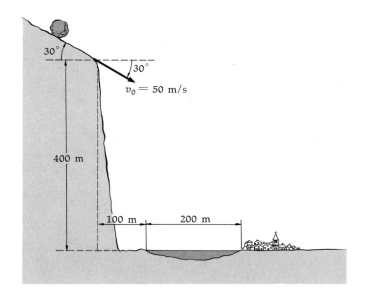

21. A projectile is fired into the air from the top of a 200-m cliff above a valley (Figure 3-32). Its initial velocity is 60 m/s at 60° to the horizontal. Neglecting air resistance, where does the projectile land?

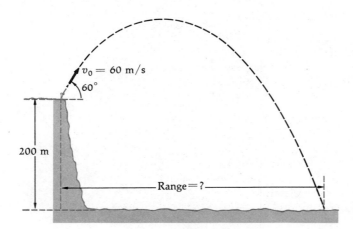

**Figure 3-32**
Problem 21.

22. A boy stands 4 m from a vertical wall and throws a ball (Figure 3-33). The ball leaves the boy's hand at 2 m above the ground with initial velocity $\mathbf{v} = 10\mathbf{i} + 10\mathbf{j}$ m/s. When the ball hits the wall, its horizontal component of velocity is reversed and its vertical component remains unchanged. Where does the ball hit the ground?

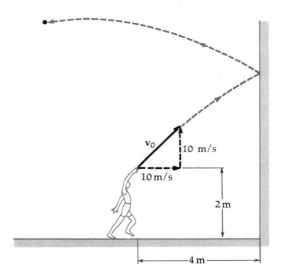

**Figure 3-33**
Problem 22.

23. A baseball just clears a 3-m wall 120 m from home plate. If it leaves the bat at 45° and 1.2 m above the ground, what must its initial velocity be (again making the unreal assumption that air resistance can be ignored)?

24. An acceleration of 31$g$ was withstood for 5 s by R. F. Gray in 1959. How many revolutions per minute does a centrifuge have to undergo to produce an acceleration of 31$g$ on somebody positioned in the arm at a radius of 5 m?

# CHAPTER 4  Newton's Laws

**Objectives**  After studying this chapter, you should:

1. Be able to discuss the definitions of force and mass and to state Newton's laws of motion.

2. Be able to distinguish between mass and weight.

3. Be familiar with the following units and know how they are defined: kilogram, newton, dyne, pound, slug.

4. Know that forces always occur in action-reaction pairs and act on different bodies, so that they never can act as balancing forces for a body.

5. Be able to discuss the law of conservation of momentum, and the "action-at-a-distance" problem.

6. Be able to apply Newton's laws in a systematic way to the solution of a variety of mechanics problems.

7. Know the meaning of an inertial reference frame.

The Bettmann Archive

Portrait of Newton.

Classical, or newtonian, mechanics is a theory of motion based on the ideas of mass and force and the laws connecting these physical concepts to the kinematic quantities—position, velocity, and acceleration—discussed in the preceding chapters. The fundamental relations of classical mechanics are contained in Newton's laws of motion. We begin by stating Newton's laws and then illustrate their application in rather simple problems involving constant forces. In the next chapter we shall discuss some more general applications.

It is interesting to read Newton's version of the laws of motion:*

> Law I. Every body continues in its state of rest, or in uniform motion in a right line unless it is compelled to change that state by forces impressed upon it.

* *Philosophiae Naturalis Principia Mathematica,* 1686, trans. Andrew Motte, 1729, University of California Press, Berkeley, 1960; reprinted by permission of the Regents of the University of California.

Law II. The change of motion is proportional to the motive force impressed; and is made in the direction of the right line in which that force is impressed.

Law III. To every action there is always opposed an equal reaction; or, the mutual actions of two bodies upon each other are always directed to contrary parts.

Corollary I. A body, acted on by two forces simultaneously, will describe the diagonal of a parallelogram in the same time as it would describe the sides by those forces separately.

A modern version of these laws is:

1. A body continues in its initial state of rest or motion with uniform velocity unless acted on by an unbalanced external force.

*Newton's laws of motion*

2. The acceleration of a body is inversely proportional to its mass and directly proportional to the resultant external force acting on it:*

$$\Sigma \mathbf{F} = \mathbf{F}_{net} = m\mathbf{a} \qquad \qquad 4\text{-}1$$

3. Forces always occur in pairs. If body $A$ exerts a force on body $B$, an equal but opposite force is exerted by body $B$ on body $A$.

Corollary. Forces obey the parallelogram law of addition; i.e., forces are vectors.

Newton's laws related the acceleration of an object to its mass and the forces acting on it. We have intuitive ideas about the words force and mass. We think of a force as being a push or a pull, like that exerted by our muscles. We visualize a massive body as something large or heavy. These intuitive notions are all right for everyday conversation but not for the applications of Newton's laws to problems in physics or even for a precise statement of the laws. To understand Newton's laws fully and be able to apply them we must define these words carefully. This we do by outlining methods for their measurement, in what is called an *operational definition*. We shall find that Newton's second law follows directly from the definitions of force and mass.

## 4-1   Force and Mass

Consider an object, say a block of wood or metal, resting on a horizontal surface, say a table. We observe that if the body is at rest (relative to the table), it remains at rest unless we push on it or pull on it. If we project the body along the table, it slides along for a way, but eventually the speed decreases and the object comes to rest. We attribute the decrease in speed to the *frictional force* exerted on the body by the table. If we polish the surfaces of the table and body, the body slides farther and its decrease in velocity in a given time is smaller. If we support the body by a thin cushion of air (this is possible with an air table or with a glider on an air track), the body will glide for a considerable time and distance with almost no perceptible change in its velocity. We extrapolate this experience to an *ideal frictionless* surface which in no way impedes the movement of a body, and we state that on such a surface the velocity of a body will not change. We thus *define* the situation in which there are *no* (horizontal) forces acting on the body. If there are no forces acting on

---

* The Greek capital letter sigma ($\Sigma$) is used to indicate a summation. $\Sigma\mathbf{F}$ therefore means the sum of all the forces acting on the body, i.e., the net or resultant force acting on the body.

the body, the velocity of the body remains constant. This is Newton's
first law, the *law of inertia*.

*Law of inertia*

If the velocity of a body is not constant, we conclude that a net force
is acting on the body. Our next problem is to develop a quantitative
measure of force. We do this by *defining the magnitude and direction of a
given force in terms of the acceleration it produces on a particular object
which we call our standard body*. The international standard body is a cylinder of platinum carefully preserved at the International Bureau of
Weights and Measures at Sèvres, France. The mass of the standard body
is 1 kg, the SI unit of mass. (We shall discuss the definition and measurement of mass below.) The force required to produce an acceleration
of 1 m/s² on the standard body is defined to be 1 newton (N). Similarly,
the force that produces an acceleration of 2 m/s² is defined to be 2 N. We
thus define force in terms of the acceleration it produces on a standard
body.

*Force of one newton defined*

A convenient agent for exerting forces on bodies is a spring. It takes a
push or a pull to compress or extend a spring from its natural length; the
greater the push or pull, the greater the compression or extension. Consider a particular spring attached to our standard body on a frictionless
horizontal surface, as shown in Figure 4-1. If the spring is extended

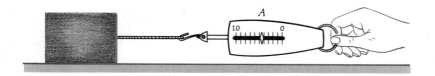

from its natural length, the body accelerates; up to a certain limit, depending on the spring, the greater the extension, the greater the acceleration. By noting the extension needed to produce a particular acceleration of our standard body and using our definition of force (that an
acceleration of $x$ m/s² means that the spring is exerting a force of $x$ N) we
can calibrate our spring in units of force. A plot of force versus extension for a typical spring is shown in Figure 4-2. For common springs,
the force exerted is proportional to the extension for small extensions.

**Figure 4-1**
A horizontal force is applied
to a body by the extended
spring. The spring can be calibrated by noting the extension produced by a given
force as measured by the
acceleration produced.

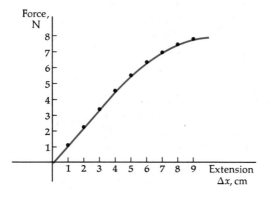

**Figure 4-2**
Calibration curve for a spring
scale like that in Figure 4-1.

This is known as *Hooke's law*. Using the same standard body, we can calibrate other springs in a similar way. We can then do experiments to see
how several forces combine. Consider two calibrated springs attached
to our standard body and extended so that they exert forces in different
directions, as in Figure 4-3. We observe experimentally that the forces
add as vectors; i.e., the resultant acceleration is found by adding the
vector accelerations each force would produce if it were acting alone.

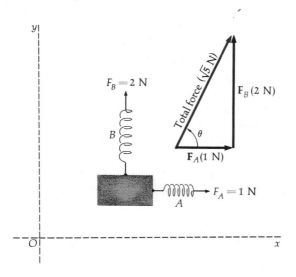

**Figure 4-3**
Forces obey the vector-addition rule. Forces of 1 N in the $x$ direction and 2 N in the $y$ direction combine as shown to give a resultant force of $\sqrt{5}$ N at an angle $\theta$ to the $x$ axis given by tan $\theta$ = 2.

For example, if spring $A$ exerts a force of 1 N along the $x$ axis and spring $B$ exerts a force of 2 N along the $y$ axis, as in Figure 4-3, the acceleration observed has the magnitude $a = \sqrt{1^2 + 2^2}$ m/s² = $\sqrt{5}$ m/s² and makes with the $x$ axis an angle $\theta$ given by tan $\theta$ = 2. This observation is known as the *parallelogram law of addition of forces; forces are vectors. We have listed this experimental result as a corollary to Newton's laws. In particular, if two equal forces act in opposite directions on a body, the acceleration is zero.

*Parallelogram law of addition of forces*

We can use our calibrated springs to measure other types of forces. For example, the force exerted by the gravitational attraction of the earth for an object is called its *weight*. When the weight of a body is the only force acting on it, the acceleration of the body is 9.81 m/s² toward the earth. The weight of our standard body is thus 9.81 N. If we hang our standard body by one of our calibrated springs, as in Figure 4-4, the spring is stretched by an amount corresponding to 9.81 N of force to balance the downward gravitational force exerted by the earth.

Now that we know how to compare forces quantitatively by comparing the accelerations they produce on our standard body, we investigate the effect of a given force on different bodies. Let us use one of our calibrated springs to produce a given force on a different body. We note that the acceleration produced for a given spring extension is not the same, in general, as it was for our standard body. If our second body is "more massive" (according to our everyday use of this term), the acceleration produced by a force of 1 N is observed to be less than 1 m/s². If it is less massive, the acceleration produced is greater. For example, if we connect two identical bodies, the acceleration produced by a given force is exactly half that produced by the same force acting on just one of the bodies. This suggests that we quantify the concept of mass by considering the acceleration a given force will produce in different bodies.

Specifically, we define the ratio of the mass of one body to that of another body to be the *inverse* ratio of the accelerations produced in those two bodies by the *same force*. If a given force produces an acceleration $a_1$ when it acts on body 1 and an acceleration $a_2$ on body 2, the ratio of the masses of the two bodies is defined to be

$$\frac{m_2}{m_1} = \frac{a_1}{a_2} \qquad \text{same force} \qquad\qquad 4\text{-}2$$

*Mass defined*

We find the ratio of the masses of any two bodies by applying the same force to each and comparing their accelerations. We find that the ratio

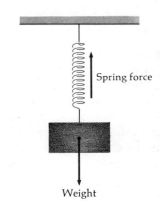

Spring force

Weight

**Figure 4-4**
Forces on a body suspended from a spring. The upward force due to the spring balances the downward force due to the gravitational attraction of the earth (the weight of the body).

of the accelerations $a_2/a_1$ produced by the same force on two bodies is independent of the magnitude or direction of the force. It is also independent of the kind of force used, i.e., whether the force is due to springs, the pull of gravity, electric or magnetic attraction or repulsion, etc. We also find that if mass $m_2$ is found to be twice mass $m_1$ by direct comparison and if a third mass $m_3$ is found to be 4 times the mass $m_1$, $m_3$ will be twice the mass of $m_2$ when compared directly. We can therefore set up a mass scale by choosing one particular body to be a standard body and assigning it a mass of 1 unit. As mentioned above, the standard body is kept at the International Bureau of Weights and Measures in France and is assigned the mass of 1 kg. (This mass was originally intended to be equal to the mass of 1000 cm$^3$ = 1 litre of water, but the relationship proved to be inexact by a very small amount.) By using this standard body secondary standards can be found by direct comparison, and the mass of any other body can then be found by comparing the acceleration produced by a given force with one of the secondary standards. As we shall see in the next section, the weight of any body is directly proportional to its mass, so an easier method of comparing masses is to weigh them at the same location.

Note that we have defined the concepts of force and mass so that Newton's second law $\Sigma\mathbf{F} = m\mathbf{a}$ follows directly from the definitions. These definitions agree with our intuitive notions of the meaning of these words. They are useful because they allow us to describe a wide variety of physical phenomena using just a few relatively simple force laws. For example, with the addition of Newton's law of gravitational attraction between two bodies we can calculate and explain such phenomena as the motion of the moon, the orbits of all the planets around the sun, the orbits of artificial satellites, the variation in the acceleration of gravity $g$ with latitude due to the rotation of the earth, the variations in the acceleration of gravity due to the presence of mineral deposits, the paths of ballistic missiles, and many other motions.

**Questions**

1. If a body has no acceleration, can you conclude that no forces act on it?

2. If only a single force acts on a body, must it be accelerating? Can it ever have zero velocity?

3. Is there a net force acting (*a*) when a body moves at constant speed along a circle, (*b*) when a body that is moving in a straight line slows down, (*c*) when a body moves at constant speed in a straight line?

4. Is it possible for an object to round any curve without a force being impressed upon it?

5. If a single known force acts on a body, can you tell *in which direction* the body will *move* from this information alone?

6. If several forces of different magnitudes and directions are applied to a body initially at rest, how can you predict the direction in which it will move?

7. Can you judge the mass of an object by its size? If *A* is twice as big as *B*, does that mean that $m_A = 2m_B$?

8. Could a body have a different mass for electric forces than for gravitational forces?

9. Can the mass of a body be negative?

10. How do we know that we can add masses as scalars? That is, if one body has a mass of 2 kg and another of 3 kg, how do we know that the two bodies taken together have a mass of 5 kg?

11. Mass is sometimes said to be the measure of the quantity of matter in a body. How does this unscientific definition compare with the definition discussed above?

## 4-2    The Force Due to Gravity: Weight

The force most common in our everyday experience is the force of gravitational attraction of the earth for an object. This force is called the *weight* of the object. If we drop an object near the surface of the earth and we can neglect air resistance so that the only force acting on the object is the force due to gravity, the object accelerates toward the earth with acceleration 9.81 m/s². At a given point in space this acceleration is the same for all objects independent of their mass. Since the acceleration of an object is the resultant force divided by the object's mass, we can conclude that the force due to gravity on an object is proportional to the mass of the object:

$$\mathbf{F}_g = \mathbf{w} = m\mathbf{g} \qquad \qquad 4\text{-}3$$

The vector $\mathbf{g}$ in Equation 4-3 is called the *gravitational field* of the earth. It is the force per unit mass exerted by the earth on any object. It is equal to the acceleration of gravity, i.e., the free-fall acceleration experienced by an object when the only force acting on it is the gravitational force of the earth. Near the surface of the earth the gravitational field of the earth and the acceleration of gravity have the value

$$g = 9.81 \text{ N/kg} = 9.81 \text{ m/s}^2$$

*Gravitational field g*

Careful measurements of $g$ at various places show that it does not have the same value everywhere. *The force of attraction of the earth for an object varies with location.* Thus weight, unlike mass, is not an intrinsic property of an object. In particular, at points above the surface of the earth, the force due to gravity varies inversely as the square of the distance of the object from the center of the earth. Thus a body weighs slightly less at very high altitudes than it does at sea level. The gravitational field also varies slightly with latitude because the earth is not exactly spherical but is flattened at the poles.*

Near the surface of the moon, the gravitational attraction of the moon is much stronger than that of the earth. The force exerted on the body by the moon is usually called the weight of the body when it is near the moon. Note again that the mass of a body is the same whether it is on the earth, on the moon, or somewhere in space. Mass is a property of the body itself, whereas weight depends on the nature and distance of other objects which exert gravitational forces on the body.

Since at any particular location the weight of a body is proportional to its mass, we can conveniently compare the mass of one body with that of another by comparing their weights as long as we determine the weights at the same place.

---

* When the free-fall acceleration of an object is measured relative to a point on the surface of the earth, there is also a variation with latitude because the reference point has an acceleration due to the rotation of the earth. This variation, discussed in detail in Section 4-6, is not a variation in the gravitational field of the earth but a variation in the measured acceleration introduced because it is measured relative to the surface of the earth, which itself has an acceleration varying with latitude.

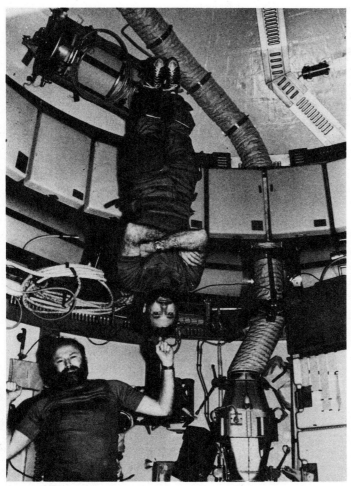

Weightlessness in a space capsule. The astronauts are in free fall accelerating toward the earth with the acceleration of gravity.

NASA

Our sensation of our own weight usually comes from other forces which balance it. For example, sitting on a chair, we feel the force exerted by the chair which balances our weight. When we stand on a spring scale, our feet feel the force exerted on us by the scale. The scale is calibrated to read the force it must exert (by compression of its springs) to balance our weight. The force which balances our weight is called our *apparent weight*. It is the apparent weight that is given by a spring scale. If there is no force to balance your weight, as in free fall, your apparent weight is zero. This condition, called *weightlessness*, is experienced by astronauts in orbiting satellites. Consider a satellite in a circular orbit near the surface of the earth with a centripetal acceleration $v^2/r$, where $r$ is the orbit radius and $v$ is the speed. The only force acting on the satellite is its weight. Thus it is in free fall with the acceleration of gravity. The astronaut is also in free fall. The only force on him is his weight, which produces the acceleration $g = v^2/r$. Since there is no force balancing the force of gravity, the astronaut's apparent weight is zero.

*Apparent weight*

*Weightlessness*

**Questions**

12. From our definitions of mass and weight, would it be conceivable to use the same units for both?

13. Suppose an object were sent far out in space away from galaxies, stars, or other bodies. How would its mass change? Its weight?

14. How would an astronaut in a condition of weightlessness be aware of his mass?

15. We commonly compare the masses of objects by comparing their weights. Would this be possible if $g$ depended on the kind of material an object is made of?

16. Under what circumstances would your apparent weight be greater than your true weight?

# 4-3  Units of Force and Mass

The definition of the unit of mass, the kilogram, as the mass of a particular standard body completes our definition of the three fundamental units of mechanics in the mks system. In this system, the unit of force, the newton (that force which produces an acceleration of 1 m/s² on a 1-kg mass) is a derived unit; i.e., it can be expressed in terms of the three fundamental units:

$$1 \text{ N} = 1 \text{ kg·m/s}^2 \qquad 4\text{-}4$$

As mentioned in Chapter 1, the mks system of mechanics units is a subset of the International System of units (SI), which also includes units of temperature, electric current, and luminous intensity. These units are used almost exclusively throughout the world except in the United States, where they will also eventually become standard. Although we generally use SI units in this book, we need to know about two other systems, the cgs system, the metric system based on the centimetre, gram, and second, which is closely related to the mks system and used by many scientists, and the U.S. customary system, based on the foot, the second, and a force unit (the pound), which is still used today in the United States.

The unit of time, the second, is common to all three systems. In the cgs system, the unit of length is the centimetre (cm), now defined to be one-hundredth the length of the metre:

$$1 \text{ cm} = 10^{-2} \text{ m} \qquad 4\text{-}5$$

The unit of mass, the gram (g), is now defined to be exactly one-thousandth the mass of the standard kilogram:

$$1 \text{ g} = 10^{-3} \text{ kg} \qquad 4\text{-}6$$

The gram was originally chosen to be the mass of one cubic centimetre of water at standard pressure and temperature. The unit of force in the cgs system, called the *dyne*, is the force which applied to a one-gram mass produces an acceleration of one centimetre per second squared:

$$1 \text{ dyn} = 1 \text{ g·cm/s}^2 \qquad 4\text{-}7$$

Because the units are so small, the cgs system is less convenient than the mks system for practical work. For example, the mass of a penny is about 3 g. Since the free-fall acceleration of gravity is 981 cm/s², the weight of a penny in the cgs system is about

$$w = mg = (3 \text{ g})(981 \text{ cm/s}^2) = 2.94 \times 10^3 \text{ dyn}$$

The dyne is a very small unit of force. The relation between the dyne and the newton is

The standard kilogram, kept at the International Bureau of Weights and Measures, Sèvres, France. (*Courtesy of the Science Museum, London.*)

*Dyne*

$$1 \text{ dyn} = \frac{1 \text{ g·cm}}{\text{s}^2} \frac{1 \text{ kg}}{10^3 \text{ g}} \frac{1 \text{ m}}{10^2 \text{ cm}} = 10^{-5} \text{ kg·m/s}^2$$

or

$$1 \text{ dyn} = 10^{-5} \text{ N} \qquad\qquad 4\text{-}8$$

The U.S. customary system differs from both the mks and cgs systems in that a unit of force is chosen as a fundamental unit rather than a unit of mass. (Another difference is that it is not a decimal system.) The pound was originally defined as the weight of a particular standard body at a point where the acceleration due to gravity is exactly 9.80665 m/s² = 32.1740 ft/s². It is now defined in terms of the standard kilogram. One pound is defined to be the weight of a body of mass 0.45359237 kg at a point where the acceleration due to gravity has the value given above. The relation between the standard pound and the newton is thus

*Definition of pound*

$$1 \text{ lb} = (0.45359237 \text{ kg})(9.80665 \text{ m/s}^2) = 4.448222 \text{ N}$$

or

$$1 \text{ lb} \approx 4.45 \text{ N}$$

Since 1 kg weighs 9.81 N, its weight in pounds is

$$9.81 \text{ N} \times \frac{1 \text{ lb}}{4.45 \text{ N}} = 2.20 \text{ lb}$$

The unit of mass in the U.S. customary system is that mass which will be given an acceleration of one foot per second squared when a force of one pound is applied to it. This unit is called a *slug.* From its definition

*Definition of slug*

$$1 \text{ slug} = \frac{1 \text{ lb}}{1 \text{ ft/s}^2} = 1 \text{ lb·s}^2/\text{ft} \qquad\qquad 4\text{-}9$$

the weight of a slug near sea level is about

$$w = mg = (1 \text{ lb·s}^2/\text{ft})(32.2 \text{ ft/s}^2) = 32.2 \text{ lb}$$

In practice, it is often convenient to work problems in this system by writing the mass as $w/g$, where $w$ is the weight and $g$ is the acceleration of gravity.

---

**Example 4-1** The net force acting on a 10.0-lb body is 3.00 lb. What is its acceleration?

The acceleration is the force divided by the mass:

$$a = \frac{F}{m} = \frac{F}{w/g} = \frac{3.00 \text{ lb}}{(10.0 \text{ lb})/(32.2 \text{ ft/s}^2)} = 9.66 \text{ ft/s}^2$$

---

Although the weight of an object varies from place to place because of changes in $g$, this variation is too small to be noticed in most practical applications. Thus, in our everyday experience, the weight of a body appears to be as much a constant characteristic of the body as its mass.*

---

* This fact has led to everyday use of two other units, which is often confusing. One is a unit of force, the kilogram force, which is the weight of a 1-kg mass. A kilogram force is equal to 9.81 N or 2.20 lb. A second practical unit is the pound mass, which is the mass of a body which weighs 1 lb. Since by definition a pound force is the weight of 0.454 kg, a pound mass is equivalent to 0.454 kg. We shall not use these practical but confusing units.

**Questions**

17. What is your weight in newtons?

18. What is your mass in kilograms? In slugs?

19. What would your weight be in pounds on the moon, where objects fall freely with acceleration of about $5\frac{1}{3}$ ft/s²?

## 4-4   Newton's Third Law and Conservation of Momentum

Newton's third law can be called the *law of interaction*. The important property of forces it describes is that they always occur in pairs. For each force exerted on some body $A$ there must be some external agent, say another body $B$, exerting the force. The third law states that body $A$ exerts an equal but opposite force on the agent $B$. For example, the earth exerts a gravitational force $\mathbf{F}_g$ on a projectile, causing it to accelerate toward the earth with acceleration $\mathbf{g} = \mathbf{F}_g/m = \mathbf{w}/m$, where $m$ is the mass of the projectile. According to the third law, the projectile in turn exerts a force on the earth equal in magnitude and opposite in direction. Thus the projectile exerts a force $\mathbf{F}_g'$ on the earth toward the projectile. If this were the only force acting on the earth, the earth would have an acceleration toward the projectile of magnitude $a = F_g'/M_E$, where $M_E$ is the mass of the earth. (Because of the large mass of the earth, the acceleration it experiences due to this force is negligible and unobserved.)

In discussions of Newton's third law the words "action" and "reaction" are frequently used. If the force exerted on body $A$ is called the *action* of $B$ upon $A$, then the force body $A$ exerts back on body $B$ is called the *reaction* of $A$ upon $B$. It does not matter which force in such a pair is called the action and which the reaction. The important point is that forces always occur in action-reaction pairs and that the reaction force is equal in magnitude and opposite in direction to the action force.

Note that the action and reaction forces can never balance each other because they act on different objects. This is illustrated in Figure 4-5, which shows two action-reaction pairs of forces for a block resting on a table. The force acting downward on the block is the weight $\mathbf{w}$ due to the attraction of the earth. An equal and opposite force $\mathbf{w}'$ is exerted by the block *on the earth*. These are an action-reaction pair. If they were the only forces acting, the block would accelerate down because it would have only a single force acting on it. However, the table in contact with

*Law of interaction*

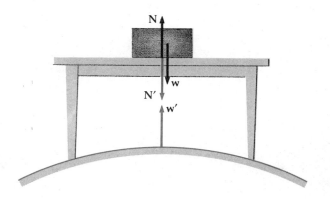

**Figure 4-5**
Action-reaction forces. The weight $\mathbf{w}$ is the force exerted on the block by the earth. The equal and opposite reaction force is $\mathbf{w}'$, exerted on the earth by the block. Similarly, the table exerts a force $\mathbf{N}$ on the block, and the block exerts an equal and opposite force $\mathbf{N}'$ on the table. Action-reaction forces are exerted on different objects and cannot balance.

the block exerts an upward force **N** on it. This force balances the weight of the block. The block also exerts a force **N′** downward on the table. The forces **N** and **N′** are also an action-reaction pair.

---

**Example 4-2** A horse refuses to pull a cart. The horse reasons, "according to Newton's third law, whatever force I exert on the cart, the cart will exert an equal and opposite force on me so the resultant force will be zero and I will have no chance of accelerating the cart." What is wrong with this reasoning?

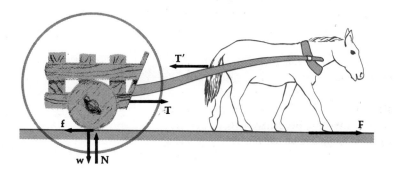

**Figure 4-6**
Horse pulling a cart. The cart will accelerate to the right if the force **T** exerted on it by the horse is greater than the frictional force **f** exerted on the cart by the ground. The force **T′** is equal and opposite to **T**, but because it is exerted *on* the *horse* it has no effect on the motion of the *cart*.

Figure 4-6 is a sketch of a horse pulling the cart. Since we are interested in the motion of the cart, we have circled it and indicated the forces acting on it. The force exerted by the horse is labeled **T**. Other forces on the cart are its weight **w**, the vertical support force of the ground **N**, and the horizontal force exerted by the ground labeled **f** (for friction). The vertical forces **w** and **N** balance. (We know this because we know the cart does not accelerate vertically.) The horizontal forces are **T** to the right and **f** to the left. The cart will accelerate if **T** is greater than **f**. Note that the reaction force to **T**, which we call **T′**, is exerted on the *horse*, not the cart. It has no effect on the motion of the cart. It does affect the motion of the horse. If the horse is to accelerate to the right, there must be a force **F** (to the right) exerted by the ground on the horse's feet that is greater than **T′**. This example illustrates the importance of a simple diagram in solving mechanics problems. Had the horse drawn a simple diagram, he would have seen that he need only push back hard against the ground so that the ground would push him forward.

---

There is a simple but important consequence of the third law for two objects isolated from their surroundings so that the only forces on them are the ones they exert on each other. Let $\mathbf{F}_{12}$ be the force exerted (by body 2) on body 1, which has mass $m_1$ and velocity $\mathbf{v}_1$, and $\mathbf{F}_{21} = -\mathbf{F}_{12}$ be that exerted on body 2, which has mass $m_2$ and velocity $\mathbf{v}_2$. Applying Newton's second law to each body, we have

$$\mathbf{F}_{12} = m_1\frac{d\mathbf{v}_1}{dt} \qquad \mathbf{F}_{21} = m_2\frac{d\mathbf{v}_2}{dt}$$

Adding these two equations and using $\mathbf{F}_{21} = -\mathbf{F}_{12}$, we obtain

$$0 = m_1\frac{d\mathbf{v}_1}{dt} + m_2\frac{d\mathbf{v}_2}{dt} = \frac{d}{dt}(m_1\mathbf{v}_1 + m_2\mathbf{v}_2)$$

or

$$m_1\mathbf{v}_1 + m_2\mathbf{v}_2 = \text{constant} \qquad\qquad 4\text{-}10$$

*Conservation of momentum*

The dynamic quantity $m\mathbf{v}$ is called the *momentum* of the particle. For two bodies subject only to their mutual interactions, the sum of the momenta of the bodies remains constant in time. This result is equivalent to Newton's third law. In fact, Newton seems to have arrived at his statement of action-reaction by studying the momentum of two bodies before and after collisions. When two bodies collide, they exert large forces on each other during the short time they are in contact. Even if there are other forces on the bodies, they are usually much smaller than these contact forces and can be neglected. From the careful measurements made by his predecessors Newton knew that no matter what kind of collision occurs, the sum of the momenta of the two colliding bodies is the same after the collision as before. Starting from Equation 4-10, which describes this result, he arrived at his statement that action equals reaction.

Equation 4-10 is known as the *law of conservation of momentum*. We have shown that it applies to two bodies each experiencing only a force exerted by the other; i.e., the total momentum of the two bodies is constant (conserved) if they are isolated from outside influences. By a straightforward generalization the same law can be shown to apply to any isolated system of bodies, no matter how great their number. We shall discuss this generalization of the conservation of momentum in more detail in later chapters.

It is convenient to write Newton's second law in terms of the momentum of a particle. We have

$$\Sigma\mathbf{F} = \frac{d\mathbf{p}}{dt} \qquad\qquad 4\text{-}11$$

where

$$\mathbf{p} = m\mathbf{v} \qquad\qquad 4\text{-}12$$

is the momentum of the particle. In classical mechanics the mass of a particle is always constant, and Equations 4-1 ($\Sigma\mathbf{F} = m\mathbf{a}$) and 4-11 are equivalent. However, when a particle moves with a speed near the speed of light (about $3 \times 10^8$ m/s), the ratio of the force to the acceleration depends on the speed. As we shall see in Chapter 35, for high-speed particles classical mechanics must be modified according to Einstein's theory of special relativity. In this theory Equation 4-11 holds if the expression for momentum is taken to be

$$\mathbf{p} = \frac{m\mathbf{v}}{\sqrt{1 - v^2/c^2}} \qquad\qquad 4\text{-}13$$

where $c$ is the speed of light in vacuum. When the speed of the particle is much less than that of light, the quantity $\sqrt{1 - v^2/c^2}$ is very nearly equal to 1 and the relativistic and classical expressions for momentum (Equations 4-12 and 4-13) are approximately equal. We shall also see that Newton's second law in the form of Equation 4-11 is easier to use when we generalize from a single particle to a system of particles in Chapter 8.

The extrapolation of the action-reaction principle for forces exerted by bodies in contact to bodies far apart presents conceptual difficulties of which Newton was well aware. We have seen that the statement that action equals reaction is equivalent to the statement that the rate at which one body gains momentum equals the rate at which the second body loses momentum. This is easily imagined if the bodies are in contact, but if they are widely separated, it implies that momentum is instantaneously transmitted from one to the other across the intervening space. This concept, called *action at a distance*, is difficult to ac-

*Action at a distance*

**Figure 4-7**
Action-reaction forces on
widely separated bodies.
Modern theory treats the
action-at-a-distance problem
by introducing the concept of
a field.

cept. For example, applied to the earth-sun system, it suggests that the
momentum lost by one travels instantly across the 150 million km between them to be taken up by the other (Figure 4-7). Newton justified
his extension of the third law to action-at-a-distance forces because the
assumption enabled him to calculate the orbits of the planets correctly
from the law of gravitation. He perceived action at a distance as a flaw
in his theory but avoided giving any other hypothesis. In 1692 Newton
made a famous comment* about the concept of action at a distance:

> It is inconceivable that inanimate, brute matter should, without the
> mediation of something else, which is not material, operate upon, and
> affect other matter without mutual contact, as it must be if gravitation,
> in the sense of Epicurus, be essential and inherent in it. And this is
> one reason why I desired you would not ascribe innate gravity to me.
> That gravity should be innate, inherent, and essential to matter, so that
> one body may act upon another at a distance through a vacuum,
> without the mediation of anything else, by and through which their
> action and force may be conveyed from one to another, is to me so
> great an absurdity that I believe no man who has in philosophical
> matters a competent faculty of thinking can ever fall into it.

Today we treat the problem of action at a distance by introducing the
concept of a *field*. For example, we consider the attraction of the earth by
the sun in two steps. The sun creates a condition in space which we call
the *gravitational field*. This field produces a force on the earth. The field
is thus the intermediary agent. Similarly, the earth produces a gravitational field which exerts a force on the sun. If the earth suddenly moves
to a new position, the field of the earth is changed. This change is not
propagated through space instantly but with the velocity $c = 3 \times 10^8$ m/s $= 1.86 \times 10^5$ mi/s, which is also the velocity of light. If we can
neglect the time it takes for propagation of the field, we can ignore this
intermediary agent and treat the forces as if they were exerted by the
sun and the earth directly on each other. For example, during the 8 min
it takes for propagation of the gravitational field from the earth to the
sun, the earth moves only a small fraction of its total orbit around the
sun. (The angular displacement of the earth after 8 min is only $9.6 \times 10^{-5}$ rad $= 5.5 \times 10^{-3}$ deg.)

*Field concept*

The third law is only an approximate law for two separated bodies. It
holds if we can neglect the time of propagation of momentum between
the interacting bodies. The law of conservation of momentum for *two
bodies* is also only approximate: it takes time for the momentum to be
transferred from one body to another. However, the conservation of
momentum can be rephrased as an exact law by introducing the idea
that the field itself can have momentum. Then during the time of

---

* Isaac Newton, *Third Letter to Bentley* (Feb. 25, 1692), R. and J. Dodsley, London, 1756.

transit, the momentum lost by the two bodies is carried by the field. It can be demonstrated in the analogous case of the electromagnetic force between two separated charges that the electromagnetic field can indeed transport momentum. It is more difficult to demonstrate this for the gravitational field.

Newton's extension of the action-reaction law to separated bodies generated the conceptual difficulties we have discussed, but the difficulties were indeed conceptual and not practical. The action-reaction law is generally an exceptionally good approximation and is very useful in practical problems.

---

**Example 4-3** A 2-kg box slides along a frictionless horizontal table at 5 m/s and collides with a stationary 6-kg box. The two boxes stick together and move away with speed $v'$. Find $v'$.

The only horizontal forces acting on the boxes are the large contact forces during the collision. These forces constitute an action-reaction pair and are equal and opposite. The horizontal momentum of the two-box system is therefore conserved. Before the collision the momentum of the first box is $p_1 = m_1 v_1 = (2 \text{ kg})(5 \text{ m/s}) = 10 \text{ kg·m/s}$ in the direction of motion of the box. After the collision each box has speed $v'$, so that the total momentum of the system has the magnitude $p_2 = (2 \text{ kg})v' + (6 \text{ kg})v' = (8 \text{ kg})v'$. Setting this equal to the initial momentum gives

$$(8 \text{ kg})v' = 10 \text{ kg·m/s}$$
$$v' = \tfrac{10}{8} \text{ m/s} = 1.25 \text{ m/s}$$

---

### Questions

20. Why can momentum conservation be applied to the collision of two cars even when there are other forces acting, such as friction?

21. When an object absorbs or reflects light, it experiences a small but measurable force. Consider the following case. A supernova (stellar explosion) occurs in which the star becomes millions of times brighter than normal for a few weeks. Light from the star reaches the earth centuries after the explosion and causes the deflection of a delicately balanced mirror. Does the action-reaction law apply to the interaction of the star and the mirror?

## 4-5   Applications to Problem Solving: Constant Forces

Newton's laws are applied in two ways: (1) to determine the acceleration, velocity, and position of a particle as functions of time given all the forces acting on the particle, and (2) to determine the forces acting on a particle given the acceleration, velocity, or position of the particle as a function of time. If all the forces acting on a particle are known, the acceleration is found from $\mathbf{a} = \Sigma \mathbf{F}/m$. On the other hand, if the position or velocity of a particle is known as a function of time, the acceleration can be determined by differentiation and the resultant force from $\Sigma \mathbf{F} = m\mathbf{a}$. In this section we illustrate the application of Newton's laws to problem solving by considering some simple examples of motion under constant forces. Careful study of these simple examples will make you aware of the content of newtonian mechanics and how it is applied.

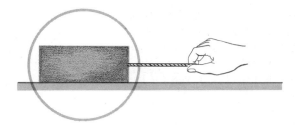

**Figure 4-8**
A block on a frictionless horizontal surface with a horizontal force exerted on it through a string. The first step in solving the problem is to isolate the system to be analyzed. In this case the circle isolates the block from its surroundings.

Practical problems are generally more complex than these examples, but the methods of solving them are natural extensions of those illustrated here. Some more general examples of the application of Newton's laws will be discussed in Chapter 5.

Consider a block resting on a frictionless horizontal table and pulled with a force applied to a light string, as shown in Figure 4-8. To find the motion of the block we need to find the resultant force acting on it. The first step is to choose the object whose acceleration is to be determined and upon which the forces to be considered act. In the figure, a circle is drawn around the block to help us isolate it mentally from its surroundings. There are two classes of forces that can act on a body:

1. *Contact forces*, exerted by objects such as strings, surfaces, etc., in direct contact with the body.

*Two classes of forces*

2. *Action-at-a-distance* forces, i.e., forces which act over a space between the body considered and the object exerting the force. The force of gravity or weight of the body is the most familiar example. Electric and magnetic forces are also of this kind. To determine whether such forces act we must know whether there is any body nearby which can exert significant gravitational force or electric charges nearby which can exert electric or magnetic forces.

Three significant external forces act on the block in this example. They are indicated in a *free-body diagram* in Figure 4-9:

*Free-body diagram*

**w**   the weight, or force due to gravity, pulling down on the block. This is an action-at-a-distance force.

**N**   the contact force exerted by the table. In general, a surface like the table exerts a force which has both a tangential component, called the *frictional force*, and a perpendicular component, called the *normal force* (the word "normal" means perpendicular). In this problem we are assuming that the frictional force is negligible compared with other forces in the problem. Thus the force exerted by a frictionless surface is perpendicular, or normal, to the surface.

*Frictional force*
*Normal force*

**T**   the contact force exerted by the string, called the *tension* in the string. We shall prove later that this force is equal to the force **F** applied by the hand to the string.

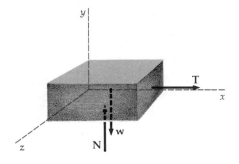

**Figure 4-9**
A free-body diagram for the block of Fig. 4-8. The three significant forces acting on the block are the force exerted by the earth **w**, the normal force exerted by the table **N**, and the force exerted by the string **T**.

A convenient coordinate system is also indicated in Figure 4-9. Note that the normal force **N** and the weight **w** are drawn with equal magnitude. We know these forces have equal magnitude because the block does not accelerate vertically. Since the resultant force is in the $x$ direction and has the magnitude $T = F$, Newton's second law gives for the acceleration $T = ma_x$; thus $a_x = T/m = F/m$.

Even in this simple example, both kinds of applications of Newton's laws are used: the horizontal acceleration is found in terms of the given force $F$, and the vertical force $N$ exerted by the table was found from the fact that the block remains on the table and thus $a_y = 0$. Information which limits the kind of motion possible for the block is called a *constraint*.

*Constraint*

According to Newton's third law, forces always act in pairs. In Figure 4-9 we have only three forces. What are the action-reaction pairs in this example? Figure 4-10 shows three forces not shown in Figure 4-9, the gravitational force **w'** exerted *by the block on the earth*, the force **N'** exerted *by the block on the table*, and the force **T'** exerted *by the block on the string*. None of these forces is exerted *on* the block. Therefore they have

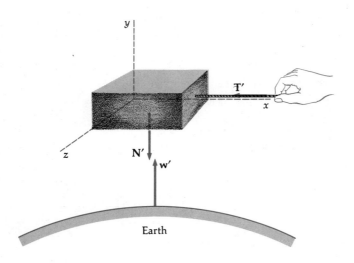

**Figure 4-10**
The reaction forces corresponding to the three forces shown in Figure 4-9. Note that these forces *do not* act on the block: **T'** acts on the string, **N'** on the table, and **w'** on the earth.

nothing to do with the motion of the block and must be omitted in the application of the second law to the motion of the block. The result that the tension in the string equals the magnitude of the force exerted by the hand on the string is obtained by considering the motion of the string. Figure 4-11 shows a free-body diagram for the string. The forces are **F** and **T'**, the force exerted by the block on the string. (We have neglected the weight of the string. If this is not neglected, the string will sag slightly and the forces **F** and **T'** will have vertical components.) If $m_s$ is the mass of the string, Newton's second law applied to the string gives

$$F - T' = m_s a_x$$

**Figure 4-11**
Free-body diagram for the string. If the string is light, so that its mass can be neglected, the forces **F** and **T'** are equal in magnitude; i.e., the tension is constant throughout the string.

We see that if the mass of the string $m_s$ is very small, the forces $F$ and $T'$ are equal. *A light string connecting two points has a tension which has constant magnitude throughout and which acts in the direction of the string at any point.* This result also holds for a string that passes over a frictionless peg or a pulley of negligible mass as long as there are no tangential forces on the string between the two points considered.

This simple example illustrates a general method of attack for problems using Newton's laws, which consists of the following steps:

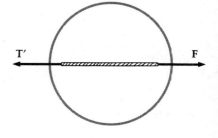

1. Draw a neat diagram.

*General methods for problem solving*

2. Isolate the body (particle) of interest and draw a free-body diagram, indicating every external force acting on the body.* Do this for each body if there is more than one in the problem, drawing a separate diagram for each.

3. Choose a convenient coordinate system for each body and apply Newton's law $\Sigma\mathbf{F} = m\mathbf{a}$ in component form.

4. Solve the resulting equations for the unknowns using whatever additional information is available, e.g., constraints. The unknowns generally will include the components of both the acceleration and some of the forces.

5. Finally, inspect the results carefully, checking whether they correspond to reasonable expectations. Particularly valuable is to determine what your solution predicts when variables in the solution are assigned extreme values. In this way you can check your work for errors.

We now give a variety of examples.

---

**Example 4-4** Find the acceleration of a block of mass $m$ which moves on a frictionless fixed surface inclined at an angle $\theta$ to the horizontal.

There are only two forces acting on the block, the weight $\mathbf{w}$ and the force $\mathbf{N}$ exerted by the incline (see Figure 4-12). We neglect air resistance, and we are instructed that there is no friction at the contact with

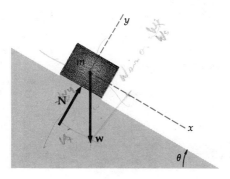

**Figure 4-12**
Forces acting on a block of mass $m$ on a frictionless incline. It is convenient to choose the $x$ axis parallel to the incline.

the incline. Since the two forces are not in the same direction, they cannot add to zero and the block must therefore accelerate. Again, we have a constraint: the acceleration is along the incline. It is convenient for this problem to choose a coordinate frame with one axis parallel to the incline and the other perpendicular, as shown in Figure 4-12. Then the acceleration has only one component, $a_x$. For this choice, $\mathbf{N}$ is in the $y$ direction, and the weight $\mathbf{w}$ has the components

$$w_x = w \sin \theta = mg \sin \theta$$

$$w_y = -w \cos \theta = -mg \cos \theta$$

4-14

where $m$ is the mass and $g$ is the acceleration due to gravity (Figure 4-13).

---

* We assume that all conceivable forces capable of acting on a body can be conveniently separated into two classes: (1) a small number of dominant forces which can be enumerated and (2) an unspecified number of other forces, e.g., air resistance, the gravitational attraction of the moon or other bodies, etc., which are too small to affect the motion of the body and can reasonably be neglected. Only the dominant forces are indicated in the free-body diagram.

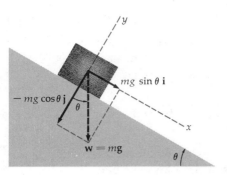

**Figure 4-13**
The weight of the block can be replaced by its components $mg \sin \theta$ parallel to the incline and $mg \cos \theta$ perpendicular to the incline. The component $mg \cos \theta$ is balanced by the normal force **N** (not shown).

The resultant force in the $y$ direction is $N - mg \cos \theta$. From Newton's second law and the fact that $a_y = 0$,

$$\Sigma F_y = ma_y = N - mg \cos \theta = 0$$

and thus

$$N = mg \cos \theta \qquad\qquad 4\text{-}15$$

Similarly, for the $x$ components,

$$\Sigma F_x = ma_x = mg \sin \theta$$

$$a_x = g \sin \theta \qquad\qquad 4\text{-}16$$

The acceleration down the incline is constant and equal to $g \sin \theta$. It is useful to check our results at the extreme values of inclination, $\theta = 0$ and $\theta = 90°$. At $\theta = 0$, the surface is horizontal. The weight has only a $y$ component, which is balanced by the normal force $N = mg \cos 0° = mg$. The acceleration is of course zero; $a_x = g \sin 0° = 0$. At the opposite extreme, $\theta = 90°$, the incline is vertical. There the weight has only an $x$ component along the incline, and the normal force is zero; $N = mg \cos 90° = 0$. The acceleration is $a_x = g \sin 90° = g$ since the block is in free fall.

When friction is neglected, the body slides down the incline with acceleration $g \sin \theta$. (*Courtesy of the Museum of Modern Art/Film Archives, New York.*)

**Example 4-5** An 8-N picture is supported by two wires of tension $T_1$ and $T_2$, as shown in Figure 4-14a. Find the tension in the wires.

This is a problem in static equilibrium. Since the picture does not accelerate, the net force acting on it must be zero. The three forces acting on the picture, its weight $m\mathbf{g}$, the tension $\mathbf{T}_1$, and the tension $\mathbf{T}_2$, must therefore sum to zero. Since the weight has only a vertical component $mg$ downward, the horizontal components of the tensions $\mathbf{T}_1$ and $\mathbf{T}_2$ must be equal in magnitude and the vertical components of the tensions must balance the weight:

$$\Sigma F_x = T_1 \cos 30° - T_2 \cos 60° = 0$$

$$\Sigma F_y = T_1 \sin 30° + T_2 \sin 60° - mg = 0$$

Using $\cos 30° = \sqrt{3}/2 = \sin 60°$ and $\sin 30° = \frac{1}{2} = \cos 60°$ and solving for the tensions, we obtain

$$T_1 = \tfrac{1}{2}mg = 4 \text{ N}$$

$$T_2 = \sqrt{3}\, T_1 = \frac{\sqrt{3}}{2} mg = 6.93 \text{ N}$$

**Example 4-6** A block hangs by a string which passes over a frictionless peg and is connected to another block on a frictionless table. Find the acceleration of each block and the tension in the string.

Figure 4-15 shows the important elements of this problem. The string tensions $\mathbf{T}_1$ and $\mathbf{T}_2$ have equal magnitude because the string is assumed to be massless and there are no tangential forces acting on it (the peg is frictionless). For block 1 on the table, the vertical forces, $\mathbf{N}$ and $\mathbf{w}_1$, have equal magnitudes because of the constraint that the vertical acceleration is zero for $m_1$. Newton's second law applied to the horizontal components gives

$$T = m_1 a_1 \qquad\qquad 4\text{-}17$$

where $a_1$ is the acceleration of $m_1$ along the horizontal surface.

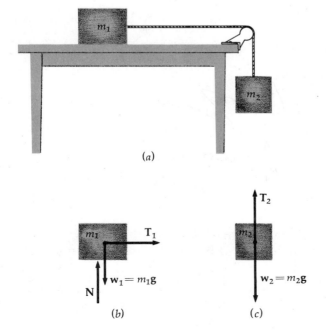

(a)

(b)          (c)

**Figure 4-14**
(a) Picture supported by two wires in Example 4-5. (b) Choice of coordinate system and resolution of the forces into $x$ and $y$ components.

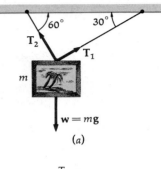

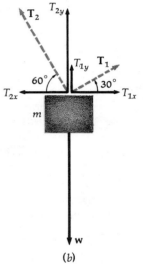

**Figure 4-15**
(a) The two blocks of Example 4-6. Free-body diagrams for (b) $m_1$ and (c) $m_2$.

If we take the downward direction to be positive for the acceleration $a_2$ of block 2, the equation of motion for $m_2$ is

$$m_2g - T = m_2a_2 \qquad\qquad 4\text{-}18$$

We can simplify these equations by noting that if the connecting string does not stretch, the accelerations $a_1$ and $a_2$ are both positive and equal in magnitude (but not in direction). Let us call this magnitude $a$. We then have

$$T = m_1a \qquad\qquad 4\text{-}19$$

$$m_2g - T = m_2a \qquad\qquad 4\text{-}20$$

Solution of these equations gives

$$a = \frac{m_2}{m_1 + m_2}g \qquad\qquad 4\text{-}21$$

$$T = \frac{m_1m_2}{m_1 + m_2}g \qquad\qquad 4\text{-}22$$

Note that although the result for $a$ appears to be the same as that for a mass $m = m_1 + m_2$ acted on by a force $m_2g$, this is not the case because the magnitude of the acceleration of each mass is the same but the direction is not.

---

**Example 4-7** *Apparent Weight in an Accelerating Elevator* A block rests on a scale in an elevator, as in Figure 4-16. What is the scale reading when the elevator is accelerating (*a*) up and (*b*) down?

(*a*) Since the block is at rest relative to the elevator, it is also accelerating up. The forces on the block are **N**, exerted by the scale platform on which it rests, and **w**, the force of gravity. If we call the upward acceleration **a**, Newton's second law gives

$$\mathbf{N} + \mathbf{w} = m\mathbf{a}$$

or in terms of the upward components,

$$N - w = ma \qquad N = w + ma = mg + ma \qquad 4\text{-}23$$

The force **N′** exerted by the block on the scale determines the reading on the scale, the apparent weight. Since **N′** and **N** are an action-reaction pair, they are equal in magnitude. Thus when the elevator accelerates up, the apparent weight of the block is greater than its true weight by the amount $ma$.

(*b*) Let us call the acceleration **a′**. Again Newton's second law gives

$$\mathbf{N} + \mathbf{w} = m\mathbf{a'}$$

but in this case the acceleration is downward. In terms of the magnitudes, we have

$$w - N = ma'$$

or

$$N = w - ma' = mg - ma' \qquad 4\text{-}24$$

Again, the scale reading, or apparent weight, equals $N$. In this case, the apparent weight is less than $mg$. If $a' = g$, as it would if the elevator were in free fall, the block would be apparently weightless. What if the acceleration of the elevator is greater than $g$? Assuming that the surface of the scale is not sticky, the scale cannot exert a force down on the block. (We assume that the scale is fastened to the floor of the elevator.)

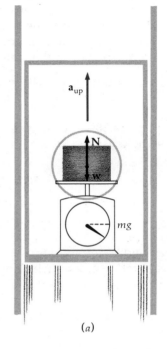

(a)

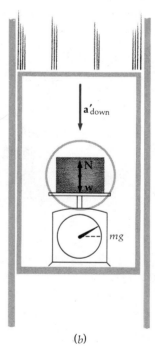

(b)

**Figure 4-16**
Block on a scale in an accelerating elevator. The scale indicates the *apparent weight N*, which is greater than *mg* when the acceleration is upward and less than *mg* when the acceleration is downward.

Since the downward force on the block cannot be greater than $w$, the scale will soon leave the block. The block will have the acceleration $g$, which is less than that of the elevator, so eventually the block will hit the ceiling of the elevator. Then if the ceiling is strong enough, it can provide the force downward necessary to give the block the acceleration $a'$.

---

## Questions

22. A picture is supported by two wires as in Example 4-5. Do you expect the tension to be greater or less in the wire that is more nearly vertical?

23. A weight is hung on a wire that is originally horizontal. Can the wire remain horizontal? Explain.

24. Give an example in which the tension in a string does not have the same magnitude throughout.

25. What effect does the velocity of the elevator have on the apparent weight of the block in Example 4-7?

## 4-6 Reference Frames

In Newton's first law no distinction is made between a particle at rest and one moving with constant velocity. If no external forces are acting, the particle will remain in its initial state—either at rest or moving with its initial velocity. Consider a particle at rest relative to you with no forces acting on it. According to the first law, the particle will remain at rest. Now consider the same particle from the point of view of a second observer moving with constant velocity relative to you. From his "frame of reference" both you and the particle are moving with constant velocity. Newton's first law also holds for him. How can we distinguish whether you and the particle are at rest and the second observer is moving or the second observer is at rest and you and the particle are moving? In this section we shall show that such a distinction cannot be made from Newton's laws. There is no way to determine the "absolute" velocity of a particle. This result is known as the *principle of relativity*.

*Principle of relativity*

To measure the velocity and acceleration of a particle, we need a coordinate system, as illustrated in Figure 4-17. Consider another coor-

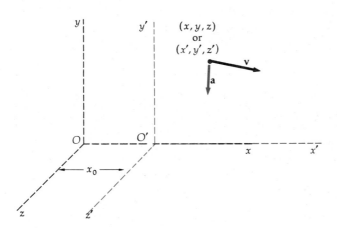

**Figure 4-17**
Two coordinate systems at rest relative to each other. A particle has the same velocity **v** and the same acceleration **a** in both systems. These coordinate systems are said to be in the same reference frame.

dinate system $x'y'z'$ *at rest* relative to $xyz$. For convenience, we have assumed the axes parallel and the origin $O'$ to be on the $x$ axis at a distance $x_0$ from $O$. A particle at $x$ with velocity **v** and acceleration **a** relative to $O$ will have the same velocity $\mathbf{v}' = \mathbf{v}$ and acceleration $\mathbf{a}' = \mathbf{a}$ relative to $O'$, but its coordinate will be different, $x' = x - x_0$. *The set of coordinate systems at rest relative to a given system is called a reference frame.* Clearly, any one coordinate system in a reference frame is as good as any other for describing the motion of a particle, for **v** and **a** are the same in all such coordinate frames.

Now consider two coordinate systems in motion relative to each other. Let axes $x'$, $y'$, $z'$ move with *constant velocity* $\mathbf{v}_0$ relative to axes $x$, $y$, $z$. (For example, the axes $x'$, $y'$, $z'$ might be fixed to a train moving at constant velocity relative to the track.) For convenience, let the origins coincide at $t = 0$ and take $\mathbf{v}_0$ along the $x$ or $x'$ axes, as in Figure 4-18. We

*Reference frame defined*

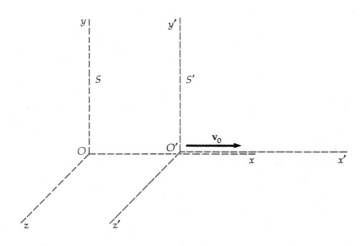

**Figure 4-18**
Two coordinate systems moving relative to each other are in different reference frames. The reference frame $S'$ moves to the right with speed $v_0$ relative to reference frame $S$. A particle has different positions and velocities but the same acceleration in the two reference frames.

now have two different reference frames, which we shall call $S$ and $S'$. $S'$ moves with velocity $v_x = v_0$ relative to $S$ whereas $S$ moves with velocity $v'_x = -v_0$ relative to $S'$. A particle at rest in one frame is not at rest in the other. The coordinates are related by

$$x' = x - v_0 t \qquad 4\text{-}25$$

The velocities of the particle measured in the two reference frames are related by

$$v'_x = \frac{dx'}{dt} = \frac{dx}{dt} - v_0 = v_x - v_0 \qquad 4\text{-}26$$

Differentiating again, we find that the accelerations in the two frames are equal since the relative velocity $v_0$ was assumed to be constant:

$$a'_x = \frac{dv'_x}{dt} = \frac{dv_x}{dt} = a_x \qquad 4\text{-}27$$

Suppose we have a body of mass $m$ attached to a spring which is extended by an amount $\Delta x$ and thus exerts a force $F_x$ in reference frame $S$. The acceleration will be $a_x = F_x/m$ assuming that Newton's law holds in this frame.

In frame $S'$ the position $x'$ and velocity $v'_x$ of the body are not the same, but the acceleration $a'_x$ is the same as $a_x$. Furthermore, the extension of the spring is the same. Thus the force is the same, and if $\mathbf{F}_{\text{net}} = m\mathbf{a}$ holds in frame $S$, it also holds in frame $S'$.

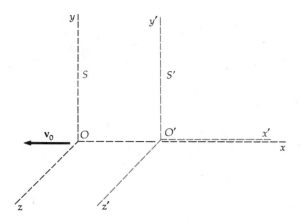

**Figure 4-19**
The two coordinate systems of Figure 4-18 from the point of view of reference frame $S'$. In this frame $S'$ is at rest and frame $S$ moves to the left with speed $v_0$. Only the relative speed $v_0$ between the reference frames is important.

We could just as well have made the drawing from the point of view of $S'$ and had $S$ moving with speed $v_0$ to the left (Figure 4-19). Is there any way to determine which frame is really at rest and which frame is moving? That is, is it possible to determine absolute velocity? Since only acceleration and not velocity appears in Newton's laws, there are no mechanics experiments available for distinguishing absolute motion. This *principle of relativity* was discussed in the fourteenth century, but it became well known only with the work of Galileo early in the seventeenth century. It is interesting to read Galileo's remarks on the impossibility of observing absolute motion:*

> Shut yourself up with some friend in the main cabin below decks on some large ship, and have with you there some flies, butterflies, and other small flying animals. Have a large bowl of water with some fish in it; hang up a bottle that empties drop by drop into a wide vessel beneath it. With the ship standing still, observe carefully how the little animals fly with equal speed to all sides of the cabin. The fish swim indifferently in all directions; the drops fall into the vessel beneath; and, in throwing something to your friend, you need throw it no more strongly in one direction than another, the distances being equal; jumping with your feet together, you pass equal spaces in every direction. When you have observed all these things carefully (though there is no doubt that when the ship is standing still everything must happen in this way), have the ship proceed with any speed you like, so long as the motion is uniform and not fluctuating this way and that. You will discover not the least change in all the effects named, nor could you tell from any of them whether the ship was moving or standing still. In jumping you will pass on the floor the same spaces as before, nor will you make larger jumps toward the stern than toward the prow even though the ship is moving quite rapidly, despite the fact that during the time you are in the air the floor under you will be going in a direction opposite to your jump. In throwing something to your companion, you will need no more force to get it to him whether he is in the direction of the bow or the stern, with yourself situated opposite. The droplets will fall as before into the vessel beneath without dropping toward the stern, although while the droplets are in the air the ship runs many spans. The fish in their water will swim toward the front of their bowl with no more effort than toward the back, and will go with equal ease toward bait placed anywhere around the edges of

---

* Galileo Galilei, *Dialogue Concerning Two Chief World Systems—Ptolemaic and Copernican*, pp. 186–187, trans. Stillman Drake, University of California Press, Berkeley, 1953; reprinted by permission of the Regents of the University of California.

Midair refueling. Although the velocity of each plane is very great relative to the earth, the velocity of one plane relative to the other is approximately zero.

the bowl. Finally the butterflies and flies will continue their flights indifferently toward every side, nor will it ever happen that they are concentrated toward the stern, as if tired out from keeping up with the course of the ship, from which they have been separated during long intervals by keeping themselves in the air. And if smoke is made by burning some incense, it will be seen going up in the form of a little cloud, remaining still and moving no more toward one side than the other.

In the late nineteenth century it was thought that a careful measurement of the velocity of light relative to the earth would reveal absolute motion of the earth through space. Such an observation would contradict the principle of relativity, which was not yet accepted as a fundamental law of nature. Careful experiments by Michelson and Morley, however, gave a null result for the absolute velocity of the earth: they could not detect motion of the earth by measuring the velocity of light. The principle of relativity was again enunciated as a universal law by Poincaré and Einstein: absolute uniform motion cannot be detected by any experiment. The consequences of this principle, along with the assumption, based on experiment, that the speed of light is the same for all observers independent of their motion relative to the light source, were investigated by Einstein in 1905. His revolutionary results are known as the *special theory of relativity*. We shall study Einstein's theory in detail in Chapter 35.

A reference frame in which Newton's laws hold is called an *inertial reference frame*. All reference frames moving with constant velocity relative to an inertial frame are also inertial reference frames. Consider now the case in which $S$ is an inertial reference frame and $S'$ is *accelerating* relative to $S$. $S'$ might be, for example, a coordinate frame at rest in a railroad car which is accelerating with acceleration $\mathbf{a}_0$ relative to the tracks. We find that Newton's laws do not hold in the railroad car. For example, a body on a smooth table does not remain at rest or move with constant velocity relative to the car; instead it has an acceleration $-\mathbf{a}_0$. Figure 4-20 shows a body suspended by a cord from the ceiling of the car. There is an unbalanced force to the right, but there is no acceleration (relative to the car). Since Newton's laws do not hold in $S'$, this is not an inertial reference frame.

*Inertial reference frame*

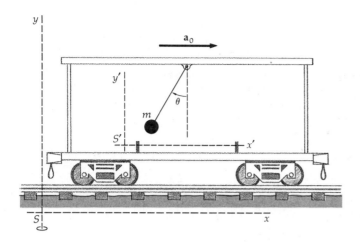

**Figure 4-20**
Body suspended by a cord from the ceiling of an accelerating car. The horizontal component of the tension in the cord is unbalanced, causing the body to accelerate to the right relative to the inertial frame $S$ attached to the tracks. Frame $S'$, attached to the car, is not an inertial reference frame because Newton's laws do not hold in this frame. Although there is an unbalanced force to the right, the body remains at rest and does not accelerate relative to frame $S'$.

It is important to note that in order to classify a reference frame as inertial or not, we need to know all the possible forces acting on a body. We can identify the possible forces by finding the agent, i.e., the other body, responsible for each force. If we suspect the existence of a force because of an observed acceleration of the body but cannot find any agent responsible for it, we must conclude that the observed acceleration results because our reference frame is noninertial.

Consider a body dropped in the accelerating railroad car. Its acceleration (relative to the car) is

$$a_y = -g \qquad a_x = -a_0$$

We describe the $y$ component of acceleration as being due to the pull of gravity of the earth. Might not the $x$ component be due to a similar cause, i.e., a large mass to the left of the car? To determine whether the car is in an inertial frame we must look outside for possible forces. Because there is no apparent agent (such as a large massive object to the left of the car) responsible for the observed acceleration to the left, we label the reference frame in the car as noninertial.

Two questions now arise: Can absolute acceleration be detected? Is there any natural inertial reference frame? Newton gave the following demonstration of the absoluteness of acceleration. Consider a pail of water suspended by a rope. The surface of the water is level. Now give the pail a twist. At first the pail rotates, but the water does not. Because of viscosity, the water begins to rotate, and the surface is no longer level but assumes a parabolic shape. Eventually the water and pail have the same angular velocity and thus no relative velocity or acceleration. If the pail is now suddenly stopped, the water continues to rotate. The shape of the water surface does not depend on the relative motion of the water and pail but only on the absolute rotation of the water, or the rotation relative to the "fixed stars." Suppose, however, that we could rotate *all* the surroundings of the water including the fixed stars while leaving the water at "rest." What shape would the surface of the water take? If it remained flat, we would have to believe in absolute acceleration. If instead it assumed the parabolic shape, as when it was rotating, we would conclude that only relative acceleration is observable.

We can state this question in terms of our accelerating boxcar at rest and give all the stars in the universe an acceleration $-\mathbf{a_0}$ relative to the boxcar. Would a body dropped in the boxcar accelerate toward the earth, i.e., obey Newton's laws, or would it accelerate also to the left, as when we accelerated only the boxcar? These questions cannot be answered with certainty, for no such experiment or its equivalent has been performed. We may, with Newton, believe in absolute acceleration.

A reference frame attached to the surface of the earth is certainly useful in describing the motion of everyday objects, but it is not an inertial frame. Because of the earth's rotation, such a frame has an acceleration directed toward the axis of rotation. At the equator, this acceleration is

$$a = \frac{v^2}{R_E} = \frac{(2\pi R_E/T)^2}{R_E} = \frac{4\pi^2 R_E}{T^2} \approx 3.4 \times 10^{-2} \text{ m/s}^2 \qquad 4\text{-}28$$

using $T \approx 86{,}400$ s for the period of rotation, 1 day (d), and $R_E = 6.4 \times 10^6$ m for the radius of the earth.* When this acceleration is very small compared with others, it can be neglected and the surface of the earth taken as approximately an inertial frame. Alternatively we can compensate for the earth's rotation by choosing a frame which rotates *relative to the earth* with a period of 1 d. Even this frame is accelerating, however, because of the revolution of the earth about the sun. Since $T = 1$ y $\approx 3.16 \times 10^7$ s for the period and $r \approx 1.49 \times 10^{11}$ m for the radius of the orbit, this acceleration is

$$a_{\text{rev}} \approx 5.9 \times 10^{-3} \text{ m/s}^2 \qquad 4\text{-}29$$

Suppose we choose a reference frame at rest relative to the sun and other fixed stars. To Newton, such a choice seemed a likely possibility for an inertial reference frame. Today we know more about the motion of the fixed stars. The acceleration of the sun corresponding to the rotation of the galaxy is about $10^{-10}$ m/s$^2$, which is small enough to be neglected for most purposes. There is apparently no simple physical system such as the earth or the sun that is exactly at rest in an inertial reference frame. We can of course set up an inertial reference frame without dependence on this result. We need merely to take a first approximation, such as a rectangular coordinate system at rest relative to the sun, project masses along the axis, and measure any deviation from constant velocity. We can then correct for any observed acceleration by choosing a second coordinate system accelerating relative to the first. In practice, we need only find a frame in which $\mathbf{F}_{\text{net}} = m\mathbf{a}$ holds within the accuracy of our measuring apparatus.

**Questions**

26. Suppose you are inside a closed railroad car (no sound or light from outside reaches you). How can you decide whether the car is moving? How can you tell whether it is speeding up? Slowing down? Rounding a curve?

27. According to the principle of relativity you cannot detect your own constant velocity. Yet even a deaf and blind person riding at constant velocity along a freeway can tell that the car is not at rest. How?

28. Should it be possible to tell that the earth itself is moving by any experiment conducted in a closed room? Can you think of any possible experiments?

---

* The free-fall acceleration of a body *relative to the surface of the earth* is thus less than $g$ by 3.4 cm/s$^2$ at the equator.

# Isaac Newton (1642–1727)

I. Bernard Cohen
*Harvard University*

When Isaac Newton was once asked how he had made his great discoveries, he replied, "By always thinking unto them." He is also reported to have said, "I keep the subject constantly before me and wait till the first dawnings open little by little into the full light." This ability to concentrate is a particular quality of Newton's genius, and it fits in well with his character and personality. For he was a solitary man, without close and intimate friends or confidants. He never married, and spent his early boyhood deprived of father (who died before young Isaac was born on Christmas day in 1642) and of mother (who remarried within 2 years and left Isaac to be reared by an aged grandmother).

A lonesome man, he developed "unusual powers of continuous concentrated introspection." In these words, his biographer Lord Keynes, the economist, epitomized Newton's "power of holding continuously in his mind a purely mental problem . . . for hours and days and weeks until it surrendered to him its secret." And then, in keeping with this character of inwardness, he was satisfied to keep his discoveries to himself, neither rushing into print, as his contemporary fellow scientists were wont to do, nor even communicating to his associates what he had accomplished. Accordingly, it has been said that every discovery of Newton's had two phases; Newton made the discovery, and then others had to find out that he had done so.

In 1684, Edmund Halley (the astronomer, after whom Halley's comet is named) went from London to Cambridge, where Newton was a professor, to ask him about a fundamental problem that had baffled the major scientists of the Royal Society (the world's oldest existing scientific organization), including Robert Hooke (of Hooke's law) and Christopher Wren (architect as well as scientist). What "curve would be described by the Planets," Halley asked Newton, "supposing the force of attraction towards the Sun to be reciprocal to the square of their distance from it?" Newton "replied immediately that it would be an Ellipsis," i.e., an ellipse. Halley, "struck with joy and amazement asked him how he knew it. Why, saith he, I have calculated it." In these final four words, Newton revealed that he had solved the major scientific problem of the century: to find the law of force that holds the solar system together, that causes the planets to move around the sun according to Kepler's laws, and that equally produces the same keplerian motion in satellites moving around planets.

Once Christiaan Huygens, in 1673, had published the law of "centrifugal force" (as he named it), one might easily compute that in uniform *circular motion* the central force is inversely proportional to the square of the distance from the center; and so, in fact, it was no great feat to *guess,* as Halley, Hooke, and Wren had done, that the same inverse-square law might apply to the *elliptical* orbits of planets and their satellites. However, Newton had done far more. He had not only proved rigorously that an inverse-square force produces a keplerian elliptical orbit; he had also found the dynamical significance of Kepler's two other laws. He then went on to show that it is one and the same universal gravitational force that keeps the planets in their observed orbits around the sun and the satellites in their orbits around the planets, that (by the pull of the moon) produces tides in the oceans, and that causes bodies to fall to earth with the observed acceleration. And he computed the law of this force: that it is directly proportional to the product of the masses of any gravitating bodies and inversely proportional to the distance between them. His actual reply to Halley ("I have calculated it") reveals the character of the man, since he had been satisfied merely to have made the calculations and had not rushed to make known the primary scientific discovery of the age: the law of universal gravitation.

The seeds of Newton's great achievements in science go back to a period of some 18 months after graduation from college (1665–1667), when fear of the plague had caused the university to be shut and Newton returned to the family farm in Woolsthorpe, Lincolnshire, where he had been born. Here he worked out his most fundamental contributions to mathematics, the methods of the differential and integral calculus (an innovation for which he must share the credit with the German philosopher and mathematician Leibniz, an independent codiscoverer). During this same time, he found the law of the inverse square and tested it by a rough calculation of the moon's motion. In this work he independently discovered the law of centrifugal force: that in uniform circular motion along a circle of radius $r$ at speed $v$ this force or acceleration must be proportional to $v^2/r$. He knew that for any system of bodies moving about a central body (such as the planets going around the sun or the set of moons around Jupiter) there holds a form of Kepler's third law: that in each such system $r^3/T^2$ is a constant $k$, where $T$ is the period of revolution. Thus, considering the orbits to be circular (an approximation, but a rather good one at a first stage), the speed $v$ is $2\pi r/T$, and so $v^2/r = 4\pi^2 r^2/T^2 r$. A clever man would see at once that multiplying by $r/r$ gives

$$\frac{v^2}{r} = \frac{4\pi^2 r^3}{T^2 r^2} = \frac{4\pi^2}{r^2}\frac{r^3}{T^2} = \frac{4\pi^2 k}{r^2}$$

That is, $v^2/r$ is equal to a constant multiplied by $1/r^2$; or the force is inversely proportional to the square of the distance. But is it? Newton's test was to note that the moon is about 60 times as far from the center of the earth as an apple. Accordingly, the "force" or acceleration of the moon, as it "falls" toward the earth and constantly "falls" away from straight-line inertial motion, should be just $1/60^2$, or $1/3600$ of the acceleration of free fall of an apple. The calculations, Newton said, agreed "pretty nearly." In his own words, "I thereby compared the force requisite to keep the Moon in her orb with the force of gravity at the surface of the Earth, and found them answer pretty nearly."

Later, Newton had to show that this inverse-square law also holds in elliptical orbits, and he had to take account of the modification required of Kepler's laws in an earth-moon system, because he proved that each of these two bodies (and any other pair) will move about their common center of mass. And he had to use his mathematical prowess to prove the fundamental theorem, that a uniform sphere (or a sphere made up of concentric uniform shells) will attract gravitationally as if all its mass were concentrated at its geometric center. Eventually, Newton cast out the misleading concept of centrifugal force, and introduced a new kind of force which he named *centripetal*.

A third set of discoveries made by Newton during his period of retreat during the plague years was in the field of optics. Newton is celebrated not only for his work in pure mathematics and for having founded the science of celestial mechanics; he is also known as an experimental scientist, primarily for his discoveries relating to light and color and that class of effects known today as *interference phenomena*. He bought a triangular glass prism, he tells us,

Manuscript drawing in Newton's papers, showing the "crucial experiment." Sunlight enters a darkened room at the left-hand side from $S$, producing a spectrum on passing through the first prism $A$. A hole at $Y$ in the board $DE$ permits light of a single color to pass through the second prism $F$, with the result of a deviated beam of light, without any further alteration in color. (*Courtesy of Cambridge University Library.*)

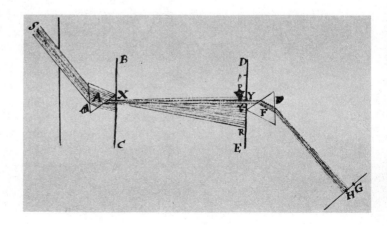

to try "the celebrated phenomenon of colours." But where others had played around with the colored spectrum produced by such a prism, Newton proceeded to analyze the phenomena of dispersion and the composition of white light and was led to devise and construct a new form of telescope, which in the succeeding three centuries has been astronomy's most powerful tool.

Newton's experiments on color showed that "white" light, or sunlight, is composite, a mixture of light of all the different colors. He demonstrated that the prism produces a spectrum, or separates out the different colors, because each color has its own index of refraction or degree of being turned or deviated from the original path. The violet light he found to be deviated the most, the red light the least. In order to prove that this is so, he devised what he called a "crucial experiment." He allowed a narrow beam of sunlight to enter a darkened room through a small hole in a window shutter. This beam then passed through a prism and produced a spectrum. Using an opaque board with a small hole in it, Newton could separate out of this spectrum a beam of light of a single color—red, orange, green, blue—and allow this monochromatic light to pass through a second prism. If prisms produce a spectrum by "altering" the incident light in some way, this effect should be apparent in the way the second prism would alter or affect an incident beam of monochromatic light. When monochromatic light entered the second prism, it emerged with no change of color; the second prism had produced only a further deviation, or change of path by refraction.

These investigations not only led Newton to an understanding of why objects appear to have the colors they display, in terms of the colors of the light they absorb or transmit or reflect, but also showed him that telescopes with simple or single lenses have definite limitations. Since a biconvex lens can be thought of as a pair of prisms base to base (Figure 1), we can see at once that there will be a different focus for each color, or a poor image due to *chromatic aberration*. Newton accordingly proceeded to invent a wholly new kind of telescope, a reflector rather than a refractor, which by using the front surface of a concave mirror to form the image avoided chromatic aberration.

A most notable set of optical investigations centered on the effects produced by thin plates, or films, as in the familiar colored patterns seen on films of oil. Newton studied the alternating fringed rings of darkness and light that are produced by the thin layer of air between a flat glass surface and the convex side of a plano-convex lens of large focal length, a phenomenon known today as *Newton's rings*. His measurements were so precise that they were used a hundred years later by Thomas Young to compute the wavelength of light.

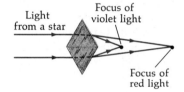

**Figure 1**
Two prisms, one on top of the other, act like a lens.

The alternating circles of brightness and darkness in monochromatic light, known as *Newton's rings*. (*From Newton's* Opticks, *1704.*)

Newton's explanation of the phenomenon of Newton's rings, produced by the thin "film" of air trapped between a plane glass surface *AB* and the convex side of the plano-convex lens *CED*. This diagram shows Newton's conception of the alternation of transmission and reflection of the light from the curved surface. (*From Newton's* Opticks, *1704.*)

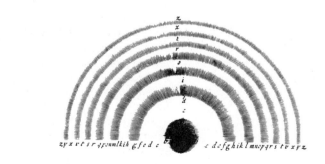

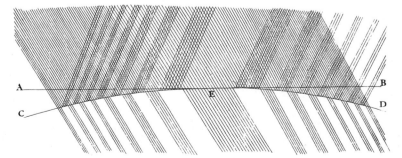

This is a remarkable set of discoveries for any lifetime, and we can only contemplate in amazement how in so short a time a young man barely out of college should have made such fundamental contributions to pure mathematics, to theoretical celestial mechanics and dynamics, and to experimental physics. As Newton himself later said, "All this was in the two plague years of 1665 and 1666, for in those days I was in the prime of my age for invention, and minded mathematics and philosophy* more than at any time since."

The remainder of Newton's scientific life was devoted to elaboration of the fundamental discoveries we have described. He became Lucasian Professor of Mathematics at Cambridge, and in 1672 his paper on light and color, together with a description of the newly invented telescope, was read to the Royal Society. After some aspects of his work had been misunderstood and criticized, he decided that he would publish no more. His mathematical innovations were distributed only in manuscript until later in his life when some were printed. Then, in 1684, after Halley's visit—and with Halley's urging—he wrote up his discoveries in dynamics and celestial mechanics in the *Philosophiae Naturalis Principia Mathematica* (London, 1687), which may be translated as *The Mathematical Principles of Celestial Mechanics.* Here he not only set forth and applied the law of universal gravitation but produced the three famous laws of motion that bear his name and that were the "axioms" on which his system of dynamics was constructed. In this work, too, he set forth for the first time a clear concept of mass and momentum and set up a series of beautiful experiments to show that mass (as measured by inertia) is proportional to weight.

In the 1690s, Newton became bored or disenchanted with the cloistered life of a university professor and moved to London, where he became director of the Mint, a post he held until his death in 1727. He became president of the Royal Society and ruled British science with a firm control. During the London years, he brought out two revised editions of his *Principia* (1713, 1726), and published his book of optics (1704 and later editions) and various tracts on mathematics. He died in 1727 and was buried with state honors in Westminster Abbey.

From early manhood, Newton devoted only a small fraction of his creative intellectual life to orthodox scientific pursuits: pure and applied mathematics, astronomy and celestial mechanics, dynamics, experimental physics, and optics. During all these years he was an ardent student of theology, reading and taking innumerable notes, and writing tracts and even whole books on religious subjects. Some of these dealt with fundamental questions of interpretation of theological doctrines, others with the meaning of the prophetic books of the Bible, and yet others with the problem of unraveling Church history. He also developed a whole new system of world chronology, in part based on astronomy, that proved to be of little value. And his chief subject of study was alchemy—reading extensively, copying out whole sections of books, making experiments, all to what end we do not know.

Through his readings of mystical philosophers, theologians, and speculative alchemists, Newton may have gained a vision of a unified system of knowledge that would embrace both physical and divine science, revealing both the laws of the created world and the plan of its Creator. Various hints of this aspect of Newton's thought appear in his writings, e.g., "And thus to discourse about God from phenomena does belong to experimental philosophy." It is this vision of a knowledge which he never attained that may have caused him to devalue his monumental scientific achievements. For shortly before his death he said, "I do not know what I may appear to the world; but to myself I seem to have been only like a boy, playing on the sea-shore, and diverting myself in now and then finding a smoother pebble or a prettier shell than ordinary, whilst the great ocean of truth lay all undiscovered before me."

---

* The word "philosophy" as used at that time meant "natural philosophy," or science.

## Review

A.  Define, explain, or otherwise identify:

Force, 80
Newton, 81
Mass, 82
Kilogram, 83
Weight, 84
Apparent weight, 85
Gram, 86
Dyne, 86
Pound, 87
Slug, 87

Momentum, 90
Action at a distance, 90
Contact force, 93
Free-body diagram, 93
Normal force, 93
Tension, 93
Constraint, 94
Principle of relativity, 99
Inertial reference frame, 102

B.  Write out the steps involved in the general method of attack for solving problems using Newton's laws of motion.

C.  True or false:

1.  If there are no forces acting on a body, the body will not accelerate.

2.  If a body is not accelerating, there must be no forces acting on it.

3.  The motion of a body is always in the direction of the resultant force.

4.  Action-reaction forces never act on the same body.

5.  The mass of a body depends on its location.

6.  The weight of a body depends on its location.

7.  Action equals reaction only if the bodies are not accelerating.

8.  Newton's laws hold only in inertial reference frames.

9.  Action-at-a-distance forces can be neglected when determining the acceleration of an object.

## Exercises

### Section 4-1, Force and Mass

1.  An object experiences an acceleration of 4 m/s² when a certain force $\mathbf{F}_0$ acts on it. (a) What is its acceleration when the force is doubled? (b) A second object experiences an acceleration of 8 m/s² under the influence of the force $\mathbf{F}_0$. What is the ratio of the masses of the two objects? (c) If the two objects are tied together, what acceleration will the force $\mathbf{F}_0$ produce?

2.  A mass is pulled in a straight line along a level, frictionless surface with a constant force. The increase in its speed in a 10-s interval is 5 km/h. When a second constant force is applied in the same direction in addition to the first force, the speed increases by 15 km/h in a 10-s interval. How do the magnitudes of the two forces compare?

3.  A force $\mathbf{F}_0$ causes an acceleration of 5 m/s² when acting on an object of mass $m$. Find the acceleration of the same mass when acted on by the forces shown in Figure 4-21a and b.

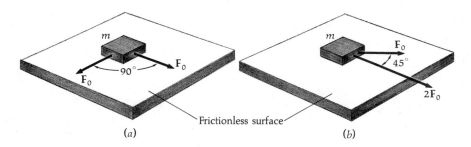

Figure 4-21
Forces acting on object for Exercise 3.

4. An electric field gives a charged object an acceleration of $6 \times 10^6$ m/s². Another field gives the same object an acceleration of $9 \times 10^6$ m/s². (*a*) If the force exerted by the first field is $\mathbf{F}_0$, what is the force exerted by the second field? What is the acceleration of the object (*b*) if the two fields act together on the same object in the same direction, (*c*) if the two fields act in opposite directions on the object, and (*d*) if the two fields are perpendicular to each other?

5. Figure 4-22 shows the path taken by an automobile. It consists of straight lines and arcs of circles. The automobile starts from rest at point $A$ and accelerates until it reaches point $B$. It then proceeds at constant speed until it reaches point $E$. From point $E$ on it slows down, coming to rest at point $F$. What is the direction of the net force, if any, on the automobile at the midpoint of each section of the path?

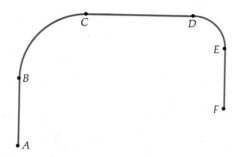

**Figure 4-22**
Path taken by automobile in Exercise 5.

6. The graph in Figure 4-23 shows a plot of $v_x$ versus $t$ for an object of mass 10 kg moving along a straight line. Make a plot of the net force on the object as a function of time.

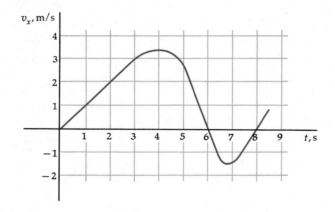

**Figure 4-23**
Graph of velocity $v_x$ versus time $t$ for Exercise 6.

7. Figure 4-24 shows the position versus time of a particle moving in one dimension. During what periods of time is there a net force acting on the particle? Give the direction ($+$ or $-$) of the net force for these times.

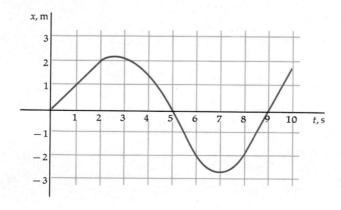

**Figure 4-24**
Graph of position $x$ versus time $t$ for Exercise 7.

8. A certain force applied to a particle of mass $m_1$ gives it an acceleration of 20 m/s². The same force applied to a particle of mass $m_2$ gives it an acceleration of 30 m/s². If the two particles are tied together and the same force is applied to the combination, find the acceleration.

9. A force of 15 N is applied to a mass $m$. The mass moves in a straight line with its speed increasing by 10 m/s every 2 s. Find the mass $m$.

10. A force of $\mathbf{F} = 6\mathbf{i} - 3\mathbf{j}$ N acts on a mass of 2 kg. Find the acceleration $\mathbf{a}$. What is the magnitude $a$?

11. A single force of 10 N acts on a particle of mass $m$. The particle starts from rest and travels in a straight line a distance of 18 m in 6 s. Find its mass.

12. A 3-kg body is acted on by a single force $\mathbf{F}_0$ perpendicular to the velocity of the body. The body travels in a circle of radius 2 m. It makes one complete revolution every 3 s. (*a*) What is the magnitude of the acceleration? (*b*) What is the magnitude of $\mathbf{F}_0$?

13. In order to drag a 100-kg log along the ground at constant velocity you have to pull on it with a force of 300 N (horizontally). (*a*) What is the resistive force exerted by the ground? (*b*) What force must you exert if you want to give the log an acceleration of 2 m/s²?

**Section 4-2, The Force Due to Gravity: Weight; and Section 4-3, Units of Force and Mass**

14. Find the weight of a 50-kg girl in newtons and pounds.

15. Find the mass of a 175-lb man in (*a*) slugs, (*b*) kilograms, and (*c*) grams.

16. A newspaper reports that 6000 lb of marijuana was found in a ship docked at Los Angeles. What is its mass in kilograms?

17. Find the weight of a 50-g body in dynes and newtons.

18. The gravitational field near the surface of the earth has the value 9.81 N/kg. What is its value in pounds per slug? What is an equivalent unit in terms of feet and seconds of 1 lb/slug?

19. The gravitational field of the earth at any height $h$ above the earth's surface can be written $g(h) = g(0)R_E{}^2/(R_E + h)^2$, where $R_E$ is the radius of the earth (about 6370 km) and where $g(0)$ is the acceleration due to gravity at the earth's surface. (*a*) Find the weight of an 80-kg man in newtons and in pounds at the earth's surface. (*b*) Find the weight of the man (in newtons and in pounds) at a height of 300 km above the earth's surface. (*c*) What is the mass of this man at this altitude?

**Section 4-4, Newton's Third Law and Conservation of Momentum**

20. Show that a newton-second is a unit of momentum.

21. A 3-kg body moves with speed 4 m/s in the $x$ direction. (*a*) What is its momentum? (*b*) If a 3-N force also in the $x$ direction acts on the body, what will its momentum be after 2 s? (*c*) If this force acts in the negative $x$ direction, what will the momentum of the body be after 2 s? After 5 s? (Assume that it is moving at 4 m/s initially in the positive $x$ direction.)

22. A 2-kg body hangs at rest from a string attached to the ceiling. (*a*) Draw a diagram showing the forces acting on the body and indicate each reaction force. (*b*) Do the same for the forces acting on the string.

23. A hand pushes two bodies on a frictionless horizontal surface, as shown in Figure 4-25. The masses of the bodies are 2 and 1 kg. The hand exerts a force of 5 N on the 2-kg body. (*a*) What is the acceleration of the system? (*b*) What is the

**Figure 4-25**
Exercise 23.

acceleration of the 1-kg body? What force is exerted on it? What is the origin of this force? (c) Show all the forces acting on the 2-kg body. What is the net force acting on this body?

24. A box slides down a rough inclined plane. Draw a diagram showing the forces acting on the box. For each force in your diagram, indicate the reaction force.

25. A 1-kg body and a 2-kg body are moving toward each other with equal speeds. They collide head on and rebound, moving away from each other with unequal speed. Compare (a) the change in momentum and (b) the change in velocity of the two bodies.

26. A $10^4$-kg freight car rolls along a horizontal track at 2 m/s with negligible friction. It hits a second stationary car of mass $1.5 \times 10^4$ kg. They couple together and roll with speed $v$. Find (a) the initial momentum of the $10^4$-kg car and (b) the speed $v$.

27. A box of mass 0.3 kg slides along a frictionless horizontal table with speed 2 m/s. It collides with a second box of unknown mass $m$. The two boxes stick together and move with speed 0.40 m/s after the collision. Find the mass of the second box.

### Section 4-5, Applications to Problem Solving: Constant Forces

28. A 5-kg object is pulled along a frictionless horizontal surface by a horizontal force of 10 N. (a) If the object is at rest at $t = 0$, how fast is it moving after 3 s? (b) How far does it travel from $t = 0$ to $t = 3$ s? (c) Indicate on a diagram all the forces acting on the object.

29. A 10-kg object is subjected to the two forces $F_1$ and $F_2$, as shown in Figure 4-26. (a) Find the acceleration **a** of the object. (b) A third force $F_3$ is applied so that the object is in static equilibrium. Find $F_3$.

30. In Figure 4-27 the bodies are attached to spring balances calibrated in newtons. Give the readings of the balances in each case, assuming the strings to be massless and the incline to be frictionless.

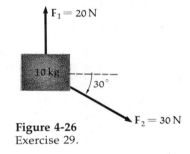

**Figure 4-26**
Exercise 29.

**Figure 4-27**
Exercise 30.

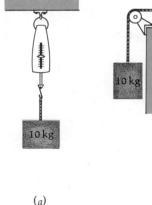

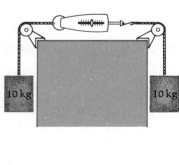

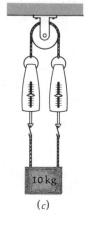

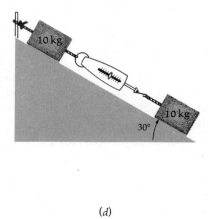

(a)                    (b)                    (c)                    (d)

31. A 4-kg object is subjected to two forces, $\mathbf{F_1} = 2\mathbf{i} - 3\mathbf{j}$ N and $\mathbf{F_2} = 4\mathbf{i} + 11\mathbf{j}$ N. The object is at rest at the origin at time $t = 0$. (*a*) What is the object's acceleration? (*b*) What is its velocity at time $t = 3$ s? (*c*) Where is the object at $t = 3$ s?

32. A vertical force $\mathbf{T}$ is exerted on a 5-kg body near the surface of the earth, as shown in Figure 4-28. Find the acceleration of the body if (*a*) $T = 5$ N; (*b*) $T = 10$ N; (*c*) $T = 100$ N.

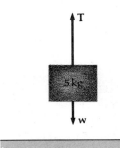

**Figure 4-28**
Exercise 32.

33. A 2-kg picture is hung by two wires of equal length which make an angle $\theta$ with the horizontal, as shown in Figure 4-29. (*a*) If $\theta = 30°$, find the tension in the wires. (*b*) Find the tension for general values of $\theta$ and weight $w$ of the picture. For what angle $\theta$ is $T$ the least? The greatest?

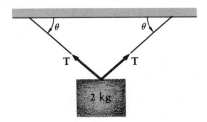

**Figure 4-29**
Exercise 33.

34. A body is held in position by a cable along a smooth incline (Figure 4-30). (*a*) If $\theta = 60°$ and $m = 50$ kg, find the tension in the cable and the normal force exerted by the incline. (*b*) Find the tension as a function of $\theta$ and $m$ and check your result for $\theta = 0°$ and $\theta = 90°$.

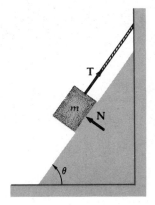

**Figure 4-30**
Mass held on frictionless incline for Exercise 34.

35. Two 5-kg bodies are connected by a light string, as shown in Figure 4-31. The table is frictionless, and the string passes over a frictionless peg. Find the acceleration of the masses and the tension in the string.

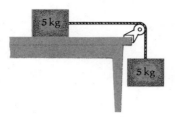

**Figure 4-31**
Exercise 35.

36. Two bodies are connected by a light string, as shown in Figure 4-32. The incline and peg are frictionless. Find the acceleration of the bodies and the tension in the string for (a) $\theta = 30°$ and $m_1 = m_2 = 5$ kg; (b) for general values of $\theta$, $m_1$, and $m_2$.

**Figure 4-32**
Exercise 36.

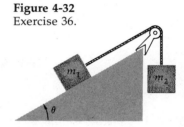

37. A body of mass $m_1$ rests on one of mass $m_2$, which rests on a frictionless horizontal surface. A force **F** is exerted on $m_2$, as shown in Figure 4-33. Consider $m_1 = 2$ kg, $m_2 = 4$ kg, and $F = 3$ N. (a) If the surface between the bodies is frictionless, find the acceleration of each body. (b) If the surface between the bodies is rough enough to ensure that $m_1$ does not slide on $m_2$, find the acceleration of the two bodies. (c) Find the resultant force acting on each body in (b). (d) What is the magnitude and direction of the contact force exerted by the upper body on the lower one?

**Figure 4-33**
Exercise 37.

38. The arrangement shown in Figure 4-34, called *Atwood's machine*, is used to measure the acceleration of gravity $g$ by measuring the acceleration of the bodies. Assuming the string to be massless and the peg to be frictionless, show that the magnitude of the acceleration of either body and the tension in the string are

$$a = \frac{m_1 - m_2}{m_1 + m_2} g \quad \text{and} \quad T = \frac{2m_1 m_2 g}{m_1 + m_2}$$

**Figure 4-34**
Exercise 38.

39. A 2-kg body hangs from a spring balance (calibrated in newtons) attached to the ceiling of an elevator (Figure 4-35). What does the balance read (*a*) when the elevator is moving up with constant velocity of 30 m/s, (*b*) when the elevator is moving down with constant velocity of 30 m/s, (*c*) when the elevator is accelerating upward at 10 m/s²? (*d*) From $t = 0$ to $t = 2$ s the elevator moves upward at 10 m/s. Its velocity is then reduced uniformly to zero in the next 2 s, so that it is at rest at $t = 4$ s. Describe the reading on the balance during the time $t = 0$ to $t = 4$ s.

40. A man holding a 10-kg body on a cord made to withstand 150 N steps into an elevator. When the elevator starts up, the line breaks. What was the minimum acceleration of the elevator?

41. A 2.8-kg box slides along a frictionless table. It is attached to a 0.2-kg box by a light string, as shown in Figure 4-36. Find the time for the 0.2-kg mass to fall 2 m to the floor if the system starts from rest.

42. A 60-kg girl weighs herself by standing on a scale in an elevator. What does the scale read when (*a*) the elevator is descending at a constant rate of 10 m/s; (*b*) the elevator is accelerating downward at 2 m/s²; (*c*) the elevator is ascending at 10 m/s but its speed is decreasing by 2 m/s in each second? (Take $g = 10 \text{ m/s}^2$.)

**Section 4-6, Reference Frames**

43. Two coordinate frames $S$ and $S'$ are coincident at $t = 0$. The origin $O'$ is moving in the $x$ direction with speed 3 m/s relative to $O$. A 2-kg body is moving along the $x$ axis with velocity $v'_x$ relative to the origin $O'$. (*a*) What is the velocity relative to $O$? (*b*) A force of 5 N in the $x$ direction acts on the body. If it starts from rest in frame $S'$ at $t = 0$, find its velocity $v'_x$ at times $t = 1$ s, $t = 2$ s, and $t = 3$ s. (*c*) Use the result from part (*a*) to find the velocity $v_x$ in $S$ at these times. (*d*) What is the acceleration of the body in frame $S'$? In frame $S$?

44. Water in a river flows in the $x$ direction at 4 m/s. A boat crosses the river with speed 10 m/s relative to the water. Set up a reference frame with its origin fixed relative to the water and another with its origin fixed relative to the shore. Write expressions for the velocity of the boat in each reference frame.

**Problems**

1. The acceleration versus spring length observed when a 0.5-kg mass is pulled along a frictionless table by a single spring is

| L, cm | 4 | 5 | 6 | 7 | 8 | 9 | 10 | 11 | 12 | 13 | 14 |
|-------|---|---|---|---|---|---|----|----|----|----|----|
| a, m/s² | 0 | 2.0 | 3.8 | 5.6 | 7.4 | 9.2 | 11.2 | 12.8 | 14.0 | 14.6 | 14.6 |

(*a*) Make a plot of the force exerted by the spring versus length L. (*b*) If the spring is extended to 12.5 cm, what force does it exert? (*c*) By how much is the spring extended when the mass is at rest suspended from it near sea level where $g = 9.81$ N/kg?

2. In a tug of war, two boys pull on a rope, each trying to pull the other over a line midway between their original positions. According to Newton's third law, the forces exerted by each boy on the other are equal and opposite. Show with a force diagram how one boy can win.

3. A bullet of mass $1.8 \times 10^{-3}$ kg moving at 500 m/s embeds itself in a large fixed piece of wood and travels 6 cm before coming to rest. Assume that the deceleration of the bullet is constant and find the force exerted by the wood on the bullet.

4. A 1-tonne (1000-kg) load is being moved by a crane. Find the tension in the cable that supports it as (*a*) the load is accelerated upward at 2 m/s², (*b*) the load is lifted at constant speed, (*c*) the load moves upward but its speed decreases 2 m/s in each second.

**Figure 4-35**
Exercise 39.

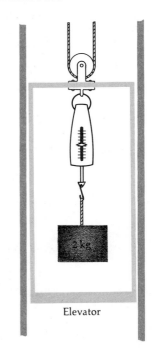

Elevator

**Figure 4-36**
Exercise 41.

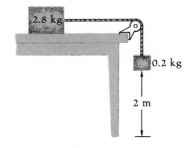

2.8 kg

0.2 kg

2 m

5. Your car is stuck in a mudhole. You are alone but have a long, strong rope. Having studied physics, you tie the rope taut to a tree and pull on it sideways, as in Figure 4-37. Find the force exerted by the rope on the car when the angle $\theta$ is 3° and you are pulling with a force of 400 N but the car does not move. How strong must the rope be if it takes a force of 600 N at an angle of $\theta = 3°$ to move the car?

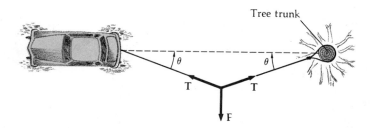

Tree trunk

**Figure 4-37**
Problem 5.

6. A box of mass $m_1$ is pulled by a force **F** exerted at the end of a rope which has a much smaller mass $m_2$, as shown in Figure 4-38. The box slides along a smooth horizontal surface. (a) Find the acceleration of the rope and box. (b) Find the tension in the rope at the point where it is attached to the box. (c) The diagram in Figure 4-38 with the rope horizontal is not quite correct for this situation. Correct the diagram and state how this correction affects your solution.

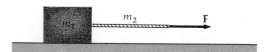

**Figure 4-38**
Problem 6.

7. A 2-kg body rests on a smooth wedge which has an inclination of 60° (Figure 4-39) and an acceleration $a$ to the right such that the mass remains stationary relative to the wedge. Find $a$. What would happen if the wedge were given a greater acceleration?

**Figure 4-39**
Problem 7.

8. A rifle bullet of mass 9 g starts from rest and moves 0.6 m in the rifle barrel. The speed of the bullet as it leaves the barrel is 1200 m/s. Assuming the force to be constant on the bullet while it is in the barrel, find (a) the average speed of the bullet in the barrel and the time it spends in the barrel and (b) the change in momentum of the bullet and the force exerted on the bullet.

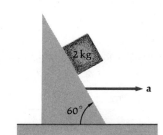

9. A car traveling 90 km/h comes to a sudden stop to avoid an accident. Fortunately the driver is wearing a seatbelt. Using reasonable values for the mass of the driver and the time it takes to come to a stop, estimate the force (assumed constant) exerted on the driver by the seat belt.

10. A 100-kg mass is pulled along a frictionless surface by a force **F** so that its acceleration is 6 m/s² (see Figure 4-40). A 20-kg mass slides along the top of the 100-kg mass and has an acceleration of 4 m/s². (It thus slides back relative to the 100-kg mass.) (a) What is the frictional force exerted by the 100-kg mass on the 20-kg mass? (b) What is the net force on the 100-kg mass? What is the force **F**? (c) After the 20-kg mass falls off the 100-kg mass, what is the acceleration of the 100-kg mass?

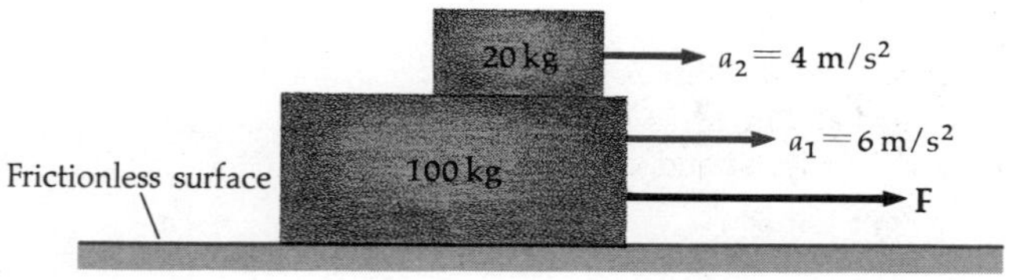

**Figure 4-40**
Problem 10.

20 kg $\quad a_2 = 4$ m/s²

$a_1 = 6$ m/s²

Frictionless surface    100 kg    **F**

11. Two 100-kg boxes are dragged along a frictionless surface with a constant acceleration of 1 m/s². Each rope has a mass of 1 kg (Figure 4-41). Find the force **F** and the tension in the ropes at points *A*, *B*, and *C*.

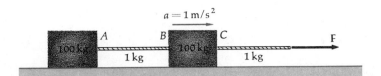

$a = 1\,\text{m/s}^2$

A

100 kg

1 kg

B    100 kg    C

1 kg

F

**Figure 4-41**
Problem 11.

12. A pulley is attached to a beam 12.0 m above the ground. A construction worker of mass 90 kg pulls an empty barrel up on a rope over the pulley and ties the rope at ground level. He climbs up and fills the barrel with spare bricks to give the loaded barrel a total mass of 180 kg. Then he climbs down and unties the rope but holds onto it. Find his speed as his hard hat hits the beam, assuming that he hangs onto the rope and rises vertically 12.0 m. How does this speed compare with the speed with which he hits the ground after he lets go of the rope upon hitting his head?

13. A man stands on scales in an elevator which has an upward acceleration *a*. The scales read 960 N. When he picks up a 20-kg box, the scales read 1200 N. Find the mass of the man, his weight, and the acceleration *a*.

14. A student has to escape from his girl friend's dormitory from a window that is 15.0 m above the ground. He has a heavy rope 20 m long, but it will break when the tension exceeds 360 N and he weighs 600 N. The student will be injured if he hits the ground with speed greater then 10 m/s. (*a*) Show that he cannot safely slide down the rope. (*b*) Find a strategy using the rope that will permit the student to reach the ground safely.

15. Two climbers on an icy (frictionless) slope and tied together on a 30-m rope are in the predicament indicated in Figure 4-42. At time *t* = 0 the speed of each is zero, but the top climber, Paul (mass 52 kg) has taken one step too many and his friend Jay (mass 74 kg) has dropped his pick. Find the tension in the rope as Paul falls and his speed just before he hits the ground. If Paul unties his rope after hitting the ground, find Jay's speed as he hits the ground.

**Figure 4-42**
Problem 15.

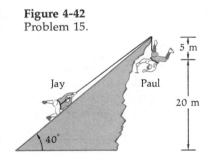

Jay

Paul

5 m

20 m

40°

16. A 60-kg house painter stands on a 15-kg aluminum platform. A rope attached to the platform and passing over an overhead pulley allows the painter to raise herself and the platform (Figure 4-43). (*a*) To get started, she accelerates herself and the platform at a rate of 0.8 m/s². With what force must she pull on the rope? (*b*) After 1 s she pulls so that she and the platform go up at a constant speed of 1 m/s. What force must she exert on the rope? (Ignore the mass of the rope.)

**Figure 4-43**
Problem 16.

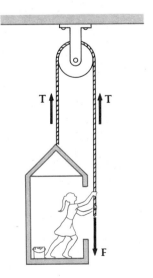

T        T

F

17. Two boxes slide on frictionless inclines, as shown in Figure 4-44. (*a*) Find the acceleration and the tension in the rope. (*b*) The two masses are replaced by two other masses $m_1$ and $m_2$ such that there is no acceleration. Find whatever information you can about the two masses.

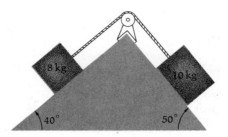

**Figure 4-44**
Problem 17.

18. In the Atwood's machine of Figure 4-34 (Exercise 38) the acceleration of gravity $g$ is determined by measuring the time $t$ for the larger body of mass $m_1$ to fall a distance $L$. (*a*) Find an expression for $g$ in terms of $m_1$, $m_2$, $L$, and $t$. (*b*) Show that if there is a small error in the time measurement $dt$, this leads to an error in the determination of $g$ by an amount $dg$ given by $dg/g = -2dt/t$. Find the values $m_1$ and $m_2$ each less than 1 kg such that $g$ can be measured to 5 percent with a watch that is accurate to 0.1 s if $L = 3$ m. Assume that the only important uncertainty in the measurement is the time of fall.

19. (*a*) Show that a point on the surface of the earth at latitude $\theta$ has an acceleration relative to the center of the earth of magnitude $3.4 \cos \theta$ cm/s². What is the direction of this acceleration? (*b*) Discuss the effect of this acceleration on the apparent weight of an object near the surface of the earth. (*c*) The free-fall acceleration of an object at sea level measured *relative to the earth's surface* has the value 978 cm/s² at the equator and 981 cm/s² at latitude $\theta = 45°$. What are the values of the gravitational field $g$ at these points?

20. A small rocket with a mass of 10 kg is initially moving horizontally near the earth's surface with speed 10 m/s. (*a*) What is the initial momentum of the rocket? (*b*) The rocket engine produces a force of 300 N on the rocket at an angle of 30° above the horizontal. What is the resultant force on the rocket (use $g = 10$ m/s²)? (*c*) Find the speed and momentum of the rocket 10 s after the engine is turned on (assume that the mass of the rocket is constant).

21. A simple accelerometer can be made by suspending a small body from a string attached to a fixed point in the accelerating object, e.g., from the ceiling of a passenger car. When there is an acceleration, the body will deflect and the string will make some angle with the vertical. (*a*) How is the direction in which the suspended body deflects related to the direction of the acceleration? (*b*) Show that the acceleration $a$ is related to the angle $\theta$ the string makes by $a = g \tan \theta$. (*c*) Suppose the accelerometer is attached to the ceiling of an automobile which brakes to rest from 50 km/h in a distance of 60 m. What angle will the accelerometer make? Will the mass swing forward or back?

22. A 65-kg girl weighs herself by standing on a scale mounted on a skateboard which rolls down an incline as shown in Figure 4-45. Assume no friction, so that the force exerted by the incline on the board is perpendicular to the incline. What is the reading on the scale if $\theta = 30°$?

**Figure 4-45**
Problem 22.

23. Two bodies of mass $m_1$ and $m_2$ rest on a horizontal frictionless table as shown in Figure 4-46. A force **F** is applied as shown. (*a*) If $m_1 = 2$ kg, $m_2 = 4$ kg, and $F = 3$ N, find the acceleration of the bodies and the contact force **F**_c exerted by one body on the other. (*b*) Find the contact force for general values of the masses of the bodies and show that if $m_2 = nm_1$, $F_c = nF/(n + 1)$.

**Figure 4-46**
Problem 23.

24. A constant force **F** is exerted on a smooth peg of mass $m_1$. Two bodies of mass $m_2$ and $m_3$ are connected to a light string which goes around the peg as shown in Figure 4-47. Assuming that $F$ is greater than $2T$, (*a*) find the acceleration of each of the bodies and the tension in the string if $m_1 = m_2 = m_3$, and (*b*) find the acceleration of each body if $m_1 = m_2$ and $m_3 = 2m_1$.

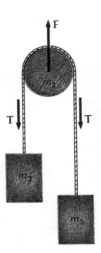

**Figure 4-47**
Problem 24.

CHAPTER 5    Applications of Newton's Laws

---

**Objectives**    After studying this chapter you should:

1. Be able to list the four basic forces found in nature.

2. Be able to discuss the similarities and differences between Newton's law of gravity and Coulomb's law for the electrostatic force.

3. Know the general order of magnitude of the range of the strong nuclear force.

4. Know that the maximum static friction force and the kinetic friction force are proportional to the normal force between the surfaces involved.

5. Be able to apply Newton's laws to work problems involving friction.

6. Be able to apply Newton's laws to circular-motion problems.

7. Be able to state the two conditions for static equilibrium and apply them to work problems.

8. Be able to discuss qualitatively motion with a velocity-dependent retarding force.

9. Be able to discuss how pseudo forces arise in accelerated frames.

10. Be able to discuss the similarities and differences between centrifugal forces and Coriolis forces.

---

In Chapter 4 we discussed Newton's laws of motion and their application to motion of a particle under the influence of constant forces. We now consider some more general applications. The methods of problem solving discussed in Section 4-5 have wide applicability to many problems in mechanics and should be reread and learned well. In particular, you should develop the habit of approaching a problem by first drawing a picture and then indicating the important forces acting on each object in a free-body diagram.

To apply Newton's laws, we must know what kinds of forces may be acting on a particular object. We therefore begin this chapter with a brief discussion of the basic forces or interactions occurring between elementary particles. We then discuss molecular forces briefly and look at some empirical descriptions of the common everyday forces such as those exerted by springs and strings, support forces, and friction, which result from the basic interactions. In Section 5-4 we apply Newton's laws to problems involving circular motion. In Section 5-5 when we consider the static equilibrium of rigid bodies, we shall see that for extended bodies (rather than just point particles) the line of action of a force is important as well as the magnitude and direction. In Section 5-6 we examine the motion of a body under the influence of a velocity-dependent retarding force, e.g., the forces that result from the viscous drag of a fluid such as air or water acting on a moving body. Finally, we conclude the chapter with a look at pseudo forces, which, unlike real forces, are not exerted by any agent or traceable to any of the basic interactions; they are fictitious forces which must be postulated in order to use Newton's second law in an accelerated, noninertial reference frame.

## 5-1   The Basic Interactions

All the different forces observed in nature can be explained in terms of four basic interactions occurring between elementary particles:

1.  The gravitational force

2.  The electromagnetic force

3.  The strong nuclear force (also called the hadronic force)

4.  The weak nuclear force

Most of the everyday forces we observe between macroscopic objects, e.g., the contact forces of support and friction or those exerted by springs and strings, are the result of molecular forces exerted by the molecules of one body on those of another; these molecular forces are themselves complicated manifestations of the basic electromagnetic force. In this section we shall take a brief look at the basic interactions. We shall study the gravitational and electromagnetic interactions in more detail in Chapters 12 and 20, respectively.

### The Gravitational Force

All particles exert a gravitational force of attraction upon each other. The magnitude of the force exerted by a particle of mass $m_1$ on another particle of mass $m_2$ a distance $r$ away is given by Newton's law of gravitation

$$F_{12} = \frac{Gm_1 m_2}{r^2} \qquad\qquad 5\text{-}1$$

*Law of gravity*

where the constant $G$, called the *universal gravitational constant*, has the value

$$G = 6.67 \times 10^{-11} \ \text{N·m}^2/\text{kg}^2 \qquad\qquad 5\text{-}2$$

*Universal gravitational constant*

The direction of the force on $m_2$ is toward $m_1$. An equal but opposite force is exerted by mass $m_2$ on mass $m_1$ in accordance with Newton's third law; i.e., the two particles exert forces of attraction on each other.

The inverse-square nature of the gravitational force was suspected by Hooke and others during Newton's time. Newton deduced the full law from his laws of motion and Kepler's empirical laws describing planetary motion. We shall study this in detail in Chapter 12.

It is not readily apparent how to apply the law of gravity for point particles to find the forces exerted by extended bodies on each other. After considerable effort, Newton was able to prove that the force exerted by the whole earth on a body on or outside its surface is the same as if all the earth's mass were concentrated at its center. The proof involves integral calculus, a tool developed by Newton for this problem, among others. We shall assume the result for now and consider the proof later. The force on a body of mass $m$ near the surface of the earth is then directed toward the center of the earth and has magnitude

$$F = \frac{GM_E m}{R_E^2}$$

where $R_E$ is the radius of the earth and $M_E$ is its mass. The acceleration of a body in free fall near the earth's surface is then

$$g = \frac{F}{m} = \frac{GM_E}{R_E^2} \qquad\qquad 5\text{-}3$$

Since the gravitational field $g = 9.81\ \text{N/kg} = 9.81\ \text{m/s}^2$ can be determined by measuring the acceleration of gravity and since the radius of the earth is known, Equation 5-3 can be used to determine either the constant $G$ or the mass of the earth $M_E$ if one of these quantities is known. Newton estimated the value of $G$ from an estimate of the earth's density. The gravitational constant was first determined accurately about 100 years later (1798) by Cavendish, who measured the force between two small spheres of known mass and separation directly. (Details of this measurement are given in Chapter 12.) We can use the known value of $G$ to compute the gravitational attraction between two ordinary objects.

---

**Example 5-1** Find the force of attraction between two balls each of mass 1 kg when their centers are 10 cm apart.

We can treat each sphere as if its total mass were at a point at its center. The magnitude of the force on either one exerted by the other is then

$$F = \frac{(6.67 \times 10^{-11}\ \text{N·m}^2/\text{kg}^2)(1\ \text{kg})(1\ \text{kg})}{(10^{-1}\ \text{m})^2}$$

$$= 6.67 \times 10^{-9}\ \text{N}$$

---

This example demonstrates that the gravitational force exerted by one object of ordinary size on another such object is extremely small. For example, the weight of a 1-g object is $9.81 \times 10^{-3}$ N, more than a million times the force computed in Example 5-1. We can usually neglect the gravitational force between objects compared with other forces acting on them. The gravitational attraction can be noticed only if the objects are extremely massive, e.g., mountains, or if great care is taken to eliminate other forces on the objects, as Cavendish did in determining $G$.

**Example 5-2** What is the free-fall acceleration of a body 200 km above the earth's surface?

The acceleration due to gravity at a distance $r$ from the center of the earth is given by $F/m$, where $F$ is the gravitational force given by Equation 5-1 with $m_1 = M_E$ and $m_2 = m$:

$$g(r) = \frac{F}{m} = \frac{GM_E}{r^2}$$

From Equation 5-3 we have $GM_E/R_E^2 = 9.81$ m/s$^2$. The acceleration at distance $r$ is then

$$g(r) = (9.81 \text{ m/s}^2) \frac{R_E^2}{r^2}$$

*Variation of g with r*

Using $R_E = 6.37$ Mm $= 6370$ km and $r = R_E + 200$ km $= 6570$ km, we obtain

$$g(r) = (9.81 \text{ m/s}^2) \left(\frac{6370 \text{ km}}{6570 \text{ km}}\right)^2 = 9.22 \text{ m/s}^2$$

**Electromagnetic Forces**

Forces exerted by one particle on another because of the electric charges of the particles are called *electromagnetic forces*. Their description is considerably more complicated than that of gravitational forces. For one thing, the electromagnetic force can be either attractive or repulsive whereas the gravitational force is always attractive. The fact that charges can either attract or repel other charges can be described by postulating the existence of two kinds of charges. Like charges repel each other and unlike charges attract each other. The two kinds of charges are labeled positive and negative, following the suggestion of Benjamin Franklin. A more serious complication arises because when charges are in motion, part of the electromagnetic force, the magnetic force, depends on both the magnitudes and directions of the velocities of the particles and is not generally along the line joining them. We avoid this complication here and discuss only the electrostatic force, i.e., the electromagnetic force between charges at rest. (Magnetism will be studied in Chapters 26–29.)

The law which gives the electrostatic force between two point charges is known as *Coulomb's law*. Like Newton's law of gravity, Coulomb's law is an inverse-square law. If there are two point charges $q_1$ and $q_2$ separated by a distance $r_{12}$, the force exerted on charge $q_2$ by charge $q_1$ is

*Coulomb's law of electric force*

$$\mathbf{F}_{12} = k \frac{q_1 q_2}{r_{12}^2} \hat{\mathbf{r}}_{12} \qquad\qquad 5\text{-}4$$

where $\hat{\mathbf{r}}_{12}$ is the unit vector pointing from $q_1$ to $q_2$ (Figure 5-1) and $k$ is a constant, known as the Coulomb constant. The SI unit of charge is the coulomb (C), which is the amount of charge flowing past a point in a wire in one second when the current in the wire is one ampere. The unit of current, the ampere (A), is defined in terms of a magnetic-force measurement (Chapter 27). It is the practical unit of current used in everyday electrical work.

*Coulomb constant*
*Fundamental charge unit*

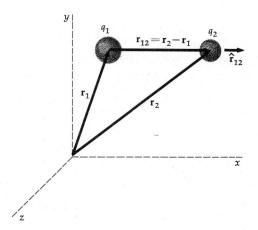

**Figure 5-1**
The electrostatic force exerted by charge $q_1$ on charge $q_2$ is in the direction of the unit vector $\hat{\mathbf{r}}_{12}$ defined by $\hat{\mathbf{r}}_{12} = \mathbf{r}_{12}/r_{12}$.

The fact that like charges repel and unlike charges attract is included in Equation 5-4. If both charges have the same sign, the product $q_1 q_2$ is positive and the force on $q_2$ is directed along $\hat{\mathbf{r}}_{12}$ away from $q_1$. If the charges have opposite signs, the force on $q_2$ is toward $q_1$. The value of the Coulomb constant $k$ is

$$k = 8.99 \times 10^9 \text{ N·m}^2/\text{C}^2 \approx 9 \times 10^9 \text{ N·m}^2/\text{C}^2 \qquad 5\text{-}5$$

The smallest magnitude of charge found in nature is the charge of the electron or proton, designated $e$. The value of $e$ in coulombs is

$$e = 1.602 \times 10^{-19} \text{ C} \qquad 5\text{-}6$$

The proton has charge $+e$ and the electron $-e$. All charges are found to have magnitudes $Ne$, where $N$ is an integer. Electric charge is therefore said to be *quantized*. This quantization of charge is usually not noticed in macroscopic experiments because the fundamental unit $e$ is so small. It takes a very large number of electrons or protons to produce a charge of magnitude 1 C, as can be seen from Equation 5-6.

Since atoms normally contain an equal number of positively charged protons and negatively charged electrons (and no other charged particles), atoms are electrically neutral, i.e., have no net electric charge. An atom can sometimes lose one of its electrons or gain an additional electron to become a charged particle called an *ion*. Macroscopic objects usually contain a large number of neutral atoms and are therefore electrically neutral. Electrons, however, can often be transferred from one macroscopic object to another, leaving one with an excess positive charge and the other with an excess negative charge. For example, when a glass rod is rubbed with a silk cloth, electrons are transferred from the rod to the cloth. With no knowledge of the process of transfer of charge, Franklin arbitrarily labeled the charge of the glass rod positive and that of the cloth negative. It is for this reason that the electron charge is now taken to be negative and the proton charge positive rather than the other way around. Excess charges of the order of several microcoulombs can be obtained by putting certain objects in intimate contact, often by simply rubbing the surfaces together. Such charges produce relatively large forces, as we shall see in Example 5-3.

---

**Example 5-3** Two point charges of 1 $\mu$C each are separated by 10 cm. Find the force between the charges and the number of fundamental units of charge in each charge.

From Coulomb's law, the magnitude of the force is

$$F = \frac{(9 \times 10^9 \text{ N·m}^2/\text{C}^2)(10^{-6} \text{ C})(10^{-6} \text{ C})}{(10^{-1} \text{ m})^2} = 0.90 \text{ N}$$

The number of electrons that must be transferred to produce such a charge is found from $Ne = Q$, $N = Q/e = (10^{-6} \text{ C})/(1.6 \times 10^{-19} \text{ C}) = 6.25 \times 10^{12}$.

---

Because both the electrostatic force and the gravitational force between two particles vary inversely with the square of the separation, the ratio of these forces is independent of the separation. We can thus compare the relative strengths of these forces for elementary particles such as two protons, two electrons, or an electron and a proton. For two protons an elementary calculation (see Exercise 4) gives $F_e/F_g = 10^{36}$ for the ratio of the electrostatic force to the gravitational force. Since electrons have a mass about $\frac{1}{1836}$ times that of protons, the ratio for two electrons is even greater. We see that the gravitational force between two elementary particles is so much smaller than the electrostatic force that it can be neglected in describing their interaction. It is only because such large masses as the earth contain almost exactly equal numbers of positive and negative charges that the gravitational force is important at all. If the positive and negative charges in bodies did not almost exactly cancel each other, the electric forces between them would be much greater than the gravitational attraction.

### Nuclear Forces

Besides the gravitational and electromagnetic force, the two other basic forces are the strong nuclear force and the weak nuclear force. It is clear from the stability of atomic nuclei that there must be some strong attractive forces besides electromagnetic and gravitational forces. Consider, for example, the helium nucleus, ⁴He, consisting of two neutrons and two protons, a total of four particles contained in a volume with a radius of about 1 fm (1 fm = $10^{-15}$ m). The repulsive electric force between protons this close together is quite large, and we have seen that by comparison the attractive gravitational force is negligible. Yet ⁴He is very stable. The force which binds the neutrons and protons together in a nucleus is called the strong nuclear force. It acts between two protons, two neutrons, or a proton and neutron but only if the particles are very close together. When two nucleons* are within about 1 fm of each other, the strong nuclear attractive force is about 10 times stronger than the repulsive electric force of two protons at this separation. However, this nuclear force has a short range; i.e., its strength decreases very rapidly with separation, much more rapidly than the inverse-square decrease in the electric force. At a separation greater than about 15 fm the nuclear force is much less than the electric force and can be neglected.

There is no macroscopic experiment which can be performed to investigate the properties of this force. Instead, information is obtained by bombarding neutrons, protons, and nuclei with neutrons, protons, and other particles and observing the results. Such *scattering experiments* have yielded much information about the nuclear force but have not produced a unique or simple formula like Coulomb's law or

---

* A nucleon is either a proton or a neutron.

Newton's law of gravitation. Some of the known characteristics of the strong nuclear force are:

1. *Charge independence*    The nuclear force between two neutrons is the same as between two protons or between a neutron and a proton.

2. *Saturation*    In a nucleus containing many nucleons, each nucleon bonds to only a few of the other nucleons. Thus the force needed to tear a neutron from a nucleus is approximately the same whether the nucleus contains five or six nucleons or a hundred nucleons.

3. *Short range*    At a distance of about 1 fm the strong nuclear force is attractive and about 10 times the electric force between two protons. The force decreases rapidly with increasing distance, becoming completely negligible at about 15 times this separation. When two nucleons are within about 0.4 fm of each other, the strong nuclear force becomes repulsive. Thus nuclei do not collapse.

A large class of particles, called *hadrons,* interact with each other via the strong nuclear force. Hadrons include *mesons* of many types and *baryons,* which in turn include nucleons and other particles. Particles in a much smaller class, the *leptons,* do not participate in the strong nuclear force at all. Leptons include electrons, positrons, muons, and neutrinos.

The *weak nuclear force* acts between all leptons and hadrons. It is also a short-range force. Its effect is to produce a degree of instability in nuclei and elementary particles. It is responsible for the type of radioactivity known as *beta decay,* and it explains, for example, why $^{12}$C, a carbon nucleus with six protons and six neutrons, is stable while $^{11}$C, a carbon nucleus with six protons and five neutrons, is not.

**Questions**

1. The greater the mass of an object, the greater the force of gravity acting on it. Why then does a heavy body fall no faster than a light one?

2. How can you tell whether the force acting on a body is gravitational or electric?

3. If all atoms are made up of electrically charged particles, and if electric forces are so much stronger than gravitational forces, why are we so aware of gravity in everyday life and so unfamiliar with electric forces?

## 5-2    Molecular Forces; Springs and Strings

The forces which bind two or more atoms together to make a molecule are electromagnetic in origin, but the details of this bonding are extremely complicated. Consider one of the simpler cases, the NaCl crystal. When neutral sodium and chlorine atoms come close to each other, the sodium atom has a tendency to give up one of its electrons to the chlorine atom because of the structure of these atoms. We then have two ions, Na$^+$ with a net charge of $+e$, and Cl$^-$ with a net charge of $-e$. These ions attract each other (much like point charges), but at a certain separation the force of attraction becomes zero and at smaller separations the ions repel each other. This repulsion is a complicated phe-

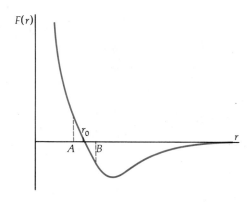

**Figure 5-2**
Force exerted by one ion on
another as a function of the
separation distance $r$. Near
the equilibrium separation $r_0$
the force is approximately
linear, like that of a spring
(see Figure 5-4).

nomenon; it cannot be understood even approximately in terms of classical mechanics or electromagnetism but only in terms of quantum
mechanics.* We can, however, give a qualitative sketch of the force
between the ions as a function of separation, as in Figure 5-2, where repulsive forces are indicated as positive and attractive forces as negative.
At the distance $r = r_0$ in this figure, the force is zero. This is called the
*equilibrium separation* for the atoms, and for NaCl it is of the order of
$10^{-10}$ m $= 0.1$ nm, the same order as the distance from the center of the
atom to its outermost electrons. For small displacements from equilibrium (between points $A$ and $B$ in the figure) the force curve is linear and
resembles the force exerted by a spring when compressed or extended
(see Figure 5-4). In its solid state, NaCl exists in a regular array of $Na^+$
and $Cl^-$ ions, each ion bound by ions of the other type into a lattice, as
illustrated in Figure 5-3. Such a structure, called a *crystal,* is quite rigid.
When we try to deform the solid by pushing or pulling on it, in an effort
to change the separation distances between the ions, the strong forces
between the ions resist compression or extension.

A more common type of molecular bonding, called *covalent bonding,*
occurs, for example, between two hydrogen atoms in the $H_2$ molecule,
between the carbon and oxygen atoms in CO, or between two carbon

---

* Early in this century it became clear that newtonian mechanics is inadequate for
treating the motions of particles in atoms. By 1926 a new and more general theory of mechanics, known as *quantum mechanics,* was developed which successfully describes the
details of subatomic motions. Quantum mechanics applied on a macroscopic scale reduces to newtonian mechanics.

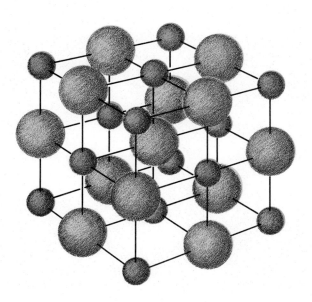

**Figure 5-3**
Arrangement of $Na^+$ and $Cl^-$
ions in an NaCl crystal.

atoms in many organic compounds. In this type of bonding the atoms do not give up or accept electrons to form ions; instead, some of the outer atomic electrons are shared by the atoms. In covalent bonding, both the attraction and the repulsion of the atoms are complicated quantum-mechanical phenomena. As with ionic bonding, the force-versus-separation curve looks like Figure 5-2.

There are other kinds of molecular bonds, e.g., metallic bonds, responsible, say, for the bonding of copper atoms in a solid piece of copper. In general, all forces between atoms and molecules are complex manifestations of the basic electromagnetic interaction, the details of which can be understood only within the framework of quantum mechanics. Nevertheless, we can often find empirical descriptions of these forces which are quite useful in determining the motion of macroscopic objects. At this level we must be content with such empirical descriptions.

As an example of such an empirical description, consider a familiar coil spring, made by winding a piece of stiff wire into a helix. The force exerted by the spring when it is compressed or extended is the result of the complicated intermolecular forces in the spring, but an empirical description of the macroscopic behavior of the spring is sufficient for most applications. If the spring is compressed or extended and released, it returns to its original, or natural, length, provided the displacement is not too great. There is a limit to such displacements beyond which the spring does not return to its original length but remains permanently deformed. If we allow only displacements below this limit, we can calibrate the extension or compression $\Delta x$ in terms of the force needed to produce this extension or compression, as outlined in Section 4-1. A typical calibration curve is shown in Figure 5-4. The sudden change in the slope of this curve at large compressions is at the point where the coils are completely collapsed so that adjacent turns are touching. Beyond this point, it takes a very large force to produce any noticeable compression. We see from this curve that for small $\Delta x$ the force exerted by the spring is approximately proportional to $\Delta x$. This result, known as *Hooke's law,* can be written

$$F_x = -k(x - x_0) = -k\,\Delta x \qquad\qquad 5\text{-}7$$

*Hooke's law*

where the empirically determined constant $k$ is called the *force constant* of the spring. It is equal to the negative of the slope of the straight-line

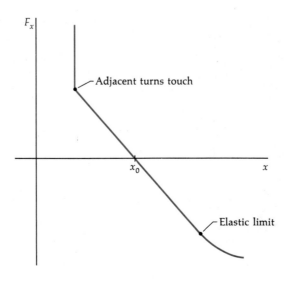

**Figure 5-4**
Force exerted by a spring versus length $x$ of the spring. The force is linear near the equilibrium length $x_0$.

portion of the curve in Figure 5-4. The distance $x$ is the coordinate of the free end of the spring or of any object attached to that end of the spring. The constant $x_0$ is the value of this coordinate when the spring is un-stretched from its equilibrium position.

The curve in Figure 5-4 is similar to that for the force between two atoms in a molecule in the region near equilibrium (Figure 5-2 from $A$ to $B$). It is often useful to visualize the atoms in a molecule as being connected by springs. For example, if we slightly increased the separation of the atoms in the molecule and released them, we would expect the atoms to oscillate back and forth as if they were two masses connected by a spring.

If we pull on a flexible string, the string stretches slightly and pulls back with an equal but opposite force (unless the string breaks). We can think of a string as a spring with such a large force constant that the extension of the string is negligible. If the string is flexible, however, we cannot exert a force of compression on it. When we push on a string, it merely flexes or bends.

## 5-3  Contact Forces; Support Forces and Friction

When two bodies are in contact with each other, they exert forces on each other due to the interaction of the molecules in one body with those of the other. Again, we describe these forces empirically.

Consider a block resting on a horizontal table. The weight of the block pulls the block downward, pressing it against the table. Because the molecules in the table have a great resistance to compression, as discussed previously, the table exerts a force upward on the block perpendicular, or normal, to the surface with no noticeable compression of the table. (Careful measurement would show that the supporting surface always does bend slightly in response to a load.) Similarly, the block exerts a downward force on the table. Note that this normal force exerted by one surface on the other can vary over a wide range of values. For example, unless the block is so heavy that the table breaks, the table will exert an upward support force on the block exactly equal to the weight of the block no matter how large or small the weight is. Of course, if you press down on the block, the table will exert a support force greater than the weight of the block to prevent it from accelerating downward.

Under certain circumstances, bodies in contact will exert forces on each other tangential to the surfaces in contact. The tangential component of the contact force exerted by one body on another is called a *frictional force*. Suppose, for example, that we exert a horizontal force on the block resting on the horizontal table in Figure 5-5. The block does not move if the horizontal force is not large enough. The table evidently exerts a horizontal force equal and opposite to the force we exert as long as the force is small enough. This force parallel to the surface of the table is the *frictional force*. The block, of course, exerts an equal and opposite frictional force on the table, tending to drag it in the direction of the horizontal force we exert. This frictional force is due to the bonding of the molecules of the block and the table at the places where the surfaces are in very close contact.

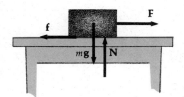

**Figure 5-5**
When a horizontal force **F** is applied to a block on a horizontal table, the table exerts a frictional force **f** on the block. The friction force will balance **F** if $F$ is less than the maximum possible value of $f$, which is $\mu_s N$.

We might expect the maximum force of friction to be proportional to the area of contact between the two surfaces. However, experimentally, to a good approximation, it is independent of this area and simply proportional to the normal force exerted by one surface on the other. Figure 5-6, an enlarged picture of the contact between the block and the table, shows that the actual microscopic area of contact at which the molecules can bond together is just a small fraction of the macroscopic area. The following is a possible model of static friction consistent with our intuition and with empirical results. The maximum force of friction *is* proportional to this microscopic area of contact, as expected, but this microscopic area is proportional to the total area and to the force per unit area exerted between the surfaces. Thus it does not depend on the total area of contact. Consider, for example, a 1-kg block with a side area of 60 cm² and an end area of 20 cm². When it is on its side on the table, a small fraction of the total 60 cm² is actually in microscopic contact. When it is placed on end, the fraction of the total area actually in microscopic contact is increased by a factor of 3 because the force per unit area is 3 times as great. However, since the area of the end is one-third that of the side, the actual microscopic area of contact is unchanged. The maximum force of static friction $f_{s,\,\text{max}}$ is thus just proportional to the normal force between the surfaces,

$$f_{s,\,\text{max}} = \mu_s N \qquad\qquad 5\text{-}8$$

where $\mu_s$, called the *coefficient of static friction*, depends on the nature of the surfaces of the block and table. If we exert a smaller horizontal force on the block, the frictional force will just balance this horizontal force. In general we can write

$$f_s \le \mu_s N \qquad\qquad 5\text{-}9$$

If we push the block hard enough, the static-friction force cannot prevent its motion. Then, as the block slides along the surface of the table, molecular bonds are continually made and broken and small pieces of the surfaces are broken off. The result is a force of sliding, or

*Static friction*

**Figure 5-6**
The microscopic area of contact between block and table is only a small fraction of the macroscopic area of the block. This fraction is proportional to the normal force exerted between the surfaces.

Drawing from one of Leonardo da Vinci's notebooks showing that he understood that the frictional force is independent of the macroscopic area of contact. (*Courtesy of the Deutsches Museum, Munich.*)

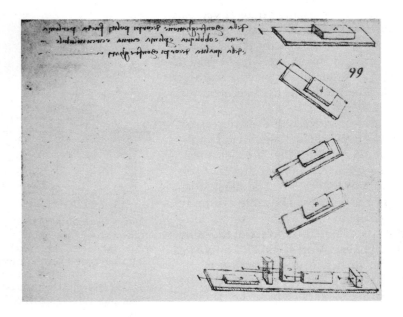

kinetic, friction which opposes the motion. Kinetic friction, like static friction, is a complicated phenomenon which even today is not completely understood. The coefficient of kinetic friction $\mu_k$ is defined as the ratio of the magnitudes of the frictional force $f_k$ and the normal force $N$. Then

*Kinetic friction*

$$f_k = \mu_k N \qquad\qquad\qquad 5\text{-}10$$

Experimentally it is found that:

1. $\mu_k$ is less than $\mu_s$.

2. $\mu_k$ depends on the relative speed of the surfaces, but for speeds in the range from about 1 cm/s to several metres per second, $\mu_k$ is approximately constant.

3. $\mu_k$ depends on the nature of the surfaces but is independent of the macroscopic area of contact.

We shall neglect any variation in $\mu_k$ with speed and assume that it is a constant which depends only on the nature of the surfaces.

We can measure $\mu_s$ and $\mu_k$ for two surfaces simply by placing a block on a plane surface and inclining the plane until the block begins to slide. Let $\theta_c$ be the critical angle at which sliding starts. For angles of inclination less than this, the block is in static equilibrium under the influence of its weight $m\mathbf{g}$, the normal force $\mathbf{N}$, and the force of static friction $\mathbf{f}_s$ (see Figure 5-7). Choosing the $x$ axis parallel to the plane and the $y$ axis perpendicular to the plane, we have

$$\Sigma F_y = N - mg \cos \theta = 0$$

and

$$\Sigma F_x = mg \sin \theta - f_s = 0$$

Eliminating the weight $mg$ from these two equations gives

$$f_s = mg \sin \theta = \frac{N}{\cos \theta} \sin \theta = N \tan \theta$$

At the critical angle $\theta_c$ the static friction is limiting, and we can replace $f_s$ by $\mu_s N$. Then

$$\mu_s = \tan \theta_c \qquad\qquad\qquad 5\text{-}11$$

The coefficient of static friction equals the tangent of the angle of inclination at which the block just begins to slide.

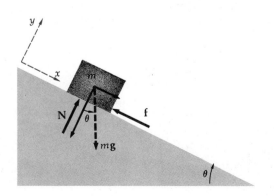

**Figure 5-7**
Forces on a block on a rough inclined plane. At angles less than the critical angle $\theta_c$ the frictional force balances the component $mg \sin \theta$ down the incline. At angles greater than $\theta_c$ the block slides down the incline. The critical angle is related to the coefficient of friction by $\tan \theta_c = \mu_s$.

Cross-country skiers use a special ski wax to maximize static friction (so that they can walk up an incline) and minimize kinetic friction.

Peter Miller/Photo Researchers

At angles greater than $\theta_c$, the block slides down the incline with acceleration $a_x$. In this case, the frictional force is $\mu_k N$, and we have

$$F_x = mg \sin \theta - \mu_k N = ma_x$$

Substituting $mg \cos \theta$ for $N$, we obtain for the acceleration

$$a_x = g(\sin \theta - \mu_k \cos \theta)$$

A measurement of the acceleration $a_x$ will then determine $\mu_k$ for these surfaces.

---

**Example 5-4** A 3-kg box rests on a horizontal table. The coefficients of static and kinetic friction between the box and the table are $\mu_s = 0.6$ and $\mu_k = 0.5$. The box is pulled by a string at an angle of 30°, as shown in Figure 5-8. Find the frictional force and the acceleration of the box if the tension in the string is (a) 10 N and (b) 20 N.

**Figure 5-8**
Example 5-4.

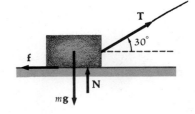

(a) The vertical and horizontal components of the tension are

$$T_y = T \sin 30° = (10 \text{ N})(0.5) = 5 \text{ N}$$

and

$$T_x = T \cos 30° = (10 \text{ N})(0.866) = 8.66 \text{ N}$$

Since there is no vertical acceleration, the net vertical force must be zero. Then $\Sigma F_y = N + T_y - mg = 0$. The normal force is therefore $N = mg - T_y = (3 \text{ kg})(9.81 \text{ m/s}^2) - 5 \text{ N} = 29.4 - 5 \text{ N} = 24.4 \text{ N}$. The maximum possible static frictional force is

$$f_{s,\,\text{max}} = \mu_s N = 0.6(24.4) = 14.7 \text{ N}$$

Since the applied horizontal force $T_x$ is less than the maximum force of static friction, the box will not accelerate horizontally. The frictional force is therefore 8.66 N to the left to balance the applied horizontal force $T_x = 8.66$ N to the right. There are two important points to note about this example: (1) the normal force is *not* equal to the weight of the box because the vertical component of the tension helps lift the box off the table and (2) the force of static friction is *not* equal to $\mu_s N$. It is less than this maximum possible limiting value.

(b) When the tension is increased to $T = 20$ N, its vertical and horizontal components are $T_y = (20$ N$)(\sin 30°) = 10$ N and $T_x = (20$ N$)(\cos 30°) = 17.3$ N. The normal force is then

$$N = mg - T_y = 29.4 \text{ N} - 10 \text{ N} = 19.4 \text{ N}$$

and the maximum possible force of static friction is

$$f_{s,\max} = \mu_s N = 0.6(19.4 \text{ N}) = 11.6 \text{ N}$$

Since this maximum static frictional force is less than the applied horizontal force, the box will slide. The frictional force on the box will thus be due to kinetic friction and will have the value $f_k = \mu_k N = 0.5(19.4$ N$) = 9.7$ N. The resultant force in the horizontal direction is thus

$$\Sigma F_x = T_x - f_k = 17.3 \text{ N} - 9.7 \text{ N} = 7.6 \text{ N}$$

and the acceleration is $a_x = \Sigma F_x/m = (7.6$ N$)/(3$ kg$) = 2.53$ m/s$^2$.

---

### Questions

4. The friction between two surfaces is first reduced by polishing both, but if the surfaces are polished until they are extremely smooth and flat, friction increases again. Explain.

5. Various objects lie on the floor of a truck. If the truck accelerates, what force acts on them to cause them to accelerate?

6. Any object resting on the floor of a truck will slip if the acceleration of the truck is too great. How does the critical acceleration at which a small box slips compare with that at which a much heavier object slips?

## 5-4 Circular Motion

If a particle moves with speed $v$ in a circle of radius $r$, it has an acceleration of magnitude $v^2/r$ directed toward the center of the circle. This centripetal acceleration is related to the change in the direction of the velocity of the particle, as discussed in Chapter 3. As with any acceleration, there must be a resultant force in the direction of the acceleration to produce it. This resultant force is called the *centripetal force*. It is important to understand that the centripetal force is *not* a new basic interaction or even a new empirical force we have not yet studied. It is merely a name for the resultant inward force that must be present to provide the centripetal acceleration needed for circular motion. The centripetal force may be due to a string, spring, or contact force such as a normal force or friction; it may be an action-at-a-distance type of force such as a gravitational force or an electrostatic force; or it may be a combination of any of these familiar forces. We shall illustrate with some examples.

It is often convenient to describe the motion of a particle moving in a circle with constant speed in terms of the time required for one complete revolution $T$, called the *period*. If the radius of the circle is $r$, the particle travels a distance $2\pi r$ during one period, so its speed is related to the radius and the period by

$$v = \frac{2\pi r}{T} \qquad \qquad 5\text{-}12$$

The reciprocal of the period is called the frequency $f$:

$$f = \frac{1}{T} = \frac{v}{2\pi r}$$    5-13

*Frequency and period in circular motion*

The frequency is usually given in revolutions per second (rev/s). Like the radian, the revolution is a dimensionless unit.

Figure 5-9 shows a portion of a circle of arc length $s$ and radius $r$. The angle $\theta$ swept out by the radius vector is $\theta = s/r$. The rate of change of the angle is called the *angular speed* $\omega$:

$$\omega = \frac{d\theta}{dt} = \frac{1}{r}\frac{ds}{dt}$$    5-14

*Angular speed*

Since $ds/dt$ is the linear speed $v$, the linear and angular speeds are related by

$$v = r\omega$$    5-15

Equations 5-14 and 5-15 hold only if the angle $\theta$ is measured in radians. The units of angular speed are radians per second (rad/s). Comparing Equations 5-15 and 5-13, we see that the angular speed is related to the frequency and period of revolution by

$$\omega = 2\pi f = \frac{2\pi}{T}$$    5-16

The centripetal acceleration can be conveniently written in terms of the angular speed:

$$a_c = \frac{v^2}{r} = \frac{(r\omega)^2}{r} = r\omega^2$$    5-17

*Centripetal acceleration*

If the speed of a particle is changing as it moves in a circle, the acceleration of the particle has a component $dv/dt$ tangent to the circle as well as an inward radial component $v^2/r$. From Equation 5-15 we see that the tangential acceleration can be written

$$a_t = \frac{dv}{dt} = r\frac{d\omega}{dt} = r\alpha$$    5-18

*Tangential acceleration*

where $\alpha = d\omega/dt$ is called the *angular acceleration*.

**Example 5-5** A particle of mass $m$ is suspended from a string of length $L$ and travels at constant speed $v$ in a horizontal circle of radius $r$. The string makes an angle $\theta$ given by $\sin \theta = r/L$, as shown in Figure 5-10. Find the tension in the string and the speed of the particle.

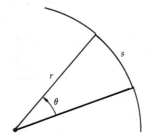

**Figure 5-9**
Portion of a circle. The arc length is related to the angle by $\theta = s/r$. The angular speed is $\omega = d\theta/dt = (1/r)\,ds/dt = v/r$.

**Figure 5-10**
(*a*) Conical pendulum for Example 5-5; (*b*) free-body diagram for the pendulum bob.

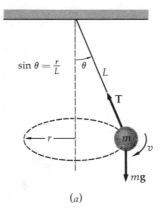

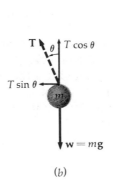

(*a*)                    (*b*)

This is known as a *conical pendulum*. The two forces acting on the particle are that due to the earth's attraction $mg$, acting vertically downward, and the tension $\mathbf{T}$, which acts along the string. In this problem we know that the acceleration is horizontal, toward the center of the circle, and of magnitude $v^2/r$. Thus the vertical component of the tension must balance the weight $mg$. The horizontal component of the tension is the resultant centripetal force. The vertical and horizontal components of $\Sigma\mathbf{F} = m\mathbf{a}$ therefore give

$$T\cos\theta - mg = 0$$

$$T\sin\theta = ma = \frac{mv^2}{r}$$

The tension is found directly from the first equation since $\theta$ is given. We can find the speed $v$ in terms of the known quantities $r$ and $\theta$ by dividing one equation by the other to eliminate $T$. We obtain

$$\tan\theta = \frac{v^2}{rg} \quad \text{or} \quad v = \sqrt{rg\tan\theta}$$

---

**Example 5-6** A pail of water is whirled in a vertical circle of radius $r$. If its speed is $v_t$ at the top of the circle, find the force exerted on the water by the pail. Find also the minimum value of $v_t$ for the water to remain in the pail.

The forces on the water at the top of the circle are shown in Figure 5-11. They are the force of gravity $mg$ and the force $\mathbf{N}$ exerted by the pail. Both these forces act downward. The acceleration, which is toward the center of the circle, is also downward at this point. Newton's second law gives

$$N + mg = m\frac{v_t^2}{r}$$

The force exerted by the pail is therefore $N = m(v_t^2/r) - mg$. Note that both the force of gravity and the contact normal force exerted by the pail contribute to the necessary centripetal force.

If we increase the speed of the pail, the bottom of the pail will exert a larger force on the water to keep it moving in a circle. If we decrease the speed, $N$ will decrease. Since the pail cannot exert an upward force on the water, the minimum speed the water can have at the top of the circle occurs when $N = 0$. Then

$$mg = m\frac{v_{t,\min}^2}{r}$$

or

$$v_{t,\min} = \sqrt{rg} \tag{5-19}$$

When the water is moving at this minimum speed, its acceleration at the top of the path is $g$, the free-fall acceleration of gravity, and the only force acting on the water is the gravitational attraction of the earth, the weight $mg$.

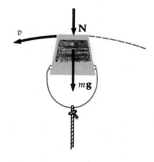

**Figure 5-11**
Pail of water whirled in a vertical circle. At the top the forces on the water are its weight $mg$ and the normal force exerted by the pail bottom $\mathbf{N}$. Both these forces act downward at this point, toward the center of the circle.

---

**Example 5-7** A block on a string moves in a vertical circle. When the string makes an angle $\theta$ with the vertical, the speed is $v$. Find the acceleration and the tension in the string.

From Figure 5-12 we see that the tension and weight are not along the same line. It is convenient to consider the components of the forces and accelerations in the radial and tangential directions. The only force

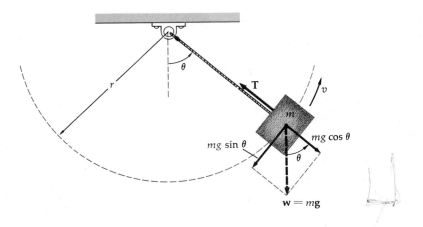

**Figure 5-12**
Forces on a block moving in a vertical circle for Example 5-7. The resultant force has a tangential component $mg \sin \theta$ and a component toward the center of the circle $T - mg \cos \theta$. The acceleration therefore also has a tangential component $dv/dt = g \sin \theta$ and a component toward the center of the circle $a_c = v^2/r$.

that has a component tangential to the circle is that due to gravity, $m\mathbf{g}$. The component of $m\mathbf{g}$ tangential to the circle has magnitude $mg \sin \theta$. If we choose the direction of increasing $\theta$ to be positive, the component is in the negative direction. The tangential acceleration is thus given by

$$-mg \sin \theta = ma_t \qquad a_t = -g \sin \theta$$

The tangential acceleration varies with angle $\theta$. The rate of change of the speed is

$$\frac{dv}{dt} = a_t = -g \sin \theta$$

The radial forces are the tension $T$ inward and $mg \cos \theta$ outward. Since the radial acceleration is toward the center of the circle, the radial component of Newton's second law is

$$T - mg \cos \theta = ma_c = m \frac{v^2}{r}$$

This result can be compared with that in Example 5-6 with $T$ replacing $N$. At the top of the circle, $\theta = 180°$, $\cos \theta = -1$, and $T = mv^2/r + mg \times \cos \theta = mv^2/r - mg$. The minimum speed for the string to remain taut is found by setting $T = 0$, resulting in $v = \sqrt{rg}$, as in Example 5-6. At the bottom of the circle, $\theta = 0$, $\cos \theta = 1$, and $T = mv^2/r + mg$. At this point the centripetal force is $T - mg$.

---

**Example 5-8** A car travels on a horizontal road in a circle of radius 30 m. If the coefficient of static friction is $\mu_s = 0.6$, how fast can the car travel without slipping?

Figure 5-13 shows the free-body diagram for the car. The normal force $\mathbf{N}$ balances the downward force due to gravity $m\mathbf{g}$. The only horizontal force is due to friction. Its maximum value is $f_{s,\,max} = \mu_s N = \mu_s mg$. In this case the frictional force is the centripetal force. The maximum speed of the car $v_{max}$ occurs when the frictional force equals its maximum value. $\mathbf{\Sigma F} = m\mathbf{a}$ then gives $f_{s,\,max} = \mu_s mg = m(v_{max}^2)/r$, or $v_{max} = \sqrt{\mu_s gr} = \sqrt{(0.6)(9.81 \text{ m/s}^2)(30 \text{ m})} = 13.3$ m/s = 47.8 km/h ($= 29.7$ mi/h). If the car travels at a speed greater than 13.3 m/s, the force of static friction will not be great enough to provide the acceleration needed for the car to travel in a circle and the car will slide out away from the center of the circle; i.e., it will tend to travel in a straight line.

If the road is not horizontal but banked, the normal force of the road will have a component inward toward the center of the circle, which

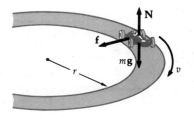

**Figure 5-13**
Car traveling in a horizontal circle. The normal force balances the force due to gravity, and the frictional force provides the centripetal acceleration.

This racetrack is banked so that the normal force exerted by the track on the car has a horizontal component toward the center of the circle to provide the centripetal force.

Ken Regan/Camera 5

will contribute to the centripetal force. The banking angle can be chosen in such a way that, for a given speed, no friction is needed for the car to make the curve (see Problem 21).

---

**Example 5-9** A satellite orbits the earth in a circular orbit. Find its period if (*a*) the satellite is just above the surface of the earth (assume it is high enough for air resistance to be neglected) and (*b*) if the satellite is at an altitude of 300 km.

Since the only force acting on the satellite is the force of gravitational attraction exerted by the earth, this force must be the centripetal force. Newton's law of gravitation combined with Newton's second law of motion then gives

$$\frac{GM_E m}{r^2} = m\frac{v^2}{r} = mr\omega^2 = mr\left(\frac{2\pi}{T}\right)^2$$

where we have used $v = r\omega = r(2\pi/T)$ because we are interested in obtaining an equation for the period $T$. Rearranging, we obtain

$$\frac{T^2}{r^3} = \frac{4\pi^2}{GM_E} \qquad\qquad 5\text{-}20$$

We see that the ratio $T^2/r^3$ is the same for all satellites in circular orbits about the earth. This result also holds for planets orbiting the sun if we replace the mass of the earth $M_E$ with the mass of the sun. It was discovered empirically for planets by Kepler (about 1600) and later derived by Newton (see Chapter 12). To evaluate Equation 5-20 it is convenient to replace $GM_E$ by $R_E^2 g$ from Equation 5-3. We then have $T^2 = (4\pi^2/gR_E^2)r^3$. If the satellite is just above the surface of the earth, $r = R_E$ and

$$T = 2\pi\sqrt{\frac{R_E}{g}} = 2\pi\sqrt{\frac{6.37 \times 10^6 \text{ m}}{9.81 \text{ m/s}^2}} = 5.06 \times 10^3 \text{ s} = 84.4 \text{ min}$$

At an altitude of 300 km above the earth's surface, $r = 6.67 \times 10^6$ m. Since $T$ is proportional to $r^{3/2}$, we can find $T$ at this distance from $T = (84.4 \text{ min})[(6.67 \times 10^6)/(6.37 \times 10^6)]^{3/2} = 90.4 \text{ min}$.

---

**Questions**

7. Explain why the following statement is incorrect: In the circular motion of a ball on a string, the ball is in equilibrium because the outward force of the ball due to its motion is balanced by the tension in the string.

8. Under what conditions might the strong nuclear force be a centripetal force?

9. Discuss the following statement: The centripetal force arises because of the circular motion of a body.

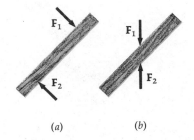

(a)                    (b)

**Figure 5-14**
(a) The two forces $\mathbf{F}_1$ and $\mathbf{F}_2$ are equal and opposite, but the stick is not in static equilibrium because these forces tend to rotate it clockwise. (b) Here the two forces have the same line of action and have no tendency to produce rotation.

# 5-5    Static Equilibrium of a Rigid Body

If a particle is in static equilibrium, i.e., is at rest and remains at rest, the resultant force acting on it must be zero. We used this condition in Example 4-5 to obtain information about the tension in wires supporting a picture. We now consider the static equilibrium of extended rigid bodies instead of point particles. Although we still must have zero resultant force for static equilibrium, this condition alone is not sufficient, as can be seen from Figure 5-14a. In this figure, a uniform stick rests on a horizontal table (assumed frictionless); the stick is subjected to two forces of equal magnitude and opposite direction, but the forces are not along the same line. (We need not consider the vertical forces $m\mathbf{g}$ downward and the normal force $\mathbf{N}$ upward, which balance.) Although the center of the stick will remain at rest, the forces $\mathbf{F}_1$ and $\mathbf{F}_2$ will produce a rotation in the clockwise direction (as viewed from above the table). The stick is clearly not in static equilibrium. On the other hand, if the forces are applied along the same line, as in Figure 5-14b, they will not produce rotation and the stick will be in equilibrium. When we consider the application of forces to bodies of finite size, the line of action of a force, as well as its magnitude and direction, is important.

In Figure 5-15 a body is pivoted at point $O$ so that point $O$ is fixed in space but the body is free to rotate. A force $\mathbf{F}$ acts at an angle $\theta$ to the radius vector $\mathbf{r}$ from $O$ to the point of application of the force. We have resolved the force into its components $F_\perp = F \sin \theta$ and $F_\parallel = F \cos \theta$, respectively perpendicular and parallel to $\mathbf{r}$. The parallel component does not tend to produce any rotation of the body about point $O$. The effect of the perpendicular component depends both on the magnitude of $F_\perp$ and on the distance $r$ from $O$. We define the *moment of the force*, or *torque*, about point $O$ to be the product of $F_\perp$ and $r$.

$$\tau = F_\perp r = (F \sin \theta)r \qquad \text{5-21}$$

A torque may be either clockwise or counterclockwise, depending on the sense of the rotation it tends to produce. The torque due to the force $\mathbf{F}$ in Figure 5-15 is clockwise.

In the figure we have indicated the line of action of the force with a dashed line. The perpendicular distance from the point $O$ to the line of action of the force is $\ell = r \sin \theta$. The distance $\ell$ is called the *lever arm* of the force. From Equation 5-21 we see that the torque about $O$ can also be written

$$\tau = F_\perp r = F\ell \qquad \text{5-22}$$

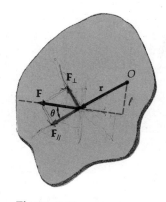

**Figure 5-15**
Force $\mathbf{F}$ exerted on a body pivoted at point $O$. The rotational effect of this force is measured by the torque $\tau$ defined as $\tau = F_\perp r = (F \sin \theta)r = F\ell$.

*Torque defined*

*Lever arm*

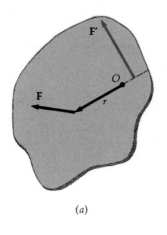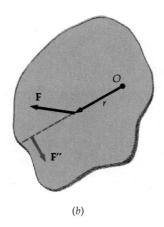

(a)                                         (b)

**Figure 5-16**
The torque produced by the
force **F** about O in Figure 5-15
can be balanced by an equal
torque produced by a larger
force **F'** at a smaller distance
from O, as in (a), or by a
smaller force **F"** at a larger
distance from O, as in (b).

If **F** is the only force applied to the body, the body will rotate about
point O. Point O remains fixed because of the force exerted by the
pivot. We can prevent the rotation of the body by applying a second
force **F'** such that it produces a torque of equal magnitude but opposite
sense, as indicated in Figure 5-16a. Note that **F'** is not necessarily equal
to **F**. In Figure 5-16a, **F'** is considerably greater in magnitude, but its
lever arm is smaller, so that $F'\ell' = F\ell$. Rotation could also be prevented
by applying a smaller force **F"** with a greater lever arm $\ell''$ such that
$F''\ell'' = F\ell$, as in Figure 5-16b.

We can summarize our results by stating the second condition for
static equilibrium:

*The net clockwise torque about any point must equal the net counter-
clockwise torque about that point.*

If we call the counterclockwise torques positive and clockwise torques
negative, this is equivalent to saying that the algebraic sum of the
torques about any point must be zero.

We have two conditions for the static equilibrium of a rigid body:

1. The resultant external force must be zero:

$$\Sigma \mathbf{F} = 0 \qquad\qquad 5\text{-}23a$$

2. The resultant external torque about any point must be zero:

$$\Sigma \tau = 0 \qquad\qquad 5\text{-}23b$$

*Conditions for static
equilibrium*

Note that we have stated our second condition for equilibrium in
terms of torques about *any point*. Although we have been considering
torques about the pivot point in Figures 5-15 and 5-16, we could have
chosen any point. If a body is stationary and not rotating about one particular point, it is not rotating about *any* point. This fact is often useful
in solving problems because we can choose the point for computing
torques at our convenience.

**Example 5-10** A 3-m board of negligible weight rests with its ends on
scales, as shown in Figure 5-17. A small 60-N weight rests on the board
2.5 m from the left end and 0.5 m from the right end, as shown. Find
the readings on the scales.

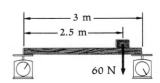

**Figure 5-17**
Board of negligible weight
resting on scales for Example
5-10.

Figure 5-18 shows the free-body diagram for the board. The force $F_L$ is that exerted by the scale at the left end of the board. Since the board exerts an equal but opposite force on the scale, the magnitude of $F_L$ is the reading on the left scale. From our first condition of equilibrium (the resultant force must be zero) we know that $F_L + F_R = 60$ N. We get a second relation between $F_L$ and $F_R$ by considering torques. If we consider the point at the weight to be the "pivot" point, we have two torques, a clockwise torque of magnitude $F_L(2.5$ m$)$ and a counterclockwise torque of magnitude $F_R(0.5$ m$)$. Equating these torques gives $0.5\, F_R = 2.5\, F_L$. Solving for $F_L$ and $F_R$, we obtain $F_L = 10$ N and $F_R = 50$ N, which are the scale readings. The scale on the right supports the greater weight, as expected.

Although there is nothing incorrect with the above analysis, there is an easier way to solve the problem without having to solve two equations for two unknowns. If we compute torques about a point on the line of action of one of the unknown forces, that force will not enter into the equation because its lever arm will be zero. We first consider torques about the left scale. The weight produces a clockwise torque of magnitude $(60$ N$)(2.5$ m$) = 150$ N·m, and $F_R$ produces a counterclockwise torque $F_R(3$ m$)$. Setting their magnitudes equal, we get $F_R(3$ m$) = 150$ N·m, $F_R = 50$ N. We can then find $F_L$ immediately from $F_L = 60$ N $- 50$ N $= 10$ N or by considering torques about the right scale. We have then $F_L(3$ m$) = (60$ N$)(0.5$ m$) = 30$ N·m, $F_L = 10$ N. As usual, whenever there are two ways of solving a problem, it is a good idea to use one method to check the results of the other.

In general, if we have several unknown forces, we can reduce the work involved in solving the problem by computing torques about a point on the line of action of one of the unknown forces so that that force does not enter into the equation.

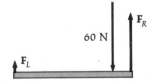

**Figure 5-18**
Free-body diagram for the board in Example 5-10.

This cube can balance on one corner if the center of gravity of the cube lies directly over the point of support.

Figure 5-19 shows an extended body which we have divided up into many smaller bodies. If we make the divisions small enough, we can consider the smaller bodies to be point particles. The force of attraction of the earth on each of these particles is indicated by the vectors $\mathbf{w}_i$.

**Figure 5-19**
The weights of all the particles of a body can be replaced by the total weight $\mathbf{W}$ acting at the center of gravity, which is the point about which the resultant torque due to the forces $\mathbf{w}_i$ is zero.

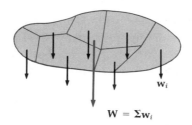

$\mathbf{w}_i$

$\mathbf{W} = \Sigma \mathbf{w}_i$

There is one point about which these parallel weight vectors produce zero torque. This point is called the *center of gravity*. For a uniform body, say a uniform sphere or a uniform stick, the center of gravity is at the body's geometric center. We can represent the force of attraction of the earth for a body as a single vector $\mathbf{W}$, the weight of the body acting at the center of gravity.*

*Center of gravity*

We can often locate the center of gravity of an object such as a stick by balancing it on a pivot. The point of the pivot at which the stick will balance is the center of gravity of the stick.

---

* Unless the gravitational field of the earth varies appreciably over the body, a very rare case, the center of gravity coincides with the center of mass of a body. In Chapter 8 we shall show that the acceleration of the center of mass of a body is given by $\mathbf{a}_{CM} = \Sigma \mathbf{F}/M$, where $\Sigma \mathbf{F}$ is the resultant force acting on the body and $M$ is the body's total mass.

If we suspend a body from a pivot (not at the center of gravity) so the body is free to rotate about the pivot under the action of gravity, the center of gravity will lie directly below the pivot when the body is in equilibrium. At any other position the attraction of the earth will produce a torque about the pivot. We can use this result to find the center of gravity of a plane figure, as illustrated in Figure 5-20.

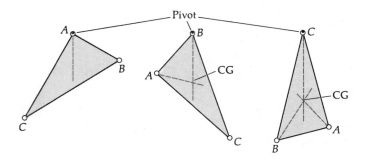

**Figure 5-20**
The center of gravity of an irregularly shaped object lies directly below any point about which it is pivoted. This property can be used to locate the center of gravity.

---

**Example 5-11** A uniform 5-m ladder weighs 12 N and leans against a frictionless vertical wall (Figure 5-21). The foot of the ladder is 3 m from the wall. What is the minimum coefficient of friction necessary between the ladder and the floor if the ladder is not to slip?

The forces acting on the ladder are the force due to gravity $w$ acting downward at the center of gravity, the force $F_1$ exerted horizontally by the wall (since the wall is frictionless, it exerts only a normal force), and the force exerted by the floor, which consists of a normal force $N$ and a horizontal force of friction $f$. From the first condition of equilibrium we have

$$N = w = 12 \text{ N} \quad \text{and} \quad F_1 = f$$

Since we know neither $f$ nor $F_1$, we must use the second condition of equilibrium and compute torques about some convenient point. We choose the point of contact between the ladder and the floor because both $N$ and $f$ act at this point and will therefore not appear in our torque equation. The torque exerted by the force of gravity about this point is clockwise with magnitude 12 N times the lever arm 1.5 m. The torque exerted by $F_1$ about the point of contact of the ladder and the floor is counterclockwise with magnitude $F_1$ times the lever arm 4 m. The second condition of equilibrium thus gives

$$F_1(4 \text{ m}) - (12 \text{ N})(1.5 \text{ m}) = 0$$

$$F_1 = 4.5 \text{ N}$$

This equals the magnitude of the frictional force. Since the frictional force $f$ is related to the normal force by

$$f \leqslant \mu_s N$$

we have

$$\mu_s \geqslant \frac{f}{N} = \frac{4.5 \text{ N}}{12 \text{ N}} = 0.375$$

where $\mu_s$ is the coefficient of static friction.

Another way of solving this problem may be easier. Let $F_2 = f + N$ be the force exerted by the floor. The ratio $f/N$ is just the cotangent of the angle between the force $F_2$ and the horizontal. We can find this angle as follows. If we extend the lines of action of the forces $w$ and $F_1$,

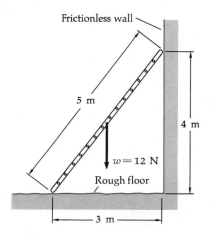

Frictionless wall

5 m

4 m

$w = 12$ N

Rough floor

3 m

**Figure 5-21**
Ladder on rough floor leaning against a frictionless wall (Example 5-11).

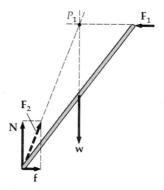

**Figure 5-22**
Free-body diagram for the ladder in Example 5-11. The lines of action of the three forces $\mathbf{F}_1$, $\mathbf{w}$, and $\mathbf{F}_2$ must all intersect at a common point $P_1$.

they meet at point $P_1$ in Figure 5-22. The torques exerted by these two forces about this point are zero. Since $\mathbf{F}_2$ is the only other force acting, it must also exert zero torque about point $P_1$. Thus its line of action must also pass through this point. From the figure one sees that if this is so, the cotangent of the angle made by $\mathbf{F}_2$ with the horizontal is $(1.5 \text{ m})/(4 \text{ m}) = 0.375 = f/N \leqslant \mu_s$. We note that if a body is in static equilibrium under the influence of three nonparallel coplanar forces, the lines of action of these forces must intersect at one point.

When solving problems in statics it is often convenient to replace a set of parallel forces by a single force equal to the resultant of the parallel forces (so that it has the same effect on the translational equilibrium of the body), and acting at such a point as to produce the same rotational effect. We find this point by requiring the resultant force to give the same torque about any point as the original parallel forces. This idea is illustrated in Figure 5-23 for two parallel forces. We have already used this idea when we replaced the forces of gravity acting on various parts of a body by a single force, the weight of the body acting at the center of gravity. However, two forces equal in magnitude but opposite in direction and having different lines of action tend to produce rotation, but since their resultant force is zero, they cannot be replaced by a single force having the same effect on both translation and rotation. Such a pair of equal and opposite forces is called a *couple*. Consider the couple shown in Figure 5-24. The torque produced by these forces about point $O$ is

$$\tau = Fx_2 - Fx_1 = F(x_2 - x_1) = FD \qquad 5\text{-}24$$

where $F$ is the magnitude of either force and $D = x_2 - x_1$ is the distance between them. This result does not depend on the choice of the point $O$. The torque produced by a couple is the same about all points in space.

Two unequal antiparallel forces such as those shown in Figure 5-25 can be replaced by a single force equal to the resultant force acting at the center of gravity plus a couple whose torque equals the torque about the center of mass of the original forces. In general, any number of parallel and antiparallel forces can be replaced by a single resultant force and a couple. Since the resultant force exerted by a couple is zero, the only way it can be balanced is by a second couple exerting an equal but opposite torque. For example, the forces $\mathbf{N}$ and $\mathbf{w}$ in Figure 5-22 form a couple of torque $(12 \text{ N})(1.5 \text{ m}) = 18 \text{ N·m}$. It is balanced by the forces $\mathbf{f}$ and $\mathbf{F}_1$, which also form a couple. Since the separation of these forces is $D = 4 \text{ m}$, their magnitude must be $(18 \text{ N·m})/(4 \text{ m}) = 4.5 \text{ N}$, as found in Example 5-11.

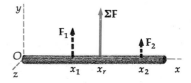

**Figure 5-23**
The two parallel forces $\mathbf{F}_1$ and $\mathbf{F}_2$ can be replaced by a single resultant force $\Sigma\mathbf{F}$ which has the same effect. The point of application of $\Sigma\mathbf{F}$ is $x_r$, given by $|\Sigma F|x_r = F_1x_1 + F_2x_2$.

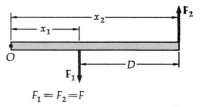

$$F_1 = F_2 = F$$

**Figure 5-24**
Two equal and opposite forces constitute a couple. A couple exerts a torque which has the same value $FD$ about any point in space.

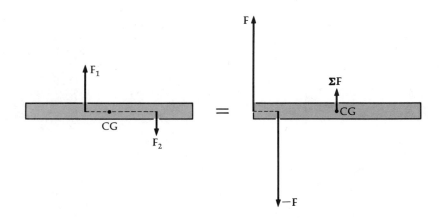

**Figure 5-25**
Two antiparallel forces of
unequal magnitude can be re-
placed by a single force $\Sigma\mathbf{F}$
acting through the center of
gravity plus a couple of
torque equal to the torque exerted by the original forces
about the center of gravity.

## Questions

10. Must there be any matter at the location of the center of gravity of a body?

11. Can a ladder in equilibrium stand on a frictionless horizontal floor leaning against a rough vertical wall? Why or why not?

## 5-6 Motion with a Retarding Velocity-Dependent Force

When an object moves through a fluid such as air or water, the fluid exerts a retarding or drag force which tends to reduce the speed of the object. This drag force depends on the shape of the object, on the properties of the fluid, e.g., viscosity, defined in Chapter 13, and on the speed of the object relative to the fluid. Like the force of friction, this drag force is very complicated; in general, however, as the speed of the object increases, the drag force increases. For small speeds the drag force is approximately proportional to the speed of the object; for higher speeds it is more nearly proportional to the square of the speed.

Consider an object dropped from rest and falling under the influence of the force of gravity, which we assume to be constant, and a retarding force of magnitude $bv^n$, where $b$ and $n$ are constants. We then have a downward constant force $mg$ and an upward force $bv^n$. If we take the downward direction to be positive, we obtain from Newton's second law

$$F = mg - bv^n = m\frac{dv}{dt} \qquad\qquad 5\text{-}25$$

Equation 5-25 is a differential equation relating the speed $v$ and its time rate of change $dv/dt$. It can be solved by standard methods of calculus, but since these methods may not be familiar to you yet, we shall consider the solution qualitatively. In practical examples, the "constants" $b$ and $n$ are not really constant but depend on the speed $v$, so that numerical methods of solution are necessary. Without solving the equation by any method at all, we can describe the important features of the motion.

At $t = 0$, when the object is dropped, the speed is zero, so that the retarding force is zero and the acceleration is $g$ downward. As the speed of the object increases, the drag force increases and the acceleration is

When the retarding force equals his weight, the sky diver is in equilibrium and falls with constant velocity. The terminal velocity (without parachute) is approximately 200 km/h.

Andy Keech

less than $g$. Eventually the speed is great enough for the drag force $bv^n$ to equal the force of gravity $mg$, and the acceleration is zero. The object then continues moving at a constant speed $v_t$, called its *terminal speed*. Setting $dv/dt$ equal to zero in Equation 5-25, we obtain for the terminal speed

*Terminal speed*

$$v_t = \left(\frac{mg}{b}\right)^{1/n} \qquad\qquad 5\text{-}26$$

Figure 5-26 shows a plot of $v$ versus $t$ for $n = 1$ and for $n = 2$. Note that in each case the initial slope of the $v$-versus-$t$ curve is $g$ since the drag force is zero initially. The dashed line in this figure is drawn with constant slope $g$. It intersects the horizontal line $v = v_t$ after a time $t_c$, given by

$$t_c = \frac{v_t}{g} \qquad\qquad 5\text{-}27$$

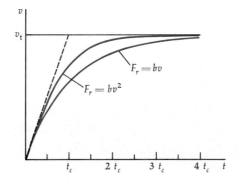

**Figure 5-26**
Plots of speed versus time for a particle under the influence of a constant force of gravity and a velocity-dependent retarding force $bv^n$ for $n = 1$ and $n = 2$. The constants $b$ have been chosen so that the terminal speed $v_t$ will be the same in each case for easy comparison. The characteristic time $t_c$ is given by $v_t = gt_c$.

The time $t_c$, called the *characteristic time* for the motion, is the time for the speed $v$ to reach an appreciable fraction of its terminal value. For $n = 1$, the speed is $0.63v_t$ after a time $t_c$. The larger the constant $b$, the smaller the terminal speed and the sooner an appreciable fraction of the terminal speed is reached. The constant $b$ depends on the shape of the object. A parachute is designed to make $b$ large so that the terminal speed will be small. On the other hand, cars are designed with $b$ small to reduce the effect of wind resistance.

For a second example of motion under the influence of a retarding force, let us assume that a particle moves with initial speed $v_0$ and that the only force acting is a drag force of magnitude $bv$ (we assume $n = 1$ for simplicity). Newton's second law gives

$$-bv = m \frac{dv}{dt}$$

or

$$\frac{dv}{dt} = -\frac{b}{m} v \qquad \qquad 5\text{-}28$$

In this case the acceleration is always negative, meaning that the speed always decreases. Initially, the acceleration is $-(b/m)v_0$, but as the speed decreases, the acceleration also decreases. A plot of $v$ versus $t$ for this problem is shown in Figure 5-27. The black dashed line in this figure has a constant slope $-bv_0/m$. It intersects $v = 0$ at the characteristic time $t_c = m/bv_0$. This is the time when the speed would equal zero if the acceleration were constant and equal to its initial maximum value. Actually, the speed decreases to $0.37v_0$ after one characteristic time $t_c$.

Equation 5-28 states that the rate of change of $v$ is proportional to $v$. The function whose rate of change is proportional to itself is the exponential. The solution of Equation 5-28 is

$$v = v_0 e^{-bt/m} = v_0 e^{-t/t_c} \qquad \qquad 5\text{-}29$$

That this is the actual solution of Equation 5-28 can be easily checked by direct differentiation of Equation 5-29, which is left as an exercise.

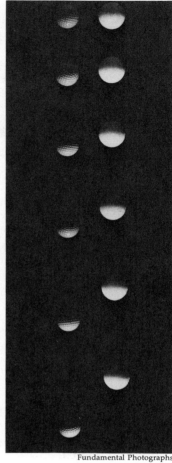

Fundamental Photographs

A golf ball and a styrofoam ball falling in air. The air resistance is negligible for the heavier golf ball, which falls with essentially constant acceleration. The styrofoam ball reaches terminal velocity quickly, as indicated by the nearly equal spacing at the bottom.

**Figure 5-27**
Plot of speed versus time for a particle moving under the influence of a single retarding force of magnitude $bv$. The characteristic time $t_c$ is the time for the speed to reach zero if its deceleration were constant and equal to $bv_0$.

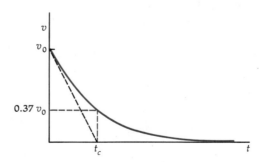

**Questions**

12. How does a sky diver vary the value of $b$?

13. How would you expect the value of $b$ for air resistance to depend on the density of air?

*Optional*

# 5-7  Pseudo Forces

Newton's laws hold only in inertial reference frames. When the acceleration of an object is measured relative to a reference frame which itself is accelerated relative to an inertial frame, the resultant force on the object does not equal the mass of the object times its acceleration. In some cases an object will be at rest relative to the noninertial frame even though there is obviously an unbalanced force acting on it. In other cases, the object has no forces acting on it but is accelerated relative to that frame. However, even in such an accelerated reference frame, we can use Newton's second law, $\Sigma\mathbf{F} = m\mathbf{a}$, if we introduce *fictitious*, or *pseudo, forces* which depend on the acceleration of the reference frame. These forces are not exerted by any agent. They are merely fictions introduced to make $\Sigma\mathbf{F} = m\mathbf{a}$ work when the acceleration $\mathbf{a}$ is measured relative to the noninertial frame. To observers in the noninertial frame the pseudo forces appear as real as other forces. The most familiar pseudo force is the centrifugal force encountered in rotating reference frames. Unfortunately, the concept of centrifugal force is probably used incorrectly more often than correctly.

Let us first consider a railroad car moving in a straight line along a horizontal track with constant acceleration $\mathbf{a}$ relative to the track, which we assume to be in an inertial reference frame. If we drop an object in the car, it does not fall straight down but toward the back of the car. Relative to the car it has vertical acceleration $\mathbf{g}$ and horizontal acceleration $-\mathbf{a}$ (Figure 5-28). If an object is placed on a smooth table so that the resulting force is zero, it accelerates toward the back of the car. Of course, from the point of view of an observer in an inertial frame on the tracks, the object does not accelerate. Instead, the car and table accelerate under the object. We can use Newton's second law in the reference frame of the car if we introduce a pseudo force $-m\mathbf{a}$ acting on any object of mass $m$. Consider, for example, a lamp hanging by a cord from the ceiling of the car. The description of the acceleration of the lamp and forces acting on it from the inertial and noninertial frames is shown in

**Figure 5-28**
An object is dropped inside a railway car which is initially at rest but has acceleration $\mathbf{a}$ to the right. (a) An observer on the ground in an inertial reference frame sees the ball fall straight down. (b) An observer in the accelerated car sees the ball fall down and toward the left of the car. He attributes the backward acceleration to a pseudo force $-m\mathbf{a}$.

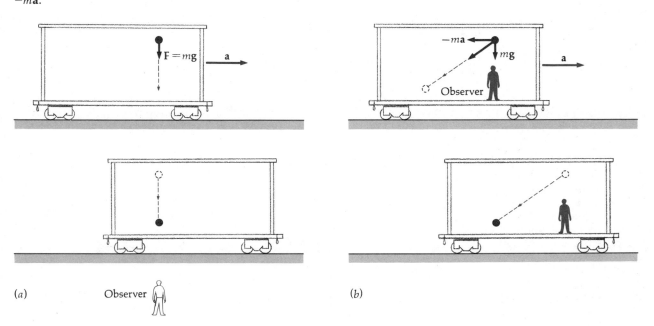

(a)     Observer     (b)

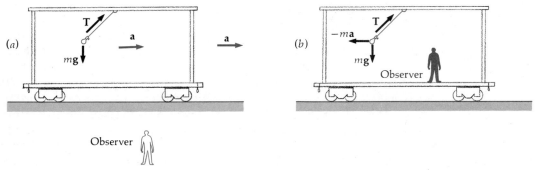

**Figure 5-29**
A lamp is hung by a cord from the roof of an accelerating car. (*a*) According to an observer in an inertial reference frame, the lamp accelerates to the right because of the unbalanced force, the horizontal component of the tension **T**. (*b*) In the accelerated frame, the lamp appears to be at rest and not accelerating. The forces are balanced by the pseudo force $-m\mathbf{a}$, which must be introduced in this frame so that Newton's second law can be used.

*Centrifugal force*

Figure 5-29. The vertical component of the tension of the cord equals the weight of the lamp according to each observer. In the inertial frame of the track, the lamp is accelerating. This acceleration is provided by the resultant force due to the horizontal component of the tension in the cord. In the frame of the car, the lamp is at rest and therefore has no acceleration. This is explained by the fact that the horizontal component of the tension balances the pseudo force $-m\mathbf{a}$ observed on all objects in the car.

Figure 5-30 shows another noninertial frame, a rotating platform. Since each point on the platform is moving in a circle, it has centripetal acceleration. Thus a frame attached to the platform is a noninertial frame. In Figure 5-30 a block at rest relative to the platform is attached to the center post by a string. According to observers in an inertial frame, the block is moving in a circle with speed $v$. It is accelerating toward the center of the circle. This centripetal acceleration $v^2/r$ is provided by the unbalanced force due to the string tension **T**. However, according to an observer on the platform, the block is at rest and not accelerating. In order to use $\Sigma\mathbf{F} = m\mathbf{a}$, he must introduce a pseudo force of magnitude $mv^2/r$ acting radially outward to balance the string tension. This fictitious outward force, called the *centrifugal force*, appears quite real to the observer on the platform. If he wants to stand "at rest" on the platform, an inward force of this magnitude must be exerted on him (by the floor) to "balance" the outward centrifugal force. We have occasion to use this pseudo force *only* in a rotating frame. Consider a satellite near the surface of the earth and observed in an inertial frame attached to the earth (we neglect the earth's rotation here). People often say that the satellite does not fall because the gravitational attraction of the earth "is balanced by the centrifugal force." This is incorrect. Pseudo forces such as the centrifugal force appear only in accelerated reference frames. In the earth frame, the satellite does "fall" toward the earth with acceleration $v^2/r$, produced by the single unbalanced force of gravity acting on

**Figure 5-30**
A block is tied to the center post of a rotating platform by a string. (*a*) An inertial observer sees the block moving in a circle with centripetal acceleration provided by the unbalanced force **T**. (*b*) According to a noninertial observer on the platform, the block is not accelerating. Newton's second law can be used only if a pseudo force $mv^2/r$ acting outward is introduced to balance the tension **T**.

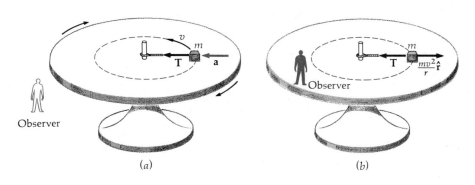

**Figure 5-31**
A boy at the center of a rotating platform throws a ball
toward a friend on the edge
of the platform. (*a*) In an inertial frame the ball travels in a
straight line and misses the
receiver because the receiver
moves with the platform. (*b*)
In the rotating frame of the
platform, the receiver is at
rest, and the ball deflects to
the right. The pseudo force
which deflects the ball from a
straight line in this frame is
called the Coriolis force.

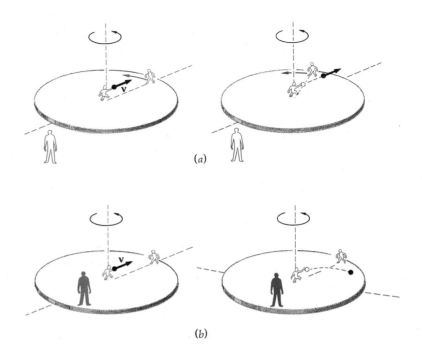

(a)

(b)

*Coriolis force*

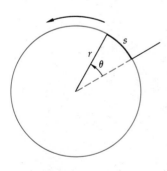

**Figure 5-32**
Diagram for defining the
angular speed $\omega = d\theta/dt$ of a
rotating platform.

it. However, an observer *in the satellite* who considers the satellite to be
at rest can use $\Sigma \mathbf{F} = m\mathbf{a}$ only by introducing an outward centrifugal
force to balance gravity.

A second pseudo force which depends on the velocity of a particle
must be introduced in a rotating frame in order to use $\Sigma \mathbf{F} = m\mathbf{a}$ in that
frame. Called the *Coriolis force*, it is perpendicular to the velocity of a
particle (relative to the rotating frame) and causes a sideways deflection. Consider two observers standing along a radial line on a rotating
platform and playing catch. If the ball is thrown outward along the
radial line, they will see it deflect to the right and miss the receiver (Figure 5-31). In an inertial frame, the ball travels in a straight line after
leaving the thrower and misses because the receiver is moving. The
path of the ball relative to the rotating platform is the curved line shown
in the figure. The ball must be thrown to the left of the receiver to take
into account this sideways deflection.

We can calculate the magnitude of the Coriolis force from a simple
special case considered in an inertial frame. Let $\theta$ be the angle between
a radial line fixed on the platform and a fixed line in space shown in Figure 5-32. This angle changes with time because the platform is rotating.
The rate of change of $\theta$ with respect to time is the *angular speed* $\omega$ of the
platform:

$$\omega = \frac{d\theta}{dt}$$

The angular speed $\omega$ describes the rotation of the platform. Any point
fixed on the platform moves in a circular arc. The distance $s$ moved is related to the angle $\theta$ and to the radial distance $r$ of the point by

$$s = r\theta$$

Thus the speed of a point fixed on the platform is

$$\frac{ds}{dt} = r\frac{d\theta}{dt} = r\omega$$

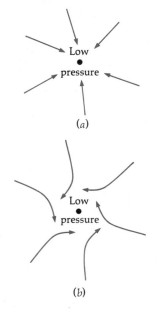

(a)

(b)

**Figure 5-33**
If the earth were not rotating, the winds would blow inward toward a low-pressure center as in (a). The Coriolis force deflects the winds to the right as in (b), setting up a counterclockwise pattern in the northern hemisphere. (In the southern hemisphere, the pattern is clockwise.)

Now consider a ball thrown from the origin with speed $v$ along a radial line. After a time $t$ the ball has moved a radial distance $r = vt$. But the point on the platform at a distance $r$ has moved along a circular arc a distance $s$ given by

$$s = r\omega t = (vt)\omega t$$

The distance $s$ is just the distance the ball is deflected by the pseudo force, the Coriolis force, in the rotating reference frame. In the rotating frame the deflection is to the right (looking along the velocity vector) and has the magnitude

$$s = \omega v t^2 = \tfrac{1}{2}(2\omega v)t^2 = \tfrac{1}{2}a_c t^2$$

where

$$a_c = 2\omega v \qquad\qquad 5\text{-}30$$

is called the *Coriolis acceleration*. The Coriolis force is

$$F_c = ma_c = 2m\omega v \qquad\qquad 5\text{-}31$$

This result, i.e., that there is a pseudo force perpendicular to the velocity vector with magnitude $2m\omega v$, is not limited to motions in the radial direction but applies to motion in any direction (perpendicular to the axis of rotation). The direction of the force is to the right looking along the velocity vector for counterclockwise rotation as seen from above.

These two pseudo forces, the centrifugal force and the Coriolis forces for a rotating reference frame, have direct application to reference frames attached to the earth because of the earth's rotation. In particular, Coriolis forces are important for understanding weather. For example, these forces are responsible for the fact that cyclones (viewed from above) are counterclockwise in the northern hemisphere and clockwise in the southern hemisphere (see Figure 5-33).

Hurricane Ginger, September 13, 1971, showing the counterclockwise rotation observed in the Northern Hemisphere.

Review

A. Define, explain, or otherwise identify:

Quantization of charge, 124
Frictional force, 129
Coefficient of friction, 130
Centripetal force, 133
Angular speed, 134
Torque, 138

Center of gravity, 141
Couple, 143
Terminal speed, 145
Pseudo force, 147
Centrifugal force, 148
Coriolis force, 149

B. List the four basic interactions.

C. Assign a letter to each basic interaction you listed in part B. After each term below, write the letter of the basic interaction (if any). If there is no basic interaction associated with the term, explain why.

1. Molecular forces

2. Force holding the nucleus together

3. Centripetal force

4. Friction

5. Your weight

6. Centrifugal force

7. Force of a spring

8. Coriolis force

9. Beta decay

10. Force holding the moon in its orbit

D. True or false:

1. The force of gravity is always attractive.

2. Protons and electrons have charges of equal magnitude.

3. Two protons always repel each other.

4. Friction is related to molecular forces.

5. Pseudo forces appear only in accelerated reference frames.

6. The terminal speed of an object depends on its shape.

7. Centripetal force is a pseudo force.

8. In static equilibrium the resultant torque about any point is zero.

9. A couple cannot be balanced by a single force.

10. The center of gravity is always at the geometric center of a body.

Exercises

**Section 5-1, The Basic Interactions**

1. Calculate the mass of the earth using the values $G = 6.67 \times 10^{-11}$ N·m²/kg², $g = 9.81$ N/kg, and $R_E = 6.37$ Mm.

2. (a) A particle is dropped from a height of 6.37 Mm above the surface of the earth. What is its initial acceleration? (b) At what distance above the earth's surface is the acceleration of gravity half its value at sea level?

3. Find the gravitational force which attracts a 65-kg boy to a 50-kg girl when they are 0.5 m apart. (Assume they are point masses.)

4. Calculate the ratio of the electrostatic force to the gravitational force between two protons.

5. Two small spheres, each with a charge of $-1$ $\mu$C, are 20 cm apart. (*a*) What is the number of excess electrons on each sphere? (*b*) Find the magnitude of the electric force exerted by one sphere on the other.

6. Two small electrified objects separated by 5 cm attract each other with a force of 4 N. Find the force of attraction when their separation is (*a*) 7 cm; (*b*) 10 cm; (*c*) 50 cm; (*d*) 0.5 cm.

7. Four equal charges $q = 50$ nC are situated at corners of a square of length 4 cm. Find the resultant electric force exerted on any one of the charges by the other three charges and indicate its direction in your diagram.

8. A penny has a mass of 3 g. Each copper atom contains 29 electrons, and there are $6.02 \times 10^{23}$ (Avogadro's number) atoms in 64 g of copper. (*a*) Find the number of atoms, the number of electrons, and the total negative charge in coulombs in a penny of pure copper. (*b*) How long would it take for this much charge to flow through a wire at a rate of 1 A $= 1$ C/s? (*c*) If 10 percent of the negative charge on one penny could be transferred to another penny, calculate the electrostatic force exerted by one penny on the other when they are separated by 20 cm (assume them to be point charges).

9. Calculate the magnitude of the strong nuclear force between two protons separated by a distance of 1 fm assuming that this force is 10 times the electric force between them. Compare this force with that between the electron and proton in the hydrogen atom, in which the separation is 0.053 nm.

### Section 5-2, Molecular Forces; Springs and Strings

10. The equilibrium separation of the $Na^+$ and $Cl^-$ ions in NaCl is about 0.24 nm. Find the electric force of attraction between the ions at this separation assuming that they are point charges. Compare your result with the results of Exercise 9 for nuclear forces and for atomic forces.

11. A spring has a force constant $k = 200$ N/m. A 5-kg object is suspended motionless from the spring. Find (*a*) the numerical values of all the forces acting on the object and (*b*) the extension of the spring from its equilibrium position.

12. A 100-kg block is pulled with constant acceleration along a frictionless table by a cable that stretches 0.3 cm. The block starts from rest and moves 4 m in 4 s. Assuming that the cable obeys Hooke's law, find (*a*) its force constant and (*b*) the stretching of the cable if the block is suspended vertically from the cable at rest.

13. A 2-kg box rests on a frictionless incline of angle 30° supported by a spring (Figure 5-34). The spring stretches by 3 cm. (*a*) Find the force constant of the spring. (*b*) If the box is pulled down the incline 5 cm from its equilibrium position and released, what will its initial acceleration be?

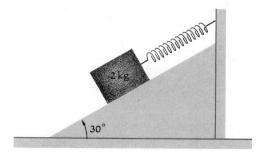

**Figure 5-34**
Exercise 13.

14. A body of mass $m$ is attached to two springs along a line, as shown in Figure 5-35. Each spring is stretched from its equilibrium position. The force constants of the springs are $k_1$ and $k_2$. (*a*) Find the ratio of the amount of stretching of the springs. (*b*) Show that if the body is displaced a small distance $x$ from equilibrium, the net restoring force is the same as if the body were attached to a single spring of force constant $k = k_1 + k_2$.

**Figure 5-35**
Exercise 14.

**Section 5-3, Contact Forces; Support Forces and Friction**

15. The coefficients of friction between the tires of a car and the road are $\mu_s = 0.6$ and $\mu_k = 0.5$. (a) If the resultant force on the car is the force of static friction exerted by the road, what is the maximum acceleration of the car? (b) What is the least distance in which the car can stop if it is initially traveling at 30 m/s?

16. The force that accelerates a car along a flat road is the frictional force between the road and the car tires. (a) Explain why the acceleration can be greater when the wheels do not spin. (b) If a car is to accelerate from 0 to 90 km/h in 12 s at constant acceleration, what is the minimum coefficient of friction needed between the road and tires?

17. The coefficient of static friction between the floor of a truck and a box resting on it is 0.30. The truck is traveling at 80 km/h. What is the least distance in which the truck can stop if the box is not to slide?

18. A chair is sliding across a polished floor with initial speed 3 m/s. It comes to rest after sliding 2 m. What is the coefficient of kinetic friction between the floor and chair?

19. A 50-kg box must be moved across a level floor. The coefficient of static friction between the box and the floor is 0.6. One method is to push down on the box at an angle $\theta$ with the horizontal. Another method is to pull up on the box at an angle $\theta$ with the horizontal. (a) Explain why one method is better than the other. (b) Calculate the force necessary to move the box by each method if $\theta = 30°$ and compare these results with that for $\theta = 0°$.

20. The coefficient of friction between box A and the cart in Figure 5-36 is 0.6. The box has a mass of 2 kg. (a) Find the minimum acceleration $a$ of the cart and box if the box is not to fall. (b) What is the magnitude of the frictional force in this case? (c) If the acceleration is greater than this minimum, will the frictional force be greater than in part (b)? Explain. (d) Show in general that for a box of any mass, the box will not fall if the acceleration is $a \geq g/\mu_s$, where $\mu_s$ is the coefficient of static friction.

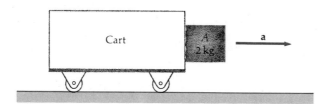

**Figure 5-36**
Exercise 20.

**Section 5-4, Circular Motion**

21. A 2-kg stone attached to a string is whirled in a horizontal circle of radius 40 cm, as in Figure 5-10. The string makes an angle of 30° with the vertical. Find the tension in the string and the speed of the stone.

22. A model airplane of mass 500 g flies in a horizontal circle of radius 6 m attached to a horizontal string. (The weight of the plane is balanced by the upward "lift" force of the air on the wings of the plane.) The plane makes 1 rev every 4 s. (a) What is the angular speed of the plane? (b) What is the acceleration of the plane? (c) Find the tension in the string.

23. A penny is placed on a record that is gradually accelerated to 78 rev/min. The penny stays on the record if it is placed within 8 cm of the center but slides off if it is placed more than 8 cm from the center. What is the coefficient of friction between the penny and the record?

24. A car rounds an unbanked curve with radius of curvature 40 m. The coefficient of friction between the tires and the road is 0.6. What is the maximum speed the car can travel without slipping?

25. A curve of radius 30 m is banked as shown in Figure 5-37 so that a car can round the curve at 48 km/h even if the road is frictionless. Show in a force diagram that a component of the normal force exerted by the road on the car can provide the centripetal force necessary and calculate the banking angle $\theta$ for these conditions.

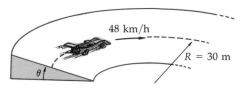

**Figure 5-37**
Car on banked road (Exercise 25).

26. A pilot comes out of a vertical dive in a circular arc such that her upward acceleration is $9g$. (a) If the mass of the pilot is 50 kg, what is the magnitude of the force exerted by the airplane seat on her at the bottom of the arc? (b) If the speed of the plane is 320 km/h, what is the radius of the circular arc?

27. Find the orbit radius of an earth satellite in a circular orbit if the period is 1 d.

28. In a carnival ride the passenger sits on a seat in a compartment that rotates in a vertical circle of radius 5 m (see Figure 5-38). Find the minimum frequency of rotation if the seat belt exerts no force on the passenger at the top of the ride.

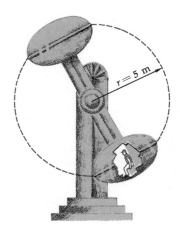

**Figure 5-38**
Exercise 28.

### Section 5-5, Static Equilibrium of a Rigid Body

29. A seesaw consists of a 4-m board pivoted at the center. A 28-kg child sits on one end of the board. Where should a 40-kg child sit to balance the seesaw?

30. Each of the objects shown in Figure 5-39 is suspended from the ceiling by a thread attached to the point marked × on the object. Describe with a diagram the orientation of each suspended object.

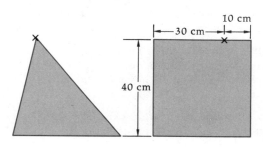

**Figure 5-39**
Exercise 30.

31. A 3-m board of mass 5 kg is hinged at one end. A force **F** is applied verti-cally at the other end to lift a 60-kg box, which rests on the board 80 cm from the hinge, as shown in Figure 5-40. (*a*) Find the magnitude of the force **F** needed to hold the board stationary at $\theta = 30°$. (*b*) Find the force exerted by the hinge at this angle. (*c*) Find the force *F* and the force exerted by the hinge if **F** is exerted perpendicular to the board when $\theta = 30°$.

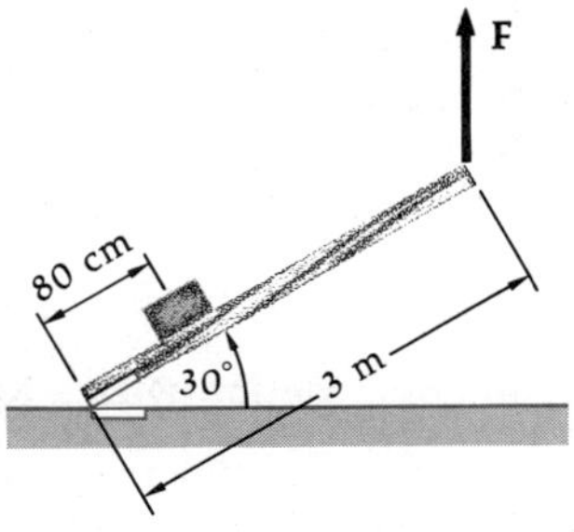

32. Indicate in a diagram the location of the center of gravity of an equilateral triangle made of three sticks of equal weight and equal length joined at their ends.

33. A 90-N board 12 m long rests on two supports each 1 m from the end of the board. A 360-N block is placed on the board 3 m from one end, as shown in Fig-ure 5-41. Find the force exerted by each support on the board.

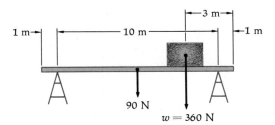

34. An 80-N weight is supported by a cable attached to a strut hinged at point *A* (Figure 5-42). The strut is supported by a second cable under tension $T_2$, as shown. The mass of the strut is negligible. (*a*) What are the three forces acting on the strut? (*b*) Show that the vertical component of the tension $\mathbf{T_2}$ must equal 80 N. (*c*) Find the force exerted on the strut by the hinge.

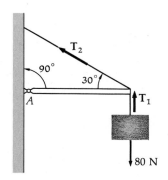

35. Find the force exerted by the hinge *A* on the strut for the arrangement in Figure 5-43 if (*a*) the strut is weightless and (*b*) the strut weighs 20 N.

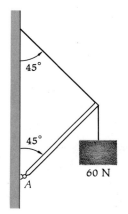

**Figure 5-40**
Exercise 31.

**Figure 5-41**
Exercise 33.

**Figure 5-42**
Exercise 34.

**Figure 5-43**
Exercise 35.

36. In Figure 5-44 the forearm is at 90° to the upper arm so that the force exerted by the muscle $F_m$ is vertical. The hand holds a 5-kg mass. Find the force exerted by the muscle $F_m$ assuming the weight of the forearm to be negligible.

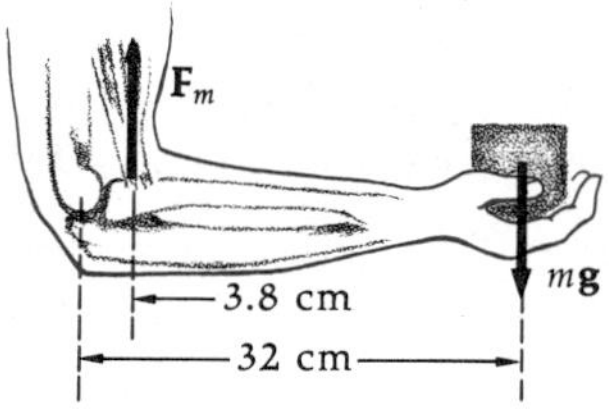

**Figure 5-44**
Exercise 36.

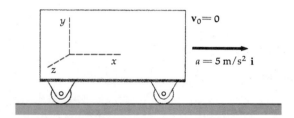

### Section 5-6, Motion with a Retarding Velocity-Dependent Force

37. What are the dimensions and SI units of the constant $b$ in the retarding force $bv^n$ if (a) $n = 1$, (b) $n = 2$?

38. A sky diver of mass 60 kg can slow herself to a constant speed of 90 km/h by adjusting her form. If the retarding force is $bv^2$, what is the value of $b$? What is the critical time $t_c$ for this motion?

### Section 5-7, Pseudo Forces

*In Exercises 39 to 43, the situations described take place in a boxcar which has initial velocity $v = 0$ but acceleration $\mathbf{a} = (5 \text{ m/s}^2)\mathbf{i}$ (Figure 5-45). Work the exercises in the frame of the boxcar using pseudo forces and in an inertial frame using only real forces. Assume that $g = 10 \text{ m/s}^2$.*

**Figure 5-45**
Box car initially at rest with constant acceleration 5 m/s² to the right for Exercises 39 to 43.

39. A 2-kg object is slid along the frictionless floor with initial velocity $(10 \text{ m/s})\mathbf{i}$. (a) Describe the motion of the object. (b) When does the object reach its original position relative to the boxcar?

40. A 2-kg object is slid along the frictionless floor with initial transverse velocity $(10 \text{ m/s})\mathbf{k}$. Describe the motion.

41. A 2-kg object is slid along a rough floor (coefficient of sliding friction 0.3) with initial velocity $(10 \text{ m/s})\mathbf{i}$. Describe the motion of the object assuming that the coefficient of static friction is greater than 0.5.

42. A 2-kg object is suspended from the ceiling by a massless unstretchable string. (a) What angle does the string make with the vertical? (b) Indicate all the forces acting on the object in each frame.

43. A 6-kg object is attached to the ceiling by a (massless) spring of force constant 1000 N/m and unstretched length 0.50 m. By how much is the spring stretched?

### Problems

1. Two small spheres of mass $m$ are suspended from a common point by threads of length $L$. When each sphere carries a charge $q$, each thread makes an angle $\theta$ with the vertical, as shown in Figure 5-46. Show that the charge $q$ is given by

$$q = 2L \sin \theta \sqrt{(mg \tan \theta)/k},$$

where $k$ is the Coulomb constant. Find $q$ if $m = 10$ g, $L = 50$ cm, and $\theta = 10°$.

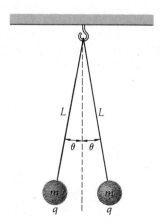

**Figure 5-46**
Problem 1.

2. A 2-kg block sits on a 4-kg block resting on a frictionless table (Figure 5-47). The coefficients of friction between the blocks are $\mu_s = 0.3$ and $\mu_k = 0.2$. (a) What is the maximum $F$ than can be applied if the 2-kg block is not to slide on the 4-kg block? (b) If $F$ is twice this maximum, find the acceleration of each block. (c) If $F$ is half this maximum, find the acceleration of each block and the friction force acting on each block.

**Figure 5-47**
Problem 2.

3. A block is on an incline whose angle can be varied. The angle is gradually increased from 0°. At 30° the block starts to slide down the incline. It slides 3 m in 2 s. Calculate the coefficients of static and kinetic friction between the block and incline.

4. The earth orbits the sun in a nearly circular orbit of radius $1.50 \times 10^{11}$ m. Its period is 1 y. Use these data to calculate the mass of the sun.

5. A platform scale calibrated in newtons is placed on the bed of a truck driven at constant speed of 14 m/s. A box weighing 500 N is placed on the scale. Find the reading on the scale if (a) the truck passes over the crest of a hill with radius of curvature 100 m and (b) the truck passes through the bottom of a dip with radius of curvature 80 m.

6. Show that if the distance of a small object from the center of the earth is changed slightly from $r$ to $r + \Delta r$, the change in the gravitational attraction of the earth is given approximately by $\Delta F/F \approx 2\, \Delta r/r$. If a man weighs 800 N at sea level, what is his approximate weight at the top of a mountain of height 400 m?

7. A horizontal plank 8.0 m long is used by pirates to make their victims walk the plank. A victim has mass 63 kg, and a pirate of mass 105 kg is to stand on the shipboard end of the plank to prevent it from tipping. Find the maximum distance the plank can overhang as the victim walks to the end if (a) the mass of the plank is negligible and (b) the mass of the plank is 25 kg.

8. In an amusement-park ride, participants stand against the wall of a spinning cylinder as the floor falls away and are held up by friction. If the radius of the cylinder is 4 m, find the minimum frequency in revolutions per minute necessary when the coefficient of friction between a rider and the wall is 0.4.

9. In Figure 5-48 the mass $m_2 = 10$ kg slides on a frictionless table. The coefficients of static and kinetic friction between $m_2$ and $m_1 = 5$ kg are $\mu_s = 0.6$ and $\mu_k = 0.4$. (a) What is the maximum acceleration of $m_1$? (b) What is the maximum value of the $m_3$ if $m_1$ moves with $m_2$ without slipping? (c) If $m_3 = 30$ kg, find the acceleration of each body and the tension in the string.

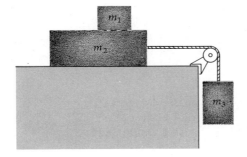

**Figure 5-48**
Problem 9.

10. A 2-kg mass is attached to a spring having a natural length of 30 cm and a force constant of 2000 N/m. The mass rotates in a circle (supported by a frictionless horizontal surface) at $v = 4$ m/s (Figure 5-49). (a) Write an exact equation for the amount $x$ the spring stretches in terms of the mass $m$, the force constant $k$, the natural length $x_0$, and the speed $v$. (b) Solve this quadratic equation for $x$ for the values given. (c) Show that if the change in the radius of the circle because of the stretching of the spring is neglected, the stretching is given approximately by $x = mv^2/kx_0$. Find $x$ for the values given in this approximation and compare your result with the exact value found in part (b). (d) A better approximate value can be found by correcting the radius of the circle using the first approximation for $x$. Compute $x = mv^2/k(x_0 + x_1)$, where $x_1$ is the value found in part (c), and compare your result with the exact value.

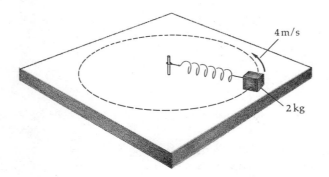

4 m/s

2 kg

**Figure 5-49**
Problem 10.

11. A small block of mass $m$ slides on a frictionless circular track in a vertical circle of radius $R$, as shown in Figure 5-50. (a) Show that the speed of the block cannot be constant. (b) If the speed of the block at the top of the track is $v$, find the force exerted on the block by the track. (c) What is the minimum value of $v$ for the block to stay on the track? If the speed is less than this value, describe the path of the block.

**Figure 5-50**
Problem 11.

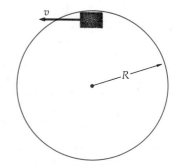

12. A 400-N box rests on a horizontal table. The coefficient of static friction is 0.6. The box is pulled by a massless rope with a force **F** at an angle $\theta$, as shown in Figure 5-51. The minimum value of the force needed to move the box depends on the angle $\theta$. Discuss qualitatively how you would expect this force to depend on $\theta$. Compute the force for the angles $\theta = 0, 10, 20, 30, 40, 50,$ and $60°$ and make a plot of $F$ versus $\theta$. From your plot, at what angle is it most efficient to apply the force to move the box?

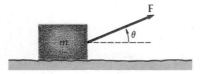

F

$\theta$

m

**Figure 5-51**
Problems 12 and 13.

13. A box of mass $m$ rests on a horizontal table. The coefficient of friction is $\mu$. A force **F** is applied at an angle $\theta$ as in Figure 5-51. (a) Find the force $F$ needed to move the box as a function of angle $\theta$. (b) At the angle $\theta$ for which this force is minimum, the slope $dF/d\theta$ of the curve $F$ versus $\theta$ is zero. Compute $dF/d\theta$ and show that this derivative is zero at the angle $\theta$ that obeys $\tan \theta = \mu$. Compare this general result with that obtained in Problem 12.

14. Show with a force diagram how a motorcycle can ride in a circle on a vertical wall. Assume reasonable parameters (coefficient of friction, radius of the circle, mass of the motorcycle, or whatever required) and calculate the minimum speed needed.

15. A uniform door, 2.0 m high by 0.8 m wide and mass 18 kg, is hung from two hinges that are 20 cm from the top and 20 cm from the bottom respectively. If each hinge supports half the weight of the door, find the magnitude and direction of the two horizontal components of the forces exerted by the hinges on the door.

16. A body is attached to a spring of force constant $k_1$, which in turn is connected to a second spring of force constant $k_2$, as shown in Figure 5-52. When the springs are unstretched, the body is at $x_0$. Show that when the body is at point $x$, the force on it is $F_x = -k(x - x_0)$, where the effective spring constant $k$ is given by $1/k = 1/k_1 + 1/k_2$.

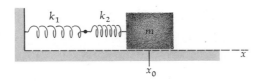

**Figure 5-52**
Problem 16.

17. A block of mass $m_1$ is attached to a cord of length $L_1$, which is fixed at one end. The mass moves in a horizontal circle supported by a frictionless table. A second block of mass $m_2$ is attached to the first by a cord of length $L_2$ and also moves in a circle, as shown in Figure 5-53. If the angular speed is $\omega$, find the tension in each cord.

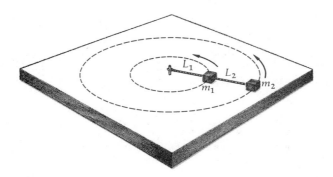

**Figure 5-53**
Problem 17.

18. A 900-N boy sits on top of a ladder of negligible weight which rests on a frictionless floor. There is a cross brace halfway up the ladder (Figure 5-54). The angle at the apex is $\theta = 30°$. (a) What is the force exerted by the floor on each leg of the ladder? (b) Find the tension in the cross brace. (c) If the cross brace is moved down toward the bottom of the ladder (with the same angle $\theta$), will its tension be greater or less?

**Figure 5-54**
Problem 18.

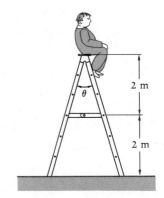

19. Show that the angle made by the string with the vertical in a conical pendulum is given by $\cos\theta = g/L\omega^2$, where $L$ is the length of the string and $\omega$ is the angular speed.

20. Two painters are working from a board 5.0 m long suspended from the top of a building by two ropes attached to the ends of the plank. Either rope will break when the tension exceeds 1 kN. Painter $A$ (mass 80 kg) is working at a distance of 1.0 m from one end. Find the range of positions available to painter $B$ if his mass is 60 kg and the plank has mass 20 kg. (Use $g \approx 10$ m/s².)

21. A road is banked so that a car traveling 40 km/h can round a curve of radius 30 m even if the road is so icy that the coefficient of friction is approximately zero. Find the range of speeds at which a car can travel around this curve without skidding if the coefficient of friction between the road and the tires is 0.3.

22. A 4-kg block rests on a 30° incline attached to a cord which passes over a smooth peg and is attached to a second block of mass $m$, as shown in Figure 5-55. The coefficient of static friction between the block and the incline is 0.4. (a) Find the range of possible values of $m$ such that the system will be in static equilibrium. (b) If $m = 1$ kg, the system will be in static equilibrium. What is the frictional force on the 4-kg block in this case?

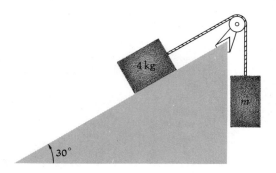

**Figure 5-55**
Problem 22.

23. A body of mass $m$ is suspended from the middle of a spring of natural length $L$ and force constant $k$, as shown in Figure 5-56. Show that the angle $\theta$ between the spring and the horizontal is given by $\tan \theta - \sin \theta = mg/2kL$. Solve this equation numerically (by trial and error) to find the distance the spring sags if $m = 10$ kg, $L = 30$ cm, and $k = 10^4$ N/m.

**Figure 5-56**
Problem 23.

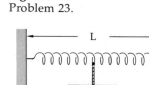

24. A light ladder rests on a rough floor and leans against a frictionless vertical wall at an angle $\theta$ with the horizontal. Show that if $L$ is the length of the ladder, a person can climb no further than $\mu_s L \tan \theta$ before the ladder slips, where $\mu_s$ is the coefficient of friction between the floor and the ladder and it is assumed that the weight of the ladder is negligible compared with that of the climber.

25. A wheel of mass $M$ and radius $R$ rests on a horizontal surface against a step of height $h$ ($h < R$). The wheel is to be raised over the step by a horizontal force $F$ applied to the axle of the wheel. Find the force $F$ necessary to raise the wheel over the step.

26. Newton showed that the air resistance on a falling object with circular cross section should be approximately $\frac{1}{2}\rho\pi r^2 v^2$, where $\rho = 1.2$ kg/m³ is the density of air. Find the terminal speed for a 56-kg sky diver assuming that his cross-sectional area is equivalent to that of a disk of radius 0.30 m.

27. A box 2 by 1 by 1 m of uniform mass is placed on end on a rough hinged plank, as shown in Figure 5-57. The plank is inclined at an angle $\theta$, which is slowly increased. The coefficient of friction is great enough to prevent the box from sliding before it tips over. Find the greatest angle that can be applied without tipping the box over. (Hint: First show that the center of gravity must be above some part of the base for the conditions of static equilibrium to be met.) What is the minimum value of $\mu_s$ if the box tips before it slides?

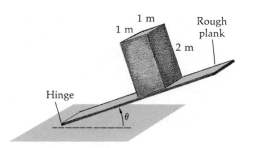

**Figure 5-57**
Problem 27.

28. Atoms are electrically neutral because they contain an equal number of positively charged protons and negatively charged electrons and because the magnitudes of the electron charge and proton charge are exactly equal. Suppose each atom contained an equal number $Z$ of electrons and protons but the charge of the electron was slightly greater in magnitude than that of the proton, so that the atom had a net charge $-fZe$, where $f$ is the fractional charge difference. Estimate the order of magnitude of the fractional difference in charge that would cause the electric repulsion between the earth and the sun to equal the gravitational attraction between them. (*Hint:* Since an atom has approximately equal numbers of protons and neutrons, assume its mass to be $2Zm_p$, where $m_p$ is the proton mass. It is sufficient to do the calculation for just two atoms. Why?)

29. A frictionless platform rotates counterclockwise (as seen from above) with angular speed $\omega$. A small object of mass $m$ rests on the platform at a distance $r$ from the center. The object is at rest relative to an inertial frame (the platform merely rotates under the object). (*a*) Describe the motion of the object relative to the platform. What are the magnitude and direction of the resultant pseudo force acting on the object in that frame? (*b*) Show that in this case the Coriolis force has twice the magnitude of the centrifugal force and is oppositely directed.

30. A space station has two compartments, as shown in Figure 5-58. The station rotates at $B$ rev/min. (*a*) A mass $m$ rests on the floor of one of the compartments a distance $r$ from the center of rotation as shown. What is the normal force exerted by the floor on the mass? (*b*) The mass is now dropped from the ceiling of the compartment. Describe its motion relative to the compartment. What forces (including pseudo forces) act on the mass while it is falling? (*c*) Explain qualitatively why the mass falls to the floor from the point of view of an inertial reference frame in which there are no forces acting on the mass.

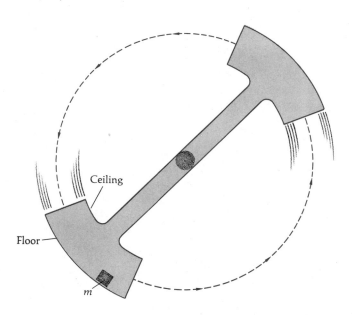

**Figure 5-58**
Problem 30.

**CHAPTER 6**      Work and Energy

---

**Objectives**   After studying this chapter, you should:

1.  Know the definitions of work, kinetic energy, potential energy, and power.
2.  Be able to derive the work-energy theorem.
3.  Be able to distinguish between conservative and nonconservative forces and know the criterion for a force to be conservative.
4.  Be able to calculate the potential-energy function associated with a given conservative force.
5.  Be able to work problems using the work-energy theorem.
6.  Be able to find the force $F_x$ from the potential-energy function $U(x)$.
7.  Be able to locate equilibrium points and discuss their stability from a graph of the potential-energy function $U(x)$.

---

In most of our illustrations applying the laws of motion, to simplify the analysis we chose situations in which the forces are constant in magnitude. When the forces are constant, the position and velocity functions are determined by the constant-acceleration formulas once the acceleration has been determined from Newton's second law. In most physical situations the forces between particles are not constant but depend on the positions of the particles. For example, the law of gravitation (and Coulomb's law of electrostatics) gives the force exerted by one particle on another as an inverse-square function of the separation of the particles. Thus in the planetary problem, we know the force on the planet exerted by the sun as a function of the distance from the planet to the sun. This distance varies as the planet moves along its elliptical orbit, and the force varies in both magnitude and direction. Similarly, in the empirical description of complex forces we often find the force on a particle given as a function of position, such as $F_x = -kx$ for the force on a body attached to a spring which obeys Hooke's law.

In this chapter we develop general methods for solving problems in which the force depends on the position of the particle. In the process we introduce the concept of the *work* done by a force as the particle on which it acts moves from one place to another. We shall see that this concept is closely related to the concept of *energy,* which plays a central role in our lives and in the description of the physical universe. In succeeding chapters we shall investigate the concept of energy more fully and see how to apply the law of conservation of energy to various problems. We begin by studying some special cases in which the motion is restricted to one dimension.

## 6-1    One-Dimensional Motion with Constant Forces

For the special case of a constant force acting on a particle which moves in one dimension, we define the work done by the force as the product of the component of the force in the direction of motion times the displacement:

$$W = F_x \, \Delta x \qquad \text{constant force} \qquad 6\text{-}1$$

*Work by a constant force*

The work is positive if the motion is in the same direction as the force and negative if it is in the opposite direction. The dimensions of work are those of force times distance. In the SI the unit of work and energy is the joule (J):

$$1 \text{ J} = 1 \text{ N·m} = 1 \text{ kg·m}^2/\text{s}^2 \qquad 6\text{-}2$$

*The joule defined*

where we have used the fact that $1 \text{ N} = 1 \text{ kg·m/s}^2$. In the U.S. customary system the unit of work is the foot-pound. The relation between these units is easily found using the relations between pounds and newtons and between metres and feet. The result is

$$1 \text{ J} = 0.738 \text{ ft·lb} \qquad 6\text{-}3$$

Work is being done by these horses because they exert a force through a distance as they pull the harrow. Despite the fact that the Amish farm boy shown in this photograph is working according to the everyday use of the word, he does very little work according to the scientific definition because he exerts little force on the harrow; rather he guides its direction by use of the reins.

Jane Latta/Photo Researchers

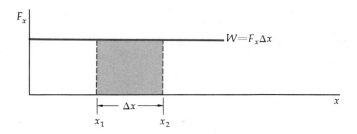

**Figure 6-1**
The work done by a constant force is interpreted geometrically as the area under the $F_x$-versus-$x$ curve.

In Figure 6-1 work by a constant force is interpreted graphically as the area under the force-versus-position curve. We note that for work to be done, the force must act through a distance. Consider a person holding a weight a distance $h$ off the floor, as in Figure 6-2. In everyday usage, we might say that it takes work to do this, but in our scientific definition, no work is done by a force acting on a stationary object. We could eliminate the effort of holding the weight by merely tying the string to some object and the weight would be supported with no help from us.

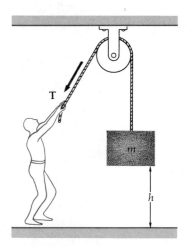
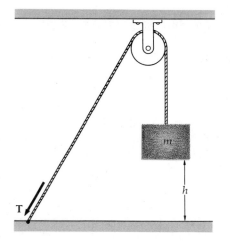

**Figure 6-2**
No work is done by the man holding the weight at a fixed position. The same task could be accomplished by tying the rope to a fixed point.

We shall see that work as we have defined it is closely related to the concept of energy. If a force is to do work, energy must be supplied from somewhere. For example, if we wish to lift the weight to a greater height, we must pull the string through a distance, doing work with our muscles. Internal chemical energy of our bodies is used, and we must eventually replenish this energy by taking in nourishment. If we attach the string to an electric motor to raise the weight, we spend electric energy to perform this work.

## 6-2 Work Done by the Resultant Force: Kinetic Energy

If several forces are acting on a particle, we can compute the work done by each from Equation 6-1. The total or net work done on the particle by all the forces acting on it is just the algebraic sum of the work done by each of the individual forces. The net work is just the work done by the resultant or net force. For example, if there are three forces $F_{x1}$, $F_{x2}$, and

Atlas. There is no work done without motion.

The work done by the rope equals the force times the distance raised. (*Courtesy of the Museum of Modern Art/Film Archives, New York.*)

$F_{x3}$, the total work done on a particle that moves from $x_1$ to $x_2$ through a distance $\Delta x = x_2 - x_1$ is

$$W_{\text{net}} = F_{x1}\,\Delta x + F_{x2}\,\Delta x + F_{x3}\,\Delta x$$
$$= (F_{x1} + F_{x2} + F_{x3})\,\Delta x = \Sigma F_x\,\Delta x \qquad \text{6-4}$$

There is an important relation between the net work done on a particle and the speed of the particle at the original and final positions. This relation is obtained by using Newton's second law, relating the resultant force to the acceleration of the particle $\Sigma \mathbf{F} = m\mathbf{a}$. For a constant force, the acceleration is constant and we can relate the distance moved to the initial and final speed by the constant-acceleration formula (Equation 2-12). If the initial speed is $v_1$ and the final speed $v_2$, we have

$$v_2^2 = v_1^2 + 2a\,\Delta x$$

Substituting $ma$ for $\Sigma F_x$, and $(1/2a)\,(v_2^2 - v_1^2)$ for $\Delta x$ in Equation 6-4, we have

$$W_{\text{net}} = ma\,\Delta x = \tfrac{1}{2}mv_2^2 - \tfrac{1}{2}mv_1^2 \qquad \text{6-5}$$

The quantity $\tfrac{1}{2}mv^2$ is called the *kinetic energy* $E_k$ of the particle. It is a scalar quantity which depends on the particle's mass and speed:

$$E_k = \tfrac{1}{2}mv^2 \qquad \text{6-6}$$

*Kinetic energy*

The quantity on the right side of Equation 6-5 is the change in the kinetic energy of the particle, i.e., the kinetic energy $\tfrac{1}{2}mv_2^2$ at the end of the interval when the particle is at $x_2$ minus the initial kinetic energy $\tfrac{1}{2}mv_1^2$ at the beginning of the interval when the particle was at $x_1$. The net work done by the resultant force is therefore equal to the change in the kinetic energy of the particle:

$$W_{\text{net}} = \Delta E_k = \tfrac{1}{2}mv_2^2 - \tfrac{1}{2}mv_1^2 \qquad \text{6-7}$$

*Work-energy theorem*

We shall see in the next sections that this result also holds when we generalize our concept of work to the case of a force that varies as the particle moves and to the case of general motion in three dimensions.

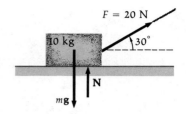

**Figure 6-3**
A constant force of 20 N applied at 30° to the horizontal to a 10-kg box for Example 6-1.

**Example 6-1** A 10-kg mass rests on a frictionless horizontal surface. It is pulled by a 20-N force making an angle of 30° with the horizontal, as shown in Figure 6-3. Find the work done by the force of the string and the final speed of the box after it moves 3 m.

The forces are shown in the figure. The vertical forces are the force due to gravity, $mg = 98.1$ N, the upward component of the force of the string, 20 N sin 30° = 10 N, and the vertical support force of the table, which must equal $98.1 - 10$ N $= 88.1$ N because there is no vertical acceleration. The only horizontal force is 20 N cos 30° = 17.3 N. The work done by the force **F** is the product of the component in the direction of motion (17.3 N) and the distance traveled (3 m). $W = F \cos \theta\, x =$ (20 N) (0.866) (3 m) = 52.0 J. Since the horizontal component of the force **F** is the resultant force, the work done by this force equals the change in kinetic energy of the block. If it starts from rest, its kinetic energy after traveling 3 m is therefore 52.0 J. Its speed can be found from

$$\tfrac{1}{2}mv^2 = E_k = 52.0 \text{ J}$$

$$v = \sqrt{\frac{2E_k}{m}} = \sqrt{\frac{2(52.0 \text{ J})}{10 \text{ kg}}} = 3.22 \text{ m/s}$$

Of course, we could have also found the speed by first finding the acceleration and then using the constant-acceleration formulas.

## 6-3   Work Done by a Force That Varies with Position

We now extend our concept of work to the case in which the force varies with position. Figure 6-4 shows a general force as a function of position $x$. In the figure we have divided the interval from $x_1$ to $x_2$ into a set of smaller intervals $\Delta x_i$. If each interval is small enough, we can approximate the varying force by a series of constant forces as shown. For each interval, the work done by the constant force is the area of the rectangle shown. The sum of the rectangular areas is the sum of the work done by the set of constant forces that approximates the varying force. As seen

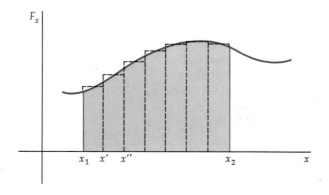

**Figure 6-4**
A general force $F_x$ versus $x$. This force can be approximated by a series of constant forces over small intervals. The work done by each constant force is the rectangular area indicated. The total work as the particle moves from $x_1$ to $x_2$ is the total area under the curve for this interval.

from the figure, this area is approximately equal to the area under the curve. We therefore define the work done by a varying force to be the area under the $F$-versus-$x$ curve.

In calculus, the area under a curve is defined to be the limit of the sum of rectangular areas as we take more and more intervals of smaller and smaller size. This limit is called the *definite integral* and written

$$\lim_{\Delta x_i \to 0} \Sigma F_x \, \Delta x_i = \int_{x_1}^{x_2} F_x \, dx \qquad\qquad 6\text{-}8$$

(The integral sign $\int$ is an elongated S denoting a sum.) The work done by a varying force is therefore the integral of the force over the displacement from $x_1$ to $x_2$:

$$W = \int F_x \, dx = \text{area under the } F_x\text{-versus-}x \text{ curve} \qquad 6\text{-}9$$

*Work by a varying force*

We now show that Equation 6-7 holds for the case of a varying force. Consider Figure 6-4. For each rectangular area, the force is constant and we can apply Equation 6-5. The area of each rectangle equals the change in kinetic energy for that interval. For example, the area of the first rectangle for the inverval $\Delta x_1 = x' - x_1$ equals $\frac{1}{2}mv'^2 - \frac{1}{2}mv_1^2$, where $v'$ is the speed of the particle at position $x'$. Similarly, the area of the second rectangle for the interval $\Delta x_2 = x'' - x'$ equals $\frac{1}{2}mv''^2 - \frac{1}{2}mv'^2$. The sum of these two areas is $\frac{1}{2}mv''^2 - \frac{1}{2}mv_1^2$; that is, the sum of the changes in the kinetic energy is just the net change for the interval from $x_1$ to $x''$. The total area under the curve therefore equals the sum of the changes in the kinetic energy for each subinterval or the net change in the kinetic energy for the complete interval $\frac{1}{2}mv_2^2 - \frac{1}{2}mv_1^2$. This is the same as Equation 6-7.

**Example 6-2** A 4-kg block resting on a frictionless table (Figure 6-5$a$) is attached to a horizontal spring which obeys Hooke's law and exerts a force $F_x = -kx$, where $x$ is measured from the equilibrium length of the

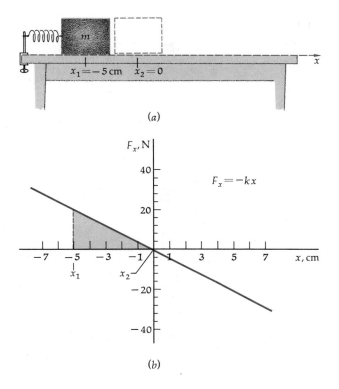

(a)

(b)

**Figure 6-5**
($a$) Block on a spring for Example 6-2. The spring is compressed from equilibrium and released. ($b$) $F_x$ versus $x$ for the block on the spring. The work done by the spring as the block moves from $x_1$ to $x_2$ is the shaded area indicated.

spring and the force constant is $k = 400$ N/m. The spring is compressed to $x_1 = -5$ cm. Find (a) the work done by the spring as the block moves from $x_1 = -5$ cm to its equilibrium position $x_2 = 0$ and (b) the speed of the block at $x_2 = 0$.

(a) Figure 6-5b is a sketch of this force versus distance. The work done as the block moves from $x_1$ to $x_2$ is by definition equal to the area under this curve between these limits, as indicated in the figure. This area is one-half the base times the height of the triangle. The base is 5 cm $= 0.05$ m, and the height is the value of the force at $x_1$, which is $F_x = -(400$ N/m$) (-0.05$ m$) = +20$ N. The work done is thus $W = \frac{1}{2}(0.05$ m$) (20$ N$) = 0.500$ N·m $= 0.500$ J and is positive because the force is in the direction of motion. This is indicated in the figure by the fact that the area is above the $x$ axis.

(b) As in Example 6-1, we find the speed of the block from the fact that its kinetic energy is 0.500 J. The speed is then $v = \sqrt{2E_k/m} = \sqrt{2(0.500 \text{ J})/4 \text{ kg}} = 0.50$ m/s $= 50$ cm/s. Note that we could *not* have found this by first finding the acceleration and using the constant-acceleration expressions because the acceleration is not constant here since the force varies with position.

---

**Example 6-3** Find the speed of the block in Example 6-2 when the spring is at its equilibrium position if the coefficient of kinetic friction between the table and block is 0.200.

In this case the work done by the spring is not the net work done on the block because the frictional force also does (negative) work. Since the frictional force is constant, the work done by it is just the force times the distance. This work is negative because the force $f_x = -\mu_k N = -\mu_k mg$ is opposite the direction of motion. The work done by the frictional force is

$$W_f = -\mu_x mgx = -(0.200) \, (4.00) \, (9.81) \, (0.05) = -0.392 \text{ J}$$

The net work done on the block is then

$$W_{net} = 0.500 \text{ J} - 0.392 \text{ J} = 0.108 \text{ J}$$

The net work equals the change in kinetic energy, which is $\frac{1}{2}mv^2$:

$$W_{net} = 0.108 \text{ J} = \tfrac{1}{2}mv^2 = \tfrac{1}{2}(4.00)v^2$$

$$v^2 = 0.054 \text{ m}^2/\text{s}^2$$

$$v = 0.232 \text{ m/s} = 23.2 \text{ cm/s}$$

---

## Questions

1. A block is pulled up an inclined plane a certain distance. Then it is pulled down the same distance. How does the work done by the friction as the block moves up compare with that done by friction as the block moves down?

2. How does the work done by gravity as the block in Question 1 moves up compare with the work done by gravity as it moves down?

3. Is it possible to exert a force which does work on a body without increasing its kinetic energy? If so, give an example.

4. How does the kinetic energy of a car change when its speed is doubled?

## 6-4 Work and Energy in Three Dimensions

In our definition of work we specified that it is the product of the displacement and the *component* of the force in the direction of the displacement. The component of force in the direction of the displacement is important because it is related to the change in the speed of the particle. In our one-dimensional examples of horizontal motion the vertical components of the forces merely helped balance the weight of the particles. Figure 6-6 shows a particle moving along a general curve in three dimensions. We consider a small displacement along the curve. The force **F** makes an angle $\phi$ with the displacement. The tangential component $F_s$ is related to the rate of change of speed of the particle by Newton's second law $\Sigma F_s = m\, dv/dt$. The perpendicular component $F_\perp$ does not affect the speed of the particle. Instead, this component changes the direction of the velocity. It is related to the centripetal acceleration $v^2/r$, where $r$ is the radius of curvature of the path at this point.

(a)  (b)

**Figure 6-6**
(a) A particle moving along a curve in space. (b) The component of force $F_\perp$ affects the direction of motion but not the speed. The component $F_s$ equals the mass times the tangential acceleration $dv/dt$. Only this component does work on the particle.

The work done by the force for the small displacement $\Delta s$ is

$$\Delta W = F_s\, \Delta s$$

To find the total work done by the force as the particle moves along the curve from point 1 to point 2 we compute the product $F_s\, \Delta s$ for each element of the path and sum. In the limit as we take smaller and smaller displacement elements, this sum becomes an integral:

$$W = \int_1^2 F_s\, ds$$

As in the case of one-dimensional motion, we can show that the net work done by the resultant force equals the change in the kinetic energy. From Newton's second law we have

$$\Sigma F_s = m\frac{dv}{dt}$$

We can now think of the speed as a function of the distance $s$ measured along the curve and apply the chain rule for derivatives (rule 3 in Appendix E):

$$\frac{dv}{dt} = \frac{dv}{ds}\frac{ds}{dt} = v\frac{dv}{ds}$$

where we have used the fact that $ds/dt$ is just the speed $v$. The work done by the resultant force is thus

$$W_{net} = \int_1^2 \Sigma F_s \, ds = \int_1^2 m \frac{dv}{dt} \, ds = \int_1^2 mv \frac{dv}{ds} \, ds = \int_1^2 mv \, dv$$

or

$$W_{net} = \int_1^2 \Sigma F_s \, ds = \tfrac{1}{2}mv_2^2 - \tfrac{1}{2}mv_1^2 \qquad\qquad 6\text{-}10$$

*Work-energy theorem in three dimensions*

Equation 6-10 along with its one-dimensional counterpart, Equation 6-7, is known as the *work-energy theorem:*

*The net work done by the resultant force equals the change in the kinetic energy of the particle.*

This theorem follows directly from the definition of work and kinetic energy and from Newton's second law of motion.

If $\phi$ is the angle between the force **F** and the displacement $\Delta$**s**, the component of the force parallel to $\Delta$**s** is $F_s = F \cos \phi$. The work done by the force during a small displacement is then

$$\Delta W = F_s \, \Delta s = (F \cos \phi) \, \Delta s$$

This type of scalar combination of two vectors and the cosine of the angle between them occurs often in physics. It is called the *scalar product* of the vectors. The scalar product of two general vectors **A** and **B** is written **A · B** and defined

$$\mathbf{A \cdot B} = AB \cos \phi \qquad\qquad 6\text{-}11$$

*Scalar product*

where $\phi$ is the angle between **A** and **B**. Because of this notation, the scalar product is also called the *dot product.* The dot product **A · B** can be thought of as the product of $A$ and the component of **B** in the direction of **A**, $B \cos \phi$, or, alternatively, as the product of $B$ and the component of **A** in the direction of **B**, $A \cos \phi$ (Figure 6-7).

*Dot product*

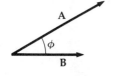 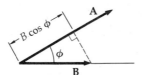 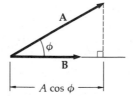

**Figure 6-7**
Geometric interpretation of the dot product **A · B**. We can think of this product as $A$ times $B \cos \phi$ or as $B$ times $A \cos \phi$.

If **A** and **B** are perpendicular, their dot product is zero because $\phi = 90°$. Conversely, if **A · B** = 0, either **A** = 0, or **B** = 0, or **A** and **B** are mutually perpendicular. If **A** and **B** are parallel vectors, the dot product is just the product of their magnitudes. The dot product of a vector with itself is the square of the magnitude of the vector:

$$\mathbf{A \cdot A} = A^2$$

It follows from the definition that the scalar product is independent of the order of multiplication, **A · B** = **B · A**. This is known as the *commutative rule of multiplication*. The scalar product also obeys the distributive rule of multiplication

$$(\mathbf{A + B}) \cdot \mathbf{C} = \mathbf{A \cdot C} + \mathbf{B \cdot C}$$

The dot product can be conveniently written in terms of the rectangular components of the two vectors. Consider the dot product of the vector $\mathbf{A} = A_x\mathbf{i} + A_y\mathbf{j} + A_z\mathbf{k}$ with the vector $\mathbf{B} = B_x\mathbf{i} + B_y\mathbf{j} + B_z\mathbf{k}$. Since the unit vectors $\mathbf{i}$, $\mathbf{j}$, and $\mathbf{k}$ are mutually perpendicular, the dot product of two different unit vectors is zero; that is, $\mathbf{i} \cdot \mathbf{j} = \mathbf{i} \cdot \mathbf{k} = \mathbf{j} \cdot \mathbf{k} = 0$. Also, the dot product of a unit vector with itself is 1, $\mathbf{i} \cdot \mathbf{i} = \mathbf{j} \cdot \mathbf{j} = \mathbf{k} \cdot \mathbf{k} = 1$. We therefore have

$$\mathbf{A} \cdot \mathbf{B} = A_x B_x + A_y B_y + A_z B_z \qquad 6\text{-}12$$

In terms of the dot-product notation, the work done by a force $\mathbf{F}$ during a small displacement $\Delta\mathbf{s}$ is written

$$\Delta W = \mathbf{F} \cdot \Delta\mathbf{s} \qquad 6\text{-}13$$

and the work done as the particle moves from point 1 to point 2 is written

$$W = \int_1^2 \mathbf{F} \cdot d\mathbf{s} \qquad 6\text{-}14$$

---

**Example 6-4** A particle of mass $m$ slides down a frictionless plane inclined at angle $\theta$ from the horizontal. It begins at a height $h$ above the bottom of the incline moving down with speed $v_0$. Find the work done by all the forces and the speed of the particle when it reaches the bottom.

The forces acting on the particle are the force of gravity $m\mathbf{g}$ and the contact force $\mathbf{N}$ exerted by the plane, as indicated in Figure 6-8. Since the plane is frictionless, $\mathbf{N}$ is perpendicular to the plane and to the motion of the particle. It therefore does no work on the particle. The only force that does work is $m\mathbf{g}$, which has the component $mg \cos \phi = mg \sin \theta$ in the direction of motion (Figure 6-8b; note that the angle $\phi$ between $m\mathbf{g}$ and the plane is the complement of the angle of inclination $\theta$). When the particle moves a distance $\Delta s$ down the incline, the earth does work $(mg \sin \theta) \Delta s$. Since the force exerted by the earth is constant, the total work done when the particle moves a distance $s$ down the incline is merely $(mg \sin \theta)s$. We see from Figure 6-8 that the total distance $s$ measured along the incline is related to the initial height $h$ by $h = s \sin \theta$, so that the work done by the earth is $(mg \sin \theta)s = mgh$. Since this is the total work done by all the forces, the work-energy theorem gives

$$W_{\text{net}} = mgh = \Delta E_k = \tfrac{1}{2}mv^2 - \tfrac{1}{2}mv_0^2$$

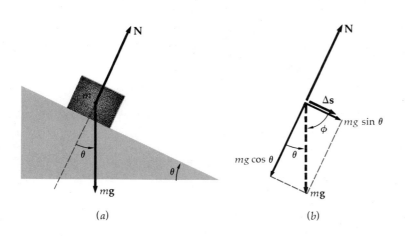

(a)                                        (b)

**Figure 6-8**
(a) Block on an inclined plane for Example 6-4. (b) Free-body diagram for the block. The resultant force is $mg \sin \theta$, which is the component of the force of gravity in the direction of the displacement $\Delta\mathbf{s}$.

The speed at the bottom of the incline is thus given by

$$v^2 = v_0^2 + 2gh$$

This result is the same as if we dropped the particle from a distance $h$ above the surface of the earth with initial downward speed $v_0$. The work done by the earth on the particle is $mgh$, independent of the angle of the incline. If the angle $\theta$ were increased, the particle would travel a smaller distance $s$ to drop the same vertical distance $h$, but the component of $m\mathbf{g}$ parallel to the motion $mg \sin \theta$ would be greater, making the work done, $(mg \sin \theta)s$, the same.

---

The results of Example 6-4 can be generalized. Consider a particle sliding down a curve of any shape under the influence of gravity. Figure 6-9 shows a small displacement $\Delta s$ parallel to the curve. The work done by the earth during this displacement is $(mg \cos \phi) \Delta s$, where $\phi$ is the angle between the displacement and the downward force of gravity.

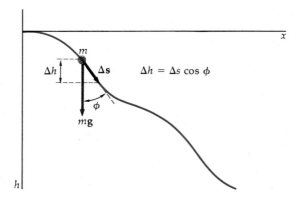

**Figure 6-9**
Particle sliding down a curve of arbitrary shape. The work done by the earth when the displacement is $\Delta s$ is $(mg \cos \phi) \Delta s = mg \Delta h$, where $\Delta h$ is the vertical component of the displacement.

The quantity $\Delta s \cos \phi$ is just $\Delta h$, the vertical distance dropped. As the particle slides down the curve, the angle $\phi$ varies, but for each displacement $\Delta s$, the downward component of displacement parallel to the weight is $\Delta s \cos \phi = \Delta h$ and the work done by the earth is $mg \Delta s \cos \phi = mg \Delta h$. Thus the total work done by the earth is $mgh$, where $h$ is the total vertical distance the particle descends. If the curve is frictionless, the weight is the only force that does work. In this case the speed of the particle after descending a vertical distance $h$ is obtained from $\frac{1}{2}mv^2 - \frac{1}{2}mv_0^2 = mgh$, where $v_0$ is the initial speed. If the curve is not frictionless, the frictional force will do work (this work will be negative because the frictional force is in the direction opposite the motion). The work done by the friction force depends on the length and shape of the curve and the coefficient of friction.

**Questions**

5. A body moves in a circle at constant speed. Does the force that accounts for its acceleration do work on it? Explain.

6. Suppose there is a net force acting on a body but it does no work. Can the body be moving in a straight line?

## 6-5   Potential Energy

Often the work done by a force applied to an object produces no increase in the kinetic energy of the object because other forces do an equal amount of negative work. For example, consider a block being pulled slowly up a smooth incline with constant velocity by an applied force $F_{app}$ which just balances the component of the weight of the block parallel to the incline, $F_{app} = mg \sin \theta$.

The work done by the applied force in moving the block a distance $s$ is

$$W = F_{app}s = (mg \sin \theta)s = mgh$$

where $s = h \sin \theta$ is the height the block has been raised above its initial level. In this case there is no increase in kinetic energy of the block because the earth does an equal but negative amount of work:

$$W_E = -(mg \sin \theta)s = -mgh$$

However, we can convert the work done by the applied force into a change in kinetic energy by merely releasing the block and letting it slide back down the incline. The earth will then do a positive amount of work $(mg \sin \theta)s = mgh$, which equals the increase in kinetic energy because it is the only work done on the block.

We can use the gravitational attraction of the earth for the block to store the work we do on the block for later use in giving the block kinetic energy. We say that the block at the height $h$ has *potential energy* $mgh$ relative to the original position. The work done by the applied force increases the potential energy of the block. When the block slides back down under the influence of gravity alone, the work done by the earth decreases the potential energy while increasing the kinetic energy by an equal amount. Here potential energy is converted into kinetic energy. Since the loss of potential energy equals the gain in kinetic energy for each part of the downward motion, the sum of potential energy and kinetic energy is constant as the block slides down the incline. This is an example of the *conservation of energy*.

If we project an object up a frictionless incline with initial velocity $v_0$, it will travel until the negative work done by the earth decreases the kinetic energy to zero. Since this negative work equals the increase in potential energy $mgh$, the maximum height attained by the object is given by $mgh = \frac{1}{2}mv_0^2$ (Figure 6-10). (Again, this height is independent of the angle of inclination of the plane.) The object stops momentarily at this height and then slides back down, gaining kinetic energy and losing potential energy. When it reaches its starting point, its kinetic energy is again $\frac{1}{2}mv_0^2$.

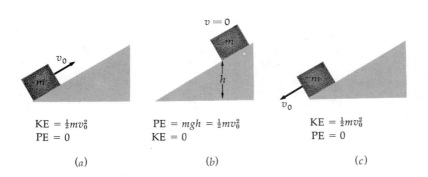

KE = $\frac{1}{2}mv_0^2$
PE = 0

(a)

PE = $mgh$ = $\frac{1}{2}mv_0^2$
KE = 0

(b)

KE = $\frac{1}{2}mv_0^2$
PE = 0

(c)

**Figure 6-10**
A particle projected up a frictionless incline moves until its potential energy $mgh$ equals its original kinetic energy $\frac{1}{2}mv_0^2$. It then slides back down and has its original kinetic energy when it reaches its starting point.

Gaston Rebuffat/Photo Researchers

This climber does work in increasing his gravitational potential energy.

There are many other kinds of potential energy. In Examples 6-2 and 6-3 we found the work needed to compress a block against a spring and then used this result to find the kinetic energy of the block when the spring was released. In this case the work we did to compress the spring was stored in the spring as potential energy.

Although we have talked about the potential energy of an object at some height above the surface of the earth or of an object attached to a spring as if the potential energy were associated with a single particle, this description is a simplification. More generally, the concept of potential energy applies to a *system of interacting particles*. Consider, for example, two bodies which exert gravitational forces of attraction on each other. If we apply an external force on each body equal but opposite to the force of interaction and pull the bodies apart with zero acceleration, we must do work on the system. The work we do by our externally applied forces can be recovered; if we release the bodies, they will accelerate toward each other and their kinetic energy will increase. The work done by our external applied forces thus increases the potential energy of the two-body system. When the bodies were released so that they moved under the influence of their mutual attraction, the work done by the forces of interaction decreased the potential energy of the system and increased the kinetic energy of the system. However, if one

*Potential energy applies to system of particles*

The waterfall in this 1961 lithograph by the Swiss artist M. C. Escher violates the law of the conservation of energy. When the water falls, part of its potential energy is converted into the kinetic energy of the waterwheel. How does the water get back to the top of the waterfall? (*Courtesy of the Escher Foundation, Haags Gemeentemuseum, The Hague.*)

of the bodies is much larger than the other, which is the case when one is a block of ordinary size and the other is the earth, the motion of the larger body is negligible. We separate such a pair of bodies by "lifting" the block, and we release them by "dropping" the block. (It doesn't matter whether we apply a force to the earth or not; its motion is completely negligible compared with that of the block.) In this case, we can simplify our description of the potential energy of the system by thinking of it as being associated with the position of the smaller body, the block.

In some situations we do work that is not readily recovered. If we push a block along a rough horizontal surface at constant velocity, we must do work against the force of sliding friction. When we stop pushing and release the block, the block slides to a stop. The work we did does not reappear as kinetic energy of the block. (We shall see later that since this work appears as thermal energy of the surroundings, the work is not lost completely; but it is lost in the sense that it is not easily reconverted into mechanical energy. We shall study this problem in more detail when we study thermodynamics in Chapters 16 to 19.) The force of friction is called a *nonconservative force*. Forces such as gravity and that of a spring for which the work done against them is stored as potential energy are called *conservative forces*.

*Conservative and nonconservative forces*

An important property of a conservative force is that the work it does on an object is independent of the path taken by the object and depends only on its initial and final positions.* We saw an example of this when we found that the work done by the force of gravity on an object sliding along a curve of any shape is just $mgh$, where $h$ is the total vertical distance dropped. The work for that case thus depends only on the initial and final elevations of the object. We use this property to define the potential-energy function $U$ associated with a conservative force. It is defined so that the work done by a conservative force equals the decrease in the potential-energy function:

$$W = \int_1^2 F_s \, ds = -\Delta U$$

or

$$\Delta U = U_2 - U_1 = -\int_1^2 F_s \, ds \qquad\qquad 6\text{-}15$$

Since only the change in the potential energy is defined, the value of the function $U$ at any point is not specified by the definition. We are free to choose the value of $U$ at any one point arbitrarily. We usually choose $U$ to be zero at some reference point. The potential energy at any other point is then the difference between the potential energy at that point and at the reference point. For example, if the gravitational potential energy of an object near the surface of the earth is chosen to be zero at sea level, its value at a distance $h$ above sea level is $mgh$. Or we could choose the potential energy to be zero at the level of a table top in our

* We discuss the properties of conservative and nonconservative forces more fully in Section 6-7.

A slingshot is an example of elastic potential energy. The work done in stretching the rubber is stored as potential energy and then converted into kinetic energy of the missile.

Culver Pictures

room, in which case its value at any other point would be $mgy$, where $y$ is measured from the table top.

When we consider a system of two particles which interact via conservative forces, we find that the work done by the forces of interaction depends only on the initial and final separation of the particles. Again, the potential energy of such a system is defined only in terms of its change.

*The change in potential energy of a two-particle system is the negative of the work done on both particles by the force of interaction when the separation of the particles changes from some initial to some final value.*

For this case we choose the value of the potential energy to be zero at some convenient separation.

The work done by nonconservative forces generally depends on things other than the initial and final positions of the object (or objects in a two-particle system). For example, the work done by frictional forces or by viscous forces may depend on the speed of the object, on the total distance traveled, or on the particular path in space taken by the object as it moves from one position to another. A potential-energy function cannot be defined for a nonconservative force.

**Questions**

7. A heavy box is dragged across the floor from point $A$ to point $B$. Will the work done by friction differ if the box is pulled along a zigzag path instead of along the shortest path from $A$ to $B$?

8. How can a man who can exert a maximum force of 400 N raise an 800-N box from the ground to a shelf a height 2 m above the ground? Could a boy who can exert only 200 N raise the box to the shelf? How much work would each have to do?

9. When you climb a mountain, is the work done on you by gravity different if you take a short steep trail instead of a long gentle trail? If not, why would you find one trail easier than the other?

10. Estimate the amount by which your gravitational potential energy changes when you walk up a flight of stairs.

# 6-6    The Conservation of Energy

The negative sign in the definition of potential energy in Equation 6-15 is introduced so that the work done by a conservative force equals the *decrease* in potential energy. If this force is the only force acting, or if it is the only force that does work on the particle, the work done by the force also equals the *increase* in *kinetic energy:*

$$W_{\text{net}} = \int \Sigma F_s \, ds = -\Delta U = +\Delta E_k$$

Hence

$$\Delta E_k + \Delta U = \Delta(E_k + U) = 0 \qquad\qquad 6\text{-}16$$

or

$$E_k + U = \text{constant} \qquad\qquad 6\text{-}17$$

*Conservation of mechanical energy*

The sum of the kinetic and potential energies of a particle is called the *total mechanical energy*. Equation 6-17 states that if only conservative forces do work on a particle, the total mechanical energy remains constant during the motion. This is known as the conservation of (mechanical) energy and is the origin of the expression *conservative* force. The principle of conservation of energy is one of the most important in physics. In the next chapter we shall show how to use energy conservation to solve many mechanics problems. In our study of thermodynamics we shall generalize this concept of energy to include thermal energy.

## 6-7    Conservative and Nonconservative Forces

The criterion for a force to be conservative can be simply stated:

*A force is conservative if the total work it does on a particle is zero when the particle moves around any closed path returning to its initial position.*

Criterion for a force to be conservative

We have already seen that the force of gravity has this property. If we project an object up an inclined plane with some initial speed, the force of gravity does work $-mgh$ while the object moves up and $+mgh$ while it moves down, where $h$ is the maximum vertical distance the object rises. If this conservative force is the only force acting, the object has the same kinetic energy when it arrives back at its starting point that it had originally. On the other hand, the force of sliding friction is always opposite to the direction of motion, so that it always does negative work. When an object makes a round trip ending at its original position, the total work done on it by sliding friction is negative. For example, if we project an object up a rough inclined plane with some initial speed, the frictional force does work $-\mu_k N s$ on the way up and $-\mu_k N s$ on the way down, where $\mu_k$ is the coefficient of friction, $N$ the normal force, and $s$ the distance traveled up the plane. The total round trip work done by sliding friction in this case is not zero but $-2\mu_k N s$.

An alternative statement of the criterion for a force to be conservative is that

*The work done on a particle by a conservative force is independent of how the particle moves from one point to another.*

Figure 6-11 shows several paths connecting points $P_1$ and $P_2$. If the work done by a conservative force is $W$ when the particle moves from $P_1$ to $P_2$ along one of the paths, it must be $-W$ when the particle returns along any of the paths, because the total round-trip work must be zero.

In one dimension a sufficient condition for force to be conservative is that the force depends only on the position of the particle. Examples are the force of gravity, which is constant (near the surface of the earth), and the force exerted by a spring, $F_x(x) = -kx$. Even though the force of sliding friction may be approximately constant independent of the speed of the object, this force depends on the direction of motion; it is always opposite to the direction of motion. Since this force is not merely a function of position, it is not conservative. Another nonconservative force is that applied by a human agent such as a push or pull.

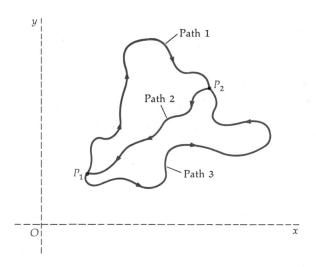

**Figure 6-11**
Several paths connecting points $P_1$ and $P_2$. If the work done by a conservative force along path 1 from $P_1$ to $P_2$ is $W$, it must be $-W$ on the return trip along path 2 because the round-trip work is zero. It follows that the work done as a particle goes from point $P_1$ to $P_2$ is the same along any path connecting the two points.

Suppose you decide to lift an object through a vertical distance $h$. You may apply a vertical force equal to the weight of the object to just balance the downward force of gravity so that the object does not accelerate, or you may apply a greater force so that the object accelerates upward. The force you apply clearly does not depend only on the position of the object and is therefore not a conservative force.

In three dimensions it is necessary but not sufficient that a force be a function only of position to be conservative. An example of a nonconservative force in three dimensions which depends only on position is one which is directed tangentially to a circle, as shown in Figure 6-12. Such a force arises in a betatron, a device for accelerating electrons. Consider the work done by a constant force tangent to the circle as the particle makes one complete revolution. Since the force is always in the direction of motion, the total work done is positive and equal to the product of the force $F$ and the total distance $2\pi r$. Since the particle is back to its original position and the net work done is not zero, this force is not conservative.

A force more common in three-dimensional motion is one directed toward a fixed point in space, e.g., the gravitational force exerted by the sun on a planet, which is toward the sun, or the electrostatic force exerted by a heavy nucleus on an orbiting atomic electron, which is directed toward the nucleus. Such a force is called a *central force*. If the magnitude of such a force depends only on the distance from the force center, as in the above examples, the force is conservative. We can represent such a force in general by

$$\mathbf{F} = f(r)\,\hat{\mathbf{r}}$$

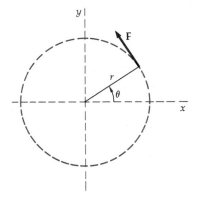

**Figure 6-12**
A force tangential to a circular path around the origin. If the particle moves once around the circle counterclockwise, the work done by **F** cannot be zero because **F** is always in the direction of motion. Such a force is therefore not conservative.

where $f(r)$ is any function of $r$ and $\hat{\mathbf{r}}$ is a unit vector pointing in the radial direction away from the origin. This general case includes the force of gravity, for which $f(r) = -Gm_1m_2/r^2$, and the electrostatic force, for which $f(r) = kq_1q_2/r^2$. We now show that such central forces are conservative no matter what the function $f(r)$ may be.

Figure 6-13 shows an arbitrary path taken by a particle from point $P_1$ to point $P_2$. Consider a small part of the displacement $\Delta\mathbf{s}$ which makes an angle $\phi$ with $\hat{\mathbf{r}}$. The work done by a force $f(r)\hat{\mathbf{r}}$ during this displacement is

$$\Delta W = \mathbf{F} \cdot \Delta\mathbf{s} = f(r)\,\hat{\mathbf{r}} \cdot \Delta\mathbf{s} = f(r)\,\Delta s \cos\phi$$

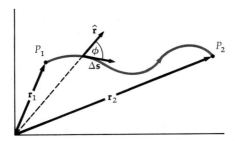

**Figure 6-13**
An arbitrary path connecting
points $P_1$ and $P_2$. The work
done by a central force $f(r)\hat{\mathbf{r}}$
for the displacement $\Delta \mathbf{s}$ is
$f(r)\,\hat{\mathbf{r}} \cdot \Delta \mathbf{s} = f(r)\,\Delta r$. The total
work from $P_1$ to $P_2$ thus de-
pends only on the initial and
final values of $r$.

But $\Delta s \cos \phi = \Delta r$, the component of the displacement in the radial
direction. The total work done by the force when the particle moves
from point $P_1$ at a distance $r_1$ from the origin to point $P_2$ at distance $r_2$ is

$$W = \int_1^2 \mathbf{F} \cdot d\mathbf{s} = \int_{r_1}^{r_2} f(r)\,dr = G(r_2) - G(r_1)$$

where $f(r) = dG(r)/dr$. Since this work depends only on the initial and
final positions and not on the path taken, the force is conservative.

There is a general test of whether a given force in three dimensions
is conservative or nonconservative, involving methods of vector calculus,
which we shall not pursue. Fortunately, nearly all of the forces we shall
be concerned with in three-dimensional motion are central forces, which
are conservative.

### Questions

11.  The force exerted by a spring which obeys Hooke's law is a conservative force. Would the force of the spring stretched beyond its elastic
limit be conservative?

12.  When you jump into the air, the forces exerted by your leg muscles
do work on your body. Are these forces conservative?

## 6-8    Finding the Potential-Energy Function

If we know a particular conservative force as a given function of position, we can calculate the potential-energy function associated with this
force directly from Equation 6-15, which states that the work done by
the conservative force equals the decrease in the potential energy. It is
usually easier to keep track of the signs if we think of applying an external force equal but opposite to the conservative force, so that the particle moves (without acceleration) in the direction opposite that of the
conservative force. For example, we lift an object up against the force of
gravity. Then the work *we* do *against* the conservative force equals the
*increase* in the potential energy of the particle.

As our first example, we consider the potential-energy function associated with the constant force of gravity directed down toward the surface of the earth. Let $y$ be the vertical axis and let the object of mass $m$ be
originally at $y = 0$. If we lift the object through a distance $y$ by exerting
a force equal to the force of gravity $mg$, we do work $mgh = mgy$. This
work equals the increase in potential energy:

This boulder, located in the
Arches National Park in Utah,
has gravitational potential energy relative to ground level.

$$\Delta U = mgy$$

If $U_0$ is the initial potential energy at $y = 0$, we have

$$U(y) = U_0 + mgy \qquad\qquad 6\text{-}18$$

*Gravitational potential energy*

As we have mentioned, potential energy is defined only in terms of its change. We can choose the constant $U_0$ to be anything we wish. If we choose $U_0 = 0$, we are choosing the potential energy to be zero at $y = 0$. Then $U(y) = mgy$. For example, if we are moving objects up and down from the surface of a table in a room, we may wish to choose $U$ to be zero at the table top. Then if the object is below the table top, its potential energy will be negative. This does not matter since only changes in potential energy are important.

For our next example, we find the potential energy associated with a spring which exerts a force $F(x) = -kx$. Here we have chosen $x = 0$ at the equilibrium length of the spring. If we apply a force $F = +kx$ and stretch the spring from $x = 0$ to some point $x = x_1$, the work we do is the area under the $F$-versus-$x$ curve in Figure 6-14. This area is $\frac{1}{2}(kx_1)x_1 = \frac{1}{2}kx_1^2$. The work that we must do to stretch the spring from $x = 0$ to some general position $x$ is then $\frac{1}{2}kx^2$. Then

$$\Delta U = \tfrac{1}{2}kx^2 \qquad \text{or} \qquad U = U_0 + \tfrac{1}{2}kx^2$$

If we choose the potential energy to be zero at the equilibrium length $x = 0$, the potential-energy function at any other length $x$ is

$$U = \tfrac{1}{2}kx^2 \qquad\qquad 6\text{-}19$$

*Potential energy of spring-mass system*

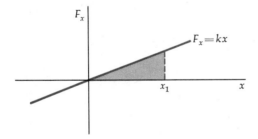

**Figure 6-14**
The work done to stretch a spring from $x = 0$ to $x = x_1$ is the area indicated, which is $\frac{1}{2}(kx_1)x_1 = \frac{1}{2}kx_1^2$.

---

**Example 6-5** A 1.5-kg box rests on a frictionless horizontal table and is attached to a spring of force constant $k = 300$ N/m. The spring is compressed 4 cm, and the box is released. How fast is it moving when the spring is at its natural length?

We choose $x = 0$ for the box when the spring is at its natural length and $U = 0$ at $x = 0$. We compress the spring 4 cm by moving the box from $x = 0$ to $x = -4$ cm. The work we must do is the potential energy stored in the spring, which is $U = \frac{1}{2}kx^2 = \frac{1}{2}(300 \text{ N/m}) (-0.04 \text{ m})^2 = 0.24$ J. Since the box is at rest at this point, its kinetic energy is zero and its total energy is 0.24 J. When we release the box, it accelerates as the spring decompresses. When $x = 0$, the potential energy is zero, and so the kinetic energy of the box is $E_k = \frac{1}{2}mv^2 = 0.24$ J. Solving for the speed gives

$$v = \left(\frac{0.48 \text{ J}}{1.5 \text{ kg}}\right)^{1/2} = 0.566 \text{ m/s}$$

---

## 6-9    Potential Energy and Equilibrium in One Dimension

Figure 6-15 shows a plot of the potential-energy function of Equation 6-19 versus $x$ for an object on a spring. The potential-energy function is related to the force by

$$dU(x) = -F_x \, dx \qquad\qquad 6\text{-}20$$

The force is therefore the negative derivative of the potential-energy function

$$F_x = -\frac{dU}{dx} \qquad\qquad 6\text{-}21$$

*Force and potential energy*

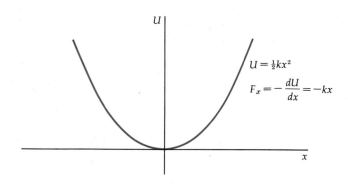

$$U = \tfrac{1}{2}kx^2$$

$$F_x = -\frac{dU}{dx} = -kx$$

**Figure 6-15**
Potential-energy function $U(x)$ for an object on a spring. The force $F_x = -dU/dx$ is the negative slope of this function. It is zero at the minimum in the curve at $x = 0$. At positive $x$ the force $F_x$ is negative; at negative $x$ the force is positive. A minimum in the potential-energy curve is a point of stable equilibrium because for displacements on either side of this point the force is toward the equilibrium position.

We can check this general relation for the case of the spring by differentiating the function $U(x) = \tfrac{1}{2}kx^2$. We obtain

$$F_x = -\frac{dU}{dx} = -kx$$

At the position on the potential-energy curve where the slope is zero, the force is zero and the object is in equilibrium; i.e., if $F_x$ is the only force acting on it, the object will remain at rest in this position if it is placed there at rest. This occurs at $x = 0$ when the spring is unstretched. For $x$ slightly greater than zero the slope is positive; thus $F_x$ is negative and the object would accelerate toward the equilibrium point $x = 0$. For $x$ slightly less than zero the slope is negative; thus $F_x$ is positive, and the object again would accelerate toward $x = 0$. The equilibrium is thus stable; i.e., if an object at rest at the equilibrium point $x = 0$ is displaced slightly, the force accelerates it back toward the equilibrium point. Figure 6-16 shows a potential-energy curve with a maximum at $x = 0$ rather than a minimum. (Such a curve could arise, for example, if we have two equal positive charges fixed on the $y$ axis at equal distances above and below the $x$ axis and a third positive charge free to move along the $x$ axis.) At $x = 0$, $dU/dx = -F_x = 0$, so that a particle placed at this point is also in equilibrium. However, this equilibrium is unstable. For $x$ slightly greater than zero, the slope is negative and $F_x = -dU/dx$ is positive. Thus if a particle in equilibrium is given a small positive displacement, it will experience a positive force $F_x$ and will accelerate away from the equilibrium position. Similarly, for $x$ slightly less than zero, the slope is positive, indicating that $F_x$ is negative. If the

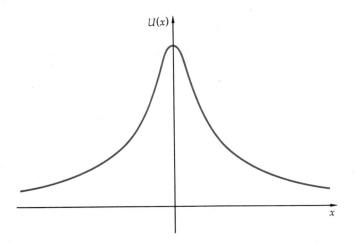

**Figure 6-16**
For this potential-energy curve, $F_x = -dU/dx$ is zero at $x = 0$, but this equilibrium is unstable because for displacements in either direction the force is away from the equilibrium position.

particle is displaced slightly to the left of its equilibrium position, it will accelerate to the left, away from equilibrium. Both maximum and minimum points on a potential-energy curve are positions of equilibrium; at minimum points the equilibrium is stable, and at maximum points it is unstable. We also note that the *conservative force always urges a particle toward lower potential energy.*

### Question

13. You are shown a graph of $U(x)$ versus $x$. At some places the curve $U(x)$ is steep and at others less steep. There are maximum and minimum points. At what points is the force associated with $U(x)$ large? Small? Zero?

## 6-10  Power

The rate at which a force does work is called the *power input P* of the force. Consider a particle with instantaneous velocity **v**. In a short time interval $dt$ the particle has a displacement $d\mathbf{s} = \mathbf{v}\, dt$. The work done by a force **F** acting on the particle during this time interval is

$$dW = \mathbf{F} \cdot d\mathbf{s} = \mathbf{F} \cdot \mathbf{v}\, dt$$

The power input of the force **F** during this interval is

$$P = \frac{dW}{dt} = \mathbf{F} \cdot \mathbf{v} \qquad\qquad 6\text{-}22 \qquad \textit{Power}$$

The SI unit of power, one joule per second, is called a watt (W). When you buy energy from a power company, you are usually charged by the kilowatt-hour (kW·h). A kilowatt-hour of energy is

$$1 \text{ kW·h} = (10^3 \text{ W}) \times (3600 \text{ s}) = 3.6 \times 10^6 \text{ W·s} = 3.6 \text{ MJ} \qquad 6\text{-}23$$

In the U. S. customary system, the unit for energy is the foot-pound, and for power it is the foot-pound per second. A common multiple of this unit is called a horsepower:

$$1 \text{ horsepower} = 550 \text{ ft·lb/s} = 746 \text{ W}$$

**Example 6-6** A small motor is used to power a lift that raises a load of bricks weighing 800 N to a height of 10 m in 20 s. What is the minimum power motor needed?

Assuming that the bricks are lifted without acceleration, the upward force equals the force of gravity 800 N. The speed of the bricks is (10 m)/(20 s) = 0.5 m/s. Since the velocity and applied force are in the same direction, the power input of the force is

$$P = Fv = (800 \text{ N}) (0.5 \text{ m/s}) = 400 \text{ N·m/s} = 400 \text{ J/s} = 400 \text{ W}$$

If there are no energy losses, e.g., to frictional forces, the motor must have a power output of 400 W.

# Energy Resources

Laurent Hodges
*Iowa State University*

Directly or indirectly, energy is essential for everything that exists or is done in modern society: coal in extracting metals from their ores, petroleum fuels for ground and air transportation, natural gas or fuel oil for heating buildings, electricity for industrial machinery and home appliances.

*Primary energy resources,* i.e., resources from the environment which are the ultimate sources of our energy, include the fossil fuels (coal, oil, and gas) and three types of electricity: hydroelectricity generated by falling water, electricity from nuclear power plants (ultimately from the energy of fission of $^{235}$U), and electricity generated by geothermal power (the energy of steam and hot water from beneath the earth). These are really only the so-called *commercial* energy sources. Although such energy sources as wood, peat, and cow or camel dung are absent from international trade and account for perhaps only 3 percent of world energy consumption, they may represent up to about half the energy consumption in certain developing countries. Wood was the leading energy source in the United States throughout most of the nineteenth century, but coal displaced it in the 1880s and the fossil fuels have been predominant ever since.

Figure 1 shows the consumption of commercial energy by the world and the United States in the years 1950 and 1980. During that period coal declined in relative importance while crude oil and natural gas grew in popularity because of their greater convenience in transportation and use and their lesser environmental problems. Nuclear power was first used during that period but still made only a small contribution in the early 1980s. Nuclear energy and natural gas are relatively sophisticated fuels used mainly in the developed countries.

The energy actually used is often in a modified form, referred to as a *secondary energy resource.* For example, about 80 percent of the electricity used in the United States is generated by the combustion of fossil fuels; in this case the fossil fuel is the primary energy resource and the electricity is a secondary resource. Similarly the hydrogen fuel of any future "hydrogen economy" will be a secondary resource obtained from one or more primary resources via the electrolysis of water.

Figure 2 shows the per capita energy consumption in the world and different parts of the world. United States per capita consumption is about 2

**Figure 1**
Commercial energy consumption in the United States and the world in 1950 and 1980.

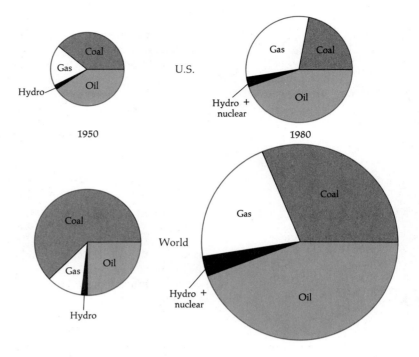

1950

U.S.

1980

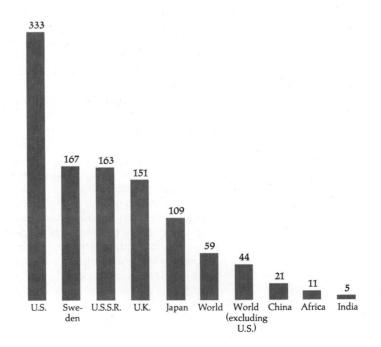

World

times that in the other developed countries (Canada, Europe, Australia, New Zealand, Japan, Israel, Soviet Russia) and 25 to 30 times that in the developing countries. It is often stated that the United States cannot cut its energy consumption significantly without returning to some unspecified "dark age." Actually a drastic 50 percent cut would only put United States per capita energy consumption at the level of that in the other developed countries, where the standard of living is certainly decent and civilized. A less drastic but supposedly intolerable cut of 30 percent would return us to consumption levels of about 20 years ago, which few of us remember as a dark age.

Despite conservation trends, the increase in United States per capita consumption in the 1970s about equaled the total per capita consumption in developing countries. This is an unfortunate trend. It would be more equitable for the consumption increases to come in the developing countries, especially since they are providing most of the world's energy production increases. In the early

**Figure 2**
Per capita energy consumption (GJ) in 1980.

1950s the United States was producing approximately as much energy as it consumed, but by 1980 it was importing about 15 percent of the energy it consumed, most of the imports being produced in developing countries.

About 40 percent of United States energy consumption is by the industrial sector; 25 percent is by ground, air, and water transportation; 20 percent by residences; and 15 percent by commercial establishments (stores, offices, restaurants, etc.). Figure 3 shows the breakdown into end uses. All the small electrical appliances in use—clocks, radios, televisions, power tools, hair dryers—account for much less than 1 percent of total energy consumption in the United States.

The fossil fuels that constitute the bulk of our energy sources are a finite nonrenewable resource and cannot meet our energy needs forever. Although the recoverable reserves of these fuels can be estimated only imperfectly and future energy demands are uncertain, it seems clear that oil and gas can continue to be a major resource for only a few more decades at most and coal only a few centuries. Nuclear energy in the form of $^{238}$U or $^{232}$Th used in breeder reactors or deuterium or other light isotopes used in controlled thermonuclear fusion reactors (should they prove feasible) could supply projected energy needs for much longer—thousands or even millions of years assuming a leveling off of the growth of energy consumption.

**Figure 3**
Approximate energy use in the United States as of the early 1980s.

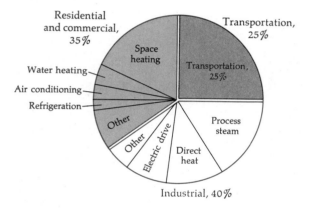

Other renewable or continuously available energy resources which are of negligible importance today may make significant contributions to our energy supply in the future: e.g., solar energy, wind power, organic wastes, tidal energy. Solar energy is especially abundant, though its low power density may make necessary the collection of solar radiation over large land areas.

Extensive use of renewable energy resources may be required if society is to have abundant and environmentally tolerable energy. Society has become increasingly concerned with the environmental costs of energy production and use. Some of the major environmental costs are land devastation by the strip mining of coal, high acid concentrations in the streams of coal mining areas, particulate matter and sulfur oxide air pollution from coal combustion, water pollution from oil-field operations, air pollution from petroleum refining, sulfur oxide pollution from petroleum combustion, carbon monoxide and hydrocarbon air pollution from the use of gasoline in motor vehicles, nitrogen oxide air pollution from any high-temperature combustion process, and radiation releases from uranium mining and from nuclear power plants.

Unfortunately much of the energy used today is wasted. For this reason energy conservation has become an important policy to implement. If we can use energy more efficiently, if we can devise new products with lower energy requirements (think of pocket calculators!), if we can change our life styles to reduce energy consumption, we can slow the rate of depletion of our energy resources and simultaneously decrease the adverse environmental effects of energy production and consumption. It will be especially advantageous to increase the use of renewable resources and thereby conserve fossil fuels for their other valuable uses, such as the manufacture of chemicals and drugs.

## Review

A. Define, explain, or otherwise identify:

Work, 163                          Conservative force, 175
Joule, 163                         Total mechanical energy, 178
Foot-pound, 163                    Stable equilibrium, 185
Kinetic energy, 165                Unstable equilibrium, 185
Work-energy theorem, 165           Power, 186
Dot product, 170                   Watt, 186
Scalar product, 170                Horsepower, 186
Potential energy, 173

B. True or false:

1. Only the resultant force acting on an object can do work.

2. No work is done on a particle which remains at rest.

3. Work is the area under the force-versus-time curve.

4. A force which is always perpendicular to the velocity of a particle does no work on the particle.

5. A kilowatt-hour is a unit of power.

6. If **A** and **B** are in opposite directions, **A** · **B** is zero.

7. Only conservative forces can do work.

8. There is a potential-energy function associated with every force.

9. If only conservative forces act, the kinetic energy of a particle does not change.

10. The work done by a conservative force decreases the potential energy associated with that force.

11. When a particle takes a round trip, the total work done by each conservative force is zero.

## Exercises

### Section 6-1, One-Dimensional Motion with Constant Forces

1. A 5-kg body is raised at constant velocity a distance of 10 m by a force $\mathbf{F}_0$. Find the work done on the body (a) by the force $\mathbf{F}_0$ and (b) by the earth. What is the net work done by all the forces acting?

2. A 10-kg box rests on a horizontal table. The coefficient of friction between the box and table is 0.4. A force $F_x$ pulls the box at constant velocity a distance of 5 m. Find the work done (a) by $F_x$ and (b) by friction.

3. The box described in Exercise 2 is pulled by a force $F_x$ so that the box has an acceleration of 2 m/s². (a) Find the work done by $F_x$ while the box is pulled a distance of 5 m. (b) What is the work done by the frictional force in this case ($\mu = 0.4$)?

4. (a) From the conversion factors for newtons to pounds and metres to feet derive the relation between the joule and the foot-pound expressed in Equation 6-3. (b) The cgs unit of work is the erg, defined as a dyne-centimetre. Find the number of ergs in a joule.

### Section 6-2, Work Done by the Resultant Force: Kinetic Energy

5. A 10-g bullet has a velocity of 1 km/s. (a) What is its kinetic energy in joules? (b) If the speed is halved, what is its kinetic energy?

6. A 2-kg block moves with speed 10 m/s. (a) What is its kinetic energy in joules? (b) What is its kinetic energy in foot-pounds?

7. A 2-kg body and a 200-kg body have equal momenta of 40 kg·m/s. Find the kinetic energy of each body.

8. A 5-kg box initially at rest on a frictionless table is pulled by a constant horizontal force of 10 N for a distance of 6 m. Find (*a*) the final kinetic energy of the box and (*b*) the final speed of the box.

9. A 5-kg mass is raised a distance of 4 m by a vertical force of 80 N. Find (*a*) the work done by the force and the work done by the earth and (*b*) the final kinetic energy of the mass if it was originally at rest.

10. A 2-kg box is initially at rest on a horizontal table. The coefficient of friction between the box and table is 0.4. The box is pulled along the table a distance of 3 m by a horizontal applied force of 10 N. (*a*) Find the work done by the applied force. (*b*) Find the work done by friction. (*c*) Find the change in kinetic energy of the box. (*d*) Find the speed of the box after it has traveled 3 m.

11. A 5-kg object is lifted by a rope exerting a force $T$ so that it accelerates upward at 5 m/s². The object is lifted 3 m. Find (*a*) the tension $T$, (*b*) the work done by $T$, (*c*) the work done by the earth, and (*d*) the final kinetic energy of the object after being lifted 3 m, if it was originally at rest.

12. A 2000-kg car moves with initial speed of 25 m/s. It is stopped in 60 m by a constant frictional force. (*a*) What is the initial kinetic energy of the car? (*b*) How much work is done by the frictional force in stopping the car? (*c*) What is the coefficient of friction?

**Section 6-3, Work Done by a Force That Varies with Position**

13. A 2-kg particle is moving with speed 3 m/s when it is at $x = 1$ m. It is subjected to a single force $F_x$, which varies with position as shown in Figure 6-17. (*a*) What is the kinetic energy of the particle when it is at $x = 1$ m? (*b*) How much work is done by the force as the particle moves from $x = 1$ m to $x = 4$ m? (*c*) What is the speed of the particle when it is at $x = 4$ m?

14. A 4-kg particle is initially at rest at $x = 0$. It is subjected to a single force $F_x$, which varies with $x$ as shown in Figure 6-18. (*a*) Find the work done by the force as the particle moves from $x = 0$ to $x = 3$ m and (*b*) as it moves from $x = 3$ m to $x = 6$ m. Find the kinetic energy of the particle when it is at (*c*) $x = 3$ m and (*d*) $x = 6$ m.

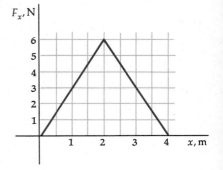

**Figure 6-17**
$F_x$ versus $x$ for Exercise 13.

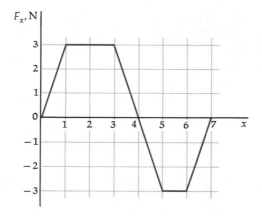

**Figure 6-18**
$F_x$ versus $x$ for Exercise 14.

15. A force $F_x$ is related to the position of a particle by the formula $F_x = Cx^3$, where $C$ is a constant. Find the work done by this force acting on the particle when the particle moves from $x = 0$ to $x = 3$ m.

**Section 6-4, Work and Energy in Three Dimensions**

16. A 5-kg box is pulled a distance of 6 m along a frictionless horizontal surface by a 20-N force which makes an angle of 30° with the horizontal. Find (*a*) the work done by the force, (*b*) the change in kinetic energy of the box, and (*c*) the final speed if the box is initially at rest.

17. A block of mass 6 kg slides down a rough incline starting from rest. The coefficient of friction is 0.2, and the angle of the incline is 60°. (*a*) List all the forces acting on the block and find the work done by each force when the block slides 2 m (measured along the incline). (*b*) What is the net work done on the block? (*c*) What is the velocity of the block after it has slid 2 m?

18. A 2-kg body attached to a string moves on a frictionless horizontal surface in a circle of radius 3 m. The speed of the body is 1.5 m/s. (*a*) Find the tension in the string. (*b*) List the forces acting on the body and find the work done by each force during 1 rev.

19. A 100-kg cart is lifted up a 1-m step by rolling the cart up an incline formed when a plank of length $L$ is laid from the lower level to the top of the step. (Assume the rolling is equivalent to sliding without friction.) (*a*) Find the force parallel to the incline needed to push the cart up without acceleration for values of $L$ of 3, 4, and 5 m. (*b*) Calculate directly from Equation 6-14 the work needed to push the cart up the incline for each of these values of $L$. (*c*) Since the work found in (*b*) is the same for each value of $L$, what advantage if any is there in choosing one length over another?

20. A 2-kg box slides down a rough incline of length 10 m and angle of inclination 30°. The coefficient of friction is 0.3. (*a*) List all the forces acting on the box and find the work done by each. (*b*) Find the kinetic energy of the box at the bottom of the incline assuming that it began at the top with speed $v = 2$ m/s.

21. Two vectors **A** and **B** have magnitudes of 6 m and make an angle of 60° with each other. Find **A** · **B**.

22. Find **A** · **B** for the following vectors: (*a*) **A** = 2**i** − 6**j**, **B** = −3**i** + 2**j**; (*b*) **A** = 4**i** + 4**j**, **B** = 2**i** − 3**j**; (*c*) **A** = 3**i** + 4**j**, **B** = 4**i** − 3**j**.

23. Find the angle $\theta$ between the vectors **A** and **B** given in Exercise 22.

24. If **A** · **B** = 0 and **A** = 2**i** − 3**j**, what can be said about **B**?

25. The scalar product of two vectors **A** and **B** is 4 units. Their magnitudes are $A = 4$ and $B = 8$. Find the angle between the vectors.

**Section 6-5, Potential Energy; Section 6-6, The Conservation of Energy; and Section 6-7, Conservative and Nonconservative Forces**

26. A man of mass 80 kg climbs up a stairway of height 5 m. (*a*) What is the increase in potential energy of the man? (*b*) What is the least amount of work the man must do to climb the stairway? (*c*) Explain why the work done by the man may be more than this minimum amount.

27. State which of the following forces are conservative and which are nonconservative: (*a*) the frictional force exerted on a sliding box, (*b*) the force exerted by a linear spring obeying Hooke's law, (*c*) the force of gravity, (*d*) the wind resistance on a moving car, (*e*) the force exerted by a nonlinear spring obeying $F_x = -kx + Cx^2$, (*f*) the force exerted by a boy on a ball when he throws it, (*g*) the force exerted by a floor on a dropped egg.

28. A 2-kg body is held at a height of 20 m above the ground and released at time $t = 0$. (*a*) What is the original potential energy of the body relative to the ground? (*b*) From Newton's laws find the distance the body falls in 1 s and its speed at $t = 1$ s. (*c*) Find the potential energy, the kinetic energy, and the total mechanical energy of the body at $t = 1$ s. (*d*) Find the kinetic energy and the speed of the body just before it hits the ground.

29. A 2-kg box slides along a long frictionless incline of angle 30°. It starts from rest at time $t = 0$ at the top of the incline a height of 20 m above the ground. (*a*) What is the original potential energy of the box relative to the ground? (*b*) From Newton's laws, find the distance the body travels in 1 s and its speed at $t = 1$ s. (*c*) Find the potential energy, the kinetic energy, and the total mechanical energy of the box at $t = 1$ s. (*d*) Find the kinetic energy and the speed of the box just as it reaches the bottom of the incline.

30. (*a*) Make a plot of the potential energy of a 10-kg body versus its height above the ground for heights ranging from 0 to 50 m. Use the ground as the reference level at which $U = 0$. (*b*) On the same graph plot the potential energy of the body, using the height 25 m as the point of zero potential energy.

31. How high must a body be lifted to gain an amount of potential energy equal to the kinetic energy it has when moving at speed 20 m/s?

32. A 0.5-kg ball is thrown straight up with initial speed of 10 m/s. (*a*) Find the initial kinetic energy of the ball and total mechanical energy, assuming that the initial potential energy is zero. (*b*) Find the kinetic energy, potential energy, and total mechanical energy of the ball at the top of its flight. (*c*) Use the law of conservation of total mechanical energy to find the maximum height of the ball. (Take $g = 10$ m/s² for convenience.) (*d*) Compare this method of finding the maximum height of the ball with that in Chapter 2 using the constant-acceleration formulas. Can you find the time needed for the ball to reach its maximum height from the conservation of mechanical energy?

33. A constant force $F_x = -6$ N acts in the negative $x$ direction. Find the potential-energy function associated with this conservative force (*a*) if $U = 0$ at $x = 0$, (*b*) if $U = 0$ at $x = 5$ m, (*c*) if $U = 50$ J at $x = 0$.

34. A constant force is given by $F_x = 4$ N. (*a*) Find the potential-energy function associated with this force for an arbitrary choice of zero potential energy. (*b*) Find $U(x)$ such that $U(x)$ is zero at $x = 6$ m. (*c*) Find $U(x)$ such that $U = 12$ J at $x = 6$ m.

35. A man places a 2-kg block against a horizontal spring of force constant $k = 300$ N/m and compresses it 9 cm. (*a*) Find the work done by the man and the work done by the spring. (*b*) The spring is released from its compression of 9 cm and returns to its original position. Find the work done by the spring and the kinetic energy of the block at this point (assuming that only the spring does work on the block).

36. A spring obeys Hooke's law with force constant $k = 10^4$ N/m. How far must it be stretched for its potential energy to be (*a*) 100 J and (*b*) 50 J?

### Section 6-8, Finding the Potential-Energy Function
*There are no exercises for this section.*

### Section 6-9, Potential Energy and Equilibrium in One Dimension

37. Figure 6-19 shows a potential-energy function $U(x)$ versus $x$. (*a*) At each point indicated, state whether the force $F_x$ is positive, negative, or zero. (*b*) At which point does the force have the greatest magnitude? (*c*) Identify any equilibrium points and state whether the equilibrium is stable or unstable.

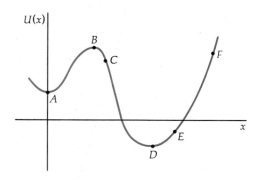

**Figure 6-19**
Exercise 37.

38. Repeat Exercise 37 for the potential-energy function $U(x)$ shown in Figure 6-20.

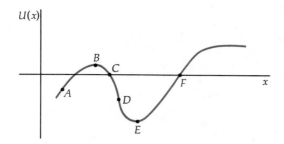

Figure 6-20
Exercise 38.

39. (a) Find the force $F_x$ associated with the potential-energy function $U(x) = Ax^4$, where $A$ is a constant. (b) At what point or points is the force zero?

40. A potential-energy function is given by $U(x) = C/x$, where $C$ is a positive constant. (a) Find the force $F_x$ as a function of $x$. (b) Is this force toward the origin or away from it? (c) Does the potential energy increase or decrease as $x$ increases? (d) Answer parts (b) and (c) if $C$ is a negative constant.

41. In the potential-energy curve $U(y)$ shown in Figure 6-21 the segments $AB$ and $CD$ are straight lines. Sketch the force $F_y$ versus $y$.

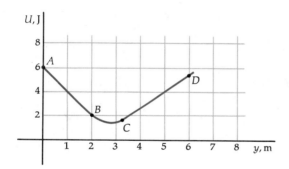

Figure 6-21
Exercise 41.

### Section 6-10, Power

42. A 4-kg mass is lifted by a force $\mathbf{F}_0$ equal to the weight of the mass. The mass moves upward at a constant velocity of 2 m/s. (a) What is the power input of the force $\mathbf{F}_0$? (b) What is the power input of the force exerted by the earth? (c) How much work is done by the force $\mathbf{F}_0$ in 2 s?

43. A constant horizontal force $F_0 = 3$ N drags a box along a rough horizontal surface at a constant velocity $v_0$. The force does work at the rate of 5 W. (a) What is the velocity $v_0$? (b) How much work is done by $F_0$ in 3 s?

44. A single force of 5 N acts in the $x$ direction on a 10-kg body. (a) If the body starts at rest at position $x = 0$ at time $t = 0$, find the velocity $v$ and position $x$ as a function of time $t$. (b) What is the power input of the force at time $t = 2$ s? (c) Write an expression for the power input of the force as a function of time $t$.

45. A constant force of 4 N acts at an angle of 30° above the horizontal on a box of mass 2 kg resting on a rough horizontal surface. The box is dragged with a constant speed of 50 cm/s. (a) Find the normal force exerted by the table on the box and the coefficient of friction. (b) What is the power input of the applied force? (c) How much work is done by the frictional force in 3 s?

46. An engine is rated at 400 horsepower. How much work is done by it operating at full power for 1 min? Express your answer in foot-pounds and in kilowatt-hours.

47. Find the power input of a force $\mathbf{F}$ acting on a particle which moves with velocity $\mathbf{v}$ for (a) $\mathbf{F} = 3\mathbf{i} + 4\mathbf{k}$ N, $\mathbf{v} = 2\mathbf{i}$ m/s; (b) $\mathbf{F} = 5\mathbf{i} - 6\mathbf{j}$ N, $\mathbf{v} = -5\mathbf{i} + 4\mathbf{j}$ m/s; (c) $\mathbf{F} = 2\mathbf{i} - 4\mathbf{j}$ N, $\mathbf{v} = 6\mathbf{i} + 3\mathbf{j}$ m/s.

## Problems

1. Figure 6-22 gives the force $F_x$ on a particle as a function of its distance $x$ from the origin. (a) Calculate from the graph the work done by the force when the particle moves from $x = 0$ to the following values of $x$: $-4$, $-3$, $-2$, $-1$, 0, 1, 2, 3, and 4 m. (b) How do you know that this force is conservative? (c) Plot the potential energy versus $x$ for the range of $x$ from $-4$ to $+4$ m, assuming that $U = 0$ at $x = 0$.

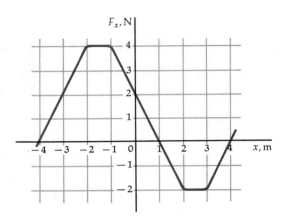

**Figure 6-22**
Problem 1.

2. Repeat Problem 1 for the force $F_x$ shown in Figure 6-23.

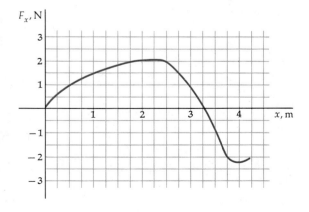

**Figure 6-23**
Problem 2.

3. A 3-kg body is moving with speed 1.00 m/s in the $x$ direction. When it passes the origin, it is acted on by a single force $F_x$, which varies with $x$ as shown in Figure 6-24. (a) Find the work done by the force from $x = 0$ to $x = 2$ m. (b) What is the kinetic energy of the body at $x = 2$ m? (c) What is the speed of the body at $x = 2$ m? (d) Find the work done on the body from $x = 0$ to $x = 4$ m. (e) What is the velocity of the body at $x = 4$ m?

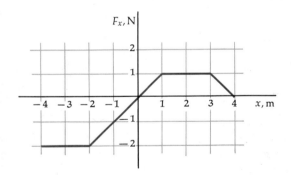

**Figure 6-24**
Problem 3.

4.  A 5-kg block is held against a spring of force constant 20 N/cm, compressing it 3 cm. The block is released and the spring expands, pushing the block along a rough horizontal surface. The coefficient of friction between the surface and the block is 0.2. (*a*) Find the work done on the block by the spring as it extends from its compressed position to its equilibrium position. (*b*) Find the work done on the block by friction while it moves the 3 cm to the equilibrium position of the spring. (*c*) What is the speed of the block when the spring is at its equilibrium position? (*d*) If the block is not attached to the spring, how far will it slide along the rough surface before coming to rest? (*e*) Suppose the block is attached to the spring so that the spring is extended when the block slides past the equilibrium point. By how much will the spring be extended before the block comes momentarily to rest? Describe the subsequent motion.

5.  A cyclist is climbing a hill inclined at 6° to the horizontal at speed 2.8 m/s. If the frictional resistance force to the motion is 0.8 percent of the weight and the mass of the cyclist plus bike is 85 kg, find the power delivered by the cyclist.

6.  A T-bar tow is required to pull a maximum of 80 skiers up a 600-m slope inclined at 15° to the horizontal at a speed of 2.5 m/s. The coefficient of friction is 0.06. Find the motor power required if the mass of the average skier is 75 kg.

7.  A car of mass 1500 kg traveling at 24 m/s is at the foot of a hill 2.0 km long and rising 120 m. At the top of the hill the speed of the car is 10 m/s. If the frictional effects are negligible, (*a*) find the work done on the car and (*b*) the average power delivered by the car motor.

8.  The force on a particle moving along the x axis is given by $F_x = -ax^2$, where *a* is a constant. Calculate the potential-energy function $U(x)$ relative to $U = 0$ at $x = 0$ and sketch a graph of $U(x)$ versus $x$.

9.  A ski jumper starts from rest at *A* (Figure 6-25), and his speed is 30.0 m/s at *B* and 23.0 m/s at *C*, where the distance *BC* is 30 m. (*a*) Find the greatest height attained by the jumper above the level of *C*. (*b*) Find the coefficient of friction between skis and snow.

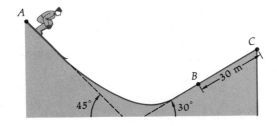

**Figure 6-25**
Problem 9.

10.  A train with a total mass of 300 Mg rises 707 m over a travel distance of 62 km at an average speed of 15.0 km/h. If the frictional force is 0.8 percent of the weight, (*a*) find the kinetic energy of the train, (*b*) the total change in potential energy, (*c*) the work done against friction, and (*d*) the power input of the engines of the train.

11.  A particle of mass *m* moves in a horizontal circle of radius *r* on a rough table. It is attached to a string fixed at the center of the circle. The speed of the particle is initially $v_0$. After completing one full trip around the circle, the speed of the particle is $\frac{1}{2}v_0$. (*a*) Find the work done by friction during that 1 rev in terms of *m*, $v_0$, and *r*. (*b*) What is the coefficient of sliding friction? (*c*) How many more revolutions will the particle make before coming to rest?

12.  A 2-kg box is projected with initial speed of 3 m/s up a rough plane inclined at 60° to the horizontal. The coefficient of friction is 0.3. (*a*) List all the forces acting on the box and find the work done by each as the box slides up the plane. (*b*) How far does the box slide along the plane before it stops momentarily? (*c*) Find the work done by each force acting on the box as it slides back down the plane. (*d*) Find its speed when it reaches its initial position.

13. When you walk along a horizontal surface, you lift part of your weight a small amount in each step and therefore do work against the force of gravity. (*a*) Estimate the work done in each step and the total work you do in lifting yourself when you walk 1 km along a horizontal surface. (*b*) If you walk 1 km in 20 min, what is your power output?

14. A force is given by $F_x = +A/x^3$, where $A = 8$ N·m³. (*a*) For positive $x$, does the potential energy associated with this force increase or decrease with increasing $x$? Convince yourself of the correct answer to this question by imagining a particle placed at rest at some point $x$ and released. (*b*) Find the potential-energy function $U(x)$ associated with this force such that $U$ approaches zero as $x$ approaches infinity. Sketch $U(x)$ versus $x$.

15. A particle is acted on by a force $\mathbf{F} = 4y\mathbf{i} + 5\mathbf{j}$, where $F$ is in newtons and $y$ is in metres. Show that this force is not conservative by computing the work done by the force in moving from the origin to the point ($x = 2$ m, $y = 2$ m) along two paths. (*a*) The first path is along the $x$ axis to point (2 m, 0) and then parallel to the $y$ axis along the line $x = 2$ m to the final point (2 m, 2 m). Find the component of the force parallel to the displacement for each of these parts of the path, the work done along each part of this path, and the total work. (*b*) The second path is first along the $y$ axis to the point (0, 2 m) and then parallel to the $x$ axis. Find the work done along each part of this path and the total work.

16. Figure 6-26 shows the potential-energy function $U(r)$ versus $r$ for a diatomic molecule, where $r$ is the separation of the atoms. (*a*) Indicate the equilibrium point on the graph and discuss its stability. (*b*) From this graph sketch a graph of the force $F_r$ exerted by one atom on the other versus separation distance $r$.

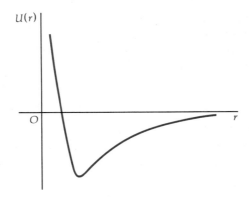

**Figure 6-26**
Problem 16.

17. A force acting in the $xy$ plane is given by $\mathbf{F} = 10\mathbf{i} + 3x\mathbf{j}$, where $F$ is in newtons and $x$ is in metres. Suppose the force acts on a particle as it moves from initial position $x = 4$ m, $y = 1$ m to a final position $x = 4$ m, $y = 4$ m. (*a*) Show that this force is not conservative by computing the work done by the force for at least two different paths. (*b*) Find a path for which the work done by the force is zero.

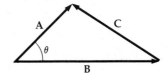

18. Vectors **A**, **B**, and **C** form a triangle, as shown in Figure 6-27. The angle between **A** and **B** is $\theta$; the vectors are related by $\mathbf{C} = \mathbf{A} - \mathbf{B}$. Compute $\mathbf{C} \cdot \mathbf{C}$ in terms of $A$, $B$, and $\theta$ and derive the law of cosines, $C^2 = A^2 + B^2 - 2AB \cos \theta$.

**Figure 6-27**
Problem 18.

19. A straight rod of negligible mass is mounted on a frictionless pivot, as shown in Figure 6-28. Masses $m_1$ and $m_2$ are suspended at distances $l_1$ and $l_2$ as shown. (*a*) Write the gravitational potential energy of the masses as a function of the angle $\theta$ made by the rod and the horizontal. (*b*) For what angle $\theta$ is the potential energy a minimum? Is the statement "systems tend to move toward minimum potential energy" consistent with your result? (*c*) Show that if $m_1 l_1 = m_2 l_2$, the potential energy is the same for all values of $\theta$. (When this holds, the system will balance at any angle $\theta$. This result is known as *Archimedes' law of the lever*.)

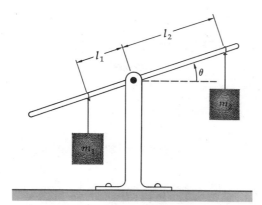

**Figure 6-28**
Problem 19.

20. The force between two atoms in a diatomic molecule can be represented approximately by the potential-energy function

$$U(x) = U_0\left[\left(\frac{x_0}{x}\right)^{12} - 2\left(\frac{x_0}{x}\right)^6\right]$$

where $x_0$ and $U_0$ are constants. (a) Find the force $F_x$. (b) At what value of $x$ is the potential energy zero? (c) At what value of $x$ is the potential energy a minimum? What is the value of this minimum potential energy? (This minimum potential energy is approximately equal to the dissociation energy of the molecule.) (d) Sketch $U(x)$ versus $x$.

21. A theoretical form for the potential energy associated with the nuclear force between two protons, two neutrons, or a neutron and proton is the *Yukawa potential*:

$$U = -U_0\left(\frac{a}{x}\right)e^{-x/a}$$

where $U_0$ and $a$ are constants. (a) Sketch $U$ versus $x$ using the parameters $U_0 = 4$ pJ and $a = 2.5$ fm. (b) Find the force $F_x$. (c) Compare the magnitudes of the forces at the separations $x = a$, $x = 2a$, and $x = 5a$.

**CHAPTER 7**     Conservation of Energy

---

**Objectives**   After studying this chapter, you should:

1.  Be able to derive the modified work-energy theorem (Equation 7-3).

2.  Be able to work mechanics problems using the work-energy theorem and the law of conservation of mechanical energy.

3.  Know the expression $U = -GM_E m/r$ for the gravitational potential energy at distance $r$ from the center of the earth.

4.  Be able to derive the expression $v_E = \sqrt{2gR_E}$ for the escape speed from a point on the surface of the earth.

5.  Know that for a circular orbit in a force field varying as $1/r^2$ the kinetic energy equals one-half the magnitude of the potential energy and that the total energy is negative and equal to one-half the potential energy.

6.  Be able to discuss qualitatively whether a system is bound or unbound and describe its general motion for various values of the total energy from a graph of the potential energy.

---

The potential energy of a system of particles is defined so that the work done on the system by conservative forces equals the decrease in the potential energy. When only conservative forces do work, the total mechanical energy of the system remains constant because the increase in kinetic energy equals the corresponding decrease in potential energy of the system. The conservation of energy is one of the most important principles in physics.

In the macroscopic world, nonconservative forces are always present, the most common being frictional forces. Another type of nonconservative force is involved in large deformations of bodies. For example, if a spring is stretched beyond its elastic limit, it becomes permanently deformed and the work done in the stretching is not recovered when the spring is released. When nonconservative forces do work, the sum of kinetic and potential energies of the system is not constant. Because some kind of frictional force is nearly always present in the motion of a macroscopic body, the importance of energy and its conservation was not realized until the nineteenth century. Then it was discovered that the disappearance of *macroscopic mechanical energy* is accompanied by the appearance of *internal energy*, usually indicated by an increase in temperature. We now know that, on the microscopic scale, this internal energy of the surroundings consists of kinetic and potential energy of molecular motion. When the concept of energy is generalized to include this internal energy, the total energy of an object plus its surroundings does not change even when friction is present.

If we carefully define a system, e.g., several bodies and their local surroundings (as we did in solving force and acceleration problems), we find that even when internal energy is included, the total energy of the system does not always remain constant. However, we are always able to account for the increase or decrease in the system energy by the appearance or disappearance of energy somewhere else. For example, energy of a system is often decreased because of some form of radiation, e.g., sound waves from a collision of two objects, water waves produced by a ship, or electromagnetic waves produced by accelerated charges in a radio antenna. We can write a generalized statement of the conservation of energy in the following way. Let $E_{sys}$ be the total energy of a given system and $P_{in}$ be the power input, i.e., the rate at which energy is put into the system. Then

$$P_{in} = \frac{dE_{sys}}{dt} \qquad\qquad 7\text{-}1$$

where $P_{in}$ is negative if energy is flowing out of the system. The significance of this statement is that if the energy concept is sufficiently generalized, changes in the total energy of the system are always precisely accounted for by energy flow into or out of the system. Energy is never created or destroyed in the system, though it may change from one form to another.

In this chapter we concentrate on systems in which a single particle moves in an external force field produced by other fixed bodies such as the earth, so that we can describe the potential energy as being associated with the particle. We shall consider only *mechanical energy*, the kinetic energy $\frac{1}{2}mv^2$, and the various forms of potential energies associated with conservative forces acting on the particle. Consider a particle acted on by a nonconservative force $\mathbf{F}_{nc}$ and several conservative forces $\mathbf{F}_i$, so that the resultant force is

$$\Sigma\mathbf{F} = \mathbf{F}_{nc} + \mathbf{F}_1 + \mathbf{F}_2 + \cdots$$

The work done by the resultant force, according to the general work-energy theorem, equals the increase in kinetic energy of the particle:

$$
\begin{aligned}
W_{net} &= \int \Sigma\mathbf{F} \cdot d\mathbf{s} \\
&= \int (\mathbf{F}_{nc} + \mathbf{F}_1 + \mathbf{F}_2 + \cdots) \cdot d\mathbf{s} \\
&= W_{nc} + W_1 + W_2 + \cdots = \Delta E_k \qquad 7\text{-}2
\end{aligned}
$$

where $W_{nc}$ is the work done by the nonconservative force, $W_1$ is that done by force $F_1$, etc. For each conservative force $F_i$, we define a potential-energy function $U_i$ in the usual way:

$$W_i = -\Delta U_i$$

Then Equation 7-2 can be written

$$W_{nc} - \Delta U_1 - \Delta U_2 + \cdots = \Delta E_k$$

or

$$W_{nc} = \Delta(E_k + U_1 + U_2 + \cdots) = \Delta E_{\text{total}} \qquad 7\text{-}3$$

*Generalized work-energy theorem*

where the total mechanical energy of the particle is

$$E_{\text{total}} = E_k + U_1 + U_2 \cdots \qquad 7\text{-}4$$

*The work done by nonconservative forces equals the change in the total mechanical energy of the particle.*

This modified form of the work-energy theorem is the most convenient in many applications, since explicit calculation of work done along the path is required only for the nonconservative forces. When the only forces that do work are conservative, the left side of Equation 7-3 is zero. Then there is no change in the total mechanical energy of the particle:

$$E_{\text{total}} = E_k + U_1 + U_2 + \cdots = \text{constant} \qquad 7\text{-}5$$

In this chapter we consider various examples of the use of the work-energy theorem and the conservation of energy expressed in Equations 7-3 and 7-5. Since we derived these equations from Newton's laws, we can never obtain more information about a mechanics problem than is contained in Newton's laws. Any problem that can be

The action of a pole vaulter demonstrates several kinds of energy. First he converts internal chemical energy into kinetic energy when he begins to run. Some of this energy is then converted into elastic potential energy of deformation of the pole and eventually into gravitational potential energy, which in turn changes into kinetic energy as he drops. The mechanical energy is finally lost as heat when he reaches the ground. (*Courtesy of Harold E. Edgerton.*)

solved by using conservation of energy can also be solved directly from Newton's laws. In spite of this, energy conservation and the work-energy theorem are extremely useful tools for the analysis of problems. Since the kinetic energy is $\frac{1}{2}mv^2$ and the potential-energy functions are functions of position, Equation 7-5 relates the speed of a particle to its position. In any problem in which one of these quantities is to be found, the energy-conservation equation is generally simpler to use than a direct application of $\Sigma \mathbf{F} = m\mathbf{a}$. We shall illustrate this in Section 7-1. In many cases, examining the potential-energy function offers a qualitative understanding of the possible motion of a system without a detailed solution of the equations of motion.

## 7-1   Some Illustrative Examples

**Example 7-1** A block of mass 3 kg is on a rough plane inclined at an angle $\theta = 20°$. The coefficient of sliding friction between the block and plane is $\mu_k = 0.4$. An external force $T = 40$ N is applied parallel to the plane. How fast is the block moving after it travels 5 m up the plane?

This problem is readily solved by direct application of Newton's second law of motion to the block and then finding $v$ from the constant-acceleration formulas. We use the work-energy theorem instead to illustrate this method of analyzing a problem.

There are four forces which act on the block as it moves up the plane (see Figure 7-1). The normal force $\mathbf{N}$ does no work because it is always perpendicular to the motion. It balances the component of the force of gravity $mg \cos \theta$ perpendicular to the plane. The force of gravity $m\mathbf{g}$ is a conservative force with the associated potential-energy function $mgh = mgs \sin \theta$, where $s$ is the distance traveled along the incline, and we choose the potential energy to be zero when $s = 0$. The applied force $\mathbf{T}$ is a nonconservative force (it depends on the action of whatever agency exerts it and not merely on the position of the block) and does work $W_T = Ts$. The force of friction $f = \mu_k mg \cos \theta$ is also a nonconservative force. Since $f$ is opposed to the motion, the work done by it is $W_f = -fs = -(\mu_k mg \cos \theta)s$.

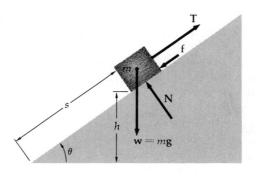

**Figure 7-1**
Block pulled up a rough inclined plane for Example 7-1.

The mechanical energy of the block consists of potential energy $mgs \times \sin \theta$ and kinetic energy $\frac{1}{2}mv^2$. Both are zero at the bottom of the incline ($s = 0$). Setting the work done by the nonconservative forces equal to the change in mechanical energy, we obtain

$$Ts - \mu_k mgs \cos \theta = \tfrac{1}{2}mv^2 + mgs \sin \theta$$

Solving for $v$ gives

$$v = \left[2s\left(\frac{T}{m} - \mu_k g \cos\theta - g \sin\theta\right)\right]^{1/2} \qquad\qquad 7\text{-}6$$

Plugging in the given data leads to

$$v = \left\{2(5 \text{ m})\left[\frac{40 \text{ N}}{3 \text{ kg}} - (0.4)(9.81 \text{ N/kg}) \cos 20° \right.\right.$$
$$\left.\left. - (9.81 \text{ N/kg}) \sin 20°\right]\right\}^{1/2}$$
$$= 7.93 \text{ m/s}$$

Note that the quantity in the brackets in Equation 7-6 must be positive for its square root to have a meaningful value for the speed $v$. This requires that $T$ be greater than $\mu_k mg \cos\theta + mg \sin\theta = f + mg \sin\theta$. This, of course, means only that the applied force $T$ must be greater than the opposing forces $f$ and $mg \sin\theta$ down the incline.

Equation 7-6 can also be obtained from the constant-acceleration formula $v^2 = 2as$, where the acceleration is the resultant force divided by the mass.

---

**Example 7-2** For the block in Example 7-1, what is the power input of the applied force if the block is to move up the incline with a constant speed of 2 m/s?

Since the speed is constant, the kinetic energy does not change. The work-energy theorem then gives

$$Ts - fs = mgh = mgs \sin\theta \qquad \text{or} \qquad Ts = fs + mgs \sin\theta$$

where we have written $f$ for the frictional force for simplicity. We obtain the rate of doing work by differentiating with respect to time. Using $ds/dt = v$ for the speed, we obtain

$$Tv = fv + mgv \sin\theta$$

The left side of this equation is the power input of the applied force. Putting in the numbers from Example 7-1 gives $Tv = 42.3$ W.

---

**Example 7-3** A pendulum consists of a bob of mass $m$ attached to a string of length $L$. It is pulled aside so that the string makes an angle $\theta_0$ with the vertical and is released from rest. What is the speed $v$ at the bottom of the swing and the tension in the string there?

The two forces acting on the bob (neglecting air resistance) are the force of gravity $m\mathbf{g}$, which is conservative, and the tension $\mathbf{T}$, which is perpendicular to the motion and does no work. The mechanical energy of the bob is therefore conserved in this problem.

Let us choose the gravitational potential energy to be zero at the bottom of the swing. At its original position, a height $h$ above the bottom, the mass has potential energy $mgh$ and no kinetic energy because it is at rest. As it swings down, the potential energy is converted into kinetic energy. Conservation of energy gives us for the speed $v$ at the bottom

$$\tfrac{1}{2}mv^2 = mgh$$

According to Figure 7-2, the distance $h$ is related to the original angle $\theta_0$ and the length of the pendulum $L$ by

$$h = L - L \cos\theta_0 = L(1 - \cos\theta_0)$$

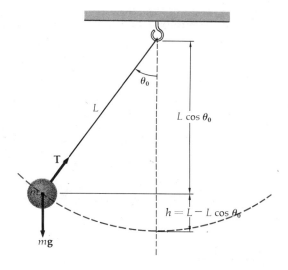

**Figure 7-2**
Simple pendulum for Example 7-3. The tension is perpendicular to the motion and does no work. The speed at the bottom is found from energy conservation, $\frac{1}{2}mv^2 = mgh$, where the initial height $h$ above the bottom is related to the initial angle by $h = L - L \cos \theta_0$.

Thus

$$v^2 = 2gL(1 - \cos \theta_0) \qquad\qquad 7\text{-}7$$

The tension at the bottom is found as in Example 5-7. At this point the mass has centripetal acceleration $v^2/r = v^2/L$. The forces acting are $T$ up and $mg$ down. Setting the resultant force equal to the mass times the acceleration gives

$$T - mg = \frac{mv^2}{r} = \frac{mv^2}{L} = 2mg(1 - \cos \theta_0)$$
$$T = mg + 2mg(1 - \cos \theta_0) \qquad\qquad 7\text{-}8$$

We note that if the bob is released from $\theta_0 = 90°$, the tension at the bottom is 3 times the force of gravity $mg$.

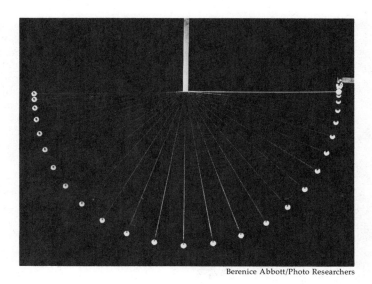

Berenice Abbott/Photo Researchers

Multiflash photograph of a simple pendulum. As the bob descends, gravitational potential energy is converted into kinetic energy and the speed increases, as indicated by the increased spacing of the recorded positions. The speed decreases as the bob moves up, and the kinetic energy is changed into potential energy.

This problem can also be solved using Newton's laws (see Problem 24), but the solution is difficult because the acceleration tangential to the circle is not constant. Instead, it varies with $\theta$ and therefore with time, so that the constant-acceleration formulas do not apply.

**Example 7-4** A skier skis down a frictionless hill of height $h$ and of arbitrary shape (see Figure 7-3). Find her speed at the bottom of the hill.

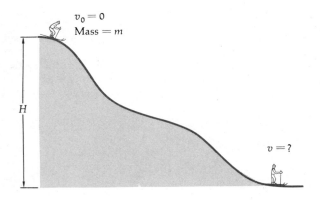

$v_0 = 0$
Mass $= m$

$H$

$v = ?$

**Figure 7-3**
Skier on a frictionless hill. The speed at the bottom can be found from energy conservation without any knowledge of the shape of the hill.

We have already considered this type of problem in Chapter 6. We repeat it here to contrast it with the pendulum problem. The only force that does work is the force of gravity, which is conservative. The normal force exerted by the snow on the skis acts to constrain the skier to follow a certain path, but it does no work. Taking the potential energy to be zero at the bottom of the hill and assuming the initial speed at the top of the hill to be zero, we obtain from conservation of energy

$$(E_k + U)_{\text{top}} = (E_k + U)_{\text{bottom}}$$

$$0 + mgh = \tfrac{1}{2}mv^2 + 0$$

or

$$v^2 = 2gh$$

If there were no friction, the kinetic energy of the skiers would equal the decrease in gravitational potential energy, and their speed after descending through a given vertical distance would be the same as if they had fallen through that distance. In practice, frictional forces decrease the mechanical energy.

This is the same as the result for the pendulum bob of Example 7-3, but in this case, it is *not* possible to obtain this result from Newton's second law because not knowing the curve along which the skier travels, we cannot compute the acceleration at each point. This example demonstrates the great power of the principle of conservation of energy. Even though we do not know the acceleration at any point, we can find the final speed quite easily.

---

**Example 7-5** A 2-kg block is pushed against a spring of force constant 500 N/m, compressing it 20 cm. It is then released, and the spring projects the block along a frictionless horizontal surface and then up a frictionless incline of angle 45°, as shown in Figure 7-4. (*a*) What is the speed of the block when it leaves the spring? (*b*) How far up the incline does the block travel?

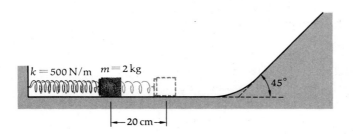

**Figure 7-4**
Example 7-5.

(*a*) After the spring is released, the only forces that do work are the force exerted by the spring and the force of gravity. Since both these forces are conservative forces, the total mechanical energy of the block-spring-earth system is conserved. In this case the total mechanical energy consists of kinetic energy of the block $\frac{1}{2}mv^2$ and two kinds of potential energy, that of the spring $\frac{1}{2}kx^2$, where $x$ is the amount the spring is compressed, and gravitational potential energy $mgh$.

When the block leaves the spring, its kinetic energy equals the original potential energy of the spring:

$$E_k = \tfrac{1}{2}mv^2 = \tfrac{1}{2}kx^2$$

$$= \tfrac{1}{2}(500 \text{ N/m}) (20 \text{ cm})^2 \left(\frac{1 \text{ m}}{100 \text{ cm}}\right)^2 = 10 \text{ J}$$

The speed is then

$$v = \sqrt{\frac{2E_k}{m}} = \sqrt{\frac{2(10)}{2}} = 3.16 \text{ m/s}$$

(*b*) The block slides up the incline until its original kinetic energy is completely converted into gravitational potential energy. The height it reaches is given by

$$mgh = \tfrac{1}{2}mv^2 = 10 \text{ J}$$

or

$$h = \frac{10 \text{ J}}{(2 \text{ kg}) (9.81 \text{ m/s}^2)} = 0.51 \text{ m}$$

The distance up the incline is found from $\sin 45° = h/s = 0.707$, giving $s = 0.721$ m.

**Example 7-6** Two bodies of mass $m_1$ and $m_2$ are attached to a light string which passes over a light, frictionless pulley, as shown in Figure 7-5. Find the speed of either body when the heavier one falls a distance $h$, and the acceleration of either body.

This device, called *Atwood's machine*, was developed in the eighteenth century to measure the acceleration of gravity $g$. As we shall see, if the values of $m_1$ and $m_2$ are not too different, the acceleration of either body is a small fraction of $g$. It could be easily measured with the rather crude timing devices available in the eighteenth century, whereas a direct measurement of $g$ was difficult if not impossible.

If the friction in the pulley bearings is negligible and the pulley mass is so small that the work done to speed it up is negligible, the tension in the string is uniform. Then the net work done by the tension is zero because when the lighter body moves upward in the direction of **T**, the heavier body moves downward the same distance in the opposite direction. The only force that does net work on the two-body system is the force of gravity. Therefore, the total mechanical energy of the system is conserved.

Let us assume that $m_2$ is greater than $m_1$ and choose the potential energy of the system to be zero at the initial position of the masses. Since they are at rest at this position, the total energy is zero. Let $v$ be the speed of $m_1$ after it has moved up a distance $y$. Then its kinetic energy is $\frac{1}{2}m_1 v^2$, and its potential energy is $m_1 g y$. Since the connecting string does not stretch, $m_2$ must move down the same distance $y$ and acquire the same speed $v$. Its kinetic energy is $\frac{1}{2}m_2 v^2$, and its potential energy is $-m_2 g y$. Conservation of energy gives

$$E = \tfrac{1}{2}m_1 v^2 + m_1 g y + \tfrac{1}{2}m_2 v^2 - m_2 g y = 0$$

or

$$\tfrac{1}{2}(m_1 + m_2)v^2 = (m_2 - m_1)g y \qquad\qquad 7\text{-}9$$

Solving this equation for $v^2$ gives

$$v^2 = \frac{2(m_2 - m_1)}{m_1 + m_2} g y \qquad\qquad 7\text{-}10$$

which relates the speed of either mass to the distance it moves. Comparing this with the constant-acceleration equation $v^2 = 2ay$, we see that the acceleration is

$$a = \frac{m_2 - m_1}{m_1 + m_2} g \qquad\qquad 7\text{-}11$$

This problem can also be solved by applying Newton's second law to each of the bodies (see Exercise 38, Chapter 4). If the masses have the values $m_1 = 0.49$ kg and $m_2 = 0.51$ kg, the acceleration is $a = (0.02 \text{ kg}/1.00 \text{ kg})g = 0.02g = 0.196$ m/s², which is considerably easier to measure than $g = 9.81$ m/s². In the laboratory experiment, $a$ is measured, and the unknown value of $g$ is obtained from $g = a(m_1 + m_2)/(m_2 - m_1)$.

In Example 7-6 we found the acceleration from the expression for $v^2$ using the constant-acceleration equation. In general, when a single conservative force acts, we can find the acceleration from the conservation-of-energy equation by differentiation. Consider such a one-dimensional problem in which a particle has initial energy $E_0$. At

*Atwood's machine*

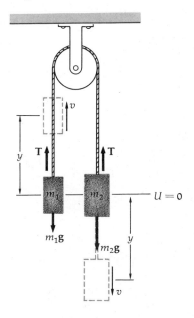

**Figure 7-5**
Atwood's machine for Example 7-6.

some time later it has kinetic energy $\frac{1}{2}mv^2$ and potential energy $U(x)$. Conservation of energy gives us

$$\frac{1}{2}mv^2 + U(x) = E_0 \qquad\qquad 7\text{-}12$$

We can obtain the acceleration from Equation 7-12 by differentiating with respect to time and using the chain rule

$$\frac{1}{2}m(2v)\frac{dv}{dt} + \frac{dU}{dx}\frac{dx}{dt} = 0$$

Since $dx/dt$ is just the speed $v$, we can divide out these terms. Then

$$m\frac{dv}{dt} = -\frac{dU}{dx} \qquad\qquad 7\text{-}13$$

This is just Newton's second law, since the negative derivative of $U$ with respect to $x$ is the force $F_x$. We could have obtained Equation 7-11 by differentiating Equation 7-10 (using $dy/dt = v$) without using the fact that the acceleration is constant.

### Questions

1. Suggest two forces that may be important in skiing and that are neglected in Example 7-3. Are these forces conservative? Is the work they do positive or negative? How would their inclusion affect the result $v^2 = 2gh$?

2. If the mass of the pulley in Atwood's machine is not negligible, the pulley has kinetic energy of rotation (to be studied in Chapter 9). Do you expect the inclusion of this kinetic energy to increase or decrease the value of the speed $v$ for a given distance $y$?

3. Discuss the advantages and disadvantages of solving mechanics problems by energy methods compared with using Newton's laws.

## 7-2   Qualitative Description of Motion Using Energy Conservation

In many situations in physics a detailed solution of the equations of motion found from Newton's laws is very difficult, but a qualitative description of the motion is both useful and easily obtained from energy considerations. Often we can get considerable information from a simple sketch of the potential-energy function for a particle moving in some force field.

In this section we give a simple example, the motion of an object attached to a light spring of force constant $k$ and moving on a frictionless horizontal table. We displace the object a distance $A$ from equilibrium and release it (Figure 7-6a). We wish to describe the motion of the object qualitatively without solving Newton's laws. A more detailed description of the motion of an object on a spring will be given in Chapter 11.

The total mechanical energy is conserved since the only force that does work is the conservative force of the spring. If we choose $x = 0$ at the natural length of the spring, the potential energy of the spring is

$$U = \frac{1}{2}kx^2$$

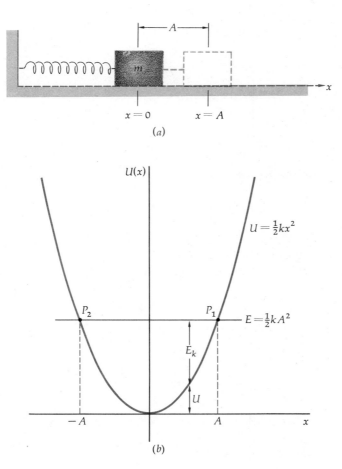

(a)

(b)

**Figure 7-6**
(*a*) Object of mass *m* attached to a horizontal spring and moving along a frictionless plane. The object is displaced a distance *A* and released. (*b*) Potential-energy function $U(x)$ for an object on a spring. The total energy $E = \frac{1}{2}kA^2$ is indicated by the horizontal line. The kinetic energy $E_k = E - U$ is represented by the vertical distance between the total-energy line and $U$. The turning points $P_1$ and $P_2$ are the intersections of $E$ and $U(x)$. At these points $E_k = 0$, and the object stops and reverses its motion.

Figure 7-6*b* shows a plot of $U(x)$ versus $x$. The minimum value of $U$ occurs at $x = 0$, the equilibrium point where the slope $dU/dx = -F_x$ is zero. Since the mass begins at $x = A$ with zero velocity, its total energy at this point is its potential energy

$$E = \tfrac{1}{2}kA^2$$

This total energy, which is constant, is represented in Figure 7-6*b* by a horizontal line. The speed $v$ of the particle is related to its position $x$ through the conservation-of-energy equation

$$\tfrac{1}{2}mv^2 + \tfrac{1}{2}kx^2 = E = \tfrac{1}{2}kA^2 \qquad\qquad 7\text{-}14$$

For any value of $x$, the kinetic energy $\frac{1}{2}mv^2 = E_k = E - U(x)$ is represented by the vertical distance between the total-energy line and the potential-energy curve $U$, as shown in the figure. Since the kinetic energy cannot be negative, the total-energy line must be above the potential-energy curve for physically possible motions. In Figure 7-6*b* the curve $U(x)$ and the total-energy line intersect at $x = A$ (point $P_1$) and $x = -A$ (point $P_2$). For this value of total energy, the motion is confined to the values of $x$ between these points.

The points $x = +A$ and $x = -A$ are called *turning points* for the motion of the mass. At these points the velocity of the mass decreases to zero and reverses. In this example, we start the mass from rest at $x = +A$. The slope of $U(x)$ at this point is positive. The force $F_x = -dU/dx$ is negative, and the particle accelerates to the left, decreasing its potential energy and increasing its kinetic energy. When it reaches the equilibrium point $x = 0$, its potential energy is zero and its kinetic energy equals its total energy. Therefore its speed is maximum at this point.

After it passes the equilibrium point, the slope of $U(x)$ is negative, indicating a positive force. This force decreases the speed since it is in the direction opposite the motion. The mass continues to the left, its potential energy increasing and its kinetic energy decreasing, until it reaches point $P_2$ at $x = -A$. Here it has no kinetic energy and is momentarily at rest. The force is still to the right, and so the particle turns and moves back toward the equilibrium point $x = 0$. Again, when it reaches this point, its kinetic energy is maximum and it continues moving to the right until it comes momentarily to rest at the starting point $P_1$. After a full cycle of motion is completed, the mass begins a new identical cycle.

The symmetry of the potential-energy curve in Figure 7-6b makes the turning points equidistant from the equilibrium point. This distance $A$, called the *amplitude of the motion*, depends only on the total energy. It is also clear from the symmetry of $U(x)$ that the motion from $x = 0$ to $x = A$ and back to $x = 0$ is exactly like the motion from $x = 0$ to $x = -A$ and back except for reversal in direction; i.e., these two parts of the cycle of motion must take the same time.

We have been able to describe most of the features of the motion of a mass on a spring by using the energy diagram of Figure 7-6b. We shall use this type of analysis often in the next section. In Chapter 11 we give a complete detailed description of the motion of an object on a spring, including the position $x$, velocity $v$, and acceleration $a$ as functions of time.

## 7-3  Gravitational Potential Energy, Escape Speed, and Binding Energy

In our discussion of the motion of an object on a spring, we found that the object is confined to a certain region of space bounded by the turning points at $x = \pm A$, where $A$, which is the amplitude, is related to the total energy by $A = \sqrt{2E/k}$. If we increase the total energy by initially pulling the object out farther (and thus doing a greater amount of work on it), the turning points change. However, for this system, with potential energy given by Figure 7-6b, no matter how great the energy, the object is confined because the total-energy line always intersects the potential energy at two values of $x$. Such a system is called a *bound system*. In this section, we calculate the gravitational potential-energy function for an object a distance $r$ from the center of the earth for the general case in which $r$ is not small compared with the radius of the earth, so the gravitational force cannot be assumed constant. We shall find that the potential energy increases with increasing distance (it takes work to lift the object up to a greater and greater height), but instead of increasing without limit, the potential-energy function never becomes greater than a certain value. If the total energy is greater than this value, the total-energy line does not intersect the potential-energy curve at large values of the separation $r$. Such a system is called an *unbound system*. We shall apply this to finding the escape speed at the surface of the earth and then consider other important systems showing similar behavior.

*Bound system*

The easiest way to calculate the gravitational potential energy of an object of mass $m$ a distance $r$ from the earth is to calculate the work we must do to lift the object from the surface of the earth to the distance $r$ by exerting a force $GM_Em/r^2$ to just balance the gravitational attraction of the earth. If we start from the distance $r = R_E$, the radius of the earth, the work we must do to lift the object to the distance $r$ with zero acceleration is

$$W = \int_{R_E}^{r} \frac{GM_E m}{r^2} \, dr = -\frac{GM_E m}{r}\bigg]_{R_E}^{r} = -\frac{GM_E m}{r} + \frac{GM_E m}{R_E}$$

This work equals the increase in the potential energy, $U(r) - U_0$. The potential energy is therefore

$$U(r) = U_0 - \frac{GM_E m}{r} + \frac{GM_E m}{R_E} \qquad 7\text{-}15$$

The constant $U_0$ is arbitrary and depends on the choice of zero potential energy. If we choose the potential energy to be zero when the mass is on the surface of the earth ($r = R_E$), the constant $U_0$ is zero. We then have

$$U(r) = \frac{GM_E m}{R_E} - \frac{GM_E m}{r} \qquad U = 0 \text{ at earth's surface} \qquad 7\text{-}16$$

*Gravitational potential energy*

We can compare this potential-energy function with that obtained previously for a mass near the surface of the earth. We have

$$U = \frac{GM_E m}{R_E} - \frac{GM_E m}{r} = \frac{GM_E m}{R_E r} (r - R_E) \qquad 7\text{-}17$$

But $r - R_E = y$, the height above the surface of the earth. Equation 7-17 can therefore be written

$$U = \frac{GM_E m}{R_E r} y = m\left(\frac{GM_E}{R_E^2}\right) y \, \frac{R_E}{r} \qquad 7\text{-}18$$

We recognize the term in parentheses as the acceleration of gravity at the surface of the earth, $g = GM_E/R_E^2$. The potential energy is thus $mgy$ times the quantity $R_E/r$. Near the surface of the earth, $r$ and $R_E$ are approximately equal and the result is approximately $mgy$, as obtained by assuming the gravitational force to be constant.

Figure 7-7 shows this function. The shaded region is inside the earth. At very large separations $r \rightarrow \infty$, the potential energy approaches $GM_E m/R_E$. This is the work we must do to lift the mass from the surface

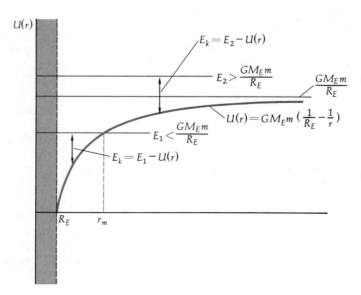

**Figure 7-7**
The potential-energy function $U(r) = GM_E m/R_E - GM_E m/r$ for a particle a distance $r$ from the center of the earth, where $U = 0$ is chosen at the earth's surface $r = R_E$. For $r$ just slightly greater than $R_E$, $U(r)$ is approximately proportional to $h = r - R_E$, but for large values of $r$, $U$ approaches its limiting value $GM_E m/R_E$. A particle with total energy $E_1$ less than this maximum value of $U$ is confined to the region between $R_E$ and $r_m$ indicated. If the total energy is greater than $GM_E m/R_E$, as in $E_2$, the total energy line does not intersect $U(r)$, indicating that the particle is not confined.

of the earth to infinity. If we project the mass from the surface of the earth with kinetic energy less than $GM_Em/R_E$, the mass will rise to some maximum separation $r_m$, stop, and then fall back to the earth. The maximum separation $r_m$ is that value of $r$ where the total-energy line intersects the potential-energy curve as shown in the figure. If we project the mass with kinetic energy greater than $GM_Em/R_E$, the mass will not return. The speed corresponding to a kinetic energy of $GM_Em/R_E$ is independent of the mass of the particle and is called the *escape speed*. The value is obtained from

$$\tfrac{1}{2}mv_E^2 = \frac{GM_Em}{R_E} \qquad\qquad 7\text{-}19a$$

$$v_E = \sqrt{\frac{2GM_E}{R_E}} \qquad\qquad 7\text{-}19b$$

Instead of using the values of $G$ and the mass of the earth $M_E$, we can write these quantities in terms of the free-fall acceleration of gravity $g = GM_E/R_E^2$ (Equation 5-3). Then $GM_E = gR_E^2$ and Equation 7-19b becomes

$$v_E = \sqrt{2gR_E} \qquad\qquad 7\text{-}20$$

Using $g = 9.81$ m/s$^2$ and $R_E = 6.37$ Mm, we obtain

$$v_E = \sqrt{2(9.81 \text{ m/s}^2)\,(6.37 \times 10^6 \text{ m})} = 11.2 \text{ km/s}$$

This is about $6.95$ mi/s $\approx 25{,}000$ mi/h $\approx 40{,}200$ km/h.

The energy $GM_Em/R_E$ may be called the *binding energy*. If the kinetic energy is less than the binding energy at the earth's surface, the mass will not leave the earth but will rise to some maximum separation $r_m$ and then fall back to earth. If the kinetic energy is greater than the binding energy, the object will continue forever without returning. The escape speed is just that speed corresponding to a kinetic energy equal to the binding energy. The earth-mass system is said to be bound or unbound according to whether the kinetic energy at the earth's surface is less or greater than the binding energy.

For small heights of a mass above the surface of the earth, it is convenient to choose the zero of potential energy at the earth's surface. Then the potential-energy function given by Equation 7-16 is approximately the same as $mgh$, as we have shown. However, for general situations in which the separation may be large compared with the earth's radius, the potential-energy function can be simplified by choosing the zero to be at an infinite separation. This is the customary choice. From Equation 7-15 we see that if we set $r = \infty$ and $U(\infty) = 0$, the arbitrary constant $U_0$ must have the value $U_0 = -GM_Em/R_E$. With this choice of zero potential energy, the potential-energy function is

$$U(r) = -\frac{GM_Em}{r} \qquad U = 0 \text{ at } r = \infty \qquad\qquad 7\text{-}21$$

*Gravitational potential energy*

and is always less than zero because $r$ is intrinsically positive. It seems peculiar to have a potential-energy function that is always negative, but there are certain advantages, as we shall see. Since only changes in potential energy are important, it does not matter what the actual value of the potential-energy function is. Figure 7-8 is a plot of $U(r)$ versus $r$ for the choice $U = 0$ at $r = \infty$. This function begins at the negative value $U = -GM_Em/R_E$ at the earth's surface and increases as $r$ increases, approaching zero at infinite $r$. Two possible values for the total energy $E$ are indicated on this graph: $E_1$, which is negative, and $E_2$, which is posi-

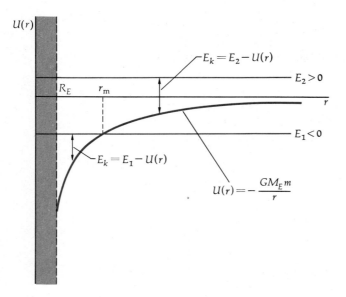

**Figure 7-8**
The same potential-energy
curve as in Figure 7-7 except
the zero of $U(r)$ is chosen at
$r = \infty$. For this choice, a particle is confined if its total energy is less than zero and not
confined if $E$ is greater than
zero.

tive. A negative total energy merely means that the kinetic energy at the
earth's surface is less than $GM_Em/R_E$, so that it is never greater in magnitude than the negative potential energy. From the figure we see that if
the total energy is negative, the total-energy line intersects the
potential-energy curve at some maximum separation $r_m$ and the system
is bound. On the other hand, if the total energy is positive, there is no
such intersection and the system is unbound. The criterion for escape is
then that the total energy be greater than or equal to zero. With this
choice of zero potential energy, conditions for a bound or unbound
system are simply stated:

If $E < 0$, the system is bound

$$7\text{-}22$$

If $E > 0$, the system is unbound

The binding energy is, of course, the same regardless of the choice of
zero for the potential energy.

---

**Example 7-7** *Energy of an Orbiting Satellite*    Consider a particle of mass
$m$ orbiting the earth in a circular orbit above the surface of the earth. Its
total energy is greater than $-GM_Em/r$ because it already has some
kinetic energy. From Newton's law $\Sigma\mathbf{F} = m\mathbf{a}$ for circular motion we
have

$$\frac{GM_Em}{r^2} = ma = \frac{mv^2}{r} \qquad 7\text{-}23$$

Thus the kinetic energy for a circular orbit is

$$\tfrac{1}{2}mv^2 = \frac{\tfrac{1}{2}GM_Em}{r} = \tfrac{1}{2}|U(r)| \qquad 7\text{-}24$$

---

This result (that the kinetic energy is half the magnitude of the potential
energy) holds for any circular orbit in an inverse-square force field. On
the energy diagram of Figure 7-8, the total energy of an orbiting satellite
is thus halfway between $-GM_Em/r$ and zero. The minimum additional
energy an orbiting satellite must be given to escape from the earth is
$\tfrac{1}{2}GM_Em/r$. The kinetic energy of the orbiting satellite must be doubled.

---

**Example 7-8** *Flight of Apollo 11*    As a practical example of escape from the earth, consider the flight of Apollo 11 in July 1969, during which man first set foot on the moon. The spacecraft was launched from the earth into a nearly circular orbit around the earth at an altitude of 191 km and speed of 28,000 km/h. From this orbit the engine of the craft was fired to give it a speed of 39,000 km/h. This speed, which nearly doubled the kinetic energy of the craft, is less than the escape velocity we calculated because the moon exerts a gravitational attraction, which we neglected. The spacecraft then traveled toward the moon, with its speed decreasing, to a point 38,000 km from the moon, where the attraction of the moon equals the attraction of the earth. Beyond this point the craft accelerated toward the moon. As the craft passed the moon, its kinetic energy was greater than that needed to escape from the moon. On the far side of the moon, retarding rockets were fired to slow it down and put it into an orbit around the moon. (This was an elliptical orbit of maximum altitude of 314 km and minimum altitude of 113 km.) From this orbit the lunar module Eagle, containing astronauts Armstrong and Aldrin, separated from the main craft and dropped into an orbit of about 15 km altitude. From this orbit the module was flown to a landing on the surface of the moon.

July 16, 1969 blast-off of Saturn V rocket carrying Apollo 11 astronauts on man's first voyage to the moon.

NASA

**Example 7-9** *Ionization of the Hydrogen Atom*    In the Bohr model of the hydrogen atom, the electron follows a circular orbit around the proton. In the lowest energy state (called the *ground state*) the radius of the orbit is $0.529 \times 10^{-10}$ m. The force exerted by one charge $q_1$ on another $q_2$ is given by Coulomb's law, $F_r = kq_1q_2/r^2$, where $k$ is the Coulomb constant. If the charges have the same sign, the force is one of repulsion. For the hydrogen atom, the charges have opposite signs but the same magnitude. That of the proton is $q_1 = +e$, and that of the electron is $q_2 = -e$, where $e$ is the magnitude of the fundamental unit of electronic charge. The force is then $F_r = -ke^2/r^2$. This is the same form as the gravitational force between two masses; i.e., it is attractive and decreases as the square of the separation distance. The electrostatic potential-energy function is found by computing the work we must do to change the separation of the charges. The calculation is the same as that for gravitational potential energy. The result with the zero potential energy chosen at infinite separation is

$$U(r) = -\frac{ke^2}{r} \qquad \text{7-25}$$

Putting in the values $k = 9 \times 10^9$ N·m²/C² and $e = 1.6 \times 10^{-19}$ C, we find for the potential energy at $r = 0.529 \times 10^{-10}$ m

$$U = -4.36 \times 10^{-16} \text{ J}$$

In atomic physics, energy is usually expressed in units of electronvolts (eV). We shall discuss the origin of this unit when we study electricity. For now, we need only the relation between the electronvolt and the joule:

$$1 \text{ eV} = 1.6 \times 10^{-19} \text{ J} \qquad \text{7-26}$$

(You will recognize that the conversion factor equals the magnitude of the fundamental unit of electric charge.) The potential energy of the electron-proton system (the hydrogen atom) when the electron is in the first Bohr orbit of radius $r = 0.529 \times 10^{-16}$ m is then

$$U = (-4.36 \times 10^{-16} \text{ J})\frac{1 \text{ eV}}{1.6 \times 10^{-19} \text{ J}} = -27.2 \text{ eV}$$

*The electronvolt energy unit defined*

The total energy of the electron is thus

$$E = \tfrac{1}{2}mv^2 + U = 13.6 \text{ eV} - 27.2 \text{ eV} = -13.6 \text{ eV} \qquad 7\text{-}27$$

To escape from the proton, the electron must have a total energy greater than or equal to zero. Thus to remove the electron, we must give it an additional energy of 13.6 eV. This is the *binding energy* of the atom. Since removing an electron from an atom is called *ionization*, the binding energy is also called the *ionization energy*.

*Binding or ionization energy*

**Example 7-10** *Molecular Oscillations and Dissociation*   The force between two atoms or ions in a molecule, discussed briefly in Chapter 5 (see Figure 5-2), is similar to that exerted by a spring connecting the two atoms, except that at great separation distances the force becomes zero. The potential energy $U$ as a function of separation distance $r$ is shown in Figure 7-9. This curve might apply to the two hydrogen atoms in the $H_2$ molecule or to the $Na^+$ and $Cl^-$ ions in an NaCl molecule. At the point $r = r_0$ the slope of the potential-energy curve is zero, and hence the force is zero. This is a stable equilibrium point since $U(r)$ increases on either side of this point. Very near this point the curve is similar to that for a mass on a spring. If the total energy is $E_1$, as shown in Figure 7-9, the atoms will oscillate about the equilibrium point like two masses joined by a spring. If the total energy is the larger value $E_2$, the motion will be an oscillation but not the simple, symmetric oscillation of two masses connected by a spring. The two turning points at $r_1$ and $r_2$ are different distances from the equilibrium point $r_0$. The two "halves" of the oscillation would take different times. Also, the average separation for this energy is slightly greater than $r_0$.

If the total energy is positive, e.g., the value $E_3$ in Figure 7-9, there is only one turning point. In this case the two particles are not bound. If initially the two particles were approaching, their separation would decrease until this single turning point was reached. Then they would move apart. Since there is no other turning point, they would continue to move apart.

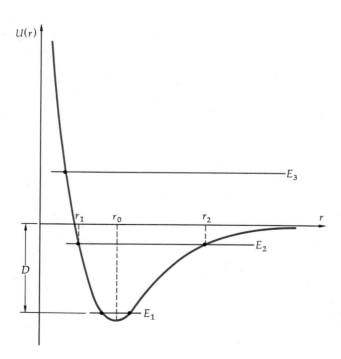

**Figure 7-9**
Potential energy versus separation for a diatomic molecule. Three possible values of the total energy are indicated by the horizontal lines. At the lowest energy $E_1$, which is typical for molecules in materials at ordinary temperatures, the atoms perform small oscillations about their equilibrium separation $r_0$. At greater energies such as $E_2$ the oscillations are of greater amplitude, and are not symmetric about the equilibrium separation. The minimum energy for dissociation $D$ is also called the binding energy. If the total energy is positive, as in $E_3$, the atoms or ions are not bound in a molecule.

The normal state of a molecule is for its parts to be nearly at rest relative to each other at the equilibrium separation $r_0$. If the molecule is given a greater energy, such as $E_1$, an oscillating motion ensues. This occurs, for example, when the body of which the molecule is a part is heated. If the heating is enough to increase the molecular energy to a value like $E_2$, the average separation of the particles is greater than $r_0$. The greater the energy $E_2$, the greater the average separation. If the molecules are in a solid, this requires the solid to expand as $E_2$ is increased. In fact, this model accounts for the familiar phenomenon of the thermal expansion of solids.

We also can see from Figure 7-9 that there is a minimum energy that must be supplied to the molecule to make its parts separate completely. This energy, called the *binding* or *dissociation energy*, is evidently nearly equal to the magnitude of the potential energy at the equilibrium separation $r_0$. (It is actually slightly less because the atoms are always oscillating with some total energy slightly greater than the minimum potential energy.)

**Question**

4. A comet enters the solar system, orbits around the sun, and leaves again. When it is near the sun, its gravitational potential energy (relative to infinity) plus its kinetic energy is less than zero. Will the comet ever return?

# Atmospheric Evaporation

Richard Goody
*Harvard University*

The expression for escape speed of a particle near the earth's surface (Equation 7-20) can be generalized to apply to the escape of a particle from the surface of any planet or the moon. If $R$ is the radius of the planet and $g$ the acceleration of gravity at the surface, the escape speed is given by

$$v_e = \sqrt{2gR}$$

This result is independent of the mass of the particle and therefore equally valid for a spacecraft, a projectile, or a molecule. The application to the escape of molecules from the surface of a planet leads to an important concept in planetary physics: using this equation and some results from the kinetic theory of gases, we can understand why some planets have atmospheres and others do not. We look to this mechanism to explain, for example, why the earth appears to have lost hydrogen to space following decomposition of water molecules, leaving an oxygen-rich atmosphere, while, to a first approximation, Jupiter is a gigantic ball of hydrogen.

Several conditions must be met before a molecule (or atom) can escape from the surface of a planet. First, the magnitude of the molecule's velocity must be greater than the escape speed $v_e$ for the planet. Table 7-1 lists escape speed

**Table 7-1**
Escape speeds for the moon and planets

| Planet | Gravitational acceleration, m/s² | Radius, km | Escape speed $v_e$, km/s |
|---|---|---|---|
| Mercury | 3.76 | 2,439 | 4.3 |
| Venus | 8.88 | 6,049 | 10.3 |
| Earth | 9.81 | 6,371 | 11.2 |
| Moon | 1.62 | 1,738 | 2.3 |
| Mars | 3.73 | 3,390 | 5.0 |
| Jupiter | 26.2 | 69,500 | 60 |
| Saturn | 11.2 | 58,100 | 36 |
| Uranus | 9.75 | 24,500 | 22 |
| Neptune | 11.34 | 24,600 | 24 |

for the moon and planets in the solar system. A second condition is that the direction of the velocity of the molecule must be away from the planet rather than toward it. Finally, the molecule must not collide with other molecules on its outward journey.

The chance that a molecule will collide with another in a particular region of space depends on the density of molecules in that region. The molecular density falls off rapidly with height above the surface of the planet. There must therefore be a level, called the *escape level,* which varies with the planet and depends upon the atmospheric temperature, above which the collision probability is so small that a molecule heading away from a planet with velocity greater than $v_e$ leaves the planet never to return. Figure 1 shows the trajectories of molecules above the escape level. Molecules with speeds greater than $v_e$ escape, while those with speeds less than $v_e$ fall back toward earth. These trajectories differ greatly from those of gas molecules near the earth's surface, where a molecule can travel only a short distance before colliding with another. For earth, the escape level averages about 500 km above the ground with considerable diurnal and seasonal variation because of atmospheric temperature changes. For Venus, the escape level is about 200 km above the surface.

Although escape speed does not depend on the mass of the molecule, the distribution of molecular speeds depends strongly on the mass of the molecule and on the temperature: in thermal equilibrium it is given by the Maxwell-Boltzmann distribution (Figure 2). In this figure, $f(v)\ dv$ is the fraction of all the molecules that have speeds in some interval $dv$. This fraction for an interval $dv$ centered about some speed $v$ is indicated by the shaded area in the

**Figure 1**
Molecules or atoms whose speeds exceed the escape speed $v_e$ can escape from the planet. These may form only a minute fraction of the total number at the critical level.

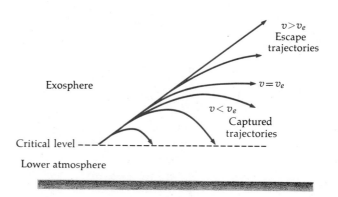

**Table 7-2**
Most probable speeds in kilometres per second

| Atom | Atomic mass, u | Temperature, K | | |
|---|---|---|---|---|
| | | 300 | 600 | 900 |
| H | 1 | 2.24 | 3.16 | 3.87 |
| He | 4 | 1.12 | 1.58 | 1.94 |
| O | 16 | 0.56 | 0.79 | 0.97 |

**Figure 2**
The Maxwell-Boltzmann speed distribution. The fraction of molecules having speeds in some interval $dv$ equals the area under the curve for that interval. The speed $v_0$ for which $f(v)$ is maximum is the most probable speed. The mean speed $\bar{v}$ and the root-mean-square speed $v_{rms}$ differ slightly from the most probable speed.

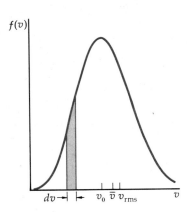

figure. The most probable speed is related to the mass of the molecule $m$ and the absolute temperature $T$ by

$$v_0 = \sqrt{\frac{2kT}{m}}$$

where $k$ is a constant called Boltzmann's constant.*

Some typical values of $v_0$ are shown in Table 7-2. Comparing Tables 7-1 and 7-2, we see that $v_0$ is greater than $v_e$ only for hydrogen on the surface of the moon at temperatures greater than 300 K. Since lunar daytime temperatures exceed 300 K, any hydrogen would evaporate into space almost instantaneously.

Even when the escape speed $v_e$ is much greater than the most probable speed $v_0$, there will be some molecules or atoms that have speeds greater than $v_e$ and therefore have a chance of escaping. From Figure 2 we discern that most molecules have speeds not much different from $v_0$ but that a few have speeds $2v_0$, $3v_0$, or even $100v_0$, though the fraction with speeds much greater than $v_0$ is very small and decreases rapidly with increasing $v$. Whether a certain type of molecule or atom is likely to be found in the atmosphere of a planet depends on the rate of escape of these molecules, which in turn depends critically on the fraction of molecules or atoms above the escape level with speeds greater than $v_e$. This fraction is shown in Figure 3 as a function of the ratio $v_e/v_0$. When this ratio is greater than about 2, this fraction is given approximately by

$$f_{v>v_e} = \frac{4}{\pi} \frac{v_e}{v_0} e^{-(v_e/v_0)^2}$$

* Because of the asymmetry of the curve for $f(v)$, the most probable speed $v_0$, the mean speed $\bar{v}$, and the root-mean-square speed $v_{rms}$ are slightly different, though they all have the same dependence on the temperature and molecular mass. In Chapter 16, it will be shown that the root-mean-square speed is given by $v_{rms} = \sqrt{3kT/m}$.

**Figure 3**
The fraction of atoms or molecules with speeds greater than $v_e$ in terms of the ratio $v_e/v_0$. The two parts of the figure differ only in scale.

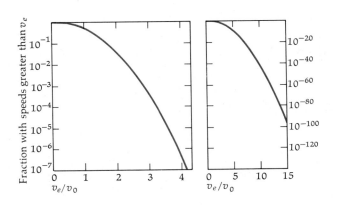

The temperature of the escape level on earth is close to 600 K, and $v_0$ is therefore 0.79 km/s for atomic oxygen and about 3.16 km/s for atomic hydrogen. Using $v_e = 11.2$ km/s, we obtain for the fraction of hydrogen atoms with speeds greater than the escape speed:

$$f_{v>v_e} (H) = \frac{4}{\pi} \frac{11.2}{3.16} e^{-(11.2/3.16)^2} \approx 1.6 \times 10^{-5}$$

This result tells us that out of every million hydrogen atoms at the escape level, about 16 will have speeds greater than $v_e$, and since half, on the average, will be moving away from the earth, 8 will escape. This fraction, seemingly small, is large enough to account for the almost complete absence of hydrogen from the earth's atmosphere now, approximately $4 \times 10^9$ years after formation of the earth.

For atomic oxygen, on the other hand, the fraction with speeds greater than $v_e$ is

$$f_{v>v_e} (O) = \frac{4}{\pi} \frac{11.2}{0.79} e^{-(11.2/0.79)^2} \approx 9.3 \times 10^{-87}$$

This fraction for oxygen atoms is much smaller than that for hydrogen atoms. If we want to evaluate the evaporation rates of hydrogen and oxygen numerically, our model of the escape process must be elaborated; nevertheless, the ratio of the numbers of hydrogen and oxygen atoms with speeds high enough for escape (more than $10^{81}$ to 1) is so large that the simultaneous retention of oxygen and escape of hydrogen is plausible. Because of the huge mass of Jupiter, its escape speed of 60 km/s enables it to retain hydrogen as easily as oxygen is retained on earth. We can therefore see in a general way why the two planets have such markedly different atmospheres.

---

## Review

A. Define, explain, or otherwise identify:

Work-energy theorem, 198    Ionization energy, 212
Escape speed, 209    Dissociation energy, 213
Binding energy, 209

B. True or false:

1. The total mechanical energy of a particle is always conserved.

2. When only conservative forces do work on a particle, its kinetic energy depends only on its position.

3. The time it takes a particle to move from one point to another can be found directly from the energy-conservation equation.

4. The kinetic energy needed to escape from the earth depends on the choice of reference point for the zero of potential energy.

## Exercises

### Section 7-1, Some Illustrative Examples

1. A 5-kg block is pushed along a horizontal rough surface by a constant horizontal force $F_0 = 25$ N. The coefficient of friction is 0.2. The block is pushed 3 m. (a) What is the work done on the block by the force $F_0$? (b) How much work is done on the block by friction? (c) What is the net increase in kinetic energy of the block? (d) What is the speed of the block after it has traveled 3 m?

2. A block is projected along a horizontal surface with initial velocity of 10 m/s. The coefficient of friction between the block and the surface is 0.2. How far does the block slide before coming to rest?

3. A 2-kg block slides down a frictionless curved ramp from rest at a height of 1 m (Figure 7-10). It slides 6 m on a rough horizontal surface before coming to rest. (*a*) What is the speed of the block at the bottom of the ramp? (*b*) How much work is done by friction on the block? (*c*) What is the coefficient of friction between the block and the horizontal surface?

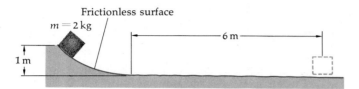

**Figure 7-10**
Exercise 3.

4. A 1-kg object is released from rest at a height of 5 m on a curved frictionless ramp. At the foot of the ramp is a spring of spring constant 400 N/m (Figure 7-11). The object slides down the ramp and onto the spring, compressing it a distance $x$ before coming momentarily to rest. (*a*) Find $x$. (*b*) What happens to the object after it comes to rest?

**Figure 7-11**
Exercise 4.

5. A 3-kg body slides along a frictionless horizontal surface with speed 5 m/s. After sliding a distance of 2 m, it encounters a frictionless ramp inclined at an angle of 30° to the horizontal. How far up the ramp does the body slide before coming momentarily to rest?

6. Work Exercise 5 if the surfaces are not frictionless and the coefficient of sliding friction between the body and the surfaces is 0.30. Find (*a*) the speed of the body when it reaches the ramp and (*b*) the distance up the ramp the body slides before coming momentarily to rest.

7. A 6-kg block slides down a 30° incline starting from rest. After sliding 10 m along the incline, its speed is 7 m/s. (*a*) What work did gravity do on the block in this descent? (*b*) What was the increase in kinetic energy of the block? (*c*) By how much did the total energy of the block increase or decrease? (*d*) How much work was done by friction on the block?

8. For the Atwood's machine shown in Figure 7-12 find the speed of the objects when they are at the same height. The system is at rest when the lower string is cut.

9. A 200-g ball is thrown with initial speed of 25 m/s at an angle of 53° above the horizontal. (*a*) What is the total mechanical energy initially? (*b*) Write the initial kinetic energy in terms of the velocity components $v_x$ and $v_y$. What is the minimum kinetic energy during the flight of the ball? At what point does the ball have this minimum kinetic energy? (*c*) What is the maximum potential energy of the ball during its flight? (*d*) Find the maximum height of the ball.

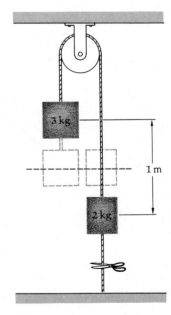

**Figure 7-12**
Exercise 8.

10. Two men stand at the edge of a 10-m cliff. Simultaneously they throw balls at initial speed of 10 m/s, one straight up and the other straight down. (a) If air resistance is neglected, how do the speeds of the balls compare when they reach the bottom of the cliff? What are their speeds there? (b) How much later does one ball reach the bottom than the other? (c) Discuss qualitatively how your answers would differ if air resistance were taken into account.

11. In Figure 7-13, $m_1 = 4$ kg, $m_2 = 2$ kg, and there is no friction. Find the speed of the masses when $m_2$ has fallen 2 m.

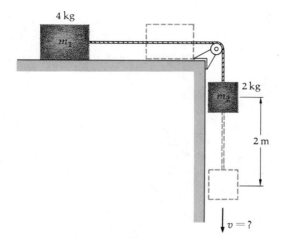

**Figure 7-13**
Exercise 11 and Problems 8 and 16.

12. A particle slides along the frictionless track shown in Figure 7-14. Initially it is at point $P$ headed downhill with speed $v_0$. Describe the motion in as much detail as you can if (a) $v_0 = 7$ m/s and (b) $v_0 = 12$ m/s. (c) What is the least speed needed by the mass to get past point $Q$?

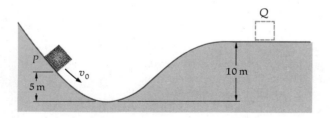

**Figure 7-14**
Exercise 12.

13. A pendulum of length $L$ has a bob of mass $m$. It is released from some angle $\theta$. The string hits a peg at a distance $x$ directly below the pivot itself (Figure 7-15) and wraps itself around the peg, shortening the length of the pendulum. Show that the maximum height the pendulum reaches equals its initial height. (Assume that initially the bob is below the height of the peg.)

14. A block of mass $m$ is attached to a spring of force constant $k$ on a horizontal smooth surface. Its equilibrium position is $x = 0$. The block is given a displacement $x = A$ and released. (a) What is the initial potential energy of the system? (b) Show that when the block is at position $x$ at any later time, its speed is given by $v = \sqrt{k/m}\,\sqrt{A^2 - x^2}$.

**Figure 7-15**
Exercise 13 and Problem 19.

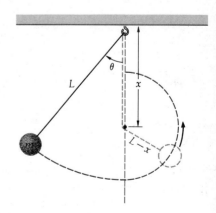

**Section 7-2, Qualitative Description of Motion Using Energy Conservation, and Section 7-3, Gravitational Potential Energy, Escape Speed, and Binding Energy**

15. (a) Find the potential energy (relative to zero at infinity) of a 100-kg mass at the surface of the earth. (Take the radius of the earth to be $R_E = 6.37$ Mm.) (b) Find the potential energy of the same mass at a height above the earth's

surface equal to the earth's radius. (c) If the mass is dropped from this height, find its speed when it hits the surface of the earth. (Neglect air resistance, although in practice air resistance is of primary importance in this type of problem.)

16. A particle is projected from the surface of the earth with speed equal to twice the escape speed. When the particle is very far from the earth, what is its speed?

17. Find the escape speed for a rocket leaving the moon. (The acceleration of gravity on the moon is 0.166 times that on earth, and the moon's radius is $0.273R_E$.)

18. The planet Saturn has a mass 95.2 times that of the earth, and a radius 9.47 times that of the earth. Find the escape speed for Saturn.

19. A space probe is to be sent from the earth so that it has a speed of 50 km/s when it is very far from the earth. What speed is needed for the probe at the surface of the earth?

20. (a) Calculate the energy in joules necessary to send a 1-kg mass away from the earth with the escape speed. (b) Convert this energy to kilowatt-hours. (c) If energy can be obtained at 10 cents per kilowatt-hour, what is the minimum cost of giving an 80-kg astronaut enough energy to escape the earth's gravitational field?

21. A satellite orbits the earth in a circular orbit. The orbit is then changed to another circular one of larger radius. Discuss the change in potential energy, total energy, and kinetic energy; i.e., whether each increases, decreases, or remains constant.

## Problems

1. A pendulum consists of a bob of mass $m$ attached to a light rod of negligible mass and of length $L$. The other end of the rod is fixed at a frictionless pivot. The pendulum is released from rest when the bob is directly over the pivot. Find the force exerted by the rod on the bob when (a) the bob is at its lowest point, (b) the rod is at an angle of 30° below the horizontal, and (c) the rod is at an angle of 30° above the horizontal.

2. A pendulum bob is released from rest at an initial angle $\theta_0$ measured from $\theta = 0$ at the lowest point. Prove that the tension in the string at the bottom is greater than the initial tension when the bob is at rest by the amount $3E_k/L$, where $L$ is the length of the string and $E_k$ is the maximum kinetic energy of the bob.

3. A child is swinging from a suspended rope 4.0 m long which will break when the tension becomes twice the weight of the child. (a) What is the greatest angle $\theta_0$ the rope can make with the vertical during the swing if the rope is not to break? (b) What is the speed of the child when the rope breaks if the greatest angle is slightly greater than that found in (a)?

4. An elastic string 25 cm long obeys Hooke's law $F_x = -kx$, where $x$ is the extension from equilibrium. When a 150-g object is suspended from the string, the string stretches 5 cm. If the object is attached to the end of the string and dropped from the point of support of the string, find the distance it falls before first coming to rest.

5. An elastic string has natural length $a$ and spring constant $k$. When an object of mass $m$ is hanging from it vertically, the string stretches by $x_0$. One end of the string is attached to the top of a frictionless plane inclined at 30° to the horizontal. With the string lying down the incline the object of mass $m$ is attached to it and released with the string in its unstretched condition. How far does the object slide down the plane before coming to rest for the first time?

6. A small mass $m$ slides without friction along the loop-the-loop track shown in Figure 7-16. The circular loop has radius $R$. The mass starts from rest at point $P$ a distance $h$ above the bottom of the loop. (a) What is the kinetic energy of $m$ when it reaches the top of the loop? (b) What is its acceleration at the top of the loop assuming that it stays on the track? (c) What is the least value of $h$ if $m$ is to reach the top of the loop without leaving the track? (d) Assuming that $h$ is greater than this least value, write an expression for the normal force exerted by the track on the mass.

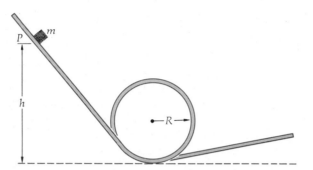

**Figure 7-16**
Problem 6.

7. A 2-kg mass is released on a frictionless incline 4 m from a spring of constant $k = 100$ N/m. The spring is fixed along the plane inclined at $\theta = 30°$ (Figure 7-17). (a) Find the maximum compression of the spring, assumed to be massless. (b) If the incline is not frictionless but the coefficient of friction between it and the mass is 0.2, find the maximum compression. (c) For the rough incline, how far up the incline will the mass travel after leaving the spring? (d) Describe the subsequent motion of the mass for the rough incline.

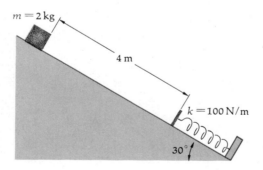

**Figure 7-17**
Problem 7.

8. For the arrangement in Figure 7-13 with general masses $m_1$ and $m_2$, (a) find an expression for the speed of the masses after $m_2$ has fallen a distance $y$. (b) Differentiate this expression to obtain an expression for the acceleration of the masses.

9. A Superball can bounce to 90 percent of its original height. (a) How much energy is lost after a 30-g ball is bounced once from an initial height of 3 m? (b) A ball dropped from an original height of $H$ makes $N$ bounces. Find a general expression for the maximum height of the ball after $N$ bounces for a ball of mass $m$ dropped from height $H$. (c) About how many bounces are required if the maximum height after the $N$th bounce is 1 percent of the original height?

10. A particle of mass $m$ is dropped from a height $h$ which is not necessarily small compared with the radius of the earth $R_E$. Show that if air resistance is neglected, the speed of the particle when it reaches the surface of the earth is given by $v = \sqrt{2gh}\sqrt{R_E/(R_E + h)}$.

11. A hole is drilled from the surface of the earth to its center. The gravitational force on a particle of mass $m$ which is inside the earth at a distance $r$ from the center ($r \leq R_E$) has the magnitude $mgr/R_E$ and points toward the center of the earth. (a) How much work is required to lift the particle from the center of the earth to the earth's surface? (b) If the particle is dropped from rest at the surface of the earth, what is its speed when it reaches the center of the earth?

12. Sketch the gravitational potential energy of a particle of mass $m$ as a function of the distance $r$ from the center of the earth using the result of part (a) of Problem 11 for $r \leq R_E$. Choose the potential to be zero at $r = \infty$. *Hint:* At $r = R_E$ both the potential energy $U(r)$ and its slope are continuous.

13. A skier starts from rest at height $H$ above the center of a rounded hummock of radius 4.0 m (Figure 7-18). There is negligible friction. Find the maximum value of $H$ for which the skier remains in contact with the snow at the peak of the hummock.

**Figure 7-18**
Problems 13 and 14.

14. A skier of mass 70 kg starts with a small initial speed from the top of the rounded hummock of Problem 13 (Figure 7-18). If friction can be neglected, find (a) his speed $v$ as a function of the angle $\theta$ and (b) the angle $\theta$ at which he loses contact with the slope.

15. In Example 7-5 find the distance the block moves up the incline if the beginning of the incline is 40 cm from the point where the block leaves the spring (the natural length of the spring) and the coefficient of friction between the block and both the horizontal surface and the incline is $\mu_k = 0.20$.

16. Work Problem 8 if the coefficient of friction between block 1 and the horizontal surface is $\mu_k$ (see Figure 7-13).

17. A 15-g ball is shot from a spring gun whose spring has a force constant of 600 N/m. The spring can be compressed 5 cm. (a) How high can the ball be shot if the gun is aimed vertically? (b) What is the greatest possible horizontal range of the ball for this compression?

18. The rate of energy loss of a certain system at a given time is directly proportional to the total energy of the system at that time. (a) Write a differential equation relating the energy $E$ to the rate of change of energy $dE/dt$ expressing the above property. Let the proportionality constant be $C$. (b) Solve your equation to obtain a general relationship for the energy of the system $E$ as a function of time $t$. (c) Find the constant $C$ in your result when 10 percent of the energy is lost in 10 s. (d) How long does it take for half the energy of the system to be lost when $C$ has the value found in part (c)?

19. In the problem of the pendulum string hitting a peg (Exercise 13) the pendulum is released from rest at $\theta = 90°$. (a) Find the speed of the bob in terms of $g$, $L$, and $x$ when it is directly above the peg and moving in a circle of radius $L - x$. (b) What is the least value of its acceleration at that point if the string is not to go slack? (c) Show that the bob will not reach this point with the string still taut unless $x$ is greater than or equal to $3L/5$.

20. A particle moves under the influence of a conservative force along the $x$ axis with total energy $E$. The potential energy associated with the force is $U(x)$. (a) Show that the speed of the particle is given by $v = \sqrt{2/m}\sqrt{E - U}$. (b) Show that the distance $dx$ traveled during the time interval $dt$ is given by

$$\frac{dx}{\sqrt{E - U(x)}} = \sqrt{\frac{2}{m}}\, dt$$

(c) For a mass on a spring the potential energy is $U = \frac{1}{2}kx^2$, where $k$ is the force constant, and the total energy is $E = \frac{1}{2}kA^2$, where $A$ is the maximum displacement. Integrate your result in part (b) to find the total time taken for a mass on a spring to move from $x = -A$ to $x = +A$.

21. (a) Show that the total energy of a satellite in a circular orbit of radius $r$ is given by $E = -GM_E m/2r$, where $M_E$ is the mass of the earth and $m$ the mass of the satellite, and that the speed $v$ is given by $v = \sqrt{GM_E/r}$. (b) If you wish to decrease the radius of the circular orbit of a satellite, must you increase or decrease the kinetic energy of the satellite? How do the potential energy and total energy change if the radius is decreased? (c) Using the differential approximation $\Delta v \approx dv$ and $\Delta r \approx dr$, show that a small change in radius must be accompanied by a small change in speed given by

$$\frac{\Delta v}{v} = -\frac{1}{2}\frac{\Delta r}{r}$$

22. A space probe launched from the surface of the earth accelerates for a short distance $\Delta r$ and then moves under the influence of gravity only. Since it is already a short distance $\Delta r$ above the surface of the earth, its escape speed is slightly less than that at the earth's surface by amount $\Delta v$. (a) Find $\Delta v$ in terms of $\Delta r$ using the differential approximation $\Delta v \approx dv$ and $\Delta r \approx dr$. (b) Find the escape speed for $\Delta r = 300$ km.

23. In calculating the escape speed the rotation of the earth was neglected. The speed that must be given to a body relative to the ground will be less if the body is launched in the direction of the earth's rotation and more if it is launched away from it. Take the earth's rotation into account for a body launched on the equator in the direction of motion of the earth's surface (horizontally eastward). (a) What is the speed relative to the center of the earth of an object at rest on the equator? (b) Relative to the earth's surface, what speed must be given to a body for it to escape the earth? (c) By what percentage is the work required to accelerate the body reduced because of the earth's rotation?

24. The bob of a pendulum of length $L$ is pulled aside so the string makes an angle $\theta_0$ and released. In Example 7-3 energy conservation was used to obtain the speed at the bottom (Equation 7-7). In this problem you are to obtain this result using Newton's second law. (a) Show that the tangential component of Newton's second law gives $dv/dt = -g \sin \theta$, where $v$ is the speed. (b) Show that $v$ can be written

$$v = L\frac{d\theta}{dt}$$

where $\theta$ is the angle made by the string and the vertical. (c) Use your results from (a) and (b) and the chain rule

$$\frac{dv}{dt} = \frac{dv}{d\theta}\frac{d\theta}{dt}$$

to obtain

$$v\, dv = -gL \sin \theta\, d\theta$$

(d) Integrate the left side of this equation from $v = 0$ to the final speed $v$ and the right side from $\theta = \theta_0$ to $\theta = 0$ to obtain Equation 7-7.

# CHAPTER 8    Systems of Particles

---

**Objectives**    After studying this chapter, you should:

1. Be able to find the center of mass of a discrete system of particles and of a continuous body.

2. Know how to find the acceleration of the center of mass of any system of particles.

3. Know when the momentum of a system is conserved.

4. Be able to discuss the various kinds of energy of a system of particles.

5. Know that the kinetic energy of a system can be written as the sum of the kinetic energy of center-of-mass motion and the kinetic energy of motion relative to the center of mass.

6. Be able to solve elastic and inelastic collisions in one dimension.

7. Know the definitions of coefficient of restitution, impulse, time average of a force, and impact parameter.

8. Be able to derive an equation (8-33) for the change in velocity of a rocket.

---

So far we have been concentrating on the motion of a single particle. We now turn to more complicated problems dealing with systems of two or more particles. Although the detailed description of the complete motion of a system of particles is quite complicated, some general methods which follow directly from Newton's laws are useful in dealing with a wide variety of problems.

In Section 5-5 we studied the static equilibrium of a rigid body, a special case of a system of particles. We found there that the force of gravity acting on the various parts of a body can be represented by a single force, the total weight of the body $Mg$ acting at the center of gravity, the point at which the torques due to gravity balance. In this chapter we shall show that in a uniform gravitational field the center of gravity of a body coincides with the center of mass of the body. The center of mass of a body or of a system of particles is a more general concept than the center of gravity in that its definition is independent of

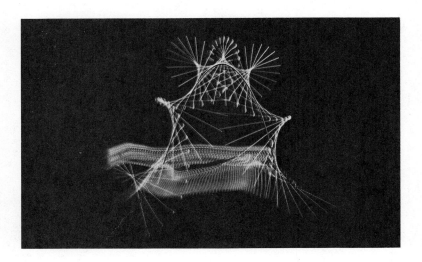

**Figure 8-1**
Multiflash photograph of a baton. One point, the center of mass, follows the same simple parabolic path as if it were a single point particle. (*Courtesy of Harold E. Edgerton.*)

the force of gravity. The center of mass of a system has the important property that its acceleration equals the resultant external force acting on the system divided by the total mass of the system. This result is illustrated by Figure 8-1, a multiflash photograph of a baton thrown into the air. Although the motion of the baton is complicated, the center of mass (and center of gravity, which is the same point) follows a parabolic path, as indicated. Since the resultant external force on the baton is the gravitational force $Mg$, the acceleration of the center of mass is $g$ downward.

# 8-1    Motion of the Center of Mass of a System

We consider a system of particles of total mass $M = m_1 + m_2 + m_3 + \cdots = \sum_i m_i$, where $m_i$ is the mass of the $i$th particle. Let $\mathbf{r}_i$ be the position vector for the $i$th particle relative to some arbitrary origin. We define the location of the center of mass of the system by its position vector $\mathbf{r}_{CM}$, which is defined by

$$M\mathbf{r}_{CM} = m_1\mathbf{r}_1 + m_2\mathbf{r}_2 + \cdots = \sum_i m_i\mathbf{r}_i \qquad \text{8-1}$$

We first show that the center of mass as defined by Equation 8-1 is the same point as the center of gravity of a body in a uniform gravitational field. For simplicity, we consider a plane figure located in the $xy$ plane, where $y$ is the vertical, as in Figure 8-2. If we choose the origin at the center of mass, the position vector $\mathbf{r}_{CM}$ relative to this origin is zero. The $x$ component of Equation 8-1 with the origin at the center of mass is then

$$0 = m_1x_1 + m_2x_2 + \cdots = \sum_i m_ix_i$$

If we multiply each term in this equation by the acceleration of gravity $g$, we have

$$0 = m_1gx_1 + m_2gx_2 + \cdots = \sum_i m_igx_i \qquad \text{8-2}$$

*Center of mass*

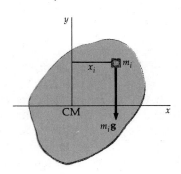

**Figure 8-2**
The torque exerted by the force of gravity on the particle of mass $m_i$ is $mgx_i$. The sum of these torques is zero if the origin is at the center of gravity, which is also the center of mass.

Since $m_i g$ is the force of gravity exerted on the $i$th particle and $x_i$ is the lever arm from the line of action of this force to the origin at the center of mass, the quantity $m_i g x_i$ is the torque exerted by the force of gravity on the $i$th particle about the center of mass. Equation 8-2 thus states that the total torque exerted by the force of gravity about the origin is zero. The origin is therefore the center of gravity as well as the center of mass. The only requirement for the center of gravity and center of mass to coincide is that the acceleration of gravity $g$ be the same at each particle in the system. This will be the case for all systems considered in this book.

We obtain the velocity of the center of mass by differentiating each term in Equation 8-1 with respect to time:

$$M\frac{d\mathbf{r}_{CM}}{dt} = m_1\frac{d\mathbf{r}_1}{dt} + m_2\frac{d\mathbf{r}_2}{dt} + \cdots = \sum_i m_i \frac{d\mathbf{r}_i}{dt}$$

or

$$M\mathbf{v}_{CM} = m_1\mathbf{v}_1 + m_2\mathbf{v}_2 + \cdots = \sum_i m_i\mathbf{v}_i \qquad \text{8-3}$$

*Velocity of the center of mass*

The quantity $m_i\mathbf{v}_i$ is the momentum of the $i$th particle. The right side of Equation 8-3 is therefore the total momentum of the particles in the system. We therefore have the important result that

*The total momentum of a system of particles equals the product of the total mass M and the velocity of the center of mass* $\mathbf{v}_{CM}$.

$$\mathbf{P} = \sum_i m_i\mathbf{v}_i = M\mathbf{v}_{CM} \qquad \text{8-4}$$

*Total momentum of a system*

We find the acceleration of the center of mass by again differentiating each term in Equation 8-3 with respect to time:

$$M\mathbf{a}_{CM} = m_1\mathbf{a}_1 + m_2\mathbf{a}_2 + \cdots = \sum_i m_i\mathbf{a}_i \qquad \text{8-5}$$

According to Newton's second law, the mass of each particle times its acceleration equals the resultant force acting on the particle. We can therefore replace the quantity $m_i\mathbf{a}_i$ with $\mathbf{F}_i$, the resultant force acting on the $i$th particle. The forces acting on a particle can be separated into two categories, internal forces due to interactions with other particles within the system and external forces due to agents outside the system:

$$\mathbf{F}_i = m_i\mathbf{a}_i = \mathbf{F}_{i,\text{int}} + \mathbf{F}_{i,\text{ext}}$$

Substituting this into Equation 8-5, we obtain

$$M\mathbf{a}_{CM} = \sum_i \mathbf{F}_{i,\text{int}} + \sum_i \mathbf{F}_{i,\text{ext}} \qquad \text{8-6}$$

According to Newton's third law, for each internal force acting on one particle there is an equal but opposite force acting on another particle. For example, if particle $m_1$ exerts a force on particle $m_2$, particle $m_2$ exerts an equal but opposite force on $m_1$. Thus the internal forces occur in pairs of equal and opposite forces. When we sum over all the particles in the system, the internal forces cancel out and we are left with only the external forces. Equation 8-5 then becomes

$$M\mathbf{a}_{CM} = \sum_i \mathbf{F}_{i,\text{ext}} \qquad \text{8-7}$$

This equation states that the total mass $M$ times the acceleration of the center of mass $\mathbf{a}_{CM}$ equals the resultant external force acting on the system. This has the same form as Newton's second law for a single particle of mass $M$ located at the center of mass and under the influence of the resultant external force $\mathbf{F}_{i,\text{ext}}$.

*The center of mass moves like a particle of mass $M = \sum_i m_i$ under the influence of the resultant external force on the system.*

*Motion of the center of mass*

This theorem is important because it shows us how to describe the motion of one point, the center of mass, for any system of particles, no matter how extensive the system may be. The center of mass of the system behaves just like a single particle subject only to the external forces. The individual motions of the members of the system generally are much more complex. For example, even if air resistance is negligible, the motion of something like a pair of masses connected by a spring and thrown into the air is quite involved. The masses tumble and turn as they move and oscillate along the line joining them. But the center of mass of the system, which lies between the two masses, moves just as if it were a single particle: it follows a simple parabolic trajectory.

The justification for our earlier treatment of large objects as point particles is actually this theorem on center-of-mass motion. All large objects can be considered to be made up of many small masses whose motions are governed by Newton's three laws. The motion of such a system generally includes rotations and oscillations relative to the center of mass. To work out all details of the motion would be very difficult, but often we are satisfied with finding the motion of the center of mass. Our theorem states that this can be treated as a problem of the motion of a single point particle.

For the baton thrown into the air (Figure 8-1) the only external force acting on the object is the force of gravity. Thus the center of mass follows the parabolic path of a particle in projectile motion predicted by Equation 8-7. This equation does not give a complete description of the complicated motion of the baton. We have described, so far, only the motion of one point, the center of mass of the system.

---

**Example 8-1** Two particles of equal mass $m$ are connected to a (massless) spring and rest on a frictionless horizontal table. If an external force $\mathbf{T}$ is applied to one of them as shown in Figure 8-3, describe the motion of the system.

Our two-particle system is indicated by the shading in the figure. The internal forces are those exerted by the spring and the gravitational attraction between the two particles, which is so small that we can

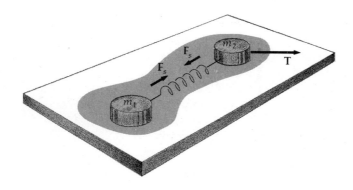

**Figure 8-3**
Two particles connected by a spring. The forces exerted by the spring are internal forces, which sum to zero. External forces on this system are the applied force $\mathbf{T}$, the force of gravity, and the normal force of the table.

neglect it. The external forces on this system are the force of gravity exerted on each particle, the normal force exerted by the table on the particles, which balances the force of gravity, and the applied force **T**. When we sum over all the forces acting on the system, the internal forces cancel (the force exerted by the spring on one particle is equal and opposite to that exerted on the other particle). Since the resultant external force is the applied force **T**, the acceleration of the center of mass is given by

$$\mathbf{a}_{CM} = \frac{\mathbf{T}}{m + m} \qquad\qquad 8\text{-}8$$

Since the masses are equal, the center of mass lies halfway between the particles. Its acceleration is given by Equation 8-7. We note that we cannot give a complete detailed description of the motion unless we know the force constant of the spring, and even then such a description is complicated. However, the description of the motion of the center of mass is simple.

---

**Example 8-2** Consider the same example with no spring, i.e., just two equal masses on a smooth table with a force **T** exerted on one of them. It is of course easy to describe the motion of each separately. One remains at rest, and the other has acceleration $\mathbf{a}_2 = \mathbf{T}/m$. However, the center-of-mass motion is the same as in the previous example, $\mathbf{a}_{CM} = \mathbf{T}/(m + m) = \mathbf{T}/2m$. We include this simple example to show that the particles of a system need not be interacting and that external forces may be acting on only some of the particles in the system. Remember that the motion of the center of mass is only part of the description of the motion of a system of particles. Equation 8-7 says nothing about the motion of the particles relative to the center of mass. In this example the motion is quite different from that in the previous one, though the motion of the center of mass is the same.

---

**Example 8-3** A projectile explodes into two equal pieces, each of mass $m$, at the top of its flight. One of the pieces drops straight down from rest after the explosion while the other moves off horizontally, so that they land simultaneously. Where does the second piece land?

Considering the projectile to be the system (whether it is in one piece or two), the only external force before the masses hit the ground is that due to gravity. The forces exerted in the explosion are internal forces, which do not affect the motion of the center of mass. After the explosion the center of mass traces out the rest of the parabola just as if there had been no explosion. Since the center of mass is halfway between equal masses and we know that one mass drops straight down, the other must land at an equal distance from the center of mass, as shown in Figure 8-4.

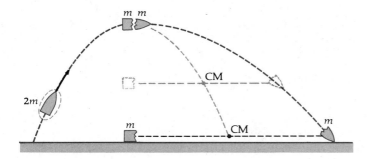

**Figure 8-4**
A projectile exploding into two equal fragments at the top of its flight so that the fragments land simultaneously. The internal forces of the explosion have no effect on the motion of the center of mass.

**Example 8-4** Consider a cylinder of mass $M$ resting on a rough piece of paper on a table. A force **f** is applied to the cylinder by pulling the paper to the right, as in Figure 8-5. Which way does the cylinder move?

The center of mass of a uniform cylinder is at its geometric center, i.e., on the axis, halfway between the faces. Taking the cylinder as our system, we have

$$\mathbf{f} = M\mathbf{a}_{\mathrm{CM}}$$

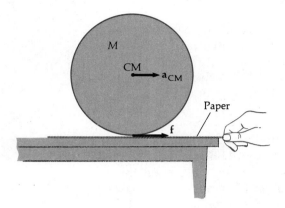

**Figure 8-5**
Cylinder rolling on a paper pulled to the right. Since the net external force on the cylinder is the frictional force to the right, the center of mass of the cylinder must accelerate to the right. The cylinder rolls backward relative to the paper because the acceleration of the paper is greater than that of the cylinder.

The acceleration of the center of mass of the cylinder is in the direction of the force **f** and has the magnitude $f/M$. Since **f** is in the direction the paper is pulled, $\mathbf{a}_{\mathrm{CM}}$ is to the right for the case illustrated in Figure 8-5. This result may be somewhat surprising at first because the paper also accelerates and the cylinder appears to move back in the direction opposite **f**. The forward acceleration of the cylinder is easily demonstrated, however, with a piece of chalk if the initial position of the chalk is marked on the *table*. The chalk accelerates back *relative to the paper*, because the acceleration of the paper is greater than that of the chalk. Relative to the table, the chalk accelerates forward in the direction of **f**. Note again that Equation 8-7 does not describe the rotation of the cylinder but only the motion of its center of mass.

## Questions

1. Two masses are connected by a light string passing over a pulley to form an Atwood's machine. Consider the two masses and the string as a system. What are the external forces? What are the internal forces?

2. A self-made man is said to have pulled himself up by his bootstraps. Discuss this from the point of view of internal and external forces and the motion of the center of mass of a system.

3. If only external forces can cause the center of mass of a system to accelerate, how can an automobile be accelerated by its motor?

4. A man is at rest at the center of a large, frictionless ice rink. Can he get off it? How?

5. A rocket ship initially is at rest (relative to the distant stars) in empty space when its engines are turned on. What happens to the center of mass of the rocket and its original contents? How can the rocket move?

# 8-2  Finding the Center of Mass

Since in a uniform gravitational field the center of mass coincides with the center of gravity, we can locate the center of mass experimentally using the methods of balance discussed in Chapter 5. For example, the center of mass of a stick can be located by finding the point at which the stick will balance on a pivot. Similarly, if we suspend an object from a pivot about which it is free to rotate, the center of mass will lie directly under the pivot when the object is in stable equilibrium. Suspending the object from a second pivot will then allow us to find the center of mass.

It is clear from our arguments about balance that the center of mass of a regular object must lie on any line of symmetry the object may have. For example, the center of mass of a uniform stick lies on the axis of the stick halfway between the ends. The center of mass of a uniform cylinder lies on its axis midway between the faces.

To find the center of mass of a system consisting of several bodies we first find the center of mass of the individual bodies. Consider for example two uniform sticks. The center of mass for this system is given by Equation 8-1:

$$M\mathbf{r}_{CM} = \sum_{\substack{\text{both} \\ \text{sticks}}} m_i \mathbf{r}_i$$

where the sum is over all the mass elements of the two sticks. We can write this sum in two parts:

$$\sum_{\substack{\text{both} \\ \text{sticks}}} m_i \mathbf{r}_i = \sum_{\text{stick 1}} m_i \mathbf{r}_i + \sum_{\text{stick 2}} m_i \mathbf{r}_i$$

The first sum on the right is just $M_1 \mathbf{r}_{CM_1}$, where $M_1$ is the mass of stick 1 and $\mathbf{r}_{CM_1}$ is the position vector to its center of mass, which is at the center of the stick since it is uniform. Similarly, the second term is $M_2 \mathbf{r}_{CM_2}$, and

$$M\mathbf{r}_{CM} = M_1 \mathbf{r}_{CM_1} + M_2 \mathbf{r}_{CM_2}$$

The center of mass of the two-stick system lies on the line joining the centers of mass of the separate sticks, as shown in Figure 8-6. We see from this type of reasoning that we can treat any part of a complicated system as a point mass located at the center of mass of the part. If the center-of-mass coordinates of a continuous body are to be calculated, the sum $\Sigma m_i x_i$ must be replaced by the integral $\int x \, dm$, where $dm$ is an element of mass. We then have

$$Mx_{CM} = \int x \, dm \qquad 8\text{-}9$$

In most cases, one or more of the coordinates can be found by symmetry without calculation. We give two examples here of finding the center of mass of a continuous body.

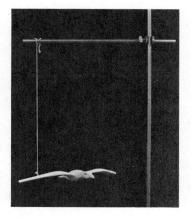

This bird looks symmetrical, but evidently its mass distribution is not. Where is its center of mass? (*Courtesy of Larry Langrill, Oakland University.*)

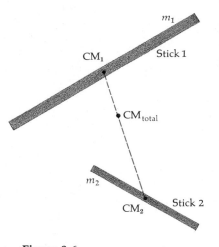

**Figure 8-6**
The center of mass of a system of two sticks can be found by treating each stick as a point particle at its individual center of mass.

---

**Example 8-5** *Center of Mass of a Uniform Stick of Mass M and Length L*
This very simple example (Figure 8-7), in which the result is known from symmetry beforehand, illustrates the technique of setting up the integration indicated in Equation 8-9. Let us call the linear density, or mass per unit length, $\lambda$. Since the density is uniform, $\lambda = M/L$.

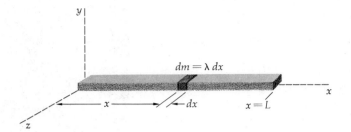

**Figure 8-7**
Calculation of the center of
mass of a uniform stick. The
mass element *dm* at a distance
*x* is treated as a point particle.

In Figure 8-7 we have chosen the origin at one end of the stick. We need
only compute $x_{CM}$. A mass element of length $dx$ is indicated. Its mass is
$dm = \lambda\, dx$, and it is a distance $x$ from the origin. Thus

$$Mx_{CM} = \int x\, dm = \int_0^L x\lambda\, dx = \frac{\lambda x^2}{2}\Big]_0^L = \frac{\lambda L^2}{2}$$

Using $\lambda = M/L$, we obtain the expected result

$$x_{CM} = \frac{\lambda L^2}{2M} = \frac{M}{L}\frac{L^2}{2M} = \tfrac{1}{2}L$$

---

**Example 8-6** *Center of Mass of a Semicircular Hoop*    Figure 8-8 shows our
choice of coordinate system for this calculation. We have chosen the origin to be on a line of symmetry of the figure. Clearly, $x_{CM} = 0$ because
for every mass element at $+x$ there is an equal one at $-x$. However, $y_{CM}$
is not zero. The center of mass is not at the origin, which is at the center

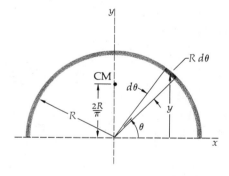

**Figure 8-8**
Geometry for the calculation
of the center of mass of a
semicircular hoop. The center
of mass lies on the *y* axis.

of curvature of the hoop. All the mass is at positive $y$; thus $\int y\, dm$ cannot
be zero. In the figure we have indicated a mass element of length $R\, d\theta$ at
height $R \sin \theta$. The mass of this element is $dm = \lambda R\, d\theta$, where $\lambda =
M/\pi R$ is the mass per unit length. We thus have

$$My_{CM} = \int y\, dm = \int_0^\pi R \sin \theta\, \lambda R\, d\theta = R^2\lambda \int_0^\pi \sin \theta\, d\theta = 2R^2\lambda$$

since

$$\int_0^\pi \sin \theta\, d\theta = -\cos \theta\Big]_0^\pi = 2$$

Using $\lambda = M/\pi R$, we have

$$My_{CM} = 2R^2 \frac{M}{\pi R} \qquad \text{or} \qquad y_{CM} = \frac{2R}{\pi}$$

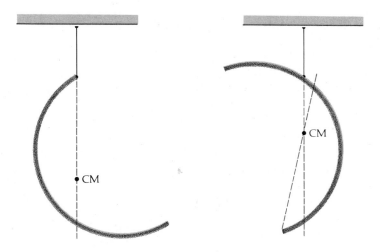

**Figure 8-9**
The center of mass of a semi-circular hoop found experimentally by suspending it from two points.

The center of mass is not within the body in this case. Figure 8-9 shows how to locate the center of mass by suspending a semicircular hoop from one end and then from some other point.

**Questions**

6. Must there be mass at the center of mass of a system? Explain.

7. Must the center of mass of a solid body be in the interior of the body? If not, give examples.

## 8-3  Conservation of Linear Momentum

An important special case of the motion of a system of particles occurs when the resultant external force is zero. From Equation 8-7, with $\Sigma\mathbf{F}_{i,\text{ext}} = 0$, we have

$$M\mathbf{a}_{\text{CM}} = M\frac{d\mathbf{v}_{\text{CM}}}{dt} = 0$$

and

$$M\mathbf{v}_{\text{CM}} = \mathbf{P} = \Sigma m_i\mathbf{v}_i = \text{constant} \qquad \text{8-10}$$

The total momentum of the system and the velocity of the center of mass are constant. This important result is known as the *law of conservation of linear momentum*. (The quantity $\mathbf{p} = m\mathbf{v}$, which we have been calling momentum, is often also called linear momentum to distinguish it from angular momentum, discussed in Chapter 10.)

*If the resultant external force on a system is zero, the velocity of the center of mass of the system is constant and the total momentum of the system is constant (conserved).*

*Conservation of momentum*

This result is the generalization to systems of particles of the conservation-of-momentum law for two particles discussed in Chapter 4. This law is one of the most important in physics. It applies, for example, to any system isolated from its surroundings so that there are no external forces acting on it. It is more generally applicable than the law of conservation of mechanical energy because internal forces exerted by one particle in the system on another are often not conservative. Thus they can change the total mechanical energy of the system, but since these internal forces always occur in pairs, they cannot change the total linear momentum of the system. The conservation of momentum is particularly useful in studying collisions (Section 8-5).

---

**Example 8-7** A man of mass 70 kg and a boy of mass 35 kg are standing together on a smooth ice surface for which friction is negligible. If after they push each other apart, the man moves away with speed 0.3 m/s relative to the ice, how far apart are they after 5 s (Figure 8-10)?

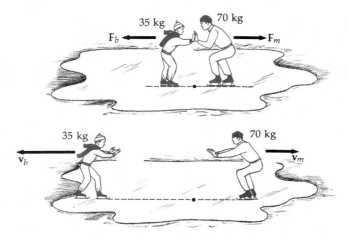

**Figure 8-10**
The resultant external force on the man-boy system is zero, and so the total momentum of the system is zero. Since the mass of the man is twice that of the boy, the speed of the man is half that of the boy.

We take the man and boy together as the system. The force of gravity on each is balanced by a corresponding normal force of the ice. Since there is no friction, the resultant force on the system is zero. The force exerted by the man on the boy is equal and opposite to that exerted by the boy on the man. The total momentum of the system is zero since they are standing at rest initially. Therefore, after they push each other, they must have equal and opposite momenta. Since the man has twice the mass of the boy, the boy must have twice the speed of the man. Since the man moves in one direction with speed 0.3 m/s, the boy moves in the opposite direction with speed 0.6 m/s. After 5 s the man has moved 1.5 m and the boy 3 m and they are 4.5 m apart. The center of mass of this system remains at rest in its original position, the point where they originally stood. Note that the mechanical energy of this system is not conserved. The force exerted by each on the other is not conservative. In this case the mechanical energy is increased, since the kinetic energy was initially zero and the potential energy did not change.

---

**Example 8-8** A bullet of mass 10 g moves horizontally with speed 400 m/s and embeds itself in a block of mass 390 g initially at rest on a frictionless table. What is the final velocity of the bullet and block (Figure 8-11)?

Since there are no horizontal forces on the bullet-block system, the horizontal component of the total momentum of the system is conserved. (There is a small vertical resultant force on the system, the weight of the bullet before it strikes the block. The bullet accelerates toward the earth before it strikes the block. We shall ignore this slight vertical motion.)

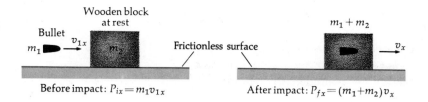

Before impact: $P_{ix} = m_1 v_{1x}$          After impact: $P_{fx} = (m_1 + m_2)v_x$

**Figure 8-11**
Bullet striking block in Example 8-8. Since there is no net external force on the bullet-block system, the momentum of this system is constant.

The total initial horizontal momentum $P_{ix}$ before the bullet strikes the block is just that of the bullet:

$$P_{ix} = m_1 v_{1x} = (10\ \text{g})\ (400\ \text{m/s}) = 4000\ \text{g·m/s} = 4\ \text{kg·m/s}$$

Afterward the bullet and block move together with a common velocity $v_x$. The total final momentum $P_{fx}$ is

$$P_{fx} = (m_1 + m_2)v_x = (10\ \text{g} + 390\ \text{g})v_x = (0.4\ \text{kg})v_x$$

Since the total momentum is conserved, the final momentum equals the initial momentum:

$$(0.4\ \text{kg})v_x = 4\ \text{kg·m/s}$$

$$v_x = 10\ \text{m/s}$$

Since the bullet and block both move together with this velocity, the center of mass must move with this velocity. We could have found the velocity of the center of mass (which is constant) before the collision from Equation 8-4:

$$Mv_{\text{CM},x} = \Sigma m_i v_{ix} = m_1 v_{1x} + m_2(0)$$

$$v_{\text{CM},x} = \frac{m_1 v_{1x}}{m_1 + m_2} = \frac{4\ \text{kg·m/s}}{0.4\ \text{kg}} = 10\ \text{m/s}$$

This is the same calculation used to find $v_x$. A simple computation of the initial and final energies shows that again mechanical energy is not conserved. The original kinetic energy is $\frac{1}{2}m_1 v_{ix}^2 = \frac{1}{2}(0.01\ \text{kg}) \times (400\ \text{m/s})^2 = 800\ \text{J}$, and the final energy is $\frac{1}{2}(0.4\ \text{kg})(10\ \text{m/s})^2 = 20\ \text{J}$. In this case most of the original kinetic energy (780 J out of 800 J) is lost because large nonconservative forces between the bullet and the block deform the bodies. A bullet embedding itself in a block is an example of an inelastic collision. We shall study such collisions in more detail later in this chapter.

# 8-4   Energy of a System of Particles

Although the total momentum of a system of particles must be constant if the resultant external force on the system is zero, the total mechanical energy of the system may change. As we saw in Examples 8-7 and 8-8, internal forces which cannot change the total momentum may be nonconservative and thus change the total mechanical energy of the system. In Example 8-8 mechanical energy is lost; some of it goes into thermal energy, some of it goes into producing permanent deformation of the block and bullet, and a small amount goes into sound waves produced when the bullet strikes the block. In Example 8-7 mechanical energy is gained; internal chemical energy of the man and boy is converted into mechanical energy as they push each other apart. Another example of an increase in mechanical energy is the explosion of a bomb, in which internal chemical energy is converted into kinetic energy of the fragments.

The total mechanical energy of a system consists of potential energy and kinetic energy. The potential energy can be separated into two kinds, the potential energy of interaction associated with conservative forces internal to the system, i.e., the forces of interaction of one particle with another, and the potential energy associated with external conservative forces such as the force of gravity due to the earth on the particles of a system. Let us consider the second kind first, for the special case of a system of particles near the surface of the earth. Let $y_i$ be the height of the $i$th particle above some reference level. If we choose the gravitational potential energy to be zero when $y_i = 0$, the potential energy of the $i$th particle is $m_i g y_i$. The total potential energy of the system is then

$$U = \sum_i m_i g y_i$$

When we compare this expression with Equation 8-1 defining the position of the center of mass, we note that $\Sigma m_i y_i = M y_{CM}$, where $M$ is the total mass of the system and $y_{CM}$ is the height of the center of mass. The total gravitational potential energy of interaction of the system with the earth is then

$$U = M g y_{CM} \qquad\qquad\qquad 8\text{-}11$$

This is the same as that of a particle of mass $M$ located at the center of mass of the system.

The second kind of potential energy of a system is that associated with internal conservative forces.

*This potential energy is defined as the negative of the work done by the particles on each other as the particles are brought from some reference position (usually infinitely separated) to their final positions.*

This is equivalent to the definition for the potential energy of a two-particle system we stated in Chapter 6; we choose the zero of potential energy to be when the particles are infinitely separated. An alternative, equivalent statement is that the potential energy of a system is the work *we* must do to bring the particles from some reference position such as that of infinite separation to their final positions without acceleration.

With these definitions, the potential energy of the system is chosen to be zero when the particles are infinitely separated. If the particles attract each other, the potential energy in their final configuration will be negative and they will tend to fall toward each other, increasing their kinetic energy and decreasing their potential energy. If the particles repel each other, the potential energy in their final configuration will be positive; to push them together from an initially infinite separation, we must do work increasing their potential energy.

The kinetic energy of a system of particles is the sum of the kinetic energies of the individual particles:

$$E_k = \sum_i \tfrac{1}{2} m_i v_i^2 \qquad\qquad\qquad 8\text{-}12$$

This kinetic energy can be written as the sum of two terms: (1) the kinetic energy of bulk motion of the system, i.e., that associated with the motion of the center of mass of the system, and (2) the kinetic energy of motion of the particles relative to the center of mass. To derive this important result we write the velocity of the $i$th particle $\mathbf{v}_i$ as the sum of the velocity of the center of mass $\mathbf{v}_{CM}$ and the velocity of the particle relative to the center of mass $\mathbf{u}_i$:

$$\mathbf{v}_i = \mathbf{v}_{CM} + \mathbf{u}_i \qquad\qquad 8\text{-}13$$

The kinetic energy of the system is then

$$E_k = \sum_i \tfrac{1}{2} m_i v_i^2 = \sum_i \tfrac{1}{2} m_i (\mathbf{v}_{CM} + \mathbf{u}_i) \cdot (\mathbf{v}_{CM} + \mathbf{u}_i)$$

$$= \sum_i \tfrac{1}{2} m_i v_{CM}^2 + \sum_i \tfrac{1}{2} m_i u_i^2 + \mathbf{v}_{CM} \cdot \sum_i m_i \mathbf{u}_i$$

where in the last term we have removed $\mathbf{v}_{CM}$ from the sum. The quantity $\Sigma m_i \mathbf{u}_i$ is zero. This quantity is the total momentum of the system as seen from an origin at the center of mass (since $\mathbf{u}_i$ is the velocity relative to the center of mass). But relative to an origin at the center of mass, the center of mass is at rest and the total momentum of the system is zero. Setting this term equal to zero, we obtain our desired result:

$$E_k = \tfrac{1}{2} M v_{CM}^2 + E_{k,\text{rel}} \qquad\qquad 8\text{-}14$$

where $M = \Sigma m_i$ is the total mass of the system and

$$E_{k,\text{rel}} = \sum_i \tfrac{1}{2} m_i u_i^2 \qquad\qquad 8\text{-}15$$

is the kinetic energy of motion relative to the center of mass.

*The kinetic energy of a system of particles can be written as the sum of two terms: (1) the energy associated with the center-of-mass motion, $\tfrac{1}{2}Mv_{CM}^2$, where M is the total mass of the system, and (2) the energy of motion relative to the center of mass, $\Sigma\tfrac{1}{2}m_i u_i^2$, where $u_i$ is the velocity of the ith particle relative to the center of mass.*

*Kinetic energy of a system*

We shall find many occasions to use this result in the following chapters. For example, the kinetic energy of a rolling ball will be written as the sum of $\tfrac{1}{2}Mv_{CM}^2$ and the energy of relative motion, which in this case is the energy of rotation.

If there are no external forces acting on the system, the velocity of the center of mass is constant and the first term in Equation 8-14 does not change. Only the relative energy can increase or decrease in an isolated system.

---

**Example 8-9** In Example 8-8, where a 10-g bullet moves with initial speed of 400 m/s and embeds itself in a 390-g block initially at rest, find the energy associated with center-of-mass motion and the relative energy.

The speed of the center of mass was found in Example 8-8 to be $v_{CM} = 10$ m/s. Since the total mass of the two-particle system is 400 g = 0.40 kg, the energy associated with motion of the center of mass is $\tfrac{1}{2}Mv_{CM}^2 = \tfrac{1}{2}(0.4\text{ kg})(10\text{ m/s})^2 = 20$ J. Since the original kinetic energy of the particles computed in Example 8-8 is 800 J, the kinetic energy relative to the center of mass must be $E_{k,\text{rel}} = 800\text{ J} - 20\text{ J} = 780$ J. We can compute this result directly by finding the velocity of each particle relative to the center of mass. These velocities are (from Equation 8-13)

$$u_1 = v_1 - v_{CM} = 400\text{ m/s} - 10\text{ m/s} = 390\text{ m/s}$$

and

$$u_2 = v_2 - v_{CM} = 0\text{ m/s} - 10\text{ m/s} = -10\text{ m/s}$$

The corresponding kinetic energies relative to the center of mass are

$$E_{k1,\text{rel}} = \tfrac{1}{2}m_1u_1^2 = \tfrac{1}{2}(0.01 \text{ kg}) (390 \text{ m/s})^2 = 760.5 \text{ J}$$

and

$$E_{k2,\text{rel}} = \tfrac{1}{2}m_2u_2^2 = \tfrac{1}{2}(0.39 \text{ kg}) (-10 \text{ m/s})^2 = 19.5 \text{ J}$$

The total energy relative to the center of mass is the sum of these, which is 780 J. After the collision, the particles stick together and move with the velocity of the center of mass, so that there is no relative energy after they collide.

---

## 8-5    Collisions in One Dimension

When two bodies come together and interact strongly for a short time, the event is called a *collision*. Often the forces exerted by one body on the other are much stronger than any other external forces present, and the time of the collision is so short that the bodies do not move appreciably during the interaction. When the external forces are negligible, the total momentum of the two-body system is conserved, the velocity of the center of mass is constant, and the part of the kinetic energy associated with center-of-mass motion $\tfrac{1}{2}Mv_{\text{CM}}^2$ is also constant. However, since the internal forces may be nonconservative, $E_{kr}$ may change. If the forces are conservative, $E_{kr}$ is also constant, the total kinetic energy remains unchanged in the collision, and the collision is said to be *perfectly elastic*. On the other hand, if all the energy of relative motion is lost in the collision, the collision is said to be *perfectly inelastic*. Since there is no motion relative to the center of mass after a perfectly inelastic collision, the bodies move together after the collision with a common velocity which is the velocity of the center of mass. The bullet embedding itself in a block in Example 8-8 is a typical example of a perfectly inelastic collision. In most collisions only part of the kinetic energy of motion relative to the center of mass is lost; these collisions are neither perfectly elastic nor perfectly inelastic.

*Elastic and inelastic collisions*

Figure 8-12 shows a body of mass $m_1$ moving with initial velocity $v_{1i}$ toward a second body of mass $m_2$ moving with initial velocity $v_{2i}$ (which we assume to be less than $v_{1i}$ so that the bodies collide). Let $v_{1f}$ and $v_{2f}$ be the final velocities of the bodies after the collision. Conservation of momentum gives

$$m_1v_{1f} + m_2v_{2f} = m_1v_{1i} + m_2v_{2i} \qquad\qquad 8\text{-}16$$

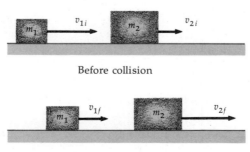

Before collision

After collision

**Figure 8-12**
General collision of two bodies in one dimension.

Equation 8-16 gives one relation between the unknown velocities $v_{1f}$ and $v_{2f}$ (we assume that the initial velocities are given). To find each of these unknowns we must have a second relation. The second relation comes from energy considerations. If the collision is perfectly elastic, the initial and final kinetic energies are equal. We then have

$$\tfrac{1}{2}m_1v_{1f}^2 + \tfrac{1}{2}m_2v_{2f}^2 = \tfrac{1}{2}m_1v_{1i}^2 + \tfrac{1}{2}m_2v_{2i}^2 \qquad 8\text{-}17$$

Equations 8-16 and 8-17 are sufficient to determine the final velocities of the two bodies. With a slight rearrangement of these two equations we can write

$$m_2(v_{2f}^2 - v_{2i}^2) = m_1(v_{1i}^2 - v_{1f}^2)$$

or

$$m_2(v_{2f} - v_{2i})\,(v_{2f} + v_{2i}) = m_1(v_{1i} - v_{1f})\,(v_{1i} + v_{1f})$$

and

$$m_2(v_{2f} - v_{2i}) = m_1(v_{1i} - v_{1f})$$

Dividing the middle equation above by the last one, we obtain

$$v_{2f} + v_{2i} = v_{1i} + v_{1f}$$

which can be written

$$v_{2f} - v_{1f} = -(v_{2i} - v_{1i}) \qquad 8\text{-}18$$

The relative velocity $v_2 - v_1$ is the velocity of body 2 as seen from body 1. Equation 8-18 states the important result that

*For perfectly elastic collisions, the relative velocity of recession after the collision equals the relative velocity of approach before the collision.*

*Elastic collisions*

In solving a perfectly elastic collision problem it is often easiest to use Equations 8-16 and 8-18 to find the final velocities, thus avoiding the quadratic terms in the conservation-of-energy equation 8-17.

**Example 8-10** A 4-kg block moving to the right at 6 m/s makes a perfectly elastic collision with a 2-kg block moving to the right at 3 m/s (Figure 8-13). Find the final velocities $v_{1f}$ and $v_{2f}$.
    Conservation of momentum gives

$$(4 \text{ kg})v_{1f} + (2 \text{ kg})v_{2f} = (4 \text{ kg})\,(6 \text{ m/s}) + (2 \text{ kg})\,(3 \text{ m/s})$$
$$= 30 \text{ kg·m/s}$$

or

$$4v_{1f} + 2v_{2f} = 30 \text{ m/s}$$

The velocity of block 2 relative to block 1 before the collision is

$$v_{2i} - v_{1i} = 3 - 6 = -3 \text{ m/s}$$

Equation 8-18 then gives

$$v_{2f} - v_{1f} = -(-3 \text{ m/s}) = +3 \text{ m/s}$$

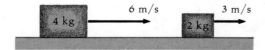

6 m/s          3 m/s

4 kg          2 kg

**Figure 8-13**
Collision of two bodies for Example 8-10.

Solving the two equations for the two final velocities, we obtain

$$v_{2f} = 7 \text{ m/s} \quad \text{and} \quad v_{1f} = 4 \text{ m/s}$$

As a check on our results, we compute the initial and final kinetic energies of the blocks:

$$E_{ki} = \tfrac{1}{2}(4 \text{ kg})(6 \text{ m/s})^2 + \tfrac{1}{2}(2 \text{ kg})(3 \text{ m/s})^2$$
$$= 72 \text{ J} + 9 \text{ J} = 81 \text{ J}$$

and

$$E_{kf} = \tfrac{1}{2}(4 \text{ kg})(4 \text{ m/s})^2 + \tfrac{1}{2}(2 \text{ kg})(7 \text{ m/s})^2$$
$$= 32 \text{ J} + 49 \text{ J} = 81 \text{ J}$$

Our results are consistent with conservation of energy.

___

**Example 8-11** A very large mass $m_1$ moving with speed $v_{1i}$ collides elastically with a very small mass $m_2$ initially at rest (Figure 8-14). What is the speed of the small mass after the collision assuming that $m_1$ is much greater than $m_2$?

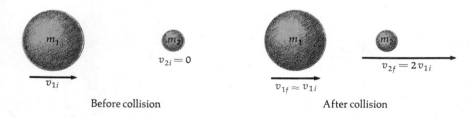

Before collision          After collision

**Figure 8-14**
Perfectly elastic collision between a very large body and a very small body. Since the relative velocity of recession after the collision must equal the relative velocity of approach before the collision and the velocity of the very large mass is essentially unchanged, the final velocity of the small body must be approximately twice that of the large body.

It should be intuitively clear that the large mass will not be affected very much by a collision with a very small mass. A cannonball will hardly be slowed down if it collides with a stationary beachball. Before the collision, the relative velocity of approach is $v_{1i}$. Then after the collision the velocity of separation must be $v_{1i}$. For a first approximation, we neglect any change in the velocity of $m_1$. Since it continues to move with velocity $v_{1i}$, the velocity of the small mass $m_2$ must be $2v_{1i}$.

An example of such a collision is that of a golf club with a golf ball. The effective mass of the club is large because it is held by the player, who is in firm contact with the ground. Careful measurement of the distance between successive flashes in Figure 8-15 shows that the initial velocity of the ball is indeed twice that of the club.

**Figure 8-15**
Multiflash photograph of a golfer hitting a ball. Comparison of the distance traveled by the club and ball between successive flashes shows that the speed of the ball after the collision is approximately twice that of the club. (*Courtesy of Harold E. Edgerton.*)

The energy transferred is just the energy of the small mass after the collision, $\frac{1}{2}m_2(2v_{1i})^2 = 2m_2v_{1i}^2$. We can use the conservation of momentum to get a correction to our first approximation for the final velocity of the large mass. Since the small mass gains momentum $2m_2v_{1i}$ to the right, the large mass must lose this much momentum. If $v_{1f}$ is the final velocity of the large mass, we have $m_1v_{1f} = m_1v_{1i} - 2m_2v_{1i}$, or $v_{1f} \approx v_{1i} - (2m_2/m_1)v_{1i}$.

*Inelastic collisions*

For perfectly inelastic collisions the second relation between the final velocities is that they are equal to each other and to the velocity of the center of the mass. This result combined with momentum conservation then gives

$$(m_1 + m_2)v_{\text{CM}} = m_1v_{1i} + m_2v_{2i} \qquad 8\text{-}19$$

This is of course the same as Equation 8-4. Example 8-8 is a typical example of an inelastic-collision problem. For the special case of one particle initially at rest we can relate the initial and final energies in a simple way by writing the kinetic energies in terms of the momentum $mv$. Let $v_{1i}$ be the initial velocity of the incoming particle and $v_{2i} = 0$ for the second particle initially at rest. After the collision they both move with velocity $v_{\text{CM}} = m_1v_{1i}/(m_1 + m_2)$. The initial kinetic energy is

An elastic collision of two equal cannonballs from Johann Marcus Marci, *De Proportione Motus*, 1639. All of the kinetic energy of cannonball *a* is transferred to cannonball *b*. (*Courtesy of the Deutsches Museum, Munich.*)

$$E_{ki} = \tfrac{1}{2}m_1v_{1i}^2 = \frac{(m_1v_{1i})^2}{2m_1} = \frac{p^2}{2m_1} \qquad 8\text{-}20$$

where $p = m_1v_{1i}$ is the initial momentum. Since momentum is conserved, the final momentum $(m_1 + m_2)v_{\text{CM}}$ is also equal to $p$. The final kinetic energy can similarly be written as the square of the momentum divided by twice the mass, where the final momentum is the same as the initial momentum but the mass is now $m_1 + m_2$. We thus have

$$E_{kf} = \frac{p^2}{2(m_1 + m_2)} \qquad 8\text{-}21$$

From a comparison of Equations 8-20 and 8-21 it is clear that the final energy is less than the initial energy. The ratio of the final and initial energies is

$$\frac{E_{kf}}{E_{ki}} = \frac{m_1}{m_1 + m_2} \qquad 8\text{-}22$$

This result holds only if the collision is perfectly inelastic and if the particle with mass $m_2$ is initially at rest.

In general, a collision is somewhere between the extreme cases of perfectly elastic, in which case the relative velocities are reversed (Equation 8-18), and perfectly inelastic, in which case there is no relative velocity after the collision. The coefficient of restitution $e$ is defined as the ratio of the relative velocity of recession and the relative velocity of approach:

*Coefficient of restitution*

$$v_{2f} - v_{1f} = -e(v_{2i} - v_{1i}) \qquad 8\text{-}23$$

For a perfectly elastic collision $e = 1$; for a perfectly inelastic collision $e = 0$. The coefficient of restitution thus measures the elasticity of the collision.

**Example 8-12** A bullet is fired with initial speed $v_{1i}$ into a large block suspended as shown in Figure 8-16. The bullet-block system rises to height $h$. Find $v_{1i}$.

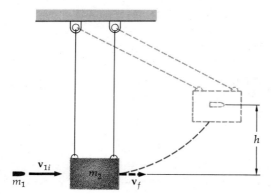

**Figure 8-16**
Ballistic pendulum. The height
$h$ is related to the speed of
the bullet-block system $v_f$ by
conservation of energy. The
speed $v_f$ can be found by
conservation of momentum of
the bullet-block system.

This arrangement is called a *ballistic pendulum*. It can be used to measure the initial speed of the bullet by measuring the height $h$ and the masses of the bullet and block. Since the time of collision is very small (we shall estimate typical collision times in Section 8-7), we can neglect any motion of the block during the collision. Since this is an inelastic collision with one object initially at rest, we can use Equation 8-22 to relate the energy after the collision to that before. If $m_1$ is the mass of the bullet and $m_2$ that of the block, we have

$$E_{kf} = \frac{m_1}{m_1 + m_2} E_{ki} = \frac{m_1}{m_1 + m_2} \tfrac{1}{2} m_1 v_{1i}^2$$

During the collision mechanical energy is not conserved, but after the collision mechanical energy is conserved. The height $h$ is that height for which the original (post-collision) kinetic energy of the bullet-block system is converted into gravitational potential energy. We then have

$$(m_1 + m_2)gh = E_{kf} = \frac{m_1}{m_1 + m_2} \tfrac{1}{2} m_1 v_{1i}^2$$

or

$$v_{1i} = \frac{m_1 + m_2}{m_1} \sqrt{2gh}$$

---

**Example 8-13** The coefficient of restitution for steel on steel is measured by dropping a steel ball onto a steel plate rigidly attached to the earth. If the ball is dropped from height $h_i$ and rebounds to height $h_f$, what is the coefficient of restitution?

The speed of the ball just before it strikes the plate is

$$v_i = \sqrt{2gh_i}$$

For the ball to reach height $h_f$ the speed just after the collision must be $v_f = \sqrt{2gh_f}$. Since the plate is attached to the earth, we can neglect its rebound velocity. The relative speeds in this case are thus $v_i$ and $v_f$:

$$v_f = ev_i$$
$$\sqrt{2gh_f} = e\sqrt{2gh_i}$$
$$e = \sqrt{\frac{h_f}{h_i}}$$

---

**Question**

8. Under what conditions can *all* the initial kinetic energy be lost in a collision?

# 8-6 Collisions in Three Dimensions

For collisions in three dimensions, the conservation of linear momentum is a vector equation. Inelastic collisions present no special difficulty. The initial momentum is found by adding the initial momentum vectors of the two bodies. Since the particles stick together and their final momentum equals the initial momentum, the particles move off in the direction of the resultant total momentum with velocity $\mathbf{v}_{CM}$, given by $\mathbf{P}/(m_1 + m_2)$.

Collisions in three dimensions which are not perfectly inelastic are more complicated. Figure 8-17 shows a particle of mass $m_1$ moving with velocity $\mathbf{v}_{1i}$ along the $x$ axis toward a particle of mass $m_2$ initially at rest.

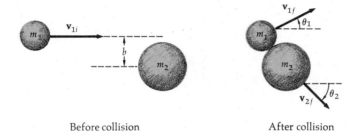

Before collision       After collision

**Figure 8-17**
Off-center collision of two bodies. The distance $b$ is called the impact parameter.

The "closeness" of the collision is measured by the distance between the centers measured perpendicular to the line of $\mathbf{v}_{1i}$. This distance $b$, called the *impact parameter*, is indicated in the figure. After the collision, particle 1 moves off with velocity $\mathbf{v}_{1f}$, making an angle $\theta_1$ with its initial velocity, and particle 2 moves with velocity $\mathbf{v}_{2f}$, making an angle $\theta_2$ with $\mathbf{v}_{1i}$. Conservation of momentum requires that $\mathbf{v}_{2f}$ lie in the plane formed by $\mathbf{v}_{1i}$ and $\mathbf{v}_{1f}$, which we shall take to be the $xy$ plane. Assuming the initial velocity $\mathbf{v}_{1i}$ to be known, we have four unknowns to find: the $x$ and $y$ components of each velocity, or alternatively, the two speeds $v_{1f}$ and $v_{2f}$, the angle of deflection $\theta_1$ of particle 1, and the angle of recoil $\theta_2$ of particle 2. The $x$ and $y$ components of the conservation-of-momentum equation give us two of the needed relations between these quantities:

*Impact parameter*

$$m_1 v_{1i} = m_1 v_{1f} \cos \theta_1 + m_2 v_{2f} \cos \theta_2 \qquad \text{8-24}$$

and

$$0 = m_1 v_{1f} \sin \theta_1 - m_2 v_{2f} \sin \theta_2 \qquad \text{8-25}$$

A third relation comes from energy considerations. If the collision is perfectly elastic, we have

$$\tfrac{1}{2} m_1 v_{1i}^2 = \tfrac{1}{2} m_1 v_{1f}^2 + \tfrac{1}{2} m_2 v_{2f}^2 \qquad \text{8-26}$$

To solve for the four unknowns, we need another relation. The fourth relationship depends on the impact parameter $b$ and on the kind of interaction force exerted by the particles on each another. If the impact parameter is zero, for example, the collision is "head on" and can be treated as a one-dimensional collision. (The angle $\theta_2$ is 0, and $\theta_1$ is either 0 or $-180°$, depending on the relative masses of the particles.) If the particles are hard spheres which interact only when in contact, there will be no collision unless $b$ is less than the sum of the radii of the spheres.

In this case, since the force exerted by one sphere on the other must be along the line of centers when the spheres are in contact (the surfaces must be frictionless if the collision is to be perfectly elastic), the angle $\theta_2$ is the angle between the line of centers and $v_{1i}$, which can easily be found from a simple diagram (see Exercise 36). In practice, the fourth relationship is often found experimentally, by measuring the angle of deflection or the angle of recoil. Such a measurement can then give information about the force of interaction of the particles.

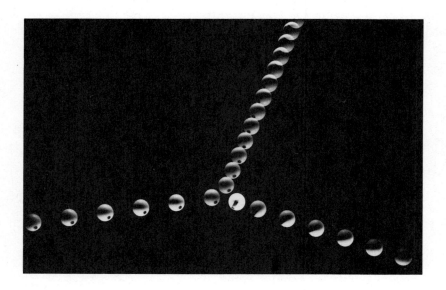

Multiflash photograph of an elastic off-center collision of two balls of equal mass. The dotted ball is incident from the left and strikes the striped ball initially at rest. The final velocities of the two balls are perpendicular to each other. (*From* PSSC Physics, *2d ed., p. 384, D. C. Heath and Company, Lexington, Mass., 1965.*)

We omit further discussion of collisions in three dimensions except for a special interesting case of a perfectly elastic collision between particles of equal mass when one of the particles is initially at rest. Figure 8-18 shows the geometry of the collision. If $\mathbf{v}_{1i}$ and $\mathbf{v}_{1f}$ are the initial and final velocities of particle 1 and $\mathbf{v}_{2f}$ is the final velocity of particle 2, conservation of momentum gives

$$m\mathbf{v}_{1i} = m\mathbf{v}_{1f} + m\mathbf{v}_{2f}$$

or

$$\mathbf{v}_{1f} + \mathbf{v}_{2f} = \mathbf{v}_{1i} \qquad\qquad 8\text{-}27$$

The final velocity vectors add to form the triangle shown in Figure 8-18b. Conservation of energy for this collision gives

$$\tfrac{1}{2}mv_{1i}^2 = \tfrac{1}{2}mv_{1f}^2 + \tfrac{1}{2}mv_{2f}^2$$

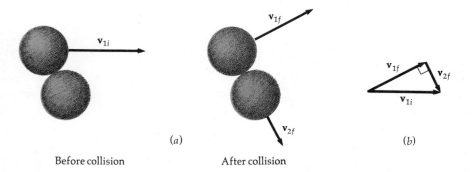

(a)

Before collision

After collision

(b)

**Figure 8-18**
(*a*) Off-center elastic collision of two spheres of equal mass when one sphere is initially at rest. After the collision the spheres move at right angles to each other. (*b*) The velocity vectors for this collision form a right triangle.

or

$$v_{1f}^2 + v_{2f}^2 = v_{1i}^2 \qquad \text{8-28}$$

Equation 8-28 is the pythagorean theorem for a right triangle formed by the vectors $\mathbf{v}_{1f}$, $\mathbf{v}_{2f}$, and $\mathbf{v}_{1i}$, the hypotenuse of the triangle being $v_{1i}$. Thus the final velocity vectors $\mathbf{v}_{1f}$ and $\mathbf{v}_{2f}$ are perpendicular to each other. We see that for the special case of a perfectly elastic collision between particles of equal mass with one particle initially at rest, the final velocities of the particles are perpendicular.

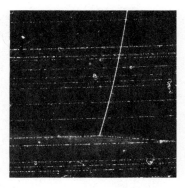

Proton-proton collision in a liquid-hydrogen bubble chamber. The incident proton enters from the left and interacts with a stationary proton in the chamber. The two particles move off at right angles after the collision. The slight curvature of the tracks is due to a magnetic field. (*Courtesy of Brookhaven National Laboratory.*)

## 8-7   Impulse and Time Average of a Force

In our study of collisions we have said nothing about the forces of interaction of the bodies except that they are usually very large and act for a short time. Figure 8-19 shows the time variation of the magnitude of a typical force exerted by one body on another during a collision. Before ·time $t_i$ the bodies are apart and the force is zero. When the bodies come into contact, the force rises steeply and then falls to zero again at time $t_f$, when the bodies separate. The time of contact $\Delta t = t_f - t_i$ is usually very small, perhaps only about 0.001 s. The area under the $F$-versus-$t$ curve is called the magnitude of the *impulse* of the force. The impulse $\mathbf{I}$ of a force is a vector defined by

$$\mathbf{I} = \int_{t_i}^{t_f} \mathbf{F}\, dt \qquad \text{8-29}$$

*Impulse defined*

Assuming that $\mathbf{F}$ is the resultant force and using Newton's second law $\mathbf{F} = d\mathbf{p}/dt$, we see that the impulse equals the total change in momentum during the time interval $t_f - t_i$:

$$\mathbf{I} = \int_{t_i}^{t_f} \mathbf{F}\, dt = \int_{t_i}^{t_f} \frac{d\mathbf{p}}{dt}\, dt = \mathbf{p}_f - \mathbf{p}_i = \Delta\mathbf{p} \qquad \text{8-30}$$

From Equation 8-30 the units of impulse are newton-seconds (kilogram-metres per second).

For a general force $\mathbf{F}$, the impulse depends on the choice of times $t_i$ and $t_f$, but the forces which occur in collisions are zero except during a small time, as shown in Figure 8-19. For these forces, the impulse does not depend on the time interval as long as $t_i$ is any time before the force occurs and $t_f$ is any time after the force has decreased to zero. It is for this type of force that the concept of impulse is most useful.

Consider a golf club hitting a ball initially at rest. The change in momentum of the ball is just its final momentum:

$$\Delta\mathbf{p} = \mathbf{I} = \mathbf{p}_f$$

The impulse exerted by the ball on the club is equal in magnitude and opposite in direction to that exerted by the club on the ball since the forces are equal and opposite according to Newton's third law.

**Figure 8-19**
Typical time variation of the force exerted by one object on another during a collision. The force becomes very large but acts only for a very short time. The area under the $F$-versus-$t$ curve is called the impulse. The impulse exerted on an object equals its change in momentum.

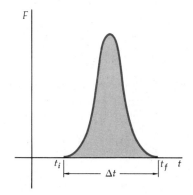

**Example 8-14** A 10-g bullet traveling at 400 m/s embeds itself in a 390-g block initially at rest. What is the impulse exerted by the bullet on the block during this inelastic collision?

The final velocity of the bullet-block system is found from momentum conservation. We have

$$(10 \text{ g} + 390 \text{ g})v_{CM} = (10 \text{ g})(400 \text{ m/s})$$

$$v_{CM} = 10 \text{ m/s}$$

Since the block is initially at rest with zero momentum, its change in momentum is just its final momentum, which is

$$p_{2f} = m_2 v_{CM} = (390 \text{ g})(10 \text{ m/s}) = 3.9 \text{ kg·m/s} = 3.9 \text{ N·s}$$

in the direction of the initial velocity of the bullet. The impulse exerted by the block on the bullet is equal and opposite; it is thus $-3.9$ N·s (in the direction opposite the original velocity of the bullet). We can check this by computing the change in momentum of the bullet:

$$p_{1f} - p_{1i} = (10 \text{ g})(10 \text{ m/s}) - (10 \text{ g})(400 \text{ m/s}) = -3.9 \text{ kg·m/s}$$
$$= -3.9 \text{ N·s}$$

The time-average force over the interval $\Delta t = t_f - t_i$ is defined by

$$\mathbf{F}_{av} = \frac{1}{\Delta t} \int_{t_i}^{t_f} \mathbf{F} \, dt = \frac{\mathbf{I}}{\Delta t} \qquad\qquad 8\text{-}31$$

*Time average of a force*

$\mathbf{F}_{av}$ is the constant force which gives the same impulse in the time interval $\Delta t$. $F_{av}$ is indicated in Figure 8-20. It is often useful to compute the average force in a collision in order to compare it with such other forces as frictional forces or the force of gravity.

**Figure 8-20**
The average force $F_{av}$ for some time interval is the constant force which gives the same impulse during that interval. The rectangular area $F_{av}\,\Delta t$ is the same as that under the $F$-versus-$t$ curve.

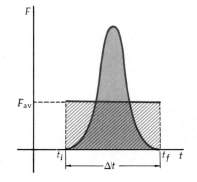

**Example 8-15** What are reasonable magnitudes for the impulse $I$, average force $F_{av}$, and collision time $\Delta t$ for a golf club hitting a golf ball?
    Let us take the mass of a typical golf ball as $m = 45$ g. The velocity of the ball when it leaves the club is related to the range by

$$R = \frac{v_0^2}{g} \sin 2\theta_0$$

where $\theta_0$ is the angle of the projectile and air resistance is neglected. Taking $\theta_0 = 45°$ for simplicity (corresponding to the maximum range) and $R = 160$ m as a reasonable range (this is about 175 yd), we find

$$v_0^2 = Rg \approx 160(10) = 1600$$

$$v_0 = 40 \text{ m/s}$$

where $g \approx 10$ m/s² has been used to simplify the calculation. The magnitude of the impulse is thus

$$I = \int F \, dt = \Delta p = mv_0 = (45 \times 10^{-3} \text{ kg})(40 \text{ m/s})$$
$$= 1.8 \text{ kg·m/s} = 1.8 \text{ N·s}$$

We can estimate the time of the collision as follows. The magnitude of the average acceleration is

$$a_{av} = \frac{F_{av}}{m} = \frac{I}{m \, \Delta t}$$

If the acceleration were constant and equal to the average acceleration, the velocity $v_0$ would be related to the distance traveled by the ball *during the collision* by the constant-acceleration expression

$$v_0^2 = 2a_{av}x$$

A reasonable estimate for the distance traveled by the ball while it is in contact with the golf club is the radius of the golf ball, about $x = 2$ cm. Then the average acceleration is

$$a_{av} = \frac{v_0^2}{2x} = \frac{1600}{2(0.02)} = 40,000 \text{ m/s}^2$$

The average force is thus estimated to be

$$F_{av} = ma_{av} = (0.045 \text{ kg}) (40,000 \text{ m/s}^2) = 1800 \text{ N}$$

The estimated collision time is then

$$\Delta t = \frac{I}{F_{av}} = \frac{1.8 \text{ N·s}}{1800 \text{ N}} = 0.001 \text{ s}$$

We could also have estimated $\Delta t$ from $v_0 = a_{av} \Delta t$. Then

$$\Delta t = \frac{v_0}{a_{av}} = \frac{40 \text{ m/s}}{40,000 \text{ m/s}^2} = 0.001 \text{ s}$$

We see that the average force exerted by the club, 1800 N, is much larger than any other forces acting on the ball. For example, the weight of the ball is only about 0.45 N, and the frictional forces exerted by the grass or tee on the ball are even less, assuming the coefficient of friction to be less than 1.

Golf club hitting a golf ball. Note the deformation of the golf ball due to the large force exerted by the club during the brief time of contact. As the ball leaves the club, it springs back to its original shape, converting the elastic potential energy of deformation into kinetic energy. (*Courtesy of Harold E. Edgerton.*)

**Example 8-16** A machine gun fires $R$ bullets per second. Each bullet has mass $m$ and speed $v$. The bullets strike a fixed target, where they stop. What is the average force exerted on the target over a time that is long compared with the time between bullets?

Figure 8-21 shows a graph of the instantaneous force on the target.

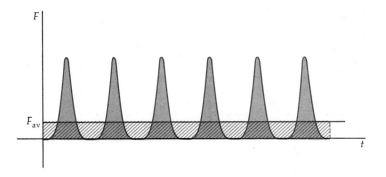

**Figure 8-21**
Instantaneous and average force exerted by bullets striking a target.

The average force is also indicated. The area under each impulse curve equals the change in momentum of each bullet. This change is just the initial momentum $mv$ since the final momentum is zero. In time $\Delta t$ the number of bullets that hit the target is $R \Delta t$. Each bullet exerts an impulse $mv$ on the target. The total impulse exerted in time $\Delta t$ by the $R \Delta t$ bullets is consequently $R \Delta t \ mv$. The average force is thus

$$F_{av} = \frac{R \Delta t \ mv}{\Delta t} = Rmv \qquad\qquad 8\text{-}32$$

Equation 8-32 also gives the average recoil force exerted by the bullets on the gun. Each bullet receives an impulse of magnitude $mv$ from the gun, and, by Newton's third law, each bullet exerts an impulse back on the gun of magnitude $mv$. Since in time $\Delta t$, $R \Delta t$ bullets leave the gun, the total impulse exerted on the gun in time $\Delta t$ is $mvR \Delta t$ and the average recoil force is $Rmv$.

## 8-8   Systems with Varying Mass: Rocket Motion

Although the mass of a particle in classical mechanics is always constant, we sometimes wish to describe a system of particles in which the mass is varying. The most common example is the motion of a rocket. In such motion fuel is burned in the rocket and exhaust gas is expelled out the back of the rocket. The force exerted by the exhaust gas on the rocket (which is equal and opposite to the force exerted by the rocket on the gas to expel it) propels the rocket forward. Consider a rocket in free space with no external forces. If we include the mass of the exhaust gas in our system, the total mass of the system does not change and the center of mass remains at rest (or moving with its initial constant velocity). We are usually not interested in this total system including the exhaust gas; instead, we want to describe the motion of the rocket itself. We now develop an equation for the motion of a rocket. The easiest way is to compute the change in momentum of the total system (including the exhaust gas) for some time interval and set this change equal to the impulse exerted on the system by the external forces acting. Let $F_{ext}$ be the resultant external force acting, $m$ be the mass of the rocket (plus fuel inside not yet burned), and $v$ be its speed at time $t$. At a later time $t + \Delta t$ the rocket has expelled some gas of mass $|\Delta m|$ (Figure 8-22). (We use absolute-value signs because we later wish to relate this small amount of mass to the rate of change of mass of the rocket $dm/dt$, which is intrinsically negative.) At time $t + \Delta t$ the rocket then has mass $m - |\Delta m|$ and is moving at speed $v + \Delta v$. If the gas is exhausted at speed $u_{ex}$ relative to the rocket, its velocity at time $t + \Delta t$ is $v - u_{ex}$. The initial momentum of the system at time $t$ is

$$P_i = mv$$

The momentum of the system at time $t + \Delta t$ is

$$
\begin{aligned}
P_f &= (m - |\Delta m|)\,(v + \Delta v) + |\Delta m|\,(v - u_{ex}) \\
&= mv + m\,\Delta v - v|\Delta m| - |\Delta m|\,\Delta v + v|\Delta m| - u_{ex}|\Delta m| \\
&\approx mv + m\,\Delta v - u_{ex}|\Delta m|
\end{aligned}
$$

where we have dropped the term $|\Delta m|\,\Delta v$, the product of two very small quantities, and a negligible term compared with the others if the time interval $\Delta t$ is very small. Computing the change in momentum and setting it equal to the impulse, we have

$$\Delta P = m\,\Delta v - u_{ex}|\Delta m| = F_{ext}\,\Delta t$$

We now divide by the time interval and take the limit as $\Delta t$ approaches zero. The term $\Delta v/\Delta t$ approaches the derivative $dv/dt$, which is the acceleration, and the term $|\Delta m|/\Delta t$ approaches $|dm/dt|$, the absolute value of the rate of change of the mass of the rocket. We then have

$$m\frac{dv}{dt} = u_{ex}\left|\frac{dm}{dt}\right| + F_{ext} \qquad\qquad 8\text{-}33$$

*Rocket equation*

Equation 8-33 is the *rocket equation*. The quantity $u_{ex}|dm/dt|$ is called the *thrust* of the rocket. When the rocket moves near the surface of the earth, the external force $F_{ext}$ is the weight of the rocket; it is negative in Equation 8-33 because it is directly opposite to the direction of the velocity (assuming the rocket is moving upward). The thrust must then be greater than the weight of the rocket if the rocket is to accelerate. To solve Equation 8-33 we must be given the exhaust speed relative to the

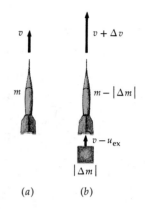

**Figure 8-22**
(*a*) Rocket moving with initial velocity $v$. (*b*) After a time interval $\Delta t$ the rocket has mass $m - |\Delta m|$ and moves with velocity $v + \Delta v$. Gas exhausted from the rocket at speed $u_{ex}$ relative to the rocket is moving with velocity $v - u_{ex}$. The change in momentum of the rocket-exhaust system equals the impulse $F_{ext}\,\Delta t$.

rocket $u_{ex}$, the rate the rocket burns fuel to obtain $dm/dt$, and the external force. The solution of this equation is complicated by the fact that $m$ is not a constant but a function of time. For example, if the rocket burns fuel at a constant rate $R$, the mass of the rocket at any time is $m = m_i - Rt$, where $m_i$ is the initial mass. We shall consider the special case of a rocket in free space with no external force. Since $dm/dt$ is negative, we write $-dm/dt = |dm/dt|$. Equation 8-33 is then

$$m\frac{dv}{dt} = -u_{ex}\frac{dm}{dt} \qquad \text{or} \qquad dv = -u_{ex}\frac{dm}{m}$$

Integrating gives

$$v_f - v_i = -u_{ex}\ln\frac{m_f}{m_i} = +u_{ex}\ln\frac{m_i}{m_f} \qquad\qquad \text{8-34}$$

Equation 8-34 gives the change in velocity of the rocket in terms of the exhaust speed and the ratio of the initial and final mass. The mass of the rocket without any fuel is called the *payload*. If the payload is just 10 percent of the total initial mass, i.e., if 90 percent of the initial mass is fuel, the ratio $m_i/m_f$ when the fuel is spent will be 10. Then if $v_i = 0$, the final speed will be

$$v_f = u_{ex}\ln 10 = 2.3u_{ex}$$

The logarithmic nature of Equation 8-34 severely restricts the final speeds attainable. For example, if we increase the fuel so that the payload is only 1 percent of the initial mass, the final velocity is $4.6u_{ex}$, just twice that for a 10 percent payload.

---

**Example 8-17** A rocket has initial mass of 20,000 kg, of which 20 percent is the payload. It burns fuel at a rate of 200 kg/s and exhausts gas at a relative speed of 2 km/s. Find the thrust of the rocket and its final speed when the fuel is spent, assuming no external forces.

The thrust is

$$F_{th} = u_{ex}\left|\frac{dm}{dt}\right| = (2 \text{ km/s})\,(200 \text{ kg/s}) = 4 \times 10^5 \text{ kg·m/s}^2$$
$$= 4 \times 10^5 \text{ N} \approx 90,000 \text{ lb}$$

Since the final mass is the payload, which is 20 percent of 20,000 kg or 4000 kg, the ratio of the initial and final masses is $m_i/m_f = (20,000 \text{ kg})/(4000 \text{ kg}) = 5$. The final speed assuming it starts from rest is

$$v = u_{ex}\ln\frac{m_i}{m_f} = (2 \text{ km/s})\,(\ln 5) = 3.22 \text{ km/s}$$

---

*Optional*

## 8-9  The Center-of-Mass Reference Frame

As we have seen, the velocity of the center of mass remains unchanged by internal forces within the system such as those exerted in a collision of two particles. It is often convenient to look at a collision in a reference frame attached to the center of mass. In such a reference frame the center of mass is at rest, and according to Equation 8-4, the total momentum of the system is zero. This reference frame is called the *center-of-mass frame* or the *zero-momentum frame*. For example, in an inelastic collision in the center-of-mass frame, the two bodies originally move toward each other with equal but opposite momenta and come to rest after the collision. We shall now show that in a one-dimensional perfectly elastic collision in the center-of-mass frame the velocity of each body is reversed by the collision. Let $\mathbf{u}_{1i}$ and $\mathbf{u}_{2i}$ be the initial velocities of the bodies. Their initial momenta are then $\mathbf{p}_{1i} = m_1\mathbf{u}_{1i}$ and $\mathbf{p}_{2i} = m_2\mathbf{u}_{2i}$. (We are using vector notation, even though we are considering only one-dimensional collisions here, in order to keep track of the signs. Of course if $\mathbf{u}_{1i}$ is to the right, $\mathbf{u}_{2i}$ must be to the left because the total momentum is zero in this frame.) Since the total momentum is zero in this frame, $\mathbf{p}_{1i} = -\mathbf{p}_{2i}$, and the magnitudes $p_{1i}$ and $p_{2i}$ are equal. The total kinetic energy $E_{ri}$ is

$$E_{ri} = \tfrac{1}{2}m_i u_{1i}^2 + \tfrac{1}{2}m_2 u_{2i}^2 = \frac{p_{1i}^2}{2m_1} + \frac{p_{2i}^2}{2m_2} = p_{1i}^2\left(\frac{1}{2m_1} + \frac{1}{2m_2}\right) \qquad 8\text{-}35$$

After the collision the energy is

$$E_{rf} = \frac{p_{1f}^2}{2m_1} + \frac{p_{2f}^2}{2m_2} = p_{1f}^2\left(\frac{1}{2m_1} + \frac{1}{2m_2}\right) \qquad 8\text{-}36$$

where $p_{1f}$ and $p_{2f}$ are the magnitudes of the final momenta of the bodies. We have used the fact that the total momentum after the collision is still zero, and hence the two masses still have momenta of equal magnitude though opposite direction. This is true to the extent that we can neglect any external force on the system during the collision. Setting the final energy equal to the initial energy for a perfectly elastic collision, we obtain

$$p_{1f}^2 = p_{1i}^2 \qquad 8\text{-}37$$

or

$$\mathbf{p}_{1f} = \pm\mathbf{p}_{1i} \qquad \text{and} \qquad \mathbf{p}_{2f} = \pm\mathbf{p}_{2i} \qquad 8\text{-}38$$

The plus sign in Equation 8-38 corresponds to no collision at all, as would be the case if the masses were initially heading away from each other. Since only conservation of momentum and conservation of energy were used to derive Equation 8-38, it is not surprising that the result includes the case of no collision. If the bodies are originally heading toward each other and collide, Equation 8-38 gives for each body

$$\mathbf{p}_{1f} = -\mathbf{p}_{1i} \qquad \mathbf{p}_{2f} = -\mathbf{p}_{2i} \qquad 8\text{-}39$$

Since the velocity of each body is just its momentum divided by its mass, Equation 8-39 also implies that the final velocity of each body is related to its initial velocity by

$$\mathbf{u}_{1f} = -\mathbf{u}_{1i} \qquad \text{and} \qquad \mathbf{u}_{2f} = -\mathbf{u}_{2i} \qquad 8\text{-}40$$

The one-dimensional perfectly elastic collision is very simple in the zero-momentum reference frame. Each body is merely turned around and leaves with the speed and energy it had before the collision.

We now consider another reference frame which is moving to the left with speed $v_{CM}$ relative to the center-of-mass frame. In this second frame, the center of mass is moving to the right with speed $v_{CM}$. We can obtain the velocities of each particle in this new frame both before and after the collision by simply adding the velocity $\mathbf{v}_{CM}$ to each one. Conversely, if we are viewing a system originally in some frame in which the center of mass is moving with velocity $\mathbf{v}_{CM}$, we can transfer into the center-of-mass frame by subtracting the velocity $\mathbf{v}_{CM}$ from each body.

As we have seen, an elastic collision does not look as simple in any general reference frame. Even though the total kinetic energy is conserved, some or all of the kinetic energy of one body is transferred in the collision to the other body. Let us calculate the energy transferred in a perfectly elastic collision from a moving body to one that is stationary before the collision. The reference frame in which one body is originally stationary is called the *laboratory frame* because experiments are most often done by bombarding a stationary particle with other moving particles. Consider a body of mass $m_1$ and initial velocity $v_{1i}$ moving toward a body of mass $m_2$ initially at rest and assume the collision to be perfectly elastic. The velocity of the center of mass in this, the laboratory, frame is

$$v_{CM} = \frac{m_1 v_{1i}}{m_1 + m_2} \qquad\qquad 8\text{-}41$$

We need only calculate the final energy of the second body, which we do by first transferring into the center-of-mass frame and then transferring back into the original laboratory frame. In the center-of-mass frame the second body is originally moving to the left with speed $v_{CM}$ given by Equation 8-41. After the collision it is moving to the right with this same speed. To transfer into the original laboratory frame, we add $v_{CM}$ to each body. The speed of the body originally at rest is then $2v_{CM}$ after the collision, and its kinetic energy is

$$E_{k2} = \tfrac{1}{2} m_2 v_{2f}^2 = \tfrac{1}{2} m_2 \left( \frac{2 m_i v_{1i}}{m_1 + m_2} \right)^2 = \frac{4 m_1 m_2}{(m_1 + m_2)^2} \tfrac{1}{2} m_1 v_{1i}^2$$

The energy transferred is proportional to the original energy of the incident body. The energy gained by particle 2 equals that lost by particle 1. If we call the energy transferred $\Delta E$, the fractional energy loss of the incident particle is

$$\frac{\Delta E}{E} = \frac{4 m_1 m_2}{(m_1 + m_2)^2} \qquad\qquad 8\text{-}42$$

The fractional energy transferred has a maximum value of 1 if the masses are equal. In this case, the body originally moving is stopped, transferring all its original energy to the second body. If the masses of the particles are very different, the fractional energy transfer is small. In the limit of one mass much larger than the other, the fractional energy transfer approaches 4 times the ratio of the small mass to the large one, as can be seen from Equation 8-42. For example, if $m_1$ is much larger than $m_2$, we neglect $m_2$ in the denominator and obtain $\Delta E/E \approx 4 m_1 m_2 / m_1^2 = 4 m_2 / m_1$.

The result that the energy transfer in an elastic collision is large only if the masses of the two bodies are about the same has an important application in slowing down neutrons in a nuclear reactor. The probability of inducing fission in $^{235}$U is large only if the neutron is moving very slowly, whereas the neutrons emitted in the fission are fast-moving. In order to sustain a chain reaction, these fast neutrons must be slowed down before they have a chance to leak out of the reactor. The main process for slowing neutrons down is elastic collision. From our calculations we have seen that efficient transfer of neutron energy by elastic collision can occur only in collisions with particles having masses near that of the neutron. Neither an electron in the reactor, with a mass about $\frac{1}{2000}$ times that of a neutron, nor a nucleus of nonfissionable $^{238}$U (which makes up about 99 percent of natural uranium), with a mass about 238

Culver Pictures

This collision, which will probably be inelastic, is observed by Laurel in the laboratory frame.

times the mass of a neutron, can efficiently absorb energy from a neutron in a collision. Thus some material containing light nuclei is placed in a reactor to slow the neutrons down. This material, called a *moderator*, is usually water (containing hydrogen and oxygen nuclei) or carbon.

---

**Example 8-18** Work Example 8-10 by first transferring into the center-of-mass reference frame.

In this example a 4-kg block moving to the right at 6 m/s makes an elastic collision with a 2-kg block moving at 3 m/s to the right. As calculated in that example, the total momentum of the system is $(4\text{ kg}) \times (6\text{ m/s}) + (2\text{ kg})(3\text{ m/s}) = 30\text{ kg·m/s}$. Dividing this by the total mass, we obtain $v_{CM} = 5$ m/s for the velocity of the center of mass. We now transfer to the center-of-mass frame (Figure 8-23) by subtracting 5 m/s from each velocity. We obtain for the velocities in the center-of-mass frame

$$u_{1i} = 6\text{ m/s} - 5\text{ m/s} = 1\text{ m/s}$$

$$u_{2i} = 3\text{ m/s} - 5\text{ m/s} = -2\text{ m/s}$$

In this frame the bodies are merely turned around by the collision. After the collision the velocities are

$$u_{1f} = -1\text{ m/s} \qquad u_{2f} = +2\text{ m/s}$$

We now transfer back into the original frame by adding $v_{CM} = 5$ m/s to each velocity. In the original frame the final velocities are

$$v_{1f} = -1\text{ m/s} + 5\text{ m/s} = 4\text{ m/s}$$

$$v_{2f} = +2\text{ m/s} + 5\text{ m/s} = 7\text{ m/s}$$

**Figure 8-23**
Calculation of final velocities after an elastic collision by considering how the collision looks in the center-of-mass reference frame. The center-of-mass velocity $v_{CM} = 5$ m/s found in the original frame is subtracted from each body to obtain the velocities relative to the center-of-mass frame, as in (b). In this frame, the objects are merely reversed in an elastic collision. Adding $v_{CM} = 5$ m/s transforms the problem back into the original frame, as in (d).

(a) Initial conditions

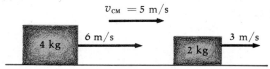

(b) Step 1: Convert to CM frame by subtracting $v_{CM}$

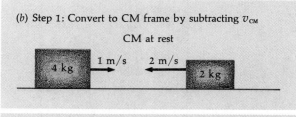

(c) Step 2: Solve collision

(d) Step 3: Reconvert to original frame by adding $v_{CM}$

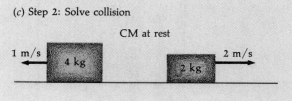

This is the same result we obtained in Example 8-10. This method is probably the easiest method of working problems involving perfectly elastic collisions in one dimension.

*Optional*

## 8-10   Reaction Threshold

When two atoms or molecules collide and the objects emerging from the collision are different from the original ones, the collision is called a *reaction.* Atomic and molecular reactions are characterized by an exchange between kinetic energy and internal energy, and energy is sometimes radiated away from the system.

In this section we consider certain reactions as inelastic collisions. Consider the reaction in which a hydrogen atom is ionized by being struck by a proton, written

$$p + {}^1\text{H} \rightarrow p + p + e + Q \qquad\qquad 8\text{-}43$$

*Exothermic and endothermic reactions*

where $Q$ is the excess kinetic energy in the zero-momentum reference frame of the final particles (those on the right) over the initial particles (those on the left). If $Q$ is positive, the reaction is called *exothermic;* energy is given off, and the reaction may occur even if there is no initial kinetic energy (in the zero-momentum frame). If $Q$ is negative, initial kinetic energy of the particles on the left side of the reaction is needed to make the reaction occur and the reaction is *endothermic.* For the reaction in Equation 8-43, $Q = -13.6$ eV; it takes 13.6 eV to remove the electron from the hydrogen atom.

In the zero-momentum frame the proton and ${}^1$H atom approach each other with equal and opposite momenta. If the total kinetic energy is less than 13.6 eV, the ${}^1$H atom cannot be ionized. If the total energy is greater than 13.6 eV, the atom can be ionized and the excess energy appears as kinetic energy of the final particles (two protons and one electron). If the energy is just 13.6 eV, the atom can be ionized with the particles relatively at rest after the reaction. Thus the minimum energy needed to produce this reaction in the zero-momentum reference frame is 13.6 eV. In the laboratory frame, where the hydrogen atom is initially at rest, the minimum kinetic energy of the incoming proton must be greater than 13.6 eV because not all of the original kinetic energy of the proton can be lost in ionizing the hydrogen atom. As we have seen, only the kinetic energy of motion relative to the center of mass can be lost. The minimum kinetic energy needed to produce the reaction in the laboratory frame is called the *reaction-threshold energy.* The reaction cannot take place in the laboratory frame at energies below the reaction threshold. We can use our results for perfectly inelastic collisions to calculate the reaction threshold, since at threshold the reaction products have no kinetic energy of relative motion but move together like the particles in a perfectly inelastic collision. Let the incoming object have mass $m_1$ and the stationary one have mass $m_2$. The energy loss in a perfectly inelastic collision can be obtained from Equation 8-22. Dropping the subscripts $k$ from the energies in Equation 8-22, we have for the final energy $E_f$ in terms of the initial energy $E_i$

*Reaction-threshold energy*

$$E_f = \frac{m_1}{m_1 + m_2} E_i$$

The energy loss $E_i - E_f$ is then

$$E_i - E_f = \left(1 - \frac{m_1}{m_1 + m_2}\right) E_i$$

$$= \frac{m_2}{m_1 + m_2} E_i$$

In our case, the initial energy is the threshold energy $E_{th}$, and the energy loss equals the magnitude of $Q$. Thus

$$\frac{m_2}{m_1 + m_2} E_{th} = |Q|$$

$$E_{th} = \frac{m_1 + m_2}{m_2} |Q| \qquad \text{for } Q \text{ negative} \qquad\qquad 8\text{-}44$$

Of course if $Q$ is positive, there is no threshold energy.

---

**Example 8-19** Find the threshold energy for the reaction of Equation 8-43.

Since the mass of the electron is $\frac{1}{1836}$ times that of the proton, the masses of a hydrogen atom and a proton are approximately equal: $m_1 \approx m_2$. Using $Q = -13.6$ eV gives

$$E_{th} = \frac{m + m}{m} \, 13.6 \text{ eV} = 2(13.6 \text{ eV}) = 27.2 \text{ eV}$$

In the laboratory frame, only half the initial kinetic energy can be used to ionize the hydrogen atom because half the energy is associated with center-of-mass motion, which cannot change. If the initial velocity of the proton is $v_{1i}$, the velocity of the center of mass is $v_{CM} = \frac{1}{2}v_{1i}$ and the center-of-mass energy is

$$\tfrac{1}{2}(2m)v_{CM}^2 = \tfrac{1}{2}(2m)\left(\tfrac{1}{2}v_{1i}\right)^2 = \tfrac{1}{2}\left(\tfrac{1}{2}mv_{1i}^2\right)$$

Thus in order for the proton to lose 13.6 eV to ionize the hydrogen atom, it must have an initial energy of 27.2 eV in the laboratory frame. The threshold energy is thus 27.2 eV.

---

**Example 8-20** A hydrogen atom at rest in the laboratory is ionized by bombarding it with electrons. What is the threshold energy?

For this case, $m_1$ is very small compared with $m_2$. We have

$$E_{th} = \frac{m_1 + m_2}{m_2} |Q| = \left(1 + \frac{m_1}{m_2}\right) |Q|$$

$$= \left(1 + \frac{1}{1836}\right) |Q|$$

We can show in another way that the threshold energy is just slightly greater than 13.6 eV. Because the mass of the hydrogen atom is so much greater than that of the electron, the center of mass is practically at the hydrogen atom and moves with the very low velocity of

$$v_{CM} \approx \frac{m_e}{m_p} v_{1i}$$

where $m_e$ is the electron mass, $m_p$ the proton mass, and $v_{1i}$ the initial velocity of the electron. Since $v_{CM}$ is so small, the laboratory frame and the center-of-mass frame are almost the same; thus nearly all the initial kinetic energy is available for the reaction.

**Example 8-21** What is the threshold in the laboratory frame for producing the nuclear reaction

$$n + {}^3\text{He} \rightarrow {}^2\text{H} + {}^2\text{H} + Q$$

by bombarding stationary ${}^3$He atoms with neutrons?

The $Q$ value of this reaction is $Q = -3.27$ MeV. (The inverse reaction, the combination of two deuterons to produce ${}^3$He plus a neutron, ${}^2\text{H} + {}^2\text{H} \rightarrow {}^3\text{He} + n + 3.27$ MeV, is an important fusion reaction which produces energy.) Since the mass of ${}^3$He is about 3 times that of the neutron, the quantity $(m_1 + m_2)/m_2$ in Equation 8-44 is $\frac{4}{3}$ and the threshold energy is

$$E_{\text{th}} = \frac{m_1 + m_2}{m_2} |Q| = \tfrac{4}{3}(3.27 \text{ MeV}) = 4.36 \text{ MeV}$$

In this case only one-fourth the energy of the neutron is wasted in center-of-mass motion. Again, if the particle initially at rest in the laboratory is much more massive than the bombarding particle, the laboratory frame is nearly the same as the center-of-mass frame and nearly all the initial energy is available for the reaction.

## Review

A. Define, explain, or otherwise identify:

B. True or false: If the statement is true only under certain conditions, write false and state the conditions under which it is true.

1. The center of mass of a two-particle system is always closer to the more massive particle.

2. The total momentum of a system equals the product of the total mass and the velocity of the center of mass.

3. The momentum of a system can be conserved even if the mechanical energy is not.

4. Internal forces do not affect the motion of the center of mass of the system.

5. In a perfectly inelastic collision all the kinetic energy of the particles is lost.

6. In a perfectly elastic collision the energy of each particle is the same before and after the collision.

7. In a perfectly elastic collision the relative velocity of recession after the collision equals the relative velocity of approach before the collision.

8. In a reference frame in which the center of mass is at rest, the total momentum of a system is zero.

9. In a reference frame in which the center of mass is at rest, it is possible for all the kinetic energy to be lost in a collision.

Exercises

### Section 8-1, Motion of the Center of Mass of a System, and Section 8-2, Finding the Center of Mass

1. A 2-kg block is moving to the right at 4 m/s, and a 4-kg block is moving to the left at 5 m/s. (a) What is the velocity of the center of mass of this system? (b) What is the total momentum of this system? (c) If at $t = 0$ the 2-kg block is at $x = 3$ m and the 4-kg block is at $x = -6$ m, where is the center of mass at this time? (d) Where is each block and the center of mass at $t = 3$ s if there are no forces acting on the system?

2. Three point masses have the following velocities along the $x$ axis: a 1-kg mass is moving to the right with $v_x = 5$ m/s, a 2-kg mass is moving to the left with $v_x = -3$ m/s, and a 4-kg mass is moving to the right with $v_x = 4$ m/s. Find (a) the total momentum of the system and (b) the velocity of the center of mass.

3. Three point masses are located in the $xy$ plane as follows: a 1-kg mass is at the origin, a second 1-kg mass is at $x = 4$ m on the $x$ axis, and a 2-kg mass is at the point $x = 2$ m, $y = 2$ m. Find the center of mass.

4. Find the center of mass of the system shown in Figure 8-24.

5. Two 3-kg masses have velocities $\mathbf{v}_1 = 2\mathbf{i} + 3\mathbf{j}$ m/s and $\mathbf{v}_2 = 4\mathbf{i} - 6\mathbf{j}$ m/s. Find (a) the velocity of the center of mass and (b) the total momentum of the system.

6. A 5-kg mass is at rest at the origin, and a second 5-kg mass is initially at rest at $x = 0$, $y = 4$ m. A constant force $\mathbf{F} = (10\text{ N})\mathbf{i}$ is applied to this second mass while no force acts on the first mass. (a) Find the acceleration of the second mass and its position at the times $t = 0, 1, 2,$ and 3 s. (b) On a diagram indicate these positions and the position of the center of mass at these times. (c) Find the acceleration of the center of mass and from it the position of the center of mass at these times.

7. A constant force $\mathbf{F} = 24\mathbf{i}$ N is applied at $t = 0$ to the two-particle system of Exercise 5. (a) Find the velocity of the center of mass at $t = 5$ s. (b) If the center of mass is at the origin at $t = 0$, where is it at $t = 5$ s?

8. Find the center-of-mass coordinates $x_{CM}$ and $y_{CM}$ for the object in Figure 8-25.

9. A nonuniform thin rod of length $L$ lies along the $x$ axis with one end at the origin. It has a linear mass density $\lambda$ kg/m, given by $\lambda = \lambda_0(1 + x/L)$. The density is thus twice as great at one end as at the other. (a) Use $M = \int dm$ to find the total mass. (b) Find the center of mass of the rod.

10. A 3-kg object slides along a frictionless horizontal plane in the $x$ direction at 4 m/s. It explodes into two pieces, one of mass 2 kg and the other 1 kg. The 1-kg piece moves along the $y$ axis at 8 m/s after the explosion. (a) What is the velocity of the 2-kg piece after the explosion? (b) What is the velocity of the center of mass after the explosion?

### Section 8-3, Conservation of Linear Momentum

11. A cart of mass 10 kg is rolling along a level floor at a speed of 5 m/s. A 4-kg mass dropped from rest lands in the cart. (a) What was the momentum of the cart before the mass was dropped? (b) What is the momentum of the cart and mass after the mass is dropped into the cart? (c) What is the speed of the cart and mass?

12. Two railroad cars have masses of $6 \times 10^4$ and $4 \times 10^4$ kg. They are initially rolling along the track in the same direction with the lighter car in front, moving at 0.5 m/s, and the heavier car trailing but moving at 1.0 m/s. They collide and couple. (a) What is the total momentum of the two-car system before the collision? (b) What is the total momentum of the two-car system after the collision?

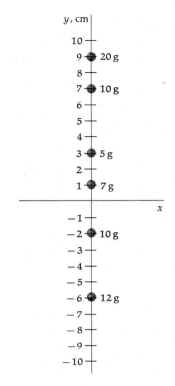

**Figure 8-24**
Exercise 4.

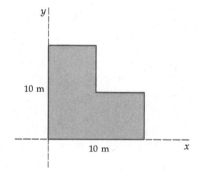

**Figure 8-25**
Exercise 8.

(c) What is the speed of the two cars after the collision? (d) Find the total kinetic energy before and after the collision.

13.  An open-topped freight car of mass 20 Mg is rolling along a track at 5 m/s. Rain is falling vertically downward into the car. After the car has collected 2 Mg of water, what is its speed?

14.  Two masses are held at rest initially on a horizontal frictionless surface. They are pushed together, compressing a small spring between them which is not attached to either mass. When the masses are released, the spring accelerates both, giving mass $m_1$ a velocity 5 m/s to the left and $m_2$ a velocity 15 m/s to the right. (a) What is the total momentum of the system before the masses are released? After they are released? (b) What is the ratio $m_1/m_2$?

15.  A 20-g bullet is fired horizontally with velocity 250 m/s from a 1.5-kg gun. If the gun were held loosely in the hand, what would its recoil velocity be?

### Section 8-4, Energy of a System of Particles

16.  (a) Find the total kinetic energy of the two blocks in Exercise 1. (b) Find the velocity of each block relative to the center of mass and use it to compute the kinetic energy of relative motion $E_{k,\text{rel}}$. (c) Show that the difference between your answers to (a) and (b) is $\frac{1}{2}Mv_{\text{CM}}^2$.

17.  Find the potential energy for the system in Exercise 4 directly by summing $m_i g y_i$ for all the particles and show that the result equals $Mgy_{\text{CM}}$.

18.  Describe how a solid ball can move so that (a) its total kinetic energy is just the energy of center-of-mass motion and (b) its total kinetic energy is energy of motion relative to the center of mass.

19.  A 3-kg body moves at 5 m/s to the right. It is chasing a second 3-kg body moving at 1 m/s also to the right. (a) Find the total kinetic energy of the two bodies in this frame. (b) Find the velocity of the center of mass. (c) Find the velocities of each body relative to the center of mass. (d) Find the kinetic energy of motion relative to the center of mass. (e) Show that your answer for part (a) is greater than that for part (d) by the amount $\frac{1}{2}Mv_{\text{CM}}^2$, where $v_{\text{CM}}$ is the velocity of the center of mass and $M$ is the total mass.

20.  Two bodies are connected by a spring. Describe how the bodies can move so that (a) the total kinetic energy is just the energy of center-of-mass motion and (b) the kinetic energy is entirely energy of motion relative to the center of mass.

21.  Repeat Exercise 19 for a 2-kg body moving to the right with speed 5 m/s toward a stationary body of mass 3 kg.

### Section 8-5, Collisions in One Dimension

22.  Find the velocity of each body if the bodies in Exercise 19 make (a) a perfectly inelastic collision; (b) a perfectly elastic collision.

23.  Find the velocity of each body if the bodies in Exercise 21 make (a) a perfectly inelastic collision; (b) a perfectly elastic collision.

24.  A 4-kg fish is swimming at 1 m/s to the right. He swallows a $\frac{1}{8}$-kg fish swimming toward him at 3 m/s to the left (Figure 8-26). What is the velocity of the larger fish immediately after his lunch?

1 m/s                   3 m/s

4 kg                    $\frac{1}{8}$ kg

**Figure 8-26**
Inelastic collision of two fish for Exercise 24.

25. Two blocks are sliding toward each other on a frictionless surface. One block of mass 10 kg is coming from the left at 5 m/s, and the other of mass 6 kg is coming from the right at −3 m/s. Find the velocity of each block after the collision if the collision is (a) perfectly inelastic; (b) perfectly elastic.

26. A 70-kg baseball player jumps vertically into the air to catch a 140-g baseball traveling horizontally at 40 m/s. If the velocity of the baseball player is 15 cm/s upward just before he catches the ball, what is his velocity just after he catches the ball?

27. A 10-g bullet is fired with velocity of 300 m/s into a pendulum bob which has a mass of 990 g. How high does the pendulum bob (plus bullet) swing after the collision?

28. A 15-g bullet is fired into the bob of a ballistic pendulum which has a mass of 1.5 kg. With the bob at maximum height the strings make an angle of 60° with the vertical. The length of the pendulum is 2 m. Find the velocity of the bullet.

29. A 20-Mg railway car stands on a hill with its brakes set. The brakes are released, and the car rolls down to the bottom of the hill 9 m below its original position (Figure 8-27). It collides with a 10-Mg car resting at the bottom of the track (with brakes off). The two cars couple together and roll up the track to a height H. Find H.

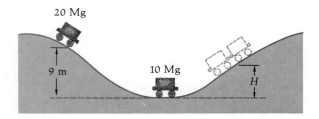

20 Mg

9 m

10 Mg

H

**Figure 8-27**
Inelastic collision of railway cars for Exercise 29.

30. A 1-kg block of wood is attached to a spring of force constant 200 N/m and rests on a smooth surface, as shown in Figure 8-28. A 20-g bullet is fired into the block, and the spring compresses 13.3 cm. (a) Find the original velocity of the bullet before the collision. (b) What fraction of the original mechanical energy is lost in this collision?

20 g

k = 200 N/m

1 kg

**Figure 8-28**
Exercise 30.

31. A 4-kg object moving at 5 m/s makes a perfectly elastic collision with a 1-kg object initially at rest. (a) Find the final velocities of each object. (b) Find the energy transferred to the 1-kg object.

32. A 10-kg object traveling at 5 m/s collides elastically with a 100-g object initially at rest. (a) Find the velocities of each object after the collision. (b) Find the fraction of the initial kinetic energy lost by the 10-kg object.

33. A ball bounces to 80 percent of its original height. (a) What fraction of its mechanical energy is lost in each bounce? (b) Where does this energy go? (c) What is the coefficient of restitution of the ball-floor system?

34. A 2-kg object moving at 3 m/s to the right collides with a 3-kg object moving at 2 m/s to the left. The coefficient of restitution is 0.6. Find the velocity of each object after the collision.

## Section 8-6, Collisions in Three Dimensions

35.  A small 1-Mg car traveling north at 80 km/h collides at an intersection with a 3-Mg truck traveling east at 60 km/h. The car and truck stick together. (*a*) What is the total momentum of the car-truck system before the collision? (*b*) Find the magnitude and direction of the velocity of the combined wreckage just after the collision.

36.  A sphere of radius $r_1$ moving parallel to the *x* axis collides with a stationary sphere of radius $r_2$. Show that when the spheres are in contact, the angle made by the line of centers of the spheres with the *x* axis is given by $\sin \theta = b/(r_1 + r_2)$, where *b* is the impact parameter.

37.  Figure 8-29 shows the results of a collision of two objects of unequal mass. (*a*) Find the speed $v_2$ of the larger mass after the collision and the angle $\theta_2$. (*b*) Show that this collision is perfectly elastic.

Before                                    After

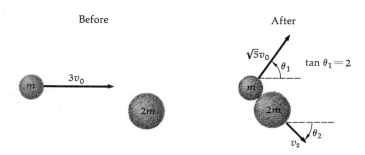

**Figure 8-29**
Exercise 37.

## Section 8-7, Impulse and Time Average of a Force

38.  A 0.4-kg particle initially moving at 20 m/s is stopped by a constant force of 50 N, which lasts for a short time $\Delta t$. (*a*) What is the impulse of this force? (*b*) Find the time interval $\Delta t$.

39.  Find the impulse necessary to stop a 1500-kg car traveling at 90 km/h.

40.  A 0.5-kg ball is dropped from a height of 3 m above the floor. It rebounds to a height of 2 m. (*a*) Find the impulse exerted by the floor on the ball. (*b*) If the ball was in contact with the floor for 1 ms, find the average force exerted by the floor on the ball during that time.

41.  A girl can exert an average force of 200 N on her machine gun. Her gun fires 20-g bullets at 1000 m/s. How many bullets can she fire per minute?

42.  Figure 8-30 shows the time variation of a force $F_x$ acting on a 2-kg particle initially at rest. Find (*a*) the impulse of this force, (*b*) the final velocity of the particle, and (*c*) the average value of the force for the time interval $t_1 = 1$ s to $t_2 = 4$ s.

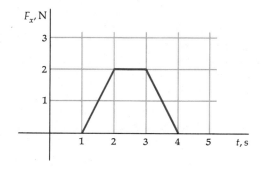

**Figure 8-30**
Exercise 42.

43. Figure 8-31 shows the time variation of a force which acts on a 3-kg particle. (*a*) Find the average value of the force for the time interval from 1 to 5 s. (*b*) If the particle is moving in the *x* direction with initial speed of 2 m/s at time *t* = 0, find its speed at *t* = 5 s.

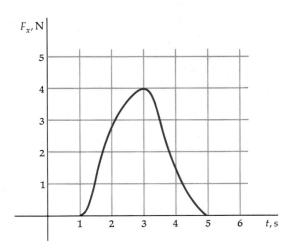

**Figure 8-31**
Exercise 43.

## Section 8-8, Systems with Varying Mass: Rocket Motion

44. A rocket burns fuel at a rate of 0.2 Mg/s and exhausts it at a relative speed of $u_{ex}$ = 6 km/s. Find the thrust of this rocket.

45. Find the ratio of the fuel to payload of the rocket in Exercise 44 if the final velocity of the rocket is to be (*a*) 6 km/s, (*b*) 12 km/s, (*c*) 30 km/s. Neglect gravity and assume that the rocket starts from rest.

46. The payload of a rocket is 5 percent of its total mass, the rest being fuel. If it starts from rest, what is its final velocity if the fuel exhaust velocity is $u_{ex}$ = 5 km/s?

## Section 8-9, The Center-of-Mass Reference Frame

47. (*a*) In Exercise 12 find the velocity of the center of mass of the two-railroad-car system. (*b*) Find the velocity of each car relative to the center of mass before the collision. What are their velocities relative to the center of mass after the collision? (*c*) Find the kinetic energies of the cars relative to the center of mass before and after the collision.

48. In Exercise 14 let $m_1$ be 12 kg and $m_2$ be 4 kg. (*a*) Find the kinetic energy of the masses after they are released from the spring. (*b*) Find the velocity of each mass in a reference frame in which the center of mass is moving to the right with velocity 3 m/s. (*c*) Find the total kinetic energy before and after the masses are released in the reference frame of part (*b*). How much energy do the masses gain from the spring in this frame?

49. A 2-kg body moving to the right with speed 6 m/s collides elastically with a stationary body of mass 4 kg. (*a*) Find the velocity of each body relative to the center of mass before and after the collision. (*b*) Find the velocity of each body in the original frame after the collision. (*c*) How much energy was transferred to the 4-kg body?

50. Work Exercise 25 by first transferring into the center-of-mass reference frame to find the velocities before and after the collision and then transferring back into the original frame.

### Section 8-10, Reaction Threshold

51. The energy that must be supplied in the center-of-mass frame to separate a deuteron into its constituents, a neutron and a proton, is 2.2 MeV. What is the reaction threshold for the reaction $n + d \rightarrow n + n + p + 2.2$ MeV, where $n$ stands for a neutron and $d$ for the deuteron at rest in the laboratory frame?

52. The energy needed to separate the two protons in an $H_2$ molecule is 4.48 eV. This process of separation is called *dissociation.* Suppose it is to be accomplished by a collision of $H_2$ with a proton. (*a*) Which takes more energy, bombarding stationary protons with $H_2$ molecules or bombarding stationary $H_2$ molecules with protons? (*b*) Calculate the threshold energy for each of these reactions.

53. The $Q$ value for the ionization of a lithium atom is $Q = -5.39$ eV. Find the threshold energy for the ionization of a lithium atom by bombardment with protons. (The mass of lithium is 7 times that of a proton.)

### Problems

1. The ratio of the mass of the earth to the mass of the moon is $M_E/m_m = 81.3$. The radius of the earth is about 6370 km, and the distance to the moon is about 384,000 km. (*a*) Locate the center of mass of the earth-moon system relative to the surface of the earth. (*b*) What external forces act on the earth-moon system? In what direction is the acceleration of the center of mass of this system? (*c*) Assume that the center of mass of this system moves in a circular orbit around the sun. How far must the center of the earth move in the radial direction (toward or away from the sun) during the 14 d between the time the moon is farthest from the sun (full moon) and when it is closest to the sun (new moon)?

2. A projectile is launched at 20 m/s at an angle of 30° with the horizontal. In the course of its flight it explodes, breaking into two parts, one of which has twice the mass of the other. The two fragments land simultaneously. The lighter fragment lands 20 m from the launch point in the direction the projectile was fired. Where does the other fragment land?

3. A man of mass 80 kg is riding on a small cart of mass 40 kg which is rolling along a level floor at speed of 2 m/s. He jumps off the back of the cart so that his speed relative to the ground is 1 m/s in the direction opposite to the motion of the cart. (*a*) What is the speed of the center of mass of the man-cart system before and after he jumps? (*b*) What is the speed of the cart after the man jumps? (*c*) What is the speed of the center of mass of the system after the man hits the ground and comes to rest? (*d*) What force is responsible for the change in $v_{CM}$? (*e*) How much energy did the man expend in jumping?

4. A 40-kg boy gets on his 10-kg wagon on level ground with two 5-kg bricks. He throws the bricks horizontally off the back of the wagon one at a time at a speed of 7 m/s relative to himself. How fast does he go after throwing the second brick? How fast would he go if he threw both bricks at the same time at 7 m/s relative to himself?

5. A circular plate of radius $r$ has a circular hole cut out with radius $\frac{1}{2}r$ (Figure 8-32). Find the center of mass of the plate. *Hint:* The hole can be represented by two disks superimposed, one of mass $m$ and the other of mass $-m$.

6. Using the hint from Problem 5, find the center of mass of a solid sphere of radius $r$ which has a spherical hole of radius $\frac{1}{2}r$, as shown in Figure 8-33.

7. A 6-kg projectile is launched at $\theta = 30°$ with initial speed of 35 m/s. At the top of its flight, it explodes into two parts of mass 2 and 4 kg. The fragments move horizontally just after the explosion, and the 2-kg piece lands back at the initial launch point. Where does the 4-kg piece land? What is the energy of the explosion?

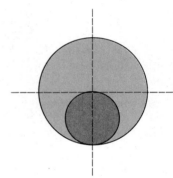

**Figure 8-32**
Problem 5.

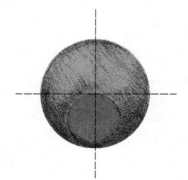

**Figure 8-33**
Problem 6.

8. A massless spring of force constant 1 kN/m is compressed a distance of 20 cm between masses of 8 and 2 kg. The spring is released on a smooth table. Since the masses are not attached to the spring, they move away with speeds $v_1$ and $v_2$. (a) Show that the kinetic energy of the masses when they leave the spring can be written $E_k = \frac{1}{2}p^2(1/m_1 + 1/m_2)$, where $p$ is the momentum of either mass. Use conservation of energy to find $p$. (b) Find the velocity of each mass as it leaves the spring. (c) Find the velocity of each mass if the system is given an initial velocity of 4 m/s perpendicular to the spring, as shown in Figure 8-34a. (d) What is the energy of center-of-mass motion in this case? What is the energy of motion relative to the center of mass? (e) Find the velocity of each mass if the system is given an initial velocity of 4 m/s in the direction along the spring, as shown in Figure 8-34b. What is the energy of center-of-mass motion in this case? What is the energy of motion relative to the center of mass?

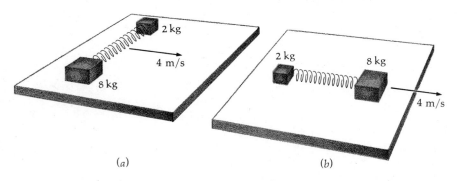

(a)                          (b)

**Figure 8-34**
Problem 8.

9. A 2-kg block moves at 6 m/s and makes a head-on collision with a 4-kg block initially at rest. After the collision, the 2-kg block is observed to be moving back at 1 m/s. (a) Find the velocity of the 4-kg block after the collision. (b) Find the energy lost in the collision. (c) What is the coefficient of restitution $e$ for this collision?

10. Show that in a one-dimensional collision between two particles the fractional relative-energy loss is related to the coefficient of restitution by

$$-\frac{\Delta E_{k,\text{rel}}}{E_{k,\text{rel}}} = 1 - e^2$$

11. A boy on a stationary skateboard (combined mass of 36 kg) throws a ball of mass 0.50 kg horizontally forward. Find the initial recoil speed of the boy if the speed of the ball is such that it would rise to a height of 15 m if thrown vertically upward.

12. A hammer of mass 0.8 kg is used to drive nails of mass 30 g into wood. When the hammer has an impact speed of 5.0 m/s, the nail penetrates 2.0 cm at a single blow. Find (a) the common speed of the hammer and nail immediately after impact, assuming the collision to be perfectly inelastic; (b) the time the nail is in motion, assuming that the initial speed is acquired in a negligible time, after which there is uniform deceleration; (c) the average resisting force of the wood against the nail as it penetrates.

13. You throw a 150-g ball to a height of 40 m. Use a reasonable value for the distance the ball moves while it is in your hand to calculate the average force exerted by your hand and the time the ball is in your hand while you are throwing it. Is it reasonable to neglect the weight of the ball while it is being thrown?

14. A cannon of mass $M$ rests on wheels on a hard horizontal surface. Its barrel makes an angle $\theta_0$ with the horizontal. It fires a cannonball of mass $m$ and recoils freely. Show that the initial angle made by the velocity of the ball with the horizontal is related to $\theta_0$ by $\tan\theta = (1 + m/M)\tan\theta_0$.

15. A ballistic pendulum consists of a pumpkin of mass 1.5 kg suspended on a 1-m string. A bullet of mass 10 g has an initial speed of 400 m/s and passes through the pumpkin. After the collision, the pendulum swings up to a max-

imum angle of 24°. Find the final speed of the bullet after the collision and the energy lost in this collision.

16. A 2000-kg car traveling at 90 km/h crashes into a concrete wall which does not give at all. Estimate the time of collision, assuming that the center of the car travels halfway to the wall with constant deceleration. (Use any reasonable length for the car.) Estimate the average force exerted by the wall on the car.

17. A ball moving at 10 m/s makes an off-center perfectly elastic collision with another ball of equal mass initially at rest. The incoming ball is deflected at an angle of 30° from its original direction of motion. Find the velocity of each ball after the collision.

18. Show that for a perfectly elastic collision between two particles of equal mass with one particle initially at rest, the energy transferred to the originally stationary particle is $(\sin^2 \theta)E_0$, where $E_0$ is the initial energy and $\theta$ is the angle of deflection of the incoming particle.

19. Use integration to find the center of mass of a uniform plate of material cut in the shape of a half circle. (Choose the origin at the center of curvature of the plate with the $y$ axis on the bisector.)

20. Use integration to find the center of mass of the right isosceles triangle shown in Figure 8-35.

21. Use integration to find the center of mass of a half-circle plate for which the mass density is proportional to the distance $r$ from the center of curvature.

22. A 2-m 80-kg man steps off a ledge and falls 5 m to the ground, making an inelastic collision with it. (a) How fast was he traveling just before he hit? (b) What was the impulse exerted on the man by the ground? (c) Estimate the time of collision, assuming that the force on the man is constant and that he travels 3 m during the collision. (d) Use your result for part (c) to find the average force exerted by the ground on the man.

23. A handball of mass 300 g is thrown straight against a wall with a speed of 8 m/s. It rebounds with the same speed. (a) What impulse was delivered to the wall? (b) If the ball was in contact with the wall for 0.003 s, what average force was exerted on the wall? (c) The ball is caught by a player who brings it to rest. In the process his hand moves back 0.5 m. What is the impulse received by the player? (d) What was the average force exerted on the player?

24. A stream of glass beads comes out of a horizontal tube at 100 per second and strikes a balance pan, as shown in Figure 8-36. They fall a distance of 0.5 m to the balance and bounce back to the same height. Each bead has mass 0.5 g. How much mass $m$ must be placed in the other pan of the balance to keep the pointer reading zero?

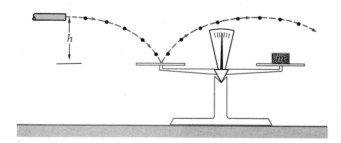
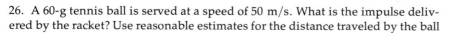

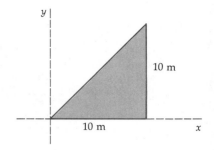

**Figure 8-35**
Problem 20.

**Figure 8-36**
Problem 24.

25. A 13-kg block is at rest on a level floor. A 400-g glob of putty is thrown at the block so that it travels horizontally, hits the block, and sticks to it. The block and putty slide 15 cm along the floor. If the coefficient of friction is 0.4, what was the original speed of the putty?

26. A 60-g tennis ball is served at a speed of 50 m/s. What is the impulse delivered by the racket? Use reasonable estimates for the distance traveled by the ball

while in contact with the racket to estimate the collision time and the average force exerted by the racket.

27. A bullet of mass $m_1$ is fired with speed $v$ into the bob of a ballistic pendulum, which has mass $m_2$. The bob is attached to a very light rod of length $L$, which is pivoted at the other end. The bullet is stopped in the bob. Find $v$ such that the bob will swing through a complete circle.

28. A bullet of mass $m_1$ is fired with speed $v$ into the bob of a ballistic pendulum of mass $m_2$. Find the maximum height attained by the bob if the bullet passes through the bob and emerges with speed $\frac{1}{4}v$.

29. Two particles, $m_1 = 4$ kg and $m_2 = 6$ kg, slide without friction along the $x$ axis. Initially $m_1$ has velocity 10 m/s, and $m_2$ has velocity 5 m/s. Particle $m_1$ overtakes $m_2$. The coefficient of restitution is 0.5. (a) What is the velocity of the center of mass? (b) What is the initial momentum of each particle in the center-of-mass reference frame? (c) What is the final momentum of each particle in the center-of-mass frame? (d) What is the final speed of each particle in the original frame? (e) How much kinetic energy was lost in this collision?

30. A particle of mass $m_1$ moves with speed $v_{1i}$ and collides head on with a stationary particle of mass $m_2$. After the collision, the velocities of the particles are $v_{1f}$ and $v_{2f}$. Prove that the condition for $v_{1f}$ to be positive, i.e., for the first particle to continue moving in the same direction, is $m_1/m_2 > e$, where $e$ is the coefficient of restitution.

31. A bowling ball of mass 3.2 kg and speed 2.5 m/s makes a direct hit with a pin of mass 1.0 kg. After the impact the forward speed of the ball is 1.4 m/s. Find the coefficient of restitution and the fraction of the kinetic energy lost in the impact.

32. A rocket is fired vertically upward near the surface of the earth where the free-fall acceleration of gravity is $g$. (a) Show that the acceleration is given by $dv/dt = (u_{ex}/m)\, dm/dt - g$. (b) Show that if the rocket starts from rest, its final velocity is given by $v = u_{ex} \ln (m_i/m_f) - g\, \Delta t$, where $\Delta t$ is the time for the fuel to burn. (c) If $dm/dt = -R$, show that the acceleration of the rocket is given by

$$\frac{dv}{dt} = \frac{u_{ex}R}{m_i - Rt} - g$$

33. A rocket has a payload of 5000 kg and a fuel supply of 20,000 kg. It starts from rest, and it burns fuel at a rate of 0.2 Mg/s and exhausts it at $u_{ex} = 6$ km/s. (a) Find its final velocity if it is in free space with no gravity. (b) Find the velocity after the fuel burns up if it moves against a uniform gravitational field $g$ [see part (b) of Problem 32]. (c) In (b), if the rocket starts from the earth's surface, is it reasonable to neglect the variation in $g$ with height above the earth's surface?

34. A particle has speed $v_0$ originally. It collides with a second particle at rest and is deflected through angle $\phi$. Its speed after the collision is $v$. The second particle recoils, its velocity making an angle $\theta$ with the initial direction of the first particle. Show that

$$\tan \theta = \frac{v \sin \phi}{v_0 - v \cos \phi}$$

Do you have to assume that this collision was elastic or inelastic to get this result?

35. A neutron of mass $m$ makes an elastic collision with a carbon atom of mass $12m$. The carbon atom is initially at rest. (a) What is the energy of the neutron after the collision if its original energy is $E_1$? (b) How many collisions must a neutron make to reduce its energy from $E_1$ to $10^{-6}E_1$? (Assume head-on collisions.)

36. A ball is thrown with speed 19.6 m/s at 15° to the horizontal and rebounds off a frictionless wall 4.9 m away. The ball returns to its initial point of projection. Find the coefficient of restitution between the ball and the wall.

**CHAPTER 9**     Rotation of a Rigid Body about a Fixed Axis

We now turn to the study of a special kind of motion of a system of particles, rotational motion. Consider, for example, the rotation of a wheel about its axis. We cannot analyze this motion by considering the wheel as a single particle because different parts of the wheel have different velocities and accelerations. We must treat the wheel as a system of particles; i.e., we consider the wheel to consist of a large number of parts, each so small that we can neglect the variation in velocity or acceleration over it. Thus each part can be treated as a particle.

No matter how the wheel moves, the distance between any two particles within it remains constant. Such a system is called a *rigid body*.     *Rigid body*
Although all objects in nature are deformable to some degree, the rigid-body approximation is often very good and it greatly simplifies the analysis of the motion of the system. In general, the change in position of a rigid body can be considered to consist of a translation of the center of mass plus a rotation of the body about an axis through the center of mass, as illustrated in Figure 9-1. We have already seen how to treat the motion of the center of mass of any system; its acceleration equals the resultant force on the system divided by the total mass. The rotational motion of a body is usually very complicated because, in general, the axis of rotation changes direction as the body moves.

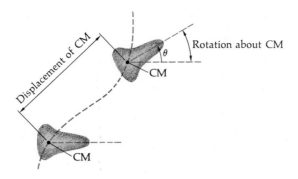

**Figure 9-1**
The general motion of a rigid body consists of translation of the center of mass plus rotation about the center of mass.

In this chapter we concentrate on the simplest kind of rotation, rotation of a rigid body about a fixed axis. In the next chapter we shall extend our discussion to systems in which the distance between any two particles can vary, and consider the situation when the axis of rotation is not fixed in space.

## 9-1  Angular Velocity and Angular Acceleration

Consider one particle of a wheel rotating about its axis of symmetry, which is fixed in space. We can specify the position of the particle $P_i$ by the distance $r_i$ from the center of the wheel and the angle $\theta_i$ between a line from the center to the particle and a reference line fixed in space, as shown in Figure 9-2. In a small time $dt$, the particle moves along the arc of a circle a distance $ds_i$, given by

$$ds_i = v_i\, dt \qquad\qquad 9\text{-}1$$

where $v_i$ is the speed of the particle. The angle swept out by the line from the center to the particle, in radians, is this distance divided by the radius $r_i$:

$$d\theta = \frac{ds_i}{r_i} = \frac{v_i\, dt}{r_i} \qquad\qquad 9\text{-}2$$

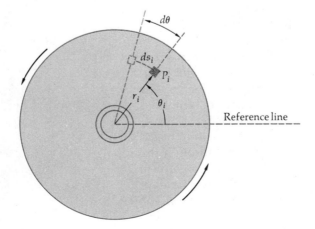

**Figure 9-2**
A wheel rotating about a fixed axis through its center. The distance $ds_i$ moved by the $i$th particle in some time interval depends on $r_i$, but the angular displacement $d\theta$ is the same for all particles on the wheel.

Although the distance $ds_i$ varies from particle to particle, the angle $d\theta$ swept out in a given time is the same for all the particles. For example, when one particle moves through a complete circle, so do all the other particles and $\Delta\theta = 2\pi$ rad for all the particles. The rate of change of angle with respect to time, $d\theta/dt$, is the same for all the particles of the wheel. It is called the *angular velocity* $\omega$ of the wheel,

$$\omega = \frac{d\theta}{dt}$$ 9-3     *Angular velocity*

From Equation 9-2 the speed of the $i$th particle is related to its radius $r_i$ and the angular velocity of the wheel by

$$v_i = r_i \frac{d\theta}{dt} = r_i\omega$$ 9-4

The angular velocity $\omega$ is positive or negative depending on whether $\theta$ is increasing or decreasing. The units of angular velocity are radians per second. Since a radian is a dimensionless unit of angle, the dimensions of angular velocity are those of reciprocal seconds ($s^{-1}$). Although angular velocity is frequently expressed in units other than radians per second, it is important to remember that expressions like Equations 9-3 and 9-4 and the other results we obtain for rotational motion are valid only when angles are expressed in radians.

Since angles are often expressed in degrees, angular velocity is sometimes expressed in degrees per second or degrees per minute. In such cases, the angular velocity can be reexpressed in radians by using

$$2\pi \text{ rad} = 360° \qquad \text{or} \qquad \frac{\pi \text{ rad}}{180°} = 1$$

Angular velocity is also commonly expressed in revolutions per second and revolutions per minute. Conversion to radians is accomplished by using

$$2\pi \text{ rad} = 1 \text{ rev} \qquad \text{or} \qquad \frac{2\pi \text{ rad}}{1 \text{ rev}} = 1$$

The rate of change of angular velocity with respect to time is called the *angular acceleration* $\alpha$. For rotation about a fixed axis

$$\alpha = \frac{d\omega}{dt} = \frac{d^2\theta}{dt^2}$$ 9-5     *Angular acceleration*

The units of angular acceleration are radians per second squared. The relation between the *tangential* linear acceleration of the $i$th particle of the wheel and the angular acceleration is obtained by taking the derivative of the speed $v_i$ in Equation 9-4 with respect to $t$:

$$a_{it} = \frac{dv_i}{dt} = r_i \frac{d\omega}{dt}$$

Thus

$$a_{it} = r_i\alpha$$ 9-6

Each particle of the wheel also has a *radial* linear acceleration, the centripetal acceleration which points inward along the radial line, and has the magnitude

$$a_{ic} = \frac{v_i^2}{r_i} = r_i\omega^2$$ 9-7

The three quantities angular displacement $\theta$, angular velocity $\omega$, and angular acceleration $\alpha$ are analogous to the linear displacement $x$, linear velocity $v_x$, and linear acceleration $a_x$ in one-dimensional motion (Chapter 2). Because of the similarity of the definitions of the rotational and linear quantities, much of what we learned in Chapter 2 will be useful in treating problems of rotation about a fixed axis. For example, if the angular acceleration is constant, we can develop constant-angular-acceleration expressions (Equations 2-9 to 2-12). If, for example, we have

$$\alpha = \frac{d\omega}{dt} = \alpha_0 = \text{constant}$$

then

$$\omega = \omega_0 + \alpha_0 t \qquad\qquad 9\text{-}8$$

where $\omega_0$ is the angular velocity at $t = 0$, and

$$\theta = \theta_0 + \omega_0 t + \tfrac{1}{2}\alpha_0 t^2 \qquad\qquad 9\text{-}9$$

*Constant-angular-acceleration equations*

As with the constant-linear-acceleration formulas, we can eliminate time from these equations to obtain an equation relating the angular displacement, angular velocity, and angular acceleration:

$$\omega^2 = \omega_0^2 + 2\alpha_0(\theta - \theta_0) \qquad\qquad 9\text{-}10$$

**Example 9-1**  A wheel rotates with constant angular acceleration of $\alpha_0 = 2$ rad/s². If the wheel starts from rest, how many revolutions does it make in 10 s?

The number of revolutions is related to the angular displacement by the fact that each revolution is an angular displacement of $2\pi$ rad. Thus we need to find the angular displacement $\theta - \theta_0$ in radians for a time 10 s and multiply by the conversion factor $(1 \text{ rev})/(2\pi \text{ rad})$.

Star tracks in a time exposure of the night sky. How can you decide whether these circular tracks are due to rotation of the earth or motion of the stars? (*Courtesy of Yerkes Observatory.*)

Equation 9-9 relates the angular displacement to the time. We are given $\omega_0 = 0$ (the wheel starts from rest). Let us choose $\theta_0 = 0$. We have

$$\theta = \tfrac{1}{2}\alpha_0 t^2 = \tfrac{1}{2}(2 \text{ rad/s}^2)(10 \text{ s})^2 = 100 \text{ rad}$$

The number of revolutions is thus

$$100 \text{ rad } \frac{1 \text{ rev}}{2\pi \text{ rad}} = \frac{50}{\pi} \text{ rev} = 15.9 \text{ rev}$$

It is convenient to assign directions to the rotational quantities $\omega$ and $\alpha$. We cannot describe the rotation of a wheel by any direction in the plane of the wheel because, by symmetry, all directions in this plane are equivalent. The direction in space uniquely associated with the rotation is the direction of the axis of rotation. We therefore choose the direction of $\omega$ to be along the axis of rotation. Similarly, since we are considering only rotation about axes which are fixed in space, the rate of change of $\omega$, which is the angular acceleration $\alpha$, must also be in this direction.

*Direction of $\omega$ and $\alpha$*

Consider a wheel rotating in the clockwise sense, as in Figure 9-3a. We have chosen the direction for $\omega$ along the axis of rotation, but we must still decide the sense of $\omega$, that is, for this rotation, whether $\omega$ points into or out of the diagram. The choice is made by convention. If the rotation is clockwise, as in the figure, $\omega$ is inward. If it is counterclockwise, $\omega$ is outward. This arbitrary decision, called the *right-hand rule*, is illustrated in Figure 9-3a. When the axis of rotation is grasped by the right hand with the fingers following the rotation, the extended thumb points in the direction of $\omega$. The direction of $\omega$ is also that of the advance of a rotating right-handed screw, as shown in Figure 9-3b. The sense of the angular acceleration (for rotation about a fixed axis) depends on whether $\omega$ is increasing or decreasing. If $\omega$ is increasing, $\alpha$ is parallel to $\omega$. If $\omega$ is decreasing, $\alpha$ is antiparallel to $\omega$.

In the next chapter when we consider general rotational motion without the restriction that the axis of rotation be fixed, we shall see that the quantities $\boldsymbol{\omega}$ and $\boldsymbol{\alpha}$ with their direction so defined are vectors; i.e., they obey the parallelogram law of vector addition. For this reason, we have used the term *angular velocity* rather than angular speed for $\omega$. When we consider only rotation about a fixed axis, there are only two possible directions for these vectors, which we can describe as positive or negative. This is the same kind of description we used for linear velocity and linear acceleration when we studied motion in one dimension.

**Figure 9-3**
(a) The right hand-rule for determining the direction of the angular velocity $\omega$. (b) The direction of $\omega$ is also the direction of advance of a rotating right-handed screw.

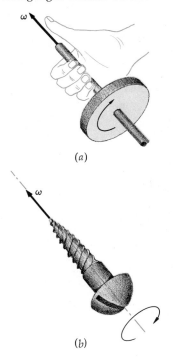

(a)

(b)

**Question**

1. Two points are on a wheel turning at constant angular velocity, one point on the rim and the other halfway between the rim and the axis. Which moves the greater distance in time $\Delta t$? Which turns through the greater angle? Which has the greater speed? The greater angular velocity? The greater acceleration? The greater angular acceleration?

## 9-2 Torque and Moment of Inertia

In our discussion of the static equilibrium of a rigid body in Chapter 5 we found that a necessary condition for a body not to rotate is that the resultant torque about any point be zero. Although this condition is *necessary*, it is not a sufficient condition for a body to be static. If a body is set rotating about an axis and there is no external torque acting, the body will continue to rotate with constant angular velocity. (This is the same as the case of linear motion. A particle can be at rest and remain at rest only if the resultant force is zero, but this is not a sufficient condition for being at rest. If the resultant force is zero, the particle could also be moving with constant velocity.) We shall now show that if there is a resultant external torque acting about the point of rotation of a rigid body, the angular velocity of the body does not remain constant but changes with an angular acceleration which is proportional to the external torque.

Figure 9-4 shows a force $\mathbf{F}_i$ acting on the *i*th particle of mass $m_i$ of a wheel which is pivoted about an axis through its center. As discussed in Chapter 5, the component $F_{ir}$ parallel to the radius vector $\mathbf{r}_i$ has no effect on the rotation of the body. The tangential component $F_{it}$, which is perpendicular to $\mathbf{r}_i$, does affect the rotation of the body. The torque produced by the force $\mathbf{F}_i$ about the center of the wheel is given by

$$\tau_i = F_{it}r_i \qquad\qquad 9\text{-}11 \qquad \textit{Torque}$$

If $\mathbf{F}_i$ is the resultant force acting on the *i*th particle, the tangential acceleration of that particle is given by Newton's second law:

$$F_{it} = m_i a_{it} = m_i r_i \alpha \qquad\qquad 9\text{-}12$$

where we have used Equation 9-6 relating the tangential acceleration of the *i*th particle to the angular acceleration of the body. Substituting this result for $F_{it}$ into Equation 9-11, we obtain

$$\tau_i = m_i r_i^2 \alpha$$

If we now sum over all the particles in the body, we obtain

$$\Sigma \tau_i = \Sigma m_i r_i^2 \alpha \qquad\qquad 9\text{-}13$$

The quantity $\Sigma \tau_i$ is the resultant torque acting on the body, which we denote by $\Sigma \tau$. The sum $\Sigma m_i r_i^2$ is a property of the wheel called the *moment of inertia I*. (The angular acceleration is the same for all particles and can therefore be taken out of the sum.) The moment of inertia depends on the mass distribution relative to the axis of rotation of the wheel:

$$I = \Sigma m_i r_i^2 \qquad\qquad 9\text{-}14 \qquad \textit{Moment of inertia}$$

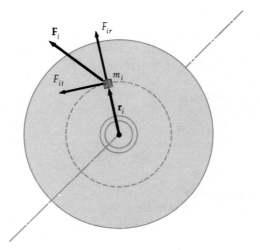

**Figure 9-4**
A force $\mathbf{F}_i$ acting on the *i*th particle of a wheel pivoted about its center. The radial component $F_{ir}$ does not affect the rotation of the wheel. The torque exerted by the force $\mathbf{F}_i$ is $F_{it}r_i$.

Culver Pictures

Torque exerted by a wrench on a nut is proportional to the force and the lever arm. Charlie could exert a greater torque with the same force if he held the wrenches nearer their ends. What is the direction of the torque exerted by each wrench when he pulls on it?

In Section 9-3 we shall illustrate how the moment of inertia is calculated for various bodies. In terms of the moment of inertia, Equation 9-13 becomes

$$\Sigma\tau = I\alpha \qquad\qquad\qquad 9\text{-}15$$

When we consider more general rotations in the next chapter, we shall show that torque, like angular velocity and angular acceleration, is a vector. The direction of the torque is along the axis of rotation in the direction of the angular acceleration it produces. As with $\boldsymbol{\omega}$ and $\boldsymbol{\alpha}$, when we consider rotation about a fixed axis, there are only two possible directions for torque, positive and negative. (This is equivalent to calling these directions clockwise and counterclockwise, as we did in Chapter 5.) In vector form, Equation 9-15 is written

$$\Sigma\boldsymbol{\tau} = I\boldsymbol{\alpha} \qquad\qquad\qquad 9\text{-}16$$

*Newton's second law for rotation*

Equation 9-16 is the rotational analog of Newton's second law $\Sigma\mathbf{F} = m\mathbf{a}$ for linear motion. The resultant torque is analogous to the resultant force; the moment of inertia is analogous to the mass; and the angular acceleration is analogous to the linear acceleration.

Since the moment of inertia of a rigid body is constant, we can write Equation 9-16 in an alternative way:

$$\Sigma\boldsymbol{\tau} = I\boldsymbol{\alpha} = I\frac{d\boldsymbol{\omega}}{dt} = \frac{d(I\boldsymbol{\omega})}{dt}$$

This looks like the linear equation $\mathbf{F} = d(m\mathbf{v})/dt = d\mathbf{p}/dt$, where $\mathbf{p}$ is the linear momentum. In Chapter 10 we shall give a general definition of the *angular momentum* of a system of particles and show that in general the resultant torque acting on a system equals the rate of change of the angular momentum. For the special case in which the axis of rotation is a symmetry axis of a body or parallel to a symmetry axis, the angular momentum of a body $\mathbf{L}$ is given by

$$\mathbf{L} = I\boldsymbol{\omega} \qquad\qquad\qquad 9\text{-}17$$

*Angular momentum*

The rotation of this nonrigid body, a galaxy in Ursa Major, is indicated by its spiral shape. (*Courtesy of Hale Observatories.*)

The rotational analog of Newton's second law can then be

$$\Sigma \boldsymbol{\tau} = \frac{d\mathbf{L}}{dt} \qquad \text{9-18}$$

Equation 9-18 is actually more general than Equation 9-16. It applies, for example, to a system of particles in which the moment of inertia is not constant, whereas Equation 9-16, obtained for a rigid body, does not. However, we must remember that Equation 9-17 is not a general definition of angular momentum; if the axis of rotation is not a symmetry axis or parallel to one, the angular momentum is not given by Equation 9-17 and in fact, as we shall see in the next chapter, $\mathbf{L}$ and $\boldsymbol{\omega}$ are not parallel in this case.

When a rotating wheel turns through a small angular displacement $\Delta\theta$, the $i$th particle moves through a distance $\Delta s_i = r_i \, \Delta\theta$. The work done by the force $\mathbf{F}_i$ is

$$\Delta W = F_{it} \, \Delta s_i = \tau_i \, \Delta\theta$$

In general, the work done by a torque $\tau$ when a body turns through a small angle $\Delta\theta$ is

$$\Delta W = \tau \, \Delta\theta \qquad \text{9-19}$$

Equation 9-19 is analogous to the similar result for linear motion in one dimension: $\Delta W = F_x \, \Delta x$. The rate of doing work is the power input of a torque:

$$P = \frac{dW}{dt} = \tau \frac{d\theta}{dt} = \tau\omega$$

In vector notation, the power input of a torque $\boldsymbol{\tau}$ is given by

$$P = \boldsymbol{\tau} \cdot \boldsymbol{\omega} \qquad \text{9-20} \qquad \textit{Power}$$

Equation 9-20 is the rotational analog of $P = \mathbf{F} \cdot \mathbf{v}$.

The net work done on a system equals the change in the kinetic energy of the system. For a wheel rotating about an axis through its center of mass, the kinetic energy of the wheel is the kinetic energy relative to the center of mass, $E_{k,\text{rel}}$, discussed in Chapter 8. This kinetic energy is just the sum of the kinetic energies of the individual particles in a body. For a rotating wheel we have

$$E_k = \Sigma \tfrac{1}{2} m_i v_i^2 = \Sigma \tfrac{1}{2} m_i (r_i \omega)^2 = \tfrac{1}{2}(\Sigma m_i r_i^2)\omega^2$$

or

$$E_k = \tfrac{1}{2} I \omega^2 \qquad\qquad 9\text{-}21$$

Equation 9-21 is the rotational analog of $E_k = \tfrac{1}{2} mv^2$ for linear motion. Table 9-1 lists some of the rotational equations we have developed in this chapter beside the analogous equations for linear motion.

**Table 9-1**
Comparison of linear motion and rotational motion

| Linear motion | | Rotational motion | |
|---|---|---|---|
| Displacement | $\Delta x$ | Angular displacement | $\Delta \theta$ |
| Velocity | $v = \dfrac{dx}{dt}$ | Angular velocity | $\omega = \dfrac{d\theta}{dt}$ |
| Acceleration | $a = \dfrac{dv}{dt} = \dfrac{d^2 x}{dt^2}$ | Angular acceleration | $\alpha = \dfrac{d\omega}{dt} = \dfrac{d^2\theta}{dt^2}$ |
| Constant-acceleration equations | $v = v_0 + at$ $\Delta x = v_{\text{av}} \Delta t$ $v_{\text{av}} = \tfrac{1}{2}(v_0 + v)$ $x = x_0 + v_0 t + \tfrac{1}{2} a t^2$ $v^2 = v_0^2 + 2a\,\Delta x$ | Constant-angular-acceleration equations | $\omega = \omega_0 + \alpha t$ $\Delta\theta = \omega_{\text{av}} \Delta t$ $\omega_{\text{av}} = \tfrac{1}{2}(\omega_0 + \omega)$ $\theta = \theta_0 + \omega_0 t + \tfrac{1}{2}\alpha t^2$ $\omega^2 = \omega_0^2 + 2\alpha\,\Delta\theta$ |
| Mass | $m$ | Moment of inertia | $I$ |
| Momentum | $\mathbf{p} = m\mathbf{v}$ | Angular momentum | $\mathbf{L} = I\omega$ |
| Force | $\mathbf{F}$ | Torque | $\tau$ |
| Power | $P = \mathbf{F} \cdot \mathbf{v}$ | Power | $P = \tau \cdot \omega$ |
| Newton's second law | $\Sigma \mathbf{F} = \dfrac{d\mathbf{p}}{dt} = m\mathbf{a}$ | Newton's second law | $\Sigma \tau = \dfrac{d\mathbf{L}}{dt} = I\alpha$ |

In Equation 9-14 for the moment of inertia of a wheel, the distance $r_i$ is the distance from the $i$th particle to the axis of rotation. In general, this distance is not the same as the distance from the $i$th particle to the origin, though for a wheel with the origin at the center on the axis of rotation these distances are the same. To avoid confusion, we shall use $\rho_i$ for the perpendicular distance from the $i$th particle to the axis of rotation. The moment of inertia of a general body about a given axis of rotation is then defined as

$$I = \Sigma m_i \rho_i^2 \qquad\qquad 9\text{-}22$$

**Example 9-2** Four particles of mass $m$ are connected by massless rods to form a rectangle of sides $2a$ and $2b$, as shown in Figure 9-5. The system rotates about an axis in the plane of the figure through the center. Find

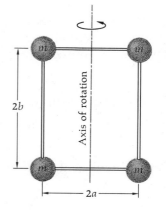

**Figure 9-5**
Four particles of equal mass connected by massless rods and rotating about an axis through the plane of the particles and through the center of mass.

the moment of inertia about this axis and the kinetic energy of rotation if the angular velocity is $\omega$.

From the figure we see that the distance from each particle to the axis of rotation is $a$. The moment of inertia of each particle about this axis is therefore $ma^2$, and since there are four particles, the total moment of inertia of the body is $I = 4ma^2$. The kinetic energy is thus $E_k = \frac{1}{2}I\omega^2 = \frac{1}{2}(4ma^2)\omega^2 = 2ma^2\omega^2$. We could also have computed the kinetic energy of each particle directly from $E_k = \frac{1}{2}mv^2 = \frac{1}{2}m(a\omega)^2 = \frac{1}{2}ma^2\omega^2$ since the linear speed of each particle is $v = a\omega$.

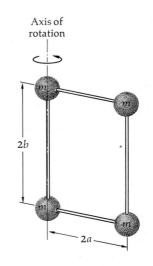

Axis of rotation

---

**Example 9-3** Find the moment of inertia of the system of Example 9-2 for rotation about an axis parallel to the first axis but passing through two of the masses, as shown in Figure 9-6.

For this rotation, two of the masses are at a distance $2a$ from the axis of rotation and two are at zero distance. The moment of inertia is thus

$$I = \Sigma m_i \rho_i^2 = m(0)^2 + m(0)^2 + m(2a)^2 + m(2a)^2 = 8ma^2$$

The moment of inertia is larger about this axis than the one parallel to it through the center of mass.

**Figure 9-6**
System of Figure 9-5 rotating about an axis through two of the particles (Example 9-3).

---

**Example 9-4** A string is wound around the rim of a uniform disk pivoted to rotate without friction about a fixed axis through its center. The mass of the disk is 3 kg, its radius is $R = 25$ cm, and its moment of inertia about this axis is $9.38 \times 10^{-2}$ kg·m². (In Section 9-3 we shall show that the moment of inertia of a uniform disk about an axis perpendicular to the disk and through its center is $\frac{1}{2}MR^2$.) The string is pulled with a force of 10 N (Figure 9-7). If the disk is initially at rest, what are its angular velocity and angular momentum after 5 s? Find also the power input of the force at 5 s.

$T = 10\,N$

**Figure 9-7**
String wrapped around a disk (Example 9-4).

Since the direction of the string as it leaves the rim of the disk is always tangent to the disk, the lever arm of the force it exerts is just $R$. The applied torque is thus

$$\tau = TR = (10\ \text{N})\,(0.25\ \text{m}) = 2.5\ \text{N·m}$$

From Equation 9-16,

$$\alpha = \frac{\Sigma\tau}{I} = \frac{2.5\ \text{N·m}}{0.0938\ \text{kg·m}^2} = 26.7\ \text{rad/s}^2$$

Since $\alpha$ is constant, we find $\omega$ after 5 s from Equation 9-8, setting $\omega_0 = 0$,

$$\omega = \omega_0 + \alpha t = 0 + (26.7\ \text{rad/s}^2)\,(5\ \text{s}) = 133\ \text{rad/s}$$

The angular momentum after 5 s is

$$L = I\omega = (0.0938\ \text{kg·m}^2)\,(133\ \text{rad/s}) = 12.5\ \text{kg·m}^2/\text{s}$$

The dimensionless unit, the radian, is usually omitted in the unit of angular momentum. The power input at $t = 5$ s is

$$P = \tau\omega = (2.5\ \text{N·m})\,(133\ \text{rad/s}) = 332.5\ \text{W}$$

Note that to compute the angular momentum and power we must express $\omega$ in radians per second, not revolutions per minute or some other unit.

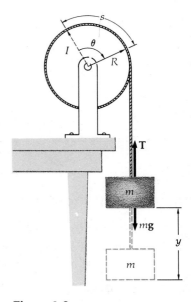

**Figure 9-8**
Body of mass $m$ tied to a string wrapped around a rotating wheel (Example 9-5). When the wheel turns through an angle $\theta$, a length of string $R\theta$ unwraps and the body drops a distance $y = R\theta$.

---

**Example 9-5** A body of mass $m$ is tied to a light string wound around a wheel of moment of inertia $I$ and radius $R$ (Figure 9-8). The wheel bearing is frictionless. Find the tension in the string, the acceleration of the body, and its speed after it has fallen a distance $h$ from rest.

The only force exerting a torque on the wheel is the tension $T$ in the string. It has lever arm $R$. Hence

$$TR = I\alpha \qquad\qquad 9\text{-}23$$

Two forces act on the suspended body, the upward tension $T$ and the downward force of gravity $mg$. Hence $\Sigma\mathbf{F} = m\mathbf{a}$ for this body gives

$$mg - T = ma \qquad\qquad 9\text{-}24$$

(where we have taken down as the positive direction).

There are three unknowns, $T$, $a$, and $\alpha$, in these two equations, but the string provides a constraint by which $a$ and $\alpha$ can be related. As the wheel turns through angle $\theta$, an amount of string of length $s = R\theta$ unwraps and the mass $m$ falls an equal distance $y = R\theta$. Thus we have

$$y = R\theta \qquad\qquad 9\text{-}25$$

$$\frac{dy}{dt} = R\frac{d\theta}{dt} \qquad \text{or} \qquad v = R\omega \qquad\qquad 9\text{-}26$$

Differentiating again gives

$$a = R\alpha \qquad\qquad 9\text{-}27$$

Thus, Equation 9-23 can be written

$$TR = I\frac{a}{R} \qquad \text{or} \qquad Ia = TR^2 \qquad\qquad 9\text{-}28$$

Solving Equations 9-24 and 9-28 for $a$ and $T$, we obtain

$$a = \frac{m}{m + I/R^2}\,g \qquad T = \frac{I/R^2}{m + I/R^2}\,mg \qquad\qquad 9\text{-}29$$

Since the acceleration is constant, the speed after the body has moved down a distance $h$ from rest can be computed from the constant-acceleration relationship

$$v^2 - v_0^2 = 2as$$

or

$$v^2 - 0 = 2\,\frac{m}{m + I/R^2}\,gh$$

$$v^2 = \frac{2m}{m + I/R^2}\,gh \qquad\qquad 9\text{-}30$$

Energy is conserved for this system, since there is no friction. We can therefore obtain the same result using the conservation-of-energy principle. As the body moves down the distance $h$, the potential energy of the system decreases by an amount $mgh$. This must be balanced by the simultaneous increase in kinetic energy of the system. Since the system starts at rest, the increase in kinetic energy is the final kinetic energy:

$$E_k = \tfrac{1}{2}mv^2 + \tfrac{1}{2}I\omega^2 = \tfrac{1}{2}mv^2 + \tfrac{1}{2}I\frac{v^2}{R^2} = \frac{1}{2}\left(m + \frac{I}{R^2}\right)v^2$$

Equating this to $mgh$ gives

$$\tfrac{1}{2}\left(m + \frac{I}{R^2}\right) v^2 = mgh$$

or

$$v^2 = \frac{2m}{m + I/R^2}\, gh$$

**Questions**

2. Can an object rotate if there is no torque acting?

3. Can a given rigid body have more than one moment of inertia?

4. Does an applied resultant torque always increase the rotational kinetic energy of a rigid body?

5. If the angular velocity of a body is zero at some instant, does this mean that the resultant torque on the body must be zero?

6. A wheel is said to be rotating about its axis with its angular velocity in the $+x$ direction. In what way is this description better than saying its angular velocity is clockwise or counterclockwise?

# 9-3    Calculating the Moment of Inertia

We can often simplify the calculation of the moments of inertia for various bodies by using general theorems relating the moment of inertia about one axis of the body to that about another axis. *Steiner's theorem*, or the *parallel-axis theorem*, relates the moment of inertia about an axis through the center of mass of a body to that about a second parallel axis. Let $I_{CM}$ be the moment of inertia about an axis through the center of mass of a body and $I$ be that about a parallel axis a distance $h$ away. The parallel-axis theorem states that

$$I = I_{CM} + Mh^2 \qquad\qquad 9\text{-}31 \qquad \textit{Parallel-axis theorem}$$

where $M$ is the total mass of the body. Examples 9-2 and 9-3 illustrated a special case of this theorem with $h = a$, $M = 4m$, and $I_{CM} = 4ma^2$.

We can prove the theorem using the result (developed in Chapter 8) that the kinetic energy of a system of particles is the sum of the kinetic energy of center-of-mass motion plus the energy of motion relative to the center of mass:

$$E_k = \tfrac{1}{2}Mv_{CM}^2 + E_{kr} \qquad\qquad 9\text{-}32$$

Consider a rigid body rotating with angular velocity about an axis a distance $h$ from a parallel axis through the center of mass, as in Figure 9-9. When the system rotates through an angle $\Delta\theta$ measured about the axis of rotation, it rotates through the same angle $\Delta\theta$ measured about any other parallel axis. The motion of the body relative to the center of mass is thus a rotation about the center-of-mass axis with the same angular velocity. The kinetic energy of motion relative to the center of mass is thus

$$E_{kr} = \tfrac{1}{2}I_{CM}\omega^2$$

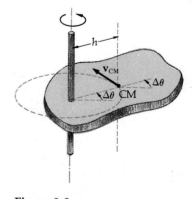

**Figure 9-9**
Body rotating about an axis parallel to an axis through the center of mass and a distance $h$ from it.

The velocity of the center of mass relative to any fixed point on the axis of rotation is

$$v_{CM} = h\omega$$

The kinetic energy of motion of the center of mass is thus

$$\tfrac{1}{2}Mv_{CM}^2 = \tfrac{1}{2}M(h\omega)^2 = \tfrac{1}{2}M\omega^2h^2$$

When the total kinetic energy of rotation is written $\tfrac{1}{2}I\omega^2$, Equation 9-32 becomes

$$E_k = \tfrac{1}{2}I\omega^2 = \tfrac{1}{2}M\omega^2h^2 + \tfrac{1}{2}I_{CM}\omega^2 = \tfrac{1}{2}(Mh^2 + I_{CM})\omega^2$$

Thus

$$I = Mh^2 + I_{CM}$$

The *plane-figure theorem* relates the moments of inertia about two perpendicular axes in a plane figure to the moment of inertia about a third axis perpendicular to the figure. If $x$, $y$, and $z$ are perpendicular axes for a figure which lies in the $xy$ plane, the moment of inertia about the $z$ axis equals the sum of the moments of inertia about the $x$ and $y$ axes. This is not difficult to prove. Figure 9-10 shows a figure in the $xy$ plane. The moment of inertia about the $x$ axis is

$$I_x = \Sigma m_i y_i^2$$

where the sum is taken over all the elements of mass $m_i$. (In practice, as we shall see, the sum is often done by integration.) The moment of inertia about the $y$ axis is

$$I_y = \Sigma m_i x_i^2$$

The moment of inertia about the $z$ axis is

$$I_z = \Sigma m_i \rho_i^2$$

But for each element $m_i$, $\rho_i^2 = x_i^2 + y_i^2$. Thus

$$I_z = \Sigma m_i \rho_i^2 = \Sigma m_i(x_i^2 + y_i^2) = \Sigma m_i x_i^2 + \Sigma m_i y_i^2$$

or

$$I_z = I_y + I_x \qquad\qquad\qquad 9\text{-}33$$

*Plane-figure theorem*

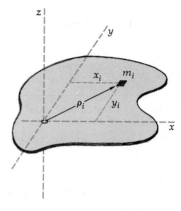

**Figure 9-10**
Plane figure with $z$ axis perpendicular to the plane. The moment of inertia about the $z$ axis equals the sum of the moments of inertia about the $x$ and $y$ axes.

**Example 9-6** What is the moment of inertia about an axis through one of the masses and perpendicular to the plane of the masses shown in Figure 9-11?

The moment of inertia calculated about the $y$ axis in this figure was found in Example 9-3 to be $8ma^2$. By similar reasoning, the moment of inertia about the $x$ axis in this figure is $8mb^2$. Thus $I_z = I_x + I_y = 8ma^2 + 8mb^2$. This result is easily checked by direct calculation.

For continuous bodies the sum in Equation 9-22 defining the moment of inertia is replaced by an integral. Let $dm$ be a mass element a distance $x$ from the axis of rotation. The moment of inertia for this axis is then

$$I = \int x^2\, dm \qquad\qquad\qquad 9\text{-}34$$

Some examples will demonstrate how to calculate moments of inertia using Equation 9-34 and the two theorems discussed above (Equations 9-31 and 9-33). Moments of inertia of uniform bodies of varied shapes are given in Table 9-2.

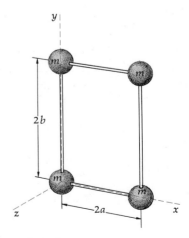

**Figure 9-11**
Example 9-6.

**Table 9-2**
Moments of inertia of uniform bodies of various shapes

| | | |
|---|---|---|
| Cylindrical shell about axis | | $I = MR^2$ |
| Solid cylinder about axis | | $I = \tfrac{1}{2}MR^2$ |
| Hollow cylinder about axis | | $I = \tfrac{1}{2}M(R_1^2 + R_2^2)$ |
| Cylindrical shell about diameter through center | | $I = \tfrac{1}{2}MR^2 + \tfrac{1}{12}ML^2$ |
| Solid cylinder about diameter through center | | $I = \tfrac{1}{4}MR^2 + \tfrac{1}{12}ML^2$ |
| Thin rod about perpendicular line through center | | $I = \tfrac{1}{12}ML^2$ |
| Thin rod about perpendicular line through one end | | $I = \tfrac{1}{3}ML^2$ |
| Thin spherical shell about diameter | | $I = \tfrac{2}{3}MR^2$ |
| Solid sphere about diameter | | $I = \tfrac{2}{5}MR^2$ |
| Solid rectangular parallelepiped about axis through center perpendicular to face | | $I = \tfrac{1}{12}M(a^2 + b^2)$ |

**Example 9-7** Find the moment of inertia of a uniform stick about an axis perpendicular to the stick through one end.

The mass element $dm$ is shown in Figure 9-12. Since the total mass $M$ is uniformly distributed along the length $L$, $dm = (M/L)\,dx$. The moment of inertia about the $y$ axis is

$$I_y = \int_0^L x^2 \frac{M}{L}\,dx = \frac{M}{L} \int_0^L x^2\,dx = \frac{M}{L}\frac{1}{3} x^3 \Big]_0^L$$

$$= \frac{M}{L}\frac{L^3}{3} = \tfrac{1}{3}ML^2 \qquad\qquad 9\text{-}35$$

The moment of inertia about the $z$ axis is also $\tfrac{1}{3}ML^2$, and that about the $x$ axis is zero, assuming that all the mass is right on the $x$ axis.

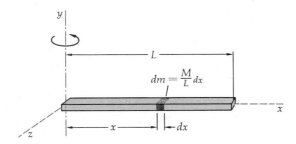

**Figure 9-12**
Calculation of $I$ for a uniform stick with axis through one end.

**Example 9-8** Find the moment of inertia of a uniform stick about the $y'$ axis through the center of mass (Figure 9-13).

Since the axis is through the center of mass of the stick, the integration extends from $-L/2$ to $+L/2$. Thus

$$I_{CM} = \int_{-L/2}^{+L/2} x'^2 \frac{M}{L}\,dx' = 2 \int_0^{L/2} x'^2 \frac{M}{L}\,dx'$$

$$= \frac{2M}{L}\frac{(L/2)^3}{3} = \tfrac{1}{12}ML^2 \qquad\qquad 9\text{-}36$$

This is smaller than that about the axis through the end, as would be expected, because, on the average, the mass is closer to this axis than to the end. This result can also be obtained from the result of Example 9-7 and the parallel-axis theorem. In this case $h = \tfrac{1}{2}L$, and

$$I = I_{CM} + M(\tfrac{1}{2}L)^2 = \tfrac{1}{3}ML^2$$

or

$$I_{CM} = \tfrac{1}{3}ML^2 - \tfrac{1}{4}ML^2 = \tfrac{1}{12}ML^2$$

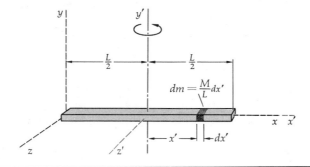

**Figure 9-13**
Calculation of $I$ for a uniform stick rotating about an axis through the center of mass.

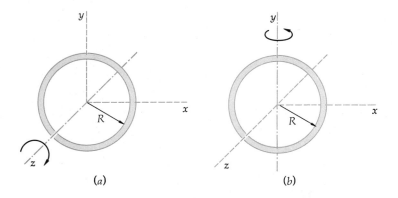

*(a)*                                    *(b)*

**Figure 9-14**
(*a*) Hoop rotating about an axis perpendicular to the plane of the hoop through its center. Since all the mass of the hoop is at a distance $R$ from the axis, the moment of inertia is $MR^2$. (*b*) Hoop rotating about an axis through its center and in the plane of the hoop. The moment of inertia is $\frac{1}{2}MR^2$, found from the plane-figure theorem.

---

**Example 9-9** Find the moment of inertia of a hoop about an axis through the center and perpendicular to the plane of the hoop (Figure 9-14*a*).

Since all the mass is at a distance $R$ from this axis, the moment of inertia is simply $I = MR^2$.

---

**Example 9-10** Find the moment of inertia of a hoop about a diameter of the hoop (Figure 9-14*b*).

Instead of solving this problem directly we use the plane-figure theorem. If we take the hoop to be in the $xy$ plane with the center at the origin, by symmetry we have $I_x = I_y$. Since we have already found $I_z$ to be $MR^2$, we have

$$I_z = I_x + I_y = 2I_x = MR^2$$

Therefore

$$I_x = I_y = \tfrac{1}{2}MR^2$$

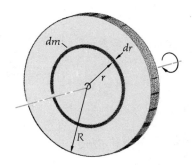

**Figure 9-15**
Calculation of $I$ for a uniform disk about an axis through its center and perpendicular to the plane of the disk.

---

**Example 9-11** Find the moment of inertia of a uniform disk about the axis through its center and perpendicular to the plane of the disk.

We expect that $I$ will be smaller than $MR^2$ since all the mass is less than the distance $R$ from the axis. We can calculate $I$ by taking mass elements as shown in Figure 9-15. Each mass element is a hoop with moment of inertia $r^2\,dm$. Since the area of each element is $2\pi r\,dr$, the mass of the element is $dm = (M/A)2\pi r\,dr$, where $A = \pi R^2$ is the area of the disk. Thus we have

$$I = \int r^2\,dm = \int_0^R r^2 \frac{M}{A}\, 2\pi r\,dr = \frac{2\pi M}{\pi R^2}\int_0^R r^3\,dr$$

$$= \frac{2M}{R^2}\frac{R^4}{4} = \tfrac{1}{2}MR^2$$

9-37

---

**Example 9-12** Find the moment of inertia of a cylinder about its axis (Figure 9-16).

We can consider the cylinder to be a set of disks each with mass $m_i$ and moment of inertia $\frac{1}{2}m_i R^2$. Then the moment of inertia of the complete cylinder is

$$I = \Sigma\tfrac{1}{2}m_i R^2 = \tfrac{1}{2}R^2\Sigma m_i = \tfrac{1}{2}MR^2$$

where $M = \Sigma m_i$ is the total mass of the cylinder.

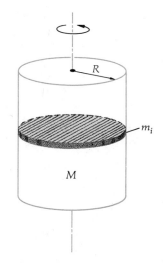

**Figure 9-16**
A cylinder rotating about its axis can be considered as a stack of disks of mass $m_i$. Since each disk has moment of inertia $\frac{1}{2}m_i R^2$, the moment of inertia of the cylinder is $\frac{1}{2}MR^2$.

**Example 9-13** Find the moment of inertia of a sphere about an axis through its center.

We can calculate this moment of inertia by treating the sphere as a set of disks. Consider the disk element in Figure 9-17 at height $x$ above the center. The radius of the disk is

$$r = \sqrt{R^2 - x^2}$$

The mass of the disk is

$$dm = \frac{M}{V} \pi r^2 \, dx = \frac{M}{V} \pi(R^2 - x^2) \, dx$$

where $M$ is the total mass of the sphere and

$$V = \tfrac{4}{3}\pi R^3$$

is the volume of the sphere. The moment of inertia of the disk is

$$dI = \tfrac{1}{2} \, dm \, r^2 = \frac{1}{2}\frac{M}{V} \pi(R^2 - x^2)^2 \, dx$$

The total moment of inertia of the sphere is

$$I = \int_{-R}^{+R} \frac{1}{2}\frac{M}{V} \pi(R^2 - x^2)^2 \, dx = \frac{1}{2}\frac{M}{V} \pi 2 \int_{0}^{R} (R^2 - x^2)^2 \, dx$$

$$= \frac{8\pi M R^5}{15V} = \tfrac{2}{5}MR^2 \qquad\qquad 9\text{-}38$$

using $V = \tfrac{4}{3}\pi R^3$.

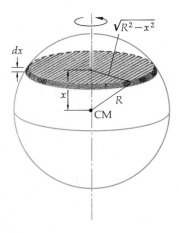

**Figure 9-17**
Calculation of the moment of inertia of a uniform sphere rotating about a diameter (Example 9-13). The sphere can be thought of as a set of disks of varying radii.

---

The moment of inertia is often given in handbooks in terms of the *radius of gyration k*, defined by

$$k^2 = \frac{I}{M} \qquad\qquad 9\text{-}39 \qquad \textit{Radius of gyration}$$

where $M$ is the total mass of the body. The radius of gyration is thus the distance from the axis at which a point mass $M$ would have the same moment of inertia about the axis. For example, the moment of inertia about an axis through one end of a stick is

$$Mk^2 = \tfrac{1}{3}ML^2$$

or

$$k = \frac{L}{\sqrt{3}}$$

The radius of gyration is a measure of the distribution of mass of a body relative to a given axis of rotation. A large radius of gyration means that, on the average, the mass is relatively far from the axis.

**Questions**

7. Through what point in a body must the axis pass if the moment of inertia is to be a minimum?

8. A disk, a hoop, and a sphere have the same radius and are uniform in density. Which has the greatest and which the least radius of gyration for an axis through the center of each and perpendicular to the plane of the disk and hoop? (Answer without looking up the moments of inertia.)

Review

A. Define, explain, or otherwise identify:

Rigid body, 263
Angular velocity, 265
Angular acceleration, 265
Moment of inertia, 268
Angular momentum, 269
Parallel-axis theorem, 274
Radius of gyration, 279

B. True or false:

1. Angular velocity and linear velocity have the same dimensions.

2. All parts of a rotating wheel have the same angular velocity.

3. All parts of a rotating wheel have the same angular acceleration.

4. The moment of inertia of a body depends on the location of the axis of rotation.

5. The moment of inertia of a body depends on the angular velocity of the body.


Exercises

### Section 9-1, Angular Velocity and Angular Acceleration

1. A particle moves in a circle of radius 100 m with constant speed of 20 m/s. (a) What is its angular velocity in radians per second about the center of the circle? (b) How many revolutions does it make in 30 s?

2. Find a formula to convert radians per second into revolutions per minute.

3. A 30-cm-diameter record rotates at $33\frac{1}{3}$ rev/min. (a) What is its angular velocity in radians per second? (b) Find the speed and (linear) acceleration of a point on the rim of the record.

4. A wheel starts from rest and has a constant angular acceleration of 2 rad/s². (a) What is its angular velocity after 5 s? (b) Through what angle has the wheel turned after 5 s? (c) How many revolutions has it made in 5 s? (d) After 5 s, what is the speed and acceleration of a point 0.3 m from the axis of rotation?

5. A turntable rotating at $33\frac{1}{3}$ rev/min is shut off. It brakes with constant angular acceleration and comes to rest in 2 min. (a) Find the angular acceleration. (b) What is the average angular velocity of the turntable? (c) How many revolutions does it make before stopping?

6. A disk of radius 10 cm rotates about its axis from rest with constant angular acceleration of 10 rad/s². At $t = 5$ s what are (a) the angular velocity of the disk and (b) the tangential and centripetal acceleration $a_t$ and $a_c$ of a point on the edge of the disk?

7. A spot on a record rests on the radial line $\theta = 0°$ (fixed in space). At $t = \frac{1}{4}$ s after the turntable is turned on, the spot has advanced to $\theta = 10°$. Assuming constant angular acceleration, how long will it be before the record is rotating at 33.3 rev/min?


### Section 9-2, Torque and Moment of Inertia

8. The bodies in Figure 9-18 are connected by a very light rod whose moment of inertia may be neglected. They rotate about the $y$ axis with angular velocity of 2 rad/s. (a) Find the speed of each body and use it to calculate the kinetic energy of this system directly from $\Sigma\frac{1}{2}m_iv_i^2$. (b) Find the moment of inertia about the $y$ axis and calculate the kinetic energy from $E_k = \frac{1}{2}I\omega^2$.

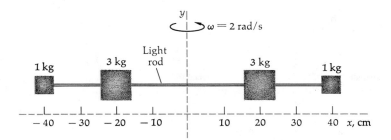

**Figure 9-18**
Exercise 8.

9. Four particles are at the corners of a square connected by massless rods, $m_1 = m_3 = 3$ kg and $m_2 = m_4 = 4$ kg. The length of the side of the square is $L = 2$ m (Figure 9-19). (a) Find the moment of inertia about an axis perpendicular to the plane of the particles, and passing through $m_4$. (b) How much work is required to produce a rotation of 2 rad/s about this axis?

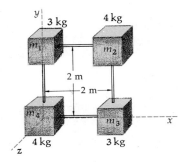

**Figure 9-19**
Exercise 9.

10. Four 2-kg particles are located at the corners of a rectangle of sides 3 and 2 m (Figure 9-20). (a) Find the moment of inertia of this system about an axis perpendicular to the plane of the masses and through one of the masses. (b) The system is set rotating about this axis with kinetic energy of 184 J. Find the number of revolutions the system makes per minute.

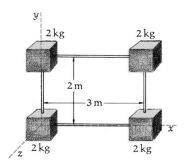

**Figure 9-20**
Exercise 10.

11. An irregularly shaped object rotates about an axis at 50 rev/min. The object does 300 J of work as it stops. What is the moment of inertia of the object?

12. A disk-shaped grindstone of mass 2 kg and radius 7 cm is spinning at 700 rev/min. When the power is shut off, a man continues to sharpen his ax by holding it against the grindstone for 10 s until the grindstone stops rotating. (a) Find the angular acceleration of the grindstone, assuming it to be constant. (b) What is the torque exerted by the ax on the grindstone? (Assume no other frictional torques.)

13. A uniform disk of radius 0.12 m and mass 5 kg is pivoted so that it rotates freely about its axis. A string wrapped around the disk is pulled with a force of 20 N (Figure 9-21). (*a*) What is the torque exerted on the disk? (*b*) What is the angular acceleration of the disk? (*c*) If the disk starts from rest, what is its angular velocity after 3 s? (*d*) What is its kinetic energy after 3 s? (*e*) What is its angular momentum after 3 s? (*f*) Find the total angle $\theta$ the disk turns through in 3 s, and (*g*) show that the work done by the torque $\tau\theta$ equals the kinetic energy.

**Figure 9-21**
Exercise 13.

14. In order to start a playground merry-go-round rotating, a rope is wrapped around it and pulled. A force of 200 N is exerted on the rope for 10 s. During this time the merry-go-round, which has a radius of 2 m, makes one complete rotation. (*a*) Find the angular acceleration of the merry-go-round, assuming it to be constant. (*b*) What torque is exerted by the rope on the merry-go-round? (*c*) What is the moment of inertia of the merry-go-round?

15. A rope is wrapped around a 3-kg cylinder of radius 10 cm which is free to turn about its axis. The rope is pulled with a force of 15 N. The cylinder is initially at rest at $t = 0$. (*a*) Find the torque exerted by the rope, the moment of inertia of the cylinder, and the angular acceleration of the cylinder. (*b*) Find the angular velocity and the angular momentum of the cylinder at time $t = 4$ s. What is the power input of the force at this time?

16. A 2000-kg block is lifted by a steel cable which passes over a pulley to a motor-driven winch (see Figure 9-22). The radius of the winch drum is 30 cm, and the moment of inertia of the pulley is negligible. (*a*) What force must be exerted by the cable to lift the block at a constant velocity of 8 cm/s? (*b*) What torque does the cable exert on the winch drum? (*c*) What is the angular velocity of the winch drum? (*d*) What power must be developed by the motor to drive the winch drum?

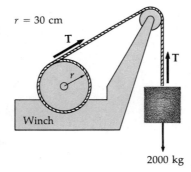

**Figure 9-22**
Exercise 16.

17. An engine develops 500 N·m of torque at 3500 rev/min. Find the power developed by the engine.

### Section 9-3, Calculating the Moment of Inertia

18. Use the parallel-axis theorem and the result of Exercise 9 to find the moment of inertia of the four-particle system in Figure 9-19 about an axis perpendicular to the plane of the masses and through the center of mass. Check your result by direct computation.

19. (a) Find the moment of inertia $I_x$ for the four-particle system of Figure 9-19 about the $x$ axis which passes through $m_3$ and $m_4$. (b) Find $I_y$ for that system about the $y$ axis which passes through $m_1$ and $m_4$. (c) Use Equation 9-33 to calculate the moment of inertia $I_z$ about the $z$ axis through $m_4$ and perpendicular to the plane of the figure.

20. (a) Use the parallel-axis theorem to find the moment of inertia about an axis parallel to the $z$ axis and through the center of mass of the system in Figure 9-20. (b) Let $x'$ and $y'$ be axes in the plane of the figure through the center of mass and parallel to the sides of the rectangle. Compute $I_{x'}$ and $I_{y'}$ and use your results and Equation 9-33 to check your result for part (a).

21. Find the moment of inertia of a disk of mass $M$ and radius $R$ about an axis through the edge of the disk parallel to the axis of the disk (see Figure 9-23).

22. Use Equation 9-33 to find the moment of inertia of a disk of radius $R$ and mass $M$ about an axis in the plane of the disk through its center (Figure 9-24).

23. Use the parallel-axis theorem and Equation 9-38 to find the moment of inertia of a solid sphere about an axis tangent to the sphere (Figure 9-25).

**Figure 9-23**
Exercise 21.

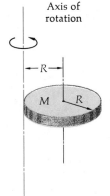

**Figure 9-24**
Exercise 22.

**Figure 9-25**
Exercise 23.

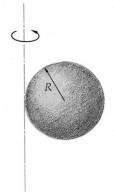

24. Find the moment of inertia about an axis tangent to a hoop of mass $M$ and radius $R$.

25. Find the moment of inertia about an axis intersecting a hoop and parallel to the axis of the hoop.

## Problems

1. A wheel mounted on a horizontal axle has a radial line painted on it which is used to determine its angular position relative to the vertical. The wheel has a constant angular acceleration. At time $t = 0$ the line points straight up. At $t = 1$ s, the line is again vertical, having turned through one full revolution. At $t = 2$ s, the line points straight down, the wheel having turned through $1\frac{1}{2}$ rev more since $t = 1$ s. (a) What is the angular acceleration of the wheel? (b) What was its angular velocity at $t = 0$?

2. Calculate the kinetic energy of rotation of the earth and compare it with the kinetic energy of the earth's center-of-mass motion. Assume the earth to be a homogeneous sphere of mass $6.0 \times 10^{24}$ kg and radius $6.4 \times 10^6$ m. The radius of the earth's orbit is $1.5 \times 10^{11}$ m.

3. A flywheel has a mass of 100 kg and a radius of gyration of 0.5 m and rotates with angular velocity of 1200 rev/min. (a) A constant tangential force is applied at a radial distance of 0.5 m. What work must this force do to stop the wheel? (b) If the wheel is brought to rest in 2 min, what torque does the force produce? What is the magnitude of the force? (c) How many revolutions does the wheel make in these 2 min?

4.  A wheel mounted on an axis that is not frictionless is at rest initially. A constant external torque of 50 N·m is applied to the wheel for 20 s. At the end of the 20 s the wheel has an angular velocity of 600 rev/min. The external torque is then removed, and the wheel comes to rest after 120 s more. (*a*) What is the moment of inertia of the wheel? (*b*) What is the frictional torque (assumed to be constant)?

5.  A typical car engine delivers about 2 MJ of mechanical energy per kilometre on the average. A car is designed to use the energy stored in a flywheel in a vacuum container. If the mass of the flywheel is not to exceed 100 kg and its angular velocity is not to exceed 400 rev/s, find the smallest radius of the flywheel such that the car can travel 300 km without recharging the flywheel. (Assume the flywheel to be a uniform cylinder.)

6.  A 1-Mg car is being unloaded by a winch, as shown in Figure 9-26. At this moment, the winch gearbox shaft breaks, and the car falls from rest. The moment of inertia of the winch drum is 320 kg·m² and that of the pulley is 4 kg·m²; the radius of the winch drum is 0.80 m and that of the pulley 0.30 m. Find the speed of the car as it hits the water.

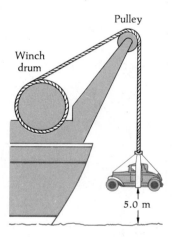

**Figure 9-26**
Problem 6.

7.  The system in Figure 9-27 is released from rest. The 30-kg body is 2 m above the floor. The pulley is a uniform disk with a radius of 10 cm and mass 5 kg. Find (*a*) the speed of the 30-kg body just before it hits the floor and the angular speed of the pulley at that time, (*b*) the tensions in the strings, and (*c*) the time it takes for the 30-kg body to reach the floor.

8.  A metrestick is pivoted at one end so it can swing freely in a vertical plane. It is released from rest in a horizontal position. (*a*) What is the angular velocity of the stick when it is vertical? (*b*) What force is exerted by the pivot when the stick is vertical (answer in terms of the mass *m* of the stick)?

9.  A uniform disk of mass *M* and radius *R* is pivoted so that it can rotate freely about an axis through its center and perpendicular to the plane of the disk. A small particle of mass *m* is attached to the rim of the disk at the top directly above the pivot. The system is given a gentle start and the disk begins to rotate. (*a*) What is the angular velocity of the disk when the particle is at its lowest point? (*b*) At this point, what force must be exerted on the particle by the disk to keep it on the disk?

10.  A vertical grinding wheel is a uniform disk of mass 50 kg and radius 40 cm. It has a handle of radius 60 cm and of negligible mass. A load of mass 20 kg is attached to the handle when it is in the horizontal position. Find (*a*) the initial angular acceleration of the wheel and (*b*) the maximum angular velocity of the wheel.

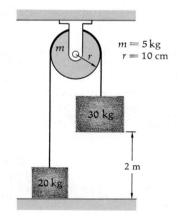

**Figure 9-27**
Problem 7.

11. A uniform cylinder of mass $m_1$ and radius $R$ is pivoted on frictionless bearings. A string wrapped around the cylinder connects to a mass $m_2$, which is on a frictionless incline of angle $\theta$, as shown in Figure 9-28. This system is released from rest with $m_2$ a height $h$ above the bottom of the incline. (a) What is the acceleration of $m_2$? (b) What is the tension in the string? (c) What is the total energy of the system when $m_2$ is at height $h$? (d) What is the total energy when $m_2$ is at the bottom of the incline and has speed $v$? (e) What is the speed $v$? (f) Evaluate your answers for the extreme cases of $\theta = 0°$, $\theta = 90°$, and $m_1 = 0$.

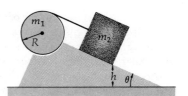

**Figure 9-28**
Problem 11.

12. An Atwood's machine has two masses, $m_1 = 500$ g and $m_2 = 510$ g, connected by a string of negligible mass which passes over a frictionless pulley (Figure 9-29). The pulley has mass 50 g and radius of gyration $k = 3$ cm. The radius at which the string passes over the pulley is 4 cm. The string does not slip on the pulley. (a) Find the acceleration of the masses. (b) What is the tension in the string supporting $m_1$? In the string supporting $m_2$? By how much do they differ? (c) What would your answers have been if you had neglected the motion of the pulley?

**Figure 9-29**
Problem 12.

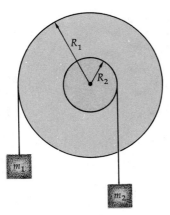
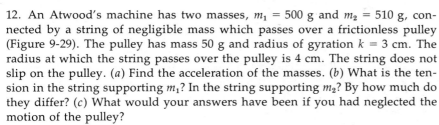

13. Two objects are attached to ropes attached to wheels on a common axle, as in Figure 9-30. The total moment of inertia of the two wheels is 40 kg·m². The radii are $R_1 = 1.2$ m and $R_2 = 0.4$ m. (a) If $m_1 = 24$ kg, find $m_2$ such that the system is in equilibrium. (b) If 12 kg is gently added to the top of $m_1$, find the angular acceleration of the wheels and the tension in the ropes.

**Figure 9-30**
Problem 13.

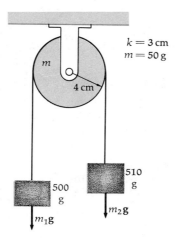

14. A uniform boom 5.0 m long and having a total mass of 150 kg is connected to the ground by a hinge at the bottom and is supported by a horizontal cable, as shown in Figure 9-31. (a) What is the tension in the cable? (b) What is the angular acceleration of the boom the instant the cable is cut? (c) If the cable is cut, what is the angular velocity of the boom when it is horizontal?

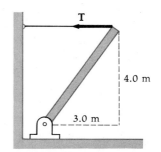

**Figure 9-31**
Problem 14.

15. Figure 9-32 shows a pair of uniform spheres each of mass 500 g and radius 5 cm. They are mounted on a uniform rod which has length $L = 30$ cm and mass 60 g. (*a*) Calculate the moment of inertia of this system about an axis perpendicular to the rod through the center of the rod, using the approximation that the two spheres can be treated as point particles a distance 20 cm from the axis of rotation and that the mass of the rod is negligible. (*b*) Calculate the moment of inertia exactly and compare your result with your approximate value.

16. A uniform rectangular plate has mass $m$ and sides $a$ and $b$. (*a*) Show by integration that its moment of inertia about an axis perpendicular to the plate and through one corner is $\frac{1}{3}m(a^2 + b^2)$. (*b*) What is the moment of inertia about an axis at the center of mass and perpendicular to the plate?

17. Show that the moment of inertia of a spherical shell of radius $R$ and mass $m$ is $\frac{2}{3}mR^2$. This can be done by direct integration or more easily by finding the increase in the moment of inertia of a solid sphere when the radius changes. To do this, first show that the moment of inertia of a solid sphere of density $\rho$ is $I = \frac{8}{15}\pi\rho R^5$. Then compute the change $dI$ in $I$ for a change $dR$, and use the fact that the mass of the shell is $m = 4\pi R^2\rho\, dR$.

18. A hollow cylinder has mass $m$, outside radius $R_2$, and inside radius $R_1$. Show that its moment of inertia about its symmetry axis is given by $I = \frac{1}{2}m(R_2^2 + R_1^2)$.

19. A uniform sphere of mass $M$ and radius $R$ is free to rotate about a horizontal axis through its center. A string is wrapped around the sphere and attached to a body of mass $m$, as shown in Figure 9-33. Find the acceleration of the body and the tension in the string.

20. A uniform stick of length $L$ and mass $M$ is hinged at one end and released from rest at an angle $\theta_0$ with the vertical. Show that when the angle with the vertical is $\theta$, the hinge exerts a force $F_r$ along the stick and $F_t$ perpendicular to the stick given by

$$F_r = \tfrac{1}{2}Mg(5\cos\theta - 3\cos\theta_0) \quad \text{and} \quad F_t = \tfrac{1}{4}Mg\sin\theta$$

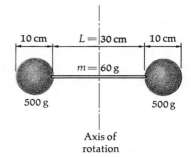

10 cm    $L = 30$ cm    10 cm

$m = 60$ g

500 g    500 g

Axis of rotation

**Figure 9-32**
Problem 15.

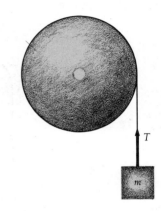

$T$

$m$

**Figure 9-33**
Problem 19.

# CHAPTER 10 Rotation in Space

---

**Objectives**  After studying this chapter, you should:

1. Know the definitions of cross product, torque, angular momentum, precession, and nutation.

2. Know how to write the angular momentum of a system of particles as the sum of the angular momentum relative to the center of mass and angular momentum due to motion of the center of mass.

3. Be able to apply the conservation of angular momentum to problems in which the moment of inertia changes.

4. Know the condition for rolling without slipping and be able to apply it in working problems involving translation and rotation.

5. Be able to describe the motion of a gyroscope qualitatively.

---

In the general motion of a body, the axis of rotation is not fixed in space. The motion is complicated because the directions as well as the magnitudes of the angular velocity and angular acceleration change. Another complication arises for systems which are not rigid because the moment of inertia may change in time. In this chapter we begin by defining torque and the angular momentum of a single particle relative to a single point in space. In doing so we shall introduce the idea of the vector product, also called the cross product. (These more general definitions of torque and angular momentum reduce to those used in the previous chapter in the special case of motion of a rigid body about a fixed axis which is a symmetry axis of the body.) We shall then discuss the third important conservation law in mechanics (in addition to the conservation of energy and of linear momentum), the conservation of angular momentum for an isolated system. After applying these ideas to the relatively simple motion of translation and rotation that occurs for rolling objects, we discuss qualitatively the more complicated motions of the gyroscope and the rotation of a rigid body about an axis which is not a symmetry axis. This latter problem is related to the problem of dynamic imbalance. Finally, we look more closely at the vector nature of rotation and show that although we can assign a magnitude and direction to a finite rotation in space, this quantity is not a vector because it does not obey the law of vector addition.

# 10-1   Torque as a Vector Product

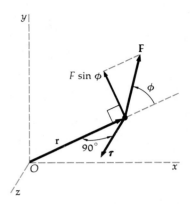

We begin by considering a single particle of mass $m$ moving with velocity $\mathbf{v} = d\mathbf{r}/dt$ and momentum $\mathbf{p} = m\mathbf{v}$, where $\mathbf{r}$ is the position vector of the particle from the origin $O$. We assume that the point $O$ is in an inertial reference frame so that Newton's law $\Sigma\mathbf{F} = d\mathbf{p}/dt$ holds, where $\Sigma\mathbf{F}$ is the resultant force. For the moment let us consider a single force $\mathbf{F}$ acting on the particle (Figure 10-1). The torque exerted by $\mathbf{F}$ relative to the point $O$ is defined to be a vector of magnitude $Fr \sin \phi$, where $\phi$ is the angle between $\mathbf{F}$ and $\mathbf{r}$, and direction perpendicular to the plane formed by $\mathbf{F}$ and $\mathbf{r}$ (Figure 10-1). This is the same definition of torque we used in Chapters 5 and 9. Here we emphasize that torque is defined *relative to a point in space* (and not relative to an axis, like the moment of inertia of a body). The torque is conveniently written as the *vector product*, or *cross product*, of $\mathbf{r}$ and $\mathbf{F}$:

**Figure 10-1**
The torque exerted by a force $\mathbf{F}$ at position $\mathbf{r}$ is perpendicular to both $\mathbf{F}$ and $\mathbf{r}$ and has the magnitude $Fr \sin \phi$.

$$\boldsymbol{\tau} = \mathbf{r} \times \mathbf{F} \qquad\qquad 10\text{-}1$$

In general,

*the cross product of two vectors* $\mathbf{A}$ *and* $\mathbf{B}$ *is defined to be a vector whose magnitude equals the area of the parallelogram formed by the two vectors (Figure 10-2a), and whose direction is perpendicular to the plane containing* $\mathbf{A}$ *and* $\mathbf{B}$ *in the sense given by the right-hand rule, as* $\mathbf{A}$ *is rotated into* $\mathbf{B}$ *through the smaller angle between the vectors.*

If $\phi$ is the angle between the two vectors and $\hat{\mathbf{n}}$ is a unit vector perpendicular to each and in the sense described, the cross product of $\mathbf{A}$ and $\mathbf{B}$ is

$$\mathbf{A} \times \mathbf{B} = AB \sin \phi \, \hat{\mathbf{n}} \qquad\qquad 10\text{-}2 \qquad\qquad \textit{Cross product}$$

The relation between the vectors $\mathbf{A}$, $\mathbf{B}$, and $\mathbf{A} \times \mathbf{B}$ is shown in Figure 10-2.

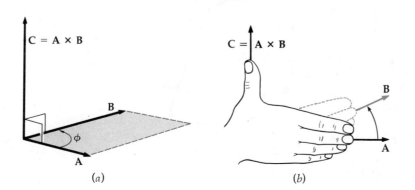

(a)                          (b)

**Figure 10-2**
(a) The vector product $\mathbf{A} \times \mathbf{B}$ is a vector $\mathbf{C}$ perpendicular to both $\mathbf{A}$ and $\mathbf{B}$ and of magnitude $AB \sin \phi$, which equals the area of the parallelogram shown. (b) The direction of $\mathbf{A} \times \mathbf{B}$ is the direction of rotation when $\mathbf{A}$ is rotated into $\mathbf{B}$ through the angle $\phi$.

If **A** and **B** are parallel, **A** × **B** is zero. From the definition, Equation 10-2, it follows that

$$\mathbf{A} \times \mathbf{A} = 0 \qquad\qquad 10\text{-}3$$

and

$$\mathbf{A} \times \mathbf{B} = -\mathbf{B} \times \mathbf{A} \qquad\qquad 10\text{-}4$$

It should be especially noted that the order in which the two vectors are multiplied is significant. Unlike the multiplication of ordinary numbers, changing the order of two vectors in a cross product changes the result. In fact, as indicated in Equation 10-4, changing the order of the factors in a cross product changes the sign of the result.

Some properties of the cross product of two vectors follow:

1. Distributive law:

$$\mathbf{A} \times (\mathbf{B} + \mathbf{C}) = \mathbf{A} \times \mathbf{B} + \mathbf{A} \times \mathbf{C} \qquad\qquad 10\text{-}5$$

2. Derivative: If **A** and **B** are functions of some variable such as $t$, the derivative of **A** × **B** follows the usual product rule for derivatives,

$$\frac{d}{dt}(\mathbf{A} \times \mathbf{B}) = \mathbf{A} \times \frac{d\mathbf{B}}{dt} + \frac{d\mathbf{A}}{dt} \times \mathbf{B} \qquad\qquad 10\text{-}6$$

We must keep the order straight here since, for example, $\mathbf{B} \times d\mathbf{A}/dt = -(d\mathbf{A}/dt) \times \mathbf{B}$.

From Equation 10-1 we see that the magnitude of the torque $Fr \sin \phi$ can be thought of as the product of $F$ and the distance $r \sin \phi$, which is the perpendicular distance from the line of the force to the origin, called the *lever arm*. Alternatively, we can think of it as the product of $r$ and $F \sin \phi$, which is the component of the force perpendicular to **r**.

**Questions**

1. Will a force acting on a body produce a torque if the body is not mounted on an axle?

2. A body is mounted on pivots so that it is free to turn about the $z$ axis. A force **F** is applied at point $(x, y, z)$. Does the $z$ component of the torque depend on what value $z$ has? On what value $F_z$ has?

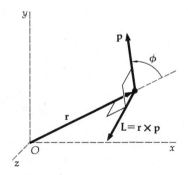

**Figure 10-3**
A particle with momentum **p** at position **r** has angular momentum, relative to the origin, of $\mathbf{L} = \mathbf{r} \times \mathbf{p}$, which is perpendicular to both **r** and **p**.

*Angular momentum of a particle*

# 10-2  Angular Momentum of a Particle

We define the angular momentum **L** of a particle relative to the origin $O$ to be the cross product of the position vector **r** and the linear momentum **p** (see Figure 10-3):

$$\mathbf{L} = \mathbf{r} \times \mathbf{p} \qquad\qquad 10\text{-}7$$

Like torque, angular momentum is defined relative to a point in space. The magnitude is the magnitude of the momentum $p$ times $r \sin \phi$, which is the perpendicular distance from the line of motion to the origin. Alternatively, $L$ is the distance $r$ times $p \sin \phi$, which is the component of the momentum perpendicular to **r**. We shall now show that

Newton's second law implies that the rate of change of angular momentum equals the resultant torque acting on the particle. If we have a number of forces acting on a particle, the resultant torque relative to the point $O$ is the sum of the torques due to each force:

$$\Sigma\boldsymbol{\tau} = \mathbf{r} \times \mathbf{F}_1 + \mathbf{r} \times \mathbf{F}_2 + \cdots = \mathbf{r} \times \Sigma\mathbf{F}$$

According to Newton's law, the resultant force equals the rate of change of linear momentum, $d\mathbf{p}/dt$. Thus

$$\Sigma\boldsymbol{\tau} = \mathbf{r} \times \Sigma\mathbf{F} = \mathbf{r} \times \frac{d\mathbf{p}}{dt} \qquad \text{10-8}$$

We can compute the rate of change of the angular momentum using the product rule for derivatives:

$$\frac{d\mathbf{L}}{dt} = \frac{d}{dt}(\mathbf{r} \times \mathbf{p}) = \frac{d\mathbf{r}}{dt} \times \mathbf{p} + \mathbf{r} \times \frac{d\mathbf{p}}{dt}$$

The first term on the right of this equation is zero because

$$\frac{d\mathbf{r}}{dt} \times \mathbf{p} = \mathbf{v} \times m\mathbf{v} = 0$$

Thus

$$\frac{d\mathbf{L}}{dt} = \mathbf{r} \times \frac{d\mathbf{p}}{dt} \qquad \text{10-9}$$

Comparing Equations 10-8 and 10-9, we have

$$\Sigma\boldsymbol{\tau} = \frac{d\mathbf{L}}{dt} \qquad \text{10-10}$$

Equation 10-10 is the rotational analog of Newton's second law $\Sigma\mathbf{F} = d\mathbf{p}/dt$. It follows directly from that law and from the definitions of torque and angular momentum. We shall give two examples of the calculation of the angular momentum for a single particle.

**Example 10-1** Find the angular momentum for a particle of mass $m$ moving in a circle of radius $r$.

The speed of the particle $v$ and the magnitude of its angular velocity $\omega$ are related by $v = r\omega$ (see Equation 9-4). Choosing the $xy$ plane to be the plane of the circle (Figure 10-4), we have for the angular momentum relative to the center of the circle.

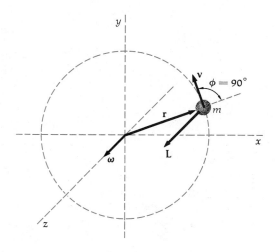

**Figure 10-4**
Particle moving in a circle (Example 10-1). The angular momentum is along the $z$ axis parallel to the angular velocity $\boldsymbol{\omega}$ and has magnitude $mvr = mr^2\omega = I\omega$.

$$L = r \times p = r \times mv = rmv \sin 90° \, k = rmvk = mr^2\omega k = mr^2\omega$$

The angular momentum is in the same direction as the angular velocity. Since the quantity $mr^2$ is the moment of inertia $I$ for a single particle about the $z$ axis, we have

$$L = mr^2\omega = I\omega \qquad\qquad 10\text{-}11$$

If the only force acting on the particle is the centripetal force directed toward the origin, there will be no torque about the origin. Then, according to Equation 10-10, the angular momentum will be constant.

---

**Example 10-2** Find the angular momentum, relative to the origin, of a particle moving in a straight line a perpendicular distance $b$ from the origin with constant speed.

This situation is shown in Figure 10-5. The angular momentum is

$$L = r \times p = -rmv \sin \phi \, k$$

The distance $r \sin \phi$ is the perpendicular distance $b$ from the line of motion to the origin, as can be seen from the figure. Thus

$$L = -mvb\mathbf{k} \qquad\qquad 10\text{-}12$$

---

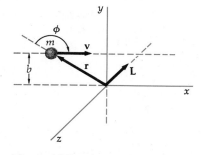

**Figure 10-5**
Particle moving with constant speed in a straight line parallel to the $x$ axis and a distance $b$ from it, for Example 10-2. The angular momentum relative to the origin is $L = r \times mv = -mvb\mathbf{k}$.

Although we usually associate angular momentum with rotational motion, Example 10-2 illustrates that even a particle moving in a straight line has angular momentum about a point not on the line. We shall use this result in Examples 10-3 and 10-5.

### Questions

3. What is the angle between a particle's linear momentum **p** and its angular momentum **L**?

4. Two particles have the same momentum **p**. Are their angular momenta also equal? Is it possible for two particles with the same momentum **p** also to have the same angular momentum if they are not at the same place?

5. A particle moves along a straight line at constant speed. How does its angular momentum about any point vary in time?

6. Two particles, $A$ and $B$, move along parallel lines a distance $D$ apart. They have the same speed $v$ but travel in opposite directions. What is the magnitude of $A$'s angular momentum about a point on its own path? About a point on $B$'s path? About particle $B$?

7. A particle moving at constant velocity has zero angular momentum about a particular point. Show that the particle either has passed through that point or will pass through it.

## 10-3  Angular Momentum of a System of Particles

The total angular momentum of a system of particles is the sum of the angular momenta of the individual particles:

$$L = \Sigma L_i$$

If the system of particles is a body rotating about a symmetry axis passing through its center of mass, the angular momentum about the center of mass is given by

$$\mathbf{L} = I\boldsymbol{\omega} \qquad 10\text{-}13$$

*Angular momentum of a system*

—the expression used in Chapter 9 for rigid bodies. We shall illustrate this relation by calculating the angular momentum of a disk rotating about an axis through its center of mass and perpendicular to the plane of the disk, as shown in Figure 10-6. In this figure we indicate a mass element $m_i$ a distance $r_i$ from the center of the disk and moving with speed $v_i = r_i\omega$. Since this particle is moving in a circle, our results of Example 10-1 apply. Its angular momentum is

$$\mathbf{L}_i = m_i r_i^2 \boldsymbol{\omega}$$

When we sum over all such mass elements, we obtain the total angular momentum of the disk,

$$\mathbf{L} = \Sigma \mathbf{L}_i = (\Sigma m_i r_i^2)\boldsymbol{\omega} = I\boldsymbol{\omega}$$

Equation 10-13 holds whether or not the body is rigid, as long as the body is rotating about a symmetry axis through its center of mass.

An example of motion for which Equation 10-13 does *not* hold is the wobbly motion of a wheel rotating about an axis inclined at a small angle to the symmetry axis of the wheel. Such motion occurs for automobile wheels that are not dynamically balanced. In such cases, the angular momentum of the body is not parallel to the angular velocity of rotation. We shall look at this problem in Section 10-7. The description of the general motion of a body rotating about any arbitrary axis can be very complicated. Except for our brief discussion of dynamic imbalance in Section 10-7 we shall be concerned only with systems for which the angular momentum and angular velocity are related by Equation 10-13.

Equation 10-13 also holds for rotation about an axis that is parallel to a symmetry axis passing through the center of mass of a body, such as that shown in Figure 10-7. We can use the parallel-axis theorem to relate the angular momentum relative to $O'$ in Figure 10-7 to that about the center of mass $O$. According to this theorem, the moment of inertia $I'$ about axis $z'$ is related to that about the $z$ axis through the center of mass a distance $h$ away, by

$$I' = I_{\text{CM}} + Mh^2$$

where $M$ is the total mass of the body. If we multiply each term by $\boldsymbol{\omega}$, we obtain

$$I'\boldsymbol{\omega} = I_{\text{CM}}\boldsymbol{\omega} + Mh^2\boldsymbol{\omega} \qquad 10\text{-}14$$

The quantity $I'\boldsymbol{\omega}$ is the angular momentum about the point $O'$, whereas $I_{\text{CM}}\boldsymbol{\omega}$ is the angular momentum about the center of mass. The quantity $Mh^2\boldsymbol{\omega}$ is the angular momentum of a particle of mass $M$ moving with angular velocity $\boldsymbol{\omega}$ in a circle of radius $h$ about the point $O'$. Equation 10-14 is thus

$$\mathbf{L} = \mathbf{L}_{\text{CM}} + \mathbf{r}_{\text{CM}} \times M\mathbf{v}_{\text{CM}} \qquad 10\text{-}15$$

This result, obtained by considering rotation of a disk about two parallel axes, holds in general.

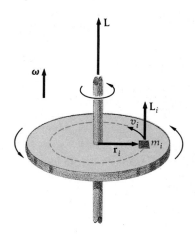

**Figure 10-6**
Disk rotating about its axis. The angular momentum is $\mathbf{L} = I\boldsymbol{\omega}$.

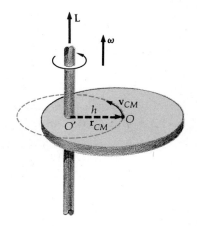

**Figure 10-7**
Disk rotating about an axis perpendicular to the disk but at a distance $h$ from the symmetry axis. The angular momentum is $\mathbf{L} = I'\boldsymbol{\omega}$, where $I'$ is the moment of inertia of the disk about the axis of rotation.

*The angular momentum about any point in space O' is the sum of the angular momentum about the center of mass of the system plus the angular momentum associated with center-of-mass motion about the point O'.*

This theorem follows directly from the definition of angular momentum and the properties of the center of mass.

One result of Equation 10-15 is that if the only motion of a body is rotation about an axis through its center of mass, the angular momentum of the body about any point in space is $\mathbf{L}_{CM}$. This follows from the fact that in this case $\mathbf{v}_{CM} = 0$. The angular momentum of a body about its center of mass is often called the *spin* of the body. For example, in the motion of a gyroscope (Section 10-6) the angular momentum consists of two parts, $\mathbf{L}_{CM}$, due to the spinning of the wheel, and $\mathbf{r}_{CM} \times m\mathbf{v}_{CM}$, due to motion of the center of mass of the gyroscope. Equation 10-15 is also used to find the angular momentum of an electron in an atom. The electron has an intrinsic angular momentum, called *spin*, just as if it were a tiny spinning ball. The total angular momentum of the electron is the sum of this spin angular momentum and the angular momentum associated with the orbital motion of the electron around the nucleus.

The resultant torque acting on a system of particles is the sum of the individual torques acting on the particles. The generalization of Equation 10-10 to a system of particles is then

$$\Sigma \tau_i = \Sigma \frac{d\mathbf{L}_i}{dt} = \frac{d}{dt}(\Sigma \mathbf{L}_i) = \frac{d\mathbf{L}}{dt} \qquad \text{10-16}$$

In Equation 10-16 the sum of the torques may include internal torques as well as those due to forces external to the system. We shall now show that if the internal forces between any two particles act parallel to the line joining the particles, the sum of the internal torques is zero. Then

$$\Sigma \tau_{\text{ext}} = \frac{d\mathbf{L}}{dt} \qquad \text{10-17}$$

*The rate of change of the angular momentum of a system equals the net torque acting on the system due to external forces.*

Consider two particles as shown in Figure 10-8. Although $\mathbf{F}_1 = -\mathbf{F}_2$ by Newton's third law, it is not immediately obvious that the internal torques cancel. The sum of the two torques in this figure is

$$\begin{aligned}\boldsymbol{\tau}_1 + \boldsymbol{\tau}_2 &= \mathbf{r}_1 \times \mathbf{F}_1 + \mathbf{r}_2 \times \mathbf{F}_2 = \mathbf{r}_1 \times \mathbf{F}_1 + \mathbf{r}_2 \times (-\mathbf{F}_1) \\ &= (\mathbf{r}_1 - \mathbf{r}_2) \times \mathbf{F}_1\end{aligned}$$

The vector $\mathbf{r}_1 - \mathbf{r}_2$ is along the line joining the two particles. Thus, if the force $\mathbf{F}_1$ acts parallel to the line joining $m_1$ and $m_2$, $\mathbf{F}_1$ and $\mathbf{r}_1 - \mathbf{r}_2$ are either parallel or antiparallel and

$$(\mathbf{r}_1 - \mathbf{r}_2) \times \mathbf{F}_1 = 0$$

If this is true for all the internal forces, the internal torques cancel in pairs.

Equation 10-17 is the rotational analog of the result for linear motion (obtained in Chapter 8) that the resultant external force equals the rate of change of the total linear momentum.

We conclude this section with an example of the use of Equation 10-17.

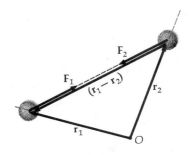

**Figure 10-8**
The internal forces $\mathbf{F}_1$ and $\mathbf{F}_2$ produce no torque about $O$ if they act along the line joining the particles.

**Example 10-3** Two masses $m_1$ and $m_2$ in an Atwood's machine are connected by a string passing over a pulley, as shown in Figure 10-9. The pulley has radius $R$ and moment of inertia $I$ about its axis of rotation. Find the acceleration of the masses and the angular acceleration of the pulley.

Let us compute the angular momentum about the center of the pulley. We note that although the masses are not rotating, they do have angular momentum. In Example 10-2 we saw that relative to some point $O$ the angular momentum of a particle moving with velocity $v$ along a line is $mvb$, where $b$ is the distance from the line to the point. In this case $b = R$, the radius of the pulley. The angular momentum of each mass is in the same direction, the direction of the angular velocity of the pulley, which is into the page. The magnitude of the total angular momentum of the two masses plus pulley is

$$L = m_1vR + m_2vR + I\omega$$

where $v$ is the speed of the masses. If we assume that the string does not slip on the pulley, the speed of the string $v$ must equal the speed of the rim of the pulley $R\omega$. We eliminate $\omega$ from the equation:

$$L = (m_1 + m_2)vR + I\frac{v}{R}$$

The only external forces acting on this system are the weights of the masses and pulley and the supporting force at the axis of the pulley. This force and the weight of the pulley do not exert any torque about the center of the pulley. The resulting external torque on this system is thus $m_2gR - m_1gR$. The direction of the torque is into the page, parallel to the angular momentum. Therefore Equation 10-17 is

$$\Sigma\tau_{\text{ext}} = (m_2 - m_1)gR = \frac{dL}{dt} = \left[(m_1 + m_2)R + \frac{I}{R}\right]\frac{dv}{dt}$$

and the acceleration is

$$a = \frac{dv}{dt} = \frac{m_2 - m_1}{m_1 + m_2 + (I/R^2)}g$$

The effect of taking the rotation of the pulley into account is to reduce the acceleration of the masses. The angular acceleration of the pulley is

$$\alpha = \frac{d\omega}{dt} = \frac{1}{R}\frac{dv}{dt} = \frac{a}{R}$$

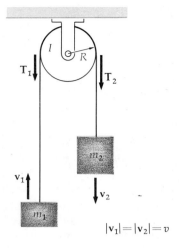

$$|\mathbf{v}_1| = |\mathbf{v}_2| = v$$

**Figure 10-9**
Atwood's machine (Example 10-3). The angular momentum of each body and that of the pulley about the center of the pulley is into the page. The rate of change of the total angular momentum equals the resultant torque about this point, which is $m_2gR - m_1gR$.

**Questions**

8. If the total angular momentum of a system of particles is zero, are all the particles at rest?

9. If the total angular momentum of a system is constant, can we conclude that no net force acts on the system?

10. If no net force acts on a system, will there be no torque on the system?

11. Two external forces act on an extended body and in general will produce a net torque on the body. Are there any points about which the total torque of these forces is zero?

## 10-4  Conservation of Angular Momentum

When there is no resultant external torque acting on a system, we have from Equation 10-17

$$\frac{d\mathbf{L}}{dt} = 0$$

or

$$\mathbf{L} = \text{constant} \qquad\qquad 10\text{-}18$$

Equation 10-18 is a statement of the *law of conservation of angular momentum.*

*If the resultant external torque acting on a system is zero, the total angular momentum of the system is constant.*

*Conservation of angular momentum*

If a system is isolated from its surroundings, there can be no external forces or torques. We therefore have a third conservation law for isolated systems. Energy, linear momentum, and angular momentum are conserved. The law of conservation of angular momentum is a fundamental law of nature. Even on a microscopic scale in atomic and nuclear physics, where newtonian mechanics does not hold, the angular momentum of an isolated system is constant in time.

We now give several examples of how the law of conservation of angular momentum is used.

Multiflash photograph of a diver. The center of mass moves in a parabolic path after the diver leaves the board. The angular momentum is provided by the initial torque due to the force of the board. This force does not pass through the center of mass if the diver leans forward as he jumps. If the diver wanted to undergo $1\frac{1}{2}$ rev in the air, he would draw in his arms and legs, decreasing his moment of inertia to increase his angular velocity. (*Courtesy of Harold E. Edgerton.*)

**Example 10-4** A disk with moment of inertia $I_1$ is rotating with angular velocity $\omega_i$ about a frictionless shaft. It drops onto another disk with moment of inertia $I_2$ initially at rest (see Figure 10-10). Because of surface friction, the two disks eventually attain a common angular velocity. What is it?

Each disk exerts a torque on the other, but there is no torque external to the two-disk system. The angular momentum of the first disk about its center of mass is $I_1\omega_i$. When both disks are rotating together, the total angular momentum is

$$I_1\omega_f + I_2\omega_f = (I_1 + I_2)\omega_f = I_1\omega_i$$

Thus the final angular velocity is

$$\omega_f = \frac{I_1}{I_1 + I_2}\,\omega_i$$

This interaction of the disks is analogous to an inelastic collision of two masses in one dimension. Energy is not conserved in this collision. We can see this best by writing the energy in terms of the angular momentum. The initial kinetic energy is

$$E_{ki} = \tfrac{1}{2}I_1\omega_i^2 = \frac{(I_1\omega_i)^2}{2I_1} = \frac{L_i^2}{2I_1}$$

The final energy is

$$\tfrac{1}{2}(I_1 + I_2)\omega_f^2 = \frac{[(I_1 + I_2)\omega_f]^2}{2(I_1 + I_2)} = \frac{L_f^2}{2(I_1 + I_2)}$$

Since $L_f = L_i$, the final kinetic energy is less than the initial energy by the factor $I_1/(I_1 + I_2)$.

**Example 10-5** A playground merry-go-round is at rest, pivoted about a frictionless axis. A child runs along a path tangential to the rim with initial speed $v_i$ and jumps onto the merry-go-round (Figure 10-11). What is the angular velocity of the merry-go-round and child?

We cannot expect that energy will be conserved because the child makes an inelastic collision with the rim of the merry-go-round. Linear momentum is not conserved either. The pivot of the merry-go-round exerts an impulse during the collision, but because it is frictionless, it cannot exert any torque. Thus, angular momentum about the pivot is conserved. This problem again demonstrates that we need not have circular motion to have angular momentum. Figure 10-11 shows that the initial angular momentum of the child about the pivot point is

$$\mathbf{L}_i = \mathbf{r} \times m\mathbf{v}_i = mv_iR\mathbf{k}$$

where $R$ is the perpendicular distance between the pivot and the line of motion of the child, and is also the radius of the merry-go-round, and $m$ is the mass of the child. The direction of the angular momentum is shown in Figure 10-11. After the child is on the merry-go-round, the angular momentum is

$$mv_fR\mathbf{k} + I\omega\mathbf{k} = (mR\omega R + I\omega)\mathbf{k} = (mR^2 + I)\omega\mathbf{k}$$

where we have used $v_f = R\omega$ because the child is at rest relative to the rim of the merry-go-round after having jumped on. Since $L_f = L_i$, we have

$$(mR^2 + I)\omega\mathbf{k} = mv_iR\mathbf{k} \qquad \text{or} \qquad \omega = \frac{mv_iR}{mR^2 + I}$$

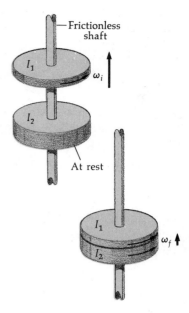

**Figure 10-10**
Inelastic rotational collision (Example 10-4). Since the only torques acting are internal to the system, the final angular velocity of the disks can be found by conservation of angular momentum.

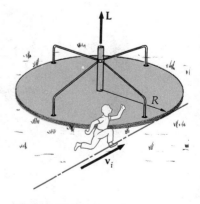

**Figure 10-11**
Example 10-5. A child runs tangentially to the rim of a merry-go-round and jumps on. If the pivot is frictionless, angular momentum is conserved.

**Example 10-6** Show that a planet moving around the sun sweeps out equal areas in equal times.

Figure 10-12 shows a planet moving in an orbit about the sun (here we are not using the approximation of a circular orbit). Since the force on the planet is directed toward the sun, there is no torque about the sun and the angular momentum about the sun is conserved. (We shall assume the sun to be at rest.) In time $\Delta t$ the planet moves a distance $v\,\Delta t$ and sweeps out the area indicated in Figure 10-12. This is half the area of the parallelogram formed by the vectors $\mathbf{r}$ and $\mathbf{v}\,\Delta t$, which is $\mathbf{r} \times \mathbf{v}\,\Delta t$.

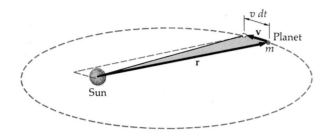

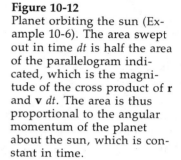

**Figure 10-12**
Planet orbiting the sun (Example 10-6). The area swept out in time $dt$ is half the area of the parallelogram indicated, which is the magnitude of the cross product of $\mathbf{r}$ and $\mathbf{v}\,dt$. The area is thus proportional to the angular momentum of the planet about the sun, which is constant in time.

Thus the area $A$ swept out in $\Delta t$ by the radius vector $\mathbf{r}$ is

$$A = \tfrac{1}{2}|\mathbf{r} \times \mathbf{v}\,\Delta t| = \frac{1}{2m}\,|\mathbf{r} \times m\mathbf{v}|\,\Delta t = \frac{1}{2m}\,L\,\Delta t$$

or

$$\frac{\Delta A}{\Delta t} = \frac{L}{2m} \qquad\qquad 10\text{-}19$$

where $L$ is the angular momentum of the planet. Since $\mathbf{L}$ is a constant, the rate at which the radius vector of the planet sweeps out area is also constant. Expressed differently, the radius vector sweeps out equal areas in equal times. This is Kepler's second law of planetary motion (Chapter 12), deduced from astronomical observation. It is a consequence of the fact that the force of gravity is a central force directed along the line from the planet to the sun. We see also that since $\mathbf{L}$ is constant in direction as well as magnitude, the motion of a planet must remain in a plane.

**Questions**

12. A man sits on a spinning piano stool with his arms folded. If he extends his arms, what happens to his angular velocity?

13. Does a particle moving at constant speed along a straight line sweep out equal areas in equal times with respect to any arbitrary point?

14. A particle moves under the influence of a central force $\mathbf{F} = f(r)\hat{\mathbf{r}}$. Will its angular momentum about the center of force be constant? Would Kepler's law of equal areas apply to this motion?

15. It is said that a cat always lands on its feet. If a cat starts falling feet up, how can it land on its feet without violating the law of conservation of angular momentum?

Focus on Sports

Because the torque exerted by the ice is small, the angular momentum of the skater is approximately constant. When she reduces her moment of inertia by drawing in her arms, her angular velocity increases.

## 10-5 Translation and Rotation

In general, the motion of a system of particles is very complicated when the axis of rotation is not fixed. As we have mentioned, we can consider any such motion as a combination of a translation of the center of mass plus a rotation about the center of mass. The description of the translation is not difficult. The center of mass moves as a point particle under the influence of the net external force. The analysis of the rotational motion is simplified by an important theorem concerning the angular momentum relative to the center of mass:

*The resultant torque about the center of mass equals the rate of change of angular momentum relative to the center of mass, no matter how the center of mass is moving:*

$$\Sigma \boldsymbol{\tau}_{\text{CM}} = \frac{d\mathbf{L}_{\text{CM}}}{dt} \qquad\qquad 10\text{-}20$$

Equation 10-20, stating that the resultant torque equals the rate of change of the angular momentum, is the same as Equation 10-17, except that now the torques and angular momentum are taken relative to the center of mass. Since we used Newton's second law to derive Equation 10-17, our derivation is valid only if the torques and angular momentum are taken relative to a point fixed in an inertial reference frame. This theorem states that the result is also valid if the torques and angular momentum are taken relative to the center of mass even if the center of mass is accelerating. We shall not prove this result. A proof is given in most intermediate-level mechanics books.[*]

Fundamental Photographs

Path of a point on the rim of a rolling checker. The motion is simpler when viewed in the center-of-mass reference frame, which moves with the center of the checker. In this frame the point on the rim moves in a circle.

In this section we apply this theorem to a special class of translational and rotational motion which can be easily treated: motion of a rigid body rotating about an axis of symmetry which moves so that it remains parallel to a line fixed in space; e.g., a rolling ball or cylinder.

Consider a cylinder rolling down an inclined plane, as in Figure 10-13. The forces on the cylinder are its weight, the normal force of the incline, and friction exerted by the incline. (If there were no friction, a cylinder released from rest at the top with no rotation would slide down the plane without rotating.) As for the general motion of a system of particles, the resultant force equals the total mass times the acceleration of the center of mass of the cylinder. If the cylinder rolls without slipping, the linear distance $s$ moved by the center of mass is related to the angle of rotation of the cylinder $\phi$ by

$$s = R\phi \qquad\qquad 10\text{-}21$$

[*] See for example, G. R. Fowles, *Analytical Mechanics*, pp. 201–203, Holt, New York, 1977.

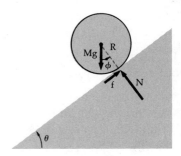

**Figure 10-13**
Forces on a cylinder rolling down an incline.

(see Figure 10-14). Thus the linear velocity of the center of mass is related to the angular velocity of rotation by

$$v_{CM} = \frac{ds}{dt} = R\frac{d\phi}{dt} = R\omega \qquad \text{10-22}$$

*Rolling without slipping*

and the acceleration of the center of mass is

$$a_{CM} = \frac{dv_{CM}}{dt} = R\frac{d\omega}{dt} = R\alpha \qquad \text{10-23}$$

where $\alpha$ is the angular acceleration. As the cylinder accelerates down the incline, the angular velocity of rotation must increase if it is to roll without slipping. Thus the angular momentum of the cylinder about the center of mass increases. This increase is due to the torque exerted by the frictional force on the cylinder. (The weight $M\mathbf{g}$ and the normal force $\mathbf{N}$ act through the center of mass and therefore exert no torque about this point.) The magnitude of the torque exerted by friction is $fR$. According to Equation 10-20, we can set this torque equal to the rate of change of angular momentum about the center of mass even though the center of mass is accelerating. The magnitude of the angular momentum about the center of mass is

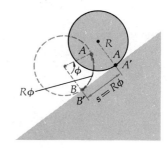

**Figure 10-14**
When the cylinder rolls without slipping, the point of contact moves the distance $s = R\phi$, which is the distance moved by the center of mass.

$$L_{CM} = I_{CM}\omega \qquad \text{10-24}$$

where $I_{CM}$ is the moment of inertia about the axis of rotation, which is constant. Since the direction of the axis of rotation does not change (the axis remains parallel to its original direction), only the magnitude of the angular momentum changes. Thus

$$\frac{dL_{CM}}{dt} = I_{CM}\frac{d\omega}{dt} = I_{CM}\alpha \qquad \text{10-25}$$

Setting the torque equal to the rate of change of the angular momentum gives

$$fR = I_{CM}\alpha \qquad \text{10-26}$$

The linear acceleration of the center of mass of the cylinder is parallel to the incline. The resultant force in this direction is $Mg\sin\theta - f$. Thus $\Sigma\mathbf{F} = M\mathbf{a}_{CM}$ gives

$$Mg\sin\theta - f = Ma_{CM} \qquad \text{10-27}$$

We can solve Equations 10-26 and 10-27 for any of the unknowns $\alpha$, $a_{CM}$, and $f$ using the condition for rolling without slipping (Equation 10-23). We first eliminate $\alpha = a_{CM}/R$:

$$fR = I_{CM}\frac{a_{CM}}{R}$$

$$f = \frac{I_{CM}}{R^2}a_{CM} \qquad \text{10-28}$$

Putting this into Equation 10-27 gives

$$Mg\sin\theta - \frac{I_{CM}}{R^2}a_{CM} = Ma_{CM}$$

$$a_{CM} = \frac{Mg\sin\theta}{M + I_{CM}/R^2} \qquad \text{10-29}$$

For a cylinder, $I_{CM} = \frac{1}{2}MR^2$. Then

$$a_{CM} = \frac{Mg\sin\theta}{M + \frac{1}{2}M} = \tfrac{2}{3}g\sin\theta \qquad \text{10-30}$$

We can use this to find the force of friction and the angular acceleration:

$$f = \frac{\frac{1}{2}MR^2}{R^2}\, a_{CM} = \tfrac{1}{3}Mg\,\sin\theta \qquad\qquad\qquad 10\text{-}31$$

$$\alpha = \frac{a_{CM}}{R} = \tfrac{2}{3}\frac{g}{R}\,\sin\theta \qquad\qquad\qquad 10\text{-}32$$

The frictional force in Equation 10-31 was found from the total mass, the moment of inertia, and the angle of inclination without any mention of the coefficient of friction. Since the cylinder is rolling without slipping, the surfaces of the cylinder and plane in contact are always instantaneously at rest relative to each other. Thus the friction is static friction, which in general is *not* equal to its maximum, limiting value $\mu_s N$. If $\mu_s$ is the coefficient of static friction, we have

$$f \leq \mu_s N = \mu_s Mg\,\cos\theta$$

Using $f = \tfrac{1}{3}Mg\,\sin\theta$ from Equation 10-31 for rolling without slipping, we have

$$\tfrac{1}{3}Mg\,\sin\theta \leq \mu_s Mg\,\cos\theta \qquad \text{or} \qquad \tan\theta \leq 3\mu_s$$

If the incline is too steep, or if the coefficient of static friction between it and the cylinder is too small, the cylinder will slip as it moves down the incline.

The linear acceleration is less than $g\,\sin\theta$, of course, because of the frictional force directed up the incline. Since the acceleration is constant, we can find the velocity of the cylinder at the bottom using the constant-acceleration formulas. If the original height of the center of mass of the cylinder is $h$, the total distance traveled is $s = h/(\sin\theta)$. The velocity at the bottom is then given by

$$v_{CM}^2 = 2a_{CM}s = 2(\tfrac{2}{3}g\,\sin\theta)\,\frac{h}{\sin\theta} = \tfrac{4}{3}gh$$

$$v_{CM} = \sqrt{\tfrac{4}{3}gh} \qquad\qquad\qquad 10\text{-}33$$

We can also obtain this result from energy considerations. We note that since the friction is static, there is no mechanical-energy dissipation. Thus the total mechanical energy, the kinetic energy plus the potential energy, is conserved. At the top of the incline the total energy is the potential energy $Mgh$. At the bottom the total energy is kinetic energy, which in general can be written $\tfrac{1}{2}Mv_{CM}^2 + E_{kr}$. Here, the energy relative to the center of mass is just the rotational energy $\tfrac{1}{2}I_{CM}\omega^2$. Conservation of energy gives

$$\tfrac{1}{2}Mv_{CM}^2 + \tfrac{1}{2}I_{CM}\omega^2 = Mgh \qquad\qquad\qquad 10\text{-}34$$

Using the rolling condition $v_{CM} = R\omega$, we can eliminate either $v_{CM}$ or $\omega$. Eliminating $\omega$ gives

$$\tfrac{1}{2}Mv_{CM}^2 + \tfrac{1}{2}I_{CM}\left(\frac{v_{CM}}{R}\right)^2 = Mgh$$

$$v_{CM}^2 = \frac{2Mgh}{M + I_{CM}/R^2}$$

For $I_{CM} = \tfrac{1}{2}MR^2$ for a cylinder we obtain

$$v_{CM}^2 = \frac{2Mgh}{M + \tfrac{1}{2}M} = \tfrac{4}{3}gh$$

in agreement with the result found from the acceleration.

**Question**

16. A wheel rolls along a level surface at speed $v_0$ without slipping. Relative to the surface, what is the speed of the bottom of the wheel? Of the center of the wheel? Of the top of the wheel? Is there any frictional force acting on the wheel?

## 10-6 Motion of a Gyroscope

We now consider an example of a gyroscope or symmetric top, in which the axis of rotation changes direction. Such motions are generally very complicated. Figure 10-15 shows such a system consisting of a bicycle wheel free to turn on an axle. The axle is pivoted at a point a distance $D$ from the center of the wheel but free to turn in any direction.

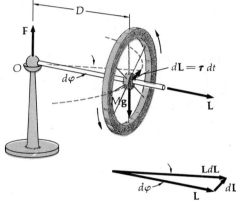

**Figure 10-15**
Gyroscope. The force of gravity $Mg$ produces a torque about the pivot into the paper, which causes a *change* in the angular momentum in that direction. If the wheel is initially spinning with **L** along its axle, the change is perpendicular to **L** and the axle moves in the direction of the torque. This motion is called precession.

When the axle is held horizontal and released, if the wheel is not spinning it simply falls. The torque about the point $O$ is $MgD$ in the direction into the page. Since the center of mass accelerates downward, the upward force $F$ exerted by the support at $O$ is evidently less than $Mg$.

Now assume that the wheel is spinning. The angular momentum relative to point $O$ is the angular momentum relative to the center of mass, in this case the spin, plus the angular momentum due to motion of the center of mass. It is quite easy in practice to make the spin angular momentum very large so that to a first approximation we can neglect the contribution due to motion of the center of mass. From Figure 10-15 we see that the torque is perpendicular to the angular momentum. In order for the angular momentum to change in the direction of the torque, the axle must move in the horizontal plane, as shown. In time $dt$ the angular-momentum change has the magnitude

$$dL = \tau\, dt = MgD\, dt$$

The angle $d\phi$ through which the axle moves is

$$d\phi = \frac{dL}{L} = \frac{MgD\, dt}{L}$$

The movement of the axle is called *precession*. The angular velocity of precession is

$$\omega_p = \frac{d\phi}{dt} = \frac{MgD}{L} \qquad\qquad 10\text{-}35 \qquad \textit{Precession}$$

Since the center of mass does not drop, the support point evidently exerts an upward force equal to $Mg$.

The observation that the wheel moves in a horizontal plane instead of falling is at first surprising. We are more familiar with situations like a falling stick, where there is no initial angular momentum, so that the direction of the change in angular momentum is also the direction of the angular momentum. There is an analogous situation regarding the linear momentum of an object in circular motion, e.g., that of the moon about the earth. The earth exerts a force on the moon toward the earth. Why doesn't the moon move toward the earth and hit it? If the moon were held with no initial momentum and released, a change in momentum (from zero) toward the earth would result in the moon's moving toward the earth; but since the moon has an initial momentum perpendicular to the line joining the earth, a change in momentum toward the earth merely results in the moon's being deviated from straight-line motion into a circular arc. Thus though $d\mathbf{p}$ is always toward the earth, $\mathbf{p}$ is tangential to the circle. For the gyroscope, if there is no initial angular momentum, the torque (into the page) causes an angular momentum (into the page) associated with center-of-mass motion as it falls; but if there is a large initial angular momentum along the axis of the wheel, this same torque merely deflects the angular momentum into the page. In this case the pivot exerts an upward force sufficient to prevent the center of mass from falling.

If the axle is not horizontal but makes an angle $\theta$ with the vertical, as in Figure 10-16, the torque about the pivot is $MgD \sin \theta$. The angle of precession in time $\Delta t$ is now

$$d\phi = \frac{dL}{L \sin \theta} = \frac{MgD \sin \theta \, dt}{L \sin \theta} = \frac{MgD}{L} dt \qquad \text{10-36}$$

The angular velocity of precession $\omega_p$ is thus independent of angle $\theta$.

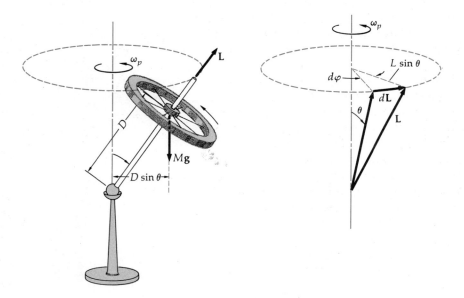

**Figure 10-16**
The rate of precession of a gyroscope is independent of the angle its axis makes with the vertical.

A careful observation of the motion of a gyroscope reveals that if the axle is held horizontally and released from rest, the motion of the axle is not confined to the horizontal plane. It dips down at first, and as the gyroscope precesses, there is a small vertical oscillation called *nutation*.    *Nutation* This effect was neglected in our previous discussion because we neglected the contribution of the center-of-mass motion to the total angular momentum about the pivot. If the spin angular momentum is much greater than that due to the motion of the center of mass during precession, the nutation is very small. The precessional motion of the center of mass results in a small component of angular momentum $MD^2\omega_p$ in the vertical direction. However, there is no torque in this direction. In order for the axle to precess in the horizontal plane without nutation, it must be given an angular impulse $MD^2\omega_p$ as it is released. If this is not done, the vertical component of the total angular momentum must remain zero. Then as the axle begins to precess, it must dip down so that there is a downward component of *spin* angular momentum to cancel the vertical angular momentum due to center-of-mass motion. We can analyze the motion qualitatively from the moment of release.

Just before the axle is released, the force of support at $O$ is $Mg/2$ and that at the hand is $Mg/2$. At the time of release, therefore, the center of mass must accelerate downward. It thus overshoots its equilibrium position, the force at $O$ becomes greater than $Mg$, and the center of mass eventually stops its downward motion and moves up again until it is horizontal.

*Optional*

## 10-7 Static and Dynamic Imbalance

In all the examples of rotation we have considered so far, the angular-momentum vector **L** has been parallel to the angular-velocity vector **ω**. This result is not general but holds only for rotation about certain axes called *principal axes*. In this section we consider some examples in which the angular-momentum and angular-velocity vectors are not parallel.

Figure 10-17 shows a simple dumbbell consisting of equal masses connected by a light rod rotating about an axis through the center of mass but not perpendicular to the line of centers. Each mass is moving in a circle of radius $r \sin \theta$ with speed $r \sin \theta \, \omega$, where $\theta$ is the angle between the rod and the axis of rotation and $\omega$ is the angular velocity. The angular momentum of the mass $m_1$ is $\mathbf{L}_1 = \mathbf{r}_1 \times m_1\mathbf{v}_1$. Its direction is in the plane of the paper perpendicular to the rod, and its magnitude is $L_1 = mr^2 \sin \theta \, \omega$. The angular momentum of mass $m_2$ is $\mathbf{L}_2 = \mathbf{r}_2 \times m_2\mathbf{v}_2$. Since $\mathbf{r}_2 = -\mathbf{r}_1$ and $\mathbf{v}_2 = -\mathbf{v}_1$, $\mathbf{L}_2 = \mathbf{L}_1$. The total angular momentum of the system is thus $2mr^2 \sin \theta \, \omega$ in the direction perpendic-

**Figure 10-17**
Particles of equal mass rotating about an axis through their center of mass when it is not a symmetry axis. The angular momentum **L** is perpendicular to the line joining the particles and makes an angle $\phi$ with the angular velocity **ω**. Since **L** changes direction as the particles rotate, a torque must be exerted by the bearings.

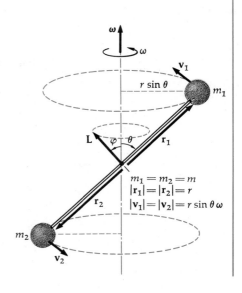

$$m_1 = m_2 = m$$
$$|\mathbf{r}_1| = |\mathbf{r}_2| = r$$
$$|\mathbf{v}_1| = |\mathbf{v}_2| = r \sin \theta \, \omega$$

ular to the rod, as shown. It is not parallel to $\boldsymbol{\omega}$, which is along the axis of rotation. As the dumbbell rotates, the angular-momentum vector rotates with its tip moving in a circle, as shown in Figure 10-18. If the angular velocity is constant, the angular momentum has a constant magnitude but its direction changes in time. Thus even if the angular velocity is constant, the angular-momentum vector changes in time and there must be a resultant torque on the system. This torque is exerted by the bearings supporting the system. At the instant shown in Figure 10-17 the torque must be out of the page since that is the direction of the change in **L**. We can calculate the magnitude of this torque by calculating the rate of change of the angular momentum.

Let us write the angular momentum vector **L** as the sum of a vector $L_z\mathbf{k}$ parallel to the axis of rotation and a vector $L_\rho\hat{\boldsymbol{\rho}}$ perpendicular to this axis, where $\mathbf{k}$ and $\hat{\boldsymbol{\rho}}$ are unit vectors. From Figure 10-18 we have for the magnitudes of these vectors

$$L_z = L \cos \phi = L \sin \theta = 2mr^2 \sin^2 \theta \, \omega$$

$$L_\rho = L \sin \phi = L \cos \theta = 2mr^2 \sin \theta \cos \theta \, \omega$$

The vector $L_z\mathbf{k}$ is constant in magnitude and direction if $\boldsymbol{\omega}$ is constant.

**Figure 10-18**
The angular-momentum vector for the system in Figure 10-17 rotates with its tip moving in a circle.

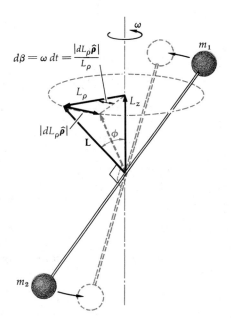

$$d\beta = \omega \, dt = \frac{|dL_\rho\hat{\boldsymbol{\rho}}|}{L_\rho}$$

The vector $L_\rho\hat{\boldsymbol{\rho}}$ is constant in magnitude, but its direction changes because the direction of the unit vector changes as the system rotates. From Figure 10-18 we have for the change in $L_\rho\hat{\boldsymbol{\rho}}$ in the time $dt$

$$\frac{|dL_\rho\hat{\boldsymbol{\rho}}|}{L_\rho} = d\beta \qquad \text{where } d\beta = \omega \, dt$$

Then

$$\frac{|dL_\rho\hat{\boldsymbol{\rho}}|}{dt} = \omega L_\rho = 2mr^2 \sin \theta \cos \theta \, \omega^2 \qquad \text{10-37}$$

Since $L_z\mathbf{k}$ is constant, the rate of change of **L** is just that of $L_\rho\hat{\boldsymbol{\rho}}$. Equation 10-37 thus gives the magnitude of $d\mathbf{L}/dt$, which in turn equals the magnitude of the torque. Note that the rate of change of angular momentum is proportional to $\omega^2$ and is zero when $\theta = 0$, corresponding to rotation about the line joining the masses, and when $\theta = 90°$, corresponding to rotation about an axis perpendicular to the line joining the masses. The torque which produces this change in angular momentum is exerted by the bearings. Its direction is perpendicular to the axis of rotation, and its effect is to maintain a constant direction for the angular velocity $\boldsymbol{\omega}$ as the direction of the angular-momentum vector changes. The rotating system exerts an equal and opposite torque on the support bearings, causing wear. Such a system is said to be *dynamically imbalanced*. The greater the angular velocity, the greater the wear on the bearings. Note that there is no static imbalance. If we pivot the dumbbell at the center of mass, it will be in static equilibrium at any orientation. If we have a system of unknown mass distribution, e.g., an automobile wheel, we cannot detect or correct dynamic imbalance by static-balance methods.

We can correct the dynamic imbalance of our two-mass system by making the angle $\theta$ equal to $90°$, as in Figure 10-19, or by adding masses, as in Figure 10-20. Then $\mathbf{L}$ and $\boldsymbol{\omega}$ will be parallel, and no torque will be required to maintain a constant angular velocity.

*Dynamic imbalance*

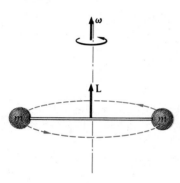

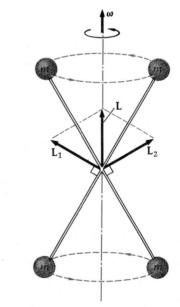

**Figure 10-19**
This arrangement of two rotating particles of equal mass is dynamically balanced because $\mathbf{L}$ does not change as the system rotates.

**Figure 10-20**
This arrangement of four particles of equal mass is dynamically balanced. The total angular-momentum vector $\mathbf{L}$ is now parallel to $\boldsymbol{\omega}$ and does not change as the system rotates.

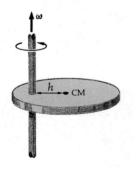

**Figure 10-21**
This system is dynamically balanced because $\mathbf{L}$ and $\boldsymbol{\omega}$ are parallel, but it is statically imbalanced. As the system rotates, the bearings exert no torque, but they must exert a force to accelerate the center of mass.

Figure 10-21 shows a wheel rotating about an axis that is off center but parallel to the symmetry axis through the center of mass, and a distance $h$ from it. The angular momentum about a point on the axis of rotation in the plane of the wheel is the sum of the angular momentum about the center of mass $I_{CM}$ and the angular momentum associated with motion of the center of mass, both parallel to the axis of rotation. The angular momentum is just $I\boldsymbol{\omega}$, where $I$ is the moment of inertia about the axis of rotation. If the wheel rotates with constant angular velocity, the angular momentum is constant and no torque is required. However, since the center of mass moves in a circle of radius $h$, there must be a net *force* on the wheel of magnitude $Mv_{CM}^2/h = M(h\omega)^2/h = Mh\omega^2$ and directed toward the axis of rotation from the center of mass. This force is exerted by the bearings. Again, the equal but opposite force exerted on the bearings causes wear on the bearings, particularly if $\omega$ is large. This *static imbalance* can be detected and corrected by pivoting the wheel on a horizontal axle and balancing it so that the axis of rotation passes through the center of mass.

## 10-8   The Vector Nature of Rotation

All vectors have both magnitude and direction, but the converse is not true: all quantities with both magnitude and direction are not necessarily vectors. A further requirement for a quantity to be a vector is that it obey the parallelogram law of vector addition. The significant property of this law of addition for this discussion is that the sum of two vectors be independent of the order in which they are added:

$$\mathbf{A} + \mathbf{B} = \mathbf{B} + \mathbf{A}$$

This is certainly true for displacements: the order in which we make two successive displacements has no effect on the resultant displacement. It is also true for velocity, acceleration, force, momentum, angular velocity, etc.—all the quantities we have assumed to be vectors. However, we shall now show that finite angular displacements do not obey the parallelogram law of addition. The final position of an object after two successive angular displacements *does* depend on the order of the displacements. Angular displacement is probably the only quantity you will encounter that has magnitude and direction but is not a vector.

Figure 10-22*a* shows a 90° rotation of a book about the *x* axis followed by a 90° rotation about the *y* axis. In Figure 10-22*b* the book is first rotated 90° about the *y* axis and then 90° about the *x* axis. The final position of the book is certainly not the same in these figures, demonstrating that the final position depends upon the order in which the two rotations occur. This proves that angular displacement is not a vector quantity because it does not obey the addition law for vectors. Figure 10-23 shows the same object undergoing much smaller successive rotations of about 20°. The difference in the final position for the two sequences is much less than for the 90° rotations. As the magnitude of the successive changes in the angle approaches zero, the difference in the

**Figure 10-22**
(*a*) The book is rotated 90° about the *x* axis and then 90° about the *y* axis. (*b*) The book is rotated first about the *y* axis and then about the *x* axis. The result of the two rotations depends on the order.

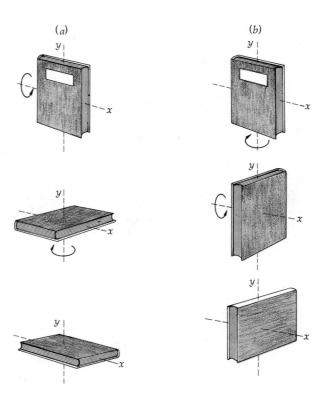

**Figure 10-23**
The same process as in Figure 10-22 except that the angle of each rotation is small. (*a*) A rotation about the *x* axis followed by a rotation about the *y* axis. (*b*) A rotation about the *y* axis followed by a rotation about the *x* axis. The result is nearly the same for the two cases.

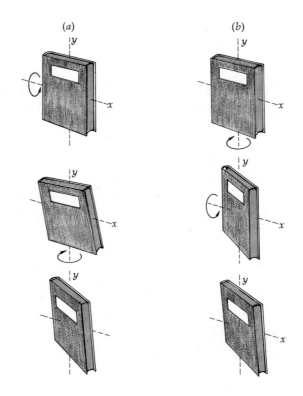

(*a*)    (*b*)

final positions also approaches zero. Thus in the limit as $\Delta\theta \to 0$, the angular displacement obeys the vector-addition law. For general rotations the angular velocity is defined by

$$\boldsymbol{\omega} = \frac{d\boldsymbol{\theta}}{dt} = \lim_{\Delta t \to 0} \frac{\Delta\boldsymbol{\theta}}{\Delta t}$$

(Here we have written $\Delta\boldsymbol{\theta}$ in boldface to indicate that it has a direction even though it is not a vector.) Since $\Delta\boldsymbol{\theta} \to 0$ as $\Delta t \to 0$, angular velocity does obey the vector-addition law and is a vector quantity. Similarly, we define the angular acceleration for general rotations by

$$\boldsymbol{\alpha} = \frac{d\boldsymbol{\omega}}{dt}$$

Since the derivative of a vector is also a vector, $\boldsymbol{\alpha}$ is a vector.

Review

A. Define, explain, or otherwise identify:

B. True or false:

1. $\mathbf{A} \times \mathbf{B}$ is the same as $\mathbf{B} \times \mathbf{A}$.

2. Angular momentum and torque are defined relative to a point in space.

3. A particle must move in a circle to have angular momentum.

4. If the resultant torque on a body is zero, its angular momentum must be zero.

5. The angular momentum of a particle is always perpendicular to its velocity.

6. If the angular momentum of a body does not change, the resultant torque on the body must be zero.

7. In general, when a body rolls down an incline without slipping, the frictional force has the magnitude $\mu_s N$.

Exercises

### Section 10-1, Torque as a Vector Product

1. A force of magnitude $F$ is applied horizontally in the negative $x$ direction to the rim of a disk of radius $R$, as shown in Figure 10-24. Write $\mathbf{F}$ and $\mathbf{r}$ in terms of the unit vectors $\mathbf{i}$, $\mathbf{j}$, and $\mathbf{k}$ and compute the torque produced by the force about the origin at the center of the disk.

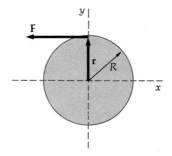

**Figure 10-24**
Exercise 1.

2. A 0.5-kg particle falls under the influence of gravity. (a) It is at $y = 10$ m and $x = 2$ m at time $t_1$. What is the torque about the origin exerted by gravity on the particle at this time? (b) At some later time the particle is at $y = 0$, $x = 2$ m. What is the torque about the origin at this time?

3. Compute the torque about the origin for the force $\mathbf{F} = -mg\mathbf{j}$ and $\mathbf{r} = x\mathbf{i} + y\mathbf{j}$ and show that this torque is independent of the coordinate $y$.

4. If $\mathbf{A} \times \mathbf{B} = 0$, what can be said about the vectors $\mathbf{A}$ and $\mathbf{B}$?

5. Find $\mathbf{A} \times \mathbf{B}$ for (a) $\mathbf{A} = 5\mathbf{i}$ and $\mathbf{B} = 5\mathbf{i} + 5\mathbf{j}$, (b) $\mathbf{A} = 5\mathbf{i}$ and $\mathbf{B} = 5\mathbf{i} + 5\mathbf{k}$, (c) $\mathbf{A} = 2\mathbf{i} + 2\mathbf{j}$ and $\mathbf{B} = -2\mathbf{i} + 2\mathbf{j}$.

6. Under what conditions is the magnitude of $\mathbf{A} \times \mathbf{B}$ equal to $\mathbf{A} \cdot \mathbf{B}$?

7. Show that for general vectors $\mathbf{A} = A_x\mathbf{i} + A_y\mathbf{j} + A_z\mathbf{k}$ and $\mathbf{B} = B_x\mathbf{i} + B_y\mathbf{j} + B_z\mathbf{k}$,

$$\mathbf{A} \times \mathbf{B} = \begin{vmatrix} \mathbf{i} & \mathbf{j} & \mathbf{k} \\ A_x & A_y & A_z \\ B_x & B_y & B_z \end{vmatrix}$$

$$= (A_yB_z - A_zB_y)\mathbf{i} + (A_zB_x - A_xB_z)\mathbf{j} + (A_xB_y - A_yB_x)\mathbf{k}$$

### Section 10-2, Angular Momentum of a Particle

8. A 2-kg particle moves with constant speed of 3 m/s in the $xy$ plane in the $y$ direction along the line $x = 5$ m. (a) Find the angular momentum $\mathbf{L}$ relative to the origin. (b) What torque about the origin is needed to maintain this motion?

9. A body of mass 3 kg moves at constant speed of 4 m/s around a circle of radius 5 m. (a) What is its angular momentum about the center of the circle? (b) What is its moment of inertia about an axis through the center of the circle and perpendicular to the plane of the motion? (c) What is the angular velocity of the particle?

10. A 3-kg body moves at constant speed of 4 m/s along a straight line. (a) What is its angular momentum about a point 5 m from the line? (b) Describe qualitatively how its angular velocity about that point varies with time.

11. A particle travels in a circular path. (a) If its linear momentum $p$ is doubled, how is its angular momentum affected? (b) If the radius of the circle is doubled but the speed is unchanged, how is the angular momentum of the particle affected?

12. A particle of mass $m$ moves with speed $v$ in a circle of radius $r$. Show that the kinetic energy of the particle can be written $E_k = L^2/2mr^2 = L^2/2I$, where $L$ is the angular momentum of the particle and $I = mr^2$ is its moment of inertia.

13. A 2-kg mass moves in a circle of radius 3 m. Its angular momentum relative to the center of the circle depends on time according to $L = 4t$, where $t$ is in sec-

onds and $L$ in kg·m²/s. (*a*) Find the torque acting on the particle. (*b*) Find the angular velocity as a function of time.

14. A planet moves in an elliptical orbit about the sun with the sun at one focus of the ellipse. (*a*) What is the torque produced by the gravitational force of attraction of the sun for the planet? (*b*) At position *A* in Figure 10-25 the planet is a distance $r_1$ from the sun and is moving with speed $v_1$ perpendicular to the line from the sun to the planet. At position *B* it is at distance $r_2$ moving with speed $v_2$, again perpendicular to the line from the sun to the planet. What is the ratio of $v_1$ and $v_2$ in terms of $r_1$ and $r_2$?

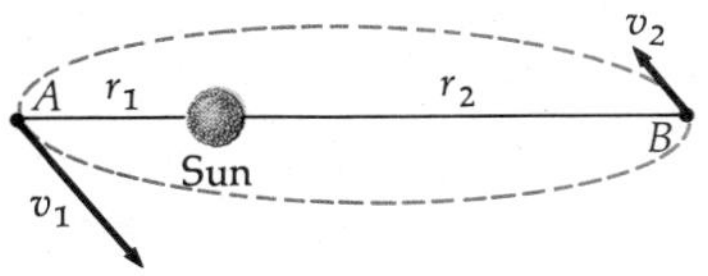

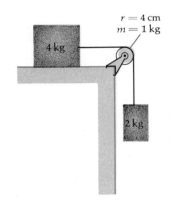

**Figure 10-25**
Exercise 14.

### Section 10-3, Angular Momentum of a System of Particles

15. Calculate the angular momentum of the earth spinning about its axis and compare it with the angular momentum the earth has about the sun. (Assume the earth to be a homogeneous sphere of mass $6.0 \times 10^{24}$ kg and radius $6.4 \times 10^6$ m.) The radius of the earth's orbit is $1.5 \times 10^{11}$ m.

16. A homogeneous cylinder of mass 100 kg and radius 0.3 m is mounted so that it turns without friction on its fixed axis of symmetry. It is rotated by a drive belt that wraps around its perimeter and exerts a constant torque. At time $t = 0$, its angular velocity is zero. At time $t = 30$ s, the angular velocity is 600 rev/min. (*a*) What is its angular momentum at that time? (*b*) At what rate is the angular momentum increasing? What is the torque acting on the cylinder? (*c*) What is the magnitude of the force acting on the rim of the cylinder?

17. A 15-g coin of diameter 1.5 cm is spinning about a vertical diameter at a fixed point on a tabletop at 10 rev/s. (*a*) What is the angular momentum of the coin about its center of mass? (*b*) What is its angular momentum about a point on the table 10 cm from the coin?

18. The coin in Exercise 17 spins about a vertical diameter at 10 rev/s but also travels in a straight line across the tabletop at 5 cm/s. (*a*) What is the angular momentum of the coin about a point on the line of motion? (*b*) What is the angular momentum of the coin about a point 10 cm from the line of motion? (There are two answers to this question. Explain why and give both.)

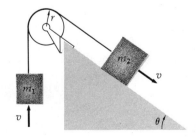

19. A 4-kg mass rests on a frictionless horizontal table. It is connected to a second mass of 2 kg by a string which passes over a frictionless pulley, as shown in Figure 10-26. The pulley is a uniform cylinder of radius 4 cm and mass 1 kg. (*a*) What is the net external torque on the system (two masses plus pulley) about the center of the pulley? (*b*) If the masses are moving with speed $v$ with the pulley turning at angular velocity $\omega = v/r$, what is the total angular momentum of the system about the center of the pulley? (Assume that the center of mass of the 4-kg mass is at the height of the string.) (*c*) Find the acceleration of the masses by differentiating your result for $L$ in part (*b*) and setting $dL/dt$ equal to the resultant torque.

**Figure 10-26**
Exercise 19.

20. (*a*) Assuming that the incline is frictionless in Figure 10-27 and the string passes through the center of mass of $m_2$, find the resultant torque acting on the system about the center of the pulley. (*b*) Write an expression for the total angular momentum of the system about the center of the pulley when the masses are moving with speed $v$. Assume the pulley has a moment of inertia $I$ and radius $r$. (*c*) Find the acceleration of the masses from your results of parts (*a*) and (*b*) by setting the resultant torque equal to the rate of change of the angular momentum of the system.

21. A uniform stick of length $l$ and mass $M$ lies on a smooth table. It rotates with angular velocity about an axis perpendicular to the table and through one end of the stick. (*a*) What is the angular momentum of the stick about the end? (*b*) What is the speed of the center of mass of the stick? (*c*) What is the angular momentum of the stick about the center of mass? (*d*) Show that Equation 10-15 is satisfied for this situation, where $L$ is the angular momentum about the end of the stick.

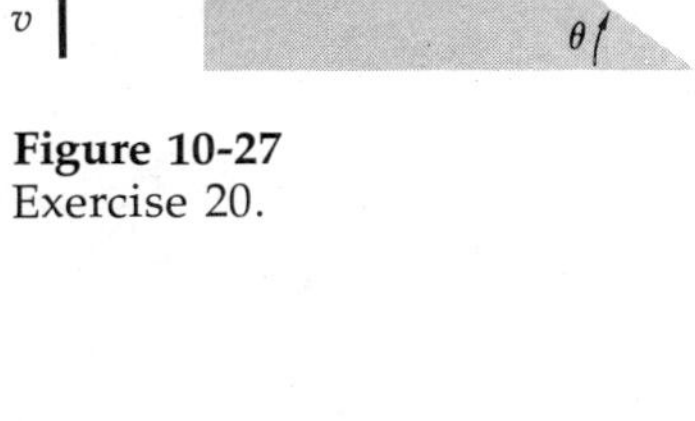

**Figure 10-27**
Exercise 20.

**Section 10-4, Conservation of Angular Momentum**

22. A merry-go-round of radius 2 m and moment of inertia 500 kg·m² is rotating without friction at 0.25 rev/s. A child of mass 25 kg sitting at the center crawls out to the rim. Find (*a*) the new angular velocity of the merry-go-round and (*b*) the initial and final kinetic energy.

23. A disk is rotating freely at 1800 rev/min about a vertical axis through its center. A second disk mounted on the same shaft above the first is initially at rest. The moment of inertia of the second disk is twice that of the first. The second disk is dropped onto the first one, and the two eventually rotate together with a common angular velocity. (*a*) Find the new angular velocity. (*b*) Show that kinetic energy is lost during the "collision" of the two disks.

24. A circular platform is mounted on a vertical frictionless axle. Its radius is $r = 2$ m, and its moment of inertia is $I = 200$ kg·m². It is at rest initially. A 70-kg man stands on the edge of the platform and begins to walk along the edge at speed $v_0 = 1.0$ m/s *relative to the ground*. (*a*) What is the angular velocity of the platform? (*b*) When the man has walked once around the platform so that he is at his original position on it, what is his angular displacement relative to the ground?

25. A man stands at the center of a circular platform holding his arms extended horizontally with a 4-kg block in each hand. He is set rotating about a vertical axis at 0.5 rev/s. The moment of inertia of the man plus platform is 1.6 kg·m², assumed constant. The blocks are 90 cm from the axis of rotation. He now pulls the blocks in toward his body until they are 15 cm from the axis of rotation. Find (*a*) his new angular velocity and (*b*) the initial and final kinetic energy of the man and platform. (*c*) How much work must the man do to pull in the blocks?

26. A particle is traveling with a constant velocity **v** along a line which is a distance $b$ from the origin $O$. Let $dA$ be the area swept out by the position vector from $O$ to the particle in time $dt$ (Figure 10-28). Show that $dA/dt$ is constant in time and equal to $\frac{1}{2}L/m$, where $L$ is the angular momentum of the particle about the origin.

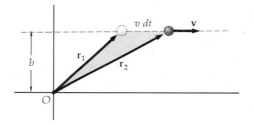

**Figure 10-28**
Exercise 26.

27. Explain why a helicopter with just one main rotor has a second smaller rotor mounted on a horizontal axis at the rear. Describe the resultant motion of the helicopter if this rotor fails during a flight.

Helicopter with two rotors (Exercise 27).

## Section 10-5, Translation and Rotation

28. A homogeneous cylinder of radius 15 cm and mass 50 kg is rolling without slipping along a horizontal floor at 6 m/s. How much work is needed to stop the disk?

29. Work Exercise 28 for a uniform sphere of the same mass, radius, and speed.

30. Find the percentage of the total kinetic energy associated with rotation and with translation for rolling without slipping if the object is (a) a uniform sphere, (b) a uniform cylinder, (c) a hoop.

31. A hoop of radius 0.50 m and mass 0.8 kg is rolling without slipping at a speed of 20 m/s toward an incline of slope 30°. How far up the incline will the hoop roll (assuming it rolls without slipping)?

32. A hoop rolls without slipping down an incline of slope 30°. Prove that its acceleration is $g/4$.

33. A ball rolls without slipping down an incline at angle $\theta$. Find (a) the acceleration of the ball, (b) the force of friction, (c) the maximum angle of the incline in terms of the coefficient of friction $\mu_s$ such that the ball can roll without slipping.

34. A ball rolls without slipping along a horizontal plane. Show that the frictional force on the ball must be zero. *Hint:* Consider a possible direction for a frictional force and what effect such a force would have on the velocity of the center of mass and on the angular velocity.

## Section 10-6, Motion of a Gyroscope

35. The angular momentum of the propeller of a small airplane points forward. (a) As the plane takes off, the nose lifts up and the airplane tends to veer to one side. Which side and why? (b) If the plane is flying horizontally and suddenly turns to the right, does the nose of the plane tend to move up or down? Why?

36. A bicycle wheel is mounted in the middle of a 60-cm-long axle. The tire and rim weigh 36 N and have a radius of 30 cm. The wheel is spun at 10 rev/s, and the axle is then placed in a horizontal position with one end resting on a pivot. (a) What is the angular momentum due to the spinning of the wheel? (Treat the wheel as a hoop.) (b) What is the angular velocity of precession? How long does it take for the axle to swing through 360° around the pivot? (c) What is the angular momentum associated with the motion of the center of mass, i.e., due to the precession? In what direction is this angular momentum?

37. A uniform disk of mass 2 kg and radius 6 cm is mounted in the center of a 10-cm axle and spun at 900 rev/min. The axle is then placed in a horizontal position with one end resting on a pivot. The other end is given an initial horizontal velocity so that the precession is smooth with no nutation. (a) Find the angular velocity of precession. (b) What is the speed of the center of mass during the precession? (c) What are the magnitude and direction of the acceleration of the center of mass? (d) What are the vertical and horizontal components of the force exerted by the pivot?

## Section 10-7, Static and Dynamic Imbalance

38. A uniform disk of radius 30 cm, thickness 3 cm, and mass 5 kg rotates at $\omega = 10$ rad/s about an axis parallel to the symmetry axis but 0.5 cm from that axis. (a) Find the net force on the bearings due to this imbalance. (b) Where should a 100-g mass be placed on the disk to correct this problem?

39. A 2-kg mass attached to a string of length 1 m moves in a horizontal circle as a conical pendulum (Figure 10-29). The string makes an angle $\theta = 30°$ with the vertical. (a) Show that the angular momentum of the mass about the point of support $P$ has a horizontal component toward the center of the circle as well as a vertical component, and find these components. (b) Find the magnitude of $d\mathbf{L}/dt$ and show that it equals the magnitude of the torque exerted by gravity about the point of support.

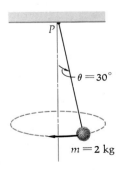

**Figure 10-29**
Exercise 39.

**Section 10-8, The Vector Nature of Rotation**

*There are no exercises for this section.*

Problems

1. A particle moves in a circle of radius $r$ with angular velocity $\boldsymbol{\omega}$. ($a$) Show that its velocity is $\mathbf{v} = \boldsymbol{\omega} \times \mathbf{r}$. ($b$) Show that its centripetal acceleration is $\mathbf{a}_r = \boldsymbol{\omega} \times \mathbf{v} = \boldsymbol{\omega} \times (\boldsymbol{\omega} \times \mathbf{r})$.

2. A particle located at $\mathbf{r} = x\mathbf{i} + y\mathbf{j} + z\mathbf{k}$ has momentum $\mathbf{p} = p_x\mathbf{i} + p_y\mathbf{j} + p_z\mathbf{k}$. A force $\mathbf{F} = F_x\mathbf{i} + F_y\mathbf{j} + F_z\mathbf{k}$ acts on it. ($a$) Calculate the components of the angular momentum of the particle and of the torque about the origin in terms of the components of these three vectors. ($b$) Compute the time derivative of the $z$ component of angular momentum and show that $dL_z/dt = v_x p_y - v_y p_x + xF_y - yF_x$, where $v_x$ and $v_y$ are the $x$ and $y$ components of the velocity of the particle. Is this expression the same as that for the $z$ component of the torque? Explain.

3. A sphere, a disk, and a hoop made of homogeneous materials have the same radius (10 cm) and mass (3 kg). They are released from rest at the top of a 30° incline and roll down without slipping through a vertical distance of 2 m. ($a$) What are their speeds at the bottom? ($b$) Find the frictional force $f$ in each case. ($c$) If they start together at $t = 0$, at what time does each reach the bottom?

4. A 100-kg uniform disk of radius 0.60 m is placed flat on some smooth ice. Two skaters wind ropes around the disk in the same sense. Each skater thus pulls on his rope and skates away, exerting constant forces of 40 N and 60 N respectively for 5 s (Figure 10-30). Describe the motion of the disk; i.e., what are the acceleration, velocity, and the position of the center of mass as functions of time, and what are the angular acceleration and angular velocity as functions of time?

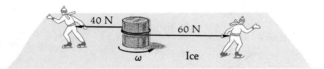

**Figure 10-30**
Problem 4.

5. A uniform ball of radius $r$ rolls without slipping along the loop-the-loop track in Figure 10-31. It starts at rest at height $h$ above the bottom of the loop. If the ball is not to leave the track at the top of the loop, what is the least value $h$ can have (in terms of the radius $R$ of the loop)? What would $h$ have been if the ball were to slide along a frictionless track instead of rolling?

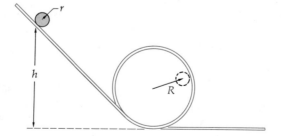

**Figure 10-31**
Problem 5.

6. A mass $m$ is attached to a light string which passes through a small hole in a frictionless tabletop. Initially the mass is sliding with speed $v_0$ in a circle of radius $r_0$ about the hole. A man under the table now begins to pull the string in slowly. ($a$) Show that when the mass is moving in a circle of radius $r$, the tension in the string is $T = L_0^2/mr^3$, where $L_0 = mv_0 r_0$ is the initial angular momentum. ($b$) The string is pulled in until the radius of the circular orbit is $r_f$.

Using the result of part ($a$), calculate the work done by integrating $T\,dr$ and show that the work done is $(L_0^2/2m)\,(r_f^{-2} - r_0^{-2})$. ($c$) What is the velocity $v_f$ when the radius of the circle is $r_f$? Show that the work done as calculated in part ($b$) equals the change in kinetic energy $\frac{1}{2}mv_f^2 - \frac{1}{2}mv_0^2$. ($d$) If $m = 0.5$ kg, $r_0 = 1$ m, $v_0 = 3$ m/s, and the maximum tension the string can withstand without breaking is 200 N, find the minimum value of $r_f$ before the string breaks.

7. A wheel of radius $R$ rolls without slipping at speed $V$. ($a$) Show that the $x$ and $y$ coordinates of point $P$ in Figure 10-32 are $r_0 \cos \theta$ and $R + r_0 \sin \theta$. ($b$) Show that the total velocity $\mathbf{v}$ of point $P$ has components $v_x = V + (r_0 V \sin \theta)/R$ and $v_y = -(r_0 V \cos \theta)/R$. ($c$) Show that $\mathbf{v}$ and $\mathbf{r}$ are perpendicular to each other by calculating their scalar product. ($d$) Show that $v = r\omega$, where $\omega = V/R$ is the angular velocity of the wheel. ($e$) These results demonstrate that in the case of rolling without slipping, the motion is the same as if the rolling object were instantaneously rotating about the point of contact with angular speed $\omega = V/R$. Calculate the kinetic energy of the wheel assuming that it is in pure rotation about point $O$, and show that the result is the same as calculated from the sum of the translational kinetic energy of the center of mass and the rotational kinetic energy of rotation about the center of mass.

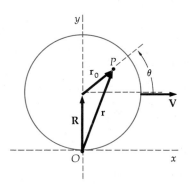

**Figure 10-32**
Problem 7.

8. A car is powered by energy stored in a flywheel. Suppose that there is just a single flywheel with angular momentum $\mathbf{L}$. Discuss problems that would arise for various orientations of $\mathbf{L}$ and various maneuvers of the car, e.g., if $\mathbf{L}$ is vertically upward and the car travels over a hilltop or through a valley; or if $\mathbf{L}$ points forward or to one side and the car attempts to turn to the left or the right. In each case consider the direction of the torque exerted on the car by the road.

9. A uniform cylinder of mass $M$ and radius $R$ has a string wrapped around it. The string is held fixed, and the cylinder falls vertically, as in Figure 10-33. ($a$) Show that the acceleration of the cylinder is downward with magnitude $a = 2g/3$. ($b$) Find the tension in the string.

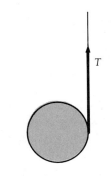

**Figure 10-33**
Problem 9.

10. In Problem 9 replace the cylinder with a uniform sphere of mass $M$ and radius $R$ with the string wound around it. Find both the acceleration of the sphere and the tension in the string.

11. A bowling ball of mass $M$ and radius $R$ is thrown so that at the instant it touches the floor it is moving horizontally with speed $v_0$ and not rotating. It slides for a time $t_1$ and a distance $s_1$ before it rolls without slipping. If $\mu_k$ is the coefficient of sliding friction between the ball and the floor, find $s_1$, $t_1$, and the final speed of the ball. Evaluate these quantities for $v_0 = 8$ m/s, and $\mu_k = 0.4$.

12. A cue ball of radius $R$ is initially at rest on a horizontal pool table. It is struck by a horizontal cue stick which delivers an impulse of magnitude $P_0 = F_{av}\,\Delta t$. (We use $P_0$ for the impulse rather than $I_0$ to avoid confusion with the moment of inertia $I$.) The stick strikes the ball at a point $h$ above the point of contact on the table (Figure 10-34). Show that the initial angular velocity $\omega_0$ is related to the initial linear velocity of the center of mass $v_0$ by $\omega_0 = 5v_0(h - R)/2R^2$. At what point should you strike the cue ball if you want it to begin rolling without slipping immediately after it is struck?

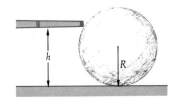

**Figure 10-34**
A cue ball is struck by a horizontal cue stick (Problem 12).

13. A uniform spherical ball is set rotating about a horizontal axis with angular speed $\omega_0$ and placed at rest on the floor. If the coefficient of sliding friction between the ball and the floor is $\mu_k$, find the speed of the center of mass of the ball when it begins to roll without slipping.

14. A uniform rod of mass $M$ and length $L$ is pivoted at one end and hangs as in Figure 10-35 so that it is free to rotate about its pivot without friction. It is struck by a horizontal force which delivers an impulse $P_0 = F_{av}\,\Delta t$ at a distance $x$ below the pivot as shown. ($a$) Show that the initial speed of the center of mass of the rod is given by $V_0 = 3P_0x/2ML$. ($b$) Find the impulse delivered by the pivot and show that this impulse is zero if $x = 2L/3$. This point is called the center of percussion of the rod.

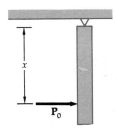

**Figure 10-35**
Problem 14.

15. The rod in Problem 14 is struck by a piece of putty of mass $m = M/6$ moving horizontally with initial speed $v_0$. The putty strikes the end of the rod and sticks to it. Find the initial angular velocity of the rod with putty attached just after the collision. What fraction of the initial energy of the putty is lost in this inelastic collision?

16. Assuming the earth to be a homogeneous sphere of radius $r$ and mass $m$, show that the period $T$ of rotation about its axis is related to its radius by $T = (4\pi m/5L)r^2$, where $L$ is the angular momentum of the earth due to its rotation. Suppose that the radius $r$ changes by a very small amount $\Delta r$ due to some internal effect, e.g., thermal expansion. (a) Show that the fractional change in the period $T$ is given approximately by $\Delta T/T = 2\,\Delta r/r$. Hint: Use differentials $dr$ and $dT$ to approximate the changes in these quantities. (b) By how many kilometres would the earth need to expand for the period to change by $\frac{1}{4}$ d/y, so that leap year would not be needed?

17. A heavy homogeneous cylinder has mass $m$ and radius $R$. It is accelerated by a force **T**, which is applied through a rope wound around a light drum of radius $r$ attached to the cylinder (Figure 10-36). The coefficient of static friction is sufficient for the cylinder to roll without slipping. (a) Find the friction force. (b) Find the acceleration $a$ of the center of the cylinder. (c) Is it possible to choose $r$ so that $a$ is greater than $T/m$? How? (d) What is the direction of the friction force in the circumstances of part (c)?

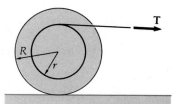

**Figure 10-36**
Problem 17.

18. The sun rotates with a period of 25.3 d. Estimate its new period of rotation if it collapses to a neutron star of radius 5 km and no mass is lost.

19. The polar ice caps contain about $2.3 \times 10^{19}$ kg of ice. This contributes essentially nothing to the moment of inertia of the earth because it is located at the poles, close to the axis of rotation. Estimate the change in the length of the day to be expected if the polar ice caps melt, distributing the water uniformly over the surface of the earth. (The moment of inertia of a spherical shell of mass $m$ and radius $r$ is $\frac{2}{3}mr^2$.)

20. A particle moves with constant speed $v$ in the $xy$ plane along a line parallel to the $x$ axis a distance $b$ from this axis. Let $r$ be the radius vector from the origin to the particle at some instant, and let $\theta$ be the angle between $r$ and the $x$ axis. (a) Let $v_r$ be the component of the velocity along the radius vector and $v_t$ be the component perpendicular to the radius vector. Show that $v_r = v\cos\theta$ and $v_t = v\sin\theta$. (b) Show that the angular velocity of the particle about the origin is $\omega = v_t/r = (v/b)\sin^2\theta$. Using this result and the moment of inertia of the particle about the $z$ axis, $I = mr^2$, show that $mbv$ and $I\omega$ are equivalent expressions for the angular momentum of the particle about the origin.

21. Prove that $\mathbf{A} \times (\mathbf{B} + \mathbf{C}) = \mathbf{A} \times \mathbf{B} + \mathbf{A} \times \mathbf{C}$ by writing out both sides in terms of their rectangular components and the unit vectors **i**, **j**, and **k**.

22. Prove that $d(\mathbf{A} \times \mathbf{B})/dt = \mathbf{A} \times d\mathbf{B}/dt + (d\mathbf{A}/dt) \times \mathbf{B}$ by writing out $\mathbf{A} \times \mathbf{B}$ in terms of the components of the vectors and the unit vectors **i**, **j**, and **k** and differentiating directly.

# CHAPTER 11 Oscillations

Motion which repeats itself is called *periodic*, the time required for each repetition being the *period*. Systems displaying periodic motion occur frequently, e.g., the motion of a mass on a spring, the motion of the moon about the earth, the oscillation of a pendulum, the oscillations of atoms in a molecule, the vibration of a string of a violin. The study of periodic motion has many applications. In addition to the examples listed above, periodic motion occurs in many types of wave motion.

## 11-1 Simple Harmonic Motion

We shall concentrate on the simplest form of oscillation, called simple harmonic motion. It occurs whenever the restoring force on a particle displaced from equilibrium is proportional to the displacement.

Common examples are a body on a spring and a simple pendulum (for small displacements from equilibrium). Figure 11-1 shows the displacement $x$ versus time of a small body attached to a spring which obeys Hooke's law, $F_x = -kx$. The origin has been chosen so that $x = 0$ is the equilibrium position. The displacement $x$ varies sinusoidally with time and can be described by the equation

$$x = A \cos (\omega t + \delta) \qquad\qquad 11\text{-}1$$

*Position function for simple harmonic motion*

where $A$, $\omega$, and $\delta$ are constants. The maximum displacement $A$ is called the *amplitude*. The quantity $\omega t + \delta$ is called the *phase* of the motion and the constant $\delta$ is called the *phase constant*. When the phase increases by $2\pi$ from its value at time $t$, the particle again has the same position and velocity as at time $t$, since $\cos (\omega t + \delta + 2\pi) = \cos (\omega t + \delta)$. During the time the phase increases by $2\pi$, the particle performs a full cycle of the motion. We can determine the *period T* of the motion by noting that the phase at time $t + T$ is just $2\pi$ plus the phase at time $t$:

*Amplitude, phase, phase constant*

$$\omega(t + T) + \delta = 2\pi + \omega t + \delta \qquad \text{or} \qquad \omega T = 2\pi$$

and

$$T = \frac{2\pi}{\omega} \qquad\qquad 11\text{-}2$$

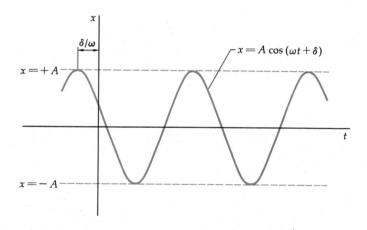

**Figure 11-1**
Plot of displacement versus time for an object attached to a spring.

The number of vibrations per unit time $1/T$ is called the *frequency f*:

$$f = \frac{1}{T} = \frac{\omega}{2\pi} \qquad\qquad 11\text{-}3$$

*Frequency and period*

The constant $\omega = 2\pi f$ is the *angular frequency*. It has the units radians per second, the same units as angular velocity, also designated $\omega$. In Section 11-2 we shall find that angular frequency is closely related to the angular velocity of a circular motion that is associated with simple harmonic motion.

*Angular frequency*

In terms of the frequency or period, Equation 11-1 is

$$x = A \cos (2\pi f t + \delta) = A \cos \left(\frac{2\pi t}{T} + \delta\right) \qquad\qquad 11\text{-}4$$

The phase constant $\delta$ depends on our choice of zero time. If we choose $t = 0$ when $x = A$, the phase constant is zero. On the other hand, if we choose $t = 0$ when $x = 0$, $\delta$ is either $\pi/2$ or $3\pi/2$, depending on whether $x$ is increasing or decreasing at $t = 0$. For example, if $\delta = 3\pi/2$,

$$x = A \cos \left( \omega t + \frac{3\pi}{2} \right) = A \sin \omega t$$

Then $x$ is increasing at $t = 0$, as can be seen from a graph of $x = A \sin \omega t$ versus $t$.

We can find the general relation between the initial position $x_0$ and the constants $A$ and $\delta$ by setting $t = 0$ in Equation 11-1:

$$x_0 = A \cos \delta \qquad\qquad 11\text{-}5$$

The phase constant is not very important for the description of one-dimensional simple harmonic motion of a single particle since we can always choose our zero time to make $\delta = 0$. If we have two particles undergoing simple harmonic motion, or if we consider the combination of simple harmonic motion of a particle in two dimensions, we have two phase constants. If they are not equal, we can make only one of them equal to zero by our choice of zero time. Then the other one, the relative phase constant of the two motions, is important in the description of the motion.

The velocity of a particle undergoing simple harmonic motion is

$$v_x = \frac{dx}{dt} = -A\omega \sin (\omega t + \delta) = A\omega \cos \left( \omega t + \delta + \frac{\pi}{2} \right) \qquad 11\text{-}6$$

The phase of the velocity differs from that of the position by $\pi/2$ rad, or 90°. When $\cos (\omega t + \delta) = 1$, $\sin (\omega t + \delta) = 0$; so when $x$ is its maximum or minimum value, the velocity is zero. Similarly when $\sin (\omega t + \delta) = 1$, $\cos (\omega t + \delta) = 0$. The speed is maximum when the particle is at its equilibrium position $x = 0$. According to Equation 11-6, the initial velocity is related to $A$ and $\omega$ by

$$v_0 = -A\omega \sin \delta \qquad\qquad 11\text{-}7$$

Since Equations 11-5 and 11-7 can be solved for $A$ and $\delta$ in terms of the initial position $x_0$ and initial velocity $v_0$, the amplitude $A$ and the phase constant $\delta$ are completely determined by the initial conditions.

The acceleration of the particle is

$$a_x = \frac{d^2x}{dt^2} = \frac{dv_x}{dt} = -\omega^2 A \cos (\omega t + \delta)$$

or

$$a_x = -\omega^2 x \qquad\qquad 11\text{-}8$$

*Acceleration is proportional to displacement*

*In simple harmonic motion the acceleration is proportional to the displacement and in the opposite direction.*

If the acceleration of a particle is known, the position $x(t)$ is completely determined by the initial conditions, e.g., the initial position and velocity. As we have seen, they are related to the constants $A$ and $\delta$ in Equation 11-1. Thus we could begin with Equation 11-8, relating the acceleration to the position, and work backward to obtain Equation 11-1 (see Problem 28). That is, the general solution of the differential equation 11-8 is of the form of Equation 11-1; the two equations are equivalent. We can therefore take either Equation 11-1 or 11-8 as the defining equation of simple harmonic motion.

The acceleration of a particle is just the resultant force divided by the mass. Therefore, we conclude that whenever a force is proportional to the displacement (and in the opposite direction), the motion is simple harmonic. For example, if a mass $m$ is acted upon by a spring of force constant $k$, the force is $-kx$ and the acceleration is

*Conditions for simple harmonic motion*

$$a_x = \frac{F_x}{m} = -\frac{k}{m} x \qquad\qquad 11\text{-}9$$

This equation is the same as Equation 11-8 with $k/m$ replacing $\omega^2$. The solution of this equation is Equation 11-1 with $\omega = \sqrt{k/m}$ and the constants $A$ and $\delta$ determined by the initial position and velocity of the mass. Note that the angular frequency $\omega$ and thus the period $T = 2\pi/\omega$ are independent of the amplitude $A$. This is an important property of simple harmonic motion.

*In simple harmonic motion the frequency and period are independent of the amplitude.*

We can write the general position function $x(t)$ for simple harmonic motion in terms of the initial position and velocity, rather than in terms of the amplitude and phase constant as in Equation 11-1. We use the trigonometric identity for the cosine of the sum of two angles,

$$\cos (\theta + \phi) = \cos \theta \cos \phi - \sin \theta \sin \phi$$

Then Equation 11-1 can be written

$$A \cos (\omega t + \delta) = A \cos \delta \cos \omega t - A \sin \delta \sin \omega t$$

But according to Equations 11-5 and 11-7, $A \cos \delta = x_0$ and $-A \sin \delta = v_0/\omega$. Then

$$x(t) = A \cos (\omega t + \delta) = x_0 \cos \omega t + \frac{v_0}{\omega} \sin \omega t \qquad\qquad 11\text{-}10$$

## Questions

1. What is the phase constant $\delta$ in Equation 11-1 if the position of the oscillating particle at time $t = 0$ is (a) 0, (b) $-A$, (c) $A$, (d) $A/2$?

2. How far does a particle oscillating with amplitude $A$ move in one full period?

3. How far apart are the extreme positions of a particle oscillating with amplitude $A$?

4. If you know that the speed of an oscillator of amplitude $A$ is zero at certain times, can you say exactly what its displacement is at those times? Can you say what direction it will be moving in a moment? If not, how close can you come?

5. If you are told that the speed of an oscillator of amplitude $A$ has its maximum value at a certain time, can you say exactly what its displacement is? Can you say in what direction it is moving?

6. Can the acceleration and displacement of a simple harmonic oscillator ever be in the same direction? The acceleration and the velocity? The velocity and the displacement? Explain.

7. What is the magnitude of the acceleration of an oscillator of amplitude $A$ and angular frequency $\omega$ when its speed is a maximum? When its displacement is a maximum?

## 11-2   Circular Motion and Simple Harmonic Motion

There is a simple relation between circular motion with constant angular velocity and simple harmonic motion. Consider a particle moving in a circle of radius $A$ with constant angular velocity $\omega$, as shown in Figure 11-2. The angular displacement of the particle relative to the $x$ axis is given by

$$\theta = \omega t + \delta \qquad\qquad 11\text{-}11$$

where $\delta$ is the angular displacement at $t = 0$. From Figure 11-2 we see that the $x$ component of the position of the particle is given by

$$x = A \cos \theta = A \cos (\omega t + \delta) \qquad\qquad 11\text{-}12$$

Thus the projection on a straight line of uniform circular motion at angular velocity $\omega$ is simple harmonic motion with angular frequency $\omega$. The frequency and period of the circular motion are the same as the corresponding frequency and period of the projected simple harmonic motion. The relation between circular motion and simple harmonic motion can be demonstrated with a turntable and a mass on a spring (Figure 11-3).

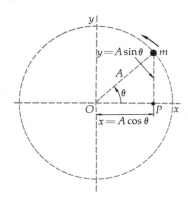

**Figure 11-2**
A particle of mass $m$ moving in a circle with constant angular speed $\omega$. The angle $\theta$ increases with time $\theta = \omega t + \delta$. The projection on the $x$ axis, point $P$, moves with simple harmonic motion $x = A \cos \theta = A \cos (\omega t + \delta)$.

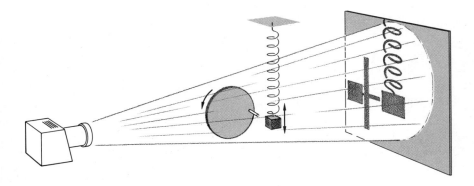

**Figure 11-3**
The shadow of a peg on a turntable is projected on a screen along with the shadow of an object on a spring. When the period of rotation of the turntable equals the period of oscillation of the object on the spring, the shadows move together.

The projection of the circular motion on the $y$ axis is $y = A \sin \theta = A \sin (\omega t + \delta) = A \cos (\omega t + \delta - \pi/2)$ (Figure 11-2). We can therefore consider the circular motion of a particle to be the combination of perpendicular simple harmonic motions having the same amplitude and frequency but having a relative phase difference of $\pi/2$.

**Question**

8. At what points does the particle traveling on the circle in Figure 11-2 have the same displacement as the oscillating particle? The same speed? The same acceleration?

## 11-3   An Object on a Spring

If an object of mass $m$ is attached to a spring stretched a distance $x$ from its equilibrium length, the force on the object is $F_x = -kx$, where $k$ is called the *force constant* of the spring. Newton's second law for the motion of the object is

$$F_x = m \frac{d^2x}{dt^2} = -kx$$

or

$$\frac{d^2x}{dt^2} = a_x = -\frac{k}{m} x \qquad\qquad 11\text{-}13$$

This is the same as Equation 11-8 with

$$\omega^2 = \frac{k}{m} \qquad\qquad 11\text{-}14$$

The object oscillates with simple harmonic motion. Its position is given by either Equation 11-1 or 11-10:

$$x = A \cos (\omega t + \delta) = x_0 \cos \omega t + \frac{v_0}{\omega} \sin \omega t$$

The velocity of the object is

$$v_x = \frac{dx}{dt} = -\omega A \sin (\omega t + \delta) = -\omega x_0 \sin \omega t + v_0 \cos \omega t$$

The potential energy is

$$U(x) = \tfrac{1}{2}kx^2 = \tfrac{1}{2}m\omega^2 x^2 \qquad\qquad 11\text{-}15$$

where we use $\omega^2 = k/m$. This is the parabolic potential-energy function discussed in Section 7-2. It is sketched again in Figure 11-4.

The total energy is

$$E = U(x) + \tfrac{1}{2}mv_x^2 = \tfrac{1}{2}m\omega^2 x^2 + \tfrac{1}{2}m \left(\frac{dx}{dt}\right)^2$$

Using Equations 11-1 and 11-6 for $x$ and $dx/dt$, we obtain

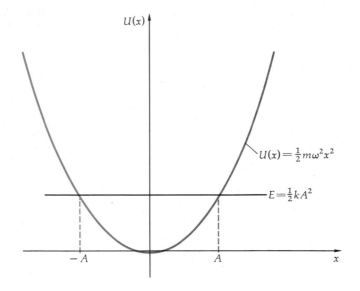

**Figure 11-4**
The potential-energy function $U(x) = \tfrac{1}{2}kx^2 = \tfrac{1}{2}m\omega^2 x^2$ for an object of mass $m$ on a spring of force constant $k$. The horizontal line represents the total energy $E = \tfrac{1}{2}kA^2$ for amplitude $A$.

$$E = \tfrac{1}{2}m\omega^2 A^2 \cos^2(\omega t + \delta) + \tfrac{1}{2}m\omega^2 A^2 \sin^2(\omega t + \delta)$$

$$= \tfrac{1}{2}m\omega^2 A^2[\cos^2(\omega t + \delta) + \sin^2(\omega t + \delta)]$$

or

$$E = \tfrac{1}{2}m\omega^2 A^2 = \tfrac{1}{2}kA^2 \qquad\qquad\qquad \text{11-16}$$

The total energy is thus constant and equal to the maximum potential energy or the maximum kinetic energy. The total energy is shown by the horizontal line in Figure 11-4. As discussed in Chapter 7, the kinetic energy is represented on this graph by the vertical distance between the total energy and the potential energy. It is easily seen from the figure that the two turning points are at equal distances from the equilibrium point $x = 0$ and that the maximum velocity occurs at $x = 0$.

---

**Example 11-1** A 2-kg object stretches a spring 10 cm when it hangs vertically in equilibrium. The object is then attached to the spring resting on a smooth table and fixed at one end, as shown in Figure 11-5. The object is held a distance 5 cm from the equilibrium position and released. Find the frequency $f$, angular frequency $\omega$, period $T$, amplitude $A$, and phase constant $\delta$ for the resulting simple harmonic motion. What is the maximum speed of the object, and when does it occur?

Equilibrium

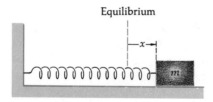

**Figure 11-5**
An object of mass $m$ attached to a spring and sliding along a frictionless horizontal surface.

The force constant of the spring is determined from the first measurement. For the vertical-equilibrium case, $ky = mg$, where $y$ is the amount the spring is stretched. Using $g = 9.81$ m/s², $m = 2$ kg, and $y = 10$ cm $= 0.1$ m, we have

$$k = \frac{mg}{y} = \frac{(2 \text{ kg})(9.81 \text{ m/s}^2)}{0.1 \text{ m}} = 196 \text{ N/m}$$

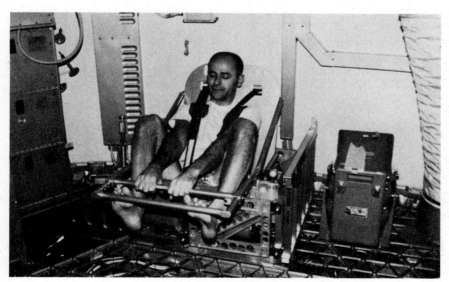

Astronaut Alan L. Bean measuring his body mass during the second Skylab mission. The total mass of the astronaut plus the apparatus is related to his frequency of vibration by Equation 11-14.

We are given that the initial position of the body is $x_0 = 5$ cm $= 0.05$ m and that the initial velocity is zero. Then from Equations 11-5 and 11-7 we have $\delta = 0$ and $A = 0.05$ m. The equation of motion of the object is

$$x = (0.05 \text{ m}) \cos \omega t$$

where $\omega = \sqrt{k/m} = \sqrt{196/2} = 9.90$ rad/s. The period, found from $\omega T = 2\pi$, is

$$T = \frac{2\pi}{9.9} = 0.63 \text{ s}$$

The reciprocal of the period is the frequency. The units of frequency are reciprocal seconds, often written cycles per second or vibrations per second for easier reading. The unit cycles per second is now called the hertz (Hz). Thus

$$f = \frac{1}{T} = 1.58 \text{ cycles/s} = 1.58 \text{ Hz}$$

The velocity of the object is

$$\begin{aligned} v_x = \frac{dx}{dt} &= -\omega A \sin \omega t \\ &= -(9.9 \text{ rad/s}) (0.05 \text{ m}) \sin \omega t \\ &= -(0.5 \text{ m/s}) \sin \omega t \end{aligned}$$

The maximum speed is 0.5 m/s. This occurs first when $\omega t_1 = \pi/2$ or $t_1 = \frac{1}{4}T = 0.16$ s. At this time, $x = 0$ and $v_x$ is negative, indicating that the motion is to the left. The speed is maximum again one-half cycle later, when $x = 0$ but the mass is moving to the right. The functions $x(t)$ and $v_x(t)$ are sketched in Figure 11-6.

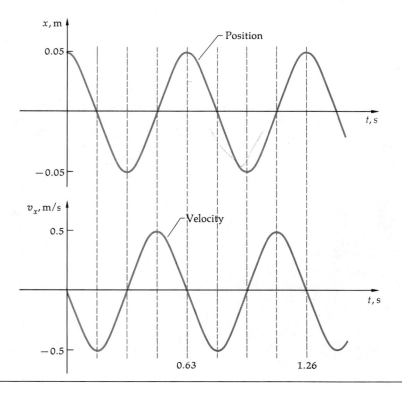

**Figure 11-6**
Position and velocity of the object on a spring for Example 11-1.

**Example 11-2** A second spring identical to that in Example 11-1 is attached to a second object, also of mass 2 kg. This spring is stretched a distance of 10 cm from equilibrium and released at the same time as the first, which is stretched to 5 cm. Which object reaches the equilibrium position first?

Figure 11-7 shows the initial positions of the objects, and Figure 11-8 is a sketch of the two position functions $x_1(t)$ and $x_2(t)$. Since both the springs have the same force constant and the masses are equal, the angular frequencies and periods of the motions are the same. Only the amplitudes differ. *Thus they both reach the equilibrium position at the same time.* Object 2 has twice as far to go to reach equilibrium as object 1, but it begins with twice the initial acceleration. The equations of motion of the two objects are

$$x_1 = (0.05 \text{ m}) \cos \omega t \qquad x_2 = (0.10 \text{ m}) \cos \omega t$$

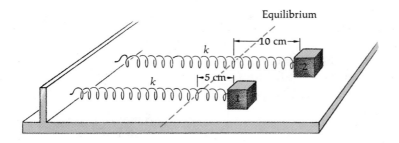

**Figure 11-7**
Two objects on identical springs released simultaneously with different amplitudes (Example 11-2).

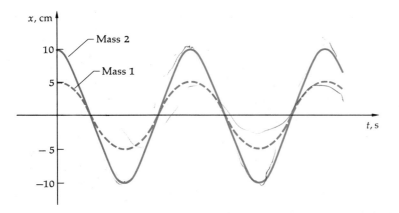

**Figure 11-8**
Plots of position versus time for the objects in Figure 11-7.

This example illustrates that the angular frequency and period in simple harmonic motion are independent of the amplitude. The time $\frac{1}{4}T$ to reach the equilibrium position is the same for each particle. Of course the energies of the two particles are different: more work is required to stretch the second spring 10 cm than to stretch the first one 5 cm. The energies are

$$E_1 = \tfrac{1}{2}kA_1^2 = \tfrac{1}{2}(200)\,(0.05)^2 = 0.25 \text{ J}$$

$$E_2 = \tfrac{1}{2}kA_2^2 = \tfrac{1}{2}(200)\,(0.1)^2 = 1.0 \text{ J}$$

You should show that at the equilibrium position the velocity of the second object is twice that of the first.

**Example 11-3** An object of mass $m$ attached to a spring at rest at $x = 0$ is struck a blow with a hammer, which delivers an impulse $I$. What is the subsequent motion?

We choose $t = 0$ just after the impulse. The object is thus at $x = 0$ with an initial velocity $v_0 = p_0/m = I/m$, where $p_0$, the initial momentum, is equal to the impulse. The form of Equation 11-10 is useful here since $x_0$ is zero. Thus

$$x = \frac{v_0}{\omega} \sin \omega t = \frac{I}{m\omega} \sin \omega t$$

When an object hangs from a vertical spring, as in Figure 11-9, there is a force $mg$ downward in addition to the force of the spring $-ky$, where we are assuming that $y$ is measured downward from the unstretched position of the spring. The effect of the gravitational force is merely to shift the position of equilibrium. When the object hangs in equilibrium from the spring, the spring is stretched by $y_0$, where $ky_0 = mg$. When the object is displaced from this equilibrium position by the amount $y' = y - y_0$, the unbalanced force is $-ky'$ and the object oscillates with (angular) frequency $\omega = \sqrt{k/m}$, the same as for a horizontal spring. We shall show below that when the object is displaced from equilibrium by the amount $y'$, the total change in the potential energy, including the gravitational potential energy, is $\frac{1}{2}ky'^2$. We can therefore ignore the effect of the gravitational force of the earth completely if we measure the displacement of the body from its equilibrium position.

*Object on a vertical spring*

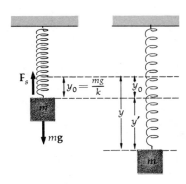

**Figure 11-9**
An object on a vertical spring. The object oscillates about its equilibrium position, in which the spring is stretched by $y_0 = mg/k$.

*Optional*

For an object hanging on a vertical spring which is stretched by the amount $y$ (measured vertically downward), Newton's second law gives

$$m\frac{d^2y}{dt^2} = -ky + mg \qquad\qquad 11\text{-}17$$

This differs from Equation 11-13 because of the constant term $mg$. We handle this extra term by changing to a new variable $y' = y - y_0$, where $y_0 = mg/k$ is the amount the spring is stretched when the object hangs in equilibrium. Since $y_0$ is a constant, we have $dy'/dt = dy/dt$ and $d^2y'/dt^2 = d^2y/dt^2$.

Multiplying $y'$ by $k$ gives

$$ky' = k(y - y_0) = ky - ky_0 = ky - mg$$

Thus $ky = ky' + mg$, and Equation 11-17 becomes

$$m\frac{d^2y'}{dt^2} = -ky' \qquad\qquad 11\text{-}18$$

which has the familiar solution $y' = A \cos(\omega t + \delta)$ with $\omega^2 = k/m$. The mass oscillates about the equilibrium point, where the total force on the mass is zero.

The spring's potential energy is not $\frac{1}{2}ky'^2$ but $\frac{1}{2}ky^2 = \frac{1}{2}k(y' + y_0)^2$. When the spring is stretched from $y = y_0$ to some value $y$, the potential energy of the *spring* is increased by

$$\Delta U_{sp} = \frac{1}{2}ky^2 - \frac{1}{2}ky_0^2 = \frac{1}{2}ky'^2 + ky_0y' + \frac{1}{2}ky_0^2 - \frac{1}{2}ky_0^2$$

$$= \frac{1}{2}ky'^2 + mgy' \qquad\qquad 11\text{-}19$$

but the potential energy due to the earth's gravity is decreased by $mgy'$. When the spring is stretched from $y_0$ to $y'$, the total potential-energy change is

$$\Delta U = \frac{1}{2}ky^2 - \frac{1}{2}ky_0^2 - mgy' = \frac{1}{2}ky'^2 \qquad\qquad 11\text{-}20$$

That is, when the spring is stretched by amount $y'$ from equilibrium, the total change in potential energy, *including gravitational potential energy*, is $\frac{1}{2}ky'^2$. Thus if we measure from the equilibrium position, we can forget about the effect of the earth. The potential-energy functions $\frac{1}{2}ky^2$ and $-mgy$ are shown as dashed curves in Figure 11-10. Their sum is the solid curve.

**Figure 11-10**
Potential energy of spring, $\frac{1}{2}ky^2$ (dashed colored line), and that of gravity, $-mgy$ (dashed black line), for object on vertical spring with $y$ positive downward. The total potential energy can be shown to be $\frac{1}{2}ky'^2 - \frac{1}{2}mgy_0$, where $y'$ is measured from equilibrium. If $\Delta U$ is measured from the constant value $-\frac{1}{2}mgy_0$, $\Delta U = \frac{1}{2}ky'^2$.

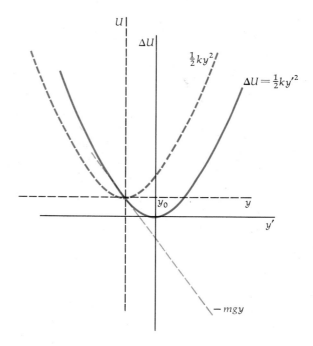

**Questions**

9. An object of mass $m$ oscillates at frequency $f$ when it is attached to a horizontal spring as in Figure 11-5. It is released from rest when its displacement is $A$. If the mass is doubled, how will the frequency be changed? The angular frequency? The period? The amplitude? The maximum speed? The maximum acceleration? The maximum force exerted by the spring?

10. Explain how each of the following quantities will change if the impulse $I$ in Example 11-3 is doubled: the amplitude, the frequency, the period, the maximum speed, the total energy, the phase constant.

11. The effect of the mass of a spring on the motion of an object attached to it is usually neglected. Describe qualitatively its effect when it is not neglected.

## 11-4  The Simple Pendulum

Another important example of periodic motion is that of a simple pendulum. If the angle made by the string with the vertical is not too great, the motion of the bob of the pendulum is simple harmonic.

Consider an object of mass $m$ on a string of length $L$, as shown in Figure 11-11. The forces on the body are the force of gravity $m\mathbf{g}$ and the tension $\mathbf{T}$ in the string. The tangential force has magnitude $mg \sin \theta$ and is in the direction of decreasing $\theta$. Let $s$ be the arc length measured from the bottom of the circle. The arc length is related to the angle measured from the vertical by

$$s = L\theta \qquad\qquad 11\text{-}21$$

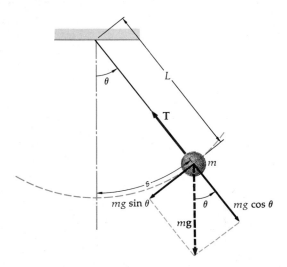

**Figure 11-11**
Forces on a simple pendulum.

The tangential acceleration is $d^2s/dt^2$. The tangential component of $\Sigma\mathbf{F} = m\mathbf{a}$ is

$$\Sigma F_t = -mg \sin \theta = m \frac{d^2s}{dt^2}$$

or

$$\frac{d^2s}{dt^2} = -g \sin \theta = -g \sin \frac{s}{L} \qquad\qquad 11\text{-}22$$

If $s$ is much less than $L$, the angle $\theta = s/L$ is small and we can approximate $\sin \theta$ by the angle $\theta$. Using $\sin (s/L) \approx s/L$ in Equation 11-22, we have

$$\frac{d^2s}{dt^2} = -\frac{g}{L} s \qquad\qquad 11\text{-}23$$

We see that for small angles for which the approximation $\sin \theta \approx \theta$ is valid, the acceleration is proportional to the displacement. The motion of the pendulum is simple harmonic for small displacement. If we write $\omega^2$ for $g/L$, Equation 11-23 becomes

$$\frac{d^2s}{dt^2} = -\omega^2 s$$

$$\omega^2 = \frac{g}{L} \qquad\qquad 11\text{-}24$$

The solution of this equation is

$$s = s_0 \cos (\omega t + \delta) \qquad\qquad 11\text{-}25$$

where $s_0$ is the maximum displacement measured along the arc of the circle. The period of the motion is

$$T = \frac{2\pi}{\omega} = 2\pi\sqrt{\frac{L}{g}} \qquad \text{11-26}$$

It is often more convenient to measure the displacement of the pendulum in terms of the angle made with the vertical, $\theta = s/L$. Equations 11-22, 11-23, and 11-25 can be written in terms of the angular displacement and the angular acceleration $d^2\theta/dt^2$ by dividing each side by $L$, giving

$$\frac{d^2\theta}{dt^2} = -\frac{g}{L}\sin\theta \qquad \text{11-22}a$$

which for small $\theta$ becomes

$$\frac{d^2\theta}{dt^2} = -\frac{g}{L}\theta \qquad \text{11-23}a$$

The solution of this equation is

$$\theta = \theta_0 \cos(\omega t + \delta) \qquad \text{11-25}a$$

where $\theta_0 = s_0/L$ is the maximum angular displacement. The criterion for simple harmonic motion stated in terms of these angular quantities is that the angular acceleration be proportional to the angular displacement, as in Equation 11-23$a$.

---

**Example 11-4** What is the period of a pendulum 1 m long when $g = 9.81$ m/s²?

The angular frequency is given by Equation 11-24:

$$\omega = \sqrt{\frac{g}{L}} = \sqrt{\frac{9.81 \text{ m/s}^2}{1 \text{ m}}} = 3.13 \text{ rad/s}$$

The period is

$$T = \frac{2\pi}{\omega} = \frac{2\pi}{3.13} = 2.01 \text{ s}$$

A time of this magnitude is easily measured accurately. For example, with a watch which can measure time intervals to ±1 s, we can measure such a period to ±1 part in 200 by measuring the time for 100 vibrations.

---

The potential energy of the simple pendulum is the gravitational potential energy $mgy$, where $y$ is the height measured from some reference level. It is convenient to choose $y = 0$ at the lowest point of the object at $\theta = 0$. Then $y$ is related to $\theta$ by

$$y = L - L\cos\theta = L(1 - \cos\theta)$$

as can be seen from Figure 11-12. The potential energy is then

$$U = mgL(1 - \cos\theta) \qquad \text{11-27}$$

and the total energy is

$$E = \tfrac{1}{2}mv^2 + U = \tfrac{1}{2}mv^2 + mgL(1 - \cos\theta) \qquad \text{11-28}$$

For small angles, we can approximate $\cos\theta$ by

$$\cos\theta = 1 - \frac{\theta^2}{2} + \cdots \approx 1 - \frac{\theta^2}{2} \qquad \text{11-29}$$

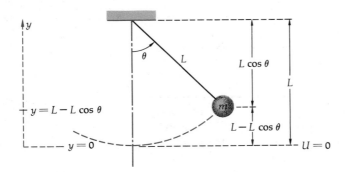

**Figure 11-12**
The height of the pendulum
bob above its equilibrium position is $y = L - L \cos \theta$, and
so its gravitational potential
energy is $U = mgy =$
$mgL(1 - \cos \theta)$ measured
from $U = 0$ at $\theta = 0$.

Then $1 - \cos \theta \approx \frac{1}{2}\theta^2$, and the potential energy is

$$U(\theta) \approx \frac{1}{2}mgL\theta^2 \qquad\qquad 11\text{-}30a$$

According to Equation 11-15, the potential energy for an object on a
spring can be written $U = \frac{1}{2}m\omega^2 x^2$. If we use $\theta = s/L$ and $\omega^2 = g/L$, we
have for the potential energy of a simple pendulum

$$U = \frac{1}{2}mgL \left(\frac{s}{L}\right)^2 = \frac{1}{2}m\frac{g}{L}s^2 = \frac{1}{2}m\omega^2 s^2 \qquad\qquad 11\text{-}30b$$

The simple pendulum was used for early determinations of the acceleration of gravity because both the period and the length are easily
measured. Direct measurement by observation of free fall, in contrast,
is difficult because the time of fall over reasonable distances is too short
for easy measurement. From Equation 11-26 we have, in terms of $L$ and
$T$,

$$g = \frac{4\pi^2 L}{T^2} \qquad\qquad 11\text{-}31$$

The motion of a simple pendulum is simple harmonic only if the
angular displacement is small so that $\sin \theta \approx \theta$ is valid. Figure 11-13
shows a plot of $\theta$ and $\sin \theta$. Even for the rather large angle $\theta = \pi/6 =$
0.523 (which is 30°), the difference between $\theta$ and $\sin \theta = \sin 30° =$
0.500 is only 4.6 percent. If the angular displacement is large, the angular acceleration given by Equation 11-22a is not proportional to the
angular displacement. The motion of the pendulum for large angles is
not simple harmonic, but the motion is periodic, though the period is
not independent of the amplitude, as in simple harmonic motion. From

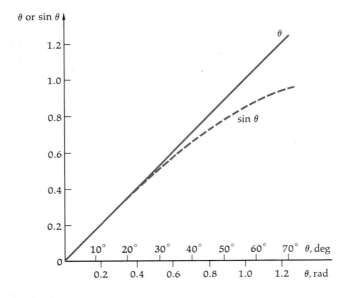

**Figure 11-13**
Comparison of $\theta$ and $\sin \theta$.
For sufficiently small angles,
the curves are approximately
the same.

Figure 11-13 we see that at large angles sin $\theta$ is less than $\theta$. The force which accelerates the bob back toward equilibrium has the magnitude $mg \sin \theta$; this is less than $mg\theta$, which would produce simple harmonic motion. Thus the acceleration at large angles is less than it must be for simple harmonic motion, and the period is slightly longer. We can see this also from Figure 11-14, comparing the actual potential energy, which is proportional to $1 - \cos \theta$, with the approximation proportional to $\frac{1}{2}\theta^2$. For a given total energy $E$, the pendulum travels farther between a turning point and equilibrium in the actual potential energy than it would in the parabolic potential-energy function for simple harmonic motion. The period is therefore slightly longer. We omit the details of the difficult solution of Equation 11-22a for large angles. The period can be expressed in the power series

$$T = T_0 \left(1 + \frac{1}{2^2} \sin^2 \tfrac{1}{2}\theta_0 + \frac{1}{2^2}\frac{3}{4^2} \sin^4 \tfrac{1}{2}\theta_0 + \cdots \right) \qquad 11\text{-}32$$

*Period of a pendulum for large amplitude*

where $\theta_0$ is the maximum angular displacement and $T_0 = 2\pi\sqrt{L/g}$ is the period in the limit of small angles.

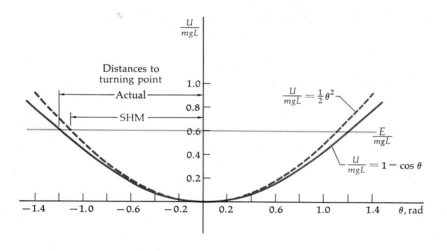

**Figure 11-14**
Comparison of the potential-energy function $U = mgL(1 - \cos \theta)$ and the simple-harmonic-motion (SHM) approximation $U = \frac{1}{2}mgL\theta^2$.

Because the actual period of a simple pendulum does vary with amplitude $\theta_0$, the simple pendulum makes a poor clock unless it is modified to compensate for energy loss due to friction and air resistance.

---

**Example 11-5** A simple pendulum clock is calibrated to keep accurate time at an amplitude of $\theta_0 = 10°$. When the amplitude has decreased so that it is very small, $\theta_0 \ll 10°$, how much time does the clock gain in 1 d?

The original period is approximately

$$T \approx T_0 \left(1 + \frac{1}{4} \sin^2 5°\right)$$

since $\sin^4 \tfrac{1}{2}\theta_0 \ll \sin^2 \tfrac{1}{2}\theta_0$ in Equation 11-32. After the amplitude has decreased, the period is just $T_0$. Since this is less, the frequency is greater and the clock gains time. The fractional change in the period is

$$-\frac{\Delta T}{T} = \frac{T - T_0}{T_0} = \frac{1}{4} \sin^2 5° = \frac{1}{4} (0.0872)^2 \approx 2 \times 10^{-3}$$

The number of minutes in 1 d is

$$\frac{24 \text{ h}}{1 \text{ d}} \frac{60 \text{ min}}{1 \text{ h}} = 1440 \text{ min/d}$$

Thus a fractional change of $2 \times 10^{-3}$ corresponds to an accumulated error in 1 d of $(2 \times 10^{-3})$ 1440 min = 2.88 min $\approx$ 3 min. Though the fractional change in period is only $2 \times 10^{-3}$, corresponding to a percentage change of 0.2 percent, this results in an intolerable error in time-keeping by today's standards for clocks. For this reason pendulum clocks are designed to maintain constant amplitude.

**Question**

12. The string or wire supporting a pendulum increases slightly in length when its temperature is raised. How would this affect a clock operated by the pendulum?

## 11-5   The Physical Pendulum

Any rigid body suspended from some point other than the center of mass will oscillate when displaced from its equilibrium position. Consider a plane figure suspended from a point a distance $D$ from the center of mass, as shown in Figure 11-15. The torque about the suspension point is $MgD \sin \theta$ in a direction tending to decrease $\theta$. The angular acceleration of the body is related to the torque by

$$\Sigma \tau = I\alpha = I\frac{d^2\theta}{dt^2}$$

where $I$ is the moment of inertia about the point of suspension. Therefore

$$-MgD \sin \theta = I\frac{d^2\theta}{dt^2} \qquad \text{11-33}$$

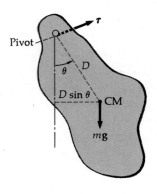

**Figure 11-15**
Physical pendulum. The torque about the pivot due to gravity is $mgD \sin \theta$, tending to decrease $\theta$.

For a simple pendulum this is the same as Equation 11-22a since $I = ML^2$ and $D = L$. Once again, the motion is simple harmonic only if the angular displacements are small, so that the approximation $\sin \theta \approx \theta$ holds. For this case, we have

$$\frac{d^2\theta}{dt^2} = -\frac{MgD}{I}\theta = -\omega^2\theta \qquad \text{11-34}$$

where $\omega^2 = MgD/I$. The period is

$$T = \frac{2\pi}{\omega} = 2\pi\sqrt{\frac{I}{MgD}} \qquad \text{11-35}$$

For large angles, the period is given by Equation 11-32, where $T_0$ is now given by Equation 11-35. The result we have just obtained can be used to measure the moment of inertia of a plane figure. The center of mass can be located by suspending the body from two different points, as discussed previously. To find the moment of inertia about some point $P$ we suspend the body from point $P$ and measure the period of oscillation. The moment of inertia is then obtained from

$$I = \frac{MgDT^2}{4\pi^2} \qquad\qquad 11\text{-}36$$

---

**Example 11-6** What is the period of oscillation of a stick pivoted at one end if the amplitude of oscillation is small?

The moment of inertia of a uniform stick of length $L$ about an axis at one end was shown in Chapter 9 to be $I = \frac{1}{3}ML^2$. The distance from the pivot to the center of mass is $\frac{1}{2}L$. Substituting these values in Equation 11-35 gives

$$T = 2\pi \sqrt{\frac{\frac{1}{3}ML^2}{Mg(\frac{1}{2}L)}} = 2\pi \sqrt{\frac{2L}{3g}}$$

If $L = 1$ m, the value of $T$ obtained from this result is 1.64 s.

---

# 11-6  General Motion near Equilibrium

Whenever a particle is displaced from a position of stable equilibrium, the motion of the particle is simple harmonic if the displacements are small enough. Figure 11-16 shows some arbitrary resultant force as a function of position. At positions $x_1$ and $x_2$ the force is zero; these are positions of equilibrium. However, the equilibrium at $x_2$ is unstable, for if the particle is displaced slightly in the positive $x$ direction, the force is positive, while if it is displaced slightly in the negative $x$ direction, the force is negative. In both cases, the force accelerates the particle away from equilibrium. The position $x_1$ is one of stable equilibrium. If the particle is displaced slightly in either direction from equilibrium, the force accelerates it back toward equilibrium. The particle thus oscillates about the equilibrium point. As we can see from Figure 11-16$b$, the plot of force versus displacement is approximately a straight

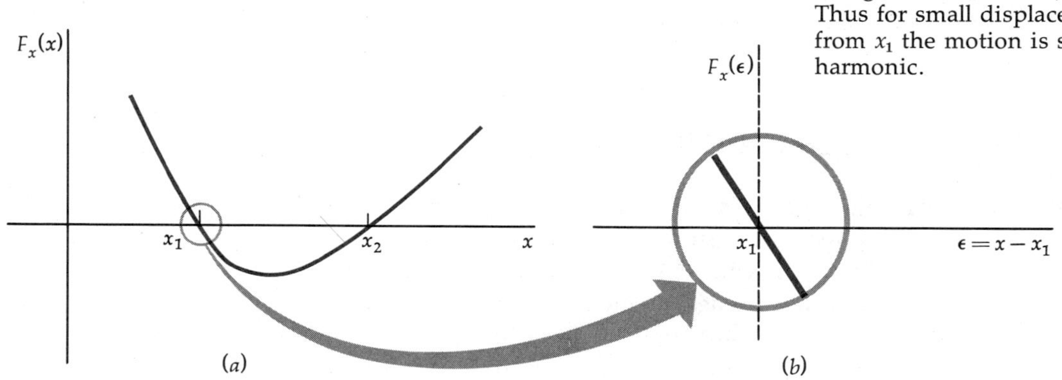

**Figure 11-16**
($a$) An arbitrary force $F_x$ versus $x$. The force is zero at the equilibrium points $x_1$ and $x_2$. At $x_1$ the equilibrium is stable because for small displacements away from $x_1$ the force is toward $x_1$. At $x_2$ the equilibrium is unstable because for small displacements from $x_2$ the force is away from $x_2$. ($b$) Near $x_1$ the force can be approximated by a straight line $F_x = -k(x - x_1)$. Thus for small displacements from $x_1$ the motion is simple harmonic.

($a$)                    ($b$)

line if we consider only small displacements from the equilibrium point $x_1$. Let $\epsilon$ measure the displacement from equilibrium

$$\epsilon = x - x_1$$

The equation for the force for small $\epsilon$ is

$$F_x = -k\epsilon$$

where $k$ is the magnitude of the slope of $F_x$ versus $x$ near $x_1$. Since the force is proportional to the displacement, the motion will be simple harmonic.

We can also examine the motion from the point of view of the potential-energy function $U(x)$ associated with the force. Figure 11-17 shows $U(x)$ versus $x$. As discussed in Chapter 6, the maximum at $x_2$ corresponds to unstable equilibrium, whereas the minimum at $x_1$ corresponds to stable equilibrium. The dashed curve in this figure is a parabolic curve which approximately fits $U(x)$ near the stable equilibrium point $x_1$. As long as the displacements from point $x_1$ are not too great, $U(x)$ is approximately parabolic and the motion is simple harmonic.

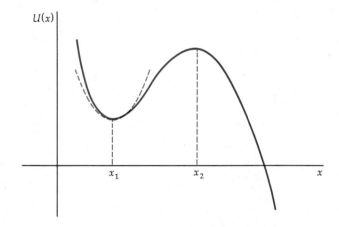

**Figure 11-17**
Potential-energy function $U(x)$ corresponding to the force in Figure 11-16. The minimum at $x_1$ indicates stable equilibrium, while the maximum at $x_2$ indicates unstable equilibrium. Near $x_1$ the curve is approximately parabolic, as in simple harmonic motion.

**Questions**

13. Give several examples of familiar motions which are either exactly or approximately simple harmonic motions.

14. The force on a particle is represented by the potential-energy function $U(x)$. The particle may execute approximate simple harmonic motion about a point $x_1$ only if $dU/dx = 0$ at that point. Explain why. The motion will be approximately simple harmonic for small displacements from $x_1$ only if $d^2U/dx^2$ is positive at $x_1$. Explain.

# 11-7   Damped Oscillations

In all real oscillatory motion, mechanical energy is dissipated because of some kind of frictional force. Left to itself, an object on a spring or a pendulum eventually stops oscillating. When the mechanical energy of oscillatory motion decreases with time, the motion is said to be *damped*. If the frictional or damping forces are small, the motion is nearly periodic except that the amplitude decreases slowly with time, as in Figure

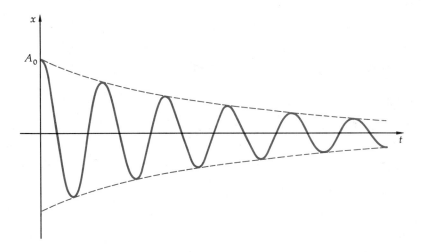

**Figure 11-18**
Displacement versus time for
a slightly damped oscillator.
The amplitude decreases
slowly with time.

11-18. In Figure 11-19 the oscillation of the body is damped because of
the motion of the plunger in the liquid. The rate of energy loss can be
varied by changing the size of the plunger or the viscosity of the liquid.
Although the detailed analysis of such a damping force is usually complicated, we can often represent such a force by an empirical expression
for which the mathematics is comparatively simple and which is in reasonable accord with experiment. The simplest and most common representation of such a force is one proportional to the velocity of the object
but in the opposite direction:

$$\mathbf{F}_d = -b\mathbf{v} \qquad\qquad 11\text{-}37$$

where $b$ is a constant which describes the degree of damping. Many
rather complicated phenomena can be approximately described with
such a force. This force is clearly nonconservative because it depends on
the velocity. Since it is always directed opposite the direction of motion, the work done by the force is always negative; thus it always decreases the mechanical energy of the system. The rate of change of the
total mechanical energy equals the power input of the damping force,

$$P = \frac{dE}{dt} = \mathbf{F}_d \cdot \mathbf{v} = -bv^2 \qquad\qquad 11\text{-}38$$

Newton's second law $\Sigma \mathbf{F} = m\mathbf{a}$ applied to the motion of the object of
mass $m$ on a spring of force constant $k$ and with damping force $-bv$ is

$$F_x = -kx - bv = m\frac{dv}{dt} \qquad\qquad 11\text{-}39$$

We can use Equation 11-38 to get a qualitative understanding of the
behavior of a damped oscillator without solving Equation 11-39 in detail. If the damping is small, we expect the body to oscillate with a frequency $\omega'$ which is approximately equal to the undamped frequency $\omega_0$:

$$\omega' \approx \omega_0 = \sqrt{\frac{k}{m}} \qquad\qquad 11\text{-}40$$

and we expect the amplitude to decrease slowly. The total mechanical
energy in simple harmonic motion oscillates between potential and
kinetic energy. The average values of the potential and kinetic energies
are equal during each cycle, and the total energy equals twice the
average value of either. We can therefore write

$$E = 2(\tfrac{1}{2}mv^2)_{\mathrm{av}} = m(v^2)_{\mathrm{av}}$$

**Figure 11-19**
Physical arrangement of a
damped oscillator. The motion is damped by the
plunger immersed in the
liquid.

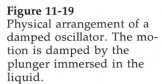
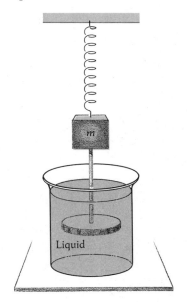

If we replace $v^2$ by its average value $(v^2)_{av} = E/m$, Equation 11-38 becomes

$$\frac{dE}{dt} = -\frac{b}{m}E \qquad\qquad 11\text{-}41$$

The function whose derivative is proportional to the function itself is the exponential. The solution of Equation 11-41 is

$$E(t) = E_0 e^{-(b/m)t} = E_0 e^{-t/t_c} \qquad\qquad 11\text{-}42$$

where $E_0$ is the energy at time $t = 0$ and the time constant

$$t_c = \frac{m}{b} \qquad\qquad 11\text{-}43$$

is the time for the energy to decrease by a factor of $1/e$. If the damping is small, the time constant will be much greater than the period of oscillation; i.e., the oscillator will lose only a very small fraction of its energy during one oscillation. In this case we can obtain the energy loss per period from Equation 11-41 by replacing the differentials $dE$ and $dt$ by differences $\Delta E$ and $\Delta t$ and setting $\Delta t = T$ for the period. We have then for the fractional energy loss per period

$$-\frac{\Delta E}{E} = \frac{b}{m}T = \frac{T}{t_c} \qquad\qquad 11\text{-}44$$

The damping of an oscillator is usually described by a dimensionless quantity $Q$ called the *quality factor* or *Q factor*, defined as the product of the angular frequency $\omega$ and the time constant $t_c$:

$$Q = \omega t_c = \frac{2\pi t_c}{T} = \frac{\omega m}{b} \qquad\qquad 11\text{-}45 \qquad \text{Q factor}$$

In terms of the $Q$ factor the fractional energy loss per period is

$$-\frac{\Delta E}{E} = \frac{2\pi}{Q} \qquad\qquad 11\text{-}46$$

When we study resonance in the next section, we shall see that the $Q$ factor is a measure of the sharpness of the resonance.

---

**Example 11-7** A bell of frequency 200 Hz can be heard for about a minute or more after it is struck. Estimate the $Q$ factor for the bell.

The period of the bell is $T = 1/f = \frac{1}{200}$ s = 5 ms. From the information given we cannot determine the time constant $t_c$ with any precision; in fact, when we study sound in Chapter 14, we shall see that the relationship between hearing and sound energy is complicated. If the sound can be heard for several minutes, however, we may guess that $t_c$ is of the order of 1 s or more. If we take $t_c \approx 2$ s, we obtain $Q = 2\pi t_c/T = 2\pi[(2\text{ s})/(5\text{ ms})] \approx 2500$.

---

We can use Equation 11-42 to find the time dependence of the amplitude, since the energy is proportional to the square of the amplitude. If $A(t)$ is the amplitude at time $t$ and $A_0$ is that at $t = 0$, we have

$$\frac{E(t)}{E_0} = \frac{A^2(t)}{A_0^2}$$

Thus from Equation 11-42

$$\frac{A^2(t)}{A_0^2} = e^{-(b/m)t} \qquad\qquad 11\text{-}47$$

or

$$A(t) = A_0 e^{-(b/2m)t} = A_0 e^{-t/2t_c} \qquad\qquad 11\text{-}48$$

The exact solution of Equation 11-39 can be found by standard methods for solving differential equations. The solution for the case of small damping, $b < 2m\omega_0$, is

$$x = A_0 e^{-(b/2m)t} \cos(\omega't + \delta) \qquad\qquad 11\text{-}49$$

where $A_0$ is the original amplitude and the frequency $\omega'$ is the angular frequency with which the oscillator passes through its equilibrium position ($x = 0$); $\omega'$ is given by

$$\omega' = \omega_0 \sqrt{1 - \left(\frac{b}{2m\omega_0}\right)^2} = \omega_0 \sqrt{1 - \frac{1}{4Q^2}} \qquad\qquad 11\text{-}50$$

in which $\omega_0 = \sqrt{k/m}$ is the angular frequency without damping. We see that our qualitative observations were correct. For small damping, the frequency is very nearly equal to the undamped frequency, and the time dependence of the amplitude is given by Equation 11-48, as expected. Equation 11-50 is given only to illustrate that for all practical purposes, the frequency of a slightly damped oscillator is the same as that of the undamped oscillator. For example, for a $Q$ factor of 10, $\omega_0$ and $\omega'$ differ by only 0.12 percent.

Figure 11-18 shows a plot of $x$ given by Equation 11-49 versus $t$. This solution is valid if $b$ is less than $2m\omega_0$. The value

$$b_c = 2m\omega_0 \qquad\qquad 11\text{-}51$$

*Critical damping condition*

is called the *condition for critical damping.* If $b$ is greater than or equal to this critical value, the system does not oscillate at all but merely returns to its equilibrium position. The greater the damping, the longer it takes for the system to return to equilibrium. When $b = b_c$, the system is said to be *critically damped.* The mass returns to its equilibrium position in the shortest possible time with no oscillation. When $b$ is greater than $b_c$, the system is *overdamped.* Figure 11-20 shows the displacement versus time for a critically damped and overdamped oscillator.

In many practical applications critical damping or nearly critical damping is used to avoid oscillations and yet have the system return to equilibrium quickly. One example is the use of shock absorbers to damp the oscillations of an automobile on its wheels. Such a system is usually slightly underdamped ($b$ slightly less than the critical value); you can see this by pushing down on the front or back of the car and observing that one or two oscillations occur before the system comes to rest.

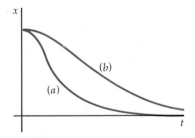

**Figure 11-20**
Displacement versus time for
(a) critically damped and (b)
overdamped oscillator.

## 11-8   The Driven Oscillator; Resonance

In the previous section we saw that if a system is set into oscillation, the oscillations gradually die down as a result of energy dissipation of some kind. Both the amplitude and the energy of the oscillations decrease exponentially with time. To keep a system oscillating, energy must be fed into the system in some way. When this is done, the oscillator is said to be forced or driven. One way to drive a system of an object on a vertical spring is to move the point of support up and down, as shown in Figure 11-21. Similarly, a simple pendulum can be driven by moving the point of support back and forth. You should try some simple demonstrations with one of these systems to familiarize yourself with the phenomena to be discussed. If the point of support of an object on a spring or a simple pendulum is moved back and forth with simple harmonic motion of small amplitude and with frequency $\omega$, the system will begin oscillating. At first the motion is complicated, but eventually a steady state is reached in which the system oscillates with the same frequency $\omega$ as that of the driver. The amplitude and therefore the energy of the system in the steady state depend not only on the amplitude of the driver but on its frequency. Even when the amplitude of the driver is very small, the system will oscillate with a large amplitude if the driving frequency is equal (or approximately equal) to the natural frequency of the system. This phenomenon is called *resonance*. When the driving frequency equals the natural frequency of the oscillator, the energy absorbed by the oscillator is maximum. Figure 11-22 shows a plot of the average power delivered to an oscillator as a function of driving frequency for two different values of damping characterized by the $Q$ factor. These curves are called *resonance curves*. The maximum power input occurs at the resonance frequency $\omega = \omega_0$. When the damping is small (high $Q$), the power input at resonance is large and the resonance is sharp; i.e., the resonance curve is narrow, indicating that the power input is large only near the resonance frequency. When the damping is large (low $Q$), the resonance curve is broad. We shall show below that the ratio of the resonance frequency $\omega_0$ to the full width at half maximum $\Delta\omega$ equals the $Q$ factor:

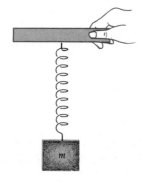

**Figure 11-21**
A body on a vertical spring can be driven by moving the point of support up and down.

$$Q = \frac{\omega_0}{\Delta\omega} \qquad\qquad 11\text{-}52$$

*Q factor measures sharpness of resonance*

The $Q$ factor is thus a direct measure of the sharpness of the resonance.

We treat the driven oscillator mathematically by assuming that in addition to a restoring force and a damping force, an oscillator is subject to an external force which varies harmonically with time:

$$F_{\text{ext}} = F_0 \sin \omega t$$

where $\omega$ is the angular frequency of the force, which is generally not related to the natural* angular frequency of the system $\omega_0$. An object of mass $m$ attached to a spring of force constant $k = m\omega_0^2$ subject to a damping force $-bv$ and to an external force $F_0 \sin \omega t$ then obeys the equation of motion given by

$$\Sigma F = -m\omega_0^2 x - bv + F_0 \sin \omega t = m\frac{dv}{dt}$$

or

$$m\frac{d^2x}{dt^2} + b\frac{dx}{dt} + m\omega_0^2 x = F_0 \sin \omega t \qquad\qquad 11\text{-}53$$

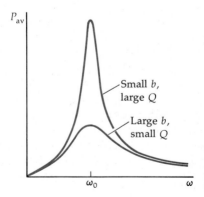

**Figure 11-22**
Average power of a driven oscillator versus angular frequency of the driving force for different values of damping. Resonance occurs when the frequency of the driving force equals the natural frequency of the system. The resonance is sharp if the damping is small.

---

* We need not distinguish between the natural frequency of a damped oscillator $\omega'$ and that of the undamped oscillator $\omega_0$ since they are very nearly equal for all but the most heavily damped oscillators.

We shall not attempt to solve Equation 11-53; instead we discuss its general solution qualitatively. The solution of Equation 11-53 consists of two parts, the transient solution $x_t$ and the steady-state solution $x_{ss}$ (Figure 11-23):

$$x(t) = x_t(t) + x_{ss}(t) \qquad \text{11-54}$$

*Transient and steady-state solutions*

The transient part of the solution is identical to that of the unforced damped oscillator given by Equation 11-49. The constants $A_0$ and $\delta'$ in this solution depend on the initial conditions. After a long time, so that $bt/2m \gg 1$, this part of the solution becomes negligible because the amplitude has decreased exponentially with time. We are then left with the steady-state solution $x_{ss}(t)$, which can be written

$$x_{ss}(t) = A \sin (\omega t - \delta) \qquad \text{11-55}$$

where the angular frequency $\omega$ is the same as that of the driving force.

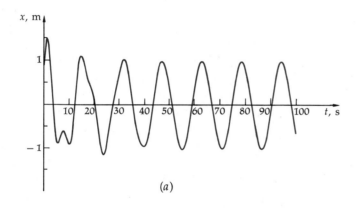

(a)

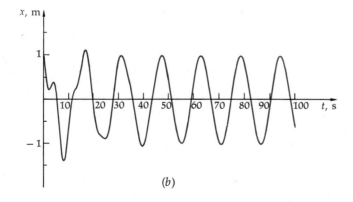

(b)

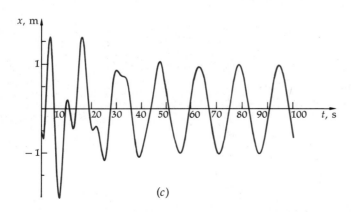

(c)

**Figure 11-23**
Three solutions of Equation 11-53 for the same driven oscillator under different initial conditions. For each solution, $\omega_0 = 1$ rad/s, $\omega = 0.4$ rad/s, and $Q = 8$. Note that the transient solutions (short times) are very different, but the steady-state solutions (long times) are all the same.

The amplitude $A$ is given by

$$A = \frac{F_0}{\sqrt{m^2(\omega_0^2 - \omega^2)^2 + b^2\omega^2}} \qquad \text{11-56}$$

and the phase constant $\delta$ by

$$\tan \delta = \frac{b\omega}{m(\omega_0^2 - \omega^2)} \qquad \text{11-57}$$

The phase angle $\delta$ is shown in Figure 11-24, from which we see that

$$\sin \delta = \frac{b\omega}{\sqrt{m^2(\omega_0^2 - \omega^2)^2 + b^2\omega^2}} = \frac{b\omega A}{F_0} \qquad \text{11-58}$$

The steady-state solution does not depend on the initial conditions. When the driving frequency equals the natural frequency, $\omega = \omega_0$, the amplitude $A$ is very large.

The velocity of the particle in the steady state is obtained from the position function (Equation 11-55) by differentiation:

$$v = \frac{dx}{dt} = A\omega \cos (\omega t - \delta) \qquad \text{11-59}$$

The power input of the driving force in the steady state is

$$P = Fv = (F_0 \sin \omega t)A\omega \cos (\omega t - \delta)$$
$$= A\omega F_0 \sin \omega t \cos (\omega t - \delta)$$

Using the identity $\cos (\omega t - \delta) = \cos \omega t \cos \delta + \sin \omega t \sin \delta$, we can write this as

$$P = \omega A F_0 \sin \delta \sin^2 \omega t + \omega A F_0 \cos \delta \sin \omega t \cos \omega t \qquad \text{11-60}$$

The power input varies with time over a cycle. During one cycle, the second term in Equation 11-60 is negative as often as it is positive, and has an average value of zero. The average value of $\sin^2 \omega t$ over one cycle is $\frac{1}{2}$ (see Figure 11-25). The average power is therefore

$$P_{av} = \tfrac{1}{2}\omega A F_0 \sin \delta$$

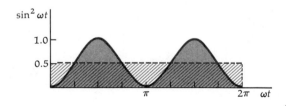

Using $\sin \delta = b\omega A/F_0$ from Equation 11-58, we can eliminate either $\sin \delta$ or $A$. We have

$$P_{av} = \tfrac{1}{2}b\omega^2 A^2 = \frac{1}{2}\frac{F_0^2}{b} \sin^2 \delta \qquad \text{11-61}$$

Alternatively, we can write either $A$ or $\sin \delta$ in terms of the driving frequency $\omega$:

$$P_{av} = \frac{1}{2}\frac{b\omega^2 F_0^2}{m^2(\omega_0^2 - \omega^2)^2 + b^2\omega^2} \qquad \text{11-62}$$

This is the function shown in Figure 11-22 for two different values of $Q$. The maximum power input occurs at the resonance frequency $\omega = \omega_0$. We can see this from Equations 11-61 and 11-57 since $\sin \delta$ has its max-

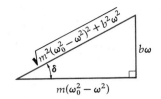

**Figure 11-24**
Phase constant $\delta$ defined by Equation 11-57.

**Figure 11-25**
Plot of $\sin^2 \omega t$ versus $\omega t$. The average value is $\frac{1}{2}$.

imum value of 1 at $\delta = 90°$. But according to Equation 11-57, when $\delta = 90°$, $\omega = \omega_0$ because $\tan 90°$ is infinite. The value of the maximum average power input is

$$P_{av} = \frac{1}{2}\frac{F_0^2}{b} \sin^2 90° = \frac{1}{2}\frac{F_0^2}{b} = \frac{1}{2}Q\frac{F_0^2}{m\omega_0} \qquad \text{11-63}$$

At resonance the displacement is 90° out of phase with the driving force, but the velocity is in phase with the driving force. Thus at resonance the particle is always moving in the direction of the driving force, as would be expected for maximum power input.

For any driving frequency the total energy is the sum of the potential and kinetic energies:

$$E = \tfrac{1}{2}kx^2 + \tfrac{1}{2}mv^2$$

Using $k = m\omega_0^2$ and Equations 11-55 and 11-59 for $x$ and $v$, we have for the total energy in the steady state

$$E = \tfrac{1}{2}m\omega_0^2 A^2 \sin^2 (\omega t - \delta) + \tfrac{1}{2}m\omega^2 A^2 \cos^2 (\omega t - \delta) \qquad \text{11-64}$$

Except at resonance, the total energy is not constant in time but varies during each cycle. The average total energy is large near resonance when the amplitude $A$ is large. At the resonance frequency $\omega = \omega_0$ the total energy is constant in the steady state and has the value

$$E = \tfrac{1}{2}m\omega_0^2 A^2 = \tfrac{1}{2}m\omega_0^2 \frac{F_0^2}{\omega_0^2 b^2} = \frac{1}{2}\frac{mF_0^2}{b^2} = \frac{Q^2}{2}\frac{F_0^2}{m\omega_0^2} \qquad \text{11-65}$$

At resonance the total energy is proportional to $Q^2$.

We can determine the width of the resonance from Equation 11-62. At resonance, the denominator of that equation has the value $b^2\omega^2 = b^2\omega_0^2$. For a sharp resonance, we can ignore the variation in $\omega$ in the numerator and in the term $b^2\omega^2$ in the denominator. Then the average power input will be half its resonance value when the denominator is twice its value at resonance, which occurs when

$$m^2(\omega^2 - \omega_0^2)^2 = b^2\omega_0^2$$

If we factor the left side of this equation and use the approximation $\omega + \omega_0 = 2\omega_0$, we have

$$m^2(\omega - \omega_0)^2 (\omega + \omega_0)^2 \approx m^2(\omega - \omega_0)^2 (2\omega_0)^2 = b^2\omega_0^2$$

or

$$\omega - \omega_0 = \pm \frac{b}{2m} = \pm \frac{\omega_0}{2Q}$$

The power input is half its maximum value at the values $\omega_1$ and $\omega_2$ given by

$$\omega_1 = \omega_0 - \frac{\omega_0}{2Q} \qquad \text{and} \qquad \omega_2 = \omega_0 + \frac{\omega_0}{2Q}$$

The full width at half maximum is therefore given by

$$\omega_2 - \omega_1 = \frac{\omega_0}{Q}$$

That is,

$$\frac{\omega_0}{\Delta\omega} = Q$$

as stated in Equation 11-52.

**Questions**

15. Give several examples of common situations that can be described as oscillators driven by sinusoidal driving forces.

16. When you push a swing, resonance occurs when the driving frequency equals the natural frequency and also when it is $\frac{1}{2}, \frac{1}{3}, \ldots$ times the natural frequency; i.e., when you push every other swing, every third swing, etc. Explain how this differs from a harmonic driving force, for which resonance occurs only at the natural frequency.

**Review**

A. Define, explain, or otherwise identify:

Periodic motion, 315
Simple harmonic motion, 315
Amplitude, 316
Phase, 316
Phase constant, 316
Period, 316
Angular frequency, 316

Damped oscillations, 332
Q factor, 334
Critical damping, 335
Resonance, 336
Transient solution, 337
Steady-state solution, 337

B. True or false:

1. All periodic motion is simple harmonic motion.

2. All simple harmonic motion is periodic motion.

3. In simple harmonic motion the period is proportional to the square of the amplitude.

4. In simple harmonic motion the total energy is proportional to the square of the amplitude.

5. In simple harmonic motion the phase constant depends on the initial conditions.

6. The motion of a simple pendulum is simple harmonic for any initial angular displacement.

7. The motion of a simple pendulum is periodic for any initial angular displacement.

8. In simple harmonic motion the acceleration is proportional to the displacement and in the opposite direction.

9. Any motion in which the acceleration is proportional to the displacement and in the opposite direction is simple harmonic motion.

10. Motion near a position of stable equilibrium is generally simple harmonic if the displacements are small.

11. The frequency of a slightly damped (unforced) oscillator is nearly equal to the undamped frequency.

12. The frequency of a damped (unforced) oscillator decreases as the amplitude decreases.

13. The amplitude of a damped (unforced) oscillator decreases exponentially with time.

14. In the steady state, the frequency of a driven oscillator equals the frequency of the driving term.

15. In the steady state the motion of a driven oscillator is independent of the initial conditions.

16. At resonance, the velocity of a driven oscillator is in phase with the driving term.

17. The resonance frequency of a driven oscillator depends on the damping.

18. The sharpness of the resonance of a driven oscillator depends on the damping.

19. In a driven oscillator, the power input of the driving term is maximum when $\omega = \omega_0$.

## Exercises
### Section 11-1, Simple Harmonic Motion

1. A particle has displacement $x$ given by $x = 3 \cos (5\pi t + \pi)$, where $x$ is in metres and $t$ in seconds. ($a$) What are the frequency $f$ and period $T$ of the motion? ($b$) What is the greatest distance the particle travels from equilibrium? ($c$) Where is the particle at time $t = 0$? At time $t = \frac{1}{2}$ s?

2. ($a$) Find an expression for the velocity of the particle whose position is given in Exercise 1. ($b$) What is the maximum velocity? ($c$) When does this maximum velocity occur?

3. A particle in simple harmonic motion is at rest a distance of 6 cm from its equilibrium position at time $t = 0$. Its period is 2 s. Write expressions for its position $x$, its velocity $v$, and its acceleration $a$ as functions of time.

4. The position of a particle traveling in one dimension is given by $x = 5 \cos 4\pi t$, where $x$ is in metres and $t$ in seconds. ($a$) Find the amplitude, angular frequency, frequency, and period of the motion. ($b$) Find an expression for the velocity of the particle at any time $t$. ($c$) What is the velocity at time $t = 0$?

5. The position of a particle is given by $x = 4 \sin 2t$, where $x$ is in metres and $t$ is in seconds. ($a$) What is the maximum value of $x$? What is the first time after $t = 0$ at which this maximum value occurs? ($b$) Find an expression for the velocity of the particle as a function of time. What is the velocity at time $t = 0$? ($c$) Find an expression for the acceleration of the particle as a function of time. What is the acceleration at time $t = 0$? What is the maximum value of the acceleration?

6. A particle oscillates with simple harmonic motion with period $T = 2$ s. It is initially at its equilibrium position and moving with speed 4 m/s in the direction of increasing $x$. Write expressions for its position $x$, its velocity $v$, and its acceleration $a$ as functions of time.

### Section 11-2, Circular Motion and Simple Harmonic Motion, and Section 11-3, An Object on a Spring

7. A particle travels in a circle in the $xy$ plane with center at the origin. The radius of the circle is 40 cm, and the speed of the particle is 80 cm/s. ($a$) What is the angular velocity of the particle? ($b$) What are the frequency $f$ and period $T$ of the circular motion? ($c$) Write the $x$ and $y$ components of the position vector $\mathbf{r}$ as functions of time.

8. A 2-kg object is attached to a spring of force constant $k = 5$ kN/m. The spring is stretched 10 cm from equilibrium and released. ($a$) Find the angular frequency, the frequency $f$, and the amplitude of oscillation. ($b$) Find $x(t)$, $v(t)$, and $a(t)$. ($c$) Write expressions for the kinetic and potential energies as functions of time. What is the total energy?

9. A 3-kg object is attached to a spring and oscillates with amplitude of 10 cm and frequency $f = 2$ Hz. ($a$) What is the force constant of the spring? ($b$) What is the total energy of the motion? ($c$) Write an equation $x(t)$ describing the position of the object relative to its equilibrium position. Can the phase constant be determined from the information given?

10. A particle is attached to a spring. At $t = 0$ it is in its equilibrium position and has speed 5 cm/s toward the negative side of the $x$ axis. Its frequency is $f = 3$ Hz. (a) At what value of $t$ does it next come to rest? (b) Where is it at that time? (c) What is its acceleration at that time? (d) Write an expression for its position $x(t)$ at any time $t$.

11. A 100-g object executes simple harmonic motion with a frequency of 20 Hz and amplitude 0.5 cm. (a) What is the constant $k$ for the force acting on it? (b) What is the maximum acceleration? (c) What is the total energy of the motion?

12. An object of mass 500 g executes simple harmonic motion with a period of 0.5 s. Its total energy is 5 J. (a) What is the amplitude of the oscillation? (b) What is the maximum speed? (c) What is the maximum acceleration?

13. When the displacement of a body oscillating on a spring is half its amplitude, what fraction of its total energy is its kinetic energy? At what displacement are its kinetic and potential energies equal?

14. A 100-g particle is attached to a spring and oscillates with a frequency of 10 Hz and total energy of 1.25 J. (a) What is the force constant of the spring? (b) What is the amplitude of the oscillation?

15. A 2-kg object is suspended from a vertically hanging spring of force constant $k = 350$ N/m. (a) Find the distance $y_0$ the spring is stretched when the object hangs in equilibrium, and the potential energy of the spring relative to the unstretched length. (b) The object is then pulled downward a distance $y' = 3$ cm below its equilibrium point. Find the change in the potential energy of the spring, the change in the gravitational potential energy, and the total change in potential energy. Show that the total change in potential energy equals $\frac{1}{2}ky'^2$. (c) The object is then released. Find the period, frequency, and amplitude of the subsequent oscillation.

16. A 3-kg object oscillates on a vertical spring of force constant $k = 500$ N/m. Its maximum kinetic energy is 5 J. (a) If the total potential energy is chosen to be zero at equilibrium, what is the maximum potential energy of oscillation? (b) What are the amplitude and period of oscillation? (c) What is the maximum stretching of the spring from its natural length?

## Section 11-4, The Simple Pendulum

17. Find the length of a simple pendulum if the period of the pendulum is 1 s at a point where $g = 9.81$ m/s$^2$.

18. What would be the period of the pendulum in Exercise 17 on the moon, where the acceleration of gravity is one-sixth that on earth?

19. If the period of a pendulum 70 cm long is 1.68 s, what is the value of $g$ at the location of the pendulum?

20. A pendulum set up in the stairwell of a 10-story building consists of a heavy weight suspended on a 34.0-m wire. What is its period of oscillation ($g = 9.81$ m/s$^2$)?

21. Derive Equation 11-23a by setting the torque about the support point equal to $I\alpha$.

## Section 11-5, The Physical Pendulum

22. A thin disk of mass 5 kg and radius 20 cm is suspended by an axis perpendicular to the disk through its rim. It is displaced slightly from equilibrium and released. Find the period of the subsequent simple harmonic motion.

23. A circular hoop of radius 50 cm is hung on a narrow horizontal rod and allowed to swing in the plane of the hoop. What is the period of its oscillation, assuming that the amplitude is small?

24. A 3-kg plane figure is suspended at a point 10 cm from its center of mass. When it is set oscillating with small amplitude, the period of oscillation is 2.6 s. Find the moment of inertia $I$ about an axis perpendicular to the plane of the figure and through the pivot point.

### Section 11-6, General Motion near Equilibrium

*There are no exercises for this section.*

### Section 11-7, Damped Oscillations

25. A 2-kg object oscillates on a spring of force constant $k = 400$ N/m with initial amplitude of 3 cm. (a) Find the period and the total initial energy. (b) If the energy decreases by 1 percent per period, find the damping constant $b$ and the $Q$ factor.

26. Show that the ratio of amplitudes for two successive oscillations is constant in a damped oscillator.

27. An oscillator has a period of 3 s. Its amplitude decreases by 5 percent each cycle. (a) By how much does its energy decrease in each cycle? (b) What is the time constant $t_c$? (c) What is the $Q$ factor?

28. Use Equation 11-42 to show that the time $t_{1/2}$ for the energy to decrease to half its initial value is related to the time constant by $t_{1/2} = t_c \ln 2$.

29. A damped oscillator loses 2 percent of its energy each cycle. (a) How many cycles elapse before half its original energy is dissipated? (b) What is the $Q$ factor?

30. An oscillator has a $Q$ factor of 20. (a) By what fraction does the energy decrease in each cycle? (b) Use Equation 11-50 to find the percentage difference between $\omega'$ and $\omega_0$. *Hint:* Use the approximation $(1 + x)^{1/2} \approx 1 + \frac{1}{2}x$ for small $x$.

31. For a child on a swing, the amplitude drops by a factor of $1/e$ in about eight periods when no energy is fed in. Estimate the $Q$ of this system.

### Section 11-8, The Driven Oscillator; Resonance

32. Find the resonance frequencies for each of the three systems shown in Figure 11-26.

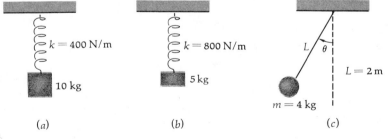

Figure 11-26
Exercise 32.

(a) $k = 400$ N/m, 10 kg
(b) $k = 800$ N/m, 5 kg
(c) $L = 2$ m, $m = 4$ kg

33. A 2-kg object oscillates on a spring of force constant $k = 400$ N/m. The damping constant is $b = 2.00$ kg/s. It is driven by a sinusoidal force of maximum value 10 N and angular frequency 10 rad/s. (a) What is the amplitude of the oscillations? (b) If the driving frequency is varied, at what frequency will resonance occur? (c) Find the amplitude of vibrations at resonance. (d) What is the width $\Delta\omega$ of the resonance?

34. In steady-state oscillations, the average power delivered by the driving force must equal the average power dissipated in the damping material. Show explicitly that this is true at all frequencies by computing $(-bv^2)_{av}$ and $(F_0v)_{av}$ and showing that they are equal.

35. A child on a swing oscillates with a period of 3 s. The child and swing have a mass of 15 kg. The swing is pushed once each cycle by a patient father so that the amplitude maintains a steady 30°. If $Q$ is 20, find the average power delivered by the father. *Note:* Pushing a swing is usually not done sinusoidally, but to maintain a steady amplitude the energy lost per cycle due to damping must be replaced by an external energy source. For a child on a swing this energy may come from the child's "pumping" or from someone else's pushing the swing.

36. Show that at resonance ($\omega = \omega_0$, $\delta = \pi/2$) the driving force is equal in magnitude to the damping force at all times and 180° out of phase.

## Problems

1. The period of an oscillating particle is 8 s. At $t = 0$ it is at its equilibrium position. (a) How does the distance it travels in the first 4 s compare with the distance it travels in the second 4 s? (b) The distance traveled in the first 2 s and the second 2 s? (c) The first second and the second second?

2. A block of wood slides on a frictionless horizontal surface. It is attached to a spring and oscillates with a period of 0.3 s. A second block rests on top of it. The coefficient of static friction between the blocks is $\mu_s = 0.25$. (a) If the amplitude of oscillation is 1 cm, will the block on top slip? (b) What is the greatest amplitude of oscillation for which the blocks will not slip?

3. An object of mass $m$ is supported by a vertical spring with force constant 1800 N/m. When pulled down 2.5 cm and released, it oscillates at 5.5 Hz. (a) Find $m$. (b) Find the equilibrium stretching of the spring. (c) Find expressions for $x(t)$, $v(t)$, and $a(t)$.

4. (a) A small particle of mass $m$ slides without friction in a spherical bowl of radius $r$. Show that the motion of the mass is the same as if it were attached to a string of length $r$. (b) A mass $m_1$ is displaced a small distance $s_1$ from the bottom of the bowl, where $s_1$ is much smaller than $r$. A second mass $m_2$ is displaced in the opposite direction a distance $s_2 = 3s_1$ ($s_2$ is also much smaller than $r$) (Figure 11-27). If the masses are released at the same time, where do they meet? Explain.

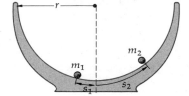

**Figure 11-27**
Problem 4.

5. It has been stated that the vibrating earth has a resonance period of 54 min and a $Q$ of about 400, and that after a large earthquake, the earth "rings" (continues to vibrate) for 2 months. By what factor has the amplitude of vibrations decreased in that time?

6. A damped oscillator loses one-twentieth of its energy during each cycle. (a) How many cycles elapse before half the original energy is dissipated? (b) During this time by how much is the amplitude reduced? (c) What is the $Q$ factor? (d) If the natural frequency is $f = 100$ Hz, what is the width of the resonance when the oscillator is driven?

7. A mass $m_1$ sliding on a frictionless horizontal surface is attached to a spring of force constant $k$. It oscillates with amplitude $A$. When the spring is at its greatest extension and the mass is instantaneously at rest, a second mass $m_2$ is placed on top of $m_1$. (a) What is the smallest value the coefficient of static friction $\mu_s$ can have without permitting $m_2$ to slip on $m_1$? (b) Explain how the total energy $E$, the amplitude $A$, the angular frequency $\omega$, and the period $T$ are changed by placing $m_2$ on $m_1$ in this way, assuming no slipping. Check your results with the expression for the total energy, $E = \frac{1}{2}m\omega^2A^2$.

8. A block is suspended from a spring and set into a vertical oscillation with a frequency of 4 Hz and an amplitude of 7 cm. A very small bit of rock is placed on top of the oscillating block just as it reaches its lowest point. (a) At what distance above the equilibrium position does the rock lose contact with the block? (b) What is the velocity of the rock when it leaves the block? (c) What is the greatest distance above the equilibrium position reached by the rock? (Assume that the rock has no effect on the oscillation.)

9. A spring is hanging vertically. An object of unknown mass is hung on the end of the unstretched spring and released from rest. If it falls 3.42 cm before first coming to rest, find the period of motion.

10. An object of mass 2.0 kg is placed on top of a vertical spring which is attached to the floor. The unstretched length of the spring is 8.0 cm and the object comes to rest 5.0 cm from the floor. When it is at equilibrium, the object is given a downward impulse with a hammer so that its initial speed is 0.3 m/s. (a) What is the maximum height (above the floor) the object will rise? (b) How long does it take for the object to first reach its maximum height? (c) Does the spring ever become unstretched? What minimum initial velocity must be given to the object for the spring to be unstretched at some time?

11. A spring with force constant $k = 150$ N/m is suspended from a rigid support. A 1-kg mass is attached at the bottom end and released from rest when the spring is unstretched. (a) How far down does the mass move before it starts up again? (b) How far below the starting point is the equilibrium position for the 1-kg mass? (c) What is the period of the oscillation? (d) Relative to this equilibrium position, what is the total energy of the mass-spring system? (e) What is the speed of the mass when it first reaches the equilibrium position? How long after starting does the mass have this speed? (f) If instead of being attached to the spring the mass had been dropped, when would it have reached the (previous) equilibrium level and what would its speed be?

12. If the earth is assumed to have constant mass density, it can be shown that the gravitational force on a particle of mass $m$ at a distance $r$ from the center ($r \leq R_E$) is directed toward the center of the earth and has magnitude $GM_E mr/R_E^3$, where $M_E$ is the mass of the earth and $R_E$ is its radius. Let the particle slide along a frictionless tunnel dug through the earth, as shown in Figure 11-28. (a) Show that when the particle is at a distance $x$ from the midpoint of the tunnel, its acceleration is $a_x = -(GM_E/R_E^3)x$ and therefore the motion is simple harmonic. (b) Show that the period of motion is given by $T = 2\pi\sqrt{R_E/g}$, and find its value in minutes. (This is the same period as that of a satellite orbiting near the surface of the earth, and is independent of the length of the tunnel.)

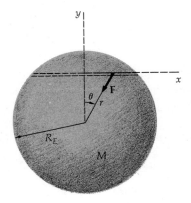

13. A 0.5-kg mass oscillates on a spring of force constant $k = 300$ N/m. During the first 10 s it loses 0.5 J to friction. If the initial amplitude was 15 cm, (a) how long before the energy is reduced to 0.1 J? (b) What is the frequency of oscillation?

**Figure 11-28**
Particle sliding along a frictionless tunnel through earth for Problem 12.

14. Show that for a weakly damped oscillator, the maximum value of the damping force in any cycle is approximately $1/Q$ times the maximum value of the restoring force during that cycle.

15. A simple pendulum 50 cm long is suspended from the roof of a cart accelerating in the horizontal direction with $a = 7$ m/s² (Figure 11-29). Find the period of small oscillations of the pendulum about its equilibrium angle.

**Figure 11-29**
Problem 15.

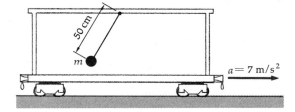

16. A pendulum clock that has run down to a very small amplitude gains 5 min/d. What is the proper (angular) amplitude for the pendulum to keep correct time?

17. A simple pendulum of length $L$ is released from rest from an angle of $\theta_0$. (a) Assuming the motion to be simple harmonic, find the velocity of the pendulum as it passes through $\theta = 0$. (b) Using conservation of energy, find this velocity exactly. (c) Show that your results for parts (a) and (b) are the same when $\theta_0$ is small. (d) Find the difference in your results for $\theta_0 = 10°$ and $L = 1$ m.

18. A 1-kg mass attached to a certain spring is critically damped by an external viscous force. Describe the resultant motion if the same viscous force damps a 2-kg mass attached to the same spring.

19. A 100-g object oscillates on a spring with a period of 5 s. Its first oscillation has an amplitude of 10 cm, and its sixth oscillation has an amplitude of 9 cm. Find (a) the time constant $t_c$, (b) the $Q$ factor, (c) the fraction of the oscillator's energy dissipated each cycle, (d) the amplitude and energy of the oscillator in its hundredth oscillation.

20. Show that for the situations in Figure 11-30a and b the mass oscillates with angular frequency $\omega = \sqrt{k_{eff}/m}$, where $k_{eff}$ is given by (a) $k_{eff} = k_1 + k_2$, and (b) $1/k_{eff} = 1/k_1 + 1/k_2$.

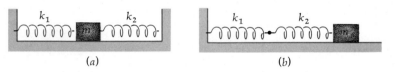

(a)               (b)

**Figure 11-30**
Problem 20.

21. A spider (mass 0.36 g) sits in the middle of its horizontal web, which sags 3.0 mm under its weight. Estimate the vertical vibration frequency of this system.

22. Show that the value of the driving frequency for which the amplitude $A$ is maximum is $\omega = \sqrt{\omega_0^2 - b^2/2m^2}$.

23. A 3-kg sphere dropped in air has a terminal velocity of 25 m/s (assume the drag force is $-bv$). It is attached to a spring of force constant $k = 400$ N/m and oscillates in air with an initial amplitude of 20 cm. (a) What is $Q$? (b) When will the amplitude be 10 cm? (c) How much energy will have been lost when the amplitude is 10 cm?

24. An object has moment of inertia $I$ about its center of mass. When pivoted at point $P_1$ (Figure 11-31), it oscillates about the pivot with period $T$. There is a second point $P_2$ on the opposite side of the center of mass about which the object can be pivoted and for which the period of oscillation is also $T$. Show that $h_1 + h_2 = gT^2/4\pi^2$.

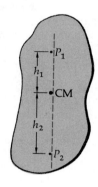

**Figure 11-31**
Problem 24.

25. A 2-kg mass is attached to a spring of force constant 600 N/m and rests on a smooth horizontal surface. A second mass of 1 kg slides along the surface toward the first at 6 m/s. (a) Find the amplitude of oscillation if the masses make a perfectly inelastic collision and remain together on the spring. What is the period of oscillation? (b) Find the amplitude and period of oscillation if the collision is perfectly elastic. (c) For each case, write down the position $x$ as a function of time $t$ for the mass attached to the spring, assuming that the collision occurs at time $t = 0$. What is the impulse given to the 2-kg mass in each case?

26. A physical pendulum consists of a spherical bob of radius $r$ and mass $m$ suspended from a string of length $L = r$ (Figure 11-32). (The distance from the center of the sphere to the point of support is $L$.) When $r$ is much less than $L$, this situation is often treated as a simple pendulum of length $L$. (a) Show that the period can be written

$$T = 2\pi \sqrt{\frac{L}{g}} \sqrt{1 + \frac{2r^2}{5L^2}} = T_0 \sqrt{1 + \frac{2r^2}{5L^2}}$$

where $T_0 = 2\pi\sqrt{L/g}$ is the period of a simple pendulum of length $L$. (b) Show that for $r$ much smaller than $L$, the period is $T \approx T_0(1 + r^2/5L^2)$. (c) If $L$ is 1 m and $r = 2$ cm, find the error made in using the approximation $T = T_0$ for this pendulum. How large must the radius of the bob be for the error to be 1 percent?

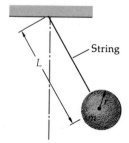

**Figure 11-32**
Problem 26.

27. Suppose that the amplitude of an oscillator decreases by a factor $R$ each cycle. That is, if $A_1$ is the amplitude after the first cycle, $A_2 = RA_1$ is that after the second cycle, etc. A driving force $F = kB \sin \omega t$ is applied to the oscillator, where $k$ is the force constant of the oscillator. Show that if $\omega$ is the resonance frequency, the amplitude is given by

$$A = \frac{\pi B}{\ln (1/R)}$$

28. In this problem you are asked to derive Equation 11-1 from the application of Newton's laws of motion and the conservation of energy to the problem of a mass $m$ which is influenced by a single force $-kx$, where $k$ is a constant. (a) From Newton's second law write down the acceleration $d^2x/dt^2$ as a function of displacement from equilibrium $x$. (b) The solution of the differential equation found in part (a) requires two integrations. The first results in an equation which is equivalent to the conservation of energy. From the conservation of energy, write an equation relating the speed $v$ to the displacement $x$ and the total energy $E$. Show that the total energy can be written $E = \frac{1}{2}m\omega^2A^2$, where $\omega = \sqrt{k/m}$ and $A$ is the maximum value of $x$. (c) In terms of the quantities defined in part (b) show that the speed $v$ can be written $v = \omega\sqrt{A^2 - x^2}$. Using $v = dx/dt$, obtain the equation

$$\frac{dx}{\sqrt{A^2 - x^2}} = \omega \, dt$$

(d) Using the substitution $x = A \sin \theta$, show that the equation obtained in part (c) becomes $d\theta = \omega \, dt$, which has the solution $\theta = \omega t + \theta_0$. Compare your solution for $x(t)$ with Equation 11-1.

29. A damped oscillator oscillates with frequency $\omega'$, which is 10 percent less than its undamped frequency. By what factor is its amplitude decreased in each cycle? By what factor is its energy reduced in each cycle?

30. Show by direct substitution that Equation 11-49 is an exact solution of Equation 11-39.

31. The acceleration of gravity $g$ varies with location on the earth because of the earth's rotation and because the earth is not exactly spherical. This was first discovered in the seventeenth century, when it was noted that a pendulum clock carefully adjusted to keep correct time in Paris lost about 90 s/d near the equator. (a) Show that a small change in $g$ produces a small change in the period of a pendulum $T$, given by

$$\frac{\Delta T}{T} \approx -\frac{1}{2}\frac{\Delta g}{g}$$

(Use differentials to approximate $\Delta T$ and $\Delta g$.) (b) How great a change in $g$ is needed to account for a change in period of 90 s/d?

# CHAPTER 12     Gravity

After studying this chapter, you should:

1.  Be able to state Kepler's three empirical laws of planetary motion.

2.  Be able to derive Kepler's law $T^2 \propto r^3$ for circular orbits.

3.  Be able to discuss gravitational and inertial mass.

4.  Be able to calculate the gravitational field and the gravitational potential for various mass distributions.

5.  Be able to draw lines of force for various mass distributions.

6.  Know how the gravitational field $g_r$ depends on $r$ for a spherical-shell mass distribution and for a solid-sphere mass distribution.

Gravity is one of the four basic interactions we discussed briefly in Chapter 5. Although of negligible importance in the interactions of elementary particles, gravity is of primary importance in the interactions of large objects. It is gravity that binds us to the earth and holds the earth and the other planets in the solar system. The gravitational force plays an important role in the evolution of stars and in the behavior of galaxies. In a sense, it is gravity that holds the universe together.

In this chapter we shall study the force of gravity in some detail. In particular we shall show how Newton's law describing the gravitational force exerted by one particle on another can be applied to the description of the force exerted by one extended body on another. In doing this, we shall introduce the concepts of a field and of lines of force, concepts used again later in the discussion of electric and magnetic forces.

Castle in the Pyrenees, *by René Magritte. (Collection of Harry Torczyner, New York; photograph by G. D. Hackett.)*

## 12-1  Kepler's Laws

The magnitude of the gravitational force exerted by a particle of mass $m_1$ on another particle of mass $m_2$ a distance $r$ away is given by Newton's law of gravity

$$F = G\frac{m_1 m_2}{r^2} \qquad\qquad 12\text{-}1$$

where the constant $G$, the universal gravitational constant, has the value

$$G = 6.67 \times 10^{-11} \ \text{N}\cdot\text{m}^2/\text{kg}^2 \qquad\qquad 12\text{-}2$$

The constant $G$ was first determined experimentally by Cavendish, as will be discussed in Section 12-3.

Newton justified his law of gravitation primarily by demonstrating that with it he could derive Kepler's laws of planetary motion, which summarized the kinematics of the motion of the planets about the sun. These laws were deduced between 1600 and 1620 by Johannes Kepler, who used the precise astronomical observations of the planets made in the late 1500s by Tycho Brahe. Kepler's laws are:

1. All planets move in elliptical orbits with the sun at one focus.

2. A line joining any planet to the sun sweeps out equal areas in equal times.

3. The square of the period of any planet is proportional to the cube of the planet's mean distance from the sun.

*Kepler's laws*

We have already discussed the relation between law 2 and angular momentum in Chapter 10. The rate of area swept out by a planet is proportional to the angular momentum of the planet about the sun. Thus Kepler's second law is equivalent to the statement that the angular momentum of a planet about the sun is constant. This implies that there is no torque exerted on the planet about the sun, i.e., that any force exerted on the planet is along the radial line from the planet to the sun, as with Newton's law of gravitation.

Newton showed that the general path of a planet moving under the influence of an inverse-square law of force is an ellipse with the center of force at one focus if the path is closed. He also showed that the path of a body which is not bound, e.g., a body that does not return, is a parabola or hyperbola. We shall consider the simpler special case of a circular orbit to show that Kepler's third law implies that the force varies inversely with the square of the distance.

Consider a planet moving with speed $v$ about the sun in a circle of radius $r$. During one period $T$, the planet travels a distance of $2\pi r$. Thus

$$T = \frac{2\pi r}{v} \qquad\qquad 12\text{-}3$$

The Granger Collection

Johannes Kepler (1571–1630).

For a circular orbit, the mean distance from the sun is just the radius $r$. (For a general elliptical path with the sun at one focus, the mean distance equals the semimajor axis of the ellipse.) Kepler's third law for this case is then

$$T^2 = Cr^3 \qquad\qquad 12\text{-}4$$

where $C$ is a proportionality constant which may depend on the properties of the sun but is the same for all planets. Substituting $2\pi r/v$ for the period, we obtain

$$T^2 = \left(\frac{2\pi r}{v}\right)^2 = Cr^3 \qquad\qquad 12\text{-}5$$

or

$$v^2 = \frac{4\pi^2}{Cr} \qquad\qquad 12\text{-}6$$

Since the planet is in a circular orbit, it has centripetal acceleration $v^2/r$. If we call the mass of the planet $m_p$, Newton's second law applied to the planet is

$$F = m_p a = \frac{m_p v^2}{r} \qquad\qquad 12\text{-}7$$

where $F$ is the central force exerted on the planet by the sun. Substituting $v^2$ from Equation 12-6 into Equation 12-7, we obtain

$$F = \frac{m_p(4\pi^2/Cr)}{r} = \frac{4\pi^2}{C}\frac{m_p}{r^2} \qquad\qquad 12\text{-}8$$

The force on the planet is proportional to the mass of the planet and inversely proportional to the square of the distance from the sun. By the law of action and reaction, this equals in magnitude the force exerted on the sun by the planet. Since the force exerted by the planet on the sun is proportional to the mass of the planet, symmetry suggests that the force exerted by the sun on the planet is proportional to the mass $M_s$ of the sun.* Writing $4\pi^2/C = GM_s$, where $G$ is a constant independent of the mass of the sun or the planets, we have

$$F = G\frac{m_p M_s}{r^2} \qquad\qquad 12\text{-}9$$

which is Newton's law of gravity.

### Question

1. The sun as seen from the earth moves more rapidly against the background of stars in the winter than in the summer. On the basis of this fact and Kepler's laws, what can you say about the relative distances of the earth from the sun during these two seasons?

---

* If the law of gravity derived from these laws of planetary motion were a special law applying only to the force between the sun and each planet, there would be no reason for the force to be proportional to $M_s$. Newton, however, argued that the same law should apply to any two masses in the universe. He supported this idea by noting that Kepler's third law also applies to such motions as the moons of Jupiter, except that the constant of proportionality is different. This of course would be the case if the gravitational force is proportional to the masses of both bodies and $G$ is a universal constant of nature.

Earth rise as seen from the moon.

NASA

# 12-2   The Acceleration of the Moon and Other Satellites

As a check on the validity of the law of gravity, Newton compared the acceleration of the moon in its orbit with the acceleration of objects near the surface of the earth (such as the legendary apple), assuming that both accelerations are due to the same cause, the gravitational attraction of the earth. Let us consider this calculation in some detail.

The moon travels in a circular orbit about the earth of radius $r_m = 3.84 \times 10^8$ m with period $T = 27.3$ d $= 2.36 \times 10^6$ s. Its speed $v$ is therefore $v = 2\pi r_m/T$, and its centripetal acceleration is

$$a_m = \frac{v^2}{r_m} = \frac{4\pi^2 r_m}{T^2} = \frac{4\pi^2(3.84 \times 10^8 \text{ m})}{(2.36 \times 10^6 \text{ s})^2}$$
$$= 2.72 \times 10^{-3} \text{ m/s}^2$$

Assuming the moon and earth to be approximately point masses, the force exerted by the earth on the moon is directed toward the earth and has the magnitude $GM_E m_m/r_m^2$, where $M_E$ and $m_m$ are the respective masses of the earth and moon. The acceleration of the moon according to Newton's second law is then

$$a_m = \frac{F}{m_m} = G\frac{M_E}{r_m^2} \qquad\qquad 12\text{-}10$$

As mentioned in Chapter 5, Newton was able to show that the force exerted by the earth on a body on or outside its surface is the same as if all the earth's mass were concentrated at its center. The acceleration of gravity of a particle near the earth's surface is then

$$g = \frac{GM_E}{R_E^2} \qquad\qquad 12\text{-}11$$

where $R_E$ is the radius of the earth. According to Equations 12-10 and 12-11, the ratio of the acceleration of gravity near the surface of the earth and the acceleration of the moon is

$$\frac{g}{a_m} = \frac{r_m^2}{R_E^2} = \frac{(3.84 \times 10^8 \text{ m})^2}{(6.37 \times 10^6 \text{ m})^2} = 3.63 \times 10^3$$

This result is in good agreement with the ratio of the measured accelerations:

$$\frac{g}{a_m} = \frac{9.81 \text{ m/s}^2}{2.72 \times 10^{-3} \text{ m/s}^2} = 3.61 \times 10^3 \qquad\qquad 12\text{-}12$$

We can relate the period of any satellite orbiting the earth (whether it is the moon or an artificial satellite) to the radius of its orbit (assuming it to be in a circular orbit) by an equation equivalent to Kepler's third law. If the period is $T$ and the radius of the orbit is $r$, we have

$$a = \frac{GM_E}{r^2} = \frac{v^2}{r} = \frac{4\pi^2 r}{T^2}$$

or

$$T^2 = \left(\frac{4\pi^2}{GM_E}\right) r^3 \qquad\qquad 12\text{-}13$$

*Kepler's third law for satellites*

Equation 12-13, which we have derived for circular orbits, also applies to any elliptical orbit if we interpret $r$ as the mean distance from the satellite to the earth, which is equal to the semimajor axis of the ellipse.

The equation also applies to the orbits of the moons of any planet if we replace $M_E$ with the mass of the planet. We note that since $G$ is known, we can determine the mass of a planet by measuring the period $T$ and radius $r$ of a moon orbiting about it.

---

**Example 12-1** Mars has a satellite with a period of 460 min and mean orbit radius of 9.4 Mm. What is the mass of Mars?

Replacing $M_E$ in Equation 12-13 with the mass of Mars $M$ and using $r = 9.4 \times 10^6$ m, $T = 460(60)$ s, and $G = 6.67 \times 10^{-11}$ N·m²/kg², we obtain

$$M = \frac{4\pi^2 r^3}{GT^2} = \frac{4\pi^2(9.4 \times 10^6)^3}{(6.67 \times 10^{-11})[460(60)]^2} = 6.45 \times 10^{23} \text{ kg}$$

---

**Question**

2. Two of Jupiter's moons have orbit radii which differ by a factor of 2. How do their periods compare?

## 12-3 The Cavendish Experiment

The first measurement of the gravitational constant $G$ was made by Henry Cavendish in 1798. Figure 12-1 shows a schematic sketch of the apparatus he used to measure the gravitational force between two small bodies.

The two small bodies of mass $m_2$ are at the ends of a light rod suspended by a fine fiber. A torque is required to turn the two masses through the angle $\theta$ from their equilibrium position because the fiber must be twisted. Careful measurement shows that the torque required to turn the fiber through a given angle is proportional to the angle. The constant of proportionality can be determined, and the fiber and the

Henry Cavendish (1731–1810).

Original drawing of Cavendish's apparatus. (*Courtesy of the Deutsches Museum, Munich.*)

suspended masses can be used to measure very small torques. This arrangement, called a *torsion balance*, was invented in the eighteenth century by John Michell. The French physicist, Charles Augustin de Coulomb, used a torsion balance in 1785 to determine the law of electric force known by his name. Cavendish used a refined and especially sensitive torsion balance in his determination of G.

In Cavendish's experiment two large masses $m_1$ are placed near the small masses $m_2$, as shown in Figure 12-1. The apparatus is allowed to come to equilibrium in this position; because the apparatus is so sensitive and the gravitational force so small, this takes hours. Instead of measuring the deflection angle directly, Cavendish reversed the positions of the large masses, as shown by the dashed lines in the figure. If the balance is allowed to come to equilibrium again, it will turn through angle $2\theta$ in response to the reversal of the torque. From the measurement of the angle and of the torsion constant, the forces between the masses $m_1$ and $m_2$ can be determined. When their masses and their separations are known, G can be calculated. Cavendish obtained a value for G within about 1 percent of the present value as given by Equation 12-2.

The very small magnitude of G means that the gravitational force exerted by one object of ordinary size on another such object is extremely small. For example, the attractive force between two objects each of mass 1 kg separated by 1 m is only $6.67 \times 10^{-11}$ N. Such forces can be observed only if extreme care is taken to eliminate all other forces on the objects, as must be done in the Cavendish experiment to measure G.

**Questions**

3. Cavendish presented his result in a paper entitled "Weighing the Earth." Explain how the earth's mass can be determined from a knowledge of G.

4. Explain how the mass of the sun can be calculated once G is known.

## 12-4 Gravitational and Inertial Mass

The property of a body responsible for the gravitational force it exerts on another body is called its *gravitational mass*. On the other hand, the property of a body that measures its resistance to acceleration is called its *inertial mass*. We have used the same symbol $m$ for these two properties because experimentally the gravitational and inertial masses of a body are equal. The fact that the gravitational force exerted by a body is proportional to its inertial mass is a characteristic unique to the force of gravity among all the forces we know, and a matter of considerable interest. One consequence is that all objects near the earth fall with the same acceleration if air resistance is negligible. This fact has seemed surprising to all since ancient times. The well-known story of how Galileo demonstrated it by dropping objects from the Leaning Tower of Pisa is just one example of the excitement this discovery aroused in the sixteenth century.

We could easily imagine that the gravitational and inertial masses of a body were not the same. Suppose we write $m_G$ for the gravitational mass and $m$ for the inertial mass. The force exerted by the earth on an object near its surface would then be

$$F = \frac{GM_E m_G}{R_E^2} = m_G g \qquad \text{12-14}$$

**Figure 12-1**
Schematic drawing of the Cavendish apparatus for determining the gravitational constant G. (*a*) Because of the gravitational attraction of the large masses $m_1$ for the nearby small masses $m_2$, the fiber is turned through a very small angle $\theta$ from its equilibrium position. (*b*) The large masses are reversed so that they are at the same distance from the equilibrium position of the balance but on the other side. The fiber then turns through the angle $2\theta$. Measurement of this angle and of the torsion constant of the fiber makes it possible to determine the force exerted by $m_1$ on $m_2$, which in turn allows the constant G to be determined.

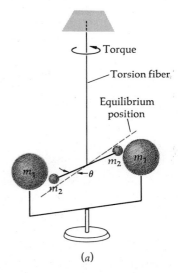

(*a*)

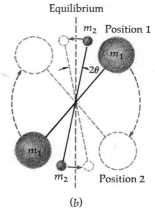

(*b*)

where $g = M_E G/R_E^2$ and $M_E$ is the gravitational mass of the earth. The free-fall acceleration of the body near the earth's surface would then be

$$a = \frac{F}{m} = \frac{m_G}{m} g \qquad \qquad 12\text{-}15$$

It might be reasonable to expect that the ratio $m_G/m$ would depend on such things as the chemical composition of the body, its temperature, or some other physical characteristics of the body. The free-fall acceleration would then be different for different objects.

The experimental fact, however, is that $a$ is the same for all bodies. This means that for every body, $m_G/m$ has the same ratio. Since this is the case, we need not maintain the distinction between $m_G$ and $m$ and can put $m_G = m$. (This amounts to choosing the constant of proportionality equal to 1, which in turn determines the magnitude and units of $G$ in the law of gravity.) We must keep in mind, however, that this is an experimental conclusion limited by the accuracy of experiment. As an experimental law, the statement that $m_G$ is equal (or proportional) to $m$ is known as the *principle of equivalence*. It is the foundation of Einstein's general theory of relativity, published in 1916. Experiments testing the equivalence principle were carried out by Simon Stevin in the 1580s (he also discovered the law of vector addition of forces). Galileo publicized this law widely, and his contemporaries made considerable improvements on the experimental accuracy with which the law was established.

*Principle of equivalence*

The most precise early experiments made to back up the principle of equivalence were those of Newton. Using simple pendulums, Newton showed that if $m_G$ and $m$ were not equal, the period of a pendulum would be given by

$$T = 2\pi \sqrt{\frac{mL}{m_G g}} \qquad \qquad 12\text{-}16$$

(The derivation of this equation is left as an exercise.) Thus one would expect that pendulums with the same length but with bobs of different masses and made of different materials would have different periods. Newton conducted careful measurements with different types of pendulum bobs and found to high accuracy that the pendulum period is independent of the mass or composition of the bob. His measurements with pendulums were much more precise than earlier experiments with falling bodies and established the equivalence of $m_G$ and $m$ to an accuracy of about 1 part in 1000. Experiments comparing gravitational and inertial mass have improved steadily over the years. The equivalence of $m_G$ and $m$ is now established to about 1 part in $10^{12}$. Thus the principle of equivalence is one of the best established of all physical laws.

The Bettmann Archive

Pen-and-ink drawing of Galileo's legendary experiment from the Leaning Tower of Pisa.

## 12-5 The Gravitational Field and Gravitational Potential

The gravitational force exerted by two bodies of mass $m_1$ and $m_2$ on a third of mass $m_0$ is just the vector sum of the forces exerted by the bodies individually; i.e., the presence of a second body of mass $m_2$ has no effect on the force exerted by $m_1$ on $m_0$. (We could imagine, for example, that the second body might shield $m_0$ from the force exerted by the first body or influence it in some other way, but no such effects are observed.) To express the gravitational force exerted by several masses on a mass $m_0$, we first express Newton's law of gravity in vector form by defining a unit vector $\hat{\mathbf{r}}_{10}$ in the direction from mass $m_1$ to mass $m_0$ (see Figure 12-2). If $\mathbf{r}_{10}$ is the vector displacement from $m_1$ to $m_0$, the unit vector is defined to be $\hat{\mathbf{r}}_{10} = \mathbf{r}_{10}/r_{10}$. In terms of the unit vector, the gravitational force exerted by mass $m_1$ on mass $m_0$ is

$$\mathbf{F}_{10} = -G\,\frac{m_1 m_0}{r_{10}^2}\,\hat{\mathbf{r}}_{10} \qquad\qquad 12\text{-}17$$

(The minus sign indicates that the force is attractive, acting toward $m_1$.)

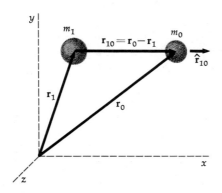

**Figure 12-2**
Definition of the unit vector $\hat{\mathbf{r}}_{10}$, which has 1 unit of length and is in the direction of $\mathbf{r}_{10} = \mathbf{r}_1 - \mathbf{r}_0$.

If there are several masses, as in Figure 12-3, the gravitational force they exert on a mass $m_0$ at some point $P$ is given by

$$\mathbf{F} = -\frac{Gm_1 m_0}{r_{10}^2}\,\hat{\mathbf{r}}_{10} - \frac{Gm_2 m_0}{r_{20}^2}\,\hat{\mathbf{r}}_{20} - \cdots - \frac{Gm_i m_0}{r_{i0}^2}\,\hat{\mathbf{r}}_{i0} - \cdots \qquad 12\text{-}18$$

where $\mathbf{r}_{i0}$ is the unit vector directed from the $i$th mass to point $P$. The force per unit mass acting on a mass at point $P$ is called the *gravitational field* at that point:

$$\mathbf{g} = \frac{\mathbf{F}}{m_0} = \sum\left(-\frac{Gm_i}{r_{i0}^2}\,\hat{\mathbf{r}}_{i0}\right) \qquad\qquad 12\text{-}19 \qquad \textit{Gravitational field}$$

The gravitational force on mass $m_0$ at point $P$ is then written

$$\mathbf{F} = m_0\mathbf{g} \qquad\qquad 12\text{-}20$$

When we have a continuous mass distribution, we calculate its gravitational field by replacing the sum in Equation 12-19 by an integral:

$$\mathbf{g} = \int -\frac{G\,dm}{r'^2}\,\hat{\mathbf{r}}' \qquad\qquad 12\text{-}21$$

where $r'$ is the distance from the mass element $dm$ to the point where

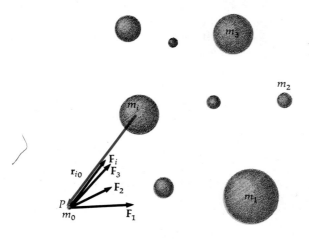

**Figure 12-3**
The force on the test mass $m_0$ is the vector sum of the forces exerted by each mass in the system. The resultant force on $m_0$ divided by $m_0$ is the gravitational field $\mathbf{g}$ at point $P$. The force on any other mass placed at point $P$ is then the product of the mass and the gravitational field $\mathbf{g}$.

the field is to be calculated and $\hat{\mathbf{r}}'$ is the unit vector pointing from $dm$ to that point. The gravitational field is a vector function of position (and of time if the positions of the other masses vary with time). By moving the test mass $m_0$ from point to point and measuring the force on it we can map out this function.

As we discussed in Chapter 4, the gravitational field is a particularly useful concept in dealing with the action-at-a-distance problem. We consider the interaction of the masses $m_i$ with the mass $m_0$ in Figure 12-3 as a two-step process. The masses $m_i$ set up a condition in space, the gravitational field, and this field in turn exerts the force on $m_0$. Similarly, $m_0$ contributes to the gravitational field at the location of each mass $m_1$. If we suddenly move one of the masses, say $m_3$, the field at point $P$ does not change immediately. There must be time for the propagation of the disturbance from $m_3$ to $m_0$. The change in the field due to a change in position of $m_3$ propagates as a gravitational wave with the speed $3 \times 10^8$ m/s, the same as the speed of light. These statements, of course, do not follow from Newton's theory of gravitation but are features of the modern theory developed by Einstein in our own century. Newton's theory can be thought of as an approximation, valid when the propagation time is negligible. In fact, it is sufficient for almost all known purposes.

In Chapter 7 we showed that the gravitational potential energy of two particles of mass $m_i$ and $m_0$ separated by a distance $r_{i0}$ is given by $U = -Gm_im_0/r_{i0}$, where the potential energy is chosen to be zero when the particles are infinitely separated. The potential energy of a particle of mass $m_0$ in a gravitational field produced by a set of particles of mass $m_1, m_2, \ldots, m_i, \ldots$ is then

$$U = -\frac{Gm_1m_0}{r_{10}} - \frac{Gm_2m_0}{r_{20}} - \cdots - \frac{Gm_im_0}{r_{i0}} - \cdots \qquad 12\text{-}22$$

The gravitational potential $V$ is defined to be the gravitational potential energy per unit mass:

$$V = \frac{U}{m_0} = \sum_i -\frac{Gm_i}{r_{i0}} \qquad 12\text{-}23 \qquad \textit{Gravitational potential}$$

The gravitational potential $V$ is a scalar function of position. For a continuous mass distribution, the gravitational potential is found from

$$V = \int -\frac{G\,dm}{r'} \qquad 12\text{-}24$$

The gravitational field $\mathbf{g}$ is related to the gravitational potential $V$ in the same way as a general conservative force $\mathbf{F}$ is related to its associated potential-energy function $U$. We shall consider only the special case of a spherically symmetric mass distribution with the origin at the center. Then both the field $\mathbf{g}$ and the potential $V$ are functions only of the radial distance $r$. Suppose we displace a test particle of mass $m_0$ through a small displacement $dr$. The work done by the gravitational force is, by definition, equal to the decrease in the potential energy. We have then

$$dW = F_r\,dr = m_0g_r\,dr = -dU = -m_0\,dV$$

Dividing by the mass $m_0$, we obtain

$$g_r\,dr = -dV \qquad 12\text{-}25a$$

or

$$g_r = -\frac{dV}{dr} \qquad 12\text{-}25b$$

**Example 12-2** Two equal point masses are located on the $y$ axis at $y = +a$ and $y = -a$, as shown in Figure 12-4. Find the gravitational field and the gravitational potential on the $x$ axis.

At a general point $(x, y, z)$ not necessarily on the $x$ axis, the gravitational field has three components, each of which is a function of $x$, $y$, and $z$: $g_x(x, y, z)$, $g_y(x, y, z)$, and $g_z(x, y, z)$; and the potential is also a function of $x$, $y$, and $z$: $V(x, y, z)$. Figure 12-4 illustrates the calculation of the field at a point in the $xy$ plane. Here the field has only two components because the $z$ component is zero. In this example, we restrict our attention to the calculation of the field and potential to points on the $x$ axis, as in Figure 12-5. By symmetry, the $y$ component of the field is zero. Since point $P$ is equidistant from each mass, the magnitudes of the contributions from each mass are equal. If we call this magnitude $g_1$, the $x$ component of the field is

$$g_x = -2g_1 \cos \theta = -2 \frac{Gm}{r^2} \frac{x}{r} = -2Gm \frac{x}{(x^2 + a^2)^{3/2}} \qquad 12\text{-}26$$

using $r^2 = x^2 + a^2$ and $\cos \theta = x/r$. The negative sign is included because for positive values of $x$, the field is toward the left.

The gravitational potential is found from Equation 12-23. Since each mass is a distance $r$ from point $P$, the potential is

$$V(x) = -\frac{Gm}{r} - \frac{Gm}{r} = -\frac{2Gm}{(x^2 + a^2)^{1/2}} \qquad 12\text{-}27$$

Figures 12-6 and 12-7 show the gravitational field $g_x$ and potential $V(x)$ versus $x$. It is left for you to show that these functions are related by $g_x = -dV/dx$ and that the field has its maximum value at $x = a/\sqrt{2}$ and approaches $-2Gm/x^2$ as $x$ becomes large.

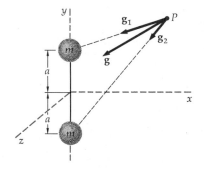

**Figure 12-4**
The gravitational field at point $P$ in the $xy$ plane due to the two masses on the $y$ axis generally has both $x$ and $y$ components.

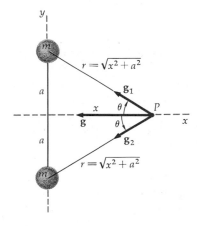

**Figure 12-5**
The gravitational field at a point $P$ on the $x$ axis due to the two masses on the $y$ axis has only an $x$ component because the $y$ components of $\mathbf{g_1}$ and $\mathbf{g_2}$ cancel.

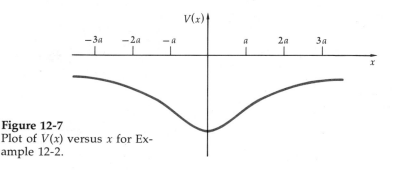

**Figure 12-6**
Plot of $g_x$ versus $x$ for Example 12-2.

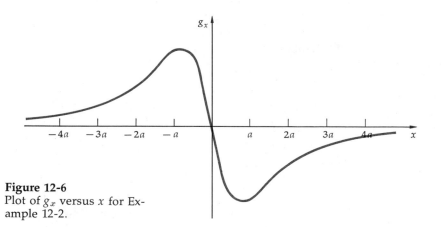

**Figure 12-7**
Plot of $V(x)$ versus $x$ for Example 12-2.

**Questions**

5. What are the SI units for the gravitational field and for the gravitational potential?

6. Is the equilibrium at $x = 0$ for the field of Example 12-2 stable or unstable?

## 12-6  Lines of Force

It is convenient to picture a field by drawing *lines of force* to indicate the direction of the field at each point. The field vector is tangent to the line at each point. Of course, there are an infinite number of points in space, so only a few representative lines are drawn. Since we could choose any point to indicate the direction of the field by a line, the lines need not be continuous, but it is both customary and useful to draw continuous lines ending at a point mass. (The lines end rather than begin at the mass because the field near a point mass points toward the mass.)

Figure 12-8 shows the lines of force for a single point mass. They are equally spaced because of spherical symmetry. As we move away from the mass, the lines become less closely spaced. The spacing of the lines is closely related to the strength (magnitude) of the gravitational field. Consider a spherical surface at a distance $r$ from the point mass. If we have a fixed number of lines, the number of lines per unit area on the sphere is inversely proportional to the area of the sphere, which is $4\pi r^2$. Thus the lines per unit area decrease with distance as $1/r^2$, just like the magnitude of the gravitational field. We can therefore use the lines to indicate both the magnitude and direction of the gravitational field.

If we have more than one point mass, we can indicate the field strength by choosing the number of lines to each mass to be proportional to the mass. Figure 12-9 shows the lines of force for two equal point masses separated by a distance $2a$. Without calculating the field at each point, we can construct this pattern as follows. Since the masses are equal, we choose an equal number of lines to each mass. At points very near one of the masses, the field is approximately due only to the nearby mass since the field varies inversely as the square of the distance. Thus the lines through the surface of a sphere of very small radius about one mass are equally spaced. At very large distances from

**Figure 12-8**
Lines of force indicating the gravitational field near a point mass. The direction of the lines shows the direction of the field, and the spacing of the lines shows the magnitude of the field.

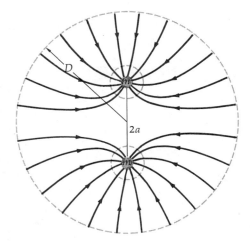

**Figure 12-9**
Lines of force for two point masses. Very near either point mass the lines are the same as if there were just the single point mass. Far from the two masses, the lines are the same as if the two masses were together as a single mass $2m$.

the two masses, the field should look like that from a single point mass of strength $2m$; the lines from the two masses are therefore equally spaced through the surface of a sphere of very large radius. We can see merely by looking at Figure 12-9 that the gravitational field in the space between the masses is weak because there are few lines in this region compared with the region just to the right or left of the masses, where the lines are more closely spaced. This information can of course also be obtained by direct calculation of the field at points in these regions.

We can apply the above reasoning to drawing the lines of force for any system of particles. We choose the number of lines entering a particle to be proportional to the mass of that particle. Very near a particle, the lines are equally spaced because the contribution to the field due to the other particles is negligible compared with that due to the nearby particle. At points very far from all the particles, the lines are equally spaced because the field must look like that of a single particle of mass equal to the total mass of the system. We note that the lines of force cannot intersect at a point in space where there is no mass because there can be only one direction for the gravitational field at such a point. Also, there can be no points like that in Figure 12-10 from which the lines of force diverge. If such a point existed, a test particle placed anywhere near that point would experience a gravitational force away from the point, implying a negative mass at that point. Negative mass which repels other masses has never been observed.

It is important to remember that this system of indicating the field strength by the spacing of the lines of force works because the area of a spherical surface increases as $r^2$ whereas the magnitude of the gravitational field due to a point mass decreases as $1/r^2$. Since the electric field of a point charge also varies inversely as the square of the distance from the charge, the concept of lines of force is also very useful in picturing the electrostatic field, as we shall see in Chapter 20.

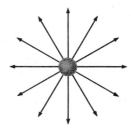

**Figure 12-10**
Lines of force diverging from a point in space. No such system is observed in nature.

## 12-7 The Gravitational Field of a Spherically Symmetric Mass Distribution

A particularly important special mass distribution is one that is spherically symmetric. To a good approximation, the earth is such a distribution. We shall consider first a spherically symmetric shell of radius $R$ and mass $M$ and investigate the properties of the gravitational field of such a system by considering the lines of force.

Let us choose the origin at the center of the shell. At very great distances compared with the shell radius $R$, the gravitational field must look like that of a point mass $M$. The lines of force are therefore radially inward and equally spaced far away from the shell. Because of the spherical symmetry of the mass distribution, the lines remain radial and equally spaced at all distances from the shell. The lines of force for this system are shown in Figure 12-11. Outside the shell the lines are exactly the same as those for a point mass $M$ at the origin. The gravitational field outside the shell is therefore given by

$$g_r = -\frac{GM}{r^2} \qquad r > R \qquad\qquad 12\text{-}28$$

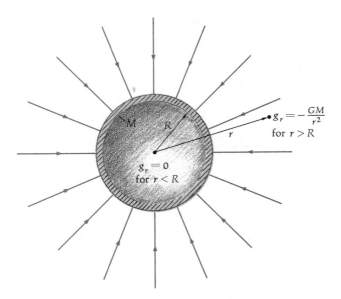

**Figure 12-11**
Lines of force for a spherical-shell mass distribution. Outside the shell the lines are the same as from a point mass at the center of the shell. Inside the shell there are no lines, indicating that the gravitational field is zero there.

$g_r = -\dfrac{GM}{r^2}$

for $r > R$

$g_r = 0$
for $r < R$

The lines end on the shell. There can be no lines inside the shell. If, for example, there were lines radially inward inside the shell, they would converge at the origin, but since there is no mass at the origin, they cannot converge there. If there were lines inside the shell radially outward, they would have to diverge from the origin, an impossibility, as we have already discussed. We thus have the remarkable result that a spherically symmetric shell of mass produces no gravitational field inside it:

$$g_r = 0 \qquad r < R \qquad\qquad\qquad\qquad 12\text{-}29$$

*A spherically symmetric shell produces no field inside it*

If we place a test mass $m_0$ anywhere inside the shell, it will experience no gravitational attraction to the shell. This is easily understood for $m_0$ at the center of the sphere since then for each mass element of the sphere there is an equal mass element at equal distance on the opposite side. Their attractions for $m_0$ would cancel. That this is also true at every other point inside the sphere is not at first obvious. It is a direct consequence of the inverse-square character of the force. As we shall see later, Franklin's discovery that the same thing occurs for the electric force led Priestley to conclude that the electric force is also an inverse-square force.

This important result for the gravitational field due to a spherical-shell mass distribution can be obtained by a direct calculation of the gravitational field from Equation 12-21 or from a direct calculation of the gravitational potential from Equation 12-24 and use of Equation 12-25b to calculate $g_r$. These calculations, which involve integral calculus and spherical coordinates, are not given here, but the identical calculations for the electric field due to a spherical-shell charge distribution appear in Chapter 20.

We now use the results of Equations 12-28 and 12-29 to find the gravitational field inside and outside a solid sphere whose density depends only on the distance $r$ from the center. We treat the solid sphere as a set of spherical shells. The gravitational field outside the sphere is just

$$g_r = -\frac{GM}{r^2} \qquad r > R \qquad\qquad\qquad 12\text{-}30$$

where $M$ is the total mass of the sphere. This follows because each shell acts as if it were a point mass at the center of the sphere. Derivation of the gravitational field inside the solid sphere is more complicated. The field inside the sphere depends upon how the density varies with $r$. We shall consider the case of constant mass density:

$$\rho = \rho_0 = \frac{M}{V} = \frac{M}{\frac{4}{3}\pi R^3} \qquad \text{12-31}$$

We wish to find the field at a point a distance $r$ from the center, where $r$ is less than $R$. As we have seen, a spherical shell produces no field at a point within the shell. Thus the only contribution to the field at this point is due to the mass $M'$, which is within the radius $r$ (Figure 12-12).

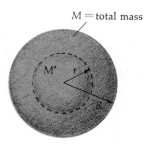

$M = $ total mass

**Figure 12-12**
Calculation of the gravitational field inside a uniform-solid-sphere mass distribution. At distance $r$ only the part of the mass $M'$ produces any gravitational field.

This mass produces a field which is the same as if the mass were at the origin. The mass within a distance $r$ is

$$M' = \rho_0(\tfrac{4}{3}\pi r^3) = \frac{M}{\frac{4}{3}\pi R^3}\,\tfrac{4}{3}\pi R^3 = M\frac{r^3}{R^3}$$

and the gravitational field at a point inside the sphere is

$$g_r = -\frac{GM'}{r^2} = -\frac{GMr^3/R^3}{r^2} = -\frac{GM}{R^3}\,r \qquad \text{12-32}$$

$g_r$ inside a spherical ball of constant density

The magnitude of the field increases with distance $r$ inside the sphere. Figure 12-13 shows $g_r(r)$ versus $r$ for a solid sphere of constant mass density.

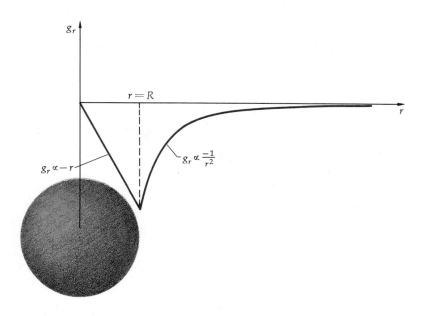

**Figure 12-13**
Plot of $g_r$ versus $r$ for a uniform-solid-sphere mass distribution. The magnitude increases with $r$ inside the sphere and decreases as $1/r^2$ outside.

**Example 12-3** A particle slides along a frictionless straight tunnel dug through the earth (Figure 12-14). Assuming the earth to have constant mass density, show that the motion is simple harmonic and find the period.

Let the $x$ axis be along the tunnel and choose the $y$ axis through the center of the earth, as shown in Figure 12-14. When the particle is at position $x$, the gravitational force acting on it is

$$\mathbf{F} = m\mathbf{g} = -m\,\frac{GM_E}{R_E^3}\,r\hat{\mathbf{r}}$$

where $M_E$ is the mass of the earth and $R_E$ is the radius of the earth. The $y$ component of this force is balanced by the normal force exerted by the tunnel. The $x$ component of this force is

$$F_x = -\frac{GM_E m}{R_E^3}\,r\sin\theta = -\frac{GM_E m}{R_E^3}\,x$$

where we have used $\sin\theta = x/r$. The acceleration in the $x$ direction is thus

$$a_x = \frac{F_x}{m} = -\frac{GM_E}{R_E^3}\,x = -\omega^2 x \qquad\qquad 12\text{-}33$$

where

$$\omega^2 = \frac{GM_E}{R_E^3}$$

Equation 12-33 is the equation for simple harmonic motion with angular frequency $\omega$. We can write $\omega$ in terms of the acceleration of gravity near the earth's surface, $g = GM_E/R_E^2$. Then

$$\omega^2 = \frac{g}{R_E}$$

The period is

$$T = \frac{2\pi}{\omega} = 2\pi\sqrt{\frac{R_E}{g}} = 2\pi\sqrt{\frac{6.37 \times 10^6 \text{ m}}{9.81 \text{ m/s}^2}}$$
$$= 5.06 \times 10^3 \text{ s} = 84.4 \text{ min}$$

The period of free fall through a straight tunnel is the same as that of a satellite orbiting near the surface of the earth. It is independent of the length of the tunnel, which means that a transit system working on this principle could provide trips from one city to any other city on earth with every trip taking about 42 min regardless of the distance involved.

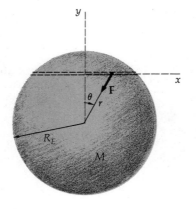

**Figure 12-14**
Tunnel through the earth for Example 12-3.

# Comets

Stephen P. Maran
*NASA-Goddard Space Flight Center*

Our present knowledge of comets indicates that if we set out to make an artificial comet and launch it into space, we might well begin with a huge lump of ice embedded with extremely fine sand. Indeed, a small version of such an experiment was contemplated for the final manned Skylab flight in 1973–1974, but not executed. It appears that the solid part of a comet, called the *nucleus,* is nothing more than a large ball of ice and other frozen gases, interspersed with rock particles, most of which are microscopic. Physical and chemical characteristics of the various parts of a comet are summarized in Table 1.

As a comet approaches the sun, solar heating sublimates the frozen matter at the surface of the nucleus, releasing gas that flows outward. This gas forms a large cloud, the *coma,* known popularly as the *head* of the comet. The gases also blow off some of the rock particles, or dust. Close to the nucleus, the gases retain the same molecular form they exhibited in the frozen state, but as these *parent molecules* diffuse outward, they are gradually broken up into *daughter products,* i.e., smaller molecules and individual atoms, by ultraviolet radiation from the sun. This radiation also ionizes some of the molecules and atoms, producing the *cometary plasma,* electrified material that is influenced by the solar wind pervading interplanetary space. The wind consists of charged particles that are constantly streaming away from the outermost region of the sun, the *corona.* The wind sweeps the cometary plasma out of the coma in a direction away from the sun, forming the *ionic,* or *plasma, tail,* which stretches for tens of millions of miles.

**Table 1**
Physical and chemical characteristics of comets

| Part of comet | Physical characteristics | Chemical composition |
|---|---|---|
| Nucleus | Typical diameter 0.3–10 km<br>Mass $10^{14}$–$10^{19}$ g | No observations but many theories; the most probable major constituents are frozen water and carbon monoxide, but other substances must be present |
| Coma | Diameter $10^4$–$10^6$ km (and $>10^7$ km when observed in the ultraviolet light emitted by hydrogen atoms of the coma) | Parent molecules include water vapor* ($H_2O$), methyl cyanide* ($CH_3CN$), and hydrogen cyanide* (HCN), all as gases<br><br>Silicate particles**<br><br>Daughter products: C, $C_2$, $C_3$, H, O, OH, CH, CN, NH, $NH_2$ |
| Tail | Width up to $10^6$ km | |
| Dust | Length up to $10^7$ km | Silicate particles** |
| Plasma | Length up to $10^8$ km | $OH^+$, $H_2O^+$, $CO^+$, $CO_2^+$, $CH^+$, $N_2^+$ |

* Substance that has been found in a comet only once.
** The presence of silicate particles is shown by infrared spectra of dust in the coma and tail.

Comet Kohoutek (1973), photographed with a wide-angle Schmidt telescope at the Joint Observatory for Cometary Research. Streaming back from the bright coma are the smooth-looking dust tail (below) and a striking plasma tail (above). Note the large disturbance in the plasma tail at the upper right of the picture; it was observed to move rapidly down the tail in the direction opposite the sun. A short "antitail," believed to consist of relatively large dust particles, extends to the left from the coma and appears (due to a perspective effect) to point toward the sun. It is faintly visible at the lower left. (*Courtesy of the Joint Observatory for Cometary Research.*)

The visible light of the sun plays a similar role; its radiation pressure exerts a significant force on the dust particles of the coma, also in a direction away from the sun. The acceleration of a dust particle away from the sun depends on its surface area and mass, whereas acceleration toward the sun, due to solar gravity, is the same for all particles. Thus the less dense dust has a large component of motion away from the sun, while the largest particles have only a small component and tend to stream out roughly back along the orbit of the nucleus. The total effect is usually to form a smooth, curved yellow dust tail, which contrasts strikingly with the kinks, wave patterns, and rayed formations of the highly structured plasma tail. The color of the dust tail comes from the reflection of sunlight. The plasma tail looks blue, due to molecular-band emission by the carbon monoxide ion ($CO^+$). The plasma tail acts as a sort of cosmic wind sock, allowing earth-bound astronomers to infer the speed and direction of the solar wind at various places in the solar system by observing the plasma tails of comets.

Since the material in the coma, dust tail, and plasma tail of a comet is not gravitationally bound to the small mass of the nucleus, it escapes into space. Thus, on each new approach to the sun, a comet evolves a new coma and tails.

Nearly all the comets that have been observed follow elliptical orbits that take them well within the orbit of Mars, and some go inside that of Mercury. Indeed, we rarely detect comets beyond Jupiter because they are so dim. When one tries to calculate how much material a comet loses each time it nears the sun and forms a coma, a problem arises. In a typical case, a bright comet like Halley's will vanish as a result of the sublimation process in a time that is very short (say, less than 1 million years) compared with the age of the solar system (5 billion years). Why then do we see any comets at all? Why weren't they all eliminated before man evolved on earth?

Two obvious possibilities are that new comets are continuously being formed, presumably on the outskirts of the solar system, or that huge numbers of comets were left over from the origin of the solar system and are stored somewhere on the outskirts. (This second hypothesis is actually the leading idea at present.) A third possibility, namely, that comets come to us from distant regions of the galaxy, seems to be excluded by the fact (old books to the contrary) that no confirmed hyperbolic orbits have been observed among comets approaching the sun.

The *Oort cloud* is the name given to the region, extending more than 1 light-year from the sun, in which perhaps 100 billion comets are thought to exist. In orbit around the sun, but usually never coming closer to it than the orbit of Jupiter, they are not significantly heated and can remain indefinitely in celestial cold storage. However, the cross-sectional area of the cloud is so great that occasionally stars must move through it on their way around the galaxy.

Indeed, roughly 3000 stars must have passed through the inner cloud (radius $7.6 \times 10^{12}$ km) since the solar system formed. The gravitation of a passing star perturbs cometary orbits. Some comets are accelerated and may escape from the cloud into interstellar space. Others are sent into the inner solar system, where, thanks to additional gravitational perturbations (especially by Jupiter), they are sometimes trapped in much smaller orbits that return them repeatedly to the vicinity of the sun. This is the source of the periodic comets well known to astronomers, such as Halley's (period of 76 years) and Encke's (3.3 years), and it presumably explains why we still see comets near the sun billions of years after solar heat ought to have dissipated them.

## Review

A. Define, explain, or otherwise identify:

Gravitational constant, 349
Kepler's laws, 349
Torsion balance, 353
Gravitational mass, 353
Inertial mass, 353

Principle of equivalence, 354
Gravitational field, 355
Gravitational potential, 356
Lines of force, 358

B. True or false:

1. Kepler's law of equal areas implies that gravity varies inversely with the square of the distance.

2. The planet closest to the sun on the average has the shortest period of revolution about the sun.

3. A spherically symmetric shell produces no gravitational field anywhere.

4. Both the magnitude and direction of the gravitational field are indicated by lines of force.

5. The gravitational field inside a solid sphere of mass of constant density is proportional to the distance from the center of the sphere.

6. There are no lines of force inside a spherically symmetric shell.

7. The gravitational field propagates through space with the speed of light.

## Exercises

### Section 12-1, Kepler's Laws

1. Suppose a small planet were discovered with a period of 5 y. What would be its mean distance from the sun?

2. Halley's comet has a period of about 76 y. What is its mean distance from the sun?

3. The comet Kohoutek has a period estimated to be at least $10^6$ y. What is its mean distance from the sun?

4. The mean distance of Jupiter from the sun is 5.22 times that of the earth. What is the period of Jupiter?

5. The radius of the earth's orbit is $1.49 \times 10^{11}$ m and that of Uranus is $2.87 \times 10^{12}$ m. What is the period of Uranus?

### Section 12-2, The Acceleration of the Moon and Other Satellites

6. A satellite orbits about the earth in a circular orbit with a period of 1 d. Find the radius of the orbit.

7. One of Jupiter's moons, Io, has a mean orbit radius of 422 Mm and a period of $1.53 \times 10^5$ s. (a) Find the mean radius of another of Jupiter's moons, Callisto, whose period is $1.44 \times 10^6$ s. (b) Use the known value of $G$ to compute the mass of Jupiter.

8. Uranus has a moon, Umbriel, whose mean orbit radius is 267 Mm and whose period is $3.58 \times 10^5$ s. (a) Find the mass of Uranus. (b) Find the period for the moon Oberon, whose mean orbit radius is 586 Mm.

9. The mass of Saturn is $5.69 \times 10^{26}$ kg. (a) Find the period of the moon Mimas, whose mean orbit radius is 186 Mm. (b) Find the mean orbit radius of the moon Titan whose period is $1.38 \times 10^6$ s.

**Section 12-3, The Cavendish Experiment**

10. Calculate the mass of the earth from the values of the period of the moon, $T = 27.3$ d, its mean orbit radius $r_m = 384$ Mm, and the known value of $G$.

11. Use the period of the earth (1 y), its mean orbit radius ($1.496 \times 10^{11}$ m), and the value of $G$ to compute the mass of the sun.

12. The masses in a Cavendish apparatus are $m_1 = 10$ kg and $m_2 = 10$ g; the separation of their centers is 5 cm; and the rod separating the two small masses is 20 cm long. What is the force of attraction between the large and small spheres? What torque must be exerted by the suspension to balance these forces?

**Section 12-4, Gravitational and Inertial Mass**

*There are no exercises for this section.*

**Section 12-5, The Gravitational Field and Gravitational Potential**

13. A point particle of mass $m$ is on the $x$ axis at $x = +a$, and a second particle of equal mass is at $x = -a$. (a) What is the gravitational field halfway between the particles at $x = 0$? (b) Find the gravitational field on the $x$ axis at points $|x| < a$.

14. At what point on the line from the earth to the moon is the gravitational field due to these two bodies zero? The moon has a mass 0.0123 times that of the earth and is a distance $3.84 \times 10^5$ km away.

15. For the mass distribution of Exercise 13, find the work required to bring a test mass $m_0$ from $x = \infty$ to $x = 0$.

16. A 2-kg point particle is moving in a circle of radius 200 m with a speed of 0.5 m/s. The only force on the particle is due to gravity. What is the gravitational field at the position of the particle due to other masses?

17. A uniform thin ring of mass $M$ and radius $r$ lies in the $yz$ plane with its center at the origin. (a) Find the gravitational potential $V(x)$ for points on the $x$ axis, which is the axis of the ring. (b) Sketch $V(x)$ versus $x$. (c) At what point does $V(x)$ have maximum magnitude? What is the gravitational field at that point?

18. For the two-particle mass distribution of Exercise 13, sketch the gravitational potential $V(x)$ versus $x$ for points on the $x$ axis. At the origin, the gravitational field is zero. Does $V(x)$ have a maximum or minimum at the origin?

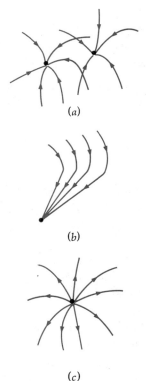

(a)

(b)

(c)

**Section 12-6, Lines of Force**

19. Two point particles of mass $m_1 = m$ and $m_2 = 2m$ are a distance $D$ apart. Sketch the lines of force for this mass distribution.

20. Explain why the lines of force shown in Figure 12-15 are not possible representations of any gravitational field.

**Section 12-7, The Gravitational Field of a Spherically Symmetric Mass Distribution**

21. A spherical shell has a radius $R = 2$ m and mass 12 kg. What is the gravitational field (a) just outside and (b) just inside the shell?

22. (a) Use Equation 12-25b to show that the gravitational potential $V(r)$ is constant inside a spherical-shell mass distribution. (b) How does $V(r)$ vary with $r$ outside a spherical-shell mass distribution? (c) Sketch $V(r)$ versus $r$ for such a distribution, using the fact that the potential must be continuous at $r = R$, where $R$ is the radius of the shell.

**Figure 12-15**
Exercise 20.

23. Find the gravitational potential $V$ just inside and just outside the spherical-shell mass distribution of Exercise 21.

24. Show that $V(r) = -GM/r$ outside a spherically symmetric mass distribution.

## Problems

1. For the mass distribution in Example 12-2, show that $g_x$ is approximately $-2GM/x^2$ when $x$ is much greater than $a$. Show also that the maximum value of $|g_x|$ occurs at the point $x = a/\sqrt{2}$.

2. The gravitational potential $V(x)$ was found for points along the $x$ axis for a ring mass in Exercise 17. (a) Calculate the gravitational field $g_x$ for points on the $x$ axis from $V(x)$. (b) At what points is the magnitude of $g_x$ maximum? (c) Calculate $g_x$ along the axis of the ring directly from Equation 12-21 by first finding $dg_x$ due to an element of the ring of mass $dm$, and then summing over all the elements.

3. Two planets of equal mass orbit a much more massive star (Figure 12-16). Planet $m_1$ moves in a circular orbit of radius $1 \times 10^8$ km with period 2 y. Planet $m_2$ moves in an elliptical orbit with closest distance $r_1 = 1 \times 10^8$ km and farthest distance $r_2 = 1.8 \times 10^8$ km, as shown. (a) Using the fact that the mean radius of an elliptical orbit is the length of the semimajor axis, find the period of $m_2$'s orbit. (b) What is the mass of the star? (c) Which planet has the greater speed at point $P$? Which has the greater total energy? (d) How does the speed of $m_2$ at point $P$ compare with that at point $A$?

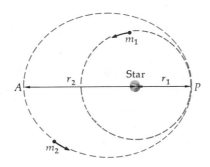

**Figure 12-16**
Problem 3.

4. The restoring torque for a torsional balance is proportional to the angular displacement $\tau = -k\theta$, where $k$ is the torque constant. (a) From the equation $\tau = I\alpha$ show that the twisting motion of the balance is simple harmonic motion with period $T = 2\pi\sqrt{I/k}$, where $I$ is the moment of inertia about the axis of rotation. (b) A Cavendish apparatus uses two 25-g metal spheres attached to the ends of a 40-cm rod with negligible mass. The rod is suspended by a fiber attached to its center. When the fiber is twisted, the period of oscillation is 1000 s. Find the torsion constant of the fiber. (c) When two large masses $m$ are placed near the smaller masses, the fiber twists through an angle $\theta = 2 \times 10^{-3}$ rad. What is the force of attraction between each large mass and the nearby 25-g mass?

5. A particle of mass $m$ slides through the tunnel discussed in Example 12-3, starting from rest at the surface of the earth. Show that the maximum speed reached is given by $v = x\sqrt{g/R_E}$, where $x$ is the initial displacement from equilibrium. Evaluate this speed for a tunnel joining two cities 300 km apart. (Use the approximation $2x = 300$ km.)

6. (a) Show that the gravitational field of a uniform ring of mass is zero at the center of the ring. (b) Figure 12-17 shows a point $P$ in the plane of the ring but not at the center. Consider two elements of the ring of length $s_1$ and $s_2$ at a distance $r_1$ and $r_2$, respectively. What is the ratio of the masses of these elements?

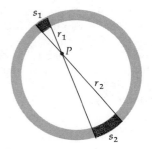

**Figure 12-17**
Problem 6.

Which produces the greater field at point $P$? What is the direction of the field at point $P$ due to these elements? (c) What is the direction of the gravitational field at point $P$? (d) Suppose that the gravitational field due to a point mass varied as $1/r$ rather than $1/r^2$. What then would be the resultant gravitational field at point $P$ due to the elements shown? (e) How would your answers to parts (b) and (c) differ if point $P$ were inside a spherical shell of mass rather than a plane circular ring?

7. A thick spherical shell has an inner radius $R_1$ and an outer radius $R_2$. It has mass $M$ and uniform density. Find the gravitational field $g_r$ as a function of $r$ for all $r$.

8. Our galaxy can be considered to be a large disk of radius $R$ and mass $M$ of approximately uniform mass density. (a) Find the gravitational potential on the axis of such a disk at a distance $x$ from it due to a ring element of radius $r$ and thickness $dr$. (b) Integrate your result in (a) to find the total gravitational potential due to the disk at distance $x$. (c) From your result in (b) find the gravitational field $g_x$ on the axis of the disk.

9. A uniform rod of mass $M$ and length $L$ lies along the $x$ axis with center at the origin. Consider an element of length $dx$ at distance $x$ from the origin. Show that this element produces a gravitational field at a point on the axis $x_0$ ($x_0$ greater than $\frac{1}{2}L$) given by $(-GM\,dx)/L(x_0 - x)^2$. Integrate this result over the rod to find the total gravitational field at the point $x_0$ due to the rod.

10. Calculate the gravitational potential $V(x)$ at point $x_0$ due to the rod of Problem 9. Use your result to find the gravitational field $g_x = -dV/dx$ and compare with your result for Problem 9.

11. A sphere of radius $R$ has its center at the origin. It has uniform mass density $\rho_0$ except that there is a spherical hole of radius $r = \frac{1}{2}R$ whose center is at $x = \frac{1}{2}R$, as shown in Figure 12-18. Find the gravitational field at points on the $x$ axis for $|x| > R$.

12. For the sphere with the hole in it of Problem 11 show that the gravitational field inside the hole is uniform, and find its magnitude and direction.

13. A straight smooth tunnel is dug through a spherical planet whose mass density $\rho_0$ is constant. The tunnel passes through the center of the planet and is perpendicular to the planet's axis of rotation, which is fixed in space. The planet rotates with an angular velocity $\omega$ so that objects in the tunnel have no acceleration relative to the tunnel. Find $\omega$.

14. A plumb bob near a large mountain is slightly deflected from the vertical by the gravitational attraction of the mountain. Estimate the order of magnitude of the angle of deflection using any assumptions you like.

15. Both the sun and the moon exert gravitational forces on oceans of the earth, causing tides. (a) Show that the ratio of the force exerted by the sun to that exerted by the moon is $M_s r_m^2/M_m r_s^2$, where $M_s$ and $M_m$ are the mass of the sun and moon and $r_s$ and $r_m$ are the distance from the earth to the sun and moon. Evaluate this ratio. (b) Even though the sun exerts a much greater force on the ocean than the moon does, the moon has a greater effect on the tides because it is the *change* in the force when the distance to the ocean changes (due to rotation of the earth) that is important. Differentiate the expression $F = Gm_1m_2/r^2$ to calculate the change in $F$ due to a small change in $r$. Show that $dF/F = (-2\,dr)/r$. (c) The largest change in distance resulting from rotation is twice the radius of the earth. Show that for a given small change in distance the change in the force exerted by the sun is related to the change in the force exerted by the moon by

$$\frac{\Delta F_s}{\Delta F_m} \approx \frac{M_s r_m^3}{M_m r_s^3}$$

and calculate this ratio.

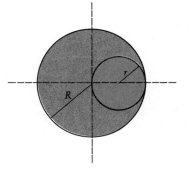

**Figure 12-18**
Problems 11 and 12.

# CHAPTER 13      Solids and Fluids

---

**Objectives**    After studying this chapter, you should:

1. Know the definitions of density, specific gravity, weight density, Young's modulus, the shear modulus, and the bulk modulus.

2. Be able to state and derive Archimedes' principle.

3. Be able to work problems involving buoyant forces on submerged or floating objects.

4. Be able to convert from one pressure unit to another, e.g., from millimetres of mercury to pascals.

5. Be able to state the assumptions leading to Bernoulli's equation and work problems using that equation.

6. Be able to state the definitions of surface tension and the coefficient of viscosity and work problems involving these quantities.

---

The states of matter in bulk can conveniently be divided into solids and fluids. Solids tend to be rigid and maintain their shape, whereas fluids do not maintain their shape but flow. Fluids include both liquids, which flow until they occupy the lowest possible regions in their container, and gases, which expand to fill their container regardless of its shape. The distinction between solids and fluids is not sharp. Glasses—even rocks under great pressure—tend to flow slightly over long periods of time. If we heat a solid, under ordinary pressure it will first become warmer, i.e., its temperature will increase, until its melting point is reached, at which point the solid will begin to change into a liquid. If we continue heating the material, the temperature does not change until all the solid has changed into a liquid. At this point the temperature will increase again until the liquid begins to vaporize, i.e., change into a gas. Again the temperature remains constant as the material is heated until all the liquid has changed to a gas, when the temperature will rise again. We shall study some of the thermal properties of matter in Chapter 18. In this chapter we look at some of the mechanical properties of solids and fluids at rest and in motion.

## 13-1  Density

An important property of a substance is the ratio of its mass to its volume, called its *density*. We usually designate the density by the greek letter rho ($\rho$):

$$\text{Density} = \frac{\text{mass}}{\text{volume}}$$

or

$$\rho = \frac{m}{V} \qquad\qquad 13\text{-}1$$

*Density defined*

Since the mass unit, the gram, was originally chosen to be the mass of 1 cm³ of water, the density of water in cgs units is 1 g/cm³. Converting to SI units of kilograms per cubic metre, we obtain for the density of water*

$$\rho = \frac{1\text{ g}}{\text{cm}^3} \frac{1\text{ kg}}{10^3\text{ g}} \left(\frac{100\text{ cm}}{\text{m}}\right)^3 = 10^3\text{ kg/m}^3$$

A convenient unit of volume is the litre (L):

$$1\text{ L} = 10^3\text{ cm}^3 = 10^{-3}\text{ m}^3$$

In terms of this unit, the density of water is 1.00 kg/L. When an object's density is greater than that of water, it will sink in water; when its density is less, it will float. In fact, we shall show in Section 13-4 that for floating objects, the fraction of the volume of the object that is submerged in any liquid equals the ratio of the density of that object to that of the liquid. For example, ice has a density of approximately 0.9 g/cm³ and floats in water with about 90 percent of its volume submerged. Since the density of water is 1 in cgs units, they are slightly more convenient than SI units. The ratio of the density of a substance to that of water is called the *specific gravity* of the substance. The specific gravity is just the magnitude of the density expressed in grams per cubic centimetre. For example, the specific gravity of ice is 0.92. Table 13-1 lists the densities of some common materials. The specific gravity of objects that sink in water ranges from 1 to about 22. Although most solids and liquids expand slightly when heated and contract slightly when subjected to an increased external pressure, these changes in volume are relatively small, so that we can say that the densities of most solids and liquids are approximately independent of temperature and pressure. On the other hand, the density of a gas depends strongly on the pressure and temperature. In fact, to a good approximation, the density of a gas is proportional to the pressure and inversely proportional to the (absolute) temperature. The temperature and pressure must therefore be specified when giving the densities of gases. In Table 13-1 the densities are given at *standard conditions*, namely, a pressure of 1 atm and a temperature of 0°C. Note that these gas densities are considerably less than those of liquids or solids; e.g., the density of water is about 800 times that of air under standard conditions.

In the U.S. customary system of units, the weight density, defined as the ratio of the weight of an object to its volume, is often used. The weight density is just $\rho g$:

$$\rho g = \frac{w}{V} = \frac{mg}{V} \qquad\qquad 13\text{-}2$$

*Weight density*

**Table 13-1**
Densities of selected substances*

| Substance | Density, kg/m³ |
|---|---|
| Aluminum | $2.70 \times 10^3$ |
| Bone | $1.7\text{–}2.0 \times 10^3$ |
| Brick | $1.4\text{–}2.2 \times 10^3$ |
| Cement | $2.7\text{–}3.0 \times 10^3$ |
| Copper | $8.96 \times 10^3$ |
| Earth (average) | $5.52 \times 10^3$ |
| Glass (common) | $2.4\text{–}2.8 \times 10^3$ |
| Gold | $19.3 \times 10^3$ |
| Ice | $0.92 \times 10^3$ |
| Iron | $7.96 \times 10^3$ |
| Lead | $11.3 \times 10^3$ |
| Wood (oak) | $0.6\text{–}0.9 \times 10^3$ |
| Alcohol (ethanol) | $0.806 \times 10^3$ |
| Gasoline | $0.68 \times 10^3$ |
| Mercury | $13.6 \times 10^3$ |
| Seawater | $1.025 \times 10^3$ |
| Water | $1.00 \times 10^3$ |
| Air | 1.293 |
| Helium | 0.1786 |
| Hydrogen | 0.08994 |
| Steam (100°C) | 0.6 |

* $t = 0$°C and $P = 1$ atm unless otherwise indicated.

* The density of water varies with temperature. Its maximum value is $10^3$ kg/m³ at 4°C.

The weight density of water is

$$\rho_w g = 62.4 \ \text{lb/ft}^3 \qquad \qquad 13\text{-}3$$

The weight density of any other material can be found by multiplying its specific gravity by $62.4 \ \text{lb/ft}^3$.

---

**Example 13-1**   A lead brick is 5 by 10 by 20 cm. How much does it weigh?

From Table 13-1 the density of lead is $11.3 \times 10^3 \ \text{kg/m}^3$. The volume of the brick is $V = (5 \ \text{cm})(10 \ \text{cm})(20 \ \text{cm}) = 1000 \ \text{cm}^3 = 10^{-3} \ \text{m}^3$. Its mass is then $m = (11.3 \times 10^3 \ \text{kg/m}^3)(10^{-3} \ \text{m}^3) = 11.3 \ \text{kg}$ and its weight is $w = mg = (11.3 \ \text{kg})(9.81 \ \text{N/kg}) = 111 \ \text{N} \approx 25 \ \text{lb}$.

---

## 13-2   Stress and Strain

If a solid object is in equilibrium but subjected to forces that tend to stretch, shear, or compress it, the shape of the object changes. If the object returns to its original shape when the forces are removed, the solid is said to be elastic. If the forces are too great, the object does not return to its original shape but is permanently deformed.

Figure 13-1 shows a solid bar subjected to a tensile force **F** to the right and an equal but opposite force to the left. In Figure 13-1b we concentrate on an element of the bar of length $\ell$ and assume the forces to be distributed uniformly over the cross-sectional area of the bar. If we increase the stretching force by a slight amount $\Delta F$, the bar stretches by a slight amount $\Delta \ell$. The ratio of the increase in the force to the cross-sectional area of the bar is called the *tensile stress:*

$$\text{Stress} = \frac{\Delta F}{A} \qquad \qquad 13\text{-}4 \qquad \textit{Tensile stress defined}$$

The fractional change in length of the bar $\Delta \ell / \ell$ is called the *strain:*

$$\text{Strain} = \frac{\Delta \ell}{\ell} \qquad \qquad 13\text{-}5 \qquad \textit{Strain defined}$$

The ratio of the stress to the strain is called *Young's modulus Y:*

$$Y = \frac{\text{stress}}{\text{strain}} = \frac{\Delta F/A}{\Delta \ell / \ell} \qquad \qquad 13\text{-}6 \qquad \textit{Young's modulus}$$

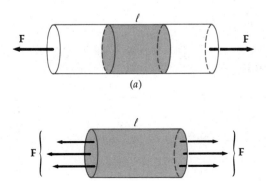

(a)

(b)

**Figure 13-1**
(a) Solid bar subjected to a stretching force **F**. (b) A section of the bar. The force per unit area is the stress $S$.

The units of Young's modulus are those of pressure, newtons per square metre or pounds per square inch. Figure 13-2 shows a graph of the strain versus the stress for a typical solid bar. The stress is proportional to the strain up to point $A$ on the graph. This is known as *Hooke's law*. (It is the same behavior as that of a coiled spring for small stretchings, but a coiled spring is more complicated because the stretching force is a combination of tensile forces and shearing forces, discussed below.) Point $B$ in Figure 13-2 is the elastic limit of the material. If the body is stretched beyond this point, it does not return to its original length but is permanently deformed. If an even greater stress is applied, the material eventually breaks.

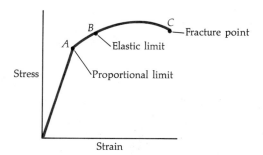

**Figure 13-2**
Stress versus strain. Up to point $A$ the strain is proportional to the stress. Point $B$, the elastic limit, is the point beyond which the bar will not return to its original length when the stress is removed. At point $C$ the bar fractures.

If the bar is subjected to forces tending to compress it rather than stretch it, the stress is called *compressive stress*. Young's modulus for compressive stress is the same as for tensile stress if $\Delta \ell$ in Equation 13-5 is taken as the decrease in the length of the bar.

In Figure 13-3 forces are applied to a bar perpendicularly to the length of the bar. Such forces are called *shear forces*. The ratio of the change in the shear force $F_s$ to the area is called the *shear stress*:

$$\text{Shear stress} = \frac{\Delta F_s}{A}$$

A shear stress tends to deform the bar as shown in Figure 13-3. The ratio $\Delta X / \ell$ is called the shear strain:

$$\text{Shear strain} = \frac{\Delta X}{\ell}$$

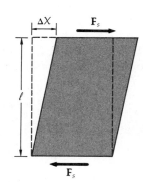

**Figure 13-3**
Change in shape of a body subjected to a shear stress.

The ratio of the shear strain to the shear stress is called the shear modulus $M_s$:

$$M_s = \frac{\text{shear stress}}{\text{shear strain}} = \frac{\Delta F / A}{\Delta X / \ell} \qquad \text{13-7}$$

*Shear modulus*

When a body is submerged in a fluid such as water, the increase in pressure due to the fluid pressing against the body tends to compress the body. The ratio of the increase in pressure to the fractional decrease in volume $(-\Delta V / V)$ is called the *bulk modulus B*:

$$B = -\frac{\Delta P}{\Delta V / V} \qquad \text{13-8}$$

*Bulk modulus*

The minus sign in Equation 13-8 is introduced to make $B$ positive since all materials decrease in volume when subjected to an increase in the external pressure. The inverse of the bulk modulus is called the compressibility $k$:

$$k = \frac{1}{B} = -\frac{\Delta V/V}{\Delta P}$$

13-9    *Compressibility*

The concept of bulk modulus and compressibility can be applied to liquids and gases as well as to solids. Solids and liquids are relatively incompressible; i.e., they have small values of compressibility $k$ or large values of the bulk modulus $B$, and these values are relatively independent of temperature and pressure. Gases, on the other hand, are easily compressed, and the values of $B$ and $k$ depend strongly on the pressure or temperature. We shall show in Chapter 14 that the speed of sound in a gas depends on the bulk modulus, and that for gases that obey the ideal-gas law (Equation 16-17) the bulk modulus is proportional to the pressure of the gas.

Table 13-2 lists approximate values of Young's modulus, the shear modulus, and the bulk modulus for various materials.

---

**Example 13-2** A 500-kg load is hung from a 3-m steel wire with a cross-sectional area of 0.15 cm². By how much does the wire stretch?

The weight of a 500-kg mass is $mg = (500 \text{ kg})(9.81 \text{ N/kg}) = 4.90 \times 10^3$ N. From Table 13-2 we find Young's modulus for steel to be about

The karate chop delivered by physicist Ronald McNair produces compressive stress at the top and tensile stress at the bottom of the patio blocks. Since concrete is weaker under tension than under compression, it breaks first at the bottom. (*Courtesy of Michael Feld, Spectroscopy Laboratory, MIT.*)

$2.0 \times 10^{11}$ N/m². The stress of the wire is

$$S = \frac{4.9 \times 10^3 \text{ N}}{0.15 \text{ cm}^2} = 3.27 \times 10^4 \text{ N/cm}^2$$

$$= 3.27 \times 10^8 \text{ N/m}^2$$

The strain is therefore

$$\frac{\Delta \ell}{\ell} = \frac{3.27 \times 10^8 \text{ N/m}^2}{2.0 \times 10^{11} \text{ N/m}^2} = 1.63 \times 10^{-3}$$

Since the wire is 300 cm long, the amount it stretches is

$$\Delta \ell = 1.63 \times 10^{-3} \, \ell = (1.63 \times 10^{-3})(300 \text{ cm}) = 0.49 \text{ cm}$$

**Question**

1. Which modulus (or moduli) would you expect to be involved in determining the force constant for a spring?

## 13-3   Pressure in a Fluid

Fluids differ from solids in being unable to support a shear stress. We define the pressure in a fluid as follows. Consider a small surface of area $A$ such as a small card submerged in the fluid. The fluid on one side of the card exerts a force $F$ on the card which is balanced by an equal but opposite force exerted by the fluid on the other side of the card. If the card is very small, so that we can neglect the difference in depth of the fluid over the surface, the force $F$ is independent of the orientation of the card. The pressure $P$ is defined as the ratio of the magnitude of the force to the area of the card:

$$P = \frac{F}{A} \qquad \text{13-10}$$

*Pressure defined*

The SI unit of pressure is the newton per square metre (N/m²), which is called the pascal (Pa):

$$1 \text{ Pa} = 1 \text{ N/m}^2 \qquad \text{13-11}$$

In the U.S. customary system, pressure is usually given in pounds per square inch (lb/in²). Another common unit is the atmosphere (atm), which is approximately air pressure at sea level. The atmosphere is now defined to be 101.325 kilopascal, which is approximately 14.70 lb/in²:

$$1 \text{ atm} = 101.325 \text{ kPa} = 14.70 \text{ lb/in}^2 \qquad \text{13-12}$$

Other units of pressure in common use will be discussed below.

The pressure in a lake or ocean increases as we go to greater and greater depths. Similarly, the atmospheric pressure decreases as we go to greater and greater altitudes. The variation in pressure in a fluid with depth or height can be easily derived. Consider a slab of a liquid such as water of area $A$ and thickness $\Delta h$, where $h$ is the depth measured from the top of the liquid (Figure 13-4). Let $P_1$ be the pressure at the top of the slab and $P_2$ be that at the bottom. The force exerted downward on the slab by the fluid above it is $P_1 A$. Similarly, the force exerted upward on the slab by the fluid below it is $P_2 A$. This upward force must be greater than the downward force to balance the weight of the slab. The

**Table 13-2**
Approximate values of Young's modulus $Y$, the shear modulus $M_s$, and the bulk modulus $B$ for various materials, GN/m²

| Material | $Y$ | $M_s$ | $B$ |
| --- | --- | --- | --- |
| Aluminum | 70 | 30 | 70 |
| Brass | 90 | 36 | 61 |
| Copper | 110 | 42 | 140 |
| Iron | 190 | 70 | 100 |
| Lead | 16 | 5.6 | 7.7 |
| Mercury | | | 27 |
| Steel | 200 | 84 | 160 |
| Tungsten | 360 | 150 | 200 |
| Water | | | 200 |

volume of the slab is $A \Delta h$, and its weight is $mg = \rho A \Delta h\, g$. Setting the upward force due to the pressure difference equal to the weight of the slab, we get

$$P_2 A - P_1 A = mg = \rho A \Delta h\, g$$

or

$$\Delta P = \rho g\, \Delta h$$

where $\Delta P$ is the pressure difference corresponding to the depth difference $\Delta h$. If we divide by $\Delta h$ and take the limit as $\Delta h$ approaches zero, the ratio $\Delta P / \Delta h$ becomes the derivative $dP/dh$ and we have

$$\frac{dP}{dh} = \rho g \qquad\qquad 13\text{-}13$$

Equation 13-13 holds for any fluid. To solve this equation for the pressure, we need to know how the density of the fluid depends on the pressure. For water and most liquids, the density is approximately independent of the pressure; i.e., liquids are nearly incompressible. Then, according to Equation 13-13, $dP/dh$ equals a constant, and $P$ must vary linearly with $h$:

$$P = P_0 + \rho g h \qquad\qquad 13\text{-}14$$

where $P_0$ is the pressure at $h = 0$. The pressure at a depth $h$ is greater than that at the top by the amount $\rho g h$. This is independent of the shape of the container and is the same at all points at the same depth. If we increase $P_0$, say by inserting a piston at the top surface and pressing down on it, the increase in pressure is the same throughout the liquid. This is known as *Pascal's principle*.

*Pressure applied to an enclosed liquid is transmitted undiminished to every point in the fluid and to the walls of the container.*

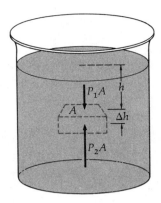

**Figure 13-4**
A slab of fluid of area $A$ and thickness $\Delta h$. The difference between the forces on the upper and lower surfaces of the slab $P_2 A - P_1 A$ must equal the weight of the slab.

*Pascal's principle*

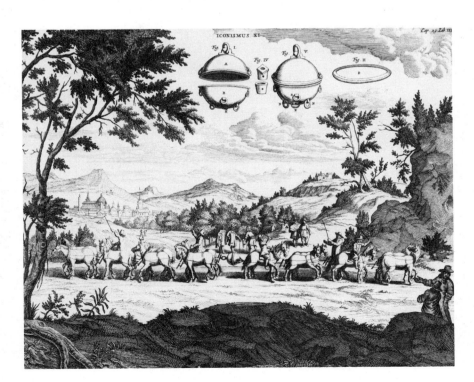

Otto von Guericke's demonstration of the large forces due to atmospheric pressure. The force exerted by the horses was not great enough to separate the two hollow hemispheres, which had been placed together and evacuated. (*Courtesy of the Deutsches Museum, Munich.*)

Figure 13-5 shows water in a container made up of parts of different shapes. At first glance, it would seem that the pressure in the largest part of the container would be greater and thus the water would be forced into the smallest part of the container to a greater height. This is known as the hydrostatic paradox. Equation 13-14 shows that the pressure depends only on the depth; the liquid should therefore be at the same height in all parts of the container, as is shown experimentally.

**Figure 13-5**
The hydrostatic paradox. The water level is the same regardless of the shape of the vessel.

Although the water in the largest part of the container weighs more than that in the smaller part, some of the weight is supported by the normal force exerted by the sides of the largest part of the container, which in this case has a component upward. In fact, the shaded portion of the water is completely supported by the sides of the container.

We can use the result that the pressure difference is proportional to the depth of the fluid to measure unknown pressures. Figure 13-6 shows the simplest pressure gauge, the open-tube manometer. The top of the tube is open to the atmosphere at pressure $P_0$. The other end of the tube is at pressure $P$ to be measured. The difference $P - P_0$ is equal to $\rho g h$, where $\rho$ is the density of the liquid in the tube. The difference between the "absolute" pressure $P$ and atmospheric pressure $P_0$ is called *gauge pressure*. The pressure you measure in your automobile tire

*Gauge pressure*

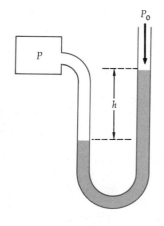

**Figure 13-6**
Open-tube manometer for measuring an unknown pressure $P$. The difference $P - P_0$ equals $\rho g h$.

is gauge pressure. When the tire is absolutely flat, the gauge pressure is zero, although the absolute pressure in the tire is atmospheric pressure. The absolute pressure is obtained from gauge pressure by adding atmospheric pressure:

$$P = P_{gauge} + P_0 \qquad 13\text{-}15$$

Figure 13-7 shows a mercury barometer used to measure atmospheric pressure. The top end of the tube has been closed off and evacuated so that the pressure there is zero. The other end is open to the atmosphere, at pressure $P_0$. The pressure $P_0$ is given by $P_0 = \rho g h$, where $\rho$ is the density of mercury.

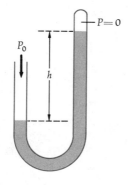

**Figure 13-7**
U-tube barometer for measuring atmospheric pressure $P_0$.

**Example 13-3** At 0°C the density of mercury is $13.595 \times 10^3$ kg/m³. What is the height of a mercury column if the pressure is 101.325 kPa? We have

$$h = \frac{P_0}{\rho g} = \frac{1.01325 \times 10^5 \text{ N/m}^2}{(13.595 \times 10^3 \text{ kg/m}^3)(9.81 \text{ N/kg})}$$

$$= 0.7597 \text{ m} \approx 760 \text{ mm}$$

In practice, pressure is often measured in millimetres of mercury (commonly called torr, after the Italian physicist Torricelli) and inches or feet of water (written $inH_2O$ or $ftH_2O$). They are related as follows:

$$1 \text{ atm} = 760 \text{ mmHg} = 760 \text{ torr} = 29.9 \text{ inHg}$$

$$= 33.9 \text{ ftH}_2\text{O} = 101.325 \text{ kPa} \tag{13-16}$$

$$1 \text{ mmHg} = 1 \text{ torr} = 1.316 \times 10^{-3} \text{ atm} = 133.3 \text{ Pa} \tag{13-17}$$

Other units commonly used on weather maps are the bar and the millibar, defined by

$$1 \text{ bar} = 10^3 \text{ millibar} = 10^5 \text{ Pa} \tag{13-18}$$

A pressure of 1 bar is just slightly less than 1 atm.

**Example 13-4** The pressure in a pipe in a water system is 50 ftH$_2$O. Express this pressure in millimetres of mercury, atmospheres, and newtons per square metre.

To say that the pressure is 50 ftH$_2$O means that the pressure is $P = \rho_w g h_w$, where $\rho_w$ is the density of water and $h_w = 50$ ft. The same pressure is related to the height $h_m$ of the mercury column by $P = \rho_m g h_m$, where $\rho_m$ is the density of mercury. Thus

$$\rho_m g h_m = \rho_w g h_w \quad \text{or} \quad h_m = \frac{\rho_w}{\rho_m} h_w$$

When we use $\rho_w = 1$ g/cm³ and $\rho_m = 13.6$ g/cm³, this gives for $h_w = 50$ ft

$$h_m = 3.68 \text{ ft} = 1120 \text{ mm}$$

Thus $P = 1120$ mmHg. We can express $P$ in other units using the conversion factors given above:

$$P = 1120 \text{ mmHg} \frac{1.316 \times 10^{-3} \text{ atm}}{1 \text{ mmHg}} = 1.47 \text{ atm}$$

and

$$P = 1120 \text{ mmHg} \frac{133.3 \text{ Pa}}{1 \text{ mmHg}} = 149 \text{ kPa}$$

Equation 13-13 also applies to gases such as air, except that we usually measure $y$ upward from the ground rather than $h$ downward. Replacing $h$ by $-y$, we obtain

$$\frac{dP}{dy} = -\rho g \tag{13-19}$$

We do not obtain a relation like Equation 13-14 because the density of a gas is not independent of the pressure. In fact, to a good approximation, the density of a gas is proportional to the pressure. In Chapter 16 we shall see that at ordinary pressures, most gases obey the ideal-gas equation (16-17). In these cases, the density of the gas is related to the pressure and temperature by

$$\rho = \frac{M}{RT} P \qquad\qquad 13\text{-}20$$

where $R$ is the *universal gas constant,* which has the value

$$R = 8.314 \text{ J/mol·K}$$

$M$ is the molecular mass, which for air has the value

$$M = 29 \times 10^{-3} \text{ kg/mol}$$

and $T$ is the absolute temperature in kelvins (K), which is related to the Celsius temperature by $T = t_C + 273$. (We discuss these quantities in more detail in Chapter 16 but include them here because they may already be familiar from your study of chemistry.)

Substituting the density relation of Equation 13-20 for an ideal gas into Equation 13-19, we obtain

$$\frac{dP}{dy} = -\frac{Mg}{RT} P = -\frac{P}{y_0} \qquad\qquad 13\text{-}21$$

where $y_0$, the *scale height,* is defined by

$$y_0 = \frac{RT}{Mg} \qquad\qquad 13\text{-}22$$

If the temperature $T$ is independent of height $y$, the scale height $y_0$ is a constant and Equation 13-21 states that the rate of change of pressure with height is proportional to the pressure. This relation is characteristic of the exponential function. The solution of Equation 13-21 for constant temperature is

$$P = P_0 e^{-y/y_0} \qquad\qquad 13\text{-}23 \qquad \textit{Law of atmospheres}$$

The pressure decreases exponentially with height. Equation 13-23 is called the *law of atmospheres.* Since the density is proportional to the pressure, the density also decreases exponentially with height. It should be noted that this result is based on the rather unreal assumption of a constant temperature.

We can compute the scale height for air at 0°C = 273 K. We have

$$y_0 = \frac{RT}{Mg} = \frac{(8.314 \text{ J/mol·K})(273 \text{ K})}{(29 \times 10^{-3} \text{ kg/mol})(9.81 \text{ N/kg})} = 7.98 \text{ km} \qquad 13\text{-}24$$

# 13-4  Archimedes' Principle

If a heavy object submerged in water is "weighed" by suspending it from a scale (Figure 13-8a), the scale reads less than when the object is weighed in air. Evidently the water exerts an upward force partially balancing the force of gravity. This force is even more evident when we submerge a piece of cork: it accelerates up toward the surface, where it floats partially submerged. The submerged cork experiences an upward force from the water greater than the force of gravity.

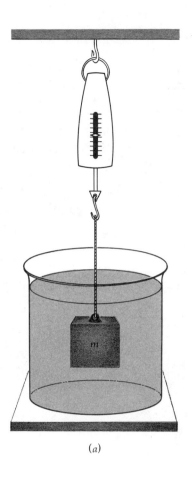

**Figure 13-8**
(*a*) Weighing a body of mass *m* submerged in a fluid. (*b*) Free-body diagram for the object. The difference $F_2 - F_1$ is the buoyant force on the object.

(*a*)

(*b*)

The force exerted by a fluid on a body submerged in it, called the *buoyant force*, depends on the density of the fluid and the volume of the body but not on the composition or shape of the body. It is equal in magnitude to the weight of the fluid displaced by the body. This result is known as *Archimedes' principle*.

*A body wholly or partially submerged in a fluid is buoyed up by a force equal to the weight of the displaced fluid.*

*Archimedes' principle*

Archimedes' principle is a direct consequence of Newton's laws applied to a fluid. In deriving this result we shall first assume that the fluid is not accelerating relative to an inertial reference frame. We shall then investigate an accelerating fluid and show that the principle still holds if we interpret weight to mean apparent weight. For example, if the fluid is in free fall, the apparent weight of the displaced fluid is zero and the buoyant force is zero.

Figure 13-8*b* shows the vertical forces acting on an object being weighed while submerged, i.e., the force of gravity **w** down, the force of the spring balance $F_s$ acting up, a force $F_1$ acting down because of the fluid pressing on the top surface of the object, and a force $F_2$ acting up because of the fluid pressing on the bottom surface of the object. Since the spring balance reads a force less than the weight, the force $F_2$ must be greater in magnitude than the force $F_1$. The difference in magnitude of these two forces is the buoyant force $B = F_2 - F_1$. The buoyant force occurs because the pressure of the fluid at the bottom of the object is greater than that at the top.

In Figure 13-9 we have replaced the submerged object by an equal volume of fluid (indicated by the dotted lines) and eliminated the spring. As we have said, the buoyant force $B = F_2 - F_1$ acting on this volume of fluid is the same as that acting on our original object. Since this volume of fluid is in equilibrium, the resulting force acting on it must be zero. The buoyant force equals the weight of the fluid in this volume:

$$B = w_f \qquad\qquad 13\text{-}25$$

We have derived Archimedes' principle.

Let $\rho_f$ be the *density* of the fluid. A volume $V$ of fluid then has mass $\rho_f V$ and weight $\rho_f g V$. If we submerge an object of volume $V$, the buoyant force on the object is $\rho_f g V$. The weight of the object can be written $\rho g V$, where $\rho$ is the density of the object. If the density of the object is greater than that of the fluid, the weight will be greater than the buoyant force and the object will sink unless supported. If $\rho$ is less than $\rho_f$, the buoyant force will be greater than the weight and the object will accelerate up to the top of the fluid unless held down. It will float in equilibrium with a fraction of its volume submerged so that the weight of the displaced fluid equals the weight of the object.

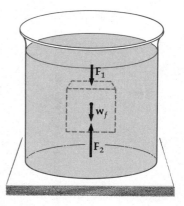

**Figure 13-9**
The same situation as in Figure 13-8 except that the object has been replaced by an equal volume of the fluid. The forces $F_1$ and $F_2$ due to the pressure of the fluid are the same as the corresponding forces on the object in Figure 13-8. The buoyant force is therefore equal to the weight of the displaced fluid.

---

**Example 13-5** A 1-N block of aluminum (density $\rho = 2.7 \times 10^3$ kg/m$^3$) is suspended by a spring scale and submerged in water ($\rho_w = 1.00 \times 10^3$ kg/m$^3$), as in Figure 13-8$a$. What is the reading on the spring scale? If the block is weighed in air of denisty $\rho_a \approx 1$ kg/m$^3$, what is the reading on the scale?

We can write the weight and buoyant force in terms of the volume of the block, $V$: $B = \rho_w g V$ and $w = \rho g V$. The spring force is the weight minus the buoyant force:

$$F_s = w - B = \rho g V - \rho_w g V = \rho g V \left(1 - \frac{\rho_w}{\rho}\right) \qquad\qquad 13\text{-}26$$

or

$$F_s = w \left(1 - \frac{\rho_w}{\rho}\right)$$

But

$$\frac{\rho_w}{\rho} = \frac{1 \times 10^3 \text{ kg/m}^3}{2.7 \times 10^3 \text{ kg/m}^3} = 0.37$$

and so

$$F_s = (1 - 0.37)w = (0.63)\,(1 \text{ N}) = 0.63 \text{ N}$$

If we replace the density of water with that of air in Equation 13-26, we obtain the force exerted by the spring scale when the block is weighed in air. For this case,

$$\frac{\rho_a}{\rho} = \frac{1 \text{ kg/m}^3}{2.7 \times 10^3 \text{ kg/m}^3} = 3.7 \times 10^{-4}$$

and

$$1 - \frac{\rho_a}{\rho} = 1 - 0.00037 \approx 1$$

showing that we can usually neglect the buoyant force of air.

---

**Example 13-6** A cork has a density of $\rho = 200$ kg/m$^3$. Find ($a$) what fraction of the volume of the cork is submerged when the cork floats in water and ($b$) the net force on the cork if it is completely submerged and then released.

Ira Rosenberg/Detroit Free Press

The buoyant force of air is not always negligible. Here it equals the weight of the balloon plus passengers, so that the balloon floats in the air. When the air in the balloon is heated more, it expands and the buoyant force is greater, causing the balloon to rise.

(*a*) Let $V$ be the volume of the cork and $V'$ be the volume that is submerged when it floats on the water. The weight of the cork is $\rho g V$, and the buoyant force is $\rho_w g V'$. Since the cork is in equilibrium, the buoyant force equals the weight. So

$$B = w \quad \text{and} \quad \rho_w g V' = \rho g V$$

$$\frac{V'}{V} = \frac{\rho}{\rho_w} = \frac{200 \text{ kg/m}^3}{1000 \text{ kg/m}^3} = \frac{1}{5}$$

Thus one-fifth of the cork is submerged.

(*b*) When the cork is held completely submerged, the buoyant force is $\rho_w g V$. This is 5 times the weight of the cork since the density of water is 5 times that of the cork. The net force upward is then

$$F = B - w = 5w - w = 4w$$

The net force upward is 4 times the weight of the cork. The upward acceleration of the cork is significantly less than $4g$ because as the cork moves upward an amount of water (which depends on the shape of the cork) must be accelerated to make room for the cork. The detailed calculation of the acceleration of the cork is a complicated problem in hydrodynamics.

Let us now consider the buoyant force in a fluid that is accelerating. In Figure 13-10*a* a container of water is falling freely with an acceleration of gravity $g$. The resultant force on any part of the water must equal $m\mathbf{g}$ because the water is accelerating with $\mathbf{a} = \mathbf{g}$. Since this resultant force equals the force of gravity, there can be no buoyant force in this case. This result can be demonstrated by placing a cork at the bottom of a long tube of water and throwing the tube into the air. (Remember that whether the tube is moving up or down, its acceleration is $\mathbf{g}$ downward; i.e., it is a projectile.) The cork does not float to the top of the tube but remains where it was when the tube was thrown.

If we submerge a cork by suspending it from the bottom of a container of water with a string and accelerate the container in a horizontal direction (Figure 13-10*b*), the somewhat surprising result is that the string inclines forward toward the direction of acceleration. Let us first consider an element of water of volume $V$ and mass $m_w = \rho_w V$. Since the water is accelerating to the left, the pressure on the right part of the water must be greater than that on the left, giving a net force to the left on the element equal to $m_w \mathbf{a}$, where $\mathbf{a}$ is the acceleration. If we now replace this element of water with a cork, the cork will experience the same force to the left; but since the mass of the cork is less than $m_w$, the acceleration of the cork will be greater than that of the water and the container. The cork will thus accelerate to the left relative to the container. If the cork is attached by a string, as in Figure 13-10*b*, the string provides a force to the right, so that the acceleration of the cork and the container are the same. How does this buoyant force to the left arise? If we look at the surface of the water, we see that it is not horizontal. The water piles up at the right, so that at a given distance from the top the water is deeper at the right of the cork than at the left. Note that the string holding the cork is perpendicular to the surface of the water.

This interesting behavior can be demonstrated by placing the container of water at the edge of a turntable and securing it so that it cannot fall off. When the table rotates, the acceleration of the container is toward the center and the string holding the cork inclines in that direction.

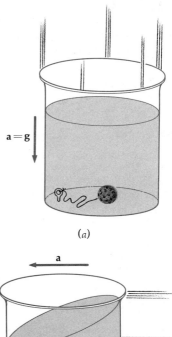

(*a*)

(*b*)

**Figure 13-10**
A cork is tied to the bottom of a container of water. (*a*) When the container is in free fall, $\mathbf{a} = \mathbf{g}$ and there is no buoyant force on the cork. (*b*) When the container accelerates to the left, the cork is deflected in the direction of acceleration and the string is perpendicular to the water surface.

Buoyant force on a submerged ball in an accelerated frame. The pendulum and water container are on a rotating platform and are therefore accelerating toward the center of rotation, which is toward the left in this photograph. Note that both strings are perpendicular to the water surface. (*Courtesy of Larry Langrill, Oakland University.*)

**Questions**

2. How can you estimate your average density at a swimming pool?

3. Smoke usually rises from a smokestack, but it may sink on a very humid day. What can be concluded about the relative densities of humid and dry air?

4. Which is denser, seawater or fresh water?

5. A certain object has a density nearly equal to that of water, so that it floats almost completely submerged; but it is much more compressible than water. What happens if the floating object is given a slight push to submerge it?

6. Fish can adjust their volume by varying the amount of oxygen and nitrogen gas (obtained from the blood) in a thin-walled sac under their spinal column called a swim bladder. Explain how this helps them swim.

7. A child riding in a car is holding a helium balloon. If the car suddenly brakes to a stop, which way does the balloon accelerate relative to the car? Why?

## 13.5  Surface Tension and Capillarity

A needle can be made to "float" on a water surface if it is placed there carefully. The forces that support the needle are not buoyant forces but are due to *surface tension*. In the interior of a liquid, a molecule is surrounded on all sides by other molecules, but at the surface there are no molecules above the surface molecules. If a surface molecule is raised slightly, the molecular bonds between it and the adjacent molecules are stretched, and there is a restoring force pulling the molecule back toward the surface. Similarly, when a needle is placed carefully on the surface, the surface molecules are depressed slightly and the adjacent molecules exert an upward restoring force on them, supporting the needle. Thus, the surface of a liquid is rather like a stretched elastic membrane. The force necessary to break the surface can be measured by lifting a thin wire out of the surface, as shown in Figure 13-11. The force needed to break the surface is found to be proportional to the length of the surface broken, which is twice the length of the wire, since there is a surface film on both sides of the wire. If the wire has a mass $m$ and length $\ell$ and it takes a force $F$ to lift it out of the surface, the coefficient of surface tension $\gamma$ is

$$\gamma = \frac{F - mg}{2\ell}$$    13-27    *Coefficient of surface tension*

The value of $\gamma$ for water is about 0.073 N/m = 73 mN/m. It is because of surface tension that small droplets of a liquid tend to be spherical; as the drop is formed, the surface tension pulls the surface together, minimizing the surface area and making the drop spherical.

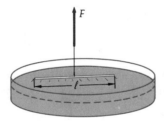

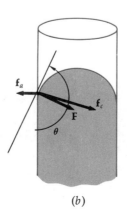

**Figure 13-11**
A wire of length $\ell$ being lifted off the surface of a liquid. Surface tension exerts a force on the wire toward the surface.

The rise of a liquid in a thin tube, called a *capillary tube,* is another surface phenomenon closely related to surface tension. The attractive forces between a molecule in a liquid and other molecules in the liquid are *cohesive forces.* The force between a liquid molecule and another substance, such as the wall of a thin tube, is an *adhesive force.* Figure 13-12a shows a molecule near the wall of a thin tube. The cohesive forces are represented by the force vector $\mathbf{f}_c$. This force points down and to the right because there are no liquid molecules above or to the left of this molecule. The adhesive force between the molecule and the wall is

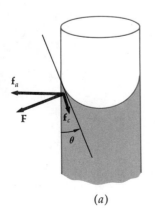

(a)            (b)

**Figure 13-12**
(a) Forces exerted on a molecule of water near the surface. The force **F** is the resultant of the adhesive force $\mathbf{f}_a$ exerted by the glass and the cohesive force $\mathbf{f}_c$ exerted by the other molecules in the liquid. (b) When the cohesive force is much greater than the adhesive force, the surface of the liquid is convex and the contact angle $\theta$ is greater than 90°.

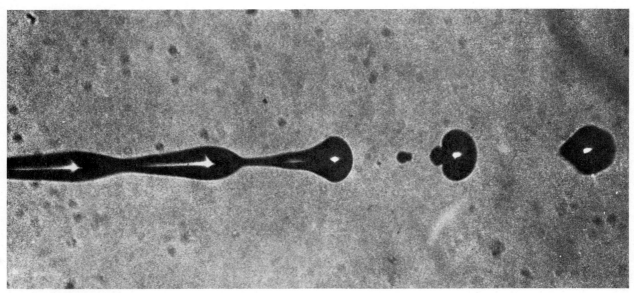

Jean Claude Chastang, I.B.M. Corporation

The formation of drops in an ink-jet printer is governed by surface tension. In the region where the jet is narrowest, the forces resulting from surface tension are greatest, and so they further reduce the diameter until the stream breaks apart into approximately spherical drops.

represented by the force vector $\mathbf{f}_a$, which is toward the wall. If the adhesive force is strong enough, the net force is down and to the left as shown. Since this force must be perpendicular to the surface (otherwise the molecule will slide along the surface), the surface curves upward, as shown in the figure. The cohesive and adhesive forces are difficult to calculate theoretically, but the angle of contact $\theta$ in Figure 13-12 can be measured. For water and glass this angle $\theta$ is about 25.5°. In Figure 13-12b the adhesive force is small compared with the cohesive force in the liquid, and the angle of contact is greater than 90°. For mercury and glass the angle of contact is about 140°.

Figure 13-13 shows a liquid that has risen to a height $h$ in a thin capillary tube of radius $r$. The tube is open to atmospheric pressure at the top. The force holding the liquid up is the vertical component of the surface tension, $F \cos \theta$. Since the length of the contact surface is $2\pi r$, this vertical force is $\gamma 2\pi r \cos \theta$. If the slight curvature of the surface is neglected, the volume of the liquid in the tube is $\pi r^2 h$. Setting the net upward force equal to the weight, we get

$$\gamma 2\pi r \cos \theta = \rho(\pi r^2 h)g$$

or

$$h = \frac{2\gamma \cos \theta}{\rho r g} \qquad\qquad 13\text{-}28$$

## Questions

8.  When the end of a glass capillary tube rests in a bowl of mercury, the level of mercury in the tube is lower than that outside. Explain the origin of the force that prevents the mercury column from reaching the same height as that outside the tube.

9.  What effect does the buoyant force have on a water bug walking on the surface of a lake?

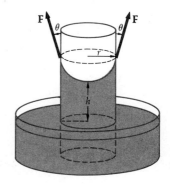

**Figure 13-13**
Fluid rising in a capillary. The upward force due to surface tension supports the weight of the column of fluid.

# 13-6 Bernoulli's Equation

The general flow of a fluid can be very complicated. Consider, for example, the rise of smoke from a burning cigarette. At first the smoke rises in a regular stream, but soon turbulence sets in and the smoke swirls irregularly. We shall be concerned only with nonturbulent steady-state flow. Instead of following the individual fluid particles as they move from point to point, we look at a fixed point in space. Let the density of the fluid at some point be $\rho$ and the velocity be $\mathbf{v}$ at some time $t$. As time passes, the fluid particles move downstream and new particles move into the region. In steady-state flow, the density and velocity of the fluid at any point are constant in time. In Figure 13-14 we have drawn lines of flow of the fluid particles. These lines are called *streamlines*. Since the particles move along these lines, there is no flow

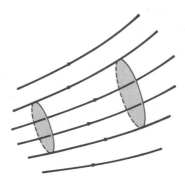

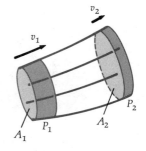

**Figure 13-14**
Streamlines of fluid flow and a tube of flow.

perpendicular to them. In the figure we have enclosed the lines to form a tube of flow. Although the density and velocity of the fluid vary from point to point in the tube, in steady-state flow, the velocity and density at one point are constant in time. Thus the total mass of the fluid in the tube remains constant in time.

In Figure 13-15 shading indicates the mass of fluid that flows into the tube from the left in some time interval $\Delta t$. The volume of this region is $v_1 \Delta t\, A_1$, where $v_1$ is the velocity at point $P_1$. If the density at that point is $\rho_1$, the mass of fluid flowing into the tube is given by

$$\Delta m = \rho_1 v_1 A_1 \, \Delta t \qquad \text{13-29}$$

The rate of fluid flow into the tube is called the *mass flow rate* $I_m$:

$$I_m = \frac{\Delta m}{\Delta t} = \rho_1 v_1 A_1 \qquad \text{13-30}$$

In steady-state flow an equal mass of fluid must flow out of the tube at point $P_2$. If the density at that point is $\rho_2$, the velocity is $v_2$, and the cross-sectional area is $A_2$, the mass flowing out of the tube is

$$\Delta m = \rho_2 v_2 A_2 \, \Delta t$$

Setting this equal to the mass flowing into the tube in the same time interval, we have

$$\rho_1 v_1 A_1 = \rho_2 v_2 A_2 \qquad \text{13-31}$$

The flow rate at points $P_1$ and $P_2$ must be equal. We therefore have the following important result:

**Figure 13-15**
In the steady state, the mass of fluid flowing in at point $P_1$ must equal the mass flowing out at point $P_2$. The flow rate in kilograms per second is $Av\rho$.

*Mass flow rate*

*In steady-state flow, the flow rate is the same at any point:*

| | |
|---|---|
| $I_m = \rho v A = \text{constant}$ | 13-32 |

Equation 13-32 is called the *equation of continuity*.

In the general steady-state flow of a fluid, layers of the fluid exert drag forces called *viscous forces* on adjacent layers. They are internal frictional forces which dissipate mechanical energy. In this section we consider nonviscous flow, in which such forces can be neglected. (We consider viscous flow in the next section.) In this case we can apply the work-energy theorem to the mass of fluid in a tube of flow. We shall also restrict our considerations to incompressible fluids, e.g., liquids for which the density is the same everywhere.

Figure 13-16*a* shows a fluid flowing through a pipe that changes in elevation and cross-sectional area. We apply the work-energy theorem to the fluid contained initially between points 1 and 2. After a short time $\Delta t$, this fluid will have moved along the pipe and will be contained between points 1' and 2', as shown in Figure 13-16*b*. If the flow is steady, the only change between Figure 13-16*a* and *b* is for the heavily shaded portions of the fluid mass. Let $\Delta m$ be this mass. From Equation 13-29,

$$\Delta m = \rho v_1 A_1 \, \Delta t = \rho v_2 A_2 \, \Delta t = \rho \, \Delta V$$

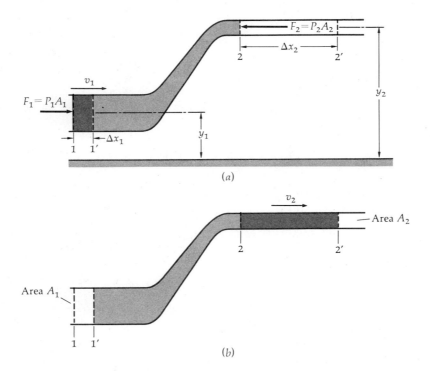

**Figure 13-16**
Fluid moving in a pipe for the derivation of Bernoulli's equation. The net work done by the forces $P_1 A_1$ and $P_2 A_2$ has the effect of raising the part of the fluid indicated by heavy shading from height $y_1$ to $y_2$ and changing its speed from $v_1$ to $v_2$.

where $\Delta V$ is the volume $vA \, \Delta t$, which is the same at each point because we are assuming the density to be constant. As the entire body of fluid being considered moves in time $\Delta t$, the fluid following it in the pipe exerts a force on it of magnitude $F_1 = P_1 A_1$, where $P_1$ is the pressure at point 1. This force does work $W_1 = F_1 \, \Delta x_1 = P_1 A_1 \, \Delta x_1 = P_1 \, \Delta V$. At the same time, the fluid preceding that under consideration exerts a force

$F_2 = P_2A_2$ to the left in the figure. This force does negative work (since it opposes the motion) $W_2 = -F_2 \, \Delta x_2 = -P_2A_2 \, \Delta x_2 = -P_2 \, \Delta V$. The net work done by these forces is

$$W = P_1 \, \Delta V - P_2 \, \Delta V = (P_1 - P_2) \, \Delta V \qquad \text{13-33}$$

This work equals the change in the kinetic energy and gravitational potential energy of the fluid under consideration, i.e., just the change in energy of the portion of mass $\Delta m = \rho \, \Delta V$. The change in potential energy of this mass is

$$\Delta U = \Delta m \, gy_2 - \Delta m \, gy_1$$

The change in kinetic energy is

$$\Delta E_k = \tfrac{1}{2}(\Delta m)v_2^2 - \tfrac{1}{2}(\Delta m)v_1^2$$

Thus the work-energy theorem gives

$$(P_1 - P_2) \, \Delta V = \Delta m \, gy_2 - \Delta m \, gy_1 + \tfrac{1}{2}(\Delta m)v_2^2 - \tfrac{1}{2}(\Delta m)v_1^2$$

If we divide each term by $\Delta V$ and write the density $\rho = \Delta m/\Delta V$, we obtain

$$P_1 - P_2 = \tfrac{1}{2}\rho v_2^2 - \tfrac{1}{2}\rho v_1^2 + \rho gy_2 - \rho gy_1$$

When we collect all the quantities with subscript 2 on one side and those with subscript 1 on the other, this equation becomes

$$P_1 + \rho gy_1 + \tfrac{1}{2}\rho v_1^2 = P_2 + \rho gy_2 + \tfrac{1}{2}\rho v_2^2 \qquad \text{13-34}$$

This result can be restated as

$$P + \rho gy + \tfrac{1}{2}\rho v^2 = \text{constant} \qquad \text{13-35}$$

*Bernoulli's equation*

meaning that this combination of quantities evaluated at any point along the pipe has the same value as at any other point. Equation 13-35 is known as *Bernoulli's equation* for steady, nonviscous flow of an incompressible fluid. To some extent Bernoulli's equation also applies to compressible fluids like gases.

A special application of Bernoulli's equation occurs for a fluid at rest. Then $v_1 = v_2 = 0$, and we obtain

$$P_1 - P_2 = \rho g(y_2 - y_1) = \rho gh$$

where $h = y_2 - y_1$ is the difference in height between points 2 and 1. This is the same as Equation 13-14. We now give some examples using Bernoulli's equation in nonstatic situations.

---

**Example 13-7** An incompressible fluid such as water flows through a horizontal pipe which has a constricted section as shown in Figure 13-17. Show that the pressure is *reduced* in the constriction.

Since both sections of the pipe are at the same elevation, we take $y_1 = y_2$ in Equation 13-34. We then have

$$P_1 + \tfrac{1}{2}\rho v_1^2 = P_2 + \tfrac{1}{2}\rho v_2^2 \qquad \text{13-36}$$

**Figure 13-17**
Constriction in a pipe carrying a moving fluid. The pressure is smaller in the narrow section of the pipe where the fluid is moving faster.

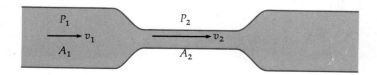

But from Equation 13-31 with constant density,

$$A_1 v_1 = A_2 v_2 \qquad \text{or} \qquad v_2 = \frac{A_1}{A_2} v_1$$

Substituting this result in Equation 13-36 gives

$$P_1 + \tfrac{1}{2}\rho v_1^2 = P_2 + \tfrac{1}{2}\rho \frac{A_1^2 v_1^2}{A_2^2}$$

or

$$P_1 - P_2 = \tfrac{1}{2}\rho \left( \frac{A_1^2}{A_2^2} - 1 \right) v_1^2 \qquad \text{13-37}$$

Since $A_1$ is larger than $A_2$, the quantity on the right is positive and $P_2$ must be less than $P_1$. Note that the only circumstance in which $P_2$ and $P_1$ can be equal, according to Equation 13-37, is when $v_1 = 0$; that is, when the fluid is at rest in the pipe.

---

**Example 13-8** A large water tower (Figure 13-18) is drained by a pipe of cross section $A$ through a valve a distance $h$ below the surface of the water in the tower. Show that the pressure in the pipe is reduced when the valve is opened.

In applying Equation 13-34 we take point 1 to be the top surface of the water in the tower, where the pressure is atmospheric pressure $P_a$. We also let $y_1 = h$. Point 2 is in the pipe just before the valve, and $y_2 = 0$. Because the surface area $A_1$ of the tower is so large compared with the area of the pipe, the speed $v_1$ with which the surface drops as the tower is drained can be neglected. Then, according to Equation 13-34,

$$P_a + \rho g h = P_2 + \tfrac{1}{2}\rho v_2^2$$

or

$$P_2 = P_a + \rho g h - \tfrac{1}{2}\rho v_2^2 \qquad \text{13-38}$$

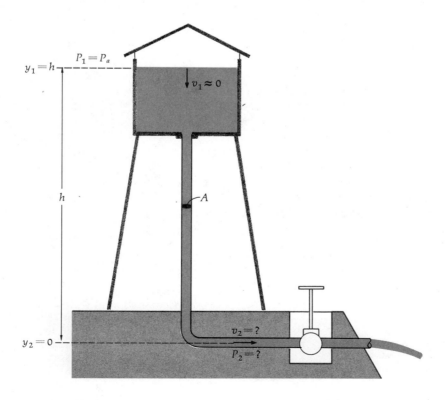

**Figure 13-18**
Water tower for Example 13-8. When the valve is opened, allowing water to flow out through the pipe, the pressure in the pipe is reduced.

If the valve is closed, the water is at rest and $v_2 = 0$. Then according to Equation 13-38, the pressure in the pipe at the valve is just $P_2 = P_a + \rho gh$. When the valve is opened, the water will move and $v_2$ will not be zero. Then, according to Equation 13-38, the pressure in the pipe will decrease by $\frac{1}{2}\rho v_2^2$.

**Example 13-9** Calculate the pressure $P_2$ in Example 13-8 if $h = 15$ m, $P_a = 101$ kPa, and $v_2 = 16$ m/s with the valve open.
   With the valve closed, $v_2 = 0$, and the pressure is

$$P_2 = P_a + \rho gh = 101 \text{ kPa} + (10^3 \text{ kg/m}^3)(9.81 \text{ N/kg})(15 \text{ m})$$
$$= 101 \text{ kPa} + 1.47 \times 10^5 \text{ N/m}^2 = 101 \text{ kPa} + 147 \text{ kPa}$$
$$= 248 \text{ kPa}$$

With the valve open, the pressure is decreased by the amount

$$\frac{1}{2}\rho v_2^2 = \frac{1}{2}(10^3 \text{ kg/m}^3)(16 \text{ m/s})^2$$
$$= 1.28 \times 10^5 \text{ kg·m/s}^2\text{·m}^2 = 128 \text{ kPa}$$

The pressure with the valve open is therefore $P_2 = 101$ kPa + 147 kPa − 128 kPa = 120 kPa.

Although Bernoulli's equation is very useful for a qualitative description of many features of fluid flow, it is often grossly inaccurate when compared quantitatively with experiment. Gases like air, of course, are hardly incompressible, and liquids like water have viscosity which invalidates the assumption of conservation of mechanical energy. In addition, it is often difficult to maintain steady-state streamline flow without turbulence.

**Question**

10. In Figure 13-17 the water entering the narrow part of the pipe is accelerated to a greater speed. What forces act on this water to produce this acceleration?

# 13-7  Viscous Flow

According to Bernoulli's equation, if a fluid flows steadily through a long narrow horizontal pipe of constant cross section, the pressure will be constant along the pipe. In practice, we observe a pressure drop as we move along the direction of the flow. This drop is due to the viscosity of the fluid. The pipe exerts a resistive drag on the fluid next to it, and the fluid layers exert a viscous drag force on adjacent layers. The fluid velocity is also not constant along a diameter of the pipe but is greatest near the center and least near the edge, where the fluid is in contact with the walls of the pipe (Figure 13-19). Let $P_1$ be the pressure at point 1 and $P_2$ be the pressure at point 2 a distance $L$ downstream from point 1. The pressure drop $\Delta P = P_1 - P_2$ is proportional to the flow rate. If we call the proportionality constant $R$, we have

$$\Delta P = P_1 - P_2 = I_V R \qquad\qquad \text{13-39}$$

where

$$I_V = vA \qquad\qquad \text{13-40} \qquad \textit{Volume flow rate}$$

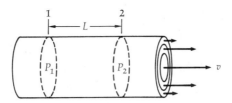

**Figure 13-19**
When a viscous fluid flows through a pipe, the speed is greater at the center of the pipe. Near the walls of the pipe, the fluid tends to remain stationary.

is the volume flow rate. (The mass flow rate $I_m$ in Equation 13-30 is just the density times the volume flow rate.) Equation 13-39 is analogous to Ohm's law (Chapter 24) for the drop in voltage along a current-carrying wire. The resistance to flow $R$ depends on the length of the pipe $L$, the radius $r$, and the viscosity of the fluid.

The coefficient of viscosity of a fluid is defined as follows. In Figure 13-20 a fluid is confined between two parallel plates, each of area $A$, separated by a distance $z_0$. The upper plate is pulled at constant velocity $v$ by a force $F$, while the bottom plate is held at rest. A force is needed to pull the upper plate because the fluid next to the plate exerts a viscous drag force opposing the motion. The velocity of the fluid between the plates is essentially $v$ near the upper plate and zero near the lower plate and varies linearly with height between the plates. The force $F$ is found to be inversely proportional to the plate separation $z_0$. The coefficient of viscosity $\eta$ is defined by

$$F = \eta \frac{vA}{z_0}$$                                    13-41    *Coefficient of viscosity*

The SI unit of viscosity is the N·s/m² = Pa·s. An older, cgs unit in common use is the dyn·s/cm², called a *poise* after the French physicist Poiseuille. These units are related by

$$1 \text{ Pa·s} = 10 \text{ poises}$$                              13-42

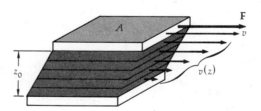

*Coefficient of viscosity*

**Figure 13-20**
Two plates of equal area $A$ with a viscous fluid between them. When the upper plate is moved relative to the lower plate, each layer of fluid exerts a drag force on adjacent layers, setting up a velocity gradient in the fluid. The force required to pull the upper plate is proportional to the speed $v$ and area $A$ and inversely proportional to the separation of the plates $z_0$.

Table 13-3 lists the coefficient of viscosity for various fluids.

In terms of the coefficient of viscosity, the resistance to fluid flow $R$ in Equation 13-39 for steady flow through a circular tube of radius $r$ can be shown to be

$$R = \frac{8\eta L}{\pi r^4}$$                                    13-43

Equations 13-39 and 13-43 can be combined to give the flow rate through a circular tube in terms of the pressure drop:

$$I_V = \frac{\pi r^4}{8\eta L} \Delta P$$                          13-44    *Poiseuille's law*

Equation 13-44, known as Poiseuille's law, can be derived similarly to Bernoulli's equation by setting the net work done by pressure forces external to the fluid equal to the work done against the viscous force exerted by the tube on the fluid. The derivation is complicated and will not be considered here. Note the inverse $r^4$ dependence of the resistance to fluid flow. If the radius of the tube is halved, the pressure drop for a given flow rate and viscosity is increased by a factor of 16. For water flowing through a long garden hose, the pressure drop is that from the water source to the open end of the hose at atmospheric pressure. The flow rate is then proportional to the fourth power of the radius. If the radius is halved, the flow rate drops by a factor of 16.

**Table 13-3**
Coefficient of viscosity of various fluids

| Fluid | $t$, °C | $\eta$, mPa·s |
|---|---|---|
| Water | 0 | 1.8 |
| | 20 | 1.00 |
| | 60 | 0.65 |
| Blood (whole) | 37 | 4.0 |
| Engine oil (SAE 10) | 30 | 200 |
| Glycerin | 0 | 10,000 |
| | 20 | 1,410 |
| | 60 | 81 |
| Air | 20 | 0.018 |

### Review

A. Define, explain, or otherwise identify:

Density, 370
Weight density, 370
Stress, 371
Strain, 371
Young's modulus, 371
Shear modulus, 372
Bulk modulus, 372
Pascal, 374
Pascal's principle, 375

Gauge pressure, 376
Buoyant force, 379
Archimedes' principle, 379
Surface tension, 382
Capillarity, 383
Equation of continuity, 386
Bernoulli's equation, 387
Coefficient of viscosity, 390
Poiseuille's law, 390

B. True or false:

1. The buoyant force on a submerged object depends on the shape of the object.

2. Young's modulus has the same dimensions as pressure.

3. If the density of an object is greater than that of water, it cannot float on the surface of water.

4. In viscous flow, the pressure drop along a pipe is proportional to the flow rate.

### Exercises

#### Section 13-1, Density

1. What is the mass of a copper sphere of radius 2 cm?

2. A station wagon can carry a maximum load of 550 kg (plus driver). How many steel rods can be transported in the wagon if each rod is 1.4 m long and has a diameter of 2.4 cm?

3. A small flask used for measuring densities of liquids (called a pycnometer) has a mass of 22.71 g. When it is filled with water, the total mass is 153.38 g; and when it is filled with milk, the total mass is 157.67 g. Find the density of milk.

4. A solid-oak door is 200 cm high, 75 cm wide, and 4 cm thick. How much does it weigh?

5. A 60-mL flask is filled with mercury at 0°C. When the temperature rises to 80°C, 1.47 g of mercury spills out of the flask. Assuming the volume of the flask to stay constant, find the density of mercury at 80°C if its density at 0°C is 13,645 kg/m$^3$.

#### Section 13-2, Stress and Strain

6. A wire 1.5 m long has a cross-sectional area of 2.4 mm$^2$. It is hung vertically and stretches 0.29 mm when an 8.5-kg block is attached to it. Find (a) the stress, (b) the strain, and (c) Young's modulus for the wire.

7. Copper wire has a breaking stress of about $3 \times 10^8$ N/m². (*a*) What is the maximum load that can be hung from a copper wire of diameter 0.42 mm? (*b*) If half this maximum load is hung from the copper wire, by what percentage of its length will it stretch?

8. What pressure is required to reduce the volume of 1 kg of water from 1.00 to 0.99 L?

9. Seawater has a bulk modulus of $23 \times 10^{10}$ N/m². Find the density of seawater at a depth where the pressure is 800 atm if the density at the surface is 1024 kg/m³.

10. As a runner's foot touches the ground, the shearing force acting on the 8-mm-thick sole is as shown in Figure 13-21. If the force of 25 N is distributed over an area of 15 cm², find the angle of shear $\theta$ shown given that the shear modulus of the sole is $1.9 \times 10^5$ N/m².

11. A length $L$ of copper wire of diameter 1.2 mm is joined to a length $2L$ of steel wire 0.8 mm in diameter, and is hung vertically. When a 10-kg load is suspended from the lower end, the total elongation is 0.65 mm. Find $L$.

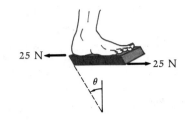

**Figure 13-21**
Shear forces on a shoe sole for Exercise 10.

### Section 13-3, Pressure in a Fluid

12. The pressure on the surface of a lake is atmospheric pressure $P_a = 101$ kPa. (*a*) At what depth is the pressure twice atmospheric? (*b*) If the pressure at the top of a deep pool of mercury is $P_a$, at what depth is the pressure $2P_a$?

13. (*a*) Find the absolute pressure at the bottom of a diving pool of depth 5.0 m. (*b*) Find the gauge pressure at the same depth.

14. A vertical U tube has a cross-sectional area of 1.40 cm² and contains 75 mL of mercury ($\rho = 13.6$ kg/L). If 25 mL of water is poured into one arm, find the difference in the level between the water-air and mercury-air surfaces.

15. A manometer is a U tube of liquid that can measure small pressure differences between the two arms. If an oil manometer ($\rho = 900$ kg/m³) can be read to ±0.05 mm, what is the smallest pressure change that can be detected?

16. What is the smallest change in altitude that can be detected by the manometer in Exercise 15? *Hint:* Equation 13-19 can be used to approximate $\Delta P / \Delta y$ because the changes are small.

### Section 13-4, Archimedes' Principle

17. A 1-kg piece of copper (specific gravity 9) is submerged in water and suspended from a spring balance (Figure 13-22). (*a*) What force does the balance read? (*b*) The whole system is dropped out of the window. Describe the initial motion of the copper block relative to the jar.

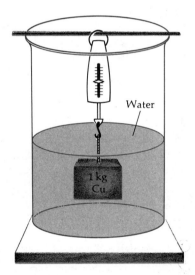

**Figure 13-22**
Exercise 17.

18. It is sometimes said that only the tip of an iceberg shows and that 90 percent lies beneath the surface. (*a*) Find the density of ice if this is true, assuming the density of seawater to be 1025 kg/m³. (*b*) What fraction of an ice cube is beneath the surface in a glass of water on the moon, where the acceleration of gravity is about one-sixth that on earth?

19. When a 60-N stone is attached to a scale and submerged in water, the scale reads 40 N. (*a*) Find the specific gravity of the stone. (*b*) What is the volume of the stone?

20. A 5-kg iron block is attached to a spring scale and submerged in a fluid of unknown density. The spring scale reads 6.16 N. What is the density of the fluid?

21. A 3- by 3-m raft is 10-cm thick and made of wood with specific gravity 0.6. How many 70-kg people can stand on the raft without getting their feet wet when the water is calm?

### Section 13-5, Surface Tension and Capillarity

22. A wire 10.0 cm long and 0.8 mm in diameter is pulled out of a water surface with its length parallel to the surface. What force in addition to the weight of the wire is needed?

23. What is the size of a steel sphere that will float on water with exactly half the sphere submerged, if the density of steel is $7.9 \times 10^3$ kg/m$^3$?

24. When a capillary with diameter 0.80 mm is dipped into methanol, the methanol rises to a height of 15.0 mm. If the angle of contact is 0°, find the surface tension of methanol (density $0.79 \times 10^3$ kg/m$^3$).

### Section 13-6, Bernoulli's Equation

25. Water flows through a 3-cm-diameter hose at 0.65 m/s. The diameter of the nozzle is 0.30 cm. (*a*) At what speed does the water pass through the nozzle? (*b*) If the pump at one end and the nozzle at the other end are at the same height and the pressure at the nozzle is atmospheric, what is the pressure at the pump?

26. A large tank of water is tapped a distance *h* below the water surface by a small pipe, as shown in Figure 13-23. (*a*) Why is the pressure the same at points *a* and *b*? (*b*) Show that the speed of the water emerging at point *b* is $\sqrt{2gh}$ in the approximation that the speed at point *a* is negligible. (This is *Torricelli's law*.) Why is this a good approximation?

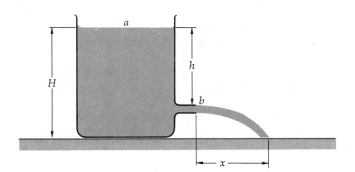

**Figure 13-23**
Exercise 26 and Problem 19.

27. Water is flowing at 3 m/s in a horizontal pipe under a pressure of 200 kPa. The pipe narrows to half its original diameter. (*a*) What is the speed of flow in the narrow section? (*b*) What is the pressure in the narrower section? (*c*) Compare the flow rates in the two sections.

28. The pressure in a section of 2-cm-diameter horizontal pipe is 142 kPa. Water flows through the pipe at 2.80 $L$/s. What should the diameter of a constricted section of the pipe be for the pressure to be atmospheric pressure (101 kPa)?

### Section 13-7, Viscous Flow

29. A horizontal tube with inside diameter 1.2 mm and length 25 cm has water flowing through it at 0.30 mL/s. Find the pressure difference required to drive this flow if the viscosity of water is 1.00 mPa·s.

30. Find the diameter of a tube that would give double the flow rate for the same pressure difference as in Exercise 29.

31. Water is flowing in a canal with vertical sides so that 2 m from the edge the flow velocity is 1.6 m/s, while the water is stationary at the edge. Find the drag force per square metre on the vertical canal wall if the viscosity of water is 1.00 mPa·s.

32. Blood flowing through a 1-mm-long capillary of the human circulating system takes about 1.0 s to pass through. If the diameter of the capillary is 7 $\mu$m and the pressure drop is 2.60 kPa, find the viscosity of the blood.

## Problems

Figure 13-24
Problem 3.

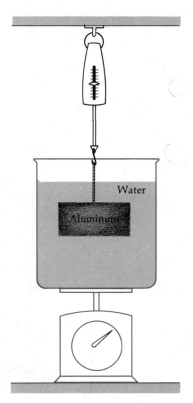

1. An object has neutral buoyancy when its density equals that of the liquid in which it is submerged, so that it neither floats nor sinks. If the average density of a human body is $0.96 \times 10^3$ kg/m³, what mass of lead should be added to give neutral bouyancy to an 85-kg diver in fresh water?

2. When you weigh yourself in air, the scale reading is less than the force exerted by gravity on you because of the bouyant force of air. Estimate the correction to the scale reading you should make to get your true weight.

3. A beaker of mass 1 kg contains 2 kg of water and rests on a scale. A 2-kg block of aluminum (specific gravity 2.70) suspended from a spring scale is submerged in water, as shown in Figure 13-24. Find the readings of both scales.

4. A rectangular dam 30 m wide supports a body of water to a height of 25 m. (a) Neglecting atmospheric pressure, find the total force due to water pressure acting on a thin strip of height $dy$ located at depth $y$. (b) Integrate your result in part (a) to find the total horizontal force exerted by the water on the dam. Why is it reasonable to neglect atmospheric pressure?

5. A water tank near the earth's surface is accelerated upward at a constant rate of $5g$. (a) Find the "buoyant force" acting on a 1-kg element of the water in the tank. (b) How does this force compare with the weight of a piece of aluminum having the same volume as 1 kg of water? Will aluminum float in this tank while the tank is accelerating up? Explain.

6. A long wire of length $L$ is stretched by a force $F$ applied to it. Show that if the wire is considered to be a spring, the force constant $k$ is given by $k = AY/L$ and the energy stored in the wire is $U = \frac{1}{2}F\,\Delta\ell$, where $\Delta\ell$ is the amount stretched.

7. When a rubber strip with a 3- by 1.5-mm cross section is suspended vertically, a student obtains the following data for the length versus load:

| Load, g | 0 | 100 | 200 | 300 | 400 | 500 |
|---|---|---|---|---|---|---|
| Length, cm | 5.0 | 5.6 | 6.2 | 6.9 | 7.8 | 10.0 |

(a) Find Young's modulus for the rubber for small loads. (b) Find the energy stored in the strip when the load is 200 g (see Problem 6).

8. The steel E string of a violin is under a tension of 53 N. The diameter of the string is 0.20 mm, and its length under tension is 35.0 cm. Find (a) the unstretched length of this string and (b) the work needed to stretch the string.

9. A steel wire of diameter 0.40 mm is stretched between rigid supports separated by a horizontal distance of 1.8 m. (a) What load suspended from the middle of the wire will produce a stress of $10^9$ N/m²? (b) Find the stored elastic energy in the wire under this load. (c) Find the drop in potential energy of the object hung from the center for this stress.

10. A fountain designed to spray a column of water 12 m into the air has a 1-cm-diameter nozzle at ground level. The water pump is 3 m below the ground. The pipe to the fountain has a diameter of 2 cm. Find the necessary pump pressure (neglect the viscosity of water).

11. The demolition of a building is to be accomplished by swinging a 400-kg steel ball on the end of a 30-m steel wire of diameter 5.0 mm hanging from a tall crane. The ball is swung through an arc from side to side, the wire making an angle of 50° with the vertical at the top of the swing. (a) Neglecting the energy stored in the stretched wire, find the amount by which the wire is stretched at the bottom of the swing. (b) Use your answer in (a) to estimate the energy stored in the stretched wire at the bottom of the swing.

12. The volume of a cone of height $h$ and base radius $r$ is $V = \frac{1}{3}\pi r^2 h$. A conical vessel of height 25 cm resting on its base of radius 15 cm is filled with water. (a) Find the volume and weight of the water in the vessel. (b) Find the force exerted by the water on the base of the vessel. Explain how this force can be greater than the weight of the water.

13. A ship sails from seawater (specific gravity 1.03) into fresh water and therefore sinks slightly. When its load of 600 Mg is removed, it returns to its original level. Assuming that the sides of the ship are vertical at the waterline, find the mass of the ship before it was unloaded.

14. The hydrometer shown in Figure 13-25 is a device for measuring liquid density. The bulb contains lead shot, and the liquid density can be read directly from the liquid level on the stem when calibrated. The bulb's volume is 20 mL, the stem is 15 cm long with diameter 5.00 mm, and the mass of the glass is 6.0 g. What mass of lead shot must be added so that the least density of liquid that can be measured is $0.90 \times 10^3$ kg/m³?

15. A hollow can with a small hole of radius 0.1 mm is pushed under water. At what depth will the water start to flow into the can through the hole if the surface tension of water is 0.073 N/m?

16. The water in the cylindrical container drains through a horizontal capillary with diameter 0.50 mm (Figure 13-26). Calculate the length of time needed for the water height in the cylinder to drop from 10.0 to 5.0 cm if the viscosity of water is 1.00 mPa·s.

17. Prove that for a spherical bubble of radius $R$ in a liquid of surface tension $\gamma$ the excess pressure inside the bubble is given by $p = 2\gamma/R$.

18. A large beer keg of height $H$ and area $A_1$ is filled with beer. The top is open to atmospheric pressure. At the bottom is a spigot opening of area $A_2$, much smaller than $A_1$. (a) Show that the speed of the beer leaving the spigot is approximately $\sqrt{2gh}$ when the height of the beer is $h$. (b) Show that in the approximation $A_2 \ll A_1$, the rate of change of height $h$ of the beer is given by $dh/dt = -(A_2/A_1)(2gh)^{1/2}$. (c) Find $h$ as a function of time if $h = H$ at $t = 0$. (d) Find the total time needed to drain the keg if $H = 2$ m, $A_1 = 0.8$ m², and $A_2 = 10^{-4}A_1$.

19. For Figure 13-23, (a) find the distance $x$ at which the water strikes the ground as a function of $h$ and $H$. (b) Show that any two values of $h$ equidistant from the point $h = \frac{1}{2}H$ give the same distance $x$. (c) Show that $x$ is a maximum when $h = \frac{1}{2}H$. What is the value of this maximum distance?

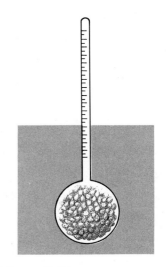

**Figure 13-25**
Problem 14.

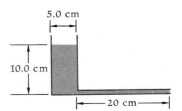

**Figure 13-26**
Problem 16.

# CHAPTER 14    Wave Motion

**Objectives**  After studying this chapter, you should:

1. Be able to state the meaning of: dispersion, transverse wave, longitudinal wave, superposition, harmonic wave, intensity, intensity level.

2. Know on what quantities the speed of a mechanical wave depends.

3. Be able to state the relationships between the speed $v$, period $T$, frequency $f$, wavelength $\lambda$, angular frequency $\omega$, and wave number $k$ for a harmonic wave.

4. Be able to derive expressions for the doppler frequency shift for a moving source or a moving receiver, and use these expressions in working problems.

5. Be able to sketch the standing-wave patterns for vibrating strings and vibrating air columns in organ pipes, and from them obtain the possible frequencies for standing waves.

6. Know how the intensity of a wave depends on its amplitude.

7. Be able to calculate the intensity level in decibels from the intensity in watts per square metre and vice versa.

Wave motion can be thought of as the transport of energy and momentum from one point in space to another without the transport of matter. In mechanical waves, e.g., water waves, waves on a string, or sound waves, the energy and momentum are transported by means of a disturbance in the medium that is propagated because the medium has elastic properties. On the other hand, in electromagnetic waves, the energy and momentum are carried by electric and magnetic fields, which can propagate through vacuum (electric field and magnetic field are defined in later chapters).

Although the variety of wave phenomena observed in nature is immense, many features are common to all kinds of waves and others are shared by a wide range of wave phenomena. In this chapter and the next we study waves on strings and sound waves. Many of the ideas and results discussed here will be applied in later chapters when we study light and other electromagnetic waves.

# 14-1  Wave Pulses

When a string stretched under tension is given a flip, as in Figure 14-1, the shape of the string changes in time in a regular way. The bump that is produced at the end by the flip and that travels down the string is called a *wave pulse*. It is a disturbance in the string, i.e., a distortion of the shape of the string from its normal, or equilibrium, shape. The pulse travels down the string at a definite speed which depends on the nature of the string and the tension. As it moves, the pulse usually changes shape, gradually spreading out. This effect, called *dispersion*, occurs to some extent in all waves (except electromagnetic waves in vacuum), but in many important examples the dispersion is negligible and the wave pulse travels with approximately the same shape. The fate of the pulse at the other end of the string depends on how the string is fastened there. If it is tied to a rigid support, the pulse will be reflected and return inverted, as in Figure 14-2. If the support is not rigid, some or all of the pulse will be absorbed. If the string is tied to another string of different mass density, part of the pulse will be transmitted and part reflected. If the second string is heavier than the first, the reflected part of the pulse will be inverted; if it is lighter, the reflected pulse will not be inverted. If the second string is very light, e.g., a thread, so that the end of the string is essentially free to move, the pulse will be totally reflected and not inverted (Figure 14-3).

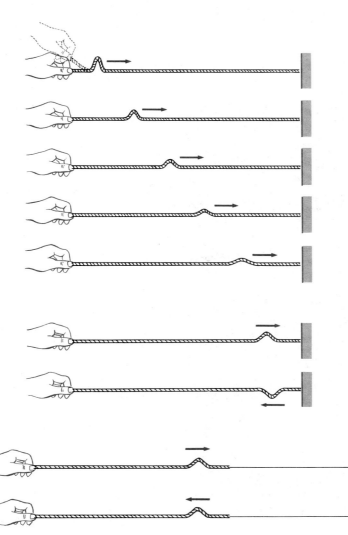

**Figure 14-1**
Wave pulse moving to the right on a stretched string. The change in shape of the pulse as it moves is called dispersion.

**Figure 14-2**
A pulse arriving at the rigid support of a stretched string is reflected and inverted.

**Figure 14-3**
A pulse arriving at a free end of the string is reflected but not inverted.

A pulse traveling down a stretched string is only one example of a wave pulse: the report of a gunshot is a wave pulse of sound; a lightning flash is a wave pulse of light; a tidal wave is a wave pulse in water. A chief characteristic of a wave pulse is that it has a beginning and an end. It is a disturbance of limited extent. At any instant only a limited region of space is disturbed. The wave pulse passes by any point in a limited time.

We can demonstrate that energy and momentum are transported by a wave pulse by hanging a weight on a string under tension, as shown in Figure 14-4, and giving the string a flip at one end. When the pulse arrives at the weight, the weight is lifted momentarily. The action of the hand in flipping one end of the string is transmitted along the string until the weight is lifted at the other end. This is usually described by saying that the momentum introduced by the hand flipping the end of the string is transmitted along the string and is received by the weight.

**Figure 14-4**
Wave pulses transmit both energy and momentum, indicated here by the upward motion of the weight when the pulse arrives.

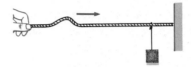

Similarly, the work done by the hand in lifting the string goes into kinetic and potential energy of the string at the pulse. (We calculate this energy in the string later.) This energy is transmitted along the string and is converted into work lifting the weight. Thus *both energy and momentum are transmitted by such a wave pulse.*

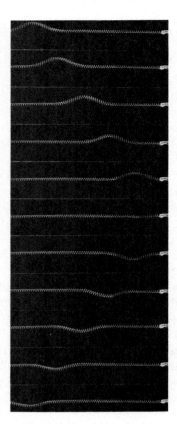

Reflection of a pulse from a fixed end. The reflected pulse is inverted. (*From* PSSC Physics, *2d ed., p. 260, D. C. Heath and Co., Lexington, Mass., 1965.*)

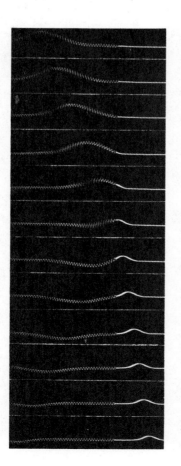
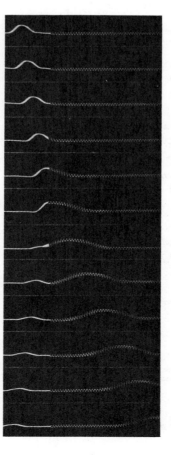

(Left) Pulse passing from a light spring to a heavy spring. The wave is partially transmitted and partially reflected. The reflected pulse is inverted. (Right) Pulse passing from a heavy spring to a light spring. The wave is partially transmitted and partially reflected. The reflected pulse is not inverted in this case. (*From* PSSC Physics, *2d ed., pp. 261 and 262, D. C. Heath and Co., Lexington, Mass., 1965.*)

It is not the mass elements of the string that are transported but the *disturbance in the shape* caused by flipping one end. The mass elements of the string, in fact, move in a direction perpendicular to the string and thus perpendicular to the direction of motion of the pulse. A wave in which the disturbance is perpendicular to the direction of propagation is called a *transverse wave*. Other examples of transverse waves are any electromagnetic wave (light, radar, radio, television, etc.) produced by a vibrating charge system which produces an alternating electric and magnetic field. The electric and magnetic field vectors are perpendicular to the direction of propagation of the electromagnetic wave. Sound waves in air, on the other hand, are not transverse waves. In sound waves, a disturbance in the pressure and density of air is set up by the vibration of a body (say a tuning fork or a violin string), and the disturbance is propagated through the air by the collisions of air molecules. The motion of the air molecules is in the same direction as the propagation. Any wave in which the disturbance is parallel to the direction of propagation is called a *longitudinal wave*. Sound consists of longitudinal waves. A longitudinal pulse in a spring analogous to a sound pulse can be produced by suddenly compressing the spring (Figure 14-5).

*Transverse wave*

*Longitudinal wave*

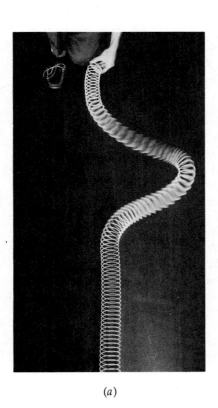 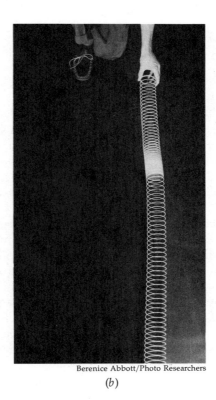

Berenice Abbott/Photo Researchers

(a)                    (b)

**Figure 14-5**
(a) Transverse wave pulse in a spring. The disturbance is perpendicular to the direction of motion of the wave. (b) Longitudinal wave pulse in a spring. The disturbance is in the direction of motion of the wave.

Water waves are neither completely transverse nor completely longitudinal, but a combination of the two. Figure 14-6 shows the motion of a water particle in a water wave. The water particle moves in a nearly circular path, its motion having both transverse and longitudinal components.

**Figure 14-6**
Surface waves on water. The water particles on the surface move in nearly circular paths having both longitudinal and transverse components.

Figure 14-7 shows a pulse on a string at time $t = 0$. The shape of the string at this time can be represented by some function $y = f(x)$. At some later time the pulse is farther down the string, so that its shape then is some other function of $x$. Let us assume that the pulse does not vary in shape, and introduce a new coordinate system with origin $O'$ which moves with the speed $v$ of the pulse. In this reference frame the pulse is stationary. The shape of the string is $y' = f(x')$ for all time. The coordinates of the two reference frames are related by

$$y = y' \quad \text{and} \quad x = x' + vt$$

Thus, the displacement of the string in frame $O$ can be written

$$y = f(x - vt) \quad \text{wave moving right} \quad 14\text{-}1$$

This same line of reasoning for a pulse moving to the left leads to

$$y = f(x + vt) \quad \text{wave moving left} \quad 14\text{-}2$$

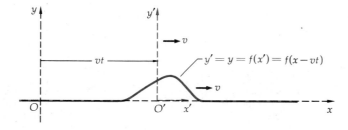

In both these expressions, $v$ is the speed of propagation of the wave. The function $y = f(x - vt)$ is called the *wave function*. For waves on a string, the wave function represents the vertical displacement of the string at point $x$ and time $t$. Hence this wave function is a function of two variables. Analogous wave functions for sound waves are the pressure $P(x - vt)$ and a related function, the displacement of gas molecules from the equilibrium position.

Consider two pulses on a string moving in opposite directions, as in Figure 14-8. The shape of the string when they meet can be found by adding the displacements produced by each pulse separately, as in the figure. In the special case where two pulses are identical except that one is inverted relative to the other (Figure 14-9) there will be one time when the pulses exactly overlap and add to zero. At this time the string is horizontal, but it is not at rest. A short time later the pulses emerge, each continuing in its original direction. Combining separate waves to produce the resultant wave is called *interference*. Interference is the characteristic property of wave motion. It occurs whenever two waves meet in the same region of space. There is no analogous situation in particle motion. Two particles never overlap or add together in this way. Interference is unique to wave motion.

If we have two pulses on a string, as in Figures 14-8 and 14-9, we cannot write the shape of the string as a simple function of either $x - vt$ or $x + vt$ alone. However, the shape is still a function of the two variables $x$ and $t$. Let us write $y(x, t)$ for the wave function of the string on which there are two pulses; that is, $y(x, t)$ is the vertical displacement of the point $x$ at time $t$. We can treat the phenomenon of interference mathematically by noting that if we call the wave function for the single pulse moving to the right $y_1(x - vt)$ and that for the pulse moving to the left $y_2(x + vt)$, the total wave function is just the algebraic sum of the individual wave functions

$$y(x, t) = y_1(x - vt) + y_2(x + vt) \quad 14\text{-}3$$

**Figure 14-7**
A wave pulse moving without change in shape in the positive $x$ direction with speed $v$ relative to origin $O$. In the primed coordinate system moving with the pulse, the wave function is $y' = f(x')$ for all time. In the unprimed system, the wave function is $y = f(x - vt)$.

*Wave function*

*Interference*

For example, if we have two pulses of the same shape but one inverted, there will be some time $t = t_1$ when they overlap. At this time, $y_1(x - vt_1) = -y_2(x + vt_1)$, so that $y(x, t_1) = 0$.

The mathematical addition of two wave functions to form the resulting wave function, as in Equation 14-3, is called *superposition*. The statement that the resultant wave function is the algebraic sum of the individual wave functions is called the *principle of superposition*. In most wave phenomena small wave pulses obey this principle. Electromagnetic waves in vacuum always obey it. For other wave phenomena, the principle of superposition does not hold for very large pulses; i.e., the wave function produced by the interference of two very large pulses is not merely the algebraic sum of the individual wave pulses. These *nonlinear waves* are not studied in this book.

If one of the two pulses is inverted relative to the other, as in Figure 14-9, the algebraic addition when the pulses meet amounts to an arithmetic subtraction. The resultant pulse is smaller than the larger pulse and perhaps smaller than either. The pulses tend to cancel each other when they overlap. This is *destructive interference*. (Complete cancellation occurs only if the shapes are identical.)

On the other hand, if the displacement of the two pulses is in the same direction, the resultant pulse when they overlap is greater than either pulse by itself. This is *constructive interference*.

#### Questions

1. Give examples of wave pulses in nature in addition to those mentioned in the text. In each case explain what kind of disturbance from equilibrium occurs.

2. Consider a long line of cars equally spaced by one car length and moving with the same speed. One car suddenly slows to avoid a dog and then speeds up until it is again one car length behind the car ahead. Discuss how the space between cars propagates back along the line. How is this like a wave pulse? Is there a transport of energy and momentum? Is the wave transverse or longitudinal? What does the speed of propagation depend on? (It may be useful to consider this situation in a reference frame where all the cars are initially at rest.)

3. When two waves moving in opposite directions interfere, does either impede the progress of the other?

## 14-2 Speed of Waves

A general property of waves is that their speed depends on the properties of the medium and is independent of the motion of the source relative to the medium. (No medium is necessary for light and other electromagnetic waves, which can propagate through a vacuum. The speed of the electromagnetic waves in vacuum is fixed at about $3 \times 10^8$ m/s.) For example, the speed of a sound wave produced by a train whistle depends only on the properties of the air and not on the motion of the train. For wave pulses on a string which do not change their shape we can derive from laws of mechanics an expression for the speed in terms of the properties of the string.

**Figure 14-8**
Two wave pulses moving in opposite directions on a string. The shape of the string when the pulses meet is found by adding the displacements of each separate pulse. This kind of wave superposition is called constructive interference.

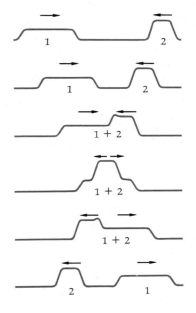

**Figure 14-9**
Superposition of pulses having opposite displacements. Here the algebraic addition of the displacements of the separate pulses amounts to subtraction of the magnitudes. This is called destructive interference.

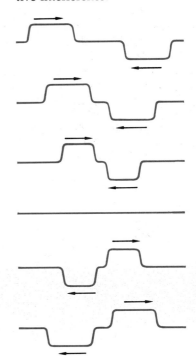

Consider a pulse moving to the right along a string with speed $v$. The speed $v$ is related to the tension $T$ in the string and the mass per unit-length $\mu$ of the string by

$$v = \sqrt{\frac{T}{\mu}}$$

14-4

The derivation we give of this result is somewhat artificial but illustrates that Equation 14-4 follows from the application of Newton's laws to the parts of the string. A more general derivation will be given in Section 14-8.

Consider the pulse in a reference frame $O'$, in which it is at rest. In this frame the string moves to the left with speed $v$, and the pulse is stationary. It is useful to imagine a glass tube of the same shape as the pulse and at rest in this frame, as in Figure 14-10$a$. We then have the remarkable phenomenon of the string passing through the tube without hitting the sides. A small segment of the string (Figure 14-10$b$) is moving in an approximately circular arc of radius $r$. The tension is just great enough to give the segment the centripetal accleration $v^2/r$. Equation 14-4 is derived by setting the resultant force on the segment due to the tension equal to the mass of the segment times its centripetal acceleration.

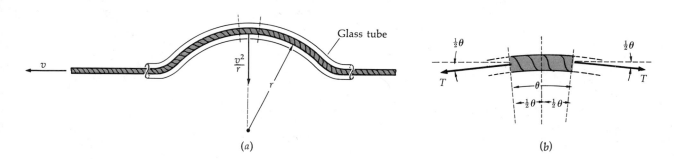

(a)                                          (b)

**Figure 14-10**
(a) A string under tension is pulled through a tube with speed $v$ so that the string does not touch the tube. (b) Forces on a string segment moving in a circular arc. The centripetal acceleration of the segment is provided by the radial components of the tension.

Let $\theta$ be the angle subtended by the segment. Figure 14-10$b$ shows that the tension $T$ at each end of the segment makes an angle $\tfrac{1}{2}\theta$ with the tangent line at the center of the segment, which is horizontal for this segment. The radial component of the tension (vertical in this figure) at each end is $T \sin \tfrac{1}{2}\theta$. Since this tension acts at both ends, the resultant radial force, which points toward the center of the circle, has the magnitude

$$F_r = 2T \sin \tfrac{1}{2}\theta \approx 2T(\tfrac{1}{2}\theta) = T\theta$$

assuming that the angle $\theta$ is small enough to permit the approximation $\sin \tfrac{1}{2}\theta \approx \tfrac{1}{2}\theta$ to be used. The length of the segment is $r\theta$, where $r$ is the radius of the circle. Hence the mass of the segment is

$$m = \mu L = \mu r\theta$$

where $\mu$ is the mass per unit length of the string (the linear mass density). Setting the resultant force $T\theta$ equal to the mass $\mu r\theta$ times the acceleration $v^2/r$, we obtain

$$T\theta = \mu r\theta \frac{v^2}{r}$$

or

$$T = \mu v^2 \qquad v = \sqrt{\frac{T}{\mu}}$$

Since this velocity is independent of $r$ and $\theta$, this result holds for all segments of the string. However, the derivation depends on the assumption that the angle $\theta$ is small. This will be true if the height of the pulse is small compared with its length or, alternatively, if the slope of the string is small.

In our original reference frame, the string is of course fixed, and the pulse moves with velocity $v = \sqrt{T/\mu}$ along the string without changing shape.

Sound waves are longitudinal waves of compression and rarefaction of a medium such as air. The velocity of a sound wave also depends on the properties of the medium through which the wave propagates. For waves in which the pressure variation is not too great, the velocity of sound $v$ is given by

$$v = \sqrt{\frac{B}{\rho_0}} \qquad\qquad 14\text{-}5$$

where $\rho_0$ is the equilibrium density of the medium and $B$ is the bulk modulus (Section 13-2). Comparing Equations 14-5 and 14-4 for the speed of waves on a string, we see that the wave speed in both depends on (1) an elastic property of the medium (the tension $T$ for string waves and the bulk modulus $B$ for compression waves) and (2) an inertial property of the medium (the linear mass density or the volume mass density). Equation 14-5 follows from the application of Newton's laws to compressions and rarefactions in a fluid or solid. Later we give a simple derivation of this relation for compression waves in one dimension. First, however, let us apply this result to an ideal gas.

At ordinary densities and pressures, air and most other gases obey the ideal-gas equation,* in which the pressure $P$, volume $V$, and absolute temperature $T$ are related by

$$PV = nRT \qquad\qquad 14\text{-}6$$

where $n$ is the number of moles and $R$ is the *universal gas constant*, with the value

$$R = 8.314 \text{ J/mol·K} \qquad\qquad 14\text{-}7$$

To compute the bulk modulus for such a gas, we need to relate the change in volume to the change in pressure. Differentiating Equation 14-6, we obtain

$$P\,dV + V\,dP = nR\,dT$$

If the compression of the gas occurs at constant temperature, $dT = 0$ and we have

$$P\,dV + V\,dP = 0 \qquad\qquad 14\text{-}8$$

or

$$dP = -\frac{P\,dV}{V}$$

A compression at constant temperature is called an *isothermal compression*. The isothermal bulk modulus for an ideal gas is

$$B_{\text{iso}} = -\frac{dP}{dV/V} = P$$

---

* This equation will be studied in some detail in Chapter 16; it is used here because it may already be familiar from your study of chemistry.

If the compressions and rarefactions are isothermal, the speed of sound through the gas is given by

$$v = \sqrt{\frac{B}{\rho_0}} = \sqrt{\frac{P}{\rho_0}}$$

We can write this in terms of the temperature using the ideal-gas equation

$$P = \frac{nRT}{V} = \frac{\rho_0 RT}{M}$$

where $M$ is the molecular mass and $\rho_0 = nM/V$ is the mass density. Then

$$v = \sqrt{\frac{RT}{M}} \qquad\qquad 14\text{-}9$$

This expression, first obtained by Newton, gives the correct temperature dependence for the speed of sound, but it yields values for $v$ that are about 20 percent too small compared with experimental measurements.*

Equation 14-9 is inaccurate because air and other gases are poor conductors and the compressions and rarefactions occurring during a sound wave are not isothermal. When a gas is compressed, its temperature tends to rise. To compress a gas isothermally, heat must be removed from the gas during the compression. If no heat is transferred out of the gas during the compression, the temperature rises and Equation 14-8 no longer holds. A compression with no heat transfer is called an *adiabatic compression*. In Chapter 17 we show that the adiabatic bulk modulus of an ideal gas is given by

$$B_{\text{adiab}} = \gamma P \qquad\qquad 14\text{-}10$$

where $\gamma$ is a constant that depends on the kind of gas; it has the value 1.4 for air. Using the adiabatic bulk modulus for a gas in Equation 14-5, we obtain for the speed of sound

$$v = \sqrt{\frac{\gamma RT}{M}} \qquad\qquad 14\text{-}11 \qquad \textit{Speed of sound waves}$$

This differs from Equation 14-9 by the factor $\gamma$ inside the radical sign, and gives values for the speed of sound about 20 percent greater than those given by Equation 14-9.

**Example 14-1** The tension in a string is provided by hanging a mass of 3 kg at one end, as in Figure 14-11. The mass density of the string is 0.02 kg/m. What is the velocity of waves on the string?

The tension in the string is

$$T = mg = 3(9.81) = 29.4 \text{ N}$$

The velocity is therefore

$$v = \sqrt{\frac{T}{\mu}} = \sqrt{\frac{29.4}{0.02}} = 38.3 \text{ m/s}$$

* Newton was not content to be within 20 percent of the experimental value for the speed of sound. He fudged his original result by introducing various "correction factors" until he obtained a result which agreed exactly with experiment, and then he claimed an accuracy of 0.1 percent. See R. S. Westfall, "Newton and the Fudge Factor," *Science*, vol. 179, p. 751, February 1973.

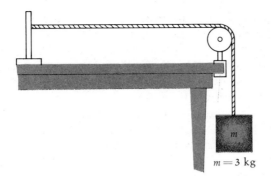

**Figure 14-11**
Example 14-1. The tension in
the string is provided by the
weight of the block.

$m = 3$ kg

**Example 14-2** Calculate the speed of sound in air at 0°C (= 32°F) and at
20°C (= 68°F).

The temperature $T$ in Equation 14-11 is the absolute temperature,
which is related to the Celsius temperature by

$$T = t_C + 273 \qquad\qquad 14\text{-}12$$

(We discuss temperature scales in Chapter 16.) The molecular mass of
air is 29.0 g/mol = $29.0 \times 10^{-3}$ kg/mol, and the value of the universal
gas constant is $R = 8.31$ J/mol·K.

We obtain for the speed of sound in SI units at 0°C

$$v = \sqrt{\frac{(1.4)(8.31)(273)}{29.0 \times 10^{-3}}} = 331 \text{ m/s}$$

To find the speed at 20°C = 293 absolute, we note that since the speed
is proportional to the square root of the absolute temperature, its value
at 293 absolute is

$$v = \sqrt{\frac{293}{273}} \ (331 \text{ m/s}) = 343 \text{ m/s}$$

**Questions**

4. Two strings are stretched between the same two posts. One weighs
twice as much as the other. How should they be adjusted so that waves
travel through both at the same speed?

5. Do you expect the speed of sound waves in helium gas to be greater
or less than that in air? Why?

6. Although the density of most solids is more than 1000 times that of
air, the speed of sound in a solid is usually greater than that in air.
Why?

*Optional*

**Derivation of Equation 14-5**

Consider a fluid of density $\rho_0$ and pressure $P$ in a long tube, as in Figure
14-12. We suddenly compress the fluid by increasing the pressure by an
amount $\Delta P$ at the left, causing the piston to move to the right for a short
time $\Delta t$. The motion of the piston is transmitted to the molecules in the
fluid by collisions of the molecules with the piston and each other. We
make the simplifying assumption that the piston moves with a constant
speed $u$ for the time $\Delta t$ and that the speed $u$ is much less than the wave
speed $v$. (It is important not to confuse these two speeds.) In time $\Delta t$ the

piston moves a distance $u \, \Delta t$, and the wave pulse moves a distance $v \, \Delta t$. We shall also assume that the action of the piston gives a speed $u$ to all the fluid from the piston to the edge of the pulse a distance $v \, \Delta t$ from the original position of the piston. The justification for this assumption is that the wave speed $v$ is the speed at which disturbances are propagated through the fluid. Thus in time $\Delta t$ the greatest distance the disturbance can penetrate into the fluid ahead of the piston is $v \, \Delta t$. The assumption that all the fluid in this region moves with constant speed $u$ amounts to the assumption of a rectangular shape for the pulse. We shall calculate the speed of the pulse by setting the change in momentum of the fluid equal to the impulse acting on the fluid due to the increased pressure acting for the time $\Delta t$.

**Figure 14-12**
Analysis of a longitudinal wave pulse produced by suddenly moving the piston to the right with speed $u$. After a short time $\Delta t$ the piston has moved a distance $u \, \Delta t$ and the wave pulse has moved a distance $v \, \Delta t$. If a rectangular pulse is assumed, the fluid in the shaded region of length $v \, \Delta t$ is moving with speed $u$.

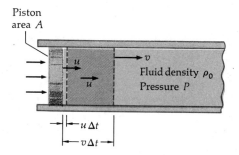

If $A$ is the area of the piston, the impulse is

$$\text{Impulse} = (A \, \Delta P) \, \Delta t$$

The mass of fluid set into motion is the density $\rho_0$ times the volume $Av \, \Delta t$. The change in momentum is this mass times the velocity $u$:

$$\text{Momentum change} = \rho_0 (Av \, \Delta t) u$$

Equating this momentum change to the impulse gives

$$A(\Delta P) \, \Delta t = \rho_0 (Av \, \Delta t) u$$

or

$$\Delta P = \rho_0 v u \qquad\qquad 14\text{-}13$$

The change in pressure is related to the decrease in volume of the fluid by the bulk modulus:

$$\Delta P = B \, \frac{-\Delta V}{V}$$

The original volume of fluid under consideration is $V = Av \, \Delta t$, and the change in volume is the volume swept out by the piston, $\Delta V = -Au \, \Delta t$. Thus

$$-\frac{\Delta V}{V} = \frac{Au \, \Delta t}{Av \, \Delta t} = \frac{u}{v} \qquad\qquad 14\text{-}14$$

and

$$\Delta P = \frac{Bu}{v} \qquad\qquad 14\text{-}15$$

Using this result for $\Delta P$ in Equation 14-13, we get

$$\frac{Bu}{v} = \rho_0 v u \qquad \text{or} \qquad v = \sqrt{\frac{B}{\rho_0}}$$

## 14-3   Harmonic Waves

If instead of flipping the end of a string, we move the end up and down in simple harmonic motion (as if it were tied to a tuning fork that is set vibrating), a sinusoidal wave train propagates down the string. Such a wave is called a harmonic wave. The shape of the string at some instant of time is that of a sine function (or cosine function, since they differ only in phase), as shown in Figure 14-13. Such a figure can be obtained

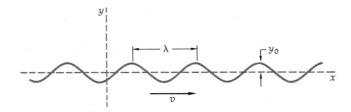

**Figure 14-13**
Harmonic wave at some instant in time; $\lambda$ is the wavelength and $y_0$ is the amplitude.

by taking a snapshot of the string. The distance between two successive crests is called the *wavelength* $\lambda$. The wavelength is the distance in space within which the wave function repeats itself. As the wave propagates down the string, each point on the string moves up and down in simple harmonic motion with the frequency $f$ of the tuning fork or whatever agent is driving the end of the string. There is a simple relation between the frequency $f$, the wavelength $\lambda$, and the velocity $v$ of the harmonic wave. Consider, for example, vibrating the end of the string with frequency $f$ for a time $t$. In this time, the number of waves generated is $N = ft$. The first wave generated travels a distance $vt$. The ratio of this distance to the number of waves in this distance is the wavelength $\lambda$. Thus

*Wavelength*

$$\lambda = \frac{vt}{N} = \frac{vt}{ft} = \frac{v}{f}$$

or

$$v = f\lambda = \frac{\lambda}{T} \qquad\qquad 14\text{-}16$$

where $T = 1/f$ is the period of vibration.

Figure 14-14 shows circular waves on the surface of water in a ripple tank generated by a point source moving up and down with simple harmonic motion. The wavelength here is the distance between successive wave crests, which are concentric circles. These circles are the curves of constant phase for these two-dimensional waves and are called *wavefronts*. If we have a point source of sound or light waves in three dimensions, the waves move out in all directions and the wavefronts are concentric spherical surfaces. The motion of the wavefront can be indicated by *rays,* directed lines perpendicular to the front (Figure 14-15).

**Figure 14-14**
Circular wavefronts diverging from a point source in a ripple tank. (*From* PSSC Physics, *2d ed., p. 272, D. C. Heath and Company, Lexington, Mass., 1965.*)

*Wavefront*

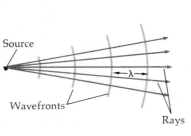

**Figure 14-15**
The motion of wavefronts can be represented by rays drawn perpendicular to the wavefronts. For a point source, the rays are radial lines diverging from the source.

For spherical waves the rays are radial lines. At a great distance from a point source, a small part of the wavefront can be approximated by a plane, and the rays are approximately parallel lines; such a wave is called a *plane wave* (Figure 14-16). The two-dimensional analog of a plane wave is a line wave, which is a small part of a circular wavefront at a great distance from the source. Such waves can also be produced in a ripple tank by a line source, as in Figure 14-17.

In Section 14-1 we stated that the wave function for a wave traveling to the right must be a function of the variables $x$ and $t$ in the combination $x - vt$, where $v$ is the wave speed. The wave function for a harmonic wave traveling to the right is written

$$y(x, t) = y_0 \sin k(x - vt) \qquad \text{14-17}$$

where $y_0$ is the amplitude and $k$ is called the *wave number*. This function can also be written

$$y(x, t) = y_0 \sin (kx - kvt)$$

or

$$y(x, t) = y_0 \sin (kx - \omega t) \qquad \text{14-18}$$

where

$$\omega = kv \qquad \text{14-19}$$

is the *angular frequency*, which is related to the frequency of vibration $f$ in the usual way:

$$\omega = 2\pi f \qquad \text{14-20}$$

Combining Equations 14-19 and 14-20 with 14-16 gives

$$2\pi f = kv = k(f\lambda)$$

or

$$k = \frac{2\pi}{\lambda} \qquad \text{14-21}$$

From our study of simple harmonic motion it should be clear that we could just as well choose a cosine function in Equation 14-18 since the sine and cosine functions differ only in phase, $\sin \theta = \cos (\theta - \pi/2)$. We also could have reversed $kx$ and $\omega t$ since $\sin (kx - \omega t) = \sin (\omega t - kx + \pi)$. We have omitted a constant phase for simplicity.

We can produce harmonic sound waves by vibrating a piston or loudspeaker cone with simple harmonic motion, causing the air molecules next to the piston or speaker to oscillate with simple harmonic motion about their equilibrium position. These molecules collide with neighboring molecules, causing them to oscillate. If we replace $y(x, t)$ and $y_0$ in Equation 14-18 with the displacement $s(x, t)$, the equation describes the displacement of the molecules from their equilibrium positions as a function of $x$ and $t$:

$$s(x, t) = s_0 \sin (kx - \omega t) \qquad \text{14-22}$$

where $s_0$ is the maximum displacement of the gas from its equilibrium position. This displacement is along the direction of motion of the wave; sound is a longitudinal wave.

**Figure 14-16**
Plane waves. At great distances from a point source the wavefronts are approximately parallel planes, and the rays are parallel lines perpendicular to the planes.

*Angular frequency*

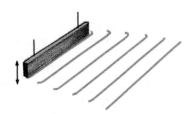

**Figure 14-17**
A two-dimensional analog of a plane wave can be generated in a ripple tank with a flat board oscillating up and down in the water, producing wavefronts which are straight lines.

The displacement of the gas molecules causes the density of the gas at a point to increase or decrease, depending on whether the gas is displaced toward the point or away from it. If the density increases (or decreases) at some point, the pressure also increases (or decreases) at that point, giving a pressure wave which is related to the displacement wave. We can use Figure 14-18a, showing a sketch of the function $s(x, t)$ at some time $t_0$, to see that the pressure wave is 90° out of phase with the displacement wave. Points $P_1$ and $P_3$ are points of zero displacement at this time. Just to the left of point $P_1$ the displacement is negative, indicating that the gas is displaced to the left away from the point. Just to the right of $P_1$ the displacement is positive, indicating that the gas is displaced to the right, again away from this point. Thus the density is a minimum at point $P_1$ because the gas is displaced away from this point on both sides. At point $P_3$ the density is a maximum because to the right of this point the displacement is negative, indicating displacement to the left toward $P_3$, and to the left of the point the displacement is positive, indicating displacement to the right, also toward $P_3$. The displacements near these points are indicated in Figure 14-18b. At point $P_2$, the displacement function is a maximum. The displacements to the left and to the right of this point are both positive and of equal magnitude. Thus the density does not change near this point. Figure 14-18c indicates the change in density for this displacement wave. Since the pressure is maximum when the density is maximum, the pressure varies with distance at this time, as indicated in Figure 14-18d.

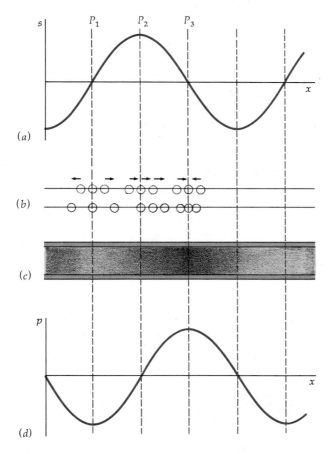

(a)

(b)

(c)

(d)

**Figure 14-18**
(a) Displacement from equilibrium of air molecules in a harmonic sound wave at some instant. At $P_1$ and $P_3$ the molecules are at their equilibrium positions. At $P_2$ they have maximum displacement. (b) Equilibrium position (top) and displacement (bottom) of three molecules near each of the points $P_1$, $P_2$, and $P_3$. (c) Density of the air at this time. The density is minimum at $P_1$ and maximum at $P_3$, points of zero displacement. (d) Pressure change versus position. The pressure and displacement are 90° out of phase with each other.

We shall show below that a displacement wave given by Equation 14-22 implies a pressure wave given by

$$p = p_0 \sin (kx - \omega t - 90°) \qquad 14\text{-}23$$

where $p$ stands for the *change* in pressure from the equilibrium pressure with no wave and the amplitude $p_0$ is the maximum value of this change. We shall also show that the pressure amplitude $p_0$ is related to the displacement amplitude $s_0$ by

$$p_0 = \rho\omega v s_0 \qquad 14\text{-}24$$

where $v$ is the speed of propagation and $\rho$ is the equilibrium density of the gas. We can think of a sound wave as either a displacement wave or a pressure wave, the displacement and pressure being 90° out of phase with each other.

*Optional*

### Derivation of Equations 14-23 and 14-24

We can relate the changes in pressure to the displacement by noting that a displacement changes the volume of a given mass of gas and that changes in pressure and volume are related by the bulk modulus. Since $p$ stands for a *change* in pressure here, we have

$$p = -B\frac{\Delta V}{V}$$

We can express the bulk modulus in terms of the wave speed $v$ and the gas density by using Equation 14-5,

$$B = \rho v^2$$

Thus the change in pressure, $p$, is related to the change in volume $V$ by

$$p = -\rho v^2 \frac{\Delta V}{V} \qquad 14\text{-}25$$

Consider a given mass of gas initially between points $x_1$ and $x_2$ occupying a volume $V = A(x_2 - x_1) = A\,\Delta x$, where $A$ is the area of the tube of gas and $\Delta x = x_2 - x_1$. This volume is shown in Figure 14-19. In general, a displacement of gas causes a change in the volume of a given mass of gas. If the displacement is the same at points $x_2$ and $x_1$, there will be no change in the volume. For example, a positive displacement at point $x_2$ increases the volume, but a positive displacement at point $x_1$ decreases the volume. There will be a change in the volume of the gas only if there is a difference in displacement at these two points. Let us call the difference in displacement $\Delta s$:

$$\Delta s = s(x_2, t_0) - s(x_1, t_0) \qquad 14\text{-}26$$

**Figure 14-19**
Change in volume of a given mass of gas due to the variation in displacement with position. The original volume is $V = A\,\Delta x$, and the change in volume is $\Delta V = A\,\Delta s$, where $A$ is the area of the tube.

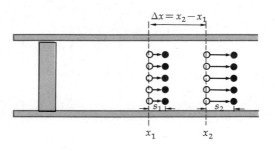

The change in volume of the gas equals this difference in displacement times the area $A$:

$$\Delta V = A \, \Delta s$$

This change in volume is related to the change in pressure $p$ by Equation 14-25:

$$p = -\rho v^2 \frac{A \, \Delta s}{A \, \Delta x} = -\rho v^2 \frac{\Delta s}{\Delta x} \qquad 14\text{-}27$$

In the limit of small $\Delta x$ the ratio $\Delta s/\Delta x$ becomes the derivative of $s$ with respect to $x$. This is a partial derivative because we are considering a single time $t_0$. Replacing $\Delta s/\Delta x$ by the notation for partial derivative $\partial s/\partial x$, we have

$$p = -\rho v^2 \frac{\partial s}{\partial x} \qquad 14\text{-}28$$

We compute $\partial s/\partial x$ from Equation 14-22:

$$s = s_0 \sin (kx - \omega t_0) \qquad \text{and} \qquad \frac{\partial s}{\partial x} = k s_0 \cos (kx - \omega t_0)$$

Hence

$$p = -\rho v^2 s_0 k \cos (kx - \omega t_0) = +k \rho v^2 s_0 \sin (kx - \omega t_0 - 90°)$$

$$= p_0 \sin (kx - \omega t_0 - 90°) \qquad 14\text{-}29$$

where

$$p_0 = k \rho v^2 s_0$$

But according to Equation 14-19, $kv = \omega$; therefore

$$p_0 = \rho \omega v s_0 \qquad 14\text{-}30$$

and Equations 14-29 and 14-30 are identical to Equations 14-23 and 14-24.

# 14-4 The Doppler Effect

If a wave source and a receiver are moving relative to each other, the frequency observed by the receiver is not the same as the source frequency. When they are moving toward each other, the observed frequency is greater than the source frequency; and when they are moving away from each other, the observed frequency is less than the source frequency. This is called the *doppler effect*. A familiar example is the change in pitch of a train whistle or car horn when the train or car is approaching or receding.

Before we calculate the doppler shift in frequency, we consider the reception of waves by a receiver at rest relative to the source. Figure 14-20 shows a source sending out waves of frequency $f_0$. This figure can represent either two-dimensional circular waves, e.g., surface water waves, or a two-dimensional drawing of three-dimensional spherical waves, e.g., waves from a point source. Let $N$ be the number of waves emitted in a time $\Delta t$. Then $N = f_0 \, \Delta t$. If $v$ is the speed of the wave relative to the medium, which we assume to be at rest relative to both the source and receiver, these waves will be contained in a distance $v \, \Delta t$.

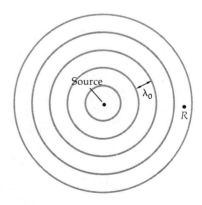

**Figure 14-20**
Wavefronts from a point source at rest. The diagram can represent either spherical surfaces in three dimensions or two-dimensional circular waves.

The wavelength of the waves is the ratio of this distance and the number of waves:

$$\lambda = \frac{v\,\Delta t}{N} = \frac{v\,\Delta t}{f_0\,\Delta t} = \frac{v}{f_0} \qquad\qquad 14\text{-}31$$

The frequency observed by the receiver is the number of waves passing per unit time. The number passing the receiver in time $\Delta t$ is the number in the distance $v\,\Delta t$, which is $v\,\Delta t/\lambda$. The frequency observed by the receiver $f_R$ is then

$$f_R = \frac{v\,\Delta t/\lambda}{\Delta t} = \frac{v}{\lambda} = f_0 \qquad\qquad 14\text{-}32$$

This is, of course, just the frequency of the wave source.

If the medium is not at rest relative to the source and receiver—if, for instance, there is a wind blowing in the case of sound waves—the wavelength changes but the frequency received is still equal to the source frequency. Let $u_w$ be the speed of the wind, which we assume to be blowing from source to receiver. The speed of the waves relative to the source or receiver is then $v' = v + u_w$. The wavelength $\lambda'$ between source and receiver is now greater than that with no wind because the same number of waves $N = f_0\,\Delta t$ is now contained in the distance $v'\,\Delta t = (v + u_w)\,\Delta t$. The wavelength is

$$\lambda' = \frac{v'\,\Delta t}{N} = \frac{(v + u_w)\,\Delta t}{f_0\,\Delta t} = \frac{v + u_w}{f_0} = \frac{v + u_w}{v}\,\lambda_0 \qquad\qquad 14\text{-}33$$

where $\lambda_0 = v/f_0$ is the original wavelength with no wind. The number of waves received in time $\Delta t$ is now the number in the distance $v'\,\Delta t$, which is just $f_0\,\Delta t$. Thus the frequency received is again $f_0$, in agreement with our common experience that the wind does not affect the observed frequency of horns, whistles, or musical instruments. Note that the wavelength and observed frequency with a wind blowing are given by Equations 14-31 and 14-32 with $v$ replaced by $v'$.

Figure 14-21 shows a point source moving with speed $u_s$ relative to a medium in which the wave speed is $v$. In front of the source the wavefronts are closer together and behind the source they are farther apart than for a stationary source. The calculation of the wavelengths is similar to Equations 14-31 and 14-32. In time $\Delta t$ the source emits $N = f_0\,\Delta t$ waves. The first wavefront travels a distance $v\,\Delta t$ while the source travels a distance $u_s\,\Delta t$. In front of the source the $N$ wavefronts occupy the

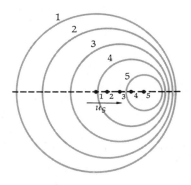

**Figure 14-21**
Successive wavefronts emitted by a point source moving with speed $u_s$ to the right. Each numbered wavefront was emitted when the source was at the correspondingly numbered position. The wavelengths are shorter in front and longer in back than wavelengths from a stationary source.

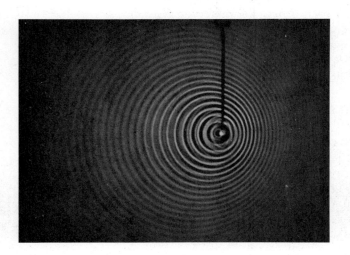

Waves in a ripple tank produced by a source moving to the right with speed less than the wave speed. (*Courtesy of Film Studio, Education Development Center, Newton, Mass.*)

distance $v \, \Delta t - u_s \, \Delta t$. The wavelength in front of the source is therefore

$$\lambda_f' = \frac{v \, \Delta t - u_s \, \Delta t}{f_0 \, \Delta t} = \frac{v - u_s}{f_0} = \frac{v}{f_0}\left(1 - \frac{u_s}{v}\right) \qquad 14\text{-}34$$

Behind the source the wavelength is

$$\lambda_b' = \frac{v \, \Delta t + u_s \, \Delta t}{f_0 \, \Delta t} = \frac{v + u_s}{f_0} = \frac{v}{f_0}\left(1 + \frac{u_s}{v}\right) \qquad 14\text{-}35$$

When we use $\lambda_0 = v/f_0$ for the wavelength of the source at rest, Equations 14-34 and 14-35 can be written

$$\lambda' = \lambda_0\left(1 \pm \frac{u_s}{v}\right) \qquad 14\text{-}36$$

The speed $v$ of the waves depends only on the properties of the medium and not on the motion of the source. The frequency at which the waves pass a point at rest relative to the medium is thus

$$f' = \frac{v}{\lambda_f'} = \frac{f_0}{1 - u_s/v} \qquad 14\text{-}37$$

for the source approaching the receiver and

$$f' = \frac{v}{\lambda_b'} = \frac{f_0}{1 + u_s/v} \qquad 14\text{-}38$$

for the source receding from the receiver.

If the source is at rest and the receiver is moving relative to the medium, there is no change in the wavelength but the frequency of the waves passing the receiver is increased when the reciever moves toward the source and decreased when he moves away from the source. The number of waves passing a stationary receiver in time $\Delta t$ is the number in the distance $v \, \Delta t$, which is $v \, \Delta t/\lambda_0$. When the receiver moves toward the source with speed $u_R$, he passes an additional number $u_R \, \Delta t/\lambda_0$ (Figure 14-22). The total number of waves passing the receiver in time $\Delta t$ is then

$$N = \frac{v \, \Delta t + u_R \, \Delta t}{\lambda_0} = \frac{v + u_R}{\lambda_0} \Delta t$$

The frequency observed is this number divided by the time interval:

$$f' = \frac{N}{\Delta t} = \frac{v + u_R}{\lambda_0} = \frac{v + u_R}{v} \frac{v}{\lambda_0}$$

or

$$f' = f_0\left(1 + \frac{u_R}{v}\right) \qquad 14\text{-}39$$

If the receiver moves away from the source with speed $u_R$, similar reasoning leads to the observed frequency

$$f' = f_0\left(1 - \frac{u_R}{v}\right) \qquad 14\text{-}40$$

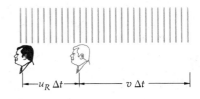

**Figure 14-22**
The number of waves passing a stationary receiver in time $\Delta t$ is the number in the distance $v \, \Delta t$, where $v$ is the wave speed. If the receiver moves toward the source with speed $u_R$, he passes the additional number in the distance $u_R \, \Delta t$. The frequency of waves received is therefore increased.

The results expressed in Equations 14-37 to 14-40 can be combined when both source and receiver are moving relative to the medium. If the medium is moving, the wave speed $v$ is replaced by $v' = v \pm u_w$, where $u_w$ is the speed of the medium.

---

**Example 14-3** The frequency of a car horn is 400 Hz. What frequency is observed if the car moves toward a stationary receiver with speed $u_s = 30$ m/s (about 67 mi/h)? Take the velocity of sound in air to be 340 m/s.

According to Equation 14-34, the wavelength in front of the car is

$$\lambda' = \frac{v - u_s}{f_0} = \frac{340 - 30 \text{ m/s}}{400 \text{ s}^{-1}} = \frac{310}{400} \text{ m} = 0.775 \text{ m}$$

The frequency observed is

$$f' = \frac{v}{\lambda'} = \frac{340 \text{ m/s}}{0.775 \text{ m}} = 439 \text{ s}^{-1} = 439 \text{ Hz}$$

---

**Example 14-4** The horn of a stationary car has a frequency of 400 Hz. What frequency is observed by a receiver moving toward the car at 30 m/s?

For a moving receiver, the wavelength does not change: the receiver merely passes more waves in a given time. The observed frequency is (Equation 14-39)

$$f' = f_0 \left( 1 + \frac{u_R}{v} \right) = (400 \text{ Hz}) \left( 1 + \frac{30}{340} \right)$$
$$= \frac{370}{340} (400 \text{ Hz}) = 435 \text{ Hz}$$

---

**Example 14-5** Work Example 14-4 in the reference frame of the receiver.

In the reference frame of the receiver, the car is moving toward the receiver with speed of 30 m/s. The situation is *not* identical to that in Example 14-3 because here the receiver observes an apparent wind of 30 m/s in the direction from the car to the receiver. The speed of sound waves relative to the receiver is $v' = v + u_w = 340 \text{ m/s} + 30 \text{ m/s} = 370 \text{ m/s}$. The wavelength of the waves in front of the car (which is moving relative to the receiver) is given by Equation 14-34 with $v'$ replacing $v$. Thus

$$\lambda_f' = \frac{v' - u_s}{f_0} = \frac{v + u_w - u_s}{f_0}$$

But the wind speed $u_w$ and the source speed $u_s$ are the same. Both are 30 m/s in the frame of the receiver. The wavelength is thus

$$\lambda_f' = \frac{v}{f_0} = \frac{340}{400} \text{ m} = 0.85 \text{ m}$$

The observed frequency is

$$f' = \frac{v'}{\lambda_f'} = \frac{370 \text{ m/s}}{0.85 \text{ m}} = 435 \text{ Hz}$$

The fact that the wavelength and frequency are the same as in Example 14-4 should not be surprising. In these two examples, identical problems were worked in two different reference frames moving with constant velocity relative to each other.

---

From the above examples we note that the doppler shift in frequency depends on whether the source or receiver moves relative to the medium. In Example 14-3 the source moved with speed 30 m/s relative to the still air, and the frequency shifted from 400 to 439 Hz. In Examples 14-4 and 14-5 the receiver moved with speed 30 m/s relative to the still air, and the frequency shifted from 400 to 435 Hz. These shifts are approximately but not exactly equal.

It is interesting to compare these doppler shifts when the speed of the source or receiver $u$ is much less than the wave speed $v$. According to Equation 14-37, the frequency observed for a source moving toward a receiver with speed $u$ is

$$f' = \frac{f_0}{1 - u/v} = f_0 \left(1 - \frac{u}{v}\right)^{-1} \qquad 14\text{-}41a$$

We can compare this with Equation 14-39 for a moving receiver by using the binomial expansion

$$(1 + x)^n = 1 + nx + \frac{n(n - 1)}{2} x^2 + \cdots$$

For $x$ much less than 1 we can approximate, using only the first few terms. With $x = -u/v$ and $n = -1$ we have

$$\left(1 - \frac{u}{v}\right)^{-1} = 1 + (-1)\left(-\frac{u}{v}\right) + \frac{(-1)(-2)}{2}\left(-\frac{u}{v}\right)^2 + \cdots$$

$$\approx 1 + \frac{u}{v} + \frac{u^2}{v^2}$$

Then Equation 14-41a is approximately

$$f' \approx f_0 \left(1 + \frac{u}{v} + \frac{u^2}{v^2}\right) \qquad 14\text{-}41b$$

On the other hand, if a receiver moves toward a stationary source with speed $u$, the observed frequency (Equation 14-39) is

$$f' = f_0 \left(1 + \frac{u}{v}\right)$$

These results differ only in terms of order $u^2/v^2$ or higher.

In many practical situations this difference can be neglected, but the difference is real and of theoretical importance because it shows that these two situations are really different. Not only is the *relative* motion of the source and receiver important but also the "absolute" motion of both relative to the medium. Thus if we can measure the doppler shift in frequency to order $u^2/v^2$, we can tell whether it is the source that is moving relative to the medium or the receiver. For sound waves in air, for example, we can also tell which is moving by noting whether it is the source or receiver that feels the wind in its own reference frame. However, a problem arises for light or other electromagnetic waves which propagate through a vacuum. Our equations for the doppler effect seem to imply that we could detect absolute motion relative to the vacuum if we could measure the doppler shift accurately enough, but this contradicts the principle of relativity. Thus either these equations are not correct, or the principle of relativity does not hold. We shall see in our study of special relativity that a small but important correction must be made to these equations describing the doppler effect. The correct expression for the doppler shift of light and other electromagnetic waves is

$$f' = \frac{\sqrt{1 - u^2/c^2}}{1 \pm u/c} f_0 \qquad 14\text{-}42$$

where $u$ is the relative velocity of source and observer and $c$ is the velocity of light. There is no way of distinguishing which is moving; only the relative velocity is important. Comparing this expression with Equations 14-37 and 14-38, we see that the expressions differ by the factor $\sqrt{1 - u^2/c^2}$. It arises in the relativistic derivation because the time interval between the emission of the first wave and the $N$th wave is not the same when measured by the source and by the receiver, though we have, of course, assumed this in our classical derivation.

In our derivation of the doppler-shift expressions we have assumed that the velocity $u$ of the source or receiver is less than the wave velocity $v$. If the source moves with velocity greater than the wave velocity, there will be no waves in front of the source. The waves behind the source will be increased in length according to Equation 14-35, but they will be confined to a cone which narrows as $u$ increases. We can easily calculate the angle of this cone. Consider the source at position $P_1$ at time $t_1$, as shown in Figure 14-23. After a time $\Delta t$, the wave emitted from this point will have traveled a distance $v\,\Delta t$. The source will have traveled a greater distance $u\,\Delta t$ and will be at point $P_2$. The line from this new position of the source to the wavefront emitted when the source was at $P_1$ makes an angle $\theta$ with the path of the source, given by

$$\sin\theta = \frac{v\,\Delta t}{u\,\Delta t} = \frac{v}{u} \qquad 14\text{-}43$$

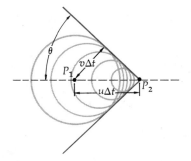

Equation 14-43 applies to electromagnetic radiation given off when a charged particle moves in a medium with speed $u$ which is greater than the speed $v$ of light in that medium.* This radiation, called Čerenkov radiation, is confined to the cone with angle given by Equation 14-43.

If the receiver moves toward the source faster than the wave speed, there is no problem. Equation 14-39 holds for the observed frequency. If the receiver moves away from the source faster than the wave speed, the waves never reach the receiver. For light waves, the relativistic expression for the doppler effect given by Equation 14-42 becomes imaginary for $u$ greater than $c$ because of the square-root term in the numerator. According to the special theory of relativity, the relative speed between the source and receiver cannot exceed the speed $c$ of light in vacuum.

**Figure 14-23**
A source moving from $P_1$ to $P_2$ with speed $u$ greater than the wave speed $v$. The envelope of the wavefronts forms a cone with the source at the apex. The angle $\theta$ of this cone is given by $\sin\theta = v/u$.

* It is impossible for a particle to move faster than $c$, the speed of light in a vacuum. In a medium such as glass, however, electrons and other particles can move faster than light can move in that medium.

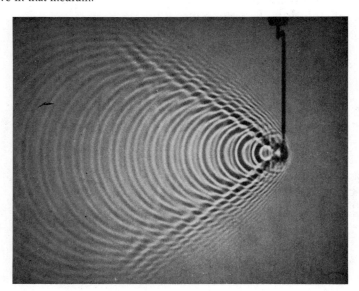

Waves in a ripple tank produced by a source moving to the right with speed greater than the wave speed. (*Courtesy of Film Studio, Education Development Center, Newton, Mass.*)

**Questions**

7. Suppose the velocity vector of a receiver makes an angle with the line between her and the source. How will the doppler shift differ from the case discussed in this section? What if the velocity vector makes a right angle with the line joining source and receiver?

8. If the source and receiver are at rest relative to each other but the wave medium is moving relative to them, will the receiver detect any wavelength or frequency shift?

## 14-5  Standing Waves on a String

When waves are confined in space, like waves on a piano string or in an organ pipe, there are reflections at both ends and therefore waves traveling in both directions. These waves combine according to the general law of wave interference. For a given string or pipe, there are certain frequencies for which this interference results in a stationary vibration pattern called a *standing wave*. The study of standing waves has many applications in the field of music and in nearly all areas of science and technology.

*Standing wave*

We shall study standing waves on strings in this section and standing sound waves in Section 14-6.

### String Fixed at Both Ends

Consider a string fixed at one end and tied at the other end to a tuning fork, which vibrates with frequency $f$ and small amplitude. The waves produced by the tuning fork travel down the string, are inverted by reflection at the fixed end, and travel back to the fork. Since the fork is vibrating with a small amplitude, it acts essentially as a fixed end as far as reflection is concerned. The waves are thus inverted once again upon reflection at the fork and start back down the string.

Let us look at a particular wave crest generated at the left end of the string by the fork. It travels to the right end of the string, is reflected back, and is reflected again at the fork. Since it has been reflected twice, it has been inverted twice and now differs from the next wave crest coming from the fork only in that the first has already traveled a distance $2L$, where $L$ is the length of the string. If this distance is exactly equal to the wavelength $\lambda$, the twice-reflected wave will be in phase with the second wave and the two will interfere constructively. The resultant wave has an amplitude twice that of either wave (assuming no loss by reflection). When this resulting wave travels the distance $2L$ to the fixed end and back and has been reflected twice, it will exactly overlap the third wave generated by the tuning fork. Thus each new wave is in phase with the waves reflected at the tuning fork, and the amplitude continues to increase as the string absorbs energy from the tuning fork. Various damping effects put a limit on the maximum amplitude that can be reached, e.g., loss of energy due to reflection or imperfect flexibility of the string. This maximum amplitude is much larger than that of the tuning fork. Thus the tuning fork is in *resonance* with the string when the tuning-fork frequency is such that the wavelength in the string equals twice the length of the string.

When the wavelength just equals the length of the string, the distance $2L$ equals 2 wavelengths. When the first wave has traveled the distance $2L$ and is reflected at the fork, it will be exactly in phase with the third wave generated at the fork. Again a large amplitude builds up in the string. Resonance will occur if the distance $2L$ is any integer times the wavelength. Thus the condition for resonance is

$$n\frac{\lambda}{2} = L \qquad\qquad\qquad 14\text{-}44$$

*Standing-wave condition for string fixed at both ends*

where $n$ is any integer. In terms of the frequency of the waves, $f = v/\lambda$, the condition for resonance is

$$2L = n\frac{v}{f}$$

or

$$f_n = n\frac{v}{2L} = nf_1 \qquad\qquad\qquad 14\text{-}45$$

where

$$f_1 = \frac{v}{2L} = \frac{1}{2L}\sqrt{\frac{T}{\mu}} \qquad\qquad\qquad 14\text{-}46$$

is the lowest resonance frequency, called the *fundamental frequency*, and we have used Equation 14-4 for the speed of waves on the string.

It is sometimes convenient to think of the resonance condition in terms of the time necessary for the first wave to travel to the end and back. Since this distance is $2L$, the time will be $2L/v$, where $v$ is the wave speed. If this time equals the period of vibration of the fork, the first wave will add constructively to the second wave. Resonance will also result if this time equals any integral number of periods. Thus we can write the resonance condition

$$\frac{2L}{v} = nT = \frac{n}{f} \qquad \text{or} \qquad f = n\frac{v}{2L}$$

which is the same as that found by fitting an integral number of wavelengths into the distance $2L$. The frequencies given by Equation 14-45 are called the *natural frequencies* of the string.

---

**Example 14-6** A string is stretched between two fixed supports 1 m apart, and the tension is adjusted until the fundamental frequency of the string is 440 Hz. What is the speed of transverse waves on the string?

From the condition for resonance (Equation 14-44) the wavelength for the fundamental, or $n = 1$, resonance frequency is $\lambda = 2L = 2$ m. Hence, the wave speed is

$$v = \lambda f = (2 \text{ m}) (440 \text{ s}^{-1}) = 880 \text{ m/s}$$

---

What happens when the frequency of the tuning fork is *not* equal to one of the natural frequencies of the string? When the first wave has traveled the distance $2L$ and is reflected from the fork, it differs in phase from the wave being generated at the fork (Figure 14-24). It combines with this wave to produce a resultant new wave whose amplitude may be greater or smaller than that of the original wave, depending on the phase difference. (We shall see how to combine two harmonic waves of different phase in the next chapter.) When this resultant wave travels a distance $2L$ and is again reflected at the fork, it will differ in phase from the next wave generated. The phase difference of the incoming waves and reflected waves depends on the number of reflections. The interference may be constructive, in which case the amplitude of the wave on the string will increase, or destructive, in which case the amplitude of the wave will decrease. On the average, the amplitude will not increase but remain that of the first wave generated; i.e., the amplitude of the resultant wave will equal that of the tuning fork, which is small. The string does not absorb energy on the average. Only when the frequency of the tuning fork equals one of the natural frequencies of the string given by Equation 14-45 will the waves add in phase and the amplitude build up.

This resonance phenomenon is analogous to the resonance of a simple harmonic oscillator with a harmonic driving force. If the frequency of the driving force equals the natural frequency of a simple harmonic oscillator, the oscillator absorbs the maximum amount of energy from the driving force. Note, however, that a string fixed at both ends has not just one natural frequency but a sequence of natural frequencies, which are integral multiples of the fundamental frequency. This is called a *harmonic series*. The second frequency $f_2 = 2f_1$ is called the second harmonic; the $n$th is called the $n$th harmonic. In the terminology often used in music, the second harmonic is called the first overtone; the third harmonic is the second overtone; etc. Thus the $n$th harmonic is also the $(n - 1)$th overtone.

When the string is vibrated at any one of its natural frequencies, it absorbs energy from the external force and the amplitude increases until it reaches a maximum value; then the rate at which energy is lost to damping effects equals the rate at which energy is absorbed, and the total energy of the vibrating string is constant. The wave produced is called a *standing wave*.

Figure 14-25 shows the standing waves on a string fixed at both ends for the fundamental and the first few harmonics. We note that in addition to the points $x = 0$ and $x = L$, there are other points which are at

**Figure 14-24**
Waves on a string produced by a tuning fork whose frequency is not in resonance with the natural frequencies of the string. The wave leaving the tuning fork for the first time (*dashed line*) and that leaving for the second time after being reflected twice (*grey line*) are not in phase and do not interfere constructively. The resultant wave (*black line*) has about the same amplitude as the individual waves.

*Harmonic series*

Winds set up standing waves in the Tacoma Narrows suspension bridge, leading to its collapse on Nov. 7, 1940, only 4 months after it had been opened for traffic.

Wide World

rest if $n$ is greater than 1. For example, in the second harmonic, the midpoint of the string is at rest. Such a point is called a *node*. In the third harmonic, there are two points that are at rest in addition to the endpoints. In general, there are $n - 1$ nodes (not counting $x = 0$ and $x = L$) for the $n$th harmonic.

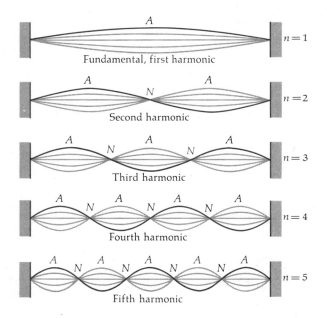

A
Fundamental, first harmonic    $n = 1$

A    N    A
Second harmonic    $n = 2$

A    N    A    N    A
Third harmonic    $n = 3$

A    N    A    N    A    N    A
Fourth harmonic    $n = 4$

A    N    A    N    A    N    A    N    A
Fifth harmonic    $n = 5$

**Figure 14-25**
Standing waves on a string fixed at both ends. The points labeled $A$ are antinodes; those labeled $N$ are nodes. In general, the $n$th harmonic has $n$ antinodes and $n - 1$ nodes (not counting the endpoints).

The points where the amplitude of vibration is maximum are called *antinodes*. There is an antinode halfway between each pair of nodes, including the ends. There are $n$ antinodes when the string is vibrating in its $n$th harmonic.

### String Fixed at One End

Standing waves can also be produced on a string with one end free instead of both ends fixed. A nearly free end results if the string is tied to a very long light thread, as in Figure 14-26. The analysis differs only slightly from that of a string fixed at both ends. When the first wave is reflected at the free end, it is not inverted; it is inverted, however, when reflected at the tuning fork. Thus if it arrives back at the fork in a time equal to *half* the period of the fork, it will add constructively to the next wave (actually the last half of the first wave). Constructive interference will also occur if the time taken for the first wave to travel twice the length of the string equals $\frac{3}{2}$ times the period, $\frac{5}{2}$ times the period, or any odd number of half periods. Thus the resonance condition for a standing wave on a string of length $L$ fixed at one end (the tuning fork) and free at the other end is

$$\frac{2L}{v} = \frac{nT}{2} \qquad \text{where } n = 1, 3, 5, 7, \ldots$$

or, in terms of the wavelength $\lambda = v/f = vT$,

$$n\frac{\lambda}{4} = L \qquad n = 1, 3, 5, 7, \ldots \qquad \text{14-47}$$

Fritz Henle/Photo Researchers

Pablo Casals producing standing waves on a cello string. The tension in the string is fixed but the effective length can be changed, thereby varying its fundamental frequency.

*Standing-wave condition for string fixed at one end*

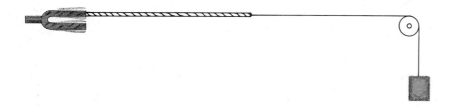

**Figure 14-26**
An approximation to a string fixed at one end and free at the other end can be produced by connecting the "free" end of the string to a very long light thread. Since the amplitude of the tuning fork is very small, the end attached to the fork is approximately fixed.

The frequencies are thus given by

$$f = n\frac{v}{4L} \qquad n = 1, 3, 5, 7, \ldots \qquad \text{14-48}$$

The natural frequencies of this system occur in the ratios $1:3:5:7$; that is, the frequencies are given by

$$f_n = nf_1 \qquad \text{where } n = 1, 3, 5, 7, \ldots \qquad \text{14-49}a$$

in which

$$f_1 = \frac{v}{4L} \qquad \text{14-49}b$$

is the fundamental frequency. The even harmonics are missing. The first few standing-wave nodes for a string fixed at one end are shown in Figure 14-27.

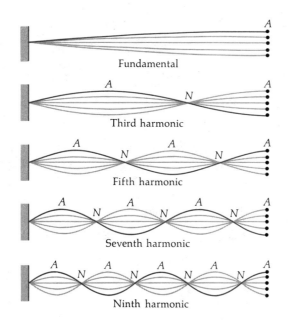

Fundamental

Third harmonic

Fifth harmonic

Seventh harmonic

Ninth harmonic

**Figure 14-27**
Standing waves on a string fixed at only one end. The free end is an antinode.

### Standing-Wave Functions

We can find the mathematical forms of the wave functions illustrated in Figures 14-25 and 14-27 and also rederive the resonance conditions, or standing-wave conditions (Equations 14-44 and 14-47), by considering the resultant wave formed by the addition of a wave traveling to the right and a wave traveling to the left. Let us call the displacement of the wave moving to the right $y_R$ and that of the wave moving to the left $y_L$

and assume that the amplitudes are equal. Then

$$y_R = y_0 \sin (kx - \omega t) \qquad \text{and} \qquad y_L = y_0 \sin (kx + \omega t)$$

where $k = 2\pi/\lambda$ is the wave number and $\omega = 2\pi f$ is the angular frequency. The sum of these two waves is

$$y(x, t) = y_R + y_L = y_0 \sin (kx - \omega t) + y_0 \sin (kx + \omega t)$$

We can simplify this by using the trigonometric identity

$$\sin A + \sin B = 2 \sin \tfrac{1}{2}(A + B) \cos \tfrac{1}{2}(A - B) \qquad\qquad 14\text{-}50$$

If we let $A = kx + \omega t$ and $B = kx - \omega t$, we have

$$\tfrac{1}{2}(A + B) = kx \qquad \text{and} \qquad \tfrac{1}{2}(A - B) = \omega t$$

Then*

$$y(x, t) = 2y_0 \cos \omega t \sin kx \qquad\qquad 14\text{-}51$$

If the string is fixed at $x = 0$ and $x = L$, we have the following *boundary conditions* on the wave functions:

$$y(x = 0, t) = 0 \qquad\qquad 14\text{-}52$$

and

*Boundary conditions*

$$y(x = L, t) = 0 \qquad\qquad 14\text{-}53$$

for all times $t$. The first boundary condition is automatically met because $\sin kx = 0$ at $x = 0$. The second boundary condition is met only for those particular values of the wave number which satisfy

$$\sin kL = 0 \qquad\qquad 14\text{-}54$$

The values $k_n$ which satisfy this equation are given by

$$k_n L = n\pi \qquad\qquad 14\text{-}55$$

where $n$ is any integer. In terms of the wavelength $\lambda = 2\pi/k$, Equation 14-55 is

$$\frac{2\pi}{\lambda_n} L = n\pi$$

or

$$n\frac{\lambda_n}{2} = L \qquad\qquad 14\text{-}56$$

This is the same condition derived earlier. The length of the string must equal an integral number of half wavelengths.

The standing-wave functions are sine functions whose wavelengths are given by Equation 14-56. The easiest way to remember the condition for standing waves is to draw a line of length $L$ and fit sine waves on it so that the displacement is zero at each end.

For a string with one end fixed and the other end free, the boundary conditions are that $y = 0$ at $x = 0$, which occurs automatically, and that $y$ be a maximum or minimum at $x = L$. That is, the point $x = L$ is an antinode. This occurs if

$$\sin k_n L = \pm 1$$

---

* It might seem logical to subtract the two waves since the wave is inverted by reflection. It makes no difference whether we add or subtract the waves. Using $\sin A - \sin B = 2 \sin \tfrac{1}{2}(A - B) \cos \tfrac{1}{2}(A + B)$, we have $y_L - y_R = 2y_0 \sin \omega t \cos kx$. This differs from Equation 14-51 only in the phases of the functions. Equation 14-51 is more convenient because we do not have to add a phase constant in order to meet the boundary conditions.

Organ pipes of St. George's Church in New York City. The fundamental frequency of each pipe is determined by its length and by whether the pipe is open or closed.

The Bettmann Archive

An organ pipe is a familiar example of the use of standing waves in air columns. In the flue-type organ pipe (Figure 14-30) a stream of air is directed against the sharp edge of an opening (point A in the figure). The complicated swirling motion of the air near the edge sets up vibrations in the air column. The resonance frequencies of the pipe depend on the length of the pipe and on whether the other end is closed or open.

In a closed organ pipe there is a pressure antinode at the closed end and a pressure node near the opening (point A in Figure 14-30). The resonance frequencies for such a pipe are therefore those for a tube open at one end and closed at the other. The wavelength of the fundamental is approximately 4 times the length of the pipe, and only odd harmonics are present.

In an open organ pipe, there is a pressure node near both ends of the pipe. The resonance frequencies for a pipe open at both ends are the same as for one closed at both ends except that there is an end correction at each end. The wavelength of the fundamental is 2 times the effective length of the pipe, and all harmonics are present.

**Figure 14-30**
Flue-type organ pipe. A stream of air is blown against the edge, causing a swirling motion of the air near A, which excites standing waves in the pipe. There is a pressure node near point A, which is open to the atmosphere. The resonance frequencies of the pipe depend on the length of the pipe and on whether the other end is open or closed.

#### Question

11. How do the resonance frequencies of an organ pipe change when the air temperature increases?

## 14-7  Energy and Intensity of Harmonic Waves

An important property of waves is that they transport energy and momentum. The transport of energy in a wave ordinarily is described in terms of the *wave intensity*, defined to be the average rate at which the wave transmits energy per unit area normal to the direction of propagation. That is, the intensity at any point in a wave is the average incident

*Intensity*

energy per unit time per unit area. Since the energy per unit time is the power, the intensity is the average power incident per unit area*:

$$I = \frac{(\Delta E/\Delta t)_{av}}{A} = \frac{P_{av}}{A} \qquad\qquad 14\text{-}59$$

If a point source emits power $P$ uniformly in all directions, the intensity will decrease with the square of the distance from the source. We can see this from the fact that at a distance $r$ from the source the power will be uniformly spread over a spherical surface of area $4\pi r^2$. In this case, the intensity will be

$$I = \frac{P}{4\pi r^2} \qquad\qquad 14\text{-}60$$

If, on the other hand, the energy is radiated uniformly into a hemi-sphere, e.g., a sound source on the ground, the power will be distributed over a hemispherical area of $2\pi r^2$ and the intensity will be given by $I = P/2\pi r^2$.

There is a simple relation between the intensity of a wave and the energy per unit volume in the medium carrying the wave. Suppose that a wave traveling to the right on a string has just reached point $P_1$ at time $t_1$, as shown in Figure 14-31. The part of the string to the left of $P_1$ contains energy because each segment is oscillating with simple harmonic motion. At this time $t_1$ there is no energy to the right of $P_1$ because the wave has not yet reached that part of the string. After a time $\Delta t$ the wave moves past point $P_1$ a distance $v\,\Delta t$, where $v$ is the wave velocity. The total energy of the string is thus increased by the amount of energy in the string past point $P_1$. This increase in energy can be written

$$\Delta E = \eta A v\, \Delta t$$

where $\eta$ is the average energy per unit volume of the string, $A$ the cross-sectional area, and $Av\,\Delta t$ the volume of the string past point $P_1$, which now contains energy.** The rate of increase of energy is the power passing point $P_1$. (The source of this energy is of course at the left end of the string, where there must be an external force doing work on the end of the string.) Thus the average incident power is

$$P_{av} = \frac{\Delta E}{\Delta t} = \eta A v$$

The intensity of the wave at a point is found by dividing the incident power by the area of the string:

$$I = \frac{P_{av}}{A} = \eta v \qquad\qquad 14\text{-}61$$

*The intensity equals the product of the wave velocity $v$ and the average energy per unit volume, or the average energy density $\eta$.*

This result is applicable to all waves.

---

* For waves on a string, the area is just the cross-sectional area of the string. We use the term power per unit area even though it is somewhat artificial for a string, because our discussion can then be carried over to any kind of waves.

** We assume that the amplitude of the wave is small so that $v\,\Delta t$ is approximately the length of string containing the wave beyond point $P_1$.

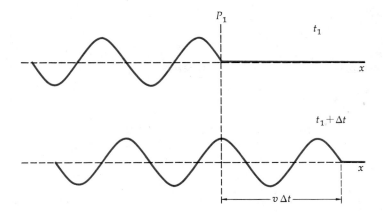

**Figure 14-31**
Energy transmitted by a wave
on a string. At time $t_1$ the
wave has just reached point
$P_1$. In an additional time $\Delta t$
the wave travels a distance
$v\,\Delta t$. The energy transmitted
past point $P_1$ is $Av\eta\,\Delta t$, where
$\eta$ is the average energy per
unit volume and $A$ is the
cross-sectional area of the
string.

To calculate the average energy density $\eta$ for waves on a string, consider a harmonic wave of angular frequency $\omega$ and amplitude $y_0$ traveling along a stretched string. Each element of the string is undergoing simple harmonic motion of angular frequency $\omega$ and amplitude $y_0$. The energy of a segment of the string of mass $\Delta m$ is of the same form as Equation 11-16 for the energy of a mass oscillating on a spring:

$$\Delta E = \tfrac{1}{2}(\Delta m)\omega^2 y_0^2 \qquad\qquad 14\text{-}62$$

The mass of the segment is just the mass density $\rho$ times the volume of the segment $\Delta V$. Substituting $\Delta m = \rho\,\Delta V$ into Equation 14-62, we have

$$\Delta E = \tfrac{1}{2}\rho\omega^2 y_0^2\,\Delta V$$

The average energy density

$$\eta = \frac{\Delta E}{\Delta V} = \tfrac{1}{2}\rho\omega^2 y_0^2 \qquad\qquad 14\text{-}63$$

is proportional to the square of the frequency and to the square of the amplitude. Substituting this result into Equation 14-61, we have for the intensity of a wave on a string,

$$I = \eta v = \tfrac{1}{2}\rho\omega^2 y_0^2 v \qquad\qquad 14\text{-}64a$$

The result that *the intensity of a wave is proportional to the square of the amplitude* is a general property of all harmonic waves. Equation 14-63 can be applied to a sound wave if we replace $y_0$ by the displacement amplitude $s_0$. We can also write the intensity of a sound wave in terms of the pressure amplitude $p_0$, which is related to the displacement amplitude. Replacing $y_0$ by $s_0$ and using $s_0 = p_0/\rho\omega v$ from Equation 14-24, we have

$$I = \tfrac{1}{2}\rho v\omega^2 s_0^2 = \frac{1}{2}\frac{p_0^2}{\rho v} \qquad\qquad 14\text{-}64b$$

The human ear can accommodate a rather large range of sound-wave intensities from about 1 pW/m², which is usually taken to be the threshold of hearing, to about 1 W/m², which produces a sensation of pain in most people. The pressure amplitudes for these extreme sound intensities can be easily calculated from Equation 14-64b. Taking $\rho = 1.29$ kg/m³ for the density of air and $v = 331$ m/s for the speed of sound, we find for the pressure amplitude corresponding to the threshold of hearing intensity of 1 pW/m² = $10^{-12}$ W/m²:

$$p_0^2 = 2\rho v I = 2(1.29)(331)(10^{-12}) = 8.54 \times 10^{-10}$$

and

$$p_0 = 2.92 \times 10^{-5} \text{ N/m}^2 = 2.92 \times 10^{-5} \text{ Pa}$$

Similarly for an intensity of 1 W/m² at the pain threshold, the pressure amplitude is

$$p_0 = \sqrt{2(1.29)(331)(1)} = 29.2 \text{ Pa}$$

These very small changes in pressure are superimposed on the normal constant atmospheric pressure level of about 101 kPa.

Because of the enormous range of intensities to which the ear is sensitive, and because the psychological sensation of loudness varies with intensity not directly but more nearly logarithmically, a logarithmic scale is used to describe the intensity level of a sound wave. The intensity level $\beta$ measured in *decibels* (dB) is defined by

$$\beta = 10 \log \frac{I}{I_0} \qquad\qquad 14\text{-}65 \qquad \textit{Intensity level in decibels}$$

where $I$ is the intensity corresponding to the level $\beta$ and $I_0$ is a reference level which we shall take to be the threshold of hearing:

$$I_0 = 10^{-12} \text{ W/m}^2 \qquad\qquad 14\text{-}66$$

**Table 14-1**
Intensity and intensity level of some common sounds, $I_0 = 10^{-12}$ W/m²

| Source | $I/I_0$ | dB | Description |
|---|---|---|---|
| | $10^0$ | 0 | Hearing threshold |
| Normal breathing | $10^1$ | 10 | Barely audible |
| Rustling leaves | $10^2$ | 20 | |
| Soft whisper (at 5 m) | $10^3$ | 30 | Very quiet |
| Library | $10^4$ | 40 | |
| Quiet office | $10^5$ | 50 | Quiet |
| Normal conversation (at 1 m) | $10^6$ | 60 | |
| Busy traffic | $10^7$ | 70 | |
| Noisy office with machines; average factory | $10^8$ | 80 | |
| Heavy truck (at 15 m); Niagara Falls | $10^9$ | 90 | Constant exposure endangers hearing |
| Old subway train | $10^{10}$ | 100 | |
| Construction noise (at 3 m) | $10^{11}$ | 110 | |
| Rock concert with amplifiers (at 2 m); jet takeoff (at 60 m) | $10^{12}$ | 120 | Pain threshold |
| Pneumatic riveter; machine gun | $10^{13}$ | 130 | |
| Jet takeoff (nearby) | $10^{15}$ | 150 | |
| Large rocket engine (nearby) | $10^{18}$ | 180 | |

On this scale, the threshold of hearing is

$$\beta = 10 \log \frac{I_0}{I_0} = 10 \log 1 = 0 \text{ dB}$$

Similarly, the pain threshold is

$$\beta = 10 \log \frac{1}{10^{-12}} = 10 \log 10^{12} = 10(12) = 120 \text{ dB}$$

Table 14-1 lists the intensity level of some common sounds.

---

**Example 14-9** A dog barking delivers about 1 mW of power. If this power is uniformly distributed over a hemispherical area, what is the sound-level intensity at a distance of 5 m? What would the level be if 5 dogs barking at the same time each delivered 1 mW of power?

The intensity at a distance of 5 m is the power divided by the area, which is $2\pi r^2 = 2\pi(5 \text{ m})^2 = 157 \text{ m}^2$. The intensity is therefore $I = P/A = (10^{-3} \text{ W})/(157 \text{ m}^2) = 6.37 \times 10^{-6} \text{ W/m}^2$. The intensity level at this distance is then

$$\beta = 10 \log \frac{I}{I_0} = \frac{10 \log (6.37 \times 10^{-6})}{10^{-12}}$$

$$= 10 \log (6.37 \times 10^{6})$$

$$= 10(\log 6.37 + \log 10^{6})$$

$$= 10(0.80 + 6) = 68 \text{ dB}$$

If there are 5 dogs barking at the same time, the intensity will be 5 times as great. Then $I/I_0 = 5 \times 6.37 \times 10^6 = 31.8 \times 10^6 = 3.18 \times 10^7$ and $\beta = 10(\log 3.18 + 7) = 75.0 \text{ dB}$.

---

**Questions**

12. If the amplitude of harmonic waves is doubled, how does the rate at which energy is used to generate the waves change?

13. Transverse waves of the same frequency and amplitude are sent along two parallel stretched wires. The wires are of the same kind except that one has a greater tension than the other. Over a long time, which wire transmits the greater energy? For which wire is the wave intensity greater?

14. A source of sound waves oscillates with a fixed amplitude and frequency. Assuming the atmospheric pressure to remain constant, how does the rate at which sound waves carry energy away from the source change if the air temperature rises?

15. Two tuning forks produce sound waves in air of the same amplitude. One has frequency 256 Hz and the other frequency 512 Hz. Which makes the louder sound, i.e., produces sound with the greater intensity?

16. How does the amplitude of a sound wave vary with distance $r$ from a point source?

# Sonic Booms

Laurent Hodges
*Iowa State University*

One of the most common examples of a shock wave resulting from a source traveling faster than the wave-propagation speed is the sonic boom generated by an airplane whose speed exceeds the speed of sound. A common misconception is that a sonic boom is produced when the airplane first accelerates past the speed of sound ("crashes through the sound barrier"). In fact the shock wave exists during the whole time that the airplane is traveling super-sonically, and a boom is heard every time the shock wave sweeps over a person with good hearing.

The airplane actually has two shock waves, associated with its front and back (Figure 1). These bow and tail shock waves have nearly the shape of cones (*Mach cones*) with apex at the aircraft's bow and tail, respectively. (Some deviation from perfect cone shape results from variations in sound speeds in different parts of the air.) The cones have a half angle $\theta$ given by $\sin \theta = v/u_a$, where $v$ is the speed of sound and $u_a$ is the speed of the airplane.

**Figure 1**
Cross-sectional represen-
tation of a supersonic transport in flight, showing bow
and tail shock waves, Mach
cone angle $2\theta$, and a plot of
the pressure at ground level.
The deviations from normal
atmospheric pressure occur in
the region of overpressure
between the two shock
waves.

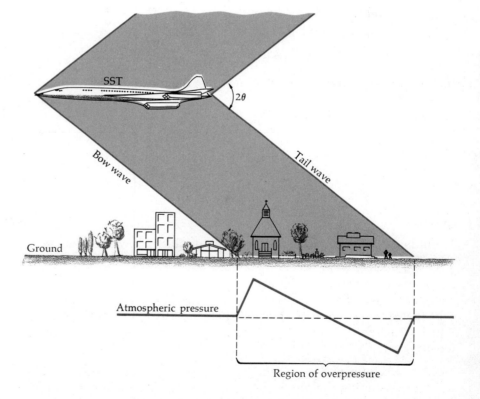

The air pressure in the region of overpressure between the two shock waves differs from normal atmospheric pressure, the bow and tail waves being associated with pressures higher and lower than normal, respectively. This pressure deviation produces the sensation of sound in an observer. As the bow wave sweeps over the observer, the pressure increases by the *overpressure* $\Delta P$ in a rise time $\tau$. It then decreases to approximately $\Delta P$ below normal atmospheric pressure before suddenly returning to normal at the tail wave. A plot of the time dependence of the pressure at a point has the shape of a slanted N and is therefore called the *N signature* (Figure 2).

**Figure 2**
A plot of the N signature, or time dependence, of the pressure at a given point where a sonic boom occurs. Shown are the overpressure $\Delta P$, the rise time $\tau$, and the time duration $T$.

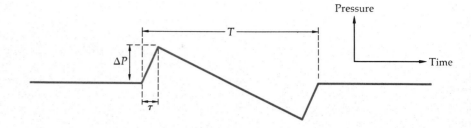

The major pressure changes experienced at the ear occur as the bow and tail shock waves reach the observer, each producing an explosive sound. This explains the double crack of a sonic boom. A supersonic bullet also has bow and tail shock waves, but its two booms are so close together in time that only one crack is heard.

The startling nature of a sonic boom and the resulting annoyance are a function of the rise time, the total duration of the N signature, and the overpressure. The rise time is of the order of 3 ms, and the time duration varies from about 0.1 s for a supersonic fighter to about 0.4 s for the British-French Concorde supersonic transport (SST). A typical overpressure for an SST 20 km overhead would be 100 N/m² = 100 Pa. The overpressure is less at places on either side of the plane's path, but booms can typically be heard over a zone 80 km wide.

The shock waves in air produced by a supersonic .22-caliber bullet. (*Courtesy of Harold E. Edgerton.*)

It is not possible to assign a decibel reading to a sonic boom because of its short duration and its highly nonsinusoidal form. A sinusoidal sound wave with an rms pressure of 100 Pa would correspond to 134 dB, but the overpressure is not the only relevant parameter. Shorter rise times, for example, are associated with "louder" booms. U.S. Air Force studies have shown that an overpressure of 100 Pa corresponds to 110 to 120 on the perceived-noise decibel scale. (This modification of the decibel scale incorporates the psychological effects of the frequency spectrum of aircraft noises.)

There have been several systematic studies of the effects of sonic-boom exposures, including some carried out by the U.S. Air Force, which has tried to make the sonic boom psychologically acceptable by referring to it as the "sound of freedom." Besides startling people and animals, sonic booms can rattle dishes, shatter glass, and even lead to structural damage to some buildings. One study has indicated that approximately $600 in damage claims would result from every million person-booms,[*] and another researcher has estimated that each SST might generate 1,000 claims totaling $500,000 annually.[**]

Some people are not bothered much by sonic booms. Many others will presumably learn to tolerate them if supersonic flight becomes commonplace, but even they will suffer the adverse physiological effects that result from any loud noise. Perhaps 25 to 50 percent of the United States population will not be able to adapt to sonic booms, and frequent supersonic flights will doubtless encounter stiff public opposition.

SST Concorde taking off from John F. Kennedy airport in New York.

Soviet Russia and a British-French consortium built SSTs, but only the state-owned airlines of the three producing countries ever purchased the aircraft. The SST program ended in the United States in 1971 after Congress cut off federal funding, although there are occasional attempts to revive the project. Many countries have instituted restrictions, e.g., no supersonic flights over land, that may well make the SST highly uneconomical, so that its future is highly uncertain.

[*] Karl D. Kryter, "Sonic Booms from Supersonic Transport," *Science*, vol. 163, p. 359, 1969.
[**] William A. Shurcliff, *S/S/T and Sonic Boom Handbook*, Ballantine Books, New York, 1970. For further information see Harvey H. Hubbard, "Sonic Booms," *Physics Today*, vol. 21, p. 31, February 1968; and Herbert A. Wilson, Jr., "Sonic Boom," *Scientific American*, vol. 206, p. 36, January 1962.

*Optional*

## 14-8 The Wave Equation

A general wave function $y(x, t)$ is a solution of a differential equation called the *wave equation*. The wave equation relates the second derivative of the wave function with respect to $x$ to the second derivative with respect to $t$. Because there are two variables, these are partial derivatives. We can obtain the wave equation by recalling that the function

$$y(x, t) = y_0 \sin (kx - \omega t)$$

is a particular solution for harmonic waves. It is useful to write the angular frequency $\omega$ in terms of the velocity $v$ and wave number $k$; that is, $\omega = kv$. Then the harmonic wave function is

$$y(x, t) = y_0 \sin (kx - kvt) \qquad \text{14-67}$$

The derivative with respect to $x$, holding $t$ constant, is

$$\frac{\partial y}{\partial x} = ky_0 \cos (kx - kvt)$$

Similarly, the second derivative with respect to $x$ is

$$\frac{\partial^2 y}{\partial x^2} = -k^2 y_0 \sin (kx - kvt) = -k^2 \, y(x, t) \qquad \text{14-68}$$

The derivative of $y(x, t)$ with respect to $t$, holding $x$ constant, is

$$\frac{\partial y}{\partial t} = -kvy_0 \cos (kx - kvt)$$

and

$$\frac{\partial^2 y}{\partial t^2} = -k^2 v^2 y_0 \sin (kx - kvy) = -k^2 v^2 \, y(x, t) \qquad \text{14-69}$$

Combining Equations 14-68 and 14-69, we obtain

$$\frac{\partial^2 y}{\partial x^2} = \frac{1}{v^2} \frac{\partial^2 y}{\partial t^2} \qquad \text{14-70}$$

*Wave equation*

—the wave equation. If $y$ is the displacement of a vibrating string, this equation describes string waves. If we interpret $y$ as the increase or decrease in the pressure or density of a gas, this equation describes sound waves. The same equation also describes electromagnetic waves, in which $y$ is the electric or magnetic field.

Equation 14-70 is satisfied by any wave in one dimension that is propagated without dispersion or change of shape. We showed earlier that in general such a wave has a wave function which can be expressed as a function of either $x + vt$ or $x - vt$. We can easily show that any function of $x - vt$ or of $x + vt$ satisfies Equation 14-70. Let $\theta = x - vt$ and consider any wave function

$$y = y(x - vt) = y(\theta)$$

The derivative of $y$ with respect to $\theta$ we shall call $y'$. Then by the chain rule for derivatives,

$$\frac{\partial y}{\partial x} = \frac{\partial y}{\partial \theta} \frac{\partial \theta}{\partial x} = y' \frac{\partial \theta}{\partial x} \qquad \text{and} \qquad \frac{\partial y}{\partial t} = \frac{\partial y}{\partial \theta} \frac{\partial \theta}{\partial t} = y' \frac{\partial \theta}{\partial t}$$

Since $\partial \theta / \partial x = 1$ and $\partial \theta / \partial t = -v$, we have

$$\frac{\partial y}{\partial x} = y' \qquad \text{and} \qquad \frac{\partial y}{\partial t} = -vy'$$

Taking the second derivatives, we obtain

$$\frac{\partial^2 y}{\partial x^2} = y''$$

$$\frac{\partial^2 y}{\partial t^2} = -v \frac{\partial y'}{\partial t} = -v \frac{\partial y'}{\partial \theta} \frac{\partial \theta}{\partial t} = +v^2 y''$$

Thus again

$$\frac{\partial^2 y}{\partial x^2} = \frac{1}{v^2} \frac{\partial^2 y}{\partial t^2}$$

Of course other equations relating the derivatives of these wave functions can be obtained by taking derivatives of the functions.

Equation 14-70 is important because it is a direct consequence of Newton's second law, $\Sigma \mathbf{F} = m\mathbf{a}$, applied to a segment of a string. For sound waves, the identical equation (with the reinterpretation of $y$ mentioned) is derived from Newton's laws applied to fluids. Similarly, an equation of the same form can be derived for electromagnetic waves from Maxwell's equations for electric and magnetic fields. We discuss the case of string waves only, to illustrate that the wave equation is a consequence of newtonian mechanics.

In Figure 14-32 we have isolated one segment of a string. Our derivation will apply only if the wave is small enough in amplitude for the angle between the string and the horizontal (the original direction of the string with no wave) to be small, in which case the length of the segment is approximately $\Delta x$ and the mass is $\mu \, \Delta x$. The string moves vertically with acceleration $\partial^2 y/\partial t^2$. The net vertical force is

$$\Sigma F = T \sin \theta_2 - T \sin \theta_1 \qquad\qquad 14\text{-}71$$

where $\theta_2$ and $\theta_1$ are the angles indicated in Figure 14-32 and $T$ is the tension in the string. Again, since the angles are assumed to be small, we can approximate $\sin \theta$ by $\tan \theta$. The tangent of the angle made by the string with the horizontal is just the slope of the curve formed by the string:

$$\tan \theta = S = \frac{\partial y}{\partial x}$$

Thus the net force on the string segment can be written

$$F = T(\sin \theta_2 - \sin \theta_1) \approx T(\tan \theta_2 - \tan \theta_1)$$

$$= T(S_2 - S_1) = T \, \Delta S$$

where $S = \tan \theta = \partial y/\partial x$ is the slope and $\Delta S$ is the change in slope. Setting this resultant force equal to the mass times the acceleration gives

$$T \, \Delta S = \mu \, \Delta x \frac{\partial^2 y}{\partial t^2} \qquad T \frac{\Delta S}{\Delta x} = \mu \frac{\partial^2 y}{\partial t^2}$$

In the limit $\Delta x \to 0$, we have

$$\lim_{\Delta x \to 0} \frac{\Delta S}{\Delta x} = \frac{\partial S}{\partial x} = \frac{\partial}{\partial x}\left(\frac{\partial y}{\partial x}\right) = \frac{\partial^2 y}{\partial x^2}$$

Thus

$$\frac{\partial^2 y}{\partial x^2} = \frac{\mu}{T} \frac{\partial^2 y}{\partial t^2} \qquad\qquad 14\text{-}72$$

It is important to realize that this wave equation for a stretched string holds only for small angles and thus for small displacements $y(x, t)$.

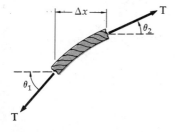

**Figure 14-32**
Segment of a stretched string used for the derivation of the wave equation. The net vertical force on the segment is $T \sin \theta_2 - T \sin \theta_1$, where $T$ is the tension. The wave equation is derived by applying Newton's second law to the segment.

Comparing Equations 14-70 and 14-72, we see again that the velocity of the wave is

$$v = \sqrt{\frac{T}{\mu}}$$

An important property of the wave equation 14-70 is that it is linear; i.e., the function $y(x, t)$ and its derivatives occur only to the first power. There are no terms like $y^2$, $(\partial y/\partial x)^2$, $y\,\partial^2 y/\partial t^2$, or $(\partial^2 y/\partial t^2)^2$. An important property of linear equations is that if $y_1(x, t)$ and $y_2(x, t)$ are two solutions of the equation, the linear combination

$$y_3(x, t) = C_1\, y_1(x, t) + C_2\, y_2(x, t) \qquad\qquad 14\text{-}73$$

where $C_1$ and $C_2$ are any constants, is also a solution. This is easily shown by direct substitution of $y_3$ into the equation (see Exercise 60).

Equation 14-73 is the mathematical statement of the superposition principle. If any two waves satisfy the wave equation 14-70, their sum also satisfies the same wave equation.

The superposition principle for waves on a string holds only if the amplitudes of the waves are small, so that the approximation used in deriving the wave equation $\sin \theta \approx \tan \theta$ holds. If the amplitudes are large, this approximation does not hold and the resulting equation relating the time derivatives and spatial derivatives of $y(x, t)$ is not linear. In that case, the sum of two solutions is not a solution and the principle of superposition does not hold.

A similar small-amplitude approximation must be made to derive the wave equation for sound waves from Newton's laws for a fluid. Thus the principle of superposition for sound waves is also limited to waves in which the amplitude of the pressure or displacement variations is small.

## Review

A. Define, explain, or otherwise identify:

Dispersion, 397
Transverse wave, 399
Longitudinal wave, 399
Wave function, 400
Interference, 400
Superposition, 401
Harmonic wave, 407
Wavelength, 407
Ray, 407
Plane wave, 408
Amplitude, 408
Wave number, 408

Doppler effect, 411
Standing wave, 417
Harmonic series, 419
Overtone, 419
Node, 420
Antinode, 420
Boundary conditions, 422
Intensity, 425
Intensity level, 428
Decibel, 428
Wave equation, 433

B. Sketch the wave functions for the fundamental and first three overtones for a string fixed at both ends. From your sketches write down the wavelengths for these harmonics. Do the same for a string fixed at one end and free at the other end.

C. Sketch both the displacement and pressure standing-wave functions for the fundamental and first three overtones in an organ pipe open at both ends. Do the same for a pipe open at only one end.

D. True or false:

1. Wave pulses on strings are transverse waves.

2. Sound waves in air are transverse waves.

3. The speed of sound at 20°C is twice that at 5°C.

4. When a wave pulse is reflected, it is always inverted.

5. The frequency of the fifth harmonic is 5 times the frequency of the fundamental.

6. In a pipe open at one end and closed at the other, the even harmonics are not excited.

7. Because of the end correction, the pressure antinodes are not separated by $\frac{1}{2}$ wavelength in an open pipe.

8. The energy in a harmonic wave is proportional to the square of the amplitude.

9. A 60-dB sound has twice the intensity of a 30-dB sound.

10. The units of intensity are watts per square metre.

11. The doppler shift of sound waves depends only on the relative velocity of the source and receiver.

## Exercises

*Unless otherwise specified, take the speed of sound to be 340 m/s in air and 1500 m/s in water.*

### Section 14-1, Wave Pulses

1. A pulse moving through a medium in which there is energy absorption but no dispersion gets smaller, but its relative shape stays the same, i.e., does not spread out. Sketch such a wave pulse at several equally spaced times.

2. Figure 14-33 shows a wave pulse at time $t = 0$. The pulse moves to the right at 1 cm/s, and it disperses in such a way that the area under the pulse stays approximately constant. Sketch the shape of the string at times $t = 1, 2,$ and 3 s.

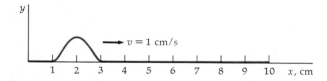

**Figure 14-33**
Pulse on string for Exercises 2 to 4.

3. Assume that the wave pulse on a string shown in Figure 14-33 is moving to the right without changing shape. At this particular time, which segments of the string are moving up? Which are moving down? Is there any segment of the string at the pulse that is instantaneously at rest? Answer these questions by sketching the pulse at a slightly later time and a slightly earlier time to see how the string segments are moving.

4. Make a sketch of the velocity of each string segment versus position for the pulse shown in Figure 14-33.

### Section 14-2, Speed of Waves

5. Show that the units of $\sqrt{T/\mu}$ are metres per second if $T$ is in newtons and $\mu$ in kilograms per metre.

6. A steel piano wire is 0.7 m long and has a mass of 5 g. It is stretched with a tension of 500 N. (*a*) What is the speed of transverse waves in the wire? (*b*) To reduce the wave speed by a factor of 2 without changing the tension, what mass of copper wire would have to be wrapped around the steel wire?

7. A common lecture demonstration of wave pulses uses a piece of rubber tubing tied at one end to a fixed post and passed over a pulley to a suspended weight at the other end. Suppose that the distance from the fixed support to the pulley is 10 m, the mass of this length of tubing is 0.7 kg, and the suspended weight is 110 N. If the tubing is given a transverse blow at one end, how long will it take the resulting pulse to reach the other end?

8. A long glass tube has a semicircular bend of radius 8 cm. A string of mass density 0.04 kg/m under tension of 20 N is pulled through the tube. (a) At what speed should the string be pulled if it is to pass through the tube without touching the sides? (b) What is the acceleration of the string segments passing through the semicircular bend in this case? (c) Draw a diagram indicating the forces acting on a string segment in the bend to provide this acceleration.

9. A steel wire 7 m long has a mass of 100 g. It is under tension of 900 N. What is the speed of a transverse wave pulse on this wire?

10. (a) What are the SI units of bulk modulus? (b) Show that $\sqrt{B/\rho}$ has the units of metres per second if B has the correct SI units and $\rho$ is in kilograms per cubic metre.

11. Find the speed of sound in air at $T = 300$ K in kilometres per hour.

12. The bulk modulus for water is $2.0 \times 10^9$ N/m$^2$. Use it to find the speed of sound in water.

13. The bulk modulus for aluminum is $7.0 \times 10^{10}$ N/m$^2$. The density of aluminum is $2.7 \times 10^3$ kg/m$^3$. Find the speed of sound in aluminum.

14. Calculate the speed of sound waves in helium gas at $T = 300$ K (take $M = 4$ g/mol and $\gamma = 1.67$).

15. The speed of sound in mercury is 1410 m/s. What is the bulk modulus for mercury ($\rho = 13.6 \times 10^3$ kg/m$^3$)?

## Section 14-3, Harmonic Waves

16. The ear is sensitive to sound frequencies in the range of about 20 to 20,000 Hz. (a) What are the wavelengths in air corresponding to these frequencies? (b) What are the wavelengths in water?

17. (a) Middle C in the musical scale has a frequency of 262 Hz. What is the wavelength of this note in air? (b) The frequency of the C an octave above middle C is twice that of middle C. What is the wavelength of this note?

18. Equation 14-18 expresses the displacement of a harmonic wave as a function of x and t in terms of the wave parameters k and $\omega$. Write equivalent expressions which instead of k and $\omega$ contain the following pairs of parameters: (a) k and v, (b) $\lambda$ and f, (c) $\lambda$ and T, (d) $\lambda$ and v, (e) f and v.

19. Equation 14-16 applies to any type of wave, including light waves in vacuum, which travel at $3 \times 10^8$ m/s. The range of wavelengths of light to which the eye is sensitive is about $4 \times 10^{-7}$ to $7 \times 10^{-7}$ m. What are the corresponding frequencies of these light waves?

20. A wave travels to the right along a string with speed 10 m/s. Its frequency is 60 Hz, and its amplitude is 0.02 m. (a) Write a suitable wave function for this wave. (b) Is this wave function the only function that could describe this wave? Explain.

21. The wave function for a harmonic wave on a string is $y(x, t) = 0.001 \sin (62.8x + 314t)$, where y and x are in metres and t is in seconds. (a) In what direction does this wave travel, and what is its speed? (b) Find the wavelength, frequency, and period of this wave. (c) What is the maximum displacement of any string segment?

22. What is the displacement amplitude for a sound wave of frequency 100 Hz and pressure amplitude $10^{-4}$ atm?

23. The displacement amplitude for a sound wave of frequency 300 Hz is $10^{-7}$ m. What is the pressure amplitude for this wave?

24. (a) Find the displacement amplitude for a sound wave of frequency 1000 Hz at the pain-threshold pressure amplitude of 29 Pa. (b) Find the displacement amplitude for a sound wave with the same pressure amplitude but of frequency 500 Hz.

25. A typical loud sound wave with a frequency of 1 kHz would have a pressure amplitude of about $10^{-4}$ atm. (a) At $t = 0$, the pressure is maximum at some point $x_1$. What is the displacement at that point at $t = 0$? (b) What is the maximum value of the displacement at any time and place? (Take the density of air to be 1.29 kg/m$^3$.)

26. (a) Find the displacement amplitude for a sound wave of frequency 500 Hz at the threshold-of-hearing pressure amplitude of $2.9 \times 10^{-5}$ Pa. (b) Find the displacement amplitude for a wave of the same pressure amplitude but of frequency 1 kHz.

### Section 14-4, The Doppler Effect

27. This exercise is a doppler-effect analogy. A conveyor belt moves to the right with speed $v = 300$ m/min. A very fast pieman puts pies on the belt at a rate of 20 per minute, and they are received at the other end by a pie eater. (a) If the pieman is stationary, find the spacing $\lambda$ between the pies and the frequency $f$ with which they are received by the stationary pie eater. (b) The pieman now walks with speed 30 m/min toward the receiver while continuing to put pies on the belt at 20 per minute. Find the spacing of the pies and the frequency with which they are received by the stationary pie eater. (c) Repeat your calculations for a stationary sender and a pie eater who moves toward him at 30 m/min.

28. For the situation described in Exercise 27 derive general expressions for the spacing of the pies and the frequency with which they are received, in terms of the speed of the belt $v$, the speed of the sender $u_S$, the speed of the receiver $u_R$, and the frequency $f_0$ of the pieman.

*In Exercises 29 to 34 the source emits sounds of frequency* 200 Hz *moving through still air at* 340 m/s.

29. The source moves with speed 80 m/s relative to still air toward a stationary listener. (a) Find the wavelength of the sound between source and listener. (b) Find the frequency heard by the listener.

30. Consider the situation in Exercise 29 from the reference frame in which the source is at rest. In this frame the listener moves toward the source with speed 80 m/s, and there is a wind blowing at 80 m/s from the listener to the source. (a) What is the speed of the sound from source to listener in this frame? (b) Find the wavelength of the sound between source and listener. (c) Find the frequency heard by the listener.

31. The source moves at 80 m/s away from the stationary listener. (a) Find the wavelength of the sound waves between the source and listener. (b) Find the frequency heard by the listener.

32. The listener moves at 80 m/s relative to still air toward a stationary source. (a) What is the wavelength of the sound between the source and listener? (b) What is the frequency heard by the listener?

33. Consider the situation in Exercise 32 in a reference frame in which the listener is at rest. (a) What is the wind velocity in this frame? (b) What is the speed of sound from source to listener in this frame, i.e., relative to the listener? (c) Find the wavelength of the sound between the source and listener in this frame. (d) Find the frequency heard by the listener.

34. The listener moves at 80 m/s relative to the still air away from a stationary source. Find the frequency heard by the listener.

35. A whistle of frequency 500 Hz moves in a circle of radius 1 m making 3 rev/s. What are the maximum and minimum frequencies heard by a stationary listener?

## Section 14-5, Standing Waves on a String

36. A 5-g steel wire 1 m long is under tension of 968 N. (a) Find the speed of transverse waves in the wire. (b) Find the wavelength and frequency of the fundamental. (c) Find the frequency of the second and third harmonics.

37. Middle C on the equal-temperament scale used by modern instrument makers has a frequency of 261.63 Hz. If this is the fundamental frequency of a 7-g piano wire 80 cm long, what should the tension in the wire be?

38. A piano string without windings has a fundamental frequency of 200 Hz. When it is wound with wire, its linear mass density is doubled. What is its fundamental frequency then?

39. A string fixed at both ends is 3 m long. It resonates in its second harmonic at a frequency of 60 Hz. What is the speed of transverse waves in the string?

40. The length of the B string on a certain guitar is 60 cm. It vibrates at 247 Hz. (a) What is the speed of transverse waves on the string? (b) If the linear mass density is 0.01 g/cm, what should the tension be when it is in tune?

41. A string 3 m long fixed at both ends is vibrating in its third harmonic. The maximum displacement of any point on the string is 4 mm. The speed of transverse waves on this string is 50 m/s. (a) What are the wavelength and frequency for this wave? (b) Write the wave function for this wave.

42. The wave function for a certain standing wave on a string fixed at both ends is $y(x, t) = 0.30 \sin 0.20x \cos 300t$, where $y$ and $x$ are in centimetres and $t$ is in seconds. (a) What are the wavelength and frequency of these waves? (b) What is the speed of transverse waves on this string? (c) If the string is vibrating in its fourth harmonic, how long is it?

43. The wave function for a certain standing wave on a string fixed at both ends is $y(x, t) = 0.5 \sin 0.025x \cos 500t$, where $y$ and $x$ are in centimetres and $t$ is in seconds. (a) Find the speed and amplitude of the two traveling waves that result in this standing wave. (b) What is the distance between successive nodes in the string? (c) What is the shortest possible length of the string?

44. A string 2.51 m long has the wave function given in Exercise 43. (a) Sketch the position of the string for the times $t = 0$, $t = \frac{1}{4}T$, $t = \frac{1}{2}T$, and $t = \frac{3}{4}T$, where $T = 1/f$ is the period of the vibration. (b) Find $T$ in seconds. (c) When the string is horizontal, what has become of the energy in the wave?

45. A 160-g rope 4 m long is fixed at one end and tied to a light string at the other end. Its tension is 400 N. (a) What are the wavelengths of the fundamental and the first two overtones? (b) What are the frequencies of these standing waves?

46. A string fixed at one end only is vibrating in its fundamental mode. The wave function is $y(x, t) = 0.02 \sin 2.36x \cos 377t$, where $y$ and $x$ are in metres and $t$ is in seconds. (a) What is the wavelength of the wave? (b) What is the length of the string? (c) What is the speed of transverse waves on the string?

47. A string 5 m long fixed at one end only is vibrating in its fifth harmonic with frequency 400 Hz. The maximum displacement of any segment of the string is 3 cm. (a) What is the wavelength of this wave? What is the wave number $k$? (b) What is the angular frequency? (c) Write the wave function for this standing wave.

48. Three successive resonance frequencies for a certain string are 75, 125, and 175 Hz. (a) Find the ratios of each pair of successive resonance frequencies. (b) How can you tell that these frequencies are for a string fixed at one end only, rather than for a string fixed at both ends? (c) What is the fundamental fre-

quency? (d) Which harmonics are these resonance frequencies? (e) If the speed of transverse waves on this string is 400 m/s, find the length of the string.

### Section 14-6, Standing Sound Waves

49. Calculate the fundamental frequency for a 10-m organ pipe that is open at both ends, and for one that is closed at one end.

50. The normal range of hearing is about 20 to 20,000 Hz. What is the greatest length of an organ pipe that would have its fundamental note in this range (a) if it is closed at one end; (b) if it is open at both ends?

51. The shortest pipes used in organs are about 7.5 cm long. (a) What is the fundamental frequency of a pipe this long which is open at both ends? (b) For such a pipe, what is the highest harmonic that is within the audible range (see Exercise 50)?

52. The space above the water in the tube shown in Figure 14-29 is 120 cm long. Near the open end is a loudspeaker driven by an audio oscillator whose frequency can be varied from 10 to 5000 Hz. (a) What is the lowest frequency of the oscillator that will resonate with the tube? (b) What is the highest frequency that will resonate? (c) How many different frequencies of the oscillator will produce resonance?

### Section 14-7, Energy and Intensity of Harmonic Waves

53. (a) Show that the power passing a point on a string equals the product of the energy per unit length and the wave speed. (b) Waves of frequency 200 Hz and amplitude 1 cm move along a 20-m string which has a mass of 0.06 kg and is under tension 50 N. What is the total energy of the waves in the string? (c) Find the power transmitted past a given point on the string.

54. Two parallel tubes with the same diameter are filled with gas at the same pressure and temperature. One tube contains $H_2$, and the other contains $O_2$. (a) If sound waves traveling in the tubes have the same displacement amplitude and frequency, how do the intensities compare? (b) If the waves have the same frequency and pressure amplitude, how do the intensities compare? (c) If the waves have the same frequency and intensity, how do the pressure amplitudes and displacement amplitudes compare?

55. A piston at one end of a long tube filled with air at room temperature and normal pressure oscillates with frequency 500 Hz and amplitude 0.1 mm. The area of the piston is 100 cm². (a) What is the pressure amplitude of the sound waves generated in the tube? (b) What is the intensity of the waves? (c) What average power is required to keep the piston oscillating (neglecting friction)?

56. What is the intensity level in decibels for a sound wave (a) of intensity of $10^{-10}$ W/m², (b) of intensity of $10^{-2}$ W/m²?

57. Find the intensity if (a) $\beta = 10$ dB and (b) $\beta = 3$ dB. (c) Find the pressure amplitude for sound waves in air at standard temperature and pressure for each of these intensities.

58. Show that if the intensity is doubled, the intensity level increases by 3.0 dB.

59. What fraction of the acoustic power of a noise would have to be eliminated to lower its sound intensity level from 90 to 70 dB?

### Section 14-8, The Wave Equation

60. Show that the function $y_3(x, t)$ given by Equation 14-73 satisfies the wave equation if $y_1(x, t)$ and $y_2(x, t)$ do.

61. Show explicitly that the following functions satisfy the wave equation: (a) $y(x, t) = (x + vt)^3$; (b) $y(x, t) = Ae^{ik(x-vt)}$, where $A$ and $k$ are constants and $i = \sqrt{-1}$; (c) $y(x, t) = \ln k(x - vt)$.

62. (a) Show that the function $y = A \sin kx \cos \omega t$ satisfies the wave equation. (b) Use the trigonometric identity $\sin A + \sin B = 2 \sin \frac{1}{2}(A + B) \cos \frac{1}{2}(A - B)$ to show that the wave function in part (a) is the sum of a wave traveling to the right and a wave traveling to the left.

## Problems

1. A common rule of thumb for finding the distance to a lightning flash is to begin counting when the flash is observed and stop when the thunder clap is heard. The number of seconds counted is then divided by 3 to get the distance in kilometres. Why is this justified? What is the velocity of sound in km/s? How accurate is this procedure? Is the correction for the time taken for the light to reach you important? (The speed of light is about $3 \times 10^8$ m/s.)

2. A method for measuring the speed of sound using an ordinary watch (with a second hand) is to stand some distance from a large flat wall and clap your hands rhythmically in such a way that the echo from the wall is heard halfway between each two claps. Show that the speed of sound is given by $v = 4LN$, where $L$ is the distance to the wall and $N$ is the number of claps per unit time. What is a reasonable value for $L$ for this experiment to be feasible? (If you have access to a flat wall outdoors somewhere, try this method and compare your result with the standard value.)

3. A man drops a stone from a high bridge and hears it strike the water below exactly 4 s later. (a) Estimate the distance to the water on the assumption that the travel time for the sound to reach the man is negligible. (b) Improve your estimate by using your result from part (a) for this distance to estimate the time for sound to travel this distance. Then calculate the distance the rock falls in 4 s minus this time. (c) Calculate the exact distance and compare with your previous estimates.

4. A coiled spring such as a Slinky is stretched to a length $L$. It has a force constant $k$ and mass $m$. Show that the velocity of longitudinal compression waves along the spring is given by $v = L\sqrt{k/m}$. Show that this is also the velocity of transverse waves along the spring if the natural length of the spring is much less than $L$.

5. A tuning fork attached to a stretched wire generates transverse waves. The vibration of the fork is perpendicular to the string. Its frequency is 440 Hz, and its amplitude of oscillation is 0.50 mm. The wire has linear mass density of 0.01 kg/m and is under a tension of 1 kN. (a) Find the period and frequency of waves in the wire. (b) What is the speed of the waves? (c) What are the wavelength and wave number? (d) Write a suitable wave function for the waves on the wire. (e) Calculate the maximum speed and acceleration of a point in the wire. (f) At what average rate must energy be supplied to the fork to keep it oscillating at a steady amplitude?

6. (a) Show that if the relative velocity of source and receiver $u$ is much less than $v$, the doppler shift in frequency and wavelength can be written

$$\frac{\Delta f}{f} = \pm \frac{u}{v} \quad \text{and} \quad \frac{\Delta \lambda}{\lambda} = \mp \frac{u}{v}$$

(b) Show that these expressions also hold for the relativistic doppler shift given by Equation 14-42. (c) In the light received from a certain galaxy, the hydrogen spectrum is present, but all the wavelengths are greater than those emitted by excited hydrogen atoms in the laboratory by 1 percent. This is known as the red shift. Calculate the speed with which the galaxy is moving away from the earth.

7. A car moves with speed 17 m/s toward a stationary wall. Its horn emits 200-Hz sound waves moving at 340 m/s. (a) Find the wavelength of the sound

in front of the car and the frequency with which the waves strike the wall. (b) Since the waves reflect off the wall, the wall acts as a source of sound waves at the frequency found in part (a). What frequency does the driver of the car hear reflected from the wall?

8. Work Problem 7 in the reference frame in which the car is stationary and the wall moves toward the car at 17 m/s.

9. Two wires of different densities are soldered together end to end and then stretched under tension $T$ (the same in both wires). The wave speed in the first wire is twice that in the second wire. When a harmonic wave traveling in the first wire is reflected at the junction of the wires, the reflected wave has half the amplitude of the transmitted wave. (a) Assuming no loss in the wire, what fraction of the incident power is reflected at the junction and what fraction is transmitted? (b) If the incident-wave amplitude is $A$, what are the reflected- and transmitted-wave amplitudes?

10. Power is to be transmitted along a stretched wire by means of transverse harmonic waves. The wave speed is 100 m/s, and the linear density is 0.01 kg/m. The power source oscillates with amplitude of 0.50 mm. (a) What average power is transmitted along the wire if the frequency is 400 Hz? (b) The power transmitted can be increased by increasing the tension in the wire, the frequency of the source, or the amplitude of the waves. If just one of these quantitites is changed, how would each quantity have to be changed to effect an increase in power by a factor of 100? (c) Which of the changes would probably be accomplished most easily?

11. The G string on a violin is 30 cm long. When played without fingering, it vibrates at frequency 196 Hz. The next higher notes on the scale are A (220 Hz), B (247 Hz), C (262 Hz), and D (294 Hz). How far from the end of the string must a finger be placed to play these notes?

12. A steel piano wire is 40 cm long, has a mass of 2 g, and is under tension of 600 N. (a) What is the fundamental frequency? (b) What is the wavelength in air of sound produced when the wire vibrates at its fundamental frequency? (c) If the highest frequency a certain listener can hear is 14 kHz, what is the highest harmonic produced by the wire that she can hear?

13. An early method for determining the speed of sound in different gases is as follows. A cylindrical glass tube is placed horizontally, and a quantity of light powder or fine wood shavings is spread along the bottom of the tube. One end is closed by a piston which can be attached to an oscillator of known frequency $f$ (such as a tuning fork). The other end is closed by a piston whose position can be varied. While the opposite piston is made to oscillate at frequency $f$, the movable piston's position is adjusted until resonance occurs. When this happens, the powder collects in piles equally spaced along the bottom of the tube. (This is known as *Kundt's method.*) (a) Explain why the powder collects in this way. (b) Derive a formula which gives the speed of sound in the gas in terms of $f$ and the distance between piles of powder. (c) Give suitable values for the frequency $f$ and the distance between piles of powder. (d) Give suitable values of the frequency $f$ and the length $L$ of the tube for which the speed of sound could be measured using either air or helium.

14. A string with mass density of $4 \times 10^{-3}$ kg/m is under tension of 360 N and is fixed at both ends. One of its resonance frequencies is 375 Hz. The next higher resonance frequency is 450 Hz. (a) What is the fundamental resonance frequency? (b) Which harmonics are the ones given? (c) What is the length of the string?

15. A string fastened at both ends has successive resonances with wavelengths of 0.54 m for the $n$th harmonic and 0.48 m for the $(n + 1)$th harmonic. (a) Which harmonics are these? (b) What is the length of the string? (c) What is the wavelength of the fundamental?

16. Three successive resonance frequencies in an organ pipe are 1310, 1834, and

2358 Hz. (*a*) Is the pipe closed at one end or open at both ends? (*b*) What is the fundamental frequency? (*c*) What is the length of the pipe?

17. A wire of mass 1 g and length 50 cm is stretched with a tension of 440 N. It is placed near the open end of the tube in Figure 14-29 and stroked with a violin bow so that it oscillates with its fundamental frequency. The water level in the tube is lowered until a resonance is first obtained, at 18 cm below the top of the tube. Use these data to determine the speed of sound in air. Why is this method not very accurate?

18. A 40-cm-long wire of mass 0.01 kg vibrates in its second harmonic. When it is placed near the open end of the tube in Figure 14-29, resonance with the fundamental of the tube occurs if the water level is 1 m below the top of the tube. Assuming the speed of sound in air to be 340 m/s, find (*a*) the frequency of oscillation of the air column in the tube, (*b*) the speed of waves along the wire, and (*c*) the tension in the wire.

19. With a tuning fork of frequency 500 Hz held above the tube in Figure 14-29, resonances are found when the water level is at distances 16, 50.5, 85, and 119.5 cm. Use these data to calculate the speed of sound in air. How far outside the open end of the tube is the pressure node?

20. In a lecture demonstration of standing waves, a string is attached to a tuning fork which vibrates at 60 Hz and sets up transverse waves of that frequency on the string. The other end of the string passes over a pulley, and the tension is varied by attaching weights to that end. The string has approximate nodes at the tuning fork and at the pulley. If the string has mass density 8 g/m and is 2.5 m long (from tuning fork to pulley), what must the tension be for the string to oscillate in its fundamental mode? Find the tensions for vibration in the first three overtones.

21. (*a*) For the wave function given in Exercise 46, find the velocity of a string segment at some point $x$ as a function of time. (*b*) Which point has the greatest speed at any time? What is the maximum speed of this point? (*c*) Find the acceleration of a string segment at some point $x$ as a function of time. Which point has the greatest acceleration? What is the maximum acceleration of this point?

22. A 2-m string is fixed at one end and vibrating in its third harmonic. The greatest displacement of any segment of the string is 3 cm. The frequency of vibration is 100 Hz. (*a*) Write the wave function for this vibration. (*b*) Write an expression for the kinetic energy of a segment of the string of length $dx$ at point $x$ at some time $t$. At what time is this kinetic energy maximum? What is the shape of the string at this time? (*c*) Find the maximum kinetic energy of the string by integrating your expression over the total length of the string. (*d*) Find the potential energy of a segment of the string and compute the maximum potential energy of the string by integration. *Hint:* Remember that the potential energy of a mass $m$ in simple harmonic motion of angular frequency $\omega$ is $\frac{1}{2}m\omega^2y^2$, where $y$ is the displacement.

23. Show that if the tension in a string fixed at both ends is changed by a small amount $\Delta T$, the frequency of the fundamental is changed approximately by $\Delta f$, where $\Delta f/f = \frac{1}{2}\Delta T/T$. (Use the differential approximation.) Does this result apply to all harmonics? Use this to find the percentage change in tension needed to raise the frequency of the fundamental of a piano wire from 260 to 262 Hz.

24. Show that if the temperature changes by a small amount $\Delta T$, the fundamental frequency of an organ pipe changes by approximately $\Delta f$, where $\Delta f/f = \frac{1}{2}\Delta T/T$. Suppose an organ pipe closed at one end has a fundamental frequency of 200 Hz when the temperature is 20°C. What will its fundamental frequency be when the temperature is 30°C? (Ignore any change in the length of the pipe due to thermal expansion.)

25. An automobile has an acoustic noise output of 0.10 W. If the sound is radiated isotropically, but confined to the upper hemisphere, (*a*) what is the in-

tensity at a distance of 30 m? (b) What is the sound intensity level in decibels at this distance? (c) At what distance is the sound intensity level 40 dB?

26. (a) Compute the derivative of the speed of a wave on a string with respect to the tension $dv/dT$ and show that the differentials $dv$ and $dT$ obey $dv/v = \frac{1}{2}dT/T$. (b) A wave moves with speed 300 m/s on a wire which is under tension of 500 N. Using $dT$ to approximate the change in tension, find how much the tension must be changed to increase the speed to 312 m/s.

27. The ratio of the frequencies of one note to those of the semitone above it on the diatonic scale is $15:16$. Find the speed of a car such that the tone of its horn drops a semitone as it passes you.

28. Compute the derivative of the velocity of sound with respect to the absolute temperature and show that the differentials $dv$ and $dT$ obey $dv/v = \frac{1}{2}dT/T$. Use this to compute the percentage change in the velocity of sound when the temperature changes from 0 to 27°C. If the speed of sound is 331 m/s at 0°C, what is it (approximately) at 27°C? How does this approximation compare with an exact calculation?

29. A loudspeaker diaphragm 30 cm in diameter is vibrating at 1 kHz with an amplitude of 0.020 mm. Assuming that the air molecules in the vicinity have the same vibration amplitude, find (a) the pressure amplitude immediately in front of the diaphragm, (b) the sound intensity, and (c) the acoustic power being radiated.

30. A stationary destroyer is equipped with sonar that sends out pulses of sound at 40 MHz. Reflected pulses are received from a submarine directly below with a time delay of 80 ms and at a frequency of 39.958 MHz. If the speed of sound in seawater is 1.54 km/s, find (a) the depth of the submarine and (b) its vertical speed.

31. A loudspeaker at a rock concert generates $10^{-2}$ W/m² at 20 m at a frequency of 1 kHz. Assume that the speaker spreads its energy uniformly in the forward hemisphere and that none is reflected from the ground or elsewhere. (a) What is the intensity level at 20 m? (b) What is the total acoustic power output of the speaker? (c) At what distance will the intensity be at the pain threshold of 120 dB? (d) What is the intensity level at 30 m?

32. When a pin of mass 0.1 g is dropped from a height of 1 m, 1 percent of its energy is converted into a sound pulse of duration 0.1 s. (a) Estimate the range at which the dropped pin can be heard if the minimum audible intensity is 10 pW/m². (b) Your result in (a) is much too large in practice because of the background noise. If instead you assume that the intensity level must be at least 40 dB to be heard, estimate the range at which the dropping pin can be heard. (In both parts assume that the intensity is $P/2\pi r^2$.)

33. Everyone at a cocktail party is speaking equally loudly. If only one person were talking, the sound level would be 72 dB. Find the sound level when all 38 people are talking.

34. The noise level in an empty examination hall is 40 dB. When 100 students are writing an exam, the sounds of heavy breathing and pens traveling rapidly over paper cause the noise level to rise to 60 dB (not counting groans). Assuming that each student contributes an equal amount of noise power, find the noise level to the nearest decibel when 50 students have left.

35. A heavy rope 3 m long is attached to the ceiling and allowed to hang freely. (a) Show that the speed of transverse waves in the rope is independent of its mass and length but does depend on the distance $y$ from the bottom according to the formula $v = \sqrt{gy}$. (b) If the bottom end of the rope is given a sudden sideways displacement, how long does it take the resulting wave pulse to go to the ceiling, reflect, and return to the bottom of the rope?

36. If a loop of chain is spun at high speed, it will roll like a hoop without collapsing. Consider a chain of linear mass density $\mu$ which is rolling without slip-

ping at a high speed $v_0$. (*a*) Show that the tension in the chain is $T = \mu v_0^2$. (*b*) If the chain rolls over a small bump, a transverse wave pulse will be generated in the chain. At what speed will it travel along the chain? (*c*) How far around the loop (in degrees) will a transverse wave pulse travel in the time the hoop rolls through one complete revolution?

37. Normal speech gives a sound intensity level of about 65 dB at 1 m. Estimate the power in human speech.

38. Three noise sources produce intensity levels of 70, 73, and 80 dB when acting separately. When acting together, the intensities of the sources add. (There is no interference between the amplitudes from the different sources because the relative phase changes randomly.) Find the sound intensity level in decibels when the three sources act at the same time. Discuss the usefulness of eliminating the two least intense sources in order to reduce the intensity level of the noise.

39. An article on noise pollution claims that the sound intensity level in large cities has been increasing by about 1 dB annually. What percentage increase in intensity does this correspond to? Does this increase seem reasonable? In about how many years would the intensity of sound double if it increased at 1 dB annually?

**CHAPTER 15**     Interference and Diffraction

---

**Objectives**     After studying this chapter, you should:

1. Be able to explain what is meant by the terms diffraction, constructive interference, destructive interference, coherent sources, nondispersive medium, and Fourier analysis and synthesis.

2. Be able to sketch the intensity pattern produced by two coherent sources.

3. Be able to discuss when diffraction is important and when it is not, and compare the propagation of a wave with that of a beam of particles.

4. Know that the beat frequency equals the difference between the two frequencies producing the beats.

5. Be able to state the conditions under which a wave packet travels without change in shape and explain why it does.

---

When two or more waves meet, they interfere and produce a resultant wave whose properties can be calculated using the principle of superposition. In this chapter we consider some phenomena related to the interference of harmonic waves. We first consider the superposition of two harmonic waves of the same frequency and wavelength but differing in phase, and apply our results to the resultant intensity pattern from two point sources. We then discuss qualitatively the bending of waves around openings and obstacles, called diffraction, which is related to the interference of many individual waves. Finally we consider such phenomena as beats and wave packets, which result from the superposition and interference of waves of difference frequencies and wavelengths.

# 15-1   Superposition and Interference of Harmonic Waves

Consider two harmonic waves of the same frequency and wavelength. If the waves are in phase, the amplitude of the resultant wave is just the sum of the amplitudes of the two waves, and we have constructive interference. On the other hand, if the waves differ in phase by 180°, the resultant wave amplitude is the difference between the amplitudes of the two waves, and we have destructive interference. To treat the general case of waves differing in phase by some arbitrary amount $\delta$, we consider the special case of waves of equal amplitude.

Let $y_1$ and $y_2$ be the wave functions for the two original waves:

$$y_1 = y_0 \sin (kx - \omega t) \qquad y_2 = y_0 \sin (kx - \omega t + \delta)$$

The resultant wave is the sum

$$y_3 = y_1 + y_2 = y_0 \sin (kx - \omega t) + y_0 \sin (kx - \omega t + \delta)$$

We can simplify this by using the trigonometric identity

$$\sin A + \sin B = 2 \sin \tfrac{1}{2}(A + B) \cos \tfrac{1}{2}(A - B)$$

For this case $A = kx - \omega t$ and $B = kx - \omega t + \delta$, so that

$$\tfrac{1}{2}(A + B) = kx - \omega t + \tfrac{1}{2}\delta \qquad \text{and} \qquad \tfrac{1}{2}(A - B) = -\tfrac{1}{2}\delta$$

Thus

$$y_3 = 2y_0 \cos \tfrac{1}{2}\delta \sin (kx - \omega t + \tfrac{1}{2}\delta) \qquad\qquad \text{15-1}$$

where we have used $\cos (-\tfrac{1}{2}\delta) = \cos \tfrac{1}{2}\delta$. We see that the resultant of these two waves is another harmonic wave with the same frequency and wavelength. It differs in phase from both the original waves, and its amplitude is $2y_0 \cos \tfrac{1}{2}\delta$.

If $\delta = 0$, the two waves are in phase, $\cos \tfrac{1}{2}\delta = 1$, and the resultant amplitude is just twice that of either wave. If $\delta = \pi$, $\cos \tfrac{1}{2}\delta = 0$ and the two waves cancel. These two extremes are *perfectly constructive interference* and *perfectly destructive interference*, respectively.

*Constructive and destructive interference*

A common cause of phase difference between two waves is a difference in the path length traveled by the waves. In Figure 15-1 we assume that the wave splits equally at point $A$; that is, the amplitudes of the waves in the two tubes are equal. When they recombine at point $B$, there will be a phase difference due to the longer path traveled by the wave in the lower tube.

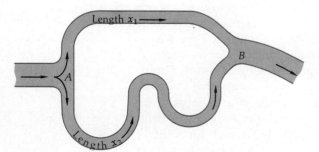

**Figure 15-1**
A wave enters from the left and splits equally at point $A$. At point $B$, where the waves combine, there is a phase difference because of the path difference. If $\lambda$ is the wavelength and $\Delta x$ the path difference, the phase difference is $\delta = 2\pi \, \Delta x / \lambda$.

If the path difference is exactly 1 wavelength, the phase difference between the two waves will be $2\pi$, which is the same as no phase difference, and the two waves will interfere constructively. Perfectly constructive interference will also occur if the path difference is any integer times the wavelength. On the other hand, if the path difference is exactly $\frac{1}{2}$ wavelength, $\frac{3}{2}$ wavelength, or any odd number of half wavelengths, the phase difference will be $\pi$ and the interference will be destructive. For any other path difference $\Delta x$ the phase difference will be

$$\delta = 2\pi \frac{\Delta x}{\lambda} \qquad\qquad 15\text{-}2$$

*Phase difference due to path difference*

## 15-2   Interference of Waves from Two Point Sources

Figure 15-2 shows the wave pattern produced by two point sources $S_1$ and $S_2$ a distance $d$ apart oscillating in phase, each producing circular waves of frequency $f$ and wavelength $\lambda$.

**Figure 15-2**
Interference of two point sources in a ripple tank. (*From* PSSC Physics, *2d ed., p. 285, D. C. Heath and Company, Lexington, Mass., 1965.*)

We can construct a similar pattern (Figure 15-3) with a compass by drawing circular arcs representing wave crests from each source at some particular time. At the points where the crests from each source overlap, the waves add constructively. At these points the paths for the waves from the two sources are either equal in length or differ by an integral number of wavelengths. These path differences are indicated in Figure 15-3. The line through a set of interference maxima for which the path difference is some integral number of wavelengths is a hyperbola. (A hyperbola is in fact defined as the locus of points whose distances from two fixed points differ by a constant amount.) Between each set of interference maxima are interference minima, where the path difference is an odd number of half wavelengths. These lines along which the waves completely cancel, called *nodes* or *nodal lines,* are also hyperbolas.

At any other point between the maxima and minima, the amplitude of the resultant wave is given by $A = 2A_0 \cos \frac{1}{2}\delta$, where $A_0$ is the amplitude of each wave separately and the phase difference $\delta$ is related to the difference in path from the point to the two sources by Equation 15-2.

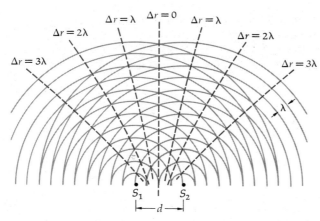

**Figure 15-3**
Geometric construction of the
interference pattern similar to
Figure 15-2. The waves add
constructively at the points of
intersection. Such points
occur whenever the path
lengths from the two sources
differ by an integral number
of wavelengths.

The interference of two sound sources can be demonstrated by driving
two separated speakers with the same amplifier fed by an audio-signal
generator. Moving about the room enables one to detect by ear the positions of constructive or destructive interference. The sound intensity
will not actually be zero at the points of destructive interference of the
direct sound waves because of sound reflections from the walls and
other objects in the room.

In optics we often observe the intensity pattern on a screen placed
parallel to the line of the sources and far away from the sources compared with their separation $d$. This is called *Young's experiment*, after
Thomas Young, who in 1801 used the interference pattern produced by
two light sources to demonstrate the wave nature of light. We can calculate the intensity pattern observed when the screen is far from the
sources.

At very large distances from the sources, the lines from the two
sources to some point $P$ on the screen are approximately parallel, and
the path difference is approximately $d \sin \theta$, as shown in Figure 15-4.
We thus have interference maxima at angles given by

$$d \sin \theta = m\lambda \qquad\qquad 15\text{-}3$$

and minima at

$$d \sin \theta = (m + \tfrac{1}{2})\lambda \qquad\qquad 15\text{-}4$$

where $m$ is any integer (0, 1, 2, . . . ). The phase difference at a point $P$
is $2\pi/\lambda$ times the path difference $d \sin \theta$:

$$\delta = \frac{2\pi}{\lambda} d \sin \theta \qquad\qquad 15\text{-}5$$

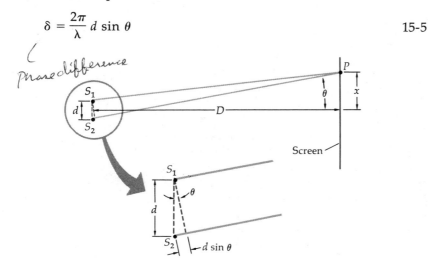

**Figure 15-4**
Geometry for calculation of
interference pattern observed
at a great distance $D$ from
two sources in phase and
separated by a much smaller
distance $d$. The rays from the
sources to some point $P$ are
approximately parallel, and
the path difference is approximately $d \sin \theta$. Constructive
interference occurs when this
path difference equals an integral number of wave-
lengths.

The distance $x$ measured along the screen from the central point to point $P$ is related to $\theta$ by $x = D \tan \theta$. We are usually interested in points for which $\theta$ is very small, so that $\sin \theta \approx \tan \theta \approx \theta$. Then $\sin \theta \approx x/D$. Substituting this into Equation 15-5, we have for the phase difference at distance $x$

$$\delta = \frac{2\pi x d}{\lambda D} \qquad\qquad 15\text{-}6$$

The amplitude at this distance is

$$2A_0 \cos \tfrac{1}{2}\delta = 2A_0 \cos \frac{\pi x d}{\lambda D}$$

The intensity is proportional to the square of the amplitude and thus is proportional to

$$4A_0^2 \cos^2 \frac{\pi x d}{\lambda D}$$

Figure 15-5 is a plot of the intensity versus $\sin \theta$, which is equivalent to a plot of intensity versus $x$ for small $\theta$, since $\sin \theta \approx x/D$.

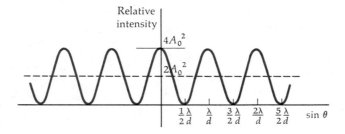

**Figure 15-5**
Plot of relative intensity versus $\sin \theta$ observed far away from two point sources in phase and separated by a small distance $d$. The maximum intensity is proportional to $4A_0^2$, where $A_0$ is the amplitude of the wave from either source. This is 4 times the intensity due to each source separately. The dashed line indicates the average relative intensity $2A_0^2$.

Since the average value of a cosine-squared function over a cycle is $\tfrac{1}{2}$, the average intensity is proportional to $2A_0^2$, indicated by the dashed line.* This is just the intensity which would arise from the two sources acting separately without interference. We see thus that as a result of the interference of waves from the two sources, the energy is redistributed in space. At a maximum point, the energy is 4 times that from a single source, whereas at a minimum point there is no energy at all; but if we average over many interference maxima and minima, we get the same energy as if the sources acted separately without interference. This is what we should expect from the conservation of energy.

Two or more sources need not be in phase to produce an interference pattern. Figure 15-6 shows the pattern produced by two sources 180° out of phase. The pattern is the same as in Figure 15-5 except that the maxima and minima are interchanged. Points equidistant from the sources or those for which the distance differs by an integral number of wavelengths are nodes because the waves are 180° out of phase. At points where the path distance differs by $\tfrac{1}{2}$ wavelength (or $\tfrac{3}{2}\lambda, \tfrac{5}{2}\lambda, \ldots$), the waves are in phase because the 180° phase difference of the source is offset by the 180° phase difference due to the path difference. It should be evident that similar interference patterns will be produced no matter what the phase difference between the sources is as long as the phase difference is constant in time. Two sources which are in phase or have a constant phase difference are called *coherent sources*. Coherent sources of water waves in a ripple tank are easy to produce by driving both

*Coherent sources*

---

* For any angle $\alpha$ the identity $\sin^2 \alpha + \cos^2 \alpha = 1$ holds. It is easily seen from a plot of $\sin^2 \alpha$ and $\cos^2 \alpha$ that the average values of these functions over a complete cycle are equal. Thus the average value of either $\sin^2 \alpha$ or $\cos^2 \alpha$ over a cycle is $\tfrac{1}{2}$.

sources by the same motor. There are many examples of sound and light sources whose phase difference is not constant but varies randomly from 0 to $2\pi$. For two sources of equal strength, the intensity at any instant is proportional to $4A_0^2 \cos^2 \frac{1}{2}\delta$, where now the phase difference $\delta$ is a function of time. If $\delta$ varies rapidly, only the time average $4A^2(\cos^2 \frac{1}{2}\delta)_{av} = 2A_0^2$ will be observed. There is then no observable interference pattern. Such sources are said to be *incoherent*.

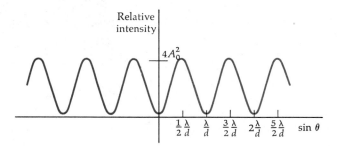

**Figure 15-6**
Relative-intensity pattern far from two sources which differ in phase by 180°. The pattern is the same as in Figure 15-5 except that it is shifted by $\frac{1}{2}\lambda/d$. The effect of a constant phase difference between two sources is merely to shift the intensity pattern.

As an example of incoherent sources, consider two candles. The light from each candle is the result of millions of independent atomic transitions. Each source, which we can consider to be a point on the macroscopic scale, is in reality the sum of millions of microscopic sources, the individual atoms. Since it is impossible to predict the exact time a particular atom will make a transition, the phase fluctuates randomly. Many fluctuations occur during a time interval as short as $10^{-8}$ s. The eye or other detection device cannot respond to such rapid intensity changes as would be indicated by the time variation of $4A_0^2 \cos^2 \frac{1}{2}\delta$. Thus only the average value of $2A_0^2$ is seen, and there is no possibility of observing an interference pattern with two independent light sources. It is, in fact, rather difficult to obtain coherent light sources except by splitting a single source into duplicates in ways to be discussed later.

---

**Example 15-1** Two loudspeakers on the $y$ axis at $y = +0.5$ m and $y = -0.5$ m are driven in phase by an amplifier whose frequency is 2000 Hz. A student stands on the $x$ axis at $x = 10$ m and walks along a line parallel to the $y$ axis. Where is the first point at which she hears no sound from the two speakers, and where is the first point beyond $y = 0$ where the interference is constructive? Assume the speed of sound to be 340 m/s.

The wavelength of the sound is $\lambda = v/f = (340 \text{ m/s})/(2000 \text{ Hz}) = 0.17$ m. The first point of destructive interference occurs at an angle $\theta$ such that the path difference $d \sin \theta = \frac{1}{2}\lambda = 0.085$ m. The angle is thus $\theta = \sin^{-1} 0.085 = 4.88°$. If $y_1$ is the student's distance from the $x$ axis, we have $\tan 4.88° = y_1/D = 0.0853$ and $y_1 = D \tan \theta = 0.853$ m = 85.3 cm. Note that if we had used the approximation $\sin \theta = \tan \theta$ we would have obtained $y_1 = 85.0$ cm. For the first position of constructive interference after $y = 0$, the path difference is $\lambda$ and $\sin \theta = \lambda/d = 0.17$, $\theta = 9.79°$. The distance $y_2$ for constructive interference is then $y_2 = D \tan \theta = (10 \text{ m}) (\tan 9.79°) = 1.72$ m.

---

**Question**

1. If two nearby sources have the same amplitude but different frequencies, is it possible for them to produce an interference pattern with fixed points of complete destructive interference?

# 15-3  Diffraction

When a portion of the wave is cut off by an obstruction, the propagation of the wave is more complicated. The portion of the wavefront which is not obstructed does not simply propagate in the direction of the straight rays, as might be expected. The situation is illustrated in Figure 15-7, which shows plane waves in a ripple tank meeting a barrier with a small opening. The waves to the right of the barrier are not confined to the narrow angle of the rays from the source that can pass through the opening, but are circular waves, as if there were a source at the opening. We can understand this by noting that the motion of the water at the opening due to the incoming waves is no different from that produced by a point source at the opening (assuming the opening to be very small).

The propagation of a wave is thus quite different from the propagation of a stream of particles. If the rays shown in Figure 15-8a indicate the direction of particles streaming out of a source, the barrier will stop any particles hitting it. Those getting through the opening will be confined to the narrow angle indicated. On the other hand, if the rays indicate the direction of propagation of a wave (Figure 15-8b), the rays appear to bend around the edges of the barrier, as in Figure 15-7. This bending of the rays, which always occurs to some extent when part of the wavefront is limited, is called *diffraction*.

**Figure 15-7**
Plane waves in a ripple tank meeting a barrier with a small opening. The waves to the right of the barrier are circular waves concentric about the opening, just as if there were a point source at the opening. (*Courtesy of Film Studio, Education Development Center, Newton, Mass.*)

*Diffraction*

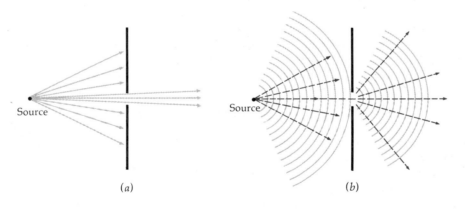

(a)                              (b)

**Figure 15-8**
Comparison of how (a) a beam of particles and (b) a wave are transmitted through a narrow opening in a barrier. In (a) the transmitted particles are confined to a narrow angle. In (b) the opening acts as a point source of circular waves which are radiated to the right through a much wider angle than in (a). Wave bending resulting from limiting the wavefront at the barrier is called diffraction.

In Figure 15-7 the opening in the barrier was chosen to be much smaller than the wavelength and could be considered to be a point. Figure 15-9 shows plane waves hitting a barrier with an opening much larger than the wavelength. The wave beyond the barrier is similar to a plane wave in the region far from the edges of the opening. Near the edges the wavefront is distorted, and the wave appears to bend slightly. In the limit of an infinitely large opening, the situation is the same as if there were no barrier. If a wave is limited by a barrier with an opening of the order of the size of the wavelength, the situation is more complicated: the transmitted wave is neither approximately a plane wave nor a circular wave from a point source (Figure 15-10).

Although the detailed calculation of the wave pattern of an obstructed wave is much too complicated to consider here, we can describe the propagation of the wave using a geometric method discovered by Christiaan Huygens about 1678. Huygens' method considers each point on a given wavefront to be a point source of waves. The new wavefront at some time later is then the surface enveloping all the little spherical wavelets emitted by these point sources, which have expanded during the time interval. Figure 15-11 illustrates the Huygens construction for the propagation of unobstructed plane waves and spherical waves. Of course, if each point on a wavefront were really a point source, there would be waves in the backward direction. These waves were ignored by Huygens. (In the refinement of Huygens' method derived by G. R. Kirchhoff in the nineteenth century, the intensity of the wavelets depends on angle and is zero in the backward direction.) If the wavefront is limited by an aperture or obstacle, the resulting wave pattern can be found by calculating the interference pattern of the line of point sources on the unobstructed portion of the wavefront. We shall use this method to calculate the diffraction pattern of a single slit in Chapter 34 when we study optics. Here we merely state the important and general result known as the ray approximation.

*If the aperture or obstacle is large compared with the wavelength, the bending of the wavefront is not noticeable and the wave propagates in straight lines or rays, much as a beam of particles does.*

Because the wavelength of audible sound ranges from a few centimetres to several metres and is often large compared with apertures and obstacles, diffraction of sound is a common experience. On the other hand, the wavelength of visible light ranges from about $4 \times 10^{-7}$ to $8 \times 10^{-7}$ m. Because these wavelengths are so small compared with the size of ordinary objects and apertures, diffraction of light is not easily noticed. It was for this reason that Newton and others thought that light was a beam of particles rather than a wave motion. We shall discuss this in more detail when we study light in Chapter 32.

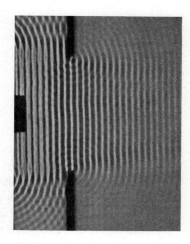

**Figure 15-9**
Plane waves in a ripple tank meeting a barrier with an opening large compared with the wavelength. The effect of the barrier is noticeable only near the edges. (*Courtesy of Film Studio, Education Development Center, Newton, Mass.*)

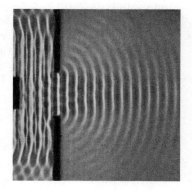

**Figure 15-10**
Plane waves in a ripple tank meeting a barrier with an opening 5 times as wide as the wavelength. When the opening is neither small nor large compared with the wavelength, the wave pattern to the right is much more difficult to analyze. (*Courtesy of Film Studio, Education Development Center, Newton, Mass.*)

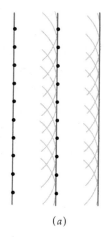

(a)

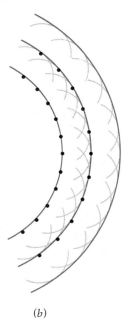

(b)

**Figure 15-11**
Huygen's construction for the propagation of (a) plane waves to the right and (b) outgoing spherical or circular waves.

An important result of the calculation of the diffraction pattern of plane waves incident on a small circular aperture is that at a great distance from the aperture, the wave intensity is essentially confined to a cone of half angle $\theta$, which is related to the aperture diameter $D$ and the wavelength by

$$\sin \theta \sim \frac{\lambda}{D} \qquad\qquad 15\text{-}7$$

(where $\sim$ is read "is of the order of"). If the wavelength is large, waves spread out as if from a point source; whereas if it is small (compared with $D$), the waves are confined to the forward direction. This result also applies to sound waves generated by a circular piston such as a speaker cone. For this reason, the high-frequency waves from a speaker, which have short wavelengths ($\lambda = v/f$), tend to be concentrated in the forward direction more than the lower-frequency waves, which spread out in all directions.

## 15-4  Beats

We now consider the superposition of waves of different frequencies. If the waves are in phase at some time, the interference will be constructive and the intensity large; but at some later time, because the frequencies are different, the waves will be out of phase and the interference will be destructive. A familiar example is the beats produced by two sound waves of nearly equal frequency, e.g., those produced by two tuning forks or two guitar strings of nearly equal but not identical frequency. What we hear is a tone whose intensity varies alternately between loud and soft. The frequency of this variation in intensity is called the *beat frequency*. As we shall see, the beat frequency equals the difference in the frequencies of the two sources.

Let us consider two waves of angular frequencies $\omega_1$ and $\omega_2$ with the same amplitude $p_0$. Since we are interested in the resultant wave at a single point in space, we can neglect the spatial part of the wave, which merely contributes a phase constant, and consider only the time dependence. The pressure variation at the ear due to either wave acting alone will be a simple harmonic function of the type

$$p_1 = p_0 \sin \omega_1 t \qquad \text{and} \qquad p_2 = p_0 \sin \omega_2 t \qquad\qquad 15\text{-}8$$

where we have chosen sine functions for convenience and assumed the waves to be in phase at time $t = 0$. Using Equation 14-50 again for the sum of two sine functions, we obtain for the resultant pressure variation

$$p = p_0 \sin \omega_1 t + p_0 \sin \omega_2 t$$

$$= 2p_0 \cos \tfrac{1}{2}(\omega_1 - \omega_2)t \, \sin \tfrac{1}{2}(\omega_1 + \omega_2)t$$

We can simplify the notation somewhat by writing $\bar{\omega} = \tfrac{1}{2}(\omega_1 + \omega_2)$ for the average angular frequency and $\Delta\omega = \omega_1 - \omega_2$ for the difference in angular frequencies. The resultant pressure variation is then

$$p = 2p_0 \cos (\tfrac{1}{2}\Delta\omega \, t) \sin \bar{\omega}t = 2p_0 \cos (2\pi \tfrac{1}{2}\Delta f \, t) \sin 2\pi \bar{f}t \qquad 15\text{-}9$$

Figure 15-12 shows this pressure variation as a function of time. The waves are originally in phase and add constructively at $t = 0$. At some later time $t_1$ the waves are 180° out of phase because of the difference in frequencies, and the waves interfere destructively. At time $t_2 = 2t_1$ they

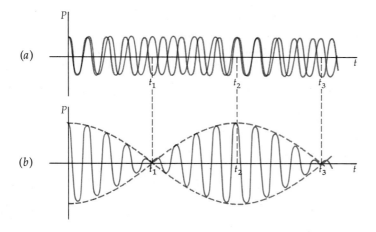

(a)

(b)

**Figure 15-12**
(a) Two waves of different frequencies that are in phase at $t = 0$. At time $t_1$ they are 180° out of phase; at $t_2$ they are back in phase; and at $t_3$ they are out of phase again. (b) The resultant of the two waves shown in (a). The frequency of the rapid oscillation is about the same as that of the original waves, but the amplitude is modulated, as indicated by the dashed envelope. The amplitude is maximum at times 0 and $t_2$ and zero at times $t_1$ and $t_3$.

are in phase again. The greater the difference in frequencies, the shorter the time $t_1$ in which they become out of phase. From Equation 15-9 we see that the time $t_1$ is that time for which the argument of the cosine function is $\frac{1}{2}\pi$, since $\cos \frac{1}{2}\pi = 0$. We therefore have

$$\pi \, \Delta f \, t_1 = \tfrac{1}{2}\pi$$

or

$$t_1 = \frac{1}{2 \, \Delta f} \qquad\qquad 15\text{-}10$$

The ear hears the average frequency $f = \frac{1}{2}(f_1 + f_2)$ with the amplitude $2y_0 \cos (2\pi \frac{1}{2}\Delta f \, t)$. Since the energy of the wave is proportional to the square of the amplitude, the sound is loud whenever the amplitude is either maximum or minimum, i.e., whenever $\cos 2\pi \frac{1}{2}\Delta f \, t$ is $+1$ or $-1$. Since the maximum amplitude occurs with frequency $\frac{1}{2}\Delta f$, the frequency of maximum and minimum amplitudes is just twice this, or $\Delta f$. For example, if the tuning forks have frequencies 241 and 243 Hz, the ear will hear the frequency 242 Hz and the sound will be loud twice each second.

*Beat frequency*

The phenomenon of beats is often used to compare an unknown frequency with a known frequency, as in tuning a piano with a tuning fork. The ear can detect beats up to about 10 per second. Above this the fluctuations in loudness are too rapid to be heard. Beats are also used to detect small frequency changes like those produced in a radar beam reflected from a moving car. The shift in frequency of the reflected beam is due to the doppler effect. The velocity of the car is found by measuring the change in frequency, determined by measuring the beats produced by the reflected beam and the original radar source. There are many interesting related phenomena, such as the moiré pattern in Figure 15-13, which can be thought of as beats produced by two sets of parallel lines when the spacing of one set differs slightly from that of the other.

**Figure 15-13**
Moiré pattern showing beats produced by two sets of parallel lines when the spacing of one set differs slightly from that of the other.

## Questions

2. Two violinists are practicing the same piece of music together. Can a listener hear the beats as they play? If beats are heard, what can be done to prevent them?

3. When musical notes are sounded together to make chords, beats are produced. Why are they not noticed as such?

4. Discuss how the phenomenon of beats is used to tune musical instruments.

5. Pianos generally have more than one wire for each note. The wires are struck simultaneously by the same hammer when the note is sounded. What is the effect if these wires are slightly out of tune?

## 15-5 Harmonic Analysis and Synthesis

Many complex periodic waves are mixtures of harmonic waves of several frequencies. Figure 15-14 shows the pressure variation $P(t)$ produced by a tuning fork, clarinet, and cornet playing the same musical note. All the waves have the same period because they correspond to the same note or pitch, but the waveforms are quite different. The difference in tone quality is related to the difference in the waveforms. A useful mathematical result called *Fourier's theorem* allows us to analyze any periodic function in terms of sinusoidal functions. According to Fourier's theorem, any periodic function can be represented with arbitrary accuracy by a sum of sinusoidal functions. Consider some periodic function $P(t)$ which might describe the pressure variation at some point in space due to some sound source. Fourier's theorem states that this function can be written

$$P(t) = \Sigma A_n \sin (\omega_n t + \phi_n) \qquad \text{15-11}$$

where $A_n$ and $\phi_n$ are constants that are determined by the function $P(t)$, that is, by the waveform. If the function $P(t)$ is given, the constants $A_n$ and $\phi_n$ are determined by *Fourier analysis*.

**Figure 15-14**
Waveforms of (*a*) a tuning fork, (*b*) a clarinet, and (*c*) a cornet, each at frequency of 440 Hz and approximately the same intensity. (*Redrawn from Charles A. Culver, Musical Acoustics, 4th ed., p. 103. Copyright © 1956 by McGraw-Hill, Inc.*)

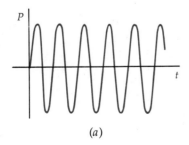

(*a*)

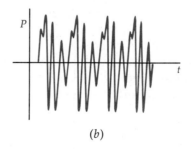

(*b*)

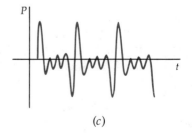

(*c*)

The lowest angular frequency in Equation 15-11, $\omega_1$, corresponds to the period of the function $P(t)$; that is, the lowest frequency is $\omega_1 = 2\pi/T$, where $T$ is the period. The other frequencies are integral multiples of the lowest or fundamental frequency:

$$\omega_n = n\omega_1 = n\frac{2\pi}{T} \qquad \text{15-12}$$

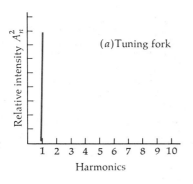

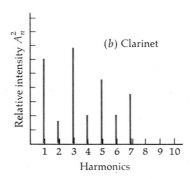

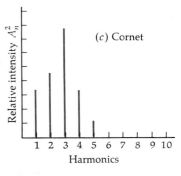

**Figure 15-15**
Relative intensities of the harmonics in the waveforms shown in Figure 15-14 for the tuning fork, clarinet, and cornet. (*Redrawn from Charles A. Culver,* Musical Acoustics, *4th ed., p. 104. Copyright © 1956 by McGraw-Hill, Inc.*)

The value of $A_n^2$ is proportional to the intensity of the harmonic component of the function. Figure 15-15 illustrates the harmonic analysis of the waveforms shown in Figure 15-14. Both figures show that the clarinet, for example, is much richer in harmonics than the tuning fork.

The construction of an arbitrary periodic waveform from its component harmonics is called *synthesis*. Figure 15-16 shows a square wave that can be produced by an electronic square-wave generator, and its approximate synthesis using only three harmonics. This waveform can be synthesized with only odd harmonics. The relative values of the constants $A_n$ are shown in Figure 15-17. The more harmonics used, the better the approximation to the actual waveform.

*Fourier synthesis*

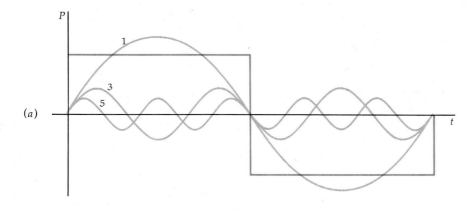

**Figure 15-16**
(*a*) A square wave and the first three odd harmonics used to synthesize the wave. (*b*) Synthesis of the square wave using the first three odd harmonics.

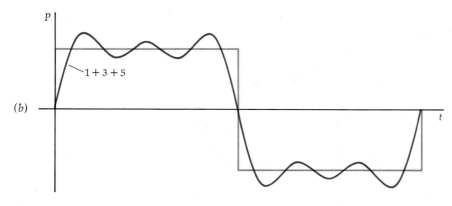

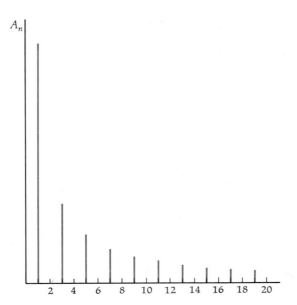

**Figure 15-17**
Relative amplitudes of the harmonics needed to synthesize the square wave shown in Figure 15-16. For even $n$, $A_n = 0$, whereas for odd $n$, $A_n$ is proportional to $1/n$. The relative energy in the $n$th harmonic is proportional to $A_n^2$.

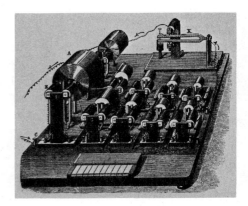

(*Right*) Synthesizer designed by Heinrich von Helmholtz and constructed by Rudolph Koenig. (*From Adolphe Ganot, Elementary Treatise on Physics, 9th ed., William Wood and Publishers, Ltd., Great Britain, 1910.*) Each electrically driven tuning fork has a resonator which can be mechanically opened or closed to vary the intensity of that harmonic. (*Left*) In the more modern Moog synthesizer, shown here with Robert Moog, the harmonics are produced and mixed electronically.

## 15-6   Wave Packets

The complicated waveforms discussed in the previous section are periodic in time. Functions which are not periodic, e.g., pulses, can also be represented by harmonic (sinusoidal) functions, but a continuous distribution of frequencies is required. The sum in Equation 15-11 then becomes an integral over frequency, and the constants $A_n$ and $\phi_n$ become functions of frequency $A(\omega)$ and $\phi(\omega)$. Often $A(\omega)$ is nearly zero except in a certain frequency range $\Delta\omega$.

The characteristic feature of a wave pulse that distinguishes it from a periodic wave or harmonic wave of a single frequency is that the pulse has a beginning and an end, whereas a harmonic wave of a single frequency repeats over and over. To send a signal with a wave, we need some kind of pulse rather than a harmonic wave of a single frequency. There is an important relation between the distribution of frequencies of the harmonic functions which make up a pulse, and the time duration of the pulse. If the duration $\Delta t$ is very short, the range of frequencies $\Delta \omega$ is very large. The general relation between $\Delta t$ and $\Delta \omega$ is

$$\Delta \omega \, \Delta t \sim 1 \qquad\qquad 15\text{-}13$$

The exact value of this product depends on just how the quantities $\Delta t$ and $\Delta \omega$ are defined. For any reasonable definitions, $\Delta \omega$ is of the order of $1/\Delta t$.

The wave pulse produced by a source of short duration $\Delta t$ has a narrow width in space, $\Delta x = v \, \Delta t$, where $v$ is the wave speed. Each harmonic wave of frequency $\omega$ has a wave number $k = \omega/v$. A range of frequencies $\Delta \omega$ implies a range of wave numbers $\Delta k = \Delta \omega/v$ in the resulting wave, which is called a *wave packet*. (For simplicity here we assume that $v$ is the same for each harmonic wave component.) Substituting $\Delta k \, v$ for $\Delta \omega$ in Equation 15-13 gives

*Wave packet*

$$\Delta k \, v \, \Delta t \sim 1$$

or

$$\Delta k \, \Delta x \sim 1 \qquad\qquad 15\text{-}14$$

where again the exact value for the product depends on the precise definition of $\Delta k$ and $\Delta x$.

The relations expressed by Equations 15-13 and 15-14 are important characteristics of wave pulses which apply to all types of waves. They are of particular importance in communications theory and the theory of quantum mechanics. Since information cannot be transported by a harmonic wave which has no beginning or end in time, the transmission of short pulses depends on the ability to transmit a wide range of frequencies. In quantum mechanics the position of a particle is described by a *wave packet* whose width reflects the uncertainty in the location of the particle. The possible values for the measurement of the momentum of the particle are proportional to the wave numbers $k$ of the harmonic waves making up the wave packet. Thus a particle which is well localized in space, as evidenced by a narrow wave packet, must contain a wide range of possible momentum values, as evidenced by the large range of wave numbers. This property of wave packets is at the heart of the famous Heisenberg uncertainty principle.

**Example 15-2** Show that the difference in time between successive zeros of the envelope function in Figure 15-12b varies inversely with the difference in the frequencies of the two waves.

The envelope in this figure resembles a wave pulse except that the envelope repeats again and again in time. Consider a time $t_1$ when the envelope is first zero. Let $t_3$ be the next time when the envelope is zero. This next time occurs when the argument of the modulated cosine function in Equation 15-9 changes by $\pi$:

$$\tfrac{1}{2}\Delta \omega \, t_3 - \tfrac{1}{2}\Delta \omega \, t_1 = \pi$$

or

$$(t_3 - t_1) \, \Delta \omega = \Delta t \, \Delta \omega = 2\pi \qquad\qquad 15\text{-}15$$

This relation is similar to Equation 15-13 except that the product equals $2\pi$ instead of 1, and $\Delta t$ is not really the duration of the waves since the envelope repeats endlessly.

## 15-7  Dispersion

We have seen from our discussion of Fourier analysis that we can consider a pulse in time to consist of a distribution of harmonic functions of different frequencies. We can extend this argument to include the spatial dependence of waves and show that a wave pulse can be considered to be a set of harmonic waves of different wavelengths and frequencies. If the pulse is to maintain the same shape as it travels, all the component harmonic waves must travel with the same speed. This will occur if the speed of harmonic waves does not depend on the wavelength or frequency of the wave. A medium for which the wave speed is independent of wavelength or frequency is called a *nondispersive medium*. An example is a perfectly flexible string, for which the wave speed is $v = \sqrt{T/\mu}$. Since the tension and mass density are independent of the frequency or wavelength, the wave speed is the same for all frequencies and wavelengths and a pulse on such a string maintains its shape as it travels. Other examples of waves in nondispersive media are sound waves in air and light waves in vacuum.

*Nondispersive medium*

If the speed of a wave does depend on the frequency and wavelength, the shape of a pulse will change as it travels. Examples of waves in a dispersive medium are waves on a real string or wire which is not perfectly flexible and light waves in any material such as glass or water. For example, when a beam of sunlight enters a glass prism from air, the light is bent, or refracted (we shall study this phenomenon in Chapter 32). The angle of refraction depends on the relative speed of the light in glass and in air. The speed of light in glass depends slightly on the wavelength of the light, being somewhat greater for the longer (red) wavelengths than for the shorter (blue) wavelengths. The long wavelengths are bent less than the short wavelengths, and the light beam is spread out, or dispersed, into its component colors or wavelengths.

If the wave speed in a dispersive medium depends only slightly on the frequency and wavelength, a wave pulse in such a medium will change shape as it travels but it will travel a considerable distance as a recognizable pulse. However, the speed of the pulse, e.g., the speed of the center of the pulse, is not the same as the (average) speed of the individual component harmonic waves. The speed of the harmonic waves is called the *phase velocity*; that of the pulse is called the *group velocity*.

*Phase velocity and group velocity*

We can get a qualitative idea of the motion of a wave packet by considering the simple case of just two waves of nearly equal but different wavelengths and frequencies. Let the waves have the same amplitude and have wave numbers $k_1$ and $k_2$ and frequencies $\omega_1$ and $\omega_2$. The wave functions are $y_1 = y_0 \sin (k_1 x - \omega_1 t)$ and $y_2 = y_0 \sin (k_2 x - \omega_2 t)$. We combine these waves to find the resultant wave in the same way as we did in our discussion of beats. We have

$$y(x, t) = y_2 + y_1 = y_0 \sin (k_2 x - \omega_2 t) + y_0 \sin (k_1 x - \omega_1 t)$$
$$= 2y_0 \cos \left[\tfrac{1}{2}(k_2 - k_1)x - \tfrac{1}{2}(\omega_2 - \omega_1)t\right]$$
$$\times \sin \left[\tfrac{1}{2}(k_2 + k_1)x - \tfrac{1}{2}(\omega_2 + \omega_1)t\right] \qquad 15\text{-}16$$

This result is simplified if we use the notation $\Delta k$ and $\Delta \omega$ for the differences in wave number and frequency and $\bar{k}$ and $\bar{\omega}$ for their averages. Then Equation 15-16 becomes

$$y = 2y_0 \cos (\tfrac{1}{2}\Delta k\, x - \tfrac{1}{2}\Delta \omega\, t) \sin (\bar{k}x - \bar{\omega}t) \qquad 15\text{-}17$$

The resultant wave is sketched in Figure 15-18 for frequencies and wavelengths nearly equal, so that $\Delta \omega$ and $\Delta k$ are small and $\bar{\omega}$ and $\bar{k}$ nearly equal to the frequency and wave number of either wave. The result is a wave of about the same frequency and wavelength as the original waves but with the amplitude *modulated* by the factor cos ($\frac{1}{2}\Delta k\, x - \frac{1}{2}\Delta\omega\, t$). The velocity of the resultant wave $v = \bar{\omega}/\bar{k}$ is nearly the same as that of the individual waves. This velocity is the phase velocity. The envelope (dashed curve in Fig. 15-18) travels as a wave of wave number $\frac{1}{2}\Delta k$ and angular frequency $\frac{1}{2}\Delta\omega$. We can find the velocity of the envelope by considering the modulating factor in Equation 15-17:

$$\cos\left(\tfrac{1}{2}\Delta k\, x - \tfrac{1}{2}\Delta\omega\, f\right) = \cos \tfrac{1}{2}\Delta k\left(x - \frac{\Delta\omega}{\Delta k}\, t\right) = \cos \tfrac{1}{2}\Delta k\,(x - v_g t)$$

where

$$v_g = \frac{\Delta\omega}{\Delta k} \qquad\qquad 15\text{-}18$$

The velocity of the envelope, $v_g = \Delta\omega/\Delta k$, is the group velocity. In the more general case of a wave packet, the group velocity $v_g$ is the speed with which the packet as a whole moves. As we have discussed, such a packet can be thought of as a superposition of many harmonic waves with wave numbers and angular frequencies distributed over limited ranges. Within these ranges $\omega$ and $k$ do not vary independently; i.e., one is a function of the other. The group velocity of a wave packet is then defined by

$$v_g = \frac{\Delta\omega}{\Delta k} \approx \frac{d\omega}{dk} \qquad\qquad 15\text{-}19$$

where $d\omega/dk$ is evaluated at the midpoint of the range of $\omega$ or $k$. Equation 15-19 is a generalization of Equation 15-18, which was found for just two waves.

   Each harmonic wave that contributes to a wave packet has its own wave number $k$ and angular frequency $\omega$. It travels at a speed $v_p$, its *phase velocity*, given by

$$v_p = \frac{\omega}{k} = f\lambda \qquad\qquad 15\text{-}20$$

By substituting $\omega = kv_p$ into Equation 15-19 we can obtain a relationship between the group velocity of a wave packet and the phase velocities of its constituent harmonic waves:

$$v_g = \frac{d\omega}{dk} = \frac{d}{dk}\,(kv_p) = v_p + k\,\frac{dv_p}{dk} \qquad\qquad 15\text{-}21$$

We see that if the phase velocity does not depend on wave number (and therefore does not depend on wavelength or frequency), the group velocity equals the phase velocity. This is the case in a nondispersive medium. In a dispersive medium the phase velocity does depend on the wave number, and the group and phase velocities are not equal.

### Dispersion of Deep Water Waves of Long Wavelength

The angular frequency $\omega$ for water waves whose wavelength is greater than about 2 cm but much smaller than the depth of the water is related to the wave number $k$ by

$$\omega = \sqrt{gk} \qquad\qquad 15\text{-}22$$

where $g$ is the acceleration of gravity. Such a relation between the angular frequency and the wave number is called a *dispersion relation*. Given the dispersion relation for waves in a medium, we can calculate the

**Figure 15-18**
Wave packet for a simple group of just two waves of nearly equal frequency and wavelength.

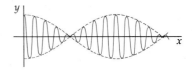

*Group velocity*

phase and group velocities of the waves. From Equation 15-20 the phase velocity is simply $\omega/k$:

$$v_p = \frac{\omega}{k} = \sqrt{\frac{g}{k}}$$

We can calculate the group velocity from either Equation 15-19 or 15-21. Using Equation 15-19, we get

$$v_g = \frac{d\omega}{dk} = \frac{1}{2}\sqrt{\frac{g}{k}} = \tfrac{1}{2}v_p$$

We thus find that the group velocity is just half the phase velocity. This phenomenon is not too difficult to observe. If a short train of crests and valleys is produced on water, the crests can be seen to move ahead faster than the group as a whole. When a crest reaches the front of the group, it disappears. New crests continuously form at the rear, move through the group, and disappear at the front. Such a group is illustrated in Figure 15-19.

For deep water waves of wavelength of the order of a few centimetres, the phase velocity is influenced by the surface tension as well as by gravity. With very short wavelengths, the phase velocity is determined mainly by the surface tension, and the phase velocity is less than the group velocity (see Exercise 26 and Problem 12).

**Question**

6. Under what conditions is group velocity greater than phase velocity? Less than phase velocity? The same?

*Optional*

**Dispersion in a Real Piano Wire**

The result that the phase velocity of a wave on a string is equal to $\sqrt{T/\mu}$ is based on the assumption that the string is perfectly flexible. When such a string is deformed by a wave, there is a restoring force proportional to the tension. The greater the tension, the greater the restoring force and the greater the phase velocity. A real piano wire is not perfectly flexible. Even if the wire is not under tension, there is a tendency for it to return to its original shape because of its stiffness. The phase velocity of a wave on a wire which is not perfectly flexible is somewhat greater than $\sqrt{T/\mu}$ because of the stiffness of the wire. The dispersion relation for a real piano wire can be written

$$\frac{\omega^2}{k^2} = \frac{T}{\mu} + \alpha k^2 \qquad\qquad 15\text{-}23$$

where $\alpha$ is a small positive constant which depends on the stiffness of the string. For a perfectly flexible string, $\alpha$ is zero. For a piano wire, $\alpha$ is usually so small that the quantity $\alpha k^2$ is much less than $T/\mu$. Using Equation 15-20, we have for the phase velocity

$$v_p = \frac{\omega}{k} = \sqrt{\frac{T}{\mu} + \alpha k^2} = \sqrt{\frac{T}{\mu}}\sqrt{1 + \frac{\alpha k^2 \mu}{T}}$$

$$\approx \sqrt{\frac{T}{\mu}}\left(1 + \frac{\alpha k^2 \mu}{2T} + \cdots\right) \qquad \text{for small } \alpha \qquad 15\text{-}24$$

We see that the phase velocity does depend on $k$. The derivative of the phase velocity with respect to wave number is

$$\frac{dv_p}{dk} = \sqrt{\frac{T}{\mu}}\frac{\alpha k \mu}{T} = \alpha k \sqrt{\frac{\mu}{T}}$$

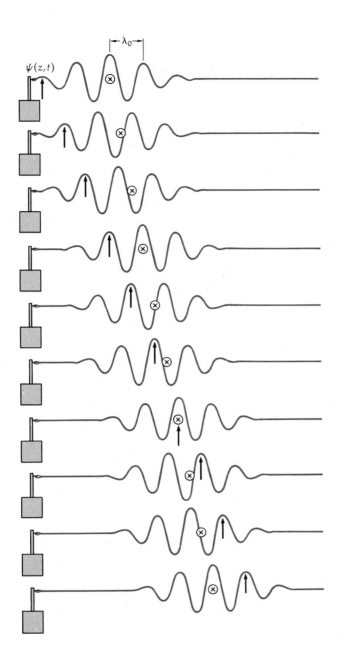

$\psi(z,t)$

$\vdash\lambda_0\dashv$

**Figure 15-19**
Wave packet for which the
group velocity is half the
phase velocity. The arrow
travels at the phase velocity
following a point of constant
phase for the dominant wavelength $\lambda_0$. The cross at the
center of the group travels at
the group velocity. (*Adapted
from Frank S. Crawford, Jr.,*
Waves and Oscillations, *p.
294, Berkeley Physics Course,
vol. 3, McGraw-Hill Book
Company, New York, 1965.
Courtesy of Education Development Center, Inc., Newton,
Mass.*)

Thus the group velocity is

$$v_g = v_p + k\frac{dv_p}{dk} = v_p + \alpha k^2 \sqrt{\frac{\mu}{T}}$$

15-25

The fact that the phase velocity of waves on a piano string depends on
the wavelength is apparently largely responsible for the distinctive tone
of a piano.* Although the wavelengths for the standing waves on a real
piano wire are still given by $\lambda_1 = 2L$, $\lambda_2 = 2L/2$, $\lambda_3 = 2L/3$, . . . , the
frequencies of the overtones are not just integral multiples of the funda-
mental because $f_n = v_p/\lambda_n$ and $v_p$ depends on the wavelength. Since the
phase velocity increases as the wavelength decreases, as indicated by
Equation 15-24 ($\lambda = 2\pi/k$), the overtones are slightly sharp compared
with those of a perfectly flexible string. The overtones are not true harmonics.

* See E. Donnell Blackham, "The Physics of the Piano," *Scientific American*, December
1965.

# 15-8  Superposition of Standing Waves

In our consideration of a perfectly flexible string fixed at both ends, we found that the standing-wave functions have the form

$$y_n(x, t) = A_n \cos (\omega_n t + \delta_n) \sin k_n x \qquad 15\text{-}26$$

where $A_n$ is the amplitude and $\delta_n$ is a phase constant which depends on the choice of zero time. The boundary conditions $y = 0$ at $x = 0$ and $x = L$ restrict the allowed frequencies and wave numbers to those obeying the standing-wave condition:

$$k_n L = n\pi \qquad \text{and} \qquad \omega_n = 2\pi f_n = n2\pi \frac{v}{2L} \qquad 15\text{-}27$$

where $v$ is the phase velocity (which is the same as the group velocity because we are considering a perfectly flexible string). If Equation 15-26 is to describe the standing-wave function for a vibrating string, the string must have the shape of a sine function at any instant. At time $t = 0$ the wave function, according to this equation, is

$$y_n(x, 0) = A_n \cos \delta_n \sin k_n x$$

For example, if the string is at rest at $t = 0$, it will vibrate in its fundamental mode with frequency $f_1$ corresponding to the wavelength $\lambda_1 = 2L$ only if the string has the initial shape indicated in Figure 15-20.

In general, a string fixed at both ends does *not* vibrate in a *single* harmonic or mode. The general vibration of a string contains a mixture of the allowed harmonic functions which satisfy the standing-wave conditions. The general wave function for waves on a perfectly flexible string fixed at both ends is a sum of harmonic wave functions:

$$y(x, t) = \sum_n A_n \sin (\omega_n t + \delta_n) \sin k_n x \qquad 15\text{-}28$$

Since each term in the sum is zero at both $x = 0$ and $x = L$, the function $y(x, t)$ also obeys the boundary conditions for any values of the amplitudes $A_n$ and the phase constants $\delta_n$. The particular values of these constants depend on the initial shape and initial velocity of the string. The velocity of any point on the string is obtained by differentiating Equation 15-28 with respect to time:

$$u_y = \frac{\partial y}{\partial t} = \sum_n -\omega_n A_n \sin (\omega_n t + \delta_n) \sin k_n x \qquad 15\text{-}29$$

At time $t = 0$, the shape of the string and the velocity of each element are given by

$$y(x, 0) = \sum_n A_n \cos \delta_n \sin k_n x = \sum_n A_n \cos \delta_n \sin \frac{n\pi x}{L} \qquad 15\text{-}30$$

and

$$u_y(x, 0) = \sum_n -\omega_n A_n \sin \delta_n \sin k_n x$$

$$= -\sum_n \omega_n A_n \sin \delta_n \sin \frac{n\pi x}{L} \qquad 15\text{-}31$$

where we have substituted $n\pi/L$ for $k_n$ from the standing-wave condition. The constants $A_n$ and $\delta_n$ are determined from the initial conditions by Fourier analysis. We shall discuss the result of such analysis for a particular example as an illustration of the general vibration of a string.

**Figure 15-20**
If a string of length $L$ fixed at both ends is to vibrate with only its fundamental frequency, it must have the initial shape shown.

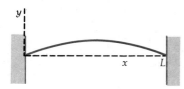

Suppose the string is initially at rest with the shape illustrated in Figure 15-21, which results from plucking the string in the center to a height $b$. The initial shape can be described by the function

$$f(x) = \begin{cases} \dfrac{2b}{L}\, x & \text{for } 0 < x < \dfrac{L}{2} \\[2mm] 2b - \dfrac{2b}{L}\, x & \text{for } \dfrac{L}{2} < x < L \end{cases}$$

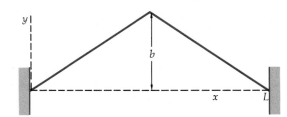

**Figure 15-21**
String plucked at the center to a height $b$.

Since the string is at rest at $t = 0$, the constants $\delta_n$ are zero from Equation 15-31. The constants $A_n$ are then determined by

$$f(x) = \sum_n A_n \sin \frac{n\pi x}{L}$$

Figure 15-22 illustrates the approximation using only the first three nonzero terms. Detailed analysis yields for these constants

$$A_1 = \frac{8b}{\pi^2} \qquad A_2 = 0 \qquad A_3 = -\frac{8b}{9\pi^2} = -0.111A_1$$

$$A_4 = 0 \qquad A_5 = \frac{8b}{25\pi^2} = 0.040A_1$$

[All the constants for even $n$ are zero because the initial shape is symmetric about the line $x = \frac{1}{2}L$, whereas $\sin(n\pi x/L)$ is antisymmetric about this line for $n$ even.]

**Figure 15-22**
Synthesis of the string plucked in Figure 15-21 using only the first three odd harmonics. The height $b$ is exaggerated in this drawing to show the relative amplitudes of the harmonics. The heavy colored line is the approximation to the original shape of the string using just these three harmonics. (*By permission from Robert M. Eisberg,* Fundamentals of Modern Physics, *p. 200. Copyright © 1961 by John Wiley & Sons, Inc.*)

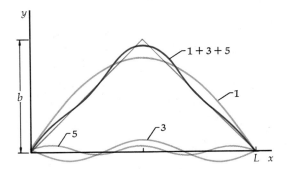

## Review

A. Define, explain, or otherwise identify:

B.  True or false:

1. The sum of two harmonic waves of the same wavelength and frequency is itself a harmonic wave.

2. The waves from two sources in phase interfere constructively everywhere in space.

3. Two sources that are out of phase by 180° are incoherent.

4. Interference patterns are observed only for coherent sources.

5. Diffraction occurs whenever the wavefront is limited.

6. Diffraction occurs only for transverse waves.

7. According to Huygens' principle, each point on a wavefront can be considered to be a source of waves.

8. The beat frequency between two sound waves of nearly equal frequency equals the difference in the frequency of the individual sound waves.

9. Information cannot be transported by a single harmonic wave.

10. Phase velocity and group velocity are never equal.

11. If the phase velocity is the same for all wavelengths, a wave pulse will maintain its shape as it propagates.

12. When a violin string is bowed, it vibrates with a single frequency equal to its fundamental frequency.

Exercises

### Section 15-1, Superposition and Interference of Harmonic Waves

1. Two waves traveling on a string in the same direction have frequency 100 Hz, wavelength 2 cm, and amplitude 0.02 m. They differ in phase by 60°. What is the amplitude of the resultant wave?

2. Two waves with the same frequency, wavelength, and amplitude are traveling in the same direction. If they differ in phase by 90° and each has amplitude 0.05 m, find the amplitude of the resultant wave.

3. Two sound sources oscillate in phase with the same amplitude. They are separated in space by $\frac{1}{3}\lambda$. What is the amplitude of the resultant wave from the two sources at a point on the line joining the sources if the amplitude due to each source separately is $A$? (Assume that the point is not between the sources.)

4. Two sound sources oscillate in phase with frequency 100 Hz. At a point 5.00 m from one source and 5.85 m from the other the amplitude of the sound from each source separately is $A$. (a) What is the difference in phase of the sound waves from the two sources at that point? (b) What is the amplitude of the resultant wave at that point?

5. Two sound speakers are separated by a distance of 6 m. A listener sits directly in front of one speaker a distance of 8 m from it so that the two speakers and listener form a right triangle. Find the two lowest frequencies for which the path difference is an odd number of half wavelengths. Why can these frequencies be heard even if the speakers are driven in phase by the same amplifier? (Use $v = 340$ m/s for the speed of sound.)

### Section 15-2, Interference of Waves from Two Point Sources

6. With a compass draw circular arcs representing wave crests for each of two point sources a distance $d$ apart for $d = 6$ cm and $\lambda = 1$ cm (see Figure 15-3). Connect the intersections corresponding to points of constant path difference and label the path difference for each line.

7. Two sound speakers are driven in phase by an audio amplifier at frequency 600 Hz. The speed of sound is 340 m/s. The speakers are on the $y$ axis, one at $y = +1.00$ m and the other at $y = -1.00$ m. A listener begins at $y = 0$ and walks along a line parallel to the $y$ axis at a very large distance $x$ away. (a) At what angle $\theta$ (between the line from the origin to the listener and the $x$ axis) will she first hear a minimum in the sound intensity? (b) At what angle will she first hear a maximum (after $\theta = 0$)? (c) How many maxima can she possibly hear if she keeps walking in the same direction?

8. Two point sources are in phase and separated by a distance $d$. The interference pattern of the sources is detected along a line parallel to that through the sources and a large distance $D$ from the sources, as in Figure 15-4. Show that the $m$th interference maximum is at a distance $x$ from the central maximum point, where $x$ is given approximately by

$$x = m\frac{D\lambda}{d}$$

9. Two sound sources, driven in phase by the same amplifier, are 2 m apart on the $y$ axis. At a point a very large distance from the $y$ axis, constructive interference is heard at the angle $\theta_1 = 8°$ and next at $\theta_2 = 16°10'$ with the $x$ axis. If the speed of sound is 340 m/s, (a) what is the wavelength of the sound waves from the sources, and (b) what is the frequency of the sources? (c) At what other angles is constructive interference heard? (d) What is the smallest angle for which the sound waves completely cancel?

10. Two violinists are standing a few feet apart and playing the same notes. Are there places in the room at which certain notes are not heard because of destructive interference? Explain.

11. Two speakers separated by some distance emit sound of the same frequency. At some point $P$ the intensity due to each speaker separately is $I_0$. The path distance from $P$ to one of the speakers is $\frac{1}{2}\lambda$ greater than that from $P$ to the other speaker. What is the intensity at $P$ if (a) the speakers are coherent and in phase; (b) the speakers are incoherent; and (c) the speakers are coherent but have a phase difference of 180°?

12. Answer the questions of Exercise 11 for the point $P'$, for which the distance to the far speaker is 1 wavelength greater than the distance to the near speaker. Again assume that the intensity at point $P'$ is $I_0$ due to each speaker separately.

13. Two speakers separated by some distance emit sound waves of the same frequency, but speaker 1 leads speaker 2 in phase by 90°. Let $r_1$ be the distance from some point to speaker 1 and $r_2$ be the distance to speaker 2. Find $r_2 - r_1$ such that the sound at the point will be (a) maximum and (b) minimum. (Express your answers in terms of the wavelength.)

## Section 15-3, Diffraction

14. Discuss the relationship between diffraction and interference. Can diffraction occur without interference? Can interference occur without diffraction?

15. If the wavelength is much larger than the diameter of a loudspeaker, the speaker radiates in all directions much like a point source. On the other hand, if the wavelength is much smaller, the sound travels approximately in a straight line in front of the speaker. Find the frequency of sound for which the wavelength is (a) 10 times and (b) one-tenth the diameter of a 30-cm speaker. Do the same for a 6-cm-diameter speaker. (Take 340 m/s for the speed of sound.)

## Section 15-4, Beats

16. Two tuning forks have frequencies 256 and 260 Hz. What is the beat frequency if the forks both vibrate at the same time?

17. When a violin string is played (without fingering) simultaneously with a tuning fork of frequency 440 Hz, beats are heard at the rate of three per second. When the tension in the string is increased slightly, the beat frequency decreases. What was the initial frequency of the violin string?

18. Two tuning forks are struck simultaneously, and four beats per second are heard. The frequency of one fork is 500 Hz. (a) What are the possible values for the frequency of the other fork? (b) A piece of wax is placed on one of the forks to lower its frequency slightly. Explain how the measurement of the new beat frequency can be used to determine which of your answers to part (a) is the correct frequency of the second fork.

### Section 15-5, Harmonic Analysis and Synthesis

*There are no exercises for this section.*

### Section 15-6, Wave Packets, and Section 15-7, Dispersion

19. Information for use by computers is transmitted along a cable in the form of short electric pulses at the rate of 100,000 pulses per second. (a) What is the maximum duration in time of each pulse if no two pulses overlap? (b) What is the range of frequencies to which the receiving equipment must respond?

20. A tuning fork of frequency $f_0$ begins vibrating at time $t = 0$ and is stopped after a time interval $\Delta t$. The sound waveform at some later time is shown as a function of $x$ in Figure 15-23. Let $N$ be the (approximate) number of cycles in this waveform. (a) How are $N$, $f_0$, and $\Delta t$ related? (b) If $\Delta x$ is the length in space of this wave group, what is the wavelength in terms of $\Delta x$ and $N$? (c) What is the wave number $k$ in terms of $N$ and $\Delta x$? (d) The number of cycles $N$ is uncertain by approximately $\pm 1$ cycle. Explain why (see Figure 15-23). (e) What is the uncertainty in the frequency $\Delta f$ due to the uncertainty in $N$? (f) Show that the uncertainty in the wave number due to the uncertainty in $N$ is $2\pi/\Delta x$.

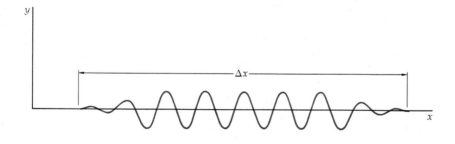

**Figure 15-23**
Exercise 20.

21. Two harmonic waves travel simultaneously along a long wire. Their wave functions are

$$y_1 = 0.002 \cos (6.0x - 600t) \quad \text{and} \quad y_2 = 0.002 \cos (5.8x - 580t)$$

where $y$ and $x$ are in metres and $t$ is in seconds. (a) What is the greatest displacement occurring in the wire? (b) Write the wave function for the resultant wave in the form of Equation 15-17. What is the phase velocity of the resultant wave? (c) What is the group velocity? (d) What is the separation in space of successive crests of the group? (e) Are these waves dispersive or nondispersive?

22. Repeat Exercise 21 for the two wave functions

$$y_1 = 0.003 \cos (8.0x - 400t) \quad \text{and} \quad y_2 = 0.003 \cos (7.8x - 380t)$$

23. Two sound waves of frequency 500 and 505 Hz travel at 340 m/s along the $x$ axis in air. The displacement amplitude of each wave is $s_0$. (a) Write wave functions $s_1(x, t)$ and $s_2(x, t)$ for each wave assuming that they are in phase at $x = 0$ and $t = 0$, and write the wave function $s(x, t)$ for the resultant wave. (b) Sketch the resultant wave function as a function of $t$ for some fixed value of $x$.

24. Show that a small spread in wave numbers $\Delta k$ implies a spread in wavelengths $\Delta \lambda$ related to $\Delta k$ by

$$|\Delta \lambda| \approx \frac{2\pi}{k^2} |\Delta k|$$

*Hint:* Differentiate $\lambda = 2\pi/k$.

25. The dispersion relation for sound waves in air is

$$\omega = \sqrt{\frac{\gamma RT}{M}}\, k$$

Find the phase velocity and the group velocity. Are these waves dispersive?

26. The dispersion relation for water waves of very short wavelength in deep water is

$$\omega^2 = \frac{\gamma}{\rho} k^3$$

where $\gamma$ is the surface tension and $\rho$ is the density. (*a*) What is the phase velocity for these waves in terms of $\gamma$, $\rho$, and $k$? (*b*) What is the group velocity? (*c*) Is the group velocity greater or less than the phase velocity?

### Section 15-8, Superposition of Standing Waves

*There are no exercises for this section.*

### Problems

1. Two loudspeakers radiate in phase at 170 Hz. An observer sits at 8 m from one speaker and 11 m from the other. The intensity level from either speaker acting alone is 60 dB. The speed of sound is 340 m/s. (*a*) Find the observed intensity $I$ when both speakers are on together. (*b*) Find the observed intensity level when both speakers are on together but one has its leads reversed so that the speakers are 180° out of phase. (*c*) Find the observed intensity level when both speakers are on and in phase but the frequency is 85 Hz.

2. Two sources have a phase difference $\delta_0$ which is proportional to time: $\delta_0 = Ct$, where $C$ is a constant. The amplitude of the wave from each source at some point $P$ is $A_0$. (*a*) Write the wave functions for each of the two waves at point $P$ assuming this point to be a distance $x_1$ from one source and $x_1 + \Delta x$ from the other. Find the resultant wave function and show that its amplitude is $2A_0 \cos \frac{1}{2}(\delta + \delta_0)$, where $\delta$ is the phase difference at $P$ due to the path difference. (*b*) Sketch the intensity at point $P$ versus time for a zero path difference. (Let $I_0$ be the intensity due to each wave separately.) What is the time average of the intensity? (*c*) Make the same sketch for the intensity at a point for which the path difference is $\frac{1}{2}$ wavelength.

3. Two identical speakers emit sound waves of frequency 680 Hz uniformly in all directions with a total audio output of 1 mW each. The speed of sound in air is 340 m/s. A point $P$ is a distance 2.00 m from one speaker and 3.00 m from the other. (*a*) Find the intensities $I_1$ and $I_2$ from each speaker at point $P$ separately. (*b*) If the speakers are driven coherently and in phase, what is the intensity at point $P$? (*c*) If they are driven coherently but out of phase by 180°, what is the intensity at point $P$? (*d*) If the speakers are incoherent, what is the intensity at point $P$?

4. A radio telescope consists of two antennas separated by a distance of 200 m. Each antenna is tuned to a particular frequency such as 20 MHz. The signals from each antenna are fed into a common amplifier, but one signal first passes through a phase adjuster, which delays the phase by an amount chosen so that the telescope can "look" in different directions. With zero phase delay, plane radio waves incident vertically produce signals which add constructively at the

amplifier. What should the phase delay be so that signals coming from an angle $\theta = 10°$ with the vertical (in the plane formed by the vertical and the line joining the antennas) add constructively at the amplifier?

5. Instructions for connecting stereo speakers to an amplifier correctly so that they are in phase are as follows: "After both speakers are connected, play a monophonic record or program with the bass control turned up and the treble turned down. While listening to the speakers, turn the balance control so that first one speaker is heard separately, then the two together, and then the other separately. If the bass is stronger when both speakers play together, they are connected properly. If the bass is weaker when both play together compared with each separately, interchange the connections on one speaker." Explain why this method works. In particular, explain why a stereo source is not used and why only the bass is compared.

*For Problems 6 and 7, review the doppler-effect discussion in Section 14-4.*

6. A radar trap radiates microwaves with a frequency of 2.00 GHz. When the waves are reflected from a moving car, the beat frequency is 293 beats per second. Find the speed of the car.

7. Hovering over the pit of hell, the devil observes that as an engineering student falls past him (with the terminal velocity), the frequency of his scream falls from 842 to 820 Hz. (*a*) Find the speed of descent of the student. (*b*) The scream generates beats when mixed with its echo from the bottom of the pit. Find the number of beats per second heard by the student. (*c*) Find the number of beats per second heard by the devil after the student has passed by.

8. Show that Equation 15-18, which gives the group velocity of the envelope of two waves exactly, leads to $v_g = v_p$ when the phase velocities of the two waves are equal.

9. A tuning fork of frequency $f_0$ can be "stopped" by looking at it in a dark room with a strobe light of the same frequency. If the strobe frequency is slightly different from $f_0$, the fork appears to vibrate in slow motion. Discuss how this phenomenon is related to beats.

10. The use of a vernier scale is related to beats. It is easiest to learn how such a scale works by constructing one. Along the edge of a card (such as a 3 by 5 index card) make a set of 9 marks equally spaced 0.9 cm apart, and number the marks (call the edge 0). Your scale can now be used to interpolate between centimetre marks on a scale that is not divided into millimetres. Place your card along a scale ruled in centimetres so that the edge is somewhere between the 2- and 3-cm marks. Then note which mark on your card is aligned with a centimetre mark on the scale. If, for example, the fourth mark on your scale is aligned with a centimetre mark, the edge of the card is at 2.4 cm. Explain how this works and how it is related to beats.

11. When two piano wires are in tune, they have the same tension $T_0$ and fundamental frequency $f_0$. What fractional change $\Delta T/T_0$ in one of the wires will produce audible beats with frequency $f_B$ when the wires vibrate simultaneously? With what accuracy must the tension in the wires be adjusted so that when the wires are tuned to middle C ($f_0 = 261$ Hz), the beat frequency will be less than one beat in 5 s?

12. The general dispersion relation for water waves can be written

$$\omega^2 = \left( gk + \frac{\gamma}{\rho} k^3 \right) \tanh kH$$

where $g$ is the acceleration of gravity, $\rho$ is the density of water, $\gamma$ is the surface tension, $H$ is the depth of the water, and the hyperbolic tangent is defined by

$$\tanh x = \frac{e^x - e^{-x}}{e^x + e^{-x}} = \frac{e^{2x} - 1}{e^{2x} + 1}$$

(a) Show that the hyperbolic tangent has the properties that for $x \gg 1$, tanh $x \approx 1$, and for $x \ll 1$, tanh $x \approx x$. Use these results to write separate dispersion relations for deep-water waves ($kH \gg 1$) and shallow-water waves ($kH \ll 1$). (b) Show that in shallow water the group velocity and the phase velocity are both equal to $\sqrt{gH}$ if the wavelength is long enough to ensure that $\gamma k^2 / \rho = 4 \pi^2 \gamma / \lambda^2 \rho \ll g$. (c) Show that for deep water the phase velocity is given by $v_p = \sqrt{g/k + \gamma k/\rho}$ and find the group velocity. (d) For water, $\rho = 10^3$ kg/m$^3$ and $\gamma = 0.075$ N/m. Evaluate $v_p$ and $v_g$ in deep water for small ripples with $\lambda = 1$ cm and for large waves with $\lambda = 1$ m. For what wavelength are the phase and group velocities equal in deep water?

**CHAPTER 16**     Temperature

---

**Objectives**  After studying this chapter, you should:

1.  Be able to explain what is meant by the terms diathermic wall, adiabatic wall, and thermal equilibrium.

2.  Be able to state the zeroth law of thermodynamics.

3.  Be able to discuss the advantages of gas thermometers over other thermometers.

4.  Be able to state how the ideal-gas temperature scale is defined.

5.  Be able to state the definitions of the Celsius temperature scale and the Fahrenheit temperature scale and convert temperatures given on one scale into those of the other.

6.  Be able to convert temperatures given on either the Celsius scale or the Fahrenheit scale into kelvins.

7.  Be able to state the equation of state for an ideal gas and give the value of the universal gas constant in joules per kelvin.

8.  Know that the average energy of a gas molecule at temperature $T$ is of the order of $kT$, where $k$ is Boltzmann's constant.

---

Thermodynamics is the study of energy transfers involving temperature, occurring between macroscopic bodies. In the following chapters we shall define the concepts of temperature, heat, and internal energy just as we carefully defined such concepts as mass, force, potential energy, etc., in our study of mechanics. Until we do, however, we shall use the words temperature and heat in their ordinary meanings: temperature is what we measure with an ordinary thermometer; heat is energy that is transferred because of a temperature difference.

# 16-1   Macroscopic State Variables

Consider a body of gas enclosed in some volume $V$ and at some pressure $P$. There are two different approaches to the description of such a system. One, the *microscopic method*, involves a description in terms of the many particles—the molecules—composing the gas. It requires numerous assumptions about the particles that are difficult to test directly. For example, we assume that the gas is composed of $N$ molecules each moving about in a random way, making elastic collisions with other molecules and with the walls of the container. Because of the huge size of $N$ (for example, 1 mol, or 32 g, of oxygen contains Avogadro's number, $N_A \approx 6 \times 10^{23}$, molecules), it is impossible to apply Newton's laws of motion separately to each molecule or even to list the coordinates of each molecule. Entering just one coordinate for each molecule of material into a computer at the rate of 1 $\mu$s per molecule would take $6 \times 10^{17}$ s $\approx 2 \times 10^{10}$ y, which is the same order of magnitude as the age of the universe. But because $N$ is so large, statistical methods of treatment are very accurate. In this chapter we shall show by a simple analysis that the pressure exerted by a gas on the wall of its container is proportional to the average kinetic energy of the molecules in the gas. This type of calculation is an example of a microscopic calculation of a macroscopic variable, the pressure $P$.

In the *macroscopic approach* characteristic of thermodynamics, the system is described by only a few variables, e.g., pressure, temperature, volume, and internal energy. These variables (except internal energy) are closely related to our senses and are easily measured. (Compare these few variables with the $6N$ coordinates and velocity components of the molecules, and with additional variables describing rotation or oscillation of molecules that, like $O_2$, contain more than one atom.) Few assumptions are made in the macroscopic description of matter. The laws of thermodynamics, which, like Newton's laws of motion, are elegant and compact generalizations of the results of experience, are therefore quite general and independent of any particular molecular assumptions made in the microscopic approach. In fact, much of thermodynamics was developed before the molecular model of matter was completely accepted.

The minimum number of macroscopic variables needed to describe a system depends on the kind of system, but it is always a small number. Usually we must specify the composition (if the system is not homogeneous; e.g., a mixture of two or more gases), the mass of each part, and only two more variables such as the pressure $P$ and the volume $V$ of a gas. If we restrict our discussion to homogeneous systems of constant mass, we usually need only two variables.

We cannot say that a gas is at pressure $P$ and temperature $t$ unless the system is in *thermal equilibrium* with itself. Consider, for example, a gas in an enclosure of volume $V$, isolated from its surroundings. If we stir the gas rapidly in one corner of the container, we cannot assign a pressure and temperature to the whole gas until the gas settles down. If we measure the temperature at various parts of the gas just after stirring, we get different results, which change with time. *When the macroscopic properties of an isolated system become constant in time, the system is in thermal equilibrium with itself.* We can then describe some property such as pressure with a single variable $P$ for the whole system.

Suppose we have a fixed mass of gas originally at some pressure $P_0$ and volume $V_0$. Let the temperature of the gas be $t_0$ (measured in any

*Thermal equilibrium*

familiar way). We now do various things to the gas—compress it, heat it, let it expand against a piston, and cool it—but finally return it to the original pressure $P_0$ and volume $V_0$. When the gas is again at equilibrium with itself at the original pressure and volume, we find that its temperature is again $t_0$. Every other macroscopic property we can measure is also the same as it was originally. The variables $P$ and $V$ thus specify a *macroscopic state* of the system. We could also specify the state of this system using the variables $P$ and $t$. For a constant-mass gas system, an equilibrium state is specified by any two macroscopic variables.

*Macroscopic state*

There are many other kinds of macroscopic systems whose states can be described by two variables (assuming constant mass). For example, a long rod can be described by its length $L$ and the pressure $P$ (which is usually atmospheric pressure unless the rod is enclosed in some way). Similarly, the state of a conducting wire can be described by the pressure and its electric resistance. Even such a complicated system as an electric cell can often be described quite well in terms of only two variables, such as the emf* of the cell and the charge. In the following chapters we shall carefully define temperature and other macroscopic variables, e.g., internal energy and entropy, which are also *state variables*. We repeat the meaning of the concept of a macroscopic equilibrium state: if a system is in a given equilibrium state at some time and after various changes it is brought back to that equilibrium state, all the macroscopic properties of the system will be the same as they were originally. Note that a *macroscopic* equilibrium state is quite different from a *microscopic state,* which is specified by the positions, velocities, and internal coordinates of all the molecules in the system. In a simple gas in thermal equilibrium with itself the microscopic state is continually changing because the molecules are changing position and making collisions which change their velocities; but in equilibrium, the macroscopic state remains the same. There are many different microscopic states which correspond to a single macroscopic state.

*Microscopic state*

## 16-2   Adiabatic and Diathermic Walls

Before we give a rigorous definition of temperature, let us consider some common experiences with properties of materials using temperature as measured by a common thermometer. Consider, for example, a given mass of gas with volume $V$ and at pressure $P$. If we heat the gas (say with a bunsen burner), the pressure or volume of the gas will change. If the volume is kept constant, the pressure increases as the temperature is raised. Alternatively, if the pressure is kept constant by allowing the gas to expand against a piston, as in Figure 16-1, the volume increases as the temperature is raised. An increase in the temperature of the gas is indicated by an increase in either the pressure $P$ or the volume $V$ when the other variable is kept constant.

Let us consider some simple experiments with constant-volume gas systems. We place two systems, originally at different temperatures, in close contact by separating them only by a thin metal wall. The temperature of each system changes until the two systems reach a common temperature between the original temperatures. The pressure of one

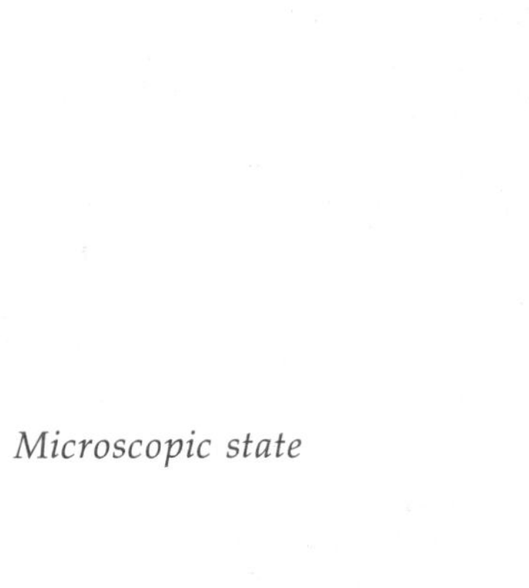

**Figure 16-1**
Heating a gas at constant pressure. As the gas is heated, the piston moves to the right against a constant force, which equals the pressure times the area.

---

* We shall discuss emf in Chapter 24.

system increases as its temperature increases, and the pressure of the other decreases as its temperature decreases, until eventually the pressures reach steady values when the systems reach their final equilibrium temperature. The final steady-state pressures of the gases are of course not generally equal, assuming that the separating wall is rigid enough to maintain constant volumes. When the pressures stop changing, the systems are in *thermal equilibrium* with each other. The time it takes to reach equilibrium depends on the masses and types of the gases, the original temperatures, and the kind of contact made between the two systems. If we replace the thin metal wall with a thick asbestos wall, the time is very long. Over rather long periods of time with this separation, the two systems remain at nearly their original pressures and temperatures. (We are assuming that both systems are otherwise isolated from other systems.) If we heat one system, stir it, or do anything to it, very little if anything happens to the other system.

We can do similar experiments with other types of systems. For example, instead of constant-volume gas systems, we might use long metal rods at constant pressure. In that case, the increase or decrease in the length of a rod indicates an increase or decrease in its temperature. From these experiments, we can define two kinds of ideal separations *without any prior reference to temperature*. Two systems are said to be separated by an *adiabatic wall* if we can arbitrarily change the variables of one system without having any effect on the other. For example, if two constant-volume gas systems are separted by an adiabatic wall, we can increase the pressure of one by heating it or by some other means, and the other system will not be affected. (The common name for an adiabatic wall is a perfectly insulating wall. An adiabatic wall does not allow heat flow. However, we do not need to define heat or temperature in order to define an adiabatic wall.) The opposite extreme is a *diathermic wall*. If two systems are connected by a diathermic wall, a change in one of the thermodynamic variables of one system will influence the variables of the other system. For example, if two constant-volume gas systems are connected by a diathermic wall, heating one gas will cause an increase in pressure in the other gas (as well as in the one heated directly). The common name for a diathermic wall is, of course, a perfect heat conductor. Systems connected by a diathermic wall are said to be in *thermal contact*. In practice thin metal walls make excellent approximations to a perfect diathermic wall. Similarly, an adiabatic wall can be approximated with thick asbestos.

*Adiabatic wall*

*Diathermic wall*

# 16-3  The Zeroth Law of Thermodynamics

We now use our definitions of adiabatic and diathermic walls to define temperature. Suppose that two systems, originally separated, are put in contact by a diathermic wall and isolated from their surroundings by adiabatic walls (Figure 16-2). In general, when the systems are put in thermal contact, the thermodynamic variables will change from their original values. For example, the pressures $P_1$ and $P_2$ of two gas systems held at constant volume will change. One will increase, and the other will decrease. When the pressures reach their final equilibrium values, the two systems are said to be in *thermal equilibrium with each other*.

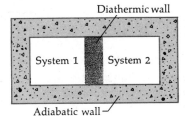

**Figure 16-2**
Systems 1 and 2, separated by a diathermic wall, are in thermal contact. In general, the thermodynamic variables of the systems will change until the systems are in thermal equilibrium with each other.

*Thermal equilibrium*

Suppose that when the two systems (originally separated) are put in contact by a diathermic wall, no change is observed in any of the variables of either system. The connected systems are in equilibrium, and they are said to be in thermal equilibrium even if they are separated. We thus generalize the concept of thermal equilibrium to include systems which are not in contact by a diathermic wall but which would be in equilibrium if they were so connected.

Suppose that system $A$ is in thermal equilibrium with system $C$ and that system $B$ is also in thermal equilibrium with $C$, as in Figure 16-3a. If we now place systems $A$ and $B$ in thermal contact, i.e., connect them by a diathermic wall (Figure 16-3b), we find that they are in thermal equilibrium with each other. This result is known as the zeroth law of thermodynamics:

*Two systems in thermal equilibrium with a third system are in thermal equilibrium with each other.*

*Zeroth law of thermodynamics*

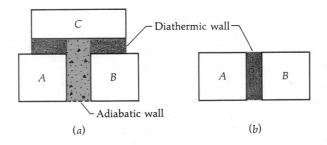

(a)                                    (b)

**Figure 16-3**
(a) Systems $A$ and $B$ are in thermal contact with system $C$. When $A$ and $B$ are in thermal equilibrium with $C$, they are in thermal equilibrium with each other. This can be checked by putting them in thermal contact with each other, as in (b). This experimental result is known as the zeroth law of thermodynamics.

The zeroth law of thermodynamics, which is an experimental law, shows that the relation between two systems implied by the statement that they are in thermal equilibrium is a *transitive relation*. That is, if $A$ is in thermal equilibrium with $C$, and $B$ is with $C$, then $A$ is in thermal equilibrium with $B$. This transitive property is needed in order to define temperature.

The first step in defining temperature is to determine a rule for saying when two systems in equilibrium have the same temperature. This resembles our procedure in mechanics, where the first law of motion defined the condition of motion without force. We adopted the rule that there is no force when velocity does not change. In the case of temperature, we shall say that two systems have the same temperature if none of their macroscopic variables change when they are connected with a diathermic wall. This definition of temperature equality is more succinctly expressed as follows:

*Two systems in thermal equilibrium with each other are at the same temperature.*

Two systems which are not in equilibrium with each other are at different temperatures. Suppose that the two systems are constant-volume gas systems and that $P_1$ decreases and $P_2$ increases when they are put in thermal contact. We could arbitrarily define either system to be at a higher temperature and still have a self-consistent concept of temperature; but since it is important to make scientific definitions of quantities as close to common usage as possible, we want a system that feels hot to the touch to be at a higher temperature than one that feels cold. (It is not always possible to distinguish the relative hotness or coldness of two objects by sense of touch, but there is a quite general agreement for a wide range of systems; e.g., ice water versus water at 30°C.) In our example of two gas systems, system 1 is originally "warmer" than system 2. We associate *decreasing pressure* of a constant-volume gas system with *decreasing temperature*, and *increasing pressure* with *increasing temperature*. If we place a constant-volume gas system in thermal contact with a metal rod when the systems are not in thermal equilibrium, we find that if the pressure of the gas increases, the length of the rod decreases and vice versa. Thus increasing rod length at constant pressure indicates that the temperature of the rod is increasing.

## Questions

1. Consider the relation expressed by "A loves B." Is this relation transitive?

2. Suppose we define two countries to be in equilibrium with each other when they are at peace with each other. Is this kind of equilibrium situation transitive?

3. Give other examples of relations between two objects and state whether they are transitive or not.

4. As we have defined it, does temperature have any meaning applied to a system not in equilibrium with itself?

5. How could you determine whether two bodies have the same temperature if it is impossible to put them into contact with each other?

## 16-4 Temperature Scales and Thermometers

We have just defined a method for comparing any two systems and deciding whether they are at the same temperature or, if not, which system is at the greater temperature. We note that if *A* is at a greater temperature than *B* and *B* is at a greater temperature than *C*, it follows that *A* is at a greater temperature than *C*. If we have *N* systems, we can order them along a line such that all those to the right of a given system have a greater temperature than that system and all those to the left have a smaller temperature. We are now in a position to define a *temperature scale*, which we can use to assign a number to each system so that greater numbers indicate greater temperatures.

We first choose a system to be our thermometer. We then define the temperature to be any monotonic function of one of the state variables of the system with the other held constant. For example, let us choose a given mass of mercury contained in a narrow glass tube. When we put this system in thermal contact with another, e.g., a constant-volume gas system, we note that the length of the mercury column increases when the temperature increases. Thus the mercury system behaves essentially the same as a rod at constant pressure. (The pressure on the mercury is usually not kept constant, but the compressibility of mercury is so small that even a rather large change in pressure has little effect on the volume or length of the mercury column.) The simplest functional relation between temperature and length of the mercury column that we

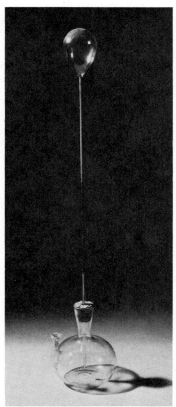

Scala/EPA

Galileo's thermoscope, one of the earliest thermometers, used by Galileo to detect temperature changes. (*Courtesy of the Science Museum, Florence.*)

can choose is a linear one. We define the temperature $t$ to be related to the length $L$ by

$$t = aL + b \qquad\qquad 16\text{-}1$$

The slope $a$ and the intercept $b$ can be determined by arbitrarily defining the temperature of two reproducible states of some standard system. Alternatively, we can choose the intercept $b$ to be zero and determine $a$ by arbitrarily defining the temperature of a single standard state.

The first method, which was universally used before 1954, is still of some practical use today. We choose for one of our states the ice point of water, i.e., water and ice in equilibrium with air at atmospheric pressure (also called the *normal melting point*), and define its temperature to be zero degrees Celsius,* written 0°C. Similarly we define the temperature of the steam point, i.e., water and water vapor in equilibrium at atmospheric pressure (also called the *normal boiling point*), to be 100°C. Let $L_i$ be the length of the mercury column when it is in thermal equilibrium with a system at the ice point and $L_s$ be that when it is in thermal equilibrium with a system at the steam point. We then have from Equation 16-1

$$t_i = aL_i + b = 0°C \qquad \text{and} \qquad t_s = aL_s + b = 100°C \qquad 16\text{-}2$$

These equations can be solved for $a$ and $b$ in terms of $L_i$ and $L_s$, giving

$$a = \frac{100°C}{L_s - L_i} \qquad \text{and} \qquad b = -\frac{(100°C)L_i}{L_s - L_i} \qquad 16\text{-}3$$

Having determined these constants, we can measure the temperature of any other system by noting the length $L$ of the mercury when the thermometer is in equilibrium with that system and using Equation 16-1. Figure 16-4 illustrates this temperature scale.

If we choose a different system to be our thermometer, we define the temperature measured by this thermometer in a similar way. For example, if we choose a constant-volume gas thermometer (Figure 16-5), we define the temperature to vary linearly with the pressure:

$$t = aP + b \qquad\qquad 16\text{-}4$$

Again we can find $a$ and $b$ in terms of the pressure $P_i$ at the ice point and $P_s$ at the steam point. We obtain equations similar to Equations 16-3:

$$a = \frac{100°C}{P_s - P_i} \qquad \text{and} \qquad b = -\frac{(100°C)P_i}{P_s - P_i} \qquad 16\text{-}5$$

Other thermometers can be similarly calibrated. For example, we can use a copper rod at constant pressure and define the temperature to be a linear function of the length, or a resistance thermometer at constant pressure, for which the temperature is defined to vary linearly with electric resistance. The constants $a$ and $b$ are different for different thermometers.

We now ask whether the various thermometers defined in this way will agree when we measure the temperature of some system other than water at the ice point or steam point. For example, if we measure the normal boiling point of sulfur (at about 444°C), we get different numbers using different thermometers. The discrepancies are particularly bad at temperatures far from the two fixed points. Similarly, if we

---

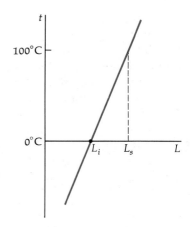

**Figure 16-4**
Temperature scale based on two fixed points. The temperature is assumed to vary linearly with the length of a mercury column. $L_i$ and $L_s$ are the lengths when the column is at the ice point and steam point of water, respectively. The scale is then determined by defining these temperatures to be 0 and 100°C.

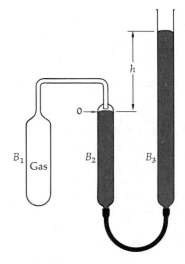

**Figure 16-5**
A constant-volume gas thermometer. The volume is maintained constant by raising or lowering the tube $B_3$ until the mercury in $B_2$ is at the zero mark. The temperature is chosen to be proportional to the pressure in the gas as determined by the height of the mercury column $h$.

measure thermal properties of some material, the results depend on the particular thermometer used.

There is one group of thermometers for which the measured temperatures are in very close agreement at all temperatures. These are the gas thermometers, either constant-volume thermometers (discussed above) or constant-pressure thermometers, for which the temperature is taken to vary linearly with volume. If we measure the temperature of a given state of some system with different gas thermometers, we get close agreement no matter what kind of gas is used as long as the pressure in the thermometer is not too high and the temperature is not too low. If we reduce the pressure in all the thermometers by using less gas, the agreement improves.

Let us consider the following procedure. We measure some temperature, say the normal boiling point of sulfur, with a constant-volume gas thermometer, choosing the amount of gas so that the pressure at the steam point of water is $P_s = 1000$ mmHg. We then remove some of the gas in the thermometer so that the steam-point pressure $P_s$ is now only 500 mmHg and again measure the temperature of the normal boiling point of sulfur. We repeat this process with smaller and smaller amounts of gas and plot the measured temperature versus $P_s$. Figure 16-6 shows such a plot for various gases used in a constant-volume thermometer. As the pressure approaches zero, all the thermometers approach the same value, $t = 444.60°C$. We obtain similar results with a constant-pressure thermometer using the volume of the gas to measure the temperature. As the mass of the gas is reduced so that the pressure becomes smaller, the temperatures measured by different gas thermometers tend to a single limit.

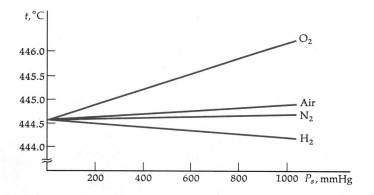

**Figure 16-6**
Temperature of the boiling point of sulfur measured with constant-volume gas thermometers filled with various gases. The pressure of the steam point of water $P_s$ is varied by varying the amount of gas in the thermometer. As the pressure is reduced, the measured temperatures approach the value 444.60°C for all thermometers.

This agreement of all gas thermometers at low pressures means that the intercept $b$ in Equation 16-4 and given by Equation 16-5 is the same for all gas thermometers. Measurements of $b$ at low pressures yield values of about $-273.15°C$. There is some disagreement between various laboratories where this number is measured because of the experimental difficulty in measuring the ice-point and steam-point temperatures. Both states are very difficult to duplicate experimentally. Because of these difficulties, a temperature scale based on a single fixed point with $b$ defined to be zero was adopted in 1954 by the International Committee on Weights and Measures. A reference state which is much more easily reproduced than either the ice point or steam point is the *triple point of water*, the single temperature and pressure at which water,

water vapor, and ice coexist in equilibrium. The triple-point pressure is 4.58 mmHg, and the temperature is 0.01°C. The temperature of this state is conventionally chosen to be 273.16° (note that this is not degrees Celsius). Setting $b$ equal to zero in Equation 16-4, we have for a constant-volume thermometer

$$t = aP \qquad\qquad 16\text{-}6$$

Let $P_3$ be the pressure when the thermometer is in equilibrium with a system containing water, water vapor, and ice; then

$$273.16° = aP_3 \qquad \text{or} \qquad a = \frac{273.16°}{P_3}$$

Thus

$$t = \frac{273.16°}{P_3} P \qquad\qquad 16\text{-}7$$

In Section 16-6 and in Chapter 17 we shall discuss the behavior of real gases at very low pressures in terms of a simple model called an *ideal gas*. The temperature scale defined by Equation 16-7 in the limit of very low pressures is called the *ideal-gas temperature T*. For a constant-volume gas thermometer, the ideal-gas temperature is defined to be

*Ideal-gas temperature*

$$T = 273.16° \lim_{P_3 \to 0} \frac{P}{P_3} \qquad \text{constant } V \qquad\qquad 16\text{-}8a$$

Similarly, for a constant-pressure thermometer the ideal-gas temperature is

$$T = 273.16° \lim_{P_3 \to 0} \frac{V}{V_3} \qquad \text{constant } P \qquad\qquad 16\text{-}8b$$

This temperature scale has the following advantages: (1) the measured temperature of any state is the same no matter what kind of gas is used, and (2) the reference state (the triple point of water) is easily reproduced throughout the world, so that temperatures measured in various laboratories are in good agreement with each other. This temperature scale depends on the properties of gases but not on the properties of any particular gas. Any substance can be used as long as it remains a gas. The lowest temperature that can be measured with a gas thermometer is about 1°, using helium for the gas. Below this temperature, even a dilute helium liquefies. All other gases liquefy at higher temperatures.

In Chapter 19 we show that we can define a temperature scale which is independent of the properties of any material, called the *Kelvin scale* or *absolute temperature scale*. We shall show that for temperatures above 1° for which the ideal-gas scale can be defined, the ideal-gas and absolute temperature scales are identical. In anticipation of this, we write* K after temperatures defined by the ideal-gas scale, and we shall use the conventional symbol $T$ for the absolute temperature scale even though we cannot define this scale until we study the second law of thermodynamics.

Figure 16-7 shows the temperature of the steam point of water measured with constant-volume gas thermometers using various gases at

---

* The name of the unit of temperature was changed by the 13th General Conference on Weights and Measures in 1967 from degree Kelvin (°K) to kelvin (K), defined to be 1/273.16 of the absolute temperature of the triple point of water.

Lord Kelvin (1824–1907).

various pressures $P_3$. As $P_3$ is reduced to zero by reducing the amount of gas in the various thermometers, the measurements approach the common value of 373.15 K, the ideal-gas temperature, which is the same as the absolute temperature. The ice-point temperature measured in a similar way is 273.15 K.

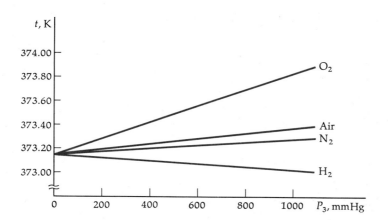

**Figure 16-7**
Temperature of the steam point of water measured with constant-volume gas thermometers filled with various gases. The temperature is plotted as a function of the triple-point pressure of water, $P_3$, which is varied by varying the amount of gas in the thermometer. As the pressure approaches zero, the measured temperatures all approach the value 373.15 K.

## 16-5  The Celsius and Fahrenheit Scales

Two other temperature scales in common use are the Celsius scale and the Fahrenheit scale. The Celsius scale has the same size degree as the Kelvin, or absolute, scale, but the zero is shifted so that the ice-point temperature is 0.00°C and the steam-point temperature is 100.00°C. In general, the Celsius temperature $t_C$ is related to the absolute temperature $T$ by

$$t_C = T - 273.15 \qquad\qquad 16\text{-}9$$

The Celsius temperature of the triple point of water (273.16 K) is 0.01°C.

It is important to remember that since the Celsius degree is the same size as the kelvin (absolute degree), temperature *differences* are the same on the two scales. That is, a temperature difference of 10 kelvins is the same as a temperature difference of 10 Celsius degrees. The difference between the Kelvin and Celsius scales lies only in the choice of zero temperature.

The Fahrenheit scale is chosen so that the ice point is 32°F and the steam point is 212°F. Since there are 180 Fahrenheit degrees between the ice point and steam point and only 100 Celsius degrees (or kelvins) between these two points, the Fahrenheit degree is smaller than the Celsius degree. A temperature difference of 1 Celsius degree equals one of $\frac{9}{5}$ Fahrenheit degrees. The zero is also shifted since the temperature 0.00°C equals the temperature 32°F. The general relation between the Celsius temperature $t_C$ and the Fahrenheit temperature $t_F$ is

$$t_F = \tfrac{9}{5}t_C + 32°F \qquad \text{or} \qquad t_C = \tfrac{5}{9}(t_F - 32°F) \qquad 16\text{-}10$$

In specifying a temperature, it is important, of course, to say what scale is used. Temperature specified on the absolute scale is customarily indicated by the letter $T$ with its numerical value followed by K (kelvins). Temperatures expressed on the Celsius or Fahrenheit scale are generally indicated by $t$ with their numerical values followed by °C (degrees Celsius) or °F (degrees Fahrenheit). For example, the steam point of water is 373.15 K = 100.00°C = 212.00°F.

Although the Celsius and Fahrenheit scales are convenient for every-day use, the absolute scale is much more convenient for scientific pur-poses, partly because many formulas are more simply expressed when the absolute scale is used, and partly because the absolute temperature can be given a more fundamental interpretation.

---

**Example 16-1** What is the Kelvin temperature corresponding to 70°F? From Equation 16-10 the Celsius temperature equal to 70°F is

$$t_C = \tfrac{5}{9}(70°F - 32°F) = \tfrac{5}{9}(38) = 21°C$$

From Equation 16-9

$$T = t_C + 273.15 \text{ K} = 21 + 273 = 294 \text{ K}$$

---

**Questions**

6. Do the temperature scales we have defined allow negative tempera-tures?

7. Can you suggest why the Celsius and Fahrenheit scales are consid-ered more convenient for ordinary nonscientific purposes?

8. Are there any circumstances in which the Kelvin and Celsius tem-peratures are not significantly different?

## 16-6  Equations of State: The Ideal Gas

If we compress a gas while keeping the temperature constant, we find that the pressure varies inversely with the volume. The experimental result that the product of the pressure and volume of a gas is constant at constant temperatures is known as *Boyle's law*. This law holds approxi-mately for all gases at low pressures. But according to Equations 16-8$a$ and $b$, the absolute temperature (or ideal-gas temperature) of a gas at low pressures is proportional to the pressure at constant volume and to the volume at constant pressure. Thus at low pressures, the product $PV$ is approximately proportional to the temperature $T$:

$$PV = CT$$

where $C$ is a constant of proportionality appropriate to a particular body of gas. The pressure exerted by a gas on the walls of its container is the result of collisions of the gas molecules with the walls. If we double the number of gas molecules in a given container, we double the number of collisions that take place in some time interval and consequently double the pressure. We therefore expect the constant $C$ in the above equation to be proportional to the number of molecules of the gas in the con-tainer and write

$$C = kN$$

where $N$ is the number of molecules of the gas and $k$ is a constant which is found experimentally to have the same value for any kind or amount of gas. We then have

$$PV = NkT \quad \text{or} \quad \frac{PV}{N} = kT \qquad\qquad 16\text{-}11$$

It is often convenient to write the amount of gas in terms of the number of moles. A *mole* (mol) of any substance is the amount of that substance that contains Avogadro's number of molecules, where Avogadro's number $N_A \approx 6.022 \times 10^{23}$ is defined to be the number of carbon atoms in 12 g of $^{12}C$. The mass of 1 mol is called the *molecular mass M*. The molecular mass of $^{12}C$ is, by definition, 12 g $= 12 \times 10^{-3}$ kg. If we have $n$ mol of a substance, the number of molecules is

*Mole*

$$N = nN_A \qquad\qquad 16\text{-}12$$

Equation 16-11 can also be written

$$\frac{PV}{n} = N_A kT = RT \qquad\qquad 16\text{-}13$$

where

$$R = kN_A \qquad\qquad 16\text{-}14$$

Figure 16-8 shows plots of $PV/nT$ versus pressure $P$ for several gases. We see that as the pressure approaches zero, the quantity $PV/nT$ approaches the same value $R$, called the *universal gas constant*, for all gases.

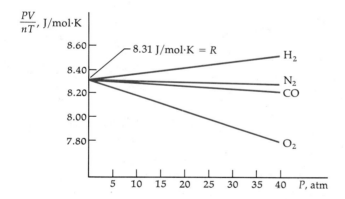

**Figure 16-8**
Plot of $PV/nT$ versus $P$ for real gases. As the pressure approaches zero, $PV/nT$ approaches the value $R = 8.31$ J/mol·K for all gases. The ideal-gas equation $pV = nRT$ is a good approximation for all real gases for low pressures up to a few atmospheres.

The numerical value of $R$ depends of course on the units chosen. In addition to the international system, in which the pressure is expressed in newtons per square metre, the volume in cubic metres, and thus $PV$ in newton-metres, i.e., joules, we sometimes express $P$ in atmospheres and $V$ in litres (L) (1 L $= 10^3$ cm$^3$ $= 10^{-3}$ m$^3$). The values of $R$ in these units are

$$R = 8.314 \text{ J/mol·K} = 0.08206 \text{ L·atm/mol·K} \qquad 16\text{-}15 \qquad \textit{Universal gas constant}$$

The constant $k$, called *Boltzmann's constant*, has the value

$$k = \frac{R}{N_A} = 1.38 \times 10^{-23} \text{ J/K} \qquad\qquad 16\text{-}16 \qquad \textit{Boltzmann's constant}$$

From Figure 16-8 we see that for real gases $PV/nT$ is very nearly constant over a rather large range of pressures. We define an *ideal gas* to be one for which $PV/nT$ is constant for all pressures. Thus for an ideal gas, the pressure, volume, and temperature are related by

$$PV = nRT \qquad\qquad 16\text{-}17$$

*Ideal-gas law*

The concept of an ideal gas is an extrapolation of the behavior of real gases at low pressures to an ideal behavior. However, as can be seen from Figure 16-8, many gases differ little from an ideal gas at reasonably low pressures. Thus Equation 16-17, called an *equation of state*, is quite useful in describing properties of real gases. By our definition of temperature, to every state $P_0V_0$ there corresponds a temperature $T_0$. Thus there must be some function $T(P, V)$ for every gas, even at high pressures. Instead of thinking of $T$ as a function of the two variables $P$ and $V$, we could equally well think of $P$ as a function of $T$ and $V$ or $V$ as a function of $P$ and $T$. There is always a functional relationship between these three variables; given any two, the third is determined. In the next section we shall take the microscopic view of an ideal gas and calculate the pressure the gas exerts on the walls of its container by considering collisions of the gas molecules with the walls. We shall see that the absolute temperature is a measure of the average kinetic energy of the molecules and that we can derive the ideal-gas law (Equation 16-17) if we assume that the gas molecules exert no forces on each other except during collisions and that the molecules are far apart on the average, which is equivalent to the assumption of molecules of negligible volume. At high pressures, these assumptions break down and the ideal-gas equation 16-17 is not a good approximation to the behavior of a real gas. At high enough pressures and low temperatures, gases liquefy, indicating that when the molecules are close together, they exert attractive forces on each other. An equation of state, known as the *van der Waals equation*, takes these considerations into account and is a better approximation for the behavior of real gases than the ideal-gas law. Van der Waals' equation for 1 mol of a gas is

*Equation of state*

$$\left(P + \frac{a}{V^2}\right)(V - b) = RT \qquad\qquad 16\text{-}18$$

*Van der Waals' equation of state*

where $a$ and $b$ are constants that are determined experimentally for a particular gas. The term $b$ corrects for the fact that the volume available to the molecules is reduced because of the finite size of the molecules. The magnitude of $b$ is the molecular volume of 1 mol of gas molecules. This correction becomes negligible at low pressures when the gas density is low and the molecules are far apart compared with their diameters. The term $a/V^2$ accounts for the reduced pressure due to the attraction of the molecules with each other when the density of the gas is high enough for the molecules to be relatively close together. This term is also negligible at low pressures when the volume $V$ is large. Thus at low pressures van der Waals' equation approaches the ideal-gas law, whereas at high pressures it is a much better description of the behavior of real gases.

Many other equations of state have been proposed for various ranges of pressure for various materials. Although no single equation has been found to be accurate over the complete possible ranges of pressure, volume, and temperature for any material, in thermodynamics it is often important just to know that such a function $T(P, V)$ exists even if the form of the equation is not known.

**Example 16-2** What volume is occupied by 1 mol of gas at standard temperature of 0°C and normal atmospheric pressure?

From Equation 16-9, the absolute temperature corresponding to 0°C is 273 K. Using this and $P = 1$ atm, we find for the volume

$$V = \frac{nRT}{P} = \frac{(1 \text{ mol}) (0.0821 \text{ L·atm/mol·K}) (273 \text{ K})}{1 \text{ atm}} = 22.4 \text{ L}$$

---

**Example 16-3** How many molecules are there in 1 cm³ of gas at 0°C and atmospheric pressure?

Here $P = 1$ atm, $V = 1$ cm³ $= 10^{-3}$ L, and $T = 273$ K. From Equation 16-17

$$n = \frac{PV}{RT} = \frac{(1 \text{ atm}) (10^{-3} \text{ L})}{(0.0821 \text{ L·atm/mol·K}) (273 \text{ K})} = 4.46 \times 10^{-5} \text{ mol}$$

From Equation 16-12

$$N = nN_A = (4.46 \times 10^{-5} \text{ mol}) (6.02 \times 10^{23} \text{ molecules/mol})$$
$$= 2.68 \times 10^{19} \text{ molecules}$$

---

**Question**

9. How does the pressure of an ideal gas change when any one of the following quantities is doubled while the others are kept fixed: (*a*) the absolute temperature, (*b*) the volume, (*c*) the number of molecules?

# 16-7 The Molecular Interpretation of Temperature

We shall now show that the absolute temperature of a gas is a measure of the average kinetic energy of the molecules of the gas. In the microscopic view a gas exerts pressure on the walls of its container because as the molecules collide with the walls, they transfer momentum to the walls. The total change in momentum per second is the force exerted on the walls by the gas. We begin by making the following assumptions:

1. The gas consists of a large number $N$ of molecules making elastic collisions with each other and with the walls of the container.

2. The molecules are separated by distances large compared with their diameters, and they exert no forces on each other except when they collide.

3. In the absence of external forces (we can neglect gravity) there is no preferred position for a molecule in the container, and there is no preferred direction for the velocity vector.

We can neglect collisions the molecules make with each other because, since momentum is conserved, such collisions have no effect on the total momentum in any direction. Let $m$ be the mass of each molecule. If we take the $x$ axis to be perpendicular to the wall, the $x$ component of momentum of a molecule is $+mv_x$ before it hits the wall and $-mv_x$ afterward. The magnitude of the change in momentum during the collision with the wall is $2mv_x$. The total change in the momentum of all the molecules in some time interval $\Delta t$ is $2mv_x$ times the number of molecules that hit the wall during this interval.

Let us consider a rectangular container of volume $V$ with the right wall having area $A$ (Figure 16-9). Let $N_i$ be the number of gas molecules with $x$ component of velocity $v_{xi}$. The number of these molecules hitting the right wall in time $\Delta t$ is the number within a distance $v_{xi} \Delta t$ traveling to the right. Since there are $N_i$ such molecules in volume $V$, the number in volume $v_{xi} \Delta t\, A$ is just $(N_i/V)\,(v_{xi} \Delta t\, A)$. If we assume for the moment that $v_{xi}$ is positive, the number hitting the right wall in time $\Delta t$ is

$$\frac{N_i}{V} v_{xi} \Delta t\, A$$

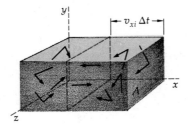

Figure 16-9
Gas molecules in a rectangular container. If a molecule has velocity component $v_{xi}$, it will hit the right face in the time interval $\Delta t$ if it is within the distance $v_{xi} \Delta t$ of the face and if $v_{xi}$ is positive. The number of molecules within this distance of the face is proportional to $v_{xi}$ and to the number of molecules per unit volume.

The impulse exerted by the wall on these molecules equals the total change in their momentum, which is $2mv_{xi}$ times the number that hit:

$$I_i = \left(\frac{N_i v_{xi} \Delta t\, A}{V}\right) 2mv_{xi} = \frac{2N_i m v_{xi}^2 A\, \Delta t}{V}$$

This also equals the magnitude of the impulse exerted by these molecules *on* the wall. We obtain the average force exerted by the molecules by dividing the impulse by the time interval $\Delta t$. The pressure is this average force divided by the area $A$. The pressure exerted by these molecules is thus

$$P_i = \frac{I_i}{\Delta t\, A} = \frac{2N_i m v_{xi}^2}{V}$$

The total pressure exerted by all the molecules is obtained by summing over all the $x$ components of velocity $v_{xi}$ that are positive. Since, on the average, at any time half the molecules will be moving to the right (positive $v_{xi}$) and half to the left (negative $v_{xi}$), we can sum over all the molecules and multiply by $\frac{1}{2}$:

$$P = \tfrac{1}{2} \sum P_i = \frac{1}{2} \sum \frac{2N_i m v_{xi}^2}{V} = \frac{m}{V} \sum N_i v_{xi}^2$$

We can write this in terms of the average value of $v_x^2$, defined to be

$$(v_x^2)_{av} = \frac{1}{N} \sum N_i v_{xi}^2$$

where $N = \Sigma N_i$ is the total number of molecules. Thus we can write for the pressure

$$P = \frac{Nm}{V} (v_x^2)_{av} \qquad \text{or} \qquad PV = 2N(\tfrac{1}{2}mv_x^2)_{av} \qquad\qquad 16\text{-}19$$

Comparing this with the equation of state for an ideal gas,

$$PV = nRT = NkT$$

we have

$$(\tfrac{1}{2}mv_x^2)_{\text{av}} = \tfrac{1}{2}kT \qquad \text{16-20}$$

The average kinetic energy associated with translational motion in the $x$ direction is $\tfrac{1}{2}kT$. Since $x$ was an arbitrary direction, we could write similar equations for $(\tfrac{1}{2}mv_y^2)_{\text{av}}$ and $(\tfrac{1}{2}mv_z^2)_{\text{av}}$:

$$(\tfrac{1}{2}mv_y^2)_{\text{av}} = \tfrac{1}{2}kT \qquad (\tfrac{1}{2}mv_z^2)_{\text{av}} = \tfrac{1}{2}kT$$

Adding these expressions and writing $v^2 = v_x^2 + v_y^2 + v_z^2$, we have

$$(\tfrac{1}{2}mv^2)_{\text{av}} = \tfrac{3}{2}kT \qquad \text{16-21}$$

The absolute temperature is thus a measure of the average translational kinetic energy of the molecules. (We include the word translational because a molecule may have other kinds of kinetic energy, e.g., rotational or vibrational. Only the translational kinetic energy comes into the calculation of the pressure exerted on the walls of the container.) The total translational kinetic energy of $n$ mol of a gas containing $N$ molecules is

$$E_k = N(\tfrac{1}{2}mv^2)_{\text{av}} = \tfrac{3}{2}NkT = \tfrac{3}{2}nRT \qquad \text{16-22}$$

The translational kinetic energy is $\tfrac{3}{2}kT$ per molecule, or $\tfrac{3}{2}RT$ per mole.

We can use these results to estimate the order of magnitude of the kinetic energies and speeds of the molecules in a gas. The average kinetic energy of translation of a molecule is $\tfrac{3}{2}kT$. At a temperature $T = 300$ K $= 27°C$ ($= 81°F$), $\tfrac{3}{2}kT$ has the value

$$\tfrac{3}{2}kT = 6.21 \times 10^{-21} \text{ J}$$

*Average molecular kinetic energy at 300 K*

In terms of the electronvolt energy unit ($1$ eV $= 1.6 \times 10^{-19}$ J) often used in atomic physics, this energy is

$$\tfrac{3}{2}kT = 6.21 \times 10^{-21} \text{ J} = 0.0388 \text{ eV}$$

Thus the average kinetic energy of a molecule in a gas at this temperature is about 0.04 eV. The square root of $(v^2)_{\text{av}}$ is the root-mean-square (rms) speed. The average value of $v^2$ is, by Equation 16-21,

$$(v^2)_{\text{av}} = \frac{3kT}{m} = \frac{3N_A kT}{N_A m} = \frac{3RT}{M}$$

where $M = N_A m$ is the mass of 1 mol, or the molecular mass. Thus the rms speed is

$$v_{\text{rms}} = \sqrt{(v^2)_{\text{av}}} = \sqrt{\frac{3kT}{m}} = \sqrt{\frac{3RT}{M}} \qquad \text{16-23}$$

For oxygen gas at $T = 300$ K we have, using $M = 32 \times 10^{-3}$ kg/mol,

$$v_{\text{rms}} = \sqrt{\frac{3(8.31 \text{ J/mol·K})(300 \text{ K})}{32 \times 10^{-3} \text{ kg/mol}}} = 483 \text{ m/s}$$

## Questions

10. How does the average translational kinetic energy of a molecule in a gas change if the pressure is doubled while the volume is kept constant? If the volume is doubled while the pressure is kept constant?

11. By what factor must the absolute temperature of a gas be increased to double the rms speed of its molecules?

12. Would you expect all the molecules in a body of gas to have the same speed?

13. Two bodies of gas have the same temperature. Do their molecules necessarily have the same rms speeds? Explain.

## Review

A. Define, explain, or otherwise identify:

Microscopic description, 473
Macroscopic description, 473
System in thermal equilibrium
    with itself, 473
Macroscopic state of a gas, 474
State variables, 474
Adiabatic wall, 475
Diathermic wall, 475
Thermal contact, 475
Thermal equilibrium, 475
Zeroth law of thermodynamics, 476
Two systems at the
    same temperature, 476

Normal melting point, 478
Normal boiling point, 478
Triple point of water, 479
Ideal-gas temperature, 480
Kelvin temperature scale, 480
Celsius temperature scale, 481
Fahrenheit temperature scale, 481
Mole, 483
Universal gas constant, 483
Boltzmann's constant, 484
Ideal-gas law, 484
Equation of state, 484

B. True or false:

1. An adiabatic wall is an insulating wall.

2. A diathermic wall is a heat-conducting wall.

3. The zeroth law of thermodynamics results from logical deduction rather than from experiment.

4. Two objects in thermal equilibrium with each other must be in thermal equilibrium with any third object.

5. All thermometers give the same result when measuring the temperature of any particular system.

6. The absolute temperature of a gas is a measure of the average translational kinetic energy of the gas molecules.

## Exercises

### Section 16-1, Macroscopic State Variables; Section 16-2, Adiabatic and Diathermic Walls; and Section 16-3, The Zeroth Law of Thermodynamics

1. A thermometer and pressure gauge are placed in the corner of a 25-L flexible container of gas. At time $t = 0$ the system begins to expand, so that at $t = 5$ s, when the volume is 30 L, the pressure reading is 1.8 atm and the thermometer reads 20.0°C. At $t = 10$ s the system begins to contract, so that at $t = 14$ s the volume is again 30 L; the reading is then 1.8 atm, and the thermometer reads $t = 19$°C. If $P$ and $V$ define a state, explain the discrepancies in the temperature readings.

2. A 75-g gaseous system is in thermal equilibrium with itself in the state $T = 30$°C, $V = 10$ L, and $P = 3$ atm. The volume is increased to 15 L while the pressure is held at 3 atm. Can the temperature be maintained at 30°C? If so, what must be done?

3. In each of the cases below, systems $A$ and $B$ are initially in thermal equilibrium with each other and are connected by a wall. Explain in each case whether the wall must be diathermic or adiabatic or if it can be either. ($a$) When $P_A$ is increased at constant volume, $P_B$ and $V_B$ do not change. ($b$) When $P_A$ is increased at constant volume, $V_B$ increases at constant pressure. ($c$) When $P_A$ is increased while $V_A$ is decreased, $P_B$ and $V_B$ do not change.

4. A class of 30 students meets in a classroom with 30 numbered seats. The teacher gives a macroscopic description of the room one day by saying that 28 of the seats are occupied. Discuss possible microscopic descriptions of the classroom on that day, and show that there are many microscopic "states" corresponding to the macroscopic state of 28 seats occupied.

5. If system $A$ is not in thermal equilibrium with $B$ and $B$ is not in thermal equilibrium with $C$, what conclusions, if any, can be drawn about the temperatures $T_A$, $T_B$, and $T_C$? In particular, can you conclude that $T_A$ and $T_C$ are not equal?

6. If system $A$ is not in thermal equilibrium with system $B$ and system $B$ *is* in thermal equilibrium with system $C$, what conclusions can be drawn about the temperatures $T_A$, $T_B$, and $T_C$?

### Section 16-4, Temperature Scales and Thermometers

7. The length of the column of a mercury thermometer is 4.0 cm when the thermometer is immersed in ice water and 24.0 cm when the thermometer is placed in boiling water. ($a$) What should the length be at room temperature of 22.0°C? ($b$) The mercury column is 25.4 cm long when the thermometer is placed in a boiling chemical solution. What is the temperature of the solution?

8. The pressure of a constant-volume gas thermometer is 0.400 atm at the ice point and 0.546 atm at the steam point. ($a$) When the pressure is 0.100 atm, what is the temperature? ($b$) What is the pressure at the boiling point of sulfur (444.6°C)?

9. A constant-volume gas thermometer reads 50 mmHg pressure at the triple point of water. ($a$) What will the pressure be when the thermometer measures a temperature of 300 K? ($b$) What ideal-gas temperature corresponds to a pressure of 68 mmHg?

10. A constant-volume gas thermometer has a pressure of 30 mmHg when it reads a temperature of 373 K. ($a$) What is its triple-point pressure $P_3$? ($b$) What temperature corresponds to a pressure of 0.175 mmHg?

### Section 16-5, The Celsius and Fahrenheit Scales

11. A certain ski wax is rated for use between −12 and −7°C. What is this temperature range on the Fahrenheit scale?

12. The boiling point of $O_2$ is −182.86°C. Express this temperature in ($a$) kelvins and ($b$) degrees Fahrenheit.

13. The boiling point of tungsten is 5900°C. Find this temperature in ($a$) kelvins and ($b$) degrees Fahrenheit.

14. The highest and lowest temperatures ever recorded in the United States are 134°F (in California in 1913) and −80°F (in Alaska in 1971). Express these temperatures on the Celsius scale.

15. What is the Celsius temperature corresponding to the normal temperature of the human body of 98.6°F?

16. In September 1933 in Portugal, the temperature rose to 70°C for 2 min. What is that temperature in degrees Fahrenheit?

17. The temperature of the interior of the sun is about $10^7$ K. What is this temperature (a) on the Celsius scale and (b) on the Fahrenheit scale?

## Section 16-6, Equations of State: The Ideal Gas

18. (a) If 1 mol of a gas occupies a volume of 10 L at a pressure of 1 atm, what is the temperature of the gas? (b) The container is fitted with a piston so that the volume can change. The gas is heated at constant pressure and expands to a volume of 20 L. What is its temperature in kelvins? In degrees Celsius? (c) The volume is now fixed at 20 L and the gas is heated at constant volume until its temperature is 350 K. What is its pressure?

19. A container fitted with a piston holds 1 mol of gas, and the initial pressure and temperature are 2 atm and 300 K. (a) What is the initial volume of the gas? (b) The gas is allowed to expand at constant temperature until the pressure is 1 atm. What is the new volume? (c) The gas is now compressed and heated at the same time until it is back to its original volume, at which time the pressure is 2.5 atm. What is its temperature then?

20. A cylindrical container with a radius of 2.5 cm and height of 20 cm is open at the top. It contains air at 1 atm pressure. A tightly fitting piston weighing 9 N is inserted and gradually lowered until the increased pressure in the container supports the weight of the piston. (a) What is the force exerted on the top of the piston due to atmospheric pressure? (b) What is the force that must be exerted by the gas in its container below the piston to keep it in equilibrium? What is the pressure in the container? (c) Assuming that the temperature of the gas in the container does not change, what is the height of the equilibrium position of the piston?

21. A gas is kept at constant pressure. If its temperature is changed from 50 to 100°C, by what factor does its volume change?

22. A pressure as low as $1 \times 10^{-8}$ mmHg can be achieved by an oil-diffusion pump. How many molecules are there in 1 cm$^3$ at this pressure if the temperature is 300 K?

23. A cubic metal box of sides 20 cm contains air at pressure of 1 atm and temperature 300 K. It is sealed so that the volume is constant, and heated to a temperature of 400 K. Find the net force on each wall of the box.

24. The mass of Avogadro's number of $^{12}$C atoms is 12 g. Each $^{12}$C atom contains six protons and six neutrons. Since the proton and neutron have approximately equal masses the mass of Avogadro's number of protons or neutrons is 1 g. The mass of the proton or neutron (in grams) is thus approximately the reciprocal of Avogadro's number. Calculate this mass and compare your answer with the mass given in Appendix Table B-1.

25. A 10-L vessel contains gas at 0°C and a pressure of 4 atm. How many moles of gas are there in the vessel? How many molecules?

## Section 16-7, The Molecular Interpretation of Temperature

26. Show that $\sqrt{3RT/M}$ has the correct units of metres per second if $R$ is in joules per mole-kelvin, $T$ is in kelvins, and $M$ is in kilograms per mole.

27. Find $v_{\text{rms}}$ of an argon atom if 1 mol of the gas is confined to a 1-L container at 10 atm. ($M = 40 \times 10^{-3}$ kg/mol.) Compare this with that of helium gas under the same conditions (the molecular mass of helium is $4 \times 10^{-3}$ kg/mol).

28. Find the total translational kinetic energy of 1 L of oxygen gas held at a temperature of 0°C and at atmospheric pressure.

29. At what temperature will the rms speed of an H$_2$ molecule equal 332 m/s?

## Problems

1. Gas is confined in a steel cylinder at 20°C and at a pressure of 5 atm. (a) If the cylinder is surrounded by boiling water and allowed to come to thermal equilibrium, how great will the gas pressure be? (b) If the gas is then allowed to escape until the pressure is again 5 atm, what fraction of the original gas (by weight) will escape? (c) If the temperature of the remaining gas in the cylinder is returned to 20°C, what will its final pressure be?

2. A constant-volume gas thermometer with a triple-point pressure of $P_3 = 500$ mmHg is used to measure the boiling point of some substance. When it is placed in thermal contact with the boiling substance, its pressure is 734 mmHg. Some of the gas in the thermometer is then allowed to escape so that its triple-point pressure is 200 mmHg. When it is placed in thermal contact with the boiling substance, its pressure is 293.4 mmHg. Again, some of the gas is removed from the thermometer so that its triple-point pressure is 100 mmHg. Its pressure when in contact with the boiling substance is then 146.65 mmHg. Find the ideal-gas temperature of the boiling substance to one place beyond the decimal point.

3. An automobile tire is filled to a gauge pressure of 200 kPa when it is at an air temperature of 20°C. After the car has been driven at high speed, the tire temperature has increased to 50°C. (a) Assuming that the volume of the tire has not changed, find the gauge pressure of the air in the tire (assume air to be an ideal gas). (b) Calculate the gauge pressure if the tire expands so that the volume increases by 10 percent.

4. (a) If 2 mol of hydrogen gas ($M = 2$ g/mol) is at atmospheric pressure and room temperature (20°C), what is the average translational kinetic energy of a hydrogen molecule in this gas? (b) What is $v_{rms}$ for the hydrogen molecules? (c) What is the total kinetic energy of translation of all the molecules in this amount of hydrogen gas? (d) If a solid object of the same mass had this kinetic energy, what would its speed be?

5. A certain gas is contained at a pressure of 300 kPa. If the density of the gas is 3.5 g/L, calculate the rms speed of the gas molecules.

6. At the surface of the sun the temperature is about 6000 K, and all materials are gaseous. From data given by the spectrum of light it is known that most elements are present on the sun. (a) What is the average kinetic energy of translation of an atom at the surface of the sun? Express your answer in both joules and electronvolts. (b) What is the range of $v_{rms}$ at the surface of the sun if the atoms present range from hydrogen ($M = 1$ g/mol) to uranium ($M = 238$ g/mol)?

7. In Chapter 7 it was shown that the escape speed for a planet of radius $R$ and acceleration of gravity $g$ at the surface is given by $v_e^2 = 2Rg$. (a) At what temperature does $v_{rms}$ for $O_2$ equal the escape speed for the earth? (b) At what temperature is $v_{rms}$ for $H_2$ the escape speed for the earth? (c) Temperatures in the upper atmosphere reach about 1000 K. Can this account for the low abundance of hydrogen in the earth's atmosphere? (d) Compute the corresponding temperatures for the moon, where $g$ has about one-sixth its value on the earth and $R = 1738$ km. Can you account in this way for the absence of an atmosphere on the moon?

8. A thermistor is a solid-state device whose resistance varies greatly with temperature. Its temperature dependence is given approximately by $R = R_0 e^{B/T}$, where $R$ is in ohms, $T$ is in kelvins, and $R_0$ and $B$ are constants which can be determined by measuring $R$ at calibration points such as the ice point and the steam point. (a) If $R = 7360$ ohms at the ice point and 153 ohms at the steam point, find $R_0$ and $B$. (b) What is the resistance of the thermistor at $t = 98.6°F$? (c) What is the rate of change of resistance with temperature ($dR/dT$) at the ice point and at the steam point? At which of these temperatures is this thermometer more sensitive?

9. The van der Waals constants for helium are $a = 0.03412$ L²·atm/mol and $b = 0.0237$ L/mol. Use these data to find the volume in cubic centimetres occupied by one helium molecule and to estimate the radius of the molecule.

10. At 127°C, 2.4 mol of a certain gas occupies a 4-L container at a density of $1.68 \times 10^{-2}$ g/cm³. (*a*) Find the total translational kinetic energy. (*b*) Find the average kinetic energy per molecule. (*c*) What is the pressure of the gas? (*d*) What is the molecular mass of the gas? (*e*) What diatomic gas could this be?

11. Ten objects have the speeds tabulated:

| Speed, m/s | 2 | 5 | 6 | 8 |
|---|---|---|---|---|
| Number of objects | 3 | 3 | 3 | 1 |

Calculate their rms speed.

ɟ

**CHAPTER 17**     Heat, Work, and the First
Law of Thermodynamics

---

**Objectives**   After studying this chapter, you should:

1.   Be able to define the terms heat capacity, adiabatic work, internal energy, and quasi-static process.

2.   Be able to work calorimetry problems.

3.   Be able to state the first law of thermodynamics and use it in solving problems.

4.   Be able to calculate the work done by a gas during various quasi-static processes and sketch the processes on a $PV$ diagram.

5.   Be able to state the Dulong-Petit law and use it to estimate the heat capacity of a given solid or to calculate the molecular mass of a solid from its specific heat.

6.   Be able to state the equipartition theorem and use it to relate the molar heat capacity of a gas to a mechanical model of the gas molecules and to derive the Dulong-Petit law.

---

When two systems at different temperatures are placed in thermal contact, heat flows from the hotter system to the colder one until the two systems reach equilibrium at a common temperature. In the seventeenth century, Galileo, Newton, and other scientists generally supported the theory of the ancient Greek atomists, who considered heat to be a manifestation of molecular motion. In the next century, methods were developed for making quantitative measurements of the amount of heat that leaves or enters the body, and it was found that often when two bodies are in thermal contact, the amount of heat that leaves one body equals the amount of heat that enters the other body. This discovery led to the development of an apparently successful theory of heat as a conserved material substance—an invisible, weightless fluid, called *caloric*, that was neither created nor destroyed but merely transferred from one body to another. The caloric theory of heat served quite well in the description of heat transfer but eventually was discarded when it was observed that caloric apparently could be created endlessly by friction with no corresponding disappearance of caloric somewhere else. In other words, the principle of conservation of caloric, which had been the experimental foundation of this theory of heat, proved to be false.

The first clear experimental observations showing that caloric cannot be conserved were made at the end of the eighteenth century by Benjamin Thompson, an American loyalist who emigrated to Europe, became the director of the Bavarian arsenal, and was given the title Count Rumford. Thompson supervised the boring of cannon for Bavaria. Because of the heat generated by the boring tool, water was used for cooling. It had to be replaced continually because it boiled away during the boring. According to the caloric theory, as the metal from the bore was cut into small chips, its ability to retain caloric was decreased. Therefore it released caloric to the water, heating it and causing it to boil. Thompson noticed, however, that even when the drill was too dull to cut the metal, the water still boiled away as long as the drill was turned. Apparently caloric was produced merely by friction and could be produced endlessly. He suggested that heat is not a substance that is conserved but some form of motion communicated from the bore to the water. He showed, in fact, that the heat produced is approximately proportional to the work done by the boring tool.

The caloric theory of heat continued to be the leading theory for some 40 years after Thompson's work, but it was gradually weakened as more and more examples of nonconservation of heat were observed. Not until the 1840s did the modern mechanical theory of heat emerge. In this view heat is another form of energy, exchangeable at a fixed rate with the various forms of mechanical energy already discussed. The most varied and precise experiments demonstrating this were performed, starting in the late 1830s, by James Joule (1818–1889), after whom the SI unit of energy is named. Joule showed that the appearance or disappearance of a given quantity of heat is always accompanied by the disappearance or appearance of an equivalent quantity of mechanical energy. The experiments of Joule and others showed that neither heat nor mechanical energy is conserved independently but that the mechanical energy lost always equals the heat produced (measured in the same units). What is conserved is the total of mechanical energy and heat energy.

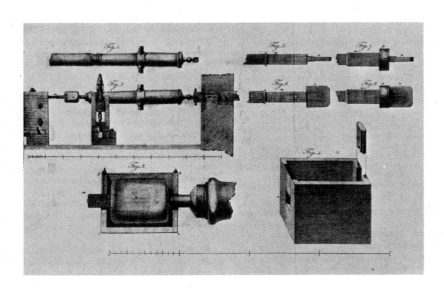

Rumford's apparatus. The drawing at bottom left labeled Figure 3 shows a close-up of the borer (faintly dotted blunt shape projecting into the box at left) within the metal being bored. The faint vertical dotted lines indicate a hole for the insertion of a thermometer. The box in Figure 4 (at the bottom right) is seen in cross section in Figure 3; it contained water and was placed around the metal and borer. It was used in later experiments to counteract the criticism that air somehow had caused the temperature increase. (*From* Philosophical Transactions of the Royal Society, *vol. 88, 1798.*)

# 17-1 Heat Capacity and Specific Heat

The amount of heat energy needed to raise the temperature of a body by 1 degree is called the heat capacity $C$ of the body. The historical unit of heat energy, the calorie, was originally defined to be the amount of heat energy needed to raise the temperature of one gram of water one Celsius degree. After careful measurement showed that this energy depends slightly on the temperature of the water, the calorie was defined more precisely to be the amount of energy needed to raise the temperature of one gram of water one degree at a specified temperature (usually 14.5 to 15.5°C). The kilocalorie is then the amount of heat energy needed to raise the temperature of one kilogram of water by one degree Celsius. (The "calorie" used in measuring the energy equivalent of foods is actually the kilocalorie.) The U.S. customary unit of heat is the Btu (for British thermal unit), which was originally defined to be the amount of energy needed to raise the temperature of one pound of water by one degree Fahrenheit (between 63 and 64°F). Since we now recognize that heat energy is just another form of energy, i.e., the energy transferred due to a temperature difference, we do not need any special units for heat that differ from other energy units. The calorie is now defined in terms of the SI unit of energy, the joule, as

$$1 \text{ cal} = 4.184 \text{ J} \qquad \qquad 17\text{-}1 \qquad \textit{Calorie defined}$$

The Btu is related to the calorie and the joule by

$$1 \text{ Btu} = 252 \text{ cal} = 1.054 \text{ kJ} \qquad \qquad 17\text{-}2$$

The heat capacity of an object is proportional to its mass. The heat capacity per unit mass* is called the specific heat $c$:

$$c = \frac{C}{m} \qquad \qquad 17\text{-}3$$

Figure 17-1 shows the specific heat of water as a function of temperature. We note that the specific heat of water varies by only about 1 percent from 0 to 100°C at a pressure of 1 atm. We can usually neglect this variation and take the specific heat of water to be 4.184 kJ/kg·K ($= 1.00$ kcal/kg·K) over the whole range from 0 to 100°C.

The heat capacity per mole is called the *molar heat capacity* $C'$. The molar heat capacity equals the heat capacity per unit mass (specific heat) times the mass $M$ of 1 mol (the molecular mass):

$$C' = Mc \qquad \qquad 17\text{-}4 \qquad \textit{Molar heat capacity}$$

---

\* In U.S. customary units the specific heat is the heat capacity per unit weight. It should be evident that the specific heat of water in these units and in the historical units is 1 cal/g·°C = 1 kcal/kg·°C = 1 Btu/lb·°F.

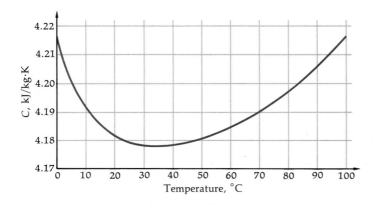

**Figure 17-1**
Specific heat of water versus temperature.

Specific heats and molar heat capacities have been determined for many substances and are listed in tables. The heat energy $Q$ which is transferred into or out of a system is related to the temperature change $\Delta T$ of the system by

$$Q = C\,\Delta T = mc\,\Delta T \qquad\qquad 17\text{-}5$$

The temperature change is positive for heat put into the system and negative for heat taken out of the system.

Because the specific heat of water varies so slightly with temperature, the heat capacity or specific heat of an object can be measured conveniently by heating it to some easily measured temperature, placing it in a water bath of known mass and temperature, and measuring the final equilibrium temperature. If the whole system is isolated from its surroundings, the heat leaving the body equals the heat entering the water. Let $m$ be the mass of the body, $c$ be its specific heat, and $T_b$ its initial temperature. Let $m_w$ be the mass of the water bath, $c_w = 4.184$ kJ/kg·K be its specific heat, $T_w$ its initial temperature, and $T_f$ the final equilibrium temperature of the system. The heat absorbed by the water is

$$Q_{in} = m_w c_w (T_f - T_w)$$

and that lost by the body is

$$Q_{out} = mc(T_b - T_f)$$

Note that we have chosen to write the temperature differences so that both $Q_{in}$ and $Q_{out}$ are positive quantities. (We know that the final temperature $T_f$ will be greater than the initial temperature of the water bath $T_w$ and less than the initial temperature of the body.) Since these amounts of heat are equal, the specific heat $c$ of the body can be calculated from

$$mc(T_b - T_f) = m_w c_w (T_f - T_w) \qquad\qquad 17\text{-}6$$

Because only differences in temperature are important in Equation 17-6, and because the kelvin and the Celsius degree are the same size, the temperatures can be measured on either the Celsius or Kelvin scale without affecting the result.

---

**Example 17-1** When 100 g of aluminum shot is heated to 100°C and placed in 500 g of water initially at 18.3°C, the final equilibrium temperature of the mixture is 21.7°C. What is the specific heat of aluminum?

Since the temperature change of the water is 21.7°C − 18.3°C = 3.4°C = 3.4 K the heat absorbed by the water is

$$Q_w = m_w c_w\,\Delta T_w = (0.5\text{ kg})\,(4.18\text{ kJ/kg·K})\,(3.4\text{ K}) = 7.11\text{ kJ}$$

The temperature change of the aluminum is 100°C − 21.7°C = 78.3 K, and the heat given off by the aluminum is

$$Q_{Al} = mc\,\Delta T_{Al} = (0.1\text{ kg})\,(c)\,(78.3\text{ K}) = 7.11\text{ kJ}$$

Solving for $c$ gives

$$c = \frac{7.11\text{ kJ}}{(0.1\text{ kg})\,(78.3\text{ K})} = 0.908\text{ kJ/kg·K}$$

The specific heat of aluminum (and of most materials) is considerably less than that of water. In practice, it is useful to choose the initial temperature of the water bath (using a rough knowledge of the specific heat to be measured) so that the final and initial temperatures of the water differ from room temperature by the same amount. Then, even if the water is not perfectly isolated from its surroundings, it will gain about as much heat from the surroundings when it is below room temperature as it loses when it is above room temperature. For Example 17-1 the initial temperature was chosen for a room temperature of 20°C (= 68°F).

In general, the amount of heat needed to raise the temperature of a body at constant pressure is not the same as that needed at constant volume. We must therefore distinguish between the specific heat at constant pressure $c_p$ and the specific heat at constant volume $c_v$. For solids and liquids this difference amounts to a few percent or less and can be ignored except in the most precise experiments. For gases, this difference is significant and will be discussed in detail in Section 17-5.

Table 17-1 lists the specific heats and molar heat capacities of some solids and liquids. Note that the molar heat capacity is about the same for all the metals, $C' \approx 3R = 24.9$ J/mol·K, a result known as the *Dulong-Petit law*. We shall discuss the significance of this law and the fact that at low temperatures the molar heat capacities of metals are all less than this value in Section 17-5. Figure 17-2 shows the temperature dependence of $C_v'$ for iron. The shape of this curve is similar for all solids. The temperature at which $C_v'$ approaches the value $3R$ differs for different materials.

*Dulong-Petit law*

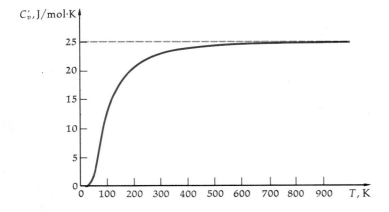

**Figure 17-2**
Molar heat capacity versus temperature for iron. All solids show this characteristic behavior, the values of the temperature scale varying from one material to another. At high temperatures the heat capacity is $3R \approx 25$ J/mol·K, the Dulong-Petit law.

**Questions**

1.  A large body of water, such as a lake or an ocean, tends to moderate the variations of temperature near it. Why? Would this be true if the specific heat of water had a value closer to that of other common substances?

2.  Explain the circumstances under which the following situations could arise. (*a*) A massive body at high temperature is brought into thermal contact with a light object at low temperature. The final equilibrium temperature is closer to the original temperature of the massive body. (*b*) A massive body at high temperature is brought into thermal contact with a light body at a low temperature. The final temperature is closer to the original temperature of the light body.

**Table 17-1**
Specific heat and molar heat capacity for various solids and liquids at 20°C

| Substance | $c$ | | $C'$, J/mol·K |
| | kJ/kg·K | kcal/kg·K or Btu/lb·°F | |
| --- | --- | --- | --- |
| Aluminum | 0.900 | 0.215 | 24.3 |
| Bismuth | 0.123 | 0.0294 | 25.7 |
| Copper | 0.386 | 0.0923 | 24.5 |
| Gold | 0.126 | 0.0301 | 25.6 |
| Ice (−10°C) | 2.05 | 0.49 | 36.9 |
| Lead | 0.128 | 0.0305 | 26.4 |
| Silver | 0.233 | 0.0558 | 24.9 |
| Tungsten | 0.134 | 0.0321 | 24.8 |
| Zinc | 0.387 | 0.0925 | 25.2 |
| Alcohol (ethyl) | 2.4 | 0.58 | 111 |
| Mercury | 0.140 | 0.033 | 28.3 |
| Water | 4.18 | 1.00 | 75.2 |

3. Body *A* has twice the mass and twice the specific heat of body *B*. If they are supplied with equal amounts of heat, how do their temperature changes compare?

## 17-2   The First Law of Thermodynamics

If we add heat to a mass of water, we raise its temperature, but we can also raise the temperature of the water by stirring it or doing work on it in some other way without adding heat. Figure 17-3 is a schematic illustration of Joule's most famous experiment to determine the amount of work required to produce a given amount of heat, i.e., the amount of work needed to raise the temperature of 1 g of water by 1°C. Once the experimental equivalence of heat and energy has been established, one can describe Joule's work as determining the size of the calorie in the usual energy units or as measuring the specific heat of water. The water is enclosed by adiabatic walls so that its surroundings cannot affect its temperature by heat conduction. The weights, falling at constant speed, turn a paddle wheel, which does work against the water. If no energy is lost through friction in the bearings, etc., the work done by the paddle wheel against the water equals the loss in mechanical energy. The latter is just the potential energy lost as the weights fall through some measured distance. The result of Joule's experiment and of many others after him is that it takes about 4.18 units of mechanical work (joules) to raise the temperature of 1 g of water by 1 Celsius degree. This result, that 4.18 J of mechanical energy is equivalent to 1 cal of heat energy,

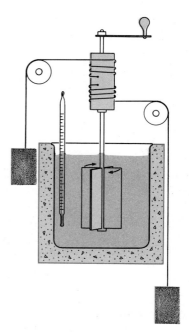

**Figure 17-3**
Schematic diagram of one of Joule's experiments to determine the amount of work required to produce a given increase in the temperature of water. The work done is determined by the loss in potential energy of the weights.

was known as the *mechanical equivalence of heat*. Historically, it was customary to express heat energy in calories and then use the mechanical equivalent of heat to convert to the standard units of mechanical energy. Today, for all forms of energy we use the same units, usually the SI unit, the joule.

There are other ways of doing work on a system. In Figure 17-4 an electric resistor is placed in a water bath and connected to a generator driven by a falling weight. (This is another of Joule's early experiments.) The water as a system is not thermally insulated because it is in thermal contact with the resistor and can absorb heat from it. If we consider the resistor plus water as the thermodynamic system, this system is shielded from its surroundings by an adiabatic wall. No heat can be conducted into this system. The electrical work needed to raise the temperature of the water by some amount is the same as the mechanical work needed to raise the temperature by the same amount.

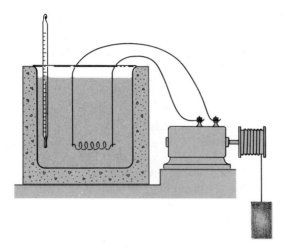

**Figure 17-4**
Another method of doing work on an adiabatically shielded system consisting of water plus resistor. Electrical work is done on the system by the generator, which is driven by the falling weight.

This type of experiment can be performed on various thermodynamic systems such as gases, solids, etc. We find that if the system is adiabatically shielded from its surroundings, the work required to change the system from state 1 to state 2 is *independent of how the work is performed* (stirring, compressing and expanding, electrical work, etc.) and depends only on the initial and final states. This experimental result expresses one form of the first law of thermodynamics.

*The adiabatic work done on a system in taking the system from state 1 to state 2 is independent of the manner of doing the work and depends only on the initial and final states of the system.*

*First law of thermodynamics*

*Adiabatic work* means work done on a system that neither gains nor loses heat to its surroundings while the work is done; i.e., it is work done on a system surrounded by adiabatic walls. Since the adiabatic work depends on the initial and final states of the system, we can define a function $U$ of the state of the system such that the change in $U$ equals the adiabatic work done. This new state function is called the *internal energy* of the system:

$$U_2 - U_1 = \text{adiabatic work done } on \text{ a system} \\ \text{to bring it from state 1 to state 2}$$

17-7

*Internal energy defined*

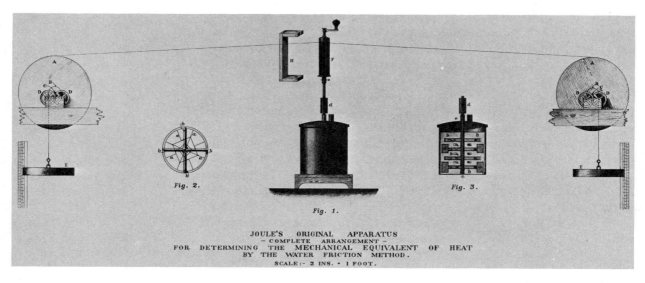

JOULE'S    ORIGINAL    APPARATUS
– COMPLETE    ARRANGEMENT –
FOR    DETERMINING    THE    MECHANICAL    EQUIVALENT    OF    HEAT
BY    THE    WATER    FRICTION    METHOD.
SCALE :– 3 INS. • 1 FOOT.

*The complete arrangement of Joule's original apparatus for demonstrating the equivalence of heat and mechanical energy. (Crown copyright; courtesy of the Science Museum, London.)*

If the system is not adiabatically shielded from its surroundings, the work done on the system as it goes from state 1 to state 2 *does* depend on the process involved. For example, a given temperature increase of the system might be accomplished with no work done, by placing the system in thermal contact with another system at a higher temperature. However, if we also measure the heat put into the system, the sum of the net heat put in plus the work done on the system going from state 1 to state 2 is the same for all processes and equal to the adiabatic work necessary to bring the system from state 1 to state 2 (assuming, of course, that the heat is expressed in energy units). Thus the sum of the heat put into a system plus the work done on the system equals the increase in internal energy of the system for any process. It is customary to write $W$ for the work done by the system on its surroundings. Then $-W$ is the work done *on* the system by the surroundings. For example, if a gas expands against a piston, $W$ is positive; if the gas is compressed, $W$ is negative. Also, the heat put *into* a system is usually written $Q$. Then we have

$$Q + (-W) = \Delta U \qquad\qquad 17\text{-}8$$

—usually written

$$Q = \Delta U + W \qquad\qquad 17\text{-}9$$

*First law of thermodynamics*

Equation 17-9 is the generalization of our first statement of the first law to include nonadiabatic work. It is the most common way of stating the first law of thermodynamics. We see that the first law is a statement of conservation of energy.

*The (heat) energy put into a system equals the sum of the work done by the system and the change in internal energy of the system.*

Equation 17-9 is sometimes taken to be the definition of heat. This point deserves elaboration because, like Newton's laws of motion, the first law of thermodynamics is often misunderstood. From experiments with adiabatic work alone we can define a state function, the internal energy of the system. We can thus prove the existence of the function $U$ with no reference to heat at all. (Recall that we have already defined what is meant by an adiabatic wall without having to define tempera-ture.) The first law of thermodynamics states that the state function *U* exists. This statement of the first law is analogous to our statement of the zeroth law, namely, that a state function called the temperature exists. Having shown that *U* exists and given a method of measuring Δ*U* for any two states (one must shield the system adiabatically from its surroundings and measure the work needed to bring the system from state 1 to state 2), we find that if the system is not adiabatically shielded, the work done on the system does not equal the change in *U*. We can then define the heat put into the system as the difference between the change in *U* and the work done on the system. We find that heat de-fined in this way has all the characteristics we have already discussed. For example, 4.18 J of heat will raise the temperature of 1 g of water by 1 Celsius degree if no work is done.

We shall usually refer to Equation 17-9 as the statement of the first law of thermodynamics. However, we should always keep in mind that the real "content" of this law is that a state function *U* exists and can be found from Equation 17-9 or by measuring adiabatic work.

Closeup view of the water-containing section of Joule's apparatus showing the paddle wheel (right) and its adiabatic container (left). (*Loaned to the Science Museum, London, by Dr. Joule, Manchester.*)

### Questions

4. The experiments of Joule discussed in this section involved the con-version of mechanical energy into heat. Can you give examples in which heat is converted into mechanical energy?

5. Can a system absorb heat with no change in its internal energy?

6. The first law of thermodynamics is sometimes stated as follows: The total energy, including heat energy, of an isolated system is conserved. How is this statement related to $Q = \Delta U + W$?

## 17-3  Work and the *PV* Diagram for a Gas

Consider a gas which can be expanded or compressed by the motion of a tightly fitting frictionless piston, as shown in Figure 17-5. When the piston moves, the volume of the gas changes and either the pressure or temperature or both must change since these three variables are related by an equation of state (such as $PV = nRT$ for an ideal gas). If the piston moves suddenly, the gas will not be in an equilibrium state just after-ward. For example, if we suddenly push the piston in to compress the gas, the pressure near the piston will be greater than that far from the piston initially. Until the gas settles down, we cannot define equilib-rium macroscopic state variables for the gas such as *T*, *P*, or *U*. How-ever, if we move the piston slowly in small steps, waiting after each step for equilibrium to be reestablished, we can compress or expand the gas in such a way that the gas is never far from an equilibrium state. This kind of a process is called a *quasi-static process*. In a quasi-static ex-pansion, the piston moves infinitely slowly with no acceleration, so that the gas moves through a series of equilibrium states. In practice, it is possible to approximate quasi-static processes fairly well.

Let us begin with a gas at a fairly high pressure and let it expand quasi-statically. The force exerted by the gas on the piston is *PA*, where *A* is the area of the piston and *P* is the gas pressure, which in general

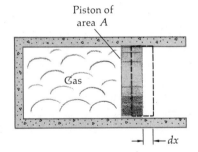

Piston of area *A*

**Figure 17-5**
Gas confined in an adiabati-cally shielded cylinder with movable piston. When the piston moves a distance *dx*, the volume changes by $dV = A\,dx$. The work done by the gas is $PA\,dx = P\,dV$, where *P* is the pressure.

*Quasi-static process*

changes as the gas expands (the pressure could be kept constant by heating the gas as it expands). Since in a quasi-static expansion the piston does not accelerate, there must be an external force pushing against the piston also equal to $PA$. The piston does work on the agent providing this external force. If the piston moves a distance $dx$, the work done *by the gas* is

$$dW = F\ dx = PA\ dx = P\ dV \qquad\qquad 17\text{-}10$$

where $dV = A\ dx$ is the increase in volume of the gas. Equation 17-10 holds also for a quasi-static compression. Then $dV$ is negative, indicating that the gas does negative work. In order to calculate the work done by the gas for an expansion from volume $V_1$ to volume $V_2$ we need to know how the pressure varies during the expansion; i.e., we cannot integrate $P\ dV$ to find the total work unless we know the function $P(V, T)$, which in general is not constant. The states of the gas can be represented on a $P$-versus-$V$ diagram as in Figure 17-6. Since any point $(P_0, V_0)$ determines the thermodynamic state of the gas, each point on this diagram corresponds to a definite state. For example, the set of states corresponding to a constant pressure of 1 atm is represented by a horizontal line parallel to the $V$ axis. *The work done by the gas in expanding a small amount $dV$ is $P\ dV$, represented by the area under the P-versus-V curve as shown.*

Consider the gas initially in the state $P_1V_1$ and expanding to the final state $P_2V_2$. Let us suppose that these two states are at the same temperature. (If the gas is ideal, $P_1V_1 = P_2V_2$.) Several possible kinds of expansion are illustrated in Figure 17-7 by different curves connecting these states. Along path $A$ (Figure 17-7$a$), the pressure is kept constant until the gas reaches the volume $V_2$, and then the volume is kept constant while the pressure is reduced to its final value $P_2$. The work done along this path is $P_1(V_2 - V_1)$ for the horizontal (constant-pressure) part of the path and zero along the constant volume. This work is indicated by the shaded area under the curve. (In order to reduce the pressure at constant volume from $P_2$ to $P_1$, the gas must be allowed to cool; i.e., heat is transferred *from* the gas.) Along path $B$ (Figure 17-7$b$), the gas is first cooled at constant volume to pressure $P_2$ and then allowed to expand at constant pressure until it reaches volume $V_2$. Clearly, the work done along this path is less than that along path $A$, as indicated by the area under the curve for this path. The work along path $B$ is $P_2(V_2 - V_1)$. Along path $C$ (Figure 17-7$c$) both the pressure and volume change for each part of the path. The work done is again the area under this curve connecting the initial and final states. We cannot compute this work unless we know the path the system takes from state 1 to state 2, which is the function $P(V)$.

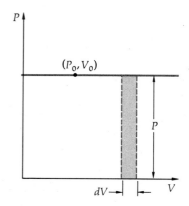

**Figure 17-6**
A particular state $P_0V_0$ is represented by a point on the $PV$ diagram. The work done by a gas as it expands is represented by the shaded area $P\ dV$.

**Figure 17-7**
Three paths in the $PV$ diagram connecting an initial state $P_1V_1$ and a final state $P_2V_2$. ($a$) Along path $A$, the gas first expands at constant pressure from volume $V_1$ to volume $V_2$, and then the pressure is reduced (by cooling) at constant volume. The work done is the shaded area. ($b$) Along path $B$, the gas is first cooled at constant volume to reduce the pressure from $P_1$ to $P_2$ and then the gas expands at constant pressure. The work done, as indicated by the shaded area, is less than that done along path $A$. ($c$) Along path $C$, both the pressure and volume vary. Again, the work done is the shaded area.

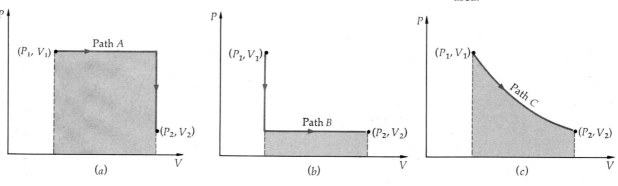

It should be clear that the work done going from state 1 to state 2 depends on how the gas goes from one state to the other. Thus, even though we have written *dW* for a small amount of work, there is no function *W* that is determined by the state. That is, *dW* does not stand for the differential of any function. In this case the work done by a system depends on the path in the *PV* diagram connecting the initial and final states. Thus the work cannot be a function of the state. Consider, for example, the system going from state 1 to state 2 along path *A* in Figure 17-8 and returning to state 1 along path *B*. Although the system has returned to its original state, the net work is not zero. It is, in fact, the difference between the area under curve *A* and the area under curve *B*. This difference is just the area enclosed by the total path of the cycle in the *PV* diagram. This work is positive if the system moves clockwise around the path.

According to the first law of thermodynamics, the internal energy of the gas *is* a state function. Thus for the cyclic path indicated in Figure 17-8, the net change in internal energy must be zero since the system is in its original state at the end. Thus an amount *Q* of heat equal to the work done must have been added to the gas during the cycle. If the gas traverses a cyclic path in the opposite direction (counterclockwise), the system does negative work and must give off an equal amount of heat. Since the work done *does* depend on the path taken, while the internal energy change does not, the heat added must also depend on the path. In our example of a cyclic path, the net heat added or subtracted was not zero but equal in magnitude to the work done. Thus heat is *not* a function of state either. When we write *dQ* for a small amount of heat added to the system, we must remember that this is not the differential of a function *Q*. No such function indicating the heat in a system in each state exists. Similarly, we can use *dW* to denote a small amount of work done by a system, but we must remember that there is no such function *W* representing the "work" in a system. To calculate either the work done by the system or the heat added to the system, we must know the kind of process, i.e., the path taken by the system on the *PV* diagram. For systems other than gases, the work done may take a different form because variables other than *P* and *V* define the state of the system. However, our discussion of work and heat for gases applies to all thermodynamic systems. Neither work nor heat is a function of the state of a system, but the internal energy *U* is a function of the state of the system. The first law says that the sum of the heat added to a system and the work done *on* a system equals the change in internal energy of the system, $\Delta U$, which does not depend on how the system goes from one state to another but only on the initial and final states.

Note that it is *not* correct to say that a system contains a lot of heat when it is hot (or that it contains a lot of work). We can increase the temperature of a system by adding heat, or by doing work on the system, or by some combination of both. It *is* correct to say that a hot system contains a lot of internal energy, because internal energy is a property of the state of a system.

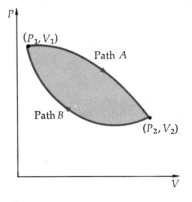

**Figure 17-8**
A gas goes from state 1 to state 2 along path *A* and returns to its original state along path *B*. The net work done for the complete trip is indicated by the shaded area. Since this work is not zero, work is not a state function. Since the net change in internal energy for the complete trip must be zero, an amount of heat equal to the work done must be added. Heat is not a state function either.

---

**Example 17-2**  One litre of an ideal gas is at a pressure of 2 atm and temperature 300 K. It expands at constant pressure until its volume is 2 L, after which it is cooled at constant volume until its pressure is 1 atm. It is then compressed at constant pressure until its volume is again 1 L. It is then heated at constant volume until it is back in its original state. Find the total work done by the gas for this cycle, and the total heat added.

This cyclic process is shown in Figure 17-9. As the gas expands from point $A$ to point $B$, it does work

$$W = \int P \, dV = P \int dV = P(V_B - V_A)$$

$$= (2 \text{ atm}) (2 \text{ L} - 1 \text{ L}) = 2 \text{ L·atm}$$

Since pressures are often given in atmospheres and volumes in litres, it is convenient to have a conversion factor between litre-atmospheres and joules:

$$1 \text{ L·atm} = (10^{-3} \text{ m}^3) (101.3 \times 10^3 \text{ N/m}^2) = 101.3 \text{ J} \qquad \text{17-11}$$

The work done by the gas as it expands at constant pressure from point $A$ to point $B$ is therefore 202.6 J. Since the gas is ideal, $PV = nRT$; so as the volume doubles at constant pressure, the absolute temperature must double from 300 to 600 K at point $B$. To calculate the heat put into the gas for this part of the cycle we would need to know the heat capacity of the gas at constant pressure. We shall discuss the heat capacities of an ideal gas in Sections 17-5 and 17-7. For now, we need only find the total heat added during the complete cycle. Since the gas ends in its original state, the net change in the internal energy for the cycle must be zero and, by the first law, the total heat added must equal the net work done by the gas.

As the gas cools from point $B$ to point $C$, the volume is constant; so no work is done. Note that since the pressure is halved at constant volume, the temperature must be halved; therefore it is again 300 K at point $C$. As the gas is compressed at constant pressure from point $C$ to point $D$, it does negative work; i.e., work must be done on it. The work done *by* the gas is

$$W = p \, dV = P(V_D - V_C) = (1 \text{ atm}) (1 \text{ L} - 2 \text{ L})$$
$$= -1 \text{ L·atm} = -101.3 \text{ J}$$

As the gas is heated at constant volume back to its original state $A$, no work is done (because $V$ is constant). The net work done by the gas for the complete cycle is therefore $202.6 - 101.3 \text{ J} = 101.3 \text{ J}$. Since the internal energy depends only on the initial and final states and the gas is back to its original state, the net change in internal energy is zero. The first law of thermodynamics then tells us that the net heat added for the complete cycle is 101.3 J.

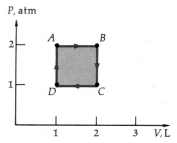

P, atm

**Figure 17-9**
Cycle for Example 17-2. The work done as the gas expands from $A$ to $B$ minus that done on the gas as it is compressed from $C$ to $D$ is the net work done by the gas for the cycle and is represented by the area enclosed by the cycle.

In Example 17-2 we calculated the net work done by the gas during a complete cycle by subtracting the work done *on* the gas during the compression from the work done *by* the gas during the expansion. This net work is represented on the $PV$ diagram by the area enclosed by the curve describing the cycle. (This result does not depend upon the fact that the pressure was held constant during the expansion and compression.) We note that the net work is positive when the cycle is traversed in the clockwise sense, and negative—i.e., work is done on the gas—when the cycle is traversed in the counterclockwise sense.

## 17-4   Internal Energy of a Gas

In Chapter 16 we saw that according to a simple molecular model of a gas, the temperature of a gas is associated with the average kinetic energy of translation of the molecules of the gas by the relation

$$PV = \tfrac{2}{3}E_k = nRT \qquad \text{or} \qquad E_k = \tfrac{3}{2}nRT$$

where $n$ is the number of moles and $E_k = N(\tfrac{1}{2}mv^2)_{av}$ is the total translational kinetic energy of the $N$ molecules. If this translational energy is taken to be the total internal energy of the gas, the internal energy will depend only on the temperature of the gas and not on the volume or pressure. Writing $U$ for $E_k$, we have

$$U = \tfrac{3}{2}nRT \qquad\qquad\qquad\qquad 17\text{-}12$$

If the internal energy of the gas includes other kinds of energy in addition to kinetic energy of translation, the internal energy will be different from that given by Equation 17-12. It may or may not depend on the pressure and volume. Consider, for example, the molecular model of a diatomic gas, such as $O_2$, known as the rigid-dumbbell model. In this model the molecule of $O_2$ is pictured as two spheres connected by a rigid (massless) rod. In addition to translational energy of the center of mass of the molecules, there can be rotational energy of the atoms about the center of mass. If we picture a gas of this kind of molecule, we can expect a kind of equilibrium between the average energy of translation and the average energy of rotation. For example, imagine the gas to have initially only translational energy with no molecules rotating. This state cannot persist for long because when two molecules collide with each other, the molecules will surely be set in rotation. The greater the speed of translation of the molecules, the greater the energy transferred to rotational energy if there is no initial rotation. However, if a molecule with a very large rotational energy collides with another molecule, it stands a very good chance of losing some of its rotational energy to translational energy. We thus expect some type of equilibrium between translational kinetic energy and rotational kinetic energy to result from molecular collisions. If we raise the temperature of the gas, the translational kinetic energy also rises. According to this discussion, we would expect the rotational kinetic energy to rise proportionally to maintain the equilibrium between these two forms of kinetic energy. It thus seems likely that rotational kinetic energy will also be just a function of the temperature, since changing the volume or pressure will change the number of collisions in a given time but will not change the nature of the energy-transfer process occurring in each collision.

If there is an attraction or repulsion between molecules in a gas even when they are far apart, we would not expect the internal energy to be a function of just the temperature. Consider the case of attraction. Since work is required to increase the separation of two molecules, the potential energy associated with the attractive force between them increases with separation. If the volume of the gas increases at constant temperature, the kinetic energy remains constant (since it depends only on the temperature) but the potential energy increases. Therefore, the total internal energy depends on the volume of the gas as well as on the temperature.

Even if there is a force of attraction between molecules of the gas and a contribution to the internal energy due to the associated potential energy, this contribution will be small if the pressure of the gas is small because then the average separation of any two molecules is very great. In the limit of very low pressures, this potential energy will be negligible. Since the equation of state for an ideal gas, $PV = nRT$, holds for real gases only at very low pressures, we place the additional restriction on the definition of an ideal gas that the internal energy be a function of temperature only:*

$$PV = nRT \qquad U = U(T) \text{ only} \qquad\qquad 17\text{-}13$$

*Ideal gas defined*

Whether or not the internal energy of a *real* gas depends on the volume or pressure was first investigated by Joule, using a process called *free adiabatic expansion of a gas*. The system shown in Figure 17-10 consists of a gas in a compartment of volume $V_1$, and an adjacent compartment that is evacuated. The compartments are connected by a stopcock initially closed. The whole system is surrounded by a rigid adiabatic wall so that no heat can be transferred into or out of the system and no work can be done by or on the system. The temperature of the gas in the left compartment is measured, and then the stopcock is opened. The gas rushes into the evacuated chamber. This expansion is not a quasi-static process. It is adiabatic because no heat can flow in or out during the expansion. Eventually the gas reaches an equilibrium state at the new volume $V_2$, which is the total volume of both compartments. The internal energy of the final state is the same as that of the initial state because no work was done by the gas and no heat was added. Joule's experiment consisted of trying to determine whether the final temperature is the same as the initial temperature. If the temperature does not change, the internal energy must depend only on the temperature and not on the volume of pressure.**

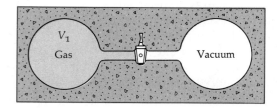

**Figure 17-10**
Free adiabatic expansion of a gas. When the stopcock is opened, the gas expands rapidly into the evacuated chamber. Since no work is done and the whole system is adiabatically insulated, the final internal energy of the gas equals its initial internal energy. If the final temperature equals the initial temperature, the internal energy does not depend on the volume but only on the temperature.

The Joule experiment is very difficult. Joule himself was unable to measure any temperature difference between the initial and final states. Later experiments of this type and others show that if the initial pressure is very high, the temperature after a free expansion is slightly lower than before the expansion. This indicates that for a real gas at high pressure, there is some attraction between the molecules. When the average separation of the molecules is increased (as in a free expansion), the potential energy increases (becomes less negative), and since the total energy does not change, the kinetic energy of translation, indicated by the temperature, must decrease. We can think of this effect by a simple model. Consider two balls which attract each other initially moving away from each other, so that their separation is increasing.

---

* This restriction is not as arbitrary as it appears. It can be shown that if the equation of state is $PV = nRT$, the first and second laws of thermodynamics (Chapter 19) imply that $U$ is a function only of temperature.

** We can see this as follows. A state of the system is determined by any two of the variables $P$, $V$, and $T$. We can thus think of the internal energy $U$ as a function of any two of these variables. Consider $U$ to be a function of the temperature $T$ and the volume $V$. Suppose that regardless of the size of the second compartment, the temperature does not change in the free expansion. Since $U$ does not change either, this means that we can change $V$ in any way and not change $U$ or $T$. Then $U$ cannot be a function of $V$. By similar reasoning, we can conclude that $U$ does not depend on $P$ either.

When they are farther apart, their velocity has decreased. Their potential energy has increased, but because of energy conservation their kinetic energy has decreased.

### Questions

7. When a bottle of compressed gas is allowed to expand into the atmosphere, the temperature of the expanding gas drops sharply. Why isn't this regarded as evidence that the internal energy of a gas depends on more than $T$?

8. In Joule's adiabatic free-expansion experiment, if the gas temperature increased on expansion, what would you conclude about the intermolecular forces in the gas? Explain.

9. At a given temperature, would you expect a mole of the diatomic gas $H_2$ to have the same, more, or less internal energy than a mole of the monatomic gas He? Explain.

## 17-5   Heat Capacities of an Ideal Gas

When a gas is heated at constant volume, no work is done, so that all the heat goes into increasing the internal energy (and therefore the temperature). On the other hand, when the gas is heated at constant pressure, it expands and does work, so that only part of the heat goes into an increase in the internal energy and the temperature. It therefore takes more heat at a constant pressure than at constant volume for a given increase in internal energy or for a given increase in temperature: $C_p$ is greater than $C_v$. In this section we shall use the equation of state for an ideal gas to develop a simple and useful expression relating these heat capacities.

Let $(dQ)_v$ be a small amount of heat put into a gas at constant volume. Since no work is done, the first law gives $dU = (dQ)_v$. The heat capacity at constant volume is then

$$C_v = \frac{(dQ)_v}{dT} = \frac{dU}{dT} \qquad\qquad 17\text{-}14$$

If $(dQ)_p$ is a small amount of heat put into a gas at constant pressure, we have for the heat capacity at constant pressure

$$C_p = \frac{(dQ)_p}{dT} \qquad\qquad 17\text{-}15$$

According to the first law, $(dQ)_p$ is related to the increase in internal energy by $(dQ)_p = dU + (dW)_p$, where $(dW)_p = P\,dV$ is the work done at constant pressure. Then

$$C_p = \frac{dU}{dT} + \frac{(dW)_p}{dT} = C_v + \frac{(dW)_p}{dT} \qquad\qquad 17\text{-}16$$

Equation 17-16 holds for any system, whether ideal gas, real gas, liquid, or solid. The work done at constant pressure $(dW)_p = P\,dV$ can be related to the change in temperature using the equation of state. For an ideal gas we have

$$PV = nRT$$

Taking the differential of both sides gives

$$P \, dV + V \, dP = nR \, dT \qquad \text{17-17}$$

At constant pressure, $dP = 0$. Then the work is

$$(dW)_p = P \, dV = nR \, dT$$

Substituting this into Equation 17-16 gives

$$C_p = C_v + \frac{nR \, dT}{dT}$$

or

$$C_p = C_v + nR \qquad \text{17-18}$$

The heat capacity of an ideal gas at constant pressure is greater than that at constant volume by the amount $nR$.

In general, for all materials that expand when heated, the heat capacity at constant pressure is greater than that at constant volume. This is because at constant volume no work is done, so that all the added heat goes into increasing the internal energy (and therefore the temperature) whereas at constant pressure some of the heat energy is used to perform work. For solids, there is very little expansion, and so little work is done when the solid is heated at constant pressure. Then, according to Equation 17-16, $C_p$ and $C_v$ are approximately equal.

# 17-6  Quasi-static Expansions of an Ideal Gas

In Example 17-2 we calculated the work done during a quasi-static expansion and compression of an ideal gas at constant pressure. A process which occurs at constant pressure is called an *isobaric process*. Two other quasi-static processes involving ideal gases are of great importance in our study of thermodynamics. One is a quasi-static *isothermal process*, i.e., one occurring at constant temperature, and the other is a quasi-static *adiabatic process*, i.e., one during which there is no heat flow into or out of the gas. In this section we shall find the curve on the *PV* diagram for each of these processes and calculate the work done during an expansion or compression of the gas.

*Isothermal process*

*Adiabatic process*

### Quasi-static Isothermal Expansion

An isothermal expansion of a gas can be carried out by placing the gas in thermal contact with a second system of very large heat capacity and at the same temperature as the gas. This second system is known as a *heat reservoir*. Since the heat capacity of a heat reservoir is very large, it can supply heat to the gas or receive heat from the gas without appreciable change in temperature. We shall assume that the heat reservoir has an infinite heat capacity so that its temperature does not change at all. In practice there are many heat reservoirs available, e.g., lakes or large conducting masses. To be effective, a heat reservoir must be able to conduct heat to and from a system quickly, so that a constant temperature is maintained in the system.

*Heat reservoir*

The curve on the *PV* diagram followed by the gas during the expansion is found from the equation of state for an ideal gas: $PV = nRT$.

Since the temperature is constant, $PV = $ constant and the curve is a hyperbola. The work done by the gas can then be calculated from

$$dW = P\ dV = \frac{nRT}{V}\ dV \qquad 17\text{-}19$$

Hence the work done by the gas as it expands isothermally from volume $V_1$ to $V_2$ is

$$W = \int_{V_1}^{V_2} P\ dV = nRT \int_{V_1}^{V_2} \frac{dV}{V} = nRT \ln \frac{V_2}{V_1} \qquad 17\text{-}20$$

$T$ has been removed from the integration because it is constant. This work is indicated by the shaded area under the $PV$ curve in Figure 17-11. Since the internal energy of an ideal gas depends only on the temperature and the temperature is constant, the internal energy does not change during an isothermal expansion of an ideal gas. Then, according to the first law of thermodynamics, the heat put into the gas is

$$Q = \Delta U + W = W$$

The heat removed from the reservoir and absorbed by the gas during an isothermal expansion equals the work done by the gas, given by Equation 17-20.

### Quasi-static Adiabatic Expansion

In a quasi-static adiabatic expansion of an ideal gas, the gas expands slowly against a piston, doing work on it. The gas is insulated from its surroundings so that no heat enters or leaves the gas. The work done by the gas equals the decrease in internal energy of the gas, and the temperature of the gas decreases. The curve representing this process on a $PV$ diagram is shown in Figure 17-12. This curve is steeper than for an isothermal expansion; i.e., when the volume increases by $dV$, the pressure in an adiabatic expansion decreases by more than it does in an isothermal expansion because the temperature decreases.

The work done by a gas undergoing a quasi-static adiabatic expansion equals the area under the curve shown in Figure 17-12. The calculation of this work is important in determining the efficiency of heat engines (Chapter 19), and for this purpose we need to know the relation between $P$ and $V$ for this curve. Although the ideal-gas equation $PV = nRT$ still applies, it is not sufficient for determing the $PV$ curve because the temperature $T$ varies during the expansion. We obtain the desired relation between $P$ and $V$ by using the equation of state and by applying the first law to the adiabatic process. From the first law we have

$$dQ = dU + P\ dV = 0 \qquad 17\text{-}21$$

According to Equation 17-14, the change in internal energy of the gas is related to the change in temperature by

$$dU = C_v\ dT \qquad 17\text{-}22$$

(Although this result was obtained by considering a constant-volume process, it relates the state variables $U$ and $T$ and is therefore valid for any process.) Then

$$C_v\ dT + P\ dV = 0 \qquad 17\text{-}23$$

We get a second relation involving the temperature change $dT$ by differentiating the equation of state $PV = nRT$. We obtain

$$P\ dV + V\ dP = nR\ dT \qquad 17\text{-}24$$

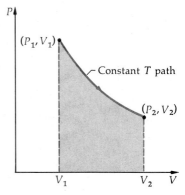

**Figure 17-11**
Isothermal expansion of an ideal gas. The curve in the $PV$ diagram for which $T$ is constant is called an isotherm. For an ideal gas $PV$ is a constant along an isotherm. The work done by the gas undergoing an isothermal expansion is indicated by the shaded area.

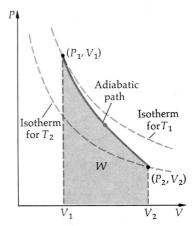

**Figure 17-12**
Quasi-static adiabatic expansion of an ideal gas. The curve on the $PV$ diagram connecting the initial and final states (called an adiabatic path) is steeper than the isotherms because the temperature drops during the adiabatic expansion.

If we multiply each term in this equation by $C_v$ and substitute $C_v \, dT = -P \, dV$ from Equation 17-23, we can eliminate the temperature. We obtain

$$C_v P \, dV + C_v V \, dP = nRC_v \, dT = nR(-P \, dV)$$

Rearranging gives

$$(C_v + nR)P \, dV + C_v V \, dP = 0$$

But $C_v + nR$ is just $C_p$. Using this fact and dividing each term by $C_v PV$, we obtain

$$\frac{C_p}{C_v} \frac{dV}{V} + \frac{dP}{P} = 0 \qquad \text{or} \qquad \gamma \frac{dV}{V} + \frac{dP}{P} = 0$$

where

$$\gamma = \frac{C_p}{C_v} \qquad\qquad\qquad 17\text{-}25$$

is the ratio of heat capacities. Integrating each term gives

$$\gamma \ln V + \ln P = \text{constant} \qquad \ln PV^\gamma = \text{constant}$$

or

$$PV^\gamma = \text{constant} \qquad\qquad\qquad 17\text{-}26$$

This is the equation relating $P$ and $V$ in a quasi-static adiabatic expansion. This equation also applies to a quasi-static adiabatic compression in which the piston does work on the gas.

We can use Equation 17-26 to calculate the adiabatic bulk modulus which we used in Chapter 14 to find the speed of sound waves in air. Differentiating Equation 17-26 we obtain

$$P\gamma V^{\gamma-1} \, dV + V^\gamma \, dP = 0$$

$$dP = -\frac{\gamma P \, dV}{V}$$

The adiabatic bulk modulus is thus

$$B_{\text{adiab}} = -\frac{dP}{dV/V} = \gamma P \qquad\qquad\qquad 17\text{-}27$$

We can use the equation of state $PV = nRT$ along with Equation 17-26 to eliminate either $P$ or $V$. For example, eliminating $P$, we obtain

$$TV^{\gamma-1} = \text{constant} \qquad\qquad\qquad 17\text{-}28$$

We now use Equation 17-26 to calculate the work done by an ideal gas during a quasi-static adiabatic expansion from an initial state of pressure $P_1$ and volume $V_1$ to a final state of pressure $P_2$ and volume $V_2$. We have

$$dW = P \, dV = \frac{K}{V^\gamma} \, dV = KV^{-\gamma} \, dV$$

where the constant $K$ is given by $K = P_1 V_1^\gamma$. The work done as the gas expands from volume $V_1$ to $V_2$ is then

$$W = \int_{V_1}^{V_2} KV^{-\gamma} \, dV = K \frac{V^{-\gamma+1}}{-\gamma + 1} \Big]_{V_1}^{V_2} = \frac{K}{1 - \gamma} (V_2^{1-\gamma} - V_1^{1-\gamma})$$

$$= \frac{K}{\gamma - 1} \left( \frac{1}{V_1^{\gamma-1}} - \frac{1}{V_2^{\gamma-1}} \right) \qquad\qquad 17\text{-}29$$

**Example 17-3** One mole of nitrogen expands isothermally at 20°C from a volume of 10 to 20 L. Assuming nitrogen to be an ideal gas, how much heat must be supplied to keep the temperature of the gas from dropping?

As mentioned, the internal energy of an ideal gas remains constant during an isothermal expansion; so, according to the first law, the heat added equals the work done by the gas. This work is given by Equation 17-20. Substituting the given values, we obtain

$$Q = W = nRT \ln \frac{V_2}{V_1} = (1 \text{ mol}) (8.31 \text{ J/mol·K}) (293 \text{ K}) \ln \frac{20 \text{ L}}{10 \text{ L}}$$

$$= 1.69 \times 10^3 \text{ J} = 1.69 \text{ kJ}$$

**Example 17-4** A quantity of air expands adiabatically and quasi-statically from an initial pressure of 2 atm and volume 2 L at room temperature (20°C) to twice its original volume. What are the final pressure and temperature? For air, $\gamma = 1.4$.

According to Equation 17-26, the quantity $PV^\gamma$ remains unchanged during a quasi-static adiabatic expansion. Thus, if $P_1$ and $V_1$ are the initial pressure and volume and $P_2$ and $V_2$ are the final pressure and volume, we have

$$P_2 V_2^\gamma = P_1 V_1^\gamma \qquad \text{or} \qquad P_2 = P_1 \left(\frac{V_1}{V_2}\right)^\gamma$$

For the given values we have

$$P_2 = (2 \text{ atm}) \left(\frac{1 \text{ L}}{2 \text{ L}}\right)^{1.4} = 0.758 \text{ atm}$$

The change in temperature is found most easily from Equation 17-28, giving

$$T_1 V_1^{\gamma-1} = T_2 V_2^{\gamma-1}$$

or

$$T_2 = T_1 \left(\frac{V_1}{V_2}\right)^{\gamma-1} = (293 \text{ K}) (\tfrac{1}{2})^{0.4} = 222 \text{ K} = -51°C$$

# 17-7   The Equipartition of Energy

If the internal energy of a gas consists of translational kinetic energy only, the internal energy of $n$ mol of the gas (Equation 17-12) is

$$U = \tfrac{3}{2}nRT \qquad\qquad 17\text{-}30$$

The heat capacities are then

$$C_v = \tfrac{3}{2}nR \qquad\qquad 17\text{-}31$$

and

$$C_p = C_v + nR = \tfrac{5}{2}nR \qquad\qquad 17\text{-}32$$

Table 17-2 lists the measured heat capacities per mole for several gases. The ideal-gas prediction $C_p - C_v = nR$ holds quite well for all gases ($R = 8.314$ J/mol·K), and the prediction $C_v = \tfrac{3}{2}nR$ holds quite well for most monatomic gases but not for other types.

**Table 17-2**
Molar heat capacities, J/mol·K, of various gases at 25°C

| Gas | $C_p'$ | $\gamma$ | $C_v'$ | $\dfrac{C_v'}{R}$ | $C_p' - C_v'$ | $\dfrac{C_p' - C_v'}{R}$ |
|---|---|---|---|---|---|---|
| **Monatomic** | | | | | | |
| He | 20.79 | 1.66 | 12.52 | 1.51 | 8.27 | 0.99 |
| Ne | 20.79 | 1.64 | 12.68 | 1.52 | 8.11 | 0.98 |
| Ar | 20.79 | 1.67 | 12.45 | 1.50 | 8.34 | 1.00 |
| Kr | 20.79 | 1.67 | 12.45 | 1.50 | 8.34 | 1.00 |
| Xe | 20.79 | 1.66 | 12.52 | 1.51 | 8.27 | 0.99 |
| **Diatomic** | | | | | | |
| $N_2$ | 29.12 | 1.4 | 20.80 | 2.50 | 8.32 | 1.00 |
| $H_2$ | 28.82 | 1.41 | 20.44 | 2.46 | 8.38 | 1.01 |
| $O_2$ | 29.37 | 1.4 | 20.98 | 2.52 | 8.39 | 1.01 |
| CO | 29.04 | 1.4 | 20.74 | 2.49 | 8.30 | 1.00 |
| **Polyatomic** | | | | | | |
| $CO_2$ | 36.62 | 1.30 | 28.17 | 3.39 | 8.45 | 1.02 |
| $N_2O$ | 36.90 | 1.30 | 28.39 | 3.41 | 8.51 | 1.02 |
| $H_2S$ | 36.12 | 1.32 | 27.36 | 3.29 | 8.76 | 1.05 |

The values of $C_p'$ and $\gamma$ are measured values taken from *The American Institute of Physics Handbook*, 3d ed., McGraw-Hill, New York, 1972. The values of $C_v'$ are calculated from $C_v' = C_p'/\gamma$.

In Section 17-4 we mentioned that we expect other kinds of kinetic energy, e.g., rotational energy, also to depend on the temperature. We discussed qualitatively how we might expect some kind of equilibrium between various kinds of energy, e.g., rotational and translational kinetic energy of a dumbbell molecule, to be established as a result of molecular collisions. Let us explore this equilibrium further.

Consider a model of a monatomic gas in which the molecules are assumed to be hard spheres with frictionless surfaces; this model is called the smooth hard-sphere model. Suppose that at some time all the molecules are moving in the $x$ direction only, that is, $v_y$ and $v_z$ are zero, and the kinetic energy of each is $\frac{1}{2}mv_x^2$. Such an unusual state is not an equilibrium state. Because of off-center collisions the molecules will acquire components $v_y$ and $v_z$. On the average, such a collision will tend at first to increase the average value of $v_y^2$ and $v_z^2$ until these quantities have the same average values as $v_x^2$. When equilibrium is established, the average values of the energies associated with each kind of translation, $\frac{1}{2}mv_x^2$, $\frac{1}{2}mv_y^2$, and $\frac{1}{2}mv_z^2$, will be equal to each other. Since we have seen that the average translational energy per molecule is $\frac{3}{2}kT$, where $k = R/N_A$ is Boltzmann's constant, each of these terms must have an average value of $\frac{1}{2}kT$ in equilibrium. (The assumption of smooth-sphere molecules rules out any exchange of rotational energy since the forces in a collision will be along the line of centers of the molecules and will not change the angular momentum of the molecules.)

The example we have just considered is somewhat artificial because it is very unlikely that all the molecules would be moving only in the $x$ direction at one time. However, there are similar situations in which the average value of the kinetic energy associated with motion in the $x$ direction is momentarily greater than that associated with motion in the $y$ or $z$ direction. Consider for example the adiabatic compression of a gas by a piston moving with a very low speed in the negative $x$ direction so that the compression is approximately quasi-static. From the molecular point of view, the energy of the gas increases because of the elastic collisions of the molecules with the moving piston. Because the piston is moving toward the molecules, the molecules rebound with a greater magnitude of $x$ component of velocity than they had before hitting the piston. Thus the moving piston directly increases the energy associated with motion in the $x$ direction but has no effect on that associated with motion in the $y$ or $z$ directions (assuming the piston to be frictionless). But immediately after colliding with the piston, molecules collide with other nearby molecules and a new equilibrium is established, $\frac{1}{2}mv_x^2$, $\frac{1}{2}mv_y^2$, and $\frac{1}{2}mv_z^2$, each having the same average value of $\frac{1}{2}kT$, which is increased by the work done on the gas by the moving piston.

Sharing the energy equally between the three terms in the translational energy is a special case of the *equipartition theorem*, a result which can be derived from more advanced statistical mechanics. Each coordinate, velocity component, angular-velocity component, etc. that appears in the expression for the energy of a molecule is called a *degree of freedom*. The equipartition theorem states that*

*In equilibrium, there is associated with each degree of freedom an average energy of $\frac{1}{2}kT$ per molecule.*   *Equipartition theorem*

For example, consider a rigid-dumbbell model of a diatomic molecule (Figure 17-13) which can translate in the $x$, $y$, and $z$ directions and can rotate about axes $x'$ and $y'$ through the center of mass perpendicular to the $z'$ axis along the line joining the two masses.** The kinetic energy for this dumbbell-model molecule is then

$$E_k = \tfrac{1}{2}mv_x^2 + \tfrac{1}{2}mv_y^2 + \tfrac{1}{2}mv_z^2 + \tfrac{1}{2}I_{x'}\omega_{x'}^2 + \tfrac{1}{2}I_{y'}\omega_{y'}^2, \qquad 17\text{-}33$$

The equipartition theorem states that the average energy associated with each of the five degrees of freedom is $\frac{1}{2}kT$, so that the total energy for $N$ molecules is

$$U = N(5)(\tfrac{1}{2}kT) = \tfrac{5}{2}NkT \qquad 17\text{-}34$$

Writing $N = nN_A$, where $n$ is the number of moles, $N_A$ is Avogadro's number, and $N_A k = R$, we have

$$U = \tfrac{5}{2}nRT$$

The heat capacity at constant volume is then

$$C_v = \frac{dU}{dT} = \tfrac{5}{2}nR \qquad 17\text{-}35$$

and

$$C_p = C_v + nR = \tfrac{7}{2}nR \qquad 17\text{-}36$$

From the observation that $C_v/n$ for nitrogen and oxygen is approximately $\frac{5}{2}R$, Clausius speculated (about 1880) that these gases must be diatomic gases which can rotate about two axes as well as translate.

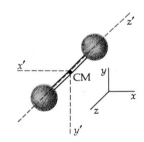

**Figure 17-13**
Rigid-dumbbell model of a diatomic gas. The motion can be described as translation of the center of mass along the $x$, $y$, or $z$ axis plus rotation about the $x'$, $y'$, or $z'$ axis. If the spheres are smooth or points, rotation about the $z'$ axis can be neglected because this energy cannot be changed by collisions.

---

* Strictly speaking, the equipartition theorem applies only to coordinates that appear squared in the expression for the energy of a molecule.

** We rule out rotation about the $z'$ axis of the dumbbell by assuming (1) that the atoms are points and that the moment of inertia about this axis is therefore zero or (2) that the atoms are hard smooth spheres, in which case rotation about this axis cannot be changed by collisions and therefore does not participate in the absorption of energy. Either of these assumptions also rules out the possibility of rotation of a monatomic molecule.

If a diatomic molecule can vibrate as well as rotate and translate, there are two more degrees of freedom, corresponding to the kinetic and potential energy of vibration along the line joining the atoms. The equipartition theorem thus predicts that the internal energy per mole will be $\frac{7}{2}RT$. According to Table 17-2, diatomic gases apparently do not vibrate. The calculation of the number of degrees of freedom for a molecule with more than two atoms is more complicated if all possible vibrations are allowed.

Although the equipartition theorem is quite successful in relating the molar heat capacities of a gas to its atomic structure, it fails in several ways. The equipartition theorem gives no hint why diatomic molecules do not vibrate or why the heat capacities vary with temperature. According to the equipartition theorem, there should be no temperature variation. The equipartition theorem fails because classical mechanics itself breaks down when applied to atomic and molecular systems and must be replaced by quantum mechanics. In the quantum-mechanical theory of motion, the possible energies of an atom or molecule are quantized. In particular there is a gap between the lowest possible energy and the next lowest. For the vibrational energies of a diatomic molecule, this energy gap is of the order of several tenths of an electronvolt, which is much higher than $kT \approx 0.026$ eV (the value at ordinary temperatures). Since a collision between molecules having energies of the order of $kT$ cannot increase their energy from the lowest possible energy to the next allowed energy, vibrational energy does not play a role in the internal energy at low temperatures.

The rotational energy of a molecule is also quantized, spacing between possible energies depending inversely on the moment of inertia. For rotation of monatomic molecules and rotation of diatomic molecules about the line joining the atoms, the moment of inertia is so small that the energy spacing is large compared with $kT$ at ordinary temperatures. For these cases, rotational energy plays no role in the internal energy at ordinary temperatures. On the other hand, the moment of inertia for rotation of diatomic molecules about axes perpendicular to the line joining the atoms is so large that the spacing of possible energies is small compared with $kT$ at ordinary temperatures. In this case, energy quantization is not important, and the predictions of the classical equipartition theory are correct.

Using the equipartition theorem and a simple model of a solid, we can understand the Dulong-Petit law for the heat capacities of a solid. (We need not distinguish between $C_v$ and $C_p$ here because for a solid they are almost equal.) Let us assume that a solid consists of a regular array of molecules each having a fixed equilibrium position and being connected by springs to each of its neighbors, as in Figure 17-14. Each molecule can oscillate in the $x$, $y$, and $z$ directions. Its total energy is thus

$$E_k = \tfrac{1}{2}mv_x^2 + \tfrac{1}{2}mv_y^2 + \tfrac{1}{2}mv_z^2 + \tfrac{1}{2}Kx^2 + \tfrac{1}{2}Ky^2 + \tfrac{1}{2}Kz^2$$

where $K$ is the effective force constant of the springs. The molecule thus has six degrees of freedom and an average energy of $6(\tfrac{1}{2}kT)$ according to the equipartition theorem. Thus the total energy of 1 mol of a solid is

$$U = 3RT \tag{17-37}$$

and the molar heat capacity is $C_v' = dU/dT = 3R$, which is the Dulong-Petit law. This result holds for many solids (but not for all) at room temperature. Again, the reason for disagreement between this prediction and experiment is the breakdown of classical physics in the atomic realm.

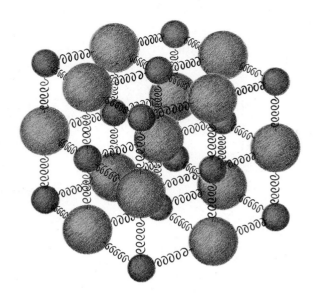

**Figure 17-14**
Model of a solid consisting of atoms connected to each other by springs. The internal energy of the solid then consists of kinetic and potential vibrational energy.

**Example 17-5** One mole of oxygen gas is heated from room temperature of 20°C and pressure of 1 atm to a temperature of 100°C. How much heat must be supplied if the volume is kept constant during the heating? How much heat must be supplied if the pressure is kept constant? How much work is done if the pressure is constant? (Assume that oxygen is an ideal gas.)

The heat capacity at constant volume of oxygen is

$$C_v = \tfrac{5}{2}nR = 20.8 \text{ J/K}$$

when we use $n = 1$ mol and $R = 8.31$ J/K. The heat added to raise the temperature from 293 to 373 K is then

$$Q_1 = C_v \, \Delta T = (20.8 \text{ J/K}) \, (80 \text{ K}) = 1.66 \text{ kJ}$$

Since no work is done when the volume is constant, the internal energy of the gas must increase by 1.66 kJ. Because the internal energy depends only on the temperature, this is the increase in internal energy when the gas temperature changes from 293 to 373 K by any process.

If we keep the pressure constant, the heat that must be added is

$$Q_2 = C_p \, \Delta T = \tfrac{7}{2}(8.31 \text{ J/K})(80 \text{ K}) = 2.33 \text{ kJ}$$

using $C_p = C_v + R = \tfrac{7}{2}R$ for 1 mol. This is greater because although the internal energy change is again 1.66 kJ, work is done when the gas expands. The work done at constant pressure is

$$W = P \, dV = P(V_2 - V_1)$$

We can find the initial and final volumes from the result of Example 16-2 that the volume of 1 mol of a gas at 1 atm and 273 K is 22.4 L. Since $PV = nRT$, the volume at constant pressure is proportional to the temperature. The volumes $V_1$ and $V_2$ are thus

$$V_1 = (22.4 \text{ L}) \, \frac{293}{273} = 24.0 \text{ L}$$

$$V_2 = (22.4 \text{ L}) \, \frac{373}{273} = 30.6 \text{ L}$$

The work done is thus

$$W = (1 \text{ atm}) (30.6 - 24.0 \text{ L}) = (6.6 \text{ L·atm}) (101.3 \text{ J/L·atm})$$
$$= 669 \text{ J} = 0.669 \text{ kJ}$$

The heat added at constant pressure, 2.33 kJ, equals the change in internal energy, 1.66 kJ, plus the work done, 0.67 kJ. These two processes are illustrated on a $PV$ diagram in Figure 17-15.

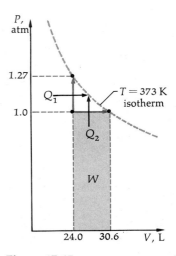

**Figure 17-15**
Two methods of raising the temperature of a gas from 293 to 373 K. The dashed gray line is the isotherm at 373 K. The internal energy of the two different final states is the same because they have the same temperature. The heat $Q_2$ added at constant pressure is therefore greater than the heat $Q_1$ added at constant volume by the amount of work done, $W$.

## Review

A. Define, explain, or otherwise identify:

Heat capacity, 495
Calorie, 495
Specific heat, 495
Molar heat capacity, 495
Adiabatic work, 499
Internal energy, 499
Heat, 500
Quasi-static process, 501

PV diagram, 502
Ideal gas, 506
Free adiabatic expansion
  of a gas, 506
Quasi-static adiabatic expansion
  of a gas, 509
Equipartition theorem, 513

B. True or false:

1. The heat capacity of a body is the amount of heat it can store at a given temperature.

2. A quasi-static process is one in which no work is done.

3. When a system goes from state 1 to state 2, the heat added is the same for all processes.

4. When a system goes from state 1 to state 2, the work done by the system is the same for all processes.

5. When a system goes from state 1 to state 2, the change in internal energy of the system is the same for all processes.

6. The internal energy of an ideal gas depends only on the temperature.

7. $C_p$ is greater than $C_v$ for any material that expands when heated.

8. In a free adiabatic expansion of an ideal gas the final state is the same as the initial state.

9. In an isothermal expansion of an ideal gas the work done by the gas equals the heat absorbed.

## Exercises

### Section 17-1, Heat Capacity and Specific Heat

1. How many joules of heat energy are required to raise the temperature of 20 kg of water from 10 to 20°C?

2. A man typically consumes food with a total energy value of 2000 kcal each day. (*a*) How many joules of energy is this? (*b*) Calculate his power output in watts assuming that this energy is dissipated at a steady rate during 24 h.

3. Find the number of Btu needed to raise the temperature of 1 gal of water from 32 to 212°F (1 gal contains 8 pints; 1 pint of water weighs 1 lb).

4. A 200-g piece of lead is heated to 90°C and dropped into 500 g of water initially at 20°C. Neglecting the heat capacity of the container, find the final temperature of the lead and water.

5. The specific heat of a certain metal is determined by measuring the temperature change which occurs when a heated piece of the metal is placed in an insulated container made of the same material and containing water. The piece of metal has mass 100 g and an initial temperature of 100°C. The container has a mass of 200 g and contains 500 g of water at an initial temperature of 17.3°C. The final temperature is 22.7°C. What is the specific heat of the metal?

### Section 17-2, The First Law of Thermodynamics

6. (a) How much work must be done on 1 kg of water to raise its temperature from 20 to 25°C, assuming that the water is adiabatically shielded from its surroundings? (b) In an actual experiment, the work done is $2.25 \times 10^4$ J. How much heat escapes to the surroundings?

7. A box of lead shot is thrown vertically into the air to a height of 4 m and allowed to fall to the floor in a lecture demonstration. The original temperature of the lead is 20°C. Five such throws are made, and the temperature is then measured. What result do you expect?

8. A lead bullet with speed 200 m/s is stopped in a block of wood. Assuming that all the energy goes into heating the bullet, find the final temperature of the bullet if the initial temperature is 20°C.

9. The water at Niagara Falls drops 50 m. (a) If all the change in potential energy goes into internal energy of the water, compute the increase in temperature. (b) Do the same for Yosemite Falls, where the water drops 740 m.

10. A body of water is heated at constant pressure from 20 to 40°C. Explain carefully why it is incorrect to say that the water at 40°C contains more heat than it did at 20°C. Is it correct to say that the water has more internal energy at 40°C than at 20°C? Why, or why not?

### Section 17-3, Work and the PV Diagram for a Gas

*In Exercises 11 to 14, 1 mol of an ideal gas is originally in state $P_1 = 3$ atm, $V_1 = 1$ L, and $U_1 = 456$ J. Its final state is $P_2 = 2$ atm, $V_2 = 3$ L, and $U_2 = 912$ J. All processes are quasi-static.*

11. The gas is allowed to expand at constant pressure to a volume of 3 L. It is then cooled at constant volume until its pressure is 2 atm. (a) Indicate this process on a PV diagram and calculate the work done by the gas. (b) Find the heat added during the process.

12. The gas is first cooled at constant volume until its pressure is 2 atm. It is then allowed to expand at constant pressure until its volume is 3 L. (a) Indicate this process on a PV diagram and calculate the work done by the gas. (b) Find the heat added during this process.

13. The gas is allowed to expand isothermally until its volume is 3 L and its pressure is 1 atm. It is then heated at constant volume until its pressure is 2 atm. (a) Indicate this process on a PV diagram and calculate the work done by the gas. (b) Find the heat added during this process.

14. The gas expands and heat is added such that the gas follows a straight-line path on the PV diagram from its initial state to its final state. (a) Indicate this process on a PV diagram and calculate the work done by the gas. (b) Find the heat added for this process.

### Section 17-4, Internal Energy of a Gas

15. An ideal gas is originally at pressure $P_1 = 2$ atm, volume $V_1 = 1$ L, and temperature $T_1 = 300$ K. It is allowed to expand at constant pressure until its volume is 4 L. (a) How much work is done during this expansion? (b) What is the temperature of the gas after this expansion? (c) The gas is now cooled at con-

stant volume until its pressure is 0.5 atm. What is its temperature now? (d) What is the net amount of heat added to the gas during the complete process of expansion and cooling?

16. A certain ionized gas is composed of ions that repel each other. The gas undergoes a free adiabatic expansion. How does the temperature of the gas change? Why?

### Section 17-5, Heat Capacities of an Ideal Gas

17. The specific heat of steam ($M = 18.0$ g/mol) is measured at constant pressure to be 2.50 kJ/kg·K. Assuming steam to be an ideal gas, what is the specific heat capacity at constant volume?

18. The specific heat of air at 0°C is listed in a handbook as having the value 1.00 J/g·K measured at constant pressure. (a) Assuming air to be an ideal gas, what is the specific heat at 0°C at constant volume? ($M = 29.0$.) (b) How much internal energy is contained in 1 L of air at 0°C?

19. (a) Show that when an ideal gas undergoes a temperature change at constant volume, its energy changes by $\Delta U = C_v \, \Delta T$. (b) Explain why the result $\Delta U = C_v \, \Delta T$ holds for an ideal gas for any temperature change no matter what the process. (c) Show explicitly that this expression holds for the expansion of an ideal gas at constant pressure by calculating the work done, showing that the work can be written $W = nR \, \Delta T$, and then subtracting the work done by the gas from the heat added, $Q = C_p \, \Delta T$.

### Section 17-6, Quasi-static Expansions of an Ideal Gas

20. Show that during a quasi-static adiabatic expansion of an ideal gas, $TV^{\gamma-1} = $ constant, where $T$ is the absolute temperature.

21. One mole of an ideal gas initially at 1 atm and 0°C is compressed isothermally and quasi-statically until its pressure is 2 atm. Find (a) the work needed to perform this compression and (b) the heat removed from the gas during the compression.

22. Show that during a quasi-static adiabatic expansion of an ideal gas, $T^{\gamma}/P^{\gamma-1} = $ constant.

23. An ideal gas initially at 20°C and 200 kPa has a volume of 4 L. It undergoes a quasi-static, isothermal expansion until its pressure is reduced to 100 kPa. Find (a) the work done by the gas and (b) the heat added to the gas during the expansion.

24. One mole of an ideal gas ($\gamma = \frac{5}{3}$) expands adiabatically and quasi-statically from a pressure of 10 atm and temperature 0°C to a final state of pressure 2 atm. Find (a) the initial and final volume and (b) the final temperature.

25. An ideal gas at a room temperature of 20°C is compressed quasi-statically and adiabatically to half its original volume. Find its final temperature if (a) $C_v = \frac{3}{2}nR$ and (b) $C_v = \frac{5}{2}nR$.

### Section 17-7, The Equipartition of Energy

26. The heat capacity at constant volume of a certain monatomic gas is 49.8 J/K. (a) Find the number of moles of the gas. (b) What is the internal energy of this gas at $T = 300$ K? (c) What is the heat capacity at constant pressure?

27. For a certain gas, the heat capacity at constant pressure is greater than that at constant volume by 29.1 J/K. (a) How many moles of the gas are there? (b) If the gas is monatomic, what are $C_v$ and $C_p$? (c) If the gas molecules are diatomic molecules which rotate but do not vibrate, what are $C_v$ and $C_p$?

28. Find the molar heat capacities $C_v'$ and $C_p'$ and the ratio $\gamma = C_p'/C_v'$ for ideal gases in which the molecules are (a) single spherical atoms, (b) rigid and dumbbell-shaped, (c) nonrigid dumbbells which vibrate and rotate.

29. For air, $\gamma = 1.40$. Which is the best model for an air molecule: (a) an elastic sphere, (b) a rigid dumbbell, or (c) a nonrigid dumbbell which vibrates and rotates? Explain.

30. For each of the four parts of the cycle in Example 17-2 calculate the heat input or output and the internal energy change, assuming the gas to be an ideal monatomic gas.

31. One mole of an ideal monatomic gas is heated at constant volume from 300 to 600 K. (a) Find the increase in internal energy, the work done, and the heat added. (b) Find these same quantities if this same gas is heated from 300 to 600 K at constant pressure. Use the first law and your result in part (a) to calculate the work done. (c) Calculate the work done in part (b) directly from $dW = P\ dV$.

32. Repeat Exercise 31 for a gas containing diatomic molecules which rotate but do not vibrate.

33. The Dulong-Petit law was used originally to determine the molecular mass of a substance from its measured heat capacities. Given that the specific heat of a certain solid is measured to be 0.447 kJ/kg·K, find the molecular mass of the substance. What element is this?

34. The specific heat of a certain solid is 0.1306 kJ/kg. Find the molecular mass of the solid assuming that it obeys the Dulong-Petit law, and state what element it is.

Problems

1. A 200-g aluminum calorimeter can contains 500 g of water at 20°C; 300 g of aluminum shot is heated to 100°C and placed in the calorimeter. Using the value of the specific heat given in Table 17-1, find the final temperature of the system (assuming no heat losses to the surroundings). If this calorimeter is to be used to make an accurate measurement of the specific heat of aluminum, what should the initial temperature of the water and container be to minimize the error due to heat transfer to and from the surroundings, if the room temperature is 20°C?

2. An ideal gas at initial pressure $P_1$ and volume $V_1$ expands quasi-statically and adiabatically to volume $V_2$ and pressure $P_2$. Show that the work done is

$$W = \frac{P_1 V_1 - P_2 V_2}{\gamma - 1}$$

Show that this is consistent with the general result that the internal-energy change is $\Delta U = C_v\ \Delta T$.

3. An adiabatically isolated system consists of 1 mol of a diatomic ideal gas at 100 K and 2 mol of a solid at 200 K separated by a rigid adiabatic wall. Find the equilibrium temperature of the system after the adiabatic wall is removed, assuming that the solid obeys the Dulong-Petit law.

4. One-half mole of helium is expanded adiabatically from an initial pressure of 5 atm and temperature of 500 K to a final pressure of 1 atm. Find the final temperature, the final volume, the work done by the gas, and the change in internal energy of the gas.

5. One mole of an ideal gas ($\gamma = 1.67$) is at 273 K and 1 atm. Find the initial and final internal energies and the work done by the gas when 500 J of heat is added (a) at constant pressure, and (b) at constant volume.

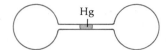

6. Prove that the slope of the adiabatic curve passing through a point on the $PV$ diagram is $\gamma$ times the slope of the isothermal curve passing through the same point.

7. Two glass spheres of equal volume are connected by a small tube containing a small amount of mercury, as shown in Figure 17-16. The spheres are sealed at 20°C with exactly 1 L of air in each side. If the cross-sectional area of the tube is 5 mm², how far will the mercury be displaced if the temperature of one sphere is raised by 0.1°C while the other is maintained at 20°C?

**Figure 17-16**
Problem 7.

8. Two glass spheres of volume 0.4 and 0.2 L are connected by a small capillary tube of negligible volume. The system is sealed with air inside at 1 atm and 20°C. If the 0.4-L sphere is heated to 80°C while the other is maintained at 20°C, what is the final pressure in the system?

9. One mole of $N_2$ ($C_v = \frac{5}{2}R$) gas is maintained at room temperature (20°C) and at a pressure of 5 atm. It is allowed to expand adiabatically and quasi-statically until its pressure equals the room pressure of 1 atm. It is then heated at constant pressure until its temperature is again 20°C. During this heating, the gas expands. After it reaches room temperature, it is heated at constant volume until its pressure is 5 atm. It is then compressed at constant pressure until it is back in its original state. (a) Construct an accurate $PV$ diagram showing each process in the cycle. (b) *From your graph* determine the work done by the gas during the complete cycle. (c) How much heat was added or subtracted from the gas during the complete cycle? (d) Check your graphical determination of the work in part (b) by calculating the work done during each process of the cycle.

10. One mole of an ideal gas is in an initial state of $P = 2$ atm and $V = 10$ L indicated by point $a$ on the $PV$ diagram of Figure 17-17. It expands at constant pressure to point $b$, where its volume is 30 L, and is then cooled at constant volume until its pressure is 1 atm at point $c$. It is then compressed at constant pressure to its original volume at point $d$ and finally heated at constant volume until it is back at its original state. (a) Find the temperature of each state $a$, $b$, $c$, and $d$. (b) Assuming that the gas is monatomic, find the heat added along each path of the cycle. (c) Calculate the work done along each path. (d) Find the internal energy of each state $a$, $b$, $c$, and $d$. (e) What is the net work done by the gas for the complete cycle? How much heat is added during the complete cycle?

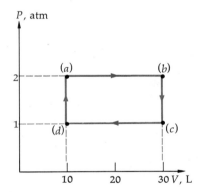

11. At very low temperatures the specific heat of a metal is given by $c = \alpha T + \beta T^3$. For copper, $\alpha = 0.0108$ J/kg·K² and $\beta = 7.62 \times 10^{-4}$ J/kg·K⁴. (a) What is the specific heat at 4 K? (b) How much heat is required to heat copper from 1 to 3 K?

**Figure 17-17**
Problem 10.

12. An ideal gas of $n$ mol is at pressure $P_1$, volume $V_1$, and temperature $T_h$. It undergoes an isothermal expansion until its pressure and volume are $P_2$ and $V_2$. It then expands adiabatically until its temperature is $T_c$ and its pressure and volume are $P_3$ and $V_3$. It is then compressed isothermally until it is at volume $V_4$, which is related to its initial volume $V_1$ by $T_c V_4^{\gamma-1} = T_h V_1^{\gamma-1}$. It is then compressed adiabatically until it is in its original state, $P_1 V_1$ and $T_h$. (a) Assuming that each process is quasi-static, plot this cycle on a $PV$ diagram (this cycle is known as a Carnot cycle for an ideal gas, to be discussed in Chapter 19). (b) Show that the heat added during the isothermal expansion at $T_h$ is $Q_h = nRT_h \ln (V_2/V_1)$. (c) Show that the heat rejected by the gas during the isothermal compression at $T_c$ is $Q_c = nRT_c \ln (V_3/V_4)$. (d) Using the result that for an adiabatic expansion or compression $TV^{\gamma-1}$ is constant, show that $V_2/V_1 = V_3/V_4$. (e) The efficiency of such a cycle is defined to be the net work done divided by the heat absorbed, $Q_h$. Show from the first law of thermodynamics that the efficiency is $1 - Q_c/Q_h$. Using your results from the previous parts of this problem, show that $Q_c/Q_h = T_c/T_h$.

# CHAPTER 18  Thermal Properties and Processes

---

**Objectives**   After studying this chapter, you should:

1. Be able to define the terms expansivity, latent heat, vapor pressure, phase diagram, critical temperature, triple point, temperature gradient, coefficient of heat conduction, thermal resistance, $R$ factor, and blackbody.

2. Be able to calculate the linear expansion and volume expansion of a substance given its temperature change.

3. Be able to work calorimetry problems which include latent heats of fusion and of vaporization.

4. Be able to sketch the liquid-vapor isotherms for a real substance and indicate the critical point on a $PV$ diagram.

5. Be able to sketch the phase diagram for a real substance.

6. Be able to calculate the thermal resistance of various heat conductors and find the rate of heat conduction for a given temperature gradient.

7. Be able to calculate the rate of heat conduction for various thermal resistors in series or in parallel.

8. Be able to state Newton's law of cooling.

9. Be able to state the Stefan-Boltzmann law of radiation and use it to calculate the power radiated by an object at a given temperature.

10. Be able to state the Wien displacement law and use it to relate the absolute temperature to the wavelength at which the power radiated by a blackbody is maximum.

---

When a body is heated, various changes take place. The temperature of the body may rise, accompanied by an expansion or contraction of the body; or the body may liquefy or vaporize with no change in temperature. In this chapter we examine some of the thermal properties of matter and some of the important processes involving thermal energy. We first look at thermal expansion and then discuss changes of phase and latent heat. Finally, we discuss the important topic of heat transfer.

# 18-1   Thermal Expansion

When the temperature of a body increases, it usually expands (one exception to this is water between 0 and 4°C, which contracts when the temperature increases). Let us first consider a long rod. Let its length be $L_0$ when the temperature is $T_0$ and $L = L_0 + \Delta L$ when the temperature is $T = T_0 + \Delta T$. We define the average coefficient of linear expansion for the interval $\Delta T$ to be the fractional change in length divided by the temperature change when the pressure is constant:

$$\bar{\alpha} = \frac{\Delta L/L_0}{\Delta T} \qquad\qquad 18\text{-}1$$

Hence $\Delta L = \bar{\alpha} L_0 \, \Delta T$. The coefficient of linear expansion at temperature $T_0$ is the limit of Equation 18-1 as $\Delta T$ approaches zero:

$$\alpha = \lim_{\Delta T \to 0} \frac{\Delta L/L_0}{\Delta T} \qquad\qquad 18\text{-}2$$

*Coefficient of thermal expansion*

The coefficient of expansion of a solid or liquid usually does not vary much with pressure, but it does vary appreciably with temperature. Thus the expression

$$\Delta L = \alpha L_0 \, \Delta T \qquad\qquad 18\text{-}3$$

where $\alpha$ at the initial temperature is used in place of $\bar{\alpha}$, holds approximately if the temperature interval is so small that $\alpha$ and its average value do not differ appreciably.

The coefficient of volume expansion $\beta$, sometimes called the *expansivity*, is defined similarly as the limit of the ratio of the fractional change in volume to the temperature interval as the temperature interval approaches zero at constant pressure:

$$\beta = \lim_{\Delta T \to 0} \frac{\Delta V/V_0}{\Delta T} \qquad\qquad 18\text{-}4$$

*Expansivity*

For a given material, the coefficient of volume expansion is just 3 times the coefficient of linear expansion. Consider a box of dimensions $L_1$, $L_2$, and $L_3$. The volume at temperature $T$ is

$$V = L_1 L_2 L_3$$

At temperature $T + \Delta T$ the volume is $V_0 + \Delta V$, where

$$
\begin{aligned}
V_0 + \Delta V &= (L_1 + \Delta L_1)(L_2 + \Delta L_2)(L_3 + \Delta L_3) \\
&= (L_1 + \alpha L_1 \, \Delta T)(L_2 + \alpha L_2 \, \Delta T)(L_3 + \alpha L_3 \, \Delta T) \\
&= L_1 L_2 L_3 (1 + \alpha \, \Delta T)^3 \\
&= V_0[1 + 3\alpha \, \Delta T + 3(\alpha \, \Delta T)^2 + (\alpha \, \Delta T)^3]
\end{aligned}
$$

Thus

$$\Delta V = V_0[3\alpha \, \Delta T + 3(\alpha \, \Delta T)^2 + (\alpha \, \Delta T)^3]$$

and

$$\frac{\Delta V/V_0}{\Delta T} = 3\alpha + 3\alpha^2 \, \Delta T + \alpha^3 (\Delta T)^2$$

Hence

$$\beta = \lim_{\Delta T \to 0} \frac{\Delta V/V}{\Delta T} = 3\alpha \qquad\qquad 18\text{-}5$$

The increase in size of any part of a body for a given temperature change is proportional to the original size of that part of the body. If we increase the temperature of a steel ruler, for example, the effect will be similar to that of a (very slight) photographic enlargement. Lines that were previously equally spaced will still be equally spaced but with correspondingly larger spaces. Values of $\alpha$ and $\beta$ for various substances are given in Table 18-1.

**Table 18-1**
Approximate Values of thermal-expansion coefficients per kelvin

| Material | $\alpha$, K$^{-1}$ | Material | $\beta$, K$^{-1}$ |
|---|---|---|---|
| Aluminum | $24 \times 10^{-6}$ | Acetone | $1.5 \times 10^{-3}$ |
| Brass | $19 \times 10^{-6}$ | Air | $3.67 \times 10^{-3}$ |
| Carbon, diamond | $1.2 \times 10^{-6}$ | Alcohol | $1 \times 10^{-3}$ |
| graphite | $7.9 \times 10^{-6}$ | Mercury | $0.18 \times 10^{-3}$ |
| Copper | $17 \times 10^{-6}$ | Water (20°C) | $0.207 \times 10^{-3}$ |
| Glass, ordinary | $9 \times 10^{-6}$ | | |
| Pyrex | $3.2 \times 10^{-6}$ | | |
| Ice | $51 \times 10^{-6}$ | | |
| Invar | $1 \times 10^{-6}$ | | |
| Steel | $11 \times 10^{-6}$ | | |

Though most materials expand when heated, the behavior of water at temperatures between 0 and 4°C is an important exception. Figure 18-1 shows the volume occupied by 1 g of water as a function of temperature. The volume is minimum and therefore the density is maximum at 4°C. Thus when water is heated at temperatures below 4°C it contracts rather than expands. This property has important consequences for the ecology of lakes. At temperatures above 4°C, as the water in a lake cools, it becomes denser and sinks to the bottom of the lake; but at temperatures below 4°C the cooler water is less dense and rises to the surface. Ice therefore forms first on the surface of a lake and, being less dense than water, remains there and acts as a thermal insulator for the water below. If the density of water were maximum at 0°C rather than at 4°C, lakes would freeze from the bottom up and would be much more likely to be completely frozen in the winter, killing the fish and other aquatic life.

**Figure 18-1**
Volume of 1 g of water at atmospheric pressure versus temperature. The minimum volume corresponding to maximum density occurs at 4°C.

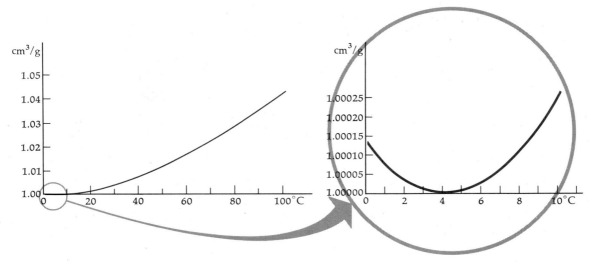

**Example 18-1** A steel bridge is 1000 m long. By how much does it expand when the temperature rises from 0 to 30°C?

From Table 18-1 the coefficient of expansion of steel is $11 \times 10^{-6}$ K$^{-1}$. The change in length for a 30-K rise in temperature is then $\Delta L = \alpha L_0 \Delta T = (11 \times 10^{-6}$ K$^{-1})$ $(1000$ m$)$ $(30$ K$) = 0.33$ m $= 33$ cm. To relieve the stresses that would occur if such expansion were not allowed, expansion joints must be included. We can calculate the stress involved if the bridge were not allowed to expand by using Young's modulus, $Y = $ stress/strain $= (\Delta F/A)/(\Delta L/L_0)$. If the bridge could not expand, the stress would be

$$\frac{\Delta F}{A} = Y \frac{\Delta L}{L_0}$$
$$= 2 \times 10^{11} \text{ N/m}^2 \frac{0.33 \text{ m}}{1000 \text{ m}}$$
$$= 6.6 \times 10^7 \text{ N/m}^2$$

where we have used the value of Young's modulus from Table 13-2.

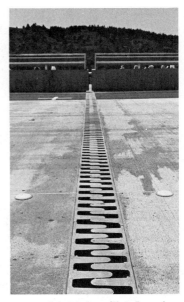

Robert A. Isaacs/Photo Researchers

An expansion grill at the end of a bridge allows the bridge to expand and contract without causing damage.

**Example 18-2** A 1-L glass flask is filled to the brim with alcohol at 10°C. If the temperature warms to 30°C, how much alcohol spills out of the flask?

From Table 18-1 we see that the coefficient of expansion for glass is negligible compared with that for alcohol; we can therefore neglect the change in volume of the glass. The increase in volume of the alcohol is

$$\Delta V = \beta V_0 \Delta T = (1.12 \times 10^{-3} \text{ K}^{-1}) (1 \text{ L}) (20 \text{ K})$$
$$= 2.24 \times 10^2 \text{ L}$$
$$= 22.4 \text{ mL}$$

**Example 18-3** Find the coefficient of expansion of an ideal gas.

For a given temperature change, a gas at constant pressure expands by much more than a liquid or solid. We can compute $\beta$ for any system if we know its equation of state. For an ideal gas we have

$$PV = nRT$$

or

$$V = \frac{nR}{P} T$$

Since $P$ is constant, we can easily find $dV/dT$:

$$\frac{dV}{dT} = \frac{nR}{P} \quad \text{and} \quad \beta = \frac{1}{V}\frac{dV}{dT} = \frac{nR}{PV} = \frac{nR}{nRT} = \frac{1}{T}$$

The expansivity equals the reciprocal of the absolute temperature.

**Example 18-4** A copper bar is heated to 300°C and then clamped rigidly between two fixed points so that it can neither expand nor contract. If the breaking stress of copper is 300 MN/m$^2$, at what temperature will the bar break as it cools?

In this example, the change in length $\Delta L$ that would occur if the bar contracted as it cooled must be offset by an equal stretching due to tensile stress in the bar. From the definition of Young's modulus (Equation 13-6), we have for the stretching $\Delta L$ caused by a tensile stress $\Delta F/A$

$$\Delta L = L \frac{\Delta F/A}{Y}$$

Setting this equal to the change in length that would occur by contraction if the bar could contract, we have

$$\Delta L = L\alpha \, \Delta T = L\alpha(300°C - t) = L\frac{\Delta F/A}{Y}$$

Solving for $\Delta T$ and using $Y = 110 \text{ GN/m}^2$ from Table 13-2, $\alpha = 17 \times 10^{-6} \text{ K}^{-1}$ from Table 18-1, and $\Delta F/A = 300 \text{ MN/m}^2$ for the breaking stress, we obtain

$$\Delta T = \frac{300 \times 10^6 \text{ N/m}^2}{(17 \times 10^{-6} \text{ K}^{-1})(11 \times 10^{10} \text{ N/m}^2)}$$
$$= 160 \text{ K}$$
$$t = 300 - 160 = 140°C$$

**Questions**

1. If a metal sheet with a hole in it expands, does the hole get larger or smaller?

2. If mercury and glass had the same coefficient of expansion, could a mercury thermometer be built?

3. Near the bottom of a deep lake, water is compressed to its maximum density by the weight of the water above it. What value do you expect for the water temperature near the bottom of the lake?

# 18-2   Change of Phase and Latent Heat

When heat is supplied to a body at constant pressure, the result usually is an increase in the temperature of the body, but sometimes a body can absorb large amounts of heat without any change in temperature. This happens when the physical condition, or *phase*, of the material is changing from one form to another, e.g., from liquid to gas, from solid to liquid, or from one crystalline form to another. Thus the effect of heat is not only to change the temperature of a body but sometimes also to change the phase without any temperature change.

These phenomena are easily understood in terms of the mechanical or molecular theory of heat. When the temperature of a body increases because heat is added, it reflects an increase of the kinetic energy of motion of the molecules. When a material changes from liquid to gaseous form, the molecules, which originally were held together by their natural attraction, are moved far apart from each other. This requires that work be done against the attractive forces; i.e., it requires that energy be supplied to the molecules to separate them even though their kinetic energies (and the temperature) do not increase in the process. From this model we would expect the change in phase from liquid to gas to require heat even though there is no rise in temperature. Similarly, other changes, e.g., from solid to liquid or from one solid form to another, are rearrangements of molecules and involve work done against the forces they exert on each other. It is not surprising that heat energy may be required to accomplish these rearrangements at constant temperature.

For a pure substance, a change of phase occurs for a given pressure only at a particular temperature. For example, pure water at atmospheric pressure changes from solid to liquid at 0°C and from liquid to gas at 100°C. The first temperature is called the normal melting point and the second the normal boiling point. These points were those chosen by Anders Celsius (1701–1744) for his temperature scale based on two fixed points.

A specific quantity of heat energy is required to change the phase of a given amount of substance. The required heat is proportional to the mass of the material:

$$Q = mL \hspace{4cm} 18\text{-}6$$

where $L$ is a constant characteristic of the substance and the kind of phase change involved. If the phase change is from solid to liquid, $L_f$ is called the *latent heat of fusion* of the substance. For water at atmospheric pressure, the latent heat of fusion is 333.5 kJ/kg = 79.7 kcal/kg. If the phase change is from liquid to gas, $L_v$ is called the *latent heat of vaporization*. For water at atmospheric pressure, the latent heat of vaporization is 2.26 MJ/kg = 540 kcal/kg. Table 18-2 gives the normal melting and boiling points and latent heats of fusion and vaporization at 1 atm for various substances.

*Latent heat*

Many substances have several different modifications of their solid or liquid phase. For example, eight different forms or phases of ice are known, each with properties different from the others. Transitions between these forms occur at definite temperatures and pressures, just like transitions between a solid and liquid phase. Helium has two very different liquid phases. Figure 18-2 shows the specific heat of helium as a function of temperature in the range 1 to 3 K. The specific heat has a sharp discontinuity at the temperature of 2.17 K. Because the shape of this curve resembles the greek letter lambda ($\lambda$), this transition temperature is called the *lambda point*. The liquid above the lambda point is called helium I, and that below the lambda point is called helium II.

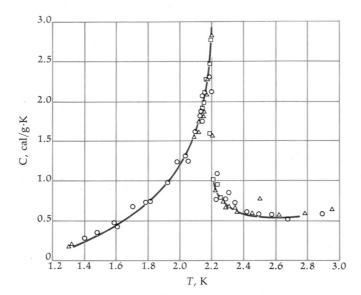

**Figure 18-2**
Specific heat of liquid helium versus temperature. Because of the resemblance of this curve to the Greek letter $\lambda$, the phase-transition point is called the lambda point. (*From F. London,* Superfluids, *New York, Dover Publications, Inc., 1964. Reprinted by permission of the publisher.*)

These two fluids are very different. For example, liquid helium II, called a superfluid because of its remarkable properties, has essentially zero viscosity and is a perfect heat conductor, whereas the viscosity and thermal properties of liquid helium I are similar to those of any other ordinary liquid.

**Table 18-2**
Normal melting point (MP), latent heat of fusion $L_f$, normal boiling point (BP), and latent heat of vaporization $L_v$ for various substances at 1 atm

| | MP, K | $L_f$, kJ/kg | BP, K | $L_v$, kJ/kg |
|---|---|---|---|---|
| Alcohol, ethyl | 159 | 109 | 351 | 879 |
| Bromine | 266 | 67.4 | 332 | 369 |
| Carbon dioxide | 215 | 108.7 | 194.6* | 573* |
| Copper | 1356 | 205 | 2839 | 4726 |
| Gold | 1336 | 62.8 | 3081 | 1701 |
| Helium | 1.76 | 2.1 | 4.2 | 21 |
| Lead | 600 | 24.7 | 2023 | 858 |
| Mercury | 234 | 11.3 | 630 | 296 |
| Nitrogen | 63 | 25.7 | 77.35 | 199 |
| Oxygen | 54.4 | 13.8 | 90.2 | 213 |
| Silver | 1234 | 105 | 2436 | 2323 |
| Sulfur | 388 | 38.5 | 717.75 | 287 |
| Water | 273.15 | 333.5 | 373.15 | 2257 |
| Zinc | 692 | 102 | 1184 | 1768 |

\* These values are for sublimation.

**Example 18-5** If 1 kg of ice at $-20°C$ is heated at atmospheric pressure until all the ice has been changed into steam, how much heat is required?

Assuming the heat capacity of ice to be constant and equal to 2.05 kJ/kg·K (Table 17-1), we find for the heat energy needed to raise the temperature of the ice from $-20$ to $0°C$

$$Q_1 = mc\ \Delta T = (1\ \text{kg})\ (2.05\ \text{kJ/kg·K})\ (20\ \text{K})$$
$$= 41\ \text{kJ}$$

The heat needed to melt 1 kg of ice is

$$Q_2 = mL_f = 334\ \text{kJ}$$

The heat needed to raise the temperature of the resulting 1 kg of water from 0 to 100°C is

$$Q_3 = mc\ \Delta T = (1\ \text{kg})\ (4.18\ \text{kJ/kg·K})\ (100\ \text{K}) = 418\ \text{kJ}$$

where we have neglected any variation in the heat capacity of water over this range of temperatures. Finally, the heat needed to vaporize 1 kg of water at 100°C is

$$Q_4 = mL_v = 2.26\ \text{MJ} = 2260\ \text{kJ}$$

and the total amount of heat required is

$$Q = Q_1 + Q_2 + Q_3 + Q_4 = 3.05\ \text{MJ}$$

Note that most of the heat supplied in this process was needed to change the phase of the water, not to raise its temperature.

## 18-3  The van der Waals Equation and Liquid-Vapor Isotherms

As mentioned in Section 16-6, the van der Waals equation of state was proposed as a correction to the ideal-gas equation to take into account the attraction of the gas molecules for each other and also their finite size. For 1 mol of a gas the van der Waals equation is (Equation 16-18)

$$\left(P + \frac{a}{V^2}\right)(V - b) = RT \qquad\qquad 18\text{-}7$$

Figure 18-3 shows the $P$-versus-$V$ isothermal curves for various temperatures for a real gas. For temperatures above some critical temperature labeled $T_c$ on the diagram, these curves are described quite accurately by the van der Waals equation and can be used to determine the constants $a$ and $b$. For example, the best-fit values of these constants for nitrogen are $a = 0.14$ Pa·m$^6$/mol$^2$ and $b = 39.1$ cm$^3$/mol. Since the constant $b$ is the actual volume of 1 mol of gas molecules, we can use it to estimate the size of a nitrogen molecule. The volume occupied by one molecule of nitrogen is

$$v = \frac{b}{N_A} = \frac{39.1 \text{ cm}^3/\text{mol}}{6.02 \times 10^{23} \text{ molecules/mol}} = 6.50 \times 10^{-23} \text{ cm}^3$$

If we assume that each molecule is a sphere of diameter $d$ occupying a cubic volume of side $d$, we obtain for $d$

$$d^3 = 6.5 \times 10^{-23} \text{ cm}^3 \quad \text{or} \quad d = 4.0 \times 10^{-8} \text{ cm}$$

which is a reasonable estimate for the diameter of a molecule.

At temperatures below $T_c$, the van der Waals equation describes the curves outside the shaded region in Figure 18-3, but inside this region the function $V(P)$ as given by Equation 18-7 is multivalued; i.e., for a given value of $P$ there are two or three values of $V$, as indicated by the dotted lines.

Suppose we have a gas at a temperature below $T_c$, initially at a low pressure and large volume, and begin to compress it isothermally. At first the pressure rises, but when we reach the dashed line on the curve in Figure 18-3, the pressure ceases to rise. As we continue to compress the gas, the pressure remains constant. What is happening here is that the gas liquefies. Along the horizontal part of the curve in the shaded region of the figure the gas and liquid are in equilibrium. As we continue to compress the gas, more and more gas liquefies until we have only liquid. Then if we try to compress further, the pressure rises sharply because a liquid is nearly incompressible. The constant pressure for which the gas and liquid are in equilibrium at that temperature is called the *saturated vapor pressure* or sometimes just the vapor pressure. It is clear from Figure 18-3 that the vapor pressure of a gas depends on temperature. If we compress the gas at a lower temperature than before, the vapor pressure is lower, as indicated by the constant-pressure horizontal line at a lower value of pressure. If the vapor pressure equals the external pressure, the liquid boils at that temperature. The temperature for which the saturated vapor pressure is 1 atm is the normal boiling point for that substance (Table 18-2). Table 18-3 gives the vapor pressure of water as a function of temperature. The vapor pressure of water is 1 atm at a temperature of 373 K = 100°C, the

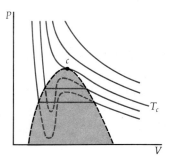

**Figure 18-3**
Isotherms on the $PV$ diagram for a real substance. For temperatures above the critical temperature $T_c$, the substance remains a gas at all pressures and is described accurately by the van der Waals equation. Below $T_c$, the van der Waals equation is not followed in the shaded region (dotted lines); instead, the gas and liquid phases are in equilibrium and the isotherm is horizontal at the saturated vapor pressure. To the left of the shaded region for $T < T_c$, the substance is liquid and nearly incompressible.

*Vapor pressure*

normal boiling point of water. At temperatures greater than the critical temperature $T_c$, the gas does not liquefy at any pressure. The critical temperature of water vapor is 647 K = 374°C. The point at which the critical isotherm intersects the dashed curve in Figure 18-3 is called the *critical point*. The critical point is a point of inflection for the critical isotherm; i.e., both $dP/dV$ and $d^2P/dV^2$ are zero at this point. The values of $P$ and $V$ at the critical point of water vapor are $P_c = 218.3$ atm and $V_c = 0.056$ m³/mol.

Consider putting a liquid such as water in an evacuated container that is sealed so that its volume is constant. At first some of the water will evaporate, and the water-vapor molecules will fill the previously empty space in the container. When there are water-vapor molecules in the container, some of them will condense again into liquid water. At first the rate of evaporation will be greater than that of condensation and the density of vapor molecules will increase, but as there are more and more water-vapor molecules in the container, the condensation rate will be greater and greater until it equals the rate of evaporation and equilibrium will be established. The pressure of the water vapor in equilibrium will be the vapor pressure for that temperature. If we now heat the container to a greater temperature, more liquid will evaporate and a new equilibrium will occur at a higher vapor pressure. Figure 18-4 is called a *phase diagram*. The portion of this diagram between points $O$ and $C$ shows the vapor pressure versus temperature for a constant volume. As we continue to heat the container, the density of the liquid decreases and the density of the vapor increases. At point $C$ in Figure 18-4 these densities are equal. This point is the critical point labeled $T_c$ on Figure 18-3. At this point and above it there is no distinction between the liquid and the gas.* Table 18-4 lists critical temperatures for various substances.

*Critical temperature*

**Table 18-3**
Vapor pressure of water versus temperature

| t, °C | P | |
|---|---|---|
| | mmHg | kPa |
| 0 | 4.581 | 0.611 |
| 10 | 9.209 | 1.23 |
| 20 | 17.535 | 2.34 |
| 30 | 31.827 | 4.24 |
| 40 | 55.335 | 7.38 |
| 50 | 92.55 | 12.3 |
| 60 | 149 | 19.9 |
| 70 | 233.8 | 31.2 |
| 80 | 355 | 47.4 |
| 90 | 526 | 70.1 |
| 100 | 760 | 101.3 |
| 110 | 1074 | 143.3 |
| 120 | 1489 | 198.5 |
| 130 | 2026 | 270.1 |

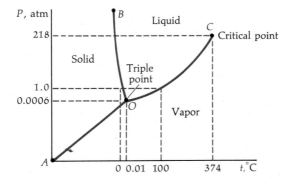

**Figure 18-4**
Phase diagram for water. The pressure and temperature scales are not linear but compressed so as to show the interesting points. Curve $OC$ is the vapor-pressure curve, curve $OA$ the sublimation curve, and curve $OB$ the melting curve.

If we now cool our container, the vapor condenses into a liquid and we move along the curve $OC$ until we reach point $O$ in Figure 18-4. At this point the liquid begins to solidify, and at this one point the vapor, liquid, and solid phases of a substance can coexist in equilibrium. This occurs at a unique temperature and pressure called the *triple point*. For water the triple-point temperature is 273.16 K = 0.01°C and the triple-point pressure is 4.58 mmHg.

At temperatures and pressures below the triple point, the liquid cannot exist. The curve $OA$ on the phase diagram of Figure 18-4 is the

---

* The word vapor is usually used if the temperature is below the critical temperature, and the word gas is used if the temperature is above the critical temperature, though there is no need for such a distinction.

locus of pressures and temperatures for which the solid and vapor coexist in equilibrium. Because the direct change from a solid to a vapor is called *sublimation,* this part of the phase-diagram curve is called the *sublimation curve.* Since the triple-point pressure of water is so low (4.58 mmHg), this is not ordinarily observed with water. The triple-point pressure and temperature for carbon dioxide ($CO_2$) are 216.55 K and 3880 mmHg. Liquid $CO_2$ can therefore exist only at pressures above 3880 mmHg = 5.1 atm. At ordinary atmospheric pressure $CO_2$ sublimates directly into gaseous $CO_2$ without going through the liquid phase; hence the name dry ice. The curve $OB$ in Figure 18-4 is the melting curve separating the liquid and solid phases. In this figure the curve is drawn for a substance such as water for which the melting temperature decreases as the pressure increases. For many other substances the melting temperature increases as the pressure increases, and the portion of the curve $OB$ for such a substance slopes upward to the right from the triple point. The fact that the melting temperature of water decreases as the pressure increases makes ice skating possible.

*Sublimation*

**Questions**

4. At high altitudes, as in the mountains, it takes longer to cook things in boiling water than at sea level. Why?

5. What is the advantage of a pressure cooker?

6. How can dry ice be stored so that it doesn't completely sublime away?

7. Why is helium so difficult to liquefy?

**Table 18-4**
Critical temperatures $T_c$ for selected substances

|  | $T_c$, °C |
| --- | --- |
| Argon | −122.3 |
| Carbon dioxide | 31 |
| Chlorine | 144 |
| Helium | −267.9 |
| Hydrogen | −239.9 |
| Neon | −228.7 |
| Nitric oxide | −93 |
| Oxygen | −118.4 |
| Sulfur dioxide | 157.8 |

## 18-4  The Transfer of Heat

Thermal energy is transferred from one place to another by three main processes: conduction, convection, and radiation. In heat conduction, thermal energy is transferred by the interactions of its molecules though there is no transport of the molecules themselves. For example, if one end of a solid bar is heated, the lattice atoms in the heated end vibrate with greater energy than those at the cooler end and, because of the interaction of these atoms with their neighbors, this energy is transported along the bar. If the solid is a metal, this transport of thermal energy is helped by free electrons which move throughout the metal, receiving and giving off thermal energy as they collide with the lattice atoms. In a gas, heat is conducted by direct collisions of the gas molecules. Molecules in the warmer part of the gas have higher energy than the average and lose energy in collisions with molecules of lower average energy from the cooler part of the gas. In convection, heat is transported by a direct mass transport. For example, warm air near the floor expands and rises because of its lower density. Thermal energy in this warm air is thus transported from the floor to the ceiling along with the mass of warm air. In heat radiation, energy is emitted and absorbed by all bodies in the form of electromagnetic radiation. If a body is in thermal equilibrium with its surroundings, it emits and absorbs energy at the same rate, but if it is warmed to a higher temperature than its surroundings, it radiates away more energy than it absorbs, thus cooling down as its surroundings warm.

## Conduction and Convection

Figure 18-5 shows a solid bar of cross-sectional area $A$. If we keep one end of the bar at a high temperature, e.g., in a steam bath, and the other end at a lower temperature, e.g., in an ice bath, thermal energy is continually conducted down the bar from the hot end to the cold end. In the steady state the temperature varies uniformly (if the bar is uniform) from the high-temperature end to the low-temperature end; i.e., there is a *temperature gradient* $\Delta T / \Delta x$ along the bar. Experimentally it is found that the rate of conduction of thermal energy is proportional to this temperature gradient and to the cross-sectional area. The proportionality constant $k$ is called the *coefficient of thermal conductivity*. Let us consider a small portion of the bar, a slab of thickness $\Delta x$, and let $\Delta T$ be the temperature difference across the slab. If $\Delta Q$ is the amount of heat energy conducted through the slab in some time interval $\Delta t$, we have for the heat current $I$

$$I = \frac{\Delta Q}{\Delta t} = kA \frac{\Delta T}{\Delta x} \qquad\qquad \text{18-8}$$

*Thermal conduction*

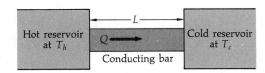

Conducting bar

**Figure 18-5**
A conducting bar between two heat reservoirs. The rate of heat conducted along the bar is proportional to the area of the bar and to the temperature difference, and inversely proportional to the length of the bar.

(This equation is often written in calculus notation with $dQ/dt$ for the heat current and $dT/dx$ for the temperature gradient; a minus sign is sometimes included because the heat energy moves in the direction of decreasing temperature. We shall omit the minus sign for simplicity in notation.) The coefficient of thermal conductivity $k$ depends on the composition of the bar. It also depends slightly on the temperature. In SI units the heat current is in watts (joules per second), and the thermal conductivity has the units of watts per metre-kelvin, though in some tables the energy may be given in calories or kilocalories and the thickness in centimetres. In practical calculations in the United States, e.g., finding the heat conducted through the walls of a room, the heat current is usually expressed in Btu per hour, the area in square feet, the thickness in inches, and the temperature in degrees Fahrenheit. The thermal conductivity is then given in Btu·in/h·ft²·°F. Table 18-5 lists values of thermal conductivity for various materials in both sets of units.

If we solve Equation 18-8 for the temperature difference, we can write

$$\Delta T = \frac{\Delta x}{kA} I = IR \qquad\qquad \text{18-9}$$

where the *thermal resistance $R$* is given by

$$R = \frac{\Delta x}{kA} \qquad\qquad \text{18-10}$$

*Thermal resistance*

Equation 18-9 is of the same form as Equation 13-39 for the viscous flow of a fluid through a pipe, except that $I$ now stands for the flow of heat

**Table 18-5**
Thermal conductivities $k$ for various materials

|  | $k$, W/m·K | $k$, Btu·in/h·ft²·°F |
|---|---|---|
| Air (27°C) | 0.026 | 0.18 |
| Ice | 2.21 | 15.3 |
| Water (27°C) | 0.609 | 4.22 |
| Aluminum | 237 | 1644 |
| Copper | 401 | 2780 |
| Gold | 318 | 2200 |
| Iron | 80.4 | 558 |
| Lead | 353 | 2450 |
| Silver | 429 | 2980 |
| Steel | 46 | 319 |
| Oak | 0.15 | 1.02 |
| Maple | 0.16 | 1.1 |
| White pine | 0.11 | 0.78 |
| Brick | 0.4–0.9 | 3–6 |
| Concrete | 0.9–1.3 | 6–9 |
| Cork board | 0.04 | 0.3 |
| Glass | 0.7–0.9 | 5–6 |
| Glass wool | 0.042 | 0.29 |
| Masonite | 0.048 | 0.33 |
| Plaster | 0.3–0.7 | 2–5 |
| Rock wool | 0.039 | 0.27 |

energy rather than the flow of a fluid, and the pressure difference $\Delta P$ is replaced by the temperature difference $\Delta T$. When we study electricity, we shall encounter a similar equation, known as Ohm's law, for the flow of electric charge. In that case we replace $\Delta T$ in Equation 18-9 by $\Delta V$, the voltage difference; $I$ becomes the electric current; and $R$ the electric resistance. (The electric resistance is also related to the thickness $\Delta x$ and cross-sectional area $A$ of the conductor by Equation 18-10, $k$ then being the electrical conductivity.)

In many practical problems we are interested in the flow of heat through two or more conductors (or insulators) in series. For example, we may wish to know the effect of adding insulating material of a certain thickness and thermal conductivity to the space between two layers of plasterboard. Figure 18-6 shows two slabs of the same area but of different material and different thicknesses. Let $T_1$ be the temperature on the warm side, $T_2$ be that at the interface between the slabs, and $T_3$ be that on the cool side. Under conditions of steady-state heat flow, the thermal current must be the same through each slab. This follows from energy conservation. If the thermal currents were not the same in both slabs, energy would be deposited or generated at the interface between the slabs. For example, if the thermal current were larger in the first slab, more energy would flow toward the interface than away from it in

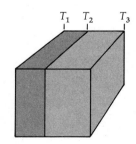

**Figure 18-6**
Two conducting slabs of the same area but different thickness and material. The conductors are in series. The equivalent thermal resistance of conductors in series equals the sum of the individual thermal resistances.

a given time. If $R_1$ and $R_2$ are the thermal resistances of the two slabs, we have from Equation 18-9 for each slab

$$T_1 - T_2 = IR_1 \quad \text{and} \quad T_2 - T_3 = IR_2$$

Adding gives

$$\Delta T = T_1 - T_3 = (R_1 + R_2)I = R_{eq}I \qquad \text{18-11}$$

where the equivalent resistance is

$$R_{eq} = R_1 + R_2 \qquad \text{18-12} \qquad \textit{Resistors in series}$$

For thermal resistors in series, the effective resistance is the sum of the individual resistances. This result can be applied to any number of resistances in series. It is the same result we shall obtain later for electric resistance in series.

In the building industry the thermal resistance per square foot in U.S. customary units is called the $R$ factor, which we shall designate by $R_f$. The $R$ factor is merely the thickness divided by the thermal conductivity:

$$R_f = \frac{\Delta x}{k} \qquad \text{18-13} \qquad \textit{R factor}$$

Table 18-6 lists $R$ factors for several materials. To find the rate of heat conduction per unit area in Btu/h·ft² through several slabs of materials, we multiply the area by the temperature difference in degrees Fahrenheit and divide by the sum of the $R$ factors.

---

**Example 18-6** A 60- by 20-ft roof is made of 1-in pine board with asphalt shingles. If the overlap in the shingles is neglected, how much heat is conducted through the roof when the inside temperature is 70°F and the outside is 40°F? By what factor is the heat loss reduced if 2 in of roof insulation is added?

From Table 18-5, since the thermal conductivity of pine board is 0.78 Btu·in/h·ft²·°F, the $R$ factor for a board of 1-in thickness is

$$R_f = \frac{\Delta x}{k} = \frac{1}{0.78} = 1.28$$

From Table 18-6 the $R$ factor for asphalt shingles is 0.44. The $R$ factor for the combination is therefore $1.28 + 0.44 = 1.72$. For a temperature difference of 30°F, the heat conducted per square foot is

$$\frac{I}{A} = \frac{\Delta T}{R_{eq}} = \frac{30}{1.72} = 17.4 \text{ Btu/h·ft}^2$$

The rate of conduction through the area of 60 by 20 ft $= 1200$ ft² is

$$I = 17.4(1200) = 21{,}000 \text{ Btu/h}$$

From Table 18-6 the $R$ factor for roof insulation is 2.8 in⁻¹, so for 2 in the $R$ factor is 5.6. Therefore the equivalent $R$ factor for the pine board plus shingles plus insulation is $R_{eq} = 1.72 + 5.6 = 7.32$. Since this is greater than 1.72 by a factor of 4.26, the rate of heat conducted through the roof is decreased by this factor. For the temperature difference and area given, the rate of heat conduction with the insulation will be (21,000 Btu/h)/4.26 = 4900 Btu/h.

---

To calculate the amount of heat leaving a room by conduction in a given time we need to know how much leaves through the walls, the windows, the floor, and the ceiling. This problem is analogous to find-

**Table 18-6**
$R$ factors $\Delta x/k$ and $1/k$ for various building materials

| Material | Thickness, in | $R_f$, h·ft²·°F/Btu | $R_f$ per in, h·ft²·°F/Btu·in |
|---|---|---|---|
| Building board | | | |
|   Gypsum or plaster board | 0.375 | 0.32 | |
|   Plywood (Douglas fir) | 0.5 | 0.62 | |
|   Plywood or wood panels | 0.75 | 0.93 | |
|   Particle board, medium density | | | 1.06 |
| Finish flooring materials | | | |
|   Carpet and fibrous pad | | | 2.08 |
|   Tile | | 0.5 | |
|   Wood, hardwood finish | 0.75 | 0.68 | |
| Insulating materials | | | |
|   Blanket and batt | ~3–3.5 | 11 | |
|   Mineral fiber | ~6–7 | 22 | |
|   Board and slabs | | | |
|     Expanded polystyrene, extruded | | 5 | |
|     Molded beads | | 3.6 | |
|   Loose fill | | | |
|     Cellulosic insulation | | | 3.1–3.7 |
|     Sawdust or shavings | | | 2.2 |
|     Vermiculite, exfoliated | | | 2.13 |
|   Roof insulation | | | 2.8 |
| Roofing | | | |
|   Asphalt roll roofing | | 0.15 | |
|   Asphalt shingles | | 0.44 | |
| Windows | | | |
|   Single-pane | | 0.9 | |
|   Double-pane | | 1.8 | |
|   Triple-pane | | 2.7 | |

ing the electric current through a system of parallel resistors. The temperature difference is the same across each conductor or resistor, but the heat current is different. The total heat current is the sum of the currents through each element. We have then

$$I_{tot} = I_1 + I_2 + \cdots = \frac{\Delta T}{R_1} + \frac{\Delta T}{R_2} + \cdots$$

or

$$I_{tot} = \frac{\Delta T}{R_{eq}} \qquad\qquad 18\text{-}14$$

where the equivalent thermal resistance in this case is given by

$$\frac{1}{R_{eq}} = \frac{1}{R_1} + \frac{1}{R_2} + \cdots \qquad\qquad 18\text{-}15 \qquad \textit{Resistors in parallel}$$

Snow patterns on the roof of a house indicate the conduction—or lack of conduction—of heat through the roof. Note the snow over the unheated garage. This new house is poorly insulated compared to the one in the background, which still has snow on its roof. (*Courtesy of A. A. Bartlett, Boulder, Colorado.*)

Although the thermal conductivity of air is relatively small, so that air makes a good insulator, the efficiency of a large amount of air—as between a storm window and the inside window—is greatly reduced because of convection. As soon as there is a temperature difference between different parts of the airspace, convection currents act quickly to equalize the temperature and the effective conductivity is greatly increased. Air gaps of about 1 to 2 cm are optimal; wider air gaps actually reduce the thermal resistance of a double-pane window because of convection. It is possible to use the insulating properties of air if it can be trapped in small pockets that are separated from each other so that convection cannot take place. For example, goose down is a good thermal insulator because it fluffs up, trapping air, which cannot circulate and transport heat by convection. Another example is styrofoam, a cellular material with tiny pockets of air separated by the poorly-conducting cell walls that prevent convection. The thermal conductivity of such a material is essentially the same as that of air.

If you touch the inside glass surface of a window when it is cold outside, you will observe that the surface is considerably colder than the inside air. The thermal resistance of a window is due mainly to thin insulating air films, which adhere to either side of the glass surface. The thickness of the glass has little effect on the overall thermal resistance. An air film typically adds an $R$ factor of about 0.45 h·ft²·°F/Btu, so the $R$ factor of a window with $N$ panes is approximately $0.9N$ h·ft²·°F/Btu. Under windy conditions, the outside air film may be greatly decreased, leading to a smaller $R$ factor for the window.

It is possible to write an equation for the thermal energy transported by convection and define a coefficient of convection, but the analysis of practical problems is very difficult and will not be treated here. To some approximation, the heat transferred from a body to its surroundings is proportional to the area of the body and to the difference in temperature between the body and the surrounding fluid.

### Radiation

The third mechanism for heat transfer is radiation in the form of electromagnetic waves. The rate at which a body radiates thermal energy is proportional to the area of the body and to the fourth power of its absolute temperature. This result, found empirically by Josef Stefan in 1879, is written

$$I = e\sigma AT^4 \qquad \text{18-16}$$

*Stefan-Boltzmann law*

where $I$ is the power radiated in watts, $A$ is the area, $e$ is a fraction between 0 and 1 called the *emissivity* of the body, and $\sigma$ is a universal constant called *Stefan's constant*, which has the value

$$\sigma = 5.6703 \times 10^{-8} \text{ W/m}^2 \cdot \text{K}^4 \qquad \text{18-17}$$

Equation 18-16 was derived theoretically by Ludwig Boltzmann about 5 years later and is now called the Stefan-Boltzmann law. When radiation falls on an opaque body, part of the radiation is reflected and part is absorbed. Light-colored bodies reflect most of the radiation, whereas dark bodies absorb most of it. The radiation absorbed is proportional to the area of the body and to the fourth power of the temperature of the surroundings. We can write for the absorption of radiation

$$I_a = a\sigma AT^4 \qquad \text{18-18}$$

where $a$ is the coefficient of absorption, which is a fraction between 0 and 1. Consider a hot body placed in an environment at a lower temperature. The body emits more radiation than it absorbs and cools down while the surroundings absorb radiation from the body and warm up. Eventually the body and surroundings are at the same temperature and are in thermal equilibrium. Then the body must absorb energy at the same rate as it emits it. The coefficient of absorption $a$ must therefore be equal to the emissivity $e$. We can write for the net power radiated by a body at temperature $T$ in an environment at temperature $T_0$

$$I_{\text{net}} = e\sigma A(T^4 - T_0^4) \qquad \text{18-19}$$

A body that absorbs all the radiation incident upon it has an emissivity equal to 1 and is called a *blackbody*. A blackbody is also an ideal radiator. Materials like black velvet or lampblack come close to being ideal blackbodies, but the best practical realization of an ideal blackbody is a small hole leading into a cavity (Figure 18-7), e.g., a keyhole in a closet door. Radiation incident on the hole has little chance of being reflected back out the hole before it is absorbed by the walls of the cavity. The ideal blackbody is important because the characteristics of the radiation emitted by such a body can be calculated theoretically.

At ordinary temperatures (below about 600°C) the thermal radiation emitted by a body is not visible; most of it is concentrated in wavelengths much longer than those of visible light. (When we study light in Chapter 32, we shall see that visible light is an electromagnetic radiation with wavelengths between about 400 and 700 nm.) As the body is heated, the quantity of energy emitted increases (Equation 18-16) and the energy radiated extends to shorter and shorter wavelengths. At about 600 to 700°C there is enough energy in the visible spectrum for the body to glow and become a dull red; at higher temperatures it becomes bright red or even "white hot." Figure 18-8 shows the power radiated by a blackbody as a function of wavelength for several different temperatures. The wavelength at which the power is a maximum varies inversely with the temperature:

$$\lambda_m \propto \frac{1}{T}$$

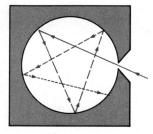

**Figure 18-7**
Cavity approximating an ideal blackbody. Radiation entering the cavity has little chance of leaving before it is completely absorbed.

*Blackbody*

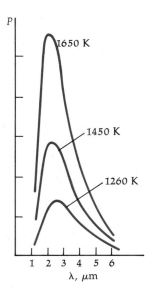

**Figure 18-8**
Spectral distribution of radiation from a blackbody for three different temperatures.

The proportionality constant has the value 2.898 mm·K. Thus

$$\lambda_m = \frac{2.898 \text{ mm·K}}{T} \qquad\qquad 18\text{-}20$$

*Wien's displacement law*

This result is known as *Wien's displacement law*. The calculation of the shape of the spectral-distribution curves shown in Figure 18-8 played an important role in the history of physics. It was the discrepancy in the theoretical calculation of these curves from classical thermodynamics and the experimental measurements that led to the first ideas of the quantization of energy by Max Planck in 1897. We shall discuss this problem further in Chapter 36.

**Example 18-7** The surface temperature of the sun is about 5000 K. If the sun is assumed to be a blackbody radiator, at what wavelength $\lambda_m$ would its spectrum peak?

From the Wien displacement law (Equation 18-20) we have

$$\lambda_m = \frac{2.898 \text{ mm·K}}{5000 \text{ K}} = 579.8 \times 10^{-9} \text{ m} = 579.8 \text{ nm}$$

This wavelength is in the middle of the visible spectrum.

**Example 18-8** Calculate the net loss in radiated energy of a naked person in a room at 20°C assuming the person to be a blackbody, the area of the body to be 1.4 m², and its surface temperature to be 33°C = 306 K. [The surface temperature of the human body is slightly less than the internal temperature (37°C) because of the thermal resistance of the skin.]

Using Equation 18-19, we have

$$I_{\text{net}} = (1)\ (5.67 \times 10^{-8} \text{ W/m}^2\text{·K}^4)\ (1.4 \text{ m}^2)\ [(306)^4 - (293)^4]$$
$$= 111 \text{ W}$$

This is slightly greater than the basic metabolism rate of about 100 W (assuming about 2000 kcal intake per day; see Exercise 17-2). We protect ourselves from such a great energy loss by wearing clothing, which, because of its low thermal conductivity, has a much lower outside temperature and therefore a much lower rate of heat radiation.

If the absolute temperature of a body is not too different from that of its surroundings, the net power radiated is approximately proportional to the temperature difference. We can see this from Equation 18-19 by noting that $T^4 - T_0^4 = (T^2 + T_0^2) \times (T^2 - T_0^2) = (T^2 + T_0^2) \times (T + T_0) \times (T - T_0)$, and also noting that in the sums we can replace either $T$ or $T_0$ by the other or by their mean value without changing the result much. Replacing $T_0$ by $T$ in these sums, we have, to a good approximation,

$$T^4 - T_0^4 \approx (T^2 + T^2)\ (T + T)\ (T - T_0) = 4T^3\ \Delta T$$

[This result can also be obtained by calculus by taking the differential of $T^4$; $d(T^4) = 4T^3\ dT$.] Thus for all mechanisms of heat transfer:

*The rate of cooling of a body is approximately proportional to the temperature difference between the body and its surroundings.*

*Newton's law of cooling*

This result is known as *Newton's law of cooling*.

**Questions**

8. In a cold room, a metal or marble table top feels much colder to the touch than a wood surface does even though they are at the same temperature. Why?

9. Which heat-transfer mechanisms are most important in the warming effect of a fire in a fireplace? In the transfer of energy from the sun to the earth?

10. In the body, thermal energy is carried from one part to another by the blood. Which heat-transfer mechanism is involved in this process?

11. Do you think that the fact that the radiant energy from the sun is concentrated in the visible range of wavelengths is an accident or not? If not, what is the reason?

---

# Solar Energy

Laurent Hodges
*Iowa State University*

Until very recently we have depended on renewable solar energy in its many forms as our sole energy resource. Sunlight and wood provided warmth, beasts of burden provided mechanical power, wind enabled ships to transport passengers and freight, and wind and water power provided mechanical energy for such purposes as grinding grain. The United States was predominantly a solar-energy economy until the 1880s, when coal displaced wood as the major energy resource. By that time, Great Britain and other European countries had already been relying on coal for some time. (Coal and the other fossil fuels are nonrenewable energy resources, although their ultimate source was also solar energy.)

### Advantages and Disadvantages of Solar Energy

Today there is renewed interest in solar energy because of its many advantages. It is available to at least some extent everywhere in the world, unlike the fossil and nuclear fuels. Solar energy itself costs nothing and is thus immune to rising energy prices. It can be used in many different ways, to provide heating, cooling, lighting, mechanical power, electricity, and transportation. Most (but not all) methods of using solar energy create few environmental problems.

Solar energy also has some disadvantages. It is not highly concentrated, although enough solar energy for some important purposes can be collected by using a modest area of land or a rooftop or wall. It is intermittent, its flow being interrupted by nights and cloudy days, but there are some good and often inexpensive ways of storing it for these periods. Many solar-energy systems require a large capital investment, but the amortization costs are often more than offset by the savings in energy prices.

### Solar Radiation

The solar radiation that lands on the earth resembles the spectral distribution of a blackbody at about 6000 K, the temperature of the sun's surface, with about 47 percent of solar energy in the visible spectrum, 45 percent in the infrared, and 8 percent in the ultraviolet.

The earth is $150 \times 10^6$ km from the sun and intercepts only a small part—about 0.5 billionths—of the total radiation from the sun. The average energy flux at this distance on a surface perpendicular to the sun's rays is about 1353 W/m²; since the earth-to-sun distance varies by about 6 percent over the course of the year, from perihelion in early January to aphelion in early July, this "solar constant" varies from about 1308 to 1398 W/m².

These values apply to the top of the earth's atmosphere. The atmosphere reduces the flux considerably, typically to 1 kW/m² on a clear day and a few watts per square metre on an overcast day. When the sun's rays intercept a surface at an angle, the flux onto the surface is further reduced for geometrical reasons.

The total solar radiation reaching the top of the atmosphere is about $1.7 \times 10^{17}$ W, only about half of which actually reaches the ground. Distributed over the whole globe, this amounts to about 170 W/m² averaged over the whole year, day and night. Thus 1 m² of the earth's surface receives about 4 kW·h daily or 1460 kW·h annually. For the land area of the United States, the average is a little higher, about 4.7 kW·h/m² daily or a total of over $10^{16}$ kW·h annually for the entire country; this is about 600 times greater than the total energy consumption in the United States.

The total amount of solar radiation falling on a horizontal surface is much greater in summer, when the days are longer and the sun's altitude is very high at midday, than it is in winter. In the Midwest, for example, a horizontal surface might receive 30 MJ/m², or a little more, on a sunny day in June but rarely over 10 MJ/m² on sunny days in December. The situation is different for a vertical south-facing surface, which receives far more in winter (when the sun is lower in the sky) than in summer; a vertical south-facing solar collector thus is ideal for collecting solar heat for winter space heating. A south-facing surface tilted up at 60° or so above the horizontal has considerably less variation over the course of a year and would thus be well suited for collecting solar heat on a year-round basis, as for water heating. These results are illustrated in Figure 1, which shows the daily insolation (solar energy per unit area) for surfaces of various inclinations.

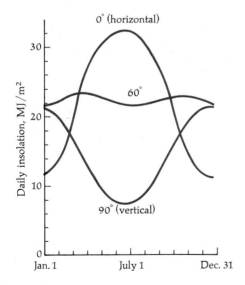

**Figure 1**
Plot of the solar energy per unit area (daily insolation) for surfaces of various inclinations.

This array of silicon solar cells mounted behind plastic lenses produces a peak electric power of 1 kW along with about 5 kW of peak thermal output. (*Courtesy of Edward L. Burgess, Sandia Corp.*)

### Forms of Solar Energy

"Solar energy" includes not only direct sunlight but also several indirect forms of solar energy, such as photosynthetic fuels and the energy from water power and wind power. These are regarded as solar energy because their energy ultimately derives from the sun: the energy of wood and other plant materials is solar energy fixed by photosynthesis; wind power is solar power because it is the heating of land, air, and water by solar radiation that produces winds; and water power is possible because sunlight drives the hydrological cycle whereby water evaporates, rains on high ground, and returns to the oceans by gravity.

### Uses of Solar Energy

Some of the possible and present uses of solar energy are listed in Table 18-7. They are currently the subject of considerable research and development by both government and industry. As discussed below, some of these uses are clearly economical today, others appear to be marginally economical, and still others need considerable development before they are competitive.

Hydroelectricity has been a significant source of electricity for many decades and is the cheapest type of electricity available today. Unfortunately, its potential is limited by the availability of suitable sites and the environmental disruption caused by hydroelectric reservoirs.

The next most economical type of solar energy appears to be solar space heating of small buildings, particularly passive solar space heating. Solar heating involves the use of (1) collectors to collect solar radiation, (2) thermal storage to store the excess energy for use at night or on cloudy days, and (3) a distribution system to move solar energy from collector to storage or living space and from storage to living space.

**Table 18-7**
Forms of solar energy

| Form | Explanation | Use |
|------|-------------|-----|
| Solar radiation and solar thermal energy | From direct sunlight | Heating buildings and water; process heat for industry and agriculture Cooling buildings Generating electricity: Photovoltaic cells Power towers Ocean thermal-energy conversion |
| Photosynthesis | Solar energy converted to chemical energy of plants and fossil fuels | Solid fuels (wood, coal) Liquid fuels (alcohols) Gaseous fuels (methane) |
| Water power | Sunlight drives hydrological cycle (water evaporates, rains on high ground, and returns to ocean by gravity) | Generating electricity (hydroelectricity) Mechanical power (water mills) |
| Wind power | Heating of land, air, and water by solar radiation produces winds | Generating electricity (wind generators) Mechanical power (windmills) Sailing vessels |

When the distribution system requires external energy inputs (such as electricity for fans or pumps), the solar-energy system is called an *active system*; when the distribution is by natural means (conduction, radiation, and natural convection), the system is called a *passive system*. Passive systems generally use large south-facing windows as the collector and large amounts of concrete, masonry, or water as thermal storage. As discussed later, a well-insulated passive solar home in the northern states can have extremely low heating bills. Passive solar space-heating systems add very little to the cost of a home and are virtually maintenance-free; in the northern states, they are the preferred solar space-heating system for new construction and can be successfully used in many existing buildings.

Active systems have also been used for homes, but they tend to be more expensive and require occasional maintenance. They are most economical for water heating, especially if the only alternative is electric heating, and are sometimes economical for the retrofit of existing homes for space heating. They are also becoming feasible for industrial and agricultural process heat applications.

Since space heating and water heating are the two largest sources of home energy consumption nationally, the fact that solar-energy systems are economical for these purposes means that solar energy can become a very important energy resource in the near future.

Wind-generated electricity is marginally economic today. Although a wind electric system for a residence costs several thousand dollars, it may be economical in locations where electricity is very expensive. In most places the high capital costs make wind-generated electricity more expensive than utility electricity; however, it may well prove to be cheaper over the lifetime of the system if electricity continues to increase in price and the wind system is reliable. Current research and development includes small wind generators for individual use and large generators for incorporation into electric utility grids.

Photovoltaic or solar cells are semiconductor devices usually (but not always) made of silicon. They produce electricity directly from solar radiation, acting much like a battery when the sun is shining on them. Although their price has dropped substantially in recent years, they are still far too expensive to be economical for home use. Some scientists are optimistic that efficient solar cells will be so cheap by the early 1990s that homeowners can install them on or near their homes to meet all or part of their electricity requirements. It appears that the most economical method of solar air conditioning in the future will be the use of ordinary electric air conditioners operating from photovoltaic cells. New buildings should be designed so that photovoltaic cells can easily be added when they have become economical. Large photovoltaic power plants in space may someday supply electric power to earth; the energy would be returned to earth as microwave energy.

There are several ways in which solar energy might be used for transportation, which accounts for about 25 percent of United States energy consumption. Solar electricity from photovoltaic cells or other solar electricity might be used with electric vehicles when they become more common. Solar electricity can also be used to generate hydrogen (by electrolysis of water), which is a possible future fuel for transportation. Biological materials can be converted into alcohol fuels, which are already being mixed 10 percent with gasoline to form gasohol. None of these methods is currently economical.

### The Future of Solar Energy

Solar energy has been neglected as an energy resource for many years. The energy crisis has revived interest in it, and its advantages are being discovered by a world that can no longer afford to waste nonrenewable energy resources. It appears that within the lifetimes of most students using this textbook, solar energy in its various forms will become an important energy resource in the United States and much of the world.

**Figure 2**
The Hodges residence.
(*Courtesy of Laurent Hodges.*)

**Passive Solar Heating: A Case Study**

The Hodges residence in Ames, Iowa (Figure 2) has a passive solar system. It consists of about 40 m² of vertical south-facing double-pane glass serving as solar collector and over 100 Mg of concrete floors and walls serving as thermal storage.

The heat transfers in the house are very simple. Solar radiation passes through the glass into the house. Much of it is directly absorbed by the surface of the concrete and then conducted into it. Part of the solar radiation heats the interior air, which comes in contact with the concrete and also transfers heat into it. When the air begins to cool down, the concrete radiates heat back into the air. This occurs naturally, without any controls or mechanical system. Since the radiant-heating system is the Rolls-Royce of heating systems, this natural radiation makes for an extremely comfortable interior. All parts of the house are within a few feet of an exposed radiant concrete surface. Natural convection in the interior of the house tends to keep the upper level a few degrees warmer, which is desirable since the upper level is the living area and the bedrooms are on the lower level.

On a sunny day the 40 m² of south-facing glass may collect 14 MJ/m², or a total of 560 MJ. The heat loss of the house is about 300 W/K, so that the heat loss would total about 500 MJ on a 0°C day and 1000 MJ on a −20°C day. Internal heat (from lights, cooking, appliances, people, etc.) contributes about 100 MJ per day, reducing the needed heat to about 400 MJ or 900 MJ, respectively. Since the thermal storage has a heat capacity of about 100 MJ/K, a change of 5 C° (9 F°) in its temperature corresponds to an exchange of 500 MJ in or out of storage. On a sunny day averaging 0°C, the 560 MJ collected would provide an excess of 160 MJ, which would be absorbed by a 1.6°C (2.9°F) temperature rise in the storage, corresponding to a 3.4°C (6.1°F) decrease.

These figures show that the house is very slow to change in temperature, since its time constant (ratio of heat capacity to heat loss) is about 4 days. Whenever the indoor temperature drops too low because of cloudiness or cold weather, auxiliary heat is needed. Much of the winter the interior temperature is in the range of 20 to 25°C, and the furnace is often not used at all for several consecutive days.

For the Hodges residence, typically about 15 MJ of auxiliary heat is needed for a whole heating season. This house requires less than one-third as much energy for space heating as the family uses for water heating.

Review

A. Define, explain, or otherwise identify:

Coefficient of linear expansion, 522     Sublimation, 530
Expansivity, 522     Thermal conductivity, 531
Phase transition, 525     R factor, 533
Latent heat of fusion, 526     Stefan-Boltzmann law, 536
Latent heat of vaporization, 526     Emissivity, 536
Saturated vapor pressure, 528     Coefficient of absorption, 536
Critical temperature, 529     Blackbody, 536
Phase diagram, 529     Wein's displacement law, 537
Triple point, 529     Newton's law of cooling, 537

B. True or false:

1. All materials expand when heated.

2. During a phase change the temperature remains constant while heat is added.

3. The temperature at which water boils depends on the pressure.

4. The vapor pressure of a gas depends on the temperature.

5. The melting temperature of a substance depends on the pressure.

6. The rate of conduction of thermal energy is proportional to the temperature gradient.

7. The emissivity and the coefficient of absorption of any body must be equal.

Exercises

### Section 18-1, Thermal Expansion

1. A steel ruler has a length of 30 cm at 20°C. What is its length at 100°C?

2. Use Table 18-1 to find the coefficient of volume expansion for steel.

3. A 100-m-long bridge is built of steel. If it is built as a single continuous structure, how much will its length change from the coldest winter days (−30°C) to the hottest summer days (40°C)?

4. The experimental value of $\beta$ for $N_2$ gas at 0°C and 1 atm is 0.003673 $K^{-1}$. Compare this with the theoretical value $\beta = 1/T$, assuming that $N_2$ is an ideal gas.

5. Define a coefficient of area expansion. Calculate it for a square and a circle, and show that it is 2 times the coefficient of linear expansion.

6. A steel tape is placed around the equator when the mean temperature is 0°C. What will the clearance between the tape and the ground (assumed constant) be if the temperature rises to 30°C? (Neglect expansion of the earth.)

7. A mercury thermometer is made from an ordinary glass tube of inside diameter 0.60 mm. The distance between the ice point and steam point is to be 20.0 cm. Find the volume of mercury needed in the bulb and tube.

8. A spherical rubber balloon filled with helium in a house at 20°C has a diameter of 36 cm. Find the diameter when the balloon is taken outside at a temperature of −10°C assuming that the balloon has negligible elastic forces so that the internal pressure is always 1.0 atm.

### Section 18-2, Change of Phase and Latent Heat

9. A 200-g piece of ice at 0°C is put into 500 g of water at 20°C. The system is in a container of negligible heat capacity and insulated from its surroundings. (a)

What is the final equilibrium temperature of the system? (b) How much of the ice melts?

10. A 50-g piece of ice at 0°C is placed in 500 g of water at 20°C, as in Exercise 9. What is the final temperature of the system assuming no heat loss to the surroundings?

11. How much heat must be removed when 100 g of steam at 150°C is cooled and frozen into 100 g of ice at 0°C? (Take the specific heat of steam to be 2.01 kJ/kg·K.)

12. A 50-g piece of aluminum at 20°C is cooled to −196°C by placing it in a large container of liquid nitrogen at that temperature. How much nitrogen is vaporized? (Assume the specific heat of aluminum to be constant and equal to 0.90 kJ/kg·K.)

13. Steam at 100°C is passed into a flask containing 100 g of ice at −10°C. Neglecting any heat loss from the flask, determine the mass of boiling water which will finally be produced in the flask.

14. A lead bullet initially at 30°C just melts upon striking a target. Assuming that all the initial kinetic energy of the bullet goes into internal energy of the bullet to raise its temperature and melt it, calculate its speed upon impact.

15. If 500 g of molten lead at 327°C is poured into a cavity in a large block of ice at 0°C, how much ice melts?

### Section 18-3, The van der Waals Equation and Liquid-Vapor Isotherms

16. (a) Calculate the volume of 1 mol of steam at 100°C and 1 atm pressure, assuming that it is an ideal gas. (b) Find the temperature at which the steam will occupy the volume found in part (a) at 1 atm if the steam obeys the van der Waals equation with $a = 0.55$ Pa·m$^6$ and $b = 30$ cm$^3$.

17. Which gases in Table 18-4 cannot be liquefied by applying pressure at 20°C?

18. Use the values in Table 18-3 to draw a graph of the vapor pressure of water versus temperature. From your graph find (a) the temperature at which water boils on a mountain where the atmospheric pressure is 70 kPa; (b) the temperature at which water will boil in a container in which the pressure is reduced to 0.5 atm; (c) the pressure at which the boiling temperature of water is 115°C.

### Section 18-4, The Transfer of Heat

19. A house has a window area of 300 ft$^2$ of glass 6.0 mm thick. Calculate the rate of heat loss per hour by conduction through the windows if the inside temperature is 68°F and the outside is −10°F.

20. A slab of insulation 20 by 30 ft has an $R$ factor of 11. How much heat in Btu per hour is conducted through the slab if the temperature on one side is 68°F and that on the other side is 30°F?

21. A copper bar 2 m long has a circular cross section of radius 1 cm. One end is kept at 100°C and the other at 0°C, and the surface is insulated so that negligible heat is lost through the surface. Find (a) the thermal resistance of the bar, (b) the thermal current $I$, (c) the temperature gradient $dT/dx$, and (d) the temperature 25 cm from the hot end.

22. Two metal cubes with 3-cm edges of copper and aluminum are arranged as shown in Figure 18-9. Find (a) the thermal resistance of each cube and the total thermal resistance of the two-cube system; (b) the thermal current $I$; and (c) the temperature $T_i$ at the interface. (d) Find the temperature at the interface if the cubes are interchanged.

23. The same metal cubes of Exercise 22 are arranged as in Figure 18-10. Find (a) the total thermal current from one reservoir to the other, (b) the effective

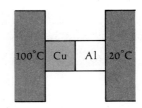

**Figure 18-9**
Two metal cubes in series for Exercise 22.

thermal resistance of the two-cube system, and (c) the ratio of the thermal current carried by the copper cube to that carried by the aluminum cube.

24. The thermal conductivity of an insulating material is measured by constructing a cubical box of side 0.5 m and thickness 2 cm from the material. When a 135-W heater is placed inside the box, the steady-state inside temperature is greater than the outside temperature by 60°C. Calculate the thermal conductivity of the material.

25. A steam pipe of radius 5 cm carries steam at 100°C. The pipe is covered by a jacket of insulating material 2 cm thick having a thermal conductivity of 0.07 W/m·K. If the temperature at the outer wall of the pipe jacket is 20°C, how much heat is lost through the jacket per metre length in an hour?

26. A 1-kW electric heater has heating wires that are "red hot" at a temperature of 900°C. Assuming that 100 percent of the heat output is due to radiation and that the wires act as blackbody radiators, what is the effective area of the radiating surface? (Assume the room temperature to be 20°C.)

27. Calculate $\lambda_m$ for a human blackbody radiator assuming the skin-surface temperature to be 33°C.

28. Calculate the temperature of a blackbody that radiates energy with a peak in the spectrum at (a) $\lambda_m = 3$ cm (microwave region) and (b) $\lambda_m = 3$ m (FM radio waves).

29. If the absolute temperature of a blackbody is doubled, by what factor does its total radiated power increase?

30. The surface temperature of the filament of an incandescent lamp is 1300°C. If the input electric power is doubled, what will the temperature become? *Hint:* Show that you can neglect the temperature of the surroundings.

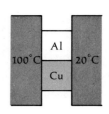

**Figure 18-10**
Exercise 23.

## Problems

1. A steel tube has an outside diameter of 3.000 cm at room temperature (20°C). A brass tube has an inside diameter of 2.997 cm at the same temperature. To what temperature must the ends of the tubes be heated if the steel tube is to be inserted into the brass tube?

2. One way to construct a device with two points whose separation remains the same in spite of temperature changes is to use rods with different coefficients of expansion, in the arrangement shown in Figure 18-11. The two rods are bolted together at one end. Show that the distance $L$ will not change with temperature if the lengths $L_A$ and $L_B$ are chosen so that $L_A/L_B = \alpha_B/\alpha_A$. If material $B$ is steel, material $A$ is brass, and $L_A = 250$ cm at 0°C, what is the value of $L$? If both rods had been made of brass, how much would $L$ have increased if the rods were heated to 100°C?

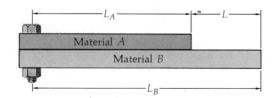

**Figure 18-11**
Problem 2.

3. A piece of ice is dropped from a height $H$. Find the minimum value of $H$ such that the ice melts when it makes an inelastic collision with the ground. Is it reasonable to neglect the variation in the acceleration of gravity in doing this problem? Comment on the reasonableness of neglecting air resistance. What effect would air resistance have on your answer?

4. An insulated cylinder with a movable piston (to maintain constant pressure) initially contains 100 g of ice at −10°C. Heat is supplied to the cylinder at a constant rate by a 100-W heater. Make a graph showing the temperature of the cylinder contents as a function of time starting at time $t = 0$, when the temperature is −10°C, and ending when the temperature is 110°C. (Use $c = 2.0$ kJ/kg·K for the average specific heat of ice from −10 to 0°C and for steam from 100 to 110°C.)

5. One mole of water at 100°C is vaporized at a constant pressure of 1 atm. (a) Find the amount of heat added. (b) Find the change in volume and the work done by the system in expanding against atmospheric pressure (treat the steam as an ideal gas). (c) Find the change in internal energy of the water. (d) If 1 mol of ice at 0°C and 1 atm is completely liquefied, is the internal-energy change greater or less than $18L_f$? Explain.

6. (a) A 200-g aluminum calorimeter can contains 500 g of water at 20°C. A 100-g piece of ice cooled to −20°C is placed in the can. Find the final temperature of the system assuming no heat losses (assume that the specific heat of ice is 2.0 kJ/kg·K). (b) A second 200-g piece of ice at −20°C is added. How much ice remains in the system after it reaches equilibrium? (c) Would your answer to part (b) be different if both pieces were added at the same time?

7. A solid material has a density $\rho$, coefficient of linear expansion $\alpha$, and mass $m$. (a) Show that at pressure $P$ the heat capacities $C_p$ and $C_v$ are related by $C_p - C_v = 3\alpha m P/\rho$. (b) Find the difference (in joules per mole-kelvin) in the molar heat capacities of aluminum at atmospheric pressure using $\alpha = 23 \times 10^{-6}$ K$^{-1}$, $M = 27 \times 10^{-3}$ kg/mol, and $\rho = 2700$ kg/m³.

8. A copper-bottomed saucepan containing 0.8 L of boiling water boils dry in 10 min. Assuming that all the heat flows through the flat copper bottom of diameter 15 cm and thickness 3.0 mm, calculate the temperature of the outside of the copper bottom while some water is still present.

9. For a boiler at a power station heat must be transferred to boiling water at the rate of 3 GW. The boiling water passes through copper pipes of wall thickness 4.0 mm and surface area 0.12 m² per metre length of pipe. Find the total length of pipe (actually there are many pipes in parallel) that must pass through the furnace if the steam temperature is 225°C and the external temperature of the pipes is 600°C.

10. In a cold room at 0°C the refrigerant is brine at −16°C circulating through copper pipes with a wall thickness of 1.5 mm. By what fraction is the heat transfer reduced when the pipes are coated with a 5-mm layer of ice?

11. A blackened copper sphere of radius 4.0 cm hangs in a vacuum in an enclosure with a wall temperature of 20°C. If the sphere is at temperature 0°C, find its rate of temperature change assuming that heat transfers by radiation only.

12. A body at initial temperature $T_i$ cools in a room at temperature $T_0$ by convection and radiation and obeys Newton's law of cooling, which can be written $dQ/dt = hA(T - T_0)$, where $A$ is the area of the body and the constant $h$ is called the surface coefficient of heat transfer. Show that the temperature $T$ at any time $t$ is given by $T = T_0 + (T_i - T_0)e^{-hAt/mc}$, where $m$ is the mass of the body and $c$ is its specific heat.

13. A 0.2-mm-diameter steel wire stretched between two rigid supports 30 cm apart is under a tension of 36 N at 25°C. If the breaking stress is 1.2 GN/m², find the temperature at which the wire will snap as it cools.

14. A piece of copper of mass 100 g is heated in a furnace at temperature $t$. The copper is then inserted into a copper calorimeter of mass 150 g containing 200 g of water. The initial temperature of the water and calorimeter is 16°C and the final temperature after equilibrium is established is 38°C. When the calorimeter and contents are weighed, 1.2 g of water is found to have evaporated. What was the temperature $t$ of the furnace?

15. Three metal cubes with 3-cm edges are made of lead (Pb), copper (Cu), and aluminum (Al), and arranged as shown in Figure 18-12. The heat reservoirs are maintained at the temperatures indicated. After steady state has been reached, find (a) the temperatures $T_1$ and $T_2$ at the interfaces and (b) the temperatures $T_1$ and $T_2$ if the lead and copper blocks are interchanged.

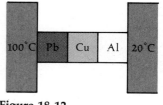

**Figure 18-12**
Problem 15.

16. Liquid helium is stored at its boiling point (4.2 K) in a spherical can; the can is separated by a vacuum space from a surrounding shield maintained at the temperature of liquid nitrogen (77 K). If the can is 30 cm in diameter and is blackened on the outside so it acts as a blackbody, how much helium boils away per hour?

17. Heat energy is transferred from a high-temperature heat source at $t_1 = 300°C$ to two lower-temperature heat reservoirs maintained at temperatures $t_2 = 50°C$ and $t_3 = 100°C$. Three identical solid steel rods, each of length $L = 1$ m and cross-sectional area $0.01$ m², are used to pipe the heat, as shown in Figure 18-13. Find the power in watts delivered by the source to each of the lower-temperature reservoirs and the temperature of the junction; consider only axial heat flow.

**Figure 18-13**
Problem 17.

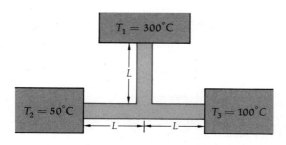

18. At the critical point on the critical isotherm, $dP/dV$ and $d^2P/dV^2$ are both zero. Show that for a van der Waals gas the critical volume is $V_c = 3b$.

19. A copper container of mass 200 g with 0.7 L of water at 60°C is linked by a copper rod of length 10 cm and cross-sectional area 1.5 cm² to a similar container maintained at 0°C. (a) Show that the temperature of the first container changes with time according to $t_c = t_{c0}e^{-t/RC}$, where $t_{c0} = 60°C$ is the initial temperature, $R$ is the thermal resistance of the rod, and $C$ is the total heat capacity of the container plus water. Evaluate $R$ and $C$ and the "time constant" $RC$. (b) Show that the total amount of heat conducted after a time $t$ is $Q = Ct_{c0} \times (1 - e^{-t/RC})$. (c) Find the time needed to reduce the temperature to 30°C.

20. The *solar constant* is the power per unit area received from the sun at the earth on an area perpendicular to the sun's rays. Its value at the upper atmosphere of the earth is about 1.35 kW/m². Calculate the effective surface temperature of the sun if it radiates like a blackbody (the radius of the sun is 696 Mm).

21. A steel bar of radius 2.2 cm and length 60 cm is jammed horizontally perpendicular to two stationary vertical concrete walls at a temperature of 20°C. With a blow torch the temperature of the bar is raised to 60°C. Find the force exerted by the bar on each wall.

22. A hollow steel sphere is filled with water at 20°C and sealed. Find the water pressure in atmospheres if the temperature rises to 25°C. (Neglect any elastic yielding of the steel sphere under the internal pressure.)

23. (a) Show that the density of a material changes with temperature by $\Delta\rho = -\rho_0\beta\,\Delta T$, where $\beta$ is the expansivity. (b) A spherical glass bulb of diameter 5.00 cm is loaded with lead shot so that its mass is 64.8 g and it floats in water at 20°C (water density 998.2 kg/m³). At what temperature will the bulb just sink, i.e., have neutral buoyancy?

**CHAPTER 19**     The Availability of Energy

---

**Objectives**     After studying this chapter you should:

---

1. Be able to give the definition of the efficiency of a heat engine and of the coefficient of performance of a refrigerator.

2. Be able to give both the Kelvin-Planck and Clausius statements of the second law of thermodynamics and illustrate their equivalence with a numerical example.

3. Be able to list the necessary conditions for a process to be reversible.

4. Be able to state the Carnot theorem and illustrate it with a numerical example.

5. Be able to give the definition of the absolute temperature scale.

6. Be able to give the expression for the Carnot efficiency of a heat engine.

7. Be able to discuss the concept of entropy, including its interpretation in terms of available energy and in terms of molecular motion.

---

Our study of the first law of thermodynamics showed that it is possible to increase the internal energy of a body either by adding heat to the body or by doing work on it. In this sense, work and heat are equivalent: 10 J of either work or heat added to a system increases the internal energy of the system by 10 J. On the other hand, if we decrease the internal energy of a system and try to use this energy to do work on another system or to add heat to another system, we find an important difference between heat and work. There is no difficulty in taking internal energy out of a system in the form of heat, but it is impossible to convert this internal energy completely into work without any other change taking place in the surroundings. There is thus a lack of symmetry in the roles played by heat and work which is not evident from the first law.

It is quite easy to convert work completely into heat without any change in the surroundings (except in the body which absorbs the heat). A common example is the work done against friction, which can go completely into heat. This process can continue indefinitely as long as there is a supply of work, but the reverse is often impossible. We cannot easily convert heat completely into work. One situation in which heat is converted into work is the isothermal expansion of an ideal gas (or an expansion of a real gas along a path of constant internal energy). If the expansion is allowed to proceed freely, the heat absorbed by the gas is completely converted into work until the pressure of the gas equals atmospheric pressure and the expansion stops. If the gas system is to be used again, however, work must be done on it to compress it.

Besides the conversion of heat into work, other conceivable processes are consistent with the first law of thermodynamics but do not occur in nature. For example, heat does not go from a cold body to a hot body by itself. Processes which conserve energy but which never occur are all related to processes which are irreversible. For example, a block *Irreversible processes* with initial kinetic energy slides along a rough table. The work against friction goes completely into heat, which increases the internal energy of the block and table. The reverse process never occurs. The internal energy of the block and table is never spontaneously converted into kinetic energy of the block, sending it sliding along the table while the table and block cool off. Similarly, heat can flow from a hot body to a cold body until the two bodies are at the same temperature. The reverse never happens. Two bodies in thermal contact at the same temperature remain at the same temperature. Heat is never spontaneously conducted from a cold body to a hot body. A third type of irreversible process not involving heat is the free adiabatic expansion of a gas. Whether the gas is ideal or not, the final energy equals the initial energy. Left to itself, a gas will never contract so that it fills only one of two connected compartments.

The explosion of the Hindenburg at Lakehurst, New Jersey on May 6, 1937 was a dramatic example of an irreversible process.

UPI

The second law of thermodynamics summarizes the fact that processes of this type do not occur. There are many different ways of stating the second law; we shall study several and show them to be equivalent. Before giving precise statements of the second law we briefly consider heat engines and refrigerators because the study of heat-engine efficiency gave rise to the first clear statements of the second law of thermodynamics.

# 19-1   Heat Engines and the Second Law of Thermodynamics

The first practical heat engine was the steam engine, invented in the seventeenth century for pumping water out of coal mines. Today the primary use of steam engines is in generating electric energy. We need not consider details of the operation of the steam engine or any other type because we can represent them all—steam engines, internal-combustion engines, diesel engines—schematically. A substance or system called the *working substance* (water in the case of the steam engine) absorbs a quantity of heat $Q_h$ from a heat reservoir at temperature $T_h$. It does work $W$ and rejects heat $|Q_c|$* to a heat reservoir at a lower temperature $T_c$. The working substance then returns to its original state. (The subscripts $h$ and $c$ indicate quantities associated with the hot and cold reservoirs.) The engine is therefore a cyclic device. At various parts of its cycle, it absorbs or rejects heat and does work, the purpose being to gain some work at the expense of heat absorbed in each cycle.

Since the initial and final states of the engine are the same, the final internal energy must equal the initial internal energy. The first law of thermodynamics thus relates the heat absorbed $Q_h$, the heat rejected $|Q_c|$, and the work done $W$ by

$$Q_h - |Q_c| = W \qquad\qquad 19\text{-}1$$

Figure 19-1 shows a schematic representation of a heat engine. Heat $Q_h$ enters from a reservoir at temperature $T_h$ (hot reservoir), and heat $|Q_c|$ is rejected to a reservoir at temperature $T_c$ (cold reservoir). Work $W$ is done by the engine. When the cycle is completed, the process repeats. The efficiency $\epsilon$ of the engine is defined to be the ratio of the work done to the heat absorbed:

$$\epsilon = \frac{W}{Q_h} = \frac{Q_h - |Q_c|}{Q_h} = 1 - \frac{|Q_c|}{Q_h} \qquad\qquad 19\text{-}2$$

Since the heat $Q_h$ is usually produced by burning coal, oil, or some other kind of fuel which must be paid for, one tries to design a heat engine with the greatest possible efficiency. We can see from Equation 19-2 that we want to reject as small a fraction of the heat absorbed as possible. For perfect efficiency ($\epsilon = 1 = 100$ percent), $|Q_c| = 0$ and no heat is rejected in a cycle. In that case, all the heat absorbed from the first reservoir would be converted into work. None would be rejected or lost to the second reservoir.

---

* The heat rejected by a system is negative according to our sign convention in the first law. Since we are interested in the magnitudes of heat absorbed or rejected, we indicate them by using absolute-value signs with $Q_c$ for heat engines and $Q_h$ for refrigerators, which reject heat to the hot reservoir.

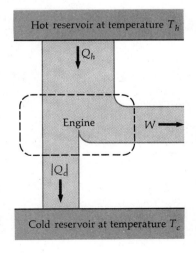

**Figure 19-1**
Schematic representation of a heat engine which removes heat energy $Q_h$ from a hot reservoir at temperature $T_h$, does work $W$, and rejects heat energy $|Q_c|$ to a cold reservoir at temperature $T_c$.

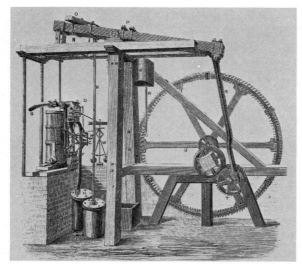

Early steam engine designed by James Watt in the mid-1700s.

The Granger Collection

Although the efficiency of heat engines has been greatly increased since the early steam engines, it is impossible to make a heat engine which is 100 percent efficient, i.e., which would reject no heat to a reservoir at a lower temperature. This experimental result is known as the *Kelvin-Planck statement of the second law of thermodynamics:*

*It is impossible for an engine working in a cycle to produce no other effect than that of extracting heat from a reservoir and performing an equivalent amount of work.*

*Second law (Kelvin-Planck statement)*

Note the words "in a cycle": it is possible to extract heat from a single reservoir and transform it completely into work. An ideal gas undergoing an isothermal expansion does just this, but after the expansion the gas is not in its original state. In order to bring the gas back to its original state, heat must be rejected to another reservoir at a lower temperature or the original work obtained must be put back into the gas as in an isothermal compression, in which case the net heat absorbed and net work done are both zero. If the Kelvin-Planck statement were not true, it would be possible to design a heat engine for a ship which would extract energy from the ocean (a convenient heat reservoir) and use it to power the ship without using a second heat reservoir at a lower temperature to receive any of the energy.

Figure 19-2 is a schematic representation of a refrigerator, which is similar to a heat engine but absorbs heat from a cold reservoir and rejects heat to a hot one. In order to do this, work $W$ must be done *on* the refrigerator. If $Q_c$ is the heat absorbed and $W$ the work done on the refrigerator, the heat rejected is

$$|Q_h| = W + Q_c$$

The purpose of a refrigerator is to transfer heat from a cold body to a hot body. It is desirable to do this with the least possible work. It is found from experience that some work $W$ must always be done. This result is the Clausius statement of the second law of thermodynamics:

*It is impossible for a refrigerator working in a cycle to produce no other effect than the transfer of heat from a colder body to a hotter body.*

*Second law (Clausius statement)*

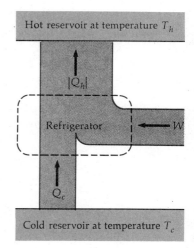

**Figure 19-2**
Schematic representation of a refrigerator which removes heat energy $Q_c$ from a cold reservoir, rejects heat energy $|Q_h|$ to the hot reservoir, and absorbs work $W$, which must be done on the refrigerator.

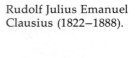

Rudolf Julius Emanuel
Clausius (1822–1888).

Brown Brothers

If the Clausius statement were not true, it would be possible in principle to cool our homes in the summer with a refrigerator which pumped heat to the outside without using any electricity or other energy.

A measure of the performance of a refrigerator is the ratio $Q_c/W$, called the *coefficient of performance* $\eta$:

$$\eta = \frac{Q_c}{W} \qquad\qquad 19\text{-}3$$

The greater the coefficient of performance, the better the refrigerator. Typical refrigerators have coefficients of performance of about 5 or 6. In terms of this ratio, the Clausius statement of the second law is that the coefficient of performance of a refrigerator cannot be infinite.

### Questions

1. What plays the role of the high-temperature reservoir in the steam engine? The low-temperature reservoir? What plays these roles in an internal-combustion engine?

2. How does friction in an engine affect its efficiency?

## 19-2 Equivalence of the Kelvin-Planck and Clausius Statements

Although the two statements of the second law we have given may seem to be quite different, they are in fact equivalent; i.e., if either statement is true, the other must also be true. We can prove this by showing that if either statement is false, the other must also be false. This is sufficient to prove equivalence.

We first assume that the Clausius statement is false, i.e., that it is possible to transfer heat from a cold reservoir to a hot reservoir without any other effects. Suppose an ordinary engine removes 100 J of energy from a hot reservoir, does 40 J of work, and exhausts 60 J of energy to the cold reservoir. This engine has an efficiency of 40 percent. If the Clausius statement were not true, we could use a perfect refrigerator to remove 60 J of energy from the cold reservoir and transfer it to the hot reservoir, doing no work in the process. The net effect of our perfect refrigerator working along with the ordinary engine would be to remove 40 J from the hot reservoir and do 40 J of work, with no energy rejected (Figure 19-3). This violates the Kelvin-Planck statement of the second law. Thus if the Clausius statement were false, the Kelvin-Planck statement would be, too, as illustrated in Figure 19-4 for general values of $|Q_h|$ and $|Q_c|$ used by the ordinary engine.

**Figure 19-3**
(a) An ordinary heat engine which removes 100 J from a hot reservoir, performs 40 J of work, and rejects 60 J to the cold reservoir. If the Clausius statement of the second law were false, a perfect refrigerator (b) could remove 60 J from the cold reservoir and reject it to the hot reservoir without requiring any work. (c) The net effect of the ordinary engine and perfect refrigerator in (a) and (b) is a perfect engine which violates the Kelvin-Planck statement of the second law by removing 40 J from the hot reservoir and converting it completely into work with no other effects.

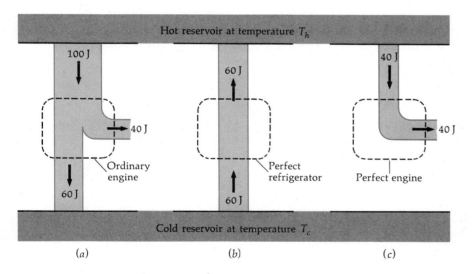

(a)                    (b)                    (c)

**Figure 19-4**
An ordinary heat engine combined with a perfect refrigerator, resulting in a perfect heat engine. The combination removes heat $Q_h - |Q_c|$ from the hot reservoir and converts it completely into work.

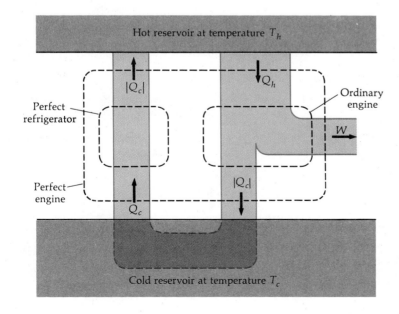

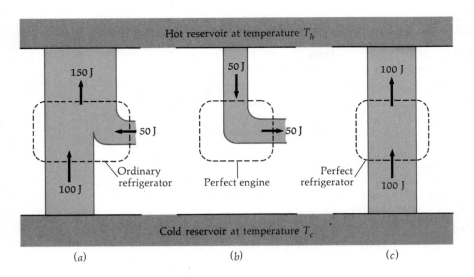

**Figure 19-5**
(*a*) An ordinary refrigerator which removes 100 J from a cold reservoir and rejects 150 J to a hot reservoir, requiring 50 J of work. If the Kelvin-Planck statement of the second law were false, a perfect engine (*b*) could remove 50 J from the hot reservoir and convert it completely into work with no other effects. (*c*) The net effect of the ordinary refrigerator and perfect engine in (*a*) and (*b*) is a perfect refrigerator which violates the Clausius statement of the second law.

We now assume that the Kelvin-Planck statement is false and show that this implies that the Clausius statement is false. Consider an ordinary refrigerator which removes 100 J of energy from a cold reservoir, uses 50 J of work, and rejects 150 J to the hot reservoir (Figure 19-5). If the Kelvin-Planck statement were false, we could remove energy from a single reservoir and convert it completely into work with 100 percent efficiency. We could thus use such a perfect engine to remove 50 J of energy from the hot reservoir and do 50 J of work (Figure 19-5*b*). The net result of the ordinary refrigerator combined with the perfect engine would be to transfer 100 J from the cold reservoir to the hot reservoir without any work being done, contradicting the Clausius statement. (Figure 19-6 illustrates this combination of an ordinary refrigerator and perfect engine for general values of $|Q_h|$, $W$, and $Q_c$ used by the refrigerator.) Hence, if the Kelvin-Planck statement were false, the Clausius statement would be, too. We have thus shown that these two statements of the second law of thermodynamics are equivalent.

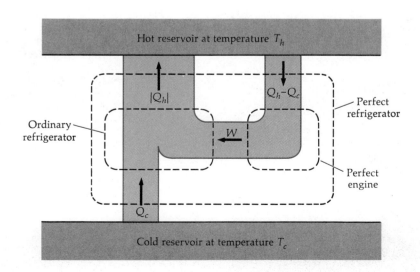

**Figure 19-6**
Ordinary refrigerator combined with a perfect heat engine, resulting in a perfect refrigerator which removes heat $Q_c$ from the cold reservoir and transfers it to the hot reservoir without requiring any work.

## 19-3 Reversibility

Consider the simple process of conduction of a certain amount of heat $Q$ from a hot reservoir to a cold reservoir. According to the Clausius statement of the second law of thermodynamics, it is impossible to transfer this heat back to the hot reservoir without other changes in the surroundings. We could use a refrigerator to remove the heat $Q$ from the lower reservoir and transfer heat $Q + W$ to the upper reservoir, where $W$ is the work done on the refrigerator, but then energy $W$ has been lost by the surroundings and an extra amount of energy $W$ is in the upper reservoir. According to the Kelvin-Planck statement, we cannot remove this energy from the hot reservoir and convert it completely into work. In any cycle some heat must be rejected to the surroundings. The process of heat conduction from a hot body to a cold body is *irreversible:* we cannot bring the system back to its original state without making a permanent change in the surroundings.

We define any process to be irreversible if the system and surroundings cannot be brought back to their initial states. Irreversibility is intimately connected with the second law of thermodynamics. We can find many irreversible processes in nature. Any process that converts mechanical energy into internal energy is irreversible; e.g., the process of a block sliding along a rough table until it stops due to friction is irreversible because we cannot extract the increased internal energy of the block and table and convert it completely back into mechanical energy. Similarly, if we drop a block of wood into a lake, mechanical energy $mgh$ is converted into internal energy of the lake (and block). We cannot reverse this process without doing work. Another irreversible process we have discussed is the free adiabatic expansion of a gas from one compartment into an evacuated compartment. Although no work is done, no heat is transferred, and the internal energy does not change; it is impossible to reverse the process.

This irreversibility is also related to the second law of thermodynamics. Let us suppose that the gas at temperature $T$ freely expands from volume $V_1$ of the original compartment to the final volume $V_2$. As we have mentioned, the final temperature is also $T$ since the internal energy of the ideal gas did not change. We can bring this gas back to its original volume by isothermal compression, in which we do work at temperature $T$. Such an isothermal compression is itself reversible, but the combination of the free expansion plus the isothermal compression puts the gas back in its original state, an amount of work $W$ being converted into energy in the heat reservoir. Since it is impossible to convert this energy back into work with no other changes, the complete process is irreversible. Thus the free adiabatic expansion is irreversible (Figure 19-7). The gas, once expanded, will never spontaneously contract to its original volume.

From these considerations and our statements of the second law of thermodynamics, we can list some conditions necessary for a process to be reversible:

1. No work must be done by friction, viscous forces, or other dissipative forces which produce heat.

2. There can be no heat conduction due to a finite temperature difference.

3. The process must be quasi-static, so that the system is always in an equilibrium state (or infinitesimally near an equilibrium state).

*Irreversibility*

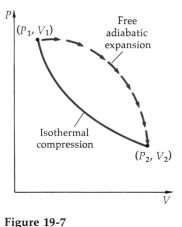

**Figure 19-7**
A free adiabatic expansion of an ideal gas is indicated by the arrows. It cannot be represented by a curve on the $PV$ diagram because the process is not quasi-static and the gas does not pass through equilibrium states. The gas can be returned to its original state by a reversible isothermal compression during which work is converted completely into heat. The combined cycle cannot be reversed because heat cannot be converted completely into work with no other change. The free adiabatic expansion is therefore irreversible.

*Conditions for reversibility*

Any process which violates any of the above conditions is irreversible. Most processes in nature are irreversible. In order to have a reversible process, great care must be taken to eliminate frictional and other dissipative forces and to make the process quasi-static. Since this can never be done completely, a reversible process seems impossible in practice. Nevertheless, one can come very close to a reversible process, and the concept is very important in theory.

Consider, for example, a quasi-static expansion of a gas from volume $V_1$ to volume $V_2$ at constant pressure, shown in Figure 19-8. If this process is to be reversible, the heat must be absorbed isothermally so that there is no heat conduction across a finite temperature difference. We can approximate the constant-pressure path by an alternate reversible path consisting of the series of isotherms and quasi-static, adiabatic paths indicated. Heat is then absorbed along each isotherm from a reservoir at the temperature of the isotherm. With a very large number of heat reservoirs, we can make this alternate path approximate the constant-pressure path as closely as we wish. Similarly, any quasi-static process represented by a curve in the $PV$ diagram can be approximated by a reversible path consisting of a series of isotherms and quasi-static, adiabatic paths.

In the next sections, we shall see how the investigation of reversible processes leads to a theoretical limit on the efficiency of a heat engine operating between two reservoirs of fixed temperatures, and to the definition of an absolute temperature scale which does not depend on the properties of any material.

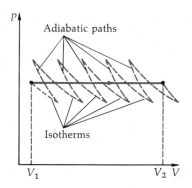

**Figure 19-8**
Quasi-static expansion at constant pressure. This process can be considered reversible because it can be approximated as closely as desired by a series of reversible isothermal and adiabatic curves as shown. Any quasi-static process can be performed approximately reversibly using a large number of heat reservoirs so that the heat is absorbed or rejected approximately isothermally.

**Questions**

3. Give several examples of irreversible processes in everyday life.

4. If a system changes from state $A$ to state $B$ by an irreversible process, does that mean it can never be returned to state $A$ again?

5. A bottle of ink is slowly, carefully, and thoroughly mixed with a tub of water. Is this a reversible or an irreversible process?

## 19-4  The Carnot Engine

In 1824, before the first law of thermodynamics was established, a young French engineer, Sadi Carnot, described an ideal reversible engine working in a simple cycle between two heat reservoirs and found the theoretical limit for the efficiency of an engine in terms of the temperatures of the heat reservoirs. Such an ideal engine is now called a *Carnot engine*, and its cycle a *Carnot cycle*. Carnot's results are given in a simple but important theorem, called the *Carnot theorem*:

*Carnot's theorem*

*No engine working between two given heat reservoirs can be more efficient than a reversible engine working between those reservoirs.*

This theorem is easily proved by using the second law of thermodynamics. Consider a *reversible* engine working between two reservoirs. Let it remove 100 J of energy from the hot reservoir, do 40 J of work, and reject 60 J of energy into the cold reservoir (Figure 19-9a). Its efficiency is thus 40 percent. Since the engine is reversible, it can be operated in reverse, absorbing 60 J of energy from the cold reservoir and rejecting 100 J into the hot reservoir while work of 40 J is done on it (Figure

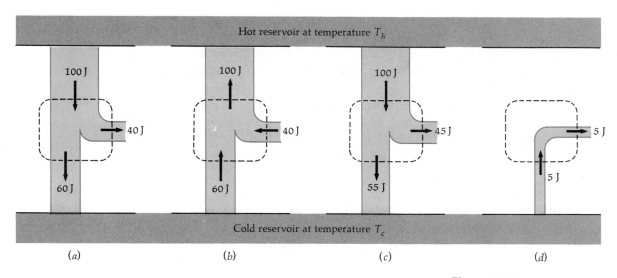

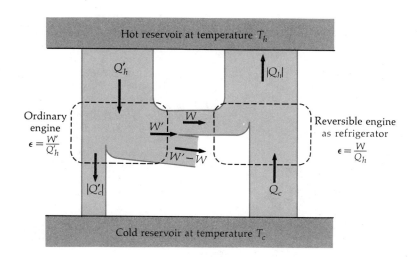

19-9b). We now consider a second engine, which may or may not be reversible, and wish to show that if its efficiency is greater than that of the first engine, the second law can be violated. Suppose we use this engine to remove 100 J of energy from the hot reservoir. Since we are assuming that its efficiency is greater than 40 percent, it does more than 40 J of work and rejects less than 60 J to the cold reservoir. Suppose it has an efficiency of 45 percent. Then 45 J of work is done and 55 J of energy is rejected (Figure 19-9c). The net effect of these two engines working together (with the reversible engine operating as a refrigerator) is that 5 J of energy has been removed from the cold reservoir and changed completely into work (Figure 19-9d). This violates the Kelvin-Planck statement of the second law. Figure 19-10 illustrates this proof for general quantities of heat removed and rejected and work done. In this figure we choose the heat removed by the second engine $Q_h'$ to be equal to that rejected by the first engine operating as a refrigerator. If the efficiency of the second engine is greater than that of the first, the work done $W'$ is greater than the work $W$ needed to operate the reversible engine as a refrigerator, and the heat rejected $|Q_c'|$ is less than that absorbed from the cold reservoir by the reversible engine. Thus the net effect of the two engines is that heat $Q_c - |Q_c'|$ is removed from the cold reservoir and changed completely into work. (On the other hand, if the efficiency of the second engine is less than that of the reversible engine, there is no contradiction of the second law of thermodynamics.)

**Figure 19-9**
Illustration of the Carnot theorem. (a) A reversible heat engine with 40 percent efficiency. (b) The same heat engine run backward as a refrigerator. (c) An assumed heat engine working between the same two reservoirs with efficiency of 45 percent. (d) The combination of the two engines with the reversible engine operating as a refrigerator results in a perfect heat engine, violating the second law.

**Figure 19-10**
Illustration of the Carnot theorem for arbitrary amounts of heat and work. The reversible engine on the right is run backward as a refrigerator, removing heat $Q_c$ from the cold reservoir and rejecting heat $|Q_h|$ to the hot reservoir while requiring work $W$. If the engine on the left has a greater efficiency than the reversible one, it can remove heat $Q_h' = |Q_h|$ and perform work $W'$ which is greater than $W$. If part of this work is used to run the refrigerator, the net effect of both engines is the removal of heat $Q_c - |Q_c'|$ from the cold reservoir and complete conversion of this heat into work.

The net result is that net work has been converted into internal energy in the cold reservoir.) We have shown that the efficiency of the reversible engine must be greater than or equal to the efficiency of any other engine working between the same reservoirs, which is the Carnot theorem.

A corollary to the Carnot theorem reads:

*All reversible engines working between the same two reservoirs have the same efficiency.*

This follows from Carnot's theorem because for any two reversible engines, the theorem requires each to have an efficiency greater than or equal to the efficiency of the other. Since neither can have an efficiency less than the other, their efficiencies must be equal.

Carnot's theorem and corollary can also be expressed in terms of refrigerators:

*No refrigerator working between two given heat reservoirs can have a greater coefficient of performance than a reversible refrigerator working between those reservoirs.*

Similarly,

*All reversible refrigerators working between the same two reservoirs have the same coefficient of performance.*

The proof that these statements follow from the second law of thermodynamics is left as an exercise.

It is not hard to see what the cycle of a Carnot engine must be. Since the cycle is reversible, heat cannot be transferred at a temperature difference, because this is always irreversible. The engine must therefore absorb and reject heat isothermally. Figure 19-11 shows a Carnot cycle for a gas used as the working substance of a Carnot engine. We assume that the gas is originally at the temperature of the hot reservoir $T_h$. (If not, we can always perform an adiabatic expansion or compression until the gas is at this temperature.) The gas first expands isothermally to position 2 on the $PV$ diagram. During this expansion the gas does some work and absorbs heat $Q_h$ from the hot reservoir. Since this heat is absorbed isothermally and quasi-statically, the expansion is reversible (assuming no friction). If the gas were ideal, the work done would

*Carnot cycle*

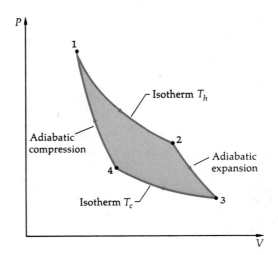

**Figure 19-11**
A Carnot cycle for a gas. The gas first expands isothermally from state 1 to state 2, absorbing heat from the reservoir at $T_h$ and performing work. It then expands adiabatically until its temperature is $T_c$ in state 3. Work is performed but no heat is absorbed or rejected. The gas is then compressed isothermally to state 4 and adiabatically back to its original state. All parts of the cycle are quasi-static. The net work performed during the cycle is indicated by the enclosed area.

equal the heat absorbed, but this is not generally true of a real gas. After the gas is removed from contact with the hot reservoir, it undergoes a quasi-static adiabatic expansion until the temperature of the gas is that of the cold reservoir $T_c$ at position 3. Some work is done, but no heat is transferred. The gas then is compressed isothermally while it is in contact with the lower-temperature reservoir to state 4. During this compression, work is done on the gas, and heat $|Q_c|$ is rejected to the reservoir. Finally, the gas is compressed adiabatically and quasi-statically until it reaches its original temperature $T_h$ at state 1, and the cycle is completed. The net work done is indicated by the enclosed area in Figure 19-11, i.e., the area defined by the two isothermal curves and two adiabatic curves, and is equal to $Q_h - |Q_c|$ by the first law of thermodynamics. The efficiency of the engine can be computed if the equation of state is known so that the work can be calculated from the curves in the $PV$ diagram. We shall compute the efficiency for an ideal gas in the next section.

## 19-5   The Absolute Temperature Scale

The efficiency is the same for all reversible engines working between the same two heat reservoirs, no matter what kind of system is used for the heat engine. As long as the engine performs in a reversible cycle, which means absorbing and rejecting the heat isothermally (and quasi-statically) and changing its temperature adiabatically (and quasi-statically), the efficiency is the same no matter what the system is. Since the only characteristic of a heat reservoir is its temperature, the efficiency of a Carnot engine must depend only on the temperatures of the two heat reservoirs. The heats absorbed and rejected by a Carnot engine are related to the efficiency by Equation 19-2, which can be written

$$\frac{|Q_c|}{Q_h} = 1 - \epsilon_R$$

where we have added the subscript $R$ to indicate that $\epsilon_R$ is the efficiency of a reversible engine. Thus $|Q_c|/Q_h$ must also be a function only of the temperatures of the reservoirs. We use this result to define an absolute temperature scale, defining the ratio of absolute temperatures of the heat reservoirs by

$$\frac{T_c}{T_h} = \frac{|Q_c|}{Q_h} \qquad\qquad\text{19-4}$$

*Absolute temperature scale defined*

To use this definition to measure the ratio of two temperatures, we would have to set up a reversible engine to operate between two reservoirs with the two temperatures and carefully measure the heats absorbed from, or rejected to, those reservoirs in each cycle of the engine.

Equation 19-4 defines only the ratio of two absolute temperatures. The temperature scale is completely determined by choosing one fixed point, as was done for the other temperature scales. We define the absolute temperature of the triple point of water to be 273.16 K. In terms of the absolute temperatures of the heat reservoirs the efficiency of a Carnot engine is

$$\epsilon_R = 1 - \frac{T_c}{T_h} \qquad\qquad\text{19-5}$$

*Carnot efficiency*

This definition of an absolute temperature scale is completely independent of the properties of any materials. Although such a scale might seem impossible to use in practice, that is not the case. At very low temperatures, where the ideal-gas scale cannot be defined, temperatures are actually measured by performing a Carnot cycle as carefully as possible between the two heat reservoirs and measuring the heats $Q_h$ and $|Q_c|$. In the temperature range in which the ideal-gas temperature scale can be defined, the absolute and ideal-gas temperature scales are identical, as will be shown in Example 19-1.

Nicolas Leonard Sadi Carnot (1796–1832) at age seventeen, in a drawing by L. L. Boilly.

---

**Example 19-1** The Carnot cycle shown in Figure 19-11 is performed using an ideal gas. Show that the ratio of the heat rejected to the heat absorbed equals the ratio of the ideal-gas temperatures of the heat reservoirs.

The equation of state of an ideal gas is $PV = nRt$, where we now use $t$ for the ideal-gas temperature to avoid assuming the result we wish to prove. We want to show that $|Q_c|/Q_h = t_c/t_h$. We first consider the isothermal expansion from state 1 to state 2. Since the internal energy of an ideal gas does not change if the temperature is constant, the work done by the gas in this isothermal expansion equals the heat absorbed, $Q_h$. The work done is

$$W_{1-2} = \int_1^2 P\, dV = \int_1^2 \frac{nRt_h}{V}\, dV = nRt_h \ln \frac{V_2}{V_1}$$

Then

$$Q_h = nRt_h \ln \frac{V_2}{V_1}$$

Similarly, the heat $|Q_c|$ rejected to the cold reservoir equals the work done *on* the gas in the isothermal compression at temperature $t_c$ from state 3 to state 4. This work is of the same magnitude as that done *by* the gas if it expands from state 4 to state 3. The heat rejected is thus

$$|Q_c| = nRt_c \ln \frac{V_3}{V_4}$$

and the ratio of these heats is

$$\frac{|Q_c|}{Q_h} = \frac{t_c \ln (V_3/V_4)}{t_h \ln (V_2/V_1)} \qquad\qquad 19\text{-}6$$

We can relate the volumes $V_1$, $V_2$, $V_3$, and $V_4$ using the equation for the adiabatic expansion and compression derived in Section 17-6. Along these curves we have

$$PV^\gamma = \text{constant}$$

If we eliminate the pressure using the equation of state, we have

$$\frac{nRt}{V} V^\gamma = \text{constant}$$

or

$$tV^{\gamma-1} = \text{constant} \qquad\qquad 19\text{-}7$$

Applying Equation 19-7 to states 2 and 3 connected by the adiabatic expansion, we have

$$t_h V_2^{\gamma-1} = t_c V_3^{\gamma-1} \qquad\qquad 19\text{-}8$$

Similarly, for the states 4 and 1 connected by the adiabatic compression,

$$t_h V_1^{\gamma-1} = t_c V_4^{\gamma-1} \qquad\qquad 19\text{-}9$$

Dividing Equation 19-8 by Equation 19-9 gives

$$\left(\frac{V_2}{V_1}\right)^{\gamma-1} = \left(\frac{V_3}{V_4}\right)^{\gamma-1} \qquad\qquad 19\text{-}10$$

and so $V_2/V_1 = V_3/V_4$ and $\ln (V_2/V_1) = \ln (V_3/V_4)$. After canceling the logarithmic terms in Equation 19-6, our result is proved:

$$\frac{|Q_c|}{Q_h} = \frac{t_c}{t_h} \qquad\qquad 19\text{-}11$$

Comparing this result with Equation 19-4, we see that the ratio of the ideal-gas temperatures of the two reservoirs is the same as the ratio of the absolute temperatures. Since both temperature scales have the same fixed value, 273.16 K for the temperature of the triple point of water, the ideal-gas and absolute temperatures are equal everywhere the ideal-gas temperature scale is defined.

---

**Example 19-2** An engine works between reservoirs at 400 and 300 K, extracting 100 J from the hot reservoir during each cycle. What is the greatest efficiency possible for this engine, and how much work can it perform during each cycle?

A Carnot engine working between these reservoirs has efficiency

$$\epsilon_R = 1 - \frac{T_c}{T_h} = 1 - \frac{300 \text{ K}}{400 \text{ K}} = 0.25 = 25\%$$

The work done by a Carnot engine in one cycle is found from $\epsilon_R = W/Q_h$. Then $W = \epsilon_R Q_h = (0.25)(100 \text{ J}) = 25 \text{ J}$. Any other engine working between these two reservoirs will have efficiency less than 25 percent because of irreversibility due to friction, non-quasi-static processes, etc.

---

**Example 19-3** What is the greatest possible coefficient of performance of a refrigerator working between reservoirs at 400 and 300 K?

From Example 19-2 we saw that a Carnot engine performed 25 J of work for each 100 J of heat removed from the upper reservoir. It must therefore reject 75 J to the lower-temperature reservoir. When this engine is run as a refrigerator, it removes 75 J from the reservoir at 300 K and rejects 100 J to the reservoir at 400 K. The coefficient of performance of a Carnot refrigerator between these reservoirs is thus

$$\eta = \frac{Q_c}{W} = \frac{75 \text{ J}}{25 \text{ J}} = 3$$

Any other refrigerator will have a smaller coefficient of performance because of irreversibility due to friction, non-quasi-static processes, etc.

---

**Questions**

6. Why do power-plant designers try to increase the temperature of the steam fed to engines as much as possible?

7. Will the efficiency of the engines in a power plant be improved if more cooling water is used so that the temperature rise in the cooling water is smaller?

## 19-6   Entropy

We used the zeroth law of thermodynamics to define a new thermodynamic state function, the temperature, which is a measure of the hotness or coldness of a system. In the molecular picture, the temperature measures the average kinetic energy of translation of the molecules. Similarly the first law allowed us to define the internal-energy function $U$. Now we consider a new thermodynamic state function, the entropy $S$, related to the second law of thermodynamics. The existence of this state function is less obvious than that of $T$ and $U$; it is also less easily related to the second law than they were to the zeroth and first laws, because the second law is more complicated than the first and zeroth laws in that it describes processes that do *not* occur. We shall first look at the special case of an ideal gas. With this simple system it is easy to show that there is a new state function which is related to the heat absorbed by the system and to the temperature at which the heat is absorbed.

Consider an arbitrary quasi-static reversible process in which an ideal gas absorbs an amount of heat $dQ$. According to the first law, $dQ^*$ is related to the internal-energy change $dU$ and the work done $dW = P \, dV$ by

$$dQ = dU + P \, dV$$

For our ideal gas we can substitute $nRT/V$ for $P$ from the equation of state and write the change in internal energy in terms of the heat capacity $dU = C_v \, dT$:

$$dQ = C_v \, dT + nRT \frac{dV}{V} \qquad \text{19-12}$$

When the system goes from state 1 to state 2, we must know the path followed in order to calculate $Q$, the total heat added. This is clear mathematically from the fact that we cannot integrate Equation 19-12. There is no problem with the first term because $C_v = dU/dT$ is a function of temperature only. The integration of this term gives the change in internal energy. The problem is with the second term, the work done by the gas. This point was discussed in Chapter 18. Since both the heat added and the work done depend on the path, neither is a state function. However, if we divide each term in Equation 19-12 by $T$, we can integrate both terms on the right-hand side:

$$\frac{dQ}{T} = C_v \frac{dT}{T} + nR \frac{dV}{V} \qquad \text{19-13}$$

Even if $C_v$ is not constant, it can depend only on $T$, so that whatever function of $T$ it is, the first term can be integrated. Thus if the system is carried from state 1 to state 2 along any quasi-static path, we can find the sum (integral) of $dQ/T$. It is in fact given by

$$\int_1^2 \frac{dQ}{T} = C_v \ln \frac{T_2}{T_1} + nR \ln \frac{V_2}{V_1} \qquad \text{19-14}$$

assuming $C_v$ to be constant. The right side of Equation 19-14 depends only on the states 1 and 2 and not on the path. If in addition the path is reversible, we get the same magnitude for the sum of $dQ/T$ by reversing the path and going from state 2 to 1, except that the sum is the negative

---

* According to our usual sign convention, $dQ$ is positive if heat enters the system and negative if it leaves the system.

of that obtained from state 1 to state 2. Thus if we travel along one reversible path from 1 to 2 and back along another reversible path from state 2 to 1, the sum of $dQ/T$ is zero. This implies that there is a state function whose change is given by $dQ/T$. We call this function the *entropy S*

$$S = S_2 - S_1 = \int_1^2 \frac{dQ_R}{T}$$

19-15

or

*Entropy*

$$dS = \frac{dQ_R}{T}$$

19-16

We have used the subscript $R$ on $dQ$ to remind us that the heat must be absorbed reversibly. The units of entropy are joules per kelvin.

A Carnot cycle is an example of a reversible closed path. During the isothermal expansion the entropy change of the gas is

$$\int_1^2 \frac{dQ_R}{T_h} = \frac{1}{T_h} \int dQ = \frac{Q_h}{T_h}$$

19-17

During the reversible adiabatic expansion and compression the entropy change is zero because no heat is absorbed or rejected by the gas. During the isothermal compression the entropy change is

$$\int_3^4 \frac{dQ_R}{T_c} = \frac{1}{T_c} \int dQ_R = \frac{Q_c}{T_c}$$

19-18

This is negative because, according to our sign convention, $dQ$ is negative when heat leaves the system.

Since the total entropy change of the gas must be zero for a complete cycle, we have

$$\frac{Q_h}{T_h} + \frac{Q_c}{T_c} = 0$$

or

19-19

$$\frac{-Q_c}{Q_h} = \frac{T_c}{T_h}$$

This is the same as the defining equation (19-4) for the ratio of absolute temperatures because the heat rejected is $Q_c = -|Q_c|$.

Although we have been considering a special sytem, the ideal gas, it can be shown that the sum of $dQ_R/T$ for any system undergoing a *reversible* change is independent of path and depends only on the initial and final states. (See the general proof in Section 19-10.) Thus Equations 19-15 and 19-16, which define the entropy function, are general. It is important to understand, though, that only for a reversible process is $\int dQ/T$ independent of path. For an irreversible process, change in the entropy of the system will not be given by $\int dQ/T$. However, since entropy is a state function, we can find the entropy change in the system undergoing an irreversible process from state 1 to state 2 by considering any *reversible* process connecting the same two states and computing $\int dQ_R/T$.

---

**Example 19-4** An ideal gas undergoes a free adiabatic expansion from volume $V_1$ to volume $V_2$. What is the change in entropy of the gas?

As we have already discussed, the final energy of the gas is the same as the initial energy, and the final temperature is the same as the initial temperature. *During the free expansion* we cannot describe the gas as having a temperature or internal energy since the gas does not pass through equilibrium states. In this process no heat is absorbed or rejected, so that $\Delta Q = 0$ for the whole process. We might expect, therefore, that there is no entropy change, but this is not correct. The change in entropy is *not* given by the sum of $dQ/T$ because this process is not reversible. We cannot even define $T$ during the process. However, changes in a state function are independent of the process. The entropy change, $S_2 - S_1$, is independent of the process connecting states 1 and 2. We can compute $S_2 - S_1$ by considering a *reversible* process connecting these two states. Any reversible process will do since the entropy change does not depend on the process. The convenient process for this example is a quasi-static isothermal expansion from volume $V_1$ to $V_2$. Since the temperature is constant for this reversible process, the entropy change is just $Q/T$, where $Q$ is the total heat absorbed. This heat absorbed equals the work done by the gas (since the internal energy is constant). Thus

$$Q = W = \int_1^2 P \, dV = nRT \int_1^2 \frac{dV}{V} = nRT \ln \frac{V_2}{V_1} \qquad \text{19-20}$$

The entropy change is

$$S_2 - S_1 = \frac{Q}{T} = \frac{nRT \ln (V_2/V_1)}{T} = nR \ln \frac{V_2}{V_1} \qquad \text{19-21}$$

This same result is obtained by integrating Equation 19-13 with $dT = 0$. The entropy of the system increases even though no heat is transferred. In order for the entropy of a gas to decrease by a free adiabatic "expansion" the final volume would have to be less than the initial volume. An adiabatic free contraction never occurs in nature.

---

**Example 19-5** The melting point of a substance is $T_0$, and its latent heat is $L$. What entropy change takes place when mass $m$ of the substance melts at this temperature?

   To compute the entropy change we must consider a reversible process for the melting and calculate the integral of $dQ_R/T$ for that process. Such a process would be the melting produced by putting the solid material into contact with a reservoir whose temperature is only very slightly higher so that melting would proceed very slowly. The process could be reversed by lowering the reservoir temperature very slightly, causing any molten material to freeze again. As we have seen, for heat conduction to be reversible the temperature difference must be vanishingly small.

   From Equation 19-15

$$S = \int \frac{dQ_R}{T} = \frac{1}{T_0} \int dQ_R = \frac{mL}{T_0}$$

The temperature $T_0$ is removed from the integral since the melting takes place at the constant temperature $T_0$. The remaining integral $\int dQ_R$ is the heat required to melt mass $m$, or $mL$.

   To illustrate our result numerically, we calculate the entropy increase when 100 g of mercury melts at $-39°C$ ($=234$ K) and atmospheric pressure. The latent heat of fusion is 11.3 J/g, and

$$S = \frac{mL}{T_0} = \frac{(100 \text{ g})(11.3 \text{ J/g})}{234 \text{ K}} = 4.83 \text{ J/K}$$

---

**Example 19-6** A substance is heated reversibly at constant pressure from temperature $T_1$ to temperature $T_2$. Show that the entropy change of the substance is $\Delta S = C_p \ln (T_2/T_1)$, where $C_p$ is the heat capacity at constant pressure, and evaluate $\Delta S$ for 1 kg of water heated from 0 to 100°C.

When a body absorbs a small amount of heat $dQ$ at constant pressure, its temperature change $dT$ is given by

$$dQ = C_p \, dT$$

and the entropy change is

$$dS = \frac{dQ}{T} = C_p \frac{dT}{T}$$

Integrating from $T_1$ to $T_2$, we obtain for the total entropy change

$$\Delta S = C_p \int \frac{dT}{T} = C_p \ln \frac{T_2}{T_1} \qquad\qquad 19\text{-}22$$

For 1 kg of water, $C_p = 4.184$ kJ/K, and the entropy change of the water when it is heated from 0°C (273 K) to 100°C (373 K) is

$$\Delta S = (4.184 \text{ kJ/K}) \ln \frac{373}{273} = 1.31 \text{ kJ/K}$$

Since entropy is a state function, its change is independent of the process. If the final pressure equals the initial pressure, the entropy change of a body when it is heated is given by Equation 19-22 no matter what the process may be. Equation 19-22 also gives the entropy change of a body that is cooled. In that case $T_2$ is less than $T_1$, and $\ln (T_2/T_1)$ will be negative, giving a negative entropy change.

---

### Questions

8. Does the entropy change in a physical system passing from state 1 to state 2 depend on the path taken between those states?

9. Is it possible to choose the value of the entropy of a system at some particular state and thereafter assign a value to the entropy of the system for every other state?

10. Give an example of a system whose entropy decreases with time.

11. A gas undergoes a quasi-static adiabatic expansion to twice its original volume. How does its entropy change?

## 19-7   Entropy Change of the Universe

We have shown that the entropy state function exists for an ideal gas,* and we have seen how to calculate changes in entropy for a system, but it is not immediately evident how this function is useful. In this section we calculate the entropy change of the universe for several processes. By "universe" we mean the system and its surroundings, e.g., heat reservoirs and other systems that interact with the system of interest. In these examples we shall often be calculating the entropy change of a heat reservoir, which is easy to do.

---

* The proof for a general system is given in Section 19-10.

The state of a heat reservoir is determined by its internal energy and its temperature, which, by definition, is constant. If a heat reservoir absorbs heat $Q$ by some reversible process, the entropy of the reservoir increases by $Q/T$, where $T$ is the temperature of the reservoir. If it absorbs the heat in an irreversible process, we need only replace the irreversible process by a reversible process to calculate the change in entropy. Again the entropy increases by $Q/T$. If the reservoir gives up heat $Q$ to the system or another reservoir, the entropy of the reservoir *decreases* by $Q/T$ since $Q$ is now leaving the reservoir.

Let us first consider the entropy change of the universe for a reversible process. A system absorbs heat $Q$ from a reservoir at temperature $T$; since the process is reversible, the absorption must occur isothermally. The system has an increase in entropy of $Q/T$, and the reservoir has a decrease in entropy of the same amount. Thus the total change in the entropy of the system plus reservoir is zero. No matter how many systems and reservoirs take part in a reversible process, for each heat absorption or rejection the entropy change of the reservoir is exactly the negative of that for the system. Thus *the entropy change of the universe is zero for a reversible process.*

---

**Example 19-7** Show that the entropy change of the universe is zero during a Carnot cycle.

The entropy change of the engine after a complete cycle must be zero because the engine is back in its original state. We thus have only to calculate the entropy change of the two heat reservoirs. The hot reservoir gives up heat $Q_h$ at temperature $T_h$; its entropy therefore decreases. The entropy change of the hot reservoir is

$$\Delta S_h = -\frac{|Q_h|}{T_h} \qquad\qquad 19\text{-}23$$

The cold reservoir absorbs heat $Q_c$ at temperature $T_c$. Its entropy change is positive, given by

$$\Delta S_c = +\frac{|Q_c|}{T_c} \qquad\qquad 19\text{-}24$$

Since this is a Carnot cycle, the heats $Q_h$ and $Q_c$ are related by Equation 19-4:

$$\frac{|Q_c|}{|Q_h|} = \frac{T_c}{T_h}$$

If we substitute $|Q_h|T_c/T_h$ for $|Q_c|$ in Equation 19-23, we have

$$S_c = +\frac{|Q_h|}{T_h}$$

Thus the sum of $\Delta S_c$ and $\Delta S_h$ is zero.

---

We now consider the entropy change of the universe for processes that are irreversible. In Section 19-10 we show that for a general irreversible process the entropy change of the universe must be positive. Here we merely investigate some typical irreversible processes and show in each case that the entropy change of the universe must be greater than zero.

Let us consider heat conduction from a hot reservoir to a cold reservoir; i.e., an amount of heat $Q$ is conducted from a reservoir at temperature $T_h$ to a reservoir at temperature $T_c$. The entropy of the hot reservoir decreases by $Q/T_h$, and that of the cold reservoir increases by $Q/T_c$. Since $T_c$ is less than $T_h$, the increase in the entropy of the cold reservoir is greater in magnitude than the decrease in the entropy of the hot reservoir. Thus the total entropy change of the universe is greater than zero and is given by

$$\Delta S_u = \frac{Q}{T_c} - \frac{Q}{T_h} > 0 \qquad\qquad 19\text{-}25$$

Next let us consider a process that is irreversible because mechanical energy is lost to internal energy due to friction or other dissipative forces—say a block with kinetic energy $E_k$ sliding along a rough table until it is stopped by friction. We assume for simplicity that the table is large enough to be considered a reservoir. (This is not important to the conclusion, but it simplifies the calculation.) Because of the work done by friction, the internal energy of the table (and block if it is not much smaller than the table) and air surroundings increases by the amount of $E_k$. The entropy change of the table and air surroundings is the same as if heat $Q = E_k$ were transferred reversibly. Thus the entropy change of the table and air is $+E_k/T$. This is the entropy change of the universe:

$$\Delta S_u = +\frac{E_k}{T} > 0 \qquad\qquad 19\text{-}26$$

Finally we consider the third kind of irreversible process discussed in Section 19-3, a non-quasi-static process, again taking the free adiabatic expansion of an ideal gas. Since no heat is transferred and no work done, the surroundings suffer no change whatsoever. We have already calculated the entropy change of the gas in Example 19-4. This is also the entropy change of the universe. According to Equation 19-21, the entropy change of the gas and universe is $nR \ln (V_2/V_1)$, which is positive because $V_2$ is greater than $V_1$.

In each of these typical irreversible processes the change in entropy of the universe is greater than zero. The reason is that a negative change in entropy of the universe would violate the second law. For example, in heat conduction, the entropy change of the universe could be negative only if the heat left the cold reservoir and entered the hot reservoir, violating the Clausius statement of the second law of thermodynamics. Similarly, in the conversion of internal energy into mechanical energy, the entropy change would be negative if heat were removed from the surroundings and converted completely into mechanical energy, violating the Kelvin-Planck statement of the second law. We have already discussed the fact that the irreversibility of the free adiabatic expansion is connected with the second law. If the entropy of the universe were to decrease in a free adiabatic "expansion," the gas would have to contract to a smaller volume by itself.

As a result we have a new statement of the second law of thermodynamics, equivalent to the other statements:

*For any process, the entropy change of the universe must be greater than or equal to zero. The entropy change is equal to zero only if the process is reversible.*

*Second law in terms of entropy*

The entropy function gives us a numerical measure of the irreversibility of a given process. Suppose, for example, we consider a real attempt to produce an isothermal quasi-static reversible expansion of a gas. We can compute the entropy change of the heat reservoir by measuring the heat leaving the reservoir. The entropy of the gas can be found in tables. If there is friction in the piston and some turbulence in the gas (because the piston has a slight acceleration or because the expansion happens too fast for the gas to remain near equilibrium states), the total entropy change will not be zero but some positive number. If we reduce the friction or make the process more nearly quasi-static, the total entropy change will be less. We thus have a numerical measure of the degree of irreversibility of any given process.

## 19-8 Entropy and the Availability of Energy

Let us consider a single heat reservoir at temperature $T$. If there are no other heat reservoirs around and no systems at a lower temperature which can be used as heat reservoirs, we cannot use any of the internal energy of the reservoir for the performance of work. The internal energy of the reservoir is unavailable. To extract some of it and do work requires a lower-temperature reservoir or system to accept the heat rejected. The second law states that we cannot remove some of the internal energy and convert it completely into work; in other words, the conversion of some of the internal energy completely into work would decrease the entropy of the universe, which is not allowed.

Whenever an irreversible process occurs, some energy that was available for work before the process took place is made unavailable for use as work. For the block sliding along the rough table, if it were not for the mechanical energy lost because of friction, the original kinetic energy of the block could be used to do some useful work, e.g., lifting a weight or turning a generator. After the block's energy has been converted into internal energy of the surroundings, all this potential for doing work is lost. In this case the entropy increase of the universe is $E_k/T = \Delta S_u$ and the energy which is made unavailable is $T \Delta S_u$, where $T$ is the temperature of the available reservoir and $\Delta S_u$ is the increase in entropy of the universe. This is a general result.

*In an irreversible process in which the entropy change of the universe is $\Delta S_u$ some energy becomes unavailable for use to perform work. The amount of energy made unavailable equals $T \Delta S_u$, where $T$ is the temperature of the lowest-temperature reservoir available.*

In another example, the free adiabatic expansion of an ideal gas, we assume that we have one reservoir at the temperature of the gas, which is the same after the expansion as before. The entropy change when the gas expands freely from volume $V_1$ to volume $V_2$ is $nR \ln (V_2/V_1)$, as calculated in Example 19-4. Since there are no effects on the surroundings, this is the entropy change of the universe. According to the result stated above, the energy made unavailable for doing work is $T$ times this entropy change, or $nRT \ln (V_2/V_1)$. This is just the work that could have been done if we had allowed the gas to expand isothermally and quasi-statically. During this reversible expansion, the gas does work

$W = nRT \ln (V_2/V_1)$, as previously calculated, while absorbing an equal amount of heat from the reservoir. Thus the result of the irreversible adiabatic free expansion is the loss of the opportunity of converting this internal energy of the reservoir into work.

In our final example the loss of ability to do work occurs in the irreversible process of conducting heat $Q$ from a hot reservoir to a cold reservoir. The entropy change of the universe for this process was computed in the previous section as

$$\Delta S_u = \frac{Q}{T_c} - \frac{Q}{T_h}$$

$Q$ is *not* the amount of energy made unavailable for work because, according to the second law, we could not have removed this heat from the hot reservoir and changed it completely into work. However, some of this energy could have been converted into work, the amount depending on the efficiency of the heat engine used. The greatest amount of work that can be obtained from the heat energy $Q$ is the amount done using a Carnot engine. The efficiency of a Carnot engine is

$$\epsilon_R = 1 - \frac{T_c}{T_h}$$

The work that can be done with heat $Q$ entering a Carnot engine is thus

$$W = \epsilon_R Q = Q - \frac{QT_c}{T_h} = T_c \left(\frac{Q}{T_c} - \frac{Q}{T_h}\right)$$

Again, the work that could have been done is just the temperature $T_c$ of the lowest-temperature heat reservoir available times the entropy change of the universe.

---

**Example 19-8** An engine working between reservoirs at 373 and 273 K extracts 100 J of energy from the hot reservoir and performs 15 J of work in each cycle. What is the efficiency of this engine? How much work could be done by a Carnot engine working between the same reservoirs? What is the entropy change of the universe in each cycle?

The efficiency of this engine is the work done divided by the energy intake $\epsilon = W/Q_{in} = (15 \text{ J})/(100 \text{ J}) = 15$ percent. This is less than the Carnot efficiency of $\epsilon_R = 1 - T_c/T_h = 1 - (273 \text{ K})/(373 \text{ K}) = 26.8$ percent. If a Carnot engine worked between these same reservoirs, it could do work of 26.8 J for each 100 J extracted. Thus in each cycle there is energy wasted because of irreversibilities of $26.8 - 15 = 11.8$ J. Since this energy equals the entropy change of the universe times the temperature of the lowest-temperature reservoir available, we can find the entropy change of the universe from $\Delta S_u = (11.8 \text{ J})/(273 \text{ K}) = 0.0432$ J/K. We can also compute the entropy change of the universe directly. When the upper reservoir loses 100 J of energy, its entropy decreases by the amount $(100 \text{ J})/(373 \text{ K}) = 0.268$ J/K. Similarly, when the lower reservoir receives the exhaust energy of $100 \text{ J} - 15 \text{ J} = 85$ J, its entropy increases by the amount $(85 \text{ J})/(273 \text{ K}) = 0.311$ J/K. The entropy change of the engine is of course zero after one complete cycle. The net entropy change of the universe is then $+0.311 - 0.268 = 0.043$ J/K, as found before.

# 19-9   Molecular Interpretation of Entropy

There is a fundamental difference between the internal energy of a system and the macroscopic mechanical energy of a system. Figure 19-12 shows a schematic illustration of the velocities of the molecules in a system whose center of mass is at rest. If the system is a solid, the molecules are vibrating about their equilibrium positions; if it is a gas, the molecules move throughout the volume. In either case, the velocity vectors $\mathbf{u}_i$ point in all directions. In a gas the velocity of any one molecule points in a direction unrelated to that of any other molecule unless the two molecules have just collided. The total linear momentum and the total angular momentum of the system are zero. The average kinetic energy associated with each degree of freedom is $\frac{1}{2}kT$ per molecule.

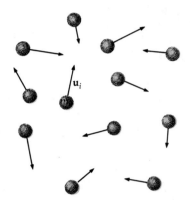

**Figure 19-12**
In a body at rest, the velocities of the molecules $\mathbf{u}_i$ are randomly directed.

Suppose we now add a velocity $\mathbf{v}$ in the $x$ direction to the velocity of each molecule. The center of mass of the system then moves with speed $v$ in the $x$ direction, and the total kinetic energy is the sum of the center-of-mass energy $\frac{1}{2}Mv^2$ and the energy relative to the center of mass:

$$E_k = \tfrac{1}{2}Mv^2 + \Sigma\tfrac{1}{2}m_iu_i^2 \qquad\qquad 19\text{-}27$$

where $M$ is the total mass. The two terms in the expression for kinetic energy are quite different. The motion giving rise to the term $\frac{1}{2}Mv^2$ is an ordered motion. Each molecule has the same component $v$ in the same direction. Thus all the molecules are moving together. The center of mass moves with velocity $\mathbf{v}$. This *mechanical kinetic energy* has nothing to do with the internal energy or the temperature. On the other hand, the second term (the energy relative to the center of mass) is the original internal kinetic energy. It is associated with the random, or *disordered*, motion of the molecules. The ordered part of the total energy of the system, the mechanical kinetic energy, can easily be changed completely into other forms, e.g., potential energy or rotational kinetic energy (we could attach a string to the moving system and use it to lift a weight or turn a wheel, which could even generate electric energy).

On the other hand, the disordered part of the total energy, the internal energy, cannot easily be converted into mechanical energy. As we have seen, only part of this energy can be extracted and converted into work, which requires that some of the random motion be converted into ordered motion. We can do this with a heat engine, say with gas as the working substance. We place the gas system in contact with the system whose internal energy we want to convert into mechanical energy, calling this system the reservoir, in keeping with previous usage. If the gas is at a slightly lower temperature, some internal energy is transferred to the gas by conduction. In the molecular picture of conduction the molecules in the walls of the gas enclosure are vibrating with less average energy than those of the walls of the reservoir because the gas temperature is lower. When the two systems are in contact, some of the kinetic energy of vibration of molecules in the reservoir walls is transferred to kinetic energy of the molecules in the gas-container walls. (This energy is quickly brought to equilibrium with the potential energy of vibration of the wall molecules.) Now the gas container is not in equilibrium with the gas molecules, and when the gas molecules collide with the container walls, energy is transferred on the average to the gas molecules. These molecules in turn collide with others and transfer kinetic energy, until the temperature of the gas and wall has increased to that of the reservoir. The pressure of the gas is thus increased. This random internal energy can now be transferred into ordered mechanical energy by letting the gas at higher pressure expand against a piston. Since the piston can move in only one direction, the transfer of energy of the gas molecules to energy of the piston amounts to transfer of random kinetic energy of the gas molecules to ordered motion of the piston.

We have seen in this chapter that ordered mechanical energy is generally more useful than unordered internal energy. The second law of thermodynamics can be stated:

*The entropy of the universe always increases (or in reversible processes remains the same).*

This can be related to the fact that molecular motion always tends toward disordered rather than ordered motion. Imagine a block with mechanical kinetic energy $\frac{1}{2}Mv^2$ moving toward a wall and making an inelastic collision, thereby changing the kinetic energy into internal kinetic energy of disordered motion of the molecules in the block, wall, and air. Our experience and intuition tell us that the reverse will not happen. Disordered motion of the molecules will never spontaneously change into ordered motion. Thus entropy increase is an increase in the disordered motion of the molecules in the universe. When the entropy of the universe reaches its maximum value, no more processes can occur (except absolutely reversible processes). Heat will not be conducted because the whole universe will be at the same temperature. No more energy will be available to perform work.

### Questions

12. When a gas is compressed isothermally, it gives up heat and its entropy decreases. In what sense is the motion of the gas molecules more orderly after compression?

13. When you pick up a spilled deck of cards and put them in a neat pile, you have increased the order of the universe. What compensating decrease in order accompanied your action?

14. One litre each of oxygen gas and nitrogen gas are initially in separate containers at room temperature and atmospheric pressure. Then the containers are connected, allowing the gases to mix. What happens to the entropy of the system consisting of the two gases?

# Power Plants and Thermal Pollution

Laurent Hodges
*Iowa State University*

The scientist's interest in a heat engine is focused primarily on the work obtained from a given heat input. Society, however, is also interested in the waste heat output. When it leads to undesirable effects, it is called *thermal pollution*.

The largest heat engines in modern society—those from which thermal pollution can be the most damaging—are the steam engines used in electric power plants to drive an ac generator and produce electricity. About 85 percent of the electricity in the United States is produced by such steam engines, the other 15 percent being generated by water power in hydroelectric plants.

The operation of a steam electric power plant is explained in Figure 1. The *heat source* is usually either the combustion of a fossil fuel (coal, oil, or natural gas) or the fission of uranium 235; occasionally geothermal power or the combustion of solid wastes is used. The *boiler* heats water, converting it into high-temperature, high-pressure steam. A nuclear power plant has a reactor core in which water is heated into steam (boiling-water reactor, or BWR) or into pressurized water (pressurized-water reactor, or PWR); in the PWR, heat is then transferred to a steam cycle in a heat exchanger. Steam expands and cools as it passes through the *turbine*, converting heat energy into the mechanical energy of a rotating shaft. A *generator* converts the mechanical energy into electric energy (alternating current). The *condenser* cools the steam exhausted from the turbine and reduces its pressure, thereby increasing the efficiency of the power plant.

Typical temperatures and efficiencies for steam electric plants are shown in Table 19-1. The average efficiency for all power plants in the United States is about 34 percent, but the best fossil-fuel plants have about 40 percent efficiency and many smaller or older plants have efficiencies of 25 percent or less. The efficiencies of BWR and PWR nuclear plants are about 34 percent, but

**Figure 1**
Major components of a steam electric power plant (schematic).

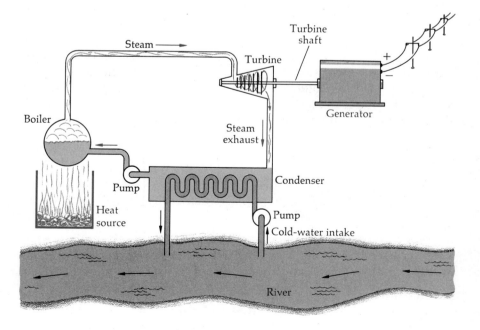

**Table 19-1**

Typical higher and lower temperatures and efficiencies for steam electric power plants

|  | Temperature, °C | | Efficiency, % | |
|---|---|---|---|---|
|  | High | Low | Carnot | Actual |
| Large fossil-fuel plant | 380 | 40 | 52 | 40 |
| Boiling-water reactor | 285 | 40 | 44 | 34 |
| Pressurized-water reactor | 315 | 40 | 47 | 34 |

those of some experimental nuclear plants (cooled by gas or liquid sodium) are closer to 40 percent.

Existing power plants must dispose of waste heat amounting to twice the electrical energy generated. A popular method, *once-through cooling,* involves passing water from a river or lake through the condenser once and then returning it at a warmer temperature. This works well only if adequate water is present. Many plants of 1000 MW$_e$ ($e$ = electric power) are being built today, sometimes several at one location. Even on a large river the resulting temperature rise may be significant: a 3000 MW$_e$–9000 MW$_t$ ($t$ = thermal power) installation using a flow of 120 m$^3$/sec (the average yearly minimum flow of the Missouri River at its mouth) would cause a 12°C rise by its 6000 MW of waste heat.

Thermal pollution is the assortment of undesirable effects resulting from waste heat. Although the temperature changes produced in a river by once-through cooling may not be as large as natural daily or seasonal fluctuations, possible adverse ecological effects include greater growth of bacteria, pathogens, or undesirable blue-green algae, or physiological and behavioral changes in aquatic life (such as fish). As an example, large-mouth bass acclimated to 30°C show a 50 percent mortality rate within 72 h if the water is warmed to 34°C.

Warmer water also contains less of the dissolved oxygen necessary for aquatic life and for the decomposition of organic wastes; e.g., the dissolved oxygen at saturation in water is 11.3 mg/L at 10°C and 6.6 mg/L at 40°C. Warmer water is also less viscous, permitting faster deposition of the sediment load, which can affect aquatic food supplies.

Most of the cooling water used in the United States is used by electric power plants. The rapid growth of electric energy consumption, averaging 7 percent annually for many decades, has led to greater use of alternatives to once-through cooling. One of these is the cooling pond, really a private lake, but the land requirements for such a pond are often too expensive to consider.

The most common alternative, the cooling tower, transfers the waste heat to the atmosphere (Figure 2). Usually these are evaporative towers, which use the heat to evaporate water. Evaporative cooling towers are either large hyperbolic types, with a natural draft, or mechanical-draft (fan) types. Nonevaporative cooling towers, which transfer heat to the air by heat exchangers, are expensive but may be needed in moist climates. At ordinary temperatures the evaporation of water requires 2.26 MJ/kg, so the water requirements are much less than in once-through cooling, although the water is consumed in the process. The evaporated water can cause fog, increased precipitation, and (in certain climates) icing of nearby roads, and the cooling tower and its plume may be blots on the landscape, so that evaporative towers are not a completely satisfactory answer to the problem of thermal pollution.

**Figure 2**
Hyperbolic natural-draft evaporative cooling towers at the Fort Martin Power Station, Fairmont, West Virginia. (*Courtesy of Marley Company, Mission, Kansas.*)

# 19-10 Proof That Entropy Exists and That the Entropy of the Universe Can Never Decrease

The general proof of the existence of the entropy function is not difficult but rather abstract. It is interesting, however, because it shows that entropy and its increase are directly related to the second law of thermodynamics. We first state and prove the *Clausius inequality*.

Let a system undergo any cycle in which it absorbs and rejects heat from any number of different heat reservoirs, performs positive or negative work, and returns to its original state. Let $\Delta Q_i$ be the heat absorbed from the $i$th reservoir at temperature $T_i$. If heat is rejected to the reservoir, $\Delta Q_i$ will be negative. The Clausius inequality states that the sum of $\Delta Q_i / T_i$ is less than or equal to zero:

$$\Sigma \frac{\Delta Q_i}{T_i} \leqslant 0 \qquad\qquad 19\text{-}28$$

To prove this inequality we use an auxiliary reservoir at temperature $T_0$ and as many Carnot cycles as we need. We run a Carnot engine between each reservoir and our auxiliary reservoir so that the original reservoir is returned to its original state. That is, if $\Delta Q_i$ was removed from the $i$th reservoir, we run a Carnot engine between it and the auxiliary reservoir to replace $\Delta Q_i$ in the $i$th reservoir. The heat extracted from the auxiliary reservoir when the engine runs between it and the $i$th reservoir is given by Equation 19-4:

$$\Delta Q_{0i} = \frac{T_0}{T_i} \Delta Q_i \qquad\qquad 19\text{-}29$$

We note that if $\Delta Q_i$ is positive, $\Delta Q_{0i}$ is positive, and if $\Delta Q_i$ is negative, $\Delta Q_{0i}$ is negative, indicating that heat was put into the auxiliary reservoir. The total heat *extracted* from the auxiliary reservoir is

$$\Delta Q_0 = \Sigma \, \Delta Q_{0i} = T_0 \, \Sigma \frac{\Delta Q_i}{T_i}$$

After this procedure all the original reservoirs are in their original state. The net result is that heat $\Delta Q_0$ has been extracted from the auxiliary reservoir and, according to the first law of thermodynamics, an equal amount of work has been done. The second law of thermodynamics tells us that $\Delta Q_0$ must be less than or equal to zero, so that either no heat was extracted or negative work was done and heat was put into the auxiliary reservoir. Thus $\Delta Q_0 \leqslant 0$ and

$$\Sigma \frac{\Delta Q_i}{T_i} \leqslant 0$$

The Clausius inequality has thus been proved. If the system undergoes a reversible cycle, absorbing heat $\Delta Q_i$ from the $i$th reservoir, the equality in Equation 19-28 holds. We can see this by merely reversing the cycle, putting $\Delta Q_i$ into each reservoir, which merely changes the sign of $\Delta Q_i$, giving

$$-\Sigma \frac{\Delta Q_i}{T_i} \leqslant 0$$

The only possibility for a reversible cycle is thus

$$\Sigma \frac{\Delta Q_i}{T_i} = 0 \qquad \text{reversible cycle} \qquad 19\text{-}30$$

We are now in a position to define entropy for a general system. We let the system move reversibly from state $A$ to state $B$ along one path and move back to state $A$ along a different *reversible* path. Any number of reservoirs can be used. According to Equation 19-30, we have

$$\sum_{\substack{A \\ \text{path 1}}}^{B} \frac{\Delta Q_i}{T_i} + \sum_{\substack{B \\ \text{path 2}}}^{A} \frac{\Delta Q_i}{T_i} = 0$$

Since each path is reversible,

$$\sum_{\substack{B \\ \text{path 2}}}^{A} \frac{\Delta Q_i}{T_i} = - \sum_{\substack{A \\ \text{path 2}}}^{B} \frac{\Delta Q_i}{T_i}$$

so that

$$\sum_{\substack{A \\ \text{path 1}}}^{B} \frac{\Delta Q_i}{T_i} - \sum_{\substack{A \\ \text{path 2}}}^{B} \frac{\Delta Q_i}{T_i} = 0 \qquad \text{or} \qquad \sum_{\substack{A \\ \text{path 1}}}^{B} \frac{\Delta Q_i}{T_i} = \sum_{\substack{A \\ \text{path 2}}}^{B} \frac{\Delta Q_i}{T_i}$$

Thus the sum $\Sigma(\Delta Q_i/T_i)$ is the same for all reversible paths connecting state $A$ to state $B$. Figure 19-13 shows an arbitrary quasi-static process for a gas. In order to make this process reversible, we must absorb heat isothermally. We do so by using a very large number of heat reservoirs at different temperatures and following the alternative path, consisting of a series of isotherms and adiabatic paths. In the limit of an infinite number of reservoirs, we can make this alternate path approach the original path as closely as desired. Thus for a general reversible path we need a large number of heat reservoirs, and the heat absorbed from each is very small. We thus write $dQ_R$ for $\Delta Q_i$ and indicate the sum by an integral sign, leaving off the subscripts $i$:

$$\int_{\text{path 1}} \frac{dQ_R}{T} = \int_{\text{path 2}} \frac{dQ_R}{T}$$

Since the integral of $dQ_R/T$ is independent of path between two states, it can depend only on the states. We thus can define a new function of state, the entropy $S$, by

$$S_B - S_A = \int_A^B \frac{dQ_R}{T}$$

where again we put a subscript $R$ on $dQ$ to remind us that the process must be reversible. For a very small change we can write

$$dS = \frac{dQ_R}{T}$$

These last two equations are the same as Equations 19-15 and 19-16.

We now consider a system which undergoes an *irreversible* change from state $A$ to state $B$ and then is brought back to state $A$ by a reversible process. This cycle is indicated in Figure 19-14, where the wiggly line indicates the irreversible process for which we cannot plot a curve because the system is not in equilibrium. According to the Clausius inequality, we have for the complete cycle

$$\int_{\substack{A \\ \text{irreversible} \\ \text{path}}}^{B} \frac{dQ}{T} + \int_B^A \frac{dQ_R}{T} < 0$$

or

$$\int_{\substack{A \\ \text{irreversible} \\ \text{path}}}^{B} \frac{dQ}{T} < - \int_B^A \frac{dQ_R}{T} = \int_A^B \frac{dQ_R}{T}$$

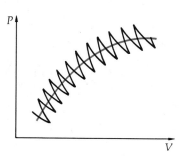

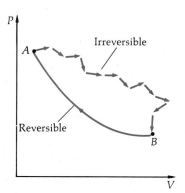

**Figure 19-13**
An arbitrary path can be approximated as accurately as desired by a series of alternating quasi-static adiabatic and isothermal paths.

**Figure 19-14**
An irreversible transition from $A$ to $B$ followed by a reversible transition from $B$ to $A$.

where we have used

$$\int_A^B \frac{dQ_R}{T} = -\int_B^A \frac{dQ_R}{T}$$

for the *reversible* path. By our definition of entropy we have

$$S_B - S_A = \int_A^B \frac{dQ_R}{T}$$

Thus

$$\int_{\substack{A \\ \text{irreversible} \\ \text{path}}}^B \frac{dQ}{T} < S_B - S_A$$

In particular, if a system is isolated from its surroundings, no heat can enter or leave. When an isolated system changes from state $A$ to state $B$, $dQ = 0$. Then if the process is irreversible,

$$0 < S_B - S_A \qquad S_B > S_A$$

We can consider the universe to be an isolated system. We have thus shown that the entropy of the universe always increases for any irreversible process. The defining equation (19-15) for entropy when applied to the universe with $dQ_R = 0$ shows that for reversible processes the entropy change of the universe is zero.

## Review

A. Define, explain, or otherwise identify:

B. True or false:

1. Work can never be converted completely into heat.

2. Heat can never be converted completely into work.

3. All heat engines have the same efficiency.

4. It is impossible to transfer a given quantity of heat from a cold reservoir to a hot reservoir.

5. The coefficient of performance of a refrigerator cannot be greater than 1.

6. All reversible processes are quasi-static.

7. All quasi-static processes are reversible.

8. All Carnot cycles are reversible.

9. The entropy of a system can never decrease.

10. The entropy of the universe can never decrease.

## Exercises

### Section 19-1, Heat Engines and the Second Law of Thermodynamics

1. An engine with 20 percent efficiency does 100 J of work in each cycle. (*a*) How much heat is absorbed in each cycle? (*b*) How much heat is rejected?

2. An engine absorbs 400 J of heat and does 120 J of work in each cycle. (*a*) What is its efficiency? (*b*) How much heat is rejected in each cycle?

3. An engine absorbs 100 J and rejects 60 J in each cycle. (*a*) What is its efficiency? (*b*) If each cycle takes 0.5 s, find the power output of this engine in watts.

4. A refrigerator absorbs 5 kJ from a cold reservoir and rejects 8 kJ. (*a*) Find the coefficient of performance of the refrigerator. (*b*) The refrigerator is reversible and is run as a heat engine ($Q_h = 8$ kJ; $Q_c = 5$ kJ). What is its efficiency?

5. An engine with an output of 200 W has an efficiency of 30 percent. It works at 10 cycles/s. How much heat is absorbed and how much rejected in each cycle?

### Section 19-2, Equivalence of the Kelvin-Planck and Clausius Statements

6. Suppose that there are two statements $A$ and $B$, each of which may be either true or false. (*a*) List the four possible combinations of truth or falsity for the two statements. (*b*) What possibilities are left if it is shown that if $A$ is true, $B$ must also be true? Does this prove that the statements are equivalent? (*c*) Suppose that it is also shown that if $A$ is false, $B$ is false. Do these two results prove that the two statements are equivalent?

7. A certain engine running at 30 percent efficiency draws 200 J of heat from a hot reservoir. Assume the Clausius statement to be false and show how this engine combined with a perfect refrigerator can violate the Kelvin-Planck statement of the second law.

8. A certain refrigerator takes in 500 J of heat from a cold reservoir and rejects 800 J to a hot reservoir. Assume that the Kelvin-Planck statement is false and show how a perfect engine working with this refrigerator can violate the Clausius statement.

### Section 19-3, Reversibility

9. Show that the reversibility of the following processes would violate the second law of thermodynamics: (*a*) an object is dropped from a height $H$ to the ground and comes to rest; (*b*) a car traveling at high speed brakes to a stop; (*c*) a hot object is placed in a cold-water bath, and the system comes to an equilibrium temperature.

### Section 19-4, The Carnot Engine

10. A reversible engine working between reservoirs at temperatures $T_h$ and $T_c$ has efficiency of 30 percent. Working as a heat engine, it rejects 140 J of heat to the cold reservoir. A second engine working between the same two reservoirs also rejects 140 J to the cold reservoir. Show that if the second engine has an efficiency of greater than 30 percent, the two engines can work together to violate the Kelvin-Planck statement of the second law.

11. A reversible engine working between reservoirs at temperatures $T_h$ and $T_c$ has an efficiency of 20 percent. Working as a heat engine, it does 100 J of work in each cycle. A second engine working between the same two reservoirs also does 100 J of work in each cycle. Show that if the efficiency of the second engine is greater than 20 percent, the two engines can work together to violate the Clausius statement of the second law.

12. A Carnot engine works between two heat reservoirs as a refrigerator. It removes 100 J from the cold reservoir and rejects 150 J to the hot reservoir during each cycle. Its coefficient of performance is $\eta = Q_c/W = 100/50 = 2$. (a) What is the efficiency of the Carnot engine when it is working as a heat engine between the same two reservoirs? (b) Show that no other engine working as a refrigerator between the same two reservoirs can have a coefficient of performance greater than 2.

## Section 19-5, The Absolute Temperature Scale

13. A Carnot engine works between two heat reservoirs at temperatures $T_h = 300$ K and $T_c = 200$ K. (a) What is its efficiency? (b) If it absorbs 100 J from the hot reservoir during each cycle, how much work does it do? How much heat does it reject during each cycle? (c) What is the coefficient of performance of this engine when working as a refrigerator between these two reservoirs?

14. An engine works between the steam point 100°C and the ice point 0°C. What is its greatest possible efficiency?

15. Which has the greater effect on increasing the efficiency of a Carnot engine, a 5-K increase in the temperature of the hot reservoir or a 5-K decrease in the temperature of the cold reservoir (in each case keeping the other reservoir fixed)?

16. A refrigerator works between 0°C and a room temperature of 20°C. (a) What is the greatest possible coefficient of performance? (b) If the refrigerator is to be cooled to $-10$°C, what is the greatest coefficient of performance assuming the same room temperature of 20°C?

17. An engine draws heat from a reservoir at a temperature of 100°C. If the engine is to be 40 percent efficient, what is the warmest possible temperature of the cold reservoir?

18. An engine is designed to exhaust heat to the atmosphere at $t = 20$°C. What is the least possible temperature of the upper reservoir if the engine is to have an efficiency of 20 percent?

## Section 19-6, Entropy, and Section 19-7, Entropy Change of the Universe

19. Two moles of an ideal gas at $T = 400$ K expands quasi-statically and isothermally from an initial volume of 40 L to a final volume of 80 L. (a) Find the entropy change of the gas. *Hint:* Use Equation 19-21. (b) What is the entropy change of the universe for this process?

20. The gas in Exercise 19 is taken from the same initial state ($T = 400$ K, $V_1 = 40$ L) to the same final state ($T = 400$ K, $V_2 = 80$ L) by a non-quasi-static process. (a) Is the entropy change of the gas the same as, greater than, or less than that found in Exercise 19? (b) Is the entropy change of the universe greater than, less than, or the same as that found in Exercise 19?

21. A system absorbs 200 J of heat reversibly from a reservoir at 300 K and rejects 100 J reversibly to a reservoir at 200 K as it moves from state $A$ to state $B$. During this process (which is quasi-static) 50 J of work is done by the system. (a) What is the change in internal energy of the system? (b) What is the change in entropy of the system? (c) What is the change in entropy of the universe? (d) If the system went from state $A$ to state $B$ by a non-quasi-static process, how would your answers to parts (a), (b), and (c) differ?

22. A system absorbs 300 J from a reservoir at 300 K and 200 J from a reservoir at 400 K. It returns to its original state, doing 100 J of work and rejecting 400 J of heat to a reservoir at temperature $T$. (a) What is the entropy change of the system for the complete cycle? (b) If the cycle is reversible, what is the temperature $T$?

23. Two moles of an ideal gas originally at $T = 400$ K and $V = 40$ L undergoes a free adiabatic expansion to twice its volume. What is (a) the entropy change of the gas and (b) the entropy change of the universe?

24. A 5-kg block is dropped from rest at a height of 6 m above the ground. It hits the ground and comes to rest. The block, ground, and atmosphere are all at 300 K initially. What is the entropy change of the universe for this process?

25. If 500 J of heat is conducted from a reservoir at 400 K to one at 300 K, what is the entropy change of the universe?

26. A 200-kg block of ice at 0°C is placed in a large lake. The temperature of the lake is just slightly higher than 0°C, and the ice melts. (a) What is the entropy change of the ice? (b) What is the entropy change of the lake? (c) What is the entropy change of the universe (ice plus lake)?

27. Calculate the entropy change of 1 kg of water at 100°C when it changes to steam under standard conditions.

### Section 19-8, Entropy and the Availability of Energy

28. A heat engine works in a cycle between reservoirs at 400 and 200 K. The engine absorbs 1000 J of heat from the hot reservoir and does 200 J of work in each cycle. (a) What is the efficiency of this engine? (b) Find the entropy change of the engine, each reservoir, and of the universe for each cycle. (c) What is the efficiency of a Carnot engine working between the same two reservoirs? How much work could be done by a Carnot engine in each cycle if it absorbed 1000 J from the hot reservoir? (d) Show that the difference in the work done by the Carnot engine and the original engine is $T_c \, \Delta S_u$, where $\Delta S_u$ is that calculated in part (b).

29. A refrigerator works between two reservoirs at 200 and 400 K. In each cycle it absorbs 200 J from the cold reservoir and rejects 600 J to the hot reservoir. (a) How much work is put into the refrigerator in each cycle? (b) What is the entropy change of the universe in each cycle? (c) If a Carnot engine is used as a refrigerator between these two reservoirs, how much heat would be rejected if it absorbed 200 J from the cold reservoir? How much work is needed? *Hint:* Use the fact that the entropy change of the universe is zero for a reversible process. (d) Show that the additional work required by the first refrigerator is $T_h \, \Delta S_u$, where $\Delta S_u$ is that found in part (b). (e) The additional work required by the first engine is not completely wasted because the energy is stored in the hot reservoir. Find how much of this additional work could be recovered using a Carnot engine between the reservoirs and show that the amount of work completely wasted by the first engine is equal to $T_c \, \Delta S_u$.

30. How much of the 500 J of heat taken from a reservoir at 400 K in Exercise 25 could have been converted into work using a cold reservoir at 300 K?

31. One mole of an ideal gas first undergoes an adiabatic free expansion from $V_1 = 12.3$ L, $T_1 = 300$ K to $V_2 = 24.6$ L, $T_2 = 300$ K. It is then compressed isothermally and quasi-statically back to its original state. (a) What is the entropy change of the universe for the complete cycle? (b) How much work is wasted in this cycle? (c) Show that the work wasted is $T \, \Delta S_u$.

32. Which process is more wasteful: (a) a block moving with 500 J of kinetic energy being slowed to rest by friction (temperature of the atmosphere = 300 K) or (b) 1 kJ of heat conducted from a reservoir at 400 K to one at 300 K? *Hint:* How much of the kJ of heat could be converted into work in an ideal situation? (c) Compute the entropy change of the universe in each case.

### Section 19-9, Molecular Interpretation of Entropy

33. Air consists mainly of diatomic molecules ($N_2$ and $O_2$), which can rotate about two axes. The total kinetic energy of 1 mol of air can be written $E_k =$

$\frac{1}{2}Mv_{CM}^2 + \frac{3}{2}RT$, where $M = 29.0$ g/mol is the molecular mass and $v_{CM}$ is the velocity of the center of mass. Which has more total energy, 1 mol of air at rest at 25°C or 1 mol of air at 20°C with a wind speed of 10 m/s (about 22.4 mi/h)? Which could be used to run a windmill?

### Section 19-10, Proof that Entropy Exists and that the Entropy of the Universe Can Never Decrease

*There are no exercises for this section.*

### Problems

1. An engine operates with a working substance which is 1 mol of an ideal gas with $C_v = \frac{3}{2}R$ and $C_p = \frac{5}{2}R$. The cycle begins at $P_1 = 1$ atm and $V_1 = 24.6$ L. The gas is heated at constant volume to $P_2 = 2$ atm. It then expands at constant pressure until $V_2 = 49.2$ L. During these two steps, heat is absorbed. The gas is then cooled at constant volume until its pressure is again 1 atm. It is then compressed at constant pressure back to its original state. During the last two steps heat is rejected. All steps are quasi-static and reversible. (*a*) Indicate this cycle on a *PV* diagram. Find the work done, the heat added, and the internal-energy change for each step of the cycle. (*b*) Find the efficiency of this cycle.

2. An engine using 1 mol of an ideal gas with $C_v = \frac{5}{2}R$ and $C_p = \frac{7}{2}R$ performs a cycle consisting of three steps: (1) an adiabatic expansion from an initial pressure of 2.64 atm and volume 10 L to a final pressure of 1 atm and volume 20 L, (2) a compression at constant pressure to its original volume of 10 L, (3) heating at constant volume to its original pressure of 2.64 atm. Find the efficiency of this cycle.

3. The behavior of a gasoline engine can be approximated by an ideal cycle called an *Otto cycle* (Figure 19-15). The steps are tabulated below. (*a*) Compute the heat input and the heat output and show that the efficiency can be written

$$\epsilon = 1 - \frac{T_e - T_b}{T_d - T_c}$$

where $T_e$ is the temperature of state $e$, etc. (*b*) Using the relation for adiabatic expansion or compression $TV^{\gamma-1} = $ a constant, show that $\epsilon = 1 - (V_1/V_2)^{\gamma-1}$, where

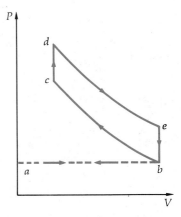

**Figure 19-15**
Otto cycle for Problem 3.

| No. | Step | Action | Comment |
|---|---|---|---|
| 1 | $a \rightarrow b$ | Air intake at constant pressure | Number of moles varies from 0 to $n$; volume varies from 0 to $V_b$, given by $P_0V_b = nRT_b$, where $T_b = $ temperature of outside air |
| 2 | $b \rightarrow c$ | Adiabatic compression to pressure $P_c$, temperature $T_c$, and volume $V_c$ | |
| 3 | $c \rightarrow d$ | Heating at constant volume | To approximate effect of explosion in gasoline engine |
| 4 | $d \rightarrow e$ | Adiabatic expansion to volume $V_e = V_b$ | The power stroke in the cycle |
| 5 | $e \rightarrow b$ | Cooling at constant volume until pressure is again atmospheric pressure $P_0$ | |
| 6 | $b \rightarrow a$ | Exhaust of gas at constant pressure | |

$V_1 = V_c = V_d$ and $V_2 = V_b = V_e$. (c) The ratio $V_2/V_1$ is called the *compression ratio*. Find the efficiency of this cycle for a compression ratio of 8 (use $\gamma = 1.5$ for ease in calculation). (Compression ratios much greater than this cannot be achieved in real engines because of preignition problems.) (d) Explain why the efficiency of a real gasoline engine might be much less than that calculated in part (c).

4. An engine using 1 mol of an ideal gas initially at $V_1 = 24.6$ L and $T = 400$ K performs in a cycle consisting of four steps: (1) isothermal expansion at $T = 400$ K to twice its volume, (2) cooling at constant volume to $T = 300$ K, (3) isothermal compression to its original volume, and (4) heating at constant volume to its original temperature of 400 K. Assume that $C_v = 21$ J/K. Sketch the cycle on a $PV$ diagram and calculate its efficiency.

5. A steam engine takes in superheated steam at 270°C and discharges condensed steam from its cylinder at 50°C. Its efficiency is 30 percent. (a) How does this efficiency compare with the best efficiency possible for these temperatures? (b) If the useful power output of the engine is 200 kW, how much heat does the engine discharge to the surroundings in 1 h?

6. One mole of an ideal gas for which $\gamma = 1.4$ is carried through a Carnot cycle. The high temperature and pressure are 400 K and 4 atm; the low temperature and pressure are 300 K and 1 atm. (a) What is the volume when $T = 400$ K and $P = 4$ atm? (b) What are the values of $PV$ and $PV^\gamma$ for the isothermal and adiabatic lines through this point? (c) What is the volume when $T = 300$ K and $P = 1$ atm? (d) What are the values of $PV$ and $PV^\gamma$ for the isothermal and adiabatic lines through this point? (e) Make an accurate plot of the Carnot cycle.

7. In the cycle shown in Figure 9-16, 1 mol of an ideal gas ($\gamma = 1.4$) is initially at 1 atm and 0°C. The gas is heated at constant volume to $t_2 = 150$°C and then expands adiabatically until its pressure is again 1 atm. It is then compressed at constant pressure back to its original state. Find (a) the temperature $t_3$ after the adiabatic expansion, (b) the heat entering or leaving the system during each process, (c) the efficiency of this cycle, and (d) the efficiency of a Carnot cycle operating between the temperature extremes of the cycle.

8. If two adiabatic curves on a $PV$ diagram intersected, a cycle could be completed using an isothermal path between the two adiabatics, as shown in Figure 19-17. Show that such a cycle could violate the second law of thermodynamics.

9. The cooling compartment of a refrigerator and its contents are at 5°C. Their heat capacity is 84 kJ/K. The refrigerator exhausts heat to the room at 25°C. What is the smallest power motor which could be used to operate the refrigerator if it is to reduce the temperature of the interior and the contents by 1°C in 1 min?

10. (a) Show that the coefficient of performance of a Carnot refrigerator working between two reservoirs at temperatures $T_h$ and $T_c$ is related to the efficiency of a Carnot engine by $\eta = T_c/\epsilon T_h$. (b) Show that the coefficient of performance of a Carnot refrigerator can be written

$$\eta = \frac{T_c}{T_h - T_c}$$

(c) Which gives the greater increase in coefficient of performance of a refrigerator, increasing the temperature of the cold reservoir by 5 K or decreasing the temperature of the hot reservoir by 5 K?

11. Equation 19-14 gives the entropy change of an ideal gas when the volume and temperature change reversibly. Use this equation and the relation $TV^{\gamma-1} = a$ constant to show explicitly that the entropy change is zero for an adiabatic expansion from state $V_1T_1$ to state $V_2T_2$.

12. (a) Show that if the Clausius statement of the second law were not true, the entropy of the universe could decrease. (b) Show that if the Kelvin-Planck statement of the second law were not true, the entropy of the universe could de-

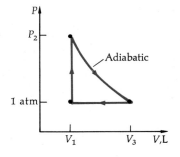

**Figure 19-16**
Problem 7.

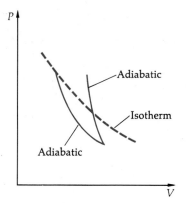

**Figure 19-17**
Problem 8.

crease. (c) An alternative statement of the second law is that the entropy of the universe cannot decrease. Have you just proved that this statement is equivalent to the Clausius and Kelvin-Planck statements?

13. One mole of an ideal, monatomic gas at initial volume $V_1 = 25$ L follows the cycle shown in Figure 19-18. All processes are quasi-static. Find (a) the temperature at each corner of the cycle, (b) the parts of the cycle where heat is absorbed, and (c) the efficiency of the cycle.

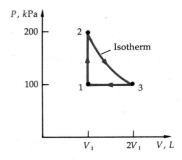

**Figure 19-18**
Problem 13.

14. Helium gas ($\gamma = 1.67$) is initially at a pressure of 16 atm, volume 1 L, and temperature 600 K. It expands isothermally until its volume is 4 L and then is compressed at constant pressure until its volume and temperature are such that an adiabatic compression will return the gas to its original state. (a) Sketch this cycle on a $PV$ diagram. Find (b) the volume and temperature after the isobaric compression, (c) the work done during the cycle, and (d) the efficiency of the cycle.

15. A heat engine which does the work of blowing up a balloon at atmospheric pressure extracts 4 kJ from a hot reservoir at 120°C. The volume of the balloon increases by 4 L, and heat is exhausted to a cold reservoir at temperature $t_c$. If the efficiency of the heat engine is half that of a Carnot engine working between the same two reservoirs, find the temperature $t_c$.

16. An ideal gas ($\gamma = 1.4$) follows the cycle shown in Figure 19-19. The temperature at point 1 is 200 K. Find (a) the temperature at the other three corners of the cycle and (b) the efficiency of the cycle.

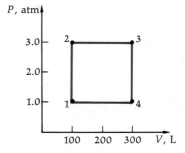

**Figure 19-19**
Problem 16.

17. A 100-g piece of ice at 0°C is placed in an insulated container with 100 g of water at 100°C. (a) When equilibrium is established, what is the final temperature of the water? (b) Find the entropy change of the universe for this process (see Example 19-6). (Ignore the heat capacity of the container.)

18. A 1-kg block of copper at 100°C is placed in a calorimeter of negligible heat capacity containing 4 L of water at 0°C. Find the entropy change of (a) the copper block, (b) the water, and (c) the universe.

19. If 2 kg of lead at 100°C is dropped into a lake at 10°C, find the entropy change of the universe.

20. Find the net change in the entropy of the universe when 10 g of steam at 100°C and 1 atm is introduced into a calorimeter of negligible heat capacity containing 150 g of water and 150 g of ice at 0°C.

21. A 1500-kg car traveling at 100 km/h crashes into a concrete wall. If the temperature of the air is 20°C, calculate the entropy change of the universe.

22. A block of metal has a heat capacity $C_p = 400$ J/K. It is to be heated from $T_1 = 200$ K to $T_2 = 400$ K. In Example 19-6 it was shown that the entropy increase of the metal is $\Delta S = C_p \ln (T_2/T_1) = 400 \ln 2 = 277$ J/K. In this problem you are to investigate the entropy change in the universe when the block is heated in several different ways. (a) The block originally at 200 K is placed in contact with a reservoir at temperature 400 K. How much heat is needed to raise the temperature of the block to 400 K? What is the entropy change in the reservoir when it loses this much heat? What is the entropy change of the universe for this process? (b) The block originally at 200 K is first placed in contact with a reservoir at 300 K until it comes to thermal equilibrium. It is then placed in contact with a reservoir at 400 K until its temperature is 400 K. Compute the entropy change of each reservoir and of the universe and compare your result with the process in part (a). (c) The block originally at 200 K is heated by first placing it in contact with a reservoir at 250 K until it comes to equilibrium. It is heated to 300 K, then to 350 K, and finally to 400 K, using successively reservoirs at 300, 350, and 400 K. Compute the entropy change of each reservoir and the entropy change of the universe for this process. Explain how reversible heating can be approximately accomplished in practice.

# CHAPTER 20    The Electric Field

---

**Objectives**    After studying this chapter you should:

1.  Be able to state Coulomb's law and use it to find the force exerted by one point charge on another.

2.  Know the value of the Coulomb constant in SI units.

3.  Know the magnitude of the electronic charge $e$ in coulombs.

4.  Be able to use Coulomb's law to calculate the electric field for both discrete and continuous charge systems.

5.  Be able to draw the lines of force for simple charge systems and to obtain information about the direction and strength of an electric field from such a diagram.

6.  Know that a spherically symmetric shell charge distribution produces zero electric field inside the shell and produces a field outside the shell the same as that of a point charge at the center of the shell.

7.  Be able to state the difference between a polar and nonpolar molecule and describe the behavior of each in a uniform electric field and in a nonuniform electric field.

---

The electrostatic force between two point charges was examined briefly in Chapter 5 in our discussion of forces in nature. We now begin a more detailed study of electricity. After a short history of the development of the concept of electric charge, we again state Coulomb's law for the force between two charges and define the electric field. We then discuss some general properties of electric fields, show how to calculate them from various charge distributions, and look at the behavior of point charges and electric dipoles in electric fields.

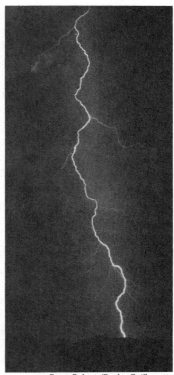

Bruce Roberts/Rapho Guillumette

A lightning discharge is a spectacular effect resulting from electromagnetic forces.

## 20-1   Electric Charge

Observations of electrical attraction can be traced back to the ancient Greeks.* The Greek philosopher Thales of Miletus (640–546 B.C.) observed that when amber is rubbed, it attracts small objects such as straw or feathers. (This attraction was often confused with the magnetic attraction of lodestone for iron.) Little more of electrical phenomena was understood until the sixteenth century, when the English physician William Gilbert (1540–1603) studied electrical and magnetic phenomena systematically. Gilbert showed that many substances besides amber acquire an attractive property when rubbed. He was one of the first to understand clearly the distinction between this attraction and the magnetic attraction, and he introduced the terms electric force, electric attraction, and magnetic pole. (The word electric comes from the Greek *elektron,* meaning amber, and the word magnetic comes from *Magnesia,* the country where magnetic iron ore was found.) Gilbert is perhaps best known for his discovery that the behavior of a compass needle results because the earth itself is a large magnet with poles at the north and the south; the orientation of a compass needle near the earth is merely an example of the general phenomenon of the attraction or repulsion of the poles of two magnets. Gilbert apparently failed to observe electric repulsion.

Around 1729 Stephen Gray, an Englishman, discovered that electric attraction and repulsion can be transferred from one body to another if the bodies are connected by certain substances, particularly metals. This discovery was of great importance, since previously experimenters could electrify an object only by rubbing it. The discovery of electric conduction also implies that electricity has an existence of its own and is not merely a property somehow brought out in a body by rubbing. The existence of two kinds of electricity was suggested by Charles François Du Fay (1698–1739), who described an experiment in which a gold leaf is attracted by a piece of glass rod previously rubbed. When the leaf is touched by the glass, it acquires the "electric virtue" and then repels the glass. "It is certain that bodies which have become electric by contact are repelled by those which have rendered them electric; but are they repelled likewise by other electrified bodies of all kinds?"** Du Fay answers this question by noting that the gold leaf, which is repelled by the glass rod, is attracted by an electrified piece of amber or resin. If the gold leaf is electrified by touching it to an electrified piece of amber, it is repelled by the amber but attracted by the electrified glass. He gave the two kinds of electricity the names vitreous and resinous and postulated the existence of two fluids which become separated by friction*** and are neutralized when they combine.

In 1747 the great American statesman and scientist Benjamin Franklin proposed a one-fluid model of electricity, described in the following experiment. If person *A* standing on wax (to insulate him from

Portrait of Charles Augustin de Coulomb.

---

* Much of this discussion follows Sir Edmund Whittaker's excellent *A History of Theories of Aether and Electricity,* Nelson, London, 1953; Torchbook edition, Harper, New York, 1960.

** *Memoir de l'Académie* (1733), as quoted by Whittaker.

*** We now know that the transfer of charge from one substance to another is not associated with friction but with the close contact of the substances achieved by rubbing them together.

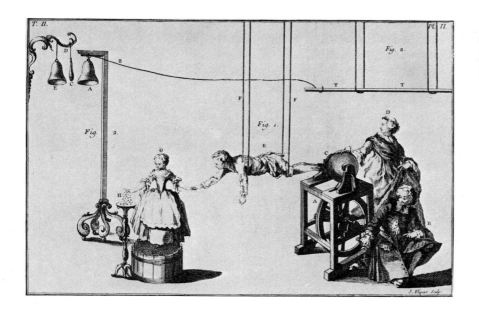

A 1745 woodcut showing electrical effects being transmitted through two persons before attracting feathers or bits of paper at table on left. (*Courtesy of the Deutsches Museum, Munich.*)

the ground) rubs a glass tube with a piece of silk cloth, and another person $B$, also standing on wax, touches the glass tube, both $A$ and $B$ become electrified. They can each give a spark to person $C$ standing on the ground. However, if $A$ and $B$ touch each other before touching $C$, the electricity of $A$ and $B$ is neutralized. Franklin proposed that every body has a "normal" amount of electricity. When a body is rubbed against another, some of the electricity is transferred from one body to the other; thus one has an excess and the other an equal deficiency. The excess and deficiency can be described with plus and minus signs: one body is plus and the other minus. An important feature of Franklin's model is the implication of conservation of electricity, now known as the *law of conservation of charge*. The electric charge is not created by the rubbing; it is merely transferred. Since Franklin chose to call Du Fay's vitreous kind of electricity positive, Du Fay's resinous electricity was merely the lack of vitreous electricity, or negative. A glass rod when rubbed acquires an excess of electricity and is positive, whereas an amber rod loses electricity when rubbed and becomes negative.

*Conservation of charge*

Franklin's choice was unfortunate because we now know that it is electrons that are transferred in the rubbing process, and according to Franklin's convention the electrons have a negative charge. Thus when we say that a (positive) charge is transferred from body $A$ to body $B$, it is really electrons that are transferred from body $B$ to body $A$. When glass is rubbed with silk, electrons are transferred from the glass to the silk, leaving the glass positive and the silk negative; when amber is rubbed with fur, electrons are transferred from the fur to the amber.

In the course of his experiments, Franklin noticed that small cork balls inside a metal cup seemed to be completely unaffected by the electricity of the cup. He asked his friend Joseph Priestley (1733–1804) to check this fact, and Priestley began experiments which showed that there is no electricity on the inside surface of a hollow metal vessel (except near the opening). From this result Priestley correctly deduced that the force between two charges varies as the inverse square of the distance between them, just like the gravitational force between two masses.

The inverse-square force law for electricity was confirmed by experiments of Charles Coulomb (1736–1806) using a torsion balance of his own invention. [The torsion balance was independently invented by the English scientist John Michell (1724–1793), who used it to show that the force between two magnetic poles also varies as the inverse square of the distance between the poles.] Coulomb's experimental apparatus was essentially the same as that described (Chapter 12) for the Cavendish experiment, with the masses replaced by small charged balls. For the magnitudes of charges easily transferred by rubbing, the gravitational attraction of the balls is completely negligible compared with the electric attraction or repulsion. Coulomb also used his torsion balance to confirm that the force between two magnetic poles varies as the inverse square of the distance.

The nineteenth century saw rapid growth in the understanding of electricity and magnetism, culminating with the great experiments of Michael Faraday (1791–1867) and the mathematical theory of James Clerk Maxwell (1831–1879). The experiments at the beginning of the twentieth century should be mentioned in this brief historical discussion of electric charge. In 1897 the English physicist J. J. Thomson showed that all materials contain particles which have the same charge-to-mass ratio. We now know that these particles, called *electrons,* are a fundamental part of the makeup of all atoms. In 1909 the American physicist Robert Millikan discovered that electric charge always occurs in integral amounts of a fundamental unit; i.e., the charge is *quantized.* Any charge can be written $q = Ne$, where $N$ is an integer and $e$ is the magnitude of the fundamental unit. The electron has charge $-e$, and the proton $+e$. The electron and the proton are very different particles. For example, the mass of the proton is nearly 2000 times that of the electron, and protons exert strong nuclear forces on neutrons, protons, pi mesons, etc., whereas electrons do not participate in the strong nuclear interaction. Yet the magnitude of the charge of the electron is exactly equal to that of the proton. Most elementary particles have either no charge (e.g., the neutron), a charge of $+e$, or a charge of $-e$. Particles made up of combinations of elementary particles may have a charge of $2e, 3e$, etc. For example, the alpha particle, which is the nucleus of the helium atom, consisting of two protons and two neutrons bound tightly together, has a charge of $+2e$. An atom with $Z$ protons in its nucleus ($Z$ is the *atomic number*) has $Z$ electrons outside the nucleus and is electrically neutral.

In this brief historical survey we have mentioned three important properties of electric charge:

1. Charge is conserved.

2. Charge is quantized.

3. The force between two point charges varies as the inverse square of the distance between the charges.

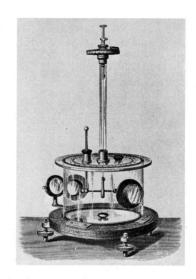

Coulomb's torsion balance. (*Courtesy of the Smithsonian Institution.*)

*Charge quantization*

**Questions**

1. Compare the properties of electric charge with those of gravitational mass. Discuss similarities and differences.

2. How might the world be different if the charge of the proton were slightly greater in magnitude than that of the electron?

## 20-2  Coulomb's Law

The experiments of Coulomb and others on the forces exerted by one point charge on another are summarized in *Coulomb's law:*

*The force exerted by one point charge on another is along the line joining the charges. It is repulsive if the charges have the same sign and attractive if the charges have opposite signs. The force varies inversely as the square of the distance separating the charges and is proportional to the magnitude of each charge.*

Coulomb's law can be written as a simple vector equation. Let $q_1$ and $q_2$ be two point charges separated by a distance $r_{12}$, which is the magnitude of the vector $\mathbf{r}_{12}$ pointing from charge $q_1$ to charge $q_2$ (Figure 20-1). The force exerted by charge $q_1$ on charge $q_2$ is then

$$\mathbf{F}_{12} = \frac{kq_1q_2}{r_{12}^2}\,\hat{\mathbf{r}}_{12} \qquad\qquad 20\text{-}1$$

*Coulomb's law*

where $k$ is a constant and $\hat{\mathbf{r}}_{12} = \mathbf{r}_{12}/r_{12}$ is the unit vector pointing from $q_1$ toward $q_2$. The force $\mathbf{F}_{21}$ exerted by $q_2$ on $q_1$ is the negative of $\mathbf{F}_{12}$ by Newton's third law. That is, $\mathbf{F}_{21}$ is equal in magnitude to $\mathbf{F}_{12}$ but opposite in direction.

Equation 20-1 includes the result that like charges repel and unlike charges attract. If both $q_1$ and $q_2$ are positive or both negative, the force $\mathbf{F}_{12}$ on $q_2$ is in the direction $\hat{\mathbf{r}}_{12}$ away from charge $q_1$ and the force $\mathbf{F}_{21}$ on $q_1$ is directed away from $q_2$. That is, if both charges have the same sign, the force is repulsive. Similarly, if one charge is positive and the other negative, the product $q_1q_2$ is negative and the forces are attractive.

The constant $k$ depends on the choice of units for charge. In SI units the unit of charge is the *coulomb* (C), which is defined as the amount of charge which flows past a point in a wire in one second when the current in the wire is one ampere. The ampere (A) is defined in terms of a magnetic force measurement which we shall describe later. In this system of units, the quantities $F_{12}$, $q_1$, $q_2$, and $r_{12}$ in Equation 20-1 are all defined independently, and the constant $k$ is determined by experiment. The measured value of $k$ is

$$k = 8.99 \times 10^9 \text{ N·m}^2/\text{C}^2 \qquad\qquad 20\text{-}2$$

In most calculations, it is convenient to use the approximation

$$k \approx 9 \times 10^9 \text{ N·m}^2/\text{C}^2 \qquad\qquad 20\text{-}3$$

*The Coulomb constant*

With his torsion balance Coulomb was able to show that the exponent of $r_{12}$ in the force equation is 2 with an uncertainty of a few percent. Although this was the most accurate measurement then, experiments similar to Priestley's can now be done with much greater precision, and we now know that the exponent is 2 with an experimental uncertainty of about 2 parts in $10^9$.

In a system of charges, each charge exerts a force on each other charge given by Equation 20-1. The resultant force on any charge is the vector sum of the individual forces exerted on that charge by all the other charges in the system.

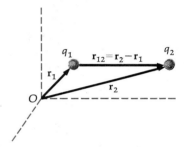

**Figure 20-1**
Charge $q_1$ at position $\mathbf{r}_1$ and charge $q_2$ at $\mathbf{r}_2$ relative to origin $O$. The force exerted by $q_1$ on $q_2$ is in the direction of the vector $\mathbf{r}_{12} = \mathbf{r}_2 - \mathbf{r}_1$ if both charges have the same sign, and in the opposite direction if the charges have opposite signs.

---

**Example 20-1** Three positive point charges lie on the $x$ axis; $q_1 = 25\ \mu\text{C}$ is at the origin, $q_2 = 10\ \mu\text{C}$ is at $x = 2$ m, and $q_3 = 20\ \mu\text{C}$ is at $x = 3$ m (Figure 20-2). Find the resultant force on $q_3$.

The force on $q_3$ due to $q_2$, which is 1 m away, is in the positive $x$ direction and has the magnitude

$$F_{23} = \frac{kq_2 q_3}{r_{23}^2} = \frac{(9 \times 10^9 \ \text{N·m}^2/\text{C}^2)(10 \times 10^{-6} \ \text{C})(20 \times 10^{-6} \ \text{C})}{(1 \ \text{m})^2}$$

$$= 1.8 \ \text{N}$$

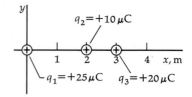

$q_2 = +10 \ \mu\text{C}$

$q_1 = +25 \mu\text{C}$    $q_3 = +20 \mu\text{C}$

**Figure 20-2**
Three point charges on the $x$ axis for Example 20-1.

The force on $q_3$ due to $q_1$, which is 3 m away, is also in the positive direction. Its magnitude is

$$F_{13} = \frac{kq_1 q_3}{r_{13}^2} = \frac{(9 \times 10^9 \ \text{N·m}^2/\text{C}^2)(25 \times 10^{-6} \ \text{C})(20 \times 10^{-6} \ \text{C})}{(3 \ \text{m})^2}$$

$$= 0.50 \ \text{N}$$

Since both forces are in the positive $x$ direction, the resultant force is also in that direction and is just the sum of the magnitudes:

$$F = F_{23} + F_{13} = 2.3 \ \text{N}$$

Note that the charge $q_2$, which is between $q_1$ and $q_3$, has no effect on the force $F_{13}$ exerted by $q_1$ on $q_3$, just as the charge $q_1$ has no effect on the force exerted by $q_2$ on $q_3$.

---

**Example 20-2** Charge $q_1 = +25 \ \mu\text{C}$ is at the origin, charge $q_2 = -10 \ \mu\text{C}$ is on the $x$ axis at $x = 2$ m, and charge $q_3 = +20 \ \mu\text{C}$ is at the point $x = 2$ m, $y = 2$ m. Find the resultant force on $q_3$.

Since $q_2$ and $q_3$ have opposite signs, the force exerted by $q_2$ on $q_3$ is attractive in the negative $y$ direction, as shown in Figure 20-3. Its magnitude is

$$F_{23} = \frac{(9 \times 10^9)(10 \times 10^{-6})(20 \times 10^{-6})}{2^2} = 0.45 \ \text{N}$$

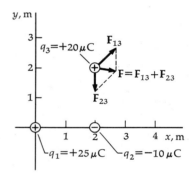

$q_3 = +20 \mu\text{C}$    $F_{13}$

$F = F_{13} + F_{23}$

$F_{23}$

$q_1 = +25 \mu\text{C}$    $q_2 = -10 \mu\text{C}$

**Figure 20-3**
Force diagram for Example 20-2. The resultant force on charge $q_3$ is the vector sum of the forces $\mathbf{F}_{13}$ due to $q_1$ and $\mathbf{F}_{23}$ due to $q_2$.

The distance between $q_1$ and $q_3$ is $2\sqrt{2}$ m. The force exerted by $q_1$ on $q_3$ is directed along the line from $q_1$ to $q_3$ and has the magnitude

$$F_{13} = \frac{(9 \times 10^9)(25 \times 10^{-6})(20 \times 10^{-6})}{(2\sqrt{2})^2} = 0.56 \ \text{N}$$

The resultant force is the vector sum of these two forces. Since $F_{13}$ makes an angle of 45° with the $x$ and $y$ axes, its $x$ and $y$ components are equal to each other and to $F_{13}/\sqrt{2} = 0.40$ N. The $x$ and $y$ components of the resultant force are therefore

$$F_x = F_{23x} + F_{13x} = 0 + 0.40 = 0.40 \ \text{N}$$

$$F_y = F_{23y} + F_{13y} = -0.45 + 0.40 = -0.05 \ \text{N}$$

---

The fundamental unit of electric charge $e$ is related to the coulomb by

$$e = 1.60 \times 10^{-19} \ \text{C} \qquad\qquad 20\text{-}4$$

*The charge of an electron*

Typical laboratory charges of 10 to 100 nC, which can be produced by rubbing, involve the transfer of many electrons. For example, the number of electrons in 10 nC is

$$10 \times 10^{-9} \ \text{C} \frac{1 \ \text{electron}}{1.60 \times 10^{-19} \ \text{C}} = 6.25 \times 10^{10} \ \text{electrons}$$

Such charges do not reveal that electric charge is quantized. A million electrons could be added to, or subtracted from, this charge without detection by ordinary instruments.

**Example 20-3** In the hydrogen atom the electron is separated from the proton by a distance of about $5.3 \times 10^{-11}$ m on the average. What is the electrostatic force exerted by the proton on the electron?

Since the proton charge is $+e$ and the electron charge $-e$, the force is attractive and has the magnitude

$$\frac{ke^2}{r^2} = \frac{(9 \times 10^9)(1.6 \times 10^{-19})^2}{(5.3 \times 10^{-11})^2} = 8.2 \times 10^{-8} \text{ N}$$

This is much greater than the negligible gravitational force between the particles, which is

$$F_G = \frac{Gm_e m_p}{r^2} = \frac{(6.67 \times 10^{-11})(9.1 \times 10^{-31})(1.67 \times 10^{-27})}{(5.3 \times 10^{-11})^2}$$
$$= 3.6 \times 10^{-47} \text{ N}$$

**Question**

3. If the sign convention for charge were changed so that the electron charge was positive and the proton charge negative, would Coulomb's law be written the same or differently?

## 20-3    The Electric Field

In Chapter 12 we defined the gravitational field due to a set of masses to be the gravitational force exerted by these masses on a test mass $m_0$ divided by $m_0$. The electric field due to a set of charges is defined analogously. If we have a set of point charges $q_i$ at various points in space and place a test charge $q_0$ at some point $P$, the force on it will be the vector sum of the forces exerted by the individual charges. Since each of these forces is proportional to the charge $q_0$, the resultant force is proportional to $q_0$. The electric field $\mathbf{E}$ at point $P$ is defined to be this force divided by $q_0$*:

$$\mathbf{E} = \frac{\mathbf{F}}{q_0} \qquad\qquad 20\text{-}5 \qquad \textit{Electric field defined}$$

The SI unit of electric field is the newton per coulomb (N/C).

**Example 20-4** When a 5-nC test charge is placed at a point, it experiences a force of $2 \times 10^{-4}$ N in the $x$ direction. What is the electric field $\mathbf{E}$ at that point?

From the definition, the electric field is

$$\mathbf{E} = \frac{2 \times 10^{-4}\mathbf{i} \text{ N}}{5 \times 10^{-9} \text{ C}} = 4 \times 10^4\mathbf{i} \text{ N/C}$$

where $\mathbf{i}$ is the unit vector in the $x$ direction.

---

\* In this discussion we assume that the presence of the test charge $q_0$ does not change the original distribution of the other charges. This is true in practice if there are no conductors present or if $q_0$ is so small that its influence on the original charge distribution is negligible.

We think of the electric field as a condition in space set up by the system of point charges. This condition is described by the vector **E**. By moving the test charge $q_0$ from point to point we can find the electric field vector **E** at any point (except one occupied by a charge $q_i$). The electric field **E** is thus a vector function of position. The force exerted on a test charge $q_0$ at any point is related to the electric field at that point by

$$\mathbf{F} = q_0 \mathbf{E} \qquad\qquad 20\text{-}6$$

---

**Example 20-5** What is the force on an electron placed at the point in Example 20-4 where the electric field is $4 \times 10^4 \mathbf{i}$ N/C?

Since the charge of the electron is $-e = -1.6 \times 10^{-19}$ C, the force is

$$\mathbf{F} = (-1.6 \times 10^{-19}\ \text{C})(4 \times 10^4 \mathbf{i}\ \text{N/C}) = -6.4 \times 10^{-15}\mathbf{i}\ \text{N}$$

---

Although it is customary to think of the test charge $q_0$ as positive, we see from Example 20-5 that the sign of $q_0$ has no effect on the definition of **E**. If $q_0$ is positive, the force on $q_0$ is in the direction of the field **E**; if $q_0$ is negative, the force is in the direction opposite to **E**. The ratio $\mathbf{F}/q_0$ is the same in either case.

Like the gravitational field, the electric field is more than a calculation device. This concept enables us to avoid the problem of action at a distance if we concede that the field is not propagated instantaneously. We thus think of the force exerted on charge $q_0$ at point $P$ as being exerted *by the field at point P* rather than by the charges, which are some distance away. Of course the field at point $P$ is produced by the other charges, but not instantaneously.

Consider a single point charge $q_1$ at the origin. If we place a test charge $q_0$ at some point given by the position vector **r**, the force on this charge will be

$$\mathbf{F} = \frac{kq_1 q_0}{r^2}\, \hat{\mathbf{r}}$$

The electric field at point **r** due to the charge $q_1$ is thus

$$\mathbf{E} = \frac{kq_1}{r^2}\, \hat{\mathbf{r}} \qquad\qquad 20\text{-}7$$

*Electric field of a point charge*

If charge $q_1$ is suddenly moved at time $t = 0$, the change in field at **r** is not instantaneous. The change in the field is propagated with the speed of light $c$, and so the test charge at **r** will not react to the change in position of the charge $q_1$ until a later time $t = r/c$, the time required for propagation of the change in the field. In this chapter we shall be concerned only with static electric fields.

The electric field due to a system of point charges can be found from Coulomb's law. Let $\mathbf{r}_{i0}$ be the vector from the $i$th charge $q_i$ to point $P$. The force on a test charge $q_0$ at $P$ is then

$$\mathbf{F} = \frac{kq_1 q_0}{r_{10}^2}\, \hat{\mathbf{r}}_{10} + \frac{kq_2 q_0}{r_{20}^2}\, \hat{\mathbf{r}}_{20} + \cdots + \frac{kq_i q_0}{r_{i0}^2}\, \hat{\mathbf{r}}_{i0} + \cdots$$

$$= q_0 \sum_i \frac{kq_i}{r_{i0}^2}\, \hat{\mathbf{r}}_{i0}$$

where $\hat{\mathbf{r}}_{i0} = \mathbf{r}_{i0}/r_{i0}$ is the unit vector pointing away from $q_i$ toward point $P$. The electric field at point $P$ due to the charges $q_i$ (but excluding the test charge $q_0$) is then

$$\mathbf{E} = \frac{\mathbf{F}}{q_0} = \sum \frac{kq_i}{r_{i0}^2}\, \hat{\mathbf{r}}_{i0} \qquad\qquad 20\text{-}8$$

**Example 20-6** A positive charge $q_i$ is at the origin, and a second positive charge $q_2$ is on the $x$ axis at $x = a$. Find the electric field at points on the $x$ axis (Figure 20-4).

In the region $x > a$ the electric field due to each charge is along the $x$ axis in the positive $x$ direction. The distance $r_{10}$ to the charge $q_1$ is $x$, and the distance $r_{20}$ to $q_2$ is $x - a$. The electric field is then

$$\mathbf{E} = \frac{kq_1}{x^2}\,\mathbf{i} + \frac{kq_2}{(x-a)^2}\,\mathbf{i} \qquad x > a$$

where $\mathbf{i}$ is the unit vector in the $x$ direction. In the region $0 < x < a$, the field due to $q_1$ is in the positive direction, but the field due to $q_2$ is in the negative $x$ direction. The distance to $q_1$ is again $x$, and that to $q_2$ is $a - x$. Since these distances are squared in calculating the field, we can write either $(a - x)^2$ or $(x - a)^2$. The electric field between the charges is thus

$$\mathbf{E} = \frac{kq_1}{x^2}\,\mathbf{i} - \frac{kq_2}{(x-a)^2}\,\mathbf{i} \qquad 0 < x < a$$

To the left of the origin, the field due to each charge is in the negative $x$ direction. The distance to $q_1$ is $-x$, and that to $q_2$ is $-x + a$ (since $x$ is a negative number). Then for negative $x$ we have

$$\mathbf{E} = -\frac{kq_1}{x^2}\,\mathbf{i} - \frac{kq_2}{(x-a)^2}\,\mathbf{i} \qquad x < 0$$

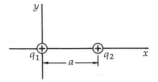

**Figure 20-4**
Point charges on $x$ axis for Example 20-6.

**Example 20-7** Find the electric field at points on the $y$ axis for the charges in Example 20-6.

The fields due to each charge at a point on the positive $y$ axis are shown in Figure 20-5. The field due to $q_1$ has magnitude $E_1 = kq_1/y^2$ and is in the $+y$ direction. The field due to $q_2$ has magnitude $E_2 = kq_2/(y^2 + a^2)$ and is in a direction making an angle $\theta$ with the $y$ axis. The resultant electric field is the vector sum of these fields, $\mathbf{E} = \mathbf{E}_1 + \mathbf{E}_2$. The $y$ component of the resultant electric field is

$$E_y = \frac{kq_1}{y^2} + \frac{kq_2}{y^2 + a^2}\cos\theta = \frac{kq_1}{y^2} + \frac{kq_2\,y}{(y^2 + a^2)^{3/2}}$$

using $\cos\theta = y/\sqrt{y^2 + a^2}$. The $x$ component of the resultant electric field is

$$E_x = -\frac{kq_2}{y^2 + a^2}\sin\theta = -\frac{kq_2\,a}{(y^2 + a^2)^{3/2}}$$

using $\sin\theta = a/\sqrt{y^2 + a^2}$.

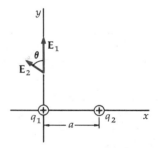

**Figure 20-5**
Example 20-7. On the $y$ axis, the electric field $\mathbf{E}_1$ due to charge $q_1$ is along the $y$ axis, and the field $\mathbf{E}_2$ due to charge $q_2$ makes an angle $\theta$ with the $y$ axis. The resultant electric field is the vector sum $\mathbf{E} = \mathbf{E}_1 + \mathbf{E}_2$.

**Example 20-8** A point charge $+q$ is at the origin and an equal but opposite charge $-q$ is on the $x$ axis a small distance away at $x = -a$ (Figure 20-6). Find the electric field at points on the $x$ axis far from the origin.

At great distances from the origin, the electric field due to the positive charge is nearly canceled by that due to the negative charge. At points on the $x$ axis, $E_y = E_z = 0$ and $E_x$ is given by

$$E_x = \frac{kq}{x^2} + \frac{k(-q)}{(x+a)^2} = kq\left[\frac{1}{x^2} - \frac{1}{(x+a)^2}\right]$$

Putting the terms in brackets over a common denominator, we get

$$\frac{1}{x^2} - \frac{1}{(x+a)^2} = \frac{(x+a)^2 - x^2}{x^2(x+a)^2} = \frac{2ax + a^2}{x^2(x+a)^2}$$

$$= \frac{2ax(1 + a/2x)}{x^4(1 + a/x)^2}$$

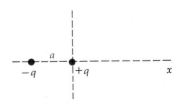

**Figure 20-6**
Equal and opposite charges separated by a distance $a$ for Example 20-8. Such a charge system is called an electric dipole.

For $x \gg a$, we can neglect $a/2x$ compared with 1 in the numerator and $a/x$ compared with 1 in the denominator. We then have

$$E_x \approx \frac{2kqa}{x^3}$$

A system of equal and opposite charges separated by a small distance is called an *electric dipole*. A vector from the negative charge to the positive charge having the magnitude $qa$ is called the dipole moment vector **p**. In terms of the dipole moment, the electric field on the axis of the dipole at a great distance is given by

$$E_x = \frac{2kqa}{x^3} = \frac{2kp}{x^3} \qquad\qquad 20\text{-}9$$

The electric field far from a dipole is proportional to the dipole moment and decreases as the cube of the distance.

## 20-4   Lines of Force

It is convenient to picture the electric field by drawing *lines of force* to indicate the direction of the field at any point, just as we did for the gravitational field in Chapter 12.* The field vector **E** is tangent to the line at each point and indicates the direction of the electric force experienced by a positive test charge placed at that point. Because there are an infinite number of points in space through which to draw a line, we choose only a few representative lines to be drawn. Since we could choose any point to indicate the direction of the field by a line, the lines need not be continuous, but it is both customary and useful to draw continuous lines which begin at positive charges and end at negative charges. At any point near a positive charge, a positive test charge is repelled from the charge; the field lines therefore diverge from a point occupied by a positive charge. Similarly, the lines converge toward a point occupied by a negative charge. (This differs somewhat from the gravitational field lines, which end at point masses but have no points from which they diverge because the gravitational field is always attractive.)

Figure 20-7 shows the lines of force, or electric field lines, of a single positive point charge. Consider a spherical surface of radius $r$ with its center at the charge. For a fixed number of lines emerging from the charge, the number of lines per unit area on the sphere is inversely proportional to the area of the sphere, $4\pi r^2$. Thus the density of lines decreases with distance as $1/r^2$, just as the magnitude of the electric field decreases. If we adopt the convention of drawing a fixed number of lines from a point charge, the number being proportional to the charge strength, and if we draw the lines symmetrically about the point charge, the field strength is indicated by the density of the lines.

The electric field near a negative point charge points inward toward the charge, so that the lines of force point toward the negative charge. If we have more than one point charge, we can indicate the electric field strength by choosing the number of lines diverging from the positive charges or converging toward the negative charges to be proportional to the charge. Figure 20-8 shows the lines of force for two equal positive point charges separated by a distance $a$.

* Much of the discussion in Chapter 12 of the gravitational field lines applies also to electric field lines. We repeat that discussion here in case Chapter 12 was omitted in your study of mechanics.

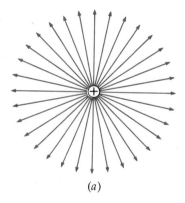

(a)

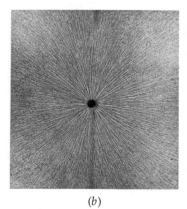

(b)

**Figure 20-7**
(a) Electric field lines, or lines of force, of a single charge.
(b) Bits of thread suspended in oil. The electric field of the charged object in the center induces opposite charges on the ends of each bit of thread, causing the thread to align itself parallel to the field.
(*Courtesy of Harold M. Waage, Princeton University.*)

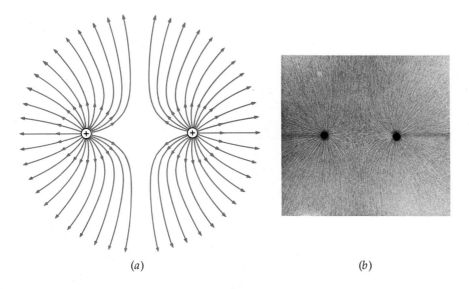

**Figure 20-8**
(*a*) Lines of force due to two positive point charges. (*b*) Electric lines of force of two equal charges of the same sign, shown by bits of thread suspended in oil. (*Courtesy of Harold M. Waage, Princeton University.*)

(*a*)                                        (*b*)

We construct this pattern without calculating the field at each point. At points near one of the charges, the field is approximately due to that charge alone because of the inverse-square-distance dependence of the field.

Thus on a sphere of very small radius ($r \ll a$) about one of the charges, the field lines are radial and equally spaced. Since the charges are equal, we draw an equal number of lines from each charge. At very large distances from the charges, the field is approximately the same as that due to a point charge of magnitude $2q$. So on a sphere of radius $r \gg a$, the lines are approximately equally spaced. (In our two-dimensional drawings, these spheres are replaced by circles.) We can see by merely looking at the figure that the electric field in the space between the charges is weak because there are few lines in this region compared with the region just to the right or left of the charges, where the lines are more closely spaced. This information can of course also be obtained by direct calculation of the field at points in these regions.

We can apply the above reasoning to drawing the lines of force for any system of point charges. Very near each charge, the field lines are equally spaced and leave or enter the charge depending on the sign of the charge. Very far from all the charges, the detailed structure of the system cannot be important, and the field lines are the same as those of a single point charge equal to the net charge of the system. For future reference, we summarize our rules for drawing lines of electric force.

1. The number of lines leaving a positive point charge or entering a negative charge is proportional to the charge.

2. The lines are drawn symmetrically leaving or entering a point charge.

3. Lines begin or end only on charges.

4. The density of lines (number per unit area perpendicular to the lines) is proportional to the magnitude of the field.

5. No two field lines can cross.

Rule 5 follows from the fact that **E** has a unique direction at any point in space (except at a point occupied by a point charge). If two lines crossed, two directions would be indicated for **E** at the point of intersection.

*Rules for drawing lines of force*

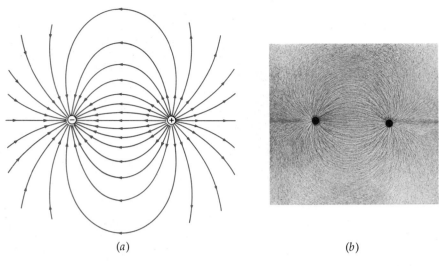

(a)                                    (b)

**Figure 20-9**
Example 20-8. (*a*) Lines of
force for an electric dipole. (*b*)
Electric lines of force of two
equal but opposite charges,
shown by bits of thread suspended in oil. (*Courtesy of
Harold M. Waage, Princeton
University.*)

Figure 20-9 shows the lines of force for an electric dipole. Very near
the positive charge the lines are radially outward. Very near the negative charge the lines are radially inward. Since the charges have equal
magnitude, the number of lines that begin at the positive charge equals
the number that end at the negative charge. In this case the field is
strongest in the region between the charges, as indicated by the high
density of field lines.

---

**Example 20-9** Sketch the lines of force for a negative charge $-q$ a distance $a$ from a positive charge $+2q$.

Since the positive charge is twice the magnitude of the negative
charge, twice as many lines leave the positive charge as enter the negative charge. At great distances from the charges the system looks like a
single charge $+q$. Half the lines beginning on the positive charge leave
the system. On a sphere of radius $r$, where $r$ is much larger than the separation of the charges $a$, these lines are approximately symmetrically
spaced and point radially outward, the same as the lines from a single
positive point charge $+q$. The other half leaving the positive charge
$+2q$ enter the negative charge. Field lines are shown in Figure 20-10.

---

It is important to realize that the *convention indicating the electric field
strength by the lines of force works only because the electric field varies
inversely as the square of the distance from a point charge.* Since the gravi-

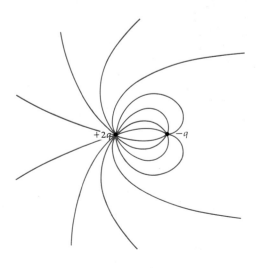

**Figure 20-10**
Example 20-9. Lines of force
for a point charge $+2q$ and a
second point charge $-q$. At
great distances from the
charges the lines are the same
as for a single charge $+q$.

tational field of a point mass also varies inversely as the square of the distance, the concept of lines of force is also useful in picturing the gravitational field. In that case, however, the lines of gravitational field enter point masses because the field points toward the mass and there are no points in space from which the gravitational field lines diverge; i.e., the gravitational force is always attractive, whereas the electric force can be either repulsive or attractive, depending on whether the charges have like or unlike signs.

As an illustration of the power of the use of lines of force, we now consider the electric field due to a spherically symmetric shell of charge of radius $R$ and negligible thickness. Let the total charge on the shell be $Q$. (Such a charge distribution can be realized in practice by placing the charge on a spherical conductor. As we shall see in the next chapter, the charge will distribute itself uniformly over the surface of such a conductor.) We choose the origin to be at the center of the shell. Because of the symmetry of the spherical shell, the only possible direction for the electric field lines is along radii, either toward the origin or away from it. Let us assume the charge to be positive. At great distances compared with the shell radius $R$, the field must look like that of a point charge $Q$. Thus the lines are radially outward and equally spaced far from the shell. The lines of force for this system are shown in Figure 20-11. Outside the shell the lines are exactly the same as those for a point charge. Thus the electric field produced outside the shell is in the radial direction and has the magnitude

$$E = \frac{kQ}{r^2} \qquad r > R \qquad \qquad 20\text{-}10a$$

The lines must begin at the charge on the shell. There can be no lines inside the shell because if there were, they would diverge from, or converge to, the center of the shell at the origin. But there is no charge at the origin, so there can be no lines converging or diverging from that point. We therefore have the remarkable result that a spherically symmetric shell of charge produces no electric field inside it:

$$E = 0 \qquad r < R \qquad \qquad 20\text{-}10b$$

If we place a test charge $q_0$ anywhere inside the shell, it will experience no electric attraction or repulsion by the shell. This is easily under-

**Figure 20-11**
Lines of force for a spherical-shell charge distribution. Outside the shell the lines are the same as from a point charge at the center of the shell. Inside the shell there are no lines, indicating that the electric field is zero there.

stood for a test charge at the center of the shell because of the symmetry, but it is also true at every other point inside the shell. This remarkable result is a consequence of the inverse-square nature of the electric force. We shall derive it directly from Coulomb's law in the next section, but the derivation is very difficult because of the difficult spherical geometry needed in setting up the problem. In Chapter 21 we shall extend the idea of lines of force and develop a mathematical equation known as Gauss' law which can be used to calculate the electric field due to highly symmetric charge distributions such as the spherical shell of charge.

## 20-5    Calculation of the Electric Field for Continuous Charge Distributions

The electric field produced by a given charge distribution can be calculated in a straightforward way from Coulomb's law. For a set of point charges, the field is obtained from Equation 20-8. We used this to calculate the field for two positive point charges and for an electric dipole in Examples 20-6 to 20-8. In this section we consider charge distributions consisting of many charges so close together that the charge can be considered to be continuously distributed over a surface or through a volume, even though electric charge is a discrete quantity microscopically. The use of a continuous charge density to describe a distribution of a large number of discrete charges is similar to the use of a continuous mass density to describe air, which in reality consists of a large number of discrete molecules. In either case, it is usually easy to find a volume element $\Delta \mathcal{V}$ large enough to contain many individual charges or molecules (billions) and yet small enough to ensure that replacing $\Delta \mathcal{V}$ by a differential $d\mathcal{V}$ and using calculus will introduce negligible error. If charge $\Delta Q$ is distributed throughout a volume $\Delta \mathcal{V}$, the *charge density $\rho$* is defined by

$$\rho = \frac{\Delta Q}{\Delta \mathcal{V}}$$
20-11    *Volume charge density*

Often charge is distributed in a thin layer on the surface of an object. (We shall show later that a static charge on a conductor always resides on the surface of the conductor.) In these cases, it is convenient to define a *surface density $\sigma$*. Let $t$ be the thickness of the layer of charge. Then in a volume element of area $\Delta A$ the charge is

$$\Delta Q = \rho t \, \Delta A = \sigma \, \Delta A$$

where $\sigma$ is the charge per unit area:

$$\sigma = \frac{\Delta Q}{\Delta A} = \rho t$$
20-12    *Surface charge density*

Similarly, if the charge is along a line, e.g., on a string of cross-sectional area $A$, we choose a volume element of length $\Delta L$, $\Delta \mathcal{V} = A \, \Delta L$, and define the charge per unit length $\lambda$ by

$$\Delta Q = \rho A \, \Delta L = \lambda \, \Delta L$$

or

$$\lambda = \frac{\Delta Q}{\Delta L} = \rho A$$
20-13    *Linear charge density*

**Example 20-10** Calculate the electric field on the axis of a uniform ring of charge.

Let $Q$ be the total positive charge, which is distributed uniformly on a ring of radius $a$. We wish to find the field at a point on the axis of the ring a distance $x$ from the center of the ring.

Figure 20-12 shows the part of the field $\Delta\mathbf{E}$ due to a portion of the charge $\Delta Q$. This field has a component $\Delta E_x$ along the axis of the ring and a component $\Delta E_\perp$ perpendicular to the axis. From the symmetry of the figure we see that the resultant field due to the entire ring must lie along the axis of the ring; i.e., the perpendicular components will sum to zero. In particular, the perpendicular component shown will be canceled by that due to another portion of the charge on the ring directly opposite the one shown.

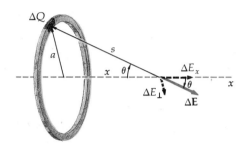

**Figure 20-12**
Charged ring of radius $a$. The electric field on the $x$ axis is in the $x$ direction.

The axial component due to the part of the charge shown is

$$\Delta E_x = \frac{k\,\Delta Q}{s^2}\cos\theta = \frac{k\,\Delta Q}{s^2}\frac{x}{s} = \frac{k\,\Delta Q\,x}{(x^2 + a^2)^{3/2}}$$

where

$$s^2 = x^2 + a^2 \qquad \text{and} \qquad \cos\theta = \frac{x}{s} = \frac{x}{\sqrt{x^2 + a^2}}$$

Since the distance from the field point at which we are calculating the field to all parts of the charge is the same, and since the angle $\theta$ is the same for all parts of the charge, the field due to the entire ring of charge is

$$E_x = \sum \frac{k\,\Delta Q\,x}{(x^2 + a^2)^{3/2}} = \frac{kx}{(x^2 + a^2)^{3/2}}\sum \Delta Q = \frac{kQx}{(x^2 + a^2)^{3/2}} \qquad \text{20-14}$$

**Example 20-11** Calculate the electric field on the perpendicular bisector of a uniform line charge.

Let $\lambda$ be the linear density. At a point on a perpendicular bisector the field $\mathbf{E}$ will not have a component parallel to the line of charge.

Figure 20-13 shows the geometry of the problem. We have chosen a coordinate system such that the origin is at the center of the line charge, the charge is on the $x$ axis, and our field point is on the $y$ axis at $P$. The magnitude of the field produced by an element of charge $\Delta Q = \lambda\,\Delta x$ is

$$|\Delta\mathbf{E}| = \frac{k\lambda\,\Delta x}{s^2}$$

The perpendicular component (in this case, the $y$ component) is

$$\Delta E_y = \frac{k\lambda\,\Delta x}{s^2}\cos\theta = \frac{k\lambda\,\Delta x}{s^2}\frac{y}{s} \qquad \text{20-15}$$

The total field $E_y$ is computed by summing over all $\Delta x$ of the line charge.

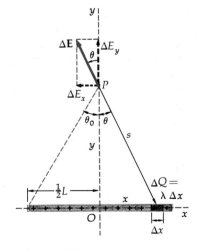

**Figure 20-13**
Line charge. The electric field on the $y$ axis which bisects the charge is in the $y$ direction.

For our case, we can sum (integrate) from $x = 0$ to $x = L/2$ and multiply by 2 because, by symmetry, each half of the line charge gives an equal contribution. The integration is somewhat simplified if we change from the variable $x$ to $\theta$ and integrate from $\theta = 0$ to $\theta_0$, defined by

$$\tan \theta_0 = \frac{\frac{1}{2}L}{y}$$

Equation 20-15 in terms of the differentials $dE_y$ and $dx$ is

$$dE_y = \frac{ky\lambda}{s^3} \, dx$$

The distance $x$ is related to the angle $\theta$ by

$$x = y \tan \theta$$

Then

$$dx = y \sec^2 \theta \, d\theta = y \left(\frac{s}{y}\right)^2 d\theta$$

and

$$dE_y = \frac{ky\lambda}{s^3} \frac{s^2}{y} \, d\theta = \frac{k\lambda}{y} \frac{y}{s} \, d\theta = \frac{k\lambda}{y} \cos \theta \, d\theta$$

The total $y$ component of the field is twice the integral of this from $\theta = 0$ to $\theta = \theta_0$:

$$E_y = 2 \int_{\theta=0}^{\theta=\theta_0} dE_y = \frac{2k\lambda}{y} \int_0^{\theta_0} \cos \theta \, d\theta = \frac{2k\lambda}{y} \sin \theta_0 \qquad \text{20-16}$$

where

$$\sin \theta_0 = \frac{\frac{1}{2}L}{\sqrt{(\frac{1}{2}L)^2 + y^2}}$$

The field produced by a line charge of infinite length is obtained from this result by setting $\theta_0 = 90°$. Then

$$E_y = \frac{2k\lambda}{y} \qquad \text{20-17}$$

---

**Example 20-12** Calculate the electric field on the axis of a uniformly charged disk.

Let the disk have a radius $R$ and contain a uniform charge per unit area $\sigma$. The electric field on the axis of the disk will be parallel to the axis. We can calculate this field by treating the disk as a set of concentric ring charges and using our result from Example 20-10. Consider a ring of radius $r$ and width $dr$, as in Figure 20-14. The area of this ring is $2\pi r \, dr$, and its charge is $dq = 2\pi\sigma r \, dr$. The field produced by this ring is given by Equation 20-14, replacing $Q$ by $2\pi\sigma r \, dr$ and $a$ by $r$. Thus

$$dE_x = \frac{kx 2\pi\sigma r \, dr}{(x^2 + r^2)^{3/2}}$$

The total field produced by the disk is found by integrating from $r = 0$ to $r = R$:

$$E_x = kx\pi\sigma \int_0^R (x^2 + r^2)^{-3/2} 2r \, dr$$

$$= kx\pi\sigma \left[\frac{(x^2 + r^2)^{-1/2}}{-\frac{1}{2}}\right]_0^R$$

$$= -2kx\pi\sigma \left(\frac{1}{\sqrt{x^2 + R^2}} - \frac{1}{x}\right) = 2\pi k\sigma \left(1 - \frac{x}{\sqrt{x^2 + R^2}}\right) \qquad \text{20-18}$$

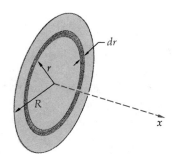

**Figure 20-14**
Disk carrying uniform surface charge density. The electric field at a point on the axis of the disk is along the axis. It can be calculated by finding the field due to a ring of radius $r$ and area $2\pi r \, dr$ and summing over the rings from $r = 0$ to $r = R$, the radius of the disk.

The interesting and important result for the case of the field near an infinite plane of charge can be obtained from this result by letting either $R$ go to infinity or $x$ go to zero. Then

$$E_x = 2\pi k\sigma \qquad\qquad 20\text{-}19$$

Thus the field due to an infinite-plane charge distribution is uniform; i.e., the field does not depend on $x$.

---

*Optional*

**Figure 20-15**
Spherical shell of radius $R$ carrying a uniform surface charge. The field at point $P$ a distance $r$ from the center of the sphere is found by first finding the field of a ring of width $R\, d\theta$ and radius $R \sin \theta$ and summing over the rings from $\theta = 0$ to $\theta = \pi$. The resulting field for $r > R$ is the same as if all the charge were at the origin. When point $P$ is inside the shell ($r < R$), the field due to the shell is zero.

**Example 20-13** Calculate the electric field due to a spherical shell of charge.

We first consider a field point $P$ outside the shell. The geometry is shown in Figure 20-15. By symmetry, the field must be radial. We can consider the spherical shell to be a set of ring elements and use our result found in Example 20-10. We thus choose for our charge element the strip shown, which has circumference $2\pi R \sin \theta$ and width $R\, d\theta$. The area of this strip is $dA = 2\pi R^2 \sin \theta\, d\theta$, and the charge is

$$dQ = \sigma\, dA = \sigma 2\pi R^2 \sin \theta\, d\theta$$

The radial component of the field due to this ring is

$$dE_r = \frac{k\, dQ}{s^2} \cos \alpha = \frac{k\sigma 2\pi R^2 \sin \theta\, d\theta}{s^2} \cos \alpha \qquad 20\text{-}20$$

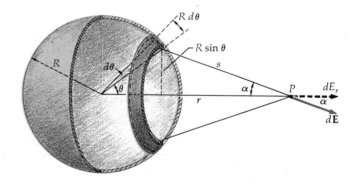

Before we integrate over the charge distribution, we must eliminate two of the three related variables $s$, $\theta$, and $\alpha$. It is convenient to write everything in terms of $s$, which varies from $s = r - R$ at $\theta = 0$ to $s = r + R$ at $\theta = 180°$. By the law of cosines, we have

$$s^2 = r^2 + R^2 - 2rR \cos \theta \qquad 20\text{-}21$$

Differentiating gives

$$2s\, ds = +2rR \sin \theta\, d\theta$$

or

$$\sin \theta\, d\theta = \frac{s\, ds}{rR} \qquad 20\text{-}22$$

An expression for $\cos \alpha$ can also be obtained from the law of cosines applied to the same triangle. We have

$$R^2 = s^2 + r^2 - 2sr \cos \alpha$$

or

$$\cos \alpha = \frac{s^2 + r^2 - R^2}{2sr} \qquad 20\text{-}23$$

Substituting these results into Equation 20-20 then gives

$$dE_r = \frac{k\sigma 2\pi R^2}{s^2} \frac{s\,ds}{rR} \frac{s^2 + r^2 - R^2}{2sr}$$

$$= \frac{k\sigma\pi R}{r^2}\left(1 + \frac{r^2 - R^2}{s^2}\right)ds \qquad\qquad 20\text{-}24$$

The field due to the entire shell of charge is found by integrating from $s = r - R$ ($\theta = 0$) to $s = r + R$ ($\theta = 180°$):

$$E_r = \frac{k\sigma\pi R}{r^2}\int_{r-R}^{r+R}\left(1 + \frac{r^2 - R^2}{s^2}\right)ds$$

$$= \frac{k\sigma\pi R}{r^2}\left[s - \frac{r^2 - R^2}{s}\right]_{r-R}^{r+R} \qquad\qquad 20\text{-}25$$

Substitution of the upper and lower limits yields $4R$ for the quantity in brackets. Thus

$$E_r = \frac{k\sigma 4\pi R^2}{r^2} = \frac{kQ}{r^2} \qquad\qquad 20\text{-}26$$

where $Q = 4\pi R^2\sigma$ is the total charge. The field is therefore the same as that due to a point charge $Q$ at the origin.

We now consider a field point inside the shell. The calculation for this case is identical except that $s$ now varies from $R - r$ to $r + R$. Thus

$$E_r = \frac{k\sigma\pi R}{r^2}\left[s - \frac{r^2 - R^2}{s}\right]_{R-r}^{r+R} \qquad\qquad 20\text{-}27$$

Substitution of these upper and lower limits yields 0. Therefore

$$E_r = 0 \qquad \text{point inside shell} \qquad\qquad 20\text{-}28$$

These are the same results as we obtained by considering the lines of force for this charge distribution.

---

**Example 20-14** Calculate the electric field due to a solid sphere of constant charge density $\rho = Q/V$, where $V = \frac{4}{3}\pi R^3$ is the volume of the sphere of charge.

We consider the sphere to consist of a set of concentric spherical shells and use the result already obtained, that the field produced by a shell is zero inside the shell and $kq/r^2$ outside the shell, where $q$ is the charge of the shell. For points outside our solid sphere of charge each shell produces a field identical to that of a point charge at the center of the sphere, and for $r > R$ we have

$$E_r = \frac{kQ}{r^2}$$

Consider now a point $P$ inside the charge, $r < R$. In Figure 20-16 the dotted line encloses a sphere of radius $r$. Let $q'$ be the portion of the charge inside this sphere. Since the charge density is uniform, $q'$ is given by

$$q' = \rho V' = \rho\tfrac{4}{3}\pi r^3 = \frac{Q}{\frac{4}{3}\pi R^3}\frac{4}{3}\pi r^3 = \frac{Qr^3}{R^3}$$

The field at $P$ due to $q'$ is the same as if $q'$ were at the origin. The rest of the charge produces no field at point $P$. This part of a charge can be con-

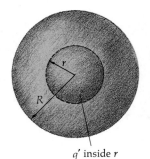

$q'$ inside $r$

**Figure 20-16**
Solid ball of charge. The electric field at a point $r < R$ inside the ball is $kq'/r^2$, where $q' = (r^3/R^3)Q$ is that part of the total charge $Q$ which is inside the sphere of radius $r$.

sidered to be a set of spherical shells, and for each shell, point $P$ is inside. The field at a distance $r < R$ is thus

$$E_r = \frac{kq'}{r^2} = \frac{kQr^3/R^3}{r^2} = \frac{kQ}{R^3}\, r \qquad\qquad 20\text{-}29$$

Figure 20-17 shows a sketch of $E_r$ versus $r$. This function is sometimes used to describe the electric field of an atomic nucleus, which can be considered to be approximately a uniform sphere of charge.

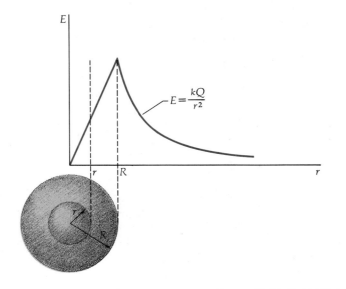

**Figure 20-17**
Plot of $E_r$ versus $r$ for a solid ball of charge of radius $R$. For $r$ less than $R$ the field increases linearly with $r$. Outside the charge, the electric field is $kQ/r^2$, the same as that due to a point charge $Q$ at the origin.

**Question**

4.  Explain why the electric field increases with $r$ rather than decreasing as $1/r^2$ as one moves out from the center inside a spherical charge distribution of constant density.

# 20-6 Motion of Point Charges in Electric Fields

When a particle with charge $q$ is placed in an electric field $\mathbf{E}$, it experiences a force $q\mathbf{E}$. If this is the only force on it, the particle has an acceleration $q\mathbf{E}/m$, where $m$ is the mass of the particle.* If the electric field is known, the charge-to-mass ratio of the particle can be determined from the measured acceleration. For example, for an electric field uniform in space and constant in time, the path of the particle is a parabola, similar to that of a projectile in a uniform gravitational field. The measurement of the deflection of electrons in a uniform electric field was used by J. J. Thomson in 1897 to demonstrate the existence of electrons and to measure their charge-to-mass ratio. (Since magnetic fields were also used in this measurement, we defer discussion of the Thomson experiment until Chapter 26.) We shall give some examples of motion of electrons in

---

* We are assuming that the speed of the particle is small enough for us to use classical mechanics and neglect special relativity. This assumption is often *not* valid for the motion of electrons in electric fields.

constant electric fields. Problems of this type can be worked using the constant-acceleration formulas from Chapter 2 or the equations for projectile motion from Chapter 3.

---

**Example 20-15**  An electron is projected into a uniform electric field $E = 1000 \text{ N/C}$ with initial speed $v_0 = 2 \times 10^6$ m/s parallel to the field. How far does the electron travel before it is brought momentarily to rest?

Since the charge of the electron is negative, the force $-e\mathbf{E}$ is in the direction opposite the field. We thus have a constant-acceleration problem in which the acceleration is opposite to the initial velocity, and we are asked to find the distance traveled. We can use the constant-acceleration expression relating the distance to the velocity,

$$v^2 = v_0^2 + 2a(x - x_0)$$

Using $x_0 = 0$, $v = 0$, $v_0 = 2 \times 10^6$, and $a = -eE/m$, we find for the distance

$$x = \frac{mv_0^2}{2eE} = \frac{(9.11 \times 10^{-31})(2 \times 10^6)^2}{2(1.6 \times 10^{-19})(1000)} = 1.14 \times 10^{-2} \text{ m}$$

---

**Example 20-16**  An electron is projected into a uniform electric field $E = 2000 \text{ N/C}$ with initial velocity $v_0 = 10^6$ m/s perpendicular to the field. Compare the weight of the electron to the electric force on it. By how much is the electron deflected after it has traveled 1 cm?

The electric force on the electron is $eE$, and the gravitational force is $mg$. Their ratio is

$$\frac{F_{elec}}{F_{grav}} = \frac{eE}{mg} = \frac{(1.6 \times 10^{-19})(2000)}{(9.1 \times 10^{-31})(9.8)} = 3.6 \times 10^{13}$$

As in most common cases, the electric force is huge compared with the gravitational force, which is wholly negligible.

It takes the electron a time

$$t = \frac{x}{v_0} = \frac{10^{-2} \text{ m}}{10^6 \text{ m/s}} = 10^{-8} \text{ s}$$

to travel the distance of 1 cm perpendicular to the field. In this time it is deflected antiparallel to the field a distance given by

$$y = \tfrac{1}{2}at^2 = \frac{1}{2}\frac{eE}{m}t^2$$

Substituting the known values of $e/m$, $E$, and $t$ gives

$$y = 1.76 \times 10^{-2} \text{ m} = 1.76 \text{ cm}$$

---

**Questions**

5.  The direction of the force on a positive charge in an electric field is, by definition, in the direction of the field line passing through the position of the charge. Must the acceleration be in this direction? The velocity? Explain.

6.  A positive charge is released from rest in an electric field. It starts out in the direction of a field line. Will it continue to move along the field line?

# 20-7  Electric Dipoles in Electric Fields

Although atoms and molecules are electrically neutral, they are affected by electric fields because they contain positive and negative charges. In some molecules the center of positive charge does not coincide with the center of negative charge. These *polar molecules* are said to have a permanent electric dipole moment. When such a molecule is placed in a uniform electric field, there is no net force on it but there is a torque, which tends to rotate the molecule. In a nonuniform electric field the molecule experiences a net force because the field at the center of the positive charge is different from that at the center of the negative charge. An example of a polar molecule is NaCl, which is essentially a positive sodium ion of charge $+e$ combined with a negative chlorine ion of charge $-e$.

Atoms and molecules for which the centers of positive and negative charge coincide are also affected by an electric field. Because the electric force on the positive charge is in the direction opposite that on the negative charge, the electric field tends to separate, or polarize, these charges. These systems then have an induced dipole moment when they are in an electric field, and they also experience a net force in a nonuniform field. The force produced by a nonuniform electric field on an electrically neutral charge system is responsible for the familiar attraction of a charged comb for uncharged bits of paper.

Figure 20-18 shows a schematic picture of a polar molecule with the centers of positive and negative charge indicated. (The center of charge is defined analogously to the center of mass, with the mass replaced by the charge.) The behavior of such a molecule can be described by a vector **p**, called the *dipole moment*. Let the total positive charge of the molecule be $q$ and the total negative charge $-q$. The dipole moment of the molecule is defined to be the product of the charge $q$ and the displacement vector **L** pointing from the center of negative charge to the center of positive charge:

$$\mathbf{p} = q\mathbf{L} \qquad\qquad 20\text{-}30$$

The diameter of an atom or molecule is of the order of $10^{-10}$ m $= 0.1$ nm. A convenient unit for electric dipole moments of atoms and molecules is the fundamental electronic charge $e$ times the distance 1 nm. For example, the dipole moment of NaCl in these units has a magnitude of about $0.2e$ nm.

We often simplify the description of the behavior of a polar molecule in an electric field by replacing the complicated charge distribution of the molecule with a simple electric dipole consisting of two charges $q$ and $-q$ separated by a distance $L$ and having the same dipole moment as the molecule.

Figure 20-19 shows a simple electric dipole whose dipole moment makes an angle $\theta$ with an external uniform electric field **E**. The forces acting on the dipole,

$$\mathbf{F}_1 = q\mathbf{E} \qquad \mathbf{F}_2 = -q\mathbf{E}$$

are opposite in direction and equal in magnitude since the field is uniform. The net force on the dipole is thus zero. However, the two forces produce a torque, which tends to rotate the dipole so that it points in the direction of the field. We found in Chapter 5 that the torque produced by two equal and opposite forces, called a couple, is the same about any point in space. From Figure 20-19 we see that the torque about the negative charge has the magnitude $F_1 L \sin \theta = qEL \sin \theta =$

*Polar molecules*

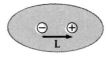

**Figure 20-18**
Schematic diagram of a polar molecule. The centers of positive and negative charge are separated by a distance $L$. The molecule behaves like an electric dipole of moment $\mathbf{p} = q\mathbf{L}$, where $q$ is the magnitude of the total positive or negative charge.

*Dipole moment defined*

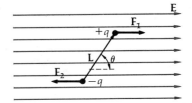

**Figure 20-19**
Electric dipole in a uniform electric field. The net force on the dipole is zero, but there is a net torque $\mathbf{L} \times \mathbf{F}_1 = q\mathbf{L} \times \mathbf{E} = \mathbf{p} \times \mathbf{E}$ which tends to align the dipole in the direction of the field.

$pE \sin \theta$. The direction of the torque is into the paper and such that it rotates the dipole moment **p** into the direction of the electric field **E**. This torque can be conveniently written as the cross product of the dipole moment **p** and the electric field **E**:

$$\boldsymbol{\tau} = \mathbf{p} \times \mathbf{E} \qquad\qquad 20\text{-}31$$

*Torque on a dipole*

If an electric dipole is placed so that it makes an angle $\theta$ with a uniform electric field, it will experience a torque of magnitude $pE \sin \theta$ tending to rotate it toward its equilibrium position $\theta = 0$. If the dipole is free, the torque will cause it to rotate so that the vector **p** oscillates about the equilibrium direction $\theta = 0$ until its energy is dissipated. We can calculate the potential energy of a dipole in a uniform field by computing the work we must do to increase the angle $\theta$. If we apply a torque of magnitude $pE \sin \theta$ and rotate the dipole through an angle $d\theta$, the work we do equals the increase in the potential energy:

$$dU = pE \sin \theta\, d\theta$$

Integrating gives

$$U = -pE \cos \theta + U_0$$

*Potential energy of a dipole in an electric field*

It is customary to choose the potential energy to be zero when the dipole is perpendicular to the electric field, i.e., when $\theta = 90°$. Then $U_0 = 0$, and the potential energy of the dipole is

$$U = -pE \cos \theta = -\mathbf{p} \cdot \mathbf{E} \qquad\qquad 20\text{-}32$$

If the electric dipole is placed in a nonuniform electric field, as in Figure 20-20, there is a net force acting on the dipole in addition to the torque tending to align the dipole with the field. The net force depends on the orientation of the dipole in the field, the dipole moment, and how rapidly the field varies in space. For the situation pictured in Figure 20-20 the net force on the dipole has a component downward and one in the direction of increasing field.

Most molecules and all atoms are nonpolar; i.e., the centers of positive and negative charge coincide, and there is no permanent electric

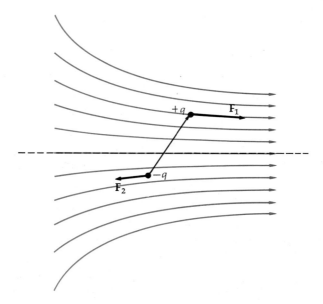

**Figure 20-20**
For a general orientation of an electric dipole in a nonuniform electric field there is both a net torque tending to align the dipole with the field and a net force on the dipole. Here, the net force has a component parallel to **E** and a small component downward. Its magnitude depends on the dipole moment **p**, its orientation, and on how rapidly the field varies in space.

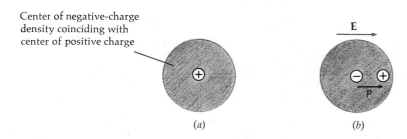

Center of negative-charge density coinciding with center of positive charge

(a)    (b)

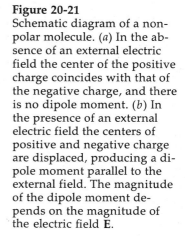

**Figure 20-21**
Schematic diagram of a nonpolar molecule. (a) In the absence of an external electric field the center of the positive charge coincides with that of the negative charge, and there is no dipole moment. (b) In the presence of an external electric field the centers of positive and negative charge are displaced, producing a dipole moment parallel to the external field. The magnitude of the dipole moment depends on the magnitude of the electric field **E**.

dipole moment. However, an external electric field produces a separation of the positive and negative charge distribution, thus inducing a dipole moment (Figure 20-21). Since the positive charge is displaced in the direction of the field and the negative charge is displaced in the opposite direction, the induced dipole moment is always in the direction of the external electric field. Thus no torque is exerted because the angle between the dipole moment **p** and the field **E** is zero. However, if the electric field is not uniform, there will be an external force acting on the dipole. Figure 20-22 shows a nonpolar molecule in an external electric field of a positive point charge $Q$. The induced dipole moment is parallel to **E** in the radial direction from the point charge. Since the field is stronger at the negative charge nearer the point charge, the force on the dipole is toward the point charge and the dipole is attracted toward the point charge. If the point charge were negative, the induced dipole would be in the opposite direction and the dipole would again be attracted to the point charge.

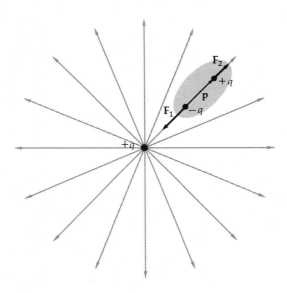

**Figure 20-22**
Nonpolar molecule in the electric field of a positive point charge. The induced dipole moment is parallel to the field **E**. Since the point charge is closer to the center of negative charge than to the center of positive charge, the dipole is attracted to the point charge.

## Question

7. A small, light conducting ball with no net electric charge is suspended from a thread. When a positive charge is brought near the ball, the ball is attracted toward the charge. How does this come about? If the ball is an insulator instead of a conductor, it still is attracted. How? How would the situation have been different if the charge brought near were negative instead of positive?

# Benjamin Franklin (1706–1790)

I. Bernard Cohen
*Harvard University*

When Franklin's contemporaries wanted to express their admiration for his scientific achievements, they could think only of comparing him to Newton. Joseph Priestley wrote that Franklin's book on electricity "bid fare to be handed down to posterity as expressive of the true philosophy of electricity; just as the Newtonian philosophy is of the true system of nature in general."

In order to appreciate Franklin's contribution, we must remember that electricity is a young branch of physics. Whereas we can trace an early history of statics and dynamics, heat and light, and even atomic theory back to the Greeks, electricity emerged as a proper subject of scientific study only in the days of Newton. How meager the information about electricity was in the early eighteenth century can be seen from the fact that the fundamental distinction between conductors and nonconductors had not yet been made, nor had it been discovered that there are two kinds of electric charge. When Franklin took up this new subject, he was almost forty, past the age when we usually think of great scientific discoveries being made. Largely self-educated, Franklin had already obtained a solid grounding in experimental physics, having studied Newton's *Opticks* and many of the primary textbooks of newtonian experimental science. His printing and newspaper business was successful enough to allow him to retire from active participation, and he eagerly seized upon the new subject and began to work at it intensively, together with a small group of coworkers.

His interest had first been sparked when he attended some public lectures on science given by a Dr. Adam Spencer, first in Boston (where Franklin was visiting his family) and then in Philadelphia (where Franklin sponsored Spencer's lectures). He purchased Spencer's apparatus in order to perform experiments himself. Soon thereafter the Library Company of Philadelphia (of which Franklin was the principal founder) received a gift of electrical apparatus, with instructions for using it, from Peter Collinson, a London merchant.

Before long Franklin and his fellow experimenters realized that they had progressed in knowledge far beyond the literature that accompanied the apparatus. Franklin periodically sent reports of his work to Collinson and to other London correspondents, which eventually were assembled into a book entitled *Experiments and Observations on Electricity, Made at Philadelphia*, first published in England in 1751.

For his pioneering research in electricity, Franklin was elected a Fellow of the Royal Society of London, with the singular distinction of being forgiven the annual dues. Shortly thereafter he was awarded the Society's Copley Gold Medal, the highest scientific honor then being awarded in England. His book on electricity was a spectacular success, going through five editions in English and being translated into French (three editions in two different translations), Italian, and German. In 1773 Franklin was elected a Foreign Associate of the French Academy of Sciences, an extraordinary honor, since, according to the terms of the Academy's foundation, there could be only eight such foreign associates at any one time. No American was similarly honored again for another century.

Franklin's discoveries included not only important new experimental evidence but a new theory of electricity which made a science of the subject. This theory was based upon the fundamental postulate that all electrostatic phenomena (charging and discharging) result from the motion or transfer of a single electrical fluid. This hypothetical fluid was made up of "particles," or atoms, of electricity that repelled one another but were attracted by the particles of "ordinary" matter. A charged body was one that had either lost or gained elec-

Title page of the first edition of Franklin's book on electricity, from the copy he presented to Harvard College. (*Courtesy of the Houghton Library, Harvard University.*)

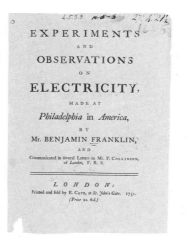

Portrait of Franklin.

The Bettmann Archive

trical fluid and accordingly was in a state that Franklin called "plus" (positive) or "minus" (negative). This postulate was closely associated with a major theoretical principle, known today as the law of conservation of charge: whatever charge is lost by one body must be gained by one or more other bodies, so that negative and positive charges always appear simultaneously or are simultaneously canceled out in equal amounts. This law implied that electrical effects were not the results of the "creation" of some mysterious entity (as had often been supposed before Franklin) but merely followed from an alteration or redistribution of the amount of electrical fluid in a body. A variety of experiments confirmed the universality of this principle of conservation of charge, which remains (along with conservation of momentum) one of the most fundamental principles of physical science.

One of Franklin's most startling experiments was the analysis of the Leyden jar, the first "condenser," or capacitor. The Leyden jar is a glass jar with a metal-foil outer coating. Inside, the jar has another metal-foil coating, or contains water or lead shot, which is in contact with a wire passing through the cork of the jar. The jar is usually charged by bringing the wire into contact with an electrostatic generator or a bit of rubbed amber, glass, or sulfur, while the outer coating is grounded. Franklin discovered that when a jar is charged, and the two conductors are then separated from the glass, neither the inner nor the outer conductor will show any sign of being charged; but when they are placed once again on the two sides of the jar, that jar will produce all the familiar phenomena of being charged. Franklin said that the whole charge "resides" in the glass, but today we refer to this phenomenon under the name of "polarization of the dielectric." Franklin also showed that such a device does not depend upon the shape of the bottle, and he invented the parallel plane capacitor.

The sentry-box experiment, from a manuscript of Franklin's with drawings. (*Courtesy of the American Academy of Arts and Sciences, Office of Charles and Ray Eames.*)

In a variety of experiments, Franklin also showed the effects of grounding and insulation. He discovered that a grounded pointed conductor can actually "draw off" the charge of a nearby charged object and that contrariwise a charged pointed conductor, however well-insulated, will "throw off" the charge through its point. This led him to study the electrical nature of the lightning discharge. Franklin's most important experiment with lightning was not made with the familiar kite but was the experiment of the sentry box. Franklin proposed that on a high building a sentry box be erected, with a long pointed conductor rising up through the roof. To determine whether the pointed conductor were charged, an experimenter would stand on an insulated stool inside the sentry box and bring up to the pointed conductor a grounded conductor set in an insulating handle. His objective was to draw a spark. If, as Franklin supposed, thunderclouds are electrically charged, then the pointed conductor would always become charged by induction when such clouds passed overhead, and a spark could be obtained. In practice, this experiment not only proved that clouds are electrically charged, so that lightning is only an ordinary electrical discharge from clouds on a large scale, but such rods also attracted a stroke of lightning and showed it to be an electrical phenomenon. Described in his book on electricity, this experiment was first performed in France. Franklin had been waiting for the completion of the spire of Christ Church in Philadelphia, where he hoped to erect a sentry box in order to perform the experiment. The kite was thought of as an alternative, an afterthought.

The lightning experiments brought fame to Franklin far and wide. Today the historical importance of these experiments is often misunderstood. In proving that lightning is an electrical discharge, Franklin showed that the electrical experiments performed in the laboratory are directly related to events in the natural world on a large scale. Thereafter, any general science of nature that did not include electricity would obviously be incomplete. Furthermore, the lightning experiments led Franklin to the invention of the lightning rod. For the first time in history, research in pure science led to a practical invention of major consequence. Bacon had indeed been correct in predicting that pure scientific knowledge would lead men to practical applications which would enable them to control their environment.

When Franklin began his research in electricity, the great French scientist Buffon (1707–1788) pointed out that electricity was not yet a science, but only a collection of bizarre phenomena subject to no single law. After Franklin produced his theory of electricity, which explained and correlated the known phenomena and also predicted verifiable new ones, it was generally agreed that electricity had indeed become a science. In fact, electricity was the first new science (or branch of science) to arise since Newton.

It is often thought that Franklin was not really a pure scientist in the ordinary sense of this expression, that he was, rather, a gadgeteer and inventor whose claim to science was aggrandized by his success as a statesman. In point of fact the opposite is true. By 1776, when Franklin was sent to France as the American representative, he had already gained an international reputation for his scientific work, a factor of the greatest importance in his diplomatic success. This was no unknown local patriot, but one of the leading figures of the scientific world. We continually pay tribute to Franklin's scientific genius whenever we use the many words he introduced into the language of electricity: plus and minus, positive and negative, electric battery, and a host of others.

When Franklin retired from business, he hoped to devote the rest of his life to the peaceful pursuit of a career in science. The demands of his community and his country, however, all too soon drew him into a life of public service. Franklin's choice between his love of science and his duty to his fellow men was expressed by him as follows: "Had Newton been Pilot but of a single common Ship, the finest of his Discoveries would scarce have excused, or atoned for his abandoning the Helm one Hour in Time of Danger; how much less if she carried the Fate of the Commonwealth."

## Review

A. Define, explain, or otherwise identify:

Charge conservation, 585          Electric dipole, 592
Charge quantization, 586          Lines of force, 592
Coulomb's law, 587                Polar molecule, 603
Electric field, 589               Dipole moment, 603

B. Without looking back at the examples, draw a careful diagram and set up the integration to find (a) the electric field on the bisector of a line charge (Example 20-11) and (b) the electric field on the axis of a uniformly charged disk (Example 20-12). Then check your diagram and integral against those in the examples.

True or false:

1. The electric field of a point charge always points away from the charge.

2. The charge of the electron is the smallest charge found.

3. Electric lines of force never diverge from a point in space.

4. Electric lines of force cannot cross at a point in space.

## Exercises

### Section 20-1, Electric Charge, and Section 20-2, Coulomb's Law

1. Find the number of electrons in a charge of (a) 1 $\mu$C; (b) 1 pC.

2. In electrolysis a quantity of electricity called the faraday $\mathscr{F}$, equal to Avogardro's number of electron charges, will deposit 1 gram ionic mass of monovalent ions, for example, 23 g of Na and 35.5 g of Cl in the decomposition of NaCl. Calculate the number of coulombs in a faraday. (Avogadro's number is $6.02 \times 10^{23}$.)

3. Two protons in the helium nucleus are about $10^{-15}$ m apart. Calculate the electrostatic force exerted by one proton on the other.

4. A charge $q_1 = 4.0$ $\mu$C is at the origin, and a charge $q_2 = 6.0$ $\mu$C is on the $x$ axis at $x = 3.0$ m. (a) Find the force on charge $q_2$. (b) Find the force on $q_1$. (c) How would your answers differ if $q_2$ were $-6.0$ $\mu$C?

5. Three point charges are on the $x$ axis; $q_1 = -6.0$ $\mu$C is at $x = -3.0$ m, $q_2 = 4.0$ $\mu$C is at the origin, and $q_3 = -6.0$ $\mu$C is at $x = 3.0$ m. Find the force on $q_1$.

6. Two equal charges of 3.0 $\mu$C are on the $y$ axis, one at the origin and the other at $y = 6$ m. A third charge $q_3 = 2$ $\mu$C is on the $x$ axis at $x = 8$ m. Find the force on $q_3$.

7. Three charges are at the corners of a square of side $L$. The two charges at the opposite corners are positive, and the other is negative. All have the same magnitude $q$. Find the force exerted by these charges on a fourth charge $+q$ placed at the remaining corner.

### Section 20-3, The Electric Field

8. A charge of 4.0 $\mu$C is at the origin. What is the magnitude and direction of the electric field on the $x$ axis at (a) $x = 6$ m and (b) $x = 10$ m? (c) Sketch the function $E_x$ versus $x$ for both positive and negative $x$. (Remember that $E_x$ is negative when $\mathbf{E}$ points in the negative $x$ direction.)

9. Two charges each $+4$ $\mu$C are on the $x$ axis, one at the origin and the other at $x = 8$ m. (a) Find the electric field on the $x$ axis at $x = 10$ m and at $x = 2$ m. (b) At what point on the $x$ axis is the electric field zero? (c) What is the direction of $\mathbf{E}$ at points on the $x$ axis just to the right of the origin? Just to the left of the origin? (d) Sketch $E_x$ versus $x$.

10. Two equal positive charges of magnitude $q_1 = q_2 = 6.0$ nC are on the $y$ axis at points $y_1 = +3$ cm and $y_2 = -3$ cm. (a) What is the magnitude and direction

of the electric field at the point on the $x$ axis at $x = 4$ cm? (b) What is the force exerted on a test charge $q_0 = 2$ nC placed on the axis at $x = 4$ cm?

11.  Charge $q_1 = +6.0$ nC is on the $y$ axis at $y = +3$ cm, and charge $q_2 = -6.0$ nC is on the $y$ axis at $y = -3$ cm. (a) What is the magnitude and direction of the electric field on the $x$ axis at $x = 4$ cm? (b) What is the force exerted on a test charge $q_0 = 2$ nC placed on the $x$ axis at $x = 4$ cm?

12.  When a test charge $q_0 = 2$ nC is placed at the origin, it experiences a force of $8.0 \times 10^{-4}$ N in the positive $y$ direction. (a) What is the electric field at the origin? (b) What would be the force on a charge $-4$ nC placed at the origin?

### Section 20-4, Lines of Force

13.  Figure 20-23 shows lines of force for a system of two point charges. (a) What are the relative magnitudes of the charges? (b) What are the signs of the charges? (c) In what regions of space is the electric field strong? In what regions is it weak?

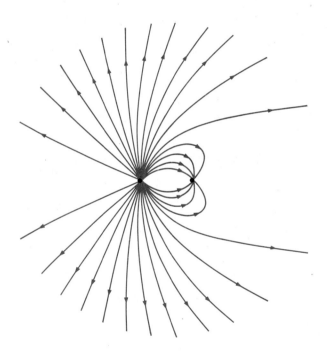

**Figure 20-23**
Exercise 13.

14.  Two charges $+q$ and $-3q$ are separated by a small distance. Draw lines of force for this system.

15.  Three equal positive point charges are situated at the corners of an equilateral triangle. Sketch the lines of force in the plane of the triangle.

### Section 20-5, Calculation of the Electric Field for Continuous Charge Distributions

16.  Show that the electric field $E_x$ on the axis of a ring charge of radius $a$ approaches that of a point charge, as expected, when the distance $x$ from the plane of the ring is much greater than $a$.

17.  A $0.5$-$\mu$C charge is uniformly distributed on a ring of radius 3 cm. Find the electric field on the axis of the ring at (a) 1 cm, (b) 2 cm, (c) 3 cm, and (d) 600 cm from the center of the ring.

18.  Show that the electric field on the perpendicular bisector of a line charge approaches that of a point charge, as expected, when the distance $y$ from the charge is much greater than the length $L$.

19. A uniform line charge of density $\lambda$ is on the $x$ axis from $x = -a$ to $x = +b$. (a) Find an expression for the $y$ component of the electric field at a point on the $y$ axis. (b) Show that your expression is the same as Equation 20-16 when $a = b = \frac{1}{2}L$.

20. A dipole consists of a charge $+q$ on the $x$ axis at $x = a$ and a charge $-q$ on the $x$ axis at $x = -a$. (a) What direction is the electric field at a point on the $y$ axis? (b) Find an expression for the electric field at a point on the $y$ axis? (c) Show that for $y$ much greater than $a$, the electric field on the $y$ axis is given by $E_x \approx -kp/y^3$, where $p = q(2a)$ is the dipole moment.

21. A line charge has density $\lambda = 2\ \mu C/m$ and a total length of 6 m. Use approximate expressions to find the electric field on the bisector of the line at distances 2 cm and 60 m.

22. (a) Using the fact that at very great distances a disk charge should look like a point charge, write down the expected form of the electric field on the axis and very far from a uniformly charged disk of radius $a$ and charge density $\sigma$. (b) Derive this result from Equation 20-18 by showing that $x/\sqrt{x^2 + a^2} = (1 + a^2/x^2)^{-1/2}$ and approximating this expression for $a/x$ much less than 1, using the binomial expansion $(1 + \epsilon)^n \approx 1 + n\epsilon$ for small $\epsilon$.

23. A disk of radius 6 cm carries a uniform charge density 30 $\mu C/m^2$. Using reasonable approximations, find the electric field on the axis of the disk at distances (a) 0.01 cm, (b) 0.02 cm, (c) 0.03 cm, and (d) 500 cm.

24. Show that the electric field just outside a spherical shell of charge is $\sigma/\epsilon_0$, where $\sigma$ is the charge per unit area.

25. A spherical shell of radius 10 cm carries a charge of 2 $\mu C$ uniformly distributed on its surface. Find the electric field at the following distances from the center of the shell: (a) 5 cm, (b) 9.99 cm, (c) 10.01 cm, (d) 20 cm, and (e) 40 cm.

26. A sphere of radius 10 cm has a charge 2 $\mu C$ uniformly distributed throughout its volume. Find the electric field at the following distances from the center of the sphere: (a) 5 cm, (b) 9.99 cm, (c) 10.01 cm, (d) 20 cm, (e) 40 cm. Compare these results with those of Exercise 25.

### Section 20-6, Motion of Point Charges in Electric Fields

27. In finding the acceleration of an electron or other charged particle the ratio of the charge to mass of the particle is important. (a) Compute $e/m$ for an electron. (b) What is the magnitude and direction of acceleration of an electron in a uniform electric field of magnitude 100 N/C? (c) Nonrelativistic mechanics can be used only if the speed of the electron is significantly less than the speed of light $c$. Compute the time it takes for an electron placed at rest in an electric field of magnitude 100 N/C to reach a speed $0.01c$. (d) How far does the electron travel in that time?

28. (a) Compute $e/m$ for a proton and find its acceleration in a uniform electric field of magnitude 100 N/C. (b) Find the time it takes for a proton initially at rest in such a field to reach the speed of $0.01c$ (see Exercise 27).

29. An electron has an initial velocity 2.0 Mm/s in the $x$ direction. It enters a uniform electric field $\mathbf{E} = 400\mathbf{j}$ N/C which is in the $y$ direction. (a) Find the acceleration of the electron. (b) How long does it take for the electron to travel 10 cm in the $x$ direction? (c) By how much and in what direction is the electron deflected after traveling 10 cm in the $x$ direction?

30. Calculate the magnitude and direction of the electric field which would be needed to balance the weight of (a) an electron, (b) a proton, (c) an oil drop which has a mass of 0.2 ng and carries a charge of $+10e$, (d) a pingpong ball of mass 25 g and charge $+0.01\ \mu C$.

31. The earth has an electric field in its atmosphere which is about 150 N/C directed downward. Compare the upward electric force on an electron with the downward gravitational force.

32. An electron is projected with an initial speed of $3.0 \times 10^5$ m/s at an upward angle of 30° with the horizontal. There is a vertically upward electric field of strength $10^5$ N/C. (a) What is the highest point reached by the electron? (b) How long does it take to reach this height? (c) What is the horizontal range of the electron projectile?

### Section 20-7, Electric Dipoles in Electric Fields

33. Two point charges $q_1 = 2.0$ pC and $q_2 = -2.0$ pC are separated by 4 $\mu$m. What is the dipole moment of this pair of charges? Sketch the pair and indicate the direction of the dipole moment.

34. A dipole of moment $0.5e$ nm is placed in a uniform electric field of strength $4.0 \times 10^4$ N/C. What is the magnitude of the torque on the dipole when (a) the dipole is parallel to the electric field, (b) the dipole is perpendicular to the electric field, (c) the dipole makes an angle of 30° with the electric field?

### Problems

1. Four charges of equal magnitude are arranged at the corners of a square of side $L$, as shown in Figure 20-24. (a) Find the magnitude and direction of the force exerted on the charge on the lower left corner by the other charges. (b) Show that the electric field at the midpoint of one of the sides of the square is directed along that side toward the negative charge and has magnitude $E$ given by

$$E = k \frac{8q}{L^2} \left(1 - \frac{\sqrt{5}}{25}\right)$$

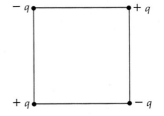

**Figure 20-24**
Problem 1.

2. Two charges $q_1$ and $q_2$ are placed a distance $L$ apart. (a) What must the relative values of $q_1$ and $q_2$ be (including sign) if the electric field is to be zero at some point on the line of the charges a distance $D$ from $q_2$ and $D + L$ from $q_1$? (b) Is there a second point anywhere on the line of the charges at which the electric field is zero?

3. Two charges $q_1$ and $q_2$ when combined give a total charge of 6 $\mu$C. When they are separated by 3 m, the force exerted by one charge on the other has the magnitude 8 mN. Find $q_1$ and $q_2$ if (a) both are positive, so that they repel each other; (b) one is positive and the other negative, so that they attract each other.

4. A positive charge $Q$ is to be divided into two positive charges $q_1$ and $q_2$. Show that for a given separation $D$ the force exerted by one charge on the other is greatest if $q_1 = q_2 = \frac{1}{2}Q$.

5. Two equal positive charges $q$ are on the $y$ axis; one is at $y = a$ and the other at $y = -a$. (a) Show that the electric field on the $x$ axis is along the $x$ axis with $E_x$ given by $E_x = 2kqx(x^2 + a^2)^{-3/2}$. (b) Show that near the origin, when $x$ is much smaller than $a$, $E_x$ is approximately $2kqx/a^3$. (c) Show that for $x$ much larger than $a$, $E_x$ is approximately $2kq/x^2$. Explain why you would expect this result even before calculating it.

6. (a) Show that the electric field for the charge distribution in Problem 5 has its greatest magnitude at the points $x = a/\sqrt{2}$ and $x = -a/\sqrt{2}$ by computing $dE_x/dx$ and setting the derivative equal to zero. (b) Sketch the function $E_x$ versus $x$ using the results of part (a) and parts (b) and (c) of Problem 5.

7. For the charge distribution in Problem 5 the electric field at the origin is zero. A test charge $q_0$ placed at the origin will therefore be in equilibrium. (a) Discuss the stability of the equilibrium for a positive test charge by considering small displacements from equilibrium along the $x$ axis and small displacements along the $y$ axis. (b) Repeat part (a) for a negative test charge. (c) Find the magnitude and sign of a charge $q_0$ which can be placed at the origin so that the net force on each of the three charges is zero. Consider what happens if any of the charges is displaced slightly from equilibrium.

8. Show that $E_x$ on the axis of a ring charge (Equation 20-14) has its maximum and minimum values at $x = +a/\sqrt{2}$ and $x = -a/\sqrt{2}$. Sketch $E_x$ versus $x$.

9. (a) Find the slope $dE_x/dx$ at the origin for $E_x$ due to a ring charge (Equation 20-14). (b) A bead of mass $m$ carrying a negative charge $-q$ slides along a stretched thread which is along the axis of a ring of radius $a$ and total charge $Q$. Show that if the bead is displaced slightly from the origin and released, the motion is simple harmonic. Find the frequency of the motion.

10. A disk of radius 30 cm carries a uniform charge density $\sigma$. Compare the approximation $E = \sigma/2\epsilon_0$ with the exact expression for the electric field (Equation 20-18) by computing the neglected term as a percentage of the field for distances $x = 0.1, 0.2$, and 3 cm. At what distance is the neglected term 1 percent of $\sigma/2\epsilon_0$?

11. A line charge of density $\lambda$ has the shape of a square of side $L$ which lies in the $yz$ plane with its center at the origin. Find the electric field on the $x$ axis at arbitrary distance $x$ and compare your result to the field on the axis of a charged ring of approximately the same size carrying the same total charge.

12. A semi-infinite line charge of uniform density $\lambda$ lies along the $x$ axis from $x = 0$ to $x = \infty$. Find both $E_x$ and $E_y$ at a point on the $y$ axis.

13. Two infinite planes are parallel to each other and separated by a distance $d$. Find the electric field to the left of the planes, to the right of the planes, and between the planes when (a) each plane contains a uniform charge density $\sigma = +\sigma_0$, and (b) the left plane has a uniform charge density $\sigma = +\sigma_0$ and the right plane $\sigma = -\sigma_0$. Draw the lines of force for each case.

14. A thin wire carries a uniform linear charge density $\lambda$ and is bent into a circular arc which subtends an angle $2\theta_0$, as shown in Figure 20-25. Show that the electric field at the center of curvature of the arc has the magnitude $E = (2k\lambda \sin \theta_0)/R$. *Hint:* Consider an element of length $dl = R\, d\theta$ at some angle $\theta$; find the component of the field along the line $OC$, and integrate from $\theta = -\theta_0$ to $\theta = +\theta_0$.

15. An electric dipole consists of two charges $+q$ and $-q$ separated by a very small distance $2a$. Its center is on the $x$ axis at $x = x_1$, and it points along the $x$ axis toward positive $x$. It is in a nonuniform electric field, which is also in the $x$ direction, given by $\mathbf{E} = Cx\mathbf{i}$, where $C$ is a constant. (a) Find the force on the positive charge and that on the negative charge and show that the net force on the dipole is $Cp\mathbf{i}$. (b) Show that, in general, if a dipole of moment $\mathbf{p}$ lies along the $x$ axis in an electric field in the $x$ direction, the net force on the dipole is given approximately by $(dE_x/dx)p\, \mathbf{i}$.

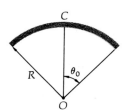

**Figure 20-25**
Problem 14.

16. A positive point charge $+Q$ is at the origin, and a dipole of moment $\mathbf{p}$ is a distance $r$ away and in the radial direction, as in Figure 20-22. (a) Show that the force exerted by the electric field of the point charge on the dipole is attractive with approximate magnitude $2kQp/r^3$ (see Problem 15). (b) Consider now the dipole at the origin and a point charge $Q$ a distance $r$ away along the line of the dipole. From your result of part (a) and Newton's third law, show that the magnitude of the electric field of the dipole along the line of the dipole a distance $r$ away is approximately $2kp/r^3$.

17. Two small spheres of mass 10 g are suspended from a common point by threads of length 50 cm. Find the angle made by either thread with the vertical when each sphere carries a charge of $2\mu C$ by (a) using the small-angle approximation $\sin \theta \approx \tan \theta$ and (b) using numerical methods.

18. A nonconducting thread forms a circle of radius $a$ and lies in the $xy$ plane with its center at the origin. It carries a nonuniform charge density $\lambda = \lambda_0 \sin \theta$, where $\theta$ is measured from $\theta = 0$ on the positive $x$ axis. Find the electric field at the origin.

# CHAPTER 21

# Gauss' Law and Conductors in Electrostatic Equilibrium

---

**Objectives**   After studying this chapter, you should:

1. Be able to state Gauss' law and use it to find the electric field produced by various symmetrical charge distributions.

2. Be able to discuss the difference between conductors and insulators.

3. Be able to prove that in electrostatic equilibrium the free charge on a conductor resides on the surface of the conductor.

4. Know and be able to derive the result that the electric field just outside the surface of a conductor has the magnitude $\sigma/\epsilon_0$.

5. Be able to discuss charging by induction.

---

Radio Times/Hulton Picture Library

Karl Friedrich Gauss, German mathematician, astronomer, and physicist (1777–1855).

In this chapter we put the idea of lines of force on a quantitative basis and develop a mathematical equation known as Gauss' law. We then show how Gauss' law can be used to find the electric field in problems with a high degree of symmetry, such as that of a uniform sphere of charge. We shall see that in such situations it is much easier to find the field using Gauss' law than it is to calculate the field directly from Coulomb's law. We then discuss electric conductors and apply Gauss' law to show that in electrostatic equilibrium the charge on a conductor resides completely on the surface.

Consider a mathematical surface enclosing the dipole shown in Figure 21-1 (the surface may be spherical or any other shape). The number of lines of force coming from the positive charge and crossing the surface going out of the enclosure depends on where the surface is drawn, but this number is exactly equal to the number of lines entering the enclosure and ending on the negative charge. If we count the number leaving as positive and the number entering as negative, the net number leaving or entering is zero. In the lines-of-force figures for other charge distributions, such as that shown in Figure 21-2, the net number of lines leaving any surface enclosing the charges is proportional to the net charge enclosed by the surface. The number is the same for all surfaces enclosing the charges. This result in its quantitative form (developed in Section 21-2) is known as Gauss' law.

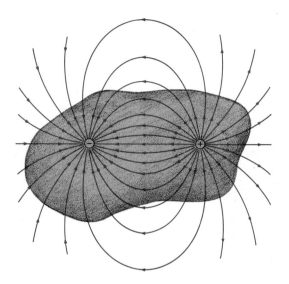

**Figure 21-1**
Surface enclosing an electric
dipole. The number of lines
leaving the surface is exactly
equal to the number of lines
entering the surface no matter
where the surface is drawn, as
long as it encloses both
charges.

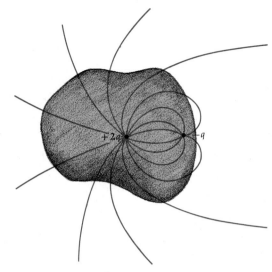

**Figure 21-2**
Surface enclosing charges $+2q$
and $-q$. The net number of
lines leaving the surface
(those leaving minus those
entering) is proportional to
the net charge inside the surface. In this drawing the
number of lines leaving or
entering a charge has been
chosen to be $8q$ (16 lines leave
charge $+2q$, and 8 enter $-q$).
The net number of lines
leaving the surface is therefore 8 no matter where the
surface is drawn, as long as it
encloses both charges.

## 21-1   Electric Flux

The mathematical quantity related to the number of lines of force
through a surface area is called the *electric flux*. Figure 21-3 shows an
electric field which is uniform in magnitude and direction over some
region. The electric flux $\phi$ through a surface of area $A$ which is perpendicular to the field is defined as the product of the field $E$ and the area $A$:

$$\phi = EA \qquad\qquad 21\text{-}1$$

The units of flux are newton-metres² per coulomb (N·m²/C). Since the
electric field is proportional to the number of lines per unit area, the
electric flux is proportional to the number of lines of force through the
surface area:

$$\phi \propto N \qquad\qquad 21\text{-}2$$

The proportionality constant depends upon the choice of the number of
lines leaving or entering a unit charge.

In Figure 21-4 the surface of area $A_2$ is not perpendicular to the electric field $\mathbf{E}$. The number of lines that cross area $A_2$ is the same as the
number that cross area $A_1$. The areas are related by

$$A_2 \cos\theta = A_1$$

where $\theta$ is the angle between $\mathbf{E}$ and the unit vector $\hat{\mathbf{n}}$ perpendicular to
surface $A_2$, as shown. The flux through a surface not perpendicular to $\mathbf{E}$
is defined to be

$$\phi = \mathbf{E}\cdot\hat{\mathbf{n}}A = EA\cos\theta = E_nA \qquad\qquad 21\text{-}3$$

where $E_n = \mathbf{E}\cdot\hat{\mathbf{n}}$ is the component of the electric-field vector perpendicular, or normal, to the surface.

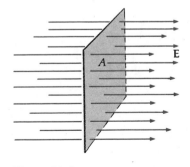

**Figure 21-3**
Lines of force for a uniform
electric field crossing an area
$A$ perpendicular to the field.
The product $EA$ is called the
electric flux.

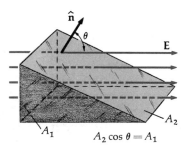

**Figure 21-4**
Lines of force for a uniform electric field which is perpendicular to the area $A_1$ but makes an angle $\theta$ with the unit vector $\hat{n}$ normal to area $A_2$. When **E** is not perpendicular to the area, the flux through the area is $E_n A$, where $E_n = E \cos \theta$ is the component of **E** perpendicular to the area. The flux through $A_2$ is then the same as that through $A_1$.

We can generalize our definition of electric flux to curved surfaces over which the electric field may vary in magnitude or direction or both, by dividing the surface up into a large number of very small elements. If each element is small enough, it can be considered to be a plane and the variation of the electric field across the element can be neglected. Let $\hat{n}_i$ be the unit vector perpendicular to such an element and $\Delta A_i$ be its area (Figure 21-5). (If the surface is curved, the unit vectors $\hat{n}_i$ will have different directions for different elements.) The flux of the electric field through this element is

$$\Delta \phi_i = \mathbf{E} \cdot \hat{n}_i \, \Delta A_i$$

The total flux through the surface is the sum of $\Delta \phi_i$ over all the elements.

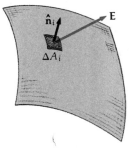

**Figure 21-5**
When **E** varies in either magnitude or direction, the area is divided into small area elements $\Delta A_i$. The flux through the area is computed by summing $E_n \, \Delta A_i$ over all the area elements.

In the limit as the number of elements approaches infinity and the area of each element approaches zero, this sum becomes an integral. The general definition of electric flux is then

$$\phi = \lim_{\Delta A_i \to 0} \sum_i \mathbf{E} \cdot \hat{n}_i \, \Delta A_i = \int \mathbf{E} \cdot \hat{n} \, dA \qquad \text{21-4}$$

*Electric flux defined*

As with a uniform electric field and a plane surface, the number of lines of force through any surface is proportional to the flux.

We are often interested in the flux of the electric field through a closed surface, i.e., a surface which separates space into two regions, one inside the surface and one outside. On a closed surface the unit normal vector $\hat{n}$ is defined to be directed outward at each point. At a point where a line of force leaves the surface, **E** is directed outward and $\mathbf{E} \cdot \hat{n}$ is positive; but at a point where a line of force enters the surface, **E** is directed inward and $\mathbf{E} \cdot \hat{n}$ is negative. The total or net flux $\phi_{net}$ through the closed surface is positive or negative depending on whether **E** is predominantly outward or inward on the surface. Since the flux through any part of the surface is proportional to the number of lines through the surface, the net flux is proportional to the *net* number of lines of force leaving the surface, i.e., the number of lines going out of the surface minus the number going into the surface. The integral over a closed surface is indicated by the symbol $\oint$. The net flux through a closed surface is therefore written

$$\phi_{net} = \oint \mathbf{E} \cdot \hat{n} \, dA = \oint E_n \, dA \qquad \text{21-5}$$

**Example 21-1** A point charge $q$ is at the center of a spherical surface of radius $R$. Calculate the net flux of the electric field through this surface.

The electric field at any point on the surface is radial, with magnitude

$$E = \frac{kq}{R^2}$$

Since $\hat{n}$ is also radial on the surface, $\mathbf{E} \cdot \hat{n} = kq/R^2$ is constant everywhere on the surface. The net flux through the spherical surface is therefore just the product of the radial component of $\mathbf{E}$ and the total area of the spherical surface, $4\pi R^2$:

$$\phi_{\text{net}} = \oint \mathbf{E} \cdot \hat{n}\, dA = \oint \frac{kq}{R^2}\, dA = \frac{kq}{R^2} \oint dA = \frac{kq}{R^2} 4\pi R^2$$

or

$$\phi_{\text{net}} = \oint \mathbf{E} \cdot \hat{n}\, dA = 4\pi kq \qquad 21\text{-}6$$

## 21-2 Gauss' Law

In Example 21-1 we calculated the flux of a point charge through a spherical surface of radius $R$ to be $4\pi kq$, independent of $R$. The number of lines of force going out through a spherical surface of radius $R$ is also proportional to the charge $q$. This is consistent with our previous observations that the net number of lines going out of a surface is proportional to the net charge inside the surface. This number of lines is the *same for all surfaces surrounding* the charge, independent of the shape of the surface. Since the number of lines and the flux are just proportional to each other, it follows that Equation 21-6 holds for the flux through any surface enclosing the point charge $q$. *The net flux through any surface surrounding a point charge $q$ equals $4\pi kq$.*

We can extend this result to systems of more than one point charge. In Figure 21-6 the surface $S$ encloses two point charges $q_1$ and $q_2$, and there is a third point charge $q_3$ outside the surface. Since the electric field at any point on the surface is the vector sum of the electric fields produced by each of the three charges, the net flux $\phi_{\text{net}} = \oint \mathbf{E} \cdot \hat{n}\, dA$ through the surface is just the sum of the fluxes due to the individual charges. The flux through the surface due to the charge $q_3$, which is outside the surface, is zero because every line of force from $q_3$ that enters the surface at one point leaves the surface at some other point. The net number of lines through the surface from a charge outside the surface is zero. The flux through the surface due to charge $q_1$ is $4\pi kq_1$, and that due to charge $q_2$ is $4\pi kq_2$. The net flux through the surface equals $4\pi k(q_1 + q_2)$, which may be positive, negative, or zero depending on the signs and magnitudes of the two charges.

*In general, for a system of charges, the net flux through any surface $S$ equals $4\pi k$ times the net charge inside the surface.*

$$\phi_{\text{net}} = \oint \mathbf{E} \cdot \hat{n}\, dA = 4\pi kq_{\text{inside}} \qquad 21\text{-}7 \qquad \textit{Gauss' law}$$

This important result is Gauss' law. Its validity depends on the fact that the electric field due to a single point charge varies inversely with the square of the distance from the charge. It was this property of the elec-

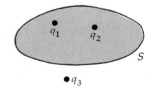

**Figure 21-6**
Three point charges $q_1$, $q_2$, and $q_3$ and a surface $S$ which encloses $q_1$ and $q_2$.

tric field that made it possible to draw a fixed number of lines of force from a charge and have the density of lines be proportional to the field strength. It is customary to write the Coulomb constant $k$ in terms of another constant $\epsilon_0$, called the *permittivity of free space:*

$$k = \frac{1}{4\pi\epsilon_0} \qquad\qquad 21\text{-}8$$

With this notation Coulomb's law and Gauss' law are written

$$\mathbf{F}_{12} = \frac{1}{4\pi\epsilon_0} \frac{q_1 q_2}{r_{12}^2} \hat{\mathbf{r}}_{12} \qquad\qquad 21\text{-}9 \qquad \textit{Coulomb's law}$$

and

$$\phi_{\text{net}} = \oint \mathbf{E} \cdot \hat{\mathbf{n}} \, dA = \frac{1}{\epsilon_0} q_{\text{inside}} \qquad\qquad 21\text{-}10 \qquad \textit{Gauss' law}$$

The value of $\epsilon_0$ in SI units is

$$\epsilon_0 = \frac{1}{4\pi k} = \frac{1}{4\pi(8.99 \times 10^9)} = 8.85 \times 10^{-12} \text{ C}^2/\text{N·m}^2 \qquad 21\text{-}11 \qquad \textit{Permittivity of free space}$$

Equation 21-7 (and its equivalent, Equation 21-10) can be derived mathematically from Coulomb's law without reference to lines of force, but the derivation is rather difficult. It is given at the end of this chapter.

Gauss' law can be used to calculate the magnitude of **E** when there is a high degree of symmetry so that the direction of **E** is known everywhere and the magnitude of **E** is constant over some simple surface.

---

**Example 21-2** Find the field of a point charge $q$.

We consider this simple example to illustrate the method and to show that Gauss' law is indeed equivalent to Coulomb's law. We choose our origin at the point charge and consider a spherical surface of radius $r$ (Figure 21-7). This mathematical surface, used for calculation only, is called a *gaussian surface*. It is clear from symmetry that **E** must be radial and that its magnitude can depend only on the distance from the charge. The normal component $\mathbf{E} \cdot \hat{\mathbf{n}} = E_r$ has the same value everywhere on our spherical surface. The flux of **E** through this surface is thus

$$\phi_{\text{net}} = \oint \mathbf{E} \cdot \hat{\mathbf{n}} \, dA = \oint E_r \, dA = E_r \oint dA$$

But $\oint dA$ is just the total area of the spherical surface, $4\pi r^2$. Since the total charge inside the surface is just the point charge $q$, Gauss' law gives

$$E_r 4\pi r^2 = \frac{q}{\epsilon_0} \quad \text{or} \quad E_r = \frac{q}{4\pi\epsilon_0 r^2} \qquad\qquad 21\text{-}12$$

We have thus derived Coulomb's law from Gauss' law. Since we originally derived Gauss' law from Coulomb's law, we have now shown the two laws to be equivalent.*

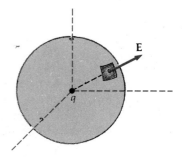

**Figure 21-7**
Calculation of the electric field of a point charge using Gauss' law. On a spherical surface surrounding the charge, the electric field **E** is perpendicular to the surface and constant in magnitude. The flux of the field through this surface is then $E_r 4\pi r^2$.

---

* Gauss' law and Coulomb's law are strictly equivalent only for electrostatic fields. Nonelectrostatic electric fields, produced by changing magnetic fields, are not described by Coulomb's law. Because the lines of force for these fields are closed lines, which do not diverge from, or converge to, any point in space, the net flux of nonelectrostatic fields through any closed surface is zero. Hence, Gauss' law holds for all electric fields.

**Example 21-3** Find the electric field due to a uniform line charge of infinite length along the $x$ axis.

Because of the symmetry of this situation, **E** must be perpendicular to the line and directed away from the line charge. Also, the magnitude of **E** can depend only on the perpendicular distance of the field point from the line.

Figure 21-8 shows a cylindrical can coaxial with the line charge. Let $r$ be the radius of the cylinder, which we use for our gaussian surface. Everywhere on the cylindrical part of the surface **E** is perpendicular to the surface and constant in magnitude. On the ends of the can **E** is parallel to the surface; thus the ends make no contribution to the flux.

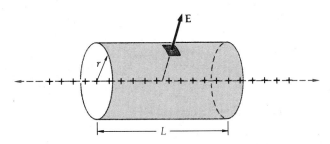

**Figure 21-8**
To calculate the electric field of an infinite line charge from Gauss' law a cylindrical surface concentric with the line charge is chosen. On the cylinder the electric field is constant in magnitude and perpendicular to the surface. The flux through the surface is $E_n 2\pi rL$, where $L$ is the length of the cylinder.

The total charge inside the gaussian surface is $\lambda L$, where $\lambda$ is the charge per unit length of the line charge and $L$ is the length of the can. Setting the flux through this surface equal to $1/\epsilon_0$ times the net charge inside, we obtain

$$\phi_{net} = \oint E_n \, dA = E_n \oint dA = \frac{\lambda L}{\epsilon_0}$$

Since the area of the cylindrical surface is $2\pi rL$,

$$E_n 2\pi rL = \frac{\lambda L}{\epsilon_0}$$

or

$$E_n = \frac{1}{2\pi\epsilon_0 L} \frac{\lambda L}{r} = 2k \frac{\lambda}{r} \qquad 21\text{-}13$$

in agreement with our result from Example 20-11.

Consider now a finite line charge which extends beyond the cylindrical gaussian surface shown (Figure 21-9). Since the net charge inside the surface is still $\lambda L$, it might seem that we would get the same result for **E** no matter how long the line charge is. However, if the line charge is not infinite, our symmetry arguments break down. One problem is that **E** is not perpendicular to the cylindrical surface except at points equidistant from the ends of the charge. Also, $E_n$ is not constant everywhere on the surface; that is, $E_n$ depends not only on $r$ but also on the distance from the center of the line charge. For this case of a finite line charge and the gaussian surface shown, Gauss' law still holds. The integral of $E_n$ over the gaussian surface (including the faces) equals $1/\epsilon_0$ times the net charge inside the surface, $\lambda L$. However, the law is *not useful* for calculating the field **E** because $E_n$ is not constant on the gaussian surface.

This example is typical of many applications of Gauss' law. Whenever there is a high degree of symmetry (infinite line charge in this example), the field **E** is obtained from Gauss' law far more easily than by direct integration using Coulomb's law. In the absence of a high degree of symmetry, Gauss' law is of no use in obtaining **E**.

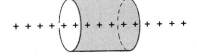

**Figure 21-9**
Cylindrical surface concentric with a line charge of finite length. Gauss' law is not useful for finding the electric field of a finite line charge because the electric field is not constant in magnitude on the surface and is not perpendicular to the surface.

**Example 21-4** Find the electric field due to an infinite plane of uniform charge density $\sigma$.

Again, we can use symmetry to argue that the field $\mathbf{E}$ must be perpendicular to the plane, must depend only on the distance from the plane to the field point, and must have the same magnitude but opposite direction at points the same distance below the plane as above it. We choose for our gaussian surface a pillbox-shaped cylinder with its axis perpendicular to the plane and with its center on the plane (Figure 21-10). Let each end of the cylinder be parallel to the plane and have area $A$. In this case, $\mathbf{E}$ is parallel to the cylindrical surface, and there is no flux through this curved surface. Since the flux out of each face is $\mathbf{E} \cdot \hat{\mathbf{n}}A = E_nA$, the total flux is $2E_nA$. The net charge inside the surface is $\sigma A$. From Gauss' law

$$\phi_{\text{net}} = 2E_nA = \frac{\sigma A}{\epsilon_0}$$

$$E_n = \frac{\sigma}{2\epsilon_0} = 2\pi k\sigma \qquad\qquad 21\text{-}14$$

This result agrees with that obtained from Coulomb's law in Example 20-12.

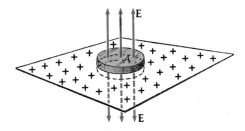

**Figure 21-10**
Gaussian surface for the calculation of the electric field due to an infinite plane of charge. On the upper and lower faces of this pillbox surface, $\mathbf{E}$ is perpendicular to the surface and constant in magnitude. The flux through this surface is $2E_nA$, where $A$ is the area of each face.

As in Example 21-3, the symmetry arguments essential for the *use* of Gauss' law depend on the fact that the plane charge distribution is infinite in extent.

**Example 21-5** Find the field due to a spherical shell of charge of radius $R$.

We choose a spherical gaussian surface of radius $r$, concentric with the charged shell. Whether $r$ is greater than $R$ or less than $R$, the symmetry of the problem implies that $\mathbf{E}$ is perpendicular to the gaussian surface and constant in magnitude on the surface. The flux through the gaussian surface is thus $4\pi r^2E_r$. For a field point outside the shell $(r > R)$,

$$\phi_{\text{net}} = 4\pi r^2E_r = \frac{Q}{\epsilon_0}$$

where $Q$ is the total charge on the shell. Then

$$E_r = \frac{1}{4\pi\epsilon_0}\frac{Q}{r^2} = \frac{kQ}{r^2} \qquad \text{outside}$$

For a field point inside the shell $(r < R)$, the total charge inside the gaussian surface is zero, so that the flux must be zero:

$$\phi_{\text{net}} = 4\pi r^2E_r = 0$$

and therefore

$$E_r = 0 \qquad \text{inside}$$

Since Newton's law of gravitation and Coulomb's law have the same inverse-square-distance dependence, the results of all of these calculations of the electric field using Gauss' or Coulomb's law can be applied directly to the analogous cases of calculation of the gravitational field. We need only replace the appropriate charge or charge density by the corresponding mass or mass density and change the sign, since the gravitational field of a point mass (which is always positive) points toward the mass, whereas the electric field of a (positive) point charge points away from the charge. Gauss' law for the flux of the gravitational field $\mathbf{g}$ is

$$\phi_{net} = \oint \mathbf{g} \cdot \hat{\mathbf{n}} \, dA = \oint g_n \, dA = -4\pi G m_{inside} \qquad \text{21-15}$$

where $m_{inside}$ is the total mass enclosed by surface $S$.

### Questions

1. If the electric field $\mathbf{E}$ is zero everywhere on a closed surface, is the net flux through the surface necessarily zero? What then is the net charge inside the surface?

2. If the net flux through a closed surface is zero, does it follow that the electric field $\mathbf{E}$ is zero everywhere on the surface? Does it follow that the net charge inside the surface is zero?

3. Is the electric field $\mathbf{E}$ in Gauss' law the part of the electric field due to the charge inside the surface, or is it the net electric field due to all charges whether they are inside or outside the surface?

4. What information is needed in addition to the total charge inside a surface to use Gauss' law to find the electric field?

## 21-3   Electric Conductors

The great difference in the electrical behavior of conductors and insulators was noted even before the discovery of electric conduction. Gilbert had classified materials according to their ability to be electrified. Objects which could be electrified he called *electrics*; those which could not (metals and some other materials) he called *nonelectrics*. After Gray discovered conduction, Du Fay showed that all materials can be electrified but that care must be taken in insulating Gilbert's nonelectrics from the ground (or the experimenter) lest the charge be quickly conducted away. Using only physiological sensation for detection, Cavendish compared the conducting abilities of many substances.*

> It appears from some experiments, of which I propose shortly to lay an account before this Society, that iron wire conducts about 400 million times better than rain or distilled water—that is, the electricity meets with no more resistance in passing through a piece of iron wire 400,000,000 inches long than through a column of water only one inch long. Sea-water, or a solution of one part of salt in 30 of water, conducts 100 times, or a saturated solution of sea-salt about 720 times, better than rain-water.

* Henry Cavendish, *Philosophical Transactions of the Royal Society of London,* vol. 66, p. 196 (1776).

Because of the enormous variation in the ability to conduct electricity, it is possible and convenient to classify most materials as conductors or insulators (nonconductors). The ability of a material to conduct electricity is measured by its conductivity. The conductivity of a typical conductor is of the order of $10^{15}$ times that of a typical insulator, whereas within the group of conductors the conductivity varies only over several orders of magnitude. We shall give a precise definition of electric conductivity (and its reciprocal, resistivity) in Chapter 24, when we study electric currents, i.e., electric charges in motion. There we also discuss *semiconductors*, which are neither conductors nor insulators.

In this chapter we are concerned only with the behavior of conductors in electrostatic equilibrium, i.e., when all electric charges are at rest. The property of a conductor that is important in studying electrostatic fields is the availability inside the conductor of charge that is free to move about. The source of this free charge is electrons that are not bound to any atom. For example, in a single atom of copper, 29 electrons are bound to the nucleus by electrostatic attraction of the positively charged nucleus. The outermost electrons are more weakly bound than the innermost electrons because of the greater distance from the positive nucleus and because of the repulsion of the inner electrons. (This is called *screening*.) When a large number of copper atoms combine to form metallic copper, the electron binding of a single atom is changed by interaction with neighboring atoms. One or more of the outer electrons in an atom are no longer bound but are free to move throughout the whole metal, much as a gas molecule is free to move about in a box. The number of free electrons depends on the particular metal but is of the order of one per atom.

In the presence of an external electric field, the free charge in a conductor moves about the conductor until it is so distributed that it creates an electric field which cancels the external field inside the conductor. Consider a charge $q$ inside a conductor. If there is a field **E** inside the conductor, there will be a force $q\mathbf{E}$ on this charge; and if it is free to move, i.e., if it is not bound to an atom or molecule by a stronger force, it will accelerate. Thus electrostatic equilibrium is impossible in a conductor unless the electric field is zero everywhere inside the conductor.

*The electrostatic field is zero inside a conductor*

Figure 21-11 shows a conducting slab placed in an external electric field $\mathbf{E}_0$. The free electrons are originally distributed uniformly throughout the slab. Since the slab is made up of neutral atoms, it is electrically neutral (assuming that no extra charge has been placed on it). If the external electric field is to the right, there will be a force on each electron $-e\mathbf{E}_0$ to the left because the electron has a negative charge, and the free electrons accordingly accelerate to the left. At the surface of the conductor, the conductor exerts forces on these electrons which balance that due to the external field, so that the electrons are bound to the conductor. (If the external field is very strong, the electrons can be stripped off from the surface. In electronics this is called *field emission*. We assume here that the external field is not strong enough to overcome the forces binding the electrons to the surface.) The result is a negative surface charge density on the left side of the slab and a positive surface charge density on the right side because of the removal of some of the free electrons from that side. Both these charge densities produce an electric field inside the slab which is opposite the external field. These two fields cancel everywhere inside the conductor, so that there is no unbalanced force on the free electrons and electrostatic equilibrium results. The behavior of the free charge in a conductor placed in an ex-

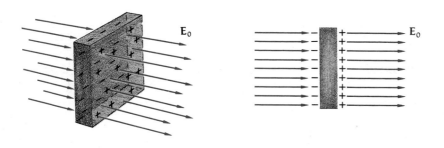

**Figure 21-11**
Conducting slab in an external electric field $E_0$. A positive charge is induced on the right face and a negative charge on the left face, so that the resultant electric field inside the conductor is zero. The field lines then end on the left face and begin again on the right face.

ternal electric field is similar no matter what the shape of the conductor may be. When an external field is applied, the free charge quickly distributes itself until an equilibrium distribution is reached such that the net electric field is zero everywhere inside the conductor. The time to reach equilibrium depends on the conductivity. For copper and other good conductors it is less than about $10^{-16}$ s. For all practical purposes, electrostatic equilibrium is reached instantaneously.

**Question**

5. Distinguish between free charge in a conductor and net charge in a conductor.

# 21-4  Charge and Field at Conductor Surfaces

In this section we use Gauss' law to show that in electrostatic equilibrium (1) any net electric charge on a conductor resides on the surface of the conductor and (2) the electric field just outside the surface of a conductor is perpendicular to the surface and has the magnitude $\sigma/\epsilon_0$, where $\sigma$ is the local surface charge density at that point on the conductor.

To obtain the first result we consider a gaussian surface just inside the actual surface of a conductor in electrostatic equilibrium, as shown in Figure 21-12. Since the electric field is zero everywhere inside the conductor, it is zero everywhere on the gaussian surface, which is chosen to be completely within the conductor. Since $E_n = 0$ at all points on the gaussian surface, the net flux $\oint E_n \, dA$ through the surface must be zero. By Gauss' law this flux equals $1/\epsilon_0$ times the net charge inside the surface. Hence there can be no net charge inside any surface lying completely within the conductor. If there is any net charge on the conductor, it must be on the conductor surface.

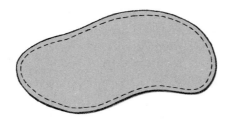

**Figure 21-12**
The dashed line indicates a gaussian surface chosen to be just inside the actual surface of the conductor. Since the electric field is zero everywhere on the gaussian surface, Gauss' law implies that there is zero net charge inside that surface. Any charge on the conductor must therefore reside on the surface of the conductor.

To find the electric field just outside the surface of a conductor, we consider a portion of the conductor surface small enough to be considered flat, with charge density $\sigma$, which has negligible variation over the portion. We construct a cylindrical pillbox gaussian surface (Figure 21-13) with one face just outside the conductor and parallel to its surface, and the other face just inside the conductor. On the surface of the conductor, in equilibrium, the electric field must be perpendicular to the surface. If there were a tangential component of **E**, the free charge on the conductor would move until this component became zero. Since one face of the pillbox surface is just outside the conductor, we can take **E** to be perpendicular to this face. The other face of the pillbox is inside the conductor, where **E** is zero. There is no flux through the cylindrical surface of the pillbox because **E** is tangent to this surface. The flux through the pillbox is thus $E_n A$, where $E_n$ is the field just outside the conductor surface and $A$ is the area of the face of the pillbox. The net charge inside the gaussian surface is $\sigma A$. Gauss' law gives

$$\phi_{net} = \oint E_n \, dA = E_n A = \frac{\sigma A}{\epsilon_0}$$

or

$$E_n = \frac{\sigma}{\epsilon_0} \qquad\qquad\qquad 21\text{-}16$$

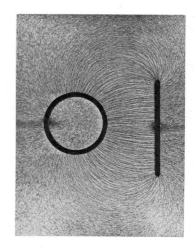

Lines of force for an oppositely charged cylinder and plate, shown by bits of fine thread suspended in oil. Note that the field lines are perpendicular to the conductors and that there are no lines inside the cylinder. (*Courtesy of Harold M. Waage, Princeton University.*)

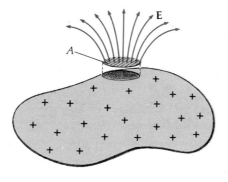

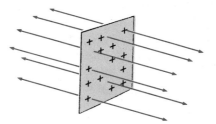

**Figure 21-13**
Pillbox gaussian surface with one face just outside and the other face just inside the surface of a charged conductor. The flux through the gaussian surface is $E_n A$, where $E_n$ is the electric field at the surface of the conductor.

This result is just twice the field produced by an infinite plane of charge. We can understand this result by comparing the flux lines for an infinite plane charge with those for a conducting slab carrying the same charge density. Figure 21-14 shows a very large charged plane sheet. The flux lines are perpendicular to the sheet and point away from it on both sides. If we construct a pillbox gaussian surface with one face to the right of the sheet and the other to the left of the sheet, the net flux through this surface is $E_n 2A$, where $A$ is the area of each face. Setting this net flux equal to $1/\epsilon_0$ times the net charge inside the pillbox $A$, we obtain $E_n = \frac{1}{2}(\sigma/\epsilon_0)$ for the electric field of an infinite sheet.

Figure 21-15 shows a large conducting slab. If we put a net charge on this conductor, it will be distributed equally on both faces. The conducting slab is equivalent to two charged sheets. The electric field due to each of these sheets is $\frac{1}{2}(\sigma/\epsilon_0)$. Between the sheets these fields are in opposite directions, so that their magnitudes subtract, giving zero net field inside the conductor. To the right or left of both sheets the two fields add, giving a magnitude of $\sigma/\epsilon_0$ for the net electric field. The flux lines from the conducting slab are twice as dense as those for a single charged sheet if the charge *densities* are the same. (To produce equal

**Figure 21-14**
Electric field lines for a charge density $\sigma$ on an infinite plane. The electric field has the magnitude $\sigma/2\epsilon_0$.

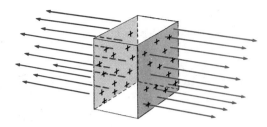

charge densities on a large conducting slab and a large plane sheet, twice as much charge is needed on the slab.)

Similar but slightly more complicated reasoning can be applied to a conductor of arbitrary shape. Consider point $P$ just outside the surface of a conductor, as shown in Figure 21-16. We can consider the charge on the surface of a conductor to consist of two parts, (1) the charge in the immediate neighborhood of point $P$ and (2) all the rest of the charge. Since point $P$ is just outside the surface, the charge in the immediate neighborhood looks like an infinite plane charge. It produces a field of magnitude $\frac{1}{2}(\sigma/\epsilon_0)$ at $P$ and a field of equal magnitude just inside the conducting surface pointing away from the surface. The rest of the charge on the conductor (or elsewhere) must produce a field $\frac{1}{2}(\sigma/\epsilon_0)$ inside the conductor pointing toward the surface, so that the net field inside the conductor is zero. Whereas the field due to this second part of the charge cancels the field *inside* the conductor, it adds to that produced by the neighboring charge just outside the conductor, giving a net field $\frac{1}{2}(\sigma/\epsilon_0) + \frac{1}{2}(\sigma/\epsilon_0) = \sigma/\epsilon_0$ just outside the conductor. (The part of the field due to the distant charges has the same magnitude and direction at points just inside and just outside the surface.) This argument was first given by Laplace in about 1800.

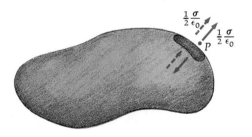

## 21-5   Charging by Induction

A simple and practical method of charging a conductor makes use of the free movement of charge in a conductor. In Figure 21-17 two uncharged metal spheres are in contact. When a charged rod is brought near the spheres, free electrons on one flow to the other. Suppose, for example, that the rod is positively charged. The rod attracts negatively charged electrons, and the sphere nearest the rod acquires electrons from the other, leaving the near sphere with a negative charge and the far sphere with an equal positive charge due to lack of electrons. If the spheres are separated before the rod is removed, they will have equal and opposite charges. A similar result is obtained, of course, with a negatively charged rod which drives electrons from the nearest sphere to the other. In each case, the spheres are charged without being touched by the rod, and the charge on the rod is undisturbed. This is called *electrostatic induction*.

**Figure 21-15**
Electric field lines for a charge density $\sigma$ on each face of a conducting slab. Inside the conductor, the electric field due to one face cancels that due to the other face. Outside the conductor, the field has the magnitude $\sigma/\epsilon_0$ because there are two planes of charge.

**Figure 21-16**
Arbitrarily shaped conductor carrying a charge on its surface. The charge in the vicinity of point $P$ looks like an infinite plane sheet of charge if $P$ is very close to the conductor. This charge produces an electric field of magnitude $\sigma/2\epsilon_0$ both inside and outside the conductor, as indicated by the solid arrows. Since the resultant field inside the conductor must be zero, the rest of the charge must produce a field of equal magnitude, indicated by the dashed arrows. Inside the conductor these fields cancel, but outside at point $P$ they add to give $E = \sigma/\epsilon_0$.

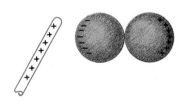

**Figure 21-17**
Charging by induction. The two spherical conductors in contact become oppositely charged because the positively charged rod attracts electrons to the left side, leaving the right side positive. If the spheres are now separated with the rod in place, they retain equal and opposite charges.

A convenient large conductor is the earth itself. For most purposes we can consider the earth to be an infinitely large conductor. When a conductor is connected to the earth, it is said to be *grounded*. This is indicated by a connecting wire and parallel horizontal lines, as in Figure 21-18*b*. We can use the earth to charge a single conductor by induction.

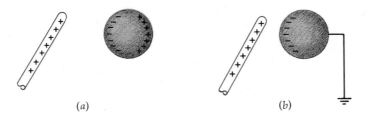

(a)                                    (b)

**Figure 21-18**
(a) The charge on a single conducting sphere is polarized because the electrons are attracted to the positively charged rod, leaving the opposite side positive. (b) When the conductor is grounded, i.e., connected to a very large conductor such as the earth, electrons from the earth neutralize the positive charge, leaving an excess negative charge. This charge remains if the ground connection is broken before the rod is removed.

In Figure 21-18*a* a positively charged rod is brought near a neutral conductor, and the conductor becomes polarized as shown. Free electrons are attracted to the side near the positive rod, leaving the other side with a positive charge. If we ground the conductor while the charged rod is still present, the conductor becomes charged oppositely to the rod because electrons from the earth travel along the connecting wire and neutralize the positive charge on the far side of the conductor. The connection to ground is broken before the rod is removed to complete the charging by induction.

**Questions**

6. An insulating rod is given a charge and then used to charge a set of conductors by induction. What practical limit is there on the number of times the rod can be used without recharging?

7. Can insulators as well as conductors be charged by induction?

*Optional*

## 21-6 Mathematical Derivation of Gauss' Law

Gauss' law can be derived mathematically using the concept of the solid angle. Consider an area element $\Delta A$ on a spherical surface. The solid angle $\Delta \Omega$ subtended by $\Delta A$ at the center of the sphere is defined to be

$$\Delta \Omega = \frac{\Delta A}{r^2}$$

where $r$ is the radius of the sphere. Since $\Delta A$ and $r^2$ both have dimensions of length squared, the solid angle is dimensionless. The unit of solid angle is the *steradian* (sr). Since the total area of a sphere is $4\pi r^2$, the total solid angle subtended by a sphere is

$$\frac{4\pi r^2}{r^2} = 4\pi \text{ sr}$$

There is a close analogy between the solid angle and the ordinary plane angle, which is defined to be the ratio of an element of arc length of a circle, $\Delta s$, divided by the radius of the circle: $\Delta \theta = \Delta s/r$ rad. The total (plane) angle subtended by a circle is $2\pi$ rad.

In Figure 21-19 the area element $\Delta A$ is not perpendicular to the radial lines from point $O$. The unit vector $\hat{\mathbf{n}}$ normal to the area element makes an angle $\theta$ with the unit radial vector $\hat{\mathbf{r}}$. In this case, the solid angle subtended by $\Delta A$ is

$$\Delta \Omega = \frac{\Delta A \, \hat{\mathbf{n}} \cdot \hat{\mathbf{r}}}{r^2} = \frac{\Delta A \cos \theta}{r^2} \qquad \text{21-17}$$

**Figure 21-19**
Area element $\Delta A$ whose normal $\hat{\mathbf{n}}$ is not parallel to the radial line from $O$ to the center of the element. The solid angle subtended by this element at $O$ is $(\Delta A \cos \theta)/r^2$.

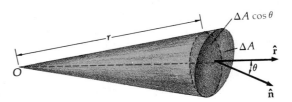

Figure 21-20 shows a point charge $q$ surrounded by a surface of arbitrary shape. To calculate the flux through this surface, we want to find $\mathbf{E} \cdot \hat{\mathbf{n}} \, \Delta A$ for each element of area on the surface and sum over the entire surface. The flux through the area element shown is

$$\Delta \phi = \mathbf{E} \cdot \hat{\mathbf{n}} \, \Delta A = \frac{kq}{r^2} \hat{\mathbf{r}} \cdot \hat{\mathbf{n}} \, \Delta A = kq \, \Delta \Omega$$

The solid angle $\Delta \Omega$ is the same as that subtended by the corresponding area element of a spherical surface of any radius. The sum of the flux through the entire surface is $kq$ times the total solid angle subtended by the closed surface, which is $4\pi$ sr:

$$\phi_{\text{net}} = \oint \mathbf{E} \cdot \mathbf{n} \, dA = kq \oint d\Omega = 4\pi kq = \frac{q}{\epsilon_0}$$

**Figure 21-20**
Point charge $q$ enclosed by surface $S$. The flux through the area element $\Delta A$ is proportional to the solid angle subtended by the area element at the charge. The net flux through the surface found by summing over all area elements is proportional to the total solid angle $4\pi$, which is independent of the shape of the surface.

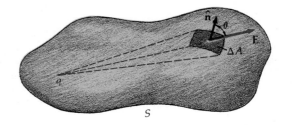

## Review

A. Define, explain, or otherwise identify:

Flux, 616
Gauss' law, 617
Permittivity of free space, 618
Gaussian surface, 618

Electrostatic equilibrium, 622
Free charge in a conductor, 622
Field emission, 622
Charging by induction, 625

B. True or false:

1. If there is no charge in a region of space, the electric field on a surface surrounding the region must be zero everywhere.

2. Gauss' law holds only for symmetric charge distributions.

3. The electric field inside a uniformly charged spherical shell is zero.

4. The electric field inside a conductor is always zero.

5. The result that $E = 0$ inside a conductor in equilibrium can be derived from Gauss' law.

6. A conductor has free electrons only if it has an excess negative charge.

7. If the net charge on a conductor is zero, the charge density must be zero at every point on the surface.

8. Half of the electric field at a point just outside the surface of a conductor is due to the charge on the surface in the immediate vicinity of that point.

Exercises

### Section 21-1, Electric Flux, and Section 21-2, Gauss' Law

1. Consider a uniform electric field $E = 2i$ kN/C. (a) What is the flux of this field through a square of side 10 cm in a plane parallel to the $yz$ plane? (b) What is the flux through the same square if the normal to its plane makes a 30° angle with the $x$ axis?

2. What is the net flux of the uniform electric field of Exercise 1 through a cube of side 10 cm oriented so that its faces are parallel to the coordinate planes?

3. A point charge $+2 \mu C$ is at the center of a sphere of radius 0.5 m. (a) Find the surface area of the sphere. (b) Find the magnitude of the electric field at points on the surface of the sphere. (c) What is the flux of the electric field due to the point charge through the surface of the sphere? (d) Would your answer to part (c) change if the point charge were moved so that it was inside the sphere but not at the center? (e) What is the net flux through a cube of side 1 m which circumscribes the sphere?

4. A single point charge $+2 \mu C$ is at the origin. A spherical surface of radius 2.0 m has its center on the $x$ axis at $x = 5$ m. (a) Sketch lines of force for the point charge. Do any lines enter the spherical surface? (b) What is the net number of lines that leave the spherical surface, counting those that enter as negative? (c) What is the net flux of the electric field due to the point charge through the spherical surface?

5. A positive point charge $q$ is at the center of a cube of side $L$. A large number $N$ of lines of force are drawn from the point charge. (a) How many of the lines pass through the surface of the cube? (b) How many lines pass through each face (assuming none are on the edges or corners)? (c) What is the net outward flux of the electric field through the cubical surface? (d) Use symmetry arguments to find the flux of the electric field through one face of the cube. (e) Which if any of your answers would change if the charge were inside the cube but not at its center?

6. Careful measurement of the electric field at the surface of a black box indicates that the net outward flux through the surface of the box is 6.0 kN·m²/C. (a) What is the net charge inside the box? (b) If the net outward flux through the surface of the box were zero, could you conclude there were no charges inside the box? Why or why not?

7. An electric field is uniform in the positive $x$ direction for positive $x$, and uniform with the same magnitude but in the negative $x$ direction for negative $x$. $E = 200i$ N/C for $x > 0$ and $E = -200i$ N/C for $x < 0$. A right circular cylinder of length 20 cm and radius 5 cm has its center at the origin and its axis along the $x$ axis so that one face is at $x = +10$ cm and the other at $x = -10$ cm. (a) What is the outward flux through each face? (b) What is the flux through the side of the cylinder? (c) What is the net outward flux through the cylindrical surface? (d) What is the net charge inside the cylinder?

8. Can you use Gauss' law to calculate the electric field on the axis of an electric dipole? If so, do it. If not, explain why not.

9. A sphere of radius $R$ has a constant volume charge density $\rho$. Use Gauss' law to derive expressions for the electric field a distance $r$ from the center of the sphere for (a) $r < R$ and (b) $r > R$.

10. A thin-walled cylindrical shell of radius $R$ and infinite length carries a uniform surface charge density $\sigma$. (a) Show that the electric field is zero for $r < R$. (b) Show that for $r > R$ the electric field has the magnitude

$$E = \frac{\sigma R}{\epsilon_0} \frac{1}{r}$$

(c) Show that your result for part (b) is the same as for an infinite line charge of the same charge per unit length.

## Section 21-3, Electric Conductors

11. A penny has a mass of 3 g and is made of copper of molecular mass 63.5 g/mol. Assume one free electron for each copper atom. (a) How many free electrons are there in a penny? (b) How much free charge (in coulombs) is there?

## Section 21-4, Charge and Field at Conductor Surfaces

12. A penny is in an external electric field of magnitude 1 kN/C and direction perpendicular to its faces. (a) Find the charge densities on each face of the penny. (b) If the radius of a penny is taken to be 1 cm, what is the total charge on one face?

13. A metal slab has square faces of side 10 cm. It is placed in an external electric field which is perpendicular to its faces. What is the magnitude of the electric field if the total charge on one of the faces of the slab is 1 nC?

14. A charge of 6 nC is placed uniformly on a square sheet of nonconducting material of side 20 cm in the $yz$ plane. (a) What is the charge density $\sigma$? (b) What is the magnitude of the electric field just to the right and just to the left of the sheet? (c) The same charge is placed on a conducting square slab of side 20 cm and thickness 1 mm. What is the charge density $\sigma$? (Assume the charge distributes itself uniformly on the large square surfaces.) (d) What is the magnitude of the electric field just to the right and just to the left of each face of the slab?

15. A nonconducting spherical shell of outer radius 15 cm carries a net charge of 2 $\mu$C uniformly distributed on its surface. (a) What is the charge density $\sigma$? What is $\mathbf{E}$ (b) just outside the surface of the sphere and (c) just inside the shell? (d) A portion of the shell is removed, leaving a small hole. What is $\mathbf{E}$ just outside and just inside the shell at the hole? Assume the rest of the charge on the shell is undisturbed.

16. An irregularly shaped conductor carries a surface charge. At some point $P$ on the surface the charge density is $\sigma = 1 \ \mu$C/m². (a) What is the electric field just outside the surface at point $P$? (b) The conductor is now replaced by an insulator of exactly the same shape and carrying exactly the same charge density $\sigma$ at each corresponding point. What now is the electric field just outside the surface at point $P$? (c) If your answers for parts (a) and (b) are different, explain why. If they are the same, in what way is the insulator different from the conductor?

## Section 21-5, Charging by Induction

17. Explain, giving each step, how a positively charged insulating rod can be used to give a metal sphere (a) a negative charge and (b) a positive charge. (c) Can the same rod be used to give one sphere a positive charge and another sphere a negative charge without recharging the rod?

18. Figure 21-21 shows a device called an electroscope, which consists of two metal-foil leaves attached to a conducting rod with a conducting knob on top. When uncharged, the leaves hang together vertically. When charged, the leaves repel each other. The divergence of the leaves indicates the amount of charge. (a) A positively charged nonconducting rod is brought close to the knob of the uncharged electroscope, and the leaves diverge. Explain why, with a diagram showing the charge distribution. (b) If the rod is removed, the leaves come back together. If the knob is momentarily grounded with the rod close, the leaves also come together but then diverge when the rod is removed. Explain with a diagram. What sign of net charge is now on the electroscope? (c) With the electroscope charged as in part (b), a negatively charged rod is brought close. Do the leaves come together or diverge more? Explain with a diagram.

**Figure 21-21**
Exercise 18.

### Section 21-6, Mathematical Derivation of Gauss' Law

*There are no exercises for this section.*

### Problems

1. A nonconducting sphere of radius $R$ carries a volume charge density proportional to the distance from the center: $\rho = Ar$ for $r \leq R$, where $A$ is a constant; $\rho = 0$ for $r > R$. (a) Find the total charge by summing the charges in shells of thickness $dr$ and volume $4\pi r^2 \, dr$. (b) Find the electric field $E_r$ both inside and outside the charge distribution and sketch $E_r$ versus $r$.

2. Repeat Problem 1 for a sphere with charge density $\rho = B/r$ for $r \leq R$ and $\rho = 0$ for $r > R$.

3. Repeat Problem 1 for a sphere with charge density $\rho = C/r^2$ for $r \leq R$ and $\rho = 0$ for $r > R$.

4. Consider two infinitely long concentric cylindrical shells. The inner shell has radius $R_1$ and carries a uniform surface charge density $\sigma_1$ while the outer shell has radius $R_2$ and carries a uniform surface charge density $\sigma_2$. (a) Use Gauss' law to find the electric field in the regions $r < R_1$, $R_1 < r < R_2$, and $R_2 < r$. (b) What should the ratio $\sigma_2/\sigma_1$ and the relative sign be for the electric field to be zero at $r > R_2$? What then is the electric field between the shells? (c) Sketch lines of force for the situation in part (b).

5. A spherical shell of radius $R_1$ carries a total charge $q_1$ uniformly distributed on its surface. A second larger spherical shell of radius $R_2$ concentric with the first carries a charge $q_2$ uniformly distributed on its surface. (a) Use Gauss' law to find the electric field in the regions $r < R_1$, $R_1 < r < R_2$, and $R_2 < r$. (b) What should the ratio of the charges $q_1/q_2$ and their relative sign be for the electric field to be zero for $r > R_2$? (c) Sketch the lines of force for the situation in part (b).

6. An infinitely long cylinder of radius $R$ has a uniform volume charge density $\rho$. Show that the electric field has the magnitude

$$E = \begin{cases} \dfrac{R^2 \rho}{2\epsilon_0} \dfrac{1}{r} & r \geq R \\[2ex] \dfrac{\rho}{2\epsilon_0} r & r \leq R \end{cases}$$

Sketch $E$ versus $r$.

7. A thick nonconducting spherical shell of inner radius $r_1$ and outer radius $r_2$ has a uniform volume charge density. Find the total charge and the electric field everywhere.

8. A ring of radius $R$ carries a uniform positive charge density $\lambda$. Figure 21-22 shows a point in the plane of the ring but not at the center. Consider two elements of the ring, of lengths $s_1$ and $s_2$ (indicated in the figure) and at distances

$r_1$ and $r_2$ from point $P$. (a) What is the ratio of the charges of these elements? Which produces the greater field at point $P$? (b) What is the direction of the field at point $P$ due to these elements? What is the direction of the total electric field at point $P$? (c) Suppose that the electric field due to a point charge varied as $1/r$ rather than $1/r^2$. What would the electric field be at point $P$ due to the elements shown? (d) How would your answers differ if point $P$ were inside a spherical shell of charge and the elements were of *area* $s_1$ and $s_2$?

9. Use Gauss' law for the flux of the gravitational field (Equation 21-15) to derive expressions for the gravitational field both inside and outside a sphere of uniform mass density, and sketch $g(r)$ versus $r$.

10. A solid conductor has a cavity as shown in Figure 21-23. A small charge $+q$ is placed in the cavity. (This can be done by placing a charge on a small piece of cork and suspending the cork with an insulating string.) (a) Prove that there is an induced charge density on the inner surface of the cavity such that the total induced charge is $q' = -q$, independent of the location of $q$. (b) Draw lines of force for this problem.

11. A spherical conducting shell with zero net charge has inner radius $a$ and outer radius $b$. A point charge $q$ is placed at the center in the cavity. (a) Use Gauss' law and the properties of conductors in equilibrium to find the electric field in each of the regions $r < a$, $a < r < b$, and $b < r$. (b) Draw lines of force for this situation. (c) Describe the charge density on the outer surface ($r = b$) of the sphere. How would this charge density be affected if the point charge in the cavity were moved away from the center? Sketch lines of force for the case in which the point charge is not at the center of the cavity.

12. The electrostatic force on a charge at some point is the product of the charge and the electric field due to all other charges. Consider a small charge on the surface of a conductor, $\Delta q = \sigma \, \Delta A$. (a) Show that the electrostatic force on the charge is $\sigma^2 \, \Delta A / 2\epsilon_0$. (b) Explain why this is just half $\Delta q \, E$, where $E = \sigma/\epsilon_0$ is the electric field just outside the conductor at that point. (c) The force per unit area is called the electrostatic stress. Find the stress when a charge of 2 $\mu$C is placed on a conducting sphere of radius 10 cm.

13. (a) Show that for both the infinite plane charge and the spherical shell, the electric field is discontinuous at the surface charge by the amount $\sigma/\epsilon_0$. (b) Prove that, in general, when there is a surface charge $\sigma$, the electric field component perpendicular to the surface is discontinuous by the amount $\sigma/\epsilon_0$. Do this by constructing a gaussian pillbox with faces on each side of the surface. Use Gauss' law to find $E_2 - E_1$, where $E_2$ is the normal component of $E$ on one side and $E_1$ is that on the other side of the surface.

14. A nonconducting sphere of radius $a$ has a spherical cavity of radius $b$ with center at the point $x = b$, $y = 0$, as shown in Figure 21-24. The sphere contains a uniform charge density $\rho$. Show that the electric field in the cavity is uniform and is given by $E_y = 0$, $E_x = \rho b/3\epsilon_0$. Hint: Replace the cavity by spheres of equal positive and negative charge densities.

15. A very long cylindrical shell of inner radius $a$ and outer radius $b$ carries a uniform charge density

$$\rho = \begin{cases} 0 & r < a \\ \rho_0 & a < r < b \\ 0 & r > b \end{cases}$$

Find the electric field everywhere.

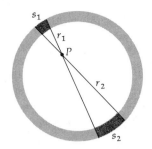

**Figure 21-22**
Problem 8.

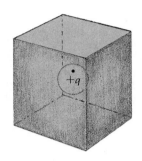

**Figure 21-23**
Problem 10.

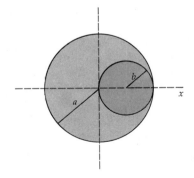

**Figure 21-24**
Problem 14.

# CHAPTER 22    Electric Potential

---

**Objectives**    After studying this chapter you should:

1.  Be able to give definitions of electric potential difference, electric potential, and electrostatic potential energy.

2.  Be able to calculate the potential difference between two points, given the electric field in the region.

3.  Be able to give the definition of the electronvolt energy unit and the conversion factor between it and the joule.

4.  Be able to calculate the electrostatic potential energy of a system of point charges.

5.  Be able to sketch the equipotential surfaces given a pattern of electric field lines.

6.  Be able to calculate the electric potential for various charge distributions.

7.  Be able to discuss the phenomena of charge sharing, dielectric breakdown, and corona discharge.

---

The concept of energy, which was so important in our study of mechanics, is also important in electricity. In Chapter 6 we discussed the fact that any central force such as the electrostatic force given by Coulomb's law is conservative; i.e., the work done on a particle by this force as the particle moves from one point to another depends only on the initial and final positions and not on the path taken. There is thus a potential-energy function associated with the force. In this chapter we discuss the concept of electrostatic potential energy and a related concept of electric potential.

# 22-1 Potential Difference

The work we must do to move a body at constant speed from one point to another in a conservative force field equals the change in the potential energy of the body. Consider a test charge $q_0$ in an electrostatic field **E** produced by some system of charges. The electric force on the charge is $q_0\mathbf{E}$. If we wish to move the charge (at constant speed), we must exert a force $-q_0\mathbf{E}$ on it. If we give the charge a small displacement $d\boldsymbol{\ell}$, the work we do, $dW = -q_0\mathbf{E} \cdot d\boldsymbol{\ell}$, equals the increase in the potential energy of the test charge. The change in the potential-energy function, $dU$, is thus defined by

$$dU = -q_0\mathbf{E} \cdot d\boldsymbol{\ell} \qquad\qquad 22\text{-}1$$

If the charge is moved from some initial point $a$ to some final point $b$, the change in its potential energy is

$$U_b - U_a = \int_a^b dU = -\int_a^b q_0\mathbf{E} \cdot d\boldsymbol{\ell} \qquad\qquad 22\text{-}2$$

The potential-energy change is proportional to the test charge $q_0$. The potential-energy change per unit charge is called the *potential difference* $\Delta V$. For an infinitesimal displacement $d\boldsymbol{\ell}$, we obtain the potential difference from Equation 22-1 by dividing by $q_0$:

*Potential difference defined*

$$dV = \frac{dU}{q_0} = -\mathbf{E} \cdot d\boldsymbol{\ell} \qquad\qquad 22\text{-}3$$

and for a finite displacement from point $a$ to point $b$,

$$\Delta V = V_b - V_a = \frac{\Delta U}{q_0} = -\int_a^b \mathbf{E} \cdot d\boldsymbol{\ell} \qquad\qquad 22\text{-}4$$

*The potential difference $V_b - V_a$ is the work per unit charge necessary to move a test charge at constant speed from point $a$ to point $b$.*

Equation 22-4 defines the change in the function $V$, which is called the *electric potential* (or sometimes just the *potential*). Like the potential energy $U$, only the *change* in the electric-potential function $V$ is significant. The value of the potential function at any point is determined by arbitrarily choosing $V$ to be zero at some convenient point. If we choose $U$ and $V$ to be zero at the same point, which is customary, the electric potential at any point equals the potential energy of the charge divided by the charge $q_0$.

*Potential*

The SI unit of potential and potential difference is the joule per coulomb, called a volt (V):

$$1 \text{ V} = 1 \text{ J/C} \qquad\qquad 22\text{-}5$$

*Volt defined*

From Equation 22-4 we note that the dimensions of potential are also those of electric field times distance. Thus the unit of electric field $E$, the newton per coulomb, is also equal to a volt per metre:

$$1 \text{ N/C} = 1 \text{ V/m} \qquad\qquad 22\text{-}6$$

If we place a positive test charge $q_0$ in an electric field **E** and release the charge, it will accelerate in the direction of **E** along the field line. The kinetic energy of the charge will increase, and its potential energy will decrease. *The electric field lines therefore point in the direction of decreasing*

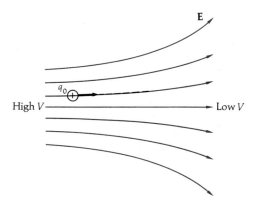

**E**

High $V$ ——————————————→ Low $V$

$q_0$

**Figure 22-1**
When a positive test charge
$+q_0$ is placed in an electric
field, it accelerates in the
direction of the field. Its
kinetic energy increases, and
its potential energy decreases.
The field lines point in the
direction of decreasing
potential.

*electric potential* (Figure 22-1). For example, if we have a constant electric field of 10 V/m in the $x$ direction, the potential decreases by 10 V in each metre in the $x$ direction.

The simplest charge distribution is a single point charge $Q$ at the origin. The electric field of such a charge is radially outward (assuming $Q$ to be positive) and has the magnitude $kQ/r^2$. We can find the electric potential at any distance from this charge from the definition of potential difference (Equations 22-3 and 22-4). Consider a test charge $q_0$ at an initial distance $r_a$ from the charge $Q$. Let us give it a small displacement in the radial direction away from the origin. Then $\mathbf{E} \cdot d\boldsymbol{\ell} = E_r\, dr$, and the change in potential energy $dU$ is

$$dU = -q_0\mathbf{E} \cdot d\boldsymbol{\ell} = -q_0 E_r\, dr = -\frac{kq_0Q}{r^2}\, dr \qquad 22\text{-}7$$

and the change in the potential $dV$ is

$$dV = -\mathbf{E} \cdot d\boldsymbol{\ell} = -E_r\, dr = -\frac{kQ}{r^2}\, dr \qquad 22\text{-}8$$

Therefore the change in the potential from point $r_a$ to point $r_b$ is

$$\Delta V = V_b - V_a = -\int_a^b \frac{kQ}{r^2}\, dr = \frac{kQ}{r_b} - \frac{kQ}{r_a}$$

If we define the potential energy and the potential to be zero at an infinite distance from the point charge ($r_b = \infty$), the potential at point $r_a$ is $V_a = kQ/r_a$. At a general distance $r$ from the point charge $Q$ the electric potential is

$$V(r) = \frac{kQ}{r} \qquad 22\text{-}9 \qquad \textit{Potential of a point charge}$$

The potential is positive or negative depending on the sign of the charge $Q$. The potential energy of the two-charge system is

$$U(r) = q_0 V(r) = \frac{kQq_0}{r} \qquad 22\text{-}10$$

If we bring a test charge $q_0$ from a very large distance from the origin ($r = \infty$) to some distance $r$ (Figure 22-2), the work we must do against the electric force equals the increase in the potential energy, which is given by Equation 22-10.

*The electric potential $V = kQ/r$ is the work per unit charge that must be done to bring a positive test charge from infinity to the distance $r$ at constant speed.*

*Potential is work per unit charge*

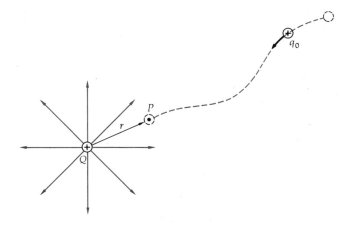

**Figure 22-2**
The work required to bring a
test charge $q_0$ at constant
speed from infinity to a point
$P$ that is a distance $r$ from a
charge $Q$ at the origin is
$kQq_0/r$. The work per unit
charge is $kQ/r$, the electric
potential at point $P$ relative to
zero potential at infinity.

**Example 22-1** What is the electric potential at the distance of the first Bohr orbit, $r = 0.529 \times 10^{-10}$ m, from a proton? What is the potential energy of the electron and proton at this separation?

The charge of the proton is $q = 1.6 \times 10^{-19}$ C. Equation 22-9 gives

$$V = \frac{kq}{r} = \frac{(9 \times 10^9)(1.6 \times 10^{-19})}{0.529 \times 10^{-10}} = 27.2 \text{ V}$$

The charge of the electron is $-e = -1.6 \times 10^{-19}$ C. In SI units the potential energy of the electron a distance $0.529 \times 10^{-10}$ m from a proton is

$$U = qV = (-1.6 \times 10^{-19} \text{ C})(27.2 \text{ V}) = -4.36 \times 10^{-18} \text{ J}$$

A convenient unit of energy is the *electronvolt* (eV), defined to be one volt times the magnitude of the charge of the electron. Since the electron has a charge of $-1.6 \times 10^{-19}$ C, and 1 C·V equals 1 J (Equation 22-5), we have

$$1 \text{ eV} = 1.6 \times 10^{-19} \text{ C·V} = 1.6 \times 10^{-19} \text{ J} \qquad \text{22-11}$$    *Electronvolt defined*

In Example 22-1 the potential energy of the electron and proton at the separation of the first Bohr orbit is 27.2 eV.

**Questions**

1. Explain in your own words the distinction between electric potential and electric potential energy.

2. If a charge is moved a small distance in the direction of an electric field, does its electric *potential energy* increase or decrease? Does your answer depend on the sign of the charge?

3. If a charge is moved a small distance in the direction of an electric field, does the electric *potential* increase or decrease? Does your answer depend on the sign of the charge?

4. What direction can you move relative to an electric field so that the electric potential does not change?

5. A positive charge is released from rest in an electric field. Will it move toward a region of greater or smaller electric potential?

## 22-2 Potential of a System of Point Charges and Electrostatic Potential Energy

If we have a system of point charges, the potential at some point (relative to zero potential at an infinite distance from the system) equals the sum of the potentials at that point due to the individual point charges. Consider, for example, two point charges $q_1$ and $q_2$ at some arbitrary positions. To bring a test charge $q_0$ from an infinite distance away to some position near these charges, we must do work against the resultant electric field. Since this field is just the vector sum of the individual electric fields due to each of the charges, this work is the sum of the work that must be done against the field due to each charge separately, which is $kq_1q_0/r_{10} + kq_2q_0/r_{20}$, where $r_{10}$ is the distance from $q_0$ to $q_1$ and $r_{20}$ is that from $q_0$ to $q_2$. The potential at some point relative to zero potential at infinity is therefore the sum of the potentials due to each charge. In general, for a system of point charges, the potential at some point $P$ is given by

$$V = \sum_i \frac{kq_i}{r_{i0}} \qquad\qquad 22\text{-}12$$

where $r_{i0}$ is the distance between the field point $P$ and the $i$th charge $q_i$.

**Example 22-2**  Three positive 2-$\mu$C point charges are at the corners of a square of side 3 m (Figure 22-3). What is the potential $V$ at the fourth, unoccupied corner of the square? How much work is needed to bring up a fourth positive charge of 2 $\mu$C and place it at the fourth corner of the square?

The unoccupied corner of the square is 3 m from two of the charges and $3\sqrt{2}$ m from the other charge. The potential at that point, according to Equation 22-12, is then

$$V = \frac{kq_1}{r_1} + \frac{kq_2}{r_2} + \frac{kq_3}{r_3}$$

$$= (9 \times 10^9) \left( \frac{2 \times 10^{-6}}{3} + \frac{2 \times 10^{-6}}{3} + \frac{2 \times 10^{-6}}{3\sqrt{2}} \right)$$

$$= 1.62 \times 10^4 \text{ V}$$

Note that an alternative expression for the units of the Coulomb constant $k$ is the volt-metre per coulomb (V·m/C).

The work needed to bring a fourth charge to the unoccupied corner of the square is the product of the charge and the potential there due to the other charges:

$$W = qV = (2 \times 10^{-6} \text{ C})(1.62 \times 10^4 \text{ V}) = 3.24 \times 10^{-2} \text{ J}$$

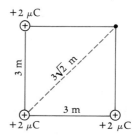

**Figure 22-3**
Three equal positive point charges at the corners of a square of side 3 m for Example 22-2.

In the above example we calculated the work needed to bring up a fourth charge to a position near a system of three other charges. Since the electrostatic force is conservative, this work does not depend on the path taken by the charged particle but only on its position relative to the other charges. We did not calculate the work needed to assemble the original three charges. This work is also independent of the paths taken by the charges and of the order of assembly; it depends only on the initial and final positions of the charges. The work needed to bring two charges $Q$ and $q_0$ from an infinite separation to a final separation $r$ is given by Equation 22-10. This work is the electrostatic potential energy of the two-charge system relative to zero potential energy when the charges are infinitely separated. In general:

*The electrostatic potential energy of a system of point charges is the work needed to bring the charges from an infinite separation to their final positions.*

*Electrostatic potential energy*

---

**Example 22-3** Find the electrostatic potential energy of three point charges $q_1$, $q_2$, and $q_3$ separated by distances $r_{12}$, $r_{23}$, and $r_{13}$ (Figure 22-4).

The work needed to bring two charges $q_1$ and $q_2$ from an infinite separation to a separation of $r_{12}$ is $kq_1q_2/r_{12}$. The work needed to bring the third charge $q_3$ to a distance $r_{13}$ from $q_1$ and $r_{23}$ from $q_2$ is $W_3 = kq_1q_3/r_{13} + kq_2q_3/r_{23}$. The total work to assemble the three charges is therefore

$$W = \frac{kq_1q_2}{r_{12}} + \frac{kq_1q_3}{r_{13}} + \frac{kq_2q_3}{r_{23}}$$

This result is independent of the order in which the charges are brought to their final positions.

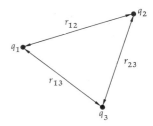

**Figure 22-4**
Three charges separated by arbitrary distances. The electrostatic potential energy of the system is the work required to bring the charges from an initial infinite separation to their final positions.

---

## 22-3  Electric Field and Potential: Equipotential Surfaces

The electric field lines point in the direction of decreasing potential. Consider a small displacement $\Delta\ell$ in an arbitrary electric field $\mathbf{E}$. The work done against the electric field in giving a test charge $q_0$ such a displacement is $\Delta U = -q_0\mathbf{E} \cdot \Delta\ell$, and the change in potential is

$$\Delta V = -\mathbf{E} \cdot \Delta\ell \qquad\qquad 22\text{-}13$$

If the displacement $\Delta\ell$ is perpendicular to the electric field, the potential does not change. The greatest change in $V$ occurs when the displacement $\Delta\ell$ is parallel or antiparallel to $\mathbf{E}$. When $\Delta\ell$ is parallel to $\mathbf{E}$, we have

$$\Delta V = -E_\ell \, \Delta\ell$$

If we divide by $\Delta\ell$ and take the limit as $\Delta\ell$ goes to zero, we have

$$E_\ell = -\frac{dV}{d\ell} \qquad d\ell \text{ parallel to } \mathbf{E} \qquad\qquad 22\text{-}14$$

A vector which points in the direction of the greatest change in a scalar function and which has a magnitude equal to the derivative of that function with respect to distance in that direction is called the *gradient* of the function. The electric field is the negative gradient of the potential. The field lines point in the direction of the greatest decrease in the potential function.

Figure 22-5 shows the field lines of the electric field due to a point charge $Q$ at the origin. If we move a test charge perpendicular to these lines, no work is done and the potential does not change. A surface on which the electric potential is constant is called an *equipotential surface*.

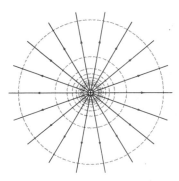

**Figure 22-5**
Lines of force and equipotential surfaces of a point charge. The equipotentials are spherical surfaces. The lines of force are everywhere perpendicular to the equipotentials.

For the potential $V = kQ/r$ produced by a point charge at the origin, the surfaces defined by $r =$ constant are equipotential surfaces. These surfaces are concentric spherical surfaces. The lines of force are always perpendicular to an equipotential surface. For a point charge at the origin, the lines of force are radial lines, and the equipotential surfaces are spheres. If we have a uniform electric field in the $x$ direction, e.g., that produced by an infinite plane charge in the $yz$ plane, the lines of force are parallel lines in the $x$ direction and the equipotential surfaces are planes parallel to the $yz$ plane. Figure 22-6 shows equipotential surfaces for the field produced by an electric dipole. In a two-dimensional drawing of lines of force, the curve representing the intersection of an equipotential surface with the plane of the paper is shown by sketching a continuous curve everywhere perpendicular to the lines of force.

Since the electric field at the surface of a conductor in equilibrium is perpendicular to the surface, the conductor surface is itself an equipotential surface; since the electric field is zero inside a conductor (in electrostatic equilibrium), the potential is constant everywhere on and inside a conductor.

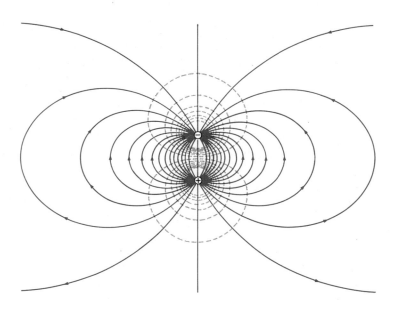

**Figure 22-6**
Lines of force and equipotential surfaces of a dipole. The lines of force are always perpendicular to the equipotential surfaces.

## 22-4 Calculation of Electric Potential

There are two ways of calculating the potential at some point due to a given charge distribution. If the electric field $\mathbf{E}$ is already known, the potential $V$ can be found directly from its definition, Equation 22-4. Consider, for example, an infinite-plane charge distribution of surface charge density $\sigma$. Let the charge be in the $yz$ plane. In Chapter 20 we found that the electric field for an infinite plane charge has magnitude $2\pi k\sigma = \sigma/2\epsilon_0$ (where $k$ and $\epsilon_0$ are related by $k = 1/4\pi\epsilon_0$), independent of the distance from the plane. For positive $x$, the electric field is

$$E_x = \frac{\sigma}{2\epsilon_0}$$

Let $V(x)$ be the potential at some distance $x$ from the $yz$ plane and $V(0)$ be that on the $yz$ plane at $x = 0$. Integrating from $x = 0$ to some general point $x$, we obtain from Equation 22-4

$$V(x) - V(0) = -\int_0^x E_x\, dx = -\int_0^x \frac{\sigma}{2\epsilon_0}\, dx = -\frac{\sigma}{2\epsilon_0} x$$

or

$$V(x) = V(0) - \frac{\sigma}{2\epsilon_0} x \qquad\qquad 22\text{-}15$$

The potential decreases linearly with distance from the plane. Since it does not approach any limiting value as $x$ approaches infinity, we cannot choose the potential to be zero at $x = \infty$. We can, however, choose $V$ to be zero at $x = 0$; then $V(0) = 0$ and $V(x) = -(\sigma/2\epsilon_0)x$. Alternatively we could choose the potential to have some other value at $x = 0$, say 100 V, in which case the constant $V(0)$ would be 100 V. Since only differences in potential are important, it does not matter what value is chosen for the potential at $x = 0$.

The second method of finding the potential due to a given charge distribution is to compute the potential directly from Equation 22-12. In general, for a system of point charges, the potential at a point $P$ is given by

$$V = \sum_i \frac{kq_i}{r_{i0}} \qquad\qquad 22\text{-}16$$

where $r_{i0}$ is the distance between the field point $P$ and the $i$th charge $q_i$. If we have a continuous charge distribution of finite size, the potential at some point $P$ can be found by treating a small element of the charge $dq$ as a point charge and integrating:

$$V = \int \frac{k\, dq}{r} \qquad\qquad 22\text{-}17$$

where $r$ is the distance from the charge element $dq$ to point $P$. This result is based on the assumption that the potential is zero at infinite distance from the charge distribution. Equation 22-17 can be used only if the charge distribution is of finite extent, so that the potential approaches a limiting value at infinity which can be chosen to be zero. Sufficiently far from any finite charge distribution, the charge distribution looks like a point charge. Then the electric field decreases as $1/r^2$, and the potential decreases as $1/r$, which approaches zero as $r$ approaches infinity. If the charge distribution is not finite, however, the electric field does not decrease as $1/r^2$ and in general the potential does not approach any limiting value at infinity. We saw an example of this for the field of an infinite plane charge with electric field independent of the distance from the plane. Another example, which we consider below, is the infinite line charge. For such charge distributions of infinite extent we cannot use Equation 22-17 to find the potential $V$. Instead we find the electric field **E** first and then find $V$ from Equation 22-4, choosing $V$ to be zero at any convenient point.

---

**Example 22-4** Find the potential on the axis of a ring of radius $a$ and charge $Q$.

An element of charge $dq$ is shown in Figure 22-7. The distance from this charge element to the field point on the axis of the ring is $s = \sqrt{x^2 + a^2}$, which is the same for all elements of charge on the ring. The potential due to the ring is thus

$$V = \int \frac{k\,dq}{\sqrt{x^2 + a^2}} = \frac{k}{\sqrt{x^2 + a^2}} \int dq = \frac{kQ}{\sqrt{x^2 + a^2}} \qquad 22\text{-}18$$

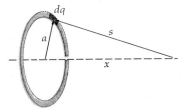

**Figure 22-7**
Geometry for the calculation of electric potential at a point on the axis of a uniformly charged ring.

The electric field can be calculated from Equation 22-14:

$$E_x = -\frac{dV}{dx} = -kQ(-\tfrac{1}{2})(x^2 + a^2)^{-3/2}2x = \frac{kQx}{(x^2 + a^2)^{3/2}}$$

which agrees with the direct calculation from Coulomb's law in Chapter 20.

---

**Example 22-5** Find the potential on the axis of a disk of uniform surface charge density $\sigma$.

Let the disk axis be the $x$ axis and consider positive $x$ only. We can treat the disk as a set of ring charges and use our result from Example 22-4. Consider a ring of radius $r$ and width $dr$ (Figure 22-8). The area of this ring is $2\pi r\,dr$, and its charge is $dq = \sigma\,dA = \sigma 2\pi r\,dr$. The potential on the axis of the disk is found by summing from $r = 0$ to $r = R$:

$$V = \int_{r=0}^{r=R} \frac{k\,dq}{\sqrt{x^2 + r^2}} = \int_0^R \frac{k\sigma 2\pi r\,dr}{\sqrt{x^2 + r^2}} = k\sigma\pi \int_0^R (x^2 + r^2)^{-1/2}2r\,dr$$

This integral is of the form $\int u^n\,du$, with $u = x^2 + r^2$ and $n = -\tfrac{1}{2}$:

$$V = k\sigma\pi \frac{(x^2 + r^2)^{+1/2}}{\tfrac{1}{2}}\Big]_0^R = 2\pi k\sigma[(x^2 + R^2)^{1/2} - x] \qquad 22\text{-}19$$

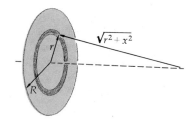

**Figure 22-8**
Geometry for the calculation of the electric potential at a point on the axis of a uniformly charged disk. The disk is divided into rings of area $2\pi r\,dr$.

Again, the electric field can be obtained from Equation 22-14:

$$E_x = -\frac{dV}{dx} = 2\pi k\sigma\left(1 - \frac{x}{\sqrt{x^2 + R^2}}\right) \qquad 22\text{-}20$$

which agrees with the result obtained directly from Coulomb's law in Chapter 20. The potential function (Equation 22-19) becomes infinite as $R$ becomes infinite. As discussed, we cannot use Equation 22-17 with the potential at infinity defined to be zero when the charge distribution is of infinite extent. The potential function for an infinite-plane charge distribution is found from the electric field, as before.

---

**Example 22-6** Find the potential due to a spherical shell of charge $Q$.

We consider a spherical shell of radius $R$ having a uniform surface charge density $\sigma$. We are interested in the potential at all points inside and outside the shell. Since this shell is of finite extent, we can use Equation 22-17. The calculation is quite difficult and is given at the end of this section. For this problem it is much easier to find the potential from the electric field, which we already know.*

Outside the spherical shell, the electric field is radial and the same as if all the charge were at the origin:

$$E_r = \frac{kQ}{r^2} \qquad r > R$$

where $Q = 4\pi R^2 \sigma$ is the total charge on the shell. The work needed to bring a test charge from infinity to a point a distance $r$ from the center of the shell for $r > R$ is the same as that for a point charge at the origin; the potential is therefore given by

$$V(r) = \frac{kQ}{r} \qquad r > R$$

where we have chosen the potential to be zero at infinite $r$. Inside the spherical shell the electric field is zero. It therefore takes no work to move a test charge from point to point inside the shell, and the potential inside the shell is constant. As $r$ approaches $R$ from outside the shell, the potential approaches $kQ/R$. Hence the constant value of $V$ inside must be $kQ/R$ to make $V(r)$ continuous. Thus

$$V(r) = \begin{cases} \dfrac{kQ}{R} & r \leq R \\[2mm] \dfrac{kQ}{r} & r \geq R \end{cases} \qquad\qquad 22\text{-}21 \qquad \textit{Potential of a spherical shell}$$

This potential function is sketched in Figure 22-9.

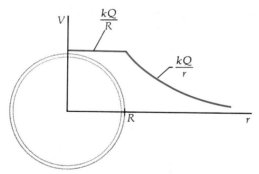

**Figure 22-9**
Electric potential of a uniformly charged spherical shell of radius $R$ as a function of distance $r$ from the center of the shell. Inside the shell the potential has the constant value $kQ/R$. Outside the shell the potential is the same as that due to a point charge at the center of the sphere.

A common mistake is to think that the potential must be zero inside a spherical shell because the electric field is zero. Actually zero electric field merely implies that the potential does not change. Consider a spherical shell with a small hole so that we can move a test charge in and out of the shell. If we move the test charge from an infinite distance to the shell, the work per charge we must do against the electric force is $kQ/R$. Inside the shell there is no electric field, and so it takes no work to move the test charge inside the shell. The total amount of work per charge it takes to bring the test charge from infinity to any point inside the shell is just the work it takes to bring it up to the shell radius $R$: $kQ/R$. The potential is therefore $kQ/R$ everywhere inside the shell.

---

* The direct calculation of the electric field for a spherical shell is even more difficult (see Example 20-13); however, we were able to infer the properties of this electric field from the consideration of symmetry and the lines of force. If neither the electric field nor the electric potential were known, it would be easiest to find the potential directly, as shown at the end of this section, and then find $E$ from $E_r = -dV/dr$.

**Example 22-7** Find the potential due to a spherical ball of uniform charge density.

Let the radius of the ball be $R$ and the total charge be $Q = \frac{4}{3}\pi R^3 \rho$, where $\rho$ is the charge per unit volume. We found the electric field for this charge distribution in Example 20-14. Outside the ball the electric field is $kQ/r^2$, the same as that due to a point charge. Thus the potential outside is $kQ/r$. Inside the charged ball the electric field is

$$E_r = \frac{kQ}{R^3} r$$

Since the electric field inside the ball is not zero, the potential will not be constant; and since the electric field points radially outward inside the ball, the potential must increase as we move inward; i.e., work must be done on a test charge to move it from the outside edge of the ball inward. The potential therefore has its greatest value at the origin. We can find the potential inside from Equation 22-4. Integrating from $r = 0$ to a general point $r$ inside the ball gives

$$V(r) - V(0) = -\int_0^r E_r \, dr = -\int_0^r \frac{kQ}{R^3} r \, dr = -\frac{1}{2}\frac{kQ}{R^3} r^2 \qquad 22\text{-}22$$

The potential at the origin, $V(0)$, cannot be chosen to be zero because we have already chosen the potential to be zero at $r = \infty$. We find $V(0)$ by requiring the potential to be continuous at $r = R$. At this point at the edge of the ball the potential is $kQ/R$. Setting $r = R$ in Equation 22-22, we have

$$\frac{kQ}{R} - V(0) = -\frac{1}{2}\frac{kQ}{R^3} R^2 = -\frac{1}{2}\frac{kQ}{R}$$

or

$$V(0) = \frac{3}{2}\frac{kQ}{R}$$

The potential inside the ball is then

$$V = \frac{3kQ}{2R} - \frac{kQr^2}{2R^3} = \frac{kQ}{2R}\left(3 - \frac{r^2}{R^2}\right) \qquad 22\text{-}23$$

The complete potential function is sketched in Figure 22-10.

**Example 22-8** Find the potential due to a uniform infinite line charge.

Let the charge per unit length be $\lambda$. Since this charge distribution extends to infinity, we cannot use Equation 22-17 to find the potential. The potential does not approach a limiting value at infinity and therefore cannot be chosen to be zero there. We use Equation 22-4. Let $E_r$ be the component of the electric field vector that points away from the line charge. $E_r$ is most easily found using Gauss' law, as in Chapter 21. The result is $E_r = 2k\lambda/r$. If $V(a)$ is the potential at some distance $a$, the potential at distance $r$ is obtained from Equation 22-4:

$$V(r) - V(a) = -\int_a^r E_r \, dr = -\int_a^r \frac{2k\lambda}{r} \, dr = -2k\lambda \ln\frac{r}{a}$$

For a positive line charge, the electric field lines point away from the line and the potential decreases with increasing distance from the line charge. At large values of $r$ the potential decreases without limit. The potential therefore cannot be chosen to be zero at $r = \infty$. Instead, it is

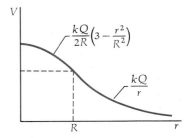

**Figure 22-10**
Plot of the electric potential $V$ of a ball of charge of uniform density versus distance $r$ from the center of the ball. $V$ has its maximum value at the center of the ball. Outside the ball, $V$ is the same as that due to a point charge at the center of the sphere.

chosen to be zero at any convenient point $r = a$. Then $V(a) = 0$, and the potential is given by

$$V(r) = -2k\lambda \ln \frac{r}{a} \qquad\qquad 22\text{-}24$$

## Questions

6. In Example 22-4, does it matter whether the charge $Q$ is uniformly distributed around the ring? Would either $V$ or $E_x$ be different if it were not?

7. If the electric potential is constant throughout a region of space, what can you say about the electric field in that region?

*Optional*

### Direct Calculation of $V(r)$ for a Spherical Shell

This calculation is similar to the calculation of the electric field given in Example 20-13, but slightly easier. The geometry is shown in Figure 22-11. We first consider a point $P$ outside the shell. By symmetry, the potential can depend only on the distance $r$ from the center of the shell. We choose for our charge element the strip shown, which has circumference $2\pi R \sin \theta$ and width $R\, d\theta$. The area of this strip is $dA = 2\pi R^2 \times \sin \theta\, d\theta$, and the charge is $dq = \sigma\, dA = \sigma 2\pi R^2 \sin \theta\, d\theta$. The potential due to this element of charge is

$$dV = \frac{k\, dq}{s} = \frac{k\sigma 2\pi R^2 \sin \theta\, d\theta}{s} \qquad\qquad 22\text{-}25$$

where $s$ is the distance from this ring to the point $P$, as shown in the figure. The total potential produced by the complete shell is obtained by summing over all the circular strips on the shell, i.e., by integrating from $\theta = 0$ to $\theta = 180°$. The quantities $\theta$ and $s$ are related by the law of cosines:

$$s^2 = r^2 + R^2 - 2rR \cos \theta \qquad\qquad 22\text{-}26$$

Differentiating gives

$$2s\, ds = +2rR \sin \theta\, d\theta \qquad\qquad 22\text{-}27$$

Instead of substituting $s$ from Equation 22-26 into Equation 22-25 and integrating over $\theta$, it is easier to substitute $\sin \theta\, d\theta$ from Equation 22-27 and integrate from $s = r - R$ at $\theta = 0$ to $s = r + R$ at $\theta = 180°$. We then have

$$dV = \frac{k\sigma 2\pi R^2 (s\, ds/rR)}{s} = \frac{k\sigma 2\pi R\, ds}{r}$$

Integrating from $s = r - R$ to $s = r + R$ gives

$$V = \int_{r-R}^{r+R} \frac{k\sigma 2\pi R}{r}\, ds = \frac{k\sigma 2\pi R}{r}\left[(r + R) - (r - R)\right]$$

$$= \frac{k\sigma 2\pi R}{r}\, 2R = \frac{k(4\pi R^2 \sigma)}{r} = \frac{kQ}{r}$$

The potential at a point outside the spherical shell is the same as that due to a point charge $Q$ at the origin. The calculation of the potential at a point inside the shell is the same except that now $s$ varies from

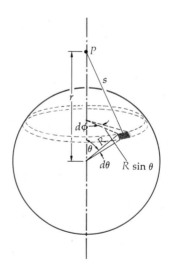

**Figure 22-11**
Geometry for calculating the potential of a spherical shell of charge. The area of the element indicated is $dA = 2\pi R^2 \sin \theta\, d\theta$, and its distance from the field point $P$ is related to $r$, $R$, and $\theta$ by the law of cosines: $s^2 = r^2 + R^2 - 2rR \cos \theta$.

$s = R - r$ at $\theta = 0$ to $s = R + r$ at $\theta = 180°$. The potential at a point inside the shell is thus given by

$$V = \int_{R-r}^{R+r} \frac{k\sigma 2\pi R}{r} \, ds = \frac{k\sigma 2\pi R}{r} [(R + r) - (R - r)]$$

$$= \frac{k\sigma 2\pi R}{r} (2r) = \frac{k\sigma 4\pi R^2}{R} = \frac{kQ}{R}$$

Inside the sphere the potential is constant and independent of the distance $r$ and has the value $kQ/R$.

## 22-5  Charge Sharing

In general, two conductors which are separated in space will not be at the same potential. The potential difference between the conductors depends on the geometrical shapes of the conductors, their separation, and the net charge on each conductor. When two conductors are brought into contact, the charge on the conductors distributes itself so that in electrostatic equilibrium the electric field is zero inside both conductors. In this situation the two conductors in contact may be considered a single conductor. In equilibrium each conductor has the same potential. The transfer of charge from one conductor to another is called *charge sharing*. Coulomb used the method of charge sharing to produce various charges of known ratios to some original charge, in his experiment to find the force law between two small (point) charges.

*Charge sharing*

Consider a spherical conductor carrying a charge $+Q$. The lines of force outside the conductor point radially outward, and the potential of the conductor relative to infinity is $kQ/R$. If we bring up a second identical but uncharged conductor, the potential and field lines will change: negative electrons on the uncharged conductor will be attracted to the positive charge $Q$, leaving the near side of the uncharged conductor with a negative charge and the far side with a positive charge (Figure 22-12). This charge separation on the neutral conductor will affect the originally uniform charge distribution on the positive conductor.

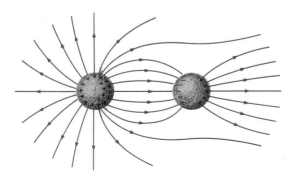

**Figure 22-12**
Electric field lines for a charged spherical conductor near an uncharged spherical conductor.

Although the detailed calculation of the charge distributions and potential in this case is quite complicated, we can see that some of the field lines leaving the positive conductor will end on the negative charge on the near side of the neutral conductor, and an equal number of lines will leave the far side of that conductor. Since the potential decreases as we move along a field line, the positively charged conductor is at a greater potential than the neutral conductor. If we put the two conductors in contact, positive charge will flow to the neutral conductor until both conductors are at the same potential. (Actually, negative electrons flow from the neutral conductor to the positive conductor. It is slightly more convenient to think of this as a flow of positive charge in the opposite direction.) By symmetry, since the conductors are identical, they will share the original charge equally. If the conductors are now separated, each will carry charge $\frac{1}{2}Q$ and both will be at the same potential.

In Figure 22-13 a small conductor carrying a positive charge $q$ is inside the cavity of a second larger conductor. In equilibrium, the electric field is zero inside the conducting material. The lines of force that leave the positive charge $q$ must end on the inner surface of the large conductor. A negative charge $-q$ must therefore be induced on the inner surface of this conductor. This must occur no matter what the charge is on the outside surface of this conductor. Regardless of the charge on the larger conductor, the small conductor in the cavity is at a greater potential because the lines of force go from this conductor to the larger conductor. If the conductors are now connected, say with a fine conducting wire, *all* the charge originally on the smaller conductor will flow to the larger one. When the connection is broken, there is no charge on the small conductor in the cavity and there are no field lines anywhere within the outer surface of the larger conductor. The positive charge transferred from the smaller conductor to the larger one resides completely on the outside surface. If we bring up a second small positively charged conductor into the cavity (through the small opening shown in the figure) and touch it to the inner surface, we again transfer all the charge to the outer conductor. This procedure can be repeated indefinitely. This method is used to produce large potentials in the Van de Graaff generator, where the charge is brought to the inner surface of a larger spherical conductor by a continuous belt (Figure 22-14). The greater the net charge on the outer conductor, the greater its potential. For example, if the conductor is a spherical shell, its potential is $kQ/R$. The maximum potential obtainable in this way is limited only by the fact that air molecules become ionized in very high electric fields and the air becomes a conductor. This phenomenon, which is called *dielectric breakdown*, occurs in air at electric field strengths of about $E_{\max} \approx$ 3 MN/C = 3 MV/m. The resulting discharge through the conducting

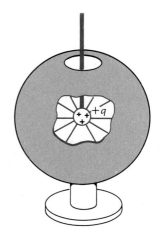

**Figure 22-13**
Small conductor carrying a positive charge inside a cavity in a larger conductor.

*Dielectric breakdown*

**Figure 22-14**
Small demonstration Van de Graaff generator. The top is removed in the photograph on the right to show the charging belt. (*Courtesy of Larry Langrill.*)

air is called *corona discharge*. Since the potential of a spherical conductor is $kQ/R$ and the electric field just outside such a conductor is $kQ/R^2$, the maximum potential is related to the field by

$$V_{max} = RE_{max}$$

A sphere of radius 1 m can therefore be raised to a potential of about 3 MV in air before breakdown.

When a charge is placed on a conductor of nonspherical shape like that in Figure 22-15, the conductor will be an equipotential surface but the charge density and the electric field $E = \sigma/\epsilon_0$ just outside the conductor will vary from point to point. Near a point where the radius of curvature is small (*A* in Figure 22-15) the charge density and electric field will be large, whereas near a point where the radius of curvature is large (*B* in Figure 22-15) the charge density and electric field will be small. For an arbitrarily shaped conductor, the potential at which dielectric breakdown occurs depends on the smallest radius of curvature of any part of the conductor. If the conductor has sharp points of very small radius of curvature, dielectric breakdown will occur at relatively low potentials.

**Figure 22-15**
Nonspherical conductor. The charge density $\sigma$ is greatest at points of small radius of curvature (near *A*) and least at points of large radius of curvature (near *B*). The electric field $E = \sigma/\epsilon_0$ at the conductor surface is greatest where the radius of curvature is smallest.

*Corona discharge*

William Vandivert

Corona discharge from overloaded power lines at General Electric test facilities in Pittsfield, Massachusetts. Because of the intense electric field near the wire, the air molecules are ionized and the air becomes conducting. The light is given off when the ions and electrons recombine.

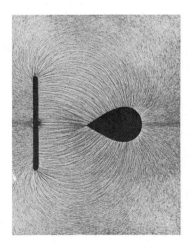

Electric field lines near a nonspherical conductor and plate carrying equal and opposite charges. The lines are shown by small bits of thread suspended in oil. The electric field is strongest near points of small radius of curvature, such as at the ends of the plate and at the pointed left side of the conductor. (*Courtesy of Harold M. Waage, Princeton University.*)

**Review**

A. Define, explain, or otherwise identify:

| | |
|---|---|
| Potential difference, 633 | Gradient, 637 |
| Electric potential, 633 | Equipotential surface, 638 |
| Volt, 633 | Charge sharing, 644 |
| Electronvolt, 635 | Dielectric breakdown, 645 |
| Electrostatic potential energy, 637 | Corona discharge, 646 |

B. True or false:

1. If the electric field is zero in some region of space, the electric potential must also be zero in that region.

2. If the electric potential is zero in some region of space, the electric field must also be zero in that region.

3. If the electric field is zero at a point, the potential must also be zero at that point.

4. Lines of electric field point toward regions of lower potential.

5. The value of the electric potential can be chosen to be zero at any convenient point in space.

Exercises

### Section 22-1, Potential Difference

1. A uniform electric field of 2 kN/C is in the $x$ direction. A point charge $Q = 3 \ \mu C$ initially at rest at the origin is released. (a) What is the kinetic energy of the charge when it is at $x = 4$ m? (b) What is the change in potential energy of the charge from $x = 0$ to $x = 4$ m? (c) What is the potential difference $V(4 \text{ m}) - V(0)$? Find the potential $V(x)$ if $V(x)$ is chosen to be (d) zero at $x = 0$, (e) 4 kV at $x = 0$, (f) zero at $x = 1$ m.

2. An infinite plane sheet of charge density $\sigma = +1 \ \mu C/m^2$ is in the $yz$ plane. (a) What is the magnitude of the electric field in newtons per coulomb? In volts per metre? What is the direction of $\mathbf{E}$ for positive $x$? (b) What is the potential difference $V_b - V_a$ when point $b$ is at $x = 20$ cm and point $a$ is at $x = 50$ cm? (c) How much work is required by an outside agent to move a test charge $q_0 = +1$ nC from point $a$ to point $b$ without acceleration?

3. A uniform electric field is in the negative $x$ direction. Points $a$ and $b$ are on the $x$ axis, $a$ at $x = 2$ m and $b$ at $x = 6$ m. (a) Is the potential difference $V_b - V_a$ positive or negative? (b) If the magnitude of $V_b - V_a$ is $10^5$ V, what is the magnitude $E$ of the electric field?

4. Two parallel conducting plates carry equal and opposite charge densities so that the electric field between them is approximately uniform. The difference in potential between the plates is 500 V, and they are separated by 10 cm. An electron is released from rest at the negative plate. (a) What is the magnitude of the electric field between the plates? Is the positive or negative plate at the higher potential? (b) Find the work done by the electric field as the electron moves from the negative plate to the positive plate. Express your answers in both electron-volts and joules. (c) What is the change in potential energy of the electron when it moves from the negative plate to the positive plate? What is its kinetic energy when it reaches the positive plate?

5. An electric field is given by $\mathbf{E} = ax\mathbf{i}$, where $\mathbf{E}$ is in newtons per coulomb, $x$ is in metres, and $a$ is a positive constant. (a) What are the SI units of $a$? (b) How much work is done by this field on a positive point charge $q_0$ when the charge moves from the origin to some point $x$? (c) Find the potential function $V(x)$ such that $V = 0$ at $x = 0$.

6. A positive charge of magnitude 2 $\mu C$ is at the origin. (a) What is the electric potential $V$ at a point 4 m from the origin relative to $V = 0$ at infinity? (b) What is the potential energy when a $+3\text{-}\mu C$ charge is placed at $r = 4$ m? (c) How much work must be done by an outside agent to bring the $3\text{-}\mu C$ charge from infinity to $r = 4$ m, assuming that the $2\text{-}\mu C$ charge is held fixed at the origin? (d) How much work must be done by an outside agent to bring the $2\text{-}\mu C$ charge from infinity to the origin if the $3\text{-}\mu C$ charge is first placed at $r = 4$ m and held fixed?

### Section 22-2, Potential of a System of Point Charges and Electrostatic Potential Energy

7. Four $2\text{-}\mu C$ point charges are at the corners of a square of side 4 m. Find the potential at the center of the square (relative to zero potential at infinity) if (a) all the charges are positive, (b) three of the charges are positive and one is negative, (c) two are positive and two are negative.

8. Three point charges are on the $x$ axis, $q_1$ at the origin, $q_2$ at $x = 3$ m, and $q_3$ at $x = 6$ m. Find the potential at the point $x = 0$, $y = 3$ m if (a) $q_1 = q_2 = q_3 = 2 \ \mu C$, (b) $q_1 = q_2 = 2 \ \mu C$ and $q_3 = -2 \ \mu C$, (c) $q_1 = q_3 = 2 \ \mu C$ and $q_2 = -2 \ \mu C$.

9. Find the electrostatic potential energy for each of the charge distributions of Exercise 7.

10. Find the electrostatic potential energy for each of the charge distributions of Exercise 8.

11. Points $A$, $B$, and $C$ are at the corners of an equilateral triangle of side 3 m. Equal positive charges of 2 $\mu$C are at $A$ and $B$. (a) What is the potential at point $C$? (b) How much work is required to bring a positive charge of 5 $\mu$C from infinity to point $C$ if the other charges are held fixed? (c) Answer parts (a) and (b) if the charge at $B$ is replaced by a charge of $-2$ $\mu$C.

### Section 22-3, Electric Field and Potential: Equipotential Surfaces

12. A point charge $q = 3.00$ $\mu$C is at the origin. (a) Find the potential $V$ on the $x$ axis at $x = 3.00$ m and at $x = 3.01$ m. (b) Does the potential increase or decrease as $x$ increases? Compute $-\Delta V/\Delta x$, where $\Delta V$ is the change in potential from $x = 3.00$ m to $x = 3.01$ m and $\Delta x = 0.01$ m. (c) Find the electric field at $x = 3.00$ m and compare its magnitude with $-\Delta V/\Delta x$ found in part (b). (d) Find the potential (to three significant figures) at the point $x = 3.00$ m and $y = 0.01$ m and compare your result with the potential on the $x$ axis at $x = 3.00$ m. Discuss the significance of this result.

13. A charge of $+3.00$ $\mu$C is at the origin, and a charge of $-3.00$ $\mu$C is on the $x$ axis at $x = 6.00$ m. (a) Find the potential on the $x$ axis at $x = 3.00$ m. (b) Find the electric field on the $x$ axis at $x = 3.00$ m. (c) Find the potential on the $x$ axis at $x = 3.01$ m and compute $-\Delta V/\Delta x$, where $\Delta V$ is the change in potential from $x = 3.00$ m to $x = 3.01$ m and $\Delta x = 0.01$ m; compare your result with your answer to part (b).

14. Sketch $V(x)$ versus $x$ for points on the $x$ axis for the charge distribution in Exercise 13.

15. In the following, $V$ is in volts and $x$ is in metres. Find $E_x$ when (a) $V(x) = 2000 + 3000x$; (b) $V(x) = 4000 + 3000x$; (c) $V(x) = 2000 - 3000x$; (d) $V(x) = -2000$ independent of $x$.

16. The electric potential in some region of space is given by $V(x) = C_1 + C_2 x^2$, where $V$ is in volts, $x$ is in metres, and $C_1$ and $C_2$ are positive constants. (a) What is the significance of the constant $C_1$? (b) Find the electric field $\mathbf{E}$ in this region. In what direction is $\mathbf{E}$?

17. An infinite sheet of charge has surface charge density 1 $\mu$C/m². How far apart are the equipotential planes whose potentials differ by 100 V?

18. A point charge $q = +\frac{1}{9} \times 10^{-8}$ C is at the origin. Taking the potential to be zero at $r = \infty$, locate the equipotential surfaces at 20-V intervals from 20 to 100 V and sketch to scale. Are these surfaces equally spaced?

19. Two equal positive charges are separated by a small distance. Sketch the lines of force and the equipotential surfaces for this system.

### Section 22-4, Calculation of Electric Potential

20. Two positive charges $+q$ are on the $x$ axis at $x = +a$ and $x = -a$. (a) Find the potential $V(x)$ as a function of $x$ for points on the $x$ axis. (b) Sketch $V(x)$ versus $x$. (c) What is the significance of the minimum in your curve?

21. Sketch $V(x)$ versus $x$ for the uniformly charged ring in the $yz$ plane of Example 22-4. At what point is $V(x)$ a maximum? What is $E_x$ at this point?

22. Consider a ball of charge of uniform charge density with radius $R = 10^{-15}$ m and total charge $+e$. (This is a model of a proton.) If the center of the charge is at the origin, find the electric field and the potential at (a) $r = R$ and (b) $r = 0$.

23. A charge of $q = +10^{-8}$ C is uniformly distributed on a spherical shell of radius 10 cm. (a) What is the magnitude of the electric field just outside and just inside the shell? (b) What is the magnitude of the electric potential just outside and just inside the shell? (c) What is the electric potential at the center of the shell? What is the electric field at that point?

24. An electric field is given by $E_x = 2.0x^3$ kN/C. Find the potential difference between the points on the $x$ axis at $x = 1$ m and $x = 2$ m.

### Section 22-5, Charge Sharing

25. Suppose that a Van de Graaff generator has a potential difference of 1 MV between the belt and the outer shell, and that charge is supplied at the rate of 200 $\mu$C/s. What minimum power is needed to drive the moving belt?

26. Sketch lines of force and equipotential surfaces both near to and far from the conductor shown in Figure 22-15, assuming that the conductor carries some charge $q$.

27. To what potential can a small Van de Graaff sphere of radius 10.0 cm be raised before dielectric breakdown?

### Problems

1. In the Bohr model of the hydrogen atom (Example 22-1) the electron moves in a circular orbit of radius $r$ around the proton. (a) Find an expression for the kinetic energy of the electron as a function of $r$ by setting the force on the electron (given by Coulomb's law) equal to $ma$, where $a$ is the centripetal acceleration. Show that at any distance $r$ the kinetic energy is half the magnitude of the potential energy. (b) Evaluate $\frac{1}{2}mv^2$ and the total energy $E = \frac{1}{2}mv^2 + U$ for $r = 0.529 \times 10^{-10}$ m. (c) How much energy (in electronvolts) must be supplied to the hydrogen atom to ionize it, i.e., to remove the electron to infinity with zero energy?

2. In a Van de Graaff accelerator, protons are released from rest at a potential of 5 MV and travel through a vacuum to a region at zero potential. (a) Find the speed of the 5-MeV protons. (b) If the potential change occurs over a distance of 2.0 m, find the accelerating electric field.

3. Consider two infinite parallel sheets of charge, one in the $yz$ plane and the other at distance $x = a$. (a) Find the potential everywhere in space with $V = 0$ at $x = 0$ if the sheets carry equal positive charge density $+\sigma$. (b) Do the same if the charge densities are equal and opposite, with the sheet in the $yz$ plane positive.

4. When uranium $^{235}$U captures a neutron, it splits into two nuclei (and emits several neutrons, which can cause other uranium nuclei to split). Assume that the fission products are equally charged nuclei with charge $+46e$ and that these nuclei are at rest just after fission and separated by twice their radius $2R \approx 1.3 \times 10^{-14}$ m. (a) Using $U = kq_1q_2/2R$, calculate the electrostatic potential energy of the fission fragments. This is approximately the energy released per fission. (b) About how many fissions per second are needed to produce 1 MW of power in a reactor?

5. Radioactive $^{210}$Po emits alpha particles with energy 5.30 MeV. Assume that just after the alpha particle is formed and escapes from the nucleus, the alpha particle with charge $+2e$ is a distance $R$ from the center of the daughter nucleus $^{206}$Pb with charge $+82e$. Calculate $R$ by setting the electrostatic potential energy of the two particles at this separation equal to 5.30 MeV.

6. Four charges are at the corners of a square centered at the origin as follows: $q$ at $x = -a$, $y = +a$ $(-a, +a)$; $2q$ at $(a, a)$; $-3q$ at $(a, -a)$; and $6q$ at $(-a, -a)$. Find (a) the electric field at the origin and (b) the potential at the origin. (c) A

fifth charge $+q$ is placed at the origin and released from rest. Find its speed when it is a great distance from the origin.

7. Find the electrostatic energy of the four point charges in Problem 6.

8. Show that Equation 22-19 for the potential on the axis of a disk charge gives the expected limit $V \rightarrow kq/x$ for $x$ much greater than $R$, where $q$ is the total charge in the disk.

9. Two large parallel metal sheets carry equal and opposite charge densities of magnitude $\sigma$. The sheets have area $A$ and are separated by a distance $d$. (a) Find the potential difference between the sheets. (b) A third sheet of the same area and thickness $a$ inserted between the original two sheets carries no net charge. Find the potential difference between the original two sheets and sketch the lines of $\mathbf{E}$ in the region between the original two sheets.

10. A point charge $+3e$ is at the origin and a second charge $-2e$ is on the $x$ axis at $x = a$. Sketch the potential function $V(x)$ versus $x$ for all $x$. At what point or points is $V(x)$ zero? Are these points equilibrium points? How much work is needed to bring a third charge $+e$ to the point $x = \frac{1}{2}a$ on the $x$ axis?

11. An electric dipole consists of a negative charge $-q$ at the origin and a positive charge $+q$ on the $z$ axis at $z = a$ (Figure 22-16). (a) Find the potential on the $z$ axis for $z > a$ and show that when $z$ is much greater than $a$, $V \approx kp/z^2$, where $p = qa$ is the dipole moment. (b) Show that this result can be obtained by considering a single charge $+q$ at the origin and finding $V(z) - V(z + \Delta z) \approx -(dV/dz)\,\Delta z$, where $\Delta z = a$.

12. For the dipole of Problem 11, show that the potential at a point off-axis a great distance $r$ from the origin is given approximately by

$$V = \frac{kaq \cos \theta}{r^2} = \frac{kp \cos \theta}{r^2} = \frac{kpz}{r^3}$$

Hint: If $r_1$ is the distance from the positive charge to the field point, as in Figure 22-16, use the law of cosines to show that $r_1 \approx r[1 - (2a \cos \theta)/r]^{1/2}$ for $r \gg a$. Then use the binomial expansion to approximate $kq/r_1 = kqr_1^{-1}$.

13. Two concentric spherical-shell conductors carry equal and opposite charges. The inner shell has radius $a$ and charge $+q$; the outer shell has radius $b$ and charge $-q$. Find the potential difference between the shells, $V_a - V_b$.

14. Two very long coaxial cylindrical-shell conductors carry equal and opposite charges. The inner cylinder has radius $a$ and charge $+q$; the outer shell has radius $b$ and charge $-q$. The length of each cylinder is $L$. Find the potential difference between the shells.

15. A hollow spherical conductor has inner radius $a$ and outer radius $b$. A positive point charge $+q$ is at the center of the sphere, and the conductor is uncharged. Find the potential $V(r)$ everywhere, assuming that $V = 0$ at $r = \infty$, and sketch $V(r)$ versus $r$.

16. A nonconducting sphere of radius $R$ has charge density $\rho = \rho_0 r/R$, where $\rho_0$ is a constant. (a) Show that the total charge is $Q = \pi R^3 \rho_0$. (b) Show that the total charge inside a sphere of radius $r < R$ is $q = Qr^4/R^4$. (c) Use Gauss' law to find the electric field $E_r$ everywhere. (d) Use $dV = -E_r\,dr$ to find the potential $V(r)$ everywhere, assuming that $V = 0$ at $r = \infty$.

17. Starting with $a + e$ charge, a line of $n$ charges is assembled, each separated by a distance $a$ from its two neighbors. The charges are equal in magnitude and they alternate in sign. (a) Find an expression for the potential at the $(n + 1)$th site when the row contains $n$ charges. (b) If $a = 0.1$ nm, find the binding energy (in eV) of a single charge in the middle of an infinite row. Note: $1 - \frac{1}{2} + \frac{1}{3} - \frac{1}{4} \cdots = \ln 2$.

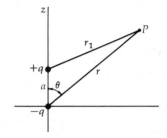

**Figure 22-16**
Electric dipole on the $z$ axis for Problem 11. The law of cosines can be used to relate the distance $r_1$ to $r$ and $\theta$.

**CHAPTER 23**     Capacitance, Electrostatic
Energy, and Dielectrics

---

**Objectives**  After studying this chapter, you should:

1.  Be able to derive expressions for the capacitance of a parallel-plate capacitor, cylindrical capacitor, and spherical capacitor.

2.  Be able to calculate the effective capacitance of systems of parallel and series capacitors.

3.  Be able to derive the expression $U = \frac{1}{2}QV$ for the energy stored in a charged capacitor.

4.  Be able to discuss the concept of electrostatic field energy.

5.  Be able to discuss the effect of a dielectric on the capacitance, charge, potential difference, and electric field in a parallel-plate capacitor.

6.  Know what is meant by bound charge and be able to discuss how bound charge arises and what effect it has.

---

A capacitor* is a useful device for storing charge and energy. It consists of two conductors insulated from each other. A typical capacitor, called a parallel-plate capacitor, consists of two large conducting plates of area $A$ separated by a small distance $d$. When the plates are connected to a charging device, e.g., a battery, as in Figure 23-1, charge is transferred from one conductor to the other until the potential difference between the conductors due to their equal and opposite charges equals the potential difference between the battery terminals. The amount of charge separated (which equals the magnitude of the charge on either conductor) depends on the geometry of the capacitor—e.g., on the area and separation of the plates in a parallel-plate capacitor—and is directly proportional to the potential difference $V$. The proportionality constant is called the *capacitance:*

$$Q = CV \qquad\qquad\qquad\qquad \text{23-1} \qquad \textit{Capacitance defined}$$

---

* A capacitor was formerly called a condenser, a term little used today.

(We write $V$ rather than $\Delta V$ for the potential difference to simplify the appearance of the formulas involving capacitance.) The capacitance depends on the size, shape, and geometrical arrangement of the conductors. For example, as we shall see below, the capacitance of a parallel-plate capacitor is proportional to the area of the plates and inversely proportional to their separation distance. The SI unit of capacitance is the coulomb per volt, called a *farad* (F):

$$1 \text{ farad} = 1 \text{ coulomb/volt} \qquad \qquad 23\text{-}2$$

Since the farad is a rather large unit, submultiples such as the microfarad (1 $\mu$F = $10^{-6}$ F) or the picofarad (1 pF = $10^{-12}$ F) are often used.

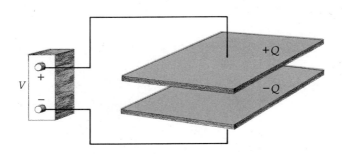

**Figure 23-1**
Capacitor consisting of two closely spaced parallel-plate conductors. When the conductors are connected to the terminals of a battery, the battery transfers charge from one conductor to the other until the potential difference between the conductors equals that of the battery terminals. The amount of charge transferred is proportional to the potential difference.

# 23-1 Calculation of Capacitance

The calculation of the capacitance of a capacitor is not difficult in principle. Given any two conductors, we find the potential difference $V$ when there is a charge $+Q$ on one conductor and $-Q$ on the other, and then calculate the capacitance $C$ from

$$C = Q/V \qquad \qquad 23\text{-}3$$

For simple geometries, e.g., the parallel-plate capacitor or the cylindrical capacitor, we can find the potential difference by first finding the electric field using either Gauss' or Coulomb's law, whichever is more convenient. The potential difference is then found by integrating the electric field along any path connecting the conductors, according to Equation 22-4. We shall illustrate this calculation for three simple arrangements of practical importance. Finding the potential difference between two conductors in more complicated arrangements, such as two nearby (nonconcentric) spherical conductors, is a problem in intermediate or advanced electrostatics and will not be considered here.

---

**Example 23-1** *The Parallel-Plate Capacitor* A parallel-plate capacitor consists of two parallel conducting plates very close together. Let each plate have area $A$, with charge $+Q$ on one plate and $-Q$ on the other. Each plate then has a surface charge density $\sigma = Q/A$, and the field will be nearly uniform between the plates, as indicated by the equally spaced field lines in Figure 23-2. The electric field in the region of space between the plates and far from the edges will essentially be that due to two infinite plane sheets of charge. The field due to each sheet has the magnitude $\sigma/2\epsilon_0$. Outside the plates these fields cancel, but in the space between the plates they add, and the electric field between the plates is

$$E = \frac{\sigma}{\epsilon_0} = \frac{Q}{\epsilon_0 A} \qquad\qquad 23\text{-}4$$

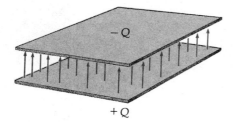

**Figure 23-2**
Parallel-plate capacitor. If the plates are closely spaced, the electric field is approximately uniform between the plates.

This result could also be obtained from Gauss' law applied to a pillbox gaussian surface with one face between the plates and the other inside one of the conductors. This result is only approximate because effects near the edges of the plates have been neglected. In practice, plates are often very close to each other. Since this field is constant in the region between the plates, the potential difference between the plates is $Ed$, where $d$ is the separation distance between the plates. Thus

$$V = Ed = \frac{Qd}{\epsilon_0 A}$$

The capacitance is

$$C = \frac{Q}{V} = \frac{Q}{Qd/\epsilon_0 A} = \frac{\epsilon_0 A}{d} \qquad\qquad 23\text{-}5$$

*Parallel-plate capacitor*

The capacitance is proportional to the area of the plates and inversely proportional to the separation distance. We see from the dimensions of Equation 23-5 that the SI unit of $\epsilon_0$ can be written as a farad per metre:

$$\epsilon_0 = \frac{1}{4\pi k} = 8.85 \times 10^{-12} \text{ F/m} \qquad\qquad 23\text{-}6$$

A numerical calculation illustrates how large the farad unit of capacitance is. Let each plate be a square of side 10 cm and let the plate separation be 1 mm. The capacitance is then

$$C = \frac{\epsilon_0 A}{d} = \frac{(8.85 \times 10^{-12} \text{ F/m}) (0.1 \text{ m}) (0.1 \text{ m})}{0.001 \text{ m}}$$

$$= 8.85 \times 10^{-11} \text{ F} \approx 90 \text{ pF}$$

**Example 23-2** The 90-pF capacitor of Example 23-1 is connected to a 12-V battery and charged to 12 V. How many electrons are transferred from one plate to the other?

From the definition of capacitance (Equation 23-1) the charge transferred is

$$Q = CV = (90 \times 10^{-12} \text{ F}) (12 \text{ V}) = 1.1 \times 10^{-9} \text{ C}$$

This is the magnitude of the charge on either plate. The number of electrons in a charge of $1.1 \times 10^{-9}$ C is

$$N = \frac{Q}{e} = \frac{1.1 \times 10^{-9} \text{ C}}{1.6 \times 10^{-19} \text{ C}} = 6.9 \times 10^9$$

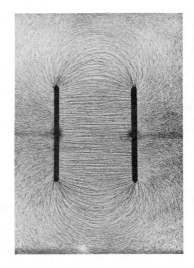

Electric field lines between plates of a parallel-plate capacitor, shown by small bits of thread suspended in oil. When the plates are very close together, the fringing of the field near the edges can be neglected. (*Courtesy of Harold M. Waage, Princeton University.*)

**Example 23-3** *The Cylindrical Capacitor* A cylindrical capacitor consists of a small cylinder or wire of radius $a$ and a larger concentric cylindrical shell of radius $b$. Let $+Q$ be the charge on the inner conductor and $-Q$ be on the outer conductor. The electric field between the conductors is found most easily from Gauss' law. Let us consider a cylindrical gaussian surface of radius $r$ and length $L_1$, as shown in Figure 23-3. According to Gauss' law, the flux through this surface equals $q/\epsilon_0$, where $q$ is the net charge inside the surface. If we assume that $L$ is much larger than $a$ or $b$ so that we can neglect edge effects, the electric field will be perpendicular to the axis of the conductors and depend only on the perpendicular distance from the axis of the conductors. Then there will be flux through only the cylindrical part of the gaussian surface; i.e., there will be no flux through the faces at each end of the surface. Since the electric field is constant in magnitude on the gaussian surface and perpendicular to this surface, the flux equals the magnitude $E_r$ times the area, which is $2\pi r L_1$. The net charge inside the gaussian surface is $q = \lambda L_1 = QL_1/L$, where $\lambda = Q/L$ is the charge per unit length of the inner conductor. Thus Gauss' law gives for the electric field

A variable parallel-plate capacitor. (*Courtesy of Larry Langrill, Oakland University.*)

$$\phi_{net} = E_r 2\pi r L_1 = \frac{q}{\epsilon_0} = \frac{\lambda L_1}{\epsilon_0}$$

$$E_r = \frac{\lambda}{2\pi\epsilon_0 r} = \frac{Q}{2\pi\epsilon_0 L}\frac{1}{r} \qquad\qquad 23\text{-}7$$

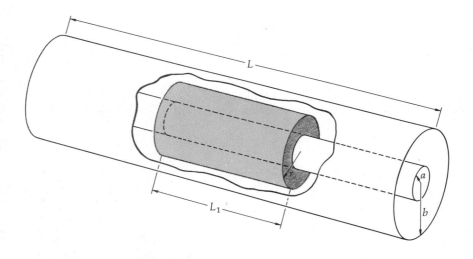

**Figure 23-3**
Cylindrical capacitor consisting of an inner wire of radius $a$ and a concentric outer shell of radius $b$. The electric field is calculated from Gauss' law using a cylindrical gaussian surface of radius $r$ and length $L_1$ between the conductors.

This result is the same as that for an infinite line charge. The potential difference between the conductors is found from Equation 22-4. Let $V_a$ be the potential of the inner conductor and $V_b$ be that of the outer conductor. Then

$$V_b - V_a = -\int_a^b E_r\,dr = -\frac{Q}{2\pi\epsilon_0 L}\int_a^b \frac{dr}{r} = -\frac{Q}{2\pi\epsilon_0 L}\ln\frac{b}{a}$$

The potential is of course greater on the inner conductor, which carries the positive charge, since the electric field lines point from this conductor to the outer conductor. The magnitude of this potential difference is

$$V = V_a - V_b = \frac{Q\ln\,(b/a)}{2\pi\epsilon_0 L} \qquad\qquad 23\text{-}8$$

and the capacitance is (Equation 23-3)

$$C = \frac{Q}{V} = \frac{2\pi\epsilon_0 L}{\ln(b/a)} \qquad 23\text{-}9$$

As expected, the capacitance is proportional to the length of the cylinders. The greater the length, the greater the amount of charge that can be put on the conductors for a given potential difference, since the electric field, and therefore the potential difference, depends only on the charge per unit length.

**Example 23-4** *The Spherical Capacitor* A spherical capacitor consists of a small inner conducting sphere of radius $R_1$ and a larger concentric spherical shell of radius $R_2$ (Figure 23-4). The inner sphere can be supported on an insulator. A small opening is made in the outer shell so that charge can be placed on the inner sphere. If this opening is small enough, it will have a negligible effect on the spherical symmetry of the capacitor. To find the capacitance we place a charge $+Q$ on the inner sphere and $-Q$ on the outer sphere and find the potential difference. The electric field between the conductors is the same as that due to a point charge $Q$ at the origin. As we have seen, the charge on the outer shell does not contribute to the electric field inside it. (We do not need to know the electric field in the region $r < R_1$ or, outside the outer shell, $r > R_2$, but it is easy to show from Gauss' law that $\mathbf{E} = 0$ in both these regions.) The electric field between the conductors is thus

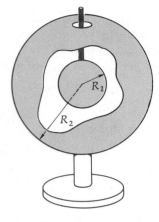

**Figure 23-4**
Spherical capacitor consisting of an inner sphere of radius $R_1$ and a concentric outer spherical shell of radius $R_2$. The electric field between the conductors is radial.

$$E_r = \frac{kQ}{r^2}$$

If $V_1$ is the potential on the inner sphere and $V_2$ on the outer sphere, the potential difference is

$$V_2 - V_1 = -\int_{R_1}^{R_2} E_r \, dr = -\int_{R_1}^{R_2} \frac{kQ}{r^2} \, dr$$

$$= \frac{kQ}{R_2} - \frac{kQ}{R_1} = -\frac{kQ(R_2 - R_1)}{R_1 R_2}$$

The potential of the inner sphere $V_1$ is greater than that of the outer sphere, as evidenced by the fact that the electric field lines point radially outward from the inner to the outer sphere. The magnitude of the potential difference is

$$V = V_1 - V_2 = \frac{kQ(R_2 - R_1)}{R_1 R_2} \qquad 23\text{-}10$$

The capacitance is

$$C = \frac{Q}{V} = \frac{R_1 R_2}{k(R_2 - R_1)} = 4\pi\epsilon_0 R_1 \frac{R_2}{R_2 - R_1} \qquad 23\text{-}11$$

As the radius of the outer sphere approaches infinity, the capacitance approaches $4\pi\epsilon_0 R_1$. We can think of this result with the outer spherical shell at infinity as the capacitance of a single isolated sphere. For a sphere of radius $R$, its capacitance, defined as the ratio of its charge to its potential (relative to zero potential at infinity), is

$$C = 4\pi\epsilon_0 R \qquad 23\text{-}12$$

The capacitance of an isolated spherical conductor is proportional to its radius. Note that in all three examples the capacitance is proportional to the constant $\epsilon_0$ and to some characteristic length of the system.

## 23-2  Parallel and Series Combinations of Capacitors

Two or more capacitors are often used in combination. In electric circuits a capacitor is indicated by the symbol ┤├. Figure 23-5 shows two capacitors connected in parallel. The upper plates of the two capacitors are connected together by a conducting wire and are therefore at the same potential. The lower plates are also connected together and are at a common potential. It is clear that the effect of adding a second capacitor connected in this way is to increase the capacitance; i.e., the area is essentially increased, allowing more charge to be stored for the same potential difference $V = V_a - V_b$. If the capacitances are $C_1$ and $C_2$, the charges $Q_1$ and $Q_2$ stored on the plates are given by

$$Q_1 = C_1 V \quad \text{and} \quad Q_2 = C_2 V$$

where $V$ is the potential difference across either capacitor, and the total charge stored is

$$Q = Q_1 + Q_2 = C_1 V + C_2 V = (C_1 + C_2)V \qquad \text{23-13}$$

The effective capacitance of two capacitors in parallel is defined to be the ratio of the total charge stored to the potential. Thus

$$C_{\text{eff}} = \frac{Q}{V} = C_1 + C_2 \qquad \text{23-14}$$

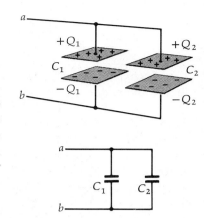

**Figure 23-5**
Two capacitors in parallel. The potential difference across the capacitors is the same, $V_a - V_b$, for each capacitor.

*Capacitors in parallel*

The effective capacitance is that of a single capacitor which could replace the parallel combination and store the same amount of charge for a given potential difference $V$. This reasoning can be extended to three or more capacitors connected in parallel, as in Figure 23-6. The effective capacitance equals the sum of the individual capacitances.

In Figure 23-7 two capacitors $C_1$ and $C_2$ are connected in series with a potential difference $V = V_a - V_b$ between the upper plate of the first capacitor and the lower plate of the second. Such a situation can be realized in practice by connecting points $a$ and $b$ to the terminals of a battery. If a charge $+Q$ is placed on the upper plate of the first capacitor, there will be an equal negative charge $-Q$ induced on its lower plate. This charge comes from electrons drawn from the upper plate of the second capacitor. Thus there will be an equal charge $+Q$ on the upper plate of the second capacitor and $-Q$ on its lower plate. The potential difference across the upper capacitor is $V_a - V_c = Q/C_1$. Similarly, the potential difference across the second capacitor is $V_c - V_b = Q/C_2$. The

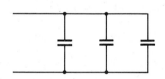

**Figure 23-6**
Three capacitors in parallel. The effect of adding a parallel capacitor is to increase the effective capacitance.

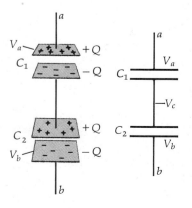

**Figure 23-7**
Two capacitors in series. The potential difference $V_a - V_b$ is the sum of the potential differences across the capacitors. The charge on each capacitor is the same.

potential difference across the two capacitors in series is the sum of these potential differences:

$$V_a - V_b = (V_a - V_c) + (V_c - V_b) = \frac{Q}{C_1} + \frac{Q}{C_2}$$

Calling this potential difference $V$, we have

$$V = Q\left(\frac{1}{C_1} + \frac{1}{C_2}\right) = \frac{Q}{C_{eff}} \qquad 23\text{-}15$$

where $C_{eff} = Q/V$ is the ratio of the charge to the total potential difference across the two capacitors connected in series. From Equation 23-15,

$$\frac{1}{C_{eff}} = \frac{1}{C_1} + \frac{1}{C_2} \qquad 23\text{-}16 \qquad \textit{Capacitors in series}$$

or

$$C_{eff} = \frac{C_1 C_2}{C_1 + C_2} \qquad 23\text{-}17$$

The effective capacitance of two capacitors in series is less than that of either of the individual capacitors. For example, in Equation 23-17, $C_{eff}$ can be considered the product of $C_1$ and $C_2/(C_1 + C_2)$, which is less than 1. Equation 23-16 can be generalized to three or more capacitors in series:

$$\frac{1}{C_{eff}} = \frac{1}{C_1} + \frac{1}{C_2} + \frac{1}{C_3} + \cdots \qquad 23\text{-}18$$

## 23-3  Electrostatic Energy in a Capacitor

While a capacitor is being charged, positive charge is transferred from the negatively charged conductor to the positively charged conductor. Since the positive conductor is at a greater potential than the negative conductor, the potential energy of the charge being transferred is increased. For example, if a small amount of charge $\Delta q$ is transferred through a potential difference $V$, the potential energy of the charge is increased by the amount $\Delta q\, V$ (by definition, potential difference is the potential-energy difference per unit charge). Work must therefore be done to charge a capacitor. Some of this work (or all of it, depending on the charging process) is stored as electrostatic potential energy. (If the capacitor is charged by connecting it to a battery, for example, the battery does twice as much work as is stored as electrostatic potential energy. Half the work done by the battery is wasted as heat in the connecting wires and the battery itself. We shall discuss this in detail in Chapter 25.)

Consider the charging of a parallel-plate capacitor. Since only the potential *difference* between the plates is important, we are free to choose the potential to be zero at any point. It is convenient to choose the potential of the negative plate to be zero. At the beginning of the charging process, neither plate is charged. There is no electric field, and both plates are at the same potential. After the charging process a charge $Q_0$ has been transferred from one plate to the other, and the potential difference is $V_0 = Q_0/C$, where $C$ is the capacitance. We then have a negative charge $-Q_0$ on one plate, which we have chosen to be at zero potential, and $+Q_0$ on the other plate at potential $V_0$. One might expect that the work needed to accomplish this would be just the charge $Q_0$ times the potential energy per unit charge $V_0$, but only the last bit of charge must be raised by the full potential difference $V_0$. The potential difference between the plates increases from zero originally to its final value $V_0$. The average value of the potential difference during the charging process is just $\frac{1}{2}V_0$, and the work needed is $\frac{1}{2}Q_0V_0$. We can see this as follows.

Let $q$ be the charge that has been transferred at some time during the process. The potential difference is then $V = q/C$. If a small amount of charge $dq$ is now transferred from the plate with charge $-q$ at zero potential to the plate with charge $q$ at potential $V$, its potential energy is increased by

$$dU = V \, dq = \frac{q}{C} \, dq$$

The total increase in potential energy in charging from $q = 0$ to $q = Q_0$ is the energy stored in the capacitor (Figure 23-8):

$$U = \int dU = \int_0^{Q_0} \frac{q}{C} \, dq = \frac{1}{C} \int_0^{Q_0} q \, dq = \frac{1}{2} \frac{Q_0^2}{C}$$

Using $C = Q_0/V_0$, we can write this in a variety of other ways:

$$U = \frac{1}{2} \frac{Q_0^2}{C} = \frac{1}{2} Q_0 V_0 = \frac{1}{2} C V_0^2 \qquad \text{23-19}$$

The derivation of Equation 23-19 was carried out for the parallel-plate capacitor, but a review of the steps will show that the geometry of the capacitor plays no role in the argument. The expression $C = \epsilon_0 A/d$ was never used. The argument is applicable to any capacitor, and Equation 23-19 is a general expression for the energy stored in a charged capacitor as electrostatic potential energy.

**Questions**

1. The potential difference of a capacitor is doubled. By what factor does its stored electric energy change?

2. Half the charge is removed from a capacitor. What fraction of its stored energy was removed along with the charge?

# 23-4  Electrostatic Field Energy

In the process of charging a capacitor, an electric field is created. For example, in a parallel-plate capacitor, there is no electric field when the plates are uncharged. When they have their final charge $Q_0$, there is a field $E_0 = \sigma/\epsilon_0 = Q_0/\epsilon_0 A$, where $A$ is the area of the plates. The work done in charging the capacitor is therefore the same as the work needed to create an electric field. Consider again the charging of a parallel-plate capacitor. When one plate has charge $+q$ and the other $-q$, the field between the plates is

$$E = \frac{\sigma}{\epsilon_0} = \frac{q}{\epsilon_0 A} \qquad \text{23-20}$$

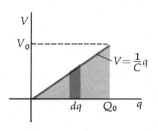

**Figure 23-8**
The work needed to charge a capacitor is the integral of $V \, dq$ from the original charge $q = 0$ to the final charge $q = Q_0$. This work is the area $\frac{1}{2}Q_0V_0$ under the curve.

*Energy in a charged capacitor*

The work needed to transfer an additional charge $dq$ is force $dq\,E$ times the distance $s^*$, since the field (and therefore the force) is constant. This is the work done to push $dq$ against the field from $-q$ to $+q$,

$$dW = dq\,Es \qquad\qquad 23\text{-}21$$

From Equation 23-20

$$dE = \frac{dq}{\epsilon_0 A}$$

Substituting $dq = \epsilon_0 A\,dE$ into Equation 23-21 gives

$$dW = (\epsilon_0 As)E\,dE \qquad\qquad 23\text{-}22$$

Therefore the work done to increase the field from $E = 0$ to $E = E_0 = Q_0/\epsilon_0 A$ is

$$W = \epsilon_0 As \int_0^{E_0} E\,dE = \tfrac{1}{2}\epsilon_0 E_0^2 As$$

This work appears as electrostatic potential energy $U$:

$$U = \tfrac{1}{2}\epsilon_0 E_0^2 As \qquad\qquad 23\text{-}23$$

This result could have been obtained directly from Equation 23-19 using $Q_0 = \epsilon_0 AE_0$ and $C = \epsilon_0 A/s$. It is convenient to think of the energy as being stored in the electric field. Note that, neglecting the edge effects, the field is constant in the region between the plates and zero outside. The volume $\mathcal{V}$ between the plates is $As$. The energy per unit volume in the field, called the *energy density* $\eta$, is thus

$$\eta = \frac{U}{\mathcal{V}} = \tfrac{1}{2}\epsilon_0 E^2 \qquad\qquad 23\text{-}24$$

*Electric field energy density*

Although we have been considering the simple case of a parallel-plate capacitor, the result (Equation 23-24) for the energy per unit volume in our electrostatic field is generally valid. We shall illustrate the generality of Equation 23-24 by using it to calculate the energy in an electrostatic field that is not constant in space.

**Example 23-5** *Energy of a Single Isolated Spherical Conductor of Radius R Carrying a Charge Q* We noted in Example 23-4 that a single spherical conductor can be thought of as a spherical capacitor with the outer shell at infinity, where the potential is chosen to be zero. The potential of this conductor is $kQ/R$, and its capacitance is $C = 4\pi\epsilon_0 R = R/k$, where $k$ is the Coulomb constant. According to Equation 23-19, the electrostatic energy is

$$U = \tfrac{1}{2}QV = \tfrac{1}{2}Q\,\frac{kQ}{R} = \frac{1}{2}\frac{kQ^2}{R} \qquad\qquad 23\text{-}25$$

A direct derivation of this result by considering the work done in bringing charge from infinity and placing it on the conductor is given as an exercise. Here we give a derivation using the result that the energy per unit volume in an electric field is $\tfrac{1}{2}\epsilon_0 E^2$, as in Equation 23-24. At a distance $r$ from the center of the sphere the electric field is

$$E_r = \frac{kQ}{r^2}$$

---

* We use $s$ here for the plate separation rather than $d$ to avoid confusion with the $d$ used for the differentials $dq$ and $dE$.

The energy per unit volume is then

$$\eta = \tfrac{1}{2}\epsilon_0 E^2 = \tfrac{1}{2}\epsilon_0 \frac{k^2 Q^2}{r^4} \qquad\text{23-26}$$

Consider a spherical shell volume element of radius $r$, thickness $dr$, and volume $d\mathcal{V} = 4\pi r^2\, dr$ (Figure 23-9). The energy in this volume element is

$$dU = \eta\, d\mathcal{V} = \tfrac{1}{2}\epsilon_0 k^2 \frac{Q^2}{r^4} 4\pi r^2\, dr$$

$$= 2\pi\epsilon_0 k^2 \frac{Q^2}{r^2}\, dr = \frac{1}{2}\frac{kQ^2}{r^2}\, dr \qquad\text{23-27}$$

using $k = 1/4\pi\epsilon_0$. We calculate the total energy by integrating from $r = R$ to $r = \infty$ since $E = 0$ for $r < R$. Thus

$$U = \int_R^\infty \frac{1}{2}\frac{kQ^2}{r^2}\, dr = \frac{1}{2}\frac{kQ^2}{R}$$

in agreement with Equation 23-25.

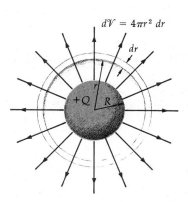

**Figure 23-9**
Calculation of the electrostatic energy of a spherical conductor with charge Q. The volume of space between $r$ and $r + dr$ is $d\mathcal{V} = 4\pi r^2\, dr$. The electrostatic field energy in this volume element is $\eta\, d\mathcal{V}$, where $\eta = \tfrac{1}{2}\epsilon_0 E^2$ is the energy density.

## 23-5    Dielectrics

A nonconducting material, e.g., glass or wood, is called a *dielectric*. Faraday discovered that when the space between the two conductors of a capacitor is occupied by a dielectric, the capacitance is increased. If the space (between the plates of a parallel-plate capacitor, for example) is completely filled by the dielectric, the capacitance increases by a factor $K$ which is characteristic of the dielectric and called the *dielectric constant*.

Suppose a capacitor of capacitance $C_0$ is connected to a battery, which charges it to a potential difference $V_0$ by placing a charge $Q_0 = C_0 V_0$ on the plates. If the battery is disconnected and a dielectric is inserted, filling the space between the plates, the potential difference decreases to a new value,

$$V = \frac{V_0}{K} \qquad\text{23-28}$$

Since the original charge $Q_0$ is still on the plates, the new capacitance is

$$C = \frac{Q_0}{V} = \frac{KQ_0}{V_0} = KC_0 \qquad\text{23-29}$$

If, on the other hand, the dielectric is inserted while the battery is still connected, the battery must supply more charge to maintain the original potential difference. The total charge on the plates is then $Q = KQ_0$. In either case, the capacitance is increased by the factor $K$.

Since the potential difference between the plates of a parallel-plate capacitor equals the electric field between the plates times the separation $d$, the effect of the dielectric (with the battery disconnected) is to decrease the electric field by the factor $K$. If $E_0$ is the original field without the dielectric, the new field $E$ is

$$E = \frac{E_0}{K} \qquad\text{23-30}$$

*Dielectric constant*

*Dielectric reduces electric field*

We can understand this result in terms of molecular polarization of the dielectric. If the molecules of the dielectric are polar molecules, i.e., have permanent dipole moments, these moments are originally randomly oriented. In the presence of the field between the capacitor plates, these dipole moments experience a torque, which tends to align them in the direction of the field (Figure 23-10). The amount of alignment depends on the strength of the field and the temperature. At high temperatures, the random thermal motion of the molecules tends to counteract the alignment. In any case, the alignment of the molecular dipoles produces an additional electric field due to the dipoles which is in the direction opposite the original field. The original field is thus weakened. Even if the molecules of the dielectric are nonpolar, they will have induced dipole moments in the presence of the electric field between the plates. The induced dipole moments are in the direction of the original field. Again, the additional electric field due to these induced moments weakens the original field.

A dielectric which has electric dipole moments predominantly in the direction of the external field is said to be *polarized* by the field, whether the polarization is due to alignment of permanent dipole moments of polar molecules or to the creation of induced dipole moments in nonpolar molecules. The net effect of the polarization of a homogeneous dielectric is the creation of a surface charge on the dielectric faces near the plates. Figure 23-11 shows a rectangular slab of homogeneous dielectric which is in a uniform electric field to the right. The molecular dipole moments are indicated schematically. As can be seen from the figure, the net effect of the polarization is to produce a positive surface charge density on the right face.* The charge densities on the dielectric faces are due to the displacement of positive and negative molecular charges near the faces. This displacement is due to the external electric field of the capacitor. The charge on the dielectric, called *bound charge,* is not free to move about like the ordinary free charge on the conducting capacitor plates. Although it disappears when the external electric field disappears, it produces an electric field just like any other charge. We shall now relate the bound charge density $\sigma_b$ to the dielectric constant $K$ and to the surface charge density $\sigma_f$ on the capacitor plates, which we call the free charge density because it is free to move about the conductor rather than being bound to the molecules, as in the case of the bound charge density.

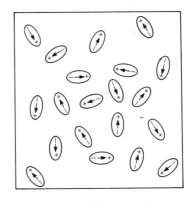
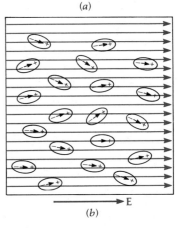

**Figure 23-10**
(a) Randomly oriented electric dipoles in the absence of an external electric field. (b) In the presence of an external field the dipoles are partially aligned parallel to the field.

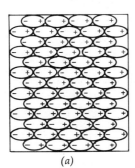
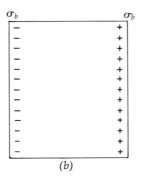

**Figure 23-11**
The net effect of alignment of electric dipole moments distributed uniformly in a volume (a) is a positive bound surface charge density on one side of the dielectric and a negative bound surface charge density on the other side, as shown in (b). The result is to weaken the electric field in the dielectric.

The electric field inside the dielectric slab due to the bound charge densities $+\sigma_b$ on the right and $-\sigma_b$ on the left is just the field due to two

* If the dielectric is not homogeneous or the field is not uniform, there may be a volume charge density created inside the dielectric because the amount of positive charge pushed out the right side of some volume element may not equal the negative charge pushed out the left side of the element. We shall neglect such complications.

infinite-plane charge densities (assuming that the slab is very thin; i.e., the plates of the capacitor are close together). The field $E'$ thus has the magnitude

$$E' = \frac{\sigma_b}{\epsilon_0} \qquad\qquad 23\text{-}31$$

This field is to the left and subtracts from the electric field due to the ordinary free charge density on the capacitor plates. The original field $E_0$ has the magnitude

$$E_0 = \frac{\sigma_f}{\epsilon_0} \qquad\qquad 23\text{-}32$$

The magnitude of the resultant field $E$ is the difference of these magnitudes. It also equals $E_0/K$:

$$E = E_0 - E' = \frac{E_0}{K} \qquad\qquad 23\text{-}33$$

or

$$E' = E_0 \left(1 - \frac{1}{K}\right) = \frac{K - 1}{K} E_0 \qquad\qquad 23\text{-}34$$

Writing $\sigma_b/\epsilon_0$ for $E'$ and $\sigma_f/\epsilon_0$ for $E_0$, we have

$$\sigma_b = \frac{K - 1}{K} \sigma_f \qquad\qquad 23\text{-}35$$

*Bound and free charge densities*

The bound charge density $\sigma_b$ is always less than the free charge density $\sigma_f$ on the capacitor plates and is zero if $K = 1$, which is the case of no dielectric. Figure 23-12 shows the lines of force for a parallel-plate capacitor with a dielectric of constant $K = 2$ which does not completely fill the space between the plates. According to Equation 23-35, the bound charge density on the dielectric is half the free charge density on the plates. Half the lines that leave the positive charge on the lower plate end on the negative bound charge on the dielectric. Similarly, half the lines which end on the negative charge on the upper plate begin on the positive bound charge on the dielectric. The density of field lines in this diagram shows that the electric field inside the dielectric is just half that outside. The effect of the dielectric on the capacitance depends on how much space is filled.

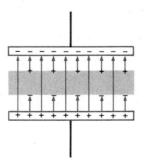

**Figure 23-12**
Electric field lines for a dielectric of constant $K = 2$ between the plates of a parallel-plate capacitor. The electric field is reduced to half its original strength, as indicated by the fact that half the original field lines end on one side of the dielectric and begin again on the other side.

**Example 23-6** A slab of material of dielectric constant $K$ has the same area as the plates of a parallel-plate capacitor but has a thickness $\frac{3}{4}d$, where $d$ is the separation of the plates. How is the capacitance changed when the slab is inserted between the plates?

Let $E_0 = V_0/d$ be the electric field between the plates when there is no dielectric and the potential difference is $V_0$. If the dielectric is inserted, the electric field in the dielectric will be $E = E_0/K$ and the potential difference will be

$$V = E_0(\tfrac{1}{4}d) + \frac{E_0}{K}\left(\frac{3}{4}\,d\right) = E_0 d\left(\frac{1}{4} + \frac{3}{4K}\right) = V_0\,\frac{K+3}{4K}$$

The potential difference decreases by the factor $(K + 3)/4K$ while the free charge $Q_0$ on the plates remains unchanged. The capacitance thus increases:

$$C = \frac{Q_0}{V} = \frac{4K}{K+3}\,\frac{Q_0}{V_0} = \frac{4K}{K+3}\,C_0$$

In addition to increasing the capacitance, a dielectric has two other functions in a capacitor: (1) it provides a mechanical means of separating the two conductors, which must be very close together in order to obtain a large capacitance; (2) the dielectric strength is increased because the dielectric strength of a dielectric is usually greater than that of air. We have already mentioned that the dielectric strength of air is 3 MV/m = 3 kV/mm. Fields greater than this magnitude cannot be maintained in air because of dielectric breakdown; i.e., the air becomes ionized and conducts. Many materials have dielectric strengths greater than that of air, allowing greater potential differences between the conductors of a capacitor.

*Three functions of a dielectric*

An example of these three dielectric functions is a parallel-plate capacitor made from two sheets of metal foil of large area (to increase the capacitance) separated by a sheet of paper. The paper increases the capacitance because of its polarization; that is, $K$ is greater than 1. It also provides a mechanical separation so that the sheets can be very close together without being in electrical contact. (A small separation is important because the capacitance varies inversely with separation.) Finally, the dielectric strength of paper is greater than that of air, so that greater potential differences can be attained without breakdown. Table 23-1 lists the dielectric constant and dielectric strength of some dielectrics.

**Table 23-1**
Dielectric constant and strength of various materials

| Material | Dielectric constant $K$ | Dielectric strength, kV/mm |
|---|---|---|
| Air | 1.00059 | 3 |
| Bakelite | 4.9 | 24 |
| Glass (Pyrex) | 5.6 | 14 |
| Mica | 5.4 | 10–100 |
| Neoprene | 6.9 | 12 |
| Paper | 3.7 | 16 |
| Paraffin | 2.1–2.5 | 10 |
| Plexiglas | 3.4 | 40 |
| Polystyrene | 2.55 | 24 |
| Porcelain | 7 | 5.7 |
| Transformer oil | 2.24 | 12 |
| Water (20°C) | 80 | |

# Electrostatics and Xerography

Richard Zallen
*Xerox Research Laboratories, Webster, N.Y.*

There are many important and beneficial technological applications which could be included in a discussion of uses of electrostatic phenomena. For example, a powerful air-pollution preventer is the electrostatic precipitator, which years ago made life livable near cement mills and ore-processing plants, and which is currently credited with extracting better than 99 percent of the ash and dust from the gases about to issue from chimneys of coal-burning power plants. The basic idea of this very effective antipollution technique is shown in Figure 1. The outer wall of a vertical metal duct is grounded, while a wire running down the center of the duct is kept at a very large negative voltage. In this concentric geometry (which corresponds to the cylindrical capacitor in Example 23-3) a very nonuniform electric field is set up, with lines of force directed radially inward toward the negative wire electrode. Close to the wire the field attains enormous values, large enough to produce an electrical breakdown of air, and the normal placid mixture of neutral gas molecules is replaced by a turmoil of free electrons and positive ions. The electrons from this corona discharge are driven outward from the wire by the electric field. Most of them quickly become attached to oxygen molecules to produce negative $O_2^-$ ions, which are also accelerated outward. As this stream of ions passes across the hot waste gas rising in the duct, small particles carried by the gas become charged by capturing ions and are pulled by the field to the outer wall. If the noxious particles are solid, they are periodically shaken down off the duct into a hopper; if they are liquid, the residue simply runs down the wall and is collected below.

Besides electrostatic precipitation, other technological examples include electrocoating with spray paints and the electrostatic separation of granular mixtures used for the removal of rock particles from minerals, garlic seeds from wheat, even rodent excreta from rice. However, the application which is

**Figure 1**
Schematic diagram of the use of a corona discharge in an electrostatic precipitator.

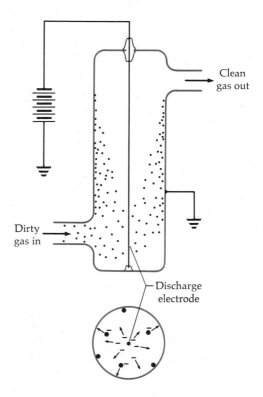

Clean
gas out

Dirty
gas in

Discharge
electrode

the main focus of this essay is xerography, the most widely used form of electrostatic imaging, or electrophotography. This is the most familiar use of electrostatics in terms of the number of people who have occasion to use plain-paper copying machines in offices, libraries, and schools, and it also provides a fine example of a process utilizing a sequence of distinct electrostatic events.

The xerographic process was invented in 1937 by Chester Carlson. The term xerography, literally "dry writing," was actually adopted a bit later to emphasize the distinction from wet chemical processes. Carlson's innovative concept did not find early acceptance, and a practical realization of his idea became available only after a small company (in a famous entrepreneurial success story) risked its future in its intensive efforts to develop the process.

Four of the main steps involved in xerography are illustrated in Figure 2. In the interest of clarity the process has been oversimplified, and several subtleties (as well as gaps in our understanding) have been suppressed. Electrostatic imaging takes place on a large thin plate of a photoconducting material supported by a grounded metal backing. A photoconductor is a solid which is a good insulator *in the dark* but which becomes capable of conducting electric current when exposed to light. The unilluminated, insulating state is indicated by shading in Figure 2. In the dark, a uniform electrostatic charge is laid down on the surface of the photoconductor. This charging step (Figure 2a) is accomplished by means of a positive corona discharge surrounding a fine wire held at about $+5000$ V. This corona (a miniature version of, and opposite in sign to, the intense precipitator corona of Figure 1) is passed over the photoconductor surface, spraying positive ions onto it and charging it to a potential of the order of $+1000$ V. Since charge is free to flow within the grounded metal backing, an equal and opposite induced charge (see Chapter 21) develops at the metal-photoconductor interface. In the dark the photoconductor contains no mobile charge, and the large potential difference persists across this dielectric layer, which is only 0.005 cm thick.

The photoconductor plate is next exposed to light in the form of an image reflected from the document being copied. What happens now is indicated in Figure 2b. Where light strikes the photoconductor, light quanta (photons) are absorbed, and pairs of mobile charges are created. Each photogenerated pair consists of a negative charge (an electron) and a positive charge (a hole; crudely, a missing electron). Photogeneration of this free charge depends not only on the photoconductor used and on the wavelength and intensity of the incident light but also on the electric field present. This large field ($1000$ V/0.005 cm $= 2 \times 10^5$ V/cm $= 2 \times 10^7$ V/m) helps to pull apart the mutually attracting electron-hole pairs so that they are free to move separately. The electrons then move under the influence of the field to the surface, where they neutralize positive charges, while the holes move to the photo-

**Figure 2**
Steps in the xerographic process: (a) charging, (b) exposure, (c) development, and (d) transfer.

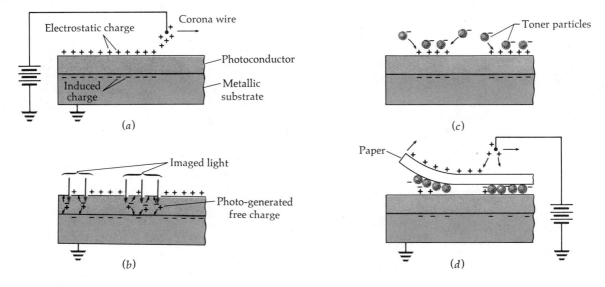

conductor-substrate interface and neutralize negative charges there. Where intense light strikes the photoconductor, the charging step is totally undone; where weak light strikes it, the charge is partially reduced; and where no light strikes it, the original electrostatic charge remains on the surface. The critical task of converting an optical image into an electrostatic image, which is now recorded on the plate, has been completed. This latent image consists of an electrostatic potential distribution which replicates the light and dark pattern of the original document.

To develop the electrostatic image, fine negatively charged pigmented particles are brought into contact with the plate. These *toner particles* are attracted to positively charged surface regions, as shown in Figure 2*c*, and a visible image appears. The toner is then transferred (Figure 2*d*) to a sheet of paper which has been positively charged in order to attract them. Brief heating of the paper fuses the toner to it and produces a permanent photocopy ready for use.

Finally, to prepare the photoconductor plate for a repetition of the process, any toner particles remaining on its surface are mechanically cleaned off, and the residual electrostatic image is erased, i.e., discharged, by flooding with light. The photoconductor is now ready for a new cycle, starting with the charging step. In high-speed duplicators the photoconductor layer is often in the form of a moving continuous drum or belt, around the perimeter of which are located stations for performing the various functions of Figure 2. The speed of xerographic printing technology is presently on the order of a few copies per second.*

---

* For further information on electrostatics in xerography, consult J. H. Dessauer and H. E. Clark (eds.), *Xerography and Related Processes,* Focal Press, New York, 1965, and R. M. Schaffert, *Electrophotography,* rev. ed., Focal Press, New York, 1973. Other applications of electrostatics are discussed in A. D. Moore, *Scientific American,* March 1972.

---

## Review

A. Define, explain, or otherwise identify:

B. True or false:

1. The capacitance of a conductor is defined to be the total amount of charge it can hold.

2. The capacitance of a parallel-plate capacitor depends on the voltage difference between the plates.

3. The capacitance of a parallel-plate capacitor is proportional to the charge on the plates.

4. The effective capacitance of two capacitors in parallel equals the sum of the individual capacitances.

5. The effective capacitance of two capacitors in series is less than that of either capacitor.

6. Since $V$ is energy per charge, the total energy in a capacitor is $QV$.

7. The electrostatic energy per unit volume at some point is proportional to the square of the electric field at that point.

## Exercises

### Section 23-1, Calculation of Capacitance

1. If a parallel-plate capacitor has 0.1-mm separation, what must its area be to have a capacitance of 1 F? If the plates are square, what is the length of their sides?

2. A parallel-plate capacitor has capacitance of 2.0 $\mu$F and plate separation of 1.0 mm. (a) How much potential difference can be placed across the capacitor before dielectric breakdown of air occurs ($E_{max} = 3$ MV/m)? (b) What is the magnitude of the greatest charge the capacitor can store before breakdown?

3. A coaxial cable between two cities has an inner radius of 1.0 cm and an outer radius of 1.1 cm. Its length is $8 \times 10^5$ m (about 500 mi). Treat this cable as a cylindrical capacitor and calculate its capacitance.

4. A Geiger tube consists of a wire of radius 0.2 mm and length 12 cm with a coaxial cylindrical shell conductor of the same length and radius 1.5 cm. (a) Find the capacitance assuming that the gas in the tube has a dielectric constant of 1. (b) Find the charge per unit length on the wire when the capacitor is charged to 1.2 kV.

5. What is the capacitance of an isolated spherical conductor of radius (a) 1.8 cm, (b) 1.8 m, (c) 1.8 km?

6. Considering the earth to be a spherical conductor of radius 6400 km, calculate its capacitance.

7. (a) Find the radius of an isolated spherical conductor which has a capacitance of 1 F. (b) What is the ratio of the radius of this sphere to the radius of the earth?

8. Calculate the capacitance of a parallel-plate capacitor of plate area 1.0 m and separation 1.0 cm.

### Section 23-2, Parallel and Series Combinations of Capacitors

9. A 10.0-$\mu$F capacitor is connected in series with a 20.0-$\mu$F capacitor across a 6.0-V battery. (a) What is the equivalent capacitance of this combination? (b) Find the charge on each capacitor. (c) Find the potential difference across each capacitor.

10. A 10.0- and a 20.0-$\mu$F capacitor are connected in parallel across a 6.0-V battery. (a) What is the equivalent capacitance of this combination? (b) What is the potential difference across each capacitor? (c) Find the charge on each capacitor.

11. Three capacitors have capacitance 2.0, 4.0, and 8.0 $\mu$F. Find the equivalent capacitance (a) if the capacitors are in parallel and (b) if they are in series.

12. A 2.0-$\mu$F capacitor is charged to a potential difference of 12.0 V and disconnected from the battery. (a) How much charge is on the plates? (b) When a second capacitor (initially uncharged) is connected in parallel across this capacitor, the potential difference drops to 4.0 V. What is the capacitance of the second capacitor?

13. (a) How many 1.0-$\mu$F capacitors would have to be connected in parallel to store 1 mC of charge with a potential difference of 10 V across each? (b) What would be the potential difference across the combination? (c) If these capacitors are connected in series and the potential difference across each is 10 V, find the charge on each and the potential difference across the combination.

14. A 1.0-$\mu$F capacitor is connected in parallel with a 2.0-$\mu$F capacitor, and the combination is connected in series with a 6.0-$\mu$F capacitor. What is the equivalent capacitance of this combination?

15. A 3.0- and a 6.0-$\mu$F capacitor are connected in series, and the combination is connected in parallel with an 8.0-$\mu$F capacitor. What is the equivalent capacitance of this combination?

### Section 23-3, Electrostatic Energy in a Capacitor

16. (a) A 3-$\mu$F capacitor is charged to 100 V. How much energy is stored in the capacitor? (b) How much additional energy is required to charge the capacitor from 100 to 200 V?

17. How much energy is stored in an isolated spherical conductor of radius 10 cm charged to 3 kV?

18. (a) A 10-$\mu$F capacitor is charged to $Q = 4$ $\mu$C. How much energy is stored? (b) If half the charge is removed, how much energy remains?

19. (a) Find the energy stored in a 20-pF capacitor when it is charged to 5 $\mu$C. (b) How much additional energy is required to increase the charge from 5 to 10 $\mu$C?

20. A conducting sphere of radius $R$ carries charge $q$. (a) Show that the work needed to bring additional charge $dq$ from infinity to the sphere is $(kq/R)\,dq$. (b) Use this result to show that the work needed to increase the charge from 0 to $Q$ is $\frac{1}{2}(kQ^2/R)$.

21. A parallel-plate capacitor of area $A$ and separation $d$ is charged to a potential difference $V$ and then disconnected from the charging source. The plates are then pulled apart until the separation is $2d$. Find expressions in terms of $A$, $d$, and $V$ for (a) the new capacitance, (b) the new potential difference, and (c) the new stored energy. (d) How much work was required to change the plate separation from $d$ to $2d$?

### Section 23-4, Electrostatic Field Energy

22. Find the energy per unit volume in an electric field equal to the breakdown field in air (3 MV/m).

23. A parallel-plate capacitor with plate area 2 m² and separation 1.0 mm is charged to 100 V. (a) What is the electric field between the plates? (b) What is the energy per unit volume in the space between the plates? (c) Find the total energy by multiplying your answer to part (b) by the total volume between the plates. (d) Find the capacitance $C$, calculate the total energy from $U = \frac{1}{2}CV^2$, and compare with your answer to part (c).

### Section 23-5, Dielectrics

24. A parallel-plate capacitor is made by placing polyethylene ($K = 2.3$) between sheets of aluminum foil. The area of each sheet is 400 cm², and the spacing is 0.3 mm. Find the capacitance.

25. A parallel-plate capacitor has plates of area 600 cm² and separation 4 mm. It is charged to 100 V and disconnected from the battery. (a) Find the electric field $E$, the charge density $\sigma$, and the energy $U$. A dielectric of constant $K = 4$ is inserted, completely filling the space between the plates. (b) Find the new electric field $E$ and the potential difference $V$. (c) Find the new energy. (d) Find the bound charge density.

26. What is the dielectric constant of a dielectric on which the induced bound charge density is (a) 80 percent of the free charge density, (b) 20 percent of the free charge density, and (c) 98 percent of the free charge density?

### Problems

1. For the arrangement shown in Figure 23-13, find (a) the total effective capacitance between the terminals, (b) the charge stored on each capacitor, and (c) the total stored energy.

2. A spherical capacitor consists of two spherical shells of radii $R_1$ and $R_2$. Show that when these radii are nearly equal, the capacitance is given approximately

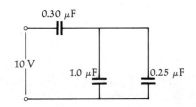

**Figure 23-13**
Problem 1.

by the expression for the capacitance of a parallel-plate capacitor, $C = \epsilon_0 A/d$, where $A$ is the area of the sphere and $d = R_2 - R_1$.

3. For the arrangement in Figure 23-14, find (a) the total effective capacitance between the terminals, (b) the charge stored on each capacitor, and (c) the total stored energy when the potential difference across the terminals is 200 V.

4. The effective capacitance of two capacitors in series can be written $C = C_1 C_2 / (C_1 + C_2)$. Show that the correct expression for the effective capacitance of three capacitors in series is $C = C_1 C_2 C_3 / (C_1 C_2 + C_2 C_3 + C_1 C_3)$.

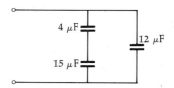

**Figure 23-14**
Problem 3.

5. A 20-pF capacitor is charged to 3.0 kV and is then removed and connected in parallel with an uncharged 50-pF capacitor. (a) What is the new charge on each capacitor? (b) Find the initial energy stored in the 20-pF capacitor and the final energy stored in the two capacitors. Is energy gained or lost in connecting the two capacitors?

6. Find all the different possible effective capacitances that can be obtained using a 1.0-, a 2.0-, and a 4.0-$\mu$F capacitor in any combination which includes all three or any two capacitors.

7. Three identical capacitors are connected so that their maximum equivalent capacitance is 15 $\mu$F. Find the three other combinations possible (using all three capacitors) and their equivalent capacitances.

8. Two capacitors of capacitance $C_1 = 4$ $\mu$F and $C_2 = 12$ $\mu$F are connected in series across a 12-V battery. They are then carefully disconnected without being discharged and connected in parallel with positive side to positive side and negative side to negative side. (a) Find the potential difference across each capacitor after they are connected in parallel. (b) Find the energy lost as they are connected in parallel.

9. Work Problem 8 for the two capacitors first connected in parallel across the 12-V battery and then connected in parallel, with the positive side of one capacitor connected to the negative side of the other.

10. A 100- and a 400-pF capacitor are both charged to 2.0 kV. They are then disconnected from the voltage source and connected together in parallel, positive side to positive side and negative side to negative side. (a) Find the resulting potential difference across each capacitor. (b) Find the energy lost as the connections are made.

11. Work Problem 10 for the capacitors connected in parallel with the positive side of one connected to the negative side of the other after they have been charged to 2.0 kV.

12. A parallel-plate capacitor has capacitance $C_0$ and plate separation $d$. Two dielectric slabs, of constants $K_1$ and $K_2$, each of thickness $\frac{1}{2}d$, and of the same area as the plates, are inserted between the plates as indicated in Figure 23-15. Show that the capacitance is then given by

$$C = \frac{2K_1 K_2}{K_1 + K_2} C_0$$

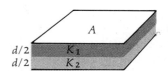

**Figure 23-15**
Problem 12.

13. A parallel-plate capacitor of area $A$ and separation $d$ is charged to potential difference $V$ and removed from the charging source. A dielectric slab of constant $K = 2$, thickness $d$, and area $\frac{1}{2}A$ is inserted, as shown in Figure 23-16. Let $\sigma_1$ be the free charge density at the conductor-dielectric surface and $\sigma_2$ be the charge density at the conductor-vacuum surface. (a) Why must the electric field have the same value inside the dielectric as in the free space between the plates? (b) Show that $\sigma_1 = 2\sigma_2$. (c) Show that the new capacitance is $3\epsilon_0 A/2d$ and the new potential difference is $\frac{2}{3}V$.

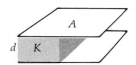

**Figure 23-16**
Problem 13.

14. A parallel-plate capacitor has plate area $A$ and separation $d$. A metal slab of thickness $t$ and area $A$ is inserted between the plates. Show that the capacitance is given by $C = \epsilon_0 A/(d - t)$ regardless of where the metal plate is placed. Show that this arrangement can be considered to be a capacitor of separation $a$ in series with one of separation $b$, where $a + b + t = d$.

15. In Figure 23-17, $C_1 = 2$ $\mu$F, $C_2 = 6$ $\mu$F, and $C_3 = 3.5$ $\mu$F. (a) Find the equivalent capacitance of this combination. (b) If the breakdown voltages of the individual capacitors are $V_1 = 100$ V, $V_2 = 50$ V, and $V_3 = 400$ V, what maximum voltage can be placed across points $a$ and $b$?

16. A parallel-plate capacitor is filled with two dielectrics of equal size, as shown in Figure 23-18. Show that the capacitance is increased by the factor $(K_1 + K_2)/2$.

17. Early capacitors, called Leyden jars, were actually glass jars coated inside and outside with metal foil. Suppose that the jar is a cylinder 40 cm high with 2.0-mm-thick walls of inner diameter 8 cm. Ignore any field fringing. (a) Find the capacitance of this jar. (b) What maximum charge can it take without collapse? The dielectric constant of the glass is 5.0, and its dielectric strength is 15 MV/m.

18. A parallel-plate capacitor of plate area $A$ and separation $x$ is given a charge $Q$ and removed from the charging source. (a) Find the electrostatic energy stored as a function of $x$. (b) Find the increase in energy $dU$ due to an increase in plate separation $dx$ from $dU = (dU/dx)$ $dx$. (c) If $F$ is the force exerted by one plate on the other, the work done to move one plate a distance $dx$ is $F$ $dx = dU$. Show that $F = Q^2/2\epsilon_0 A$. (d) Show that the force found in part (c) equals $\frac{1}{2}EQ$, where $Q$ is the charge on one plate and $E$ is the electric field between the plates. Discuss the reason for the factor $\frac{1}{2}$ in this result.

19. Design a circuit of capacitors which has a capacitance of 2 $\mu$F and breakdown voltage of 400 V using as many 2-$\mu$F capacitors as needed, each with a breakdown voltage of 100 V.

20. A parallel-plate capacitor of plate area 1.0 m$^2$ and plate separation distance 0.5 cm has a glass plate of the same area and thickness between its plates. The glass has dielectric constant 5.0. The capacitor is charged to a potential difference of 12.0 V and removed from its charging source. How much work is required to pull the glass plate out of the capacitor?

21. A spherical capacitor has an inner sphere of radius $R_1$ with charge $+Q$ and an outer concentric spherical shell of radius $R_2$ with charge $-Q$. (a) Find the electric field and the energy density at any point in space. (b) How much energy is in the volume of the spherical shell of radius $r$, thickness $dr$, and volume $4\pi r^2$ $dr$ between the conductors? (c) Integrate your expression in part (b) to find the total energy stored in the capacitor and compare your result with that obtained from $U = \frac{1}{2}QV$.

22. A cylindrical capacitor consists of a long wire of radius $R_1$ and length $L$ with positive charge $Q$ and a concentric outer cylindrical shell of radius $R_2$, length $L$, and charge $-Q$. (a) Find the electric field and the energy density at any point in space. (b) How much energy is in the cylindrical shell of radius $r$, thickness $dr$, and volume $2\pi rL$ $dr$ between the conductors? (c) Integrate your expression in part (b) to find the total energy stored in the capacitor and compare your result with that obtained from $U = \frac{1}{2}QV$.

23. A ball of charge of radius $R$ has a uniform charge density $\rho$ and total charge $Q = \frac{4}{3}\pi R^3 \rho$. (a) Find the electrostatic energy density at distance $r$ from the center of the charge for $r < R$ and for $r > R$. (b) Find the energy in a spherical shell of volume $4\pi r^2$ $dr$ for both $r < R$ and $r > R$. (c) Compute the total electrostatic energy by integrating your expressions in part (b) and show that your result can be written $U = \frac{3}{5}kQ^2/R$. Explain why this result is greater than that for a spherical conductor of radius $R$ carrying a total charge $Q$.

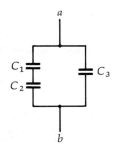

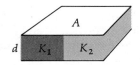

**Figure 23-17**
Problem 15.

**Figure 23-18**
Problem 16.

# CHAPTER 24      Electric Current

---

**Objectives**   After studying this chapter, you should:

1. Be able to define and discuss the concepts of electric current, current density, drift velocity, resistance, and emf.

2. Be able to state Ohm's law and distinguish between it and the definition of resistance.

3. Be able to give the definition of resistivity and describe its temperature dependence.

4. Be able to discuss the simple model of a real battery in terms of an ideal emf and an internal resistance and find the terminal voltage of a battery when it delivers a current $I$.

5. Be able to give the general relationship between potential difference, current, and power.

---

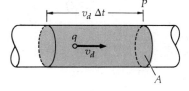

**Figure 24-1**
The current through area $A$ at point $P$ is the rate of flow of charge through that area. In time $\Delta t$ all the charge in the shaded volume $Av_d\, \Delta t$ flows past point $P$, where $v_d$ is the speed of the carriers. If there are $n$ carriers per unit volume each with charge $q$, the total charge is $\Delta Q = nqv_dA\, \Delta t$ and the current is $I = \Delta Q/\Delta t = nqv_dA$.

An electric current is the rate of flow of electric charge past a point. Although the flow of charge often takes place inside a conductor, this is not necessary; e.g., a beam of charged ions in vacuum from an accelerator constitutes a current. We define electric current as follows. Consider a small area element $A$, as in Figure 24-1, which might be the cross-sectional area of a conducting wire. The current through the area is defined as the amount of charge flowing through the area per unit time. If $\Delta Q$ is the charge that flows through the area in time $\Delta t$, the current is

---

$$I = \frac{\Delta Q}{\Delta t} \tag{24-1}$$

*Electric current defined*

---

The SI unit of current is the ampere (A):

$$1\text{ A} = 1\text{ C/s} \tag{24-2}$$

The direction of current is taken to be the direction of flow of positive charge. If the current is due to particles such as electrons which have negative charge, the direction of current is in the direction opposite the flow of the negatively charged particles. This definition of the direction of the current is purely arbitrary. In a conducting wire it is electrons which are free to move and produce the flow of charge. By our definition, the current is in the direction opposite to the motion of the electrons.

There are many examples of electric current outside of conducting wires. For example, a beam of protons from an accelerator produces a current in the direction of the motion of the positively charged protons. In electrolysis, the current is produced by the motion of both electrons and positive ions. Since these particles move in opposite directions, both produce current in the same direction. In nearly all applications, the motion of negative charges to the left is indistinguishable from the motion of positive charges to the right. [An exception is the Hall effect (Chapter 26).] We can always think of current as motion of positive charges in the direction of the current and remember (if we need to) that in conducting wires, for example, the electrons are moving in the direction opposite to the current.

## 24-1   Current and Motion of Charges

An electric current can be related to the motion of the charged particles responsible for it. Let us first consider the case of current in a conducting wire of cross-sectional area $A$. Let $n$ be the number of free charge-carrying particles per unit volume. For a conducting wire, these charge carriers are the free electrons. We shall assume that each particle carries a charge $q$ and moves with velocity $v_d$. In a time $\Delta t$ all the particles in the volume $A v_d \Delta t$, shaded in Figure 24-1, pass through the area element at point $P$. The number of particles in this volume is $n A v_d \Delta t$, and the total charge is

$$\Delta Q = q n A v_d \Delta t$$

The current at point $P$ is thus

$$I = \frac{\Delta Q}{\Delta t} = n q v_d A \qquad\qquad 24\text{-}3$$

The current per unit area is the *current density* $J$:

$$J = \frac{I}{A} = n q v_d \qquad\qquad 24\text{-}4$$

We can generalize the idea of current density to apply to any type of current whether confined to a wire or not. We define the current-density vector $\mathbf{J}$ by

$$\mathbf{J} = nq\mathbf{v}_d \qquad\qquad 24\text{-}5 \qquad \textit{Current density}$$

The current-density vector is in the direction of $\mathbf{v}_d$ if the charge $q$ is positive, and in the opposite direction if $q$ is negative. The quantity $\mathbf{v}_d$ is the average velocity of the charge carriers. For electrons in conductors this velocity is called the *drift velocity*. The current density is a fundamental quantity related by Equation 24-5 to the number density of charge carriers $n$, the charge $q$, and the average velocity of the carriers $\mathbf{v}_d$. If the current is due to more than one type of charged particle (as in electrolysis), the current density is found by summing $n_i q_i (\mathbf{v}_d)_i$ over the particles:

$$\mathbf{J} = \sum_i n_i q_i (\mathbf{v}_d)_i \qquad\qquad 24\text{-}6$$

If $\mathbf{J}$ is constant over the area $A$, the current through $A$ is

$$I = \mathbf{J} \cdot \hat{\mathbf{n}} A = J_n A \qquad\qquad 24\text{-}7$$

where $\hat{\mathbf{n}}$ is the unit vector perpendicular to the plane of the area $A$ and $J_n$ is the component of the current density parallel to $\hat{\mathbf{n}}$. If the current density is not constant, the current through a surface is found by integration:

$$I = \int \mathbf{J} \cdot \hat{\mathbf{n}} \, dA \qquad\qquad 24\text{-}8$$

We can get an idea of the order of magnitude of the drift velocity for electrons in a conducting wire by putting typical magnitudes into Equation 24-3.

---

**Example 24-1** What is the drift velocity of electrons in a typical copper wire (14 gauge) of radius 0.0814 cm carrying a current of 1 A?

If we assume one free electron per copper atom, the density of free electrons is the same as the density of atoms. Then

$$n = \frac{(6.02 \times 10^{23} \text{ atoms/mol}) \, (8.92 \text{ g/cm}^3)}{63.5 \text{ g/mol}}$$

$$= 8.46 \times 10^{22} \text{ atoms/cm}^3$$

and

$$v_d = \frac{I}{Ane} = \frac{1 \text{ C/s}}{\pi (0.0814 \text{ cm})^2 \, (8.46 \times 10^{22} \text{ cm}^{-3}) \, (1.60 \times 10^{-19} \text{ C})}$$

$$\approx 3.55 \times 10^{-3} \text{ cm/s}$$

We see that typical drift velocities are very small.

---

The actual instantaneous velocity of an electron in a metal is much larger than the drift velocity, which is the average velocity. The behavior of electrons in a metal is similar to that of gas molecules in air. In still air, the gas molecules move with large instantaneous velocities between collisions, but the average velocity is zero. When there is a breeze, the air molecules have a small drift velocity in the direction of the breeze superimposed on the much larger instantaneous velocity. We discuss this point further in Section 24-4 when we consider the classical model of electric conduction in metals.

### Question

1. Two wires of the same diameter but different materials are joined together to carry the same electric current. In one wire the density of charge carriers is twice that in the other. How do the drift velocities in the two wires compare?

## 24-2 Ohm's Law and Resistance

In our study of conductors in electrostatics we argued that the electric field inside a conductor must be zero in electrostatic equilibrium. If this were not so, the free charges inside the conductor would move about. We are now considering nonelectrostatic equilibrium situations, in which the free charge *does* move in a conductor. When a conductor carries a current, there is an electric field inside the conductor. The current is in the direction of the electric field. Figure 24-2 shows a segment of wire of length $L$ and cross-sectional area $A$ carrying a current $I$. Since there is an electric field in the wire, the potential at point $a$ is greater than that at point $b$. If the segment is short enough for us to neglect any variation of the electric field over the distance $L$, the potential difference between points $a$ and $b$ is

$$\Delta V = V_a - V_b = EL \qquad\qquad\qquad 24\text{-}9$$

where $E$ is the electric field. For most materials such as wire conductors, the potential difference between two points is found experimentally to be proportional to the current $I$ in the conductor. This experimental result is known as *Ohm's law*. The constant of proportionality is called the resistance $R$:

$$\Delta V = RI \qquad\qquad\qquad 24\text{-}10 \qquad \textit{Ohm's law}$$

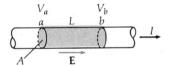

**Figure 24-2**
Segment of wire carrying current $I$. The potential difference is related to the electric field by $V_a - V_b = EL$.

The SI unit of resistance, the volt per ampere, is called an ohm ($\Omega$). The resistance of a conducting wire depends on the length of the wire, its cross-sectional area, the type of material, and the temperature, but for materials obeying Ohm's law, it does not depend on the current $I$. Such materials, which include most metals, are called *ohmic* materials. For many other, nonohmic materials the potential difference is not proportional to the current. For such materials, the resistance between two points is defined by

$$R = \frac{\Delta V}{I} \qquad\qquad\qquad 24\text{-}11 \qquad \textit{Resistance defined}$$

where $\Delta V$ is the potential difference between the points. Although the definition of resistance for nonohmic materials (Equation 24-11) is the same as that for ohmic materials, the resistance $R$ so defined is independent of the current $I$ for ohmic materials, whereas for nonohmic materials, the resistance does depend on the current. Figure 24-3 shows the potential difference $\Delta V$ versus current $I$ for ohmic and nonohmic materials. In both cases, the resistance $R$ is $\Delta V/I$, but for nonohmic materials $R$ depends on the current $I$ and for ohmic materials it does not. Ohm's law therefore is not a fundamental law of nature like Newton's laws or the laws of thermodynamics but an empirical description of a property shared by many materials.

For ohmic materials, the resistance of a conducting wire is found to be proportional to the length of the wire and inversely proportional to its cross-sectional area. The proportionality constant is called the *resistivity* $\rho$ of the conductor:*

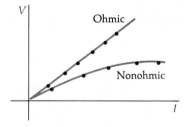

**Figure 24-3**
$V$ versus $I$ for ohmic and nonohmic materials. The resistance $R = \Delta V/I$ is independent of $I$ for ohmic materials, as indicated by the constant slope of the line.

$$R = \rho \frac{L}{A} \qquad\qquad\qquad 24\text{-}12 \qquad \textit{Resistivity defined}$$

* Care must be taken to distinguish between the resistivity $\rho$ and the volume charge density, also designated by $\rho$. It should be clear from the context which interpretation of $\rho$ is intended.

The unit of resistivity is the ohm-metre ($\Omega\cdot$m). The reciprocal of the resistivity is the conductivity $\sigma$.*

$$\sigma = \frac{1}{\rho} \tag{24-13}$$

*Resistivity and conductivity*

In terms of the conductivity, the resistance of an ohmic conducting wire is

$$R = \frac{L}{\sigma A} \tag{24-14}$$

Note that Equations 24-10 and 24-14 are the same as the equations for thermal conduction and resistance except that $\Delta T$ for thermal conduction is replaced by $\Delta V$ and the thermal conductivity $k$ is replaced by the electric conductivity $\sigma$.

The resistivity (and conductivity) of any given metal depends on the temperature. Figure 24-4 shows the temperature dependence of the resistivity for copper. Except at very low temperatures, the resistivity varies nearly linearly with temperature. The resistivity is usually given in tables in terms of its value $\rho_{20}$ at 20°C and the temperature coefficient of resistivity $\alpha$, which is the slope of the $\rho$-versus-$T$ curve. The resistivity $\rho$ at some other Celsius temperature $t$ is then given by

$$\rho = \rho_{20}[1 + \alpha(t - 20°C)] \tag{24-15}$$

(Since the Celsius and absolute temperatures differ only in the choice of zero, the resistivity has the same slope whether plotted against $t$ or $T$.)

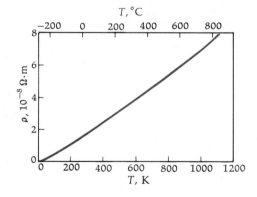

**Figure 24-4**
Plot of the resistivity $\rho$ versus absolute temperature $T$ for copper. (*By permission from D. Halliday and R. Resnick,* Fundamentals of Physics, *p. 512, John Wiley & Sons, Inc., New York, 1974.*)

The resistivity $\rho$ and temperature coefficient $\alpha$ are listed in Table 24-1 for various materials. This table shows that there is a wide range of values for the resistivity and an enormous difference between conductors and insulators.

Wires used to carry electric current are manufactured in standard sizes. The diameter of the circular cross section is indicated by a gauge number; higher numbers correspond to smaller diameters. For example, the diameter of a 10-gauge copper wire is 2.588 mm and that of 14-gauge wire is 1.628 mm. Handbooks give the combination $\rho/A$ in ohms per centimetre or ohms per foot.

---

* Again, the same symbol $\sigma$ is used in two different contexts, for conductivity and for surface charge density.

**Table 24-1**
Resistivities and temperature coefficients

| Material | Resistivity $\rho$ at 20°C, $\Omega \cdot m$ | Temperature coefficient $\alpha$ at 20°C, $K^{-1}$ |
|---|---|---|
| Silver | $1.6 \times 10^{-8}$ | $3.8 \times 10^{-3}$ |
| Copper | $1.7 \times 10^{-8}$ | $3.9 \times 10^{-3}$ |
| Aluminum | $2.8 \times 10^{-8}$ | $3.9 \times 10^{-3}$ |
| Tungsten | $5.5 \times 10^{-8}$ | $4.5 \times 10^{-3}$ |
| Iron | $10 \times 10^{-8}$ | $5.0 \times 10^{-3}$ |
| Lead | $22 \times 10^{-8}$ | $4.3 \times 10^{-3}$ |
| Mercury | $96 \times 10^{-8}$ | $0.9 \times 10^{-3}$ |
| Nichrome | $100 \times 10^{-8}$ | $0.4 \times 10^{-3}$ |
| Carbon | $3500 \times 10^{-8}$ | $-0.5 \times 10^{-3}$ |
| Germanium | $0.45$ | $-4.8 \times 10^{-2}$ |
| Silicon | $640$ | $-7.5 \times 10^{-2}$ |
| Wood | $10^{8} - 10^{14}$ | |
| Glass | $10^{10} - 10^{14}$ | |
| Hard rubber | $10^{13} - 10^{16}$ | |
| Amber | $5 \times 10^{14}$ | |
| Sulfur | $1 \times 10^{15}$ | |

The Bettmann Archive

Georg Simon Ohm (1781–1854)

**Example 24-2**  Calculate $\rho/A$ in ohms per metre for 14-gauge copper wire.
From Table 24-1 we have for the resistivity of copper

$$\rho = 1.7 \times 10^{-6} \ \Omega \cdot cm$$

The cross-sectional area of 14-gauge wire is

$$A = \frac{\pi d^2}{4} = \frac{\pi}{4} (0.163 \text{ cm})^2 = 2.09 \times 10^{-2} \text{ cm}^2$$

Thus

$$\frac{\rho}{A} = \frac{1.7 \times 10^{-6} \ \Omega \cdot cm}{2.09 \times 10^{-2} \ cm^2}$$
$$= (8.13 \times 10^{-5} \ \Omega/cm) = 8.13 \times 10^{-3} \ \Omega/m$$

This example shows that the copper connecting wires used in the laboratory have a very small resistance.

**Example 24-3**  What is the electric field in a 14-gauge copper wire that carries a current of 1 A?
According to Example 24-2, the resistance of a 1-m length of 14-gauge copper wire is $8.13 \times 10^{-3} \ \Omega$. The voltage drop across 1 m of this wire is

$$\Delta V = IR = 8.13 \times 10^{-3} \ V$$

and the electric field is

$$E = \frac{\Delta V}{L} = 8.13 \times 10^{-3} \ V/m$$

Note that the electric field in a conducting wire is very small.

Ohm's law can also be written in terms of the electric field $E$ and the current density $J$. Substituting $\Delta V = EL$ (Equation 24-9) and $I = AJ$ into Equation 24-10, we have

$$EL = R(AJ) = \rho \frac{L}{A} (AJ)$$

or

$$\mathbf{J} = \frac{1}{\rho} \mathbf{E} = \sigma \mathbf{E} \qquad\qquad 24\text{-}16$$

We have written this as a vector equation because the current density and the electric field vectors are in the same direction.

There are many metals for which the resistivity is zero below a certain temperature $T_c$, called the *critical temperature*. This phenomenon, called *superconductivity*, was discovered in 1911 by the Dutch physicist H. Kamerlingh Onnes. Figure 24-5 shows his plot of the resistance of mercury versus temperature. The critical temperature for mercury is 4.2 K. Critical temperatures for other superconductors range from less than 0.1 K for hafnium and iridium to 9.2 K for niobium. Many alloys are also superconducting. The highest critical temperature yet known (23 K) is for the alloy $Nb_3Ge$. The search continues for higher-temperature superconductors.

The conductivity of a superconductor cannot be defined since there can be a finite current density even when the electric field in the superconductor is zero. Steady currents have been observed to persist in superconducting rings for years with no electric field and no apparent loss. The phenomenon of superconductivity can be understood only with the help of quantum mechanics, which is beyond the scope of this book. The first successful theory of superconductivity was published by Bardeen, Cooper, and Schrieffer in 1957 and is known as the BCS theory.

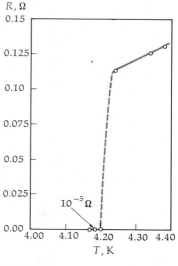

**Figure 24-5**
Plot by Kamerlingh Onnes of resistance of mercury versus temperature, showing sudden decrease at the critical temperature $T = 4.2$ K. (*By permission from C. Kittel*, Introduction to Solid State Physics, *3d ed., John Wiley & Sons, Inc., New York, 1966.*)

### Questions

2. Wire $a$ and wire $b$ have the same electric resistance and are made of the same material. Wire $a$ has twice the diameter of wire $b$. How do the lengths of the wires compare?

3. If wires $a$ and $b$ in Question 2 carry the same currents, how do their current densities compare? The voltage drops across them? The electric fields inside the wires?

4. In our study of electrostatics we concluded that there is no electric field within the material of a conductor. Why do we now find it possible to discuss electric fields inside conducting material?

## 24-3  Energy in Electric Circuits

When there is electric current in a conductor, electric energy is continually converted into thermal energy in the conductor. For example, in the simple classical model of conduction, covered in detail in Section 24-4, the electric field in the conductor accelerates the free electrons for a short time, giving them an increased kinetic energy which is quickly transferred into thermal energy of the conductor by collisions between the electrons and the lattice ions of the conductor. Thus, though the electrons continually gain energy from the electric field, this energy is

Demonstration of persistent currents. Oppositely directed superconducting current loops are set up in the ring and ball so that the magnetic force between the currents is repulsive. The ball floats above the ring, its weight balanced by the magnetic force of repulsion. (*Courtesy of Cryogenic Technology, Inc.*)

immediately transferred to thermal energy of the conductor and the electrons maintain a steady drift velocity on the average.

In general when (positive) charge flows in a conductor, it flows from high potential to low potential in the direction of the electric field (the negatively charged electrons of course move in the opposite direction). The charge thus loses potential energy. This potential-energy loss does not appear as kinetic energy of the charge carriers except momentarily, before it is transferred to the lattice ions by collisions. Consider the charge $\Delta Q$ which passes point $P_1$ in Figure 24-6 during a time $\Delta t$. If the potential at that point is $V_1$, the charge has potential energy $\Delta Q\, V_1$. During that time interval, the same amount of charge passes point $P_2$, where the potential is $V_2$. It has potential energy $\Delta Q\, V_2$, which is less than the original potential energy. The energy lost by the charge passing through this segment of the conductor is $-\Delta W$, given by

$$-\Delta W = \Delta Q(V_1 - V_2) = \Delta Q\, V$$

where $V = V_1 - V_2$ is the potential drop from point 1 to point 2. (Here we write $V$ rather than $\Delta V$ to simplify the notation.) The rate of energy loss by the charge is

$$-\frac{\Delta W}{\Delta t} = \frac{\Delta Q}{\Delta t}\, V = IV$$

where $I$ is the current. The power loss in the conductor is thus

---

$$P = IV \hspace{5cm} 24\text{-}17 \qquad \textit{Power}$$

---

If $I$ is in amperes and $V$ in volts, the power loss is in watts.

This expression for electric-power loss can easily be remembered by recalling the definitions of $V$ and $I$. The voltage drop is the potential-energy decrease per unit charge, and the current is the charge flowing per unit time. The product $IV$ is thus the energy loss per unit time, or the power loss. As we have seen, this power goes into heating the conductor.

Using the definition of resistance $R = V/I$, we can write Equation 24-17 in several other useful forms by eliminating either $V$ or $I$:

$$P = (IR)I = I^2 R \qquad \text{or} \qquad P = V\frac{V}{R} = \frac{V^2}{R} \hspace{2cm} 24\text{-}18$$

The energy put into a conductor is called *Joule heat*.

To have a steady current in a conductor we need to have a supply of electric energy. A device which supplies electric energy is called a *seat of electromotive force* or simply a *seat of emf*. It converts chemical, me- *emf* chanical, or other forms of energy into electric energy. It is often a bat- tery, which converts chemical energy into electric energy, or a genera- tor, which converts mechanical energy into electric energy. A seat of emf does work on the charge passing through it, raising the potential energy of the charge. The increase in the potential energy per unit charge is called the emf $\varepsilon$ of the seat. When a charge $\Delta Q$ flows through a seat of emf, its potential energy is increased by the amount $\Delta Q\, \varepsilon$. The unit of emf is the volt, the same as the unit of potential difference. The symbol for a seat of emf in a circuit diagram is ⊣⊢. The longer line indicates the higher-potential side. Ideally, a seat of emf maintains a constant potential difference $\varepsilon$ between its two sides independent of the rate of flow of charge, i.e., independent of the current in a circuit.

Figure 24-7 shows a simple circuit consisting of a resistance $R$ con- nected to a seat of emf. The resistance is indicated by the symbol ⌇⌇⌇.

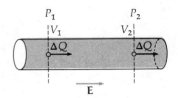

**Figure 24-6**
At point $P_1$ the charge $\Delta Q$ has potential energy $\Delta Q\, V_1$, and at point $P_2$ its potential energy is $\Delta Q\, V_2$. The energy loss in the conductor is $\Delta Q\, V$, where $V = V_1 - V_2$, and the power loss is $IV$.

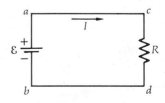

**Figure 24-7**
Simple circuit with an ideal emf, a resistance $R$, and a connecting wire that is as- sumed to be resistanceless.

Russ Kinne/Photo Researchers, Inc.

An unusual source of emf is the electric eel (*Electrophorus electricus*), which lives in the Amazon and Orinoco rivers of South America. When attacking, the eel can develop up to 600 V between its head (positive) and its tail (negative).

The straight lines in the circuit diagram indicate connecting wires of negligible resistance. Since the choice of zero potential is always arbitrary, we are interested only in the potential differences between various points in the circuit. The seat of emf maintains a constant potential difference $\varepsilon$ between points $a$ and $b$, with point $a$ at the higher potential. There is no potential difference between points $a$ and $c$ or between points $d$ and $b$ because the connecting wire is assumed to have negligible resistance. The potential difference between points $c$ and $d$ is therefore also $\varepsilon$, and the current in the resistor is given by $I = \varepsilon/R$. The direction of the current in this circuit is clockwise, as indicated in the figure. Note that *in the seat of emf the charge flows from low potential to high potential.** When charge $\Delta Q$ flows through the seat of emf, its potential energy is increased by the amount $\Delta Q\,\varepsilon$. It then flows through the conductor, where it loses this potential energy to heat. The rate at which energy is supplied by the emf is

$$P = \frac{\Delta W}{\Delta t} = \frac{\Delta Q\,\varepsilon}{\Delta t} = \varepsilon\,I \qquad\qquad 24\text{-}19$$

In this simple circuit, the power input by the seat of emf equals that dissipated in the resistor. An emf can be thought of as a sort of charge pump. It pumps the charge from low potential energy to high potential energy, analogously to a water pump pumping water from low to high gravitational potential energy.

Figure 24-8 shows a mechanical analog to the simple electric circuit discussed above. Marbles of mass $m$ roll along an inclined board with many nails in it. The marbles are accelerated by the gravitational field between collisions with the nails. They transfer the kinetic energy obtained between collisions to the nails during the collisions. Because of the many collisions, the marbles move with a small drift velocity toward the ground. When they reach the bottom, a boy picks them up and starts them again. The boy is the analog of the emf. He does work $mgh$ on each marble. The work per mass is $gh$, which is analogous to the work per charge done by the emf. The energy source in this case is the internal chemical energy of the boy.

A real battery is more than merely a seat of emf. The potential difference across the battery terminals, called the *terminal voltage*, is not simply the emf of the battery. Consider the simple circuit of a battery

(a)

(b)

**Figure 24-8**
Mechanical analog of resistance and emf. (*a*) As the marbles roll down the incline, their potential energy is converted into kinetic energy, which is quickly converted into heat because of collisions with the nails in the board. (*b*) A boy lifts the marbles up from low potential energy to high potential energy, converting his internal chemical energy into potential energy of the marbles.

*Terminal voltage*

* In some cases studied in the next chapter, such as charging one battery by another, the charge can flow in the opposite direction through a seat of emf.

Brown Brothers

The first practical method of maintaining a steady electric current was the battery, invented by Alessandro Volta in 1800. The earliest battery was a series of electric cells containing a copper disk and a zinc disk separated by a moistened pasteboard. Volta is shown here demonstrating his electric cell to Napoleon.

and resistor. Figure 24-9 shows a typical dependence of the terminal voltage of the battery on the current in the battery. The terminal voltage decreases slightly as the current increases.* As we have seen, an ideal seat of emf maintains a constant potential difference between its sides, independent of the current. Because the decrease in terminal voltage is approximately linear, we can represent a real battery by a seat of emf plus a small resistance $r$, called the *internal resistance* of the battery. Figure 24-10 shows a simple circuit containing a battery, a resistor, and connecting wires. As before, we ignore any resistance in the connecting wires. The circuit diagram for this circuit is shown in Figure 24-11. As charge passes from point $b$ to point $a$, its potential energy is first increased as it passes through the seat of emf and then decreased slightly as it passes through the internal resistance of the battery. If the current in the circuit is $I$, the potential at point $a$ is related to that at point $b$ by

$$V_a = V_b + \varepsilon - Ir$$

* This is assuming that the current through the battery is in the same direction as the emf of the battery, which is the case for a circuit with only one seat of emf. When a battery is being charged (by another seat of emf), the current is opposite to the direction of the emf and the terminal voltage increases slightly with increasing current.

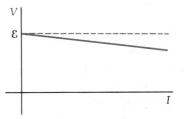

**Figure 24-9**
Terminal voltage $V$ versus current $I$ through a real battery. The dashed line shows the expected result for an ideal emf $\varepsilon$.

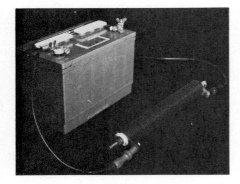

**Figure 24-10**
A simple circuit consisting of a battery, a resistor, and connecting wires. (*Courtesy of Larry Langrill, Oakland University.*)

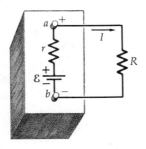

**Figure 24-11**
Circuit diagram representing the circuit with battery and resistor in Figure 24-10. The battery can be represented by an ideal emf $\varepsilon$ and a small resistance $r$.

The terminal voltage is thus

$$V_a - V_b = \varepsilon - Ir \qquad\qquad 24\text{-}20$$

The terminal voltage of the battery decreases linearly with current, as indicated in Figure 24-9. The potential drop across the resistor $R$ is the same as the terminal voltage. The current $I$ is thus given by

$$I = \frac{V_a - V_b}{R} = \frac{\varepsilon - Ir}{R}$$

or

$$I = \frac{\varepsilon}{R + r} \qquad\qquad 24\text{-}21$$

**Questions**

5.  What are several common kinds of emf's? What sort of energy is converted into electric energy by the emf in each?

6.  In a simple electric circuit like that shown in Figure 24-11 the charge outside the emf flows from positive voltage toward negative voltage, but in the same circuit the current inside the emf flows from minus to plus. Explain how this is possible.

7.  Figure 24-8 illustrates a mechanical analog to a simple electric circuit. Devise another in which the current is a flow of water instead of marbles.

8.  An emf is a device which converts chemical, mechanical, or another form of energy into electric energy. Could a device which uses electric energy as its source of energy be considered an emf? Can you think of any examples? What about an electronic power supply which serves the same function as a battery?

# 24-4   Classical Model of Electric Conduction

A classical model of electric conduction was first proposed by P. Drude in 1900 and developed by Hendrik A. Lorentz about 1909. This model successfully predicts Ohm's law and relates the conductivity and resistivity to the motion of free electrons in conductors. This classical theory is useful in understanding conduction even though it has been replaced by a modern theory based on quantum mechanics. We present the classical theory of electric conduction in this section and then discuss how some of the parameters are reinterpreted in the modern theory in Section 24-5.

In the classical model of electric conduction, a metal or other conductor is pictured as a regular three-dimensional array of atoms or ions with a large number of electrons free to move about the whole metal. In copper, for example, there is approximately one free electron per copper atom. The number of free electrons per atom can be measured using the Hall effect (Chapter 26). A copper atom with one electron missing is a copper ion with a positive charge $+e$. The regular arrangement of copper ions in metallic copper is called a *lattice*. In the absence of an electric field, the free electrons move about the metal much like gas molecules in a container. The free electrons make collisions with the ions of the lattice and are in thermal equilibrium with it. The rms speed of the electrons can be calculated from the equipartition theorem. The result is the same as that for an ideal-gas molecule with the electron mass replacing the molecular mass in Equation 16-23:

$$v_{\text{rms}} = \sqrt{\frac{3kT}{m}} \qquad\qquad 24\text{-}22$$

For example, at temperature $T = 300$ K the rms speed is

$$v_{\text{rms}} = \sqrt{\frac{3(1.38 \times 10^{-23} \text{ J/K}) (300 \text{ K})}{9.11 \times 10^{-31} \text{ kg}}} = 1.17 \times 10^5 \text{ m/s}$$

This speed is much larger than the drift velocity calculated in Example 24-1.

At first glance it is surprising that any material obeys Ohm's law. In the presence of an electric field, a free electron experiences a force $q\mathbf{E}$ which would cause an acceleration $q\mathbf{E}/m$ if it were the only force acting on the electron. But Ohm's law implies a steady drift velocity (not acceleration) of the electron which is proportional to the electric field. The current density $\mathbf{J} = nqv_d$ is proportional to the drift velocity, and according to Ohm's law, $\mathbf{J}$ is proportional to the electric field $\mathbf{E}$. Of course there are other forces acting on the electrons because they make collisions with the lattice ions. In the classical model of conduction, it is assumed that after an electron makes a collision with a lattice ion, its velocity is completely unrelated to that before the collision. The justification of this assumption is that the drift velocity is very small compared with the random thermal velocity of the electron. We thus picture the electron as being accelerated by the electric field for a short time between collisions with the lattice ions, thereby acquiring a small drift velocity. After each collision, which gives it a random velocity, the electron is again accelerated by the field. The motion of such an electron is shown in Figure 24-12, in which the drift velocity is greatly exaggerated. In Figure 24-13$b$ a small drift velocity $\mathbf{v}_d$ has been added to the electron velocities, which in Figure 24-13$a$ are random. We can relate the drift velocity to the electric field by ignoring the random thermal velocities of the electrons and assuming that the electron starts from rest after a collision. After a time $\tau$ the electron will have a velocity of magnitude equal to the acceleration $q\mathbf{E}/m$ times the time:

$$v = \frac{qE}{m} \tau \qquad\qquad 24\text{-}23$$

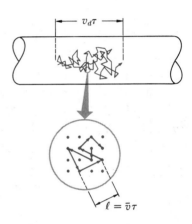

**Figure 24-12**
Electron path in a wire. Superimposed on the random thermal motion is a slow drift in the direction of the electric force $q\mathbf{E}$. The mean free path $\ell$, mean time between collisions $\tau$, and mean speed $\bar{v}$ are related by $\ell = \bar{v}\tau$.

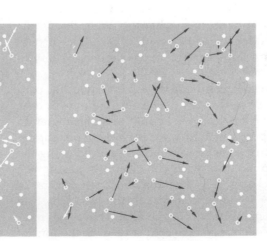

Electron    Positive ion

(a) Average electron velocity = 0          (b) Average electron velocity = ⟶

**Figure 24-13**
(a) Stationary positive ions, and electrons with a random distribution of velocities. (b) A small drift velocity to the right has been added to each electron velocity, as indicated for the electron in the lower left corner. (*Adapted from E. M. Purcell,* Electricity and Magnetism, *p. 123,* Berkeley Physics Course, *McGraw-Hill Book Company, New York, 1965. Courtesy of Education Development Center, Inc., Newton, Mass.*)

If we use this expression for the drift velocity and insert it in Equation 24-4 for the current density, we have

$$J = nqv_d = nq\,\frac{qE}{m}\,\tau = \frac{nq^2\tau}{m}\,E \qquad\qquad 24\text{-}24$$

Comparing this result with the definition of electric conductivity and its reciprocal, the resistivity, gives

$$\sigma = \frac{nq^2\tau}{m} \quad\text{and}\quad \rho = \frac{m}{nq^2\tau} \qquad\qquad 24\text{-}25$$

The average distance the electron travels between collisions is called the *mean free path* $\ell$. It is the product of the mean speed $\bar{v}$ and the mean time between collisions $\tau$:

*Mean free path*

$$\ell = \bar{v}\tau \qquad\qquad 24\text{-}26$$

In terms of the mean free path and the mean speed, the conductivity and resistivity are

$$\sigma = \frac{nq^2\ell}{m\bar{v}} \qquad \rho = \frac{m\bar{v}}{nq^2\ell} \qquad\qquad 24\text{-}27$$

According to Ohm's law, the conductivity and resistivity are independent of the electric field $E$. The quantities in Equation 24-27 that might depend on the electric field are the mean speed $\bar{v}$ and the mean free path $\ell$. As we have seen, the drift velocity is very much smaller than the rms speed of the electrons in thermal equilibrium with the lattice ions. Thus the electric field has essentially no effect on the mean speed of the electrons. The mean free path of the electron depends on the chance that in traveling some distance the electron will collide with a lattice ion. According to the classical model, this chance depends on the size of the lattice ion and on the density of ions, neither of which depends on the electric field $E$. Thus this model predicts Ohm's law with the resistivity and conductivity given by Equations 24-27.

---

**Example 24-4** Use Equation 24-23 and the results of Examples 24-1 and 24-3 to estimate the mean time between collisions for electrons in 14-gauge copper wire when the current is 1 A.

From these examples we have $v_d = 3.55 \times 10^{-3}$ cm/s $= 3.55 \times 10^{-5}$ m/s and $E \approx 8.13 \times 10^{-3}$ V/m. Using $q = 1.60 \times 10^{-19}$ C and $m = 9.11 \times 10^{-31}$ kg for the charge and mass of the electron, we have

$$\tau = \frac{mv_d}{qE} = \frac{(9.11 \times 10^{-31}\text{ kg})(3.55 \times 10^{-5}\text{ m/s})}{(1.60 \times 10^{-19}\text{ C})(8.13 \times 10^{-3}\text{ V/m})}$$

$$= 2.49 \times 10^{-14}\text{ s}$$

The time between collisions is very short.

---

**Example 24-5** Assuming that the mean speed is about $10^5$ m/s, the same order of magnitude as the rms speed calculated in this section, find the order of magnitude of the mean free path of an electron in a copper wire.

Using $\tau \approx 2.5 \times 10^{-14}$ s from Example 24-4, we have

$$\ell = \bar{v}\tau = (10^5\text{ m/s})(2.5 \times 10^{-14}\text{ s})$$
$$= 2.5 \times 10^{-9}\text{ m}$$

Although successful in predicting Ohm's law, the classical theory of conduction has several defects. The numerical magnitudes of the conductivity and resistivity calculated from Equation 24-27 using classical methods for finding the mean free path and the mean speed differ from measured values by a factor of 10 or so, depending on the temperature at which these values are measured. The temperature dependence of resistivity is given completely by the mean speed $\bar{v}$ in Equation 24-27, which is proportional to $\sqrt{T}$. Thus this calculation does not give a linear dependence on temperature. Finally, the classical model says nothing about why some materials are conductors, others insulators, and still others semiconductors, a subject we discuss briefly in Section 24-6.

## 24-5   Corrections to the Classical Theory of Conduction

The results of the modern theory of electric conduction based on quantum mechanics can be discussed briefly and qualitatively. The important features of this theory for our discussion are that rms speed $v_{rms}$ is not given by Equation 24-22, because the equipartition theorem does not hold for electrons in a metal, and that the mean free path must be calculated using the wave nature of propagation of electrons in the metal. The calculation of the mean speed $\bar{v}$ is quite complicated. The results are that $\bar{v}$ is essentially independent of temperature and equals about 1.6 Mm/s for copper, roughly 16 times that calculated at $T = 300$ K from the equipartition theorem. The wave analog of electron collisions with the lattice ions is the scattering of electron waves by the ions. A detailed calculation of electron-wave scattering by a perfectly periodic lattice of identical ions gives the result that there is no scattering; the mean free path is infinite. Thus for such a perfect crystal, the resistance is zero. Electron waves are scattered only if the lattice is not perfectly periodic. There are two main causes for lattice deviations from perfect periodicity. One is impurities. For example, if some zinc is introduced into pure copper, the previously perfect periodicity is destroyed. At very low temperatures the resistance of a metal is primarily due to impurities. The other cause of deviations is the displacement of the lattice ions due to vibrations. This effect is dominant at ordinary temperatures. Thus the mean free path is not determined by the size of the lattice ions. At very low temperatures the ions effectively look like points to the electron so far as scattering is concerned. At normal temperatures, the effective area of an ion is proportional to the square of the amplitude of vibration. This in turn is proportional to the energy of vibration, which is proportional to $T$ (except at very low temperatures). The modern theory accurately accounts for the temperature dependence of resistance.

As an illustration of these ideas it is instructive to compare the resistivity of copper and brass (an alloy of copper and zinc) at temperatures of 300 and 4 K. The resistivity of copper drops by a factor of about 50 to 100 (depending on the purity) when the temperature drops from 300 to 4 K, indicating that most of the resistivity at 300 K is due to thermal motion. However, the resistivity of brass drops by only a factor of 4 with temperatures from 300 to 4 K. A large part of the resistivity of brass is due to the temperature-independent irregularity of the copper-zinc lattice.

*Optional*

## 24-6   Conductors, Insulators, and Semiconductors

The classical model we have discussed for electric conduction with the modifications due to quantum mechanics gives a good account of Ohm's law and the temperature dependence of resistivity and conductivity, but it does not deal with the important question of why some materials are good conductors and others are insulators. As we have seen, there is an enormous variation in the conductivity and resistivity from the best conductors to the best insulators. The difference between a conductor and an insulator can be characterized empirically by the density of charge carriers $n$. For a good conductor the number of charge carriers is about one per atom; for insulators the number of charge carriers is nearly zero.

*Energy quantization*

To discuss why this is so, even qualitatively, we need two important ideas from quantum mechanics. One is that the energy of an electron in an atom cannot take on just any value but is quantized. This idea was first put forward by Bohr, who proposed that the possible energy values or levels for the hydrogen atom are given by the simple formula

$$E_n = -\frac{13.6}{n^2}\ \text{eV}$$

where $n$ is any integer, $n = 1, 2, 3, \ldots$ . The lowest possible energy for the hydrogen atom corresponds to $n = 1$; the other possible energies are given by this expression with other values of $n$. No simple formulas exist for the possible energies of electrons in other more complicated atoms, but the important point for this discussion is that for any atom there is a discrete set of possible energies.

*Pauli exclusion principle*

The other quantum-mechanical result needed to understand why some materials are conductors and others are not is the *Pauli exclusion principle*. Only two electrons (with opposite spins) can occupy a given quantum state. Consider for a moment two identical atoms very far apart. Each atom has the same set of allowed energy levels. Let us focus our attention on one such level (such as $n = 2$ for hydrogen). There is a level of the same energy for each atom. If the atoms are brought close together, the energy of this level changes because of the influence of the other atom. If the atoms are close enough together, we can no longer associate the energy levels with one of the atoms but must look at the two-atom system. The two previously identical energy levels split into two levels of slightly different energy for the two-atom system. Similarly, if we have $N$ identical atoms, a particular energy level of the isolated atom splits into $N$ different nearly equal energy levels when the atoms are sufficiently close together. The number of copper atoms in a piece of copper is very large, on the order of Avogadro's number. A single atomic energy level identified with an isolated copper atom splits into a *band* of a very large number of energy levels which are closely spaced in energy. Because the number of levels in the band is so large, they are spaced almost continuously within the band. There is a separate band of levels for each particular energy level of the isolated atom. In the isolated atom the different levels are often far apart. (For example, in hydrogen the energy for $n = 1$ is $-13.6$ eV, and for $n = 2$ it is $-13.6/4 = -3.4$ eV.) The energy bands corresponding to these individual levels for a large number of atoms in a solid may be widely separated in energy, may be close together, or may even overlap in energy, depending on the kind of atom and the type of bonding in the crystal.

Figure 24-14 shows four possible kinds of band structure for a solid. The band structure for copper is shown in Figure 24-14a. The lower bands are filled with the inner electrons of the atoms; i.e., according to the Pauli exclusion principle, no more electrons can occupy levels in these bands. The uppermost band containing electrons is only about half full. In the normal state, at low temperatures, the lower half of the energy band is filled, and the upper half is empty. At higher temperatures, a few of the electrons are in the higher energy states in that band because of thermal excitation, but there are still many unfilled energy states above the filled ones.

When an electric field is established in the conductor, the electrons in the upper band are accelerated, which means that their energy is increased. This is consistent with the Pauli exclusion principle because there are many empty energy states just above those occupied by electrons in this band. These electrons are thus the conduction electrons, and the band is called the *conduction band*.

Figure 24-14b shows the band structure for a typical insulator. At $T = 0$ K the highest energy band that contains electrons is completely full. The next energy band, containing empty energy states, is separated from the last filled band by an energy gap which is quite large compared with thermal energies (of the order of 0.02 eV). This energy gap is sometimes referred to as a *forbidden energy band*. Very few electrons can be thermally excited to the nearly empty conduction band, even at fairly high temperatures. When an electric field is established in the solid, electrons cannot be accelerated because there are no empty energy states at nearby energies. In the classical model, we describe this by saying that there are no free electrons. The small conductivity that is observed is due to the very few electrons that are thermally excited into the upper nearly empty conduction band.

*Conduction band*

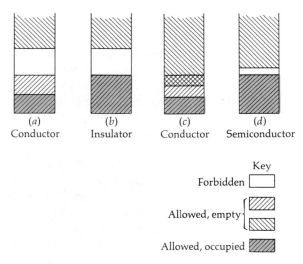

(a)
Conductor

(b)
Insulator

(c)
Conductor

(d)
Semiconductor

Key

Forbidden ☐

Allowed, empty

Allowed, occupied

**Figure 24-14**
Four possible band structures for a solid. In (a) an allowed band is only partially full, whereas in (c) two allowed bands overlap. In both cases nearby energy states are available for excitation of electrons, and so these materials are conductors. (b) The band structure of an insulator, where there is a forbidden band with a large energy gap between a filled allowed band and the next allowed band. (d) Since the energy gap between the filled band and the next allowed band is small, some electrons are excited to the next allowed band at normal temperatures, making this material a semiconductor.

These results—that the conduction band is essentially empty and that the gap between it and the lower energy band that is nearly filled is large—can be derived by the methods of quantum mechanics. In some materials the energy gap between the top filled band and the empty conduction band is very small (Figure 24-14d). At ordinary tem-

*Semiconductor*

peratures there are an appreciable number of electrons in the conduction band due to thermal excitation. Such a material is called a *semiconductor*. In the presence of an electric field, the electrons in the conduction band can be accelerated. Also, for each electron in the conduction band there is a vacancy, or hole, in the nearly filled band (this band is called the *valence band*). In the presence of a field, electrons in this band can be excited to a vacant energy level. This contributes to the electric current and is most easily described as the motion of a hole in the direction of the field and opposite the motion of the electrons. The hole thus acts like a positive charge. (An analogy is a line of cars with a space the size of one car: as the cars move to fill the space, the space moves backward, in the direction opposite the motion of the cars.) An interesting feature of a semiconductor is that as the temperature increases, the conductivity increases (and the resistivity decreases), contrary to the usual behavior of conductors. The reason is that as the temperature is increased, the number of free electrons $n$ is increased because there are more electrons in the conduction band. (The number of holes of course also increases.) This increase in $n$ outweighs the natural decrease in conductivity due to a smaller mean free path at higher temperatures.

A semiconductor can also be made from an insulator by introducing certain impurities (this procedure is called *doping*) which slightly modify the band structure. These slight modifications have a great effect on the conductivity. In one type of doped semiconductor the impurity atoms provide electrons in energy states which lie in the original forbidden band just below the empty conduction band (Figure 24-15a).

**Figure 24-15**
(a) Energy bands of an *n*-type semiconductor. Impurity atoms provide filled energy levels just below the empty conduction band and donate electrons to the conduction band. (b) Energy bands of a *p*-type semiconductor. Impurity atoms provide empty energy levels just above the filled valence band and accept electrons from it.

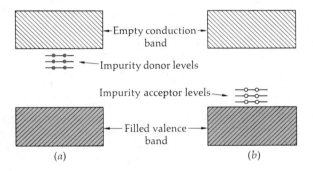

These impurity energy levels are called *donor levels*. The impurity atoms donate electrons to the conduction band. This is called an *n-type semiconductor* because the charge carriers are negative electrons. In a second kind of doped semiconductor the impurity atoms have empty energy levels just above the filled band in the original solid, as in Figure 24-15b. These levels accept electrons from the filled valence band when these electrons are thermally excited to a higher energy state. Thus a hole is created in the valence band, which is free to propagate in the direction of an electric field. This is called a *p-type semiconductor* because the charge carriers are positive holes.

# Transistors

Reuben E. Alley, Jr.
*U.S. Naval Academy*

In 1949, John Bardeen, Walter Brattain, and William Shockley, all of Bell Telephone Laboratories, initiated a revolution in electronics with the invention of the transistor. In 1956 they received a Nobel prize for their work.

Transistors, semiconductor diodes, and the integrated circuit, which combines the functions of many transistors and their associated resistors and capacitors into one small package, have almost completely replaced vacuum tubes in all but a few special applications, e.g., television picture tubes and high-power radio transmitters. These solid-state devices can be made very small, very light, and from readily available materials. They are cheap to manufacture, require little power, and operate at room temperatures. A spectacular example of reduction in size and cost is the pocket calculator, available in some models for under $10. The most expensive parts of the modern pocket calculator are said to be the control-button switches. Their size, which is limited by the size of the human finger, prevents further miniaturization by a factor of perhaps 25.

At least one company offers a programmable calculator whose memory capacity and computational abilities equal or perhaps surpass those of the best vacuum-tube computer of the early 1950s. Its cost (a few hundred dollars) is less than one-hundredth and its volume is less than one-thirty-thousandth that of the vacuum-tube counterpart. It is portable, operates on tiny batteries, and is highly reliable. In contrast, the vacuum-tube computer required an enormous power supply and air conditioning to remove the great quantities of waste heat resulting from the high operating temperature of the vacuum tubes. This high temperature seriously limited the life and reliability of the individual components.

Manufacture of diodes, transistors, and integrated circuits requires a very pure sample of one element throughout which a carefully controlled, exceedingly small amount (a few parts per million) of another element has been uniformly distributed. The basic material usually is germanium or silicon. Both these elements are semiconductors; i.e., at room temperature they have relatively few free charges (compared with the practically unlimited supply of free charge in a conductor such as copper). The second element (the *impurity* or *doping* material) must lie near silicon or germanium in the periodic table but have one less or one more valence electron than silicon or germanium. Commonly used are boron (valence 3) and arsenic (valence 5). Adding boron produces a material that acts as if it had a large number of free positive charges. Such material is called a *p*-type semiconductor. Similarly, the introduction of arsenic into the pure material produces a sample in which current is carried by flow of negative charges (an *n*-type semiconductor).

A single crystal of silicon made to contain a *p*-type region and an adjacent *n*-type region with a fairly abrupt discontinuity between them is a diode. Within a specified voltage range (below the breakdown voltage indicated in Figure 1) a diode is a good conductor for one polarity of applied voltage and a very poor conductor for applied voltage of opposite polarity.

Semiconductor diodes are used in *rectifiers*, which convert the alternating voltage supplied by all power companies into unidirectional voltage. With *filters* and *regulating circuits* one can construct a *power supply* that provides a direct voltage of constant amplitude.

Radios, television sets, sound-reproducing systems, and many other devices require a circuit that will accept a time-dependent voltage or current of very small amplitude (the *input*) and provide a greatly magnified replica of the input voltage or current as its *output*. Such a circuit, which must be supplied with energy from a battery or a power supply, is called an *amplifier*.

**Figure 1**
Diode characteristic. Note the different scales in the first and third quadrants.

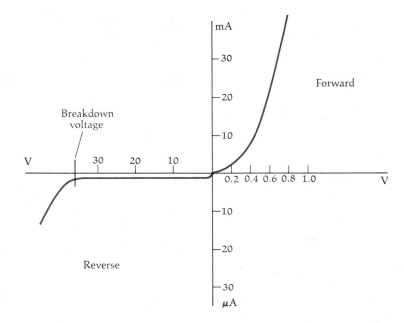

**Figure 2**

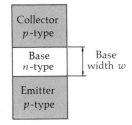

**Figure 3**
A *pnp* transistor biased for normal operation in the common-emitter connection.

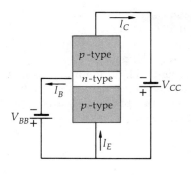

The basic element of most amplifiers is the transistor, which contains three distinct regions of semiconductor material. Figure 2 is an idealized representation of a *pnp* transistor, where the three regions are labeled with their common designations. In a properly designed transistor the level of doping is much greater in the emitter than in the other two regions, and the base width $w$ is very narrow.

The device represented in Figure 2 contains two *pn* junctions. In a practical circuit (Figure 3) voltages are supplied so that the emitter-base (*EB*) junction is forward-biased and the collector-base (*CB*) junction is reverse-biased. The difference in doping levels of the emitter and the base means that practically all the current across the *EB* junction consists of positive charges from the emitter. Because the base width is small and the *CB* junction is reverse-biased, most of the positive charges from the emitter that enter the base diffuse across its narrow width and enter the collector. The few positive charges that are not collected by the *CB* junction leave the base through the external connection. In Figure 3, therefore, $I_C$ is almost but not quite equal to $I_E$, and $I_B$ is much smaller than either $I_C$ or $I_E$. It is customary to write

$$I_C = \beta I_B$$

where $\beta$, the *current gain*, is an important parameter of the transistor. Transistors can be designed to have values of $\beta$ as low as 10 or as high as several hundred.

In a useful transistor, $\beta$ will be constant; i.e., the device will be linear, over a restricted range of base current, provided that proper biases are maintained on the two junctions. Figure 4 shows a simple amplifier. A small time-varying voltage, say from the motion of a needle in the groove of a phonograph record, is in series with a bias voltage $V_{BB}$. The base current is then the sum of a steady current $I_B$ produced by $V_{BB}$ and a varying current $i_b$ produced by $v_s$, the voltage to be amplified. Because $v_s$ may at any instant be either positive or negative, $V_{BB}$ must be included and must be large enough to ensure that there is always a forward bias on the *EB* junction. Since the device is linear, the collector current will consist of two parts, a steady current $I_C = \beta I_B$ and a time-varying current $i_c = \beta i_b$. We thus have a *current amplifier* where the time-varying output current $i_c$ is $\beta$ times the input current $i_b$. In such an amplifier, the steady currents $I_C$ and $I_B$, while essential to proper operation, are usually not of interest. (If an *npn* transistor is used, the bias voltages must be reversed and the currents will all be in the other direction. Otherwise, operation is identical with that described here.)

(a)

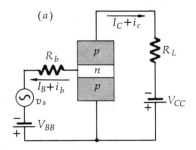

(b)

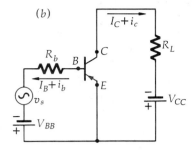

(c)

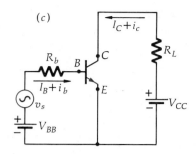

**Figure 4**
(a) Simple amplifier circuit using *pnp* transistor. (b) The same circuit with the transistor represented by a standard symbol. (c) Amplifier using *npn* transistor. Battery polarities and current directions are opposite those in part (b).

**Figure 5**
(a) Five-watt audio amplifier using an integrated circuit. The integrated circuit is partly hidden by a piece of metal 3.3 cm long. The five cylinders are capacitors. (b) Photomicrograph of an integrated circuit with more than 50 components, on a silicon chip approximately 0.25 cm by 0.2 cm.

(a)

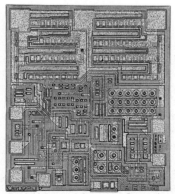

Fairchild Camera & Instrument Corporation

(b)

Ohm's law tells us that $v_s$ and $i_b$ are related by

$$i_b = \frac{v_s}{R_b + r_b}$$

where $r_b$ is the internal resistance of the transistor between base and emitter. Likewise, in the collector circuit the current $i_c$ produces a voltage $v_L$ across the resistor $R_L$,

$$v_L = i_c R_L$$

We already have

$$i_c = \beta i_b = \beta \frac{v_s}{R_b + r_b}$$

and therefore,

$$v_L = \beta \frac{R_L}{R_b + r_b} v_s$$

and the voltage gain or voltage amplification is

$$\frac{v_L}{v_s} = \beta \frac{R_L}{R_b + r_b}$$

In a practical case $\beta$ may be 100, and the ratio $R_L/(R_b + r_b)$ may be $\frac{1}{2}$. Then the voltage gain is 50. A more detailed derivation shows that $v_L$ and $v_s$ are exactly *out of phase*; i.e., when $v_s$ has its most positive value, $v_L$ has its most negative value. For a simple amplifier this phase shift is not important because all input voltages, regardless of frequency, are affected identically. The simple voltage amplifier is the basis of many circuits such as oscillators and modulators that constitute communication systems.

The complete amplifier in a record or tape player generally consists of several *stages* similar to Figure 4 connected in *cascade* so that the output of one becomes the input of the next. Thus the very small voltage produced by the motion of the needle or by the passage of magnetized tape through the player head controls the large amounts of power required to drive a system of loudspeakers. The energy delivered by the speakers is supplied by the sources of direct voltage connected to each transistor.

The resistance of a semiconductor material depends upon the impurity concentration. Reverse-biased diodes have capacitance that can be controlled by controlling the bias voltage. These two facts are exploited in the manufacture of *integrated circuits*, in which several transistors along with associated resistors and capacitors are interconnected on a single small piece of silicon. Figure 5a shows a 5-W audiofrequency amplifier that uses one integrated circuit containing 14 transistors, 3 diodes, and 10 resistors. Figure 5b shows a different integrated circuit containing even more circuit elements.

In addition to greater reductions in their size and weight, the introduction of integrated circuits has changed the design of electronics devices radically. For example, one need no longer design an amplifier but can choose from those available a complete amplifier with the desired characteristics. It is then procured and installed as a complete unit. Such possibilities have enabled small companies with little expertise in design to build rather complex electronics equipment and compete favorably in price and quality.

## Review

A.  Define, explain, or otherwise identify:

Current, 671                        Superconductivity, 677
Ampere, 671                         Joule heat, 678
Current density, 672               emf, 678
Drift velocity, 672                Terminal voltage, 679
Ohm's law, 674                     Internal resistance, 680
Resistance, 674                    Mean free path, 683
Resistivity, 674                   Energy bands, 685
Conductivity, 675                  Conduction band, 686

B.  True or false:

1.  The net motion of electrons is in the direction of the current.

2.  Current density is proportional to the drift velocity.

3.  $R = \Delta V/I$ is Ohm's law.

4.  The current in a conductor is always proportional to the potential drop across the conductor.

5.  $R = \Delta V/I$ holds for all materials for which resistance can be defined.

6.  The resistance of a metal is due to irregularities in the spacing of the lattice ions.

## Exercises

### Section 24-1, Current and Motion of Charges

1.  A wire carries a steady current of 2.0 A. (*a*) How much charge flows past a point in the wire in 5.0 min? (*b*) If the current is due to flow of electrons, how many electrons flow past a point in this time?

2.  A 10-gauge copper wire (diameter 2.59 mm) carries a current of 20 A. Assuming one free electron per copper atom, calculate the electron drift velocity.

3.  In a 3.0-cm-diameter fluorescent tube, $2.0 \times 10^{18}$ electrons and $0.5 \times 10^{18}$ positive ions (charge $+e$) flow past a point each second. (*a*) What is the current in the tube? (*b*) What is the current density?

4.  In a certain electron beam, there are $5.0 \times 10^6$ electrons per cubic centimetre. The kinetic energy of the electrons is 10.0 keV, and the beam is cylindrical with diameter 1.00 mm. (*a*) What is the velocity of the electrons? (*b*) Find the current density. (*c*) Find the beam current.

5.  A charge $+q$ moves in a circle of radius $r$ with speed $v$. (*a*) Express the frequency $f$ with which the charge passes a point in terms of $r$ and $v$. (*b*) Show that the average current is $qf$ and express it in terms of $v$ and $r$.

6.  A ring of radius $R$ has charge per unit length $\lambda$ which rotates with angular velocity $\omega$ about its axis. Find an expression for the current.

7.  A 20.0-MeV proton beam of diameter 2.0 mm in a certain accelerator constitutes a current of 1.0 mA. The beam strikes a metal target and is absorbed by it. (*a*) What is the density of protons in the beam? (*b*) How many protons strike the target in 1.0 min? (*c*) If the target is initially uncharged, express the charge of the target as a function of time.

8.  A 10-gauge wire (diameter 2.59 mm) is welded end to end to a 14-gauge wire (diameter 1.63 mm). The wires carry a current of 15 A. (*a*) What is the current density in each wire? (*b*) If both wires are copper with one free electron per atom, find the drift velocity in each wire.

## Section 24-2, Ohm's Law and Resistance

9. A 10-m wire of resistance 0.2 Ω carries a current of 5 A. (*a*) What is the potential difference across the wire? (*b*) What is the magnitude of the electric field in the wire?

10. A potential difference of 100 V produces a current of 3 A in a certain resistor. (*a*) What is its resistance? (*b*) What is the current when the potential difference is 25 V?

11. A copper wire and an iron wire have the same length and diameter and carry the same current *I*. (*a*) Find the potential drop across each wire and the ratio of these drops. (*b*) In which wire is the electric field greater?

12. A piece of carbon is 3.0 cm long and has a square cross section 0.5 cm on a side. A potential difference of 8.4 V is maintained across its long dimension. (*a*) What is the resistance of the block? (*b*) What is the current in this resistor? (*c*) What is the current density?

13. A tungsten rod is 50 cm long and has a square cross section 1.0 mm on a side. (*a*) What is its resistance at 20°C? (*b*) What is its resistance at 40°C?

14. The third (current-carrying) rail of a subway track is made of steel and has a cross-sectional area of about 55 cm². What is the resistance of 10 km of this track?

15. A wire of length 1 m has a resistance of 0.3 Ω. It is uniformly stretched to a length of 2 m. What is its new resistance?

16. What is the potential difference across one wire of a 30-m extension cord of 16-gauge copper wire (diameter 0.130 cm) carrying a current of 5.0 A?

17. At what temperature will the resistance of a copper wire be 10 percent greater than it is at 20°C?

18. How long is a 14-gauge copper wire which has a resistance of 2 Ω?

19. A cube of copper has side 2.0 cm. If it is drawn out into a 14-gauge wire, what will its resistance be?

## Section 24-3, Energy in Electric Circuits

20. What is the power dissipated in a 10.0-Ω resistor if the potential difference across it is 50 V?

21. Find the power dissipated in a resistor of resistance (*a*) 5 Ω and (*b*) 10 Ω connected across a constant potential difference of 110 V.

22. A 200-W heater is used to heat water in a cup. Assume that 90 percent of the energy goes into heating the water. (*a*) How long does it take to heat 0.25 kg of water from 15 to 100°C? (*b*) How long does it take to boil this water away after it reaches 100°C?

23. A 1-kW heater is designed to operate at 220 V. (*a*) What is its resistance, and what current does it draw? (*b*) What is the power of this resistor if it operates at 110 V?

24. A 10.0-Ω resistor is rated capable of dissipating 5.0 W. (*a*) What maximum current can this resistor tolerate? (*b*) What voltage across this resistor will produce this current?

25. If energy costs 6 cents per kilowatt-hour, (*a*) how much does it cost to operate an electric toaster for 4 min if the toaster has resistance 11.0 Ω and is connected across 100 V? (*b*) How much does it cost to operate a heater of resistance 5.0 Ω across 100 V for 8 h?

26. A battery has an emf of 12.0 V. How much work does it do in 5 s if it delivers 3 A?

27. A battery with 12-V emf has a terminal voltage of 11.4 V when it delivers 20 A to the starter of a car. What is the internal resistance $r$ of the battery?

28. (a) How much power is delivered by the emf of the battery in Exercise 27 when it delivers 20 A? How much of this power is delivered to the starter? (b) By how much does the chemical energy of the battery decrease when it delivers 20 A for 3 min in starting the car? (c) How much heat is developed in the battery when it delivers 20 A for 3 min?

29. A 12-V car battery has an internal resistance of 0.4 $\Omega$. (a) What is the current if the battery is shorted momentarily? (b) What is the terminal voltage when the battery delivers 20 A to start the car?

### Section 24-4, Classical Model of Electric Conduction

30. A current of 30 mA exists in a copper wire of cross-sectional area 2.0 mm². (a) What is the current density? (b) What is the drift velocity of the electrons? (c) What is the average time between collisions? (d) What is the mean free path of electrons in this wire (assume $\bar{v} = 10^5$ m/s)?

31. Repeat Exercise 30 for a current of 5.0 mA in a copper wire of cross-sectional area 1.0 mm².

### Section 24-5, Corrections to the Classical Theory of Conduction, and Section 24-6, Conductors, Insulators, and Semiconductors

*There are no exercises for these sections.*

### Problems

1. A 16-gauge copper wire (diameter 1.29 mm) can safely carry a maximum current of 6 A (assuming rubber insulation). (a) How great a potential difference can be applied across 40 m of such a wire? (b) Find the current density and the electric field in the wire when it carries 6 A. (c) Find the power dissipated in the wire when it carries 6 A.

2. A Van de Graaff accelerator belt carries a surface charge density of 5 $\mu$C/m². The belt is 0.5 m wide and moves at 20 m/s. (a) What current does it carry? (b) If this charge is raised to a potential of 100 kV, what is the minimum power of the motor needed to drive the belt?

3. A linear accelerator produces a beam of electrons in which the current is not constant but consists of a pulsed beam of particles. Suppose that the pulse current is 1.6 A for a 0.1-$\mu$s duration. (a) How many electrons are accelerated in each pulse? (b) What is the average beam current if there are 1000 pulses per second? (c) If the electrons acquire an energy of 400 MeV, what is the average (minimum) power input to the accelerator? (d) What is the peak power input? (e) What fraction of the time is the accelerator actually accelerating particles? (This is called the *duty factor* of the accelerator.)

4. A coil of nichrome wire is to be used as the heating element in a water boiler required to generate 8.0 g of steam per second. The wire has a diameter of 1.80 mm and is connected to a 115-V power supply. Find the length of wire required.

5. An 80.0-m copper wire 1.0 mm in diameter is joined end to end with a 49.0-m iron wire of the same diameter. The current in each is 2.0 A. (a) Find the current density in each wire. (b) Find the electric field in each wire. (c) Find the potential difference across each wire. (d) Find the equivalent resistance which would carry 2.0 A at a potential difference equal to the sum of that across the two, and compare it with the sum of the resistances of the two.

6. Two wires have the same length but different cross-sectional areas and are made from different materials. They are laid side by side and joined at each end so that the potential difference across each wire is the same. Let $R_1$ be the resistance of one wire and $R_2$ be that of the other. (a) If the potential difference across each is $V$, find the current in each and find the total current $I = I_1 + I_2$. (b) Find an expression for the resistance of a single wire which would carry the same total current when the potential difference is $V$.

7. A wire of area $A$, length $l_1$, resistivity $\rho_1$, and temperature coefficient $\alpha_1$ is connected to a second wire of length $l_2$, resistivity $\rho_2$, temperature coefficient $\alpha_2$, and the same area $A$, so that the wires carry the same current. (a) Show that if the current is $I$, the potential drop across the two wires is $IR$, where $R = R_1 + R_2$ is the sum of the resistances. (b) Show that if $\rho_1 l_1 \alpha_1 + \rho_2 l_2 \alpha_2 = 0$, the total resistance $R$ is independent of temperature for small temperature changes. (c) If one wire is made of carbon and the other copper, find the ratio of their lengths such that $R$ is approximately independent of temperature.

8. A 100-W heater resistor is designed to operate across 100 V. (a) What is its resistance and what current does it draw? (b) Show that if the potential difference across the resistor changes by a small amount $\Delta V$, the power changes by a small amount $\Delta P$, where $\Delta P/P \approx 2 \, \Delta V/V$. Hint: Approximate the changes with differentials. (c) Find the approximate power dissipated in the resistor if the potential difference increases to 115 V.

# CHAPTER 25 Direct-Current Circuits

---

**Objectives**   After studying this chapter you should:

1.   Be able to determine the equivalent resistances of series or parallel resistors to simplify various combinations of resistors.

2.   Be able to state Kirchhoff's rules and use them to analyze various dc circuits.

3.   Be able to find the time constant for an $RC$ circuit and sketch both the charge $Q$ on the capacitor and the current $I$ as functions of time for charging and discharging a capacitor.

4.   Be able to draw the circuit diagrams and calculate the proper series or shunt resistors needed to make an ammeter, voltmeter, and ohmmeter from a given galvanometer.

---

In this chapter we analyze some simple circuits consisting of batteries, resistors, and capacitors in various combinations. These circuits are called direct-current (dc) circuits because the direction of the current in any one part of the circuit does not vary. Circuits in which the current alternates in direction (ac circuits) will be discussed in Chapter 30.

## 25-1   Series and Parallel Resistors

Two or more resistors connected so that the same charge must flow through each are said to be connected in series. Resistors $R_1$ and $R_2$ in Figure 25-1a are examples of resistors in series. They must carry the

**Figure 25-1**
(a) Two resistors in series. (b) The resistors in part (a) can be replaced by a single equivalent resistance $R_{eq} = R_1 + R_2$, which gives the same total potential drop when carrying the same current as in part (a).

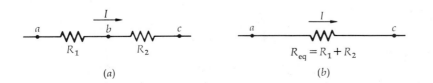

(a)

(b)

same current $I$. The potential drop across the two resistors (from point $a$ to point $c$ in Figure 25-1) is

$$V = IR_1 + IR_2 = I(R_1 + R_2)$$

We can often simplify the analysis of a circuit with resistors in series by replacing such resistors with a single equivalent resistance which gives the same total potential drop $V$ when carrying the same current $I$. The equivalent resistance for series resistances is the sum of the original resistances:

$$R_{eq} = \frac{V}{I} = R_1 + R_2$$

When there are more than two resistances in series, the equivalent resistance is

$$R_{eq} = R_1 + R_2 + R_3 + \cdots \qquad\qquad 25\text{-}1 \qquad \textit{Series resistors}$$

Two resistors connected as in Figure 25-2a so that they have the same potential difference across them are said to be connected in parallel. Let $I$ be the current from point $a$ to point $b$. At point $a$ the current splits into two parts, $I_1$ through resistor $R_1$ and $I_2$ through $R_2$. The total current is the sum of the individual currents:

$$I = I_1 + I_2 \qquad\qquad 25\text{-}2$$

Let $V = V_a - V_b$ be the potential drop across either resistor. In terms of the currents and resistances,

$$V = I_1 R_1 = I_2 R_2$$

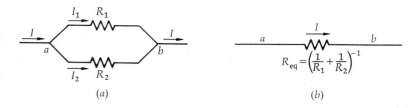

(a)

(b)

**Figure 25-2**
(a) Two resistors in parallel.
(b) The two resistors in part (a) can be replaced by a single equivalent resistance $R_{eq}$ related to $R_1$ and $R_2$ by $1/R_{eq} = 1/R_1 + 1/R_2$.

The ratio of the two currents is therefore the inverse ratio of the resistances; the smaller resistance carries the larger current and vice versa:

$$\frac{I_1}{I_2} = \frac{R_2}{R_1} \qquad\qquad 25\text{-}3$$

The equivalent resistance of a combination of parallel resistors is defined to be that resistance $R_{eq}$ for which the same total current $I$ produces the potential drop $V$:

$$R_{eq} = \frac{V}{I}$$

Equation 25-2 can then be written

$$I = \frac{V}{R_{eq}} = I_1 + I_2 = \frac{V}{R_1} + \frac{V}{R_2}$$

or

$$\frac{1}{R_{eq}} = \frac{1}{R_1} + \frac{1}{R_2} \qquad\qquad 25\text{-}4$$

This result can be generalized to any number of resistances in parallel:

$$\frac{1}{R_{eq}} = \frac{1}{R_1} + \frac{1}{R_2} + \frac{1}{R_3} + \cdots \qquad \text{25-5}$$

The equivalent resistance of a combination of parallel resistors is less than that of any resistor. Consider, for example, any one resistor such as $R_1$ carrying current $I_1$ with potential drop $V = I_1 R_1$. The effect of another resistor or other resistors in parallel with $R_1$ is to increase the total current for the same potential drop $V$. The equivalent resistance $V/I$ is thereby decreased.

The equivalent resistance for two parallel resistors can be found from Equation 25-4 by putting the two terms on the right side of the equation over a common denominator and taking the reciprocal of each side. The result is

$$R_{eq} = \frac{R_1 R_2}{R_1 + R_2} \qquad \text{25-6}$$

This result explicitly demonstrates that $R_{eq}$ is less than either $R_1$ or $R_2$ since the right side of the equation can be considered to be the product of either resistance (such as $R_1$) and a factor less than 1, such as $R_2/(R_1 + R_2)$. When there are three or more parallel resistors, it is usually easier to solve Equation 25-5 for $1/R_{eq}$ than to remember a general result analogous to Equation 25-6. For example, the equivalent resistance of three parallel resistors is given by the somewhat complicated expression (see Problem 16)

$$R_{eq} = \frac{R_1 R_2 R_3}{R_1 R_2 + R_2 R_3 + R_3 R_1} \qquad \text{25-7}$$

The current in each of two parallel resistors can be found from the fact that the potential drop across the combination is $IR_{eq}$ and is also equal to $I_1 R_1$ and to $I_2 R_2$:

$$I_1 R_1 = I_2 R_2 = IR_{eq} = I \frac{R_1 R_2}{R_1 + R_2}$$

Then

$$I_1 = \frac{R_2}{R_1 + R_2} I \quad \text{and} \quad I_2 = \frac{R_1}{R_1 + R_2} I \qquad \text{25-8}$$

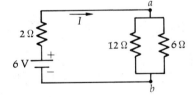

**Example 25-1** For the circuit shown in Figure 25-3 find the equivalent resistance of the parallel combination of resistors, the current carried by each, and the total current through the emf.

We first find the equivalent resistance of the 6- and 12-$\Omega$ resistors in parallel:

**Figure 25-3**
Circuit for Example 25-1.

$$\frac{1}{R_{eq}} = \frac{1}{6\ \Omega} + \frac{1}{12\ \Omega} = \frac{3}{12\ \Omega} = \frac{1}{4\ \Omega}$$

$$R_{eq} = 4\ \Omega$$

Note that this equivalent resistance is less than that of either of the parallel resistors. Figure 25-4 shows the circuit with $R_{eq}$ replacing the parallel combination. The resistances $R_{eq} = 4\ \Omega$ and $r = 2\ \Omega$ are in series. The equivalent resistance of this series combination is $R_{eq}' = R_{eq} + r = 6\ \Omega$. The current in the circuit is therefore

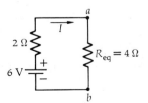

$$I = \frac{\mathcal{E}}{R_{eq}'} = \frac{6\ V}{6\ \Omega} = 1\ A$$

This is the total current in the seat of emf.

**Figure 25-4**
Simplification of the circuit in Figure 25-3 by replacing the parallel resistors by a single equivalent resistance.

The currents in the 6- and 12-$\Omega$ resistors could be found from Equation 25-8 or more simply by noting that since the total current from point $a$ to $b$ in Figure 25-4 is 1 A and the resistance $R_{eq}$ is 4 $\Omega$, the voltage drop from point $a$ to $b$ is $IR_{eq}$ = 4 V. The current in the 6-$\Omega$ resistor needed to produce a 4-V potential drop is $I_6 = V/R_1$ = (4 V)/(6 $\Omega$) = $\frac{2}{3}$ A. Similarly, the current in the 12-$\Omega$ resistor is $I_{12}$ = (4 V)/(12 $\Omega$) = $\frac{1}{3}$ A. The 6-$\Omega$ resistor carries twice as much current as the 12-$\Omega$ resistor.

**Example 25-2** Find the equivalent resistance between points $a$ and $b$ for the combination of resistors shown in Figure 25-5.

This combination of resistors may look complicated, but it is easily treated step by step. We first find the equivalent resistance of the 4- and 12-$\Omega$ resistors, which are in parallel. Using Equation 25-4, we obtain

$$\frac{1}{R_{eq}} = \frac{1}{4\ \Omega} + \frac{1}{12\ \Omega} = \frac{4}{12\ \Omega} = \frac{1}{3\ \Omega}$$

or $R_{eq}$ = 3 $\Omega$. In Figure 25-6 we have replaced the 4- and 12-$\Omega$ resistors by their equivalent, a 3-$\Omega$ resistor. Since this 3-$\Omega$ resistor is in series with a 5-$\Omega$ resistor, the equivalent resistance of the bottom branch of this combination is 8 $\Omega$. We are now left with an 8-$\Omega$ resistor in parallel with a 24-$\Omega$ resistor. The equivalent resistance of these two parallel resistors is again found from Equation 25-4. The result is $R_{eq}$ = 6 $\Omega$ for the complete combination.

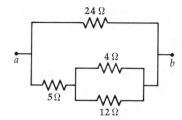

**Figure 25-5**
Combination of resistors for Example 25-2.

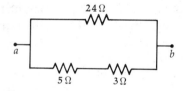

**Figure 25-6**
Figure 25-5 simplified by replacing the parallel combination of 4- and 12-$\Omega$ resistances with the equivalent 3-$\Omega$ resistor.

### Question

1. A common mistake is to generalize Equation 25-6 to $R_1R_2R_3/(R_1 + R_2 + R_3)$ for the equivalent resistance of three parallel resistors. What simple argument can you give to show that such an expression could not be correct?

## 25-2  Kirchhoff's Rules

Although the methods discussed in the previous section for replacing series and parallel combinations of resistors by their equivalent resistances are very useful for simplifying many combinations of resistors, they are not sufficient for the analysis of many simple circuits, particularly those containing more than one battery. For example, the two resistors $R_1$ and $R_2$ in the circuit of Figure 25-7 cannot be replaced by an equivalent resistance for parallel resistors because the potential difference is not the same across each *resistor*. Two simple rules, called *Kirchhoff's rules,* are applicable to any dc circuit containing batteries and resistors connected in any way:

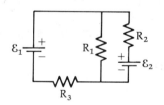

**Figure 25-7**
A simple circuit that cannot be analyzed by replacing series or parallel resistors by their equivalents. The potential differences across the resistors $R_1$ and $R_2$ are not equal because of the emf $\mathcal{E}_2$. Therefore these resistors cannot be replaced by an equivalent resistance.

*Kirchhoff's rules*

1. The algebraic sum of the increases or decreases in potential around any closed circuit loop must be zero.

2. At any junction point in a circuit where the current can divide, the sum of the currents into the junction must equal the sum of the currents out of the junction.

Rule 1, called the loop rule, follows from the simple fact that in the steady state the potential difference between any two points is constant. As we move around a circuit loop, the potential may decrease or increase as we pass through a resistor or battery, but when we have completely traversed a loop and arrive back at our starting point, the net change in the potential must be zero. There are several ways of applying this rule in practice. If we call the potential increases positive and the decreases negative, we can put all the changes on one side of an equation and set the algebraic sum of these changes equal to zero. Alternatively, we can list the potential increases separately from the potential decreases and set the sum of the increases equal to the sum of the decreases.

Rule 2 follows from the conservation of charge. This rule is needed for multiloop circuits containing points where the current can divide, as in Figure 25-8. We have labeled the currents in the three wires $I_1$, $I_2$, and $I_3$. In a time interval $\Delta t$, charge $I_1 \Delta t$ flows into the junction from the left. In the same time, charges $I_2 \Delta t$ and $I_3 \Delta t$ flow out of the junction to the right. Since there is no way that charge can originate at this point or collect there, conservation of charge implies rule 2, which for this case gives

$$I_1 = I_2 + I_3 \qquad 25\text{-}9$$

In this section we shall illustrate the application of Kirchhoff's laws to various dc circuits. We begin with a simple circuit with one battery and one loop and then give examples of more complicated circuits containing more than one battery and more than one loop.

Figure 25-9 shows a simple circuit consisting of a battery of emf $\varepsilon$, internal resistance $r$, and external resistance $R$ across the terminals of the battery. The current $I$ is in the direction shown. The connecting wires indicated by straight lines in this diagram are assumed to have negligible resistance. Any two points along such a line are therefore always at the same potential. If we start from point $a$ and move around the circuit with the current (clockwise in this case), we first encounter a potential drop $IR$ from point $b$ to point $c$. We then encounter a potential increase $\varepsilon$ from point $d$ to point $e$ across the emf of the battery and finally another potential drop $Ir$ from point $e$ to point $a$ across the internal resistance of the battery. Kirchhoff's first rule then gives

$$\varepsilon - IR - Ir = 0$$

or

$$I = \frac{\varepsilon}{r + R} \qquad 25\text{-}10$$

Note that the power output $\varepsilon I$ of the emf goes partly into Joule heating of the external resistor and partly into Joule heating of the battery because of its internal resistance. We can see this by multiplying each term in Equation 25-10 by $I$, giving

$$\varepsilon I = I^2 r + I^2 R \qquad 25\text{-}11$$

---

**Example 25-3** An 11-$\Omega$ resistance is connected across a battery of emf 6 V and internal resistance 1 $\Omega$.* Find the current, the terminal voltage of the battery, the power delivered by the emf, and the power delivered to the external resistance.

---

* Real batteries usually have internal resistance of the order of a few hundredths of an ohm. To simplify calculations we may exaggerate the value of the internal resistance, as in this example; in other examples we may ignore it completely.

The Bettman Archive

Gustav Robert Kirchhoff (1824–1887).

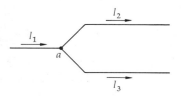

**Figure 25-8**
If charge cannot originate or collect at the junction point $a$, conservation of charge implies that the current $I_1$ into the point must equal the current $I_2 + I_3$ out of the point.

**Figure 25-9**
Simple circuit containing a battery of emf $\varepsilon$, an internal resistance $r$, and an external resistance $R$.

From Equation 25-10 the current is

$$I = \frac{6 \text{ V}}{1 \ \Omega + 11 \ \Omega} = 0.5 \text{ A}$$

The terminal voltage across the battery, $V_a - V_d$, is equal to the voltage drop $V_b - V_c$ across the external resistance, which is $IR = (0.5 \text{ A})(11 \ \Omega) = 5.5$ V. This differs from the emf of 6 V of the battery because of the potential drop across the internal resistance, $Ir = (0.5 \text{ A})(1 \ \Omega) = 0.5$ V. The power delivered by the emf is $P = (6 \text{ V})(0.5 \text{ A}) = 3$ W. The power delivered to the external resistance is $I^2R = (0.5 \text{ A})^2(11 \ \Omega) = 2.75$ W. The other 0.25 W of power is dissipated as heat in the internal resistance of the battery.

---

**Example 25-4** For a battery of given emf and internal resistance $r$, what value of external resistance $R$ should be placed across the terminals to obtain the greatest Joule heating in $R$?

The resistance $R$ is sometimes called the *load resistance*. The power into $R$ is $I^2R$, where $I$ is given by Equation 25-10. Thus the power is

$$P = I^2R = \frac{\varepsilon^2}{(r + R)^2} R \qquad \qquad 25\text{-}12$$

Figure 25-10 shows a sketch of $P$ versus $R$. The maximum value of $P$ occurs when $R = r$. Thus the maximum power is obtained when the load resistance equals the internal resistance. A similar result also occurs for more complicated ac circuits and is known as *impedance matching*. It is left as an applied calculus problem for the student to derive the result that $P$ is a maximum when $R = r$.

---

**Example 25-5** Find the current in the circuit shown in Figure 25-11.

In this circuit we have two seats of emf and three external resistors. We cannot predict the direction of the current unless we know which emf is greater, but we do not have to know the direction of the current before solving the problem. We can assume any direction and solve the problem with that assumption. If the assumption is incorrect, we shall get a negative number for the current, indicating that its direction is opposite that assumed.

Let us assume $I$ to be clockwise as indicated in the figure. For this example we shall apply Kirchhoff's first rule by finding the potential increases and setting their sum equal to the sum of the potential decreases. The potential drops and increases as we traverse the circuit in the assumed direction of the current, starting at point $a$, are given in Table 25-1. We encounter a potential drop traversing one of the emfs and an increase traversing the other. Kirchhoff's first rule gives

$$\varepsilon_1 = IR_1 + IR_2 + \varepsilon_2 + Ir_2 + IR_3 + Ir_1 \qquad 25\text{-}13$$

Solving for the current $I$, we obtain

$$I = \frac{\varepsilon_1 - \varepsilon_2}{R_1 + R_2 + R_3 + r_1 + r_2} \qquad \qquad 25\text{-}14$$

Note that if $\varepsilon_2$ is greater than $\varepsilon_1$, we get a negative number for the current $I$, indicating that we have chosen the wrong direction for $I$. For $\varepsilon_2$ greater than $\varepsilon_1$ the current is in the counterclockwise direction. On the other hand, if $\varepsilon_1$ is the greater emf, we get a positive number for $I$, indicating that its assumed direction is correct.

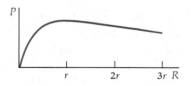

**Figure 25-10**
Plot of the power into resistor $I^2R$ versus the resistance $R$. The power is a maximum when the load resistance equals the internal resistance of the battery.

*Impedance matching*

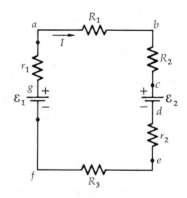

**Figure 25-11**
Circuit for Example 25-5.

**Table 25-1**

| $a \rightarrow b$ | Drop $IR_1$ |
|---|---|
| $b \rightarrow c$ | Drop $IR_2$ |
| $c \rightarrow d$ | Drop $\varepsilon_2$ |
| $d \rightarrow e$ | Drop $Ir_2$ |
| $e \rightarrow f$ | Drop $IR_3$ |
| $f \rightarrow g$ | Increase $\varepsilon_1$ |
| $g \rightarrow a$ | Drop $Ir_1$ |

Let us assume for this example that $\varepsilon_1$ is the greater emf and that the current is clockwise, as indicated. A charge $\Delta Q$ moving through battery 2 from point $c$ to point $d$ loses energy $\varepsilon_2 \Delta Q$. In this battery, electric energy is converted into chemical energy and stored. Battery 2 is *charging*. We can account for the energy balance in this circuit by multiplying each term in Equation 25-13 by the current $I$. Then the term on the left, $\varepsilon_1 I$, is the rate at which battery 1 puts energy into the circuit. This energy comes from internal chemical energy of the battery. The first term on the right, $I^2 R_1$, is the rate of production of Joule heat in resistor $R_1$. There are similar terms for each of the other resistors. The term $\varepsilon_2 I$ is the rate at which electric energy is converted into chemical energy in the second battery.

---

**Example 25-6** The quantities in the circuit of Example 25-5 have the magnitudes $\varepsilon_1 = 12$ V, $\varepsilon_2 = 4$ V, $r_1 = r_2 = 1\ \Omega$, $R_1 = R_2 = 5\ \Omega$, $R_3 = 4\ \Omega$. Find the potentials at the points indicated in Figure 25-12, assuming that the potential at point $f$ is zero, and discuss the energy balance in the circuit.

Since only potential differences are important, any point in a circuit can be chosen to have zero potential. In this example we have chosen point $f$ in Figure 25-12. The potentials of the other points are then found relative to point $f$.* This procedure often simplifies the analysis of a problem.

We first find the current in the circuit. From Equation 25-14 we have

$$I = \frac{12 - 4}{5 + 5 + 4 + 1 + 1} = \frac{8\ \text{V}}{16\ \Omega} = 0.5\ \text{A}$$

We can now find the potentials at the indicated points relative to zero potential at point $f$. Since, by definition, the emf maintains a constant potential difference $\varepsilon_1 = 12$ V between point $g$ and point $f$, the potential at point $g$ is 12 V. The potential at point $a$ is less than at $g$ by the potential drop $Ir_1 = 0.5(1) = 0.5$ V. Thus the potential at point $a$ is $12 - 0.5 = 11.5$ V. Similarly the potential drops across the 5-$\Omega$ resistors $R_1$ and $R_2$ are each $IR_1 = 0.5(5) = 2.5$ V. The potential at point $b$ is then $11.5 - 2.5 = 9$ V, and that at $c$ is 6.5 V. The potential drop across $\varepsilon_2$ is 4 V. Thus point $d$ is at a potential of 2.5 V. Since the drop across the 1-$\Omega$ resistance $r_2$ is 0.5 V, the potential at $e$ is 2 V. The potential drop across the 4-$\Omega$ resistance $R_3$ is $IR_3 = 2$ V. This gives zero for the potential at $f$, consistent with our original assumption.

The power delivered by the emf $\varepsilon_1$ is $\varepsilon_1 I = 12(0.5) = 6$ W. The power into the internal resistance of battery 1 is $I^2 r_1 = 0.5^2(1) = 0.25$ W. Thus the power delivered by the battery to the external circuit is $6 - 0.25 = 5.75$ W. This also equals $V_1 I$, where $V_1 = V_a - V_f = 11.5$ V is the terminal voltage of that battery. The total power into the external resistances in the circuit is $I^2 R_1 + I^2 R_2 + I^2 R_3 = I^2(R_1 + R_2 + R_3) = 0.5^2(5 + 5 + 4) = 3.5$ W. The power into the battery being charged is $(V_c - V_e)I = (6.5 - 2)(0.5) = 2.25$ W. Part of this, $I^2 r_2 = 0.25$ W, is dissipated in the internal resistance, and part, $\varepsilon_2 I = 2$ W, is the rate at which energy is stored in that battery.

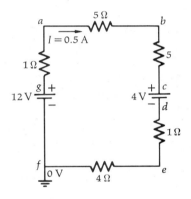

**Figure 25-12**
Circuit for Example 25-6.

---

* In more complicated circuits, we may wish to keep some point such as point $f$ at a fixed potential relative to the earth even though some of the elements of the circuit change. This can be done by connecting that point to the earth, i.e., grounding it. Since the earth is such a large conductor, it has nearly infinite capacitance. Thus its potential remains essentially constant and is usually chosen to be zero.

Note that the terminal voltage of the battery that is being charged in Example 25-6 is $V_c - V_e = 4.5$ V, which is greater than the emf of the battery. Because of its internal resistance, a battery is not completely reversible. If the same 4-V battery were to deliver 0.5 A to an external circuit, its terminal voltage would be 3.5 V (again assuming the value of 1 for its internal resistance). If the internal resistance is very small, the terminal voltage of the battery is nearly equal to its emf, whether the battery is delivering current to an external circuit or being charged. Some real batteries, such as the storage batteries used in automobiles, are nearly reversible and can easily be recharged; others (such as dry cells) are not. If you attempt to recharge a dry cell by driving current through it from its positive to its negative terminal, most, if not all, of the energy expended goes into heat rather than into chemical energy of the dry cell.

We now consider an example of a circuit which contains more than one loop. In such circuits we need to apply Kirchhoff's second rule at junction points where the current splits into two or more parts.

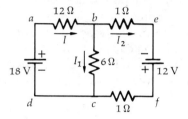

**Figure 25-13**
Circuit for Example 25-7.

---

**Example 25-7** Find the current in each part of the circuit shown in Figure 25-13.

Let $I$ be the current through the 18-V battery in the direction shown. At point $b$, this current divides into currents $I_1$ and $I_2$, as shown. Until we know the solution for the currents, we cannot be sure of their direction. For example, we need to know whether point $b$ or point $c$ is at the higher potential in order to know the direction of current through the 6-$\Omega$ resistor. The current directions indicated are merely guesses. Applying the junction rule (rule 2) to the point $b$, we obtain

$$I = I_1 + I_2$$

There are three possible loops for applying rule 1, loops $abcd$, $befc$, and $abefcd$. We need only two more equations to determine the three unknown currents. Equations for any two of the loops will be sufficient. (The third loop will then give redundant information.) Traversing the loop $abcd$ in the clockwise direction and applying Kirchhoff's first rule gives

$$(12\ \Omega)I + (6\ \Omega)I_1 = 18\ V$$

or

$$2I + I_1 = 3\ A \qquad\qquad\qquad 25\text{-}15$$

Applying rule 1 to loop $befc$, we obtain another equation,

$$(1\ \Omega)I_2 + (1\ \Omega)I_2 = 12\ V + (6\ \Omega)I_1$$

Note that in moving from $c$ to $b$ we encounter a voltage increase because the (assumed) current $I_1$ is in the opposite direction. Combining terms and substituting $I - I_1$ for $I_2$ in this equation, we obtain

$$2(I - I_1) = 12\ A + 6I_1$$

$$2I = 12\ A + 8I_1 \qquad\qquad\qquad 25\text{-}16$$

Equations 25-15 and 25-16 can be solved for the unknown currents $I$ and $I_1$. The results are

$$I = 2\ A \qquad I_1 = -1\ A$$

Then

$$I_2 = I - I_1 = 2 - (-1) = 3\ A$$

Our original guess about the direction of $I_1$ was incorrect. The current through the 6-$\Omega$ resistor is in the direction from point $c$ to point $b$.

---

**Example 25-8** Find the currents $I$, $I_1$, and $I_2$ for the circuit in Figure 25-14.

This circuit is the same as in Figure 25-7, with $\mathcal{E}_1 = 12$ V, $\mathcal{E}_2 = 5$ V, $R_1 = 4\ \Omega$, $R_2 = 2\ \Omega$, and $R_3 = 3\Omega$. We choose directions for the currents as shown in the figure, and we obtain from the junction rule:

$$I = I_1 + I_2$$

Using this to eliminate $I$ and applying the loop rule to the outer loop, we get

$$12 - 2I_1 - 5 - 3(I_1 + I_2) = 0 \qquad\qquad \text{25-17}$$

Applying the loop rule to the first loop on the left gives

$$12 - 4I_2 - 3(I_1 + I_2) = 0 \qquad\qquad \text{25-18}$$

When we solve Equations 25-17 and 25-18 for the currents $I_1$ and $I_2$, we obtain $I_1 = 0.5$ A and $I_2 = 1.5$ A. The total current through the 12-V emf is then $I = 0.5$ A $+ 1.5$ A $= 2.0$ A.

---

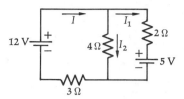

**Figure 25-14**
Circuit for Example 25-8.

## 25-3  *RC* Circuits

In this section we consider circuits containing capacitors, in which the current is not steady but varies with time. Figure 25-15 shows a capacitor with an initial charge $+Q_0$ on the upper plate and $-Q_0$ on the lower plate. It is connected to a resistor $R$ and a switch $S$, which is open to prevent the charge from flowing through the resistor. The potential difference across the capacitor is initially $V_0 = Q_0/C$, where $C$ is the capacitance. Since there is no current with the switch open, there is no potential drop across the resistor. Thus there is also a potential difference $V_0$ across the switch.

We close the switch at time $t = 0$. Since there is now a potential difference across the resistor, there must be a current in it. The initial current is $I_0 = V_0/R$. The current is due to the flow of charge from the positive plate to the negative plate through the resistor, and so after a time the charge on the capacitor is reduced. The rate of decrease of the charge equals the current. If $Q$ is the charge on the capacitor at any time, the current at that time is

$$I = -\frac{dQ}{dt} \qquad\qquad \text{25-19}$$

We can write an equation for the charge and current as functions of time by applying Kirchhoff's first rule to the circuit after the switch is closed. Traversing the circuit in the direction of the current, we encounter a potential drop $IR$ across the resistor and a potential increase $Q/C$ across the capacitor. Kirchhoff's first rule gives

$$IR = \frac{Q}{C} \qquad\qquad \text{25-20}$$

where both $Q$ and $I$ are functions of time and are related by Equation

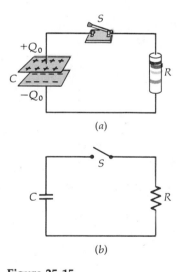

**Figure 25-15**
(*a*) Parallel-plate capacitor in series with a switch and a resistor $R$. (*b*) Circuit diagram for (*a*).

25-19. Substituting $-dQ/dt$ for $I$ in this equation, we have

$$-R\frac{dQ}{dt} = \frac{Q}{C}$$

or

$$\frac{dQ}{dt} = -\frac{1}{RC}Q \qquad\qquad 25\text{-}21$$

The function $Q(t)$ whose rate of change is proportional to itself is the exponential. The solution of Equation 25-21 is

$$Q(t) = Q_0 e^{-t/RC} = Q_0 e^{-t/t_c} \qquad\qquad 25\text{-}22$$

where the time

$$t_c = RC \qquad\qquad 25\text{-}23$$

*Time constant*

is the time in which the charge decreases to $1/e$ of its original value; it is called the *time constant* of the circuit. After a time $t_c$ the charge is $Q = Q_0 e^{-1} = 0.37 Q_0$. After a time $2t_c$, the charge is $Q = Q_0 e^{-2} = 0.135 Q_0$, and so forth. We can understand the time constant as follows. Originally, the charge on the capacitor is $Q_0$, and its rate of change as given by Equation 25-21 is $-Q_0/RC$. If this rate of change were constant in time, the charge would decrease to zero in a time $t = t_c$, as indicated by the dashed line in Figure 25-16. However, the rate of change of the charge is not constant but proportional to the charge itself. As the charge decreases, the magnitude of the rate of change decreases. The function $Q(t)$ is shown by the solid curve in Figure 25-16. After a long time both the value of $Q$ and its slope, which is proportional to $Q$, approach zero as shown.

The current is obtained from Equation 25-19 by differentiating Equation 25-22:

$$I = -\frac{dQ}{dt} = \frac{Q_0}{RC} e^{-t/RC} = I_0 e^{-t/RC} \qquad\qquad 25\text{-}24$$

where $I_0 = Q_0/RC = V_0/R$ is the initial current. The current also decreases exponentially with time and falls to $1/e$ of its initial value after a time $t_c = RC$.

Figure 25-17 shows a circuit used for charging a capacitor, which we assume to be originally uncharged. The switch, originally open, is closed at time $t = 0$. Charge immediately begins to flow into the capacitor. If the charge on the capacitor at some time is $Q$ and the current in the circuit is $I$, Kirchhoff's first rule gives

$$\mathcal{E} = V_R + V_C = IR + \frac{Q}{C} \qquad\qquad 25\text{-}25$$

In this circuit, the current equals the rate of increase of charge on the capacitor:

$$I = \frac{dQ}{dt}$$

Substituting this into Equation 25-25 gives

$$\mathcal{E} = R\frac{dQ}{dt} + \frac{Q}{C} \qquad\qquad 25\text{-}26$$

At time $t = 0$ the charge is zero and the current is $I_0 = \mathcal{E}/R$. The charge

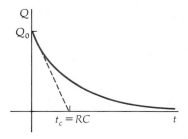

**Figure 25-16**
Plot of charge $Q$ on capacitor versus time $t$ for the circuit in Figure 25-15. The charge decreases by a factor $1/e$ in the time $t_c = RC$. The time constant $t_c$ is also the time in which the capacitor would be fully discharged if the discharge rate were constant, as indicated by the dashed line.

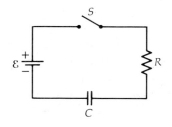

**Figure 25-17**
Circuit for charging a capacitor to a potential difference $\mathcal{E}$.

then increases and the current decreases, as can be seen from Equation 25-25. The charge reaches a maximum value of $C\varepsilon$, again from Equation 25-25, when the current $I$ equals zero. Figures 25-18 and 25-19 show the behavior of the charge versus time and the current versus time. The analytical expressions for $Q(t)$ and $I(t)$ are obtained by solving the differential equation 25-26:

$$Q(t) = C\varepsilon(1 - e^{-t/RC}) \qquad\qquad 25\text{-}27$$

$$I = \frac{dQ}{dt} = \frac{\varepsilon}{R}\, e^{-t/RC} \qquad\qquad 25\text{-}28$$

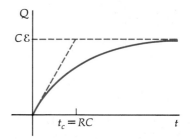

**Figure 25-18**
Plot of charge on the capacitor versus time for the circuit in Figure 25-17. After time $t = t_c$ the charge is 63 percent of its final value of $C\varepsilon$. If the charging rate were constant and equal to its initial value, the capacitor would be fully charged after time $t = t_c$, as indicated by the dashed line.

**Example 25-9** A 6-V battery of negligible internal resistance is used to charge a 2-$\mu$F capacitor through a 100-$\Omega$ resistor. Find the initial current, the final charge, and the time required to obtain 90 percent of the final charge.

The initial current is $I_0 = \varepsilon/R = (6\text{ V})/(100\ \Omega) = 0.06$ A. The final charge is $Q_f = \varepsilon C = (6\text{ V})(2\ \mu\text{F}) = 12\ \mu\text{C}$. The time constant for this circuit is $RC = (100\ \Omega)(2\ \mu\text{F}) = 200\ \mu\text{s}$. We expect the charge to reach 90 percent of its final value in a time of the order of several time constants. We can find the exact solution from Equation 25-27, using $Q = 0.9\varepsilon C$:

$$Q = 0.9\varepsilon C = \varepsilon C(1 - e^{-t/RC})$$

$$0.9 = 1 - e^{-t/RC}$$

$$e^{-t/RC} = 1 - 0.9 = 0.1$$

$$\ln e^{-t/RC} = -\frac{t}{RC} = \ln 0.1 = \ln 1 - \ln 10 = -\ln 10 = -2.3$$

since $\ln 0.1 = \ln \frac{1}{10} = \ln 1 - \ln 10$ and $\ln 1 = 0$. Thus

$$t = RC\ \ln 10 = 2.3RC = 2.3(200\ \mu\text{s}) = 460\ \mu\text{s}$$

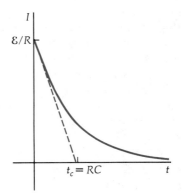

**Figure 25-19**
Plot of current versus time for the circuit in Figure 25-17. The current decreases by a factor $1/e$ in the time $t = t_c = RC$.

During the charging process a total charge $Q_f = \varepsilon C$ flows through the battery. The battery does work

$$W = Q_f\varepsilon = \varepsilon^2 C$$

The energy stored in the capacitor is just half this amount. By Equation 23-19,

$$U = \tfrac{1}{2}QV = \tfrac{1}{2}Q_f\varepsilon = \tfrac{1}{2}\varepsilon^2 C$$

We now show that the other half of the energy provided by the battery goes into Joule heat in the resistor. The rate of energy put into the resistor is

$$\frac{dW_R}{dt} = I^2 R$$

Using Equation 25-28 for the current, we have

$$\frac{dW_R}{dt} = \left(\frac{\varepsilon}{R}\, e^{-t/RC}\right)^2 R = \frac{\varepsilon^2}{R}\, e^{-2t/RC}$$

We find the total heat produced by integrating from $t = 0$ to $t = \infty$:

$$W_R = \int_{t=0}^{t=\infty} dW_R = \int_0^\infty \frac{\varepsilon^2}{R}\, e^{-2t/RC}\, dt$$

The integration can be done by substituting $x = 2t/RC$. Then

$$dt = \frac{RC}{2}\, dx$$

$$W_R = \frac{\varepsilon^2}{R}\frac{RC}{2}\int_0^\infty e^{-x}\, dx = \tfrac{1}{2}\varepsilon^2 C$$

since the integral is 1. This answer is independent of the resistance $R$. When a capacitor is charged by a constant emf, half the energy provided by the battery is stored in the capacitor and half goes into heat, independent of the resistance. (This heat energy includes power into the internal resistance of the battery.)

# 25-4  Ammeters, Voltmeters, and Ohmmeters

We turn now to the consideration of the measurement of electrical quantities in dc circuits. Devices which measure current, potential difference, and resistance are called ammeters, voltmeters, and ohmmeters, respectively. To measure the current through the resistor in the simple circuit in Figure 25-20 we place an ammeter in series with the resistor, as indicated in the figure. Since the ammeter has some resistance, the current in the circuit is changed when the ammeter is inserted. Ideally, the ammeter should have a very small resistance so that only a small change will be introduced in the current to be measured. The potential difference across the resistor is measured by placing a voltmeter across the resistor in parallel with it, as shown in Figure 25-21. The voltmeter reduces the resistance between points $a$ and $b$, thus increasing the total current in the circuit and changing the potential drop across the resistor. An ideal voltmeter has a very large resistance, to minimize its effect on the circuit.

The principal component of an ammeter or voltmeter is a galvanometer, a device which detects a small current through it. A common type, the *d'Arsonval galvanometer*, consists of a coil of wire free to turn, an indicator of some kind, and a scale. The galvanometer is designed so that the scale reading is proportional to the current in the galvanometer. The galvanometer operates on the principle that a coil carrying a current in a magnetic field experiences a torque which is proportional to the current. This torque rotates the coil until it is balanced by the restoring torque provided by the mechanical suspension of the coil. We shall study the effects of magnetic field on current-carrying wires in Chapter 26; here we merely needed to know that such a torque exists, that it is proportional to the current in the coil, and that with proper design of the magnet the torque does not depend on the angle of rotation of the coil. Figure 25-22 shows a sketch of a galvanometer coil in a magnetic field.

Since the restoring torque of the suspension is proportional to the angle of rotation of the coil, the equilibrium angle of rotation will be proportional to the current in the coil. The resistance of the galvanometer and the current needed to produce full-scale deflection are the two parameters important for the construction of an ammeter or voltmeter from a galvanometer. Typical values of these parameters for a portable pivoted-coil laboratory galvanometer are $R_g = 20\ \Omega$ and $I_g = 0.5$ mA. The voltage drop across a galvanometer with these parameters

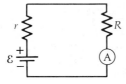

**Figure 25-20**
To measure the current in the resistor $R$ an ammeter Ⓐ is placed in series with the resistor.

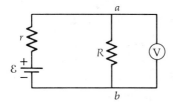

**Figure 25-21**
To measure the voltage drop across a resistor, a voltmeter Ⓥ is placed in parallel with the resistor.

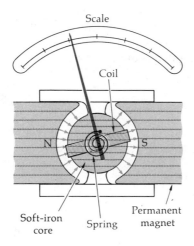

**Figure 25-22**
D'Arsonval galvanometer.
When the coil carries a current, the magnet exerts a
torque on the coil proportional to the current, causing
the coil to twist. The deflection read on the scale is proportional to the current in the
coil.

is thus $I_g R_g = 10^{-2}$ V for full-scale deflection. To construct an ammeter
from a galvanometer, we place a small resistance, called a *shunt resistor,*
in parallel with the galvanometer. Since the shunt resistance is usually
much smaller than the resistance of the galvanometer, most of the current is carried by the shunt and the effective resistance of the ammeter
is much smaller than the galvanometer resistance.

Resistors are added in series with a galvanometer to construct a voltmeter. Figure 25-23 illustrates the construction of an ammeter and voltmeter from a galvanometer. The choice of the appropriate resistors for
the construction of an ammeter or voltmeter from a galvanometer is best
illustrated by example.

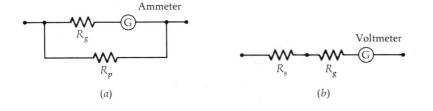

**Figure 25-23**
(*a*) An ammeter consists of a
galvanometer Ⓖ and a small
parallel resistance $R_p$, called a
shunt resistance. (*b*) A voltmeter consists of a gal-
vanometer and a large series
resistance $R_s$.

*Ammeter*

**Example 25-10** Using a galvanometer with a resistance of 20 Ω, for
which $5 \times 10^{-4}$ A gives full-scale deflection, design an ammeter which
will read full scale when the current is 5 A.

Since the total current through the ammeter must be 5 A when the
current through the galvanometer is just $5 \times 10^{-4}$ A, most of the current
must go through the shunt resistor. Let $R_p$ be the shunt resistance and
$I_p$ be the current through the shunt. Since the galvanometer and shunt
are in parallel, we have $I_g R_g = I_p R_p$ and $I_p + I_g = 5$ A or

$$I_p = 5 \text{ A} - I_g = 5 - 5 \times 10^{-4} \text{ A} \approx 5 \text{ A}$$

Thus the value of the shunt resistor should be

$$R_p = \frac{I_g}{I_p} R_g = \frac{5 \times 10^{-4}}{5} \times 20 = 2 \times 10^{-3} \text{ } \Omega$$

Since the resistance of the shunt is so much smaller than the resistance
of the galvanometer, the effective resistance of the parallel combination
is approximately equal to the shunt resistance.

*Voltmeter*

**Example 25-11** Using the same galvanometer as in Example 25-10, design a voltmeter which will read 10 V at full-scale deflection.

Let $R_s$ be the value of a resistor in series with the galvanometer. We want to choose $R_s$ so that a current of $I_g = 5 \times 10^{-4}$ A gives a potential drop of 10 V. Thus

$$I_g(R_s + R_g) = 10 \text{ V}$$

$$R_s + R_g = \frac{10 \text{ V}}{5 \times 10^{-4} \text{ A}} = 2 \times 10^4 \ \Omega$$

$$R_s = 2 \times 10^4 - R_g = 2 \times 10^4 - 20 = 19{,}980 \ \Omega \approx 20 \text{ k}\Omega$$

*Ohmmeter*

Figure 25-24 shows an ohmmeter consisting of a battery, a galvanometer, and a resistor, which can be used to measure resistance. The resistance $R_s$ is chosen so that the galvanometer reads full scale when the terminals $a$ and $b$ are shorted (touched together). Thus full scale on the galvanometer is marked zero resistance. When the terminals are connected across an unknown resistance $R$, the current is less than $I_g$ and the galvanometer reads less than full scale. The current in this case is

$$I = \frac{\mathcal{E}}{R + R_s + R_g} \qquad\qquad 25\text{-}29$$

Since this depends on $R$, the scale can be calibrated in terms of the resistance measured, from zero at full scale to infinite resistance at zero deflection. Since the calibration of the scale is far from linear and depends on the constancy of the emf of the battery, such an ohmmeter is not a high-precision instrument, but it is quite useful for making quick, rough determinations of resistance.

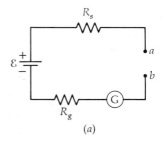

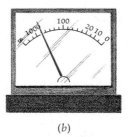

(a)                    (b)

**Figure 25-24**
(a) Ohmmeter consisting of a battery in series with a galvanometer and a resistor $R_s$ chosen so that the galvanometer reads full-scale deflection when points $a$ and $b$ are shorted. (b) Calibration of the galvanometer scale.

Some caution must be exercised in the use of an ohmmeter since it sends a current through the resistance to be measured. For example, consider an ohmmeter with a 1.5-V battery and a galvanometer similar to that in the previous examples. The series resistance needed is

$$I_g(R_s + R_g) = 1.5 \text{ V}$$

or

$$R_s = \frac{1.5}{5 \times 10^{-4}} - R_g = 3000 - 20 = 2980 \ \Omega$$

Suppose we were to use the ohmmeter to measure the resistance of a more sensitive galvanometer which gives a full-scale reading with a current of $10^{-5}$ A and has a resistance of about 20 $\Omega$. When the terminals $a$ and $b$ are placed across this more sensitive galvanometer, the current will be just slightly less than $5 \times 10^{-4}$ A because the total resistance is 3020 Ω, which is just slightly more than 3000 Ω. Such a current, about 50 times that needed to produce full-scale deflection, would ruin a sensitive galvanometer.

**Questions**

2. Under what conditions might it be advantageous to use a galvanometer which is less sensitive, i.e., requires a greater current $I_g$ for full-scale deflection, than the one discussed in Examples 25-10 and 25-11?

3. When the series resistance $R_s$ is properly chosen for a particular emf $\varepsilon$ of an ohmmeter, any value of resistance from zero to infinity can be measured. Why are there different scales on a practical ohmmeter for different ranges of resistance to be measured?

## 25-5    The Wheatstone Bridge

An accurate method of measuring resistance uses a circuit known as a *Wheatstone bridge* (Figure 25-25). In this figure $R_x$ is an unknown resistance, and $R_1$, $R_2$, and $R_4$ are presumed known. The galvanometer is used as a null detector. The resistances $R_1$ and $R_2$ are varied until there is no current in the galvanometer; the bridge is then said to be balanced. Under this condition, the potential at point $a$ is the same as that at point $b$. Thus the potential drop across $R_x$ must equal that across $R_1$:

$$I_2 R_x = I_1 R_1$$

Similarly, the drop across $R_2$ must equal that across $R_4$:

$$I_2 R_4 = I_1 R_2$$

We can eliminate the currents from these two equations by dividing one equation by the other, and then we can solve for the unknown resistance $R_x$:

$$\frac{R_x}{R_4} = \frac{R_1}{R_2}$$

or

$$R_x = R_4 \frac{R_1}{R_2} \qquad\qquad 25\text{-}30$$

In practice, a large variable resistance is placed in series with the galvanometer to reduce its sensitivity and protect it when the bridge is unbalanced. When the bridge is nearly balanced, the series resistance is decreased and eventually shorted out for maximum sensitivity of the galvanometer. Often a 1-m wire is used for the resistors $R_1$ and $R_2$ (Figure 25-26). Point $a$ is a sliding contact, which is varied until the bridge is balanced. If the wire is uniform, the ratio $R_1/R_2$ is just the ratio of the lengths $L_1/L_2$. The measurement does not depend on the emf used to supply the current.

**Question**

4. What are the advantages and disadvantages of a Wheatstone bridge compared with an ohmmeter?

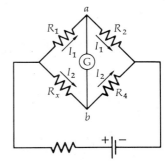

**Figure 25-25**
Wheatstone-bridge circuit for measuring an unknown resistance $R_x$ in terms of known resistances $R_1$, $R_2$, and $R_4$.

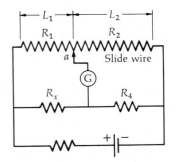

**Figure 25-26**
Slide-wire Wheatstone bridge. The resistances $R_1$ and $R_2$ are proportional to the lengths $L_1$ and $L_2$. Their ratio is varied by moving a sliding contact along the wire until the bridge is balanced.

# Exponential Growth

Albert A. Bartlett
*University of Colorado at Boulder*

If a town with a population of 10,000 has a steady annual population growth of 5 percent, how large will the population be in 10, 30, and 100 y? If I invest $100 in an account that pays 15 percent annual interest (compounded continuously), how large will the account be in 10, 30, and 100 y? These questions are real-life applications of what may be the single most important differential equation:

$$\frac{dN}{dt} = kN \qquad\qquad 1$$

This equation describes the situation in which the rate of change of a quantity $dN/dt$ is proportional to the size $N$ of the quantity. In our first example $N$ is the population of the town and $k = 0.05\ y^{-1}$; in the second example $N$ is the number of dollars in the account and $k = 0.15\ y^{-1}$. In both cases $t$ is the time in years. The meaning of $k$ can be seen by rearranging Equation 1 to read

$$k = \frac{1}{N}\frac{dN}{dt} \qquad\qquad 2$$

Thus $k$ is the fractional change in $N$, $dN/N$, per unit time $dt$. It follows that $100k$ is $P$, the percent growth per unit time. These two examples are cases of steady growth in which the quantity always increases by a fixed fraction (for example, 5 percent) in a fixed length of time (each year).

The solution of Equation 1 is the exponential function

$$N = N_0 e^{kt} \qquad\qquad 3$$

where $e$ is the base of natural logarithms (2.718 $\cdots$ ) and $N_0$ is the size of $N$ at the time $t = 0$. If $k > 0$, Equation 3 describes exponential growth (steady growth), and if $k < 0$, it describes exponential decay. Exponential decay occurs in Newton's law of cooling (Chapter 18), where $N$ is the temperature difference between an object and its surroundings; in the study of $RC$ circuits (Chapter 25), where $N$ is the charge on the capacitor; in $LR$ circuits (Chapter 28), where $N$ is the current in the circuit; and in radioactive decay (Chapter 37), where $N$ is the number of radioactive nuclei which have not yet decayed. Here we focus our attention on exponential or steady growth, where $k$ is a positive constant. Exponential growth occurs whenever the rate of increase in a quantity is proportional to that quantity.

In our two examples in the first paragraph, the town's population will be

$$N = 10{,}000\,e^{0.05t}$$

and the size of the bank account will be

$$N = 100\,e^{0.15t}$$

where $t$ is the time in years.

Using Equation 3 one can show that for a given value of $k$, a constant time $T_c$ is required for $N$ to increase in size from its present value $N_0$ to $CN_0$; where $C$ is a constant:

$$T_c = \frac{1}{k}\ln C \qquad\qquad 4$$

The time for $N$ to double in size ($C = 2$) is

$$T_2 = \frac{1}{k}\ln 2 = \frac{0.693}{k} \qquad\qquad 5$$

**Table 1**
Doubling times for various rates of steady growth

| $k$, $y^{-1}$ | $P$, %/y | $T_2$, y |
|---|---|---|
| 0.01 | 1 | 69.3 |
| 0.02 | 2 | 34.7 |
| 0.03 | 3 | 23.1 |
| 0.05 | 5 | 13.9 |
| 0.07 | 7 | 9.90 |
| 0.10 | 10 | 6.93 |
| 0.15 | 15 | 4.62 |
| 0.20 | 20 | 3.47 |
| 0.30 | 30 | 2.31 |

which can be expressed as

$$T_2 = \frac{69.3}{P} \approx \frac{70}{P} \qquad\qquad 6$$

where $P = 100k$ is the percent growth per unit time. Table 1 shows doubling times for different steady rates of growth calculated from Equation 6.

From Table 1 we see that a quantity growing steadily at 5 percent per year will double in size in 13.9 y (steady growth at a rate of 5 percent per month will cause doubling to take place in 13.9 months). A quantity growing steadily at 15 percent per year will double in 4.62 y. We should develop the habit of doing the mental arithmetic of Equation 6 every time we see a percent growth rate in the news or elsewhere. People who do not grasp the meaning of 15 percent annual inflation will understand when you tell them that this means that prices will double every 4.6 years!

Equation 3 can be rewritten in a very useful form using $T_2$:

$$N = N_0 e^{kt} = N_0 2^{t/T_2} \qquad\qquad 7$$

In this form we can estimate the answers to the questions that opened this essay. Since the doubling time for an annual growth rate of 5 percent is roughly 14 y, it can be seen that 10 y is less than one doubling time, 30 y is approximately two doubling times, and 100 y is approximately 7 doubling times. Thus in 10 y the population will increase from 10,000 to roughly 16,000; in 30 y it will have doubled twice (to 40,000); and in 100 y it will have doubled 7 times (a factor of $2^7 = 128$), and so the population will be 1,280,000. For a 15 percent annual interest rate compounded continuously ($T_2 = 4.62$ y) the figures are shown in Table 2.

**Table 2**
The value of $100 deposited at 15% annual interest (compounded continuously) after a time $t$

| $t$, y | $t/T_2$ | $2^{t/T_2}$ | Size of account |
|---|---|---|---|
| 10 | $10/4.62 = 2.16$ | 4.48 | $448 |
| 30 | $30/4.62 = 6.49$ | 90 | $9000 |
| 100 | $100/4.62 = 21.6$ | $3.27 \times 10^6$ | $327,000,000 |

We can estimate the value of large powers of 2 by noting that

$$2^{10} = 1024 \approx 10^3 \qquad\qquad 8$$

For example,

$$2^{22} = 2^{10} \times 2^{10} \times 2^2 \approx 10^3 \times 10^3 \times 4 = 4 \times 10^6$$

Some very important properties of steady growth can be illustrated by the following example. Legend has it that the game of chess was invented by a mathematician who worked for a king. As a reward the mathematician asked the king for the quantity of wheat determined by placing one grain of wheat on the first square of the chessboard, on the next square doubling the one grain to make two, on the next square doubling the two grains to make four, and continuing until the doubling has been done for all the squares (see Table 3). We can use Equation 8 to get an approximate expression for this total number of grains:

$$2^{64} \approx 10^3 \times 10^3 \times 10^3 \times 10^3 \times 10^3 \times 10^3 \times 16 = 1.6 \times 10^{19} \text{ grains}$$

The actual value of $2^{64}$ is $1.84 \times 10^{19}$. This is approximately 500 times the 1980 worldwide harvest of wheat, which may be more wheat than people have harvested in the entire history of the earth. How did we get such a large number? It was simple. We started with one grain and let the number grow until it had doubled a mere 63 times. A study of Table 3 reveals another important point.

**Table 3**
Grains of wheat on a
chessboard

| No. of square | No. on square | Total no. on board |
|---|---|---|
| 1 | 1 | 1 |
| 2 | 2 | 3 |
| 3 | 4 | 7 |
| 4 | 8 | 15 |
| 5 | 16 | 31 |
| 6 | 32 | 63 |
| . . . . . . . . . . . . . . . . . |
| 64 | $2^{63}$ | $2^{64} - 1$ |

*The growth in any doubling time is greater than the sum of all of the preceding growth.* Thus when we put 32 grains on the sixth square, the 32 is larger than the total of 31 already on the board.

In his speech on energy (April 18, 1977) President Carter said, "in each of those decades [the 1950s and 1960s] more oil was consumed [worldwide] than in all of mankind's previous history." This statement sounds incredible, but a look at Table 3 shows that this is a simple consequence of steady growth with a doubling time of 10 years, i.e., an annual growth rate of 7 percent. This was the growth rate of world oil consumption for nearly 100 years before 1970. This feature of exponential growth is extremely important. Suppose, for example, that the annual growth rate of 7 percent continues and that just as the world's supply of oil is to be exhausted, a new source of oil is discovered equal to the total amount of oil consumed worldwide in the history of mankind. If oil consumption continues to grow at the annual rate of 7 percent, this fantastic new supply will last one doubling time, or 10 years. This example makes it clear that, contrary to what prominent people often suggest, exploration and production of fossil fuels cannot meet the needs of continued growth for very long.

It is interesting to calculate the length of time a given finite resource will last if its rate of use grows steadily ($k$ = a constant) until the last of the resource is used. The rate of use of the resource $R$ will then be

$$R = R_0 e^{kt} \qquad 9$$

where $R_0$ is the rate at $t = 0$. If we have a given quantity $M_0$ of the resource at time $t = 0$, we can find the time $T_e$ when this quantity will be exhausted by integrating the rate $R\,dt$ from time $t = 0$ to $t = T_e$ to find the total amount used in $T_e$, and setting this equal to $M_0$:

$$\int_0^{T_e} R\,dt = \frac{R_0}{k}(e^{kT_e} - 1) = M_0$$

Solving for $T_e$, we obtain

$$T_e = \frac{1}{k} \ln\left(\frac{kM_0}{R_0} + 1\right) \qquad 10$$

We can use Equation 10 to estimate the lifetime $T_e$ of United States coal reserves for different steady rates of growth of coal production. The U.S. Geological Survey* estimates the Reserve Base of United States coal to be $M_0 = 379 \times 10^9$ tonnes (1 tonne = $10^3$ kg). The rate of production in 1978 was $R_0 = 0.6 \times 10^9$ tonnes/y. Using these numbers in Equation 10 we find the results shown in Table 4.

If we have zero growth of coal production, United States coal would last $T_e = M_0/R_0 = 632$ y. This is the basis for news stories which say "at present rates of consumption" United States coal will last about 600 y.** These stories go on then to stress the national urgency of achieving rapid growth of United States coal production, but the stories don't tell how growth reduces $T_e$ (see Table 4). For example, President Gerald Ford called for United States coal production to grow at an annual rate of about 10 percent ($T_e = 42$ y); President Carter called for 5 percent ($T_e = 70$ y); and President Reagan is continuing and enlarging President Carter's program of rapid growth of United States coal exports.*** Part of our problem with oil in the United States in 1981 arises from the fact that for decades the United States was a major exporter of oil. Now we seem to be repeating the process with coal.

The exponential arithmetic that arises so naturally in physics and in many other real-life situations is incredibly important—in part because it is so poorly

**Table 4**
Lifetime of United States coal
reserves for different steady
rates of growth of coal
production

| Annual growth rate, % | Expiration time of the reserve base |
|---|---|
| 0 | 632 |
| 1 | 199 |
| 2 | 131 |
| 3 | 100 |
| 5 | 70 |
| 10 | 42 |
| 15 | 30 |

* P. Averitt, Coal Resources of the United States, *U.S. Geological Survey Bulletin* 1412, Jan. 1, 1974.
** *Newsweek Magazine*, July 16, 1979, p. 23.
*** National Geographic Magazine, Feb. 1981, p. 19.

understood by the people who make our policies (both local and national), who follow the dictum that "growth is good," and who never ask how long growth can continue. One can see that the energy shortage is caused by growth in consumption; the energy crisis is caused by our failure to understand the arithmetic and consequences of this growth. In the words of Aldous Huxley, "Facts do not cease to exist because they are ignored."

Let us summarize the points we have made about the arithmetic of steady growth (the exponential function):

1. Steady growth happens whenever the rate of increase of a quantity is proportional to the quantity (Equation 1).

2. It is characterized by a fixed doubling time (Equations 5 and 6).

3. In every doubling time the amount of growth is greater than the total of all of the preceding growth.

4. Growth for a modest number of doubling times gives astronomically large numbers.

5. If consumption of a resource is growing steadily, enormous new supplies of the resource will allow only very short extensions of the period of steady growth.

It has been said that the greatest shortcoming of the human race is our inability to understand the exponential function.*

* These topics have been developed in more detail in *The American Journal of Physics,* vol. 46, September 1978, p. 876; *Journal of Geological Education,* vol. 28, January 1980, p. 4; "The Exponential Function," *The Physics Teacher,* October, November 1976; January, March, April 1977; January to March 1978; January 1979; and "The Forgotten Fundamentals of the Energy Crisis," videotape, the Media Center, University of Colorado, Boulder, Colorado.

---

## Review

A. Define, explain, or otherwise identify:

Series resistors, 696
Parallel resistors, 697
Kirchhoff's rules, 698
Load resistance, 700
Impedance matching, 700
Time constant, 704

Galvanometer, 706
Ammeter, 707
Voltmeter, 708
Ohmmeter, 708
Wheatstone bridge, 709

B. True or false:

1. The effective resistance of two series resistors is always greater than that of either resistor.

2. The effective resistance of two parallel resistors is always less than that of either resistor.

3. To measure the current through a resistor, an ammeter is placed in series with the resistor.

4. To measure the potential drop across a resistor a voltmeter is placed in parallel with the resistor.

5. The time constant $t_c = RC$ is the time needed to discharge a capacitor completely.

6. A Wheatstone bridge is used to measure emf.

**Figure 25-27**
Exercise 1.

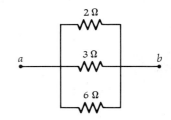

### Exercises

#### Section 25-1, Series and Parallel Resistors

1. (a) Find the equivalent resistance between points $a$ and $b$ in Figure 25-27. (b) If the potential drop between $a$ and $b$ is 12 V, find the current in each resistor.

2. Repeat Exercise 1 for the resistors shown in Figure 25-28.

3. Repeat Exercise 1 for the resistors shown in Figure 25-29.

4. Repeat Exercise 1 for the resistors shown in Figure 25-30.

**Figure 25-28**
Exercise 2.

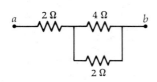

**Figure 25-29**
Exercises 3 and 5.

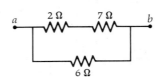

**Figure 25-30**
Exercise 4.

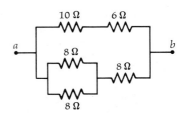

5. In Figure 25-29 the current in the 6-$\Omega$ resistor is 3 A. (a) What is the potential drop between $a$ and $b$? (b) What is the current in the 2-$\Omega$ resistor?

6. (a) Show that the equivalent resistance between points $a$ and $b$ in Figure 25-31 is $R$. (b) What would be the effect of adding a resistance $R$ between points $c$ and $d$?

7. Repeat Exercise 1 for the resistors shown in Figure 25-32.

8. Consider the equivalent resistance of two parallel resistors $R_1$ and $R_2$ as a function of the ratio $x = R_2/R_1$. (a) Show that $R_{eq} = R_1 x/(1 + x)$. (b) Sketch $R_{eq}$ as a function of $x$.

9. The battery in Figure 25-33 has negligible resistance. Find (a) the current in each resistor and (b) the power delivered by the battery.

**Figure 25-31**
Exercise 6.

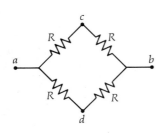

**Figure 25-32**
Exercise 7.

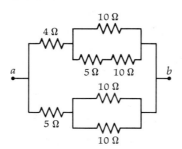

**Figure 25-33**
Exercise 9.

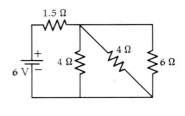

### Section 25-2, Kirchhoff's Rules

10. A 6-V (emf) battery with internal resistance 0.3 $\Omega$ is connected to a variable resistance $R$. Find the current and power delivered by the battery when $R$ is (a) 5 $\Omega$ and (b) 10 $\Omega$. (c) What is the maximum power that can be dissipated in the resistance $R$?

11. A variable resistance $R$ is connected across a potential difference $V$ which remains constant independent of $R$. At one value $R = R_1$, the current is 6.0 A. When $R$ is increased to $R_2 = R_1 + 10.0$ $\Omega$, the current drops to 2.0 A. Find (a) $R_1$ and (b) $V$.

12. A battery has emf $\varepsilon$ and internal resistance $r$. When a 5.0-$\Omega$ resistor is connected across the terminals, the current is 0.5 A. When this resistor is replaced by an 11.0-$\Omega$ resistor, the current is 0.25 A. Find the emf $\varepsilon$ and internal resistance $r$.

13. In Figure 25-34 the emf is 6 V and $R = 0.5$ $\Omega$. The rate of Joule heating in $R$ is 8 W. (a) What is the current in the circuit? (b) What is the potential difference across $R$? (c) What is $r$?

14. For the circuit shown in Figure 25-35 find (a) the current, (b) the power delivered or absorbed by each emf, and (c) the rate of Joule heat production in each resistor. (Assume the batteries have negligible internal resistance.)

15. In the circuit shown in Figure 25-34 let $\varepsilon = 18.0$ V and $r = 3.0$ $\Omega$. (a) Graph the current through $R$ as a function of $R$ for the range $R = 0$ to $R = 6.0$ $\Omega$. (b) For the same range of $R$, graph the power supplied to $R$ as a function of $R$. (c) At what value of $R$ is the power supplied to $R$ a maximum? What is the value of this maximum power?

16. In the circuit shown in Figure 25-36 the batteries have negligible internal resistance, and the ammeter has negligible resistance. (a) Find the current through the ammeter. (b) Find the energy delivered by the 12-V battery in 3 s. (c) Find the total heat produced in that time. (d) Account for the difference in your answers to parts (b) and (c).

17. In the circuit shown in Figure 25-37 the batteries have negligible internal resistance. (a) Find the current in each resistor, (b) the potential difference between points $a$ and $b$, and (c) the power supplied by each battery.

18. Repeat Exercise 17 for the circuit shown in Figure 25-38.

**Figure 25-34**
Exercises 13 and 15.

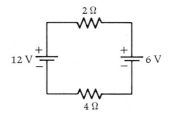

**Figure 25-35**
Exercise 14.

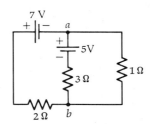

**Figure 25-36**
Exercise 16.

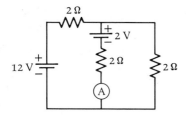

**Figure 25-37**
Exercise 17.

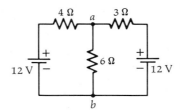

**Figure 25-38**
Exercise 18.

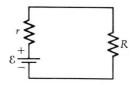

### Section 25-3, RC Circuits

19. A 6-$\mu$F capacitor is initially charged to 100 V and then connected across a 500-$\Omega$ resistor. (a) What is the initial charge on the capacitor? (b) What is the initial current just after the capacitor is connected to the resistor? (c) What is the time constant of this circuit? (d) How much charge is on the capacitor after 6 ms?

20. (a) For Exercise 19, find the initial energy stored in the capacitor. (b) Show that the energy stored in the capacitor is given by $U = U_0 e^{-2t/t_c}$, where $U_0$ is the initial energy and $t_c = RC$ is the time constant. (c) Make a careful sketch of the energy $U$ in the capacitor versus time $t$.

21. A 0.10-$\mu$F capacitor is given a charge $Q_0$. After 4 s its charge is observed to be $\frac{1}{2}Q_0$. What is the effective resistance across this capacitor?

22. A 1.0-$\mu$F capacitor, initially uncharged, is connected in series with a 10-k$\Omega$ resistor and a 5.0-V battery of negligible internal resistance. (a) What is the charge on the capacitor after a very long time? (b) How long does it take for the capacitor to reach 99 percent of its maximum charge?

23. A 1-M$\Omega$ resistor is connected in series with a 1.0-$\mu$F capacitor and a 6.0-V battery of negligible internal resistance. The capacitor is initially uncharged. After a time $t = t_c = RC$, find (a) the charge on the capacitor, (b) the rate at which the charge is increasing, (c) the current, (d) the power supplied by the battery, (e) the power dissipated in the resistor, and (f) the rate at which the energy stored in the capacitor is increasing.

24. Repeat Exercise 23 for the time $t = 2t_c$.

### Section 25-4, Ammeters, Voltmeters, and Ohmmeters

25. A galvanometer has a resistance of 100 $\Omega$. It requires 1 mA to give full-scale deflection. (a) What resistance should be placed in parallel with the galvanometer to make an ammeter that reads 2 A at full-scale deflection? (b) What resistance should be placed in series to make a voltmeter that reads 5 V at full-scale deflection?

26. Sensitive galvanometers can detect currents as small as 1 pA. How many electrons per second does this current represent?

27. A sensitive galvanometer has resistance 100 $\Omega$ and requires 1.0 $\mu$A of current to produce full-scale deflection. (a) Find the shunt resistance needed to construct an ammeter which reads 1.0 mA full scale. (b) What is the resistance of the ammeter? (c) What resistance would be required to construct a voltmeter reading 3.0 V full scale from this galvanometer?

28. A galvanometer of resistance 100 $\Omega$ gives full-scale deflection when its current is 1.0 mA. It is used to construct an ammeter whose full-scale reading is 100 A. (a) Find the shunt resistance needed. (b) What is the resistance of the ammeter? (c) If the shunt resistor consists of a piece of 10-gauge copper wire (diameter 2.59 mm), what should its length be?

29. The galvanometer in Exercise 28 is used with a 1.5-V battery of negligible internal resistance to make an ohmmeter. (a) What resistance $R_s$ should be placed in series with the galvanometer? (b) What resistance $R$ will give half-scale deflection? (c) What resistance will give a deflection of one-tenth full scale?

30. For the ohmmeter in Exercise 29, indicate how the galvanometer scale should be calibrated by representing the galvanometer scale on a straight line of some length $L$ where the end of the line ($x = L$) represents full-scale reading at $R = 0$. Divide the line into 10 equal divisions and indicate values of the resistance at each division.

31. A galvanometer of resistance 100 $\Omega$ gives full-scale reading when its current is 0.1 mA. It is to be used in a multirange voltmeter as shown in Figure 25-39, where the connections refer to full-scale readings. Determine $R_1$, $R_2$, and $R_3$.

**Figure 25-39**
Exercise 31.

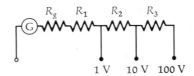

$R_g$    $R_1$    $R_2$    $R_3$

1 V    10 V    100 V

32. The galvanometer of Exercise 31 is to be used in a multirange ammeter with full-scale readings as indicated in Figure 25-40. Determine $R_1$, $R_2$, and $R_3$.

### Section 25-5, The Wheatstone Bridge

33. In a slide-wire Wheatstone bridge using a uniform wire of length 1 m, $R_1$ is proportional to the length of wire from 0 to the balance point and $R_2$ is proportional to the length from the balance point to 100 cm (the end of the wire). If the fixed resistor is $R_4 = 200$ Ω, find the unknown resistance if (a) the bridge balances at the 18-cm mark, (b) the bridge balances at the 60-cm mark, and (c) the bridge balances at the 95-cm mark.

34. In the Wheatstone bridge of Exercise 33, with $R_4 = 200$ Ω, the bridge balances at the 98-cm mark. (a) What is the unknown resistance? (b) What effect would an error of 2 mm have on the measured value of the unknown resistance? (c) How should $R_4$ be changed so that this unknown resistor will give a balance point nearer the 50-cm mark?

### Problems

1. A battery with emf $\varepsilon$ and internal resistance $r$ is connected to an external resistance $R$. Show that (a) the power $P$ supplied by the emf is given by $\varepsilon^2/(r + R)$, (b) the power $P_R$ supplied to the external resistor is $\varepsilon^2 R/(r + R)^2$, and (c) the power $P_r$ dissipated in the battery is $\varepsilon^2 r/(r + R)^2$. Sketch on the same graph $P$, $P_R$, and $P_r$ versus $R$.

2. Show that as $R$ varies in Problem 1, $P_R$ is maximum at $R = r$, and that the maximum power is $P_R = \varepsilon^2/4r$.

3. A sick car battery of emf 11.4 V and internal resistance 0.01 Ω is connected to a load of 2.0 Ω. To help the ailing battery, a second battery of emf 12.6 V and internal resistance 0.01 is connected by jumper cables to the terminals of the first battery. Draw a circuit diagram for this situation and find the current in each part of the circuit. Find the power delivered by the second battery and discuss where this power goes, assuming that both emfs and internal resistances are constant.

4. In Problem 3 assume that the first emf increases at a constant rate of 0.2 V/h, but that that of the second battery and the internal resistances are constant. Find the current in each part of the circuit as a function of time. Sketch a graph of the power delivered to the first battery as a function of time.

5. In the circuit shown in Figure 25-41 find (a) the current in each resistor, (b) the power supplied by each emf, and (c) the power dissipated in each resistor.

6. Two identical batteries with emf $\varepsilon$ and internal resistance $r$ may be connected across a resistance $R$ either in series or in parallel. Which method of connection supplies the greater power to $R$ when (a) $R < r$, (b) $R > r$?

7. In the circuit shown in Figure 25-42 find the potential difference between points $a$ and $b$.

8. The space between the plates of a parallel-plate capacitor is filled with a dielectric of constant $K$ and resistivity $\rho$. (a) Show that the charge on the plates decays with time constant $t_c = \epsilon_0 K \rho$. (b) If the dielectric is mica with $K = 5.0$ and $\rho = 9 \times 10^{13}$ Ω·m, find the time for the charge to decrease to $1/e^2 \approx 14$ percent of its initial value.

9. In the circuit shown in Figure 25-43 the battery has internal resistance 0.01 Ω. An ammeter of resistance 0.01 Ω is inserted at point $a$. (a) What is the reading of the ammeter? (b) By what percentage is the current changed because of the ammeter? (c) The ammeter is removed and a voltmeter of resistance 1 kΩ is connected from $a$ to $b$. What is the reading of the voltmeter? (d) By what percentage is the voltage drop from $a$ to $b$ changed by the presence of the voltmeter?

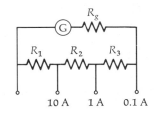
**Figure 25-40**
Exercise 32.

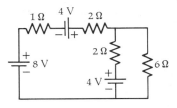
**Figure 25-41**
Problem 5.

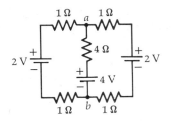
**Figure 25-42**
Problem 7.

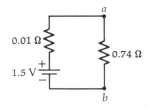
**Figure 25-43**
Problem 9.

10. If you have two batteries, one with $\varepsilon$ = 9.0 V and $r$ = 0.8 $\Omega$ and the other with $\varepsilon$ = 3.0 V and $r$ = 0.4 $\Omega$, how would you connect them to give the largest current through a resistor $R$ for (a) $R$ = 0.2 $\Omega$, (b) $R$ = 0.6 $\Omega$, (c) $R$ = 1.0 $\Omega$, and (d) $R$ = 1.5 $\Omega$?

11. In the circuit shown in Figure 25-44 the reading of the ammeter is the same with both switches open as with both closed. Find the resistance $R$.

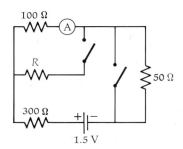

**Figure 25-44**
Problem 11.

12. In Figure 25-45 the galvanometer has resistance 10 $\Omega$ and is connected across a 90-$\Omega$ resistor. The value of the current which causes full-scale deflection can be chosen by using connections $ab$, $ac$, $ad$, or $ae$. (a) How should the 90-$\Omega$ resistor be divided so that the current causing full-scale deflection decreases by a factor of 10 for each successive connection $ab$, $ac$, etc.? (b) What should be the full-scale deflection current in the galvanometer $I_g$ so that this ammeter has ranges 1.0 mA, 10 mA, 100 mA, and 1 A?

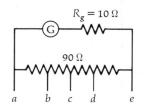

**Figure 25-45**
Problem 12.

13. Figure 25-46 shows two possible ways to use a voltmeter and ammeter to measure an unknown resistance. Assume that the internal resistance of the battery is negligible and that the resistance of the voltmeter is 1000 times that of the ammeter: $R_v$ = 1000 $R_a$. The calculated value of $R$ is taken to be $R_c = V/I$, where $V$ and $I$ are the readings of the voltmeter and ammeter. (a) Discuss which circuit is preferred for values of $R$ in the range from $10R_a$ to $0.9R_v$. (b) Find $R_c$ for each circuit if $R_a$ = 0.1 $\Omega$, $R_v$ = 100 $\Omega$, and $R$ = 0.5, 3, and 80 $\Omega$.

14. See Problem 13. For the circuits shown in Figure 25-46 show that $R_c = V/I$ is related to the true value $R$ by

$$\frac{1}{R_c} = \frac{1}{R} + \frac{1}{R_v}$$

for circuit $a$ and $R_c = R + R_a$ for circuit $b$. If $\varepsilon$ = 1.5 V, $R_a$ = 0.01 $\Omega$, and $R_v$ = 10 k$\Omega$, what range of values of $R$ can be measured such that $R_c$ is within 5 percent of $R$ using circuit $a$? Using circuit $b$?

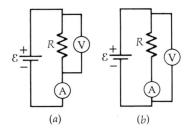

**Figure 25-46**
Problems 13 and 14.

(a)         (b)

15. Consider the circuit in Figure 25-47, where $r$ is the internal resistance of the emf and $R_a$ the resistance of the ammeter. (*a*) Show that the ammeter reading is given by

$$\mathcal{E}\left(R_2 + R_a + r + \frac{R_2 + R_a}{R_1}r\right)^{-1}$$

(*b*) Show that if the ammeter and emf are interchanged, the ammeter reading is

$$\mathcal{E}\left(R_2 + R_a + r + \frac{R_2 + r}{R_1}R_a\right)^{-1}$$

Note that if $R_a = r$, or if both are negligible, the reading is the same. (When one can neglect $R_a$ and $r$, this symmetry can be very useful in analyzing circuits with only one emf; it is not valid for more than one emf.)

16. Derive Equation 25-7 for the equivalent resistance of three parallel resistors.

17. In the circuit shown in Figure 25-48 the capacitor is originally uncharged with the switch open. At $t = 0$ the switch is closed. (*a*) What is the current supplied by the emf just after the switch is closed? (*b*) What is the current a long time after the switch is closed? (*c*) Derive an expression for the current through the emf for any time after the switch is closed. (*d*) After a long time $t'$ the switch is opened. How long does it take for the charge on the capacitor to decrease to 10 percent of its value at $t = t'$ if $R_1 = R_2 = 5$ k$\Omega$ and $C = 1.0$ $\mu$F?

18. Two batteries with emfs $\mathcal{E}_1$ and $\mathcal{E}_2$ and with internal resistances $r_1$ and $r_2$ are connected in parallel. Prove that the optimal load resistance (for maximum power delivery) $R$ connected in parallel with this combination is $R = r_1 r_2/(r_1 + r_2)$.

**Figure 25-47**
Problem 15.

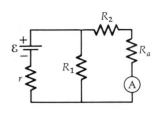

**Figure 25-48**
Problem 17.

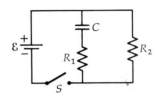

CHAPTER 26    The Magnetic Field

---

Objectives    After studying this chapter, you should:

1.  Be able to calculate the magnetic force on a current element and on a moving charge in a given magnetic field.

2.  Be able to calculate the magnetic moment of a current loop and the torque exerted on a current loop in a magnetic field.

3.  Be able to discuss the Thomson experiment for measuring $q/m$ for electrons.

4.  Be able to describe a velocity selector, a mass spectrograph, and a cyclotron.

5.  Be able to discuss the Hall effect.

---

It was known to the early Greeks more than 2000 years ago that certain stones from Magnesia, now called magnetite, attract pieces of iron.* Early experiments in magnetism were concerned chiefly with the behavior of permanent magnets. It is not known when a magnet was first used for navigation, but there is written reference to such use as early as the twelfth century. In 1269 an important discovery was made by Pierre de Maricourt, when he laid a needle on a spherical natural magnet at various positions and marked the directions taken by the needle. The lines encircled the magnet just as the meridian lines encircle the earth, passing through two points at opposite ends of the sphere. Because of the analogy with the meridian lines of the earth, he called these points the *poles* of the magnet. It was noted by many experimenters that every magnet of whatever shape had two poles, a north pole and a south pole, where the magnetic force exerted by the magnet is strongest. Like poles of two magnets repel each other, and unlike poles attract.

---

* Much of this discussion follows Sir Edmund Whittaker's excellent *A History of the Theories of Aether and Electricity*, Nelson, London, 1953; Torchbook edition, Harper, New York, 1960.

In 1600 William Gilbert discovered why a compass needle orients it-self in definite directions: the earth itself is a permanent magnet. Since the north pole of a compass needle is attracted to the north geographic pole of the earth, this north geographic pole of the earth is actually a south magnetic pole. The attraction and repulsion of magnetic poles was studied quantitatively by John Michell in 1750. Using a torsion bal-ance, Michell showed that the attraction and repulsion of the poles of two magnets are of equal strength and vary inversely as the square of the distance between the poles. These results were confirmed shortly thereafter by Coulomb. The law of force between two magnetic poles is similar to that between two electric charges, but there is an important difference: magnetic poles always occur in pairs. It is impossible to iso-late a single magnetic pole. If a magnet is broken in half, equal and op-posite poles appear at the break point, so that there are two magnets, each with equal and opposite poles. Coulomb explained this result by assuming that the magnetism is contained in each molecule of the magnet.

The connection between electricity and magnetism was not known until the nineteenth century, when Hans Christian Oersted discovered that an electric current affects the orientation of a compass needle. Sub-sequent experiments by André Marie Ampère and others showed that electric currents attract bits of iron and that parallel currents attract each other. Ampère proposed the theory that electric currents are the source of all magnetism; in ferromagnets these currents were thought to be "molecular" current loops which are aligned when the material is mag-netized. Ampère's model is the basis of the theory of magnetism today, though in many cases the "currents" in magnetic materials are related to the intrinsic electron spin, a quantum property of the electron which

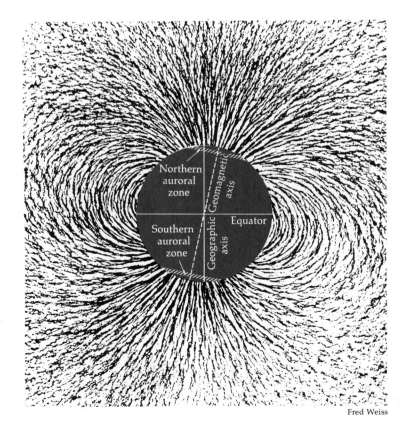

Fred Weiss

Magnetic field lines of the earth indicated by iron filings around a uniformly magne-tized sphere.

could not be envisioned in the nineteenth century. Further connections between magnetism and electricity were demonstrated by experiments of Michael Faraday and Joseph Henry showing that a changing magnetic field produces a nonconservative electric field, and by James Clerk Maxwell's theory showing that a changing electric field produces a magnetic field.

The basic magnetic interaction is one between two charges in motion. As with the electrostatic interaction, it is convenient to consider the magnetic force exerted by one moving charge on another to be transmitted by a third agent, the magnetic field. A moving charge produces a magnetic field, and that field in turn exerts a force on a second moving charge. Since a moving charge constitutes an electric current, the magnetic interaction can also be thought of as an interaction between two currents.

In this chapter we consider only the effects of given magnetic fields on moving charges and currents in wires. We postpone consideration of the origin of the magnetic field until Chapter 27.

# 26-1   Definition of the Magnetic Field **B**

We can define the magnetic field vector **B** at a point in space similar to the way we defined the electric field **E** (recall that we defined **E** at a point in space to be the force per unit charge on a test charge $q_0$ placed at that point). It is observed experimentally that when a charge has a velocity **v** in the vicinity of a magnet or a current-carrying wire, there is an additional force on it which depends on the magnitude and direction of the velocity. We can easily separate these two forces by measuring the force on the charge when it is at rest and subtracting this *electric* force from the total force acting when the charge is moving. For simplicity, let us assume that there is no electric field at the point in space we are considering. Experiments with various charges moving with various velocities at such a point in space give the following results for the magnetic force:

1. The force is proportional to the magnitude of the charge $q$.

2. The force is proportional to the speed $v$.

3. The magnitude and direction of the force depend on the direction of the velocity **v**.

4. If the velocity of the particle is along a certain line in space, the force is zero.

5. If the velocity is not along this line, there is a force which is perpendicular to this line and also perpendicular to the velocity.

6. If the velocity makes an angle $\theta$ with this line, the force is proportional to sin $\theta$.

7. The force on a negative charge is in the direction opposite that on a positive charge with the same velocity.

*Magnetic force on a moving charge*

We can summarize these experimental results by defining a magnetic field vector **B** to be along the line described in result 4, and by writing

$$\mathbf{F} = q\mathbf{v} \times \mathbf{B} \qquad\qquad 26\text{-}1$$

For historical reasons, the vector **B** is usually called the *magnetic-induction vector* or the *magnetic flux density*. (A related vector **H**, the magnetic field intensity, will be introduced in Chapter 29 when we study magnetic materials. In free space, the vectors **B** and **H** are proportional to each other. In keeping with custom we shall refer to this **B** field as the *magnetic induction*, although the phrase magnetic field **B** may also be used.)

Figure 26-1 shows the force exerted on various moving charges when the magnetic-induction vector **B** is in the vertical direction. Note that the direction of **B** can be found by experiment using Equation 26-1.

The SI unit of magnetic induction is the tesla (T).* A charge of one coulomb moving with a velocity of one metre per second perpendicular to a magnetic field of one tesla experiences a force of one newton:

$$1 \text{ T} = 1 \text{ N·s/C·m} = 1 \text{ N/A·m}$$

This unit is rather large; a common unit, derived from the cgs system, is the gauss (G), related to the tesla as follows:

$$1 \text{ T} = 10^4 \text{ G} \qquad\qquad 26\text{-}2$$

Since magnetic fields are often given in gauss, which is not an SI unit, it is important to remember to convert to teslas when making calculations.

---

**Example 26-1** The magnetic induction of the earth has a magnitude of 0.6 G and is directed downward and northward, making an angle of about 70° with the horizontal. (These data are approximately correct for the central United States.) A proton moves horizontally in the northward direction with speed $v = 10^7$ m/s. Calculate the force on the proton.

Figure 26-2 shows the directions of the magnetic induction **B** and the proton's velocity **v**. The angle between them is $\theta = 70°$. The direction of the force is $\mathbf{v} \times \mathbf{B}$, which is west for a proton moving north. The magnitude of the magnetic force is

$$F = qvB \sin \theta = (1.6 \times 10^{-19} \text{ C}) (10^7 \text{ m/s}) (0.6 \times 10^{-4} \text{ T}) (0.94)$$
$$= 9.02 \times 10^{-17} \text{ N}$$

The acceleration of the proton is

$$a = \frac{F}{m} = \frac{9.02 \times 10^{-17} \text{ N}}{1.67 \times 10^{-27} \text{ kg}} = 5.40 \times 10^{10} \text{ m/s}^2$$

---

When a wire carries a current in a magnetic field, there is a force on the wire which is the sum of the magnetic forces on the charged particles whose motion produces the current. Figure 26-3 shows a short segment of wire of cross-sectional area $A$ and length $\ell$ carrying a current $I$. If the wire is in a magnetic field **B**, the magnetic force on each charge is $q\mathbf{v}_d \times \mathbf{B}$, where $\mathbf{v}_d$ is the drift velocity of the charge carriers. The number of charges in the wire segment is the number $n$ per unit volume times the volume $A\ell$. Thus the total force **F** on the wire segment is

$$\mathbf{F} = (q\mathbf{v}_d \times \mathbf{B})nA\ell$$

\* This unit is also called a weber per square metre, because the unit for magnetic flux is called the weber (Wb).

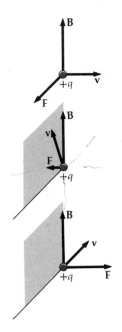

**Figure 26-1**
Direction of the magnetic force on a charged particle with velocity **v** at various orientations in a magnetic field **B**.

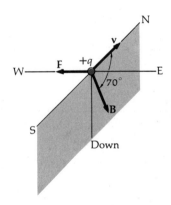

**Figure 26-2**
Force on a proton moving north in the magnetic field of the earth, which makes an angle of 70° with the horizontal north direction. The force is directed toward the west.

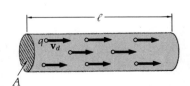

**Figure 26-3**
Current element of length $\ell$. Each charge experiences a magnetic force $q\mathbf{v}_d \times \mathbf{B}$.

By Equation 24-3, the current in the wire is

$$I = nqv_dA$$

Hence the force can be written

$$\mathbf{F} = I\boldsymbol{\ell} \times \mathbf{B} \qquad\qquad 26\text{-}3$$

where $\boldsymbol{\ell}$ is a vector whose magnitude is the length of the wire and whose direction is parallel to $q\mathbf{v}_d$, the direction of the current $I$. Equation 26-3 assumes that the wire is straight and that the magnetic induction does not vary over the length of the wire. It is easily generalized for an arbitrarily shaped wire in any magnetic field. We merely choose a very small wire segment $d\boldsymbol{\ell}$ and write the force on this segment $d\mathbf{F}$:

$$d\mathbf{F} = I\,d\boldsymbol{\ell} \times \mathbf{B} \qquad\qquad 26\text{-}4$$

*Force on a current element*

where $\mathbf{B}$ is the magnetic-induction vector at the segment. The quantity $I\,d\boldsymbol{\ell}$ is called a *current element*. We find the total force on the wire by summing (or integrating) over all the current elements using the appropriate field $\mathbf{B}$ at each element.

Equation 26-4 may be taken as an alternate definition of $\mathbf{B}$ equivalent to Equation 26-1. The magnitude and direction of $\mathbf{B}$ can be found using this definition by measuring the force on a current element $I\,d\boldsymbol{\ell}$ for various orientations of the element.

### Questions

1. Charge $q$ moves with velocity $\mathbf{v}$ through a magnetic field $\mathbf{B}$. At this instant it experiences a magnetic force $\mathbf{F}$. How would the force differ if the charge had the opposite sign? If the velocity had the opposite direction? If the magnetic field had the opposite direction?

2. For what angle between $\mathbf{B}$ and $\mathbf{v}$ is the magnetic force on $q$ greatest? Least?

3. A moving electric charge may experience both electric and magnetic forces. How could you distinguish whether a force causing a charge to deviate from a straight path is an electric or a magnetic force?

4. How can a charge move through a region of magnetic induction without ever experiencing any magnetic force?

5. Show that the force on a current element is the same in direction and magnitude regardless of whether positive charges, negative charges, or a mixture of positive and negative charges create the current.

6. A current-carrying wire passes through a magnetic field, but the wire does not experience any magnetic force. How is this possible?

## 26-2   Magnets in Magnetic Fields

When a small permanent magnet such as a compass needle is placed in a magnetic field, it tends to orient itself so that the north pole points in the direction of $\mathbf{B}$. This effect also occurs with previously unmagnetized iron filings, which become magnetized in the presence of a $\mathbf{B}$ field. Figure 26-4 shows a small magnet which makes an arbitrary angle with a magnetic field $\mathbf{B}$. There is a force $\mathbf{F}_1$ on the north pole in the direction of $\mathbf{B}$ and an equal but opposite force $\mathbf{F}_2$ on the south pole. The *pole strength* *of the magnet* $q_m$ is defined* to be the ratio of the magnitude of the force **F** on the pole and the magnitude of the magnetic induction **B**:

$$q_m = \frac{F}{B} \qquad\qquad 26\text{-}5$$

*Magnetic pole strength defined*

**Figure 26-4**
A small magnet experiences a torque in a uniform magnetic field.

Since a tesla is equal to a newton per ampere-metre, the SI unit of magnetic pole strength is the ampere-metre (A·m). If we adopt the sign convention that the north pole is positive and the south pole negative, the force on a pole can be written as a vector equation:

$$\mathbf{F} = q_m\mathbf{B} \qquad\qquad 26\text{-}6$$

From Figure 26-4 we see that there is a torque acting on a magnet in a magnetic field. If $\ell$ is a vector pointing from the south pole to the north pole with the magnitude of the distance between the poles, the torque is

$$\boldsymbol{\tau} = \boldsymbol{\ell} \times \mathbf{F} = \boldsymbol{\ell} \times q_m\mathbf{B} = q_m\boldsymbol{\ell} \times \mathbf{B} \qquad\qquad 26\text{-}7$$

The magnetic moment **m** of a magnet is defined to be

$$\mathbf{m} = q_m\boldsymbol{\ell} \qquad\qquad 26\text{-}8$$

*Magnetic moment*

The SI unit of magnetic moment is the ampere-metre$^2$ (A·m$^2$). In terms of the magnetic moment, the torque on a magnet is

$$\boldsymbol{\tau} = \mathbf{m} \times \mathbf{B} \qquad\qquad 26\text{-}9$$

These equations are completely analogous to those for the torque on an electric dipole in an electric field (Section 20-7). However, there is an important difference between the magnetic pole strength $q_m$ and the electric charge $q$. An isolated magnetic pole has never been experimentally detected. Magnetic poles always come in pairs; i.e., the fundamental unit of magnetism is the magnetic dipole. Experimentally, it is the magnetic dipole moment **m** of a magnet that can be easily measured by placing the magnet in a magnetic field of known strength and measuring the torque. The magnetic pole strength $q_m$ is then found from Equation 26-8 by dividing the magnitude of the magnetic moment by the length of the magnet.

It is customary to indicate the direction of the magnetic induction **B** by drawing lines which are parallel to **B** at each point in space and to indicate the magnitude of **B** by the density of the lines, just as for the electric field **E**. Such lines are called *lines of magnetic induction*. For a given magnetic induction **B** the lines can be found using a compass needle or iron filings since these small magnets align themselves in the direction of **B**. Figure 26-5 shows the lines of magnetic induction near a bar magnet and near a long straight current-carrying wire. External to the magnet, the lines of **B** leave the north pole and enter the south pole.

*Lines of magnetic induction*

---

* The notation for magnetic pole strength $q_m$ is used so that the magnetic equations resemble the corresponding equations for electric charges in electric fields. The subscript $m$ reminds us that $q_m$ denotes a magnetic pole and not an electric charge.

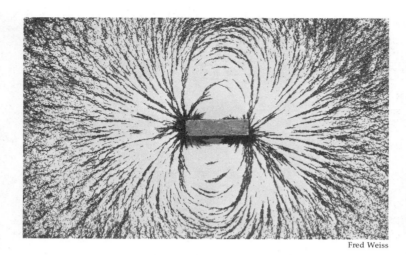

Fred Weiss

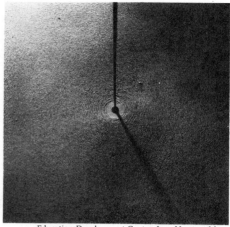

Education Development Center, Inc., Newton, Mass.

This is consistent with the fact that the force on a north pole is in the direction of **B** and two north poles repel each other. We shall investigate how to calculate **B** for these cases in later chapters.

**Figure 26-5**
Magnetic field lines indicated by the alignment of iron filings for (*left*) a magnet and (*right*) a long straight current-carrying wire.

### Questions

7.  Although the force on a north pole of a magnet is in the direction of the lines of magnetic induction, the force on an electric charge $q$ moving through a region of magnetic induction **B** is not. Is it possible to draw a set of lines which would indicate the direction of the magnetic force on a moving charge at each point in space?

8. The lines of magnetic induction near the long straight current-carrying wire shown in Figure 26-5 (*right*) are circles concentric with the wire. If a single isolated magnetic pole did exist, what would happen to it if it were placed near such a wire?

## 26-3   Torque on a Current Loop in a Uniform Magnetic Field

When a wire carrying a current $I$ is placed in a uniform magnetic field, there are forces on each part of the wire. If the wire is in the shape of a closed loop, there is no net force on the loop because the forces on the different parts of the wire sum to zero (assuming that the magnetic induction **B** is the same at all parts of the wire). In general, however, the magnetic forces produce a torque on the wire which tends to rotate the loop so that the area of the loop is perpendicular to the magnetic induction **B**.

Figure 26-6 shows a rectangular loop carrying a current $I$. The loop is in a region of uniform magnetic field parallel to the plane of the loop. The forces on each segment of the loop are indicated in the figure. There are no forces on the top or bottom of the loop because for those segments the current is parallel or antiparallel to the magnetic induction **B**, so that $I \, d\boldsymbol{\ell} \times$ **B** is zero. The forces on the sides of the loop have the magnitude

$$F_1 = F_2 = IaB$$

where $a$ is the length of the side. These forces are equal and opposite.

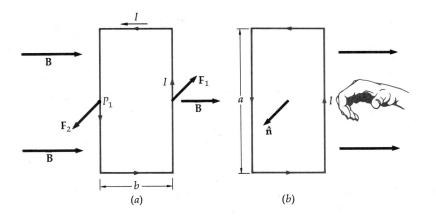

**Figure 26-6**
(*a*) Forces exerted on a rectangular current loop in a uniform magnetic field **B** parallel to the plane of the loop. (*b*) Unit vector $\hat{n}$ perpendicular to the plane of the loop. The sense of $\hat{n}$ is defined by the right-hand rule. With the fingers curling in the direction of the current, the thumb points in the direction of $\hat{n}$. The forces in (*a*) tend to twist the loop so that $\hat{n}$ rotates into **B**.

Hence they form a couple. The net force is zero, and the torque about any point is independent of the location of the point. Point $P_1$ is a convenient point about which to compute the torque. The magnitude of the torque is

$$F_1 b = IabB = IAB$$

where $b$ is the width of the loop and $A = ab$ is the area of the loop. The torque is therefore the product of the current, the area of the loop, and the magnetic induction $B$. This torque tends to twist the loop so that its plane is perpendicular to **B**. The orientation of the loop can be described conveniently by a unit vector $\hat{n}$ perpendicular to the plane of the loop. The sense of $\hat{n}$ is chosen to be that given by the right-hand rule applied to the circulating current, as illustrated in Figure 26-6. The torque thus tends to rotate $\hat{n}$ into the direction of **B**.

Figure 26-7 shows the forces exerted by a uniform magnetic field on a rectangular loop whose normal unit vector $\hat{n}$ makes an angle $\theta$ with the magnetic induction **B**. Again, the net force on the loop is zero. The torque about any point (point $P$, for example) is the product of the force $F_2 = IaB$ and the lever arm $b \sin \theta$. The torque thus has the magnitude

$$\tau = IaBb \sin \theta = IAB \sin \theta$$

where again $A = ab$ is the area of the loop. This torque can be written conveniently in terms of the cross product of $\hat{n}$ and **B**:

$$\tau = IA\hat{n} \times \mathbf{B} \qquad\qquad 26\text{-}10$$

Equation 26-10 is of the same form as Equation 26-9 for the torque on a bar magnet in a uniform magnetic field, with the quantity $IA\hat{n}$ for the loop playing the role of the magnetic moment $q_m \ell$ of the bar magnet. We thus define the magnetic moment of a current loop to be $IA\hat{n}$. Often a loop contains several turns of wire, each turn having the same area and carrying the same current. The total magnetic moment of such a loop is the product of the number of turns and the magnetic moment of each turn. The total magnetic moment of a loop of $N$ turns is then

$$\mathbf{m} = NIA\hat{n} \qquad\qquad 26\text{-}11$$

With this definition, the torque on the current loop is given by Equation 26-9:

$$\tau = \mathbf{m} \times \mathbf{B}$$

Equation 26-10, which we have derived for a rectangular loop, holds in general for a loop of any shape. The torque on any loop is the cross

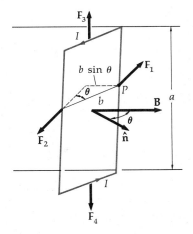

**Figure 26-7**
Rectangular current loop whose unit normal $\hat{n}$ makes an angle $\theta$ with a uniform magnetic field **B**. The torque on the loop is $IA\hat{n} \times \mathbf{B} = \mathbf{m} \times \mathbf{B}$, where $\mathbf{m} = IA\hat{n}$ is the magnetic moment of the loop.

*Magnetic moment of current loop*

product of the magnetic moment of the loop and the magnetic induction **B**, where the magnetic moment is defined to be a vector whose magnitude is the current times the area of the loop and whose direction is perpendicular to the area of the loop (Figure 26-8). A current loop thus behaves like a small bar magnet. We shall see in the next chapter that the magnetic field produced by a current loop at points far from the loop is identical to that produced by a bar magnet at points far from the magnet if the magnetic moments are the same. Since such a field is of the same form as the electric field produced by an electric dipole, the analogy of a current loop with a bar magnet is useful for understanding the behavior of the loop in an external magnetic field and for explaining the magnetic field produced by the loop. We say more about the significance of this analogy in Chapter 27.

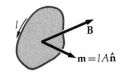

**Figure 26-8**
A current loop of arbitrary shape experiences a torque **m × B** in a magnetic field.

An important application of the torque exerted on a coil in a magnetic field is the d'Arsonval galvanometer discussed in Section 25-4 (see Figure 25-22). In this case, the magnetic field produced by a permanent magnet is radial at the coil, so that the angle $\theta$ between **B** and the normal to the plane of the coil is 90° independent of the orientation of the coil. The torque produced by the magnetic field is proportional to the current in the coil and is balanced by a restoring torque due to the spring suspension. If this restoring torque is proportional to the angular deflection of the coil, the angular deflection will be proportional to the current $I$. In any case, the angular deflection of the coil can be calibrated to indicate the current in the coil.

### Question

9. A current loop has its magnetic moment antiparallel to a uniform **B** field. What is the torque on the loop? Is this equilibrium stable or unstable?

## 26-4   Motion of a Point Charge in a Magnetic Field

An important characteristic of the magnetic force on a moving charged particle described by Equation 26-1 is that the force is always perpendicular to the velocity of the particle. The magnetic force therefore does no work on the particle, and the kinetic energy of the particle is unaffected by this force. The magnetic force changes the direction of the velocity but not its magnitude.

In the special case where the velocity of a particle is perpendicular to a uniform magnetic field, as shown in Figure 26-9, the particle moves in a circular orbit. The magnetic force provides the centripetal force necessary for circular motion. We can relate the radius of the circle to the magnetic field and the velocity of the particle by setting the resultant force equal to the mass $m$ times the centripetal acceleration $v^2/r$. The resultant force in this case is just $qvB$ since **v** and **B** are perpendicular. Thus Newton's second law gives

$$qvB = \frac{mv^2}{r}$$

or

$$r = \frac{mv}{qB} \qquad\qquad 26\text{-}12$$

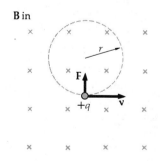

**Figure 26-9**
Charged particle moving in a plane perpendicular to a uniform magnetic field which is into the plane of the paper (indicated by the crosses). The magnetic force is perpendicular to the velocity of the particle, causing it to move in a circle of radius $r$, which can be found from $qvB = mv^2/r$.

Circular path of electrons moving in a magnetic field produced by two large coils. The electrons ionize the gas in the tube, causing it to give off a bluish glow that indicates the path of the beam. (*Courtesy of Larry Langrill, Oakland University.*)

The angular frequency of the circular motion is

$$\omega = \frac{v}{r} = \frac{qB}{m}$$  26-13

and the period is

$$T = \frac{2\pi}{\omega} = \frac{2\pi m}{qB}$$  26-14

Note that the frequency given by Equation 26-13 does not depend on the radius of the orbit or the velocity of the particle. This frequency is called the *cyclotron frequency*. Two of the many interesting applications of the circular motion of charged particles in a uniform magnetic field, the mass spectrograph and the cyclotron, will be discussed in Examples 26-3 and 26-4.

*Cyclotron frequency*

If a charged particle enters a region of uniform magnetic induction with a velocity which is not perpendicular to **B**, the path of the particle is a helix. The component of velocity parallel to **B** is not affected by the magnetic field. Consider, for example, a uniform magnetic field in the $z$ direction, and let $v_z$ be the component of velocity of the particle parallel to the field. In a reference frame moving in the $z$ direction with speed $v_z$, the particle has a velocity perpendicular to the field and moves in a circle in the $xy$ plane. In the original frame the path of the particle is a helix which winds around the lines of **B**, as shown in Figure 26-10.

The magnetic force on a charged particle moving in a uniform magnetic field can be balanced by an electrostatic force if the magnitudes and directions of the magnetic and electric fields are properly chosen. Since the electric force is in the direction of the electric field (for positive particles) and the magnetic force is perpendicular to the magnetic field, the electric and magnetic fields must be perpendicular to each other if

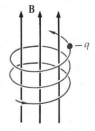

**Figure 26-10**
When a charged particle has a component of velocity parallel to a magnetic field **B**, it moves in a helical path.

Cloud-chamber photograph of the helical path of an electron moving in a magnetic field. The path of the electron is made visible by the condensation of water droplets in the cloud chamber. (*Courtesy of Carl E. Nielsen, Ohio State University.*)

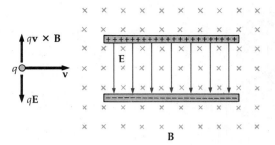

**Figure 26-11**
Crossed electric and magnetic
fields. When a positive particle moves to the right, it
experiences a downward electric force $q\mathbf{E}$ and an upward
magnetic force $q\mathbf{v} \times \mathbf{B}$, which
balance if $vB = E$.

the forces are to balance. Figure 26-11 shows a region of space between
the plates of a capacitor where there is an electric field and a perpendicular magnetic field (which can be produced by a magnet not shown).
Such an arrangement of perpendicular fields is called *crossed fields*. Consider a particle of charge $q$ entering this space from the left. If $q$ is posi-
tive, the electric force $q\mathbf{E}$ is down and the magnetic force $q\mathbf{v} \times \mathbf{B}$ is up. If
the charge is negative, each of the forces is reversed. Since $\mathbf{v}$ is perpendicular to $\mathbf{B}$, the magnitude of the magnetic force is just $qvB$. These two
forces will balance if $qE = qvB$ or

$$v = \frac{E}{B} \qquad\qquad 26\text{-}15$$

For given magnitudes of the electric and magnetic fields, the forces will
balance only for particles with the speed given by Equation 26-15. Any
particle, no matter what its mass or charge, will traverse the space undeflected if its speed is given by Equation 26-15. A particle of greater
speed will be deflected in the direction of the magnetic force; one of less
speed will be deflected in the direction of the electric force. Such an
arrangement of fields is called a *velocity selector*; its use is illustrated in
Examples 26-2 and 26-3.     *Velocity selector*

The motion of charged particles in nonuniform magnetic fields is
quite complicated. We shall discuss qualitatively some cases in which
the magnetic field $\mathbf{B}$ is nearly uniform so that the path of the particle is
approximately a helix. Consider a magnetic field indicated by the lines
of induction in Figure 26-12, which represents a three-dimensional
field axially symmetric about the horizontal axis. The field is strongest
at the ends and weakest at the center, as indicated by the line spacing.
Such a field is sometimes called a *magnetic bottle* because charged par-     *Magnetic bottle*
ticles can be trapped in it, as we shall show.

Let a positive particle have an initial velocity into the paper at point
$P_1$. If the magnetic induction $\mathbf{B}$ were horizontal at this point, the particle would move in a circle. However, $\mathbf{B}$ has a small vertical component, and the force $q\mathbf{v} \times \mathbf{B}$ has a small component to the right. The particle thus accelerates to the right as it moves in a nearly circular path
about the axis of symmetry of the field. According to Equation 26-12,

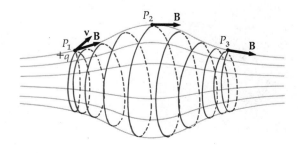

**Figure 26-12**
Magnetic bottle. The positive
particle moves back and forth
between $P_1$ and $P_3$ along the
indicated path.

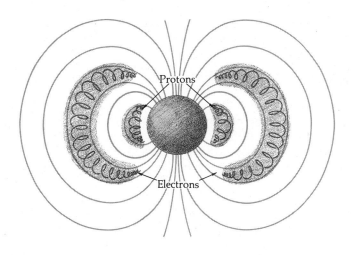

**Figure 26-13**
Van Allen belts. Protons
(inner belts) and electrons
(outer belts) are trapped in
the earth's magnetic field and
spiral along the field lines
between the north and south
poles, as in a magnetic bottle.

the radius of the circular motion is proportional to the velocity component perpendicular to **B** and inversely proportional to **B**. (Since the speed of the particle cannot be changed by the magnetic force, which is perpendicular to the velocity, an increase in the axial component of the velocity is accompanied by a slight decrease in the component perpendicular to the axis.) As the particle moves to the right, it enters a region of weaker magnetic field. The radius of its nearly circular path thus increases. At point $P_2$ the magnetic field is horizontal. The particle has no axial acceleration at this point but continues to move to the right because of its axial velocity. To the right of point $P_2$ the magnetic force has a component to the left which decreases the axial velocity until it is zero at point $P_3$, where the particle begins to move to the left. Since the field is stronger at point $P_3$, the radius of the circular path is smaller than at point $P_2$. The particle thus oscillates back and forth between points $P_1$ and $P_3$. A motion somewhat similar to this is the oscillation of charged ions back and forth between the earth's magnetic poles in the Van Allen belts (Figure 26-13).

**Example 26-2** *Thomson's Measurement of q/m for Electrons* An example of the use of a velocity selector and the measurement of the deflection of charged particles in an electric field is the famous experiment of J. J. Thomson in 1897, in which he showed that rays in cathode-ray

J. J. Thomson in his laboratory. (*Courtesy of the Cavendish Laboratory, University of Cambridge.*)

tubes can be deflected by electric and magnetic fields and therefore consist of charged particles. By observing the deflection of these rays with various combinations of electric and magnetic fields, Thomson was able to prove that all the particles had the same charge-to-mass ratio $q/m$ and to determine the ratio. He showed that particles with this charge-to-mass ratio can be obtained using any material for the cathode, which means that these particles, now called electrons, are a fundamental constituent of all matter.

Figure 26-14 shows the cathode-ray tube he used. Electrons are emitted from a cathode $C$, which is at a negative potential relative to the

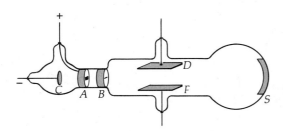

**Figure 26-14**
Thomson's tube for measuring $q/m$. Electrons from the cathode $C$ pass through the slits at $A$ and $B$ and strike a phosphorescent screen $S$. The beam can be deflected by an electric field between the plates $D$ and $F$ or by a magnetic field (not shown). [*From J. J. Thomson, Philosophical Magazine, Ser. 5, vol. 44 (1897).*]

slits $A$ and $B$. An electric field in the direction from $A$ to $C$ accelerates the electrons. They pass through the slits $A$ and $B$ into a field-free region and then encounter between the plates $D$ and $F$ an electric field that is perpendicular to the velocity of the electrons. The acceleration produced by this electric field gives the electrons a vertical component of velocity when they leave the region between the plates. They strike the phosphorescent screen $S$ at the far right side of the tube at some displacement $\Delta y$ from the point at which they strike when there is no field between the plates $D$ and $F$. When the electrons strike the screen, it glows, indicating the location of the beam. The deflection $\Delta y$ consists of two parts, $\Delta y_1$, which occurs while the electrons are between the plates, and $\Delta y_2$, which occurs after the electrons leave the region between the plates (Figure 26-15).

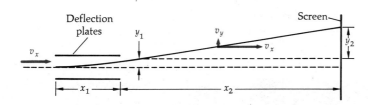

**Figure 26-15**
Deflection of the electron beam in the Thomson experiment.

Let $x_1$ be the horizontal distance across the deflection plates $D$ and $F$. If the velocity of the electron is $v_0$ when it enters the plates, the time spent between the plates will be

$$t_1 = \frac{x_1}{v_0}$$

Assuming that the electron was moving horizontally when it entered the region between the deflection plates, the vertical velocity when it leaves this region will be

$$v_y = at_1 = \frac{qE}{m}\, t_1 = \frac{qE}{m}\frac{x_1}{v_0}$$

where $E$ is the electric field between the plates. The deflection in this region will be

$$\Delta y_1 = \tfrac{1}{2}at_1^2 = \frac{1}{2}\frac{qE}{m}\left(\frac{x_1}{v_0}\right)^2$$

The electron then travels an additional horizontal distance $x_2$ in the field-free region between the deflection plates and the screen. Since there is no force on the electron (except the negligible gravitational force), the velocity of the electron is constant in this region. It therefore takes a time

$$t_2 = \frac{x_2}{v_0}$$

to reach the screen and suffers an additional vertical deflection given by

$$\Delta y_2 = v_y t_2 = \frac{qE}{m}\frac{x_1}{v_0}\frac{x_2}{v_0}$$

The total deflection at the screen is

$$\Delta y = \Delta y_1 + \Delta y_2 = \frac{1}{2}\frac{qE}{m}\left(\frac{x_1}{v_0}\right)^2 + \frac{qE}{m}\frac{x_1 x_2}{v_0^2} \qquad 26\text{-}16$$

If the initial velocity $v_0$ is known, a measurement of the deflection for a given electric field $E$ yields a value for $q/m$. The initial velocity $v_0$ is determined by introducing a magnetic field **B** between the plates perpendicular to both the electric field and the initial velocity and adjusting its magnitude so that the beam is undeflected. The velocity of the electrons is then given by Equation 26-15. The magnetic field is then turned off, and the deflection is measured.

---

**Example 26-3** *The Mass Spectrograph* The mass spectrograph, first made by Francis William Aston in 1919 and improved by Kenneth Bainbridge and others, was designed to measure the masses of isotopes. It measures the mass-to-charge ratio of ions by determining the velocity of the ions and then measuring the radius of their circular orbit in a uniform magnetic field. According to Equation 26-12, the mass-to-charge ratio is given by

$$\frac{m}{q} = \frac{Br}{v} \qquad 26\text{-}17$$

where $B$ is the magnetic induction, $r$ the radius of the circular orbit, and $v$ the speed of the particle. In Figure 26-16, a simple schematic drawing of a mass spectrograph, ions from the source are accelerated by an electric field and enter a uniform magnetic field produced by an electromagnet. If the ions start from rest and move through a potential drop $V$, their kinetic energy when they enter the magnet equals the loss in potential energy $qV$:

$$\tfrac{1}{2}mv^2 = qV \qquad 26\text{-}18$$

The ions move in a semicircle of radius $r$ given by Equation 26-17 and strike a photographic film at point $P_2$, a distance $2r$ from the point where they entered the magnet. The speed $v$ can be eliminated from Equations 26-17 and 26-18 to find $m/q$ in terms of the known quantities $V$, $B$, and $r$. The result is

$$\frac{m}{q} = \frac{B^2 r^2}{2V} \qquad 26\text{-}19$$

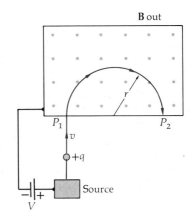

**Figure 26-16**
Schematic drawing of a mass spectrograph. The outward magnetic field is indicated by the dots.

Strictly speaking, the mass of an ion can be determined only if the charge $q$ is known. However this is not important because the charge $q$ is either one or two electron charges and the spectrograph is used to compare the masses of several isotopes which have nearly equal mass. In Aston's original spectrograph this could be done to a precision of about 1 part in 10,000. The precision is improved by the introduction of a velocity selector between the ion source and the magnet, making it possible to determine the velocity of the ions accurately and to limit the range of velocities of the incoming ions.

**Example 26-4**   *The Cyclotron*   The cyclotron was invented by E. O. Lawrence and M. S. Livingston in 1934 to accelerate particles such as protons or deuterons to high kinetic energy. The high-energy particles are then used to bombard nuclei, causing nuclear reactions, which are studied to obtain information about the nucleus. High-energy protons or deuterons are also used to produce radioactive materials and for medical purposes. The operation of the cyclotron is based on the fact that the period of motion of the charged particle in a uniform magnetic field is independent of the velocity of the particle.

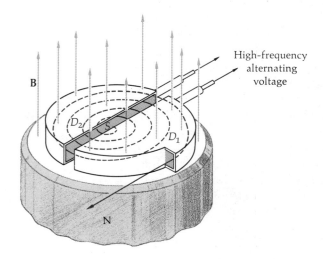

**Figure 26-17**
Schematic drawing of a cyclotron. The upper (south) pole face of the magnet has been omitted.

Figure 26-17 is a schematic drawing of a cyclotron. The particles move in two semicircular metal containers called *dees* (because of their shape). The containers lie in a vacuum chamber which is in a uniform magnetic field provided by an electromagnet. (The region in which the particles move must be evacuated so that the particle will not lose energy and be scattered in collisions with air molecules.) The dees are maintained at a potential difference $V$, which alternates in time with the period $T$, chosen to be equal to the cyclotron period given by Equation 26-14. The charged particle is initially injected with a small velocity from an ion source $S$ near the center of the magnetic field. It moves in a semicircle in one of the dees and arrives at the gap between the dees after a time $\frac{1}{2}T$, where $T$ is the cyclotron period and also the period of the alternating potential across the dees. Let us call the dee in which the particle traverses this first semicircle dee 1. The alternating potential is phased so that when the particle arrives at the gap between the dees, dee 1 is at a higher potential than dee 2 so that the particle is accelerated across the gap by the electric field across the gap and gains kinetic energy $qV$. The region inside each dee is shielded from electric fields by the metal dee. Thus the particle moves in a semicircle of larger radius in dee 2 and again arrives at the gap after a time $\frac{1}{2}T$. By this time the poten-

M. S. Livingston and E. O. Lawrence standing in front of their 27-in cyclotron in 1934. Lawrence won the Nobel prize (1939) for the invention of the cyclotron. (*Courtesy of Lawrence Radiation Laboratory, University of California, Berkeley.*)

tial between the dees has been reversed, and dee 2 is at the higher potential. Once more the particle is accelerated across the gap and gains kinetic energy $qV$. In each half revolution, the particle gains kinetic energy $qV$ and moves into a semicircular orbit of larger and larger radius until it leaves the magnetic field. In a typical cyclotron, a particle may make 50 to 100 revolutions.

### Questions

10. By observing the path of a particle how can you distinguish whether the particle is deflected by a magnetic field or an electric field?

11. A beam of positively charged particles passes undeflected from left to right through a velocity selector in which the electric field is up. The beam is then reversed so that it travels from right to left. Will the beam be deflected in the velocity selector? If so, in which direction?

A modern 83-in cyclotron at the University of Michigan. (A deflecting magnet is shown in the foreground.) (*Courtesy of University of Michigan.*)

## 26-5 The Hall Effect

In Section 26-1 we calculated the force exerted by a magnetic field on a current-carrying wire. This force is actually exerted directly on the charge carriers in the wire, the electrons. The force is transferred to the wire by the forces which bind the electrons to the wire at the surface. Since the charge carriers themselves experience the magnetic force when a current-carrying wire is in a magnetic field, the carriers are accelerated toward one side of the wire. This phenomenon, called the *Hall effect*, allows us to determine the sign of the charge on the carrier and the number density $n$ of charge carriers in a conductor. It also provides a convenient method for measuring magnetic fields.

Figure 26-18a shows a conducting strip carrying a current $I$ to the right in a magnetic field which we have chosen to be into the paper. Let us assume for the moment that the current consists of positively charged particles moving to the right. The magnetic force on these particles will be in the direction $q\mathbf{v}_d \times \mathbf{B}$, which is up in the figure. The positive particles will thus move up to the top of the strip, leaving the bottom of the strip with an excess negative charge. This charge separation causes an electrostatic field in the strip which opposes the magnetic force on the charge carriers. When the electrostatic and magnetic forces cancel, the charge carriers will no longer move upward. In this equilibrium situation, the upper part of the strip will be positively charged and be at a greater potential than the negatively charged lower part. On the other hand, if the current consists of negatively charged particles, they must be moving to the left for a current to the right. The magnetic force $q\mathbf{v}_d \times \mathbf{B}$ will again be up since we have changed the sign of both $q$ and $\mathbf{v}_d$. Again the carriers will be forced to the upper part of the strip, but the upper part of the strip now carries a negative charge (because the charge carriers are negative) and the lower part a positive charge. A measurement of the sign of the potential difference between the upper and lower part of the strip tells us the sign of the charge carriers. Experiments for the current and magnetic field in the directions indicated in the figure show that the upper part of the strip carries a negative charge and is at a lower potential than the lower part. This observation led to the discovery that the charge carriers in metallic conductors are negative.

If we connect the upper and lower portions of the strip with a wire of resistance $R$, the negative electrons will flow from the upper part of the strip through the wire to the lower part. As soon as some electrons leave the upper part of the strip, more carriers in the strip will be driven up by the magnetic force because the electrostatic force is momentarily weakened by the reduced charge separation. Thus the magnetic force maintains the potential difference across the strip. The strip is thus a seat of emf. The potential difference between the top and bottom of the strip is called the *Hall emf*. Its magnitude is not hard to calculate. The magnetic force on the charge carriers in the strip has the magnitude $qv_dB$. This magnetic force is balanced by the electrostatic force $qE$, where $E$ is the electric field due to the charge separation. Thus $E = v_dB$. If the width of the strip is $w$, the potential difference is $Ew$. The Hall emf is therefore

$$\mathcal{E}_H = Ew = v_dBw \qquad\qquad 26\text{-}20$$

We can see from Equation 26-20 that for ordinary-sized strips and magnetic fields, the Hall emf is very small since the drift velocity for ordinary currents is very small.

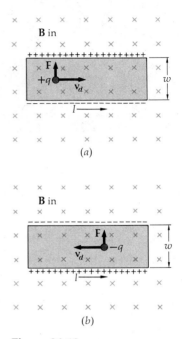

**Figure 26-18**
The Hall effect. Whether the current to the right is due to positive particles moving to the right as in (a) or to negative particles moving to the left as in (b), the magnetic force $q\mathbf{v}_d \times \mathbf{B}$ will be upward on the charge carriers. The sign of the carriers can be determined by observing whether the top of the strip is (a) positive or (b) negative.

*Hall emf*

From measurements of the size of the Hall emf for a strip of a given size carrying a known current in a known magnetic field we can determine the number of charge carriers per unit volume in the strip. By Equation 24-5 the current density is

$$J = nqv_d \qquad\qquad 26\text{-}21$$

The current density is determined by measuring the current $I$ and the cross-sectional area of the strip $A$. The quantity $q$ is just one electron charge. The drift velocity is determined from Equation 26-20 by measuring the Hall emf. Thus the number density of charge carriers $n$ is given by

$$n = \frac{J}{qv_d} = \frac{I}{Aqv_d} \qquad\qquad 26\text{-}22$$

(In Example 24-1 we estimated the drift velocity $v_d$ by assuming a value for $n$, namely one electron per atom for copper. This assumption for $n$ is justified by Hall emf measurements for copper.)

Although the Hall emf is ordinarily very small, it provides a convenient method for measuring magnetic fields. Combining Equations 26-20 and 26-21, we can write for the Hall emf

$$\mathcal{E}_H = \frac{J}{nq}\, wB = \frac{Iw}{nqA}\, B \qquad\qquad 26\text{-}23$$

A given strip can be calibrated by measuring the Hall emf for a given current in a known magnetic field. The strength of the magnetic induction $B$ of an unknown field can then be measured by placing the strip in the unknown field, sending a current through the strip, and measuring $\mathcal{E}_H$.

### Review

A. Define, explain, or otherwise identify:

Magnetic induction, 723

Magnetic flux density, 723

Tesla, 723

Gauss, 723

Current element, 724

Pole strength, 725

Magnetic moment of magnet, 725

Lines of magnetic induction, 725

Magnetic moment of current loop, 727

Cyclotron frequency, 729

Crossed fields, 730

Velocity selector, 730

Magnetic bottle, 730

Mass spectrograph, 733

Cyclotron, 734

Hall effect, 736

Hall emf, 736

B. True or false:

1. The magnetic force is always perpendicular to the velocity of the particle.

2. The torque on a magnet tends to align the magnetic moment parallel to **B**.

3. A current loop in a uniform magnetic field behaves like a small magnet.

4. The period of a particle moving in a circle in a magnetic field is proportional to the radius of the circle.

5. The drift velocity of electrons in a wire can be determined from the Hall effect.

### Exercises

#### Section 26-1, Definition of the Magnetic Field B

1. Find the magnetic force on a proton moving with velocity 4 Mm/s in the positive $x$ direction in a magnetic field of 2 T in the positive $z$ direction.

2. A charge $q = -2$ nC moves with velocity $\mathbf{v} = -3.0 \times 10^6 \mathbf{i}$ m/s. Find the force on the charge if the magnetic field is (a) $\mathbf{B} = 0.6\mathbf{j}$ T, (b) $\mathbf{B} = 0.6\mathbf{i} + 0.6\mathbf{j}$ T, (c) $\mathbf{B} = 0.8\mathbf{i}$ T, (d) $\mathbf{B} = 0.6\mathbf{i} + 0.6\mathbf{k}$ T.

3. A uniform magnetic field of magnitude 1.5 T is in the positive $z$ direction. Find the force on a proton if its velocity is (a) $\mathbf{v} = 3.0\mathbf{i}$ Mm/s, (b) $\mathbf{v} = 2.0\mathbf{j}$ Mm/s, (c) $\mathbf{v} = 8\mathbf{k}$ Mm/s, (d) $\mathbf{v} = 3.0\mathbf{i} + 4.0\mathbf{j}$ Mm/s.

4. An electron moves with velocity 5 Mm/s in the $xy$ plane at an angle of 30° to the $x$ axis and 60° to the $y$ axis. A magnetic field of 1.5 T is in the positive $y$ direction. Find the force on the electron.

5. A straight wire segment 2 m long makes an angle of 30° with a uniform magnetic field of 0.5 T. Find the force on the wire if it carries a current of 2 A.

6. A straight wire 10 cm long carrying a current of 3.0 A is in a uniform magnetic field of 1.5 T. The wire makes an angle of 37° with the direction of $\mathbf{B}$. What is the magnitude of the force on the wire?

7. A long wire parallel to the $x$ axis carries a current of 10.0 A in the direction of increasing $x$. There is a uniform magnetic field of magnitude 2.0 T in the positive $y$ direction. Find the force per unit length on the wire.

8. The wire segment in Figure 26-19 carries a current of 2.0 A from $a$ to $b$. There is a magnetic field $\mathbf{B} = 1.0\mathbf{k}$ T. Find the total force on the wire and show that it is the same as if the wire were a straight segment from $a$ to $b$.

**Figure 26-19**
Exercise 8.

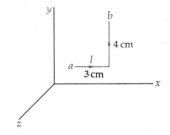

### Section 26-2, Magnets in Magnetic Fields

9. A small magnet is placed in a magnetic field of 0.1 T. The maximum torque experienced by the magnet is 0.20 N·m. (a) What is the magnetic moment of the magnet? (b) If the length of the magnet is 4 cm, what is the pole strength $q_m$?

10. Show that the SI unit of magnetic moment, ampere-metres², is the same as joules per tesla.

11. The cgs unit for magnetic moment is the dyn·cm/G (1 dyn = $10^{-5}$ N). (a) Find the conversion factor between 1 dyn·cm/G and 1 A·m² = 1 J/T (see Exercise 10). (b) A useful unit in atomic physics for magnetic moment is electronvolts per gauss. Find the conversion factor between electronvolts per gauss and ampere-metres² (= joules per tesla).

12. A small magnet of length 5 cm is placed at an angle of 45° to the direction of a uniform magnetic field of magnitude 0.04 T. The observed torque has the magnitude 0.10 N·m. (a) Find the magnetic moment of the magnet. (b) Find the pole strength $q_m$.

### Section 26-3, Torque on a Current Loop in a Uniform Magnetic Field

13. A small circular coil of 20 turns of wire lies in a uniform magnetic field of 0.5 T so that the normal to the plane of the coil makes an angle of 60° with the direction of $\mathbf{B}$. The radius of the coil is 4 cm, and it carries a current of 3 A. (a) What is the magnitude of the magnetic moment of the coil? (b) What torque is exerted on the coil?

**Figure 26-20**
Exercises 15 and 16.

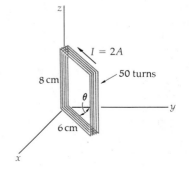

14. The SI unit for the magnetic moment of a current loop is ampere-metres². Use this to show that 1 T = 1 N/A·m.

15. A rectangular 50-turn coil has sides 6.0 and 8.0 cm long and carries a current of 2.0 A. It is oriented as shown in Figure 26-20 and pivoted about the $z$ axis. (a) If the wire in the $xy$ plane makes an angle of 37° with the $y$ axis as shown, what angle does the unit normal $\hat{\mathbf{n}}$ make with the $x$ axis? Write an expression for $\hat{\mathbf{n}}$ in terms of the unit vectors $\mathbf{i}$ and $\mathbf{j}$. (b) What is the magnetic moment of the coil? (c) Find the torque on the coil when there is a uniform magnetic field $\mathbf{B} = 1.5\mathbf{j}$ T.

16. The coil of Exercise 15 is pivoted about the $z$ axis and held at various posi-

tions in a uniform magnetic field $\mathbf{B} = 2.0\mathbf{j}$ T. Make a sketch of the coil position and find the torque exerted when the unit normal is (a) $\hat{\mathbf{n}} = \mathbf{i}$, (b) $\hat{\mathbf{n}} = \mathbf{j}$, (c) $\hat{\mathbf{n}} = -\mathbf{j}$, (d) $\hat{\mathbf{n}} = (\mathbf{i} + \mathbf{j})/\sqrt{2}$.

17. What is the maximum torque on a 500-turn circular coil of radius 0.5 cm which carries a current of 1.0 mA and resides in a uniform magnetic field of 0.1 T?

18. A particle of charge $q$ and mass $M$ moves in a circle of radius $r$ with angular velocity $\omega$. (a) Show that the average current is $I = q\omega/2\pi$ and the magnetic moment has the magnitude $m = \frac{1}{2}q\omega r^2$. (b) Show that the angular momentum of this particle has the magnitude $L = Mr^2\omega$ and that magnetic-moment and angular-momentum vectors are related by $\mathbf{m} = (q/2M)\mathbf{L}$.

### Section 26-4, Motion of a Point Charge in a Magnetic Field

19. A proton moves in a circular orbit of radius 80 cm perpendicular to a uniform magnetic field of magnitude 0.5 T. (a) What is the period for this motion? (b) Find the speed of the proton. (c) Find the kinetic energy of the proton.

20. A particle of charge $q$ and mass $m$ has momentum $p = mv$ and kinetic energy $E_k = \frac{1}{2}mv^2 = p^2/2m$. If it moves in a circular orbit of radius $r$ in a magnetic field $B$, show that (a) $p = Bqr$ and (b) $E_k = B^2q^2r^2/2m$.

21. An electron of kinetic energy 25 keV moves in a circular orbit in a magnetic field of 0.2 T. (a) Find the radius of the orbit. (b) Find the angular frequency and the period of the motion.

22. Protons, deuterons, and alpha particles of the same kinetic energy enter a uniform magnetic field $\mathbf{B}$ which is perpendicular to their velocities. Let $r_p$, $r_d$, and $r_\alpha$ be the radii of their circular orbits. Find the ratios $r_d/r_p$ and $r_\alpha/r_p$. Assume that $m_\alpha = 2m_d = 4m_p$.

23. An alpha particle travels in a circular path of radius 0.5 m in a magnetic field of 1.0 T. Find the (a) period, (b) speed, and (c) kinetic energy (in electronvolts) of the alpha particle. Take $m = 6.65 \times 10^{-27}$ kg for the mass of the alpha particle.

24. Show that the cyclotron frequency is the same for deuterons as for alpha particles and that each is half that of a proton in the same magnetic field.

25. A beam of protons moves along the $x$ axis in the positive $x$ direction with speed 10 km/s through a region of crossed fields. (a) If there is a magnetic field of magnitude 1.0 T in the positive $y$ direction, find the magnitude and direction of the electric field. (b) Would electrons of the same velocity be deflected by these fields? If so, in what direction?

26. A velocity selector has a magnetic field of magnitude 0.1 T perpendicular to an electric field of magnitude 0.2 MV/m. (a) What must the speed of a particle be to pass through undeflected? What energy must (b) protons and (c) electrons have to pass through undeflected?

27. The plates of a Thomson $q/m$ apparatus are 5.0 cm long and are separated by 1.0 cm. The end of the plates is 25.0 cm from the tube screen. The kinetic energy of the electrons is 2.0 keV. (a) If a potential of 20.0 V is applied across the deflection plates, by how much will the beam deflect? (b) Find the magnitude of a crossed $\mathbf{B}$ field which will allow the beam to pass through undeflected.

28. A singly ionized $^{24}$Mg ion (mass 24.0 u) is accelerated through a 2-kV potential and bent in a magnetic field of 50 mT in a mass spectrometer. (a) Find the radius of curvature of the orbit for the ion. (b) What is the difference in radius for $^{26}$Mg and $^{24}$Mg ions?

29. A cyclotron for accelerating protons has a magnetic field of 1.5 T and a radius of 0.5 m. (a) What is the cyclotron frequency? (b) Find the maximum energy of the protons when they emerge. (c) How do your answers change if deuterons are used instead of protons?

30. A certain cyclotron has a magnetic field of 2.0 T and is designed to accelerate protons to 20 MeV. (*a*) What is the cyclotron frequency? (*b*) What must the minimum radius of the magnet be to achieve the 20-MeV emergence energy? (*c*) If the alternating potential applied to the dees has a maximum value of 50 kV, how many orbital trips must the protons make before emerging with 20 MeV energy?

**Section 26-5, The Hall Effect**

31. A metal strip 2 cm wide and 0.1 cm thick carries a current of 20 A in a uniform magnetic field of 2.0 T, as shown in Figure 26-21. The Hall emf is measured to be 4.27 $\mu$V. (*a*) Calculate the drift velocity of the electrons in the strip. (*b*) Find the number density of charge carriers in the strip.

**Figure 26-21**
Exercises 31 to 34.

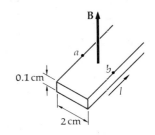

32. (*a*) In Figure 26-21 which point (*a* or *b*) is at the higher potential? (*b*) If the metal strip is replaced by a *p*-type semiconductor in which the charge carriers are positive, which point will be at the higher potential?

33. The number density of free electrons in copper is $8.5 \times 10^{22}$ electrons per cubic centimetre. If the metal strip in Figure 26-21 is copper and the current is 10 A, find (*a*) the drift velocity $v_d$ and (*b*) the Hall emf. (Assume that the magnetic field is 2.0 T, as in Exercise 31.)

34. A copper strip ($n = 8.5 \times 10^{22}$ electrons per cubic centimetre) 2 cm wide and 0.1 cm thick (like that shown in Figure 26-21) is used to measure the magnitudes of unknown magnetic fields which are perpendicular to the strip. Find the magnitude of B when $I = 20$ A and the Hall emf is (*a*) 2.00 $\mu$V, (*b*) 5.25 $\mu$V, and (*c*) 8.00 $\mu$V.

**Problems**

1. A 10.0-cm length of wire has mass 5.0 g and is connected to a source of emf by light flexible leads. A magnetic field $B = 0.5$ T is horizontal and perpendicular to the wire. Find the current necessary to float the wire; i.e., the current such that the magnetic force balances the weight of the wire.

2. A metal crossbar of mass $M$ rides on a pair of long horizontal conducting rails separated by a distance $L$ and connected to a device that supplies constant current $I$ to the circuit, as shown in Figure 26-22. A uniform magnetic field **B** is established as shown. (*a*) If there is no friction and the bar starts from rest at $t = 0$, show that at time $t$ the bar has velocity $v = (BIL/M)t$. (*b*) In which direction will the bar move? (*c*) If the coefficient of static friction is $\mu_s$, find the minimum field $B$ necessary to start the bar moving.

**Figure 26-22**
Problems 2 and 3.

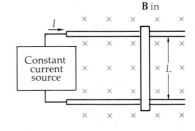

3. In Figure 26-22 assume the rails are frictionless but tilted upward so that they make an angle $\theta$ with the horizontal. (*a*) What vertical magnetic field $B$ is needed so that the bar will not slide down the rails? (*b*) What is the acceleration of the bar if $B$ is twice the value found in (*a*)?

4. Show that the radius of the orbit in a cyclotron is proportional to the square root of the number of orbits completed.

5. A mass spectrograph is preceded by a velocity selector consisting of parallel plates separated by 2.0 mm and having a potential difference of 160 V. The magnetic field between the plates is 0.42 T. The magnetic field in the mass spectrograph is 1.2 T. Find (*a*) the speed of the ions entering the mass spectrograph and (*b*) the separation distance of the peaks on the photographic plate for singly ionized $^{238}$U and $^{235}$U.

6. A wire bent into some arbitrary shape carries a current $I$ in a uniform magnetic field **B**. Show that the total force on the part of the wire from some point $a$ to some point $b$ is $\mathbf{F} = I\mathbf{L} \times \mathbf{B}$, where **L** is the vector from $a$ to $b$. *Hint:* Integrate $d\mathbf{F} = I\, d\boldsymbol{\ell} \times \mathbf{B}$ for constant **B**.

7. If you have a wire of fixed length $L$ and make a coil of $N$ turns, the larger the number of turns, the smaller the area enclosed by the wire. Show that for a wire of given length carrying current $I$ the maximum magnetic moment is achieved with a coil of just one turn and the magnitude of this magnetic moment is $IL^2/4\pi$. (You need consider only circular coils. Why?)

8. A nonconducting rod of length $\ell$ has a uniform charge per unit length $\lambda$ and is rotated with angular velocity $\omega$ about an axis through one end and perpendicular to the rod. (a) Consider a small segment of length $dx$ and charge $dq = \lambda\,dx$ at a distance $x$ from the pivot. Show that the magnetic moment of this segment is $\frac{1}{2}\lambda\omega x^2\,dx$. (b) Integrate your result to show that the total magnetic moment of the rod is $m = \frac{1}{6}\lambda\omega\ell^3$. (c) Show that the magnetic moment $\mathbf{m}$ and angular momentum $\mathbf{L}$ are related by $\mathbf{m} = (Q/2M)\mathbf{L}$, where $M$ is the mass of the rod and $Q$ is the total charge.

9. A nonconducting disk of radius $R$ has uniform surface charge density $\sigma$ and rotates with angular velocity $\omega$ about its axis. (a) Consider a ring of radius $r$ and thickness $dr$. Show that the total current in this ring is $dI = (\omega/2\pi)\,dq = \omega\sigma r\,dr$. (b) Show that the magnetic moment of the ring is $dm = \pi\omega\sigma r^3\,dr$. (c) Integrate your result for part (b) to show that the total magnetic moment of the disk is $m = \frac{1}{4}\pi\omega\sigma R^4$. (d) Show that the magnetic moment $\mathbf{m}$ and angular momentum $\mathbf{L}$ are related by $\mathbf{m} = (Q/2M)\mathbf{L}$, where $M$ is the total mass of the disk and $Q$ is its total charge.

10. Particles of charge $q$ and mass $m$ are accelerated from rest through a potential difference $V$ and enter a region of uniform magnetic field $\mathbf{B}$ perpendicular to the velocity. If $r$ is the radius of curvature of their circular orbit, show that $q/m = 2V/r^2B^2$.

11. A beam of particles enters a region of uniform magnetic field $\mathbf{B}$ with velocity $\mathbf{v}$ which makes a small angle $\theta$ with $\mathbf{B}$. Show that after a particle moves a distance $2\pi(m/qB)v\cos\theta$ measured along the direction of $\mathbf{B}$, the velocity of the particle is in the same direction as when it entered the field.

12. A small magnet of moment $\mathbf{m}$ makes an angle $\theta$ with a uniform magnetic field $\mathbf{B}$. (a) How much work must be done by an external torque to twist the magnet by a small amount $d\theta$? (b) Show that the work required to rotate the magnet until it is perpendicular to the field is $W = mB\cos\theta$. (c) Use your result for part (b) to show that if the potential energy of the magnet is chosen as zero when the magnet is perpendicular to the field, the potential energy at angle $\theta$ is $U(\theta) = -\mathbf{m}\cdot\mathbf{B}$. (d) Would any part of this problem be different if the magnet were replaced by a coil carrying current such that its magnetic moment is $\mathbf{m}$?

CHAPTER 27    Sources of the Magnetic Field

---

**Objectives**    After studying this chapter, you should:

1. Be able to state the Biot-Savart law and use it to calculate the magnetic field **B** due to a straight current-carrying wire and on the axis of a circular current loop.

2. Be able to sketch the magnetic field lines for (*a*) a long straight current, (*b*) a circular current loop, (*c*) a solenoid, and (*d*) a uniformly magnetized bar magnet.

3. Be able to state Ampère's law and discuss its uses and limitations.

4. Be able to use Ampère's law to derive expressions for $B$ due to an infinite straight current and inside a long tightly wound solenoid.

5. Be able to state the definition of magnetic flux and discuss the significance of the result that the net magnetic flux out of a closed surface is zero.

6. Be able to state the definition of Maxwell's displacement current and discuss its significance.

---

We now turn to a consideration of the origins of the magnetic field **B**. The earliest known sources of magnetism were magnets, but after Hans Christian Oersted described his finding in September 1820 that a compass needle is deflected by an electric current, many scientists investigated the properties of the magnetism associated with electric currents. One month later, Jean Baptiste Biot and Félix Savart announced the results of their measurements of the force on a magnetic pole near a long current-carrying wire and analyzed these results in terms of the magnetic field produced by each element of the current. André Marie Ampère extended these experiments and showed that current elements themselves experience a force in the presence of a magnetic field; in particular, he showed that two currents exert forces on each other.

Ampère also obtained the Biot-Savart result for the magnetic field due to a current element and developed a model explaining the behavior of permanent magnets in terms of microscopic current loops within the magnetic material. In this model it is proposed that all matter contains microscopic current loops. In nonmagnetic materials the loops are randomly oriented, producing no net effect, but in magnetic materials, current loops are aligned and produce a magnetic field with the same characteristics as that produced by aligned current loops of ordinary conduction current in wires. Ampère's model is essentially correct in that we can treat all magnetic fields as arising from an electric current of some kind. However, the microscopic currents (called *amperian currents*) are more complicated than Ampère could have foreseen, being related to the motions of atomic electrons, which cannot be fully described classically. In particular, ferromagnetism is intimately connected with electron spin, an intrinsic property of the electron.

## 27-1   The Biot-Savart Law

From their investigations of the force produced by a long wire carrying a current on a magnetic pole, Biot and Savart proposed an expression relating the magnetic induction vector **B** at a point in space to an element of the current that produces it. Let $I\, d\boldsymbol{\ell}$ be an element of current (Figure 27-1). The magnetic induction $d\mathbf{B}$ produced by this element at a field point $P$ a distance $r$ away is

$$d\mathbf{B} = k_m \frac{I\, d\boldsymbol{\ell} \times \hat{\mathbf{r}}}{r^2}$$

27-1   *Biot-Savart law*

where $\hat{\mathbf{r}}$ is the unit vector pointing from the current element to the field point and $k_m$ is a constant which depends on the units. Equation 27-1, the *Biot-Savart law*, was also deduced by Ampère. The magnetic induction due to the total current in a circuit can be found by using the Biot-Savart law for the field due to each current element and summing over all the current elements in the circuit.

The SI unit of current, the ampere, is defined so that the magnetic constant $k_m$ is exactly $10^{-7}$. This definition of the ampere will be discussed in more detail in the next section. The units of $k_m$ will be shown below to be newtons per ampere². As with the Coulomb constant, it is customary to write the magnetic constant $k_m$ in terms of another constant $\mu_0$, called the *permeability of free space*. These constants are related by

$$k_m = \frac{\mu_0}{4\pi} = 10^{-7} \text{ N/A}^2$$

$$\mu_0 = 4\pi k_m = 4\pi \times 10^{-7} \text{ N/A}^2$$

27-2

In terms of the constant $\mu_0$ the Biot-Savart law is written

$$d\mathbf{B} = \frac{\mu_0}{4\pi} \frac{I\, d\boldsymbol{\ell} \times \hat{\mathbf{r}}}{r^2}$$

27-3

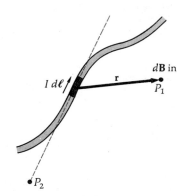

**Figure 27-1**
The magnetic field $d\mathbf{B}$ at point $P_1$ due to the current element $I\, d\boldsymbol{\ell}$ is $d\mathbf{B} = k_m (I\, d\boldsymbol{\ell} \times \hat{\mathbf{r}})/r^2$. At point $P_2$ the field due to this element is zero because $d\boldsymbol{\ell} \times \hat{\mathbf{r}}$ is zero.

The Biot-Savart law is analogous to Coulomb's law. The source of the magnetic field is the current element $I\, d\boldsymbol{\ell}$, just as the charge $q$ is the source of the electrostatic field. The magnetic field decreases as the square of the distance from the current element, like the decrease in the electric field with distance from a point charge, but the directional aspects of these fields are quite different. Whereas the electrostatic field points in the radial direction **r** from the point charge to the field point (assuming a positive charge), the magnetic field is perpendicular both to the radial direction and to the direction of the current element $I\, d\boldsymbol{\ell}$. At a point along the line of the current element, e.g., point $P_2$ in Figure 27-1, the magnetic field is zero.

By combining the Biot-Savart law with the expression for the force on a current element in a magnetic field (Equation 26-4) we can write an equation for the force exerted by one current element on another. The force on current element $I_2 \, d\ell_2$ exerted by element $I_1 \, d\ell_1$ is

$$d\mathbf{F}_{12} = I_2 \, d\ell_2 \times \left( k_m \frac{I_1 \, d\ell_1 \times \hat{\mathbf{r}}}{r^2} \right) \qquad \text{27-4}$$

From this equation we can see that if the force is in newtons, the distances $d\ell_1$, $d\ell_2$, and $r$ are in metres, and the currents are in amperes, the units of $k_m$ are newtons per ampere$^2$.

*Optional*

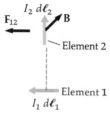

**Figure 27-2**
The forces exerted by current elements on each other are not equal and opposite. Here **B** at element 2 due to element 1 is into the paper, giving a force on element 2 to the left, but **B** due to element 2 at element 1 is zero.

This relation is remarkable in that the force exerted by element 1 on element 2 is not equal and opposite to that exerted by element 2 on element 1. That is, these forces do not obey Newton's action-reaction law, as can be demonstrated by considering the special case illustrated in Figure 27-2. Here, the magnetic field at element 2 due to element 1 is into the paper, and the force on element 2 is to the left. However, the magnetic field at element 1 due to element 2 is zero because $d\ell_2 \times \hat{\mathbf{r}}$ is zero. Thus no force is exerted on element 1. If the force $\mathbf{F}_{12}$ is the only force acting on the two-current-element system, the system will accelerate in the direction of $\mathbf{F}_{12}$ and linear momentum of the system will apparently not be conserved. We recall that it was the experimental observation of conservation of momentum in collisions that originally led Newton to the law of action and reaction. In most situations, current elements are but a part of a complete circuit. If the current elements shown are parts of complete circuits, as in Figure 27-3, there will be forces on the elements from other parts of the circuits. A detailed analysis of the total force exerted on one circuit by the other shows that the total forces *do* obey Newton's third law; i.e., the force exerted on circuit 1 by circuit 2 is equal and opposite to the total force exerted on circuit 2 by circuit 1. This analysis can be found in intermediate books on electricity and magnetism.

It is possible to produce the equivalent of isolated current elements by accelerating charges for a short time and then stopping them. Then the problem of the apparent violation of Newton's action-reaction law and of conservation of momentum is real. It is this type of situation for which our discussion of action at a distance in Chapter 4 is pertinent. We need to include in our system the electric and magnetic fields as well as the two current elements. The acceleration of electric charges necessary to produce isolated current elements also produces electromagnetic radiation, which carries momentum. A detailed analysis of a situation like that pictured in Figure 27-2 shows that indeed the current element system is accelerated to the left while the electromagnetic radiation carries momentum to the right. When we include the field and its momentum in the system, the total momentum of the system is again conserved.

**Figure 27-3**
Complete circuits containing the current elements shown in Figure 27-2. The total force which circuit 1 exerts on circuit 2 is equal and opposite to the total force which circuit 2 exerts on circuit 1.

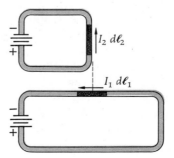

We now illustrate how the Biot-Savart law is used to calculate the magnetic induction for two important current configurations, a long straight wire and a current loop.

### Magnetic Induction of a Long Straight Current

Figure 27-4 shows the geometry needed to calculate the magnetic induction **B** at point $P$ due to the current in the long wire shown. We choose the wire to be the $x$ axis and the $y$ axis to be perpendicular to the wire through point $P$. A typical current element $I\,d\ell$ at a distance $x$ is shown.

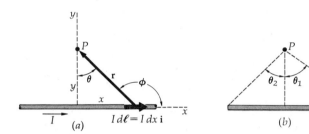

(a)     $I\,d\ell = I\,dx\,\mathbf{i}$     (b)

**Figure 27-4**
(a) Geometry for calculation from the Biot-Savart law of the magnetic field at point $P$ due to a straight current segment. Each element gives a contribution to the magnetic field at point $P$ directed out of the paper. (b) The result is given by Equation 27-6 in terms of the angles $\theta_1$ and $\theta_2$.

The vector **r** points from the element to the field point $P$. The direction of the magnetic field at $P$ due to this element is out of the paper, as determined by the direction of $I\,d\ell \times \mathbf{r}$. We note that all such current elements of the wire give contributions in this same direction, and so we need only compute the magnitude of the field. The field due to the current element shown has the magnitude

$$dB = k_m \frac{I\,dx}{r^2} \sin \phi = k_m \frac{I\,dx \cos \theta}{r^2} \qquad 27\text{-}5$$

where $\phi$ and $\theta$ are the angles shown. In order to sum over all the current elements, we need to relate the variables $\theta, r,$ and $x$. It turns out to be easiest to eliminate $x$ and $r$ in favor of $\theta$. We have

$$x = y \tan \theta$$

Then $dx = y \sec^2 \theta\, d\theta = y(r^2/y^2)\, d\theta = (r^2/y)\, d\theta$ using $\sec \theta = r/y$. Substituting this expression for $dx$ into Equation 27-5, we have

$$dB = \frac{k_m I}{r^2} \frac{r^2\, d\theta}{y} \cos \theta = \frac{k_m I}{y} \cos \theta\, d\theta$$

Let us first calculate the contribution from the current elements to the right of the point $x = 0$. We sum over these elements by integrating from $\theta = 0$ to $\theta = \theta_1$, where $\theta_1$ is the angle between the line perpendicular to the wire and the line from $P$ to the right end of the wire, as shown. We have for this contribution

$$B_1 = \int_0^{\theta_1} \frac{k_m I}{y} \cos \theta\, d\theta = \frac{k_m I}{y} \sin \theta_1$$

Similarly, the contribution from elements to the left of the point $x = 0$ is

$$B_2 = \frac{k_m I}{y} \sin \theta_2$$

We thus have for the total magnetic field due to the wire

$$B = \frac{k_m I}{y}(\sin\theta_1 + \sin\theta_2) \qquad\qquad 27\text{-}6$$

*Magnetic field of a straight wire*

This result gives the magnetic field due to any wire segment in terms of the angles subtended at the field point by the ends of the wire. If the wire is very long, these angles are nearly 90°. The result for an infinitely long wire is obtained from Equation 27-6 by using $\theta_1 = \theta_2 = 90°$:

$$B = \frac{2k_m I}{y} = \frac{\mu_0 I}{2\pi y} \qquad\qquad 27\text{-}7$$

*B due to an infinite straight wire*

The direction of **B** is such that the lines of **B** encircle the wire as shown in Figure 27-5. This direction can be remembered by a right-hand rule, as shown in that figure. This result for the magnetic field **B** due to a current in a long straight wire was found experimentally by Biot and Savart in 1820. From the analysis of this experimental result they were able to discover the expression given in Equation 27-1 for the magnetic induction due to an element of the current.

**Figure 27-5**
The magnetic field due to a very long straight wire is tangent to a circle around the wire. The direction of **B** is related to the direction of the current $I$ by the right-hand rule illustrated.

**Example 27-1** Find the magnetic induction field at the center of a square current loop of side 1 m carrying a current of 1 A.

From Figure 27-6 we see that each side of the loop contributes a field in the direction out of the paper. By the symmetry of the situation we need only calculate the field due to one side and multiply by 4. The distance between one side and the field point $y$ is $\frac{1}{2}\ell = \frac{1}{2}$m, and the angles $\theta_1$ and $\theta_2$ are 45°. Thus the field due to one segment is

$$B_1 = \frac{k_m I}{\frac{1}{2}\ell}(\sin 45° + \sin 45°) = \frac{2k_m I}{\ell}2\frac{1}{\sqrt{2}} = 2\sqrt{2}\frac{k_m I}{\ell}$$

and the total field is

$$B = 4B_1 = 8\sqrt{2}\frac{k_m I}{\ell} = 8\sqrt{2}\frac{10^{-7}\times 1}{1} = 11.3\times 10^{-7}\text{ T}$$

**Figure 27-6**
Square current loop for Example 27-1.

### Magnetic Induction on the Axis of a Circular Current Loop

The geometry needed for this calculation is shown in Figure 27-7. We first consider the current element at the top of the loop. Here $I\,d\ell$ is out of the paper and perpendicular to the vector **r**, which lies in the plane of the paper. The magnetic field due to this element is in the direction shown. The field $d\mathbf{B}$ is perpendicular to **r**, and since it is perpendicular to $I\,d\ell$, it must be in the plane of the paper as shown. The magnitude of $d\mathbf{B}$ is

$$|d\mathbf{B}| = k_m\frac{|I\,d\ell\times\hat{\mathbf{r}}|}{r^2} = k_m\frac{I\,d\ell}{x^2 + R^2}$$

since $d\ell$ and $\hat{\mathbf{r}}$ are perpendicular and $r^2 = x^2 + R^2$.

When we sum around the current elements in the loop, components of $d\mathbf{B}$ perpendicular to the axis of the loop sum to zero, leaving only the components parallel to the axis. We thus compute only the $x$ component. From the figure,

$$dB_x = dB\sin\theta = dB\frac{R}{\sqrt{x^2 + R^2}} = \frac{k_m I\,d\ell}{x^2 + R^2}\frac{R}{\sqrt{x^2 + R^2}}$$

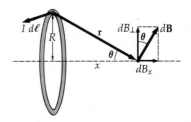

**Figure 27-7**
Geometry for the calculation of **B** on the axis of a current loop.

Since neither $x$ nor $R$ varies as we sum over the elements in the loop, this sum merely gives $\oint d\ell = 2\pi R$:

$$B_x = \frac{k_m I(2\pi R)R}{(x^2 + R^2)^{3/2}} = \frac{\mu_0}{2}\frac{IR^2}{(x^2 + R^2)^{3/2}} \qquad \text{27-8}$$

*B on axis of a circular loop*

An important special case of this result is the magnetic field at the center of the loop, which we obtain by setting $x = 0$:

$$B_x = \frac{\mu_0 I}{2R} \qquad \text{27-9}$$

At great distances from the loop, $x$ is much greater than $R$ and we can neglect $R^2$ compared with $x^2$ in the denominator of Equation 27-8. Then

$$B_x \to \frac{2k_m I(\pi R^2)}{x^3} = \frac{2k_m m}{x^3} = \frac{\mu_0}{2\pi}\frac{m}{x^3} \qquad \text{27-10}$$

*Dipole field of a current loop*

where $m = I(\pi R^2)$ is the magnetic moment of the loop. Note the similarity of this expression to that for the electrostatic field on the axis of an electric dipole of moment $p$:

$$E_x \to \frac{2kp}{x^3}$$

A current loop behaves like a magnetic dipole both in experiencing a torque $\mathbf{m} \times \mathbf{B}$ when placed in an external magnetic field and in producing a dipole field at great distances from the loop. Figure 27-8 shows the lines of $\mathbf{B}$ for a current loop. Our result that a current loop produces a dipole field is not confined to the axis of the loop.

### Questions

1. For a given distance $r$ from a current element, in what direction does the contribution of the current element to the total magnetic field have its greatest value?

2. What is the effect of replacing the single loop in Figure 27-7 with a coil of $N$ turns of the same radius, each carrying the current $I$?

3. A common method for cutting down on the magnetic fields created by leads carrying current to electrical devices is to twist together the wires carrying the incoming and outgoing currents. Explain.

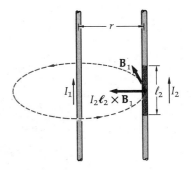

**Figure 27-8**
Lines of magnetic induction $\mathbf{B}$ due to a circular current loop indicated by iron filings. (*Avco Research Laboratory.*)

## 27-2  Definition of the Ampere and the Coulomb

One week after Ampère heard of Oersted's discovery of the effect of a current on a compass needle, he showed that two parallel currents attract each other if they are in the same direction and repel each other if they are in opposite directions. Figure 27-9 shows two long straight wires separated by a distance $r$ and carrying currents $I_1$ and $I_2$ in the same direction. We can compute the force exerted by one wire on the other from our result for the magnetic field of a long current-carrying wire and Equation 26-3 for the force exerted by a magnetic field on a current element. Let us consider a segment of the second wire of length $\ell_2$. The magnetic field $\mathbf{B}_1$ at this segment is perpendicular to the segment

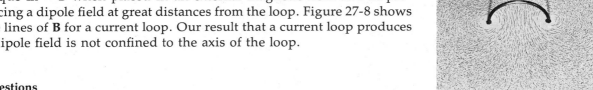

**Figure 27-9**
Two long straight wires carrying parallel currents. The magnetic field at the current element $I_2\ell_2$ due to current $I_1$ is into the paper, giving an attractive force $\mathbf{F}_2 = I_2\ell_2 \times \mathbf{B}_1$.

and has the magnitude $2k_m I_1/r$ if the segment is sufficiently close to wire 1 and wire 1 is long enough for the angles in Equation 27-6 to be approximately 90°. The force on this segment has the magnitude

$$F_2 = |I_2 \boldsymbol{\ell}_2 \times \mathbf{B}_1| = \frac{2k_m I_1 I_2 \ell_2}{r}$$

From the direction of the cross product $\boldsymbol{\ell}_2 \times \mathbf{B}_1$ we see that this force on the segment of wire 2 is toward wire 1. The force per unit length is

$$\frac{F_2}{\ell_2} = \frac{2k_m I_1 I_2}{r} \qquad\qquad 27\text{-}11$$

In Chapter 20 we deferred the definition of the coulomb as a unit of charge, mentioning that it is defined in terms of the ampere. The ampere is defined as follows:

*If two very long parallel wires one metre apart carry equal currents, the current in each is defined to be one ampere if the force per unit length on each wire is $2 \times 10^{-7}$ newton per metre.*

*Ampere defined*

This definition allows the unit of current (and therefore of electric charge) to be determined by a mechanical experiment. In practice of course, the currents are chosen to be much closer together than 1 m so that the wires need not be so long and the force is large enough to measure accurately. Figure 27-10 shows a current balance which can be used to calibrate an ammeter from the fundamental definition of the ampere. The upper conductor is free to rotate about the knife edges and is balanced so that the wires are a small distance apart. Since the conductors are wired in series to carry the same current but in opposite directions, the wires repel rather than attract. The force of repulsion can be measured by placing weights on the upper conductor until it balances again at the original separation. This definition of the ampere makes the magnetic constant $k_m = \mu_0/4\pi$ exactly equal to $10^{-7}$. The ratio of the Coulomb constant $k$ and the magnetic constant $k_m$ is

$$\frac{k}{k_m} = \frac{1/4\pi\epsilon_0}{\mu_0/4\pi} = \frac{1}{\epsilon_0\mu_0} = \frac{9 \times 10^9 \text{ N·m}^2/\text{C}^2}{10^{-7} \text{ N·s}^2/\text{C}^2} = 9 \times 10^{16} \text{ m}^2/\text{s}^2$$

This ratio equals the square of the speed of light, a fact noted by James Clerk Maxwell in 1860. He showed that the laws of electricity and magne-

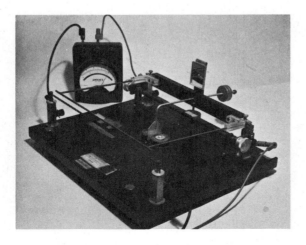

**Figure 27-10**
Current balance used in elementary physics laboratory to calibrate an ammeter. (*Courtesy of Larry Langrill, Oakland University.*)

tism imply that an accelerated charge radiates energy in the form of waves that travel with speed

$$c = \sqrt{\frac{k}{k_m}} = \frac{1}{\sqrt{\epsilon_0 \mu_0}} = 3 \times 10^8 \text{ m/s} \qquad 27\text{-}12$$

Since this speed is the same as that of light, Maxwell speculated that light is an electromagnetic wave produced by the acceleration of atomic charges.

# 27-3  Ampère's Law

We have noted that the lines of $\mathbf{B}$ for a long straight current-carrying wire encircle the wire. These lines are quite different from any lines of electric field we have studied. The electrostatic field is conservative. The work done on a test charge by the electrostatic field when the charge is moved in a complete circle is always zero. This work per charge is the sum of $\mathbf{E} \cdot d\ell$ around the path. The line integral of the electrostatic field around any closed path is zero because the electrostatic field is conservative:

$$\oint_C \mathbf{E} \cdot d\ell = 0 \qquad C \text{ any closed curve}$$

Clearly the sum of $\mathbf{B} \cdot d\ell$ around a closed path is not necessarily zero. If we take this sum along a circular path enclosing a long wire carrying a current, the magnetic-induction vector $\mathbf{B}$ is everywhere tangent to the path (Figure 27-11). Then $\mathbf{B} \cdot d\ell$ is everywhere positive if we travel in the direction of the field lines. Since $\mathbf{B}$ is in fact parallel to $d\ell$ and has the constant magnitude given by equation 27-7, we can easily compute this sum:

$$\oint \mathbf{B} \cdot d\ell = \frac{2k_m I}{r} \oint d\ell = \frac{2k_m I}{r} 2\pi r = 4\pi k_m I \qquad 27\text{-}13a$$

This integral is independent of the radius chosen for the circular path.

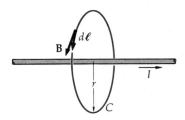

This result, known as *Ampère's law*, is more general than indicated by our calculation. It holds for the line integral of $\mathbf{B} \cdot d\ell$ around any closed path which encloses a steady current.

The line integral around any closed path equals $4\pi k_m$ times the total current which crosses any area bounded by the path. In terms of the constant $\mu_0 = 4\pi k_m$, Ampère's law is written

$$\oint_C \mathbf{B} \cdot d\ell = \mu_0 I \qquad 27\text{-}13b \qquad \textit{Ampère's law}$$

(It is because of the $4\pi$ factor in Equation 27-13a that the magnetic constant $k_m$ is written $\mu_0/4\pi$.)

Brown Brothers

André Marie Ampere (1775–1836).

**Figure 27-11**
Circular path around a long current-carrying wire. The integral of $\mathbf{B} \cdot d\ell$ around this curve is $4\pi k_m I = \mu_0 I$.

Ampère's law holds for any curve $C$ as long as the currents are steady. Like Gauss' law, it is *useful* for the calculation of the magnetic field only in situations with considerable symmetry. If the symmetry is great enough, the line integral can be written as the product of $B$ and some distance, thus relating $B$ to the current enclosed by the curve. Also, like Gauss' law, if the currents are steady, Ampère's law holds even if there is no symmetry; but in those cases it is no help in finding an expression for the magnetic field.

**Figure 27-12**
Long straight wire of radius $a$ carrying current $I$ uniformly distributed with current density $J = I/\pi a^2$. The magnetic field at distance $r$ inside the wire can be calculated from Ampère's law applied to the circle of radius $r$.

**Example 27-2** A long straight wire of radius $a$ carries a current $I$ which is uniformly distributed over the wire with current density $J = I/\pi a^2$ (Figure 27-12). Find the magnetic field both inside and outside the wire.

We can calculate **B** from Ampère's law because of the high degree of symmetry. Consider a circle of radius $r$ concentric with the axis of the wire. We expect **B** to be tangential to the circle, as with a very thin wire. By symmetry, the magnitude of **B** is constant everywhere on the circle. Thus

$$\oint \mathbf{B} \cdot d\boldsymbol{\ell} = B \oint d\ell = 2\pi r B$$

The current through $C$ depends on whether $r$ is less or greater than the radius of the wire. For $r$ greater than $a$, the total current $I$ crosses the area bounded by $C$. Then

$$\oint \mathbf{B} \cdot d\boldsymbol{\ell} = B 2\pi r = \mu_0 I$$

$$B = \frac{\mu_0 I}{2\pi r} = 2 \frac{\mu_0}{4\pi} \frac{I}{r} \qquad r > a$$

This result is the same as obtained in Equation 27-7. To find **B** inside the wire we choose $r$ less than $a$. Then the current through $C$ is

$$\pi r^2 J = \frac{r^2}{a^2} I$$

and

$$\oint \mathbf{B} \cdot d\boldsymbol{\ell} = 2\pi r B = \mu_0 \frac{r^2}{a^2} I$$

$$B = \frac{\mu_0}{2\pi} \frac{I}{a^2} r \qquad r < a \tag{27-14}$$

Figure 27-13 is a sketch of $B$ versus $r$.

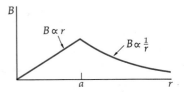

**Figure 27-13**
Plot of $B$ versus $r$ calculated in Example 27-2. Inside the wire, $B$ is proportional to $r$, whereas outside, it is inversely proportional to $r$.

*B inside wire carrying uniform current*

For an example in which Ampère's law is not useful in calculating the magnetic induction due to a steady current, consider a current loop as shown in Figure 27-14. We calculated the magnetic induction on the axis of such a loop from the Biot-Savart law (Equation 27-8). According to Ampère's law, the line integral of $\mathbf{B} \cdot d\boldsymbol{\ell}$ around a curve such as curve $C$ in Figure 27-14 equals $\mu_0$ times the current $I$ in the loop. Although Ampère's law is true for this curve, the magnetic induction **B** is not constant along any curve encircling the current, nor is it everywhere tangent to any such curve. Thus there is not enough symmetry in this situation to allow us to calculate $B$ from Ampère's law.

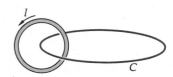

**Figure 27-14**
Ampère's law holds for curve $C$ encircling the current in the circular loop, but it is not useful in finding **B** because **B** is neither constant along the curve nor tangent to it.

### Limitations of Ampère's Law

Suppose we attempt to use Ampère's law to find the magnetic field at a point on the perpendicular bisector of a current segment of length $\ell$. Let the point be a distance $r$ from the segment.

This situation is shown in Figure 27-15. A direct application of Ampère's law again gives

$$B = \frac{\mu_0}{2\pi} \frac{I}{r}$$

This result is the same as for an infinitely long wire since we have the same symmetry arguments. The result does not agree with that obtained from the Biot-Savart law. That law gives a smaller result which depends on the length of the current segment and agrees with experiment. There are two possibilities for the discrepancy: (1) Ampère's law is correct but was incorrectly applied; (2) Ampère's law is not correct. In order to understand which explanation applies we must ask how we can obtain a current segment as shown in Figure 27-15. One possibility is shown in Figure 27-16, where the current segment is just one segment of a continuous loop carrying a steady current. For this situation, Ampère's law is correct, but our application was not. If there are other current segments in the problem, as in Figure 27-16, we do not have the symmetry we assumed to compute $\oint \mathbf{B} \cdot d\boldsymbol{\ell}$. For example, $\mathbf{B}$ is not tangential to the curve shown, and its magnitude is not constant along the curve. Then even though Ampère's law is correct, it is not useful for the calculation of the magnetic field for this situation. Figure 27-17 shows another possibility for obtaining a current segment, i.e., a source of charge at point $P_1$ and a sink at point $P_2$. For example, we might have a small spherical conductor with charge $+Q$ at the left and one with charge $-Q$ at the right. When they are connected, a current $I = -dQ/dt$ exists in the segment for a short time until the spheres are uncharged. For this case we do have the symmetry needed to assume that $\mathbf{B}$ is tangential to the curve and of constant magnitude along the curve, but the current is not a steady current. *When the current is not steady, Ampère's law does not hold.* We shall see in Section 27-7 how Maxwell generalized Ampère's law to hold for all currents, steady or not. When the generalized form is used, the magnetic field calculated for this situation agrees with that found from the Biot-Savart law.

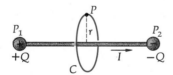

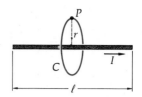

**Figure 27-15**
Application of Ampère's law to find the magnetic field on the bisector of a finite current segment gives an incorrect result.

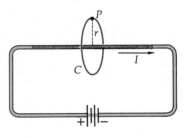

**Figure 27-16**
If the current segment in Figure 27-15 is a part of a complete circuit carrying a steady current, Ampère's law for curve $C$ is correct, but there is not enough symmetry to use it to find the magnetic field at point $P$.

**Figure 27-17**
If the current segment in Figure 27-15 is due to a momentary flow of charge from a small conductor on the left to one at the right, there is enough symmetry to compute $\oint \mathbf{B} \cdot d\boldsymbol{\ell}$ along curve $C$, but Ampère's law does not hold because the current is not steady.

## 27-4 The Magnetic Field of a Solenoid

A wire wound into a helix, as in Figure 27-18, called a *solenoid*, is used to produce a strong uniform magnetic field in a small region of space. It plays a role in magnetism analogous to that of the parallel-plate capacitor in providing a strong uniform electrostatic field between its plates. The magnetic field of a solenoid is essentially that of a set of $N$ identical coils placed side by side. Figure 27-19 shows the lines of magnetic induction for two such coils. In the space between the coils near the axis, the fields of the individual coils add; between the coils but at distances from the axis greater than the coil radius the individual fields tend to cancel. Figure 27-20 shows the field lines for a long tightly wound solenoid. Inside the solenoid, the field lines are approximately parallel to

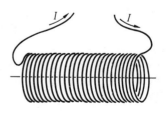

**Figure 27-18**
A solenoid produces a strong uniform magnetic field in the space inside it.

the axis and closely and uniformly spaced, indicating a strong uniform magnetic field. Outside the solenoid, the lines are much less dense. They diverge from one end and converge at the other end, like the lines of electric field from disks of positive and negative charge (Figure 27-21). Note that *inside* the solenoid the magnetic field lines are very different from electric field lines from the disk charges.

Because the magnetic field is very strong and nearly uniform inside the solenoid, as indicated by the field lines in Figure 27-20, and the field is very weak outside the solenoid, we can use Ampère's law to find a useful approximate result for the magnitude of **B**. Let the solenoid have radius $r$, length $\ell$, and $N$ turns of wire carrying a current $I$, and assume that $\ell$ is much greater than $r$. We shall assume that **B** is uniform and parallel to the axis inside the solenoid and zero outside. We apply Ampère's law to the curve $C$ (Figure 27-22), which is a rectangle with sides of length $a$ and $b$. The only contribution to the integral $\oint \mathbf{B} \cdot d\boldsymbol{\ell}$ around this curve is along side 1, since along the opposite side (outside the solenoid) the magnetic field is assumed to be zero, and along the two short sides **B** is perpendicular to $d\boldsymbol{\ell}$. Since **B** is assumed to be uniform and axial, the total line integral is merely $Bb$. The total current through this curve is the current $I$ times the number of turns of wire enclosed.

**Figure 27-19**
The magnetic fields due to two adjacent loops of a solenoid add in the region inside the solenoid and subtract in the region outside.

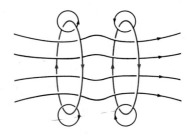

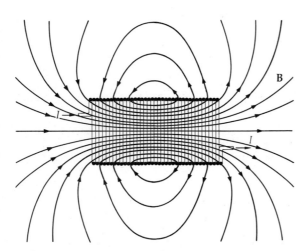

**Figure 27-20**
Magnetic field lines of a tightly wound solenoid. The field is strong and nearly uniform inside the solenoid. Outside it the field lines are like those of the electric field due to two disk charges (Figure 27-21).

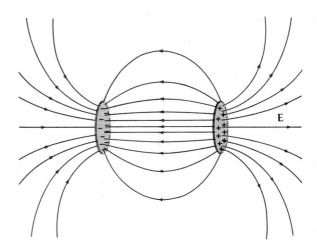

**Figure 27-21**
Electric field lines due to two oppositely charged disks. The lines outside the cylinder bounded by the disks are like the magnetic field lines of a solenoid (Figure 27-20). Between the disks the **E** lines are directed opposite to the corresponding magnetic field lines of a solenoid.

Since there are $N$ turns in a total length $\ell$, the number of turns in the length $b$ is $Nb/\ell$ and Ampère's law gives

$$\oint \mathbf{B} \cdot d\boldsymbol{\ell} = Bb = \mu_0 \frac{NbI}{\ell}$$

or

$$B = \mu_0 \frac{N}{\ell} I = \mu_0 n I \qquad\qquad 27\text{-}15 \qquad B \text{ inside solenoid}$$

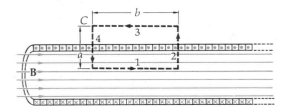

**Figure 27-22**
The magnetic field inside a solenoid can be calculated by applying Ampere's law to the curve C shown, assuming that **B** is uniform inside and zero outside.

where $n = N/\ell$ is the number of turns per unit length. Figure 27-23 shows how the magnetic field on the axis of a solenoid varies with distance from the ends. The approximation that the field is constant independent of position along the axis is quite good except very near the ends. At a point right at the end, the magnitude of **B** is about half that at the center for a long solenoid.

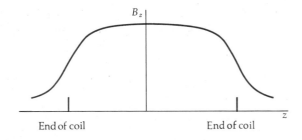

**Figure 27-23**
Plot of $B_z$ versus $z$ on the axis of a solenoid. Note that $B_z$ is nearly constant inside the solenoid except near the ends.

We shall calculate the magnetic field outside the solenoid only for the special case of a point far from the end compared with the radius of the coils. Then each coil gives a dipole field, since we are far from it compared with its radius. We can calculate the resultant field for $N$ coils by summing the fields of the individual coils. We can anticipate the result. Since each loop produces a field like that of a small bar magnet, i.e., a dipole field, we expect a set of adjacent loops to give a field like that of a set of adjacent small bar magnets, i.e., like a bar magnet of length $\ell$. This result is derived at the end of this section.

Figure 27-24 shows the magnetic field lines of a bar magnet of the same shape as the solenoid in Figure 27-18. The lines of **B** are identical for these two cases.

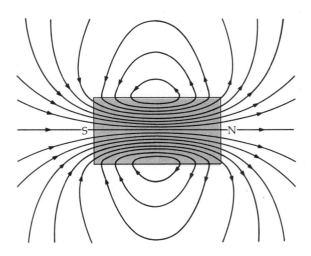

**Figure 27-24**
Magnetic field lines of a bar magnet. The lines are the same everywhere as those of a solenoid of similar shape.

**Question**

4. The magnetic field inside a long solenoid has magnitude $B$. How would the field change if the current in the solenoid were doubled? If the diameter were doubled without changing the length or number of turns? If the number of turns were doubled by making the solenoid twice as long? If the number of turns were doubled without changing the length of the solenoid?

*Optional*

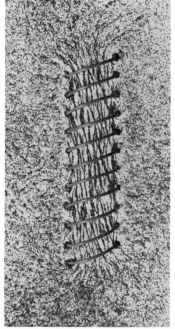

Fred Weiss

Magnetic field lines of a solenoid, indicated by iron filings.

**Figure 27-25**
Geometry for calculating the magnetic field on the axis of a tightly wound solenoid at a point far from both ends. The element of length $dx$ gives a magnetic dipole field of a current loop carrying current $di = nI\, dx$, where $I$ is the current in the solenoid.

Let us consider a point on the axis a distance $x_0$ from the center and distances $x_0 - \frac{1}{2}\ell$ from the close end and $x_0 + \frac{1}{2}\ell$ from the far end, with $x_0 - \frac{1}{2}\ell$ much greater than the solenoid radius $r$ (Figure 27-25). We first consider an element of the solenoid of length $dx$ at a distance $x$ from the center. If $n = N/\ell$ is the number of turns per unit length, there are $n\, dx$ turns of wire in this element, each carrying a current $I$. The element is thus equivalent to a loop carrying a current $di = nI\, dx$, giving a dipole field

$$dB = 2k_m \frac{di\, A}{x'^3} = 2k_m nIA \frac{dx}{x'^3}$$

where $x' = x_0 - x$ is the distance from the element to the field point and $A = \pi r^2$ is the area of the coil. We find the total field at $x_0$ due to the solenoid by summing from $x = -\frac{1}{2}\ell$ to $x = +\frac{1}{2}\ell$. Performing the integration between these limits, we obtain

$$B = k_m nIA \left( \frac{1}{r_1^2} - \frac{1}{r_2^2} \right) \qquad \text{27-16}$$

where $r_1 = x_0 - \frac{1}{2}\ell$ is the distance to the near end and $r_2 = x_0 + \frac{1}{2}\ell$ is the distance to the far end. We have obtained the expected result. We can put this into a form similar to Coulomb's law by defining a magnetic-pole strength $q_m$:

$$q_m = nIA$$

Then

$$B = k_m \frac{q_m}{r_1^2} - k_m \frac{q_m}{r_2^2} \qquad \text{27-17}$$

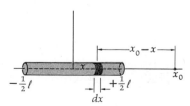

Note that the dipole moment of the solenoid is

$$q_m \ell = nIA\ell = NIA \qquad \text{27-18}$$

which is the sum of the dipole moments of the individual loops, as expected.

# 27-5  The Magnetic Field of a Bar Magnet

The first quantitative investigations of magnetic fields were made with permanent magnets. Using a torsion balance, John Michell found that the force of attraction or repulsion between two poles varies as the product of the pole strengths and inversely with the square of the distance between the poles. Since the force acting on a single pole in a magnetic induction field **B** is $q_m\mathbf{B}$, this result implies that the magnetic field due to a pole is proportional to $q_m/r^2$ and in the radial direction. According to Ampère's model, all magnetic fields are due to currents of some kind. In permanent magnets or other magnetized material these currents are due to the intrinsic motion of atomic electrons. Although these motions are very complicated, for this model we need only assume that the motions are equivalent to closed-circuit loops (Figure 27-26). The magnetic moments of the current loops shown in this figure are in the direction parallel to the axis. If the material is homogeneous, the net current at any point inside the material is zero because of cancellation of neighboring current loops. However, since there is no cancellation on the surface of the material, the result of these current loops is equivalent to a current on the surface of the material, called an *amperian current*. The surface current is similar to the real conduction current in a tightly wound solenoid. The magnetic field due to the surface current is the same as that due to an equivalent "surface" current in a solenoid. Let $M$ be the amperian current per unit length on the surface of a cylindrical bar magnet. The corresponding quantity for a solenoid is $nI$, where $n$ is the number of turns per unit length and $I$ is the current in each turn. If we replace the current per unit length $nI$ in Equation 27-18 by $M$, we have for the dipole moment of the magnet

$$q_m\ell = M(A\ell) \qquad\qquad 27\text{-}19$$

where $A$ is the cross-sectional area of the magnet and $\ell$ the length. Since $A\ell$ is the total volume of the magnet, $M$ equals the magnetic moment per unit volume (Figure 27-27). The vector **M**, whose magnitude is the magnetic moment per unit volume and whose direction is that of the dipole moment of the magnet, is called the *magnetization vector* of the magnet. The direction of **M** is from the south (negative) pole to the north (positive) pole. (Any magnetized material, whether it is a permanent magnet or some material magnetized by an external magnetic field, can be characterized by the magnetic moment per unit volume, **M**, which for non-homogeneous material may vary from point to point in the material.) The pole strength of a cylindrical bar magnet of cross-sectional area $A$ is related to its magnetization $M$ through Equation 27-19:

$$q_m = MA \qquad\qquad 27\text{-}20$$

The magnetic induction $B$ due to a bar magnet is then given by Equation 27-17 with $q_m$ equal to the magnetization times the area. In our derivation of that equation for a solenoid we assumed that the field point was far away compared with the radius of the solenoid. Thus this equation is only an approximation for the field due to a magnet also. This approximation is equivalent to the assumption that the pole strength of the magnet is concentrated at a single point at each end of the magnet. From the observation that the lines of magnetic induction outside a cylindrical bar magnet are identical to the lines of **E** for two equally but oppositely charged disks, we see that we can improve this approximation in the region just outside the magnet by replacing these

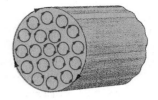

**Figure 27-26**
Model of atomic current loops when all the atomic dipoles are parallel to the axis of the cylinder. The net current at any point inside the material is zero due to cancellation of neighboring atoms. The result is a surface current similar to that of a solenoid.

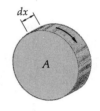

**Figure 27-27**
Disk for relating the surface current to the dipole moment. If $M$ is the current per unit length on the surface, the total current is $M\,dx$ and the dipole moment is $M\,dx\,A = M\,d\mathcal{V}$, where $d\mathcal{V}$ is the volume.

*Magnetization vector* **M**

point poles with a surface pole density $\sigma_m$ on the ends, where $\sigma_m$ is given by

$$\sigma_m = \frac{q_m}{A} = M \qquad\qquad 27\text{-}21$$

In the region inside a solenoid the magnetic induction is approximately $\mu_0 nI$, according to Equation 27-15. This approximation is good as long as the field point is not near the ends, as indicated in Figure 27-23. Replacing the current per unit length of the solenoid, $nI$, with the corresponding amperian current per unit length of a bar magnet, $M$, we have for the field inside a magnet far from the ends

$$\mathbf{B} = \mu_0 \mathbf{M} \qquad\qquad 27\text{-}22$$

We emphasize that the magnetic induction $\mathbf{B}$ for a bar magnet of magnetization $\mathbf{M}$ is identical everywhere to that of a solenoid of the same shape and carrying a current per unit length $nI$ equal to $M$.

## 27-6  Magnetic Flux

The flux of the magnetic field through a surface is defined similarly to the flux of the electric field. Let $dA$ be an element of area on the surface and $\hat{\mathbf{n}}$ be the unit vector perpendicular to the element. The magnetic flux $\phi_m$ is then defined by

$$\phi_m = \int \mathbf{B} \cdot \hat{\mathbf{n}}\, dA \qquad\qquad 27\text{-}23$$

If the surface is a plane with area $A$ and $\mathbf{B}$ is constant in magnitude and direction over the surface, making an angle $\theta$ with the unit normal vector, the flux is

$$\phi_m = BA \cos\theta \qquad\qquad 27\text{-}24$$

The unit of flux is called the weber (Wb):

$$1\ \text{Wb} = 1\ \text{T·m}^2 \qquad\qquad 27\text{-}25$$

We are often interested in the magnetic flux through a coil containing several turns of wire. Figure 27-28 shows a coil with two turns in a uniform magnetic field $\mathbf{B}$ perpendicular to the plane of the coil. The total area enclosed by the coil is twice the area enclosed by each turn. The magnetic flux through the coil is therefore the product of the field $B$, the area of each coil $A$, and the number of turns $N$, which in this case is two.

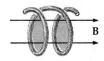

**Figure 27-28**
The area enclosed by a coil of two turns is twice the area bound by each turn. In general, the area bound by a coil of $N$ turns is $N$ times that of each turn.

**Example 27-3**  Find the magnetic flux through a long, tightly wound solenoid the length $\ell$, cross-sectional area $A$, and number of turns $N$, carrying current $I$.

We shall use the approximation that the magnetic field inside the solenoid is constant, equal to its value at the center:

$$B = \mu_0 nI = \mu_0 \frac{N}{\ell} I$$

The total area is $NA$ since there are $N$ turns, each of area $A$. Since the magnetic field is perpendicular to the area of each turn, the flux is

$$\phi_m = BA = \mu_0 \frac{N}{\ell} INA = \mu_0 \frac{N^2}{\ell} AI = \mu_0 n^2 (\ell A) I$$

where $n = N/\ell$ is the number of turns per unit length, and the quantity

$\ell A$ is the volume of the space inside the solenoid containing the magnetic field $B$.

In Chapter 21 we considered the flux of the electric field through a closed surface and found that the flux, which is proportional to the net number of lines leaving the surface, is proportional to the net electric charge within the surface. The lines of magnetic field differ from those of electric field in that the magnetic lines are continuous and have no beginning or end. Thus for any closed surface, the same number of lines must enter the surface as leave the surface. This means that the net magnetic flux through any *closed* surface is always zero:

$$\oint \mathbf{B} \cdot \hat{\mathbf{n}}\, dA = 0 \qquad\qquad 27\text{-}26$$

The magnetic field differs fundamentally from the electric field in that there are no magnetic charges or poles where the field lines originate or terminate. In Section 27-4 we defined a magnetic-pole strength $q_m$ in order to understand the magnetic field *outside* a solenoid or bar magnet better. Figure 27-29 shows a closed surface surrounding one end of a solenoid or bar magnet (the lines for both are identical). This figure should be compared with Figure 27-30 showing a similar closed surface surrounding the positive charge of an electric dipole. In the region corresponding to that outside the solenoid or bar magnet, the lines of **B** are identical to the lines of **E** for the dipole, and in each case they leave the closed surface indicated. However, inside the solenoid or magnet, the lines of **B** are different from the lines of **E** for the charge distribution. The magnetic field lines enter the closed surface inside the solenoid or magnet and there is no divergence of lines from the region of the pole. The lines which leave the surface also enter it.

Equation 27-26 is a statement of the experimental result that isolated magnetic poles do not exist. There are no regions of space in which lines of magnetic induction begin or end.

*Isolated magnetic poles have not been found experimentally*

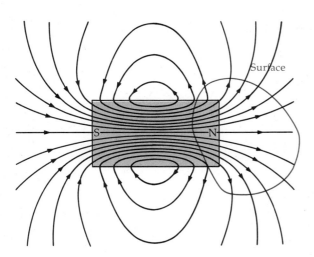

**Figure 27-29**
The net magnetic flux through a surface enclosing the north pole of a magnet is zero because each line that leaves the surface outside the magnet enters the surface inside the magnet.

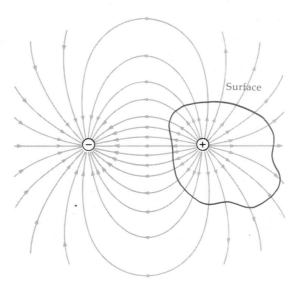

**Figure 27-30**
Electric flux through a surface enclosing a positive charge, for comparison with Figure 27-29. Here the net flux is not zero.

**Questions**

5. How can a small disk of area $A$ be placed in a magnetic field so that there is no magnetic flux through this area?

6. How should a small loop of wire be oriented in a magnetic field so that the flux of field lines through it is a maximum? A minimum?

7. How does a positive (north) magnetic pole $q_m$ as defined for a solenoid or magnet in Sections 27-4 and 27-5 differ from a true isolated positive pole? Sketch the lines of **B** near the poles for these cases.

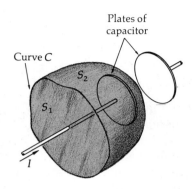

# 27-7   Maxwell's Displacement Current

When the current is not steady, Ampère's law does not hold. We can see why by considering the charging of a capacitor (Figure 27-31). Consider the curve $C$. According to Ampère's law, the line integral of the magnetic field around this curve equals $\mu_0$ times the total current through any surface bounded by the curve. Such a surface need not be a plane. Two surfaces bounded by the curve $C$ are indicated in the figure. The current through surface 1 is $I$. There is no current through surface 2 since the charge stops on the capacitor plate. There is thus an ambiguity in the phrase "current through any surface bounded by the curve." However, for steady currents, charge does not build up at any point, and we get the same current no matter which surface we choose.

Maxwell recognized this flaw in Ampère's law and showed that the law can be generalized to include all situations if the current $I$ in the equation is replaced by the sum of the true current $I$ and another term $I_d$, called *Maxwell's displacement current*. The displacement current is defined by

$$I_d = \epsilon_0 \frac{d\phi_e}{dt} \qquad\qquad 27\text{-}27$$

where $\phi_e$ is the flux of the electric field. The generalized form of Ampère's law is then

$$\oint \mathbf{B} \cdot d\boldsymbol{\ell} = \mu_0(I + I_d) = \mu_0 I + \mu_0\epsilon_0 \frac{d\phi_e}{dt} \qquad\qquad 27\text{-}28$$

We can understand the generalization of Ampère's law by considering Figure 27-31 again. Let us call the sum $I + I_d$ the generalized current. According to our arguments above, the same generalized current must cross any area bounded by the curve $C$. Thus there can be no net generalized current into or out of the closed volume. If there is a net true current $I$ into the volume, there must be an equal net displacement current $I_d$ out of the volume. In the volume pictured there is a net true current $I$ into the volume which increases the charge within the volume:

$$I = \frac{dQ}{dt}$$

The flux of the electric field out of the volume is related to the charge by Gauss' law:

$$\phi_{e \text{ net out}} = \oint \mathbf{E} \cdot \hat{\mathbf{n}} \, dA = \frac{1}{\epsilon_0} Q$$

**Figure 27-31**
Two surfaces $S_1$ and $S_2$ bounded by the same curve $C$. The current $I$ passes through surface $S_1$ but not $S_2$. Ampère's law, which relates the line integral of $B$ around the curve $C$ to the total current passing through any surface bounded by $C$, is not valid when the current is not continuous, as here, where it stops at the capacitor plate.

*Displacement current*

The rate of increase of charge is thus proportional to the rate of increase of the net flux out of the volume:

$$\epsilon_0 \frac{d\phi_{e \text{ net out}}}{dt} = \frac{dQ}{dt}$$

and the net current into the volume equals the net displacement current out of the volume.

A significant feature of Maxwell's generalization is that a magnetic field is produced by a changing electric field as well as by true electric currents. Maxwell was undoubtedly led to this generalization by the reciprocal result that an electric field is produced by a changing magnetic flux as well as by electric charges. This latter result, known as *Faraday's law*, preceded Maxwell's generalization. We shall study Faraday's law in the next chapter. These two properties of electric and magnetic fields—that a changing flux of one field produces the other—are connected with electromagnetic radiation, of which light is an example.

### Review

A. Define, explain, or otherwise identify:

Biot-Savart law, 743                    Amperian current, 755
Current balance, 748                    Magnetization vector, 755
Ampère's law, 749                       Magnetic flux, 756
Solenoid, 751                           Displacement current, 758
Pole strength of a solenoid, 754

B. True or false:

1. The magnetic field due to a current element decreases as the square of the distance from the element.

2. The magnetic field due to a current element is parallel to the element.

3. The magnetic field due to a very long wire carrying a current decreases as the square of the distance from the wire.

4. A solenoid and a uniformly magnetized magnet of the same shape and same magnetic moment produce the same magnetic field everywhere in space.

5. Lines of **B** never diverge from a point in space.

### Exercises

#### Section 27-1, The Biot-Savart Law

1. A small current element $I\, d\ell$, with $d\ell = 2\mathbf{k}$ mm and $I = 2$ A, is centered at the origin. Find the magnetic-induction field $d\mathbf{B}$ at the following points: (a) on the $x$ axis at $x = 3$ m, (b) on the $x$ axis at $x = -6$ m, (c) on the $z$ axis at $z = 3$ m, (d) on the $y$ axis at $y = 3$ m.

2. For the current element of Exercise 1 find the magnitude and indicate the direction of $d\mathbf{B}$ at the point $x = 0$, $y = 3$ m, $z = 4$ m.

3. For the current element of Exercise 1, find the magnitude of $d\mathbf{B}$ and indicate its direction on a diagram for the points (a) $x = 2$ m, $y = 4$ m, $z = 0$; (b) $x = 2$ m, $y = 0$, $z = 4$ m.

4. A long straight wire carries a current of 75 A. Find the magnitude of **B** at distances (a) 10 cm, (b) 50 cm, and (c) 2 m from the center of the wire.

*Exercises 5 to 10 refer to Figure 27-32, which shows two long straight wires in the xy plane parallel to the x axis. One wire is at y = −6 cm and the other is at y = +6 cm. The current in each wire is 20 A.*

5. If the currents in Figure 27-32 are in the negative $x$ direction, find **B** at the points on the $y$ axis at (a) $y = -3$ cm, (b) $y = 0$, (c) $y = +3$ cm, (d) $y = +9$ cm.

6. Sketch $B_z$ versus $y$ for points on the $y$ axis when both currents are in the negative $x$ direction.

7. Find **B** at points on the $y$ axis as in Exercise 5 but with the current in the wire at $y = -6$ cm in the negative $x$ direction and the current in the wire at $y = +6$ cm in the positive $x$ direction.

8. Sketch $B_z$ versus $y$ for points on the $y$ axis when the currents are in the direction opposite that in Exercise 7.

9. Find **B** at the point on the $z$ axis at $z = 8$ cm if (a) the currents are parallel, as in Exercise 5; (b) the currents are antiparallel, as in Exercise 7.

10. Find the magnitude of the force per unit length exerted by one wire on the other.

11. The current in the wire of Figure 27-33 is 8.0 A. Find **B** at point $P$ due to each wire segment and sum to find the resultant **B**.

12. Consider a circular loop of radius $R$ carrying current $I$. Show directly from the Biot-Savart law that the magnetic field at the center of the loop due to each segment of length $d\ell$ has the magnitude

$$dB = \frac{\mu_0}{4\pi} \frac{I\, d\ell}{R^2}$$

and that the resultant magnetic field at the center is $B = \mu_0 I / 2R$.

13. Find the magnetic field at point $P$ in Figure 27-34 if the current is 15 A.

14. A single-turn circular loop of radius 10.0 cm is to produce a field at its center which will just cancel the earth's field at the equator, which is 0.7 G directed north. Find the current in the wire and make a sketch showing the orientation of the loop and current.

15. A wire of length $L$ carries a current $I$. Find the magnetic field $B$ at the center when (a) the wire is bent into a square of side $L/4$ and (b) the wire is bent into a circle of circumference $L$. (c) Which gives the greater value of $B$?

16. In Figure 27-35, find the magnetic field at point $P$, which is at the common center of the two semicircular arcs.

## Section 27-2, Definition of the Ampere and the Coulomb

17. Two long straight parallel wires 10.0 cm apart carry currents of equal magnitude $I$. They repel each other with a force per unit length of 4.0 nN/m. (a) Are the currents parallel or antiparallel? (b) Find $I$.

18. A wire of length 12 cm is suspended by flexible leads above a long straight wire. Equal and opposite currents are established in the wires such that the 12-cm wire floats 1.5 mm above the long wire with no tension in its suspension leads. If the mass of the 12-cm wire is 10.0 g, what is the current?

19. In a student experiment with a current balance, the upper wire of length 30 cm is pivoted so that with no current it balances at 2 mm above a fixed parallel wire also 30 cm long. When the wires carry equal and opposite currents $I$, the upper wire again balances at its original position when a 2.4-g mass is placed on it. What is the current $I$?

20. Three long parallel straight wires pass through the corners of an equilateral triangle of side 10.0 cm, as shown in Figure 27-36, where a dot means that the current is out of the paper and a cross means that it is into the paper. If each cur-

**Figure 27-32**
Two long straight parallel wires for Exercises 5 to 10.

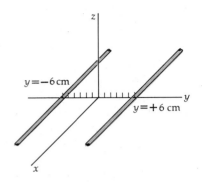

**Figure 27-33**
Exercise 11.

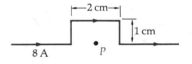

**Figure 27-34**
Exercise 13.

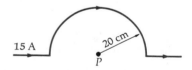

**Figure 27-35**
Exercise 16.

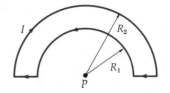

**Figure 27-36**
Exercises 20 and 21.

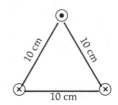

rent is 15.0 A, find (*a*) the force per unit length on the upper wire and (*b*) the magnetic field **B** at the upper wire due to the lower two wires. *Hint:* It is easier to find the force per unit length directly from Equation 27-11 and use your result to find **B** than to find **B** first and use it to find the force.

21. Work Exercise 20 with the current in the lower right corner of Figure 27-36 reversed.

### Section 27-3, Ampère's Law

22. A long straight thin-walled cylindrical shell of radius $R$ carries a current $I$. Find **B** inside and outside the cylinder.

23. In Figure 27-37, one current is 10 A into the paper; the other current is 10 A out of the paper; and each curve is a circular path. (*a*) Find $\oint \mathbf{B} \cdot d\boldsymbol{\ell}$ for each path indicated. (*b*) Which path, if any, can be used to find **B** at some point due to these currents?

**Figure 27-37**
Exercise 23.

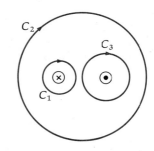

24. A very long coaxial cable has an inner wire and a concentric outer cylindrical conducting shell of radius $R$. At one end the wire is connected to the shell. At the other end the wire and shell are connected to the opposite terminals of a battery so that there is a current $I$ down the wire and back up the shell. Assume that the cable is straight and find $B$ (*a*) at points far from the ends and between the wire and shell, and (*b*) outside the cable.

25. A wire of radius 0.5 cm carries a current of 100 A uniformly distributed over the cross section. Find $B$ (*a*) at 0.1 cm from the center of the wire, (*b*) at the surface of the wire, and (*c*) at a point outside the wire 0.2 cm from the surface of the wire. (*d*) Make a graph of $B$ versus distance from the center of the wire.

26. Show that a uniform magnetic field with no fringing field as shown in Figure 27-38 is impossible because it violates Ampère's law. Do this by applying Ampère's law to the rectangular curve shown by the dashed lines.

**Figure 27-38**
Exercise 26.

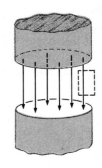

### Section 27-4, The Magnetic Field of a Solenoid

27. A solenoid has a length of 25 cm, radius 1 cm, and 400 turns, and carries 3 A of current. Find (*a*) $B$ on the axis at the center, (*b*) the magnetic moment of the solenoid, (*c*) the pole strength $q_m$, and (*d*) an approximate value for $B$ on the axis a distance 2 m from one end.

28. Work Exercise 27 for a solenoid of length 30 cm, radius 2 cm, and 800 turns, and carrying 2 A.

29. Two tightly wound solenoids each 10 cm long have 200 turns and a cross-sectional area of 0.5 cm², and carry current of 4.0 A. Each has its axis along the $x$ axis, and the currents are such that each has its north pole on the right and its south pole on the left. Their centers are 60 cm apart. (*a*) Find the pole strength of each solenoid. (*b*) Find the net force of attraction of the solenoids from Coulomb's law of attraction or repulsion between magnetic poles, $F = k_m q_{1m} q_{2m} / r^2$.

### Section 27-5, The Magnetic Field of a Bar Magnet

30. What is the magnitude of the magnetization vector **M** that gives a magnetic induction field of 5000 G at the center of a long cylindrical bar magnet?

31. A bar magnet has the same shape as the solenoid in Exercise 27. Find the magnetization **M** such that the magnetic field produced by the bar magnet is the same as that produced by the solenoid.

32. A long thin rod of iron is uniformly magnetized with magnetization $M = 2 \times 10^5$ A/m along its axis. Its length is 15 cm, and its cross-sectional area is 0.3 cm². (*a*) What is the magnitude of **B** at the center? (*b*) Find the magnetic-pole strength $q_m$ at each end. (*c*) Find the magnetic moment of the magnet.

33. An iron atom is like a small bar magnet of magnetic moment $1.8 \times 10^{-23}$ A·m². The number of atoms per unit volume in iron is $8.55 \times 10^{28}$ atoms per cubic metre. (a) If all the iron atoms in an iron bar are aligned with their magnetic moment along the axis, what is the magnitude of the magnetization $M$? (b) What is $B$ at the center of the magnet, assuming it to be a long rod?

### Section 27-6, Magnetic Flux

34. A uniform magnetic field of magnitude 2000 G is parallel to the $x$ axis. A square coil of side 5 cm has a single turn and makes an angle $\theta$ with the $z$ axis, as shown in Figure 27-39. Find the magnetic flux through the coil when (a) $\theta = 0$, (b) $\theta = 30°$, (c) $\theta = 60°$, (d) $\theta = 90°$.

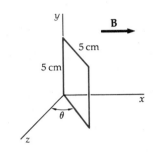

**Figure 27-39**
Exercise 34.

35. A circular coil has 25 turns and a radius of 5 cm. It is at the equator, where the earth's magnetic field is 0.7 G north. Find the magnetic flux through the coil when (a) its plane is horizontal, (b) its plane is vertical and its axis points north, (c) its plane is vertical and its axis points east, and (d) its plane is vertical and its axis makes an angle of 30° with the north.

36. Find the magnetic flux through the solenoid in Exercise 27. (Assume the magnetic field to be constant inside the solenoid.)

37. Find the magnetic flux through the solenoid in Exercise 28.

38. A circular coil of 15 turns of radius 4 cm is in a uniform magnetic field of 4000 G in the positive $x$ direction. Find the flux through the coil when the unit normal vector to the plane of the coil is (a) $\hat{n} = \mathbf{i}$, (b) $\hat{n} = \mathbf{j}$, (c) $\hat{n} = (\mathbf{i} + \mathbf{j})/\sqrt{2}$, (d) $\hat{n} = \mathbf{k}$, and (e) $\hat{n} = 0.6\mathbf{i} + 0.8\mathbf{j}$.

### Section 27-7, Maxwell's Displacement Current

39. A parallel-plate capacitor has circular plates of radius 1.0 cm separated by 1.0 mm in air. Charge is flowing onto the upper plate and off the lower plate at a rate of 5 A. (a) Find the time rate of change of the electric field between the plates. (b) Compute the displacement current between the plates and show that it equals 5 A. (c) Find the displacement current density $J_d$ between the plates and show that $J_d = \epsilon_0 \, dE/dt$.

40. For Exercise 39, show that at distance $r$ from the axis of the plates the magnetic field between the plates is given by $B = \frac{1}{2}\mu_0 J_d r$ if $r$ is less than the radius of the plates and $J_d$ is the displacement current density between the plates.

41. Show that for a parallel-plate capacitor the displacement current is given by $I_d = C \, dV/dt$, where $C$ is the capacitance and $V$ the voltage across the capacitor.

### Problems

1. Three very long parallel wires are at the corners of a square, as shown in Figure 27-40. Find the magnetic field $B$ at the unoccupied corner of the square when (a) all the currents are into the paper, (b) $I_1$ and $I_3$ are in and $I_2$ is out, (c) $I_1$ and $I_2$ are in and $I_3$ is out.

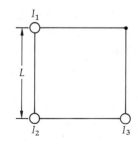

**Figure 27-40**
Problem 1.

2. Four long straight parallel wires each carry current $I$. In a plane perpendicular to the wires, the wires are at the corners of a square of side $a$. Find the force per unit length on one of the wires if (a) all the currents are in the same direction and (b) the currents in the wires at adjacent corners are oppositely directed.

3. A power cable carrying 50.0 A is located 2.0 m below the earth's surface, but its direction and precise position are unknown. Show how you could locate the cable using a compass. Assume that you are at the equator, where the earth's magnetic field is 0.7 G north.

4. A square loop of side $L$ lies in the $yz$ plane with its center at the origin. It carries a current $I$. Find the magnetic field $B$ at any point on the $x$ axis and show from your expression that for $x$ much larger than $L$

$$B_x = \frac{\mu_0}{2\pi} \frac{m}{x^3}$$

where $m = IL^2$ is the magnetic moment of the loop.

5. A very long straight wire carries a current of 20.0 A. An electron is 1.0 cm from the center of the wire and is moving with speed $5.0 \times 10^6$ m/s. Find the force on the electron when it moves (a) directly away from the wire, (b) parallel to the wire in the direction of the current, and (c) perpendicular to the wire and tangent to a circle around the wire.

6. A relatively inexpensive ammeter, called a *tangent galvanometer*, can be made using the earth's field. A plane circular coil of $N$ turns and radius $R$ is oriented such that the field $B_c$ it produces in the center of the coil is either east or west. A compass is placed at the center of the coil. When there is no current in the coil, the compass points north. When there is a current $I$, the compass points in the direction of the resultant magnetic field $B$ at an angle $\theta$ to the north. Show that the current $I$ is related to $\theta$ and the horizontal component of the earth's field $B_e$ by

$$I = \frac{2RB_e}{\mu_0 N} \tan \theta$$

7. A particle of positive charge $q$ moves in a helical path around a long straight wire carrying current $I$. Its velocity has components $v_\parallel$ parallel to the current and $v_t$ tangent to a circle around the current. Find the relationship between $v_\parallel$ and $v_t$ such that the magnetic force provides the necessary centripetal force.

8. A very long bar magnet of pole strength $q_m$ lies along the axis of a circular loop with its north pole at the center of the loop, as in Figure 27-41. The loop has radius $r$ and carries current $I$. (a) From the definition $\mathbf{F} = q_m\mathbf{B}$, find the magnitude and direction of the force exerted by the loop on the magnet. Assume that the south pole of the magnet is far enough away to be neglected. (b) Let $\mathbf{B}$ be the magnetic field at the loop due to the north pole of the magnet. Assuming that $\mathbf{B}$ points directly away from the pole, find the magnitude and direction of the total force exerted by the pole on the loop (in terms of $\mathbf{B}$, $r$, and $I$). (c) Using Newton's third law, set the magnitudes of the forces obtained in (a) and (b) equal to obtain the relation

$$B = \frac{\mu_0}{4\pi} \frac{q_m}{r^2}$$

for the magnetic field due to a magnetic pole.

**Figure 27-41**
Problem 8.

9. A very long straight conductor has a circular cross section of radius $R$ and carries a current $I$. Inside the conductor there is a cylindrical hole of radius $a$ whose axis is parallel to the axis of the conductor and a distance $b$ from it. Let the $z$ axis be the axis of the conductor, and let the axis of the hole be at $x = b$ (Figure 27-42). Find the magnetic field $B$ at the point (a) on the $x$ axis at $x = 2R$, (b) on the $y$ axis at $y = 2R$. *Hint:* Consider a uniform current distribution throughout the cylinder of radius $R$ plus a current in the opposite direction in the hole.

10. For the cylinder with the hole in Problem 9, show that the magnetic field inside the hole is uniform and find its magnitude and direction.

**Figure 27-42**
Problems 9 and 10.

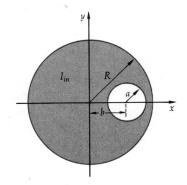

11. A disk of radius $R$ carries a fixed charge density $\sigma$ and rotates with angular velocity $\omega$. (a) Consider a circular strip of radius $r$ and width $dr$. Show that the current produced by this strip is $dI = (\omega/2\pi)\, dq = \sigma\omega r\, dr$. (b) Use your result for part (a) to show that the magnetic field at the center of the disk is $B = \frac{1}{2}\mu_0\sigma\omega R$. (c) Use your result for part (a) to find the magnetic field at a point on the axis of the disk a distance $x$ from the center.

12. A circular loop of radius $R$ carrying a current $I$ is centered at the origin with its axis along the $x$ axis. Its current is such that it produces a magnetic field in the positive $x$ direction. (a) Sketch $B_x$ versus $x$ for points on the $x$ axis. Include both positive and negative $x$. Compare this sketch with that for $E_x$ due to a charged ring of the same size. (b) A second identical loop carrying the same current in the same sense is in a plane parallel to the $xy$ plane with its center at $x = d$. Sketch the magnetic field on the $x$ axis due to each coil separately and the resultant field due to the two coils. Show from your sketch that $dB_x/dx$ is zero midway between the coils.

13. Two coils which are separated by a distance equal to their radius and carrying equal currents such that their axial fields add are called *Helmholtz coils*. A feature of Helmholtz coils is that the resultant magnetic field between the coils is very uniform. Let $R = 10$ cm, $I = 20$ A, and $N = 300$ turns for each coil. Place one coil with center at the origin and the other at $x = 10$ cm (as in Problem 12 with $d = R = 10$ cm). (a) Calculate the resultant field $B_x$ at points $x = 5$ cm, $x = 7$ cm, $x = 9$ cm, and $x = 11$ cm and from this, sketch $B_x$ versus $x$. (Note that $B_x$ is symmetric about the midpoint of the coils.) (b) Show that $dB_x/dx = 0$ and $d^2B_x/dx^2 = 0$ at $x = \frac{1}{2}R$. (c) Show that $d^3B_x/dx^3 = 0$ at $x = \frac{1}{2}R$.

14. The magnetic field of a small current-carrying coil can be found at points off axis and a great distance from the coil by considering the coil to be a small magnet of pole strength $q_m$ and length $\ell$ with $q_m\ell = IA$, where $A$ is the area of the coil and $I$ is its current. Consider a small coil at the origin with its magnetic moment in the positive $z$ direction. Show that the field at a point on the $x$ axis a great distance away is given by

$$\mathbf{B} = -\frac{\mu_0}{4\pi}\frac{IA}{x^3}\mathbf{k}$$

*Hint:* Calculate $\mathbf{B}$ from Coulomb's law, assuming a positive pole on the $z$ axis at $z = +\frac{1}{2}\ell$ and a negative pole at $z = -\frac{1}{2}\ell$. [See part (c) of Problem 8.]

15. A long straight wire carries a current of 20 A, as shown in Figure 27-43. A rectangular coil with two sides parallel to the straight wire has sides 5 and 10 cm with its near side a distance 2 cm from the wire. (a) Compute the magnetic flux through the rectangular coil. *Hint:* Calculate the flux through a strip of area $dA = (10\ \text{cm})\ dx$ and integrate from $x = 2$ cm to $x = 7$ cm. (b) If the rectangle carries a counterclockwise current of 5 A, find the net force on it.

16. A solenoid has $n$ turns per unit length and radius $R$ and carries current $I$. Its axis is at the $x$ axis with one end at $x = -x_1$ and the other at $x = +x_2$. Consider a part of the solenoid of length $dx$ at distance $x$ from the origin to be a circular loop carrying current $nI\ dx$. (a) Use Equation 27-8 for the magnetic field at the origin due to this loop and integrate from $x = -x_1$ to $x = +x_2$ to find the total magnetic field at the origin due to the solenoid. *Hint:* You will need to use the indefinite integral

$$\int \frac{dx}{r^3} = \frac{x}{R^2 r} \qquad \text{where } r = \sqrt{x^2 + R^2}$$

(b) Show that the result can be written as $B = \frac{1}{2}\mu_0 nI\ (\cos\theta_1 + \cos\theta_2)$, where $\cos\theta_1 = x_1/r_1$ and $\cos\theta_2 = x_2/r_2$. (c) Use this result to show that $B \approx \mu_0 nI$ if $R$ is much less than $x_1$ and $x_2$.

17. A large 50-turn circular loop of radius 10.0 cm carries a current 4.0 A. At the center of the loop is a small 20-turn coil of radius 0.5 cm carrying current 1.0 A. The planes of the two coils are perpendicular. Find the torque exerted by the large coil on the small coil. (Neglect any variation in $\mathbf{B}$ due to the large coil over the region occupied by the small coil.)

18. A solenoid has $n$ turns per unit length, radius $R_1$, and carries current $I$. (a) A large circular loop of radius $R_2 > R_1$ and $N$ turns encircles the solenoid at a point far away from the ends of the solenoid. Find the magnetic flux through the loop. (b) A small circular loop of radius $R_3 < R_1$ is completely inside the solenoid far from its ends with its axis parallel to that of the solenoid. Find the magnetic flux through the loop.

**Figure 27-43**
Problem 15.

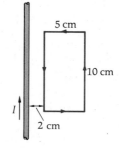

5 cm

10 cm

$I$

2 cm

These Helmholtz coils at the Kettering Magnetics Laboratory at Oakland University are used to cancel the earth's magnetic field and to provide a uniform magnetic field in a small region.

19. A long cylindrical conductor of radius $R$ carries current $I$ of uniform density $J = I/\pi R^2$. Find the magnetic flux per unit length through the area indicated in Figure 27-44.

**Figure 27-44**
Problem 19.

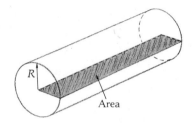

20. A long solenoid has $n$ turns per unit length and radius $R$ and carries current $I$. Its axis is the $x$ axis with one end at $x = -\frac{1}{2}L$ and the other at $x = +\frac{1}{2}L$, where $L$ is the total length of the solenoid. Show that the magnetic field $B$ at any point on the $x$ axis is given by

$$B = \tfrac{1}{2}nI \ (\cos\,\theta_1 + \cos\,\theta_2)$$

where

$$\cos\,\theta_1 = \frac{\tfrac{1}{2}L + x}{[R^2 + (\tfrac{1}{2}L + x)^2]^{1/2}} \quad \text{and} \quad \cos\,\theta_2 = \frac{\tfrac{1}{2}L - x}{[R^2 + (\tfrac{1}{2}L - x)^2]^{1/2}}$$

Note that for points outside the solenoid, $\cos\,\theta_2$ is negative (for $x$ positive). (See Problem 16.)

21. Use the results of Problem 20 to plot the magnetic field $B$ on the axis of a solenoid as a function of $x$. Let $L = 10$ cm, $R = 0.5$ cm, $I = 4$ A, and $n = 1500$ turns/m. Plot $B$ for positive $x$ ranging from $x = 0$ to $x = 20$ cm.

22. In this problem you are to show that the generalized form of Ampère's law (Equation 27-28) and the Biot-Savart law give the same result in a situation in which they both can be used. Figure 27-45 shows two charges $+Q$ and $-Q$ on the $x$ axis at $x = -a$ and $x = +a$ with a current $I = -dQ/dt$ along the line between them. Point $P$ is on the $y$ axis at $y = R$. (a) Use Equation 27-6, obtained from the Biot-Savart law, to show that the magnitude of $B$ at point $P$ is

$$B = \frac{\mu_0}{2\pi R}\,\frac{Ia}{\sqrt{a^2 + R^2}}$$

(b) Show that the electric field at any point $y$ on the $y$ axis is in the $x$ direction with magnitude

$$E_x = \frac{1}{2\pi\epsilon_0}\,\frac{Qa}{(y^2 + a^2)^{3/2}}$$

(c) Consider a circular strip of radius $r$ and width $dr$ in the $yz$ plane with center at the origin. Show that the flux of the electric field through this strip is $E_x\,dA = (Q/\epsilon_0)a(r^2 + a^2)^{-3/2}r\,dr$. (d) Use your result in part (c) to find the total flux $\phi_e$ through a circular area of radius $R$. Show that $\epsilon_0\phi_e = Q(1 - a/\sqrt{a^2 + R^2})$. (e) Find the displacement current $I_d$ and show that $I + I_d = Ia/\sqrt{a^2 + R^2}$. Then show that Equation 27-28 gives the same result for $B$ as that found in part (a).

**Figure 27-45**
Problem 22.

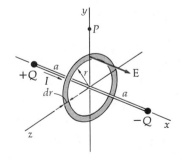

**CHAPTER 28**  Faraday's Law

Objectives  After studying this chapter you should:

1. Be able to state Faraday's law and use it to find the emf induced by a changing magnetic flux.

2. Be able to state Lenz's law and use it to find the direction of the induced current in various applications of Faraday's law.

3. Be able to discuss the various forces involved and the energy balance in an example of motional emf.

4. Be able to describe the betatron as an example of the application of Faraday's law.

5. Be able to discuss eddy currents.

6. Be able to state the definition of self- and mutual inductance and derive an expression for the self-inductance of a tightly wound solenoid.

7. Be able to apply Kirchhoff's laws to obtain the differential equations for $LR$, $LC$, and $LRC$ circuits and to discuss the general behavior of the solution for each circuit.

8. Be able to state the expression for the energy stored in a magnetic field and for the magnetic field energy density.

**Figure 28-1**
Demonstration of induced emf. When the magnet is moved away from the coil, an emf is induced in the coil, as shown by the galvanometer deflection. No deflection is observed when the magnet is stationary. (*Courtesy of Larry Langrill, Oakland University.*)

The fact that electric currents can be induced by changing magnetic fields was discovered in the early 1830s by Michael Faraday and simultaneously by the American physicist Joseph Henry. In a simple demonstration of induced currents, the ends of a coil of wire are attached to a galvanometer and a strong magnet is moved toward or away from the coil (Figure 28-1). The momentary deflection of the galvanometer *during* the motion indicates an induced electric current in the coil-galvanometer circuit. In another demonstration, a circuit containing a large electromagnet made of many turns of wire wrapped around an iron core is broken by a knife switch. As the circuit is broken, a spark jumps the gap across the switch. The changing magnetic field caused by breaking the circuit induces a large emf across the switch. It results in dielectric breakdown of the air, indicated by the spark. The results of these and many other experiments can be expressed by a single relation known as Faraday's law. Let $\phi_m$ be the flux of the magnetic field through a circuit (Figure 28-2). If this flux is changed in any way, there is an emf induced in the circuit given by

$$\varepsilon = -\frac{d\phi_m}{dt} \qquad\qquad 28\text{-}1$$

The emf is usually detected by observing a current in the circuit. In our previous discussions, the emf in a circuit has been localized in a specific region of the circuit, namely, between the terminals of the battery. The emf produced by a changing magnetic flux is not localized but must be considered to be distributed throughout the circuit. Since emf is work done per unit charge, there must be a force exerted on the charge associated with the emf. Energy is put into the circuit by the emf, and this force is therefore not conservative. The force per unit charge is a nonconservative electric field **E**; it differs from electrostatic fields given by Coulomb's law, which are conservative. The integral of $\mathbf{E} \cdot d\boldsymbol{\ell}$ around the complete circuit is the work per charge done by this nonconservative electric field, and is equal to the emf in the circuit. (This is true whether the emf is localized or not. In a battery the nonconservative electric field arises because of chemical forces inside the battery.) In terms of this nonconservative electric field, Faraday's law can be written

$$\varepsilon = \oint \mathbf{E} \cdot d\boldsymbol{\ell} = -\frac{d\phi_m}{dt} \qquad\qquad 28\text{-}2 \qquad \textit{Faraday's law}$$

The magnetic flux through a circuit can be changed in many different ways: the current producing the flux may be increased or decreased; permanent magnets may be moved toward the circuit or away from it; the circuit itself may be moved toward or away from the source of the flux; or the area of the circuit may be increased or decreased in a fixed magnetic field. In every case, an emf is induced in the circuit equal in magnitude to the rate of change of the magnetic flux. We discuss the significance of the negative sign in Faraday's law in Section 28-2.

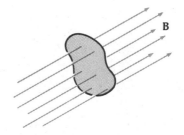

**Figure 28-2**
Flux of **B** through a circuit. If the flux changes, an emf is induced in the circuit.

# 28-1 Motional EMF

Figure 28-3 shows a conducting rod sliding along conducting rails which are connected to a resistor. A uniform magnetic field is directed into the paper. Since the magnetic flux through the circuit is changing (the area of the circuit increases as the rod moves), an emf is induced in the circuit. Let $\ell$ be the separation of the rails and $x$ be the distance from the left end to the rod at some time. The magnetic flux at this time is then

$$\phi_m = BA = B\ell x$$

The rate of change of the flux is

$$\frac{d\phi_m}{dt} = B\ell\frac{dx}{dt} = B\ell v$$

where $v = dx/dt$ is the speed of the rod. The emf induced in this circuit is therefore

$$\varepsilon = (-)\frac{d\phi_m}{dt} = (-)B\ell v \qquad\qquad 28\text{-}3$$

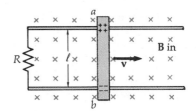

**Figure 28-3**
Rod sliding on rails in magnetic field.

An engraving of Michael Faraday (1791–1867) in his laboratory in the basement of the Royal Institution in London, where he carried out his research on electricity and magnetism. (*Courtesy of the Niels Bohr Library, American Institute of Physics.*)

(We defer the discussion of the minus sign in Faraday's law until the next section.)

The direction of the emf in this case is such as to produce a current in the counterclockwise sense. This example of Faraday's law is one in which we can understand the origin of the emf by considering the known forces acting on the electrons in the circuit. Since the current in the rod is upward from $b$ to $a$, the electrons move downward from $a$ to $b$. The velocity of a typical electron has a downward vertical component, the drift velocity, and a horizontal component $v$ equal to the speed of the rod. The electron's velocity $\mathbf{v}_e$ thus makes an angle $\theta$ with the horizontal, as shown in Figure 28-4. Since the horizontal component of the velocity is $v$, the speed of the rod, we have

$$v_e \cos \theta = v \qquad\qquad 28\text{-}4$$

The magnetic force $\mathbf{f}_m = -e\mathbf{v}_e \times \mathbf{B}$ is in the plane of the figure perpendicular to $\mathbf{v}_e$, as shown, and has the magnitude

$$f_m = ev_e B \qquad\qquad 28\text{-}5$$

If $\mathbf{f}_m$ were the only force acting on the electron, it could not stay in the rod as the rod moves to the right. The rod exerts a horizontal force $\mathbf{f}_r$ on the electron to balance the horizontal component of $\mathbf{f}_m$, which is $f_m \sin \theta$:

$$f_r = f_m \sin \theta \qquad\qquad 28\text{-}6$$

Since $\mathbf{f}_m$ is perpendicular to the motion of the electron, it does no work. The work done on the electron is done by the force $\mathbf{f}_r$. As the electron moves down the rod, the rod moves to the right; the electron therefore moves along a diagonal path of length $S$ (Figure 28-4). The lengths $S$ and $\ell$ are related by

$$\ell = S \sin \theta \qquad\qquad 28\text{-}7$$

**Figure 28-4**
Forces on an electron in the moving rod of Figure 28-3. The electron velocity $v_e$ has a horizontal component $v$, the speed of the rod, and a vertical component, its drift velocity along the rod. The magnetic force $\mathbf{f}_m$ is perpendicular to $\mathbf{v}_e$ and does no work. The rod exerts a horizontal force $\mathbf{f}_r$ on the electron with magnitude $f_m \sin \theta$. This force has a component in the direction of motion of the electron and therefore does work on it. The work per charge is shown to be equal to $BLv$.

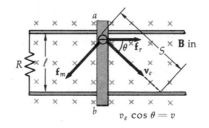

The work done on the electron as it moves down the complete length of the rod is

$$W = f_r \cos \theta \, S = (f_m \sin \theta) \cos \theta \, S = f_m \cos \theta \, \ell$$

where we have used Equation 28-6 for $f_r$ and $S \sin \theta = \ell$ from Equation 28-7. Substituting $ev_eB$ for $f_m$, we obtain for the work done on the electron:

$$W = ev_eB \cos \theta \, \ell = eB(v_e \cos \theta)\ell = eBv\ell$$

The work per unit charge is the emf $Bv\ell$, in agreement with our result from Faraday's law.

The force $\mathbf{f}_r$ is exerted by the rod on the electron. The electron exerts an equal but opposite force $-\mathbf{f}_r$ on the rod. This force is to the left in the figure. To keep the rod moving to the right with constant speed, an external agent must exert a force to the right. The work done by this external agent provides the energy that goes into the circuit. (In this case, the energy goes into Joule heat $I^2R$ in the resistor in the circuit.)

Faraday and his wife in the 1850s.

## 28-2  Lenz's Law

The direction of the induced emf and current can be found from a general statement known as *Lenz's law*.

*The emf and induced current are in such a direction as to tend to oppose the change which produced them.*

*Lenz's law*

In our statement of Lenz's law we did not specify just what kind of change it is that causes the induced emf. This was left vague to allow a variety of interpretations, all of which are valid and give the same result for the direction of the induced emf and current. A few illustrations will clarify this point.

In the example of the rod moving on the rails, above, the magnetic field is into the paper. Let us call this direction positive. The movement of the rod then tends to increase the flux in this direction. According to Lenz's law, the induced current is in the direction that will oppose this change. As we have seen, the induced emf and current are counterclockwise. The flux produced *by the induced current* is out of the paper, opposing the increase in flux through the circuit which results from the motion of the rod. If the current induced were in the clockwise direction, the induced flux would add to the increase produced by the moving rod, contrary to Lenz's law. Lenz's law is thus a statement that the induced current tends to maintain the status quo. If we try to increase the flux through a circuit, a current will be induced which tends to decrease the flux. If we move the rod to the left in our example, decreasing the flux, Lenz's law tells us that the induced current will be clockwise to help maintain the original flux through the circuit.

We can also interpret Lenz's law in terms of forces and the conservation of energy. In our example of the rod moving to the right, there is a magnetic force on the rod due to the induced current. Since the current is upward in the rod and the magnetic field is inward, the force is to the left,

$$\mathbf{F}_m = I\boldsymbol{\ell} \times \mathbf{B}$$

This force tends to oppose the motion of the rod (see Figure 28-5). If the rod is to move to the right with constant velocity, there must be an external force to the right of magnitude $I\ell B$. Lenz's law then tells us that there will be a current induced in such a direction that the magnetic force opposes this external force.

When the rod is moving with constant velocity, the power input by the external force on the rod goes into heating the resistor. The power input is $Fv$, where $F$ is the external force. Thus

$$P = Fv = I\ell Bv = I^2R \qquad\qquad 28\text{-}8$$

or

$$IR = B\ell v \qquad\qquad 28\text{-}9$$

By Kirchhoff's first rule the potential drop $IR$ must equal the emf in the circuit. Thus we find by means of energy balance that the emf in the circuit is

$$\varepsilon = B\ell v \qquad\qquad 28\text{-}10$$

If we assume the opposite of Lenz's law, energy cannot be conserved. Suppose we start with the rod at rest on the rails and give it a slight push to the right. Assuming the opposite of Lenz's law, we would have a current in the clockwise direction. Then there would be a magnetic force on the rod to the right which would accelerate the rod, increasing its velocity. The emf and induced current would thus be increased. That would further increase the force on the rod to the right. Hence, the opposite of Lenz's law gives a situation in which the rod continues to accelerate to the right and the current continues to increase without limit. With no power input to this circuit, the power $I^2R$ into the resistor would increase without limit.

The negative sign in Faraday's law (Equation 28-1 or 28-2) is the mathematical expression of Lenz's law: the sense of the induced emf is opposite to the sense of the rate of change of magnetic flux. In practice, it is usually easiest to compute the magnitude of the emf by computing the magnitude of $d\phi_m/dt$ (momentarily ignoring the sign) and then finding the direction of the emf and the induced current using whatever interpretation of Lenz's law is most convenient. We shall illustrate this in the next section.

**Figure 28-5**
The magnetic force $\mathbf{F}_m$ on the rod opposes the motion of the rod. Here the rod is moving to the right, and $\mathbf{F}_m$ is to the left.

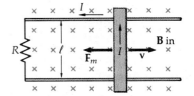

# 28-3  Applications of Faraday's Law

In most applications of Faraday's law it is not possible to understand the origin of the induced emf from our previous knowledge of magnetic forces. That is, we cannot derive Equation 28-1 by considering magnetic or electric forces on charges, as we did for the special case of the rod moving on rails. Instead, we must take Faraday's law as expressed in Equation 28-1 to be an independent law of nature expressing a wide range of experimental results. In this section we consider more examples to illustrate the use of Faraday's law and Lenz's law to find the direction of the induced current.

Figure 28-6 shows a bar magnet moving toward a loop which has a resistance $R$. Since the magnetic field from the bar magnet is to the right, out of the north pole of the magnet, the movement of the magnet toward the loop tends to increase the flux through the loop to the right. The induced current in the loop is in the direction shown, to create a flux which opposes the change.

**Figure 28-6**
When the bar magnet moves toward the loop, the induced emf in the loop produces a current in the direction shown.

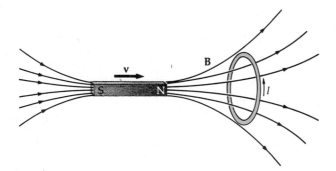

**Figure 28-7**
The magnetic moment of the loop (indicated by the outlined magnet) due to the induced current is such as to oppose the motion of the bar magnet. Here the bar magnet is moving toward the loop, and so the induced magnetic moment repels the bar magnet.

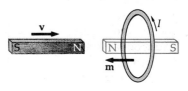

Figure 28-7 shows the induced magnetic moment of the current loop in Figure 28-6. The loop acts like a small magnet with north pole to the left and south pole to the right. Since opposite poles attract and like poles repel, the induced magnetic moment of the loop exerts a force on the bar magnet to the left to oppose its motion toward the loop. Again, we could use Lenz's law in terms of forces rather than flux. If the bar magnet is moved toward the loop, the induced current must produce a magnetic moment to oppose this change.

In Figure 28-8 the bar magnet is at rest, and the loop is moved away from it. The induced current and magnetic moment are shown in the figure. In this case the magnetic moment of the loop attracts the bar magnet, as required by Lenz's law.

In Figure 28-9 when the current in circuit 1 is changed, there is a change in the flux through circuit 2. Suppose the switch in circuit 1 is open, with no current in the circuit. When the switch is closed (Figure 28-9b), the current in circuit 1 does not reach its steady value $\varepsilon_1/R_1$ instantaneously but takes a short time to change from zero to this value. During this short time, while the current is increasing, the flux in circuit 2 is changing and there is an induced current in that circuit in the direction shown. When the current in the first circuit reaches its steady value, the flux is no longer changing and there is no induced current in circuit 2. An induced current in circuit 2 in the opposite direction appears momentarily when the switch in circuit 1 is opened (Figure 28-9c) and the current is decreasing to zero.

**Figure 28-8**
When the loop is moved away from a stationary bar magnet, the induced magnetic moment attracts the bar magnet, again opposing the relative motion.

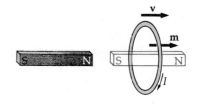

**Figure 28-9**
(a) Two adjacent circuits. (b) Just after the switch is closed, $I_1$ is increasing and B is increasing in the direction shown. The changing flux in circuit 2 induces current $I_2$. The flux due to $I_2$ opposes the increase in flux due to $I_1$. (c) As the switch is opened, $I_1$ decreases and B decreases. The induced current $I_2$ tends to maintain the flux in the circuit, opposing the decrease.

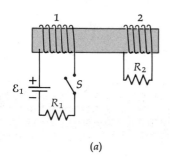

(a)

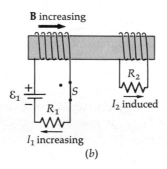

B increasing

$I_2$ induced

$I_1$ increasing

(b)

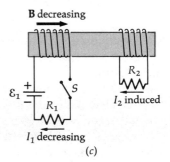

B decreasing

$I_2$ induced

$I_1$ decreasing

(c)

For our next example we consider a single isolated circuit. When there is a current in the circuit, there is magnetic flux through the circuit due to its own current. When the current is changing, the flux is changing and there is an induced emf in the circuit. This *self-induced* emf opposes the change in the current. It is because of this self-induced emf that the current in a circuit cannot jump instantaneously from zero to some finite value or from some value to zero. Henry first noticed this effect when he was experimenting with a circuit consisting of many turns of a wire (Figure 28-10), an arrangement that gives a large flux through the circuit for even a small current. Henry noticed a spark across the switch when he tried to break the circuit. This spark is due to the large induced emf which occurs when the current varies rapidly, as with opening the switch. In this case the induced emf tries to maintain the original current. The large induced emf produces a large voltage drop across the switch as it is opened. The electric field between the poles of the switch is large enough to tear the electrons from the air molecules, causing dielectric breakdown. When the molecules in the air dielectric are ionized, the air conducts electric current in the form of a spark.

For our final example, we consider a *flip coil,* a device for measuring magnetic fields. The plane of the coil in Figure 28-11 is perpendicular to a uniform magnetic induction **B**, which is to be measured. The coil is connected to a *ballistic galvanometer,* a device designed to measure the total charge passing through it in a short time. The flux through the flip coil is

$$\phi_m = NBA$$

where $N$ is the number of turns and $A$ is the area of the coil. If the coil is suddenly rotated about a diameter through 90°, the flux decreases to zero. While the flux is decreasing, there is an emf in the coil and a current in the coil-galvanometer circuit. The current is

$$I = \frac{\mathcal{E}}{R} = \frac{1}{R}\frac{d\phi_m}{dt}$$

where $R$ is the total resistance of the coil and galvanometer. The total charge that passes through the galvanometer is

$$Q = \int I\,dt = \frac{1}{R}\int d\phi_m = \frac{\phi_m}{R} = \frac{NAB}{R}$$

If the total charge is measured, the magnetic induction $B$ can be found from

$$B = \frac{RQ}{NA} \qquad\qquad 28\text{-}11$$

## 28-4  Eddy Currents

In the examples we have discussed, the currents produced by a changing flux were set up in definite circuits. Often a changing flux sets up circulating currents, called *eddy currents,* in a piece of bulk metal like the core of a transformer. Consider a conducting slab between the pole faces of an electromagnet (Figure 28-12). If the magnetic induction $B$ between the pole faces is changing with time (as it will if the current in the magnet windings is alternating current), the flux through any closed loop in the slab will be changing. For example, the flux through the

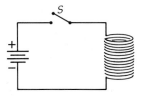

**Figure 28-10**
Circuit with many turns of wire, giving a large flux for a given current in the circuit. The emf induced in the circuit when the current changes opposes the change, trying to maintain the original current.

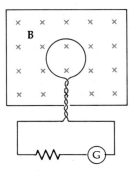

**Figure 28-11**
Flip-coil circuit for measuring the magnetic field **B**. When the coil is flipped over, the total charge flowing through the galvanometer is proportional to $B$.

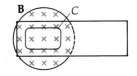

**Figure 28-12**
Eddy currents. If the magnetic field is changing, an emf is induced in any closed path in the metal, such as the curve $C$ shown. The induced emf causes a current in the circuit.

curve $C$ indicated in the figure is just the magnetic induction $B$ times the area enclosed by the curve. If $B$ varies, the flux will vary and there will be an induced emf around curve $C$. Since path $C$ is in a conductor, there will be a current given by the emf divided by the resistance of the path. In this figure we have indicated just one of the many closed paths which will contain currents if the magnetic field between the pole faces varies. Circulating, or eddy, currents are usually unwanted because the heat produced is not only a power loss but must be dissipated. The power loss can be reduced by increasing the resistance of the possible paths for the eddy currents, as in Figure 28-13. Here the conducting slab is laminated, i.e., made up of small strips glued together. Because of the resistance between the strips, the eddy currents are essentially confined to the strips. The large eddy-current loops are broken up, and the power loss is greatly reduced.

The existence of eddy currents can be demonstrated by pulling a copper or aluminum sheet between the poles of a strong permanent magnet (Figure 28-14). Part of the area enclosed by curve $C$ in this figure is in the magnetic field, and part is outside the field. As the sheet is pulled to the right, the flux through this curve decreases (assuming that the flux into the paper is positive). According to Faraday's law and Lenz's law, a clockwise current will be induced around this curve. Since this current is directed upward in the region between the pole faces, the magnetic field exerts a force on the current to the left, opposing motion of the sheet. You can feel this force on the sheet if you try to pull a sheet suddenly through a strong magnetic field. If the sheet has cuts in it, as in Figure 28-15, the eddy currents are lessened and the force is greatly reduced.

**Questions**

1. A bar magnet is dropped inside a long vertical pipe. The pipe is evacuated so that there is no air resistance, but the falling magnet still reaches a terminal velocity. Explain why.

2. A sheet of metal fixed to the end of a pivoted rod will swing like a pendulum about the pivot, but if the sheet is made to swing through the gap between two poles of a magnet, the oscillation will rapidly damp out. Why?

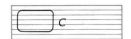

**Figure 28-13**
The eddy currents in a metal slab can be reduced by constructing the slab from small strips. The resistance of the path indicated by $C$ is now large because of the glue between the strips.

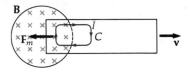

**Figure 28-14**
Demonstration of eddy currents. When the metal slab is pulled to the right, there is a magnetic force to the left on the induced current, opposing the motion.

**Figure 28-15**
If the metal slab has cuts as shown, the eddy currents are reduced because of the lack of good conducting paths.

## 28-5 The Betatron

An interesting use of Faraday's law is the betatron (Figure 28-16), invented in 1941 by Donald Kerst at the University of Illinois to accelerate electrons (also called beta particles) to high energies. In the betatron, electrons move in a circular orbit at a constant radius in an evacuated doughnut-shaped tube. They are accelerated by a tangential, nonconservative electric field, which is produced by a changing magnetic flux generated by an electromagnet through the orbit. The centripetal force needed to keep the electrons moving in a circle is provided by the same magnet. This can be done if the magnetic induction $B_0$ at the orbit radius is just half the average value of the magnetic induction over the complete area of the orbit. To accomplish this the pole pieces of the magnet are shaped as shown in Figure 28-17, so that the magnetic induction decreases with distance from the center of the circle.

Figure 28-16
D. W. Kerst with the first betatron in 1941. This betatron
was capable of accelerating
electrons to an energy of 2.3
MeV. The electron beam
strikes a target inside the
doughnut, producing x-rays
used for research in nuclear
physics. (*Courtesy of the Niels
Bohr Library, American Institute of Physics.*)

We define the average magnetic induction $B_{av}$ to be the flux divided by the area:

$$B_{av} = \frac{1}{\pi R^2} \int \mathbf{B} \cdot \hat{\mathbf{n}} \, dA = \frac{\phi_m}{\pi R^2} \qquad \text{28-12}$$

The emf around the orbit is

$$\varepsilon = \frac{d\phi_m}{dt} = \pi R^2 \frac{dB_{av}}{dt} \qquad \text{28-13}$$

This emf is equivalent to a tangential, nonconservative electric field given by Equation 28-2:

$$\varepsilon = \oint \mathbf{E} \cdot d\boldsymbol{\ell} = E 2\pi R = \pi R^2 \frac{dB_{av}}{dt}$$

or

$$E = \tfrac{1}{2} R \frac{dB_{av}}{dt} \qquad \text{28-14}$$

The tangential force $qE$ increases the magnitude of the momentum of the electron according to Newton's second law:

$$\frac{dp}{dt} = qE = \tfrac{1}{2} qR \frac{dB_{av}}{dt} \qquad \text{28-15}$$

The centripetal force which keeps the electrons moving in a circle is provided by the radial force $qvB_0$, where $B_0$ is the value of the magnetic induction at the orbit radius. The condition for circular motion is Equation 26-12, which we rewrite as*

$$p = B_0 qR \qquad \text{28-16}$$

These equations are consistent with each other if the magnetic induction at the orbit is exactly half the average value over the area

$$B_0 = \tfrac{1}{2} B_{av} \qquad \text{28-17}$$

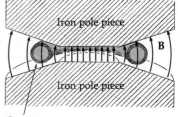

Ceramic
doughnut

**Figure 28-17**
The pole faces in a betatron
magnet are shaped as shown
so that the magnetic field at
the orbit radius of the electrons has a magnitude half
that of the average magnetic
field through the area enclosed by the orbit.

*Betatron condition*

---

* Because electrons in a betatron move with speeds nearly equal to the speed of light, newtonian mechanics does not apply and the theory of special relativity must be used. We shall see when we study this theory in Chapter 35 that for high-speed particles, Newton's second law in the form $\mathbf{F} = d\mathbf{p}/dt$ is valid relativistically if momentum is redefined in such a way that $\mathbf{p} = m\mathbf{v}$ holds in the low-speed nonrelativistic limit. If the relativistic expression for momentum is used, Equations 28-15 and 28-16 are valid both relativistically and classically.

The magnetic field is produced by coils which carry alternating current. The magnetic induction $B$ then varies as in Figure 28-18. The electrons are injected into orbit just as the magnetic field crosses zero and begins to increase. (The electrons are injected from an electron gun just outside the stable orbit and are quickly brought into the stable orbit by a magnetic field from a current pulse through some auxiliary coils.) Only the first quarter of the cycle can be used. As the flux through the orbit changes, the electrons gain energy and their momentum increases, according to Equation 28-15. They stay in their orbit because the field at the orbit $B_0$ changes at the same rate as the momentum. Typically, the electrons make several hundred thousand revolutions during the quarter cycle in which they are accelerated. When the magnetic field is near its maximum value, the electrons are ejected from their orbit by a current pulse through a set of auxiliary coils which momentarily weakens the induction at the orbit. Usually the electron beam is moved into a larger orbit and hits the injector gun, producing x-rays. The x-rays are used for nuclear research or for medical purposes.

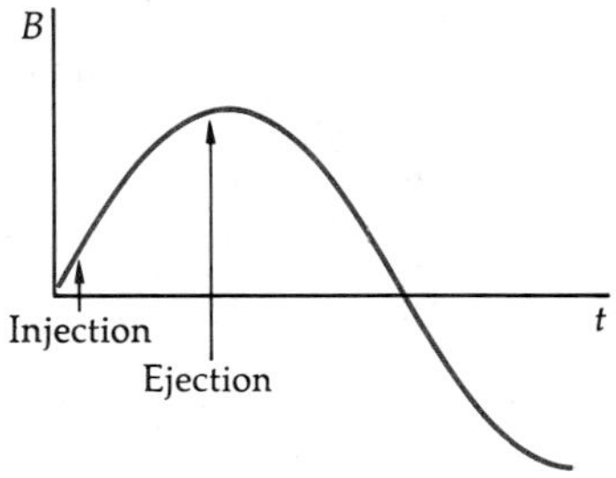

**Figure 28-18**
$B$ versus $t$ in the accelerating part of the betatron cycle. The electrons are injected just after $B$ crosses zero and ejected just before maximum $B$.

## Questions

3. Why can't the second quarter of the cycle in Figure 28-18 be used to accelerate the electrons?

4. Why can't the fourth quarter of the cycle be used? Is there any other part of the cycle that can be used?

5. The iron core of the betatron magnet-pole face is made not from one piece of iron but from many pie-shaped wedges. Why?

## 28-6 Inductance

The flux through a circuit can be related to the current in that circuit and the currents in other nearby circuits. (We shall assume that there are no permanent magnets around.) Consider the two circuits in Figure 28-19. The magnetic field at some point $P$ consists of a part due to $I_1$ and a part due to $I_2$. These fields are proportional to the currents producing them and could, in principle, be calculated from the Biot-Savart Law. We can therefore write the flux through circuit 2 as the sum of two parts; one part is proportional to the current $I_1$ and the other to the current $I_2$:

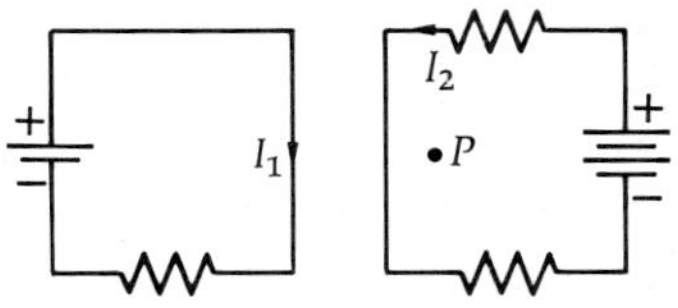

**Figure 28-19**
Two adjacent circuits. The magnetic field at point $P$ is partly due to current $I_1$ and partly due to $I_2$. The flux through either circuit is the sum of two terms, one proportional to $I_1$ and the other to $I_2$.

*Self-inductance and mutual inductance*

$$\phi_{m2} = L_2 I_2 + M_{12} I_1 \qquad 28\text{-}18$$

where $L_2$ and $M_{12}$ are constants. The constant $L_2$, called the *self-inductance* of circuit 2, depends on the geometrical arrangement of that circuit. The constant $M_{12}$, called the *mutual inductance* of the two circuits, depends on the geometrical arrangement of both circuits. In particular, we can see that if the circuits are far apart, the flux through circuit 2 due to the current $I_1$ will be small and the mutual inductance will be small. An equation similar to Equation 28-18 can be written for the flux through circuit 1:

$$\phi_{m1} = L_1 I_1 + M_{21} I_2 \qquad 28\text{-}19$$

The self-inductance $L_1$ depends only on the geometry of circuit 1, whereas the mutual inductance $M_{21}$ depends on the arrangement of both circuits. Though it is certainly not obvious, it can be shown in general that these two mutual inductances are equal:

$$M_{12} = M_{21} \tag{28-20}$$

When the circuits are fixed and only the currents change, their induced emfs are, from Faraday's law,

$$\mathcal{E}_1 = -\frac{d\phi_{m1}}{dt} = -L_1 \frac{dI_1}{dt} - M \frac{dI_2}{dt}$$

$$\mathcal{E}_2 = -L_2 \frac{dI_2}{dt} - M \frac{dI_1}{dt} \tag{28-21}$$

where we have dropped the subscripts on the mutual inductance.

The SI unit of inductance is the henry (H). From Equations 28-19 and 28-21 we see that the henry is related to other SI units by

$$1 \text{ H} = 1 \text{ T·m}^2/\text{A} = 1 \text{ V·s/A}$$

We shall see that inductance can always be written as the product of a $\mu_0$ and some characteristic length. The SI unit of the constant $\mu_0$ can therefore be conveniently expressed as henrys per metre:

$$\mu_0 = 4\pi \times 10^{-7} \text{ H/m}$$

In principle, the calculation of the self- and mutual inductances is straightforward for any given circuits and arrangement. For example, to calculate $L_1$ we need only assume some current $I_1$ and calculate the magnetic field from the Biot-Savart law at every point on some surface bounded by the circuit. This magnetic field will of course be proportional to the current $I_1$. We then calculate the flux of this field by integrating over the area. Since $B$ is proportional to $I_1$, the flux will also be proportional to $I_1$. The proportionality constant is the self-inductance $L_1$. Similarly, we calculate the mutual inductance by assuming some current $I_1$ in circuit 1 and calculating the magnetic field everywhere on a surface bounded by circuit 2, and from it the flux through that circuit. Because of the difficulty in calculating the magnetic field in general from the Biot-Savart law and in integrating this field over any surface, it is not surprising that the calculations of self-inductance and mutual inductance are extremely difficult except for a few very simple geometrical arrangements. Consider, for example, the calculation of the self-inductance for a single circular loop of radius $R$. The magnetic field at the center of the loop is $B = \frac{1}{2}\mu_0 I/R$. But in order to find the flux through the loop, we need to know the value of $B$ everywhere on the circle enclosed by the loop, not just at the center. Such a calculation is too difficult to do here. The self-inductance of a single loop depends not only on the radius of the loop but also on the radius of the conducting wire.

We now discuss some cases in which inductance can be calculated, at least approximately.

### Self-Inductance of a Solenoid

This calculation is not difficult because, to a good approximation, the magnetic field inside the solenoid is uniform and given by

$$B = \mu_0 \frac{N}{\ell} I = \mu_0 n I$$

where $n = N/\ell$ is the number of turns per unit length, $N$ the total number of turns, $\ell$ the total length, and $I$ the current in the solenoid.

The area enclosed by this circuit is the area of one turn times the number of turns, and so the flux is

$$\phi_m = BNA = \mu_0 nINA = \mu_0 n \frac{N}{\ell} \ell AI = \mu_0 n^2 (\ell A) I \qquad 28\text{-}22$$

where $A$ is the area of a single loop. As expected, the flux is proportional to the current $I$. The proportionality constant is the self-inductance:

$$L = \frac{\phi_m}{I} = \mu_0 n^2 A \ell = \mu_0 N^2 \frac{A}{\ell} \qquad 28\text{-}23$$

*Self-inductance of a solenoid*

The self-inductance is proportional to the square of the number of turns per unit length and to the volume $A\ell$. It can also be thought of as the product of $\mu_0$, $N^2$, and the characteristic length $A/\ell$.

**Example 28-1** Find the self-inductance of a solenoid of length 10 cm, area 5 cm², and 100 turns.

We can calculate the self-inductance from Equation 28-23 in henrys if we put all quantities in SI units:

$$n = \frac{N}{\ell} = \frac{100 \text{ turns}}{0.1 \text{ m}} = 10^3 \text{ turns/m}$$

$$A\ell = (5 \times 10^{-4} \text{ m}^2)(0.1 \text{ m}) = 5 \times 10^{-5} \text{ m}^3$$

$$\mu_0 = 4\pi \times 10^{-7} \text{ H/m}$$

$$L = (4\pi \times 10^{-7})(10^3)^2(5 \times 10^{-5}) = 2\pi \times 10^{-5} \text{ H}$$

**Example 28-2** At what rate must the current in the solenoid of Example 28-1 change to induce an emf of 20 V?

From Equation 28-21 with $M = 0$ we have

$$\varepsilon = -L \frac{dI}{dt} = 20 \text{ V}$$

Then

$$\frac{dI}{dt} = -\frac{\varepsilon}{L} = \frac{20 \text{ V}}{2\pi \times 10^{-5} \text{ H}} = -3.18 \times 10^5 \text{ A/s}$$

**Mutual Inductance between a Long Wire and a Rectangular Loop**

Two circuits for which the mutual inductance can be calculated are shown in Figure 28-20, where the wire on the left is assumed to be so long that the remaining part of its circuit is far from the rectangular loop and contributes negligible flux. To find the flux through the rectangular loop we must integrate. Since the magnetic field due to a long wire decreases as $1/r$ with the distance from the wire, we choose the strip-shaped area element shown in the figure. The area of this strip is $c\, dx$, and the flux through the strip is

$$d\phi_m = \frac{\mu_0 I}{2\pi x} c\, dx \qquad 28\text{-}24$$

where $I$ is the current in the long wire. The total flux is found by integrating from $x = a$ to $x = b$:

$$\phi_m = \frac{\mu_0}{2\pi} cI \int_a^b \frac{dx}{x} = \frac{\mu_0 Ic}{2\pi} \ln \frac{b}{a} \qquad 28\text{-}25$$

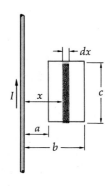

**Figure 28-20**
Long wire and circuit for calculation of mutual inductance.

The flux through the rectangular loop due to the current $I$ in the long wire is proportional to $I$, as expected. The proportionality constant is the mutual inductance:

$$M = \frac{\mu_0 c}{2\pi} \ln \frac{b}{a}$$

### Questions

6. How would the inductance of a solenoid be changed if the same length of wire were wound onto a cylinder of the same diameter but twice as long? If twice as much wire were wound onto the same cylinder?

7. Two solenoidal inductances are wound on cylinders of the same diameter. Is it possible for them to have the same inductance even though one has twice the total length of wire of the other?

## 28-7  *LR* Circuits

As we have seen, self-inductance in a circuit prevents the current from rising or falling instantaneously. Circuits containing coils or solenoids of many turns have a large self-inductance. Such a coil or solenoid is called an *inductor*. The symbol for an inductor is ⌢⌢⌢. We can often neglect the self-inductance of the rest of the circuit compared with that of an inductor.

A circuit containing batteries, resistors, and inductors is called an *LR* circuit. Since all circuits contain resistance and self-inductance, the analysis can be applied to some extent to all circuits. All circuits also have some capacitance between parts of the circuit at different potentials. We shall include the effects of capacitance in Section 28-9, when we study *LC* and *LCR* circuits. We neglect capacitance here in order to simplify the analysis and to bring out the special features of inductance in a simple situation.

Figure 28-21 shows a typical circuit. We neglect the internal resistance of the battery and treat it as an ideal seat of emf. The current in the circuit is originally zero. At $t = 0$ we close the switch. As discussed before, while the current is increasing, there is an induced emf in the inductance in the direction opposing the current increase. In the figure the current will be clockwise, as shown. We can relate the current, its rate of change $dI/dt$, and the emf of the battery by applying Kirchhoff's first rule to the circuit shown. Because of the negative sign in Equation 28-21, which reduces to $\varepsilon = -L\, dI/dt$ for this circuit, and the possibility that the $dI/dt$ itself may be positive or negative, there can be confusion in applying Kirchhoff's rule and getting the correct sign for the potential difference across an inductor. The best approach is to assume a direction for $I$, assume $dI/dt$ to be positive, and then write $L\, dI/dt$ for the potential difference across the inductor, obtaining the proper sign from Lenz's law. In Figure 28-21 there is a potential drop across the resistor from $a$ to $b$ of magnitude $IR$. If $dI/dt$ is positive, the induced emf in the inductor is so directed from $c$ to $b$ as to decrease the current. Point $b$ is therefore like the positive terminal of a battery. From $b$ to $c$ the potential drops by the amount $L\, dI/dt$. As we move from point $c$ to point $a$, the potential increases by the amount of the emf of the battery, which is $\varepsilon_0$. Kirchhoff's first rule applied to this circuit therefore gives

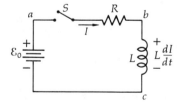

**Figure 28-21**
Typical *LR* circuit.

$$\mathcal{E}_0 = IR + L \frac{dI}{dt} \qquad\qquad 28\text{-}26$$

This equation is similar to Equation 25-26, for an *RC* circuit. Again, we can understand the general behavior of the current *I* versus time without going into the mathematical details of the solution of the equation.

Just after the switch is closed, at $t = 0$, the current is zero and the rate of change of the current is, from Equation 28-26,

$$\left(\frac{dI}{dt}\right)_0 = \frac{\mathcal{E}_0}{L} \qquad\qquad 28\text{-}27$$

The current thus increases as expected. After a short time, the current has reached some positive value, and the rate of change is given by

$$\frac{dI}{dt} = \frac{\mathcal{E}_0}{L} - \frac{IR}{L}$$

At this time the current is still increasing, but its rate of increase is less than at $t = 0$. The final value of the current can be obtained by setting $dI/dt$ equal to zero. From Equation 28-26 we see that the final value of the current is

$$I_f = \frac{\mathcal{E}_0}{R} \qquad \text{when} \quad \frac{dI}{dt} = 0 \qquad\qquad 28\text{-}28$$

Figure 28-22 shows *I* versus *t*. The time for the current to reach an appreciable fraction of its final value depends on the resistance and inductance. The mathematical expression for the current as a function of time obtained by solving Equation 28-26 is

$$I = \frac{\mathcal{E}_0}{R}\left(1 - e^{-Rt/L}\right) = I_f(1 - e^{-t/t_c}) \qquad\qquad 28\text{-}29$$

where

$$t_c = \frac{L}{R} \qquad\qquad 28\text{-}30$$

is the *time constant* of the circuit. The larger the self-inductance *L* or the smaller the resistance *R*, the longer it takes for the current to build up.

We note that the product of the initial slope $\mathcal{E}_0/L$ and the time constant $t_c = L/R$ equals the final current $\mathcal{E}_0/R$:

$$\left(\frac{dI}{dt}\right)_0 t_c = \frac{\mathcal{E}_0}{L}\frac{L}{R} = \frac{\mathcal{E}_0}{R} = I_f$$

If the current continued to increase at its initial rate, it would reach its final value after one time constant.

Figure 28-23 shows a slightly different arrangement, with an additional switch that allows us to remove the battery and an additional resistor *r* to protect the battery so that it is not shorted when both switches are momentarily closed. If both switches are open and we close switch $S_1$, the current builds up in the circuit as discussed above, except that the total resistance is now $r + R$ and the final current is $\mathcal{E}/(r + R)$. Suppose that this switch has been closed for a time long compared with the time constant, which is now $L/(R + r)$, so that the current is approximately steady at its final value, which we shall call $I_0$.

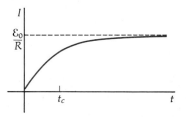

**Figure 28-22**
Graph of current *I* versus time *t* for the circuit in Figure 28-21, in which the switch is closed at $t = 0$. The current increases to its maximum value $\mathcal{E}_0/R$, reaching 63 percent of this value after time $t_c = L/R$.

*Time constant*

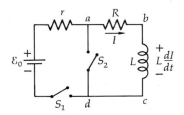

**Figure 28-23**
*LR* circuit with two switches so that the battery can be removed from the circuit. After the current in the inductor reaches its maximum value with $S_1$ closed, $S_2$ is closed and $S_1$ opened. The current then decreases in time as shown in Figure 28-24.

Switch $S_2$ is then closed and switch $S_1$ is opened, to remove the battery from consideration completely. Let us choose the time $t = 0$ when switch $S_2$ is closed. We now have a circuit with just a resistor and an inductor (loop *abcd*) with an initial current $I_0$. Again we analyze this circuit by applying Kirchhoff's rule. The potential drop from point *a* to *b* across the resistors is $IR$. Across the inductor, the potential drop from *b* to *c* is $L \, dI/dt$. (Since $dI/dt$ will turn out to be negative, this is really a potential increase. Less confusion with signs arises if we always assume that the quantities $I$ and $dI/dt$ are positive when writing the equations. The solutions of the equations then give us the correct signs of the quantities.) We are now back at our original point. The sum of the potential drops must be zero. Thus

$$IR + L\frac{dI}{dt} = 0 \qquad\qquad 28\text{-}31$$

Since the current $I$ is positive, the rate of change $dI/dt$ is negative:

$$\frac{dI}{dt} = -\frac{R}{L}I \qquad\qquad 28\text{-}32$$

The induced emf is in the direction of the current and opposite to that indicated in Figure 28-23. In agreement with Lenz's law, the induced emf is in the direction that maintains the current and tends to prevent its decrease. Equation 28-32 is similar to Equation 25-21 for the discharge of a capacitor, with $I$ replacing $Q$ and $L/R$ replacing the time constant $RC$.

Figure 28-24 shows a sketch of the current $I$ versus time $t$. The original value of the current is $I_0$, and the original value of the slope is $dI/dt = -RI_0/L$. As the current decreases, the slope decreases in magnitude until both the current and its slope are zero. The mathematical expression for the current is

$$I = I_0 e^{-Rt/L} = I_0 e^{-t/t_c} \qquad\qquad 28\text{-}33$$

where $t_c = L/R$ is the time constant.

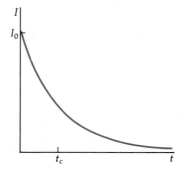

**Figure 28-24**
Current versus time after switch $S_2$ is closed in Figure 28-23. The current decreases as $e^{-t/t_c}$, where $t_c = L/R$ is the time constant.

## 28-8  Magnetic Energy

When a current is set up in a circuit like the one in Figure 28-21, only part of the energy supplied by the battery goes into Joule heat in the resistor; the rest of the energy is stored in the inductor. We can see this by multiplying each term in Equation 28-26 by the current. We have then

$$\mathcal{E}_0 I = I^2 R + LI\frac{dI}{dt} \qquad\qquad 28\text{-}34$$

The term on the left side of this equation represents the rate at which energy is supplied by the seat of emf. The first term on the right, $I^2R$, is the rate at which energy is put into the resistor. This energy appears as heat. The second term, $LI \, dI/dt$, is the rate at which energy is put into the inductor and is the term we wish to discuss. Let $U_m$ be the energy in the inductor. According to Equation 28-34, the rate of increase of this energy is

$$\frac{dU_m}{dt} = LI\frac{dI}{dt} \qquad\qquad 28\text{-}35$$

We can find the total energy in the inductor by integrating from the time $t = 0$, when the current is zero, to the time $t = \infty$, when the current has reached its final value $I_f$:

$$U_m = \int dU_m = \int LI \, dI = \tfrac{1}{2}LI_f^2 \qquad\qquad 28\text{-}36$$

*Energy in an inductor*

This energy is stored in the self-inductance of the circuit, as we can see by considering what happens when the emf is removed from the circuit, as in Figure 28-23 with $S_2$ closed and $S_1$ open. Though there is now no power input into the circuit by the emf, the current does not immediately fall to zero, so there is still some power put into the resistor. The rate at which energy is put into the resistor in this case equals the rate at which the energy in the inductor is decreased. We can see this by multiplying each term in Equation 28-31 by the current $I$:

$$I^2 R = -LI \frac{dI}{dt} \qquad\qquad 28\text{-}37$$

Whenever a current is set up in a circuit, a magnetic field is set up. We can consider the energy stored in a circuit by virtue of its self-inductance and current as energy stored in the magnetic field. When the current is decreased, decreasing the energy $\tfrac{1}{2}LI^2$ in the inductor, the magnetic field is also decreased. The idea that energy is stored in a magnetic field is similar to the idea that energy is stored in an electric field when a capacitor is charged. In Chapter 23 we showed that the electrostatic energy stored in a parallel-plate capacitor can be written

$$U_e = \tfrac{1}{2}QV = \tfrac{1}{2}\epsilon_0 E^2 A d$$

where the area of the plates $A$ times their separation $d$ is the volume in which there is an electrostatic field $E$. We then stated that this relation is more general than indicated by that example; namely, that whenever an electrostatic field is set up in some volume, there is energy stored $\tfrac{1}{2}\epsilon_0 E^2$ per unit volume. We now give an analogous argument for magnetic energy. Again, for simplicity, we consider a special case, that of a solenoid.

Let $n$ be the number of turns per unit length of a solenoid and $A\ell$ be the volume, where $A$ is the cross-sectional area and $\ell$ the length. The magnetic field inside the solenoid is

$$B = \mu_0 n I$$

and the self-inductance is

$$L = \mu_0 n^2 \ell A$$

Substituting this expression for $L$ and $I = B/\mu_0 n$ for the current into Equation 28-36 for the energy of an inductor, we have

$$U_m = \tfrac{1}{2}LI^2 = \tfrac{1}{2}\mu_0 n^2 \ell A \left(\frac{B}{\mu_0 n}\right)^2 = \frac{B^2}{2\mu_0}\ell A$$

The energy stored in a solenoid can be written as the product of the volume $\ell A$ and the term $\tfrac{1}{2}B^2/\mu_0$. Again, the result is more general than is indicated by this discussion. Whenever a magnetic field is set up in some volume $\mathcal{V}$ in space, energy is stored. The energy per unit volume $\eta_m$ is

$$\eta_m = \frac{U_m}{\mathcal{V}} = \frac{B^2}{2\mu_0} \qquad\qquad 28\text{-}38$$

*Magnetic energy density*

We see that there is energy stored in both the electric and magnetic fields. In general, if there is an electric field $E$ and magnetic field $B$ in some volume of space, the energy per unit volume associated with these fields is

$$\eta = \frac{U}{\mathcal{V}} = \tfrac{1}{2}\epsilon_0 E^2 + \frac{B^2}{2\mu_0} \qquad\qquad 28\text{-}39$$

## 28-9  *LC* and *LCR* Circuits

Figure 28-25 shows a capacitor connected to an inductor and a switch. Assume that the capacitor carries an initial charge $Q_0$ and the switch is initially open. At $t = 0$ we close the switch, and the charge flows through the inductor. For simplicity, we shall neglect any resistance in the circuit.

We have chosen the direction for the current in the circuit so that when the charge on the bottom capacitor plate is $+Q$, the current is

$$I = \frac{dQ}{dt}$$

This choice of direction is purely arbitrary. With this choice, the current will be negative just after the switch is closed. With this choice of positive sense for the current, the potential drop across the inductor from point $a$ to $b$ is $L\, dI/dt$. Across the capacitor from $c$ to $d$ there is a potential drop $Q/C$. Thus Kirchhoff's rule for this circuit gives

$$L\frac{dI}{dt} + \frac{Q}{C} = 0 \qquad\qquad 28\text{-}40$$

Substituting $dQ/dt$ for $I$ in this equation, we obtain for the charge $Q$ in the capacitor

$$L\frac{d^2Q}{dt^2} + \frac{Q}{C} = 0$$

or

$$\frac{d^2Q}{dt^2} = -\frac{1}{LC}Q \qquad\qquad 28\text{-}41$$

Equation 28-41 is of the same form as the equation for the acceleration of a mass on a spring,

$$\frac{d^2x}{dt^2} = -\omega^2 x$$

where $\omega^2 = k/m$, $k$ is the spring constant, and $m$ is the mass.

In Chapter 11 we studied the solutions of this equation for simple harmonic motion and found that we could always write the solution in the form

$$x = A\cos(\omega t + \delta)$$

where $\omega = \sqrt{k/m}$ is the angular frequency, $A$ is the maximum value of $x$ (called the amplitude), and $\delta$ is the phase constant, which depends on the initial conditions. We can put Equation 28-41 in the same form by writing $\omega^2$ for the quantity $1/LC$. Then

$$\frac{d^2Q}{dt^2} = -\omega^2 Q \qquad\qquad 28\text{-}42$$

**Figure 28-25**
*LC* circuit. The capacitor is initially charged. The charge on the plates and the current $I$ are related by $I = dQ/dt$ for the choice of current direction indicated. Just after the switch is closed, $I$ is negative.

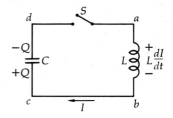

$$\omega = \frac{1}{\sqrt{LC}} \qquad\qquad 28\text{-}43 \qquad \textit{Frequency of an LC circuit}$$

The solution of Equation 28-42 is

$$Q = A \cos (\omega t + \delta) \qquad\qquad 28\text{-}44$$

The current is obtained by differentiating this solution:

$$I = \frac{dQ}{dt} = -\omega A \sin (\omega t + \delta) \qquad\qquad 28\text{-}45$$

For our initial conditions, the phase constant $\delta$ must be zero because the current is zero at time $t = 0$. Setting $t = 0$ in Equation 28-45, we have

$$I_0 = -\omega A \sin \delta = 0$$

and hence $\delta = 0$. Then our solution for the charge is

$$Q = A \cos \omega t$$

The constant $A$ is just the value of the charge at $t = 0$ since $\cos 0 = 1$. Writing $Q_0$ for this initial charge, we have

$$Q = Q_0 \cos \omega t \qquad\qquad 28\text{-}46$$

and

$$I = -\omega Q_0 \sin \omega t \qquad\qquad 28\text{-}47$$

Figure 28-26 shows graphs of $Q$ and $I$ versus time. The charge oscillates between the values $+Q_0$ and $-Q_0$ with the angular frequency $\omega = 1/\sqrt{LC}$. The current also oscillates with this frequency and is 90° out of phase with the charge. The current is maximum when the charge is zero and zero when the charge is maximum.

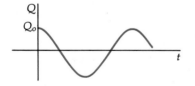 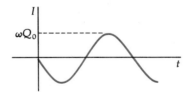

**Figure 28-26**
Graphs of $Q$ versus $t$ and $I$ versus $t$ for the $LC$ circuit in Figure 28-25.

In our study of the oscillation of a mass on a spring we found that the total energy is constant but oscillates between potential and kinetic energy. In our $LC$ circuit we also have two kinds of energy, electrostatic energy and magnetic energy. The electrostatic energy of the capacitor is

$$U_e = \tfrac{1}{2}QV = \frac{Q^2}{2C}$$

where $V = Q/C$ is the voltage drop across the capacitor. Substituting $Q_0 \cos \omega t$ for $Q$ in this equation, we have for the electrostatic energy of the capacitor

$$U_e = \frac{Q_0^2}{2C} \cos^2 \omega t \qquad\qquad 28\text{-}48$$

The electrostatic energy stored in the capacitor oscillates between its maximum value $Q_0^2/2C$ and zero. When the current in the circuit is $I$, the magnetic energy stored in the inductor is

$$U_m = \tfrac{1}{2}LI^2 \qquad\qquad 28\text{-}49$$

Substituting the value for the current from Equation 28-47 into this expression, we get

$$U_m = \tfrac{1}{2}L\omega^2 Q_0^2 \sin^2 \omega t = \frac{Q_0^2}{2C} \sin^2 \omega t \qquad \text{28-50}$$

where we have used the fact that $\omega^2$ is equal to $1/LC$. The magnetic energy also oscillates between its maximum value of $Q_0^2/2C$ and zero. The sum of the electrostatic and magnetic energies is the total energy, which is constant in time:

$$U_{\text{total}} = U_e + U_m = \frac{Q_0^2}{2C} \cos^2 \omega t + \frac{Q_0^2}{2C} \sin^2 \omega t = \frac{Q_0^2}{2C}$$

---

**Example 28-3** A 2-$\mu$F capacitor is initially charged to 20 V and then shorted across a 6-$\mu$H inductor. What are the frequency of oscillation and the maximum value of the current?

The frequency of oscillation is independent of the initial charge and depends only on the values of the capacitance and inductance. The frequency is

$$f = \frac{\omega}{2\pi} = \frac{1}{2\pi}\sqrt{\frac{1}{LC}} = \frac{1}{2\pi\sqrt{LC}}$$

$$= \frac{1}{2\pi\sqrt{(6 \times 10^{-6})(2 \times 10^{-6})}} = 4.59 \times 10^4 \text{ Hz}$$

According to Equation 28-47, the maximum value of the current is related to the maximum value of the charge by

$$I_m = \omega Q_0 = \frac{Q_0}{\sqrt{LC}}$$

The initial charge on the capacitor is

$$Q_0 = CV_0 = (2 \ \mu\text{F})(20 \text{ V}) = 40 \ \mu\text{C}$$

Thus

$$I_m = \frac{40 \ \mu\text{C}}{\sqrt{(6 \ \mu\text{H})(2 \ \mu\text{F})}} = 11.5 \text{ A}$$

---

In Figure 28-27 we include a resistor in series with the capacitor and inductor. Again we assume that the switch is initially open, with the capacitor carrying an initial charge $Q_0$, and we close the switch at $t = 0$. We need only modify Equation 28-40 by including the potential drop $IR$ across the resistor. We then have from Kirchhoff's rule

$$L\frac{dI}{dt} + \frac{Q}{C} + IR = 0$$

or

$$L\frac{d^2Q}{dt^2} + \frac{Q}{C} + R\frac{dQ}{dt} = 0 \qquad \text{28-51}$$

using $I = dQ/dt$ as before. Equation 28-51 is analogous to Equation 11-39 for a damped harmonic oscillator. If the resistance is small, the charge and current still oscillate with very nearly the same frequency $1/\sqrt{LC}$ but the oscillations are damped; i.e., the maximum values of the charge and current decrease with each oscillation. We can under-

**Figure 28-27**
LCR circuit.

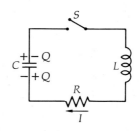

stand this qualitatively from energy considerations. If we multiply each term in Equation 28-51 by the current $I$, we have

$$IL\frac{dI}{dt} + I\frac{Q}{C} + I^2R = 0 \qquad\qquad 28\text{-}52$$

The first term in this equation is the current times the voltage across the inductor. This is the rate at which energy is put into the inductor or taken out of it, i.e., the rate of change of magnetic energy, which is positive or negative depending on whether $I$ and $dI/dt$ have the same or different signs. Similarly, the second term is the current times the voltage across the capacitor. This is the rate of change of the energy of the capacitor. Again, it may be positive or negative. The last term, $I^2R$, the rate at which energy is dissipated in the resistor as Joule heat, is always positive regardless of the sign of the current since it depends only on $I^2$. The sum of the electric and magnetic energies is not constant for this circuit because energy is continually dissipated in the resistor. Figure 28-28 shows graphs of $Q$ versus $t$ and $I$ versus $t$ for small resistance. If we increase $R$, the oscillations are more heavily damped, until a critical value of $R$ is reached for which there is not even one oscillation. Figure

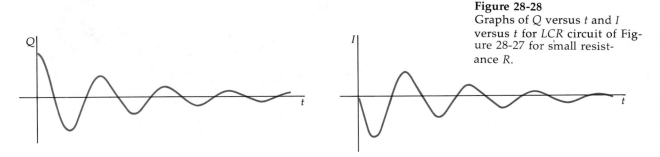

**Figure 28-28**
Graphs of $Q$ versus $t$ and $I$ versus $t$ for *LCR* circuit of Figure 28-27 for small resistance $R$.

28-29 shows the charge versus $t$ for $R$ greater than the critical damping value. The detailed solution of Equation 28-51 is exactly analogous to the solution of Equation 11-39 for the damped harmonic oscillator and will not be discussed here.

**Figure 28-29**
Graph of $Q$ versus $t$ for very large resistance $R$ in *LCR* circuit of Figure 28-27.

### Questions

8. It is not difficult to produce *LC* circuits with frequencies of oscillation of thousands of hertz or more, but it is difficult to produce *LC* circuits with small frequencies. Why?

9. How will the frequency of an *LCR* circuit be altered if the inductance in the circuit is doubled? The capacitance doubled? The voltage with which the capacitor is initially charged is doubled? The resistance in the circuit is doubled?

# Electric Motors

Reuben E. Alley, Jr.
*U.S. Naval Academy*

In Thomas Edison's machine shop the various tools were driven from a single steam engine by means of shafts, pulleys, and belts, as shown in Figure 1. Such a power-distribution system was typical of most industrial establishments around 1900. The installation of electric-power distribution facilities and the development of practical electric motors made it possible to replace the complicated and hazardous system of belts and pulleys with individual energy sources for each machine. The growth of a nationwide electric-power network and the concomitant invention of several types of electric motors resulted in a revolution in our society as profound as the electronics revolution associated with the invention and application of transistors and integrated circuits.

**Figure 1**
Interior of Thomas Edison's second machine shop, Menlo Park, New Jersey, as it appeared during a New Year's Eve lighting demonstration, 1879. (*Collections of Greenfield Village and the Henry Ford Museum, Dearborn, Michigan.*)

A single electric motor may be supplied from a battery or from a small generator, but a nationwide power network is required to supply the millions of joules of electric energy that are converted into mechanical energy in such varied applications as steel rolling mills, paper mills, subway trains, electric locomotives, automobile assembly lines, and mining machinery. Although automobiles, ships, diesel locomotives, trucks, and airplanes do not depend upon the power network, they all use electric motors. In the home, electric motors are essential parts of furnaces, air conditioners, refrigerators, and washing machines. Imagine what your everyday life would be like if there were no electric motors.

Figure 2 shows a simple *dc motor*. Current from the battery magnetizes the soft-iron *armature*, which is free to rotate about axis $AA'$ and turns to align itself with the *field* produced by the poles labeled N and S. As the armature rotates, it carries with it the *commutator*, whose two segments provide means for reversing current direction just as the armature reaches its equilibrium position. The mass of the armature ensures that it will rotate past the equilibrium position; then, because of its reversed polarity, there is a further rotation of a half revolution. Because the commutator reverses the current direction every 180°, continuous rotation is achieved. Useful work can be done by the shaft of the motor.

The motor shown in Figure 2 has several disadvantages: (1) when power is disconnected, the motor tends to stop in the equilibrium position, and so there may be no starting torque, and (2) the torque is zero twice during each revolution. If the armature is provided with additional poles and windings, and if the commutator is divided into more segments, the torque is more nearly uniform and the motor always starts.

If the field is supplied from a permanent magnet, as in model trains and most battery-powered toys, the speed depends upon supply voltage and changes with load. For applications requiring appreciable amounts of power, electromagnets supply the field. Adjustment of field current permits control of speed independent of armature current. In general, if the load on a dc motor increases, speed decreases and more armature current is required. Field current can be adjusted to maintain constant speed. Automatic controls are frequently used with dc motors.

DC motors with field windings are very flexible. They may be connected in *series,* so that the armature current is also the field current; in *parallel* (or *shunt*), so that the field current is independent of armature current; or in a *compound* arrangement using two field windings, so that one is connected in series and the other is connected in parallel. DC motors are widely used for traction (subways and electric railways) and in applications where speed control is critical, e.g., steel rolling mills.

The dc motor in Figure 2 becomes a *synchronous motor* if the commutator is replaced by *slip rings* (Figure 3) and alternating current is supplied. Although this simple device has no starting torque, it will "lock in" if it is accelerated by external means to *synchronous speed* (determined by power-line frequency). As the load is increased, armature current increases but the speed stays constant unless the load is great enough to make the machine stall. Synchronous motors designed for use with three-phase power (see below) can be self-starting. When the horsepower rating is high, field current is much less than armature current. In order that the smaller current can be supplied through slip rings, most such machines have a stationary armature while the field rotates on the shaft that drives the load.

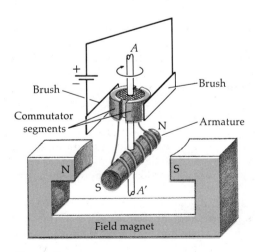

**Figure 2**
A simple dc motor. The field magnet may be a permanent magnet or a soft-iron core wrapped with many turns of wire, through which a direct current flows.

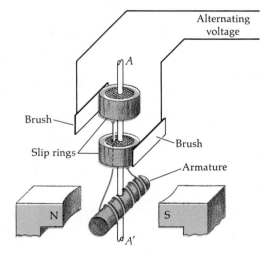

**Figure 3**
A simple synchronous motor. The field magnet is like that in Figure 2. Only the pole pieces are shown here.

The *universal series motor* has many low-power applications such as hand drills and small saws. A simple form of such a motor results if the current, before entering the commutator (Figure 2), first goes through the field winding. This motor will operate on either alternating or direct current. No-load speed depends upon the number of field poles, the number of armature poles, and the supply voltage. Many hand drills now are equipped with solid-state circuits that control speed by controlling applied voltage. These machines slow down under load, but they develop reasonably high torque and so are especially satisfactory for driving hand tools.

Practically all commercially generated electric power in the United States is three-phase, 60-Hz sinusoidal voltage. The term three-phase means that there are three sinusoidal voltages of equal amplitude and frequency whose successive peak values are separated in time by one-third of a cycle (see Figure 4). These three voltages are usually carried in the three-wire transmission lines that are a common sight around the countryside. One phase of the power appears between each pair of wires. Residences, almost without exception, are supplied with only a single phase, but all three phases are commonly supplied to industries.

**Figure 4**
Voltages in a three-phase power system. At any instant the sum of the three voltages is zero.

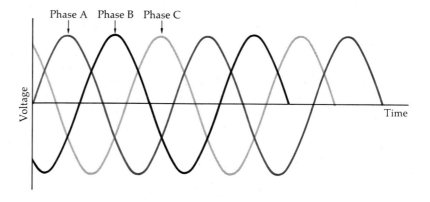

Figure 5 shows a simple three-phase *induction motor*. The winding of each of the three pairs of poles is connected to a different phase. The currents in the windings therefore reach their maximum values successively, and the result is a rotating magnetic field in the region between the poles. A stationary rotor of conducting material in this field will have currents set up in it. The magnetic fields of the *induced currents* interact with the rotating field to produce a torque. If there is no load on the motor, the torque accelerates the rotor to a speed almost equal to the synchronous speed of rotation of the field. (If the rotor actually attained synchronous speed, there would be no relative motion between field and rotor and so no induced currents or torque.) As the motor is loaded, rotor speed decreases; this results in larger induced currents and correspondingly greater torques. Speed may decrease to 75 percent of synchronous speed before the motor stalls. There are no moving electrical contacts in the three-phase induction motor, which can be built to develop hundreds of horsepower. The elimination of any possibility of electric sparks makes these motors especially attractive for applications in explosive atmospheres, e.g., mines or flour mills.

A practical induction motor usually has a cylindrical iron rotor with insulated copper bars embedded in its surface and interconnected to form good conducting paths for induced currents. In some motors the rotor windings, instead of being short-circuited, are connected to slip rings so that external resistors can be used for speed control, but the resulting moving electrical contacts remove one of the important advantages of induction motors.

If only one phase is connected in Figure 5, there is no rotating magnetic field and no starting torque. If a second winding in Figure 5 is connected in series with a large (hundreds of microfarads) capacitor and supplied from the same phase as the first winding, the currents in the two windings are suf-

**Figure 5**
Schematic cross section of a simple three-phase induction motor.

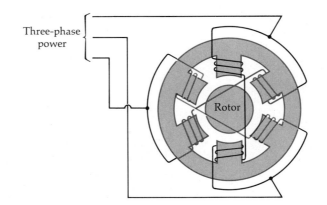

Three-phase power

Rotor

ficiently out of phase to provide a rotating field and therefore a starting torque. This fact is employed in the design of *single-phase* induction motors, used to drive refrigerators, washing machines, air conditioners, and furnaces.

Since 1965, especially in Great Britain, Germany, and Japan, serious efforts have been made to develop a high-speed mass-transit system based upon the *linear induction motor*. It is possible to achieve both acceleration and levitation (once speeds of about 30 km/h are reached) by electrical means. Such support for the vehicle eliminates most of the friction losses associated with the rolling of flanged wheels on steel rails.

Suppose the cylindrical shell that supports the poles of an induction motor is cut along an axial line and flattened (just as a tin can may be cut and unrolled). Then the poles will lie in a line and the magnetic field will move across the poles periodically. Now imagine a long strip of magnetic material with many poles equally spaced along its length. Let the first pole and every third succeeding pole be supplied from one phase of the power line; let the second pole and every third succeeding pole be supplied from the second phase; and let the third pole and every third succeeding pole be supplied from the third phase. The result will be a magnetic field that travels along the strip with a speed relative to the ground that depends upon the distance between poles and the frequency of the power. Now let a mass of conducting material be supported (by wheels running on rails, for example) so that it is free to move in the direction of the strip. When the mass is stationary, currents will be induced whose magnetic fields will interact with the moving field to produce an accelerating force along the direction of the strip (Figure 6). The mass will acquire a speed that is somewhat less than synchronous speed, just like the rotor of the induction motor. What has been described is a simple linear induction motor. Such a device was proposed in 1946 as a means of catapulting aircraft from a ship.

**Figure 6**
Schematic diagram of a simple linear induction motor.

Conducting material on wheels

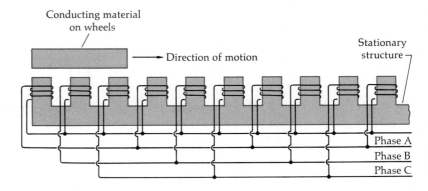

→ Direction of motion

Stationary structure

Phase A
Phase B
Phase C

An alternative design of the linear induction motor involves mounting a number of poles on the vehicle and supplying them through flexible collectors in contact with a stationary three-phase distribution system. The interaction of the fields of the poles in the vehicle with currents induced in a stationary metal strip (aluminum backed by a soft magnetic material to provide a path for magnetic flux) produces support and accelerating force for the vehicle (Figure 7).

**Figure 7**
Linear induction motor with moving poles.

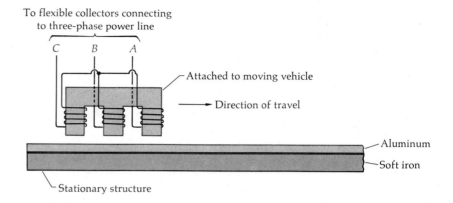

In the United States some engineers are investigating the use of a single-phase *linear synchronous motor*. Speed variation is achieved through change in frequency of the alternating voltage supplied to the vehicle. The required steady magnetic field is stationary. One proposal is to operate the field windings at superconducting temperatures, thus greatly reducing the winding resistance and thereby reducing the power loss from the direct current supplying these windings.

Although no system has yet reached the commercial stage, some prototypes have been tested. There seems to be ample evidence that it will be possible to design and build a practical transportation system that will operate at speeds of 300 km/h or more based upon the concept of a linear motor.

## Review

A. Define, explain, or otherwise identify:

Faraday's law, 767
Motional emf, 767
Lenz's law, 769
Self-induced emf, 772
Flip coil, 772
Ballistic galvanometer, 772

Eddy currents, 772
Betatron, 773
Self-inductance, 775
Mutual inductance, 775
Inductor, 778
Magnetic energy, 781

B. True or false:

1. The induced emf in a circuit is proportional to the magnetic flux through the circuit.

2. There can be an induced emf at an instant when the flux through the circuit is zero.

3. The induced emf in a circuit always decreases the magnetic flux through the circuit.

4. Faraday's law can be derived from the Biot-Savart law.

5. The energy in a magnetic field is proportional to $B^2$.

## Exercises

### Section 28-1, Motional EMF

1. A rod 30 cm long moves at 8 m/s in a plane perpendicular to a magnetic field of 500 G. Its velocity is perpendicular to the length of the rod. Find (*a*) the mag-

netic force on an electron in the rod, (b) the electrostatic field **E** in the rod, and (c) the potential difference $V$ between the ends of the rod.

2. Find the speed of the rod in Exercise 1 if the potential difference between its ends is 6 V.

3. In Figure 28-5 let $B$ be 0.8 T, $v = 10.0$ m/s, $\ell = 20$ cm, and $R = 2 \, \Omega$. Find (a) the induced emf in the circuit, (b) the current in the circuit, and (c) the force needed to move the rod with constant velocity assuming negligible friction. (d) Find the power input by the force found in part (c) and the rate of heat production $I^2R$.

4. Work Exercise 3 for $B = 1.5$ T, $v = 6$ m/s, $\ell = 40$ cm, and $R = 1.2 \, \Omega$.

### Sections 28-2, Lenz's Law, and Section 28-3, Applications of Faraday's Law

5. The two loops in Figure 28-30 have their planes parallel to each other. As viewed from $A$ toward $B$, there is a counterclockwise current in $A$. Give the direction of the current in loop $B$ and state whether the loops attract or repel each other if the current in loop $A$ is (a) increasing and (b) decreasing.

6. A bar magnet moves with constant velocity along the axis of a loop, as shown in Figure 28-31. (a) Make a qualitative sketch of the flux $\phi_m$ through the loop as a function of time $t$. Indicate the time $t_1$ when the magnet is halfway through the loop. (b) Sketch the current $I$ in the loop versus time, choosing $I$ positive when it is counterclockwise as viewed from the left.

7. Give the direction of the induced current in the circuit on the right in Figure 28-32 when the resistance in the circuit on the left is suddenly (a) increased; (b) decreased.

8. A uniform magnetic field **B** is established perpendicular to the plane of a loop of radius 5.0 cm, resistance 0.4 $\Omega$, and negligible self-inductance. The magnitude of **B** is increasing at a rate of 40 mT/s. Find (a) the induced emf in the loop, (b) the induced current in the loop, and (c) the rate of Joule heating in the loop.

9. A 100-turn coil has radius 4.0 cm and resistance 25 $\Omega$. At what rate must a perpendicular magnetic field change to produce a current of 4.0 A in the coil?

10. Show that if the flux through each turn of an $N$-turn coil of resistance $R$ changes from $\phi_{m1}$ to $\phi_{m2}$ in any manner, the total charge that passes through the coil is given by $Q = N(\phi_{m2} - \phi_{m1})/R$.

11. The flux through a loop is given by $\phi_m = (t^2 - 4t) \times 10^{-1}$ T·m², where $t$ is in seconds. (a) Find the induced emf $\varepsilon$ as a function of time. (b) Find both $\phi_m$ and $\varepsilon$ at $t = 0$, $t = 2$ s, $t = 4$ s, and $t = 6$ s.

12. For the flux given in Exercise 11, sketch $\phi_m$ versus $t$ and $\varepsilon$ versus $t$. (a) At what time is the flux maximum? What is the emf at this time? (b) At what times is the flux zero? What is the emf at these times?

13. A 100-turn circular coil has a diameter of 2.0 cm and resistance of 50 $\Omega$. The plane of the coil is perpendicular to a uniform magnetic field of magnitude 1.0 T. The field is suddenly reversed in direction. (a) Find the total charge passing through the coil. If the reversal takes 0.1 s, find (b) the average current in the circuit and (c) the average emf in the circuit.

14. A 1000-turn coil of cross-sectional area 300 cm² and resistance 15.0 $\Omega$ is aligned with its plane perpendicular to the earth's magnetic field of 0.31 G (at the equator). If the coil is flipped over, how much charge flows through it?

15. A circular coil of 300 turns and radius 5.0 cm is connected to a ballistic galvanometer. The total resistance of the circuit is 20 $\Omega$. The plane of the coil is originally aligned perpendicular to the earth's magnetic field at some point. When the coil is rotated through 90°, the charge passing through the galvanometer is measured to be 9.4 $\mu$C. Calculate the magnitude of the earth's magnetic field at that point.

**Figure 28-30**
Exercise 5.

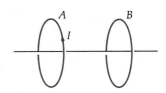

**Figure 28-31**
Exercise 6.

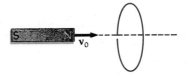

**Figure 28-32**
Exercise 7.

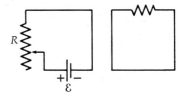

### Section 28-4, Eddy Currents

*There are no exercises for this section.*


### Section 28-5, The Betatron

**16.** Consider a plane perpendicular to a magnetic field **B** which is uniform over a circular region of radius $R$ and is essentially zero outside the circle $(r > R)$. Show that if the magnetic field is changing at the rate $dB/dt$, the induced electric field at distance $r$ from the center of the circle is tangent to the circle of radius $r$ and has the magnitude

$$E = \begin{cases} \dfrac{r}{2}\dfrac{dB}{dt} & \text{for } r < R \\[2mm] \dfrac{R^2}{2r}\dfrac{dB}{dt} & \text{for } r > R \end{cases}$$


### Section 28-6, Inductance

**17.** A coil with a self-inductance of 8.0 H carries a current of 3 A, which is changing at a rate of 200 A/s. Find (*a*) the magnetic flux through the coil and (*b*) the induced emf in the coil.

**18.** A coil with self-inductance $L$ carries a current $I$, given by $I = I_0 \sin 2\pi ft$. Find and graph the flux $\phi_m$ and the self-induced emf as functions of time.

**19.** A solenoid has length 25 cm, radius 1 cm, and 400 turns, and carries a 3-A current. Find (*a*) $B$ on the axis at the center, (*b*) the flux through the solenoid, assuming $B$ to be uniform, (*c*) the self-inductance of the solenoid, and (*d*) the induced emf in the solenoid when the current changes at 150 A/s.

**20.** A circular coil of $N_1$ turns and area $A_1$ encircles a long tightly wound solenoid of $N_2$ turns, area $A_2$, and length $\ell_2$, as shown in Figure 28-33. The solenoid carries a current $I_2$. (*a*) Show that the flux through the circular coil due to the current in the solenoid is given by $\phi_{m1} = \mu_0 N_1 (N_2/\ell_2) A_2 I_2$. (*b*) Explain why the flux depends on the area $A_2$ of the solenoid and not on that of the coil. (*c*) Find the mutual inductance of the solenoid and coil. (*d*) Explain why it is much more difficult to calculate the mutual inductance by assuming current $I_1$ in the coil and then finding the flux through the solenoid due to $I_1$.

**21.** A very small circular coil of $N_1$ turns and area $A_1$ is completely inside a long tightly wound solenoid with its plane perpendicular to the axis of the solenoid. The solenoid has $N_2$ turns, area $A_2$, and length $\ell_2$ and carries current $I_2$. (*a*) Show that the flux through the circular coil due to the current in the solenoid is given by $\phi_{m1} = \mu_0 N_1 (N_2/\ell_2) A_1 I_2$. (*b*) Find the mutual inductance of the coil and solenoid. (*c*) Explain why it is much more difficult to calculate the mutual inductance by assuming current $I_1$ in the coil and then finding the flux through the solenoid due to $I_1$.

**Figure 28-33**
Exercise 20.

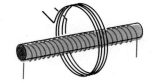


### Section 28-7, *LR* Circuits

**22.** The current in an *LR* circuit is zero at time $t = 0$ and increases to half its final value in 4.0 s. (*a*) What is the time constant of this circuit? (*b*) If the total resistance is 5 $\Omega$, what is the self-inductance?

**23.** A coil of resistance 8.0 $\Omega$ and self-inductance 4.0 H is suddenly connected across a constant potential difference of 100 V. Let $t = 0$ be the time of connection at which the current is zero. Find the current $I$ and its rate of change $dI/dt$ at times (*a*) $t = 0$, (*b*) $t = 0.1$ s, (*c*) $t = 0.5$ s, and (*d*) $t = 1.0$ s.

**24.** How many time constants must elapse before the current in an *LR* circuit which is originally zero reaches (*a*) 90 percent, (*b*) 99 percent, and (*c*) 99.9 percent of its final value?

25. The current in a coil of self-inductance 1 mH is 2.0 A at $t = 0$, when the coil is shorted through a resistor. The total resistance of the coil plus resistor is 10.0 Ω. Find the current after (a) 0.5 ms and (b) 10 ms.

26. Compute the initial slope $dI/dt$ at $t = 0$ from Equation 28-33 and show that if the current decreased steadily at this rate, it would be zero after one time constant.

27. In Exercise 25, find the time for the current to decrease to one electron per second.

### Section 28-8, Magnetic Energy

28. A solenoid of 2000 turns, area 4 cm², and length 30 cm carries a 4.0-A current. (a) Calculate the magnetic energy stored from $\frac{1}{2}LI^2$. (b) Divide your answer in part (a) by the volume of the solenoid to find the magnetic energy per unit volume in the solenoid. (c) Find $B$ in the solenoid. (d) Compute the magnetic energy density from $\eta_m = B^2/2\mu_0$ and compare with part (b).

29. In the circuit in Figure 28-21 let $\varepsilon_0 = 12.0$ V, $R = 3.0$ Ω, and $L = 0.6$ H. The switch is closed at time $t = 0$. At time $t = 0.5$ s, find (a) the rate at which the battery supplies power, (b) the rate of Joule heating, and (c) the rate at which energy is being stored in the inductor.

30. Do Exercise 29 for the times $t = 1$ s and $t = 100$ s.

31. A coil with self-inductance 2.0 H and resistance 12.0 Ω is connected across a 24-V battery of negligible internal resistance. (a) What is the final current? (b) How much energy is stored in the inductor when the final current is attained?

32. In a plane electromagnetic wave such as a light wave, the magnitudes of the electric and magnetic fields are related by $E = cB$, where $c = 1/\sqrt{\epsilon_0\mu_0}$ is the speed of light. Show that in this case the electric and magnetic energy densities are equal.

33. Find (a) the magnetic energy, (b) the electric energy, and (c) the total energy in a volume of 1.0 m³ in which there is an electric field of $10^4$ V/m and a magnetic field of 1 T.

### Section 28-9, $LC$ and $LCR$ Circuits

34. Show from the definitions of the henry and farad that $1/\sqrt{LC}$ has units of $s^{-1}$.

35. What is the period of oscillation of an $LC$ circuit consisting of a 2-mH coil and a 20-μF capacitor?

36. What inductance is needed with an 80-μF capacitor to construct an $LC$ circuit which oscillates with frequency 60 Hz?

37. An $LC$ circuit has capacitance $C_1$ and inductance $L_1$. A second circuit has $C_2 = \frac{1}{2}C_1$ and $L_2 = 2L_1$, and a third circuit has $C_3 = 2C_1$ and $L_3 = \frac{1}{2}L_1$. (a) Show that each of these circuits oscillates with the same frequency. (b) In which circuit would the maximum current be the greatest if in each case the capacitor were charged to potential $V$?

**Figure 28-34**
Problem 1.

### Problems

1. A conducting rod of length $\ell$ rotates at constant angular velocity $\omega$ about one end in a plane perpendicular to a uniform magnetic field **B** (Figure 28-34). (a) Show that the magnetic force on a charge $q$ at distance $r$ from the pivot is $Bqr\omega$. (b) Show that the potential difference between the ends of the rod is $V = \frac{1}{2}B\omega\ell^2$. (c) Draw any radial line in the plane from which to measure $\theta = \omega t$. Show that the area of the pie-shaped region between the reference line and the rod is $A = \frac{1}{2}\ell^2\theta$. Compute the flux through this area and show that $\varepsilon = \frac{1}{2}B\omega\ell^2$ follows from Faraday's law applied to this area.

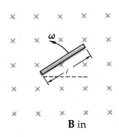

**B** in

2. A rod of length $\ell$ lies with its length perpendicular to a long wire carrying current $I$, as shown in Figure 28-35. The near end of the rod is a distance $d$ away from the wire. The rod moves with speed $v$ in the direction of the current $I$. (a) Show that the potential difference between the ends of the rod is given by

$$V = \frac{\mu_0 I}{2\pi} v \ln \frac{d + \ell}{d}$$

(b) Use Faraday's law to obtain this result by considering the flux through a rectangular area $A = \ell vt$ swept out by the rod.

3. A rectangular loop 10 by 5.0 cm with resistance 1 $\Omega$ is pulled through a region of uniform magnetic field, $B = 1.0$ T (Figure 28-36), with constant speed $v = 2.0$ cm/s. The front end of the loop enters the region of magnetic field at time $t = 0$. (a) Find and graph the flux through the loop as a function of time. (b) Find and graph the induced emf and the current in the loop as functions of time. Neglect any self-inductance of the loop and extend your graphs from $t = 0$ to $t = 16$ s.

4. Show that for two inductors $L_1$ and $L_2$ connected in series such that none of the flux of either passes through the other, the effective inductance is given by $L = L_1 + L_2$.

5. Two concentric solenoids have the same length $\ell$. The inner solenoid has $n_1$ turns per unit length and radius $r_1$, whereas the outer solenoid has $n_2$ turns per unit length and radius $r_2$ ($r_2 > r_1$). Find the mutual inductance of these solenoids.

6. Show that for two inductors $L_1$ and $L_2$ connected in parallel such that none of the flux of one passes through the other, the effective inductance is given by $1/L = 1/L_1 + 1/L_2$.

7. When the current in a certain coil is 5.0 A and is increasing at the rate of 10.0 A/s, the potential difference across the coil is 140 V. When the current is 5.0 A and decreasing at the rate of 10.0 A/s, the potential difference is 60 V. Find the resistance and self-inductance of the coil.

8. A coaxial cable consists of two very thin-walled conducting cylinders of radii $r_1$ and $r_2$ (Figure 28-37). Current $I$ goes in one direction down the inner cylinder and back up the outer cylinder. (a) Use Ampère's law to find $B$ and show that $B = 0$ except in the region between the conductors. (b) Show that the magnetic energy density in the region between the cylinders is

$$\eta_m = \frac{\mu_0 I^2}{8\pi^2 r^2}$$

(c) Find the magnetic energy in a cylindrical shell volume element of length $\ell$ and volume $d\mathcal{V} = \ell 2\pi r \, dr$ and integrate your result to show that the total magnetic energy in the volume of length $\ell$ between the cylinders is

$$U_m = \frac{\mu_0}{4\pi} I^2 \ell \ln \frac{r_2}{r_1}$$

(d) Use the result of part (c) and $U_m = \frac{1}{2} L I^2$ to show that the self-inductance per unit length is

$$\frac{L}{\ell} = \frac{\mu_0}{2\pi} \ln \frac{r_2}{r_1}$$

9. In the circuit of Figure 28-21 let $\varepsilon = 12.0$ V, $R = 3.0$ $\Omega$, and $L = 0.6$ H (as in Exercise 29). After a time $t = t_c$, find (a) the total energy supplied by the battery and (b) the energy stored in the inductor. *Hint:* Find the rates as functions of time and integrate from $t = 0$ to $t = t_c = L/R$.

10. In Figure 28-37 compute the flux through a rectangular area of sides $\ell$ and $r_2 - r_1$ between the conductors. Show that the self-inductance per unit length can be found from $\phi_m = LI$ [see part (d) of Problem 8].

**Figure 28-35**
Problem 2.

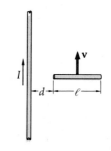

**Figure 28-36**
Problem 3.

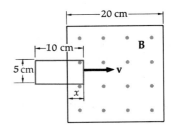

**Figure 28-37**
Problems 8 and 10.

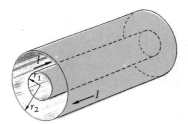

11. An ac generator rectangular loop of dimensions $a$ and $b$ has $N$ turns. The loop is connected to slip rings (Figure 28-38) and rotates with angular velocity $\omega$ in a uniform magnetic field $\mathbf{B}$. ($a$) Show that the potential difference between the two slip rings is $\varepsilon = NBab\omega \sin \omega t$. ($b$) If $a = 1.0$ cm, $b = 2.0$ cm, $N = 1000$, and $B = 2$ T, at what angular frequency $\omega$ must the coil be rotated to generate an emf whose maximum value is 110 V?

**Figure 28-38**
Problem 11.

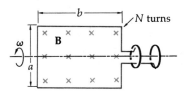

12. A conducting rod of length $\ell$, resistance $R$, and mass $m$ rides on a pair of horizontal frictionless rails of negligible resistance (Figure 28-39). A uniform magnetic field $\mathbf{B}$ is perpendicular to the plane of the rails. The rod is moving with initial speed $v_0$ when points $a$ and $b$ are connected with a conductor of negligible resistance. ($a$) Find the retarding force on the rod. ($b$) Show that the speed is given by $v = v_0 e^{-t/T}$, where $T = mR/(B\ell)^2$, assuming the connection across $ab$ is made at $t = 0$. ($c$) Find the rate of Joule heat production $I^2R$ and show that the total heat produced equals $\frac{1}{2}mv_0^2$.

**Figure 28-39**
Problems 12 and 14.

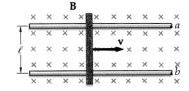

13. In the circuit in Figure 28-23 the current is $I_0 = 4.0$ A when switch $S_2$ is closed and $S_1$ opened. The inductance is 50 mH, and the resistance is 150 $\Omega$. ($a$) Find the rate of Joule heat production $I^2R$ as a function of time and ($b$) calculate the total heat produced. ($c$) Calculate the initial energy stored in the inductor and show that this equals your answer to part ($b$).

14. A battery of emf $\varepsilon$ and negligible internal resistance is connected across $ab$ in Figure 28-39, and the rod is placed at rest across the rails at $t = 0$. ($a$) Find the force on the rod as a function of speed $v$ and write Newton's second law for the rod when it has speed $v$. ($b$) Show that the rod reaches a terminal velocity and find an expression for it. ($c$) What is the current when the rod reaches its terminal velocity?

15. Show that the electric field in a charged parallel-plate capacitor cannot go abruptly to zero at the edge by evaluating $\oint \mathbf{E} \cdot d\boldsymbol{\ell}$ for the rectangular curve shown in Figure 28-40 and applying Faraday's law to this curve.

**Figure 28-40**
Problem 15.

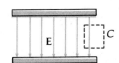

16. Show by direct substitution that Equation 28-51 is satisfied by $Q = Q_0 e^{-Rt/2L} \cos \omega't$, where $\omega' = \sqrt{(1/LC) - (R/2L)^2}$ and $Q_0$ is the charge on the capacitor at $t = 0$.

17. ($a$) Compute the current $I = dQ/dt$ from the solution of Equation 28-51 given in Problem 16 and show that

$$I = -I_0 \left( \sin \omega't + \frac{R}{2L\omega'} \cos \omega't \right) e^{-Rt/2L}$$

where $I_0 = \omega'Q_0$. ($b$) Show that this can be written

$$I = -\frac{I_0}{\cos \delta} (\cos \delta \sin \omega't + \sin \delta \cos \omega't) e^{-Rt/2L}$$

$$= -\frac{I_0}{\cos \delta} \sin (\omega't + \delta) e^{-Rt/2L}$$

where $\tan \delta = R/2L\omega'$. When $R/2L\omega'$ is small, $\cos \delta \approx 1$ and $I \approx -I_0 \sin (\omega't + \delta) e^{-Rt/2L}$.

CHAPTER 29    Magnetism in Matter

---

**Objectives**    After studying this chapter, you should:

1. Be able to list the three types of magnetic materials and discuss the origins, directions, and strengths of the magnetic effects in each.

2. Be able to state the definition of the magnetic intensity **H**.

3. Be able to state the analogous expressions of Ampère's law, Gauss' law, and Coulomb's law for the magnetic intensity **H**.

4. Be able to relate the magnetic vectors **B**, **H**, and **M** using the definitions of magnetic susceptibility and permeability.

5. Be able to derive the relation between magnetic moment and angular momentum for a charged particle moving in a circle.

6. Be able to sketch the general shape of the *B*-versus-*H* curve for a ferromagnetic material.

7. Be able to state Curie's law for paramagnetic materials.

8. Be able to describe the general temperature dependence of the magnetization in paramagnetic materials and explain its origin.

---

In studying electric fields in matter we found that the electric field is affected by the presence of electric dipoles. For polar molecules, which have a permanent electric dipole moment, the dipoles are aligned by the electric field; whereas for nonpolar molecules, electric dipoles are induced by the external field. In both cases, the dipoles are aligned parallel to the external electric field, and the alignment tends to weaken the external field.

Similar but more complicated effects occur in magnetism. Atoms have magnetic moments due to the motion of their electrons. In addition, each electron has an intrinsic magnetic moment associated with its spin. The net magnetic moment of an atom depends on the arrangement of the electrons in the atom. Unlike the situation with electric dipoles, the alignment of magnetic dipoles parallel to an external magnetic field tends to increase the field. We can see this difference by comparing the lines of **E** for an electric dipole with the lines of **B** for a magnetic dipole, e.g., a small current loop as shown in Figure 29-1. Far from the dipoles, the lines are identical, but in the region inside the dipole the lines of **B** and **E** are oppositely directed. Thus in an electrically polarized material, the dipoles create an electric field *antiparallel* to their dipole-moment vector, whereas in a magnetically polarized material, the dipoles create a magnetic field *parallel* to the magnetic-dipole-moment vectors.

We can classify materials into three categories, paramagnetic, diamagnetic, and ferromagnetic. Paramagnetic and ferromagnetic materials have molecules with permanent magnetic dipole moments. In *paramagnetism*, these moments do not interact strongly with each other and are normally randomly oriented. In the presence of an external magnetic field, the dipoles are partially aligned in the direction of the field, thereby increasing the field. However, at ordinary temperatures and ordinary external fields, only a very small fraction of the molecules are aligned because thermal motion tends to randomize their orientation. The increase in the total magnetic field is therefore very small. *Ferromagnetism* is much more complicated. Because of a strong interaction between neighboring magnetic dipoles, a high degree of alignment can be achieved even with weak external magnetic fields, thereby causing a very large increase in the total field. Even when there is no external magnetic field, ferromagnetic material may have magnetic dipoles aligned, as in permanent magnets. *Diamagnetism* is the result of an induced magnetic moment opposite in direction to the external field. The induced dipoles thus weaken the resultant magnetic field. This effect occurs in all materials but is very small and often masked by the paramagnetic or ferromagnetic effects if the individual molecules have permanent magnetic dipole moments.

In Section 27-5 we discussed the magnetic field produced by a permanent magnet, characterizing the magnet by the magnetic moment

*Three types of magnetic materials*

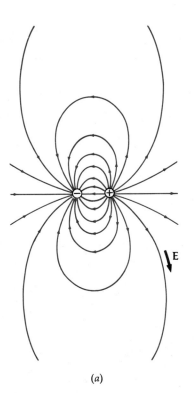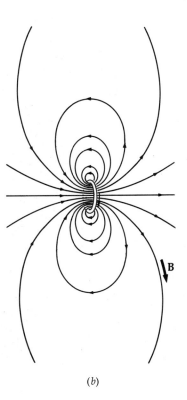

(a)    (b)

**Figure 29-1**
(a) Electric field lines of an electric dipole. (b) Magnetic field lines of a magnetic dipole. Far from the dipoles, the field patterns are identical. In the region inside the dipoles, the lines of **B** and **E** are oppositely directed.

per unit volume **M**, known as the magnetization vector. We found that the field due to the magnetized material is the same as that due to a current per unit length $M$ on the surface of the magnet, and we argued that this amperian current can be understood in terms of atomic current loops which do not completely cancel on the surface. The discussion and results of that section apply to any magnetized material, whether it is paramagnetic, diamagnetic, or ferromagnetic. We describe the magnetization of any such material by the magnetic dipole moment per unit volume **M**. Except for permanent magnets, this magnetization is caused by some external magnetic field and disappears when the external field is removed. The magnetization vector **M** is parallel to the external field that causes it for paramagnetism and ferromagnetism, and antiparallel for diamagnetism. The magnetic induction **B** due to the magnetization of the material is very small compared with the external field for paramagnetism and diamagnetism, and very large for ferromagnetism.

The calculation of the resultant magnetic-induction field **B** is complicated when magnetized material is in the presence of an external field. Although the field due to the material can be calculated from Ampère's law or the Biot-Savart law using the amperian current associated with the magnetization vector **M**, the magnetization itself depends on the resultant magnetic-induction field **B**. Because of this complication, a new vector, the magnetic intensity **H**, is defined in an attempt to separate out the external field from that due to the material. In many cases of interest, **H** can be found from the external conduction currents in wires alone without reference to any material present. The magnetization **M** can then be related to **H**, and the resultant magnetic induction field **B** can be found from **H** and **M**. We define the magnetic intensity **H** in Section 29-1 and discuss some of its properties. We then estimate the order of magnitude of the magnetic moment of an atom and use it to estimate the possible magnitude of the magnetization **M** for paramagnetic and ferromagnetic materials. After a brief look at paramagnetism and diamagnetism we shall discuss qualitatively the complicated phenomenon of ferromagnetism, which has many important practical applications.

## 29-1  Magnetic Intensity **H**

Let us consider a long solenoid with $n$ turns per unit length carrying a current $I$. The magnetic induction at the center of the solenoid is

$$B_0 = \mu_0 n I \tag{29-1}$$

This expression is good if we are far from the ends of the solenoid. Since we wish to avoid any complications from end effects, we shall assume that the solenoid is very long. (End effects can be completely eliminated in practice, using a doughnut-shaped solenoid called a *toroid*, illustrated in Figure 29-2, but we continue to use a solenoid because we have already studied it in some detail.) If we now insert some material inside the solenoid, there will be an additional contribution to the magnetic-induction field due to the magnetization of the material. If **M** is the magnetic moment per unit volume, the induction field $\mathbf{B}_m$ produced by this magnetization is given by an expression similar to Equation 29-1, with the amperian current per unit length $M$ replacing the current per unit length in the solenoid $nI$:

$$\mathbf{B}_m = \mu_0 \mathbf{M} \tag{29-2}$$

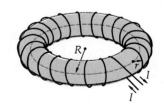

**Figure 29-2**
Toroid. The magnetic field is similar to that of a solenoid, but there are no end effects.

The total magnetic induction at the center of the solenoid is the vector sum of these separate fields:

$$\mathbf{B} = \mathbf{B}_0 + \mu_0\mathbf{M} \qquad\qquad 29\text{-}3$$

(We have written this as a vector sum because **M** is in the opposite direction to $\mathbf{B}_0$ for diamagnetic materials.) As mentioned above, it is convenient to separate out the external magnetic field from that due to the material. The *magnetic-intensity* vector **H** is defined by

$$\mathbf{H} = \frac{\mathbf{B}}{\mu_0} - \mathbf{M} \qquad\qquad 29\text{-}4a$$

*Magnetic intensity* **H** *defined*

Then

$$\mathbf{B} = \mu_0\mathbf{H} + \mu_0\mathbf{M} \qquad\qquad 29\text{-}4b$$

For a toroid or very long solenoid, for which we can neglect the effects at the ends of the material, **H** is simply the original external magnetic induction $\mathbf{B}_0$ divided by $\mu_0$. The magnitude of **H** at the center of a long solenoid is

$$H = nI \qquad\qquad 29\text{-}5$$

From Equation 29-4 we see that the dimensions of **H** are the same as those for **M**, which are amperes per metre. (Because the number of turns of wire often comes into the equation for **H**, as in Equation 29-5, the units of **H** are often stated as ampere-turns per metre.)

In situations like this where we can neglect end effects of the magnetized material, the magnetic intensity **H** is determined by the external conduction currents and is not affected by the magnetization. In these cases **H** can be obtained from Ampère's law or the Biot-Savart law in the forms

$$\oint_C \mathbf{H} \cdot d\boldsymbol{\ell} = I \qquad\qquad 29\text{-}6$$

and

$$d\mathbf{H} = \frac{1}{4\pi}\frac{I\,d\boldsymbol{\ell} \times \hat{\mathbf{r}}}{r^2} \qquad\qquad 29\text{-}7$$

It is important to realize that these equations differ from the corresponding equations for the magnetic induction **B** not only in the absence of the constant $\mu_0$ but also because the current $I$ is the macroscopic conduction current, whereas in the equations for **B** the current $I$ represents any kind of current, including the atomic or amperian current associated with the magnetization of the material.

**Example 29-1** Use Equations 29-6 and 27-13 to find **B** and **H** for a long cylinder which has a magnetization **M** and is wound with $n$ turns per unit length carrying a current $I$.

Figure 29-3 shows a rectangular path for computing the line integrals in these equations. As in Section 27-4, we assume that the tangential component of the magnetic field is constant along path $ab$ and zero along the other legs of the path. The conduction current through the rectangle is $nI\ell$, where $\ell$ is the length from $a$ to $b$. Thus Equation 29-6 gives

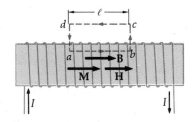

**Figure 29-3**
Path for calculation of $\oint \mathbf{H} \cdot d\boldsymbol{\ell}$ and $\oint \mathbf{B} \cdot d\boldsymbol{\ell}$ in Example 29-1. The only contribution is from $a$ to $b$.

$$\oint \mathbf{H} \cdot d\boldsymbol{\ell} = H\ell = nI\ell \qquad \text{or} \qquad H = nI$$

To calculate $B$ we need to include the atomic current, which is $M\ell$ since $M$ is the atomic current per unit length. Thus Equation 27-13 gives

$$\oint \mathbf{B} \cdot d\boldsymbol{\ell} = \mu_0 I_{\text{total}} = \mu_0(I_{\text{con}} + I_{\text{mag}})$$

or

$$B\ell = \mu_0(nI\ell + M\ell)$$

$$B = \mu_0(nI + M)$$

At points near the ends, or poles, of a magnetized material, the magnetic intensity $\mathbf{H}$ is not completely determined by the external conduction currents because of a contribution from the poles of the material. We can see this by considering the flux of $\mathbf{H}$ through any closed surface. Using the defining equation 29-4, we have

$$\oint_S \mathbf{H} \cdot \hat{\mathbf{n}}\, dA = \frac{1}{\mu_0} \oint_S \mathbf{B} \cdot \hat{\mathbf{n}}\, dA - \oint_S \mathbf{M} \cdot \hat{\mathbf{n}}\, dA$$

Since the net flux of the magnetic induction $\mathbf{B}$ is necessarily zero, we have

$$\oint_S \mathbf{H} \cdot \hat{\mathbf{n}}\, dA = - \oint_S \mathbf{M} \cdot \hat{\mathbf{n}}\, dA \qquad\qquad 29\text{-}8$$

In a homogeneous material the magnetization $\mathbf{M}$ is uniform. The flux of $\mathbf{M}$ through any closed surface within the material is then zero. There is a net flux of $\mathbf{M}$ only at the ends of the material. Let us consider the uniformly magnetized cylinder shown in Figure 29-4. The flux of $\mathbf{M}$ through the surface $S_1$ enclosing the left end of the cylinder is positive and has the magnitude $MA$, where $A$ is the cross-sectional area of the cylinder. Similarly, the net flux of $M$ through surface $S_2$ enclosing the right end is $-MA$. From our discussion in Section 27-5 we note that $MA$ is just the pole strength of the material and that the pole at the left is negative and the pole at the right is positive. In general, we can define the pole strength of a magnetized material $q_m$ by

$$q_m = - \oint_S \mathbf{M} \cdot \hat{\mathbf{n}}\, dA \qquad\qquad 29\text{-}9$$

Then the flux of the magnetization vector $\mathbf{H}$ is

$$\oint_S \mathbf{H} \cdot \hat{\mathbf{n}}\, dA = q_m \qquad\qquad 29\text{-}10 \qquad\qquad \textit{Gauss' law for } \mathbf{H}$$

This equation is similar to Gauss' law for the flux of the electric field $\mathbf{E}$, with the magnetic pole strength $q_m$ replacing the electric charge $q$. It follows from this equation that $\mathbf{H}$ varies inversely as the square of the distance from a magnetic pole according to

$$\mathbf{H} = \frac{1}{4\pi} \frac{q_m}{r^2} \hat{\mathbf{r}} \qquad\qquad 29\text{-}11 \qquad\qquad \textit{Coulomb's law for } \mathbf{H}$$

If we wish to calculate $\mathbf{H}$ at a point near a magnetic pole, we must add the contribution due to the pole given by Equation 29-11 to that due to the conduction currents, which can be calculated from Ampère's law or the Biot-Savart law. In many cases we are interested in the magnetic intensity $\mathbf{H}$ only in regions far from any poles, e.g., at the center of a long solenoid, and we can ignore the contribution given by Equation 29-11.

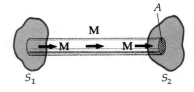

**Figure 29-4**
Surfaces $S_1$ and $S_2$ enclosing the ends of a uniformly magnetized cylinder. The net flux of $\mathbf{M}$ out of $S_1$ is $+MA$, and that out of $S_2$ is $-MA$, where $A$ is the area of the cylinder.

**Example 29-2** A permanent cylindrical bar magnet of radius $a$ and length $L = 10a$ has uniform magnetization **M** to the right parallel to its axis. Find **H** and **B** on the axis at the center of the magnet and just outside the right end.

Since there are no conduction currents in this example, **H** is due solely to the magnetic poles. If we consider the center of the magnet to be far from the ends so that the poles can be neglected, **H** = 0 and **B** = $\mu_0$**M** at that point. We can find a correction of this approximation by using Coulomb's law for **H** at the center due to the two poles. At the center, both the positive pole on the right and the negative pole on the left contribute equally to give **H** to the left in the direction opposite **M**. Using $q_m = \pi a^2 M$ and $r = \frac{1}{2}L$ in Equation 29-11, we have for the magnitude of $H$ due to either pole,

$$H = \frac{1}{4\pi} \frac{q_m}{r^2} = \frac{1}{4\pi} \frac{\pi a^2 M}{(\frac{1}{2}L)^2} = \frac{a^2}{L^2} M$$

Adding the contributions due to each pole, we obtain for $H$ at the center

$$\mathbf{H} = -\frac{2a^2}{L^2} \mathbf{M}$$

The magnetic induction **B** at the center is found from Equation 29-4$b$:

$$\mathbf{B} = \mu_0(\mathbf{H} + \mathbf{M}) = \mu_0\mathbf{M}\left(1 - \frac{2a^2}{L^2}\right)$$

We note that for $a = L/10$ the correction due to the poles is 2 percent.

Just outside the right end, the near pole can be treated as a disk of pole density $\sigma_m = M$ (Equation 27-21) and the far pole can be treated as a point pole a distance $L$ away. We found in Chapter 20 that a disk charge produces an electric field at a nearby point given by $\sigma/2\epsilon_0$. Since Equations 29-10 and 29-11 for **H** are the same as the corresponding equations for **E** except for the absence of $\epsilon_0$, the magnetic intensity **H** produced by a disk of pole density $\sigma_m$ is $\sigma_m/2$. The total **H** just outside the right end of the magnet is then

$$H = +\frac{\sigma_m}{2} = \frac{1}{4\pi} \frac{q_m}{L^2} = +\frac{M}{2} - \frac{1}{4\pi} \frac{\pi a^2 M}{L^2}$$

or

$$\mathbf{H} = \frac{\mathbf{M}}{2}\left(1 - \frac{a^2}{2L^2}\right)$$

Since there is no magnetization **M** outside the magnet, **B** is just $\mu_0$**H**,

$$\mathbf{B} = \frac{\mu_0\mathbf{M}}{2}\left(1 - \frac{a^2}{2L^2}\right)$$

If we neglect the term $a^2/2L^2$ due to the far pole, the magnetic induction $B$ is $\frac{1}{2}\mu_0M$, approximately half its value at the center of the magnet, a result which also holds for a long solenoid.

**Questions**

1. How are **B** and **H** related in a vacuum?

2. Must **B** and **H** necessarily be parallel?

## 29-2  Magnetic Susceptibility and Permeability

For paramagnetic and diamagnetic materials the magnetization **M** is proportional to the magnetic intensity **H**:

$$\mathbf{M} = \chi_m \mathbf{H} \qquad\qquad 29\text{-}12$$

The proportionality factor $\chi_m$ is called the *magnetic susceptibility*. Since **M** and **H** have the same dimensions, the susceptibility is dimensionless. It is positive for paramagnetic materials and negative for diamagnetic materials. For diamagnetic materials the susceptibility is essentially independent of temperature, but for paramagnetic materials the susceptibility decreases as the temperature increases. This temperature dependence for paramagnetic materials is the result of the tendency for thermal vibrations to reduce the alignment of the permanent dipole moments, discussed in more detail in Section 29-4.

*Magnetic susceptibility*

From Table 29-1 listing typical values of the susceptibility, we see that $\chi_m$ is much less than 1. If **M** is proportional to **H**, then **B** is also proportional to **H**. Substituting $\chi_m \mathbf{H}$ for **M** in Equation 29-4b, we can write

$$\mathbf{B} = \mu_0(\mathbf{H} + \mathbf{M}) = \mu_0(\mathbf{H} + \chi_m \mathbf{H}) = \mu_0(1 + \chi_m)\mathbf{H} \qquad 29\text{-}13$$

or

$$\mathbf{B} = \mu \mathbf{H} \qquad\qquad 29\text{-}14$$

where

$$\mu = (1 + \chi_m)\mu_0 \qquad\qquad 29\text{-}15$$

is called the *permeability* of the material. Because the susceptibility is so small, the permeability of all paramagnetic and diamagnetic material is very nearly equal to the permeability of free space $\mu_0$.

*Magnetic permeability*

**Table 29-1**
Magnetic susceptibility of various materials at 20°C

| Material | $\chi_m$ |
| --- | --- |
| Aluminum | $2.3 \times 10^{-5}$ |
| Bismuth | $-1.66 \times 10^{-5}$ |
| Copper | $-0.98 \times 10^{-5}$ |
| Diamond | $-2.2 \times 10^{-5}$ |
| Gold | $-3.6 \times 10^{-5}$ |
| Magnesium | $1.2 \times 10^{-5}$ |
| Mercury | $-3.2 \times 10^{-5}$ |
| Silver | $-2.6 \times 10^{-5}$ |
| Sodium | $-0.24 \times 10^{-5}$ |
| Titanium | $7.06 \times 10^{-5}$ |
| Tungsten | $6.8 \times 10^{-5}$ |
| Hydrogen (1 atm) | $-9.9 \times 10^{-9}$ |
| Carbon dioxide (1 atm) | $-2.3 \times 10^{-9}$ |
| Nitrogen (1 atm) | $-5.0 \times 10^{-9}$ |
| Oxygen (1 atm) | $2090 \times 10^{-9}$ |

Equation 29-14 is also written for ferromagnetic materials, but it is difficult to interpret because the magnetization **M** is not a linear function of **H**. The permeability defined by Equation 29-14 depends not only on the value of **H** but also on the previous state of magnetization of a ferromagnetic material. In such cases, **B** and **M** are multivalued functions of **H**. The maximum value of $\mu$ for ferromagnetic materials is typically several thousand times greater than $\mu_0$. We shall discuss the relation between **B**, **H**, and **M** for ferromagnetic materials in more detail in Section 29-6.

### Questions

3. Why are some values of $\chi_m$ in Table 29-1 positive and others negative?

4. Why is it more convenient to list $\chi_m$ than $\mu$ for the materials in Table 29-1?

## 29-3  Atomic Magnetic Moments

The magnetization **M** of a paramagnetic or ferromagnetic material can be related to the permanent magnetic moments of the individual atoms of the material. In this section we calculate the order of magnitude of these magnetic moments from the Bohr theory of the atom, in which the electrons move in circular orbits around the nucleus. Although this theory has severe defects, its results are in agreement with those from the correct quantum-mechanical theory of the atom in many cases and insight can often be obtained from the simpler Bohr picture. We can use the simple circular motion of a charged particle to show that there is a general connection between the angular momentum of the particle and the magnetic moment (Figure 29-5). Consider a charge moving in a circle of radius $r$ with speed $v$. The time for each revolution, the period $T$, is related to the speed and radius by

$$vT = 2\pi r$$

This motion results in a current loop and thus produces a magnetic moment. The current is the charge $q$ divided by the period $T$:

$$I = \frac{q}{T} = \frac{qv}{2\pi r}$$

The magnetic moment is the product of the current and the area $\pi r^2$:

$$m = IA = \frac{qv}{2\pi r}\,\pi r^2 = \tfrac{1}{2}qvr \qquad\qquad 29\text{-}16$$

If the mass of the particle is $M$ (we use $M$ here for the mass to avoid confusion with $m$ for the magnetic moment), the angular momentum is

$$L = Mvr = M\,\frac{2m}{q}$$

If the charge $q$ is positive, the angular momentum and magnetic moment are in the same direction. We can therefore write

$$\mathbf{m} = \frac{q}{2M}\,\mathbf{L} \qquad\qquad 29\text{-}17$$

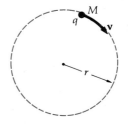

**Figure 29-5**
Particle of charge $q$ and mass $M$ moving in a circle of radius $r$. The angular momentum is into the paper with magnitude $Mvr$, and the magnetic moment is in (if $q$ is positive) with magnitude $\tfrac{1}{2}qvr$.

For the electron, with negative charge, the magnetic moment points in the direction opposite to the angular momentum. Equation 29-17 is the general classical relation between magnetic moment and angular momentum. It also holds in the quantum-mechanical theory of the atom for orbital angular momentum but not for the intrinsic spin angular momentum of the electron. For electron spin, the magnetic moment is twice that predicted by this equation. This extra factor of 2 is a quantum-mechanical result which has no analog in classical mechanics.

In both the Bohr theory of the atom and in the rigorous quantum-mechanical theory, the orbital angular momentum is quantized and takes on only values which are integral multiples $h/2\pi$, where $h$ is a fundamental constant called Planck's constant,* and has the value

$$h = 6.63 \times 10^{-34} \text{ J·s}$$

The combination $h/2\pi$ occurs often and is designated by $\hbar$, where

$$\hbar = \frac{h}{2\pi} = \frac{6.63 \times 10^{-34}}{2\pi} \text{ J·s} = 1.05 \times 10^{-34} \text{ J·s}$$

The orbital angular momentum $L$ can therefore be written

$$L = N\hbar \qquad\qquad 29\text{-}18$$

where $N$ is an integer. The spin angular momentum of the electron is also quantized and has magnitude $\frac{1}{2}\hbar$. It is therefore convenient to write the magnetic moment as

$$\mathbf{m} = \frac{q\hbar}{2M} \frac{\mathbf{L}}{\hbar} \qquad\qquad 29\text{-}19a$$

For an electron, $q = -e$ and $M - m_e$. The magnetic moment and angular momentum are then oppositely directed, and their magnitudes are related by

$$m = \frac{e\hbar}{2m_e} \frac{L}{\hbar} = m_B \frac{L}{\hbar} \qquad\qquad 29\text{-}19b$$

where

$$m_B = \frac{e\hbar}{2m_e} \qquad\qquad 29\text{-}20 \qquad \textit{Bohr magneton}$$

is called a *Bohr magneton*. It is a convenient unit with which to measure atomic magnetic moments. For example, the magnetic moment associated with the spin of an electron is 1 Bohr magneton. Although the calculation of the magnetic moment of any atom is a complicated quantum-mechanical problem, the result of theory and experiment for all atoms is that the magnetic moment is of the order of a few Bohr magnetons (or zero for atoms with closed-shell electron structure). The magnitude of the Bohr magneton in SI units is

$$m_B = \frac{(1.60 \times 10^{-19} \text{ C})(1.05 \times 10^{-34} \text{ J·s})}{2(9.11 \times 10^{-31} \text{ kg})}$$

$$= 9.27 \times 10^{-24} \text{ A·m}^2$$

The SI unit for magnetic moment, the ampere-metre², can also be expressed as a joule per tesla (see Exercise 10 in Chapter 26). The value of the Bohr magneton is therefore

$$m_B = 9.27 \times 10^{-24} \text{ A·m}^2 = 9.27 \times 10^{-24} \text{ J/T} \qquad\qquad 29\text{-}21$$

* Quantization and Planck's constant will be discussed in Chapter 36.

If all the atoms or molecules in some material have their magnetic moments aligned, the magnetic moment per unit volume of the material is just the product of the number of molecules per unit volume $n$ and the magnetic moment of each molecule. For this extreme case, the saturation magnetization vector $M_s$ is

$$M_s = nm \qquad\qquad 29\text{-}22$$

The number of molecules per unit volume is related to the molecular weight, the density, and Avogadro's number. For example, the number of iron molecules (atoms) per unit volume is

$$n = \frac{6.02 \times 10^{23} \text{ atoms/mol}}{55.8 \text{ g/mol}} \frac{7.9 \text{ g}}{1 \text{ cm}^3} \frac{10^6 \text{ cm}^3}{1 \text{ m}^3}$$
$$= 8.52 \times 10^{28} \text{ atoms/m}^3$$

If we assume that the magnetic moment of each iron atom is 1 Bohr magneton, the maximum value of the magnetization vector in iron has the magnitude

$$M_s = (8.52 \times 10^{28} \text{ atoms/m}^3)(9.27 \times 10^{-24} \text{ A·m}^2/\text{atom})$$
$$= 7.90 \times 10^5 \text{ A/m}$$

The magnetic induction on the axis inside a long iron cylinder with maximum magnetization is then

$$B = \mu_0 M_s = (4\pi \times 10^{-7} \text{ T·m/A})(7.90 \times 10^5 \text{ A/m})$$
$$= 0.993 \text{ T} \approx 1 \text{ T}$$

The measured saturation magnetic induction of annealed iron is about 2.16 T, indicating that the magnetic moment of an iron atom is slightly greater than 2 Bohr magnetons. This magnetic moment is due mainly to the spins of two unpaired electrons in the iron atom.

## 29-4   Paramagnetism

Paramagnetism occurs in materials whose atoms have permanent magnetic moments interacting with each other only very weakly. When there is no external magnetic field, these magnetic moments are randomly oriented. In the presence of an external magnetic field they tend to line up parallel to the field, but this is counteracted by the tendency for the moments to be randomly oriented due to thermal motion. The fraction of moments that line up with the field depends on the strength of the field and the temperature. At very low temperatures and high external fields nearly all the moments are aligned with the field. In this situation the contribution to the total magnetic field due to the material is very large, as indicated in the numerical estimates in the previous section. Even with the largest magnetic field obtainable in the laboratory, the temperature must be as low as a few degrees absolute in order to obtain a high degree of alignment. At higher temperatures and weaker external fields only a small fraction of the moments are aligned with the field, and the contribution of the material to the total magnetic field is very small. We can state this more quantitatively by comparing the energy of a magnetic moment in an external magnetic field with the thermal energy, which is of the order of $kT$, where $k$ is Boltzmann's constant and $T$ is the absolute temperature. The potential energy of a magnetic moment in an external magnetic field is least when the moment is parallel to the field and greatest when it is antiparallel to the field. We can find the expression for this potential energy by computing the work that must be done to rotate a magnetic dipole in an external field. Let a dipole of moment **m** make an angle $\theta$ with the magnetic induction **B** (Figure 29-6$a$). The torque exerted by the field on the dipole has the magnitude

$$\tau = mB \sin \theta$$

If we wish to rotate the dipole to increase the angle by $d\theta$, the work we must do is

$$dW = \tau \, d\theta = mB \sin \theta \, d\theta$$

The work we must do to rotate the dipole from $\theta = 0$ to $\theta = 180°$ (Figure 29-6$b$) equals the change in potential energy:

$$W = \int_0^{180°} mB \sin \theta \, d\theta = -mB \cos \theta \, \Big]_0^{180°} = 2mB = \Delta U \qquad 29\text{-}23$$

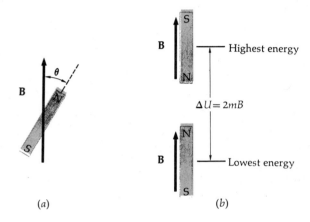

(a)                              (b)

**Figure 29-6**
(a) Magnetic dipole whose moment makes an angle $\theta$ with a magnetic field **B**. (b) When the dipole is antiparallel to **B**, the energy is greater by the amount $2mB$ than when it is parallel to **B**.

The potential energy when $\theta = 180°$ is thus greater than at $\theta = 0°$ by the amount $2mB$. For a typical magnetic moment of 1 Bohr magneton and a typical strong field of 1 T, this difference in potential energy is

$$2mB = 2(9.27 \times 10^{-24} \text{ J/T}) (1 \text{ T}) = 1.85 \times 10^{-23} \text{ J}$$

At a normal temperature of T $= 300$ K, the typical thermal energy $kT$ is $4.14 \times 10^{-21}$ J, about 200 times greater than the difference in potential energy when the dipole is aligned parallel or antiparallel with the field. Thus even in a strong field of 1 T, most of the moments will be essentially randomly oriented because of thermal motions.

Figure 29-7 shows a plot of the ratio of the magnetization $M$ to its saturation value $M_s$ versus a dimensionless parameter $x = mB/kT$. At $x = 0$, corresponding to zero external field, the magnetization is zero because the dipoles are randomly aligned. At very large values of $x$, which can occur only for high fields and very low temperatures, nearly all the dipoles are aligned with the field and $M/M_s = 1$. In the region of small $x$ the curve is approximately linear, with a slope of $\frac{1}{3}$. In this region we have

$$M = \frac{1}{3} \frac{mB}{kT} M_s \qquad 29\text{-}24$$

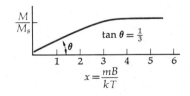

**Figure 29-7**
$M/M_s$ versus $mB/kT$. At weak external fields or high temperatures, $M/M_s$ increases linearly with $mB/kT$. When $mB$ is much greater than $kT$, $M$ reaches its saturation value $M_s$.

The result that the magnetization varies inversely with the absolute temperature was discovered experimentally by Pierre Curie and is known as *Curie's law*.

*Curie's law*

# 29-5 Diamagnetism

Diamagnetism was discovered by Faraday in 1846, when he found that a piece of bismuth is repelled by either pole of a magnet, indicating that the external field of the magnet induces a magnetic dipole in the bismuth in the direction opposite the field. We can understand this effect qualitatively using Lenz's law. Figure 29-8 shows two positive charges moving in circular orbits with the same speed but in opposite directions. Their magnetic moments are in opposite directions and cancel. (It is simpler to consider positive charges even though it is the negatively charged electrons which provide the magnetic moments in matter.) Consider now what happens when an external magnetic field **B** is turned on in the direction into the page. According to Lenz's law, currents will be induced to oppose the change in flux. If we assume that the radius of the circle is not changed, the charge on the left will be speeded up to increase its flux out of the page and the charge on the right will be slowed down to decrease its flux into the page. In each case, the *change* in magnetic moment of the charges will be in the direction out of the page, opposite that of the external field.

We can estimate the magnitude of this effect by relating the change in the speeds of the particles to the change in the centripetal force due to the external magnetic field. We assume that the radius of the orbit does not change and that the change in speed is small compared with the original speed. Both these assumptions can be justified. The original centripetal force is provided by the electrostatic force of attraction of the electron to the nucleus. Calling this force $F$ and setting it equal to the mass times the acceleration, we have

$$F = \frac{m_e v^2}{r} \qquad\qquad 29\text{-}25$$

where $m_e$ is the electron mass. In the presence of an external magnetic field there is an additional force on each particle, $q\mathbf{v} \times \mathbf{B}$. Note that on the particle on the left in Figure 29-8 this force is inward, thus increasing the centripetal force. This is necessary because, according to our argument from Lenz's law, this particle must speed up to oppose the change in flux. Similarly, the force is outward on the particle on the right, decreasing the centripetal force. Again, this is in the correct direction since this particle slows down to oppose the change in flux when the field is turned on. In each case the magnitude of the change in force is $qvB$ since **v** and **B** are perpendicular. Since this change is very small, we use the differential approximation. From Equation 29-25 we have

$$\Delta F \approx dF = \frac{2m_e v}{r}\, dv \approx \frac{2m_e v}{r}\,\Delta v$$

Substituting $qvB$ for $\Delta F$ and solving for $\Delta v$ gives

$$\Delta v = \frac{qrB}{2m_e} \qquad\qquad 29\text{-}26$$

This change in speed causes a change in the magnetic moment in the outward direction for both particles. The magnetic moment of a charge moving in a circle is related to its speed by Equation 29-16,

$$m = \tfrac{1}{2}qvr$$

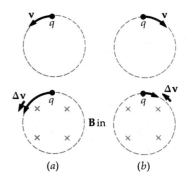

**Figure 29-8**
(*a*) Charge moving in a circle with magnetic moment out (assuming $q$ positive). In the presence of an external magnetic field the speed is increased. (*b*) Charge moving in a circle with its magnetic moment in. An external magnetic field causes a decrease in its speed. In both cases the *change* in the magnetic moment is out.

Then

$$\Delta m = \tfrac{1}{2}qr\,\Delta v = \tfrac{1}{2}qr\,\frac{qrB}{2m_e} = \frac{q^2r^2}{4m_e}B \qquad\qquad 29\text{-}27$$

When we use $R = 10^{-10}$ m for a typical radius, the values of $q$ and $m_e$ for the electron, and a strong field of $B = 1$ T, the order of magnitude of this induced magnetic moment is

$$\Delta m = \frac{(1.60 \times 10^{-19}\text{ C})^2(10^{-10}\text{ m})^2}{4(9.11 \times 10^{-31}\text{ kg})}\,(1\text{ T}) \approx 7 \times 10^{-29}\text{ A·m}^2$$

which is about $10^5$ times smaller than a Bohr magneton. Since $m$ is of the order of a Bohr magneton, $\Delta m/m$ is about $10^{-5}$. This justifies our assumption that the change in speed is very small.

Since the induced magnetic moment is antiparallel to the external field whether the original moment is parallel or antiparallel to the field, there is a contribution from each electron in the atom to the induced magnetic moment of the atom. The magnetization vector **M** is the product of the number of *electrons* per unit volume times the induced magnetic moment of each electron. If $n$ is the number of atoms per unit volume and $Z$ is the atomic number, the number of electrons per unit volume is $nZ$. (Of course, an accurate calculation of the induced dipole moment of an atom must take into account the different values of $r^2$ for the different electrons and the fact that the orbits are not perpendicular to the external field.)

Atoms with closed-shell electron structures have zero angular momentum and therefore no permanent magnetic dipole moment. They are diamagnetic. Nonferromagnetic atoms with permanent magnetic dipole moments are either paramagnetic or diamagnetic, depending on which effect is stronger. Since the diamagnetic effect does not depend on temperature, and alignment of permanent moments decreases with temperature for both paramagnetic and ferromagnetic substances, all materials are diamagnetic at sufficiently high temperatures.

## 29-6  Ferromagnetism

Ferromagnetism occurs in pure iron, cobalt, and nickel, in alloys of these metals with each other and with some other elements, and in a few other substances (gadolinium, dysprosium, and a few compounds). In these substances a small external magnetic field can produce a very large degree of alignment of the atomic magnetic dipole moments, which, in some cases, can persist even when there is no external magnetizing field. This occurs because the magnetic dipole moments of atoms of these substances exert strong forces on their neighbors so that over a small region of space the moments are aligned with each other even with no external field. These dipole forces are predicted for these substances by quantum mechanics but cannot be explained with classical physics. At temperatures above a critical temperature, called the *Curie temperature*, these forces disappear and ferromagnetic materials become paramagnetic.

The region of space over which the magnetic dipole moments are aligned is called a *domain*. The size of a domain is usually microscopic. Within the domain, all the magnetic moments are aligned, but the

*Curie temperature*

*Domain*

direction of alignment varies from domain to domain so that the net magnetic moment of a macroscopic piece of material is zero in the normal state. Figure 29-9 illustrates this situation. When an external magnetic field is applied, the boundaries of the domains (called *domain walls*) shift and the direction of alignment within a domain may change so that there is a net magnetic moment in the direction of the applied field. Since the degree of alignment is great even for a small external field, the magnetic field produced in the material by the dipoles is often much greater than the external field.

 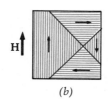 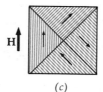

(a)  (b)  (c)

**Figure 29-9**
Schematic illustration of magnetization by domain changes: (*a*) unmagnetized; (*b*) magnetization by domain boundary changes; (*c*) magnetization by rotation of domains.

Consider magnetizing a long iron rod inside a solenoid by gradually increasing the current in the solenoid windings. Assume that the rod and solenoid are long enough to permit us to neglect end effects. The magnetic intensity **H** at the center of the solenoid is then simply related to the current by

$$H = nI = \frac{N}{\ell} I \qquad\qquad 29\text{-}28$$

where $N$ is the number of turns and $\ell$ is the length. The magnetic induction **B** is given by

$$\mathbf{B} = \mu_0 \mathbf{H} + \mu_0 \mathbf{M} \qquad\qquad 29\text{-}29$$

where **M** is the magnetization. For iron and other ferromagnetic materials, the magnetization **M** is often much greater than the magnetic intensity **H** by a factor of several thousand or more.

William Vandivert

Photomicrograph showing magnetic domains in a thin single crystal of yttrium iron garnet magnified 330 times. The photograph was made with transmitted light through the crystal between crossed Polaroids. The different directions of magnetization of the domains give rise to different colors in the transmitted light. The magnetization is pointing down into the paper in the light striped regions and out of the paper in the dark striped regions. In the large gray areas, the magnetization lies in the plane of the paper.

Figure 29-10 shows a plot of $B$ versus $H$. As $H$ is gradually increased from zero, $B$ increases from zero along the part of the curve from the origin $O$ to point $P_1$. The flattening of this curve near point $P_1$ indicates that the magnetization $M$ is approaching its saturation value $M_s$ when all the atomic dipoles are aligned. The external field needed to produce saturation in a given ferromagnetic material is called the saturation intensity $H_s$. A further increase in $H$ above $H_s$ increases $B$ only through the term $\mu_0 H$ in Equation 29-29. When $H$ is gradually decreased from point $P_1$, there is no corresponding decrease in the magnetization. The shift of the domains in a ferromagnetic material is not completely reversible, and some magnetization remains even when $H$ is reduced to zero, as indicated in Figure 29-10. This effect is called *hysteresis*, from a word which means to lag. The value of the magnetic induction at point $r$ when $H$ is zero is called the *remanent field $B_r$*. If the current in the solenoid is now reversed so that **H** is in the opposite direction, the magnetic induction $B$ is gradually brought to zero at point $c$. The value of $H$ needed to reduce $B$ to zero is called the *coercive force $H_c$*. The remaining part of the hysteresis curve is obtained by further increasing the current in the opposite direction until point $P_2$ is reached, corresponding to saturation in the opposite direction; decreasing the current to zero at point $P_3$; and increasing the current again to produce a magnetic intensity $H$ in the original direction. We can define the magnetic susceptibility $\chi_m$ formally just as we did for paramagnetic or diamagnetic materials:

$$\mathbf{M} = \chi_m \mathbf{H}$$

Then

$$\mathbf{B} = \mu_0(\mathbf{H} + \mathbf{M}) = \mu_0(1 + \chi_m)\mathbf{H} = K_m\mu_0\mathbf{H} = \mu\mathbf{H} \qquad \text{29-30}$$

where

$$K_m = 1 + \chi_m = \frac{\mu}{\mu_0} \qquad \text{29-31}$$

is called the *relative permeability*. Since, according to Figure 29-10, **B** and **H** are sometimes parallel and sometimes antiparallel and either can be zero when the other is not, $\mu$ and $K_m$ may have any value from zero to infinity and may be positive or negative depending on **H** and the history of the material. If we confine the definitions of $\mu$ and $K_m$ to that part of the magnetization curve from the origin $O$ to point $P_1$, these quantities are always positive and have maximum values characteristic of the material. Table 29-2 lists the saturation magnetization $M_s$ and the maximum value of $K_m$ for some ferromagnetic materials. Note that the maximum values of $K_m$ are much greater than 1. In a ferromagnetic material the quantity $\mu_0\mathbf{H}$ can usually be neglected compared with $\mu_0\mathbf{M}$.

*Hysteresis*

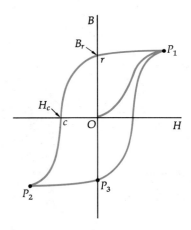

**Figure 29-10**
$B$ versus $H$ for hard ferromagnetic material. The outer curve is called a hysteresis curve.

**Table 29-2**
Maximum values of $\mu_0 M$ and $K_m$ for some ferromagnetic materials

| Material | $\mu_0 M_s$, T | $K_m$ |
| --- | --- | --- |
| Iron (annealed) | 2.16 | 5,500 |
| Iron-silicon (96% Fe, 4% Si) | 1.95 | 7,000 |
| Permalloy (55% Fe, 45% Ni) | 1.60 | 25,000 |
| Mu-metal (77% Ni, 16% Fe, 5% Cu, 2% Cr) | 0.65 | 100,000 |

The area enclosed by the hysteresis curve represents loss in energy because of the irreversibility of the process. The energy appears in the material as heat. We can see this as follows. The emf in the solenoid due to the changing flux is given by Faraday's law:

$$\varepsilon = \frac{d\phi}{dt} = NA\frac{dB}{dt} \qquad \text{29-32}$$

where $A$ is the cross-sectional area of the coil and rod. (We neglect the minus sign in Faraday's law because we are interested in magnitudes only.) The rate at which work is done against this emf is

$$\frac{dW}{dt} = \varepsilon I \qquad \text{29-33}$$

Using Equation 29-32 for the emf and Equation 29-28 for the current, we have

$$\frac{dW}{dt} = \left(NA\frac{dB}{dt}\right)\frac{H\ell}{N} = \ell AH\frac{dB}{dt}$$

or

$$dW = \ell AH\,dB \qquad \text{29-34}$$

The sum of this work for a complete hysteresis cycle is thus the area enclosed in the $B$-versus-$H$ curve times the volume $\ell A$ of the material. If the hysteresis effect is small, so that the area is small, indicating a small energy loss, the material is called *magnetically soft* (soft iron is an example). The hysteresis curve for a magnetically soft material is shown in Figure 29-11. Here the remanent field $B_r$ and the coercive force $H_c$ are nearly zero, and the energy loss per cycle is small. Magnetically soft materials are used for transformer cores to increase the induction $B$ without incurring a large energy loss when the field alternates many times per second.

On the other hand, it is desirable to have a large remanent field $B_r$ and a large coercive force $H_c$ in a permanent magnet (the large coercive force is important so that the magnetization will not be destroyed by small stray fields). Magnetically hard materials, e.g., carbon steel and the alloy Alnico 5, are used for permanent magnets.

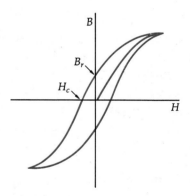

Figure 29-11
Hysteresis curve for magnetically soft material. The remanent field $B_r$ and the coercive force $H_c$ are small compared with those for a magnetically hard material (Figure 29-10).

### Review

A. Define, explain, or otherwise identify:

Paramagnetism, 797
Diamagnetism, 797
Ferromagnetism, 797
Magnetization, 798
Magnetic intensity, 799
Magnetic pole strength, 800
Gauss' law for **H**, 800
Coulomb's law for **H**, 800
Magnetic susceptibility, 802
Permeability, 802

Permeability of free space, 802
Bohr magneton, 804
Curie's law, 806
Curie temperature, 808
Magnetic domain, 808
Hysteresis, 810
Remanent field, 810
Coercive force, 810
Magnetically soft, 811
Magnetically hard, 811

B. True or false:

1. Diamagnetism occurs in all materials.

2. Diamagnetism is the result of induced magnetic dipole moments.

3. In free space, **H** is always zero.

4. Paramagnetism is the result of partial alignment of permanent magnetic dipole moments.

5. Hysteresis is associated with a loss in electromagnetic energy.

6. Magnetically hard materials are good for transformer cores.

## Exercises

### Section 29-1, Magnetic Intensity H

1. A tightly wound solenoid 20 cm long has 400 turns carrying a current of 4 A such that its axial field is in the $z$ direction. Neglecting end effects, find $\mathbf{H}$ and $\mathbf{B}$ at the center when (a) there is no core and (b) there is an iron core in the solenoid with magnetization $\mathbf{M} = 1.2 \times 10^6 \mathbf{k}$ A/m.

2. Consider a single magnetic pole of strength $q_m$ far away from any other pole. Use Gauss' law for $\mathbf{H}$ (Equation 29-10) to derive Coulomb's law for $\mathbf{H}$ (Equation 29-11).

3. In Exercise 1 the cross-sectional area of the solenoid core is 4 cm². Assuming the same current and magnetization as in Exercise 1, find (a) the magnetic pole strength $q_m$ of the core, (b) $\mathbf{H}$ at the center of the solenoid due to the magnetic poles, (c) the resultant $\mathbf{H}$ at the center due to both the magnetic poles and the current in the windings, and (d) $\mathbf{B}$ at the center.

4. For the permanent magnet in Example 29-2, show that just inside the right end of the magnet (a) $\mathbf{H} = -\frac{1}{2}M(1 + a^2/2L^2)$ and (b) $\mathbf{B} = \frac{1}{2}\mu_0 M(1 - a^2/2L^2)$. (c) How do these results compare with those in Example 29-2 for the point just outside the magnet?

### Section 29-2, Magnetic Susceptibility and Permeability

5. Which of the four gases listed in Table 29-1 are diamagnetic and which are paramagnetic?

6. If the solenoid in Exercise 1 has an aluminum core, find $\mathbf{H}$, $\mathbf{M}$, and $\mathbf{B}$ at the center, neglecting end effects.

7. Repeat Exercise 6 for a tungsten core.

### Section 29-3, Atomic Magnetic Moments

8. Nickel has density 8.7 g/cm³ and molecular mass 58.7 g/mol. Its saturation magnetization is $\mu_0 M_s = 0.61$ T. Calculate the magnetic moment in Bohr magnetons of a nickel atom.

9. Repeat Exercise 8 for cobalt, which has density 8.9 g/cm³, molecular mass 58.9 g/mol, and saturation magnetization $\mu_0 M_s = 1.79$ T.

### Section 29-4, Paramagnetism

10. Show that Curie's law predicts that the magnetic susceptibility for a paramagnetic substance is given by $\chi_m = m\mu_0 M_s/3kT$.

11. Assume that the magnetic moment of an aluminum atom is 1 Bohr magneton. The density of aluminum is 2.7 g/cm³, and its molecular mass is 27 g/mol. (a) Calculate $M_s$ and $\mu_0 M_s$ for aluminum. (b) Use the result of Exercise 10 to calculate $\chi_m$ at $T = 300$ K. (c) Explain why you expect this result to be larger than that listed in Table 29-1.

12. In a simple model of paramagnetism we can consider that some fraction $f$ of the molecules have their magnetic moment aligned with the external magnetic

field and the rest of the molecules are randomly oriented to produce no contribution to the magnetic moment. (a) Use this model and Curie's law to show that at temperature $T$ and external field $B$ this fraction aligned is $f = mB/3kT$. (b) Calculate this fraction for $T = 300$ K, $B = 1$ T, assuming $m$ to be 1 Bohr magneton.

## Section 29-5, Diamagnetism

*There are no exercises for this section.*

## Section 29-6, Ferromagnetism

13. The saturation magnetic intensity for annealed iron is $H_s = 1.6 \times 10^5$ A/m. Find the permeability $\mu$ and the relative permeability $K_m$ at saturation (see Table 29-2).

14. For annealed iron the relative permeability $K_m$ has its maximum value of about 5500 at $H = 125$ A/m. (This is well below the saturation intensity.) Find $M$ and $B$ when $K_m$ is maximum.

15. The coercive force for a certain permanent bar magnet is $H_c = 4.4 \times 10^4$ A/m. The bar magnet is to be demagnetized by placing it inside a long solenoid 15 cm long with 600 turns. What is the minimum current needed in the solenoid to demagnetize the magnet?

16. A long solenoid has 50 turns/cm and carries a current of 2 A. The solenoid is filled with iron, and $B$ is measured to be 1.72 T. (a) What is $H$ (neglecting end effects)? (b) What is $M$? (c) What is the relative permeability $K_m$ for this case?

17. When the current in the solenoid in Exercise 16 is 0.2 A, the magnetic induction is measured to be 1.58 T. (a) Neglecting end effects, what is $H$? (b) What is $M$? (c) What is the relative permeability $K_m$?

## Problems

1. A toroid has mean radius $R$ and cross-sectional radius $r$ (Figure 29-2), where $r < R$. (a) Use Equation 29-6 to show that $H = nI = IN/2\pi R$, where $n$ is the number of turns per unit length and $I$ is the current. (b) When the toroid is filled with material, it is called a *Rowland ring*. Find **H** and **B** in such a ring. Assume a magnetization **M** everywhere parallel to **H**.

2. Show that the normal component of the magnetic induction **B** is continuous across any surface. Do this by applying Gauss' law for **B** ($\oint \mathbf{B} \cdot \hat{n}\, dA = 0$) to a pillbox gaussian surface which has a face on each side of the surface.

3. Use Equation 29-10 to show that the normal component of **H** on one side of a surface differs from that on the other side of the surface by the amount $\sigma_m$, where $\sigma_m$ is the magnetic pole density on the surface (see Problem 2).

4. A Rowland ring has windings with 5 turns/cm carrying current $I = 1$ A. The magnetic induction in the ring is 1.40 T. (a) Find **H** and **M** in the ring. (b) What is $K_m$? (c) A small gap is cut in the ring. Find **B** and **H** in the gap assuming that they are unchanged in the bulk of the material (see Problems 2 and 3).

5. In our derivation of the magnetic moment induced in an atom we used the Bohr model and assumed that the radius of the orbit did not change in the presence of an external magnetic field. Show that the assumption of constant radius is justified by showing that when **B** is applied, there is an impulse which increases or decreases the speed of the electron by just the correct amount given by Equation 29-26. (a) Use Faraday's law and the expression $\varepsilon = \oint \mathbf{E} \cdot d\ell$ to show that the induced electric field is related to the rate of change of the magnetic field by $E = \frac{1}{2}r\, dB/dt$, assuming $r$ to be constant. (b) Use Newton's second law to show that the change in speed of the electron $dv$ is related to the change in $B$ by $dv = (qr/2m_e)\, dB$. Integrate to obtain Equation 29-26.

6. Equation 29-27 gives the induced magnetic moment for a single electron in an orbit which has its plane perpendicular to **B**. If an atom has $Z$ electrons, a reasonable simplifying assumption is that on the average one-third have their planes perpendicular to **B**. Show that the diamagnetic susceptibility obtained from Equation 29-27 is then

$$\chi_m = \frac{-nZq^2r^2}{12m_e}\mu_0$$

where $n$ is the number of atoms per unit volume. Use $n \approx 6 \times 10^{28}$ atoms/m³ and $r \approx 5 \times 10^{-11}$ m to estimate $\chi_m$ for $Z \approx 50$.

7. Use the values in Table 29-3 to plot $B$ versus $H$ and $K_m$ versus $H$.

8. A long solenoid of length $\ell$ and cross-sectional area $A$ has $n$ turns per unit length and carries a current $I$. It is filled with iron of relative permeability $K_m$. Find the self-inductance of the solenoid.

**Table 29-3**
For Problem 7

| $H$, A/m | $B$, T |
|---|---|
| 0 | 0 |
| 50 | 0.04 |
| 100 | 0.67 |
| 150 | 1.00 |
| 200 | 1.2 |
| 500 | 1.4 |
| 1,000 | 1.6 |
| 10,000 | 1.7 |

CHAPTER 30      Alternating-Current Circuits

---

**Objectives**    After studying this chapter, you should:

1. Be able to state the definition of rms current and relate it to the maximum current in an ac circuit.

2. Be able to state the definition of capacitive reactance, inductive reactance, and impedance.

3. Be able to draw a phasor diagram for a series $LRC$ circuit and from it relate the phase angle $\phi$ to the capacitive reactance, inductive reactance, and resistance.

4. Be able to state the definition of the $Q$ value and discuss its significance.

5. Be able to state the resonance condition for an $LRC$ circuit with generator and sketch the power versus $\omega$ for both a high-$Q$ and low-$Q$ circuit.

6. Be able to describe a step-up and a step-down transformer.

---

In Chapter 28 we looked briefly at circuits with inductance, capacitance, and resistance and saw that in an $LC$ circuit with no resistance, the energy oscillates between electric energy in the capacitor and magnetic energy in the inductor. With resistance in the circuit, the oscillations are damped as energy is dissipated as heat. In this chapter we look at similar circuits driven by a generator of sinusoidal alternating current. Much of the mathematics involved in the detailed analysis of such circuits is studied in a course in differential equations. As in the last sections of Chapter 28 and in the study of forced oscillators in Chapter 11, we shall content ourselves with a qualitative analysis of these circuits, to bring out the main features without relying on differential-equation theory.

    We begin by studying a simple ac generator which puts out a sinusoidal emf, and then we consider the results of placing a resistor, capacitor, or inductor separately across its terminals. We then look at a combination of these circuit elements in series with such a generator. There are several reasons for studying sinusoidal currents and voltages.

Although practical generators are considerably different from the simple device we shall study in Section 30-1, they are designed to put out a sinusoidal emf. We shall see that when the generator output is sinusoidal, the current in an inductor, capacitor, or resistor is also sinusoidal, though not necessarily in phase with the generator emf. When the emf and current are both sinusoidal, their maximum values can be simply related. Finally, an important reason for studying sinusoidal currents is that even when the current in some circuit is not sinusoidal, it can be analyzed in terms of sinusoidal components using Fourier analysis. The results for sinusoidal currents can therefore be applied to any kind of alternating current.

## 30-1  An AC Generator

A simple generator of alternating current is a coil rotating in a uniform magnetic field, as shown in Figure 30-1, where the unit vector $\hat{n}$ normal to the plane of the coil makes an angle $\theta$ with a uniform magnetic field **B**. The magnetic flux through the coil is

$$\phi_m = NBA \cos \theta \qquad\qquad 30\text{-}1$$

where $N$ is the number of turns and $A$ the area of the coil. If the coil is mechanically rotated, the flux through the coil will change and, according to Faraday's law, there will be an emf in the coil. Let $\omega$ be the angular velocity of the coil. Then

$$\theta = \omega t \qquad\qquad 30\text{-}2$$

where we have chosen $\theta = 0$ at time $t = 0$ for convenience. Substituting this expression for $\theta$ into Equation 30-1, we have for the flux

$$\phi_m = NBA \cos \omega t \qquad\qquad 30\text{-}3$$

The emf in the coil will then be

$$\mathcal{E} = -\frac{d\phi_m}{dt} = -NBA(-\omega \sin \omega t) = +NBA\omega \sin \omega t$$

or

$$\mathcal{E} = \mathcal{E}_{max} \sin \omega t \qquad\qquad 30\text{-}4$$

where

$$\mathcal{E}_{max} = NBA\omega \qquad\qquad 30\text{-}5$$

is the maximum value of the emf. We can thus produce a sinusoidal emf in a coil by rotating it with constant angular velocity in a magnetic field. In this type of emf source, mechanical energy is converted into electric energy. In circuit diagrams, an ac generator is represented by the symbol $\ominus$. Although practical generators are considerably more complicated, they work on the principle that there is an alternating emf in a coil rotating in a magnetic field, and they are designed so that the emf produced is sinusoidal.

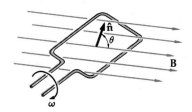

**Figure 30-1**
A coil rotating in a magnetic field with constant angular velocity generates a sinusoidal emf.

### Questions

1. Does the sinusoidal nature of the emf depend on the size or shape of the coil?

2. How could such a coil be used to generate a nonsinusoidal emf?

3. If such a generator delivers electric energy to a circuit, where does the energy come from?

## 30-2   Alternating Current in a Resistor

The simplest ac circuit consists of a generator and a resistor, as shown in Figure 30-2. We find the current in this circuit as usual by Kirchhoff's rule. Let the emf of the generator be given by

$$\mathcal{E} = \mathcal{E}_{max} \sin \omega t \qquad\qquad 30\text{-}6$$

Then Kirchhoff's rule gives

$$\mathcal{E} = IR = \mathcal{E}_{max} \sin \omega t \qquad\qquad 30\text{-}7$$

The current is proportional to the generator voltage and is given by

$$I = \frac{\mathcal{E}_{max}}{R} \sin \omega t = I_{max} \sin \omega t \qquad\qquad 30\text{-}8$$

where the maximum current is

$$I_{max} = \frac{\mathcal{E}_{max}}{R} \qquad\qquad 30\text{-}9$$

The instantaneous power dissipated in the resistor is

$$I^2R = (I_{max} \sin \omega t)^2R = I_{max}^2R \sin^2 \omega t$$

This power varies from zero to its maximum value $I_{max}^2R$, as shown in Figure 30-3. We are usually interested in the average power over one or more cycles. The average value of $\sin^2 \omega t$ over one or more cycles is $\frac{1}{2}$.* The average power dissipated in the resistor is thus

$$(I^2R)_{av} = \tfrac{1}{2}I_{max}^2R \qquad\qquad 30\text{-}10$$

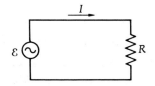

**Figure 30-2**
An ac generator in series with a resistor $R$.

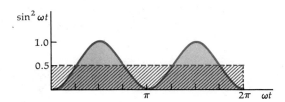

**Figure 30-3**
Plot of $\sin^2 \omega t$ versus $\omega t$. The average value over a cycle is $\frac{1}{2}$. This result can be obtained from the fact that $\cos^2 \omega t$ has the same average value over a cycle, and from $\sin^2 \omega t + \cos^2 \omega t = 1$.

Most ac ammeters and voltmeters measure root-mean-square (rms) values of current and voltage rather than the maximum or peak values. The rms value of the current, $I_{rms}$, is defined by

$$I_{rms} = \sqrt{(I^2)_{av}} \qquad\qquad 30\text{-}11$$

*rms current defined*

The average value of $I^2$ is

$$(I^2)_{av} = (I_{max} \sin \omega t)_{av}^2 = \tfrac{1}{2}I_{max}^2$$

---

* This can be seen from the identity $\sin^2 \omega t + \cos^2 \omega t = 1$, and the fact that the average values of $\sin^2 \omega t$ and $\cos^2 \omega t$ are equal.

The rms value of the current is thus

$$I_{rms} = \frac{1}{\sqrt{2}} I_{max} \qquad\qquad 30\text{-}12$$

The rms value of any quantity that varies sinusoidally equals the maximum value of that quantity divided by $\sqrt{2}$. In terms of the rms current, the average power dissipated in the resistor is

$$(I^2 R)_{av} = I_{rms}^2 R \qquad\qquad 30\text{-}13$$

The average power delivered by the generator is of course equal to that dissipated in the resistor for this simple circuit:

$$P_{av} = (\mathcal{E}I)_{av} = \mathcal{E}_{max} I_{max} (\sin^2 \omega t)_{av}$$

or

$$P_{av} = \tfrac{1}{2}\mathcal{E}_{max} I_{max} = \mathcal{E}_{rms} I_{rms} \qquad\qquad 30\text{-}14$$

where $\mathcal{E}_{rms} = (1/\sqrt{2})\, \mathcal{E}_{max}$. Since Equation 30-9 holds for the rms values of emf and current as well as the maximum values,

$$I_{rms} = \frac{\mathcal{E}_{rms}}{R} \qquad\qquad 30\text{-}15$$

The power input of the emf (Equation 30-14) equals the power dissipated in the resistor (Equation 30-13).

## 30-3   Alternating Current in a Capacitor

Figure 30-4 shows a capacitor connected across the terminals of a generator. Assuming $I$ to be positive in the clockwise direction as indicated, we have

$$I = \frac{dQ}{dt}$$

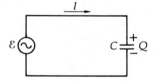

**Figure 30-4**
An ac generator in series with a capacitor $C$.

where $Q$ is the charge on the capacitor. Kirchhoff's rule for this circuit gives

$$\mathcal{E} = \frac{Q}{C} = \mathcal{E}_{max} \sin \omega t \qquad \text{or} \qquad Q = \mathcal{E}_{max} C \sin \omega t$$

The current is

$$I = \frac{dQ}{dt} = \omega \mathcal{E}_{max} C \cos \omega t = I_{max} \cos \omega t \qquad\qquad 30\text{-}16$$

where

$$I_{max} = \mathcal{E}_{max} \omega C \qquad\qquad 30\text{-}17$$

The current in this circuit is not in phase with the generator voltage, which is also the potential drop across the capacitor. Using the trigonometric identity

$$\cos \omega t = \sin (\omega t + 90°)$$

we have

$$I = I_{max} \sin (\omega t + 90°)$$

The current and voltage across the capacitor are plotted in Figure 30-5, from which we see that the maximum value of the voltage occurs 90° or one-fourth period after the maximum value of the current. The voltage drop across the capacitor is said to *lag* the current by 90°, or one-fourth period.

The relation between the maximum (or rms) current and the maximum (or rms) voltage can be written in a form similar to Ohm's law:

$$I_{max} = \omega C \mathcal{E}_{max} = \frac{\mathcal{E}_{max}}{1/\omega C} = \frac{\mathcal{E}_{max}}{X_C} \qquad 30\text{-}18a$$

or

$$I_{rms} = \frac{\mathcal{E}_{rms}}{X_C} \qquad 30\text{-}18b$$

where

$$X_C = \frac{1}{\omega C} \qquad 30\text{-}19$$

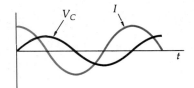

**Figure 30-5**
Plot of current $I$ and voltage $V_C$ across the capacitor versus time. The maximum in $V_C$ occurs one-fourth period after the maximum in the current. The voltage $V_C$ thus lags the current by one-fourth period, or 90°.

*Capacitive reactance*

is called the *capacitive reactance* of the circuit. Like resistance, capacitive reactance has units of ohms; the larger the reactance for a given emf, the smaller the maximum value of the current. Unlike resistance, the capacitive reactance depends on the frequency of the current: the larger the frequency, the smaller the reactance. The instantaneous power input by the generator is

$$P = \mathcal{E}I = (\mathcal{E}_{max} \sin \omega t)(I_{max} \cos \omega t)$$
$$= \mathcal{E}_{max}I_{max} \sin \omega t \cos \omega t$$

The average power input by the generator is zero. We can see this by writing $\sin \omega t \cos \omega t = \frac{1}{2} \sin 2\omega t$. This term oscillates twice during each cycle and is negative as often as it is positive.

---

**Example 30-1** A 20-$\mu$F capacitor is placed across a generator which has a maximum emf of 100 V. Find the reactance and maximum current when the frequency is 60 Hz and when it is 5000 Hz.

These frequencies correspond to angular frequencies $\omega_1$ and $\omega_2$, given by $\omega_1 = 2\pi f_1 = 2\pi(60 \text{ s}^{-1}) = 337 \text{ s}^{-1}$ and $\omega_2 = 2\pi(5000 \text{ s}^{-1}) = 3.14 \times 10^4 \text{ s}^{-1}$. The reactances at these frequencies are

$$X_1 = \frac{1}{\omega_1 C} = [377(20 \times 10^{-6})]^{-1} = 133 \ \Omega$$

and

$$X_2 = \frac{1}{\omega_2 C} = [(3.14 \times 10^4)(20 \times 10^{-6})]^{-1} = 1.59 \ \Omega$$

The maximum currents are then

$$I_{1,max} = \frac{\mathcal{E}_{max}}{X_1} = \frac{100 \text{ V}}{133 \ \Omega} = 0.754 \text{ A}$$

and

$$I_{2,max} = \frac{100 \text{ V}}{1.59 \ \Omega} = 62.8 \text{ A}$$

## 30-4 Alternating Current in an Inductor

Figure 30-6 shows an inductor across the terminals of an ac generator. Kirchhoff's rule for this circuit gives

$$\mathcal{E} = L\frac{dI}{dt} = \mathcal{E}_{max} \sin \omega t \qquad 30\text{-}20$$

**Figure 30-6**
An ac generator in series with an inductor $L$.

Solving for the current $I$ gives

$$I = -\frac{\mathcal{E}_{max}}{\omega L} \cos \omega t = \frac{\mathcal{E}_{max}}{\omega L} \sin (\omega t - 90°) \qquad 30\text{-}21$$

(We have neglected the constant of the integration because it depends on the initial conditions, which are of no interest for this discussion.) From the plot in Figure 30-7 of the current and voltage across the inductor we see that the maximum value of the voltage occurs 90° or one-fourth period before the corresponding maximum value of the current. The voltage across an inductor is said to *lead* the current by 90° or one-fourth period.

The maximum (or rms) current is related to the maximum (or rms) voltage across the inductor by

$$I_{max} = \frac{\mathcal{E}_{max}}{\omega L} = \frac{\mathcal{E}_{max}}{X_L} \qquad 30\text{-}22a$$

or

$$I_{rms} = \frac{\mathcal{E}_{rms}}{X_L} \qquad 30\text{-}22b$$

where

$$X_L = \omega L \qquad 30\text{-}23$$

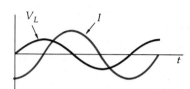

**Figure 30-7**
Plot of current $I$ and voltage $V_L$ across the inductor as functions of time. The maximum in $V_L$ occurs one-fourth period before the maximum current. The voltage $V_L$ thus leads the current by one-fourth period, or 90°.

*Inductive reactance*

The quantity $X_L$ is called the *inductive reactance*. As the frequency increases, the inductive reactance increases and the maximum current decreases. As it was for an ac generator and capacitor, the average power input of the generator into the inductor is zero. This is because the emf is proportional to $\sin \omega t$ and the current to $\cos \omega t$, and $(\sin \omega t \cos \omega t)_{av} = 0$.

## 30-5 *LCR* Circuit with Generator

An important circuit containing many of the features of most ac circuits is a series *LCR* circuit with a generator (Figure 30-8). We discussed a similar circuit without the generator in Chapter 28, finding that the current oscillates with angular frequency nearly equal to $\omega_0 = 1/\sqrt{LC}$ (assuming that the resistance is small) and that the maximum value of the current decreases exponentially with time. Kirchhoff's rule applied to this circuit with the generator gives

$$L\frac{dI}{dt} + IR + \frac{Q}{C} = \mathcal{E}_{max} \sin \omega t \qquad 30\text{-}24$$

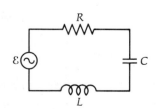

**Figure 30-8**
A series *LRC* circuit with an ac generator.

with $I = dQ/dt$. This equation is analogous to Equation 11-53 for the driven oscillator. The first term $L\,dI/dt = L\,d^2Q/dt^2$ is analogous to $m\,d^2x/dt^2$; the second term $IR = (dQ/dt)R$ is the damping term analogous to the term $bv = b\,dx/dt$ in Equation 11-53; the third term $Q/C$ is analogous to the restoring-force term $kx$. The solution of this equation for the current $I$ contains a part called the transient current, which decreases exponentially with time and is identical to the solution of the equation without the driving term on the right side, plus a steady-state current which does not decrease exponentially in time. At a sufficient time after the switch is closed, the transient current is negligible compared with the steady-state current. We shall ignore the transient current in our discussion of this circuit and concentrate on the steady-state solution. The current obtained by solving Equation 30-24 is

$$I = I_{max} \sin (\omega t - \phi) \qquad\qquad 30\text{-}25$$

where

$$\tan \phi = \frac{X_L - X_C}{R} \qquad\qquad 30\text{-}26$$

and

$$I_{max} = \frac{\mathcal{E}_{max}}{\sqrt{(X_L - X_C)^2 + R^2}} = \frac{\mathcal{E}_{max}}{Z} \qquad\qquad 30\text{-}27$$

The quantity

$$Z = \sqrt{(X_L - X_C)^2 + R^2} \qquad\qquad 30\text{-}28$$

is called the *impedance*. In terms of impedance, the current is given by

$$I = \frac{\mathcal{E}}{Z} \sin (\omega t - \phi) \qquad\qquad 30\text{-}29$$

We can obtain these results from a simple vector diagram called a *phasor diagram*. At any instant, the current is the same in each element of the series circuit. We represent this current by a vector (called a phasor) which has magnitude equal to $I_{max}$ and which makes an angle $\omega t - \phi$ with the $x$ axis. This vector rotates counterclockwise with angular frequency $\omega$. The instantaneous current is the $y$ component of this vector. If we multiply this vector by the resistance $R$, we obtain a vector $\mathbf{V}_R$ representing the voltage drop across the resistor. These vectors are parallel because the voltage drop across the resistor is in phase with the current. The voltage drop across the capacitor lags the current by 90°. This voltage is therefore represented by a vector $\mathbf{V}_C$ which has magnitude $I_{max}X_C$ and lags the vector $\mathbf{V}_R$ by 90°, as shown in Figure 30-9. The $y$ component of this vector equals the instantaneous voltage drop across the capacitor. Similarly, we represent the voltage drop across the inductor by a vector $\mathbf{V}_L$ which has magnitude $I_{max}X_L$ and which leads $\mathbf{V}_R$ by 90°. The sum of the $y$ components of these three vectors equals the instantaneous sum of the voltage drops across the resistor, capacitor, and inductor. By Kirchhoff's law, this sum equals the instantaneous emf. The sum of these $y$ components is the $y$ component of the resultant vector:

$$\mathcal{E} = \mathbf{V}_R + \mathbf{V}_L + \mathbf{V}_C \qquad\qquad 30\text{-}30$$

The vector $\mathcal{E}$ representing the emf makes an angle $\omega t$ with the $x$ axis and makes an angle $\phi$ with the current and resistance vector $\mathbf{V}_R$ (Figure 30-9b). From the right triangle in Figure 30-9a we obtain for the magnitude of the resultant vector

$$\mathcal{E}_{max} = |\mathbf{V}_R + \mathbf{V}_L + \mathbf{V}_C| = \sqrt{V_R^2 + (V_L - V_C)^2}$$

$$= I_{max} \sqrt{R^2 + (X_L - X_C)^2} = I_{max}Z$$

*Impedance*

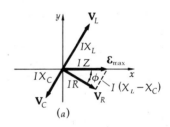

(a)

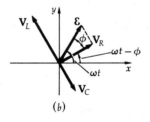

(b)

**Figure 30-9**
Phasor diagram for an *LRC* circuit with generator. The emf $\mathcal{E}$ and the voltage drops $V_R$, $V_C$, and $V_L$ are the $y$ components of the vectors or phasors representing these quantities. In (a) these vectors are shown at $t = 0$. The vector $\mathbf{V}_R$ has magnitude $IR$ and is in phase with the current. $\mathbf{V}_C$ lags the current by 90° and $\mathbf{V}_L$ leads the current by 90°. The current and $\mathbf{V}_R$ lag the emf by $\phi$, where $\tan \phi = (X_L - X_C)/R$. (The current $I$ can be either $I_{max}$ or $I_{rms}$.) In (b) each of the vectors is rotated through the angle $\omega t$.

This result is Equation 30-27. We see also from the diagram that the phase angle $\phi$ is given by Equation 30-26.

## Resonance

Since both the inductive reactance $X_L = \omega L$ and the capacitive reactance $X_C = 1/\omega C$ depend on the angular frequency $\omega$ of the applied emf, the impedance $Z$ and the rms current $I_{rms}$ also depend on $\omega$. As we increase $\omega$, the inductive reactance increases and the capacitive reactance decreases. When $X_L$ and $X_C$ are equal, the impedance $Z$ has its minimum value $Z_{min} = R$ and $I_{rms}$ has its greatest value. The value of $\omega$ for which $X_L$ and $X_C$ are equal is obtained from

$$\omega L = \frac{1}{\omega C}$$

or

$$\omega = \frac{1}{\sqrt{LC}} = \omega_0 \qquad \text{30-31} \qquad \textit{Resonance condition}$$

The impedance is minimum and the rms current maximum when the angular frequency of the emf equals the natural angular frequency $\omega_0 = 1/\sqrt{LC}$. At this frequency the circuit is said to be at *resonance*. This resonance condition in a driven *LCR* circuit is exactly the same as that in a driven simple harmonic oscillator. We note from Equation 30-26 that when $X_L = X_C$, the phase angle $\phi$ is zero; i.e., at resonance the current is in phase with the applied emf.

The instantaneous power supplied by the generator to the circuit is

$$P = \mathcal{E}I = \mathcal{E}_{max} \sin \omega t \; I_{max} \sin (\omega t - \phi)$$

The average power is most easily computed by first expanding the function $\sin (\omega t - \phi)$ using the trigonometric identity for the sine of the sum of two angles:

$$\sin (\omega t - \phi) = \sin \omega t \cos \phi - \cos \omega t \sin \phi$$

Then

$$P = \mathcal{E}_{max}I_{max} \cos \phi \sin^2 \omega t - \sin \phi \sin \omega t \cos \omega t$$

The time average of $\sin^2 \omega t$ is $\frac{1}{2}$, and that of $\sin \omega t \cos \omega t$ is zero, as we have seen before. The average power is therefore

$$P_{av} = \tfrac{1}{2}\mathcal{E}_{max}I_{max} \cos \phi = \mathcal{E}_{rms}I_{rms} \cos \phi \qquad \text{30-32}$$

As in a dc circuit, the power supplied depends on the product of the emf and the current. However, Equation 30-32 also contains the factor $\cos \phi$, called the *power factor*. The power supplied is maximum at resonance as can be seen from this equation, since $\phi = 0$ at resonance and $I_{rms}$ has its maximum value.

The power can be expressed as a function of the angular frequency $\omega$. From the triangle in Figure 30-9a we have

$$\cos \phi = \frac{R}{Z} \qquad \text{30-33}$$

When we use this result and write $I_{rms} = \mathcal{E}_{rms}/Z$, the average power is

$$P_{av} = \mathcal{E}_{rms}^2 \frac{R}{Z^2} \qquad \text{30-34}$$

From the definition of the impedance $Z$ we have

$$Z^2 = (X_L - X_C)^2 + R^2 = \left(\omega L - \frac{1}{\omega C}\right)^2 + R^2$$

$$= \frac{L^2}{\omega^2}\left(\omega^2 - \frac{1}{LC}\right)^2 + R^2 = \frac{L^2}{\omega^2}(\omega^2 - \omega_0^2)^2 + R^2$$

where we have used $\omega_0^2 = 1/LC$. Using this expression for $Z^2$, we obtain for the average power as a function of $\omega$

$$P_{av} = \frac{\mathcal{E}_{rms}^2 R\omega^2}{L^2(\omega^2 - \omega_0^2)^2 + \omega^2 R^2} \qquad\qquad 30\text{-}35$$

Equation 30-35 is the same as Equation 11-62 for the average power input in a driven oscillator with $R$ replacing the damping constant $b$, $L$ replacing the mass $m$, and $\mathcal{E}$ replacing the driving force $F_0$ (the factor of $\frac{1}{2}$ in Equation 11-62 is the result of using the maximum driving force $F_0$ rather than the rms force).

Figure 30-10 shows a plot of the average power versus generator frequency $\omega$ for two values of the resistance $R$. The smaller the resistance, the more sharply peaked the resonance. Let $\Delta\omega = \omega_2 - \omega_1$ be the width of the resonance curve, where $\omega_1$ and $\omega_2$ are the two values of $\omega$ for which $P_{av}$ is half its maximum value. For a sharply peaked resonance we can find a useful approximation for the width as follows. At resonance, the denominator of the expression on the right in Equation 30-35 is just $\omega^2 R^2$. The power will be half its maximum value when this denominator is approximately twice this value (neglecting the variation of the numerator with $\omega$). Then at the half-power points

$$L^2(\omega^2 - \omega_0^2)^2 \approx \omega^2 R^2$$

or

$$L(\omega^2 - \omega_0^2) = L(\omega - \omega_0)(\omega + \omega_0) \approx \pm\omega R$$

Since $\omega$ and $\omega_0$ are nearly equal for a sharply peaked resonance, we can replace $\omega R$ with $\omega_0 R$ and $\omega + \omega_0$ with $2\omega_0$. Then we have approximately

$$\omega - \omega_0 \approx \pm\frac{R}{2L}$$

$$\omega \approx \omega_0 \pm \frac{R}{2L}$$

That is, $\omega_2 \approx \omega_0 + R/2L$ and $\omega_1 \approx \omega_0 - R/2L$. The width is then

$$\Delta\omega \approx \frac{R}{L} \qquad\qquad 30\text{-}36$$

As we did for the resonance curve for a mechanical driven oscillator, we describe the width of the resonance by a dimensionless parameter, the $Q$ value or quality factor, which in general is defined by $Q = L\omega_0/R$. When $Q$ is greater than about 2 or 3, the resonance is sharply peaked and our approximations for the width are valid. Then $Q$ is simply the ratio of the resonance frequency to the width:

$$Q = \frac{L\omega_0}{R} = \frac{\omega_0}{\Delta\omega} \qquad\qquad 30\text{-}37 \qquad Q \ value$$

A common application of series resonance circuits is a radio receiver, where the resonance frequency of the circuit is varied by varying the capacitance. Resonance occurs when the natural frequency of the circuit

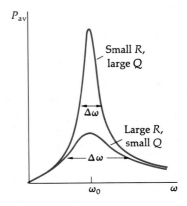

**Figure 30-10**
Plot of average power $P_{av}$ given by Equation 30-35 versus $\omega$ for small and large $R$. The $Q$ value is the ratio of the resonance frequency $\omega_0$ to the resonance width $\Delta\omega$. A large $Q$ value indicates a narrow resonance.

equals the frequency of the radio waves picked up at the antenna. At resonance, there is a relatively large current in the antenna circuit. If the $Q$ value of the circuit is sufficiently high, currents due to other stations off resonance will be negligible compared with those due to the station to which the circuit is tuned.

---

**Example 30-2** A series $LCR$ circuit with $L = 2$ H, $C = 2$ $\mu$F, and $R = 20$ $\Omega$ is driven by a generator of maximum emf 100 V and variable frequency. Find the resonance frequency $\omega_0$, the phase $\phi$, and maximum current $I_{max}$ when the generator angular frequency is $\omega = 400$ rad/s.

The resonance frequency is $\omega_0 = 1/\sqrt{LC} = 1/\sqrt{(2 \text{ H}) (2 \times 10^{-6} \text{ F})} = 500$ rad/s. When the generator frequency is 400 rad/s, it is well below the resonance frequency. The capacitive and inductive reactances at 400 rad/s are

$$X_C = \frac{1}{\omega C} = \frac{1}{(400) (2 \times 10^{-6})} = 1250 \ \Omega$$

and

$$X_L = \omega L = (400) (2) = 800 \ \Omega$$

The total reactance is $X_L - X_C = 800 \ \Omega - 1250 \ \Omega = -450 \ \Omega$. This is of much greater magnitude than the resistance, a result which always holds far from resonance. The total impedance is

$$Z = \sqrt{(X_L - X_C)^2 + R^2} \approx 450 \ \Omega$$

since $20^2$ is negligible compared with $450^2$. The maximum current is then

$$I_{max} = \frac{\mathcal{E}_{max}}{Z} = \frac{100 \text{ V}}{450 \ \Omega} = 0.222 \text{ A}$$

This is small compared with $I_{max}$ at resonance, which is $(100 \text{ V})/(20 \ \Omega) = 5$ A. The phase angle $\phi$ is given by

$$\tan \phi = \frac{X_L - X_C}{R} = \frac{-450 \ \Omega}{20 \ \Omega} = -22.5 \qquad \phi = -87°$$

From Equation 30-25 we see that a negative phase angle means that the current leads the generator voltage.

---

**Example 30-3** Find the resonance width and $Q$ value of the circuit of Example 30-2.

In that example we found the resonance frequency to be $\omega_0 = 500$ rad/s. The $Q$ value is then

$$Q = \frac{L\omega_0}{R} = \frac{2(500)}{20} = 50$$

and the width of the resonance is

$$\Delta\omega = \frac{\omega_0}{Q} = \frac{500}{50} = 10 \text{ rad/s}$$

---

**Questions**

4. In general, does the power factor depend on frequency?

5. Can the instantaneous power delivered by the generator ever be negative?

6. What is the power factor for a circuit which has inductance or capacitance or both but zero resistance?

7. What is the power factor for a circuit with resistance but no inductance or capacitance? Does this power factor depend on frequency?

## 30-6  The Transformer

In order to transport power with a minimum $I^2R$ heat loss in transmission lines, it is economical to use a high voltage and low current. On the other hand, safety and other considerations, e.g., insulation, make it convenient to use power to run motors and other electrical appliances at lower voltage and higher current. This is accomplished by using a transformer, a device for changing ac voltage and current without appreciable loss in power. If $V$ is the voltage and $I$ the current, the instantaneous power is $VI$. If the voltage is to be changed without change in power, the current must also be changed.

Figure 30-11 diagrams a simple transformer consisting of two coils of wire around a common core of soft iron. The coil carrying the input power is called the *primary*, and the other coil is called the *secondary*. Either coil of a transformer can be used for the primary or secondary. The function of the iron core is to increase the flux greatly for a given current and to guide it so that nearly all the flux through one turn of one coil goes through each turn of both coils. The iron core is laminated to reduce eddy-current losses. Other possible losses are the $I^2R$ losses in the coils, which can be reduced by using a low-resistance wire for the coils, and hysteresis losses in the core, which can be reduced by using soft iron. It is relatively easy to design a transformer for which power is transferred from the primary to the secondary with efficiency of 90 to 99 percent. We shall discuss only an ideal transformer, which has no losses.

Consider an ac generator of emf $\varepsilon$ across the primary of $N_1$ turns with the secondary coil of $N_2$ turns open. Because of the iron core, there is a large flux through each coil even with a very small magnetizing current $I_m$ in the primary circuit. We can neglect the resistance of the coil compared with the inductive reactance. The primary is then a simple circuit consisting of an ac generator and a pure inductance, as discussed in Section 30-4. The current and voltage in the primary are out of phase by $90°$, and the average power in the primary circuit is zero. We can see from Faraday's law that the voltage across the secondary is just the turns ratio $N_2/N_1$ times the voltage across the primary. The induced emf $V_1$ in the primary circuit is

$$V_1 = -\frac{d\phi}{dt}$$

The total flux $\phi$ is the flux through each turn $\phi_{\text{turn}}$ times the number of turns $N_1$:

$$V_1 = -N_1 \frac{d\phi_{\text{turn}}}{dt} \tag{30-38}$$

Assuming no flux leakage out of the iron core, the flux through each

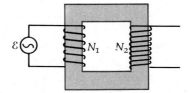

**Figure 30-11**
Transformer with $N_1$ turns in the primary and $N_2$ turns in the secondary.

turn is the same for both coils. Thus the total flux through the secondary coil is $N_2\phi_{\text{turn}}$, and the voltage across the secondary coil is

$$V_2 = -N_2 \frac{d\phi_{\text{turn}}}{dt} \qquad\qquad 30\text{-}39$$

Comparing these equations, we see that

$$V_2 = \frac{N_2}{N_1} V_1 \qquad\qquad 30\text{-}40$$

According to Kirchhoff's law for the primary circuit, the applied emf $\mathcal{E}$ is the negative of the induced emf $V_1$:

$$\mathcal{E} - \frac{d\phi}{dt} = 0$$

$$\mathcal{E} = \frac{d\phi}{dt} = -V_1 \qquad\qquad 30\text{-}41$$

In terms of the applied emf, the voltage across the secondary coil is

$$V_2 = -\frac{N_2}{N_1} \mathcal{E} \qquad\qquad 30\text{-}42$$

The transformer is called a *step-up transformer* if $N_2$ is greater than $N_1$ so that the output voltage is greater than the input voltage. If $N_2$ is less than $N_1$, it is a *step-down transformer*. There is no current in the secondary because that circuit is open. The very small current $I_m$ in the primary coil is 90° out of phase with the emf in that coil.

Consider now what happens when we put a resistance $R$, called a *load resistance*, across the secondary coil. There will then be a current $I_2$ in the secondary circuit which is in phase with the voltage $V_2$ across the resistance. This current will set up an additional flux through each turn $\phi'_{\text{turn}}$ proportional to $N_2 I_2$. This flux adds to the original flux $\phi_{\text{turn}}$ set up by the original magnetizing current in the primary $I_m$. However, the voltage across the primary coil is determined by the generator emf, which is unaffected by the secondary circuit. According to Equation 30-41, the flux in the iron core must therefore change at the original rate; i.e., the total flux in the iron core must be the same as with no load across the secondary. The primary coil thus draws an additional current $I_1$ to maintain the original flux $\phi_{\text{turn}}$. The flux through each turn produced by this additional current is proportional to $N_1 I_1$. Thus the additional current $I_1$ in the primary is related to the current $I_2$ in the secondary by

$$N_1 I_1 = -N_2 I_2 \qquad\qquad 30\text{-}43$$

The negative sign indicates that these currents are 180° out of phase because they produce counteracting fluxes. Since $I_2$ is in phase with $V_2$, the additional current in the primary $I_1$ is in phase with the applied emf.

Figure 30-12 is a phasor diagram of the phase relationships between the voltages and currents. The total current in the primary $I$ is the "vector" sum of the original magnetizing current $I_m$ and the additional current $I_1$, which is usually very much greater than $I_m$. The power delivered by the generator is the product of the rms applied emf, the rms total current in the primary $I_{\text{rms}}$, and the power factor $\cos\phi$, where $\phi$ is the phase angle between the applied emf and the total current $I$. Since $I_1$ is in phase with the applied emf, $\phi$ is the angle between $I_1$ and $I$ shown in Figure 30-12.

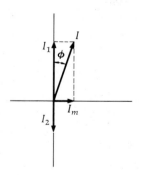

**Figure 30-12**
Phase relations between the magnetizing current $I_m$, the secondary current $I_2$, and the additional primary current $I_1$. The total primary current $I$ is nearly equal to $I_1$ because $I_m$ is usually small. Then the power factor, $\cos\phi$, is nearly 1.

We note from this figure that $I \cos \phi$ is just the additional current $I_1$, and so the power input by the primary is

$$P = \mathcal{E}_{rms} I_{rms} \cos \phi = \mathcal{E}_{rms} I_{1,rms}$$

Using Equations 30-42 and 30-43 to relate the applied emf and the additional current $I_1$ in the primary to the voltage $V_2$ and the current $I_2$ in the secondary, we have

$$\mathcal{E}I_1 = \left(-\frac{N_1}{N_2} V_2\right)\left(-\frac{N_2}{N_1} I_1\right) = V_2 I_2 \qquad 30\text{-}44$$

Thus

$$\mathcal{E}_{rms} I_{1,rms} = V_{2,rms} I_{2,rms} \qquad 30\text{-}45$$

The power input to the primary equals the power output in the secondary, as assumed for an ideal transformer with no losses.

In most cases the additional current in the primary $I_1$ is much greater than the original magnetizing current $I_m$ with no load. This can be demonstrated by putting a light bulb in series with the primary coil: it is much brighter when there is a load across the secondary than when the secondary circuit is open. If $I_m$ can be neglected, Equation 30-43 relates the total currents in the primary and secondary circuit.

The currents $I_1$ and $I_2$ can be related to the resistance $R$ across the secondary. For the secondary current we have simply

$$I_2 = \frac{V_2}{R}$$

Using Equations 30-42 and 30-43 to write $I_2$ and $V_2$ in terms of the primary current $I_1$ and the primary emf $\mathcal{E}$, we have

$$I_1 = \frac{\mathcal{E}}{(N_1/N_2)^2 R} \qquad 30\text{-}46$$

The current $I_1$ is the same as if we connected a resistance $(N_1/N_2)^2 R$ across the generator. This effect is called *impedance transformation* since in general some combination of capacitance, inductance, and resistance with impedance $Z$ is connected across the secondary coil as the load.

## Review

A. Define, explain, or otherwise identify:

| | |
|---|---|
| rms current, 817 | $Q$ value, 823 |
| Capacitive reactance, 819 | Transformer, 825 |
| Inductive reactance, 820 | Primary, 825 |
| Impedance, 821 | Secondary, 825 |
| Phasor, 821 | Step-up transformer, 826 |
| Resonance condition, 822 | Step-down transformer, 826 |
| Power factor, 822 | Impedance transformation, 827 |
| Resonance width, 823 | |

B. True or false:

1. At very high frequencies a capacitor acts like a short circuit.

2. A circuit with a high $Q$ value has a narrow resonance curve.

3. At resonance the impedance equals the resistance.

4. Off resonance, the current always lags the generator voltage.

5. At resonance, the current and generator voltage are in phase.

6. If a transformer increases the current, it must decrease the voltage.

## Exercises

### Section 30-1, An AC Generator

1. A coil has an area of 4 cm$^2$, has 200 turns, and rotates in a magnetic field of 0.5 T. (a) What must its angular frequency be to generate an emf of 10 V maximum? (b) If it rotates at 60 Hz, what is $\varepsilon_{max}$?

2. In what magnetic field should the coil of Exercise 1 rotate to generate an emf of 10 V maximum at 60 Hz?

3. (a) Show that the rms emf of a generator is related to the maximum emf by $\varepsilon_{rms} = 0.707\varepsilon_{max}$. (b) The rms value of standard house voltage is 120 V. What is the maximum voltage?

### Section 30-2, Alternating Current in a Resistor

4. A 100-W light bulb is plugged into a standard 120-V rms outlet. Find (a) $I_{rms}$, (b) $I_{max}$, and (c) the maximum power.

5. A 3-$\Omega$ resistor is placed in series with a 12.0-V (maximum) generator of frequency 60 Hz. (a) What is the angular frequency of the current? (b) Find $I_{max}$ and $I_{rms}$. What is (c) the maximum power into the resistor, (d) the minimum power, and (e) the average power?

6. A 5.0-kW rms electric clothes dryer runs on 220 V rms. Find (a) $I_{rms}$ and (b) $I_{max}$. (c) Find these same quantities for a dryer of the same power that operates at 120 V rms.

### Section 30-3, Alternating Current in a Capacitor

7. What is the reactance of a 1.0-nF capacitor at (a) 60 Hz, (b) 6 kHz, and (c) 6 MHz?

8. Find the reactance of a 10.0-$\mu$F capacitor at (a) 60 Hz, (b) 600 Hz, and (c) 6 kHz.

9. Sketch a graph of $X_C$ versus frequency $f$ for $C = 100$ $\mu$F.

10. An emf of 10.0 V maximum and frequency 20 Hz is applied to a 20-$\mu$F capacitor. Find (a) $I_{max}$ and (b) $I_{rms}$.

11. At what frequency is the reactance of a 10-$\mu$F capacitor (a) 1 $\Omega$, (b) 100 $\Omega$, and (c) 0.01 $\Omega$?

12. Show that the maximum power delivered by a generator to a capacitor is $P_{max} = \varepsilon_{rms}I_{rms}$. What is the average power?

### Section 30-4, Alternating Current in an Inductor

13. What is the reactance of a 1.0-mH inductor at (a) 60 Hz, (b) 600 Hz, and (c) 6 kHz?

14. An inductor has a reactance of 100 $\Omega$ at 80 Hz. (a) What is its inductance? (b) What is its reactance at 160 Hz?

15. At what frequency would the reactance of a 10.0-$\mu$F capacitor equal that of a 1.0-mH inductor?

### Section 30-5, *LCR* Circuit with Generator

16. Show that Equation 30-27 can be written

$$I_{max} = \frac{\omega \varepsilon_{max}}{\sqrt{L^2(\omega^2 - \omega_0^2)^2 + \omega^2 R^2}}$$

17. A series $LCR$ circuit with $L = 10$ mH, $C = 2$ $\mu$F, and $R = 5$ $\Omega$ is driven by a generator of maximum emf 100 V and variable angular frequency $\omega$. Find ($a$) the resonance frequency $\omega_0$ and ($b$) $I_{rms}$ at resonance. When $\omega = 8000$ rad/s, find ($c$) $X_C$ and $X_L$, ($d$) $Z$ and $I_{rms}$, and ($e$) the phase angle $\phi$.

18. ($a$) Show that Equation 30-26 can be written $\tan \phi = L(\omega^2 - \omega_0^2)/\omega R$. Find $\phi$ approximately ($b$) at very low frequencies and ($c$) at very high frequencies.

19. For the circuit of Exercise 17, let the generator frequency be $f = \omega/2\pi = 1$ kHz. Find ($a$) the resonance frequency $f_0 = \omega_0/2\pi$, ($b$) $X_C$ and $X_L$, ($c$) the total impedance $Z$ and $I_{rms}$, and ($d$) the phase angle $\phi$.

20. A series $LCR$ circuit in a radio receiver is tuned by a variable capacitor so that it can resonate at frequencies from 500 to 1600 kHz. If $L = 1.0$ $\mu$H, find the range of $C$ necessary to cover this range of frequencies.

21. FM radio stations have carrier frequencies which are separated by 0.20 MHz. When the radio is tuned to a station such as 100.1 MHz, the resonance width of the receiver circuit should be much smaller than 0.2 MHz so that adjacent stations are not received. If $f_0 = 100.1$ MHz and $\Delta f = 0.05$ MHz, what is the $Q$ value of the circuit?

22. ($a$) Find the power factor for the circuit in Example 30-2 when $\omega = 400$ rad/s. ($b$) At what angular frequency is the power factor $\frac{1}{2}$?

23. Find ($a$) the $Q$ value and ($b$) the resonance width for the circuit of Exercise 17. ($c$) What is the power factor when $\omega = 8000$ rad/s?

24. Show that when there is resistance and capacitance but no inductance, the power factor is given by $\cos \phi = RC\omega/\sqrt{1 + (RC\omega)^2}$. Sketch the power factor versus $\omega$.

25. An ac generator of maximum emf 20 V is connected in series with a 20-$\mu$F capacitor and an 80-$\Omega$ resistor. There is no inductance in the circuit. Find ($a$) the power factor, ($b$) the rms current, and ($c$) the average power if the angular frequency of the generator is 400 rad/s.

26. Express the power factor as a function of $\omega$ and sketch for a circuit with resistance and inductance in series with a generator.

27. A coil can be considered to be a resistance and an inductance in series. Assume that $R = 100$ $\Omega$ and $L = 0.4$ H. The coil is connected across a 120-V 60-Hz line. Find ($a$) the power factor, ($b$) the rms current, and ($c$) the average power supplied.

28. Find the power factor and the phase $\phi$ for the circuit of Exercise 17 when the generator frequency is ($a$) 900 Hz, ($b$) 1.1 kHz, and ($c$) 1.3 kHz.

### Section 30-6, The Transformer

29. A transformer has 400 turns on the primary and 8 turns on the secondary. ($a$) Is this a step-up or step-down transformer? ($b$) If the primary is connected across 120 V rms, what is the open-circuit voltage across the secondary? ($c$) If the primary current is 0.1 A, what is the secondary current, assuming negligible magnetizing current and no power loss?

30. The primary of a step-down transformer has 250 turns and is connected to a 120-V rms line. The secondary is to supply 20 A at 9 V. Find ($a$) the current in the primary and ($b$) the number of turns in the secondary, assuming 100 percent efficiency.

31. A transformer has 500 turns in its primary, which is connected to 120 V rms. Its secondary coil is tapped at three places to give an output of 2.5, 7.5, and 9 V. How many turns are needed for each part of the secondary coil?

## Problems

1. A coil with resistance and inductance is connected to a 120-V rms 60-Hz line. The average power supplied to the coil is 60 W, and the rms current is 1.5 A. Find (a) the power factor, (b) the resistance of the coil, and (c) the inductance of the coil. (d) Does the current lead or lag the voltage? What is the phase angle $\phi$?

2. Compute by direct integration the area under the curve $\sin^2 \omega t$ from $t = 0$ to $t = T = 2\pi/\omega$ and show that it equals $\frac{1}{2}T$.

3. In a certain LCR circuit $X_C = 16\ \Omega$ and $X_L = 4\ \Omega$ at some frequency. The resonance frequency is $\omega_0 = 10^4$ rad/s. (a) Find $L$ and $C$. If $R = 5\ \Omega$ and $\mathcal{E}_{max} = 26$ V, find (b) the $Q$ value and (c) the maximum current.

4. Show that the formula $P_{av} = \mathcal{E}_{rms}^2 R/Z^2$ gives the correct result for a circuit containing only a generator and (a) a resistor, (b) a capacitor, (c) an inductor.

5. In a series LCR circuit connected to an ac generator whose maximum emf is 200 V, the resistance is 60 $\Omega$ and the capacitance is 8.0 $\mu$F. The inductor can be varied from 8.0 to 40.0 mH by insertion of an iron core into a solenoid. The angular frequency is 2500 rad/s. If the capacitor voltage is not to exceed 150 V, find (a) the maximum current and (b) the range of $L$ that is safe to use.

6. Sketch the impedance $Z$ versus $\omega$ for (a) a series LR circuit, (b) a series RC circuit, and (c) a series LCR circuit.

7. When an LCR series circuit is connected to a 120-V rms 60-Hz line, the current is $I_{rms} = 11.0$ A and the current leads the emf by 45°. (a) Find the power supplied to the circuit. (b) What is the resistance? (c) If the inductance $L = 0.05$ H, find the capacitance $C$. (d) What capacitance or inductance would you add to make the power factor 1?

8. (a) Show that Equation 30-26 can be written $\tan \phi = Q(\omega^2 - \omega_0^2)/\omega\omega_0$. (b) Show that near resonance, $\tan \phi \approx 2Q(\omega - \omega_0)/\omega$. (c) Sketch $\phi$ versus $x$ where $x = \omega/\omega_0$, for a circuit with high $Q$ and for one with low $Q$.

9. Show by direct substitution that the current given by Equation 30-25 with $I_{max}$ and $\phi$ given by Equations 30-26 and 30-27 satisfies Equation 30-24. *Hint:* Use trigonometric identities for the sine and cosine of the sum of two angles and write the equation in the form $A \sin \omega t + B \cos \omega t = 0$. Since this equation must hold for all time, $A = 0$ and $B = 0$.

10. A resistor and inductor are *in parallel* across an emf $\mathcal{E} = \mathcal{E}_{max} \sin \omega t$ as in Figure 30-13. Show that (a) $I_R = (\mathcal{E}_{max}/R) \sin \omega t$, (b) $I_L = (\mathcal{E}_{max}/X_L) \sin (\omega t - 90°)$, and (c) $I = I_R + I_L = I_{max} \sin (\omega t - \phi)$, where $\tan \phi = R/X_L$ and $I_{max} = \mathcal{E}_{max}/Z$ with $Z^{-2} = R^{-2} + X_L^{-2}$.

11. A certain electrical device draws 10 A rms and has an average power of 720 W when connected to a 120-V rms 60-Hz power line. (a) What is the impedance of the device? (b) What series combination of resistance and reactance is this device equivalent to? (c) If the current leads the emf, is the reactance inductive or capacitive?

12. An ac generator is in series with a capacitor and an inductor in a circuit with negligible resistance. (a) Show that the charge on the capacitor obeys the equation

$$L \frac{d^2Q}{dt^2} + \frac{Q}{C} = \mathcal{E}_{max} \sin \omega t$$

(b) Show by direct substitution that this equation is satisfied by $Q = Q_{max} \sin \omega t$ if

$$Q_{max} = -\frac{\mathcal{E}_{max}}{L(\omega^2 - \omega_0^2)}$$

**Figure 30-13**
Parallel RL circuit and ac generator for Problem 10.

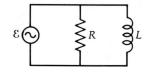

(c) Show that the current can be written $I = I_{max} \sin (\omega t - \phi)$, where

$$I_{max} = \frac{\mathcal{E}_{max}}{|\omega^2 - \omega_0^2|} = \frac{\mathcal{E}_{max}}{|X_L - X_C|}$$

and $\phi = -90°$ for $\omega < \omega_0$ and $\phi = +90°$ for $\omega > \omega_0$.

13. A method for measuring inductance is to connect the inductor in series with a known capacitance, a resistor, an ac ammeter, and a variable-frequency signal generator. The frequency of the signal generator is varied and the emf kept constant until the current is maximum. If $C = 10 \mu F$, $\mathcal{E}_{max} = 10$ V, $R = 100 \Omega$, and $I$ is maximum at $\omega = 5000$ rad/s, what is $L$? What is $I_{max}$?

14. A resistor, inductor, and capacitor are *in parallel* across an ac generator $\mathcal{E} = \mathcal{E}_{max} \sin \omega t$, as shown in Figure 30-14. Show that the total current in the generator is given by $I = I_{max} \sin (\omega t - \phi)$, where $I_{max} = \mathcal{E}_{max}/Z$ with $Z^{-2} = R^{-2} + (1/X_L - 1/X_C)^2$ and $\tan \phi = R(1/X_L - 1/X_C)$.

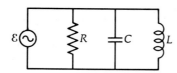

**Figure 30-14**
Parallel *LCR* circuit and ac generator for Problem 14.

15. Find the power factor for the parallel circuit of Problem 14 and show that $\phi = 0$ when $X_C = X_L$.

16. A resistor and capacitor are connected *in parallel* across a sinusoidal emf $\mathcal{E} = \mathcal{E}_{max} \sin \omega t$, as in Figure 30-15. (a) Show that the current in the resistor is $I_R = (\mathcal{E}_{max}/R) \sin \omega t$. (b) Show that the current through the capacitor branch is $I_C = (\mathcal{E}_{max}/X_C) \sin (\omega t + 90°)$. (c) Show that the total current is given by $I = I_R + I_C = I_{max} \sin (\omega t + \phi)$, where $\tan \phi = R/X_C$ and $I_{max} = \mathcal{E}_{max}/Z$ with $Z^{-2} = R^{-2} + X_C^{-2}$.

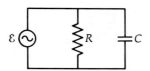

**Figure 30-15**
Parallel *RC* circuit and ac generator for Problem 16.

CHAPTER 31 Maxwell's Equations and Electromagnetic Waves

---

**Objectives** After studying this chapter, you should:

1. Be able to write down Maxwell's equations and discuss the experimental basis of each.

2. Be able to state the expression for the speed of an electromagnetic wave in terms of the fundamental constants $\mu_0$ and $\epsilon_0$.

3. Be able to state the expression for the Poynting vector and discuss its significance.

4. Be able to state the relationships between the Poynting vector, the intensity of an electromagnetic wave, and radiation pressure.

5. Be able to calculate the radiation pressure and the maximum values of $E$ and $B$ from the intensity of an electromagnetic wave.

---

About 1860 the great English physicist James Clerk Maxwell found that the experimental laws of electricity and magnetism, which we have studied in Chapters 20 to 30, could be conveniently summarized in a set of equations now known as *Maxwell's equations*. They relate the electric and magnetic field vectors **E** and **B** to their sources, which are electric charges, currents, and changing fields. Maxwell's equations express in a concise mathematical form the laws of Coulomb, Gauss, Biot-Savart, and Ampère (with the important addition of Maxwell's displacement current). They play a role in classical electromagnetism analogous to that of Newton's laws in classical mechanics. In principle, all problems in electricity and magnetism can be solved from Maxwell's equations, just as all problems in classical mechanics can be solved from Newton's laws. Maxwell's equations are considerably more complicated than Newton's laws, however, and their application to the solution of most problems involves mathematics beyond the scope of this book. Nevertheless, Maxwell's equations are of great theoretical importance. Maxwell showed that these equations could be combined to yield a wave equation for the electric and magnetic field vectors **E** and **B**. Such *elec-*

*tromagnetic waves* are caused by oscillating charges and currents. They were first produced in the laboratory by Heinrich Hertz in 1887. Maxwell showed that the speed of electromagnetic waves is predicted to be

$$c = \frac{1}{\sqrt{\mu_0 \epsilon_0}} \qquad\qquad 31\text{-}1$$

where $\epsilon_0$ is the constant appearing in Coulomb's and Gauss' laws and $\mu_0$ is that in the Biot-Savart law and Ampère's law. When the measured values of these fundamental constants are put into Equation 31-1, the speed of electromagnetic waves is found to be about $3 \times 10^8$ m/s, the same as the measured speed of light. Maxwell noted this "coincidence" with great excitement and correctly surmised that light itself is an electromagnetic wave.

In this chapter we state Maxwell's equations and relate each one to the laws of electricity and magnetism already studied. We then show that Maxwell's equations imply that the electric and magnetic field vectors obey a wave equation which describes waves propagating through free space with speed $c = 1/\sqrt{\mu_0 \epsilon_0}$.

## 31-1  Maxwell's Equations

Maxwell's equations are

$$\oint_S \mathbf{E} \cdot \hat{\mathbf{n}} \, dA = \frac{1}{\epsilon_0} Q \qquad\qquad 31\text{-}2$$

$$\oint_S \mathbf{B} \cdot \hat{\mathbf{n}} \, dA = 0 \qquad\qquad 31\text{-}3$$

$$\oint_C \mathbf{E} \cdot d\boldsymbol{\ell} = -\frac{d}{dt} \int_S \mathbf{B} \cdot \hat{\mathbf{n}} \, dA \qquad\qquad 31\text{-}4$$

$$\oint_C \mathbf{B} \cdot d\boldsymbol{\ell} = \mu_0 I + \mu_0 \epsilon_0 \frac{d}{dt} \int_S \mathbf{E} \cdot \hat{\mathbf{n}} \, dA \qquad\qquad 31\text{-}5$$

Equation 31-2 is Gauss' law; it states that the flux of the electric field through any closed surface equals $1/\epsilon_0$ times the net charge inside the surface. As discussed in Chapter 21, Gauss' law implies that the electric field due to a point charge varies inversely as the square of the distance from the charge. This law describes how the lines of **E** diverge from a positive charge and converge on a negative charge. Its experimental basis is Coulomb's law.

Equation 31-3, sometimes called Gauss' law for magnetism, states that the flux of the magnetic induction vector **B** is zero through any closed surface. This equation describes the experimental observation that the lines of **B** do not diverge from any point in space or converge to any point; i.e., isolated magnetic poles do not exist.

Equation 31-4 is Faraday's law. The integral of the electric field around any closed curve $C$ is the emf. This emf equals the (negative) rate of change of the magnetic flux through any surface $S$ bounded by the curve. (This is not a closed surface, and so the magnetic flux through $S$ is not necessarily zero.) Faraday's law describes how the lines of **E** encircle an area through which the magnetic flux is changing. Faraday's law relates the electric field vector **E** to the magnetic field vector **B**.

Equation 31-5, Ampère's law with Maxwell's displacement-current modification, states that the line integral of the magnetic induction **B** around any closed curve $C$ equals $\mu_0$ times the current through any surface bounded by the curve plus $\mu_0\epsilon_0$ times the rate of change of the electric flux through such a surface. This law describes how the magnetic-induction lines encircle an area through which a current is passing or the electric flux is changing.

The charges and currents in these equations include the bound charges which occur on dielectrics and the magnetization or atomic currents which occur in magnetized materials. To be useful when such materials are present, Equations 31-5 and 31-2 are rewritten in terms of free charge and conduction currents (plus displacement currents) only. This is done by substituting **H** for **B** in Equation 31-5 and omitting the $\mu_0$. Similarly, Equation 31-2 can be written in terms of free charge only by defining a new vector **D**, called the *displacement vector*, which is related to **E** and the electric polarization of dielectrics analogously to the way **B**, **H**, and the magnetic polarization **M** are related. Since we shall not consider the complications introduced by polarized materials, we need not be concerned with such refinements.

## 31-2  The Wave Equation for Electromagnetic Waves

In Section 14-8 we showed that the harmonic wave functions for waves on a string and for sound waves obey a partial differential equation called the *wave equation*:

$$\frac{\partial^2 y(x,\,t)}{\partial x^2} = \frac{1}{v^2}\frac{\partial^2 y(x,\,t)}{\partial t^2} \qquad\qquad 31\text{-}6$$

In this equation, $y(x,\,t)$ is the wave function, which for string waves is the displacement of the string and for sound waves can be either the pressure change or the displacement of air particles from equilibrium. The derivatives are partial derivatives because the wave function depends on both $x$ and $t$. The quantity $v$ is the velocity of the wave, which depends on the medium and on the frequency if the medium is dispersive. We also showed in Section 14-8 that the wave equation for string waves can be derived by applying Newton's laws of motion to a string under tension, and we found that the velocity of the wave is $\sqrt{T/\mu}$, where $T$ is the tension and $\mu$ the linear mass density.

The solutions of this equation, studied in Chapter 14, were harmonic wave functions of the form

$$y = y_0 \sin{(kx - \omega t)}$$

where $k = 2\pi/\lambda$ is the wave number and $\omega = 2\pi f$ is the angular frequency.

In this section we shall use Maxwell's equations to derive the wave equation for electromagnetic waves. We shall not consider how such waves arise from the motion of charges but merely show that the laws of electricity and magnetism imply a wave equation, which in turn implies the existence of electric and magnetic fields **E** and **B** propagating through space with the velocity of light $c$. We shall consider only free space, in which there are no charges or currents. We shall also assume that the electric and magnetic fields **E** and **B** are functions of time and one space coordinate only, which we shall take to be the $x$ coordinate. This assumption is equivalent to the assumption of plane waves.

We can use Equations 31-2 and 31-3 to show that the electric and magnetic fields associated with electromagnetic waves must be transverse to the direction of propagation. For the situation we are considering, i.e., free space with no charge density and no current, Equations 31-2 and 31-3 state that the net flux of either **E** or **B** through any closed surface is zero. Let us consider the flux of **E** through a cube of sides $\Delta x$, $\Delta y$, and $\Delta z$, as shown in Figure 31-1. We are assuming that **E** depends only on $x$, the direction of propagation. Consider first the cube's top and bottom faces; their area is $\Delta x\, \Delta z$. Flux through these faces depends on $E_y$. Since, by our assumption, $E_y$ does not depend on $y$, the flux of **E** into the cube through the bottom face must equal the flux out through the top face. Similar reasoning leads to the result that there can be no contribution to the net flux through the faces of area $\Delta x\, \Delta y$ because $E_z$ does not depend on $z$. The only possible contribution is through the faces of area $\Delta y\, \Delta z$. The flux out of the right face is $E_{xR}\, \Delta y\, \Delta z$, where $E_{xR}$ is the value of $E_x$ at the right face. The flux in through the left face is $E_{xL}\, \Delta y\, \Delta z$, where $E_{xL}$ is $E_x$ at the left face. Since the net flux out of the cube must be zero, $E_{xR}$ must equal $E_{xL}$; that is, $E_x$ must be the same at the left and right sides of the cube. Thus $E_x$ does not depend on $x$. If the electric field has any component at all in the $x$ direction, it must be constant in space. Such a constant field is unrelated to electromagnetic waves and can be ignored for this discussion. Similar reasoning applied to the flux of **B** through such a cube leads to the conclusion that $B_x$ must also be constant in space. Thus the part of the electric or magnetic fields which varies in time and space must be perpendicular to the direction of propagation.

To obtain the wave equation relating the time and space derivatives of either the electric field **E** or the magnetic field **B**, we first relate the time derivative of one of the field vectors to the space derivative of the other. We do this by applying the other two Maxwell equations (31-4 and 31-5) to appropriately chosen curves in space. Let us assume for simplicity that the electric field is in the $y$ direction. We can relate the spatial dependence of $E_y$ to the time dependence of $B_z$ by applying Equation 31-4 (which is Faraday's law) to the rectangular curve of sides $\Delta x$ and $\Delta y$ lying in the $xy$ plane, as shown in Figure 31-2. Assuming that $\Delta x$ and $\Delta y$ are very small, the line integral of **E** around this curve is approximately

$$\oint \mathbf{E} \cdot d\boldsymbol{\ell} = E_y(x_2)\, \Delta y - E_y(x_1)\, \Delta y = [E_y(x_2) - E_y(x_1)]\, \Delta y$$

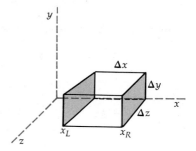

**Figure 31-1**
Volume element with sides parallel to the coordinate planes. If **E** depends only on $x$ and $t$, there is no contribution to the net flux of **E** from the top and bottom or from the front and back sides. If the net flux of **E** out of the volume is zero, $E_x$ must have the same value at $x_R$ as at $x_L$.

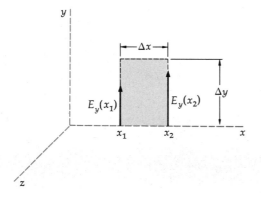

**Figure 31-2**
Rectangular curve in the $xy$ plane for derivation of Equation 31-7.

The contributions of the type $E_x \Delta x$ from the top and bottom of this curve cancel because we have assumed that **E** does not depend on $y$ (or $z$). Assuming that $\Delta x$ is very small, we can approximate the difference in $E_y$ at the points $x_1$ and $x_2$ by

$$E_y(x_2) - E_y(x_1) = \Delta E_y \approx \frac{\partial E_y}{\partial x} \Delta x$$

Then

$$\oint \mathbf{E} \cdot d\ell \approx \frac{\partial E_y}{\partial x} \Delta x \, \Delta y$$

The flux of the magnetic induction through this curve is approximately

$$\int \mathbf{B} \cdot \hat{\mathbf{n}} \, dA = B_z \, \Delta x \, \Delta y$$

Faraday's law then gives

$$\frac{\partial E_y}{\partial x} \Delta x \, \Delta y = - \frac{\partial B_z}{\partial t} \Delta x \, \Delta y$$

or

$$\frac{\partial E_y}{\partial x} = - \frac{\partial B_z}{\partial t} \qquad\qquad 31\text{-}7$$

Equation 31-7 implies that if there is a component of electric field $E_y$ that depends on $x$, there must be a component of magnetic induction $B_z$ that depends on time; or, conversely, if there exists a magnetic induction field $B_z$ that depends on time, there must be an electric field $E_y$ that depends on $x$. We can get a similar equation relating $E_y$ to $B_z$ by applying Equation 31-5 to the curve of sides $\Delta x$ and $\Delta z$ in the $xz$ plane, as shown in Figure 31-3. For the case with no currents Equation 31-5 is

$$\oint_C \mathbf{B} \cdot d\ell = \mu_0 \epsilon_0 \frac{d}{dt} \int_S \mathbf{E} \cdot \hat{\mathbf{n}} \, dA$$

We omit the details of this calculation, which is similar to that already done; the result is

$$\frac{\partial B_z}{\partial x} = -\mu_0 \epsilon_0 \frac{\partial E_y}{\partial t} \qquad\qquad 31\text{-}8$$

which relates the space variation of the magnetic induction $B_z$ and the time variation of the electric field $E_y$.

   We can eliminate either $B_z$ or $E_y$ from Equations 31-7 and 31-8 by differentiating either equation with respect to $x$ or $t$. If we differentiate both sides of Equation 31-7 with respect to $x$, we obtain

$$\frac{\partial}{\partial x}\left(\frac{\partial E_y}{\partial x}\right) = - \frac{\partial}{\partial x}\left(\frac{\partial B_z}{\partial t}\right) \qquad \text{or} \qquad \frac{\partial^2 E_y}{\partial x^2} = - \frac{\partial}{\partial t}\left(\frac{\partial B_z}{\partial x}\right)$$

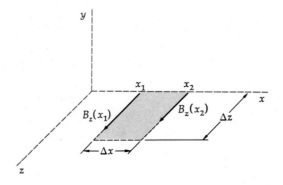

**Figure 31-3**
Rectangular curve in the $xz$ plane for the derivation of Equation 31-8.

where we have interchanged the order of the time and space derivatives on the right side. We now use Equation 31-8 for $\partial B_z/\partial x$:

$$\frac{\partial^2 E_y}{\partial x^2} = -\frac{\partial}{\partial t}\left(-\mu_0\epsilon_0\,\frac{\partial E_y}{\partial t}\right)$$

or

$$\frac{\partial^2 E_y}{\partial x^2} = \mu_0\epsilon_0\,\frac{\partial^2 E_y}{\partial t^2} \qquad\qquad 31\text{-}9 \qquad \textit{Wave equation for } E_y$$

Comparing this equation with Equation 31-6, we see that $E_y$ obeys a wave equation for waves with speed

$$v = \frac{1}{\sqrt{\mu_0\epsilon_0}}$$

which is Equation 31-1. If we had instead chosen to eliminate $E_y$ from Equations 31-7 and 31-8 (by differentiating Equation 31-7 with respect to $t$, for example), we would have obtained an equation identical to Equation 31-9 with $B_z$ replacing $E_y$. We have thus shown that both the electric field $E_y$ and the magnetic field $B_z$ obey a wave equation for waves traveling with the velocity $1/\sqrt{\mu_0\epsilon_0}$, which is the velocity of light. As we noted in discussing harmonic waves, a particularly important solution to Equation 31-9 is the harmonic wave function of the form

$$E_y = E_{y0}\sin(kx - \omega t) \qquad\qquad 31\text{-}10$$

If we substitute this solution into either Equation 31-7 or 31-8, we can show that the magnetic induction $B_z$ is in phase with the electric field $E_y$. From Equation 31-7 we have

$$\frac{\partial E_y}{\partial x} = kE_{y0}\cos(kx - \omega t) = -\frac{\partial B_z}{\partial t}$$

Solving for $B_z$ gives

$$B_z = \frac{k}{\omega}E_{y0}\sin(kx - \omega t) = B_{z0}\sin(kx - \omega t) \qquad 31\text{-}11 \qquad \textit{E and B are in phase}$$

where

$$B_{z0} = \frac{k}{\omega}E_{y0} = \frac{E_{y0}}{c} \qquad\qquad 31\text{-}12$$

and $c = \omega/k$ is the velocity of the wave.*

We have shown that Maxwell's equations imply the wave equation 31-9 for the electric field component $E_y$ and the magnetic induction component $B_z$, and that if $E_y$ varies harmonically, as in Equation 31-10, the magnetic induction $B_z$ is in phase with $E_y$ and has an amplitude related to the amplitude of $E_y$ by Equation 31-12. The electric and magnetic fields are perpendicular to each other and to the direction of the wave propagation, the $x$ direction here.

From Figure 31-4 we note that the direction of propagation (the $x$ direction in this case) is parallel to the vector $\mathbf{E}\times\mathbf{B}$. The vector $(\mathbf{E}\times\mathbf{B})/\mu_0 = \mathbf{E}\times\mathbf{H}$ is known as the *Poynting vector* (after John H. Poynting):

$$\mathbf{S} = \frac{\mathbf{E}\times\mathbf{B}}{\mu_0} = \mathbf{E}\times\mathbf{H} \qquad\qquad 31\text{-}13 \qquad \textit{Poynting vector}$$

---

* In obtaining Equation 31-11 by integration from the previous equation an arbitrary constant of integration arises, but we have omitted this constant magnetic induction field from Equation 31-11 because it plays no part in the electromagnetic waves we are interested in. Note that if any constant electric field is added to Equation 31-10, the new function still satisfies the wave equation.

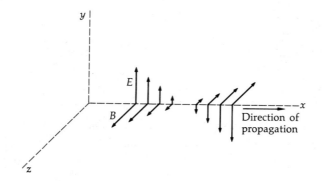

We shall now show that its average magnitude equals the intensity of the electromagnetic wave. In our discussion of the transport of energy by waves of any kind, we showed that the intensity (average energy per unit time per unit area) is in general equal to the product of the average energy density (energy per unit volume) and the speed of the wave. The energy per unit volume associated with an electric field **E** is

$$\eta_e = \tfrac{1}{2}\epsilon_0 E^2$$

For the harmonic wave given by Equation 31-10 we have

$$\eta_e = \tfrac{1}{2}\epsilon_0 E_0^2 \sin^2 (kx - \omega t)$$

For simplicity in notation, we shall drop the $y$- and $z$-component subscripts. The energy per unit volume associated with a magnetic induction field **B** is

$$\eta_m = \frac{B^2}{2\mu_0}$$

From Equations 31-11 and 31-12 we have

$$\eta_m = \frac{B_0^2}{2\mu_0} \sin^2 (kx - \omega t) = \frac{E_0^2}{2\mu_0 c^2} \sin^2 (kx - \omega t)$$

But since $c^2 = 1/\epsilon_0\mu_0$,

$$\eta_m = \tfrac{1}{2}\epsilon_0 E_0^2 \sin^2 (kx - \omega t)$$

and we see that the magnetic and electric energy densities are equal. The total electromagnetic energy per unit volume is

$$\eta = \eta_e + \eta_m = \epsilon_0 E_0^2 \sin^2 (kx - \omega t)$$

$$= \frac{E_0^2}{\mu_0 c^2} \sin^2 (kx - \omega t) \qquad\qquad 31\text{-}14$$

The energy per unit time per unit area is the product of this energy density and the speed $c$:

$$\eta c = \frac{E_0^2}{\mu_0 c} \sin^2 (kx - \omega t)$$

$$= [E_0 \sin (kx - \omega t)] \left[ \frac{E_0}{\mu_0 c} \sin (kx - \omega t) \right]$$

$$= E \frac{B}{\mu_0} = EH = |\mathbf{S}|$$

The intensity at any point $x$ is the time average of the energy per unit time per unit area. From Equation 31-14 we see that the energy density is proportional to $\sin^2 (kx - \omega t)$. The average of this quantity over one or more cycles is $\frac{1}{2}$. The average energy density is then

$$\eta_{av} = \frac{E_0^2}{2\mu_0 c^2}$$

and the intensity is

$$I = c\eta_{av} = \frac{E_0^2}{2\mu_0 c} = \frac{E_0}{\sqrt{2}} \frac{B_0}{\mu_0 \sqrt{2}} = E_{rms} H_{rms} = |\mathbf{S}|_{av} \qquad 31\text{-}15$$

where $E_{rms} = E_0/\sqrt{2}$ is the rms value of $E$ and $H_{rms} = B_{rms}/\mu_0 = B_0/(\mu_0\sqrt{2})$ is the rms value of $H$.

The magnitude of the Poynting vector is the instantaneous power per unit area. The rate at which electromagnetic energy flows through any area is the flux of the Poynting vector through that area:

$$P = \int \mathbf{S} \cdot \hat{\mathbf{n}} \, dA = \int (\mathbf{E} \times \mathbf{H}) \cdot \hat{\mathbf{n}} \, dA \qquad 31\text{-}16$$

where $P$ is the power through the area and $\mathbf{S} = \mathbf{E} \times \mathbf{H}$ is the Poynting vector.

In Chapter 14 we pointed out that traveling waves carry momentum as well as energy. The momentum carried by an electromagnetic wave equals the energy carried divided by the speed $c$. Since the intensity of a wave is the energy per unit time per unit area, the intensity divided by $c$ is the momentum carried by the wave per unit time per unit area. The SI unit of momentum per unit time is the unit of force, the newton. The intensity divided by $c$ is therefore a force per unit area, or pressure, with SI units of N/m² = Pa. This pressure is called radiation pressure $p_r$:

*Radiation pressure*

$$p_r = \frac{I}{c} \qquad 31\text{-}17$$

We can relate the radiation pressure to the electric or magnetic fields using Equations 31-15 and 31-12, giving

$$p_r = \frac{I}{c} = \frac{E_0^2}{2\mu_0 c^2} = \frac{B_0^2}{2\mu_0} = \frac{1}{c}|\mathbf{S}|_{av} \qquad 31\text{-}18$$

---

**Example 31-1** A 100-W light bulb emits electromagnetic waves uniformly in all directions. Find the intensity, the radiation pressure, and the electric and magnetic fields at a distance of 3 m from the bulb, assuming that 50 W goes into electromagnetic radiation.

At a distance $r$ from the bulb, the energy is spread uniformly over an area $4\pi r^2$. The intensity is therefore

$$I = \frac{50 \text{ W}}{4\pi r^2} = 0.442 \text{ W/m}^2 \qquad \text{at } r = 3 \text{ m}$$

The radiation pressure is the intensity divided by the speed of light: $p_r = I/c = (0.442 \text{ W/m}^2)/(3 \times 10^8 \text{ m/s}) = 1.47 \times 10^{-9}$ Pa. This is a very small pressure compared with atmospheric pressure of the order of $10^5$ Pa. The maximum value of the magnetic field is, from Equation 31-18,

$$B_0 = (2\mu_0 p_r)^{1/2} = [2(4\pi \times 10^{-7})(1.47 \times 10^{-9})]^{1/2} = 6.08 \times 10^{-8} \text{ T}$$

The maximum value of the electric field is the speed of light times $B_0$:

$$E_0 = cB_0 = 18.2 \text{ V/m}$$

The electric and magnetic fields are of the form $E = E_0 \sin(kx - \omega t)$ and $B = B_0 \sin(kx - \omega t)$ with $E_0 = 18.2$ V/m and $B_0 = 6.08 \times 10^{-8}$ T.

# James Clerk Maxwell (1831–1879)

C. W. F. Everitt
*Stanford University*

Maxwell aged about 32. (*Courtesy of the Master and Fellows of Peterhouse, Cambridge, England.*)

One day in 1877 a young Scottish undergraduate named Donald MacAlister, afterwards a distinguished physician and academic statesman, wrote home from Cambridge University that he had just had dinner with a professor who was "one of the best of our men, and a thorough old Scotch laird in ways and speech." This description of James Clerk Maxwell was accurate. He was the proprietor of an estate of 2000 acres in the southwest of Scotland and a man with all the qualities of the better kind of Victorian country gentleman: cultivated, considerate of his tenants, taking his due part in local affairs, an expert swimmer and horseman. He was a descendant through one line (the Maxwells) of a family long famous in Scottish Border history, and through another (the Clerks) of one of the leading families in Edinburgh society. His uncle, Sir George Clerk, had been a prominent Conservative politician. Few would suspect that the "Scotch laird" who seemed to MacAlister charmingly old-fashioned in 1877 was also a scientist whose writings remain astonishingly up to date in the 1980s; that he was the greatest mathematical physicist since Newton; that he had created the electromagnetic theory of light and predicted the existence of radio waves; that he had written the first significant paper on control theory; that he was just then writing a profound article on statistical mechanics, a science he and Ludwig Boltzmann jointly invented; that he had performed with his wife's aid a brilliant series of experiments on color vision and had taken the first color photograph; and that in the remaining 2 years until his death in 1879, at the age of 48, he would lay the foundations of another new subject that was to reach fruition in the twentieth century, rarefied gas dynamics.

Maxwell owed his interest in science to his father, John Clerk Maxwell, who before marrying in 1826 and moving to his estate, had lived in Edinburgh and regularly attended meetings of the Royal Society of Edinburgh. His great interest was in mechanical contrivances; he published one scientific paper, a proposal for an automatic-feed printing press, in 1831, the year Maxwell was born. The neat ingenuity and carefully thought-out details of Maxwell's own scientific apparatus bear the stamp of his father's mind. Maxwell's first 10 years were spent in the country, remote from any school, and his education was in the hands of his mother, Frances Cay, a woman of intelligence and great force of character. From her he imbibed the love of English literature that left its mark in the graceful literary style of his writings. She died when he was 8 years old, and after a disastrous 2-year experiment with a tutor, he was sent to Edinburgh to attend Edinburgh Academy, then and now one of the leading schools in Britain. There, after a slow start, where his back-country accent and odd dress earned him the nickname "Dafty," his intellect began to blossom, and through family connections he was gradually introduced to the magnificent intellectual and scientific culture then flourishing in Edinburgh.

One friend of the Clerks and Cays was D. R. Hay, a decorative artist interested in principles of design. He happened in 1845 to be seeking, not too successfully, a geometrical method of drawing oval curves, similar to the string property of the ellipse. Maxwell at age 14 discovered that where the string for the ellipse is folded back on itself $n$ times toward one focus and $m$ times towards the other, like the block and tackle of a crane, a true oval is formed, of the kind first studied in the 1680s by Réné Descartes in connection with the refraction of light. Figures 1a and 1b reproduce four ovals from Maxwell's paper, three with $m = 1$ and $n = 2$, the other $m = 2$ and $n = 3$. (Figure 1c illustrates one of the interesting optical properties of these ovals.) Simple as the idea was, it had not been thought of before. Maxwell's father showed it to James David Forbes, Professor of Natural Philosophy at Edinburgh University, and between them the two men secured its publication in the *Proceedings of the Royal Society of Edinburgh*. Thereafter Maxwell often went to meetings of the Society and soon came

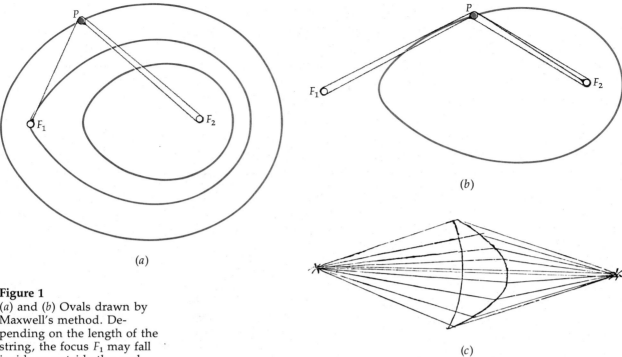

(a)

(b)

(c)

**Figure 1**
(a) and (b) Ovals drawn by
Maxwell's method. De-
pending on the length of the
string, the focus $F_1$ may fall
inside or outside the oval.
The ovals in (a) are generated
by attaching the string at the
focus $F_1$, passing it around
the pencil $P$ and the other
focus $F_2$, and fixing it at $P$.
The oval in (b) is generated
by attaching the string at $F_2$,
passing it around $P$ and $F_1$,
passing it back around $P$ and
$F_2$, and finally fixing it at $P$.
(*Redrawn from* The Scientific
Papers of James Clerk Maxwell, *Vol. I, p. 1.*) (c) Optical
properties of a Cartesian oval.
This figure reproduced from
an early manuscript of Maxwell's shows a lens with a
convex oval surface to the
right and a concave spherical
surface to the left, the center
of the sphere being on the
inner focus of the oval. A lens
of this kind has perfect
imaging properties for the
two foci: a point source of
light at either focus forms a
perfect point image at the
other.

to know many of the leading scientists in Edinburgh. One encounter, which he
afterwards described as a turning point in his life, was when his uncle John Cay
took him and a friend to see the private laboratory of William Nicol, inventor of
the Nicol polarizing prism. Nicol later presented him with a pair of prisms,
with which over the next few years he made many experiments on polarized
light, discovering among other things the *Maxwell spot*, a curious visual phenomenon arising from a previously unknown polarizing structure in the foveal
region of the human eye. The prisms are now on exhibit in the Cavendish Laboratory Museum at Cambridge.

Maxwell's undergraduate career was unusually protracted. He spent 3 years
at Edinburgh University and another $3\frac{1}{4}$ at Cambridge. Unlike Einstein, he enjoyed student life and was fortunate to gain the attention of some outstanding
teachers. At Edinburgh he was influenced by two powerful and sharply con-
trasted men, Forbes and Sir William Hamilton, the metaphysician. Forbes was
an experimentalist, who had invented the seismometer and done important
work on the polarization of infrared radiation, besides achieving fame as one of
the earliest British Alpinists. He gave Maxwell the run of his laboratory and
with his help began the experiments on color vision that led eventually to Maxwell's own work on the subject. Hamilton, who had a genius for inspiring
youth, imparted to Maxwell the ranging philosophic vision that can be seen in
the many interesting metaphysical asides in his papers.

In 1850 Maxwell went up to Cambridge. By then his mathematical bent was
clear and the Cambridge Mathematical Tripos offered the finest training in applied mathematics along with the most grueling examination system the wit of
man has devised. Like many another clever undergraduate before and since,
Maxwell worked hard while pretending not to; but he just missed the coveted
position of Senior Wrangler, having to be content with second place. His private tutor was William Hopkins, the founder of modern geophysics and
arguably the greatest teacher Cambridge has ever produced. Others who influenced him were G. G. Stokes, the mathematical physicist who held the chair
Newton had occupied, and William Whewell, the mathematician-philosopher
of whom it was remarked that "science was his forte and omniscience his
foible." Maxwell's student friendships were more with classical scholars than

scientists. He too affected omniscience. After a discussion on New Year's Eve 1854 one friend wrote of

> Maxwell as usual showing himself acquainted with every subject upon which the conversation turned. I never met a man like him. I do believe there is not a single subject on which he cannot talk, and talk well too, displaying always the most curious and out of the way information.

Maxwell's electromagnetic theory of light is rooted in the work of two men, Michael Faraday and William Thomson. Faraday's invention of the electric motor and his researches on electromagnetic induction, electrochemistry, dielectric and diamagnetic action, and magneto-optical rotation made him in Maxwell's words "the nucleus of everything electric since 1830." His contributions to theory lay in his progressively advancing ideas about lines of electric and magnetic force, in particular the geometrical relations governing electromagnetic phenomena and the idea that magnetic forces might be accounted for not by direct attractions and repulsions between elements of current but by attributing to lines of force the property of shortening themselves and repelling each other sideways (Figure 2). Thomson's role was to relate lines of force to existing theories in electrostatics and magnetostatics, to invent a number of highly ingenious analytical techniques for solving electrical problems, and to emphasize the cardinal importance of energy principles in electromagnetism. Maxwell then introduced a series of new concepts: the *electrotonic function* (vector potential), the energy density of the field, the displacement current, and the significance of the operation *curl* in the field equations; he organized the subject into a coherent structure and made in 1861 the momentous discovery of the equivalence between light and electromagnetic waves.

**Figure 2**
Faraday's explanation of forces between current-carrying wires. The two diagrams show the lines of force observed when currents are flowing in parallel wires. Faraday assumed that the lines of force tend to shorten and repel each other sideways. (*a*) For wires with currents flowing in the same direction, the lines of force pull the two wires together. (*b*) For wires with currents flowing in opposite directions, the lines of force push the wires apart.

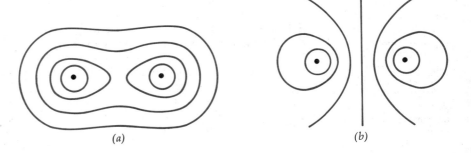

(*a*)                    (*b*)

The first part of Maxwell's paper "On Faraday's Lines of Force" (1855–1856) developed an analogy, due in essence to Thomson, between lines of electric and magnetic force and stream lines in a moving incompressible fluid. Maxwell applied this to interpret many of Faraday's observations, prefacing his paper with a luminous discussion of the significance of analogies in physics, a subject he enlarged on soon afterwards in a philosophical essay read in April 1856 to the Apostles, a famous Cambridge quasi-secret intellectual society of which he was a member. Next, still building on Faraday and Thomson, Maxwell extended the discussion to electromagnetism. He formulated a group of equations summarizing the relations of the electric and magnetic fields to the charges and currents producing them—the beginnings of what we now call Maxwell's equations. They described the phenomena with elegant precision from a point of view completely different from the then-popular action-at-a-distance theories of André-Marie Ampère and Wilhelm Weber. A nice feature was that the central theorem in all this work was one which, following Maxwell, we now call *Stokes' theorem*. It had been published in January 1854 by Stokes *as an examination question* in the Smith's Prize Examination at Cambridge, which Maxwell sat for immediately after completing his bachelor's degree.

After such a brilliant start one might have expected a rush of papers following up the new ideas. But other physicists ignored them, and Maxwell had the habit of investigating different subjects in turn, often with long intervals between successive papers in the same field. Six years elapsed before the appearance of his next paper, "On Physical Lines of Force," published in four parts in 1861–1862. During the interval Maxwell made brilliant contributions to three distinct subjects before returning to electromagnetism: color vision, the theory of Saturn's rings, and the kinetic theory of gases. He left Cambridge, became a professor at Marischal College in Aberdeen, married the daughter of the Principal of the College, and then found himself in the odd position of being forced to retire at age 29 with a life pension after his chair had been abolished when the two universities in Aberdeen were united by Act of Parliament. Fortunately the chair at King's College in London, had just fallen vacant, and so he went there.

"On Physical Lines of Force" contained Maxwell's extraordinary molecular-vortex model of the electromagnetic field. In order to account for the pattern of stresses associated with lines of force by Faraday, Maxwell investigated the properties of a medium occupying all space in which tiny molecular vortices rotate with their axes parallel to the lines of force. The closer together the lines are, the faster the rotation of the vortices. In a medium of this kind the lines of force do tend to shorten themselves and repel each other sideways, yielding the right forces between currents and magnets; the question is: what makes the vortices rotate? Here Maxwell put forward an idea as ingenious as it was weird. He postulated that an electric current consists in the motion of tiny particles that mesh like gear wheels with the vortices, and that the medium is filled with similar particles between the vortices. Figure 3 gives the picture. Maxwell remarks:

> I do not bring [this hypothesis] forward as a mode of connexion existing in nature . . . [but] I venture to say that anyone who understands [its] provisional and temporary character . . . will find himself helped rather than hindered by it in his search for the true interpretation of [electromagnetic] phenomena.

**Figure 3**
Maxwell's vortex model of the magnetic field. The rotating vortices represent lines of magnetic force. They mesh with small particles which act like gear wheels. In free space the particles are restrained from moving, except for a small elastic reaction (the displacement current), but in a conducting wire they are free to move. Their motion constitutes an electric current, which in turn sets the vortices in rotation, creating the magnetic field around the wire. A and B represent current through a wire, and p and q represent an induced current in an adjacent wire. (Redrawn from The Scientific Papers of James Clerk Maxwell, *Vol. I, fig. 2 after p. 488.*)

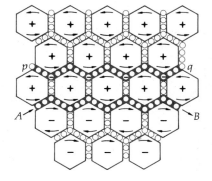

The question then was how to fit electrostatic phenomena into the model. Maxwell made the medium an elastic one. Thus magnetic forces were accounted for by rotations in the medium, and electric forces by its elastic distortion. Any elastic medium will transmit waves. In Maxwell's medium the velocity of the waves turned out to be related to the ratio of electric to magnetic forces. Putting in numbers from an experiment of 1856 by G. Kohlrausch and W. Weber, Maxwell found to his astonishment that the propagation velocity was equal to the velocity of light. With excitement manifested in italics he wrote "we can scarcely avoid the inference that *light consists in the transverse undulation of the same medium which is the cause of electric and magnetic phenomena.*"

Having made the great discovery, Maxwell promptly jettisoned his model. Instead of attempting a more refined mechanical explanation of the phenomena, he formulated a system of electromagnetic equations from which he deduced

Balance arm of apparatus with which Maxwell and Charles Hockin compared the ratio of electric to magnetic forces to determine the characteristic velocity, showing that it was equal to the velocity of light (1868).

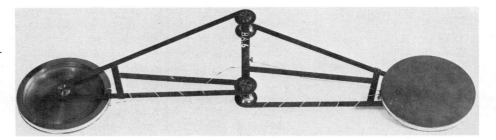

that waves of electric and magnetic force would propagate through space with the velocity of light. That is why his is called an *electromagnetic* theory of light, in contrast to the theories of the mechanical ether that preceded it. The theory appeared in two papers of 1865 and 1868, and in its most general form in the great *Treatise on Electricity and Magnetism,* published in 1873, a work of such scope that Robert Andrews Millikan, author of the famous oil-drop experiment to measure the charge on the electron, ranked it with Newton's *Principia* in considering them the two most influential books in the history of physics, "the one creating our modern mechanical world and the other our modern electrical world." Maxwell wrote it during a period of retirement from academic life after he had resigned his chair at King's College in 1865, finishing the work off somewhat hastily after his appointment in 1871 as Cavendish Professor of Experimental Physics at Cambridge.

Equally profound were Maxwell's contributions to statistical and molecular physics. They began with a paper in 1859 on the kinetic theory of gases, in which Maxwell introduced the velocity distribution function and enunciated the *equipartition theorem*, which in its original form stated that the average translational and rotational energies of large numbers of colliding molecules, whether of the same or different species, are equal. One surprising conclusion, afterwards confirmed experimentally by Maxwell and his wife, was that the viscosity of a gas should over a wide range be independent of its pressure. Another result was Maxwell's estimate of the mean free path of a gas molecule, which Loschmidt in 1865 applied to make the first serious estimates of the diameters of molecules. Later, Maxwell developed the general theory of transport phenomena, from which the Boltzmann equation is derived; invented the concept of ensemble averaging; created rarefied-gas dynamics; and conceived that "very small BUT lively being" the Maxwell demon. Brilliant as the successes of classical statistical mechanics were, its failures, as Maxwell saw, were in some ways even more striking. The equipartition theorem gave an answer for the ratio of the specific heats of gases that disagreed with experiment, while some of Boltzmann's theorems "proved too much" because they would apply to the properties of solids and liquids as well as gases. These questions remained shrouded in mystery until the emergence in 1900 of Planck's quantum hypothesis. Writing about them in 1877, Maxwell confessed his bewilderment and stated that nothing remained but to adopt the attitude of "thoroughly conscious ignorance that is the prelude to every real advance in knowledge."

Maxwell was an unusually sensitive man, with strong religious feeling and a fascinating and astonishing sense of humor.* Many of his letters reveal a delightfully sly irony. He also had some talent for writing poetry, usually light, but occasionally touching a deeper note. The last stanza of one poem to his wife, written in 1867, was

> All powers of mind, all force of will
>   May lie in dust when we are dead,
> But love is ours, and shall be still
>   When earth and seas are fled.

Apparatus with which Maxwell and his wife measured the viscosity of gases as a function of pressure and temperature (1863–1865).

* For more details of Maxwell's life and work see L. Campbell and W. Garnett, *The Life of James Clerk Maxwell,* Johnson Reprint Corp., Harcourt, Brace & Jovanovich, New York, 1970 (reprint of Oxford 1882 edition); and C. W. F. Everitt, *James Clerk Maxwell: Physicist and Natural Philosopher,* Scribner, New York, 1975.

## Review

A. Define, explain, or otherwise identify the following:

Maxwell's equations, 833

Poynting vector, 837

Wave equation, 834

Radiation pressure, 839

B. True or false:

1. Maxwell's equations apply only to fields that are constant in time.

2. The wave equation can be derived from Maxwell's equations.

3. Electromagnetic waves are transverse.

4. In an electromagnetic wave, the electric and magnetic fields are in phase.

5. In an electromagnetic wave, the electric and magnetic field vectors **E** and **B** are equal in magnitude.

6. In an electromagnetic wave, the electric and magnetic energy densities are equal.

## Exercises

### Section 31-1, Maxwell's Equations

1. Write the Maxwell equation which corresponds most closely to the law stated: (a) Faraday's law; (b) Ampère's law; (c) Gauss' law; (d) Gauss' law for magnetism.

### Section 31-2, The Wave Equation for Electromagnetic Waves

2. Show by direct substitution that the wave function $E_y = E_0 \sin (kx - \omega t) = E_0 \sin k(x - ct)$, where $c = \omega/k$ satisfies Equation 31-9.

3. An electromagnetic wave has intensity 100 W/m². Find (a) the radiation pressure $p_r$, (b) $E_{rms}$, and (c) $B_{rms}$.

4. Use the known values of $\mu_0$ and $\epsilon_0$ in SI units to compute $c = 1/\sqrt{\mu_0\epsilon_0}$ and show that its value is approximately $3 \times 10^8$ m/s.

5. The amplitude of an electromagnetic wave is $E_0 = 400$ V/m. Find (a) $E_{rms}$, (b) $B_{rms}$, (c) the intensity $I$, and (d) the radiation pressure $p_r$.

6. (a) Show that if $E$ is in volts per metre and $B$ in teslas, the units of the Poynting vector $(\mathbf{E} \times \mathbf{B})/\mu_0$ are watts per square metre. (b) Show that if the intensity $I$ is in watts per square metre, the units of radiation pressure $p_r = I/c$ are newtons per square metre.

7. (a) An electromagnetic wave of intensity 200 W/m² is incident normally on a black card of sides 20 by 30 cm which absorbs all the radiation. Find the force exerted on the card by the radiation. (b) Find the force exerted by the same wave if the card reflects all the radiation incident on it.

8. Find the force exerted by the electromagnetic wave on the reflecting card in part (b) of Exercise 7 if the radiation is incident at an angle of 30° to the normal.

## Problems

1. A laser beam has a diameter of 1.0 mm and average power of 1.5 mW. Find the intensity of the beam, $E_{rms}$, $B_{rms}$, and the radiation pressure.

2. The intensity of sunlight striking the upper atmosphere is 1.4 kW/m². (a) Find $E_{rms}$ and $B_{rms}$ due to the sun at the upper atmosphere of the earth. (b) Find the average power output of the sun. (c) Find the intensity and the radiation pressure at the surface of the sun.

3. (a) Find the intensity and radiation pressure 6 m from a 500-W light bulb, assuming it to be a point source. (b) Find $E_{rms}$ and $B_{rms}$ at this distance from the bulb. (Assume half the power goes into electromagnetic radiation.)

4. A 10- by 15-cm card has a mass of 2 g and is perfectly reflecting. The card hangs in a vertical plane and is free to rotate about a horizontal axis through one edge. The card is illuminated uniformly by an intense light so that the card makes an angle of 1° with the vertical. Find the intensity of the light.

5. A long cylindrical conductor of radius $a$ and resistivity $\rho$ carries a steady current $I$ uniformly distributed over its cross-sectional area. (a) Use Ohm's law to relate the electric field $\mathbf{E}$ in the conductor to $I$, $\rho$, and $a$. (b) Find the magnetic field $\mathbf{B}$ just outside the conductor. (c) At $r = a$ (the edge of the conductor) use the results of parts (a) and (b) to compute the Poynting vector $\mathbf{S} = \mathbf{E} \times \mathbf{H} = (1/\mu_0)\mathbf{E} \times \mathbf{B}$. In what direction is $\mathbf{S}$? (d) Find the flux of $\mathbf{S}$ through the surface of the wire of length $L$ and area $2\pi a L$ and show that the rate of energy flow into the wire equals $I^2 R$, where $R$ is the resistance.

6. (a) Using arguments similar to those given in the text, show that

$$\frac{\partial E_z}{\partial x} = \frac{\partial B_y}{\partial t} \quad \text{and} \quad \frac{\partial B_y}{\partial x} = \mu_0 \epsilon_0 \frac{\partial E_z}{\partial t}$$

(b) Show that $E_z$ and $B_y$ also satisfy the wave equation.

7. Discuss how Maxwell's equations might have to be modified in a region of space containing magnetic monopoles. (A magnetic monopole is a particle carrying a single "magnetic charge" $qm$, that is, an isolated magnetic pole. Its existence was predicted by P. A. M. Dirac in 1931. The discovery of one such particle was reported in 1975, but the evidence is not conclusive.)

CHAPTER 32      Light

Electromagnetic radiation with wavelengths in the range of about 400 to 700 nm, to which the eye is sensitive, is called light. This phenomenon has intrigued mankind for centuries. Despite the many advances in recent times, much important research in both the theory of light and its applications is still being carried out today. To gain historical perspective, we first look briefly at some early theories and experiments which help us understand light as a wave motion. We then discuss the basic phenomena of reflection, refraction, and polarization. They can all be adequately understood by using rays to describe the straight-line propagation of light and neglecting interference and diffraction effects. As discussed in Chapter 15, this ray approximation is valid for the propagation of any wave motion if the wavelength is small compared with any apertures or obstacles—often the situation in optics because of the very small wavelengths of light. In the next chapter we shall apply our knowledge of reflection and refraction to the study of image formation by mirrors and lenses, a subject known as ray optics or geometrical optics. The fascinating study of the interference and diffraction of light will be taken up in Chapter 34.

## 32-1  Waves or Particles?

The controversy over the nature of light is one of the most interesting in the history of science. Early theories considered light to be a stream of particles which emanated from a source and caused the sensation of vision upon entering the eye. The most influential proponent of this particle theory of light was Newton. Using it, he was able to explain the laws of reflection and refraction.

When light strikes a boundary between two transparent media, e.g., air and water or air and glass, part of the light energy is reflected and part transmitted. The reflected ray makes an angle with the normal to the surface equal to that of the incident ray, whereas the transmitted ray is bent toward the normal (Figure 32-1). (If the light is leaving the water or glass and entering air, the transmitted ray is bent away from the normal.) This bending of the ray is called *refraction*.

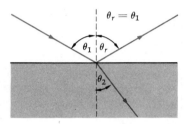

**Figure 32-1**
When a light ray strikes the boundary between two different media, e.g., air and glass, part of the energy is reflected and part is transmitted. The angle of reflection equals the angle of incidence. The angle of refraction $\theta_2$ is less than the angle of incidence for air-to-glass refraction.

The law of reflection of light from a plane boundary is easily explained by the particle theory. Figure 32-2 shows a particle bouncing off a hard plane surface. If there is no friction, the component of particle momentum parallel to the surface is not changed by the collision but the component perpendicular to the wall is reversed (assuming that the mass of the wall is much greater than that of the particle and that the collision is elastic). Thus the angle of reflection equals the angle of incidence, as is observed for light rays reflecting from a plane mirror.

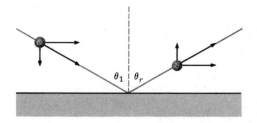

**Figure 32-2**
Reflection of a particle. If the surface is frictionless, the angle of reflection $\theta_r$ equals the angle of incidence $\theta_1$.

Figure 32-3 illustrates Newton's explanation of refraction at an air-glass or air-water surface. Newton assumed that the light particles are strongly attracted to the glass or water so that when they approach the surface, they receive a momentary impulse which increases the component of momentum perpendicular to the surface. Thus the direction of the momentum of the light particle changes, the light beam being bent toward the normal to the surface, in agreement with observation. An important feature of this theory is that the velocity of light must be greater in water or glass than in air to account for the bending of the beam toward the normal when it enters the water or glass and away from the normal when it enters air from either of those media. Experimental determination of the velocity of light in water did not come until about 1850, nearly 200 years later.

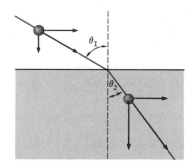

**Figure 32-3**
Refraction of a particle from
air into water or glass. According to Newton's particle
theory of light, the light particle is attracted by the water
or glass, and its component of
momentum normal to the surface is increased, bending the
path toward the normal to the
surface.

The chief proponents of the wave theory of light propagation were
Christian Huygens and Robert Hooke. Huygens, using his wavelet
construction (Section 15-3), was able to explain reflection and refraction, assuming that light travels more slowly in a glass or water medium
than in air. Newton saw the virtues of the wave theory of light, particularly as it explained the colors formed by thin films, which Newton
studied extensively. However, he rejected the wave theory because of
the observed straight-line propagation of light:

> To me the fundamental supposition itself seems impossible, namely,
> that the waves or vibrations of any fluid can, like the rays of light, be
> propagated in straight lines, without a continual and very extravagant
> spreading and bending every way into the quiescent medium, where
> they are terminated by it. I mistake if there be not both experiment and
> demonstration to the contrary.*

Because of Newton's great reputation and authority, this reluctant rejection of the wave theory of light, based on lack of evidence of diffraction, was strictly adhered to by Newton's followers. Even after evidence
of diffraction was available, they sought to explain it as scattering of
light particles from the edges of slits. Newton's particle theory of light
was accepted for more than a century.

In 1801 Thomas Young revived the wave theory of light. He was one
of the first to introduce the idea of interference as a wave phenomenon
in both light and sound. Figure 32-4 shows his drawing of the combination of waves from two sources. His observation of interference with
light was a clear demonstration of the wave nature of light. Young's
work went unnoticed by the scientific community for more than a decade, however. Perhaps the greatest advance in the general acceptance of
the wave theory of light was due to the French physicist Augustin
Fresnel (1788–1827), who performed extensive experiments on interference and diffraction and put the wave theory on a mathematical basis.

* Newton, *Opticks*, 1704. Reprint, Dover, New York, 1952.

The Science Museum, London

Portrait of Thomas Young,
British physicist, physician,
and egyptologist (1773–1829).

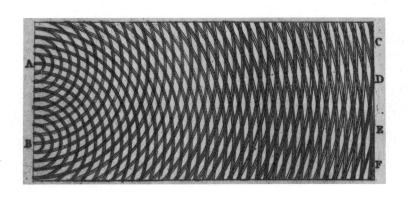

**Figure 32-4**
Thomas Young's drawing of a
two-source interference pattern. (*Courtesy of the Niels
Bohr Library, American Institute of Physics.*)

He showed, for example, that the observed rectilinear propagation of light is a result of the very short wavelength of visible light. An interesting triumph of the wave theory was an experiment suggested to Fresnel by S. Poisson, who sought to discredit the wave theory. Poisson noted that if an opaque disk is illuminated by light from a source on its axis, the Fresnel wave theory predicts that light waves bending around the edge of the disk should meet and interfere constructively on the axis, producing a bright spot in the center of the shadow of the disk. Poisson considered this to be a ridiculous contradiction of fact, but Fresnel's immediate demonstration that such a spot does in fact exist convinced many doubters that the wave theory of light is valid (see Figure 32-5).

In 1850 Jean Foucault measured the speed of light in water and showed that it is less than that in air, thus ruling out Newton's particle theory.

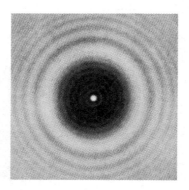

**Figure 32-5**
Diffraction of light by an opaque disk, showing the bright spot in the center of the shadow. (*From M. Cagnet, M. Françon, and J. C. Thrierr, Atlas of Optical Phenomena, plate 33, Springer-Verlag, Berlin, 1962.*)

In 1860 James Clerk Maxwell published his mathematical theory of electromagnetism, which predicted the existence of electromagnetic waves propagating with the speed $c = 1/\sqrt{\epsilon_0 \mu_0} \approx 3 \times 10^8$ m/s, the speed of light, and correctly suggested that this agreement is not accidental but indicates that light is an electromagnetic wave. Maxwell's theory was confirmed in 1886 by Hertz, who produced and detected waves in the laboratory by strictly electrical means. Hertz used a spark gap in a tuned circuit to generate the waves and another similar circuit to detect them. In the latter half of the nineteenth century, Kirchhoff and others applied Maxwell's equations to explain the interference and diffraction of light and other electromagnetic waves and put Huygens' empirical methods of construction on a firm mathematical basis.

This discussion of the wave-particle controversy of light would not be complete if we did not mention the discovery in the twentieth century that although the wave theory is generally correct in describing light and other electromagnetic waves, it fails to account for all their observed properties. One of the ironies of the history of science is that in his famous experiment of 1887, which confirmed Maxwell's wave theory, Hertz also discovered the *photoelectric effect,* which can be explained only by a particle model of light, as Einstein showed only a few years later. Hertz noticed that the passage of sparks in the receiving gap of his apparatus was facilitated by light from the generating gap. Investigating this effect in 1900, Lenard found that light falling on a metal surface ejects electrons and that the energies of these electrons do *not* depend on the intensity of the light. This result was quite surprising since the intensity is the energy per second per unit area falling on the metal surface. In 1905 Einstein demonstrated that this result can be explained by the assumption that the energy of a light wave is *quantized* into small bundles, called *photons*. The energy of a photon is propor-

The Bettmann Archive

Nineteenth-century engraving of Augustin Jean Fresnel (1788–1827).

*Photons*

tional to the frequency of the wave. According to Einstein, the energy of a photon is

$$E = hf \qquad\qquad 32\text{-}1$$

where $h = 6.63 \times 10^{-34}$ J·s is Planck's constant, and $f$ is the frequency of the light wave.

An electron ejected from a metal surface exposed to light receives its energy from a single photon. When the intensity of light of a given frequency is increased, more photons fall on the surface, ejecting more electrons, but the energy of each electron is not increased. Using this model, Einstein predicted that the maximum energy of an electron ejected in the photoelectric effect and escaping from the metal would increase linearly with the frequency of incident light, and that if the frequency of the incident light were below a certain threshold frequency $f_t$ (which depends on the kind of metal), the photoelectric effect would not occur, no matter what the intensity, because no single photon would have enough energy to eject an electron. These predictions were accurately confirmed about 10 years later in difficult experiments performed by the American physicist R. A. Millikan. Thus a particle model of light was reintroduced.

Complete understanding of this dual nature of light did not come until the 1920s, when experiments by C. J. Davisson and L. Germer and by G. P. Thompson showed that electrons (and other "particles") also have a dual nature and exhibit the wave properties of interference and diffraction besides their well-known particle properties. The behavior of fundamental quantities such as light, electrons, and other subatomic particles is correctly described by the theory of quantum mechanics worked out by E. Schrödinger, W. Heisenberg, P. A. M. Dirac, and others. Although this theory differs from both a classical wave theory and a classical particle theory, in some circumstances the quantum theory resembles a classical wave theory and in others it resembles a classical particle theory. For example, the propagation of these fundamental quantities can always be described as a wave propagation exhibiting the usual wave effects of interference and diffraction. On the other hand, the exchange of energy between these fundamental quantities, as in the photoelectric effect, is usually best described in terms of particle mechanics. As discussed in Chapter 15, when the wavelength of any wave is very small compared with the various obstacles and apertures it encounters, the ray approximation is valid and the propagation is in straight lines. The wave nature of electrons and other so-called "particles" is not easily observed because of their extremely short wavelengths—the same difficulty which prevented Newton from observing the wave character of light. On the other hand, the particle nature of the energy exchanges involving light is often not noticed because the number of photons is so enormous and the energy of each photon is so small. This difficulty is not unlike that of observing the particle nature of gas molecules which exert pressure on the walls of a container.

### Questions

1. The spreading of a beam of light after passing through a very small opening was observed before Newton's time, but it was argued that this was to be expected if the beam consisted of particles moving in not quite perfectly parallel paths. Explain.

2. Supporters of the particle theory of light were able to explain the slight diffraction of light observed, but they could not hope to explain

Young's observation of interference on the basis of the particle theory. Why?

3. According to Newton's theory of light, reflection and refraction are the result of attractive forces between the particles of light and the atoms of the refracting material. Explain why this attraction affects the path of a light beam only as it enters or leaves the refracting material and not in the center of the material.

4. Newton concluded that the force between particles of light and atoms of matter must be an attractive force. Can you see why? How would absorption of light by matter be explained in Newton's theory?

## 32-2  Electromagnetic Waves

Electromagnetic waves include light, radio, x-rays, gamma rays, microwaves, and others, all involving the propagation of electric and magnetic fields through space with speed $c = 3 \times 10^8$ m/s (in vacuum). All electromagnetic waves are generated when electric charge is accelerated. The differences between the various types of electromagnetic waves are in their frequency and wavelength. Table 32-1 shows the

**Table 32-1**

Electromagnetic spectrum

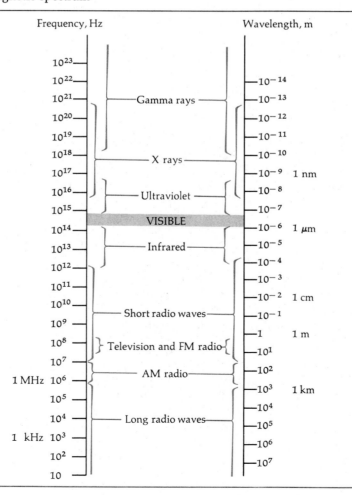

electromagnetic spectrum and the names usually associated with the various frequency and wavelength ranges. The ranges often are not well defined and sometimes overlap. For example, electromagnetic waves with wavelength of about 0.1 nm are called x-rays if their origin is atomic and gamma rays if their origin is nuclear. The human eye is sensitive to electromagnetic radiation of wavelengths from about 400 to 700 nm, the range called *visible light,* but the word light is also used for wavelengths just beyond the visible range. Ultraviolet light is electromagnetic radiation just on the shorter-wavelength side of the visible range, and infrared light is that just on the longer-wavelength side. There are no limits on the wavelengths of electromagnetic radiation: all frequencies and wavelengths are theoretically possible.

A complete mathematical description of electromagnetic waves can be derived from Maxwell's equations (Chapter 31), but such a description is much too difficult for an introductory course. As shown in Chapter 31, the oscillating electric and magnetic fields associated with an electromagnetic wave are mutually perpendicular and perpendicular to the direction of propagation of the wave, as illustrated in Figure 32-6.

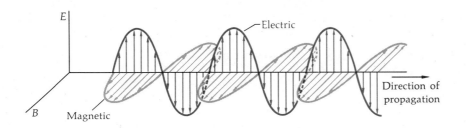

**Figure 32-6**
The electric and magnetic field vectors are in phase, perpendicular to each other, and perpendicular to the direction of propagation of the wave.

These fields are produced by charges which accelerate. In particular, according to classical electromagnetic theory, if a charge oscillates with simple harmonic motion of frequency $f$, it radiates energy in the form of electromagnetic waves of the same frequency. Figure 32-7 shows a simple and common type of radiating charge system, two equal but opposite charges whose separation varies harmonically with time. The radiation from such an oscillating electric dipole is called *electric-dipole radiation.* Some characteristics of electric-dipole radiation are important for our understanding of light:

*Electric-dipole radiation*

1. The magnitude of the electric field at large distances varies inversely as the distance and is proportional to sin $\theta$, where $\theta$ is the angle between the dipole axis and the point of observation as shown in Figure 32-7. Since the energy density and therefore the intensity is proportional to $E^2$, the intensity of electric-dipole radiation varies as $1/r^2$ and is proportional to $\sin^2 \theta$.

2. The electric field in an electromagnetic wave from an electric dipole is in the plane formed by the dipole axis and the direction of propagation. This is the plane of the paper in Figure 32-7.

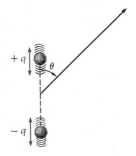

**Figure 32-7**
Electric dipole. The intensity of radiation is proportional to $\sin^2 \theta$. It is zero along the direction of the dipole and maximum perpendicular to this direction.

Many electromagnetic waves exhibit the characteristics of electric-dipole radiation. For example, radio waves are produced by currents oscillating in a straight antenna, an arrangement equivalent to an electric dipole. Light waves come from charge oscillations in atoms. Although the classical theory of electromagnetic radiation has been modified by the theory of quantum mechanics, many of the features of light and other electromagnetic waves can be understood in terms of the classical theory.

All electromagnetic waves are similar except for their wavelength and frequency, but this difference is very important. For example, it is not possible to generate light waves using ordinary circuits to produce the acceleration of the charges because the eye is sensitive only to electromagnetic waves with frequency of the order of $10^{14}$ Hz, much greater than can be attained with ordinary circuits. As we know, the behavior of waves closely depends on the relative sizes of the wavelength and of the physical objects and apertures the waves encounter. Since the wavelength of light is in the rather narrow range of about 400 to 700 nm, it is much smaller than most obstacles and apertures and the ray approximation often is valid.

**Questions**

5. What sort of radiation is an electromagnetic wave whose frequency is $10^5$ Hz? $10^{10}$ Hz? $10^{15}$ Hz? $10^{20}$ Hz?

6. A dipole antenna on top of a high broadcasting tower has a vertical axis. Describe qualitatively the variation in intensity you would observe if you walked on the ground along a straight path which passes directly under the antenna, starting a great distance from the tower and continuing to a great distance on the other side.

7. If you want to broadcast radio waves from a dipole antenna, would you place the axis of the antenna horizontally or vertically? Why?

## 32-3  The Speed of Light

The first effort to measure the speed of propagation of light was made by Galileo. He and a partner situated themselves on hilltops about a kilometre apart, each with a lantern and a shutter to cover it. Galileo proposed to measure the time for light to traverse twice the distance between the experimenters. A would uncover his lantern, and when B saw the light, B would uncover his. The time between A's uncovering his lantern and seeing the light from B's would be the time for light to travel back and forth between the experimenters. Though this method is sound in principle, the speed of light is so great that the time interval to be measured is much smaller than fluctuations in the human response time, and Galileo was unable to obtain any value for the speed of light.

The first indication of the true magnitude of the speed of light came from astronomical observations of the period of one of the moons of Jupiter. This period is determined by measuring the time between eclipses (when the moon disappears behind Jupiter). The eclipse period is about 42.5 h, but measurements made when the earth is moving away from Jupiter, as from point A to C in Figure 32-8, give a greater time for this period than measurements made when the earth is moving toward Jupiter, as from point C to A in the figure. Since these measurements differ by only about 15 s from the average value, the discrepancies were difficult to measure accurately. In 1675 the astronomer Ole Roemer attributed these discrepancies to the fact that the velocity of light is not infinite. During the 42.5 h between eclipses of Jupiter's moon, the distance between the earth and Jupiter changed, making the path for the light longer or shorter. Roemer devised the following method for measuring the cumulative effect of these discrepancies. We neglect the motion of Jupiter (it is much slower than that of the earth). When the

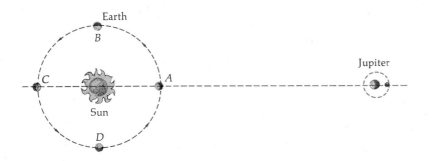

**Figure 32-8**
Roemer's method of measuring the speed of light. The time between eclipses of Jupiter's moon appears greater when the earth is moving from $A$ to $C$ than when it is moving from $C$ to $A$. The difference is due to the time it takes light to travel the distance traveled by the earth along the line of sight during one period of the moon.

earth is at the nearest point $A$, the period of the moon is measured. The time when an eclipse should occur $\frac{1}{2}$ y later, when the earth is at point $C$, is computed. The observed eclipse time is about 16 min later than predicted. This is the time it takes light to travel a distance equal to the diameter of the earth's orbit.

The first nonastronomical measurement of the velocity of light was made by the French physicist A. H. L. Fizeau in 1849. His method (Figure 32-9) was greatly improved by Foucault, who replaced the toothed wheel by a rotating mirror. In about 1850, Foucault measured the speed of light in air and in water, showing that it is less in water. Precise measurements of the speed of light using essentially the same method were performed by the American physicist A. A. Michelson from 1880 to 1930.

Another method does not involve light directly but is based on Maxwell's electromagnetic theory, which predicts the speed of light to be $c = 1/\sqrt{\epsilon_0 \mu_0}$. The electrical constant $\epsilon_0$ can be determined by accurately measuring the capacitance of a parallel-plate capacitor, and the magnetic constant $\mu_0$ can be determined by measuring the force between two current-carrying wires. (This is essentially the calibration experiment to determine the SI unit of current, the ampere.) Accurate electrodynamic measurements of the speed of light were first made by Rosa and Dorsey of the U.S. Bureau of Standards in 1906.

**Figure 32-9**
Fizeau's method of measuring the speed of light. Light from the source is reflected by mirror $B$ and transmitted through a gap in the toothed wheel to mirror $A$. The speed of light is determined by measuring what angular speed of the wheel will permit the reflected light to be transmitted by the next gap in the toothed wheel so that an image of the source is observed. (*Redrawn from Bernard Jaffe, Michelson and the Speed of Light. Copyright © 1960 by Doubleday & Company, Inc. Used by permission of the publisher.*)

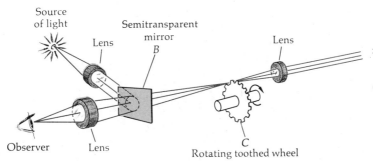

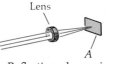
Reflecting plane mirror (5.39 mi from light source)

There is general agreement between the results of these various methods for determining the speed of light and many others which we shall not discuss. The accepted value for the speed of light is now

$$c = 2.997925 \pm 0.000003 \times 10^8 \text{ m/s}$$

The value $3 \times 10^8$ m/s is accurate enough for nearly all calculations.

### Question

8. Estimate the time required for light to make the round trip in Galileo's experiment to determine $c$.

## 32-4 Reflection

When waves of any type strike a plane barrier, new waves are generated which move away from the barrier. Experimentally it is found that the rays corresponding to the incident and reflected waves make equal angles with the normal to the barrier and that the reflected ray lies in the *plane of incidence* formed by the incident ray and the normal. This *law of reflection* can be stated by saying that *the reflected ray lies in the plane of incidence such that the angle of reflection equals the angle of incidence*. The law of reflection holds for all waves whether they are water waves, sound waves, electromagnetic waves, or others. Figure 32-10 shows ultrasonic plane waves in water reflecting from a steel plate. The law of reflection can be derived in several ways. According to Huygens' principle (Section 15-3), each point on a given wavefront can be considered to be a point source of secondary wavelets. Figure 32-11 shows a plane wavefront $AA'$ striking a barrier at point $A$. The angle $\theta_i$ between the ray corresponding to this wavefront and the normal to the barrier is called the *angle of incidence*. As can be seen from the figure, the angle of incidence $\theta_i$ equals the angle $\phi_i$ between the incident wavefront and the barrier. The position of the wavefront after a time $\Delta t$ is found by constructing wavelets of radius $v\,\Delta t$ with centers on the wavefront $AA'$. Wavelets which do not strike the barrier form the portion of the new wavefront $BB'$. Wavelets which do strike the barrier are reflected and form the portion of the new wavefront $BB''$. By a similar construction,

*Law of reflection*

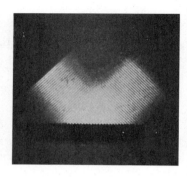

**Figure 32-10**
Ultrasonic plane waves in water reflecting from a steel plate. (*Battelle-Northwest Photography.*)

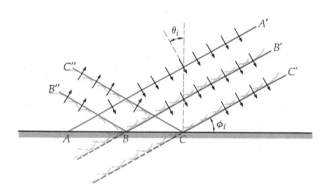

**Figure 32-11**
Plane wave reflected at a plane barrier. The angle $\theta_i$ between the incident ray and the normal to the barrier is the angle of incidence. It is the same as the angle $\phi_i$ between the incident wavefront and the barrier.

the wavefront $C''CC'$ is obtained from the Huygens' wavelets originating on the wavefront $B''BB'$. Figure 32-12 is an enlargement of a portion of Figure 32-11 showing a part of the original wavefront $AP$ which strikes the barrier in time $\Delta t$. In this time the wavelet from point $P$ reaches the barrier at point $B$, and the wavelet from point $A$ reaches point $B''$. The reflected wavefront $BB''$ makes an angle $\phi_r$ with the barrier which is equal to the angle of reflection $\theta_r$ between the reflected ray and the normal to the barrier. The triangles $ABP$ and $BAB''$ are both right triangles with a common side $AB$ and equal sides $AB'' = BP =$

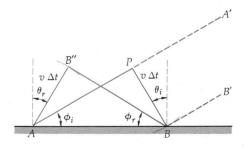

**Figure 32-12**
Geometry of Huygens' construction for the calculation of the law of reflection. The wavefront $AP$ originally intersects the barrier at point $A$. After a time $\Delta t$, the Huygens' wavelet from $P$ strikes the barrier at point $B$, and the one from $A$ reaches point $B''$. Since the triangles $ABP$ and $BAB''$ are congruent, the angles $\theta_i$ and $\theta_r$ are equal.

$v \, \Delta t$. Hence these triangles are congruent, and the angles $\phi_i$ and $\phi_r$ are equal, implying that the angle of reflection $\theta_r$ equals the angle of incidence $\theta_i$, which is the law of reflection. Having derived this law from consideration of the propagation of wavefronts and Huygens' principle, it is now much easier to consider only the rays which are perpendicular to the wavefronts.

A second derivation of the law of reflection uses an interesting principle enunciated in the seventeenth century by the French mathematician Pierre de Fermat, stated for our purposes as follows:

*The path taken by light in traveling from one point to another is such that the time of travel is a minimum when compared with nearby paths.*

*Fermat's principle*

In Figure 32-13 we assume that light leaves point $A$, strikes the plane surface, which we can consider to be a mirror, and travels to point $B$. We wish to find the path taken by light. The problem of reflection stated for the application of Fermat's principle is: At what point $P$ in Figure 32-13 must light strike the mirror so that it will travel from point $A$ to point $B$ in the least time? Since the light always travels in the same medium for this problem, the time will be minimum when the distance is minimum. In Figure 32-13 the distance $APB$ is the same as the distance $A'PB$, where $A'$ is the image point of the source $A$. Point $A'$ lies along the perpendicular from $A$ to the mirror and is equidistant behind the mirror. Obviously, as we vary point $P$, the distance $A'PB$ is least when the points $A'$, $P$, and $B$ lie on a straight line. We can see from the figure that when this occurs, the angle of incidence equals the angle of reflection.

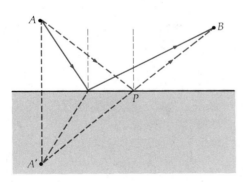

**Figure 32-13**
Geometry for deriving the law of reflection by Fermat's principle.

Although we have derived the law of reflection in several different ways, we have not discussed the detailed mechanism by which light is reflected by atoms, nor have we developed expressions for the relative intensities of the reflected and transmitted light. A complete discussion of the reflection of light taking into account the fact that light is an electromagnetic wave of very great frequency and applying the laws of electricity and magnetism to such a wave cannot be given in an elementary text. The results of such a treatment are in full accord with experiment, and we shall mention a few.

The fraction of energy reflected at a boundary between two media depends on the angle of incidence, the indexes of refraction of the two media, and the polarization of the incident light wave. The *index of refraction n* of a medium is related to the speed $v$ of waves in that medium and is defined by

$$n = \frac{c}{v}$$

32-2   *Index of refraction*

where $c$ is the speed of light in vacuum. The general expression for the fraction of energy reflected is quite complicated. When light is incident normally, the reflected intensity is

$$I = \left(\frac{n_2 - n_1}{n_2 + n_1}\right)^2 I_0 \qquad\qquad 32\text{-}3$$

where $I_0$ is the incident intensity and $n_1$ and $n_2$ are the indexes of refraction of the two media. For a typical case of reflection from an air-glass surface for which $n_1 = 1$ and $n_2 = 1.5$, Equation 32-3 gives $I = \frac{1}{25}I_0$; only about 4 percent of the energy is reflected, the rest being transmitted.

For light reflected from an air-glass surface, the reflected wave is 180° out of phase with the incident wave when the incident wave is in air, but the two waves are in phase when the incident wave is in glass. We can understand this from our discussion of the reflection of pulse waves on a string. In Chapter 14 we found that when a pulse is incident on a boundary beyond which the velocity of the pulse is slower, the pulse is inverted; but it is not inverted if the velocity of the pulse is greater beyond the boundary. If we consider a harmonic wave to be a series of pulses, an inversion of the pulse is equivalent to a 180° phase difference between the incident and reflected wave. This phase difference can be observed experimentally (see Chapter 34).

## 32-5   Refraction

When light or any other wave is incident on a boundary surface separating two media, some of the energy is reflected and some transmitted. The angle between the transmitted ray and the normal to the surface $\theta_2$, called the angle of refraction, is related to the angle of incidence $\theta_1$ by

$$n_1 \sin \theta_1 = n_2 \sin \theta_2 \qquad\qquad 32\text{-}4 \qquad \textit{Snell's law}$$

where $n_1$ and $n_2$ are the indexes of refraction of the first and second media, respectively, as defined by Equation 32-2. This result was discovered experimentally in 1621 by W. Snell and is known as Snell's law or the law of refraction. Like the law of reflection, this result can be derived either from Huygens' or Fermat's principle. In Figure 32-14 we apply Huygens' construction to find the wavefront of the transmitted wave, in much the same way as we did for reflection. $AP$ is a portion of a wavefront in medium 1 at angle of incidence $\theta_1$ which intersects the line dividing the media at point $A$.

In time $\Delta t$ the wavelet from $P$ travels the distance $v_1 \Delta t$ and reaches the point $B$ on the line $AB$ separating the two media, while the wavelet from point $A$ travels the distance $v_2 \Delta t$ into the second medium. The new wavefront $BB'$ is not parallel to the original wavefront $AP$ because the speeds $v_1$ and $v_2$ are different. From the triangle $APB$

$$\sin \phi_1 = \frac{v_1 \Delta t}{AB} \qquad \text{or} \qquad AB = \frac{v_1 \Delta t}{\sin \phi_1} = \frac{v_1 \Delta t}{\sin \theta_1}$$

using the fact that the angle $\phi_1$ equals the angle of incidence $\theta_1$. Similarly, from triangle $AB'B$,

$$\sin \phi_2 = \frac{v_2 \Delta t}{AB} \qquad \text{or} \qquad AB = \frac{v_2 \Delta t}{\sin \phi_2} = \frac{v_2 \Delta t}{\sin \theta_2}$$

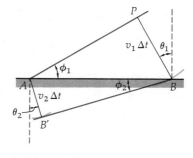

**Figure 32-14**
Application of Huygens' principle to the refraction of plane waves at the surface separating a medium in which the wave speed is $v_1$ from a medium in which the wave speed is $v_2$. For the case shown, in which $v_2$ is less than $v_1$, the rays are bent toward the normal to the surface.

where $\theta_2 = \phi_2$ is the angle between the refracted ray and the normal to the line separating the media. $\theta_2$ is the angle of refraction. Equating the two values of $AB$ gives

$$\frac{\sin \theta_1}{v_1} = \frac{\sin \theta_2}{v_2} \qquad \text{32-5}$$

If we write the wave speeds $v_1$ and $v_2$ in terms of the indexes of refraction $n = c/v$, Equation 32-5 becomes Equation 32-4, Snell's law. Note that if $v_2$ is less than $v_1$, $n_2$ is greater than $n_1$ and the angle of refraction $\theta_2$ is less than the angle of incidence $\theta_1$. The ray is then bent toward the normal. For example, when light passes from air to glass, the ray is bent toward the normal; when it passes from glass to air, the ray is bent away from the normal.

The frequency of the transmitted (and reflected) wave is the same as that of the incident wave. This can be understood in terms of Huygens' principle where we think of the incident wave at the barrier as the source of the transmitted wave. Since the speed of the transmitted wave differs from that of the incident wave, their wavelengths also differ. This can be seen in Figure 32-15, which shows plane waves in a ripple tank incident on a line barrier where the wave speed changes abruptly because the water depth differs on each side of the barrier.

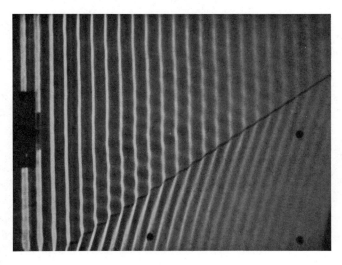

**Figure 32-15**
Refraction of plane waves in a ripple tank at a boundary at which the wave speed changes because the depth of water changes. Note that reflection also occurs at the boundary. (*Courtesy of Film Studio, Education Development Center, Newton, Mass.*)

Derivation of the law of refraction from Fermat's principle is more complicated than that of the law of reflection. Figure 32-16 shows possible paths for light traveling from point $A$ in air to point $B$ in glass. Point $P_1$ is on the straight line between $A$ and $B$, but this path is not the one for the shortest travel time because light travels with a smaller velocity in the glass. If we move slightly to the right of $P_1$, the total path length is greater but the distance traveled in the slower medium is less than for the path through $P_1$. It is not apparent from the figure which path is that of least time, but it is not surprising that a path slightly to the right of the straight-line path takes less time because the time gained by traveling a shorter distance in the glass more than makes up for the time lost traveling a longer distance in air. As we move the point of intersection of the possible path to the right of point $P_1$, the total time for the travel from $A$ to $B$ decreases until we reach a minimum point $P_{min}$. Beyond this point, the time saved by traveling a shorter distance in the glass does not compensate for the greater time required for the greater distance traveled in air. The details of the derivation of Equation 32-4 from Fermat's principle are given at the end of this section.

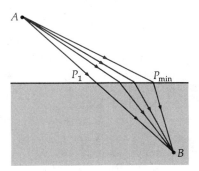

**Figure 32-16**
Geometry for deriving Snell's law of refraction from Fermat's principle. The point $P_{min}$ is the point at which light must strike the glass for travel time from $A$ to $B$ to be a minimum.

Solving Equation 32-4 for $\sin \theta_2$ gives

$$\sin \theta_2 = \frac{n_1}{n_2} \sin \theta_1 \qquad\qquad 32\text{-}6$$

If $n_2$ is greater than $n_1$ (as in the case of refraction of light entering a glass or water medium from air), this equation can be solved for the angle of refraction $\theta_2$ for any angle of incidence $\theta_1$. However, if $n_2$ is less than $n_1$ (as in the case of refraction of light entering air from glass or water), Equation 32-6 gives values for $\sin \theta_2$ greater than 1 if the angle of incidence is greater than some critical angle $\theta_c$ defined by

$$\sin \theta_c = \frac{n_2}{n_1} \qquad\qquad 32\text{-}7$$

Since the sine of any angle must be less than 1, Equation 32-6 cannot be solved for the angle of refraction for angles of incidence greater than $\theta_c$. This situation is illustrated in Figure 32-17 for light incident on a glass-to-air surface, where $n_1 = 1.5$ and $n_2 = 1$. The critical angle is then given by $\sin \theta_c = \frac{2}{3}$, $\theta_c = 41.8°$. When the angle of incidence is greater than this critical angle, there is no refracted wave in the second medium. All the energy is reflected. This phenomenon is called *total internal reflection*. An interesting application of total internal reflection is the transmission of a beam of light down a long narrow transparent glass fiber. If the beam begins approximately parallel to the axis of the fiber, it will always strike the walls of the fiber at angles greater than the critical angle, and no light energy will be lost through the walls of the fiber.

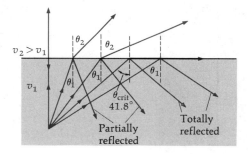

**Figure 32-17**
Total internal reflection. At the critical andlge$\theta_{\text{crit}}$, given by $\sin \theta_{\text{crit}} = v_1/v_2$, the angle of refraction is 90°, and the refracted ray emerges parallel to the boundary surface. At angles of incidence greater than the critical angle, there is no refracted ray. If the first medium is glass and the second air, $v_1/v_2$ is approximately $\frac{2}{3}$ and the critical angle is 41.8°.

A complete discussion of refraction of light based on the laws of electricity and magnetism is beyond the scope of this text. The intensity of the transmitted light can always be found by conservation of energy if the intensity of the reflected light is known. The sum of the energy transmitted and reflected must equal the energy incident.

Understanding the mechanism of refraction by atoms is also more difficult than for reflection. We can consider the transmitted wave as the resultant of the interference of the incident wave and a wave radiated from the atoms, which absorb and reradiate some of the light. For light entering glass from air, there is a phase lag between the reradiated wave and the incident wave. Thus the resultant wave also lags in phase behind the incident wave. This phase lag means that the position of a wave crest of the transmitted wave is retarded relative to the position of a wave crest of the incident wave in the medium. Therefore in a given time, the transmitted wave does not travel as far in the medium as the original incident wave; i.e., the phase velocity of the transmitted wave is less than that of the incident waves. Thus the index of refraction, defined to be the ratio of the velocity of light in vacuum to that in glass, is greater than 1.

Figure 32-18 shows the index of refraction of a glass as a function of wavelength. The slight decrease in $n$ as $\lambda$ increases means that the wave speed in a medium increases as $\lambda$ increases; i.e., the wave speed in a medium depends on the wavelength and frequency of the light. This is the phenomenon of *dispersion*, discussed for sound waves and waves on strings in Section 15-7. When a beam of light is incident at some angle on a glass surface, the angle of refraction of shorter wavelengths

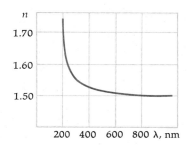

**Figure 32-18**
Typical dispersion curve for light in glass. The index of refraction decreases slightly as the wavelength increases, indicating that the speed of light in the glass depends on wavelength and frequency.

(toward the blue end of the visible spectrum) is slightly larger than that of the longer wavelengths (toward the red end of the spectrum). A beam of white light is therefore spread out or dispersed into its component colors or wavelengths (Figure 32-19). The formation of a rainbow is a familiar example of dispersion of sunlight by refraction in water droplets.

**Figure 32-19**
A beam of white light incident on a glass prism is dispersed into its component colors. The index of refraction decreases as the wavelength increases as shown in Figure 32-18, so the longer wavelengths (red) are bent less than the shorter wavelengths (blue).

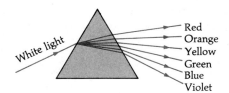

White light
Red
Orange
Yellow
Green
Blue
Violet

**Figure 32-20**
Geometry for deriving Snell's law from Fermat's principle.

*Optional*

**Derivation of Snell's Law from Fermat's Principle**

Figure 32-20 shows the geometry for finding the path of least time. If $L_1$ is the distance traveled in medium 1 with index of refraction $n_1$ and $L_2$ the distance in medium 2 with index of refraction $n_2$, the time for light to go along this path is

$$t = \frac{L_1}{v_1} + \frac{L_2}{v_2} = \frac{L_1}{c/n_1} + \frac{L_2}{c/n_2} = \frac{n_1 L_1}{c} + \frac{n_2 L_2}{c} \qquad 32\text{-}8$$

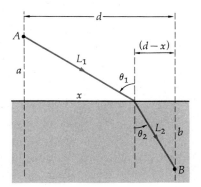

The quantity $nL$ is called the *optical-path length*. According to Fermat's principle, light will travel along the path for which the time is minimum. From Equation 32-8 this is the path for which the optical-path length is a minimum. We can derive Snell's law of refraction by consid-

Newton demonstrating the dispersion of sunlight with a glass prism.

ering the time as a function of a single parameter indicating the position of point $P_{min}$. In terms of the distance $x$ in Figure 32-20 we have

$$L_1^2 = a^2 + x^2 \quad \text{and} \quad L_2^2 = b^2 + (d - x)^2 \qquad \text{32-9}$$

Figure 32-21 shows the time $t$ as a function of $x$. At the value of $x$ for which the time is a minimum, the slope of this graph is zero, or

$$\frac{dt}{dx} = 0$$

Differentiating Equation 32-8 gives

$$\frac{dt}{dx} = \frac{1}{c} \left( n_1 \frac{dL_1}{dx} + n_2 \frac{dL_2}{dx} \right)$$

For the path of minimum time we have

$$n_1 \frac{dL_1}{dx} + n_2 \frac{dL_2}{dx} = 0 \qquad \text{32-10}$$

We can compute these derivatives from Equations 32-9. We have

$$2L_1 \frac{dL_1}{dx} = 2x \quad \text{or} \quad \frac{dL_1}{dx} = \frac{x}{L_1}$$

**Figure 32-21**
Sketch of time for light to travel from point $A$ to $B$ in Figure 32-16 versus distance $x$ measured along the refracting surface. The time is minimum at the point for which the angles of incidence and refraction obey Snell's law.

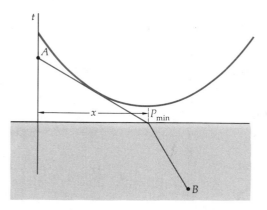

But $x/L_1$ is just $\sin \theta_1$, where $\theta_1$ is the angle of incidence. Thus

$$\frac{dL_1}{dx} = \sin \theta_1$$

Similarly

$$2L_2 \frac{dL_2}{dx} = 2(d - x)(-1)$$

or

$$\frac{dL_2}{dx} = -\frac{d - x}{L_2} = -\sin \theta_2$$

where $\theta_2$ is the angle of refraction. Hence Equation 32-10 is

$$n_1 \sin \theta_1 + n_2(-\sin \theta_2) = 0$$

or

$$n_1 \sin \theta_1 = n_2 \sin \theta_2$$

which we recognize as Snell's law.

# 32-6 Polarization

In any transverse wave, the vibration is perpendicular to the direction of propagation and can therefore be resolved into rectangular components in the plane perpendicular to that direction. For example, in an electromagnetic wave moving in the $x$ direction, the electric and magnetic fields lie in a plane parallel to the $yz$ plane and can be resolved into $y$ and $z$ components. If the electric field vector remains parallel to a line in space, the wave is said to be linearly polarized. The magnetic field then also remains parallel to a line in space.* Linear polarization is also called *plane polarization*. Electric-dipole radiation is linearly polarized.

Transverse waves can also be circularly polarized or elliptically polarized. In circularly or elliptically polarized electromagnetic waves, the electric field vector rotates in a circle or an ellipse. For example, for a wave moving in the $x$ direction, the $y$ and $z$ components of the electric field for such a wave are given by

$$E_y = E_{y0} \sin (kx - \omega t)$$

and

$$E_z = E_{z0} \sin \left( kx - \omega t + \frac{\pi}{2} \right) = E_{z0} \cos (kx - \omega t) \qquad 32\text{-}11$$

If the amplitudes $E_{y0}$ and $E_{z0}$ are equal, the wave is circularly polarized. Otherwise the polarization is elliptical.

The various kinds of polarization of transverse waves can be visualized most easily by considering mechanical waves on a string. If one end of a long string is moved back and forth along a line, the resulting waves on the string are linearly polarized, each element of the string vibrating along a line. On the other hand, if the end of the string is moved with constant speed in a circle, a circularly polarized wave will be propagated along the string with each element of the string moving in a

---

* In discussing electromagnetic waves it is simpler to consider only one of the fields, such as the electric field, since the direction of the magnetic field is then determined by the fact that $\mathbf{E} \times \mathbf{B}$ is the direction of propagation. The electric field is usually chosen because most detectors, including the human eye, are sensitive to $\mathbf{E}$ rather than to $\mathbf{B}$.

circle. Similarly, if the string is moved in an elliptical path, the resulting wave is elliptically polarized.

Most waves produced by a single source are polarized; e.g., string waves produced by the vibration of one end of a string, or electromagnetic waves produced by a single atom or a single electric dipole antenna are polarized. Waves produced by many sources are usually unpolarized. A typical light wave, for example, is produced by millions of atoms acting independently. The electric field for a light wave propagating in the $x$ direction can be resolved into $y$ and $z$ components at any instant, but these components have a phase difference which varies randomly in time because there is no correlation between the electric field vectors produced by the different atoms. Such a light wave is unpolarized.

Four phenomena produce polarized light from unpolarized light:

1. Absorption

2. Reflection

3. Scattering

4. Birefringence

*Means of polarization*

We look briefly at each in this section.

### Absorption

A common method of polarization is absorption in a sheet of commercial material called Polaroid, invented by E. H. Land in 1938. This material contains long-chain hydrocarbon molecules which are aligned when the sheet is stretched in one direction during the manufacturing process. These chains become conducting (at optical frequencies) when the sheet is dipped in a solution containing iodine. When light is incident with its electric field vector **E** parallel to the chains, electric currents are set up along the chains and the light energy is absorbed. If the electric field **E** is perpendicular to the chains, the light is transmitted. The direction perpendicular to the chains is called the *transmission axis*. We shall make the simplifying assumption that all the light is transmitted when **E** is parallel to the transmission axis and all the light is absorbed when **E** is perpendicular to the transmission axis.

Consider a light beam in the $z$ direction incident on a Polaroid which has its transmission axis in the $y$ direction. On the average, half of the incident light has its **E** vector in the $y$ direction and half in the $x$ direction. Thus half the intensity is transmitted, and the transmitted light is linearly polarized with its **E** vector in the $y$ direction.

Suppose we have a second piece of Polaroid whose transmission axis makes an angle $\theta$ with that of the first, as in Figure 32-22. The **E** vector of the light between the Polaroids can be resolved into two components, one parallel and one perpendicular to the transmission axis of the second Polaroid. If we call the direction of the transmission axis of the second Polaroid $y'$,

$$E_{y'} = E \cos \theta \quad \text{and} \quad E_{x'} = E \sin \theta$$

Only the component $E_{y'}$ is transmitted by the second Polaroid. The transmitted intensity is proportional to the square of the transmitted amplitude. Thus, if $I_1$ is the intensity between the two Polaroids, the intensity transmitted by both Polaroids is

$$I_{\text{total}} = I_1 \cos^2 \theta \qquad \qquad 32\text{-}12$$

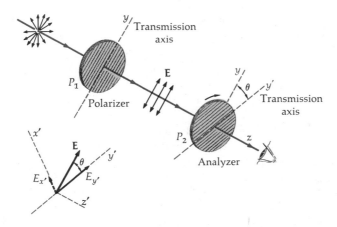

**Figure 32-22**
Two Polaroids with their
transmission directions
making an angle $\theta$ with each
other. Only the component
$E \cos \theta$ is transmitted through
the second Polaroid. If the
original intensity is $I_0$, the intensity between the Polaroids
is $\frac{1}{2}I_0$ and that transmitted by
both Polaroids is $\frac{1}{2}I_0 \cos^2 \theta$.

(The intensity incident on the second Polaroid is of course half that incident on the first.) When two polarizing elements are placed in succes-
sion in a beam of light as described here, the first is called the *polarizer*
and the second is called the *analyzer*. If the polarizer and analyzer are
crossed, i.e., have their axes perpendicular to each other, no light gets
through when the absorption is complete. In practice, usually some
weak red or blue light can be observed through crossed Polaroids, indicating that the absorption is not complete over the entire visible spectrum. Equation 32-12 is known as *Malus' law* after its discoverer, E. L.        *Malus' law*
Malus (1775–1812). It applies to any two polarizing elements whose
transmission directions make an angle $\theta$ with each other.

Polarization of electromagnetic waves by absorption can be demonstrated with microwaves (electromagnetic waves with wavelengths on
the order of centimetres). In a typical microwave generator, polarized
waves are radiated by a dipole antenna. An absorber can be made of a
screen of parallel straight wires (Figure 32-23). An electric field parallel
to the wires sets up currents in the wires, which absorb energy. If the
waves are perpendicular to the wires, no currents are set up and the
wave is transmitted. The transmission axis of this polarizer is perpendicular to the direction of the wires. When the axis is parallel to the dipole antenna (the wires are perpendicular to the dipole), the microwaves are transmitted. If the absorber is rotated 90° from this position, the waves are absorbed.

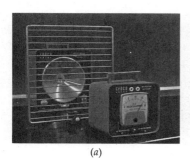

*(a)*

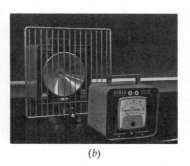

*(b)*

**Figure 32-23**
Polarization of microwaves.
The electric field of the microwaves is vertical, parallel to
the vertical dipole radiator.
(*a*) When the metal wires are
horizontal (transmission axis
vertical), the waves are transmitted, as indicated by the
high reading on the detector.
(*b*) When the wires are vertical (transmission axis horizontal), the waves are absorbed, as indicated by the
low reading on the detector.
(*Courtesy of Larry Langrill,
Oakland University.*)

### Reflection

When unpolarized light is reflected from a plane surface, e.g., that separating air and glass or air and water, the reflected light is partially polarized. The degree of polarization depends on the angle of incidence
and the indexes of refraction of the two media. When the angle of inci-

dence is such that the reflected and refracted rays are perpendicular to each other, the reflected light is completely polarized. This result was discovered experimentally by Sir David Brewster in 1812.

Figure 32-24 shows light incident at the polarizing angle $\theta_p$ for which the reflected light is completely polarized. The electric field vector **E** of the incident light can be resolved into components parallel and perpendicular to the plane of incidence. The reflected light is completely polarized with its electric field vector perpendicular to the plane of incidence. We can relate the polarizing angle $\theta_p$ to the indexes of refraction of the media using Snell's law. If $n_1$ is the index of refraction of the first medium and $n_2$ that of the second medium, we have

$$n_1 \sin \theta_p = n_2 \sin \theta_2 \qquad\qquad 32\text{-}13$$

where $\theta_2$ is the angle of refraction. From Figure 32-24 we see that the sum of the angle of reflection and the angle of refraction is 90°. Since the angle of reflection equals the angle of incidence, we have

$$\theta_2 = 90° - \theta_p$$

Then

$$n_1 \sin \theta_p = n_2 \sin (90° - \theta_p) = n_2 \cos \theta_p$$

or

$$\tan \theta_p = \frac{n_2}{n_1} \qquad\qquad 32\text{-}14 \qquad \textit{Brewster's law}$$

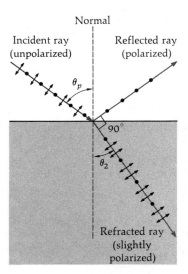

Normal

Incident ray
(unpolarized)

Reflected ray
(polarized)

$\theta_p$

90°

$\theta_2$

Refracted ray
(slightly
polarized)

**Figure 32-24**
Polarization by reflection. The incident wave is unpolarized and has components of **E** (indicated by the arrows) parallel to the plane of incidence and components (indicated by the dots) perpendicular to the plane of incidence. For incidence at the polarizing angle $\theta_p$ shown, the reflected light is completely polarized with **E** perpendicular to the plane of incidence as indicated by the dots. Then the transmitted beam is partially polarized because only a small fraction of the light is reflected. At other angles of incidence, the reflected light is partially polarized.

Equation 32-14 is known as *Brewster's law*. Although the reflected light is completely polarized when the incident angle is $\theta_p$, the transmitted light is only partially polarized because only a small fraction of the incident light is reflected. If the incident light itself is polarized with its electric field vector in the plane of incidence, there is no reflected light when the angle of incidence is $\theta_p$ (Figure 32-25). There is no simple method of deriving Brewster's law; it can be derived from the electromagnetic theory of light. We can understand the result qualitatively from Figure 32-25. If we consider the molecules of the second medium to be oscillating in the direction of the electric field of the refracted ray, they cannot radiate energy along the direction of oscillation, which would be the direction of the reflected ray.

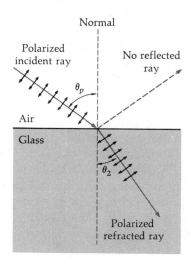

**Figure 32-25**
Polarized light incident at the polarizing angle. If the incident light has no component of **E** perpendicular to the plane of incidence, there is no reflected light.

Because of the polarization of reflected light, sunglasses made of polarizing material can be very effective in cutting out glare. If the light is reflected from a horizontal surface such as a lake or snow on the ground, the plane of incidence will be vertical and the electric field vector of the reflected light will be predominantly horizontal. Sunglasses with their transmission axis vertical will then reduce the glare by absorbing much of the reflected light.

### Scattering

The phenomenon of absorption and reradiation is called *scattering*. Reflection of light is actually a scattering of light by a large number of scattering centers closely spaced compared with the wavelength. Refraction is a similar phenomenon in which the scattered light interferes with the incident light. The term scattering, however, usually refers to the situation in which the scattering centers are separated by distances not small compared with the wavelength of light. A familiar example of light scattering is that from clusters of air molecules (due to random fluctuations in the density of air) which tend to scatter short wavelengths more than long wavelengths, giving the sky its blue color.

Scattering can be demonstrated by adding a small amount of powdered milk to a container of water. The milk particles absorb light and reradiate it as dipole radiators. Consider a beam of light in the $z$ direction (Figure 32-26). The **E** vector is therefore in the $x$ and $y$ directions. The scattering particles oscillate parallel to the **E** vector of the incident wave, i.e., in the $x$ and $y$ directions but not in the $z$ direction. If we look perpendicular to the beam, say in the $x$ direction, we see light radiated by charges in the scattering centers oscillating in the $y$ direction but we do not see radiation due to the oscillations in the $x$ direction because no light is radiated in the direction along the line of the dipoles. We also see no radiation with **E** in the $z$ direction because the scattering particles do not oscillate in that direction. The light radiated in the $x$ direction is thus polarized with its **E** vector in the $y$ direction. This is easily demonstrated by polarizing the incident beam with a Polaroid. If the incident light contains only **E** vectors in the $x$ direction, no scattered light will be observed in the $x$ direction, whereas light will be observed in the $y$ direction. Rotating the Polaroid 90° will reverse the situation: no scattered light in the $y$ direction and scattered light in the $x$ direction.

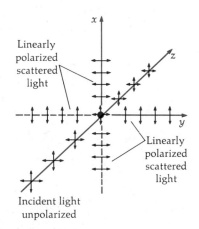

**Figure 32-26**
Polarization by scattering. Unpolarized light propagating in the $z$ direction is incident on a scattering center at the origin. The light scattered in the $x$ direction is polarized with its **E** vector in the $y$ direction, while the light scattered in the $y$ direction is polarized in the $x$ direction.

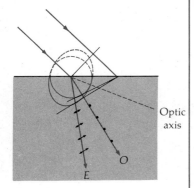

**Figure 32-27**
Huygens wavelets for the *O* and *E* rays when light is incident on a birefringent crystal whose optic axis is in the plane of incidence. The *E* ray is polarized with its electric field in the plane of incidence, whereas the *O* ray has its electric field perpendicular to the plane of incidence. The *O* ray obeys Snell's law of refraction, but the *E* ray does not.

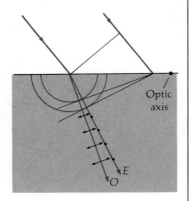

**Figure 32-28**
Huygens wavelets for the *O* and *E* rays when unpolarized light is incident on a birefringent crystal whose optic axis is perpendicular to the plane of incidence. In this case both rays obey Snell's law of refraction.

### Birefringence

Birefringence is a complicated phenomenon occurring in materials, e.g., calcite, which are anisotropic, i.e., have a preferred direction. Although most materials are isotropic (all directions are equivalent), some crystals, because of their atomic structure, have different optical properties for light traveling in different directions through the crystal. The velocity of light in calcite and some other materials is not the same in all directions. Such materials are called *double-refracting* or *birefringent*. When a light ray is incident on a calcite crystal, it may be separated into two rays which travel with different velocities and are polarized in mutually perpendicular directions. The Huygens wavelets corresponding to one ray are spherical, as in propagation through ordinary isotropic materials, and this ray is called the *ordinary* (*O*) *ray*. In the *extraordinary* (*E*) *ray* the Huygens wavelets are ellipsoids of revolution, indicating that the velocity of propagation depends on direction. Along the axis of revolution, the two rays propagate with the same speed. The Huygens wavelets are tangent to each other in this direction, called the *optic axis*. Perpendicular to the optic axis, the difference in speed of the two rays is maximum. In calcite, the extraordinary ray travels faster than the ordinary ray perpendicular to the optic axis. The index of refraction for the ordinary ray in calcite is about $n_O = 1.66$. For the extraordinary ray $n_E$ varies from 1.66 along the optic axis to 1.49 perpendicular to it.

The propagation of both the ordinary ray and extraordinary ray through a birefringent crystal can be traced by construction of the Huygens wavelets for the rays. For a general orientation of the crystal this is extremely complex. We shall restrict our discussion to the cases in which the optic axis of the crystal is either in the plane of incidence or perpendicular to it. Then both the *O* and *E* rays are in the plane of incidence, and the polarization of the rays is relatively simple. The ordinary ray is polarized with its electric field perpendicular to the plane formed by the optic axis and the direction of propagation, whereas the extraordinary ray has its electric field in the plane containing the optic axis and the direction of propagation. The Huygens wavelets and polarizations are shown in Figures 32-27 and 32-28 for the optic axis in the plane of incidence and perpendicular to the plane of incidence, respectively. Note the spatial separation of the rays.

In Figure 32-29 a light beam is incident normally on the face of a crystal whose optic axis is perpendicular to the surface. The two rays propagate along the optic axis with the same speed, and no separation of the rays is observed. In Figure 32-30 the light is incident normally on a crystal whose optic axis is not perpendicular to the surface. In this case, even for normal incidence the extraordinary ray is deflected. The two rays are separated in space and emerge from the crystal as parallel rays separated by an amount that depends on the thickness of the crystal (Figure 32-31). This effect can be demonstrated by placing a calcite crystal over a small source and observing the two images of the source through the crystal. A Polaroid analyzer can be used to show that the two rays are polarized in mutually perpendicular directions. If the crystal is rotated about the line of the incident ray, the extraordinary ray is rotated through space.

There are several interesting applications of double refraction. One is the Nicol prism, which is used as a polarizer. Figure 32-32 shows the geometry of such a prism. A crystal is cut so that the two rays are separated in space. The crystal is cut along the diagonal and cemented together with a transparent cement which has an index of refraction en-

**Figure 32-29**
Huygens wavelets for light incident normally on a birefringent crystal whose optic axis is perpendicular to the surface. Both rays propagate with the same speed and in the same direction, and so there is no separation of the rays.

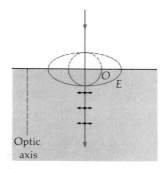

**Figure 32-30**
Huygens wavelets for light incident normally on a birefringent crystal whose optic axis is not perpendicular to the surface. The E ray is deflected even at normal incidence.

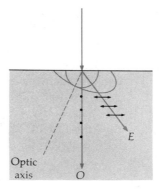

**Figure 32-31**
A narrow beam of unpolarized light incident normally on a birefringent crystal oriented as in Figure 32-30 is split into two beams (exaggerated in this figure). When the crystal is rotated in space, the E ray rotates in space.

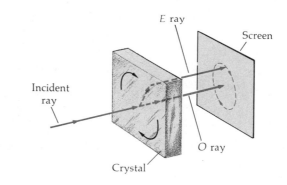

**Figure 32-32**
A Nicol prism made by cutting a birefringent crystal and gluing it back together. The O ray is totally reflected at the surface of the crystal and glue, but the E ray is not. The emerging beam is polarized.

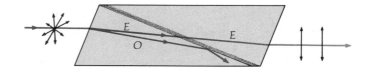

suring that one ray is totally reflected whereas the other is transmitted. The advantage of a Nicol prism over Polaroid is that the emerging light is completely polarized over the entire visible spectrum.

Double refraction can also be used to produce a rotation of the plane of polarization of linearly polarized light and to produce circularly polarized light from linearly polarized light. Figure 32-33 shows a light beam incident normally on the surface of a crystal whose optic axis is parallel to the surface. The Huygens wavelets for the O and E rays have circular cross sections but of different radii. The two rays travel in the same direction as the incident ray but with different speeds and different polarizations. In calcite, for example, the extraordinary ray travels faster than the ordinary ray so that its wavefront gets ahead of that for the O ray.

**Figure 32-33**
Huygens wavelets for unpolarized light incident normally on a birefringent crystal whose optic axis is parallel to the surface. The *O* and *E* rays propagate in the same direction but with different speeds. The two beams emerge from the crystal with a phase difference which depends on the indexes of refraction for the two rays and on the thickness of the crystal.

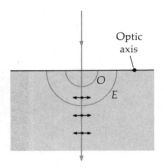

**Figure 32-34**
Rotation of the direction of polarization by a half-wave plate birefringent crystal. Light from the Polaroid is linearly polarized with its **E** vector making an angle of 45° with the optic axis, which is parallel to the surface of the crystal, as in Figure 32-33. The thickness of the crystal is such that the *E* and *O* rays differ in phase by 180° upon emergence from the crystal. The effect is to rotate the direction of polarization by 90°.

Double image produced by birefringent calcite crystal. (*From Richard T. Weidner and Robert L. Sells,* Elementary Classical Physics, *vol. II, 2d ed., fig. 36.17, p. 743. Copyright © 1973 by Allyn and Bacon, Inc. Used by permission.*)

The beams emerge with a phase difference which depends on the thickness of the plate. In a *half-wave plate* the thickness is such that the number of wavelengths of the ordinary ray is greater than the number of wavelengths of the extraordinary ray by an odd number of half wavelengths. The phase difference is thus 180°. (Since the ordinary ray travels slower, it has a smaller wavelength, $\lambda = v/f$, than the extraordinary ray. A given thickness thus has more ordinary waves in it than extraordinary waves.) If the incident light is polarized in the direction at 45° to the polarizations of the two rays, the emerging light will have its polarization rotated as shown in Figure 32-34. In a *quarter-wave plate* the thickness is such that there is a 90° phase difference between the rays when they emerge. The resulting electric field is of the form $E_x = E_0 \times \sin \omega t$, $E_y = E_0 \cos \omega t$. The **E** vector thus rotates in a circle, and the wave is circularly polarized.

Many common substances, e.g., cellophane and transparent tape, are birefringent. Interesting and beautiful patterns can be observed by placing such materials between crossed Polaroids. With no substance between the Polaroids, no light is transmitted, but a sheet of cellophane acts as a half-wave plate for light of a certain color, depending on the thickness of the cellophane. It rotates the plane of polarization of the light between the Polaroids, and some light of that color is transmitted.

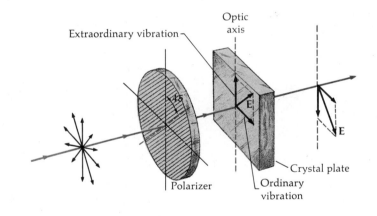

# Radar Astronomy

G. H. Pettengill
*Massachusetts Institute of Technology*

Essentially all we know about celestial objects has come from measuring the electromagnetic radiation they emit or reflect. In passing through the earth's atmosphere, however, much of this radiation is absorbed. In fact, we can observe celestial objects from the ground only at wavelengths lying within two major spectral windows, where our atmosphere is relatively transparent. The shorter-wavelength, or optical, window shows us the splendor of the night sky and has made the development of optical astronomy possible. The longer-wavelength, or radio, window has been increasingly exploited in recent years and provides the basis for radio astronomy.

Radio waves can be collected and focused by antennas using large parabolic reflecting surfaces, as shown in Figure 1. This particular antenna uses an additional reflector to redirect the incident energy collected by the paraboloid onto a feed horn, visible near the center of the primary reflector. The same antenna can be used for radiating energy by reversing the receiving process. In this case, a transmitter is attached to the feed horn, and its radiation is modified by reflection into a plane wave extending across the aperture of the large paraboloid. At great distances from the antenna, diffraction transforms this circular section of a plane wave into a spherical wavefront with a cone angle (beam width) proportional to the ratio of the radio wavelength to the reflector's diameter.

If the radio transmitter is sufficiently powerful (typically several hundred kilowatts) and the antenna sufficiently large (30 m or so in diameter), it is possible to detect radar echoes from the moon and the nearer planets. Since the time, frequency, and polarization characteristics of the coherent transmitted waveform are precisely known, the alteration of these characteristics as seen in the echo can be related to properties of the distant scattering surface and the propagating medium en route. The study of celestial objects in this way is called *radar astronomy*.

How does the distant surface scatter the incident radar energy? If the surface contains regions, or facets, which are many wavelengths across and flat

**Figure 1**
Steerable parabolic reflector, 84 ft in diameter, used for radar astronomy at 23-cm wavelength at M.I.T.'s Lincoln Laboratory, Westford, Massachusetts. [*From M. I. Skolnik (ed.), Radar Handbook, p. 33-18, McGraw-Hill Book Company, New York, 1970, by permission of the publishers.*]

**Figure 2**
Schematic representation of radar scattering and the dispersion of echo power in delay and frequency introduced by finite target size and rotation. [*After M. I. Skolnik (ed.), Radar Handbook, p. 33-3, McGraw-Hill Book Company, New York, 1970, by permission of the publishers.*]

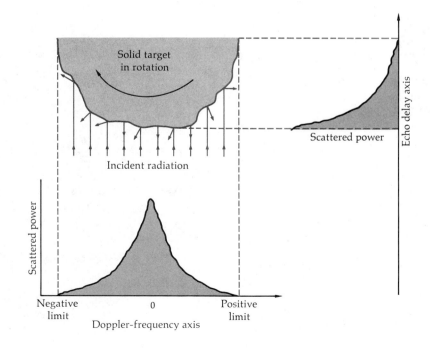

within plus or minus the wavelength, the reflection is coherent, or specular; i.e., it emerges at an angle with respect to the facet normal which is equal to the angle of incidence. Since the radar observes only backscattered radiation, only surface facets oriented at right angles to the line of sight are seen, as diagrammed schematically in Figure 2. If the surface is rough, so that its height varies rapidly by a significant fraction of the wavelength as one moves across it, the scattering is diffuse and depends only slightly on the angle of incidence to the surface.

Rotation of the target introduces a doppler frequency shift which varies linearly across the target as shown in Figure 2. Therefore, if one analyzes the echo power in frequency, one can establish a direct correspondence between the echo intensity observed in a given frequency interval and the scattering efficiency of a calculable location on the surface which imparts those particular values of frequency shift. Similarly, the distribution of echo power in time delay can be related to scattering from identifiable regions of the surface lying at different distances from the radar.

By analyzing the echo power simultaneously as a function of delay and frequency, radar scattering maps of the surface can be obtained, as shown in Figure 3. These radar maps clearly show the existence of large-scale surface features, such as craters and mountain ranges, and are of particular interest in the study of the cloud-shrouded surface of Venus, which is not otherwise accessible to study from afar. This technique can also be turned around and used to determine the planet's rotation. In this way it was discovered by radar that Mercury does not always keep one face turned to the sun, as was formerly thought, but rotates 50 percent faster than its average orbital rate.

By timing the total round-trip echo delay, the distance between the radar and the surface of the distant celestial object is obtained, since the velocity of radio waves is known. In many current experiments, the timing accuracy is better than one-millionth of a second, while the total round-trip echo delay exceeds 1000 s. Thus, we measure distance with a fractional accuracy of $1 \times 10^{-9}$, or a thousand times more accurately than is possible using optical observations alone. Not many measurements in physics achieve this level of accuracy, which is sufficient to make observable results of many extremely small perturbing effects on the orbits of the inner planets. Several of these effects are identifiable with relativistic "corrections" and have been used successfully to confirm the accuracy of Einstein's formulation of the theory of general relativity.

**Figure 3**
Radar map of the moon at 70-cm wavelength, made by T. W. Thompson using the Arecibo radar in Puerto Rico. Note the unusually strong diffuse scattering by the crater Tycho in the bottom center of the map, indicating that it has an exceptionally rough surface. (*From T. W. Thompson, Atlas of Lunar Radar Maps at 70-cm Wavelength.* The Moon, *vol. 10, no. 1, p. 54, May 1974.*)

Review

A. Identify the contributions the following people made to the understanding of light: Young, Fresnell, Foucault, Maxwell, Hertz, Einstein, Fizeau, Roemer, Fermat, Brewster.

B. Define, explain, or otherwise identify:

Photoelectric effect, 850
Photon, 850
Electric-dipole radiation, 853
Plane of incidence, 856
Law of reflection, 856
Fermat's principle, 857
Index of refraction, 857

Snell's law, 858
Polarization, 863
Malus' law, 865
Brewster's law, 866
Scattering, 867
Birefringence, 868

Exercises

### Section 32-1, Waves or Particles?

1. Show that if the wavelength of an electromagnetic wave is given in nanometres, the photon energy in electronvolts is given by $E = (1240 \text{ eV·nm})/\lambda$.

2. (a) Use the result of Exercise 1 to calculate the range of photon energies in the visible spectrum of wavelengths from about 400 to 700 nm. (b) Molecular transitions involve the absorption or emission of photons of energy less than 1 eV. Is this radiation in the infrared (long-wavelength) or ultraviolet (short-wavelength) part of the electromagnetic spectrum?

3. What is the energy of an x-ray photon whose wavelength is 0.1 nm?

4. A prominent radiation from sodium has a wavelength of 589 nm. What is the energy of the photon of this wavelength?

### Section 32-2, Electromagnetic Waves

5. What is the frequency of a 3-cm microwave?

6. What is the frequency of an x-ray of wavelength 0.1 nm?

7. Find the wavelength for a typical AM radio frequency and for a typical FM radio frequency.

8. What is the wavelength of a photon whose energy is 5 eV? In what part of the electromagnetic spectrum is it?

9. A radiating electric dipole lies along the $z$ axis. Let $I_1$ be the intensity of the radiation at a distance $r = 10$ m and at angle $\theta = 90°$. Find the intensity (in terms of $I_1$) at (a) $r = 30$ m, $\theta = 90°$; (b) $r = 10$ m, $\theta = 45°$; (c) $r = 20$ m, $\theta = 30°$.

10. (a) For the situation described in Exercise 9, at what angle is the intensity at $r = 5$ m also equal to $I_1$? (b) At what distance is the intensity equal to $I_1$ at $\theta = 45°$?

### Section 32-3, The Speed of Light

11. A light-year is the distance light travels in 1 year. (a) Find the number of kilometres in a light-year. (b) The spiral galaxy in the Andromeda constellation is about $2 \times 10^{19}$ km away from us. How many light-years is this?

12. How long does it take light to travel from the sun to the earth, a distance of about $1.5 \times 10^{11}$ m? What is the distance to the sun in light-minutes?

13. On a rocket sent to Mars to take pictures, the camera is triggered by radio waves which (like all electromagnetic waves) travel with the speed of light. What is the time delay between sending and receiving the signal from the earth to Mars? (Take the distance to Mars to be $9.7 \times 10^{10}$ m.)

**Section 32-4, Reflection**

14. Calculate the fraction of light energy reflected from water at normal incidence ($n = 1.33$ for water).

15. Light is incident normally on a slab of glass of index of refraction $n = 1.5$. Reflection occurs at both surfaces of the slab. About what percentage of the incident light energy is transmitted by the slab?

16. A physics student playing pocket billiards wants to strike his cue ball so that it hits a cushion and then hits the eight ball squarely. He chooses several points on the cushion and for each point he measures the distance from it to the cue ball and to the eight ball. He aims at the point for which the sum of these distances is least. Will his cue ball hit the eight ball? How is this method related to Fermat's principle?

17. A point source of light is 5 cm above a plane reflecting surface (such as a mirror). Draw a ray from the source which strikes the surface at an angle of incidence of about 45°; draw two more at angles slightly less than 45°; and draw the reflected ray for each. The reflected rays appear to diverge from a point called the image of the light source. Draw dotted lines extending the reflected rays back until they meet at a point behind the surface, to locate the image point.

**Section 32-5, Refraction**

18. The index of refraction of water is 1.33. Find the angle of refraction of a beam of light in air hitting the water surface at an angle of (a) 20°, (b) 30°, (c) 45°, and (d) 60° with the normal, and indicate these rays on a diagram.

19. Repeat Exercise 18 for a beam of light initially in water and incident on a water-air surface.

20. What is the critical angle for total internal reflection for light traveling from water ($n = 1.33$) to air ($n = 1.00$)?

21. Find the speed of light in water ($n = 1.33$) and in glass ($n = 1.5$).

22. A beam of monochromatic red light of wavelength 700 nm in air travels in water. What is the wavelength in water? Does a swimmer under water observe the same color or a different color for this light?

23. The critical angle for total internal reflection in diamond is about 24°. What is the index of refraction of diamond?

24. A beam of light strikes a plane glass surface at an angle of incidence of 45°. The index of refraction of the glass varies with wavelength according to the graph in Figure 32-18. How much smaller is the angle of refraction for blue light of wavelength 400 nm than for red light of wavelength 700 nm?

25. Use Figure 32-18 to calculate the critical angles for total internal reflection from a glass-air surface for blue light of wavelength 400 nm and red light of wavelength 700 nm.

26. A swimmer at $S$ in Figure 32-35 develops a leg cramp while swimming near the shore of a calm lake and calls for help. A lifeguard at $L$ hears the call. The lifeguard can run 9 m/s and swim 3 m/s. He knows physics and chooses a path that will take the least time to reach the swimmer. Which of the paths shown in Figure 32-35 does he take? Explain.

**Section 32-6, Polarization**

27. Two Polaroid sheets have their transmission directions crossed so that no light gets through. A third sheet is inserted between the two so that its transmission direction makes an angle $\theta$ with the first sheet. Unpolarized light of intensity $I_0$ is incident on the first sheet. Find the intensity transmitted through all three sheets if (a) $\theta = 45°$; (b) $\theta = 30°$.

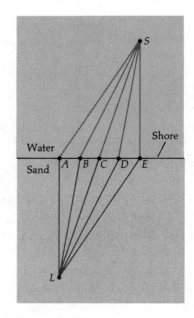

**Figure 32-35**
Swimmer at $S$ and lifeguard at $L$ for Exercise 26.

28. Two Polaroid sheets are inserted between two other Polaroid sheets which have their transmission directions crossed, so that the angle between each successive pair of sheets is 30°. Find the transmitted intensity if the original light is unpolarized with intensity $I$.

29. The polarizing angle for a certain substance is 60°. (a) What is the angle of refraction of light incident at this angle? (b) What is the index of refraction of this substance?

30. The critical angle for total internal reflection for a substance is 45°. What is the polarizing angle for this substance?

31. What is the polarizing angle for glass with $n = 1.5$?

32. A beam of linearly polarized light strikes a calcite crystal such that its electric vector makes an angle of 30° with the optic axis (Figure 32-36). What is the ratio of the intensities of the ordinary and extraordinary rays?

33. Light of wavelength $\lambda$ in air is incident on a slab of calcite so that the ordinary and extraordinary rays travel in the same direction as in Figure 32-33. Show that the phase difference between these rays after traversing a thickness $t$ is

$$\delta = \frac{2\pi}{\lambda}(n_O - n_E)t$$

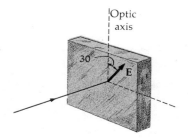

**Figure 32-36**
Exercise 32.

## Problems

1. Show that the transmitted intensity through a glass slab of index of refraction $n$ for normally incident light is approximately given by

$$I_T = I_0 \left[ \frac{4n}{(n+1)^2} \right]^2$$

2. (a) Use the result of Problem 1 to find the ratio of the transmitted intensity to the incident intensity through $N$ parallel slabs of glass for light of normal incidence. (b) Find this ratio for three slabs of glass with $n = 1.5$. (c) How many slabs of glass with $n = 1.5$ will reduce the intensity to 10 percent of the incident intensity?

3. This problem is a refraction analogy. A band is marching down a football field with a constant speed $v_1$. About midfield, the band comes to a section of muddy ground, which has a sharp boundary making an angle of 30° with the 50-yd line, as shown in Figure 32-37. In the mud, the marchers move with speed $v_2 = \frac{1}{2}v_1$. Diagram how each line of marchers is bent as it encounters the muddy section of the field so that eventually the band is marching in a different direction. Indicate the original direction by a ray and the final direction by a second ray, and find the angles between these rays and the line perpendicular to the boundary line. Is their direction of motion bent toward the perpendicular to the boundary line or away from it?

**Figure 32-37**
Problem 3.

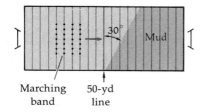

Marching band    50-yd line

4. Light is incident normally upon one face of a prism of glass where the index of refraction is $n$ (Figure 32-38). The light is totally reflected at the right side. (a) What is the minimum value $n$ can have? (b) When this prism is immersed in a liquid whose index of refraction is 1.15, there is still total reflection, but in water, whose index is 1.33, there no longer is total reflection. Use this information to establish limits for possible values of $n$.

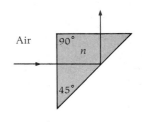

**Figure 32-38**
Problem 4.

5. Investigate the effect on the critical angle of a thin film of water on a glass surface for rays which originate in the glass. Take $n = 1.5$ for glass and $n = 1.33$ for water. (*a*) What is the critical angle for total internal reflection at the glass-water surface? (*b*) Is there any range of incident angles that are greater than $\theta_c$ for glass-to-air refraction and for which light rays will leave the glass *and* the water and pass into the air medium?

6. Show that a light ray transmitted through a glass slab emerges parallel to the incident ray but displaced from it. For an incident angle of 60°, index of refraction of glass $n = 1.5$, and slab thickness of 10 cm, find the displacement measured along the glass surface from the point at which the incident ray would emerge if there were no slab.

7. A prism is made of calcite cut so that its upper face contains the optic axis. The incident ray is unpolarized and normally incident upon the surface (Figure 32-39). (*a*) What must the angle of the prism be for the ordinary ray to be totally reflected internally? (*b*) Show that the extraordinary ray will not then be internally reflected, so that the emerging light will be plane-polarized.

**Figure 32-39**
Problem 7.

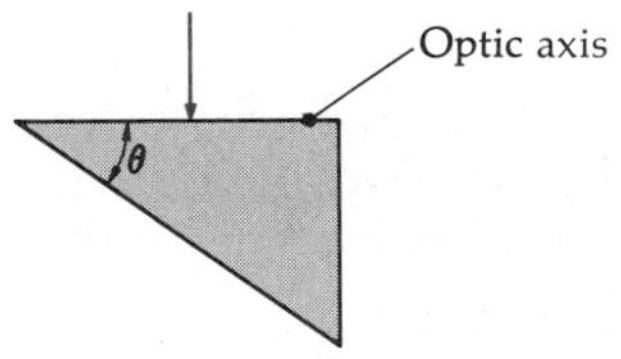

8. The indexes for ordinary and extraordinary waves given for calcite are for green light ($\lambda = 540$ nm). (*a*) What would be the minimum thickness for a quarter-wave plate? (*b*) What would be the minimum thickness for a half-wave plate? (*c*) Would you expect these to be quarter- and half-wave plates for, say, red light? Why or why not?

9. In Exercise 26, let the shoreline be the $x$ axis and the swimmer be at point $x = 20$ m, $y = 30$ m. The lifeguard is at $y = -30$ m on the $y$ axis (Figure 32-40). (*a*) Find the point $x_0$ at which the lifeguard enters the water if he can run 3 times as fast as he can swim, using the approximation $\sin \theta = \tan \theta$ for the angles made by his path and the normal to the shoreline. (*b*) Find $x_0$ to three significant figures by trial and error, using your answer from (*a*) as a first guess. (*c*) Find the time for the lifeguard to reach the swimmer if he runs at 9 m/s and swims at 3 m/s and compare with the time it would take if the guard followed a straight-line path from his initial position to that of the swimmer.

**Figure 32-40**
Problem 9.

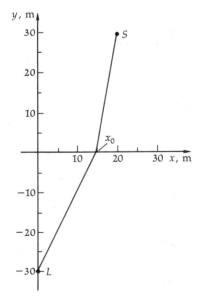

CHAPTER 33    Geometric Optics

---

**Objectives**    After studying this chapter you should:

1.  Be able to draw simple ray diagrams to locate images and determine whether they are real or virtual, erect or inverted, and enlarged or reduced for spherical mirrors and thin lenses.

2.  Be able to locate the image formed by a spherical mirror or by a thin lens using the mirror or thin-lens equation and calculate the magnification of the image.

3.  Be able to locate the image formed by a single refracting surface.

4.  Be able to discuss the origin of the various aberrations that occur with mirrors and lenses.

5.  Be able to discuss how the eye works.

6.  Be able to show with a simple diagram why close objects appear larger than far objects.

7.  Be able to describe how a simple magnifier works and calculate the angular magnification of a simple magnifier.

8.  Be able to describe with diagrams and equations how a microscope and telescope work.

---

The wavelength of light is often very small compared with the size of any obstacles or apertures it encounters, and diffraction effects can be neglected. The study of such situations, for which the ray approximation is valid and light propagates in straight lines, is known as *geometric optics*. In this chapter, we shall apply the laws of reflection and refraction discussed in Chapter 32 to study the formation of images by mirrors and lenses.

## 33-1   Plane Mirrors

Figure 33-1 shows a narrow bundle of light rays from a point source $P$ reflected from a plane mirror. After reflection, the rays diverge exactly as if they came from a point $P'$ behind the plane of the mirror. The point $P'$ is called the *image* of the object $P$. When these rays enter the eye, they cannot be distinguished from rays diverging from a source at $P'$ with no mirror. The image is called a *virtual image* because the light does not actually emanate from the image but only appears to. Geometric construction using the law of reflection shows that the image point lies on the line through the object perpendicular to the plane of the mirror and

*Virtual image*

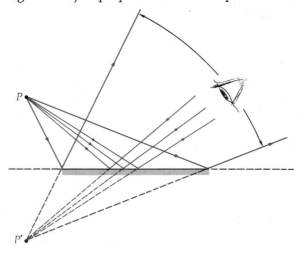

**Figure 33-1**
Image formed by a plane mirror. The rays from point $P$, which strike the mirror and enter the eye, appear to come from the image point $P'$ behind the mirror. The image can be seen by the eye anywhere in the region indicated.

at a distance behind the plane equal to that from the plane to the object. The image can be seen by an eye anywhere in the region indicated, in which a line from the image to the eye passes through the mirror. We note that the object need not be directly in front of the mirror. An image can be seen as long as the object is above the plane of the mirror.

Figure 33-2 shows an image of an extended object formed by a plane mirror. The size of the image is the same as that of the object.

Figure 33-3 illustrates the formation of multiple images by two plane mirrors making an angle with each other. Light reflected from mirror 1 strikes mirror 2 just as if it came from the image point $P_1$. The image $P_1$ is called the *object point* for mirror 2. Its image is at point $P_{12}$. This image will be formed whenever the image point $P_1$ is in front of the plane of mirror 2. The image at point $P_2$ is due to rays from the object which re-

**Figure 33-2**
Image of an extended object in a plane mirror. The image is the same size as the object.

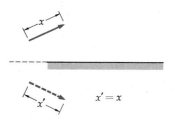

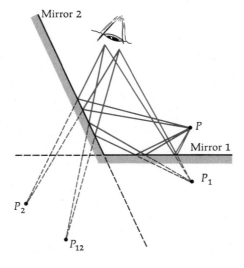

**Figure 33-3**
Images formed by two plane mirrors. $P_1$ is the image of the object $P$ in mirror 1, and $P_2$ is the image of the object in mirror 2. Point $P_{12}$ is the image of $P_1$ in mirror 2 seen when light rays from the object reflect first from mirror 1 and then from mirror 2, as shown. The image $P_2$ does not have an image in mirror 1 because it is behind that mirror.

flect directly from mirror 2. Since $P_2$ is behind the plane of mirror 1, it cannot serve as an object point for a further image in mirror 1. The number of multiple images formed by two mirrors depends on the angle between the mirrors and the position of the object.

**Question**

1. How tall must a mirror be for a standing person to see his entire reflection in it?

## 33-2 Spherical Mirrors

Figure 33-4 shows a bundle of rays from a point on the axis of a concave spherical mirror reflecting from the mirror and converging at point $P'$. The rays then diverge from this point just as if there were an object at that point. This image is called a *real image* because the light actually does emanate from the image point. It can be seen by an eye at the left of the image looking into the mirror. It could also be observed on a ground-glass viewing screen or photographic film placed at the image point. A virtual image cannot be observed on a screen at the image point because there is no light there. Despite this distinction between real and virtual images, the light rays diverging from a real image and those appearing to diverge from a virtual image are identical, so that no distinction is made by the eye when viewing a real or a virtual image.

From Figure 33-4 we see that only rays which strike the mirror at points near the axis $AO$ are reflected through the image point. Such rays are called *paraxial rays*. Because other rays converge to different points near the image point, the image appears blurred, an effect called *spherical aberration*. The image can be sharpened by reducing the size of the mirror so that nonparaxial rays do not strike it. Although the image is then sharper, its brightness is reduced because less light is reflected.

Using the law of reflection and elementary geometry, we can relate the image distance $s'$ to the object distance $s$ and the radius of curvature $r$. The geometry is shown in Figure 33-5. The result is

$$\frac{1}{s} + \frac{1}{s'} = \frac{2}{r}$$

33-1

H. M. Null/Rapho Guillumette Pictures

The lake serves as a plane mirror producing virtual images of the trees. The images are inverted and the same size as the trees.

*Spherical aberration*

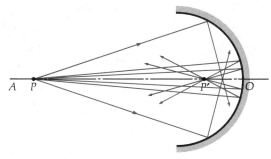

**Figure 33-4**
Rays from a point object $P$ on the axis $AO$ of a concave spherical mirror form an image at $P'$ if the rays strike the mirror near the axis. Nonparaxial rays striking the mirror at points far from $O$ are not reflected through the image point $P'$.

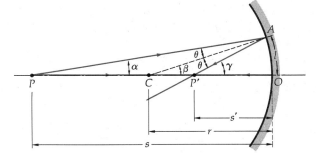

**Figure 33-5**
Geometry for calculating the image distance $s'$ from the object distance $s$ and the radius of curvature $r$. The angle $\beta$ is an exterior angle to the triangle $PAC$ and is therefore equal to $\alpha + \theta$. Similarly, from triangle $PAP'$, $\gamma = \alpha + 2\theta$. Eliminating $\theta$ from these equations gives $2\beta = \gamma + \alpha$. Equation 33-1 follows from the small-angle approximations $\alpha \approx l/s$, $\beta \approx l/r$, and $\gamma \approx l/s'$.

The derivation of this equation assumes that angles made by the incident and reflected rays with the axis are small, an assumption equivalent to that of paraxial rays.

When the object distance is much greater than the radius of curvature of the mirror, the term $1/s$ in Equation 33-1 can be neglected, resulting in $s' = \frac{1}{2}r$ for the image distance. This distance is called the *focal length $f$* of the mirror, and the image point is called the *focal point F*.

$$f = \tfrac{1}{2}r \qquad\qquad\qquad 33\text{-}2$$

In terms of the focal length $f$ the mirror equation is

$$\frac{1}{s} + \frac{1}{s'} = \frac{1}{f} \qquad\qquad\qquad 33\text{-}3$$

The focal point is the point at which parallel rays corresponding to plane waves from infinity are focused, as illustrated in Figure 33-6.

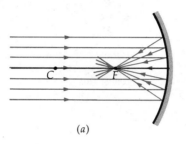 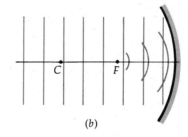

(a)                                    (b)

**Figure 33-6**
(a) Parallel rays strike a concave mirror and are reflected through the focal point at a distance $\frac{1}{2}r$. (b) The incoming wavefronts are plane waves; upon reflection they become spherical waves, which converge at the focal point.

(Again, only paraxial rays are focused at a single point.) Figure 33-7 shows rays from a point source at the focal point which strike the mirror and are reflected parallel to the axis. This illustrates a property of waves called *reversibility*. If we reverse the direction of a reflected ray, the law of reflection assures that the reflected ray will be along the original incoming ray but in the opposite direction. Reversibility holds also for refracted rays. If we have a real image of a source formed by a reflecting or refracting surface, we can place a source at the image point and the new image will be formed at the position of the original source.

A useful method of locating images is by geometric construction of a ray diagram, as illustrated in Figure 33-8, where the object is a human figure perpendicular to the axis a distance $s$ from the mirror. By a judicious choice of rays from the head of the figure we can quickly locate the image. A ray from the head parallel to the axis is reflected through the focal point a distance $\frac{1}{2}r$ from the mirror, as shown. Another ray, through the center of curvature of the mirror, strikes the mirror perpen-

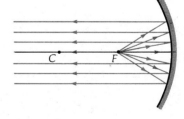

**Figure 33-7**
Illustration of reversibility. Rays diverging from a point source at the focal point of a concave mirror are reflected from the mirror as parallel rays. The rays are the same as in Figure 33-6a but in the reverse direction.

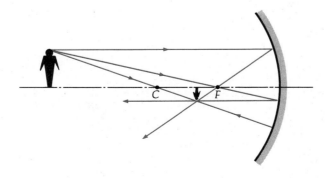

**Figure 33-8**
Ray diagram for location of image by geometric construction.

dicular to the surface and is reflected back along its original path. The intersection of these two rays locates the image point of the head. This can be checked by a third ray through the focal point, which is reflected back parallel to the axis.

We see from the figure that the image is inverted and is not the same size as the object. The magnification of the optical system (the spherical mirror in this case) is defined to be the ratio of the image size to the object size. Comparison of the triangles in Figure 33-9 shows that the magnification is just the ratio of the distances $s'$ and $s$.

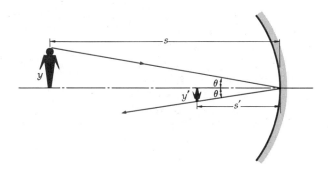

**Figure 33-9**
Geometry for finding the magnification of a spherical mirror. From the upper triangle, $\tan \theta = y/s$, and from the lower triangle, $\tan \theta = -y'/s'$, where the negative sign is introduced because $y'$ is negative. Then the magnification is $m = y'/y = -s'/s$.

When the object is between the mirror and its focal point, the rays reflected from the mirror do not converge but appear to diverge from a point behind the mirror, as illustrated in Figure 33-10. In this case the image is virtual and erect. If $s$ is less than $\frac{1}{2}r$ in Equation 33-1, the image distance $s'$ turns out to be negative. We can apply Equations 33-1 and 33-3 to this case and to convex mirrors if we adopt a convenient sign convention. Whether the mirror is convex or concave, real images can be formed only on the same side of the mirror as the object; virtual images are formed on the opposite side, where there are no actual light rays. Distances to points on the real side are taken to be positive; distances to points on the virtual side are taken to be negative. Thus for the

*Sign convention*

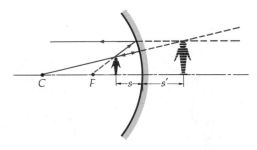

**Figure 33-10**
When the object is between the focal point and the mirror, the image is virtual and behind the mirror. Here it is located by a ray along the line from the focal point through the head of the figure and a second ray from the center of curvature through the head. The first ray is reflected parallel to the axis and the second is reflected back on itself. These rays diverge from a point behind the mirror. A third ray (not shown) could be drawn from the head parallel to the axis. It is reflected through the focal point, and its extension behind the mirror intersects the other rays at the image point.

concave mirror, $s$ and $r$ are positive, and $s'$ is positive or negative depending on whether the image is real or virtual. For a convex mirror (Figure 33-11) the center of curvature is on the virtual side, and so $r$ is taken to be negative. The focal length is also negative. For either case, Equation 33-1 gives the image distance $s'$ in terms of the object distance and radius of curvature. The lateral magnification of the image is given by

$$m = \frac{y'}{y} = -\frac{s'}{s} \qquad \qquad 33\text{-}4$$

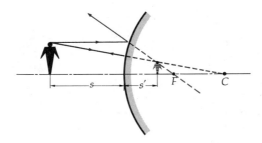

**Figure 33-11**
Ray diagram for a convex mirror. The ray parallel to the axis is reflected as if it came from the focal point to the right of the mirror, and the ray toward the center of curvature is reflected back on itself. A third ray (not shown) could be drawn toward the focal point. It would be reflected parallel to the axis, and its extension behind the mirror would intersect the other two rays of the image.

A negative magnification, which occurs when both $s$ and $s'$ are positive, indicates that the image is inverted. Although these equations with this sign convention are relatively easy to use, practical work in optics often requires only knowledge of whether the image is real or virtual and an approximate knowledge of its location. This is easiest to obtain by constructing a ray diagram. A similar method for locating images produced by a lens and various lens combinations (Section 33-4) is often of more practical use than the equations relating the image position to the object position.

---

**Example 33-1** An object 2 cm high is 10 cm from a convex mirror with a radius of curvature of 10 cm. Locate the image and find its height.

Since the center of curvature of a convex mirror is on the virtual-image side of the mirror, the focal length is negative:

$$f = \tfrac{1}{2}r = \tfrac{1}{2}(-10 \text{ cm}) = -5 \text{ cm}$$

Using Equation 33-3 to find the image distance gives

$$\frac{1}{10 \text{ cm}} + \frac{1}{s'} = \frac{1}{f} = -\frac{1}{5 \text{ cm}}$$

$$\frac{1}{s'} = -\frac{2}{10 \text{ cm}} - \frac{1}{10 \text{ cm}} = -\frac{3}{10 \text{ cm}}$$

$$s' = 3\tfrac{1}{3} \text{ cm}$$

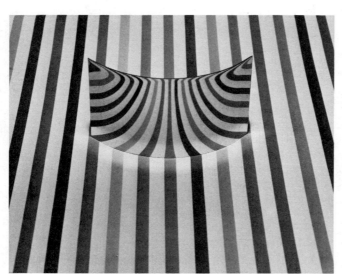

Convex mirror resting on paper with equally spaced parallel stripes. The image of each point in front of the mirror is virtual and behind the mirror. Note the reduction in size and distortion in the shape of the image and the large number of stripes that are seen.

Fundamental Photographs

The image distance is negative, indicating a virtual image behind the mirror. The magnification is

$$m = -\frac{s'}{s} = -\frac{-3\frac{1}{3}}{10} = +\frac{1}{3}$$

The image is erect and one-third the size of the object. Its size is

$$y' = my = \tfrac{1}{3}(2 \text{ cm}) = \tfrac{2}{3} \text{ cm}$$

The ray diagram for this example is similar to Figure 33-11.

**Questions**

2. Under what conditions will a concave mirror produce an erect image? A virtual image? An image smaller than the object? An image larger than the object?

3. Answer Question 2 for a convex mirror.

## 33-3 Images Formed by Refraction

Figure 33-12 illustrates the formation of an image by refraction from a spherical surface separating two media with indexes of refraction $n_1$ and $n_2$. Again, only paraxial rays converge to one point. We can derive an equation relating the image distance to the object distance, the radius of

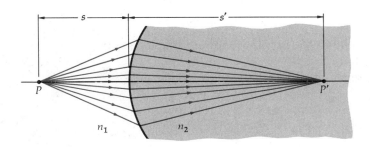

curvature, and the indexes of refraction by applying Snell's law of refraction to these rays and using the small-angle approximation. This derivation is given in Figure 33-13. The resulting equation is

$$\frac{n_1}{s} + \frac{n_2}{s'} = \frac{n_2 - n_1}{r} \qquad\qquad 33\text{-}5$$

We can use the same sign convention for this equation as we used for mirrors, but we must note that for refraction real images are formed to the right of the surface (if the object is to the left) and virtual images to the left. Thus $s'$ and $r$ are taken to be positive if the image and center of curvature lie to the right of the surface.

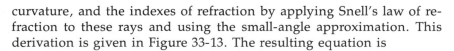
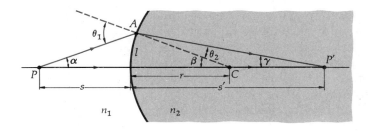

Edward Lettau, Photo Researchers

Convex mirrors are useful for wide-angle viewing when the distortion in shape is not important.

**Figure 33-12**
Image formed by refraction at a spherical surface.

**Figure 33-13**
Geometry for the derivation of Equation 33-5. The angles $\theta_1$ and $\theta_2$ are related by Snell's law, which for small angles can be written $n_1\theta_1 = n_2\theta_2$. From the triangle $ACP'$ we obtain

$$\beta = \theta_2 + \gamma = \frac{n_1}{n_2}\theta_1 + \gamma$$

or

$$n_1\theta_1 = n_2\beta - n_2\gamma$$

From triangle $PAC$ we obtain $\theta_1 = \alpha + \beta$. Eliminating $\theta_1$ gives $n_1\alpha + n_2\gamma = (n_2 - n_1)\beta$. Equation 33-5 follows from substituting the small-angle approximations $\alpha \approx l/s$, $\gamma \approx l/s'$, and $\beta \approx l/r$.

We can apply Equation 33-5 to find the apparent depth of an object under water when viewed from directly overhead. For this case the surface is a plane surface, the radius of curvature is infinite, and the image and object distances are related by

$$\frac{n_1}{s} + \frac{n_2}{s'} = 0$$

$$s' = -\frac{n_2}{n_1} s \qquad\qquad 33\text{-}6 \qquad \textit{Apparent depth}$$

The negative sign indicates that the image is virtual and on the same side of the refracting surface as the object, as shown in the ray diagram in Figure 33-14.

Figure 33-14
The image $P'$ appears at the depth $s'$, which is less than the actual depth $s$ of the object $P$. When viewed in air from directly overhead, the apparent depth equals the real depth divided by the index of refraction of water.

---

**Example 33-2** Find the apparent depth of a fish resting 1 m below the surface of water which has an index of refraction $n = \frac{4}{3}$.
    Using $n_1 = \frac{4}{3}$ and $n_2 = 1$ in Equation 33-6, we obtain

$$s' = -\tfrac{3}{4}(1 \text{ m}) = -75 \text{ cm}$$

The apparent depth is three-fourths the actual depth. Note that this result holds only when the object is viewed from directly overhead so that the rays are paraxial.

---

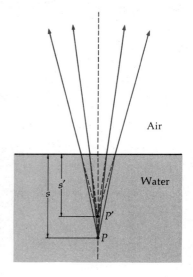

## 33-4  Lenses

The most important application of Equation 33-5 is in finding the position of the image formed by a lens. We do this by considering the refraction at each surface separately. Consider a glass lens of index of refraction $n$ with air on both sides. Let the radii of curvature of the surfaces of the lens be $r_1$ and $r_2$. If an object is at a distance $s$ from the first surface, application of Equation 33-5 gives for the distance of the image due to the refraction at the first surface

$$\frac{1}{s} + \frac{n}{s'} = \frac{n-1}{r_1} \qquad\qquad 33\text{-}7$$

This image is usually not formed (unless the lens is extremely thick) because the light is again refracted at the second surface. Assume, for example, that the image distance $s_1'$ is negative, indicating a virtual image to the left of the first surface. The light leaving this surface strikes the second surface *as if* it came from an object at the image position. If the thickness of the lens is $t$, the distance of this point from the second surface is $-s_1' + t$. (For this case $-s_1'$ is positive because $s_1'$ is negative.) We can find the final image position due to both refractions by using this distance for the object distance for the second surface. If $s_1'$ turns out to be positive, indicating a real image to the right of the first surface, the image point is to the right of the second surface if $s_1'$ is greater than the thickness $t$ and to the left of the surface if $t$ is greater than $s_1'$. In either case the distance from the surface is $-s_1' + t$. Even if this image is to the right of the second surface, we can use this distance in Equation 33-5 to find the image distance of the final image. In this case the object distance for the second refraction is negative, indicating a virtual object on the right side of the surface. Thus for all possible values of the first image distance $s_1'$, the image formed by refraction at the second surface is at a distance $s'$ from this surface given by

$$\frac{n}{-s_1' + t} + \frac{1}{s'} = \frac{1 - n}{r_2} = -\frac{n - 1}{r_2}$$   33-8

For this refraction the light is traveling from the glass medium of index $n$ into the air medium of index 1.

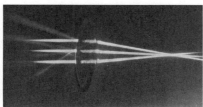

Bausch & Lomb

For a general lens of thickness $t$, it is usually easier to find the distance $s_1'$ numerically from Equation 33-7 and use this result in Equation 33-8 to find $s'$ than to eliminate $s_1'$ from these two equations. However, in many cases, the thickness $t$ is much smaller than any of the other distances involved. For such a *thin lens* we can neglect $t$ in Equation 33-8 and easily eliminate $s_1'$ from these equations. Solving for $n/s_1'$ in each equation, we obtain

$$\frac{n}{s_1'} = \frac{n - 1}{r_1} - \frac{1}{s} = \frac{1}{s'} + \frac{n - 1}{r_2}$$

or

$$\frac{1}{s} + \frac{1}{s'} = (n - 1)\left(\frac{1}{r_1} - \frac{1}{r_2}\right)$$   33-9

Equation 33-9 gives the image distance $s'$ in terms of the object distance $s$ and the properties of the thin lens—$r_1$, $r_2$, and the index of refraction $n$. As with mirrors, the focal length of a thin lens is defined to be the image distance when the object distance is very large. Setting $s$ equal to infinity and writing $f$ for the image distance $s'$, we obtain

$$\frac{1}{f} = (n - 1)\left(\frac{1}{r_1} - \frac{1}{r_2}\right)$$   33-10

and

$$\frac{1}{s} + \frac{1}{s'} = \frac{1}{f}$$   33-11   *Thin-lens equation*

Equation 33-10 is sometimes called the *lens-maker's equation* because it gives the focal length of a thin lens in terms of the properties of the lens. Equation 13-11 is the same as that for a spherical mirror. Note that one or both of the radii of curvature may be negative according to our sign convention. For example, in the double convex lens shown in Figure 33-15, the first radius $r_1$ is positive because the center of curvature lies

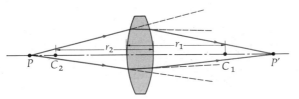

*Marginal notes:*

Parallel light rays incident on a double convex lens. The refracted rays converge at the second focal point of the lens. Reflected rays from each surface of the lens can also be seen. The first surface acts as a convex mirror producing diverging reflected rays, whereas the second surface acts as a concave mirror producing converging reflected rays.

**Figure 33-15**
Double convex lens. Both surfaces bend the light rays toward the axis. For this lens, $r_1$ is positive and $r_2$ is negative in Equation 33-10, resulting in a positive focal length.

on the right side of the surface, which is the real side, but $r_2$ is negative because the center of curvature of the second surface lies on the left or virtual side of the surface. In this case, both surfaces tend to bend a light ray toward the axis of the lens. This lens is a *converging lens*. Since both $1/r_1$ and $-1/r_2$ are positive and $n - 1$ is positive, the focal length $f$ given in Equation 33-10 must be positive, indicating that parallel light from the left is focused on the right, or real, side of the lens. A converging lens has a positive focal length and is therefore called a *positive lens*.

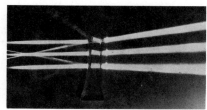

Bausch & Lomb

Parallel light rays incident on a double concave lens. The refracted rays diverge as if they came from the first focal point to the left of the lens. As in the previous photograph, reflected rays from each surface of the lens can also be seen. Here, the first surface acts as a concave mirror producing converging reflected rays, whereas the second surface acts as a convex mirror producing diverging reflected rays.

Figure 33-16 shows a double concave lens. Both surfaces tend to diverge light rays away from the lens axis. Since both $1/r_1$ and $-1/r_2$ are negative, the focal length is negative, indicating that parallel light from the left is diverged as if it came from a point on the left of the lens. A *diverging lens* has a negative focal length and is called a *negative lens*. For any other lens in which $r_1$ and $r_2$ are both positive or both negative, the lens is converging or diverging depending on which radius of curvature has the greater magnitude. If we turn any lens around, we interchange $r_1$ and $r_2$ and change their signs; i.e., the new radii of curvature for a lens which is turned around are related to the old by $r_1' = -r_2$ and $r_2' = -r_1$. We see from Equation 33-10 that the new focal length $f'$ has the same magnitude and sign as the old. Thus, for example, if parallel light strikes a double convex lens from the *right*, it is focused at a point on the left a distance $f$ from the lens. The two points on the left and right of any thin lens a distance $f$ from the lens are the first and second focal points of the lens, designated by $F$ and $F'$. Using the reversibility property of light rays, we can see that light diverging from either of the focal points and striking the lens will leave the lens as a parallel beam. If we set the object distance $s$ in Equation 33-10 equal to the focal distance $f$, we obtain $s' = \infty$, indicating that the light is not focused but emerges as a parallel beam.

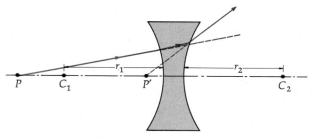

**Figure 33-16**
Double concave lens. Both surfaces bend the light rays away from the axis. Here $r_1$ is negative and $r_2$ is positive in Equation 33-10, resulting in a negative focal length.

If we have two or more thin lenses, we can find the final image produced by the system by finding the first image distance and using it along with the distance between lenses to find the object distance for the second lens. That is, we consider each image, whether it is formed or not, as the object point for the next lens.

As with images formed by mirrors, it is convenient to locate the image by graphical methods. Figure 33-17 illustrates the graphical method for a converging lens. Three convenient rays from the head of the figure are (1) a ray parallel to the axis of the lens which is bent through the focal point on the right of the lens, (2) a ray through the center of the lens (vertex) which is undeflected, and (3) a ray through the first focal point of the lens which emerges parallel to the axis. These three rays converge to the image point, as indicated. In this case the image is real and inverted. From Figure 33-17 we have $\tan \theta = y/s = -y'/s'$. The lateral magnification is then

$$m = \frac{y'}{y} = -\frac{s'}{s} \qquad\qquad 33\text{-}12$$

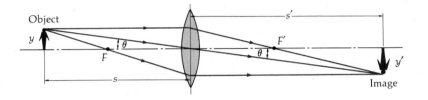

This expression is the same as that for mirrors. Again a negative magnification indicates that the image is inverted.

**Figure 33-17**
Ray diagram for a thin converging lens. For simplicity we assume all the bending to take place at a line. The ray through the center of the lens is undeflected because the lens surfaces there are nearly parallel and close together.

**Example 33-3** A double convex thin lens made of glass of index of refraction $n = 1.5$ has both radii of curvature of magnitude 20 cm. An object 2 cm high is placed 10 cm from the lens. Find the focal length of the lens, locate the image, and find its size.

We first calculate the focal length from Equation 33-10, noting that $r_1 = 20$ cm and $r_2 = -20$ cm by our sign convention. The focal length is thus

$$\frac{1}{f} = (1.5 - 1)\left(\frac{1}{20\text{ cm}} - \frac{1}{-20\text{ cm}}\right) = \frac{1}{20\text{ cm}} \quad \text{or} \quad f = 20\text{ cm}$$

Figure 33-18 shows a ray diagram for a small object 10 cm from the lens. The ray parallel to the axis is bent through the focal point, as shown. The ray through the vertex is undeflected. These rays are diverging on the right of the lens. The image is thus located by extending the rays back until they meet. These two rays are sufficient to locate the image.

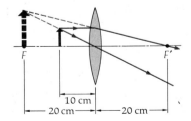

**Figure 33-18**
Ray diagram for Example 33-3. When the object is between the first focal point and a converging lens, the image is virtual and erect.

(As a check, we could draw a third ray from the object along a line through the focal point in front of the lens. This ray would then leave the lens parallel to the axis.) We can see immediately from this drawing

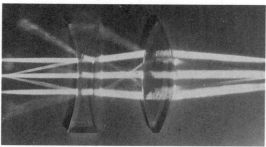

Parallel light rays incident on a diverging lens followed by a converging lens. The net effect of such a lens system can be converging rays, diverging rays, or parallel rays, depending on the lens separation and the relative size of the focal lengths of the two lenses. Here the net effect is to produce converging rays from parallel rays.

Bausch & Lomb

that the image is virtual, erect, and enlarged. It is on the same side of the lens as the object and is about twice as far away from the lens. Since it is quite easy to make an error in the calculation of the image distance from Equation 33-11, it is always a good idea to check your result with a ray diagram.

The image distance is found algebraically from Equation 33-11:

$$\frac{1}{10 \text{ cm}} + \frac{1}{s'} = \frac{1}{20 \text{ cm}}$$

$$\frac{1}{s'} = \frac{1}{20 \text{ cm}} - \frac{1}{10 \text{ cm}} = -\frac{1}{20 \text{ cm}} \quad \text{or} \quad s' = -20 \text{ cm}$$

The image distance is negative, indicating that the image is virtual and to the left of the lens. The magnification is $m = -s'/s = -(-20 \text{ cm})/(10 \text{ cm}) = +2$. The image is thus twice as large as the object, and it is erect. Since the size of the object is given as 2 cm, the size of the image is 4 cm.

**Example 33-4** A second lens of focal length $+10$ cm is placed 20 cm to the right of the lens in Example 33-3. Locate the final image.

Figure 33-19 shows a ray diagram for this example. The rays used to locate the image of the first lens are not necessarily convenient rays for the second lens. In this case one of the rays is, the one through the focal point of the first lens, which coincides with the center of the second lens. We see that the final image is real, inverted, and just outside the focal point of the second lens. We locate its position algebraically by noting that the virtual image of the first lens is 20 cm to the left of that lens and 40 cm to the left of the second lens. Using $s = 40$ cm and $f = 10$ cm, we have

$$\frac{1}{40} + \frac{1}{s'} = \frac{1}{10}$$

giving

$$s' = \tfrac{40}{3} \text{ cm} = 13.3 \text{ cm}$$

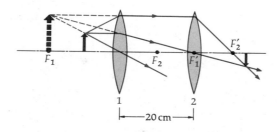

**Figure 33-19**
Ray diagram for Example 33-4. The image due to the first lens serves as the object for the second lens. The final image is located by drawing two rays from the first image. In this case one of the original rays used to locate the first image is undeflected by the second lens. A second ray from the first image parallel to the axis locates the final image.

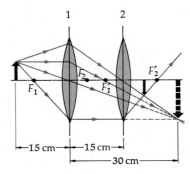

**Example 33-5** Two lenses of focal length 10 cm each are 15 cm apart. Find the final image of an object 15 cm from one of the lenses.

In the ray diagram of Fig. 33-20 the image of the first lens would be 30 cm from the lens if the second lens were not there. We calculate this using $s = 15$ cm and $f = 10$ cm in the thin-lens equation:

$$\frac{1}{15} + \frac{1}{s'} = \frac{1}{10}$$

$$\frac{1}{s'} = \frac{1}{10} - \frac{1}{15} = \frac{1}{30} \quad \text{or} \quad s' = 30 \text{ cm}$$

This image is not formed because the light rays strike the second lens before they reach the image position. We can locate the final image graphically by choosing rays which are heading toward the unformed image when they strike the lens. A ray parallel to the axis and one through the center of the second lens are sufficient. We see that the final image is between the second lens and its focal point. The final image can be located algebraically by using the first image as the object for the second lens. Since this image is not formed because it is to the right of the second lens, it is a *virtual object*. Since it is 15 cm to the right of the second lens, the object distance is $s = -15$ cm. Then

$$\frac{1}{-15} + \frac{1}{s'} = \frac{1}{f} = \frac{1}{10}$$

$$\frac{1}{s'} = \frac{1}{10} + \frac{1}{15} = \frac{5}{30} = \frac{1}{6} \quad \text{or} \quad s' = 6 \text{ cm}$$

**Figure 33-20**
Ray diagram for Example 33-5. The image due to the first lens is to the right of the second lens. The rays do not actually converge to that image because they are refracted by the second lens. That image therefore serves as a virtual object for the second lens. The final image is found by drawing rays toward the first image, as shown. Here, a ray parallel to the axis and one through the center of the second lens are used.

**Questions**

4. A fish under water is viewed from a point not directly overhead. Is its apparent depth greater or less than three-fourths its actual depth?

5. Under what conditions will the focal length of a thin lens be positive? Negative?

6. Under what conditions will the image formed by a positive thin lens be erect? Inverted? Real? Virtual? Larger than the object? Smaller than the object?

7. The focal length of a lens is different for different colors of light. Why?

## 33-5  Aberrations

When all the rays from a point object are not focused at a single image point, the resulting blurring of the images is called an *aberration*. The blurring of the image of an object point on the axis due to rays which strike a mirror or lens at points far from the axis is called *spherical aberration*. The similar blurring of the image of a point off the axis due to the different magnification of different parts of the lens or mirror is called *coma*, after the comet-shaped image observed. The imaging of a point source off axis into two mutually perpendicular lines at different locations is called *astigmatism*. The distortion in shape of the image of an extended object due to the fact that the magnification depends on the distance of the object point from the axis is called *distortion*. These aberrations, common to both mirrors and lenses, are the result of the laws of reflection and refraction applied to spherical surfaces. They are not evident in our simple equations for mirrors and lenses because we have used small-angle approximations in the derivation of these equations. Some of the aberrations can be eliminated or partially corrected by using nonspherical surfaces for the mirror or lens, but this is usually difficult and costly compared with the production of spherical surfaces. One example of a nonspherical reflecting surface is the parabolic mirror illustrated in Figure 33-21. Parallel rays incident on the parabolic surface are reflected and focused at a common point no matter how far they are from the axis. Parabolic reflecting surfaces are important in large astronomical telescopes, in which a large reflecting surface is needed to make the intensity of the image as great as possible. A parabolic surface can also be used in a searchlight to produce a parallel beam of light from a small source placed at the focus of the surface.

An important aberration of lenses not found in mirrors is *chromatic aberration* due to the variation in the index of refraction with wavelength. From Equation 33-10 we see that even for paraxial rays, the focal length of a lens depends on the index of refraction and is therefore slightly different for different wavelengths. Chromatic and other aberrations can be partially corrected by using combinations of lenses instead of a single lens. For example, a positive lens and a negative lens of greater focal length can be used together to produce a converging lens system which has much less chromatic aberration than a single lens of the same focal length.

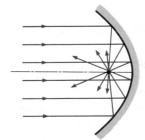

**Figure 33-21**
Spherical aberration can be eliminated by using a parabolic surface, which reflects all parallel rays through a common focal point.

## 33-6   The Eye

The optical system of prime importance is the eye, shown in Figure 33-22. Light enters the eye through a variable aperture, the pupil, and is focused by the cornea-lens system on the retina, a film of nerve fibers covering the back surface. The retina contains tiny sensing structures called *rods* and *cones*, which receive the image and transmit the information along the optic nerve to the brain. The shape of the crystalline

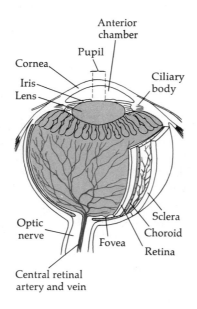

**Figure 33-22**
Cutaway view of the human eye. The amount of light entering the eye is controlled by the iris, which regulates the size of the pupil. Because the lens is surrounded by material of nearly equal index of refraction, most of the refraction occurs at the cornea. The lens thickness is controlled by the ciliary muscle. The cornea and lens together focus the image on the retina, which contains about 125 million receptors called rods and cones and about 1 million optic-nerve fibers. (*From H. Curtis,* Biology, *2d ed., p. 696. Worth Publishers, Inc., New York, 1975.*)

lens can be slightly altered by the action of the ciliary muscle. When the eye is focused on an object far away, the muscle is relaxed and the cornea-lens system has its maximum focal length, about 2.5 cm, the distance from the cornea to the retina. When the object is brought closer to the eye, the ciliary muscle tenses, increasing the curvature of the lens slightly and decreasing its focal length, so that the image is again focused on the retina. This process is called *accommodation*. If the object is too close to the eye, the lens cannot focus the light on the retina and the image is blurred. The closest point for which the lens can focus the image on the retina is called the *near point*. The distance from the eye to the near point varies greatly from one person to another and with age. At the age of 10 years, the near point may be as close as 7 cm, whereas at 60 years it may recede to 200 cm because of the loss of flexibility of the lens. The standard value taken for the near point is 25 cm.

If the relaxed eye images distant objects behind the retina, the person is said to be farsighted. Farsightedness is corrected with a converging lens (Figure 33-23). On the other hand, if light from a distant object is focused in front of the retina by the relaxed eye, the person is nearsighted. Nearsightedness is corrected with a diverging lens (Figure 33-24).

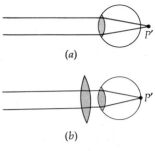

**Figure 33-23**
(*a*) The lens of a farsighted eye focuses parallel rays to form an image behind the retina. (*b*) A converging lens corrects this defect by bringing the image onto the retina.

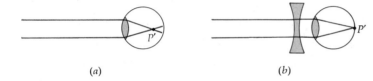

(*a*)　　　　　　　　　(*b*)

**Figure 33-24**
(*a*) The lens of a nearsighted eye focuses parallel rays to form an image in front of the retina. (*b*) A diverging lens corrects this defect.

---

**Example 33-6** By how much must the focal length of the cornea-lens system of the eye change when the object is moved from infinity to the near point at 25 cm? Assume that the distance from the cornea to the retina is 2.5 cm.

When the object is at infinity, the focal length of the cornea-lens system is 2.5 cm. When the object is at 25 cm, the focal length $f$ must be such that the image distance is 2.5 cm. From Equation 33-11 we have

$$\frac{1}{25 \text{ cm}} + \frac{1}{2.5 \text{ cm}} = \frac{1}{f}$$

$$\frac{1}{f} = \frac{1}{25 \text{ cm}} + \frac{10}{25 \text{ cm}} = \frac{11}{25 \text{ cm}}$$

$$f = \frac{25 \text{ cm}}{11} = 2.27 \text{ cm}$$

The focal length must therefore decrease by 0.23 cm.

---

(*Below, left*) Section of the retina of the human eye, magnified 500 times. (*From F. W. Sears and M. W. Zemansky,* University Physics, *4th ed., p. 582. Addison-Wesley Publishing Company, Inc., Reading, Mass., 1970.*) (*Below*) Rods (cylindrical structures extending into lower foreground) and cones (upper middle and left) as seen through a scanning electron microscope, magnified 1600 times. (*Courtesy of E. R. Lewis.*)

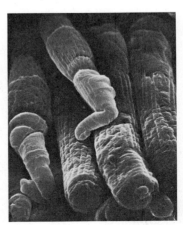

Nerve fibers

Receptor layer
Rods　Cones

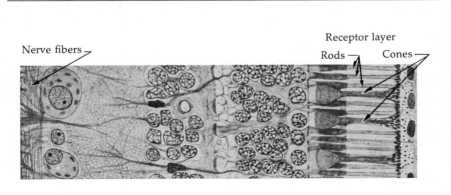

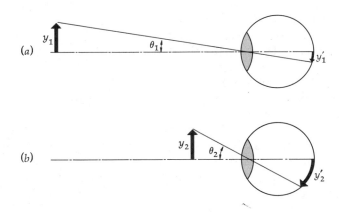
(a)
(b)

**Figure 33-25**
(a) A distant object of height $y_1$ looks small because the size of the image on the retina is small. (b) When the same object is closer, it looks larger because the retinal image is larger. The size of the image on the retina is approximately proportional to the angle $\theta$ subtended by the object.

The apparent size of an object is determined by the size of the image on the retina. The larger the image on the retina, the greater the number of rods and cones activated. From Figure 33-25 we see that the size of the image on the retina is greater when the object is close than it is when the object is far away. Even though the actual size of the object does not change, its apparent size is greater when it is brought closer to the eye. Since the near point is the closest point to the eye for which a sharp image can be formed on the retina, the distance to the near point is called the *distance of most distinct vision*. A convenient measure of the size of the image on the retina is the angle subtended by the object at the eye (Figure 33-26). When an object of height $y$ is at the near point, the angle subtended is

$$\theta_0 = \frac{y}{25 \text{ cm}} \qquad\qquad 33\text{-}13$$

where we have taken 25 cm to be the distance to the near point and have used the small-angle approximation $\theta_0 \approx \tan \theta_0$.

## 33-7 The Simple Magnifier

The apparent size of an object can be increased by using a converging lens called a *simple magnifier*. In Figure 33-26b a small object of height $y$ is at the first focal point of a thin converging lens of focal length $f$, which is less than the distance of most distinct vision (taken to be 25 cm). The image formed by the converging lens is virtual, erect, and at infinity. It subtends an angle $\theta$ given by

$$\theta = \frac{y}{f}$$

Viewed through the lens, the object looks larger because the image on the retina is larger by the factor $\theta/\theta_0$, where $\theta_0$, given by Equation 33-13, is the angle subtended by the object at the near point. The ratio $\theta/\theta_0$ is called the angular magnification or magnifying power $M$ of the lens:

$$M = \frac{\theta}{\theta_0} = \frac{25 \text{ cm}}{f} \qquad\qquad 33\text{-}14$$

The angular magnification of a simple magnifier can be increased if the object is moved closer to the magnifier, since the image moves in from infinity and the angle subtended increases. The largest usable magnifi-

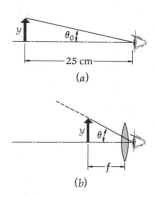
y
$\theta_0$
—25 cm—
(a)

y
$\theta$
—f—
(b)

**Figure 33-26**
(a) An object at the near point subtends an angle $\theta_0$ at the eye. (b) When the object is at the focal point of the converging lens, the image (not shown) is at infinity and can be viewed by the relaxed eye. When $f$ is less than 25 cm, the converging lens acts as a simple magnifier by allowing the object to be brought closer to the eye, increasing the angle subtended by the object to $\theta$ and thereby increasing the image size on the retina.

cation occurs when the image is at the near point of the eye, 25 cm. The angle subtended is then $y/s$, where $s$ is the object distance for the magnifier. We can calculate $s$ from the thin-lens equation 33-11. Using $s' = -25$ cm, we have

$$\frac{1}{s} + \frac{1}{-25} = \frac{1}{f}$$

$$\frac{1}{s} = \frac{1}{f} + \frac{1}{25} = \frac{25 + f}{25f} \qquad s = \frac{25f}{25 + f}$$

Then

$$\theta = \frac{y}{s} = \frac{25 + f}{25f} y \qquad \text{and} \qquad \frac{\theta}{\theta_0} = \frac{25 + f}{f} = 1 + \frac{25}{f}$$

When the image of the magnifier is viewed at the near point, the magnification is

$$M = 1 + \frac{25 \text{ cm}}{f} \qquad\qquad 33\text{-}15$$

It is left as an exercise to show that in this case the angular magnification is equal to the linear magnification $m = y/y'$.

Simple magnifiers are used as eyepieces or oculars in compound microscopes and telescopes to view the image formed by another lens or lens system. To correct aberrations, combinations of lenses with a resulting short positive focal length may be used as a simple magnifier in place of a single lens, but the principle is the same. For convenience we shall use $(25 \text{ cm})/f$ for the magnification of a simple magnifier, though the choice between Equations 33-14 and 33-15 depends on whether the image is viewed at infinity or the near point. Of course, if the distance to the near point is much greater than 25 cm, the effective magnification of a simple magnifier is greater. For example, if the near point for a person is 50 cm from the eye, a simple magnifier of focal length $f$ gives an angular magnification of $(50 \text{ cm})/f$ when the image is viewed at infinity.

# 33-8 The Compound Microscope and the Telescope

The compound microscope and the telescope are interesting applications of the principles of geometric optics. In their simplest forms each consists of two converging lenses. The lens nearest the object, called the *objective*, forms a real image of the object. The lens nearest the eye, called the *ocular* or *eyepiece*, is used as a simple magnifier to view the image formed by the objective.

The compound microscope (Figure 33-27) is used to look at very small objects at short distances. The distance between the second focal point of the objective and the first focal point of the ocular is called the *tube length l* and is usually fixed. The object is placed just outside the focal point of the objective so that an enlarged image is formed at the first focal point of the ocular a distance $l + f_o$ from the objective, where $f_o$ is the focal length of the objective. From Figure 33-27, $\tan \beta = y/f_o = -y'/l$. The lateral magnification of the objective is therefore

$$m_o = \frac{y'}{y} = -\frac{l}{f_o} \qquad\qquad 33\text{-}16$$

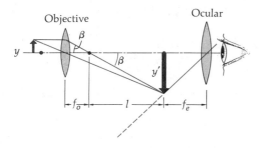

Objective

Ocular

**Figure 33-27**
Schematic diagram of a compound microscope, consisting of two positive lenses, the objective of focal length $f_o$, and the ocular, or eyepiece, of focal length $f_e$, separated by a distance $l + f_o + f_e$. The real image of the object formed by the objective is viewed by the ocular, which acts as a simple magnifier.

The overall magnification of the compound microscope is the product of the lateral magnification of the objective and the angular magnification of the ocular $M_e = (25 \text{ cm})/f_e$, where $f_e$ is the focal length of the ocular, or eyepiece:

$$m = m_o M_e = -\frac{l}{f_o}\frac{25 \text{ cm}}{f_e} \qquad\qquad 33\text{-}17$$

The astronomical telescope, illustrated schematically in Figure 33-28, is used to view objects which are far away and often large. The purpose of the objective is to form a real image at the first focal point of the ocular. Since the object is usually a great distance from the telescope, the image is at the second focal point of the objective. The objective and ocular are therefore separated by a distance $f_o + f_e$, where $f_o$ and $f_e$ are the focal lengths of the objective and ocular, respectively.

Let $\theta_o$ be the angle subtended by the object when it is viewed directly by the unaided eye. This is the same as the angle subtended by the object at the objective shown in the figure. The angle $\theta_e$ in the figure is that subtended by the final image. The angular magnification of the telescope is $\theta_e/\theta_o$. Using the small-angle approximation, we have

$$\tan \theta_o = -\frac{y'}{f_o} \approx \theta_o$$

where we have introduced the negative sign to make $\theta_o$ positive when $y'$ is negative. Similarly,

$$\tan \theta_e = \frac{y'}{f_e} \approx \theta_e$$

Since $y'$ is negative, $\theta_e$ is negative, indicating that the image is inverted. The angular magnification of the telescope is then

$$M = \frac{\theta_e}{\theta_o} = -\frac{f_o}{f_e} \qquad\qquad 33\text{-}18$$

Seventeenth-century telescope used by Galileo.

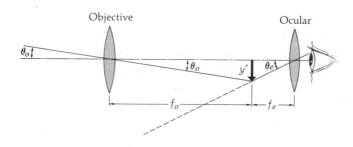

Objective

Ocular

**Figure 33-28**
Schematic diagram of an astronomical telescope. The objective forms a real image of a distant object near its second focal point, which coincides with the first focal point of the ocular. The ocular serves as a simple magnifier which views this image.

Review

A. Define, explain, or otherwise identify:

Virtual image, 878
Real image, 879
Paraxial ray, 879
Spherical aberration, 879
Focal length, 880
Focal point, 880
Reversibility, 880
Lateral magnification, 881
Apparent depth, 884
Converging lens, 886

Positive lens, 886
Diverging lens, 886
Negative lens, 886
Virtual object, 889
Chromatic aberration, 890
Accommodation, 891
Near point, 891
Angular magnification, 892
Objective, 893
Ocular, 893

B. True or false:

1. Virtual and real images look the same to the eye.

2. A virtual image cannot be displayed on a screen.

3. Aberrations occur only for real images.

4. A negative image distance implies that the image is virtual.

5. All rays parallel to the axis of a spherical mirror are reflected through a single point.

6. A diverging lens cannot form a real image from a real object.

7. The image distance for a positive lens is always positive.

8. Chromatic aberration does not occur with mirrors.

9. The image formed by the objective of a telescope is inverted and larger than the object.

10. The final image of a telescope is virtual.

Exercises

**Section 33-1, Plane Mirrors**

1. The image of the point object shown in Figure 33-29 is viewed by an eye as shown. Draw a bundle of rays from the object which reflect from the mirror and enter the eye. For this object position and mirror, indicate the region of space in which the eye can be placed to see the image.

**Figure 33-29**
Exercise 1.

2. A point object is placed between parallel mirrors separated by 30 cm. The object is 10 cm from the lower mirror and 20 cm from the upper mirror. (a) Find the distances from the upper mirror to the first four images which occur above the mirror. (b) Find the distances from the lower mirror to the first four images which occur below the lower mirror.

3. Two plane mirrors make an angle of 90°. Show that there are three images for any position of the object. Draw appropriate bundles of rays from the object to the eye for viewing each image.

4. Two plane mirrors make an angle of 60° with each other. Consider a point object on the bisector of the angle between the mirrors. Show on a sketch the location of all the images formed.

5. Repeat Exercise 4 for two mirrors at an angle of 120°.

**Section 33-2, Spherical Mirrors**

6. A concave spherical mirror has a radius of curvature of 50 cm. Draw ray diagrams to locate the image (if one is formed) for an object at a distance of (a) 100 cm, (b) 50 cm, (c) 25 cm, and (d) 10 cm from the mirror. For each case state whether the image is real or virtual; erect or inverted; and enlarged, reduced, or the same size as the object.

7. Use the mirror equations 33-2 to 33-4 to locate and describe the images for the object distances and mirror of Exercise 6.

8. Repeat Exercise 6 with a convex mirror of the same radius of curvature.

9. Repeat Exercise 7 with a convex mirror of the same radius of curvature.

10. Show that a convex mirror cannot form a real image of a real object no matter where the object is placed by showing that $s'$ is always negative for positive $s$.

11. Show that a concave mirror forms a virtual image of a real object when $s < f$ and a real image when $s > f$ (see Exercise 10).

12. A certain telescope uses a concave spherical mirror of radius 8 m as its objective. Find the location and diameter of the image of the moon, which has a diameter of $3.5 \times 10^6$ m and is $3.8 \times 10^8$ m from the mirror.

**Section 33-3, Images Formed by Refraction**

13. A sheet of paper with writing on it is protected by a thick glass plate having an index of refraction of 1.5. If the plate is 2 cm thick, at what distance beneath the top of the plate does the writing appear when viewed from directly overhead?

14. A very long glass rod has one end ground to a convex hemispherical surface of radius 5 cm. Its index of refraction is 1.5. (a) A point object is on the axis in air a distance 20 cm from the surface. Find the image and state whether it is real or virtual. (b) Repeat for an object at 5 cm from the surface. (c) Repeat for an object very far from the surface. Draw a ray diagram for this case.

15. At what distance from the rod of Exercise 14 should an object be placed so that the light rays in the rod are parallel to the axis? Draw a ray diagram for this situation.

16. Repeat Exercise 14 for a glass rod with a concave hemispherical surface of radius $(-)$ 5 cm.

17. Repeat Exercise 14 for the glass rod immersed in water of index of refraction 1.33; this means that the objects are in water.

18. Repeat Exercise 14 for a glass rod with a concave surface ($r = -5$ cm) immersed in water (index of refraction 1.33).

**Section 33-4, Lenses**

19. The following thin lenses are made of glass with an index of refraction of 1.5. Make a sketch and find the focal length in air for each: (a) double convex: $r_1 = 10$ cm, $r_2 = -21$ cm; (b) plano-convex: $r_1 = \infty$, $r_2 = -10$ cm; (c) double concave: $r_1 = -10$ cm, $r_2 = +10$ cm; (d) plano-concave: $r_1 = \infty$, $r_2 = +20$ cm.

20. Glass with an index of refraction of 1.6 is used to make a thin lens which has radii of equal magnitude. Find the radii of curvature and make a sketch if the focal length in air is to be (a) +5 cm, (b) −5 cm.

21. The following thin lenses are made of glass of index of refraction 1.5. Make a sketch and find the focal length in air for each: (a) $r_1 = 20$ cm, $r_2 = 10$ cm; (b) $r_1 = 10$ cm, $r_2 = 20$ cm; (c) $r_1 = -10$ cm, $r_2 = -20$ cm.

22. For the following object distances and focal lengths of thin lenses in air find the image distance and the magnification and state whether the image is real or virtual, erect or inverted. (a) $s = 40$ cm, $f = 20$ cm; (b) $s = 10$ cm, $f = 20$ cm; (c) $s = 40$ cm, $f = -30$ cm; (d) $s = 10$ cm, $f = -30$ cm.

23. What is meant by a negative object distance? How can it occur? Find the image distance and magnification and state whether the image is virtual or real, erect or inverted, for a thin lens in air when (a) $s = -20$ cm, $f = +20$ cm; (b) $s = -10$ cm, $f = -30$ cm.

24. An object 3.0 cm high is placed 20 cm from a thin lens of focal length 5.0 cm. Draw a careful ray diagram to find the position and size of the image and check your result using the thin-lens equation.

25. Repeat Exercise 24 for an object 1.0 cm high placed 10 cm from a thin lens of focal length 5.0 cm.

26. Repeat Exercise 24 for an object 1.0 cm high placed 10 cm from a thin lens whose focal length is $-5.0$ cm.

27. A thin converging lens of focal length 10 cm is used to obtain an image which is twice as large as a small object. (a) Where should the object be placed? (b) Where is the image? (c) Is the image real or virtual? Erect or inverted?

28. Show that a diverging lens can never form a real image from a real object. Hint: Show that $s'$ is always negative.

29. Show that a converging lens forms a real image from a real object when $s$ is greater than $f$. Hint: Show that $s'$ is then positive.

30. Show that if two thin lenses of focal lengths $f_1$ and $f_2$ are placed in contact, they are equivalent to a single thin lens of focal length $f = f_1 f_2 / (f_1 + f_2)$.

31. Two converging lenses each of focal length 10 cm are separated by 35 cm. An object is 20 cm to the left of the first lens. Find the final image using both ray diagrams and the lens equation. What is the overall lateral magnification?

### Section 33-5, Aberrations

32. A concave spherical mirror has a radius of curvature 6.0 cm. A point object is on the axis at 9.0 cm from the mirror. Construct a careful ray diagram showing rays from the object which make angles of 5, 10, 30, and 60° with the axis, strike the mirror, and are reflected back across the axis. (Use a compass to draw the mirror, and use a protractor to measure the angles needed to find the reflected rays.) What is the spread along the axis of the image for these rays?

33. For the mirror of Exercise 32, draw rays parallel to the axis at distances 0.5, 1, 2, and 4 cm and find the points at which the reflected rays cross the axis.

34. A double convex lens of radii $r_1 = +10$ cm and $r_2 = -10$ cm is made from a glass with an index of refraction of 1.53 for blue light and 1.47 for red light. Find the focal lengths of this lens for red and blue light.

### Section 33-6, The Eye

*In the following exercises take the distance from the cornea-lens system of the eye to the retina to be 2.5 cm.*

35. Suppose the eye were designed like a camera with a lens of fixed focal length $f = 2.5$ cm which could move toward or away from the retina. Approximately how far would the lens have to move to focus the image of an object 25 cm from the eye on the retina? Hint: Find the distance from the retina to the image behind it for an object at 25 cm.

36. Since the index of refraction of the lens of the eye is not very different from that of the surrounding material, most of the refraction takes place at the

cornea, where $n$ changes abruptly from 1.0 in air to about 1.4. Assuming the cornea to be homogeneous with an index of refraction of 1.4, calculate its radius if it focuses parallel light on the retina a distance 2.5 cm away.

37. If two point objects close together are to be seen as two distinct objects, the images must fall on the retina on two different cones that are not adjacent; i.e., there must be an unactivated cone between them. The separation of the cones is about 1 $\mu$m. What is the smallest angle the two points can subtend (see Figure 33-30)? How close can two points be if they are 20 m from the eye?

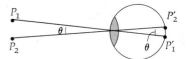

**Figure 33-30**
Exercise 37.

### Section 33-7, The Simple Magnifier

38. Show that when the image is viewed at the near point, the lateral and angular magnifications of a simple magnifier are equal.

39. What is the magnifying power of a lens of focal length 5 cm when (a) the image is viewed at infinity and (b) the image is viewed at the near point 25 cm from the eye?

40. A lens of focal length 10 cm is used as a simple magnifier with the image at infinity by one person whose near point is at 25 cm and by another whose near point is at 50 cm from the eye. What is the effective magnifying power of the lens for each person? Compare the size of the image on the retina when each looks at the same object with the magnifier.

### Section 33-8, The Compound Microscope and the Telescope

41. A crude symmetric hand-held microscope consists of two converging lenses each of focal length 5.0 cm fastened in the ends of a tube 30 cm long. (a) What is the "tube length"? (b) What is the lateral magnification of the objective? (c) What is the overall magnification of the microscope? (d) How far from the objective should the object be placed?

42. Repeat Exercise 41 for the same two lenses separated by 40 cm.

43. A simple telescope has an objective with a focal length of 100 cm and an ocular with a focal length of 5 cm. It is used to look at the moon, which subtends an angle of about 0.009 rad. (a) What is the size of the image formed by the objective? (b) What angle is subtended by the final image at infinity?

44. The objective lens of the refracting telescope at the Yerkes Observatory has a focal length of about 19.5 m. What is the diameter of the image of the moon formed by the objective? (The angle subtended by the moon is about 0.009 rad.)

### Problems

1. A slide projector is to form a real magnified image 4 m from its lens. The width of the slide (object) is 3.5 cm, and that of the image is to be 105 cm. What focal length should the projector lens have?

2. (a) Show that the focal length $f'$ for a thin lens of index of refraction $n_2$ in a medium of index $n_1$ is given by

$$\frac{1}{f'} = \frac{n_2 - n_1}{n_1}\left(\frac{1}{r_1} - \frac{1}{r_2}\right)$$

(b) Show that if $f$ is the focal length of a thin lens in air, its focal length in water is $f'$, given by

$$f' = \frac{n_1(n_2 - 1)}{n_2 - n_1}f$$

where $n_1$ is the index of refraction of water and $n_2$ is that of the lens material.

3. You wish to see an image of your face for makeup or shaving. If you want the image to be upright, virtual, and magnified 1.5 times when your face is 30 cm from the mirror, what kind of mirror should you use, convex or concave, and what should its focal length be?

4. A glass rod 96 cm long with an index of refraction of 1.6 has its ends ground to convex spherical shapes of radii 8 cm and 16 cm. A point object is in air on the axis 20 cm from the 8-cm-radius end. (*a*) Find the image distance due to refraction at the first surface. (*b*) Find the final image due to refraction at both surfaces. (*c*) Is the final image real or virtual?

5. Repeat Problem 4 for a point object in air 20 cm from the 16-cm-radius end.

6. Show that as an object moves away from a spherical mirror, its image always moves in the opposite direction. *Hint:* Compute $ds'$ in terms of $ds$ from the mirror equation.

7. Find the focal length of a thick double convex lens with an index of refraction of 1.5, thickness 4 cm, and radii +20 and −20 cm.

8. A small object is 20 cm from a thin positive lens of focal length 10 cm. To the right of the lens is a plane mirror which crosses the axis at the focal point of the lens and is tilted so that the reflected rays do not go back through the lens (see Figure 33-31). Sketch a ray diagram showing the final image. Is this image real or virtual?

**Figure 33-31**
Problem 8.

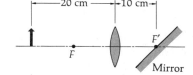

9. Find the final image for the situation in Problem 8 except that the mirror is not tilted. Assume that the image is viewed by an eye to the left of the object, looking through the lens into the mirror.

10. A horizontal concave mirror with a radius of curvature of 50 cm holds a layer of water of refractive index 1.33 and maximum depth 1 cm. At what height above the mirror must an object be placed so that its image is at the same position as the object?

11. When a bright light source is placed 30 cm in front of a lens, there is an erect image 7.5 cm from the lens. There is also a faint inverted image 6 cm in front of the lens due to reflection from the front surface of the lens. When the lens is turned around, this weaker inverted image is 10 cm in front of the lens. Find the index of refraction of the lens.

12. An object is 15 cm in front of a positive lens of focal length 15 cm. A second positive lens of focal length 15 cm is 20 cm from the first lens. Find the final image and draw a ray diagram.

13. Work Problem 12 if the second lens is a diverging lens of focal length −15 cm.

14. A glass ball of radius 10 cm has an index of refraction 1.5. The back half of the ball is silvered so that it acts as a concave mirror (Figure 33-32). Find the position of the final image seen by an eye to the left of the object and ball for an object at (*a*) 30 cm and (*b*) 20 cm to the left of the front surface of the ball.

**Figure 33-32**
Problem 14.

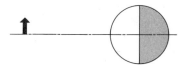

15. (a) Show that to obtain magnification of magnitude $m$ with a thin positive lens of focal length $f$, the object distance must be given by

$$s = \frac{m+1}{m} f$$

(b) A 35-mm camera has a lens with a 50-mm focal length. It is used to take a picture of a person 175 cm tall so that the image size on the film is 35 mm. How far from the camera should the person stand?

16. A convenient form of the thin-lens equation due to Newton measures the object and image distances from the focal points. Show that if $x = s - f$ and $x' = s' - f$, the thin-lens equation can be written $xx' = f^2$ and the lateral magnification is given by $m = -x'/f = -f/x$. Indicate $x$ and $x'$ on a sketch.

17. Show that a small change $dn$ in the index of refraction of a lens material produces a small change in the focal length $df$ given approximately by

$$\frac{df}{f} = -\frac{dn}{n-1}$$

Use this to find the focal length of a thin lens for blue light, for which $n = 1.53$, if the focal length for red light, for which $n = 1.47$, is 20 cm.

18. The lateral magnification of a spherical mirror or a thin lens is given by $m = -s'/s$. Show that for objects of small horizontal extent, the longitudinal magnification is approximately $-m^2$. Do this by showing that $ds'/ds = -s'^2/s^2$.

19. A disadvantage of the astronomical telescope for terrestrial use, e.g., at a football game, is that the image is inverted. A galilean telescope uses a converging lens as its objective but a diverging lens as its eyepiece. The image formed by the objective is beyond the diverging lens at its second focal point so that the final image is virtual and at infinity. Show with a ray diagram that the final image is erect. Show that the angular magnification is $M = -f_o/f_e$, where $f_o$ is the focal length of the objective and $f_e$ is the focal length of the ocular; $f_e$ is negative, making $M$ positive.

20. A galilean telescope is designed so that the final image is at the near point, 25 cm from the eye. The focal length of the objective is 100 cm and that of the ocular is −5 cm. (a) If the object distance is 30 m, where is the image of the objective? (b) What is the object distance for the ocular so that the final image is at the near point? (c) How far apart are the lenses? (d) If the object height is 1 m, what is the height of the final image? What is the angular magnification? (See Problem 19.)

CHAPTER 34 Physical Optics: Interference and Diffraction

---

**Objectives**  After studying this chapter you should:

1. Be able to work problems involving interference in thin films.

2. Be able to describe the Michelson interferometer.

3. Be able to sketch the two-slit-interference intensity pattern and locate the interference maxima and minima.

4. Be able to use the vector model to find the resultant of several harmonic waves.

5. Be able to sketch the interference pattern due to three or more equally spaced slits.

6. Be able to sketch the single-slit diffraction pattern and locate the first diffraction minimum.

7. Be able to sketch the combined interference-diffraction pattern for several slits.

8. Be able to state the Rayleigh criterion for resolution and use it to investigate the conditions for resolution of two objects.

9. Be able to discuss the use of diffraction gratings and find the resolving power of a grating.

---

Physical optics is the study of the interference and diffraction of light. These are the important phenomena that distinguish waves from particles. In Chapter 15 we discussed the interference of sound waves from two point sources and gave a brief qualitative discussion of the diffraction of sound. Since the analytical treatment of interference and diffraction is the same for all waves whether they are sound waves, waves on strings, water waves, or electromagnetic waves, you should review Chapter 15 before you begin this chapter.

When harmonic waves of the same frequency and wavelength but different phase come from two sources and combine, the resultant wave is a harmonic wave whose amplitude depends on the phase difference.

If the phase difference is 0 or an integer times $2\pi$, the waves are in phase and the interference is constructive. The resultant amplitude equals the sum of the individual amplitudes, and the intensity (which is proportional to the square of the amplitude) is maximum. If the phase difference is $\pi$ or an odd integer times $\pi$, the waves are out of phase and the interference is destructive. The resultant amplitude is then the difference between the individual amplitudes, and the intensity is a minimum. If the amplitudes are equal, the maximum intensity is 4 times that of either source and the minimum intensity is zero. A common cause of phase difference between two waves is a difference in path length traveled by the two waves. A path difference of $\Delta x$ contributes a phase difference $\delta$ given by

$$\delta = \frac{\Delta x}{\lambda} 2\pi \qquad\qquad 34\text{-}1$$

Another cause of phase difference is the $180° = \pi$ phase change a wave undergoes upon reflection from a boundary surface when the wave speed is greater in the first (incident) medium than in the second.

As mentioned in Chapter 15, interference of waves from two sources is not observed unless the sources are coherent, i.e., the phase difference between the waves from the sources is constant in time. Because a light beam is usually the result of millions of atoms radiating independently, two different light sources are usually not coherent; the phase difference between the waves from such sources fluctuates randomly many times per second. Coherence in optics is usually achieved by splitting the light beam from a single source into two or more beams, which then combine to produce an interference pattern. This splitting can be caused by (1) reflection from two nearby surfaces of a thin film (Section 34-1), (2) reflection from a mirror with the beams combined by reflection from other mirrors, as in the Michelson interferometer (Section 34-2), or (3) diffraction of the beam from two small openings or slits in an opaque barrier (Section 34-3). Coherent sources can also be obtained by using a single point source and its image in a plane mirror for the two sources, an arrangement called Lloyd's mirror.

## 34-1  Interference in Thin Films

Perhaps the most commonly observed case of interference of light is that due to reflection of light from the two surfaces of a thin film of water, air, or oil. Consider viewing, at small angles with the normal, a thin film of water (such as a soap bubble), as in Figure 34-1. Part of the light is reflected from the upper surface. Since light travels more slowly in water than in air, there is a 180° phase change in this reflection. Part of the light enters the film and is partially reflected by the bottom water-air surface. There is no phase change in this reflection. If the light is nearly perpendicular to the surfaces, both the ray reflected from the top surface and the one reflected from the bottom surface can enter the eye at point $P$ in the figure. The path difference between these two rays is $2t$, where $t$ is the thickness of the film. This path difference produces a phase difference of $2\pi(2t/\lambda')$, where $\lambda'$ is the wavelength of the light in the film; $\lambda'$ is related to the wavelength $\lambda_0$ in air by

$$\lambda' = \frac{v}{f} = \frac{c/n}{f} = \frac{\lambda_0}{n} \qquad\qquad 34\text{-}2$$

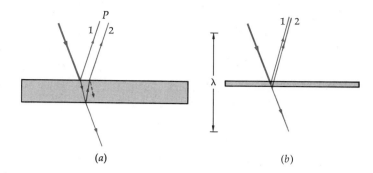

**Figure 34-1**
(a) Light rays reflected from the top and bottom surfaces of a thin film are coherent and produce interference. (b) If the film thickness is much smaller than the wavelength of light, the interference will be destructive because one of the rays undergoes a change in phase of 180° upon reflection.

where $n$ is the index of refraction. The phase difference between these two rays is thus 180° plus that due to the path difference. Destructive interference occurs when the path difference $2t$ is zero or a whole number of wavelengths (in the film). Constructive interference occurs if the path difference is an odd number of half wavelengths. We can express these conditions mathematically as

$$\frac{2t}{\lambda'} = \begin{cases} m & \text{destructive} \quad 34\text{-}3 \\ \\ m + \frac{1}{2} & \text{constructive} \quad 34\text{-}4 \end{cases} \quad m = 0, 1, 2, 3, \ldots$$

**Figure 34-2**
Interference of light reflected from a thin film of water resting on a glass surface. In this case, both rays undergo a change in phase of 180° upon reflection.

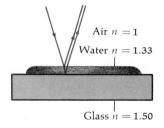

When a thin water film lies on a glass surface (Figure 34-2), the ray that reflects from the lower water-glass surface also undergoes a 180° phase change because the index of refraction of glass (about 1.5) is greater than that of water (about 1.33). Both the rays shown in the figure undergo a 180° phase change upon reflection. The phase difference between these rays is due solely to the path difference and is given by $\delta = 2\pi(2t/\lambda')$.

A common example of this type of interference occurs when there is a thin film of oil on a water-covered surface or street. Because the thickness of the film varies from point to point, constructive interference occurs for different wavelengths at different points, giving rise to colored bands.

Figure 34-3 illustrates the interference pattern observed when light is reflected from an air film between a spherical glass surface and a plane glass surface in contact. These circular interference fringes are known as *Newton's rings*. Near the point of contact, where the path difference between the ray reflected from the upper glass-air surface and the lower air-glass surface is essentially zero or at least small compared with the wavelength of light, the interference is perfectly destructive because of the 180° phase shift of the ray reflected from the lower air-glass surface. This region is therefore dark. The first bright fringe occurs at a radius such that the path difference contributes a phase difference of 180°, which adds to that due to the phase shift upon reflection and produces a total phase difference of 360°, or zero. The computation of the fringe spacing in terms of the radius of curvature of the spherical piece of glass is left as a problem.

**Figure 34-3**
Newton's rings observed with light reflected from a thin film of air between a plane glass and convex glass surface. At the center the thickness of the air film is negligible and the interference is destructive because of the phase change of one of the rays.

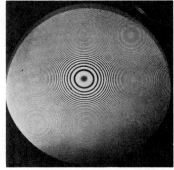

---

**Example 34-1** A wedge-shaped film of air is made by placing a small slip of paper between the edges of two flat pieces of glass. Light of wavelength 500 nm is incident normally on the glass plates, and interference fringes are observed by reflection. If the angle made by the plates is $3 \times 10^{-4}$ rad, how many interference fringes per unit length are observed?

Because of the 180° phase change of the ray reflected from the bottom plate (Figure 34-4), the first fringe near the point of contact (where the path difference is zero) will be dark. Let $x$ be the horizontal distance to the $m$th dark fringe where the plate separation is $t$ (Figure 34-4). Since the angle $\theta$ is very small, it is given approximately by

$$\theta = \frac{t}{x}$$

Using Equation 34-3 for $m$, we have

$$m = \frac{2t}{\lambda'} = \frac{2t}{\lambda}$$

since the film is an air film. Substituting $t = x\theta$ gives

$$m = \frac{2x\theta}{\lambda} \quad \text{or} \quad \frac{m}{x} = \frac{2\theta}{\lambda} = \frac{2(3 \times 10^{-4})}{5 \times 10^{-5}\ \text{cm}} = 12\ \text{cm}^{-1}$$

where we have used $\lambda = 5 \times 10^{-7}$ m $= 5 \times 10^{-5}$ cm. We therefore observe 12 dark fringes per centimetre. In practice, the number of fringes per centimetre, which is easy to count, can be used to determine the angle.

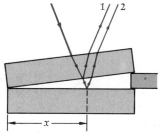

**Figure 34-4**
Light incident normally on a wedge of air between two glass plates. The path difference $2t$ is proportional to $x$. Alternate bright and dark bands are seen when viewed from above.

---

Figure 34-5 shows interference fringes produced by a wedge-shaped air film between two flat glass plates, as in Example 34-1. The straightness of the fringes indicates the flatness of the glass plates. Such plates are called *optically flat*. A similar air wedge formed by two ordinary glass plates yields the irregular fringe pattern in Figure 34-6, indicating that these plates are not optically flat.

**Figure 34-5**
Straight-line fringes from a wedge of air like Figure 34-4. The straightness of the fringes indicates the flatness of the glass plates. (*Courtesy of T. A. Wiggins.*)

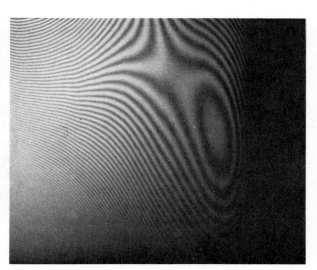

**Figure 34-6**
Fringes from a wedge of air between glass plates that are not optically flat. (*Courtesy of T. A. Wiggins.*)

**Questions**

1. Why must a film used to observe interference colors be thin?

2. Nonreflecting coatings for lenses are made by covering the lens with an extremely thin layer of material having an index of refraction greater than that of the glass. Explain how this works. Why does the coating have to be thin compared with the wavelength of visible light?

3. What effect does increasing the wedge angle in Example 34-1 have on the spacing of the interference fringes? If the angle is too large, fringes are not observed. Why?

4. The spacing of Newton's rings decreases rapidly as the diameter of the rings increases. Explain qualitatively.

## 34-2  The Michelson Interferometer

An interferometer is a device using interference fringes to make precise measurements. Figure 34-7 is a schematic diagram of a Michelson interferometer. Light from a broad source strikes a beam-splitter plate $A$, which is partially silvered. The light is partially reflected and partially transmitted by this plate. The reflected beam travels to mirror $M_2$ and is reflected back toward the eye at $O$. The transmitted beam travels through a compensating plate $B$ of the same thickness as $A$ to mirror $M_1$ and is reflected back to the eye at $O$. The purpose of the compensating plate is to make both beams pass through the same thickness of glass.

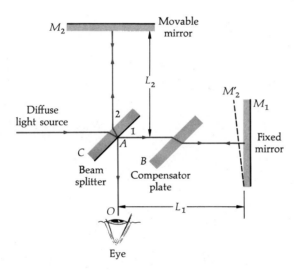

**Figure 34-7**
Michelson interferometer. The dashed line $M_2'$ is the image of mirror $M_2$ in the mirror $A$. The interference fringes observed are those of a small wedge formed by the sources $M_1$ and $M_2'$. As $M_2$ (and therefore $M_2'$) is moved, the fringes move past the field of view.

Mirror $M_1$ is fixed, but mirror $M_2$ can be moved back and forth with a fine and accurately calibrated screw adjustment. The two beams combine at $O$ and form an interference pattern. This pattern is most easily understood by considering the light sources to be mirror $M_1$ and the image of mirror $M_2$ labeled $M_2'$ in the diagram. This image is formed by the beam-splitting mirror. If the mirrors $M_1$ and $M_2$ are exactly perpendicular to each other and equidistant from the beam splitter, the image $M_2'$ will coincide with $M_1$. If not, $M_2'$ will be slightly displaced and make a small angle with $M_1$, as shown in the diagram. The interference pattern at $O$ will then be that of a thin wedge-shaped film of air between $M_1$ and $M_2'$, as discussed in Example 34-1. If mirror $M_2$ is now moved, the fringe pattern will shift. Suppose, for example, that the mirror $M_2$ is moved back $\frac{1}{4}\lambda$. The image $M_2'$ will then move an additional distance $\frac{1}{4}\lambda$ away from $M_1$, increasing the thickness of the wedge $\frac{1}{4}\lambda$ at each point. This will introduce an additional path difference of $\frac{1}{2}\lambda$ everywhere in the wedge, and the fringe pattern will move over by one-half fringe; i.e., a previously dark fringe will now be a bright fringe, etc. As mirror $M_2$ is moved, the displacement of the fringe pattern is observed. If the

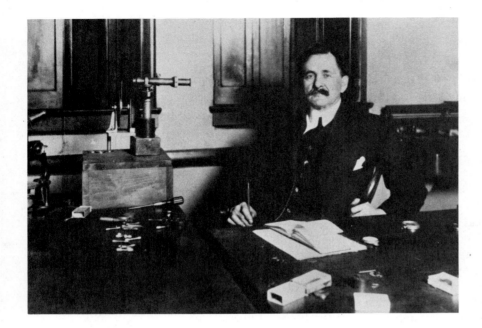

Albert A. Michelson in his laboratory. (*Courtesy of the Niels Bohr Library, American Institute of Physics.*)

distance the mirror is moved is known, the wavelength of the light can be determined. Michelson used such an interferometer to measure the wavelength of light emitted by krypton 86 in terms of the standard metre bar. This measurement was then used to redefine the standard metre in terms of this wavelength.

Another use for the Michelson interferometer is to measure the index of refraction of air (or of some other gas). One of the beams is enclosed in a container which can be evacuated. The wavelength in air $\lambda'$ is related to that in vacuum by $\lambda' = \lambda_0/n$, where $n$ is the index of refraction of air (about 1.0003). When the container is evacuated, the wavelength increases so that there are fewer waves in this distance, causing a shift in the fringe pattern. By measuring the shift the index of refraction can be determined (see Exercise 7). We shall discuss another example of the use of the Michelson interferometer in the next chapter when we discuss the famous Michelson-Morley experiment to measure the absolute velocity of the earth through space.

A student-type Michelson interferometer. The fringes are produced on a ground-glass screen by light from a laser. (*Courtesy of Dr. L. Velinsky.*)

## 34-3   The Two-Slit Interference Pattern

Interference patterns of light from two or more sources can be observed only if the sources are coherent, i.e., are in phase or have a phase difference that is constant in time. We have mentioned that the randomness of the radiation process of atoms means that two different light sources are generally incoherent. The interference of thin films discussed previously can be observed because the two beams come from the same light source but are separated by reflection. In Young's famous two-slit experiment used to demonstrate the wave nature of light, two coherent sources are produced by illuminating two parallel slits with a single source. We assume here that the width of each slit is small compared with the wavelength. (We shall treat the general case in a later section.) Then, because of diffraction, each slit acts as a line source (equivalent to a point source in two dimensions). The interference pattern is observed on a screen far from the slits. This pattern was discussed in Section 15-2. At very large distances from the slits, the lines from the two slits to some point $P$ on the screen are approximately parallel, and the path difference is approximately $d \sin \theta$, as shown in Figure 34-8. We thus have interference maxima at an angle given by

$$d \sin \theta = m\lambda \qquad \text{34-5}$$

*Two-slit interference maxima*

and minima at

$$d \sin \theta = (m + \tfrac{1}{2})\lambda \qquad \text{34-6}$$

where $m = 0, 1, 2, 3, \ldots$ . The phase difference at a point $P$ is $2\pi/\lambda$ times the path difference $d \sin \theta$:

$$\delta = \frac{2\pi}{\lambda} d \sin \theta \qquad \text{34-7}$$

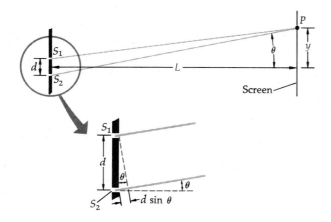

**Figure 34-8**
Two slits as coherent sources of light for the observation of interference in Young's experiment. Since the screen is very far away compared with the separation of the slits, the rays from the slits to a point on the screen are approximately parallel. The path difference between the two rays is $d \sin \theta$.

The distance $x$ measured along the screen from the central point to point $P$ is related to $\theta$ by $y = L \tan \theta$, where $L$ is the distance from the slits to the screen. For small $\theta$ we have $\sin \theta \approx \tan \theta \approx \theta$. Then $\sin \theta \approx y/L$, and Equation 34-7 becomes

$$\delta \approx \frac{2\pi}{\lambda} \frac{yd}{L} \qquad \text{34-8}$$

To calculate the intensity on the screen at a general point $P$ we need to add two harmonic wave functions that differ in phase, as we did in Chapter 15 when we discussed the interference of sound waves. The wave functions for electromagnetic waves are the electric field vectors. Let $E_1$ be the electric field at some point $P$ on the screen due to the waves from slit 1, and let $E_2$ be the electric field due to waves from slit 2. Since the angles $\theta$ of interest are small, we can assume these fields to be parallel and consider only their magnitudes. Both electric fields oscillate with the same frequency (since they result from a single source that illuminates both slits), and they have the same amplitude, assuming that the slits are the same size and are approximately equidistant from the screen. (The path difference is only of the order of a few wavelengths of light at most.) They have a phase difference $\delta$ given by Equations 34-7 and 34-8. If we represent these wave functions by

$$E_1 = E_0 \sin \omega t \qquad \text{and} \qquad E_2 = E_0 \sin (\omega t + \delta)$$

the resultant wave function is

$$E = E_1 + E_2 = E_0 \sin \omega t + E_0 \sin (\omega t + \delta)$$
$$= 2E_0 \cos \tfrac{1}{2}\delta \sin (\omega t + \tfrac{1}{2}\delta) \qquad \text{34-9}$$

where we have used a trigonometric identity for the sum of two sine functions:

$$\sin A + \sin B = 2 \cos \tfrac{1}{2}(A - B) \sin \tfrac{1}{2}(A + B) \qquad 34\text{-}10$$

(This is the same result as Equation 15-1.) The amplitude of the resultant wave is $2E_0 \cos \tfrac{1}{2}\delta$. It has its maximum value of $2E_0$ when the waves are in phase ($\delta = 0$, or an integer times $2\pi$), and is zero when they are 180° out of phase ($\delta = \pi$ or an odd integer times $\pi$). Since the intensity is proportional to the square of the amplitude, the intensity at any point $P$ is

$$I = 4I_0 \cos^2 \tfrac{1}{2}\delta \qquad 34\text{-}11$$

where $I_0$ is the intensity on the screen due to either slit acting separately. This intensity pattern is shown in Figure 34-9 as a function of $\sin \theta$, which is equivalent to a plot of intensity versus $y$ for small $\theta$. If we average over many interference maxima and minima, we obtain $2I_0$ for the average intensity since the average value of $\cos^2 \delta$ is $\tfrac{1}{2}$. This is the intensity that would arise from the two sources if they acted independently without interference. It is the intensity we would observe if the sources were incoherent, because then there would be an additional phase difference between them that fluctuates randomly so that only the average intensity could be observed.

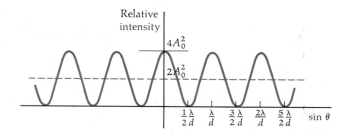

*Two-slit interference intensity*

**Figure 34-9**
Plot of relative intensity versus $\sin \theta$ observed on a screen far away from two slits as in Figure 34-8. The maximum intensity is proportional to $4A_0^2$ at $\sin \theta = m\lambda/d$, where $m$ is an integer and $A_0$ is the amplitude from each source separately. This is 4 times the intensity of either source acting separately. The average relative intensity $2A_0^2$, indicated by the dashed line, equals twice that of each source acting separately. Since $\sin \theta \approx y/L$ for small $\theta$, this is also a plot of relative intensity versus $y$.

Another method of producing the two-slit interference pattern is an arrangement shown in Figure 34-10 and known as Lloyd's mirror. A single slit is placed at a distance $\tfrac{1}{2}d$ above the plane of a mirror. Light striking the screen directly from the source interferes with that reflected from the mirror. The reflected light can be considered to come from the virtual image of the slit formed by the mirror. Because of the 180° change in phase on reflection at the mirror, the interference pattern is that of two coherent line sources which differ in phase by 180°. The pattern is the same as that shown in Figure 34-9 for two slits except that the maxima and minima are interchanged. The central fringe just above the mirror at a point equidistant from the sources is dark. Constructive interference occurs at points for which the path difference is $\tfrac{1}{2}$ wavelength or any odd number of half wavelengths. At these points, the 180° phase difference due to the path difference combines with the 180° phase difference of the sources to produce constructive interference.

*Lloyd's mirror*

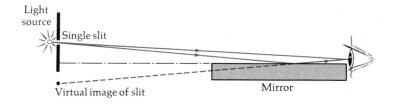

**Figure 34-10**
Lloyd's mirror for observation of double source interference with light. The two sources (light source and its image) are coherent and out of phase by 180° because of the phase change on reflection. The central interference band at points equidistant from the sources is dark.

# 34-4   The Vector Model for Addition of Harmonic Waves

To calculate the interference pattern produced by three, four, or more sources and to calculate the diffraction pattern of a single slit we need to combine several harmonic waves of the same frequency but differing in phase. A simple geometric interpretation of harmonic wave functions leads to a method of addition of harmonic waves of the same frequency by geometric construction. This method lets us find the sum of two or more harmonic waves geometrically without having to remember the trigonometric identity of Equation 34-10. It is useful even if the amplitudes of the waves are different or there are more than two waves. The method is based on the fact that the $y$ (or $x$) component of the resultant of two vectors equals the sum of the $y$ (or $x$) components of the vectors.
Let

$$E_1 = A_1 \sin (kx - \omega t) \quad \text{and} \quad E_2 = A_2 \sin (kx - \omega t + \delta)$$

be the wave functions of the two waves. We wish to add these waves at some point $x$ and some time $t$. We can simplify our notation by writing $\alpha$ for the quantity $kx - \omega t$. Our problem is then to find the sum

$$E_1 + E_2 = A_1 \sin \alpha + A_2 \sin (\alpha + \delta)$$

Consider a vector of magnitude $A_1$ making an angle $\alpha$ with the $x$ axis (Figure 34-11). The $y$ component of this vector is $A_1 \sin \alpha$, which is the wave function $E_1$. Similarly, the wave function $E_2 = A_2 \sin (\alpha + \delta)$ is the $y$ component of a vector of magnitude $A_2$ making an angle $\alpha + \delta$ with the $x$ axis. By the laws of vector addition, the sum of these components equals the $y$ component of the resultant vector, as shown in Figure 34-11. The $y$ component of the resultant vector, $A' \sin (\alpha + \delta')$, is a harmonic wave function which is the sum of the two original wave functions,

$$A_1 \sin \alpha + A_2 \sin (\alpha + \delta) = A' \sin (\alpha + \delta') \qquad 34\text{-}12$$

where $\alpha = kx - \omega t$ and $A'$ (the amplitude of the resultant wave) and $\delta'$ (the phase of the resultant wave relative to the first wave) are found by adding the vectors representing the waves, as in Figure 34-11. As time

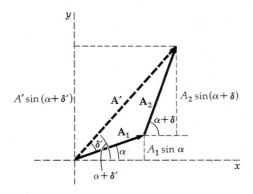

**Figure 34-11**
The wave function $A_1 \sin \theta$ is the $y$ component of the vector $\mathbf{A}_1$ making angle $\alpha$ with the $x$ axis. The wave function $A_2 \sin (\alpha + \delta)$ is the $y$ component of the vector $\mathbf{A}_2$ making angle $\alpha + \delta$ with the $x$ axis. The sum of these wave functions is $A' \sin (\alpha + \delta')$, which is the $y$ component of the resultant vector $\mathbf{A}' = \mathbf{A}_1 + \mathbf{A}_2$.

varies, $\alpha$ varies. The vectors representing the two wave functions and the resultant vector representing the resultant wave function rotate in space, but their relative positions do not change because all the vectors rotate with the same angular velocity $\omega$.

**Example 34-2** Use the vector method of addition to derive Equation 34-9 for the superposition of two waves of the same amplitude.

Figure 34-12 shows the vectors representing two waves of equal amplitude $A$ and the resultant wave of amplitude $A'$. These three vectors form an isosceles triangle in which the two equal angles are $\delta'$. Since the sum of these angles equals the exterior angle $\delta$, we have

$$\delta' = \tfrac{1}{2}\delta$$

The amplitude $A'$ can be found from the right triangles formed by bisecting the resultant vector, as shown in Figure 34-12$b$. From these triangles we have

$$\cos \tfrac{1}{2}\delta = \frac{\tfrac{1}{2}A'}{A}$$

Therefore the amplitude is given by $A' = 2A \cos \tfrac{1}{2}\delta$, and the resultant wave is

$$A' \sin (\alpha + \delta') = 2A \cos \tfrac{1}{2}\delta \sin (\alpha + \tfrac{1}{2}\delta)$$

in agreement with Equation 34-9, with $\alpha = \omega t$.

**Example 34-3** Find the resultant of the two waves

$$E_1 = 4 \sin (kx - \omega t) \quad \text{and} \quad E_2 = 3 \sin (kx - \omega t + 90°)$$

Figure 34-13 shows the vector diagram for this addition. The vectors make an angle of 90° with each other. The resultant of these two vectors has the magnitude 5 and makes an angle of 37° with the first vector, as shown. The sum of these two waves is

$$E_1 + E_2 = 5 \sin (kx - \omega t + 37°)$$

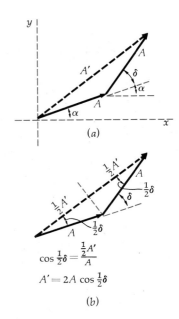

**Figure 34-12**
The vector addition for two waves having equal amplitude $A$ and phase difference $\delta$. ($a$) The vectors at a particular time at which $\alpha = \omega t$. ($b$) Construction for finding the amplitude of the resultant wave. The amplitude $A'$ is found from $\cos \tfrac{1}{2}\delta = \tfrac{1}{2}A'/A$, giving $A' = 2A \cos \tfrac{1}{2}\delta$.

## 34-5 Interference Pattern of Three or Four Equally Spaced Sources

If we have three or more sources which are equally spaced and in phase with each other, the intensity pattern on a screen far away is similar to that due to two sources, but there are important differences. The position on the screen of the intensity maxima is the same no matter how many sources we have (assuming that they are equally spaced and in phase), but these maxima have much greater intensity and are much sharper if there are many sources. In this section we shall study the pattern produced by three or four sources. The extension of this study to the problem of finding the intensity pattern of a large number of equally spaced sources is not difficult and has important applications for diffraction and the diffraction grating, discussed in the next two sections.

We first consider the case of three sources, as shown in Figure 34-14. The geometry is the same as for two sources. At a great distance from the sources, the rays from the sources to a point $P$ on the screen are approximately parallel. The path difference between the first and second source is then $d \sin \theta$, as before, and between the first and third source the path difference is $2d \sin \theta$. The wave at point $P$ is the sum of three waves. Let $\alpha = kr_1 - \omega t_0$ be the phase of the first wave at point $P$, where $t_0$ is the time of observation and $r_1$ is the distance from the first

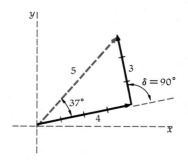

**Figure 34-13**
Vector model for the addition of the two waves in Example 34-3.

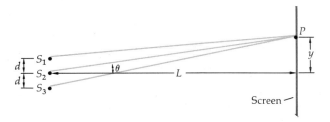

**Figure 34-14**
Geometry for calculating the intensity pattern far from three equally spaced sources in phase.

source to point $P$. We thus have the problem of addition of the three waves of the form

$$E_1 = A_0 \sin \alpha$$

$$E_2 = A_0 \sin (\alpha + \delta) \qquad\qquad 34\text{-}13$$

$$E_3 = A_0 \sin (\alpha + 2\delta)$$

where

$$\delta = \frac{2\pi}{\lambda} d \sin \theta \approx \frac{2\pi}{\lambda} \frac{yd}{L} \qquad\qquad 34\text{-}14$$

as in the two-slit problem.

It is easiest to analyze the resulting pattern in terms of the phase angle $\delta$ between the first and second sources or between the second and third sources instead of directly in terms of the space angle $\theta$. If we know the resultant amplitude due to the three waves at some point $P$ corresponding to a particular phase angle $\delta$, we can relate this phase angle to $\theta$ by Equation 34-14.

The addition of these waves is easiest with the vector model of addition. We shall be most interested in the points of perfectly constructive interference and those of perfectly destructive interference, i.e., the interference maxima and minima.

At the central maximum point $\theta = 0$, the phase angle $\delta$ is zero, and the amplitude of the resultant wave is 3 times that of each individual wave. Since the intensity is proportional to the square of the amplitude, the intensity at this central maximum is 9 times that from each source acting separately. As we move away from this central maximum, the angle $\theta$ increases and the phase angle $\delta$ also increases.

Figure 34-15 shows the vector addition of three waves for a phase angle $\delta$ of about $30° = \pi/6$ rad. (This corresponds to a point $P$ on the screen for which $\theta$ is given by $\sin \theta = \lambda\delta/2\pi d = \lambda/12d$.) The resultant amplitude is considerably less than 3 times that of each source. As the phase angle $\delta$ increases, the resultant amplitude decreases until the amplitude is zero at $\delta = 120°$. For this phase difference, the three vectors form an equilateral triangle (Figure 34-16). This first interference minimum occurs at a smaller phase angle (and therefore at a smaller angle $\theta$) than the $180°$ for only two sources. As $\delta$ increases from $120°$, the resultant amplitude increases, reaching a secondary maximum when $\delta$ is $180°$. At the phase angle $\delta = 180°$ the amplitude is the same as that from

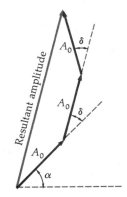

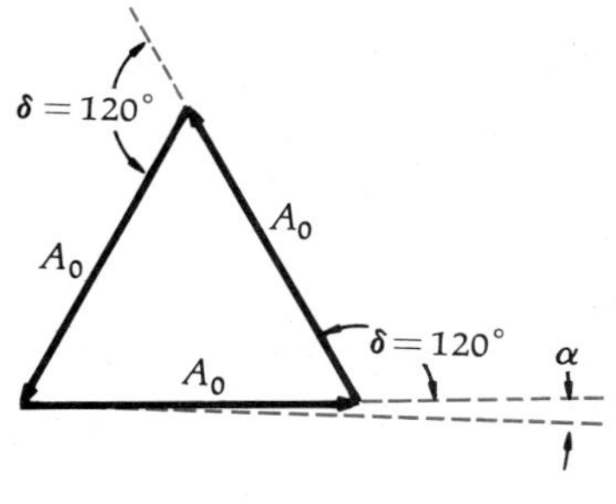

**Figure 34-15**
Vector addition to determine the resultant amplitude due to three waves, each of amplitude $A_0$, which have phase differences of $\delta$ and $2\delta$ due to path differences of $d \sin \theta$ and $2d \sin \theta$. The angle $\alpha = kx - \omega t$ varies with time but does not affect the calculation of the resultant amplitude.

**Figure 34-16**
The resultant amplitude for the waves from three sources is zero when $\delta$ is $120°$. This interference minimum occurs at a smaller angle $\theta$ than the first minimum for two sources, which occurs when $\delta$ is $180°$.

a single source since the waves from the first two sources cancel each other, leaving only the third. The intensity of the secondary maximum is one-ninth that of the central maximum. As δ increases beyond 180°, the amplitude again decreases and is zero at δ = 180° + 60° = 240°. For δ greater than 240° the amplitude increases and is again 3 times that of each source when δ = 360°. This phase angle corresponds to a path difference of 1 wavelength for the waves from the first two sources and 2 wavelengths for the waves from the first and third sources. Hence the three waves are in phase at this point. Figure 34-17 shows the intensity pattern on a screen far from three equally spaced sources. The maxima are at the same positions as for just two sources, i.e., at points corresponding to the angles $\theta$ given by

$$d \sin \theta = m\lambda \qquad\qquad 34\text{-}15$$

where $m$, the *order number*, is 0, 1, 2, . . . . These maxima are stronger and narrower than those for two sources.

These results can be generalized to more sources. For example, if we have four equally spaced sources in phase, the interference maxima are again given by Equation 34-15 but the maxima are still narrower and there are two small secondary maxima between each pair of principal maxima. At $\theta = 0$, the intensity is 16 times that from a single source.

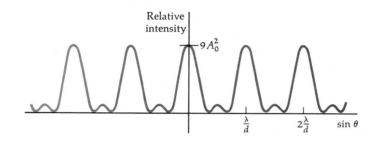

The first interference minimum occurs when δ is 90°, as can be seen by the vector diagram of Figure 34-18. The first secondary maximum is near δ = 120°, where the waves from three of the sources cancel, leaving only the wave from the fourth source. The intensity of the secondary maximum is approximately one-sixteenth that of the central maximum. There is another minimum at δ = 180°, a secondary maximum near δ = 240°, and another minimum at δ = 270° before the next principal maximum at δ = 360°. This discussion shows that as we increase the number of sources, the intensity becomes more and more concentrated in the maxima given by Equation 34-15 and the maxima become narrower. For $N$ sources, the intensity at the maxima is $N^2$ times that of a single source, and the first minimum occurs at a phase angle of δ = 360°/N. In that case the $N$ vectors form a closed polygon of $N$ sides. There is a secondary maximum between each two minimum points. These secondary maxima are very weak compared with the principal maxima.

---

**Example 34-4** Four equally spaced slits are separated by a distance $d =$ 0.1 mm and uniformly illuminated by light of wavelength 500 nm. The interference pattern is viewed on a screen at a distance of 1 m. The intensity on the screen due to each slit when the other three slits are covered is $I_0$. Find the positions of the interference maxima and compare their width with that for just two slits of the same spacing.

**Figure 34-17**
A plot of the relative-intensity pattern on a screen far from three equally spaced sources in phase. The major maxima occur when $d \sin \theta = m\lambda$ at the same position as for two sources, but for three sources the maxima are narrower and of greater intensity. Between successive major maxima there is a secondary maximum with one-ninth the intensity of the major maxima.

**Figure 34-18**
Vector diagram for the first minimum from four equally spaced sources in phase. The amplitude is zero when the phase difference of the waves from adjacent sources is 90°.

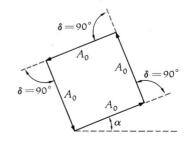

According to Equation 34-15, the maxima are at angles given by

$$\sin \theta = m \frac{\lambda}{d} = m \frac{5 \times 10^{-7} \text{ m}}{1 \times 10^{-4} \text{ m}} = 5 \times 10^{-3} m$$

where $m = 0, 1, 2, 3, \ldots$ . Since $\theta$ is small, we can approximate $\sin \theta \approx \tan \theta \approx \theta$. The distance $y$ measured along the screen from the central maximum is related to $\theta$ by

$$y = L \tan \theta \approx L\theta$$

The position of the $m$th maximum is thus

$$y_m = L\theta_m = m(1 \text{ m}) (5 \times 10^{-3}) = m(5 \text{ mm})$$

The maxima are thus separated by 5 mm on the screen. The first minimum occurs when the phase difference between two adjacent sources is $\delta = 90° = \pi/2$. This corresponds to a path difference of $\lambda/4$. The angle $\theta$ of this minimum is given by $d \sin \theta = \lambda/4$, or

$$\sin \theta = \frac{\lambda}{4d} = \frac{5 \times 10^{-7} \text{ m}}{4 \times 10^{-4} \text{ m}} = 1.25 \times 10^{-3}$$

The position $y$ of this minimum is

$$y = L\theta = (1 \text{ m})(1.25 \times 10^{-3}) = 1.25 \text{ mm}$$

We shall take for the width of the maximum the distance between the first minima on either side of the first maximum, or $2y = 2.5$ mm. If we had only two sources with this same spacing, the maxima would be at the same points but the first minimum would be at an angle $\theta$ corresponding to a path difference of $\lambda/2$. The width of this maximum would be twice as great as with four sources (Figure 34-19).

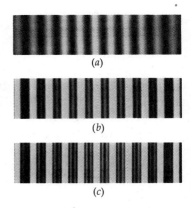

(a)

(b)

(c)

Interference patterns for (a) two, (b) three, and (c) four coherent sources. In (b) there is a secondary maximum between each pair of principal maxima, and in (c) there are two such secondary maxima. (From M. Cagnet, M. Françon, and J. C. Thrierr, Atlas of Optical Phenomena, plate 19, Springer-Verlag, Berlin, 1962.)

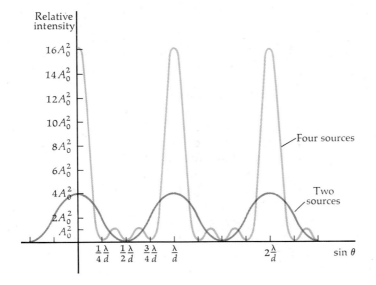

**Figure 34-19**
Comparison of relative-intensity pattern for four sources with that for two sources when the spacing $d$ is the same. The maxima for four sources are stronger and narrower than those for two sources.

**Question**

5. How many submaxima would there be between the main maxima in the interference pattern produced by eight equally spaced slits? Why would they be difficult to see?

## 34-6  Diffraction Pattern of a Single Slit

In our calculations of the interference patterns produced by two or more slits we have assumed the widths of the slits to be much less than the wavelength of the light, enabling us to consider the slits to be line sources of cylindrical waves, which in our two-dimensional diagrams are point sources of circular waves. (We can justify this assumption by our observation of circular waves in a ripple tank emanating from a small opening in a barrier, as shown in Figure 15-7.) We can also justify this assumption from Huygens' principle, according to which we can find the wave to the right of a single slit opening by assuming that each point along the line of the opening is a point source. If we replace the slit by a line of point sources separated by a distance $d$ and the total width of the slit is much less than the wavelength, $d$ will also be much less than the wavelength, and the path difference for waves from adjacent sources, $d \sin \theta$, will be much smaller than the wavelength. The waves at the screen from the different point sources will then have a negligible phase difference no matter what the angle $\theta$; therefore we might as well treat the line of sources as a single point source. However, if the slit width is *not* small compared with the wavelength of light, we must take into account the different phases of the waves from the different portions of the slit. Our calculations of the interference pattern of a line of sources in the previous section will be very useful.

In Figure 34-20 we have a single slit of width $a$, which we have divided into $N$ equal intervals. We assume that there is a point source of waves at the midpoint of each interval. If $d$ is the distance between two adjacent sources and $a$ the width of the opening, we have

$$d = \frac{a}{N}$$

For convenience we shall take the number of sources $N$ to be an even number. Since the screen on which we are calculating the intensity is very far from the sources, the rays from the sources to a point $P$ on the screen are approximately parallel. The path difference between any two adjacent sources is then $d \sin \theta$, and the phase difference is $\delta = (2\pi/\lambda)d \sin \theta$. If $A_0$ is the amplitude due to a single source, the amplitude at the central maximum point $\theta = 0$, where all the waves are in phase, is $A_{max} = NA_0$ (Figure 34-21). We find the amplitude at some other point $P$ at an angle $\theta$ by using the vector method of addition of waves. As in the addition of two, three, or four waves, the intensity is zero at a point such that the vectors representing the waves form a closed polygon. In this case the polygon has $N$ sides (Figure 34-22). At point $P$ the wave from the first source near the top of the opening and that from the source just below the middle of the opening are 180° out of phase. Consider, for example, 100 sources. These sources give zero intensity at a point $P$ when the waves from the first source and the fifty-first source are out of phase by 180° and cancel (Figure 34-23). Then

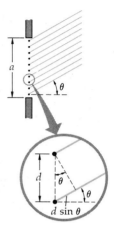

**Figure 34-20**
Diagram for the calculation of the interference pattern far from a narrow slit. The slit of width $a$ is assumed to contain a large number of in-phase point sources separated by a distance $d$. The rays from these sources to a point very far away are approximately parallel. The path difference for the waves from adjacent sources is then $d \sin \theta$.

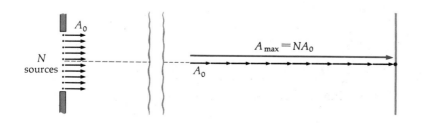

**Figure 34-21**
A single slit is represented by $N$ sources of amplitude $A_0$. At the central maximum point at $\theta = 0$ the waves from the sources add in phase, giving a resultant amplitude $A_{max} = NA_0$.

the waves from the second and fifty-second are also out of phase by 180° and cancel, as do those from the third and fifty-third sources, etc. In this case the waves from the sources near the top and bottom of the opening differ in phase by nearly 360°. (The phase difference is, in fact, $360° - 360°/N$.) Thus if the number of sources is very large, *we get complete cancellation when the waves from the first and last sources are out of phase by 360°, corresponding to a path difference of 1 wavelength.* Since this path difference is $a \sin \theta$, where $a$ is the width of the opening, the condition for the first minimum is

$$a \sin \theta = \lambda \qquad\qquad 34\text{-}16$$

We get a second minimum when the path difference for the waves from the top and bottom of the opening is 2 wavelengths. This can be seen by dividing the group of $N$ sources into two parts. The waves from the top half of the sources add to zero, as we have just noted. Similarly the waves from the bottom half of the sources also add to zero. Thus the condition for a minimum in the diffraction pattern is

$$a \sin \theta = m\lambda \qquad m = 1, 2, 3 \ldots \qquad\qquad 34\text{-}17$$

We now calculate the amplitude at a general point for which the waves from two adjacent sources differ in phase by $\delta$. Figure 34-24 shows the vector diagram for the addition of $N$ waves which differ in phase from the first wave by $\delta, 2\delta, \ldots, (N-1)\delta$. When $N$ is very large and $\delta$ is very small, the vector diagram is approximately an arc of a circle. The resultant amplitude $A$ is the length of the chord of this arc. We calculate this resultant amplitude in terms of the phase difference $\phi$ between the first and last waves. From Figure 34-24 we have

$$\sin \tfrac{1}{2}\phi = \frac{\tfrac{1}{2}A}{r} \qquad \text{or} \qquad A = 2r \sin \tfrac{1}{2}\phi \qquad\qquad 34\text{-}18$$

where $r$ is the radius of the arc. Since the length of the arc is $A_{\max} = NA_0$ and the angle subtended is $\phi$, we have

$$\phi = \frac{A_{\max}}{r} \qquad\qquad 34\text{-}19$$

or

$$r = \frac{A_{\max}}{\phi}$$

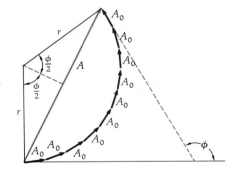

**Figure 34-22**
Vector diagram for the first minimum in the single-slit diffraction pattern. When the waves from the $N$ sources completely cancel, the $N$ vectors form a closed polygon. The phase difference between waves from adjacent sources is then $\delta = 360°/N$. When $N$ is very large, the waves from the first and last sources are approximately in phase.

*Single-slit diffraction minimum*

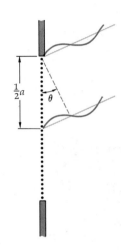

**Figure 34-23**
At the first diffraction minimum of a single slit, the waves from the source near the top and those from the source just below the middle of the slit are 180° out of phase and cancel.

**Figure 34-24**
Vector model for the calculation of the amplitude of the waves from $N$ sources in terms of the phase difference $\phi$ between the waves from the first source near the top of the slit and the last source near the bottom of the slit. For large $N$, the resultant amplitude $A$ is the chord of a circular arc of length $NA_0$.

Substituting this into Equation 34-18 gives

$$A = \frac{2A_{max}}{\phi} \sin \tfrac{1}{2}\phi = A_{max} \frac{\sin \tfrac{1}{2}\phi}{\tfrac{1}{2}\phi} \qquad\qquad 34\text{-}20$$

Since the amplitude at the central maximum point ($\theta = 0$) is $A_{max}$, the ratio of the intensity at any other point to that at the central maximum point is

$$\frac{I}{I_0} = \frac{A^2}{A_{max}^2} = \left(\frac{\sin \tfrac{1}{2}\phi}{\tfrac{1}{2}\phi}\right)^2$$

or

$$I = I_0 \left(\frac{\sin \tfrac{1}{2}\phi}{\tfrac{1}{2}\phi}\right)^2 \qquad\qquad 34\text{-}21$$

*Single-slit diffraction intensity*

The phase difference $\phi$ between the first and last waves is $2\pi/\lambda$ times the path difference $a \sin \theta$ between the top and bottom of the opening:

$$\phi = \frac{2\pi}{\lambda} a \sin \theta \qquad\qquad 34\text{-}22$$

The function $I/I_0$ is plotted in Figure 34-25. The first minimum occurs when $\phi = 360°$, at the point where the waves from the top and bottom of the opening have a path difference of $\lambda$ and are in phase. The second minimum occurs at $\phi = 720°$, where the waves from the top and bottom of the opening have a path difference of $2\lambda$.

(a)

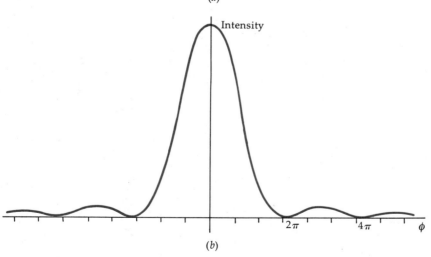

(b)

**Figure 34-25**
(a) Fraunhofer diffraction pattern of a single slit. (*From M. Cagnet, M. Françon, and J. C. Thierr,* Atlas of Optical Phenomena, *plate 18, Springer-Verlag, Berlin, 1962.*) (b) Plot of intensity versus phase difference for the pattern in (a).

   Approximately midway between these minima is a secondary maximum. Figure 34-26 shows the vector diagram for determining the relative intensity of this secondary maximum. The phase difference between the first and last waves is approximately $360° + 180°$. The vectors thus complete $1\tfrac{1}{2}$ circles. The resultant amplitude is the diameter

$A$ of a circle with a circumference $C$ which is two-thirds the total length $NA_0 = A_{max}$. Thus

$$C = \tfrac{2}{3}A_{max} = \pi A \qquad A = \frac{2}{3\pi} A_{max}$$

and

$$A^2 = \frac{4}{9\pi^2} A_{max}^2 \qquad\qquad 34\text{-}23$$

The intensity at this point is

$$I = \frac{4}{9\pi^2} I_0 = \frac{1}{22.2} I_0 \qquad\qquad 34\text{-}24$$

The intensity pattern consists of a broad central maximum which has intensity $I_0$ at $\theta = 0$ and drops to zero intensity at the first diffraction minimum at an angle $\theta$ given by

$$a \sin \theta = \lambda$$

There is then a secondary maximum of intensity approximately $I_0/22.2$ at an angle $\theta$ given approximately by

$$a \sin \theta = \tfrac{3}{2}\lambda$$

The second minimum occurs at an angle $\theta$ given by $a \sin \theta = 2\lambda$.

The intensity pattern of Figure 34-25 is called a *Fraunhofer diffraction pattern* of a single slit. The assumptions made in deriving Equation 34-21 describing the pattern were (1) that plane waves are incident normally on the slit (we assumed that the amplitudes and phases of the many Huygens sources are equal) and (2) that the pattern is observed at a great distance from the slit compared with the size of the openings. (We assumed that the rays from the sources to a point on the screen are approximately parallel to simplify the geometry.) Another condition that is usually necessary is that the width of the slit be of the order of the wavelength. If the opening, or slit, is much wider than the wavelength, the angle of the first minimum will be very small. For example, if $a = 1000\lambda$, the first minimum will occur at an angle $\theta$ given by $\sin \theta = 1/1000 \approx \theta$. Unless the screen is extremely far away, this small angle is not much different from the angle made by the rays from the top and bottom of the slit to the central maximum, rays which we assumed to be parallel in our derivation.

When the diffraction pattern is observed near the opening, it is called a *Fresnel diffraction pattern*. Because of the geometry, this pattern is much more difficult to calculate. Figure 34-27 illustrates the difference between the Fresnel and Fraunhofer patterns.[*]

When there are two or more slits, the intensity pattern on a screen far away is a combination of the single-slit diffraction pattern and the multiple-slit interference pattern we have studied. Figure 34-28 shows the intensity pattern on a screen far from two slits whose separation $d$ is 10 times the width $a$ of each slit. The pattern is the same as the two-slit pattern for a line source (Figure 34-9) except that it is modulated by the Fraunhofer single-slit diffraction pattern. The intensity can be calculated from the product of the intensity for the two-slit pattern (Equation 34-11) and the intensity due to each slit given by Equation 34-21:

$$I = 4I_0 \left(\frac{\sin \tfrac{1}{2}\phi}{\tfrac{1}{2}\phi}\right)^2 \cos^2 \tfrac{1}{2}\delta \qquad\qquad 34\text{-}25$$

[*] See Richard E. Haskell, "A Simple Experiment on Fresnel Diffraction," *American Journal of Physics*, vol. 38, p. 1039, 1970.

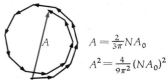

Circumference $\times \tfrac{3}{2} = NA_0 = \pi A \times \tfrac{3}{2}$

$A = \frac{2}{3\pi} NA_0$

$A^2 = \frac{4}{9\pi^2}(NA_0)^2$

**Figure 34-26**
Vector diagram for calculating the amplitude of the first secondary maximum of the single-slit diffraction pattern. This secondary maximum occurs when the $N$ vectors complete approximately $1\tfrac{1}{2}$ circles.

*Assumptions for Fraunhofer diffraction*

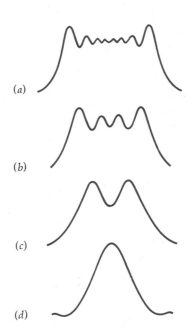

(a)

(b)

(c)

(d)

**Figure 34-27**
Diffraction pattern of a single slit for various screen distances. The Fresnel pattern observed near the slit (a) gradually merges into the Fraunhofer pattern (d) as the screen is moved to a great distance from the slit. (*Courtesy of Richard E. Haskell, Oakland University.*)

where $\phi$ is related to the width of each slit by

$$\phi = \frac{2\pi}{\lambda} a \sin \theta \qquad\qquad 34\text{-}26$$

and where $\delta$ is related to the slit separation by

$$\delta = \frac{2\pi}{\lambda} d \sin \theta \qquad\qquad 34\text{-}27$$

and $I_0$ is the intensity at $\theta = 0$ due to one slit alone. Note that in Figure 34-28 the central diffraction maximum contains 19 interference maxima, the central interference maximum and 9 maxima on either side. The tenth interference maximum on either side of the central one is at the angle $\theta$ given by $\sin \theta = 10\lambda/d = \lambda/a$, since $d = 10a$. This coincides with the position of the first diffraction minimum, and so this interference maximum is not seen. At this point the light from the two slits would be in phase and interfere constructively, but there is no light from either slit because the point is a diffraction minimum.

**Figure 34-28**
Interference-diffraction pattern of two slits with separation $d$ equal to 10 times their width $a$. The tenth interference maximum on either side of the central interference maximum is missing because it falls at the first diffraction minimum. (*From M. Cagnet, M. Françon, and J. C. Thrierr, Atlas of Optical Phenomena, plate 18, Springer-Verlag, Berlin, 1962.*)

**Questions**

6. As the width of the slit producing a single-slit diffraction pattern is slowly and steadily reduced, how will the diffraction pattern change?

7. How many interference maxima will be contained in the central diffraction maximum in the diffraction-interference pattern of two slits if the separation $d$ is 5 times the slit width $a$? How many if $d = Na$ for any $N$?

## 34-7  Diffraction and Resolution

Diffraction occurs whenever a portion of the wavefront is limited by an obstacle or aperture of some kind. The intensity of light at any point in space can be computed using Huygens' principle by taking each point on the wavefront to be a point source and computing the resulting interference pattern. Fraunhofer patterns can be observed at great distances from the obstacle or aperture so that the rays reaching any point are approximately parallel, or they can be observed using a lens to focus parallel rays on a viewing screen placed at the focal point of the lens. Fresnel patterns are observed at points close to the source. Diffraction of light is often difficult to observe because the wavelength is so small or because the light intensity is not great enough. Except for the Fraunhofer pattern of a long narrow slit, diffraction patterns are usually difficult to calculate.

Figure 34-29$a$ shows the Fresnel diffraction pattern of a straight edge illuminated by light from a point source. A graph of the intensity versus distance measured along a line perpendicular to the edge is shown in Figure 34-29$b$. The light intensity does not fall abruptly to zero in the geometric shadow but decreases rapidly and is negligible within a few wavelengths of the edge. The Fresnel diffraction pattern of a circular aperture is shown in Figure 34-30. Note the similarity of this pattern and the diffraction pattern of an opaque disk shown in Figure 32-5. The Fresnel diffraction pattern of a rectangular opening is shown in Figure 34-31. These patterns cannot be seen with broad light sources like an ordinary light bulb because the dark fringes of the pattern produced by

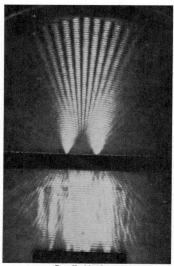

Battelle-Northwest Photography

When waves are incident on a barrier having two slits with size and separation of the order of the wavelength, both diffraction and interference can be observed. Here plane ultrasonic waves in water are incident on slits of width $2\lambda$ and separation $8\lambda$. The photograph was made using a laser and a Schlieren technique, which is sensitive to the small changes in the optical index of refraction of the water introduced by the compressions and rarefactions of the ultrasonic waves.

light from one point on the source overlap the bright fringes of the pattern produced by light from another point.

Figure 34-32 shows the Fraunhofer diffraction pattern of a circular aperture, which has important applications to the resolution of many optical instruments. The angle $\theta$ subtended by the first diffraction minimum is related to the wavelength and the diameter of the opening $D$ by

$$\sin \theta = 1.22 \frac{\lambda}{D} \qquad\qquad 34\text{-}28$$

*Diffraction by circular aperture*

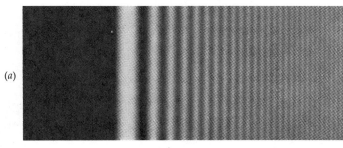

(a)

(b)

Intensity

Distance

**Figure 34-29**
(a) Fresnel diffraction of a straight edge. (*From M. Cagnet, M. Françon, and J. C. Thrierr,* Atlas of Optical Phenomena, *plate 32, Springer-Verlag, Berlin, 1962.*) (b) Intensity versus distance along a line perpendicular to the edge.

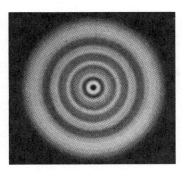

**Figure 34-30**
Fresnel diffraction of a circular aperture. Compare this pattern with Figure 32-5, the Fresnel diffraction pattern of an opaque circular disk. The patterns are complements of each other. (*From M. Cagnet, M. Françon, and J. C. Thrierr,* Atlas of Optical Phenomena, *plate 33, Springer-Verlag, Berlin, 1962.*)

**Figure 34-31**
Fresnel diffraction of a rectangular opening. (*From M. Cagnet, M. Françon, and J. C. Thrierr,* Atlas of Optical Phenomena, *plate 34, Springer-Verlag, Berlin, 1962.*)

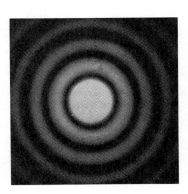

**Figure 34-32**
Fraunhofer diffraction pattern of a circular aperture. (*From M. Cagnet, M. Françon, and J. C. Thrierr,* Atlas of Optical Phenomena, *plate 16, Springer-Verlag, Berlin, 1962.*)

Equation 34-28 is similar to Equation 34-16 except for the factor 1.22; it arises from the mathematical analysis, which is similar to that for a single slit but more complicated because of the cylindrical geometry. In many applications, the angle $\theta$ is small, so that $\sin \theta$ can be replaced by $\theta$. Then

$$\theta \approx 1.22 \frac{\lambda}{D} \qquad\qquad 34\text{-}29$$

Figure 34-33 shows two incoherent point sources which subtend an angle $\alpha$ at a circular aperture far from the sources. The Fraunhofer diffraction intensity patterns are also sketched in this figure. If $\alpha$ is much greater than $1.22 \lambda/D$, there will be little overlap in the diffraction patterns of the two sources and they will be seen as two sources. However, as $\alpha$ is decreased, the overlap of the diffraction patterns increases and it becomes difficult to distinguish the two sources from one source. At the critical-angular separation

$$\alpha_c = 1.22 \frac{\lambda}{D} \qquad\qquad 34\text{-}30$$

the first minimum of the diffraction pattern of one source falls at the central maximum of the other source. These objects are said to be just resolved by the Rayleigh criterion for resolution. Figure 34-34 shows the diffraction patterns for two sources for $\alpha > \alpha_c$ and $\alpha = \alpha_c$.

*Rayleigh criterion for resolution*

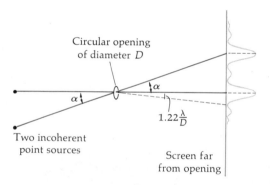

**Figure 34-33**
Two distant sources which subtend an angle $\alpha$ at a circular aperture are easily resolved if $\alpha$ is greater than $1.22\lambda/D$, where $\lambda$ is the wavelength of light and $D$ the diameter of the aperture. Then the diffraction patterns have little overlap. If $\alpha$ is not greater than $1.22\lambda/D$, the overlap of the diffraction patterns makes it difficult to distinguish two sources from one.

**Figure 34-34**
Diffraction patterns of a circular aperture and two incoherent point sources for (a) $\alpha$ greater than $1.22\lambda/D$ and (b) the limit of resolution $\alpha = 1.22\lambda/D$. (From M. Cagnet, M. Françon, and J. C. Thrierr, *Atlas of Optical Phenomena, plate 16, Springer-Verlag, Berlin, 1962.*)

(a)

**Example 34-5** What is the minimum angular separation of two point objects if they are to be just resolved by the eye? How far apart must they be if they are 100 m away? Assume that the diameter of the pupil of the eye is 5 mm and the wavelength of the light is 600 nm.

Using Equation 34-30 with $D = 5$ mm and $\lambda = 600$ nm, we have for the minimum angular separation

$$\alpha_c = 1.22 \frac{6 \times 10^{-7} \text{ m}}{5 \times 10^{-3} \text{ m}} = 1.46 \times 10^{-4} \text{ rad}$$

If the objects are separated by a distance $y$ and are 100 m away, they will be just barely resolved if $\tan \alpha_c = y/(100 \text{ m})$. Then

$$\begin{aligned} y &= (100 \text{ m}) \tan \alpha_c \approx (100 \text{ m})\alpha_c \\ &= 1.46 \times 10^{-2} \text{ m} \\ &= 1.46 \text{ cm} \end{aligned}$$

where we have used the small-angle approximation $\alpha_c \approx \tan \alpha_c$.

(b)

## 34-8  Diffraction Gratings

A useful tool for the analysis of light is the diffraction grating, which consists of a large number of equally spaced slits. Such a grating can be made by cutting parallel, equally spaced grooves on a glass or metal plate with a precision ruling machine. A grating with 10,000 slits per centimetre is not uncommon. The spacing of the slits for such a grating is $d = (1 \text{ cm})/10{,}000 = 10^{-4}$ cm. Consider a plane light wave incident normally on such a grating and assume that the width of each slit is very small, so that each slit produces a wide diffracted beam. Then the interference pattern produced on a screen a large distance from the grating is just that due to a large number of equally spaced line sources. The interference maxima are at angles $\theta$ given by

$$d \sin \theta = m\lambda \qquad\qquad 34\text{-}31$$

The position of an interference maximum does not depend on the number of sources, but the more sources there are, the sharper the maximum. We can see this by finding the angle $\theta_{\min}$ of the first minimum produced by the light from the $N$ slits after the central maximum at $\theta = 0$. The calculation of the angle of the first minimum due to $N$ sources in phase is the same as that for the first minimum of a single slit of width $a = Nd$. The minimum is therefore at the angle $\theta_{\min}$, given by

$$\sin \theta_{\min} = \frac{\lambda}{Nd}$$

For small angles we can replace $\sin \theta_{\min}$ by $\theta_{\min}$. Since the central maximum is at $\theta = 0$, the angular separation $\Delta\theta$ between the central maximum and the first minimum is approximately

$$\Delta\theta = \frac{1}{N}\frac{\lambda}{d} \qquad\qquad 34\text{-}32$$

We note that this "angular width" is approximately $1/N$ times the angular separation of the orders, which is $\lambda/d$ for small $\theta$.

Figure 34-35 shows a typical spectroscope, which uses a diffraction grating to analyze light from a source, usually a tube containing atoms of a gas, e.g., helium or sodium vapor. The atoms are excited because of bombardment by electrons accelerated by the high voltage across the tube. The light emitted by such a source does not consist of a continuous spectrum but contains only certain wavelengths characteristic of the atoms in the source. Light from the source passes through a narrow slit and is made parallel by a collimating lens. Parallel light from the lens is incident on the grating. Instead of falling on a screen a large distance

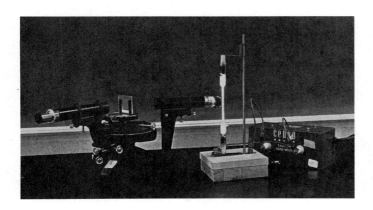

**Figure 34-35**
Typical student spectroscope. Light from a slit near a source is made parallel by a lens and falls on a grating. The diffracted light is viewed with a telescope at an angle which can be accurately measured. The wavelength of the light is found from Equation 34-31. (*Courtesy of Larry Langrill, Oakland University.*)

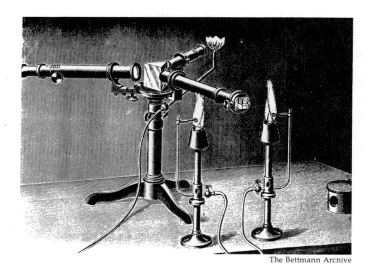

The Bettmann Archive

Spectrometer used in the late nineteenth century by Gustav Kirchhoff. In this drawing, a prism replaces the diffraction grating.

away, the parallel light from the grating is focused by a telescope and viewed by the eye. The telescope is mounted on a rotating platform, calibrated so that the angle $\theta$ can be measured. At angle $\theta = 0$ the central maximum for all wavelengths is seen. If light of a particular wavelength $\lambda$ is emitted by the source, the first interference maximum is seen at angle $\theta$, given by Equation 34-31 with $m = 1$. Each wavelength emitted by the source produces a separate image of the slit in the spectroscope and is called a *spectral line*. Thus by measuring the angle $\theta$ and knowing the spacing $d$ of the slits on the grating we can measure the wavelengths emitted by the source.

An important feature of such a spectroscope is its ability to measure light of two nearly equal wavelengths $\lambda_1$ and $\lambda_2$. For example, the two prominent yellow lines in the spectrum of sodium have wavelengths 589.00 and 589.59 nm, which can be seen as two separate wavelengths if their interference maxima do not overlap. According to the Rayleigh criterion of resolution, these wavelengths are resolved if the angular separation of their interference maxima is greater than the angular separation between one interference maximum and the first interference minimum on either side of it. Let us consider small angles so that we can replace $\sin \theta$ by $\theta$. The angular deviation of the $m$th-order maximum for light of wavelength $\lambda$ (Equation 34-31 with $\theta$ replacing $\sin \theta$) is

$$\theta_1 = m \frac{\lambda_1}{d}$$

and that for light of wavelength $\lambda_2$ is

$$\theta_2 = m \frac{\lambda_2}{d}$$

The angular separation of these maxima is then

$$\Delta\theta = \theta_2 - \theta_1 = m \frac{\Delta\lambda}{d}$$

These wavelengths will just be resolved by the Rayleigh criterion if this angular separation equals that between the maximum and minimum of one of the wavelengths, which is given by Equation 34-32:

$$m \frac{\Delta\lambda}{d} = \frac{1}{N} \frac{\lambda}{d} \qquad\qquad 34\text{-}33$$

The resolving power of a diffraction grating is defined to be $\lambda/|\Delta\lambda|$, where $|\Delta\lambda|$ is the difference between two nearby wavelengths, each approximately equal to $\lambda$. From Equation 34-33 we see that the resolving power is

$$R = \frac{\lambda}{|\Delta\lambda|} = mN \qquad\qquad\qquad 34\text{-}34$$

*Resolving power of a grating*

This result, which we have obtained using the small-angle approximation, holds in general whether the angles are small or not. A general derivation is given below.

We see from Equation 34-34 that to resolve the two yellow lines in the sodium spectrum the resolving power must be

$$R = \frac{589.00 \text{ nm}}{589.59 - 589.00 \text{ nm}} = 998$$

Thus to resolve the two yellow sodium lines in the first order ($m = 1$) we need a grating containing about 1000 slits in the area illuminated by the light.

*Optional*

### Derivation of Equation 34-34 for General Angles

The $m$th-order interference maximum corresponding to some wavelength $\lambda$ is at an angle $\theta$ given by Equation 34-31:

$$\sin\theta = \frac{m\lambda}{d}$$

If we change the wavelength by a small amount $d\lambda$, the angular change $d\theta$ is found by differentiating this expression:

$$\cos\theta \; d\theta = \frac{m}{d} \; d\lambda$$

or

$$\Delta\theta \approx \frac{m\,\Delta\lambda}{d\cos\theta} \qquad\qquad\qquad 34\text{-}35$$

This is the angular separation of the $m$th-order interference maximum for two nearly equal wavelengths. The phase difference $\phi$ between the light from two adjacent slits is given by

$$\phi = \frac{2\pi d}{\lambda}\sin\theta \qquad\qquad\qquad 34\text{-}36$$

This phase difference is $m2\pi$ for the $m$th-order interference maximum. The first interference *minimum* occurs when the vectors representing the $N$ sources form a closed polygon, as in Section 34-5. The phase difference between two adjacent sources is then $2\pi/N$. If we change the angle $\theta$ by a small amount $d\theta$, the change in phase is

$$d\phi = 2\pi d \cos\theta \, \frac{d\theta}{\lambda}$$

Setting this change in phase equal to $2\pi/N$, we obtain for the angular separation $d\theta$ between the interference maximum and the first minimum

$$2\pi d \cos\theta \, \frac{d\theta}{\lambda} = d\phi = \frac{2\pi}{N}$$

or

$$d\theta = \frac{\lambda}{Nd\cos\theta} \approx \Delta\theta \qquad\qquad\qquad 34\text{-}37$$

According to the Rayleigh criterion of resolution, these wavelengths are resolved if the angle $\Delta\theta$ between the interference maximum and minimum given by Equation 34-37 equals the angle $\Delta\theta$ between the interference maxima of the two wavelengths given by Equation 34-35. Thus the wavelengths are resolved if

$$\frac{\lambda}{Nd \cos \theta} = \frac{m \, \Delta\lambda}{d \cos \theta} \quad \text{or} \quad \frac{\lambda}{\Delta\lambda} = mN$$

## Review

A. Define, explain, or otherwise identify:

Newton's rings, 903
Interferometer, 905
Lloyd's mirror, 908
Fraunhofer diffraction pattern, 917

Fresnel diffraction pattern, 917
Rayleigh's criterion for resolution, 920
Diffraction grating, 921

B. True or false:

1. When waves interfere destructively, the energy is converted into heat energy.

2. Interference is observed only for waves from coherent sources.

3. In single-slit diffraction, the narrower the slit, the wider the diffraction pattern.

4. A circular aperture can produce both a Fraunhofer and a Fresnel diffraction pattern.

5. The ability to resolve two point sources depends on the wavelength of the light.

## Exercises

### Section 34-1, Interference in Thin Films

1. Light of wavelength 500 nm is incident normally on a film of water $10^{-4}$ cm thick. The index of refraction of water is 1.33. (a) What is the wavelength of the light in the water? (b) How many wavelengths are contained in the distance $2t$, where $t$ is the film thickness? (c) What is the phase difference between the wave reflected from the top of the film and the one reflected from the bottom after it has traveled this distance?

2. A loop of wire is dipped in soapy water and held so that the soap film is vertical. Viewed by reflection with white light, the top of the film appears black. Explain why. Below the black region are colored bands. Is the first band red or blue? Describe the appearance of the film when viewed by transmitted light.

3. A wedge-shaped film of air is made by placing a small slip of paper between the edges of two flat pieces of glass. Light of wavelength 700 nm is incident normally on the glass plates, and interference bands are observed by reflection. (a) Is the first band near the point of contact of the plates dark or bright? Why? (b) There are five dark bands per centimetre. What is the angle of the wedge?

4. A thin layer of a transparent material with index of refraction 1.30 is used as a nonreflective coating on the surface of glass with index of refraction 1.50. What should the thickness be for the film to be nonreflecting for light of wavelength 600 nm (in vacuum)?

5. The diameter of fine wires can be accurately measured by interference patterns. Two optically flat pieces of glass of length $L$ are arranged with the wire as shown in Figure 34-36. The setup is illuminated by monochromatic light, and the resulting interference fringes are detected. Suppose $L = 20$ cm and yellow

sodium light ($\lambda \approx 590$ nm) is used for illumination. If 19 bright fringes are seen along this 20-cm distance, what are the limits on the diameter of the wire? *Hint:* The 19th fringe might not be right at the end, but you do not see 20 fringes.

**Figure 34-36**
Exercise 5.

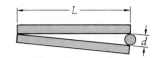

### Section 34-2, The Michelson Interferometer

6. A thin film of index of refraction $n = 1.5$ for light of wavelength 600 nm is inserted in one arm of a Michelson interferometer. (*a*) If a fringe shift of 12 fringes occurs, what is the thickness of this film? (*b*) If the illuminating light is changed to 400 nm, the fringe shift as this film is inserted becomes 16 fringes. What is the index of refraction of this film to light of wavelength 400 nm?

7. A hollow cell of length 5 cm with glass windows is inserted into one arm of a Michelson interferometer. The air is pumped out of the cell and the mirrors are adjusted to give a bright fringe at the center. As the air is gradually let back into the cell, there is a shift of 49.6 fringes when light of wavelength 589.29 nm is used. (*a*) How many waves are there in the 5.0-cm-long cell when it is evacuated? (*b*) How many waves are in the cell when it contains air? (*c*) What is the index of refraction of air as determined by this experiment?

### Section 34-3, The Two-Slit Interference Pattern

8. Two narrow slits separated by a distance $d$ are illuminated by light of wavelength $\lambda$, and the interference pattern is viewed on a screen a long distance $L$ away ($L > d$). The $m$th interference maximum is a distance $y_m$ from the central maximum measured along the screen. (*a*) Show that $y_m$ is given approximately by $y_m = m\lambda L/d$. (*b*) For $\lambda = 600$ nm, $L = 2$ m, and $d = 1$ mm, find the number of bright fringes per centimetre on the screen.

9. A long narrow horizontal slit lies 1 mm above a plane mirror. The interference pattern produced by the slit and its image is viewed on a screen a distance 1 m from the slit. The wavelength of the light is 600 nm. (*a*) Find the distance above the mirror to the first maximum. (*b*) How many dark bands per centimetre are seen on the screen?

10. (See Exercise 8.) When light of wavelength 589 nm is used and the screen is 3 m from the slits, there are 28 bright fringes per centimetre on the screen. What is the slit separation?

11. In a lecture demonstration, laser light is used to illuminate two slits separated by 0.5 mm, and the interference pattern is observed on a screen 5 m away. The distance on the screen to the 37th bright fringe is 25.7 cm. What is the wavelength of the light? [See part (*a*) of Exercise 8.]

12. Light is incident at an angle $\phi$ with the normal to a plane containing two slits of separation $d$ (Figure 34-37). Show that the interference maxima are now located at angles $\theta$ given by $\sin \theta + \sin \phi = m(\lambda/d)$.

13. Light from a helium-neon laser of wavelength 633 nm is shone normally on a plane containing two slits. The first interference maximum is 82 cm from the central maximum on a screen 12.0 m away. (*a*) Find the separation of the slits. (*b*) How many interference maxima can be observed?

**Figure 34-37**
Light incident on two slits at an angle $\phi$ with the normal for Exercise 12.

### Section 34-4, The Vector Model for Addition of Harmonic Waves, and Section 34-5, Interference Pattern of Three or Four Equally Spaced Sources

14. With a compass, draw circular arcs representing wave crests for each of three point sources a distance $d$ apart for $d = 6$ cm and $\lambda = 1$ cm. Connect the intersections corresponding to points of constant path difference and label the path difference for each line. How do the lines of constructive interference differ from those for just two sources of the same spacing?

15. Three equally spaced slits are separated by $d = 0.1$ mm and uniformly illuminated by light of wavelength 600 nm. The interference pattern is viewed on a screen 2 m away. Find the positions of the interference maxima and minima.

16. Five equally spaced slits are uniformly illuminated, and the interference pattern is viewed on a screen far away such that the approximation $\sin \theta \approx \theta$ is good for the first few maxima. (a) Find the angle $\theta_1$ between the first interference maximum and the central maximum ($\theta = 0$) and compare this angle with the angle to the first minimum. (b) Sketch the interference pattern.

17. Do Exercise 16 for six equally spaced slits.

### Section 34-6, Diffraction Pattern of a Single Slit

18. Equation 34-17, $a \sin \theta = m\lambda$, and Equation 34-5, $d \sin \theta = m\lambda$, are sometimes confused. For each equation define the symbols and explain its application.

19. Light of wavelength 600 nm is incident on a long narrow slit. Find the angle of the first diffraction minimum if the width of the slit is (a) 0.1 mm, (b) 1 mm, (c) 0.01 mm.

20. The single-slit diffraction pattern of light is observed on a screen a large distance $L$ from the slit. (a) Show that if the angle $\theta$ is small, the distance $y$ from the central maximum on the screen to the first minimum is given by $y = L\lambda/a$. Note that the width of the central maximum $2y$ varies inversely with the width of the slit $a$. Find this width $2y$ for $L = 2$ m, $\lambda = 500$ nm, and (b) $a = 0.1$ mm, (c) $a = 0.01$ mm, and (d) $a = 0.001$ mm.

21. In a lecture demonstration of diffraction, a laser beam of wavelength 700 nm passes through a vertical slit 0.5 mm wide and hits a screen 6 m away. Find the horizontal length of the principal diffraction maximum on the screen; i.e., find the distance between the first minimum on the left and the first minimum on the right of the central maximum.

22. Plane microwaves are incident on a long narrow slit of width 5 cm. The first diffraction minimum is observed at $\theta = 37°$. What is the wavelength of the microwaves?

23. A two-slit Fraunhofer diffraction-interference pattern is observed with light of wavelength 500 nm. The slits have a separation $d = 0.1$ mm and width $a$. Find the width $a$ if the fifth interference maximum is at the same angle as the first diffraction minimum. For this case, how many bright fringes will be seen in the central diffraction maximum?

24. A two-slit Fraunhofer diffraction-interference pattern is observed with light of wavelength 700 nm. The slits have a width $a = 0.01$ mm and are separated by $d = 0.2$ mm. How many bright fringes will be seen in the central diffraction maximum?

### Section 34-7, Diffraction and Resolution

25. Light of wavelength 700 nm is incident on a pinhole of diameter 0.1 mm. (a) What is the angle between the central maximum and the first diffraction minimum for Fraunhofer diffraction? (b) What is the distance between the central maximum and the first diffraction minimum on a screen 8 m away?

26. Two light sources ($\lambda = 700$ nm) are 10 m away from the pinhole in Exercise 25. How far apart must the sources be for their diffraction patterns to be resolved by the Rayleigh criterion?

27. (a) How far apart must two objects be on the moon to be resolved by the eye? Take the diameter of the pupil of the eye to be 5.0 mm, $\lambda = 600$ nm, and $L = 380,000$ km for the distance to the moon. (b) How far apart must the objects on the moon be to be resolved by a telescope which has a 5.00-m-diameter mirror?

28. Two sources of wavelength 700 nm are separated by a horizontal distance $x$. They are 5 m from a vertical slit of width 0.5 mm. What is the least value of $x$

permitting the diffraction pattern of the sources to be resolved by the Rayleigh criterion?

29. What aperture (in millimetres) is required for perfect opera glasses (binoculars) for the observer to be able to distinguish the soprano's individual eyelashes (separated by 0.5 mm) at 25 m? Assume the effective wavelength of light to be 550 nm.

30. The headlights on a small car are separated by 112 cm. At what maximum distance could you resolve them if the diameter of your pupils is 5.0 mm and the effective wavelength of the light is 550 nm?

31. You are told not to shoot until you see the whites of their eyes. If the eyes are separated by 6.5 cm and the diameter of your pupil is 5.0 mm, at what distance can you resolve the two eyes using light of wavelength 550 nm?

### Section 34-8, Diffraction Gratings

32. A diffraction grating with 2000 slits per centimetre is used to measure the wavelengths emitted by hydrogen gas. At what angles $\theta$ would you expect to find the two blue lines of wavelength 434 and 410 nm?

33. With the grating used in Exercise 32, two other lines in the hydrogen spectrum are found in first order at angles $\theta_1 = 9.72 \times 10^{-2}$ rad and $\theta_2 = 1.32 \times 10^{-1}$ rad. Find the wavelengths of these lines.

34. Repeat Exercise 32 for a grating with 15,000 slits per centimetre.

35. A grating of 2000 slits per centimetre is used to analyze the spectrum of mercury. (a) Find the angular deviation in first order of the two lines of wavelength 579.0 and 577.0 nm. (b) How wide must the beam be on the grating for these lines to be resolved?

36. What is the longest wavelength that can be observed in fifth order using a grating with 4000 slits per centimetre?

### Problems

1. Laser light falls normally on three evenly spaced slits. When one of the side slits is covered, the first-order maximum is at 0.60° from the normal. If the center slit is covered with the other two open, find (a) the angle of the first-order maximum and (b) the order number of the maximum that now occurs at the same angle the fourth-order maximum did before.

2. Two slits are separated by a distance $d$. Their interference pattern is to be observed on a screen a large distance $L$ away. Calculate the spacing $y$ of the maxima on the screen for light of wavelength 500 nm with $L = 1$ m and $d = 1$ cm. Would you expect to observe the interference of light on the screen for this situation? How close together should you place the slits for the maxima to be separated by 1 mm for this wavelength and screen distance?

3. Suppose that the central *diffraction* maximum for two slits contained 17 interference fringes for some wavelength of light. How many interference fringes would you expect in the first secondary diffraction maximum?

4. The ceiling of your lecture hall is probably covered with acoustic tile which has small holes separated by about 6.0 mm. (a) Using $\lambda = 500$ nm, how far could you be from this tile and still resolve these holes? The diameter of the pupil of your eye is about 5.0 mm. (b) Could you "see" these holes better with red or blue light?

5. The telescope on Mount Palomar has a diameter of about 5.0 m. Assuming "ideal" sky conditions, the resolution would be diffraction-limited. Suppose a double star were 4 light-years away. What would the stellar separation have to be for their images to be resolved?

6. A mica sheet 1.20 $\mu$m thick is suspended in air. In reflected light there are gaps in the visible spectrum at 421, 474, 542, and 632 nm. Find the index of refraction of mica.

7. A thin film having index of refraction 1.5 is surrounded by air. It is illuminated normally by white light and viewed by reflection. Analysis of the resulting reflected light shows that the wavelengths 360, 450, and 600 nm are the only missing wavelengths near the visible portion of the spectrum. That is, for these wavelengths there is destructive interference. (a) What is the thickness of the film? (b) What visible wavelengths would be extra bright in the reflected interference pattern? (c) If this film is supported on glass whose index of refraction is 1.6, what wavelengths in the visible spectrum will be missing from the reflected light?

8. For a ruby laser of wavelength 694 nm, the ends of the ruby crystal are the aperture that determines the diameter of the light beam emitted. If the diameter is 2.00 cm and the laser is aimed at the moon, 0.38 Gm away, find the approximate diameter of the light beam reaching the moon.

9. Sodium light of wavelength 589 nm falls normally on a 2-cm-square diffraction grating ruled with 4000 lines per centimetre. The Fraunhofer diffraction pattern is projected onto a screen at 1.5 m by a lens of focal length 1.5 m placed immediately in front of the grating. Find (a) the positions of the first two intensity maxima on one side of the central maximum, (b) the width of the central maximum, and (c) the resolution in the first order.

10. At the second secondary maximum of the diffraction pattern of a single slit, the phase difference between the waves from the top and bottom of the slit is approximately $5\pi$. The vectors used to calculate the amplitude at this point complete $2\frac{1}{2}$ circles. If $I_0$ is the intensity at the central maximum, find the intensity $I$ at this second secondary maximum.

11. White light falls at an angle of 30° to the normal of a plane containing a pair of slits separated by 2.5 $\mu$m. What visible wavelengths give a bright interference maximum in the transmitted light in the direction normal to the plane? (See Exercise 12.)

12. A Newton's-ring apparatus consists of a glass lens with radius of curvature $R$ which rests upon a flat glass plate, as shown in Figure 34-38. Usually both pieces of glass have the same index of refraction $n$, and so the thin film is air of variable thickness. The pattern is viewed by reflected light. (a) Show that for a thickness $t$ the condition for a bright (constructive) interference fringe is $t = \frac{1}{2}(m + \frac{1}{2})\lambda$, $m = 0, 1, 2, \ldots$ . (b) Show that as long as $t/R \ll 1$, the radius $r$ of a bright circular fringe is given by $r = \sqrt{(m + \frac{1}{2})\lambda R}$, $m = 0, 1, 2, \ldots$ . (c) How would the transmitted pattern look in comparison with the reflected one? (d) Use $R = 10$ m and a diameter of 4.0 cm for the lens. How many bright fringes would you see if the apparatus were illuminated by yellow sodium light ($\lambda \approx$ 590 nm) and viewed by reflection? (e) What would be the diameter of the sixth bright fringe? (f) If the glass used in the apparatus has index of refraction $n = $ 1.5 and water is placed between the two pieces of glass, what change will take place in the bright fringes?

**Figure 34-38**
Newton's-rings setup for Problem 12.

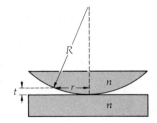

13. A *Jamin refractometer* is a device for measuring or comparing the indexes of refraction of fluids. A beam of monochromatic light is split into two parts, each directed along the axis of a separate cylindrical tube before being combined into a single beam again and viewed through a telescope. Suppose that each tube is 0.40 m long and sodium light of wavelength 589 nm is used. Both tubes are initially evacuated, and constructive interference is observed in the center of the field of view. As air is slowly allowed to enter one of the tubes, the central field of view changes to dark and back to bright a total of 198 times. (a) What is the index of refraction of air? (b) If the fringes can be counted to ±0.25 fringe, where one fringe is equivalent to one complete cycle of intensity variation at the center of the field of view, to what accuracy can the index of refraction of air be determined by this experiment?

14. A camera lens is made of glass whose index of refraction is 1.6. This lens is coated with a magnesium fluoride film ($n = 1.38$) to enhance its light transmission. This film is to produce a zero reflection for light of wavelength 540 nm. Treat the lens surface as a flat plane and the film as a uniformly thick flat film. (a) How thick must the film be to accomplish its objective in first order? (b) Would there be destructive interference for any other visible wavelengths? (c) By what factor would the reflection be reduced by this film for 400 and 700 nm? Neglect the variation in the reflected light amplitudes from the two surfaces.

15. For a diffraction grating one is interested not only in resolving power $R$, which is the ability of the grating to separate two close wavelengths, but also in the dispersion $D$ of the grating. This is defined by $D = \Delta\theta_m/\Delta\lambda$ in $m$th order. (a) Show that $D$ can be written

$$D = \frac{m}{\sqrt{d^2 - m^2\lambda^2}}$$

where $d$ is the slit spacing. (b) If a diffraction grating with 2000 slits per centimetre is to resolve the two sodium yellow lines (wavelengths 589.0 and 589.6 nm) in second order, how many slits must be illuminated by the beam? (c) What would the separation be between these resolved yellow lines if the pattern were viewed on a screen 4 m from the grating?

16. (a) Show that the positions of the interference minima on a screen a large distance $L$ away from three equally spaced sources (spacing $d$ with $d \gg \lambda$) are given approximately by

$$y = n\,\frac{\lambda L}{3d} \qquad \text{where } n = 1, 2, 4, 5, 7, 8, 10, \ldots$$

that is, $n$ is not a multiple of 3. (b) For $L = 1$ m, $\lambda = 5 \times 10^{-7}$ m, and $d = 0.1$ mm, calculate the width of the principal interference maxima (distance between successive minima) for three sources.

17. Show that the positions of the interference minima on a screen a large distance $L$ from four equally spaced sources are given by

$$y = n\,\frac{\lambda L}{4d} \qquad \text{where } n = \text{integer not a multiple of 4, and } d \geq \lambda$$

Compare the width of the principal interference maxima for four sources with that for two sources for $d = 0.1$ mm, $L = 2$ m, and $\lambda = 6 \times 10^{-7}$ m.

18. Light of wavelength $\lambda$ is diffracted through a single slit of width $a$, and the resulting pattern is viewed on a screen a long distance $L$ away from the slit. (a) Show that the width of the central maximum on the screen is approximately $2L\lambda/a$. (b) If a slit of width $2L\lambda/a$ is cut in the screen and illuminated, show that the width of its central diffraction maximum at the same distance $L$ (that is, back on the plane of the slit) is $a$ to the same approximation.

19. In a pinhole camera, the image is fuzzy because of the finite size of the pinhole and because of diffraction. As the pinhole is made smaller, the fuzziness due to its size, i.e., to rays arriving at the image point from different parts of the object, is reduced, but that due to diffraction is increased. The optimum size of the aperture for the sharpest possible image occurs when the spread due to diffraction equals that due to the aperture size. Estimate the optimum size of the aperture if the distance from the pinhole to the screen is 10.0 cm and the wavelength of the light is 550 nm.

20. Light of wavelength 480 nm falls normally on four slits; each is 2.0 $\mu$m wide and separated by 6.0 $\mu$m from the next. (a) Find the angle from the center to the first zero of the single-slit diffraction pattern. (b) Find the angles of any bright interference maxima that lie inside the central diffraction maximum. (c) Find the angular spread between the central interference maximum and the first interference minimum on either side of it. (d) Sketch the intensity as a function of angle on the distant screen.

**CHAPTER 35**     Relativity

---

**Objectives**   After studying this chapter you should:

1.   Be able to state the Einstein postulates of special relativity.

2.   Be able to discuss the results and significance of the Michelson-Morley experiment.

3.   Be able to define proper time and proper length and state the equations for time dilation and length contraction.

4.   Be able to discuss the lack of synchronization of clocks in moving reference frames.

5.   Be able to use the Lorentz transformation to work problems relating time or space intervals in different reference frames.

6.   Be able to discuss the twin paradox.

7.   Be able to state the definition of relativistic momentum and the equations relating the kinetic energy and total energy of a particle to its speed.

8.   Be able to discuss the relation between mass and energy in special relativity and compute the binding energy of various systems from the known rest masses of their constituents.

9.   Be able to state the principle of equivalence and discuss three predictions derived from it.

---

The theory of relativity consists of two rather different theories, the special theory and the general theory. The special theory, developed by Einstein and others in 1905, concerns the comparison of measurements made in different inertial reference frames moving with constant velocity relative to each other. Its consequences, which can be derived with a minimum of mathematics, are applicable in a wide variety of situations encountered in physics and engineering. On the other hand, the general theory, also developed by Einstein and others (around 1916), is concerned with accelerated reference frames and gravity. A thorough understanding of the general theory requires sophisticated mathematics, e.g., tensor analysis, and the applications of this theory are chiefly in the area of gravitation. It is of great importance in cosmology but is rarely encountered in other areas of physics or in engineering. We shall therefore concentrate on the special theory (often referred to as special relativity) and discuss the general theory only briefly in the last section of this chapter.*

The theory of special relativity can be derived from two postulates proposed by Einstein in a paper on the electrodynamics of moving bodies published when he was only 26 years old.** These two postulates, simply stated, are:

1. Absolute, uniform motion cannot be detected.

2. The speed of light is independent of the motion of the source.

*Einstein postulates*

Although each postulate seems quite reasonable, many of the implications of the two together are quite surprising and contradict what is often called common sense. For example, a direct consequence is that all observers measure the same number for the speed of light in vacuum independent of their relative motion. We shall derive this and other results later. First we shall look at an historically important experiment related to the theory of special relativity, the Michelson-Morley experiment.

## 35-1   The Michelson-Morley Experiment

From our study of wave motion we know that all mechanical waves require a medium for their propagation. The speed of such waves depends only on properties of the medium. With sound waves, for example, absolute motion, i.e., motion relative to the still air, can be detected. For example, the doppler effect for sound depends not only on the relative motion of the source and listener but on the absolute motion of each relative to the air. It was natural to expect that some kind of medium supports the propagation of light and other electromagnetic waves. Such a medium, called the *ether*, was proposed in the nineteenth century. The ether as proposed had to have unusual properties. Although it required great rigidity to support waves of such high velocity (recall that the velocity of waves on a string depends on the tension of the string), it could introduce no drag force on the planets, as their motion is fully accounted for by the law of gravitation.

It was of considerable interest to determine the velocity of the earth relative to the ether. Maxwell pointed out that in measurements of the speed of light (such as Fizeau's toothed-wheel method), the earth's speed $v$ relative to the ether appears only in the second order $v^2/c^2$, an effect then considered too small to measure. In these measurements, the time for a light pulse to travel to and from a mirror is determined. Fig-

---

* There are several good books that the interested student may want to consult to pursue the subject of relativity at the elementary level: R. Resnick, *Introduction to Special Relativity*, Wiley, New York, 1968; A. P. French, *Special Relativity*, Norton, New York, 1968; and C. Kacser, *Introduction to the Special Theory of Relativity*, Prentice-Hall, Englewood Cliffs, N.J., 1967.

** *Annalen der Physik*, vol. 17, p. 841, 1905. For a translation from the original German, see W. Perrett and G. B. Jeffery (trans.), *The Principle of Relativity: A Collection of Original Memoirs on the Special and General Theory of Relativity* by H. A. Lorentz, A. Einstein, H. Minkowski, and H. Weyl, Dover, New York, 1923.

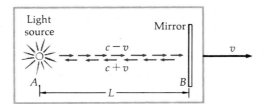

Figure 35-1
Light source and mirror
moving with speed $v$ relative
to the "ether." According to
classical theory, the speed of
light relative to the source
and mirror would be $c - v$
toward the mirror and $c + v$
away from the mirror.

ure 35-1 shows a light source and mirror a distance $L$ apart. If we assume that both are moving with speed $v$ through the ether, classical
theory predicts that the light will travel toward the mirror with speed
$c - v$ and back with speed $c + v$ (both speeds relative to the mirror and
light source). The time for the total trip will be

$$t_1 = \frac{L}{c - v} + \frac{L}{c + v} = \frac{2cL}{c^2 - v^2} = \frac{2L}{c}\left(1 - \frac{v^2}{c^2}\right)^{-1} \tag{35-1}$$

For $v$ much less than $c$, we can expand this result using the binomial expansion:

$$(1 + x)^n \approx 1 + nx + \cdots \qquad \text{for } x \ll 1$$

Then

$$t_1 \approx \frac{2L}{c}\left(1 + \frac{v^2}{c^2} + \cdots\right) \tag{35-2}$$

If we take the orbital speed of the earth about the sun for an estimate of
$v$, we have $v \approx 3 \times 10^4$ m/s $= 10^{-4}c$ and $v^2/c^2 = 10^{-8}$. Thus the correction for the earth's motion is small indeed. Michelson realized that
though this effect is too small to be measured directly, it should be possible to determine $v^2/c^2$ by a difference measurement. For this measurement he used the Michelson interferometer discussed in Section 34-2.
Let us assume that the interferometer (see Figure 34-7) is oriented so
that the beam transmitted by the beam splitter is in the direction of the
assumed motion of the earth. Equation 35-2 then gives the classical result for the round-trip time $t_1$ for the transmitted beam. Since the reflected beam travels (relative to the earth) perpendicular to the earth's
velocity, the velocity of this beam relative to earth (according to classical
theory) is the vector difference $\mathbf{u} = \mathbf{c} - \mathbf{v}$ (Figure 35-2). The magnitude
of $\mathbf{u}$ is $\sqrt{c^2 - v^2}$; so the round-trip time for this beam is

$$t_2 = \frac{2L}{\sqrt{c^2 - v^2}} = \frac{2L}{c}\left(1 - \frac{v^2}{c^2}\right)^{-1/2} \approx \frac{2L}{c}\left(1 + \frac{1}{2}\frac{v^2}{c^2} + \cdots\right) \tag{35-3}$$

where again the binomial expansion has been used. Hence there is a
time difference:

$$\Delta t = t_1 - t_2 \approx \frac{2L}{c}\left(1 + \frac{v^2}{c^2}\right) - \frac{2L}{c}\left(1 + \frac{1}{2}\frac{v^2}{c^2}\right) = \frac{Lv^2}{c^3} \tag{35-4}$$

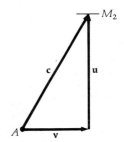

Figure 35-2
A light beam from the beam
splitter $A$ to the mirror $M_2$ in
a Michelson interferometer
moves with velocity $\mathbf{c}$ at an
angle as shown in a reference
frame in which the interferometer is moving to the right
with speed $v$. Relative to the
earth in which the interferometer is fixed, the light
beam travels with velocity
$\mathbf{u} = \mathbf{c} - \mathbf{v}$ according to classical theory. The speed of the
light beam relative to the
earth is then $u = (c^2 - v^2)^{1/2} =$
$c(1 - v^2/c^2)^{1/2}$ according to
classical theory.

The time difference is to be detected by observing the interference of
the two beams of light. Because of the difficulty of making the two paths
of equal length to the precision required, the interference pattern of the
two beams is observed and then the whole apparatus rotated 90°. The
rotation produces a time difference given by Equation 35-4 for each
beam. The total time difference of $2 \Delta t$ is equivalent to a path difference

of $2c\,\Delta t$. The interference fringes observed in the first orientation should thus shift by a number of fringes $\Delta N$ given by

$$\Delta N = \frac{2c\,\Delta t}{\lambda} = \frac{2L}{\lambda}\frac{v^2}{c^2} \qquad\qquad 35\text{-}5$$

where $\lambda$ is the wavelength of the light. In Michelson's first attempt, in 1881, $L$ was about 1.2 m and $\lambda$ was 590 nm. For $v^2/c^2 = 10^{-8}$, $\Delta N$ was expected to be 0.04 fringe. Even though the experimental uncertainties were estimated to be about this same magnitude, when no shift was observed, Michelson reported this as evidence that the earth did not move relative to the ether. In 1887, when he repeated the experiment with Edward W. Morley, he used an improved system for rotating the apparatus without introducing a fringe shift because of mechanical strains, and he increased the effective path length $L$ to about 11 m by a series of multiple reflections. Figure 35-3 shows the configuration of the Michelson-Morley apparatus. For this attempt, $\Delta N$ was expected to be about 0.4 fringe, about 20 to 40 times the minimum possible to observe. Once again, no shift was observed. The experiment has since been repeated under various conditions by a number of people, and no shift has ever been found.

The null result of the Michelson-Morley experiment is easily understood in terms of the Einstein postulates. According to postulate 1, absolute uniform motion cannot be detected. We can consider the whole apparatus and the earth to be at rest. No fringe shift is expected when the apparatus is rotated 90° since all directions are equivalent. It should be pointed out that Einstein did not set out to explain this experiment. His theory arose from his considerations of the theory of electricity and magnetism and the unusual property of electromagnetic waves, i.e., that they propagate in a vacuum. In his first paper, which contains the complete theory of special relativity, he made only a passing reference to the Michelson-Morley experiment, and in later years he could not recall whether he was aware of the details of this experiment before he published his theory.

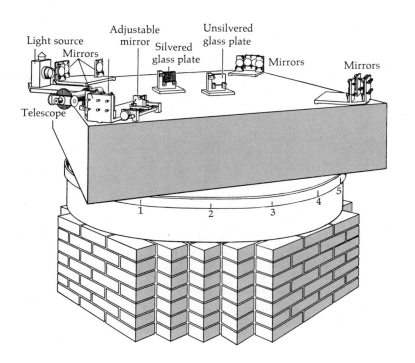

**Figure 35-3**
Drawing of Michelson-Morley apparatus used in their 1887 experiment. The optical parts were mounted on a sandstone slab 5 ft square, which was floated in mercury, thereby reducing the strains and vibrations that had affected the earlier experiments. Observations could be made in all directions by rotating the apparatus in the horizontal plane. (*From R. S. Shankland, "The Michelson-Morley Experiment," Copyright © November 1964 by Scientific American, Inc. All rights reserved.*)

## 35-2 Consequences of Einstein's Postulates

An immediate consequence of the two Einstein postulates is that

*Every observer measures the same value for the speed of light independent of the relative motion of the sources and observers.*

Consider a light source $S$ and two observers, $R_1$ at rest relative to $S$ and $R_2$ moving toward $S$ with speed $v$, as shown in Figure 35-4$a$. The speed of light measured by $R_1$ is $c = 3 \times 10^8$ m/s. What is the speed measured by $R_2$? The answer is *not* $c + v$. By postulate 1, Figure 35-4$a$ is equivalent to Figure 35-4$b$, in which $R_2$ is pictured at rest and the source $S$ and $R_1$ are moving with speed $v$. That is, since absolute motion has no meaning, it is not possible to say which is really moving and which is at rest. By postulate 2, the speed of light from a moving source is independent of the motion of the source. Thus, looking at Figure 35-4$b$, we see that $R_2$ measures the speed of light to be $c$, just as $R_1$ does.

This result—that all observers measure the same value for the speed of light—contradicts our intuitive ideas about relative velocities. If a car moves at 50 km/h away from an observer and another car moves at 80 km/h in the same direction, the velocity of the second car relative to the first car is 30 km/h. This result is easily measured and conforms to our intuition. However, according to Einstein's postulates, if a light beam is moving in the direction of the cars, observers in both cars will measure the same speed for the light beam. We shall see later that our intuitive ideas about the combination of velocities are approximations which hold only when the speeds are very small compared with the speed of light. In practice, even in a plane moving with the speed of sound, it is not possible to measure the speed of light accurately enough to distinguish the difference between the result $c$ and $c \pm v$, where $v$ is the speed of the plane. In order to make such a distinction, we must either move with a very great velocity (much greater than that of sound) or make extremely accurate measurements, as in the Michelson-Morley experiment.

Einstein recognized that his postulates have important consequences for measuring time intervals and space intervals as well as relative velocities. He showed that the size of time and space intervals between two events depends on the reference frame in which the events are observed. The changes in such measurements from one reference frame to another are called *time dilation* and *length contraction*. Instead of developing the general formalism of relativity, known as the *Lorentz transformation*, we shall derive these famous relativistic effects directly from the Einstein postulates by considering some simple special cases of measurement of time and space intervals. We shall also study the problems of clock synchronization and simultaneity, which are of central importance in understanding special relativity. In these discussions we shall be comparing measurements made by observers who are moving relative to each other. We shall use a rectangular coordinate system $xyz$ with origin $O$, called the $S$ *reference frame*, and another system $x'y'z'$ with origin $O'$, called the $S'$ *frame*, which is moving with constant velocity relative to the $S$ frame. For simplicity, we shall consider the $S'$ frame to be moving with speed $v$ along the $x$ (or $x'$) axis relative to $S$. In each frame we assume that there are as many observers as needed equipped with clocks, metresticks, etc. that are identical when compared at rest (see Figure 35-5).

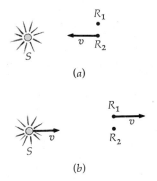

**Figure 35-4**
If absolute motion cannot be detected, the reference frame shown in ($a$), in which the receiver $R_1$ and source $S$ are stationary and receiver $R_2$ is moving toward the source with speed $v$, is equivalent to the reference frame shown in ($b$), in which $R_2$ is stationary and the source $S$ and receiver $R_1$ are moving with speed $v$. Since the speed of light does not depend on the source speed, receiver $R_2$ measures the same value for that speed as receiver $R_1$.

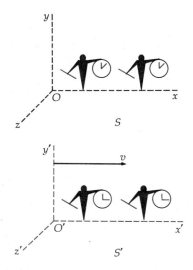

**Figure 35-5**
Coordinate reference frames $S$ and $S'$ moving with relative speed $v$. In each frame are observers with metresticks and clocks.

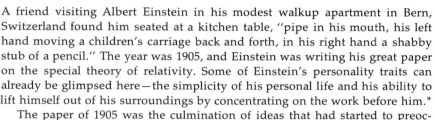

# Albert Einstein (1879–1955)

Gerald Holton
*Harvard University*

Einstein at 17. (*From Carl Seelig*, Albert Einstein: Eine Dokumentarische Biographie, *Europa Verlag, Zurich, 1954.*)

A friend visiting Albert Einstein in his modest walkup apartment in Bern, Switzerland found him seated at a kitchen table, "pipe in his mouth, his left hand moving a children's carriage back and forth, in his right hand a shabby stub of a pencil." The year was 1905, and Einstein was writing his great paper on the special theory of relativity. Some of Einstein's personality traits can already be glimpsed here—the simplicity of his personal life and his ability to lift himself out of his surroundings by concentrating on the work before him.*

The paper of 1905 was the culmination of ideas that had started to preoccupy Einstein 10 years earlier, when he was about 16. At that age, he later wrote in his autobiography,** the following paradox occurred to him. As one knows from ample experience, galilean relativity holds in mechanics; if you throw a ball forward in a moving carriage, observing the motion of the ball cannot tell you how fast you and the carriage are moving. But when it comes to optics, it seems to be different. For example, if I move along "a beam of light with the velocity $c$ [velocity of light in a vacuum], I should observe such a beam of light as a spatially oscillatory electromagnetic field." Looking back along the beam a distance of one whole wavelength, one should see that the local magnitudes of the electric and magnetic field vectors increase point by point from, say, zero to full strength, and then decrease again to zero, one wavelength away. Seeing such a curious field in free space would tell me that I am going at the speed of light with respect to absolute space, or the ether.

Einstein suspected that the imagined result of this *Gedanken experiment* (thought experiment) must somehow be in error. In any case, he said later, "from the very beginning it appeared to me intuitively clear that, judged from the standpoint of such an observer, everything would have to happen according to the same laws as for an observer who, relative to the earth, was at rest."

Einstein solved the apparent paradox in 1905 by showing that the expectation of what one would see in pursuing a light beam is false, being grounded in a wrong idea that "unrecognizedly was anchored in the unconscious" of all scientists of the time—namely the absolute character of time and of simultaneity. Instead, Einstein showed that a sound view of how physical nature operates can be gained by boldly postulating two principles.

The first principle of relativity for inertial systems generalizes galilean relativity to encompass not only mechanics but also optics and electromagnetism; the second principle says that light in a vacuum will always be found to move with the same speed $c$, regardless of the state of motion of the emitting body. The solutions to all paradoxes and experimentally puzzling observations at the time were derivable from these two principles.

At least as important to Einstein was the fact that this approach hugely simplified our view of nature in two ways: it broke down barriers between hitherto entirely separate notions, and it cleansed physics of unnecessary conceptions that had produced pseudo problems. Now electromagnetism was on the same footing as mechanics, instead of allowing "privileged" systems. Time and space were found to have interpenetrating meaning—what really existed was a space-time continuum. Electric fields and magnetic fields were at bottom the same reality perceived in different experimental conditions. Mass and energy were equivalent. Even the boldly intuitive approach to scientific

---

* See Einstein's essay "Motive of Research," pp. 224–227 in Sonja Bargmann (trans. and ed.), *Ideas and Opinions,* Crown, New York, 1954; look also at the other essays in this collection.

** Albert Einstein, *Autobiographical Notes,* in P. A. Schilpp (ed.), *Albert Einstein: Philosopher-Scientist,* Harper, New York, 1959.

discovery and the rigorously rational method were joined into one powerful approach in his paper. Moreover, the notion of the ether was at last declared to be "superfluous." It had long been embedded in physics but had required the assumption of ever more puzzling properties. While he was the first to give up the idea of an ether, to the end of his life Einstein was dedicated to the theme of the *field* (or, in general, the continuum) as the basic conceptual tool for the fundamental explanation of phenomena.

The preoccupation with the continuum, with explaining mysterious orderliness, with finding pleasing simplicity—these were conceptions in physics which characterized Einstein's work from the beginning. But as with other highly creative persons, the power of his work was derived not merely from good physical ideas. Instead it came from a fusion of his scientific interests and his characteristics as a human being, a synthesis of his life-style and his perception of the laws of nature. Much of what was most daring or novel in his great work in physics was present in Einstein's ways of everyday thinking and behaving, even as a child or a young student. Take, for example, the ability to come back, again and again, for years, to a difficult puzzle. In early life, this trait showed up as what may have seemed mere obstinacy. He was commonly reported to have been withdrawn as a child, preferring to play by himself, erecting complicated constructions. Before he was 10, with infinite patience he was making fantastic card houses that had as many as 14 floors. He was unable or unwilling to talk until the age of 3. In school, he was not an exceptional student, preferring to follow his own thoughts. Later, he stuck to his ideas with the same persistence when the experiments of others seemed to disconfirm his theories (in time, those experiments usually turned out to be wrong). And during the last 30 years of his life, he persisted in his skepticism concerning the fundamental explanatory power of quantum mechanics and in his dedication to the problems of field theory, unlike most physicists of the time.

With this single-mindedness and concentration on his own revolutionary ideas went his deep suspicion of established authority, in science no less than in daily life. The same young student who quietly challenged the established ideas in physics also abhorred the current political, religious, and social conventions. From the age of 12 years on, he confessed later, he had "suspicion against every kind of authority." One result was that his teachers were quite delighted to see him drop out of high school at the age of 15½, when he no longer could stand the regimented and militaristic way of life in his native Germany and in his school. Moving to the freer, more democratic atmosphere of Switzerland, he found at last a school to his liking, and there had a glorious year before entering the Polytechnic Institute of Zurich. It was there, too, that he had his first ideas on relativity.

Related to this trait of uncompromisingly sticking to his own identity was his search for what is really *necessary*. In his early work in physics, he said, the thought of having to explain electric and magnetic field effects as "two fundamentally different cases was for me unbearable." The highest aim was nothing less than finding the most economical, simple principles, the barest bones of nature's frame, cleansed of everything that is ad hoc, redundant, unsymmetrical. "What really interests me," he once said, "is whether God has any choice in the creation of the world." Nature does not like anything that is unnecessary. Nor did Einstein, in his personal life—in his clothing, in his manner of speech and writing, in his behavior, from his preference for the classical music of Bach and Mozart down to his preference for using the same bar of soap for washing and shaving instead of complicating life unnecessarily by facing two kinds of soap every morning.

The preference for an egalitarian democracy characterized Einstein's political life just as it did his physics. The man who declared every inertial system to be created equal before all the laws of physics was also, from youth on, fiercely opposed to every antidemocratic and narrowly nationalistic political or social system.

Einstein at 53. (*Courtesy of the Archives, California Institute of Technology.*)

Einstein kept an abiding interest in philosophy, which penetrated his work; his ideas, in turn, influenced the development of modern philosophy it-self. Here, too, he was cutting across unnecessary barriers. This is not surprising. Genius discovers itself not in splendid solutions to little puzzles but in the struggle with deep and perhaps eternal problems at the point where science and philosophy join.*

---

* For further discussion on Einstein's relativity theory, its genesis and influence, see Gerald Holton, *Thematic Origins of Scientific Thought, Kepler to Einstein,* chaps. 5–10, Harvard University Press, Cambridge, Mass., 1973. Among more recent biographies, see Banesh Hoffmann, with the collaboration of Helen Dukas, *Albert Einstein: Creator and Rebel,* Viking, New York, 1972; Jeremy Bernstein, *Einstein,* Viking, New York, 1973; or the best of the older biographies: Philipp Frank, *Einstein: His Life and Times,* Knopf, New York, 1947. For a selection of Einstein's correspondence, see *Albert Einstein, the Human Side: New Glimpses from His Archives,* edited by Helen Dukas and Banesh Hoffman, Princeton University Press, Princeton, New Jersey, 1979.

---

## 35-3 Time Dilation and Length Contraction

Since the results of measurement of the time and space intervals between events do not depend on the kind of apparatus used for the measurements or on the events, we are free to choose any events and measuring apparatus which will help us understand the application of the Einstein postulates to the results of the measurement. Convenient events in relativity are those which produce light flashes. A convenient clock is a light clock, pictured schematically in Figure 35-6. A photocell detects the light pulse and sends a voltage pulse to an oscilloscope, giving a vertical deflection of the trace on the scope. The phosphorescent material on the face of the oscilloscope tube gives a persistent light that can be observed visually or photographed. The time between two light flashes is determined by measuring the distance between pulses on the scope and knowing the sweep speed of the scope. Such a clock, which can easily be calibrated and compared with other types of clocks, is often used in nuclear-physics experiments.

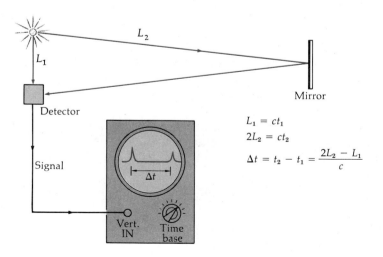

$$L_1 = ct_1$$
$$2L_2 = ct_2$$

$$\Delta t = t_2 - t_1 = \frac{2L_2 - L_1}{c}$$

**Figure 35-6**
Light clock for measuring time intervals. The time is measured by reading the distance between pulses on the oscilloscope after calibrating the sweep speed.

We first consider an observer $A'$ at rest in frame $S'$ a distance $D$ from a mirror, as shown in Figure 35-7*a*. He explodes a flash gun and measures the time interval $\Delta t'$ between the original flash and the return flash from the mirror. Since light travels with speed $c$, this time is

$$\Delta t' = \frac{2D}{c} \qquad 35\text{-}6$$

We now consider these same two events, the original flash of light and the returning flash, as observed in reference frame $S$, where observer $A'$ and the mirror are moving to the right with speed $v$. The events happen at two different places $x_1$ and $x_2$ in frame $S$ because between the original flash and the return flash observer $A'$ has moved a horizontal distance $v\,\Delta t$, where $\Delta t$ is the time interval between the events measured in $S$. In Figure 35-7*b* we see that the path traveled by the light is longer in $S$ than in $S'$. However, by Einstein's postulates, light travels with the same speed $c$ in frame $S$ as it does in frame $S'$.

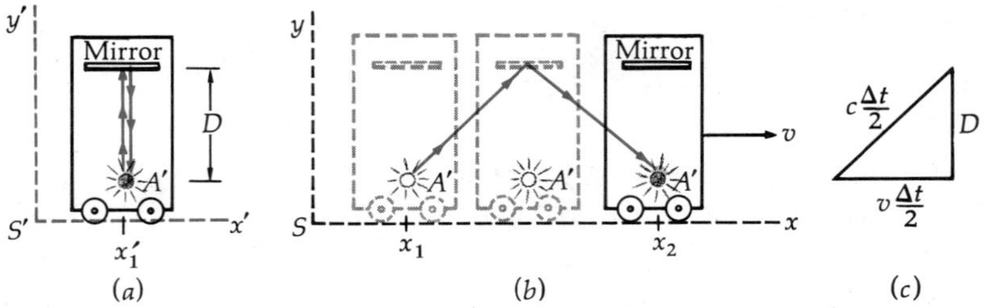


**Figure 35-7**
(*a*) $A'$ and the mirror are at rest in $S'$. The time for a light pulse to reach the mirror and return is measured to be $2D/c$ by $A'$. (*b*) In frame $S$, $A'$ and the mirror are moving with speed $v$. If the speed of light is the same in both reference frames, the time for the light to reach the mirror and return is longer than $2D/c$ in $S$ because the distance traveled is greater than $2D$. (*c*) Right triangle for computing the time $\Delta t$ in frame $S$.

Since it travels farther in $S$ at the same speed, it takes longer in $S$ to reach the mirror and return. The time interval in $S$ is thus longer than it is in $S'$. We can easily calculate $\Delta t$ in terms of $\Delta t'$. From the triangle in Figure 35-7*c* we have

$$\left(\frac{c\,\Delta t}{2}\right)^2 = D^2 + \left(\frac{v\,\Delta t}{2}\right)^2 \qquad 35\text{-}7$$

or

$$\Delta t = \frac{2D}{\sqrt{c^2 - v^2}} = \frac{2D}{c}\,\frac{1}{\sqrt{1 - v^2/c^2}}$$

Using $\Delta t' = 2D/c$, we have

$$\Delta t = \frac{\Delta t'}{\sqrt{1 - v^2/c^2}} = \gamma\,\Delta t' \quad \text{where} \quad \gamma = \frac{1}{\sqrt{1 - v^2/c^2}} \geq 1 \qquad 35\text{-}8$$

*Time dilation*

Observers in $S$ would say that the clock held by $A'$ runs slow since he claims a shorter time interval for these events.

$A'$ measures the times of the light flash and return at the same point in $S'$, while in $S$ these events happen at two different places. A single clock can be used in $S'$ to measure the time interval, but in $S$, two synchronized clocks are needed, one at $x_1$ and one at $x_2$. The time between events that happen at the *same place* in a reference frame (as with $A'$ in $S'$ in this case) is called *proper time*. The time interval measured in any other reference frame is always longer than the proper time. This expansion is called *time dilation*.

Time dilation is closely related to another phenomenon, *length contraction*. The length of an object measured in the reference frame in which the object is at rest is called its *proper length*. In a reference frame in which the object is moving, the measured length is shorter than its proper length. We can see this from our previous example using the light clocks. Suppose that $x_1$ and $x_2$ in that example are at the ends of a measuring rod of length $L_0 = x_2 - x_1$ measured in frame $S$, in which the rod is at rest. Since $A'$ is moving relative to this frame with speed $v$, the distance moved in time $\Delta t$ is $v \, \Delta t$. Since he moves from point $x_1$ to point $x_2$ in this time, this distance is $L_0 = x_2 - x_1 = v \, \Delta t$. According to observer $A'$ in frame $S'$ (Figure 35-8), the measuring rod moves with speed $v$ and takes a time $\Delta t'$ to move past him. The length of the rod in his frame is $L' = v \, \Delta t'$. Since the time interval $\Delta t'$ is less than $\Delta t$, the length $L'$ is less than $L_0$. These lengths are related by

$$L' = v \, \Delta t' = \frac{v \, \Delta t}{\gamma} = \frac{L_0}{\gamma} = \sqrt{1 - \frac{v^2}{c^2}} \, L_0 \qquad \text{35-9}$$

*Length contraction*

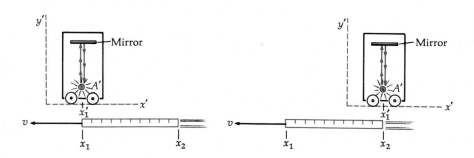

(Both $A$ and $A'$ measure the same relative velocity. Otherwise there would be a lack of symmetry, and postulate 1 would be violated; i.e., we could choose the frame with the smaller or greater relative velocity to be a preferred frame.) Thus the length of the rod is smaller when measured in a frame in which it is moving. Since this contraction is just the amount proposed by Lorentz and FitzGerald to explain the Michelson-Morley experiment, it is often called the *Lorentz-FitzGerald contraction*.

An interesting example of the observation of these phenomena is afforded by the appearance of muons as secondary radiation from cosmic rays. Muons decay according to the statistical law of radioactivity:

$$N(t) = N_0 e^{-t/T} \qquad \text{35-10}$$

where $N_0$ is the number at time $t = 0$, $N(t)$ is the number at time $t$, and $T$ is the mean lifetime, which is about 2 $\mu$s for muons at rest. Since they are created (from the decay of $\pi$ mesons) high in the atmosphere, usually several thousand metres above sea level, few muons should reach sea level. A typical muon moving with speed $0.998c$ would travel only about 600 m in 2 $\mu$s. However, the lifetime of the muon measured in the earth's reference frame is increased by the factor $1/\sqrt{1 - v^2/c^2}$, which is 15 for this particular speed. The mean lifetime measured in the earth's reference frame is therefore 30 $\mu$s, and a muon of this speed travels about 9000 m in this time. From the muon's point of view, it lives only 2 $\mu$s, but the atmosphere is rushing past it with a speed of $0.998c$. The distance of 9000 m in the earth's frame is thus contracted to only 500 m, as indicated in Figure 35-9.

It is easy to distinguish experimentally between the classical and relativistic predictions of the observation of muons at sea level. Suppose that we observe $10^8$ muons at an altitude of 9000 m in some time interval with a muon detector. How many would we expect to observe at sea level in the same time interval? According to the nonrelativistic pre-

**Figure 35-8**
Measuring the length of a moving object. The rod has length $L_0 = x_2 - x_1 = v \, \Delta t$ as measured in frame $S$, in which it is at rest. In frame $S'$ it moves past a fixed point in time $\Delta t' = \sqrt{1 - v^2/c^2} \, \Delta t$, so its length $L' = v \, \Delta t'$ is shorter in this frame.

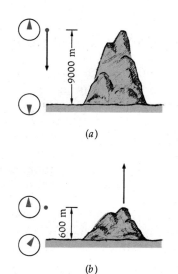

**Figure 35-9**
Although muons are created high above the earth and their mean lifetime is only about 2 $\mu$s when at rest, many appear at the earth's surface. (a) In the earth's reference frame a typical muon moving at $0.998c$ has a mean lifetime of 30 $\mu$s and travels 9000 m in this time. (b) In the reference frame of the muon, the distance traveled by the earth is only 600 m in the muon's lifetime of 2 $\mu$s.

diction, the time taken for these muons to travel 9000 m is $(9000 \text{ m})/0.998c \approx 30 \ \mu s$, which is 15 lifetimes. Putting $N_0 = 10^8$ and $t = 15T$ into Equation 35-10, we obtain

$$N = 10^8 e^{-15} = 30.6$$

We would thus expect all but about 31 of the original 100 million muons to decay before reaching sea level.

According to the relativistic prediction, the earth must travel only the contracted distance of 600 m in the rest frame of the muon. This takes only 2 $\mu s = T$. Thus the number expected at sea level is

$$N = 10^8 e^{-1} = 3.68 \times 10^7$$

Thus relativity predicts that we would observe 36.8 million muons in the same time interval. Experiments of this type have confirmed the relativistic predictions.

### Question

1.  You are standing on a corner and a friend is driving past in an automobile. Both of you note the times when the car passes two different intersections. You each determine from your watch readings the time that elapses between the two events. Which of you has determined the proper time interval?

## 35-4  Clock Synchronization and Simultaneity

At first glance, time dilation and length contraction seem contradictory not only to our intuition but to our ideas of self-consistency. If each reference frame can be considered at rest with the other moving, the clocks in the "other" frame should run slow. How can there be any self-consistency if each observer sees the clocks of the other run slow? The answer to this puzzle lies in the problems of clock synchronization and in the concept of simultaneity. We note that the time intervals $\Delta t$ and $\Delta t'$ considered in Section 35-3 were measured in quite different ways. The events in frame $S'$ happened at the same place, and the times could be measured on a single clock; but in $S$, the two events happened at different places. The time of each event was measured on a different clock and the interval found by subtraction. This procedure requires that the clocks be synchronized. We shall show in this section that:

*Two clocks synchronized in one reference frame are not synchronized in any other frame moving relative to the first frame.*

A corollary to this result is that:

*Two events that are simultaneous in one reference frame are not simultaneous in another frame moving relative to the first.*

(This is true unless the events and clocks are in the same plane perpendicular to the relative motion).

Comprehension of these facts usually resolves all relativity paradoxes. Unfortunately, the intuitive (and incorrect) belief that simultaneity is an absolute relation is difficult to correct.

How can we synchronize two clocks separated in space? The problem is not difficult but requires some thought. One obvious method is to bring the clocks together and set them to read the same time, then move them to their original positions. This method has the drawback that, as we have seen, each clock runs slowly during the time it is moved, according to the time-dilation result. Suppose the clocks are at rest at points $A$ and $B$ a distance $L$ apart in frame $S$. If an observer at $A$ looks at the clock at $B$ and sets his clock to read the same time, the clocks will not be synchronized because of the time $L/c$ it takes light to travel from one clock to another. To synchronize the clocks, the observer at $A$ must set his clock ahead by the time $L/c$. Then he will see that the clock at $B$ reads a time which is $L/c$ behind the time on his clock, but he will calculate that the clocks are synchronized when he allows for the time $L/c$ for the light to reach him. All observers except those midway between the clocks will see the clocks reading different times, but they will also compute that the clocks are synchronized when they correct for the time it takes the light to reach them. An equivalent method for the synchronization of two clocks would be for a third observer $C$ at a point midway between the clocks to send a light signal and for observers at $A$ and $B$ to set their clocks to some prearranged time when they receive the signal.

We now examine the question of simultaneity. Suppose $A$ and $B$ agree to explode bombs at $t_0$ (having previously synchronized their clocks). Observer $C$ will see the light from the two explosions at the same time, and since he is equidistant from $A$ and $B$, he will conclude that the explosions were simultaneous. Other observers in $S$ will see the light from $A$ or $B$ first, depending on their location, but after correcting for the time the light takes to reach them, they also will conclude that the explosions were simultaneous. *We shall thus define two events in a reference frame to be simultaneous if the light signals from the events reach an observer halfway between the events at the same time.* To show that two events which are simultaneous in frame $S$ are not simultaneous in another frame $S'$ moving relative to $S$ we use an example introduced by Einstein. A train is moving with speed $v$ past the station platform. We have observers $A'$, $B'$, and $C'$ at the front, back, and middle of the train. (We shall consider the train to be at rest in $S'$ and the platform in $S$.) We now suppose that the train and platform are struck by lightning at the front and back of the train and that the lightning bolts are simultaneous in the frame of the platform ($S$) (Figure 35-10). That is, an observer $C$ halfway between the positions $A$ and $B$, where the lightning strikes, observes the two flashes at the same time. It is convenient to suppose that the lightning scorches the train and platform so that the events can be easily located in each reference frame. Since $C'$ is in the middle of the train, halfway between the places on the train which are scorched, the events can be simultaneous in $S'$ only if $C'$ sees the flashes at the same time. However, $C'$ sees the flash from the front of the train before he

*Simultaneity defined*

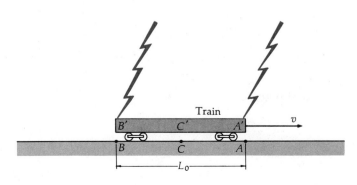

Train

$v$

**Figure 35-10**
Simultaneous lightning bolts strike the ends of a train traveling with speed $v$ in frame $S$ attached to the platform. The light from these simultaneous events reaches observer $C$ midway between the events at the same time. The distance between the bolts is $L_0$, which is also the length of the train measured in frame $S$.

sees the flash from the back. In frame $S$, when the light from the front flash reaches him, he has moved some distance toward it, so that the flash from the back has not yet reached him, as indicated in Figure 35-11. He must therefore conclude that the events are not simultaneous. The front of the train was struck before the back. As we have discussed above, all observers in $S'$ on the train will agree with $C'$ when they have corrected for the time it takes light to reach them.

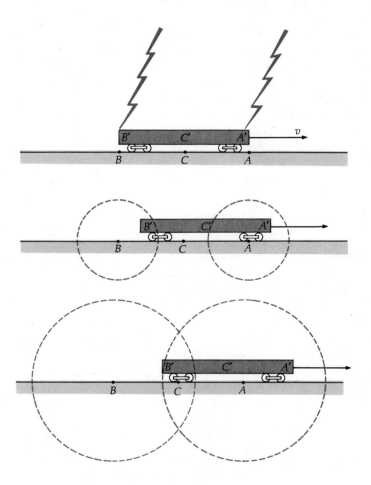

**Figure 35-11**
In frame $S$ the light from the bolt at the front of the train reaches observer $C'$ in the middle of the train before light from the bolt at the rear of the train because the train is moving. Since $C'$ is midway between the events (which occur at the front and rear of the train), these events are not simultaneous for him.

Let $L_0$ be the distance between the scorch marks on the platform, which is also the length of the train measured in frame $S$. This distance is smaller than the proper length of the train $L_T'$ because of length contraction. Figure 35-12 shows the situation in frame $S'$, in which the train is at rest and the platform is moving. In this frame the distance between the burns on the platform is contracted and is related to $L_0$ by

$$L_P' = \frac{L_0}{\gamma} \qquad\qquad 35\text{-}11$$

Similarly the train length is the proper length related to $L_0$ by

$$L_T' = \gamma L_0 \qquad\qquad 35\text{-}12$$

(The drawing of this figure has been made for $\gamma = 1.5$.) The time interval in $S'$ between these two events is the time it takes the platform to move the distance $\Delta L$, where

$$\Delta L = L_T' - L_P' = \gamma L_0 - \frac{L_0}{\gamma} = \left(1 - \frac{1}{\gamma^2}\right) \gamma L_0 = \frac{v^2}{c^2} \gamma L_0$$

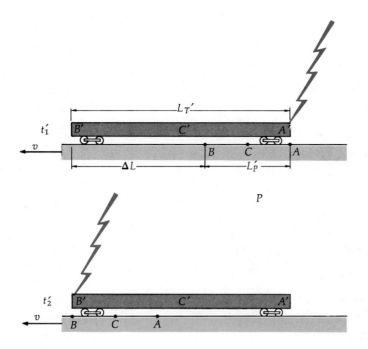

**Figure 35-12**
Lightning bolts of Figures
35-10 and 35-11 as seen in
reference frame $S'$ fixed to the
train. In this frame the distance $L'_P = BA$ along the
moving platform is con-
tracted, and the train length
$L'_T = B'A'$ is its proper
length. At time $t'_1$ lightning
strikes the front of the train
when $A'$ and $A$ are coincident. At a later time $t'_2$ the
lightning strikes the rear of
the train when $B'$ and $B$ are
coincident. During the time
between flashes the platform
moves the distance $L'_T - L'_P$.

since

$$1 - \frac{1}{\gamma^2} = 1 - \left(1 - \frac{v^2}{c^2}\right) = \frac{v^2}{c^2}$$

Thus

$$t'_2 - t'_1 = \frac{\Delta L}{v} = \frac{\gamma L_0 v}{c^2}$$

During this time (according to observers in $S'$), the clocks at $A$ and $B$
ticked off a time interval which is smaller because of time dilation

$$\Delta t_S = \frac{1}{\gamma}\,(t'_2 - t'_1) = \frac{L_0 v}{c^2} \qquad\qquad 35\text{-}13$$

This is the time interval by which the clocks in $S$ are unsynchronized
according to observers in $S'$. That is, the clock at $A$ is ahead of that at $B$
by the amount $L_0 v/c^2$. This result is worth remembering.

*If two clocks are synchronized in the frame in which they are at rest, they
will be out of synchronization in another frame. In the frame in which
they are moving, the "chasing clock" leads by an amount $\Delta t_S = L_0 v/c^2$,
where $L_0$ is the proper distance between the clocks.*

*Lack of synchronization of
moving clocks*

---

**Example 35-1** A numerical example should help clarify time dilation,
clock synchronization, and the internal consistency of these results. Let
the light clock used in Section 35-3 be moving with speed $v = 0.8c$.
Then

$$1 - \frac{v^2}{c^2} = 1 - 0.64 = 0.36$$

and

$$\gamma = \frac{1}{\sqrt{1 - v^2/c^2}} = \frac{1}{\sqrt{0.36}} = \frac{1}{0.6} = \frac{5}{3}$$

Also, let the distance $x_2 - x_1$ be 40 light-minutes, that is, the distance light travels in 40 min (a convenient notation for this unit is $c \cdot min$):

$$x_2 - x_1 = 40 \ c \cdot min$$

The time for the light clock to travel this distance is $\Delta t$ given by

$$\Delta t = \frac{x_2 - x_1}{v} = \frac{40 \ c \cdot min}{0.8c} = 50 \ min$$

(Note that the $c$ in the unit $c \cdot min$ cancels, giving the time in minutes.)
The proper time interval in $S'$ is shorter by the factor $\gamma$:

$$\Delta t' = \frac{\Delta t}{\gamma} = \frac{50 \ min}{5/3} = 30 \ min$$

(Thus we have taken $D = 15 \ c \cdot min$ for the distance from the flash gun to the mirror in Figure 35-7.) The distance traveled is

$$\Delta x' = v \ \Delta t' = (0.8c) \ (30 \ min) = 24 \ c \cdot min$$

This is just the distance $x_2 - x_1$ that is contracted in $S'$:

$$L' = \frac{L_0}{\gamma} = \frac{40 \ c \cdot min}{5/3} = 24 \ c \cdot min$$

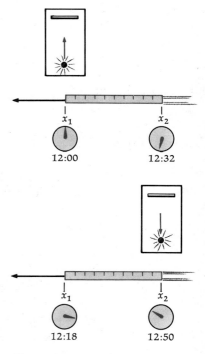

Figure 35-13 shows the situation viewed in $S'$. We assume that the clock at $x_1$ reads noon at the time of the light flash. The clocks at $x_1$ and $x_2$ are synchronized in $S$ but not in $S'$. In $S'$, the clock at $x_2$, which is chasing the one at $x_1$, leads by

$$\frac{L_0 v}{c^2} = \frac{(40 \ c \cdot min) \ (0.8c)}{c^2} = 32 \ min$$

When the light clock coincides with $x_2$, the clock there reads 50 min past noon. The time between events is therefore 50 min in $S$. Note that according to observers in $S'$, this clock ticks off $50 - 32 = 18$ min for a trip which takes 30 min in $S'$. Thus this clock runs slow by the factor $30/18 = \frac{5}{3}$.

Each observer sees clocks in the other frame run slow. According to observers in $S$ who measure 50 min for the time interval, the time interval in $S'$ is too small (30 min) because the single clock in $S'$ runs too slow by the factor $\frac{5}{3}$. According to the observers in $S'$, the observers in $S$ measure a time which is too *long* despite the fact that their clocks run too slow, because they are out of synchronization. The clocks tick off only 18 min, but the second one leads the first by 32 min, so the time interval is 50 min.

**Figure 35-13**
Example 35-1. In frame $S'$ the rod passes the light clock in a time $\Delta t' = 30$ min. During this time the clocks at each end of the rod tick off $30/\gamma$ min $= 18$ min. But the clocks are unsynchronized, with the chasing clock leading by $L_0 v/c^2 = 32$ min, where $L_0$ is the proper separation of the clocks. The time interval in $S$ is therefore 32 min + 18 min $= 50$ min, which is longer than 30 min.

**Questions**

2. Two observers are in relative motion. In what circumstances can they agree on the simultaneity of two different events?

3. If event $A$ occurs before event $B$ in some frame, might it be possible for there to be a reference frame in which event $B$ occurs before event $A$?

4. Two events are simultaneous in a frame in which they also occur at the same point in space. Are they simultaneous in other reference frames?

## 35-5   The Doppler Effect

In deriving the doppler effect for sound (Chapter 14) we found that the change in frequency for a given velocity $v$ depends on whether it is the source or receiver that is moving with that speed. Such a distinction is possible for sound because there is a medium (the air) relative to which the motion takes place, and so it is not surprising that the motion of the source or the receiver relative to the still air can be distinguished. Such a distinction between motion of the source or receiver cannot be made for light or other electromagnetic waves in vacuum. The expressions we have derived for the doppler effect cannot be correct for light. We now derive the relativistic doppler-effect equations.

We shall consider a source moving toward a receiver with velocity $v$, and work in the frame of the receiver. Let the source emit $N$ waves. If the source is moving toward the receiver, the first wave will travel a distance $c \, \Delta t_R$ and the source will travel $v \, \Delta t_R$ in the time $\Delta t_R$ measured in the frame of the receiver. The wavelength will be $\lambda' = (c \, \Delta t_R - v \, \Delta t_R)/N$. The frequency $f'$ observed by the receiver will therefore be

$$f' = \frac{c}{\lambda'} = \frac{c}{c - v} \frac{N}{\Delta t_R} = \frac{1}{1 - v/c} \frac{N}{\Delta t_R}$$

If the frequency of the source is $f_0$, it will emit $N = f_0 \, \Delta t_S$ waves in time $\Delta t_S$, measured by the source. Here $\Delta t_S$ is the proper time interval (the first wave and the $N$th wave are emitted at the same place in the source's reference frame). Times $\Delta t_S$ and $\Delta t_R$ are related by the usual time-dilation equation $\Delta t_S = \Delta t_R/\gamma$. Thus we obtain for the doppler effect for a moving source

$$f' = \frac{1}{1 - v/c} \frac{f_0 \, \Delta t_S}{\Delta t_R} = \frac{f_0}{1 - v/c} \frac{1}{\gamma} = \frac{\sqrt{1 - v^2/c^2}}{1 - v/c} f_0 \qquad 35\text{-}14a$$

which differs from our classical equation only in the time-dilation factor.

We now make the calculation in the reference frame of the source. That is, we assume that the source is at rest and the receiver moves with velocity $v$ toward the source. In time $\Delta t_S$ in the frame of the source, the receiver encounters all the waves in the distance $v \, \Delta t_S$ in addition to the waves in the distance $c \, \Delta t_S$, just as in the classical calculation. The number of waves encountered is thus

$$N = \frac{c \, \Delta t_S + v \, \Delta t_S}{\lambda} = \frac{(c + v) \, \Delta t_S}{c/f_0} = \left(1 + \frac{v}{c}\right) f_0 \, \Delta t_S$$

where we have used $\lambda = c/f_0$ for the wavelength. In the classical calculation, we need only divide $N$ by the time interval $\Delta t$ to obtain the frequency observed by the moving receiver, but here we must be careful to divide by the time interval in the receiver's frame to find the observed frequency. In this case the receiver's time interval $\Delta t_R$ is proper time, i.e., the time interval between encountering the first wave and the $N$th wave. Both events occur at the receiver. Thus $\Delta t_R = \Delta t_S/\gamma$, and the frequency observed is

$$f' = \frac{N}{\Delta t_R} = \left(1 + \frac{v}{c}\right) f_0 \frac{\Delta t_S}{\Delta t_R}$$

$$= \gamma \left(1 + \frac{v}{c}\right) f_0 = \frac{1 + v/c}{\sqrt{1 - v^2/c^2}} f_0 \qquad 35\text{-}14b$$

It is left to the student to show that this result is identical to Equation 35-14a.

When the source and receiver move away from each other with relative speed $v$, the frequency received is given by

$$f' = \frac{1 - v/c}{\sqrt{1 - v^2/c^2}} f_0 = \frac{\sqrt{1 - v^2/c^2}}{1 + v/c} f_0 \qquad\qquad 35\text{-}15$$

**Question**

5. How do the relativistic equations for the doppler effect for sound waves differ from those for light waves?

## 35-6   The Lorentz Transformation

We now consider the general relation between the coordinates $x$, $y$, $z$, and $t$ of an event as seen in reference frame $S$ and the coordinates $x'$, $y'$, $z'$, and $t'$ of the same event as seen in reference frame $S'$, which is moving with uniform velocity relative to $S$. We shall consider only the simple special case in which the origins are coincident at time $t = t' = 0$ and $S'$ is moving with speed $v$ along the $x$ (or $x'$) axis. The classical relation, called the *galilean transformation*, is

$$x = x' + vt' \qquad y = y' \qquad z = z' \qquad t = t' \qquad\qquad 35\text{-}16$$

with the inverse transformation

$$x' = x - vt \qquad y' = y \qquad z' = z \qquad t' = t$$

(For the rest of this discussion we shall ignore the equations for $y$ and $z$, which do not change for this special case of motion along the $x$ and $x'$ axes.) These equations are consistent with experiment as long as $v$ is much less than $c$. They lead to the familiar classical addition law for velocities. If a particle has velocity $u_x = dx/dt$ in frame $S$, its velocity in frame $S'$ is

$$u'_x = \frac{dx'}{dt'} = \frac{dx}{dt} - v = u_x - v \qquad\qquad 35\text{-}17$$

If we differentiate this equation again, we find that the acceleration of the particle is the same in both frames: $a_x = du_x/dt = du'_x/dt = a'_x$.

It should be clear that this transformation is not consistent with Einstein's postulates of special relativity. If light moves along the $x$ axis with speed $c$ in $S$, these equations imply that the speed in $S'$ is $u'_x = c - v$ rather than $u'_x = c$, which is consistent with Einstein's postulates and with experiment. The classical transformation equations must therefore be modified to be consistent with Einstein's postulates and to reduce to the classical equations when $v$ is much less than $c$. We shall give a brief outline of one method of obtaining the relativistic transformation, the *Lorentz transformation*. We assume that the equation for $x$ is of the form

$$x = K(x' + vt')$$

where $K$ is a constant which can depend on $v$ and $c$ but not on the coordinates. For this equation to reduce to the classical one, $K$ must approach 1 as $v/c$ approaches zero. The inverse transformation must look the same except for the sign of the velocity:

$$x' = K(x - vt)$$

Let us assume that a light pulse starts at the origin at $t = 0$. Since we have assumed that the origins are coincident at $t = t' = 0$, the pulse also starts out at the origin of $S'$ at $t' = 0$. The equation for the wavefront of the light pulse is $x = ct$ in frame $S$ and $x' = ct'$ in $S'$. Substituting these in our transformation equations gives

$$ct = K(ct' + vt') = K(c + v)t'$$

and

$$ct' = K(ct - vt) = K(c - v)t$$

We can eliminate either $t'$ or $t$ from these two equations and determine $K$. The result is $K^2 = (1 - v^2/c^2)^{-1}$, and $K$ is the same as $\gamma$:

$$K = \gamma = \frac{1}{\sqrt{1 - v^2/c^2}}$$

The transformation is therefore $x = \gamma(x' + vt')$. We can obtain equations for $t$ and $t'$ by combining this transformation with its inverse:

$$x' = x - vt$$

Using this value for $x'$ in the above equation, we get

$$x = \gamma[\gamma(x - vt) + vt']$$

which can be solved for $t'$ in terms of $x$ and $t$. The complete Lorentz transformation is

---

$$x = \gamma(x' + vt') \qquad y = y'$$

$$t = \gamma\left(t' + \frac{vx'}{c^2}\right) \qquad z = z'$$

35-18    *Lorentz transformation*

---

with the inverse

---

$$x' = \gamma(x - vt) \qquad y' = y$$

$$t' = \gamma\left(t - \frac{vx}{c^2}\right) \qquad z' = z$$

35-19

---

**Example 35-2** Two events occur at the same place $x_0$ at times $t_1$ and $t_2$ in $S$. What is the time interval between them in $S'$?

$$t_1' = \gamma\left(t_1 - \frac{vx_0}{c^2}\right) \qquad t_2' = \gamma\left(t_2 - \frac{vx_0}{c^2}\right)$$

$$t_2' - t_1' = \gamma(t_2 - t_1)$$

This is our familiar result for time dilation since $t_2 - t_1$ is the proper time interval.

---

We can obtain the velocity transformation by differentiating the Lorentz transformation equations or merely by taking differences. We shall do the latter. Suppose a particle moves a distance $\Delta x$ in time $\Delta t$ in frame $S$. Its velocity is $u_x = \Delta x/\Delta t$. Using the transformation equations, we obtain

$$\Delta x' = \gamma(\Delta x - v\,\Delta t) \qquad \text{and} \qquad \Delta t' = \gamma\left(\Delta t - \frac{v\,\Delta x}{c^2}\right)$$

The velocity in $S'$ is

$$u'_x = \frac{\Delta x'}{\Delta t'} = \frac{\gamma(\Delta x - v\,\Delta t)}{\gamma\left(\Delta t - \dfrac{v\,\Delta x}{c^2}\right)} = \frac{\dfrac{\Delta x}{\Delta t} - v}{1 - \dfrac{v}{c^2}\dfrac{\Delta x}{\Delta t}}$$

or

$$u'_x = \frac{u_x - v}{1 - vu_x/c^2} \qquad\qquad \text{35-20} \qquad \textit{Velocity addition}$$

Similarly the inverse velocity transformation is

$$u_x = \frac{u'_x + v}{1 + vu'_x/c^2} \qquad\qquad \text{35-21}$$

If a particle has components of velocity along $y$ and $z$, it is not difficult to find the components in $S'$. In these cases, $\Delta y' = \Delta y$ and $\Delta z' = \Delta z$, with the same relation between $\Delta t'$ and $\Delta t$. Thus we obtain

$$u'_y = \frac{\Delta y'}{\Delta t'} = \frac{\Delta y}{\gamma\left(\Delta t - \dfrac{v\,\Delta x}{c^2}\right)} = \frac{\Delta y/\Delta t}{\gamma\left(1 - \dfrac{v}{c^2}\dfrac{\Delta x}{\Delta t}\right)}$$

or

$$u'_y = \frac{u_y}{\gamma(1 - vu_x/c^2)} \qquad\qquad \text{35-22}$$

with a similar result for $u'_z$. The inverse is

$$u_y = \frac{u'_y}{\gamma(1 + vu'_x/c^2)} \qquad\qquad \text{35-23}$$

---

**Example 35-3** Light moves along the $x$ axis with speed $u_x = c$. What is its speed in $S'$?

$$u'_x = \frac{c - v}{1 - vc/c^2} = \frac{c(1 - v/c)}{1 - v/c} = c$$

as required by the postulates.

---

**Question**

6. The Lorentz transformation for $y$ and $z$ is the same as the classical result: $y = y'$ and $z = z'$. Yet the relativistic velocity transformation does not give the classical result $u_y = u'_y$ and $u_z = u'_z$. Explain.

# 35-7  The Twin Paradox

Homer and Ulysses are identical twins. Ulysses travels at high speed to a planet beyond the solar system and returns while Homer remains at home. When they are together again, which twin is older, or are they the same age? The correct answer is that Homer, the twin who stays at home, is older. This problem, with variations, has been the subject of spirited debate for decades, though there are very few who disagree with the answer.*

The problem is a paradox because of the seemingly symmetric roles played by the twins with the asymmetric result in their aging. The paradox is resolved when the asymmetry of the twins' roles is noted. The relativistic result conflicts with common sense based on our strong but incorrect belief in absolute simultaneity. We shall consider a particular case with some numerical magnitudes which, though impractical, make the calculations easy.

Let planet $P$ and Homer on earth be fixed in reference frame $S$ a distance $L_0$ apart, as in Figure 35-14. We neglect the motion of the earth. Reference frames $S'$ and $S''$ are moving with speed $v$ toward and away from the planet, respectively. Ulysses quickly accelerates to speed $v$, then coasts in $S'$ until he reaches the planet, when he stops and is momentarily at rest in $S$. To return he quickly accelerates to speed $v$ toward earth and coasts in $S''$ until he reaches earth, where he stops. We can assume that the acceleration times are negligible compared with the coasting times. (Given the time needed to reach the speed $v$, we can formulate the problem with $L_0$ large enough to meet this condition.) We use the following values for illustration: $L_0 = 8$ light-years and $v = 0.8c$; then $\sqrt{1 - v^2/c^2} = \frac{3}{5}$.

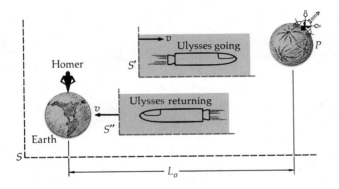

**Figure 35-14**
Twin paradox. The earth and a distant planet are fixed in frame $S$. Ulysses coasts in frame $S'$ to the planet and then coasts in $S''$ back to earth. His twin Homer stays on earth. When Ulysses returns, he is younger than his twin. The roles played by the twins are not symmetric. Homer remains in one inertial reference frame, but Ulysses must accelerate if he is to return home.

It is easy to analyze the problem from Homer's point of view. According to his clock, Ulysses coasts in $S'$ for a time $L_0/v = 10$ y and in $S''$ for an equal time. Thus Homer is 20 y older when Ulysses returns. The time interval in $S'$ between leaving earth and arriving at the planet is shorter because it is proper time; the time to reach the planet by Ulysses' clock is

$$\Delta t' = \sqrt{1 - \frac{v^2}{c^2}} \frac{L_0}{v} = \tfrac{3}{5}(10) = 6 \text{ y}$$

Since the same time is required for the return trip, Ulysses will have recorded 12 y for the round trip and will be 8 y younger than Homer.

From Ulysses' point of view, the calculation of his trip time is not difficult. The distance from earth to planet is contracted and is only

$$L_0 \sqrt{1 - \frac{v^2}{c^2}} = \tfrac{3}{5}(8) = 4.8 \text{ light-years}$$

---

* A collection of some important papers concerning this paradox can be found in *Special Relativity Theory, Selected Reprints,* American Association of Physics Teachers, New York, 1963.

At $v = 0.8c$, it takes only 6 y each way. The real difficulty in this problem is for Ulysses to understand why his twin ages 20 y during his absence. If we consider Ulysses at rest and Homer moving away, doesn't Homer's clock measure proper time? When Ulysses measures 6 y, Homer should measure only $\frac{3}{5}(6) = 3.6$ y. Then why shouldn't Homer age only 7.2 y during the round trip? This, of course, is the paradox. The difficulty with the analysis from the point of view of Ulysses is that he does not remain in an inertial frame. What happens while Ulysses is stopping and starting? To investigate this problem in detail, we would need to treat accelerated reference frames, a subject dealt with in the study of general relativity but beyond the scope of this book. However, we can get some insight into the problem by considering the lack of synchronization of moving clocks.

Suppose that there is a clock at the planet $P$ synchronized in $S$ with Homer's clock at earth. In reference frame $S'$ these clocks are unsynchronized by the amount $L_0 v/c^2$. For our example this is 6.4 y. Thus when Ulysses is coasting in $S'$ near the planet, the clock at the planet leads that at the earth by 6.4 y. After he stops, he is in the frame $S$, in which these two clocks are synchronized. Thus in the negligible time (according to Ulysses) it takes him to stop, the clock at earth must gain 6.4 y. Accordingly, his twin on earth ages 6.4 y. This 6.4 y plus the 3.6 y that Homer aged during the coasting makes him 10 y older by the time Ulysses is stopped in frame $S$. When Ulysses is in frame $S''$ coasting home, the clock at earth leads that at the planet by 6.4 y, and it will run another 3.6 y before he arrives home. We do not need to know the detailed behavior of the clocks during the acceleration in order to know the cumulative effect; the special relativity theory is enough to show us that if the clocks on earth and at the planet are synchronized in $S$, the clock on earth lags that at $P$ by $L_0/c^2 = 6.4$ y when viewed in $S'$, and the clock on earth leads that at $P$ by this amount when viewed in $S''$.

The difficulty in understanding the analysis of Ulysses lies in the difficulty of giving up the idea of absolute simultaneity. Suppose Ulysses sends a signal calculated to arrive at earth just as he arrives at $P$. If the arrival of this signal and the arrival of Ulysses at the planet are simultaneous in $S'$, they are not simultaneous in $S$; in fact, the signal arrives at earth 6.4 y before Ulysses arrives at the planet, according to observers in $S$. The roles of the twins are not symmetric because Ulysses does not remain in an inertial reference frame but must accelerate.

It is instructive to have the twins send regular signals to each other so that they can record the other's age continuously. If they arrange to send a signal once a year, the age of the other can be determined merely by counting the signals received. The arrival frequency of the signals will not be 1 per year because of the doppler shift. The frequency observed will be given by Equation 35-14 or 35-15, which for our example is 3 per year or $\frac{1}{3}$ per year, depending on whether the source is approaching or receding.

Consider the situation first from the point of view of Ulysses. During the 6 y it takes him to reach the planet (remember that the distance is contracted in his frame), he receives signals at the rate of $\frac{1}{3}$ per year, and so he receives 2 signals. As soon as he turns around and starts back to earth, he receives 3 signals per year; in the 6 years it takes him to return he receives 18 signals, giving a total of 20 for the trip. He accordingly expects his twin to have aged 20 years.

We now consider the situation from Homer's point of view. He receives signals at the rate of $\frac{1}{3}$ per year not only for the 10 years it takes Ulysses to reach the planet but also for the 8 years it takes for the last signal sent by Ulysses from $S'$ to get back to earth. (He cannot know that Ulysses has turned around until the signals reach him.) During the first 18 y Homer receives 6 signals. In the final 2 y before Ulysses arrives, Homer receives 6 signals, or 3 per year. (The first signal sent after Ulysses turns around takes 8 y to reach earth, whereas Ulysses, traveling at 0.8$c$, takes 10 y to return and therefore arrives just 2 y after Homer begins to receive signals at the faster rate.) Thus Homer expects Ulysses to have aged 12 y. In this analysis, the asymmetry of the twins' roles is apparent. Both twins agree that when they are together again, the one who has been accelerated will be younger than the one who stayed home.

The predictions of the special theory of relativity concerning the twin paradox have been tested many times using small particles which can be accelerated to such large speeds that $\gamma$ is appreciably greater than 1. Unstable particles can be accelerated and trapped in circular orbits in a magnetic field, for example, and their lifetimes compared with those of identical particles at rest. In all such experiments the accelerated particles live longer on the average than those at rest, as predicted. These predictions are also confirmed by the results of an experiment using high-precision atomic clocks flown around the world in commercial airplanes, but the analysis of this experiment is complicated by the necessity of including gravitational effects treated in the general theory of relativity.*

## 35-8   Relativistic Momentum

Our discussion of Newton's second law in Chapter 4 showed that if $\Sigma\mathbf{F} = m\mathbf{a}$ holds in one reference frame, it holds in any other reference frame moving with constant velocity relative to the first. According to the galilean transformation, the accelerations in the two frames are equal, $a'_x = a_x$, and forces, e.g., those due to stretching of springs, are also the same in the two frames. However, according to the Lorentz transformation, accelerations are not the same in two such reference frames. If a particle has acceleration $a'_x$ and velocity $u'_x$ in frame $S'$, its acceleration in $S$ obtained by computing $du_x/dt$ from Equation 35-21 is

$$a_x = \frac{a'_x}{\gamma^3(1 + u'_x v/c^2)^3}$$

Thus either the force transforms in a similar way or $\Sigma\mathbf{F} = m\mathbf{a}$ does not hold. It is reasonable to expect that $\Sigma\mathbf{F} = m\mathbf{a}$ does not hold at high speeds, for this equation implies that a constant force will accelerate a particle to unlimited velocity if it acts for a long time. However, if a particle's velocity were greater than $c$ in some reference frame $S'$, we could not transform from $S'$ to the rest frame of the particle because $\gamma$ becomes imaginary when $v > c$. We can see from the velocity transformation that if a particle's velocity is less than $c$ in some frame $S$, it is less than $c$ in all frames moving relative to $S$ with $v < c$. Thus it seems reasonable to conclude that particles never have speeds greater than $c$.

We can avoid the question of how to transform forces by considering a problem in which the total force is zero, namely, a collision of two masses. In classical mechanics, the total momentum $\Sigma m_i \mathbf{u}_i$ is conserved.

---

* The details of these tests can be found in J. C. Hafele and Richard E. Keating, "Around-the-World Atomic Clocks: Predicted Relativistic Time Gains" and "Around-the-World Atomic Clocks: Observed Relativistic Time Gains," *Science*, July 14, 1972, p. 166.

We can see by a simple example that this quantity, the classical total momentum, is not conserved relativistically. That is, the conservation of the quantity $\Sigma m_i \mathbf{u}_i$ is an approximation which holds only at low speeds.

Consider an observer in frame $S$ with a ball $A$, and one in $S'$ with a ball $B$. The balls each have mass $m$ and are identical when compared at rest. Each observer throws his ball along his $y$ or $y'$ axis with speed $u_0$ (measured in his own frame) so that the balls collide. Assuming the balls to be perfectly elastic, each observer will see his ball rebound with its original speed $u_0$. If the total momentum is to be conserved, the $y$ component must be zero because the momentum of each ball is merely reversed by the collision. However, if we consider the relativistic velocity transformations, we can see that the quantity $mu_y$ does not have the same magnitude for each ball as seen by either observer.

Let us consider the collision as seen in frame $S$ (Figure 35-15). In this frame ball $A$ moves along the $y$ axis with velocity $u_{yA} = u_0$. Ball $B$ has an $x$ component of velocity $u_{xB} = v$ and a $y$ component $u_{yB} = u'_{yB}/\gamma = -u_0\sqrt{1 - v^2/c^2}$. Here we have used the velocity-transformation equations 35-21 and 35-23 and the fact that $u'_{yB} = -u_0$ and $u'_{xB} = 0$. We see that the $y$ component of velocity of ball $B$ is smaller in magnitude than that of ball $A$. The factor $\sqrt{1 - v^2/c^2}$ comes from the time-dilation factor. The time taken for ball $B$ to travel a given distance along the $y$ axis in $S$ is greater than the time measured in $S'$ for the ball to travel this same distance. Thus in $S$, the total $y$ component of classical momentum is not zero. Since the velocities are reversed in an elastic collision, momentum as defined by $\mathbf{P} = \Sigma m\mathbf{u}$ is not conserved in $S$. Analysis of this problem in $S'$ leads to the same conclusion (Figure 35-15b). In the classical limit, $v \ll c$, momentum is conserved, of course, because $\gamma \approx 1$.

The reason for defining momentum to be $\Sigma m\mathbf{u}$ in classical mechanics was that this quantity is conserved when there are no external forces, as in collisions. We now see that this quantity is conserved only in the approximation $v \ll c$. We shall define *relativistic momentum* $\mathbf{p}$ of a particle to have the following properties:

1. $\mathbf{p}$ is conserved in collisions.

2. $\mathbf{p}$ approaches $m\mathbf{u}$ as $u/c$ approaches zero.

We shall state without proof that the quantity meeting these conditions is

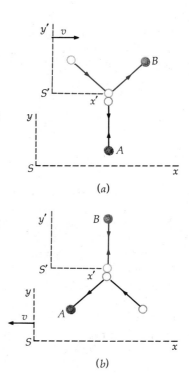

**Figure 35-15**
(a) Elastic collision of two identical balls as seen in frame $S$. The vertical component of the velocity of ball $B$ is $u_0/\gamma$ in $S$ if it is $u_0$ in $S'$. (b) The same collision as seen in $S'$. In this frame ball $A$ has vertical component of velocity $u_0/\gamma$.

$$\mathbf{p} = \frac{m\mathbf{u}}{\sqrt{1 - u^2/c^2}} \qquad 35\text{-}24$$

*Relativistic momentum defined*

where $u$ is the speed of the particle. We thus take this equation for the definition of relativistic momentum of a particle. It is clear that this definition meets our second criterion because the denominator approaches 1 when $u$ is much less than $c$. Proof that it also meets the first criterion involves much tedious algebra. From this definition, the momenta of the two balls $A$ and $B$ as seen in $S$ are

$$p_{yA} = \frac{mu_0}{\sqrt{1 - u_0^2/c^2}} \qquad p_{yB} = \frac{mu_{yB}}{\sqrt{1 - (u_{xB}^2 + u_{yB}^2)/c^2}}$$

where $u_{yB} = -u_0\sqrt{1 - v^2/c^2}$ and $u_{xB} = v$. We shall leave the details of showing that $p_{yB} = -p_{yA}$ as an exercise. Because of the similarity of the factor $1/\sqrt{1 - u^2/c^2}$ and $\gamma$ in the Lorentz transformation, Equation 35-24 is often written

$$\mathbf{p} = \gamma m \mathbf{u} \qquad \text{with } \gamma = \frac{1}{\sqrt{1 - u^2/c^2}} \qquad\qquad 35\text{-}25$$

This use of the symbol $\gamma$ for two different quantities can cause some confusion. The notation is standard, however, and simplifies many of the equations. We shall use the notation except when considering transformations between reference frames. Then, to avoid confusion, we shall write out the factor $\sqrt{1 - u^2/c^2}$ and reserve $\gamma$ for $1/\sqrt{1 - v^2/c^2}$, where $v$ is the relative speed of the frames.

One interpretation of Equation 35-24 is that the mass of an object increases with speed. The quantity

$$\frac{m}{\sqrt{1 - u^2/c^2}} = \gamma m$$

is sometimes called the *relativistic mass*, written $m(u)$. The mass of the body in its rest frame, written $m_0$, is called the *rest mass*. Although this makes the expression $m(u)\mathbf{u}$ for relativistic momentum similar to the nonrelativistic expression, the use of relativistic mass often leads to mistakes. For example, the expression $\frac{1}{2}m(u)u^2$ is not the correct relativistic expression for kinetic energy. The measurement of relativistic mass involves measuring the force needed to produce a given change in momentum. The experimental evidence often cited as verification that mass depends on velocity can also be interpreted merely as evidence of the validity of the assumption that the force equals the time rate of change of relativistic momentum. We shall avoid using a symbol for relativistic mass. The symbol $m$ in this book always refers to the rest mass.

## 35-9 Relativistic Energy

We have seen that the quantity $m\mathbf{u}$ is not conserved in collisions but that $\gamma m\mathbf{u}$ is, with $\gamma = 1/\sqrt{1 - u^2/c^2}$. Evidently Newton's law in the form $\Sigma \mathbf{F} = m\mathbf{a}$ cannot be correct relativistically, since it leads to the conservation of $m\mathbf{u}$. We can get a hint of the correct form of Newton's second law by writing it $\Sigma \mathbf{F} = d\mathbf{p}/dt$. Let us assume that this equation is correct if relativistic momentum $\mathbf{p}$ is used. The validity of this assumption can be determined only by examining its consequences, since an unbalanced force on a high-speed particle is measured by its effect on momentum and energy. We are essentially defining force by the equation $\Sigma \mathbf{F} = d\mathbf{p}/dt$. As in classical mechanics, we shall define kinetic energy as the work done by an unbalanced force in accelerating a particle from rest to some velocity. Considering one dimension only, we have

$$E_k = \int_{u=0}^{u} \Sigma F \, ds = \int_0^u \frac{d(\gamma m u)}{dt} \, ds = \int_0^u u \, d(\gamma m u) \qquad 35\text{-}26$$

using $u = ds/dt$. The computation of the integral in Equation 35-26 is not difficult but requires some messy algebra. It is left as a problem to show that

$$d(\gamma m u) = m \left(1 - \frac{u^2}{c^2}\right)^{-3/2} du$$

Substituting this into the integrand in Equation 35-26 gives

$$E_k = \int_0^u u \, d(\gamma m u) = \int_0^u m\left(1 - \frac{u^2}{c^2}\right)^{-3/2} u \, du$$

$$= mc^2\left(\frac{1}{\sqrt{1 - u^2/c^2}} - 1\right)$$

or

$$E_k = \gamma m c^2 - mc^2 \qquad\qquad 35\text{-}27$$

We can check this expression for low speeds by noting that for $u/c \ll 1$

$$\gamma = \left(1 - \frac{u^2}{c^2}\right)^{-1/2} \approx 1 + \frac{1}{2}\frac{u^2}{c^2} + \cdots$$

Thus

$$E_k \approx mc^2\left(1 + \frac{1}{2}\frac{u^2}{c^2} + \cdots -1\right) = \tfrac{1}{2}mu^2$$

The expression for kinetic energy consists of two terms. One, $\gamma m c^2$, depends on the speed of the particle (through the factor $\gamma$), and the other, $mc^2$, is independent of the speed. The quantity $mc^2$ is called the *rest energy* of the particle. The total energy $E$ is then defined to be the sum of the kinetic energy and the rest energy:

*Rest energy*

$$E = E_k + mc^2 = \gamma m c^2 = \frac{mc^2}{\sqrt{1 - u^2/c^2}} \qquad\qquad 35\text{-}28$$

*Total relativistic energy*

Thus the work done by an unbalanced force increases the energy from the rest energy $mc^2$ to $\gamma m c^2$ (or increases the mass from $m$ to $\gamma m$).

The identification of the term $mc^2$ as rest energy is not merely a convenience. The rest energy of a system of particles is related to the internal energy of the system. For example, if two particles are bound together by attractive forces so that the potential energy $U$ is negative (relative to zero potential energy at infinite separation) and the internal kinetic energy is less than $U$ in magnitude (necessary if the system is to remain bound), the total internal energy $E_{int} = E_k + U$ will be negative. Then the rest mass of this system is less by the amount $E_{int}/c^2$ than the rest mass of the two particles when separated and at rest. On the other hand, if the rest mass of a system is greater than the rest mass of its individual parts, the system is not bound. This is true in radioactive decay, e.g., a radium nucleus decaying by emission of an alpha particle. In this case the sum of the masses of the alpha particle and the final nucleus is less than the mass of the original radium nucleus by an amount $E_k/c^2$, where $E_k$ is the observed kinetic energy of the decay particles.

In practical applications the momentum or energy of a particle is often known rather than the speed. Equation 35-24 for the relativistic momentum and Equation 35-28 for the relativistic energy can be combined to eliminate the speed $u$. The result (see Exercise 24) is

$$E^2 = p^2 c^2 + (mc^2)^2 \qquad\qquad 35\text{-}29$$

If the energy of a particle is much greater than its rest energy $mc^2$, the second term on the right of Equation 35-29 can be neglected, giving the useful approximation

$$E \approx pc \qquad \text{for} \qquad E \gg mc^2 \qquad\qquad 35\text{-}30$$

Equation 35-30 is an exact relation between energy and momentum for particles with no rest mass, e.g., photons and neutrinos. If the kinetic energy of a particle is small compared with its rest energy, the term $pc$ is small compared with $mc^2$. From Equation 35-29 we then have

$$
\begin{aligned}
E &= [p^2c^2 + (mc^2)^2]^{1/2} = mc^2 \left(1 + \frac{p^2}{m^2c^2}\right)^{1/2} \\
&\approx mc^2 \left(1 + \frac{1}{2}\frac{p^2}{m^2c^2} + \cdots\right) \\
&= mc^2 + \frac{p^2}{2m} + \cdots \qquad\qquad 35\text{-}31
\end{aligned}
$$

where we have used the approximation for small $x$, $(1 + x)^{1/2} \approx 1 + \frac{1}{2}x$. This result agrees with the classical result for the kinetic energy, $E_k = E - mc^2 = p^2/2m$. The velocity of a particle is often most easily calculated from its total energy and momentum. If we multiply Equation 35-28 by $u$ and compare with Equation 35-24, we obtain the useful result

$$uE = pc^2$$

or

$$u = \frac{pc^2}{E} \qquad\qquad 35\text{-}32$$

The most convenient unit for expressing the energy of an electron or other subatomic particle is the electronvolt (eV) and its multiples, defined in Section 7-3. The electronvolt is related to the joule by

$$1 \text{ eV} = 1.602 \times 10^{-19} \text{ J}$$

---

**Example 35-4** An electron with rest energy 0.511 MeV moves with speed $u = 0.8c$. Find its total energy, kinetic energy, and momentum.
  From Equation 35-25

$$\gamma = \frac{1}{\sqrt{1 - 0.64}} = \frac{5}{3} = 1.67$$

The total energy is then

$$E = \gamma mc^2 = 1.67(0.511 \text{ MeV}) = 0.853 \text{ MeV}$$

The kinetic energy is the total energy minus the rest energy:

$$E_k = E - mc^2 = 0.853 \text{ MeV} - 0.511 \text{ MeV} = 0.342 \text{ MeV}$$

The momentum can be calculated from either Equation 35-25 or 35-32. From Equation 35-25 we have for its magnitude

$$
\begin{aligned}
p &= \gamma mu = \gamma m(0.8c) = \frac{0.8\gamma mc^2}{c} \\
&= \frac{(0.8)(1.67)(0.511 \text{ MeV})}{c} = 0.683 \text{ MeV}/c
\end{aligned}
$$

The unit MeV/$c$ is a convenient momentum unit.

**Example 35-5** An electron has a total energy 5 times its rest energy. What is its momentum and speed?

From Equation 35-29

$$p^2c^2 = E^2 - (mc^2)^2 = (5mc^2)^2 - (mc^2)^2 = 24(mc^2)^2$$

$$pc = \sqrt{24}\ mc^2 = (4.90)\ (0.511\ \text{MeV}) = 2.50\ \text{MeV}$$

$$p = 2.50\ \text{MeV}/c$$

We find its speed from Equation 35-32:

$$u = \frac{pc^2}{E} = \frac{\sqrt{24}\ mc^2}{5mc^2}\ c = 0.980c$$

# 35-10 Mass and Binding Energy

Some numerical examples from atomic and nuclear physics will illustrate changes in rest mass and rest energy. A unit of mass convenient for discussing atomic and nuclear masses is the *unified mass unit* u, defined as one-twelfth the mass of the neutral carbon atom consisting of the $^{12}$C nucleus and six electrons. (This unit replaces the older atomic mass unit based on the oxygen atom.) Since 1 mol of carbon contains Avogadro's number of atoms and has a mass of 12 g, the relation between the unified mass unit and the gram is

$$1\ u = \frac{1\ g}{6.0220 \times 10^{23}} = 1.6606 \times 10^{-24}\ \text{g}$$
$$= 1.6606 \times 10^{-27}\ \text{kg} \qquad \qquad 35\text{-}33$$

In rough calculations we can write

$$1\ u = 1.66 \times 10^{-24}\ \text{g} = 1.66 \times 10^{-27}\ \text{kg}$$

The rest energy of 1 g is

$$(1\ g)c^2 = (10^{-3}\ \text{kg})\ (3 \times 10^8\ \text{m/s})^2$$
$$= 9 \times 10^{13}\ \text{J} = 5.61 \times 10^{32}\ \text{eV} \qquad \qquad 35\text{-}34$$

The rest energy of a unified mass unit is

$$(1\ u)c^2 = 931.5\ \text{MeV} \qquad \qquad 35\text{-}35$$

The rest masses and rest energies of some elementary particles and light nuclei are given in Table 35-1, from which we can see that the mass of a nucleus is not the same as the sum of the masses of its parts.

**Example 35-6** The simplest example is that of the deuteron $^2$H, consisting of a neutron and a proton bound together. Its rest energy is 1875.63 MeV. The sum of the rest energies of the proton and neutron is 938.28 + 939.57 = 1877.85 MeV. Since this is greater than the rest energy of the deuteron, the deuteron cannot spontaneously break up into a neutron and a proton. The binding energy of the deuteron is 1877.85 − 1875.63 = 2.22 MeV. In order to break up the deuteron into a proton and a neutron, at least 2.22 MeV must be added. This can be done by bombarding deuterons with energetic particles or electromagnetic radiation.

If a deuteron is formed by combination of a neutron and proton, energy must be released. When neutrons from a reactor are incident on protons, some neutrons are captured. The nuclear reaction is $n + p \rightarrow d + \gamma$. Most of these reactions occur for the low-energy neutrons (kinetic energy less than 1 eV). The energy of the photon plus the kinetic energy of the deuteron is 2.22 MeV.

---

**Example 35-7**  A free neutron decays into a proton plus an electron plus an antineutrino:

$$n \rightarrow p + e + \bar{\nu}$$

What is the kinetic energy of the decay products?

Here rest energy is converted into kinetic energy. Before decay, $(mc^2)_n = 939.57$ MeV. After decay, $(mc^2)_p + (mc^2)_e + (mc^2)_{\bar{\nu}} = 938.28 + 0.511 + 0 = 938.79$ MeV. Thus rest energy of $939.57 - 938.79 = 0.78$ MeV has been converted into kinetic energy of the decay products.

---

**Example 35-8**  The binding energy of the hydrogen atom (energy to remove the electron from the atom) is 13.6 eV. How much mass is lost when an electron and a proton form a hydrogen atom?

The mass of a proton plus that of an electron must be greater than that of the hydrogen atom by

$$\frac{13.6 \text{ eV}}{931.5 \text{ MeV/u}} = 1.46 \times 10^{-8} \text{ u}$$

This mass difference is so small that it is usually neglected.

---

**Example 35-9**  How much energy is needed to remove one proton from a $^4$He nucleus?

Removal of one proton from $^4$He leaves $^3$H. From Table 35-1, the rest energy of $^3$H plus that of a proton is $2808.94 + 938.23 = 3747.17$ MeV. This is about 19.8 MeV greater than the rest energy of $^4$He.

---

These examples show that because atomic binding energies are so small (of the order of 1 eV to 1 keV), the mass changes are negligible in atomic (or chemical) reactions, but the nuclear binding energies are quite large and involve appreciable changes in mass.

---

**Table 35-1**

Rest energies of some elementary particles and light nuclei

| Particle | Symbol | Rest energy, MeV |
|---|---|---|
| Photon | $\gamma$ | 0 |
| Neutrino (antineutrino) | $\nu$ ($\bar{\nu}$) | 0 |
| Electron (positron) | $e$ or $e^-$ ($e^+$) | 0.5110 |
| Muon | $\mu$ | 105.7 |
| Pi meson | $\pi^0$ | 135 |
| | $\pi^{\pm}$ | 139.6 |
| Proton | $p$ | 938.280 |
| Neutron | $n$ | 939.573 |
| Deuteron | $^2$H or $d$ | 1875.628 |
| Triton | $^3$H | 2808.944 |
| Alpha | $^4$He or $\alpha$ | 3727.409 |

## 35-11 General Relativity

The generalization of relativity theory to noninertial reference frames by Einstein in 1916 is known as the general theory of relativity. It is much more difficult mathematically than the special theory of relativity, and there are fewer situations in which it can be tested. Nevertheless its importance calls for a brief qualitative discussion.

The basis of the general theory of relativity is the principle of equivalence:

*A homogeneous gravitational field is completely equivalent to a uniformly accelerated reference frame.*

*Equivalence principle*

This principle arises in newtonian mechanics because of the apparent identity of gravitational and inertial mass. In a uniform gravitational field, all objects fall with the same acceleration $\mathbf{g}$ independent of their mass because the gravitational force is proportional to the (gravitational) mass while the acceleration varies inversely with the (inertial) mass. Consider a compartment in space far from any matter and undergoing uniform acceleration $\mathbf{a}$, as shown in Figure 35-16a. If objects are dropped in the compartment, they fall to the "floor" with acceleration $\mathbf{g} = -\mathbf{a}$. If people stand on a spring scale it will read their "weight" of magnitude $ma$. No mechanics experiment can be performed *inside* the compartment that will distinguish whether the compartment is actually accelerating in space or is at rest (or moving with uniform velocity) in the presence of a uniform gravitational field $\mathbf{g} = -\mathbf{a}$.

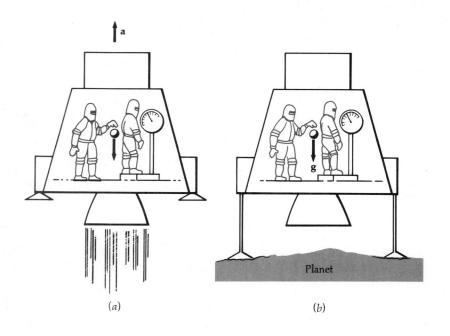

(a)    (b)

**Figure 35-16**
The results of experiments in a uniformly accelerated reference frame (*a*) cannot be distinguished from those in a uniform gravitational field (*b*) if the acceleration $\mathbf{a}$ and the gravitational field $\mathbf{g}$ have the same magnitude.

Einstein assumed that the principle of equivalence applied to all physics and not only to mechanics. In effect, he assumed that there is no experiment of any kind that can distinguish uniformly accelerated motion from the presence of a gravitational field. We shall look qualitatively at a few of the consequences of this assumption.

The first consequence of the principle of equivalence we shall discuss, the deflection of a light beam in a gravitational field, was one of

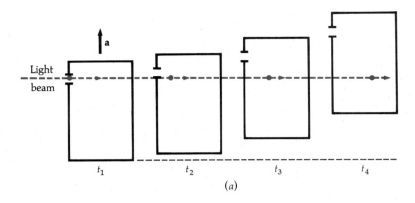

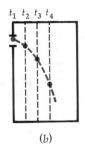

(a)

(b)

the first to be tested experimentally. Figure 35-17 shows a beam of light entering a compartment that is accelerating. Successive positions of the compartment are shown at equal time intervals. Because the compartment is accelerating, the distance it moves in each time interval increases with time. The path of the beam of light as observed from inside the compartment is therefore a parabola. But according to the equivalence principle, there is no way to distinguish between an accelerating compartment and one with uniform velocity in a uniform gravitational field. We conclude, therefore, that a beam of light, like massive objects, will accelerate in a gravitational field. For example, near the surface of the earth, light will fall with acceleration 9.81 m/s². This is difficult to observe because of the enormous speed of light. For example, in a distance of 3000 km, which takes about 0.01 s to traverse, a beam of light should fall about 0.5 mm. Einstein pointed out that the deflection of a light beam in a gravitational field might be observed when light from a distant star passes close to the sun, as illustrated in Figure 35-18. Because of the brightness of the sun, such a star cannot ordinarily be seen. Such a deflection was first observed in 1919 during an eclipse of the sun. This well-publicized observation brought instant worldwide fame to Einstein.

A second prediction from Einstein's theory of general relativity (which we shall not discuss in detail) is the excess precession of the

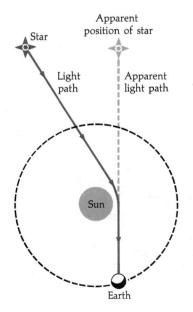

perihelion of the orbit of Mercury of about 0.01° per century. This effect had been known and unexplained for some time, so in some sense it represented an immediate success of the theory. There is, however, some difficulty in comparing the prediction of general relativity with experiment because of other effects, e.g., the perturbations due to other planets and the nonspherical shape of the sun.

A third prediction of general relativity concerns the change in time intervals and frequencies of light in a gravitational field. If $\Delta t_1$ is a time interval measured by a clock where the gravitational potential (potential energy per unit mass) is $\phi_1$ and $\Delta t_2$ is the same interval measured by a clock where the gravitational potential is $\phi_2$, general relativity predicts that the fractional difference will be approximately

$$\frac{\Delta t_2 - \Delta t_1}{\Delta t} = \frac{1}{c^2}(\phi_2 - \phi_1) \qquad 35\text{-}36$$

(Since this shift is usually very small, it does not matter by which interval we divide on the left side of the equation.) A clock in a region of low gravitational potential will therefore run slower than one in a region of high potential. Since a vibrating atom can be considered to be a clock, the frequency of vibration in a region of low potential (such as near the sun where the gravitational potential $-GM_s/r$ is negative) will be lower than that of the same atom on earth. This shift toward lower frequency and therefore longer wavelength is called the *gravitational red shift*.

*Gravitational red shift*

As our final example of the predictions of general relativity, we mention black holes, first predicted by Oppenheimer and Snyder in 1939. According to the general theory of relativity, if the density of an object such as a star is great enough, the gravitational attraction will be so great that nothing can escape, not even light or other electromagnetic radiation. A remarkable property of such an object is that nothing happening inside it can be communicated to the outside world. As is often true in physics, a simple but incorrect calculation gives the correct results for the relation between the mass and the critical radius of a black hole. In newtonian mechanics, the speed needed for escape of a particle from the surface of a planet or star is found by requiring the kinetic energy $\frac{1}{2}mv^2$ to be equal in magnitude to the potential energy $-GMm/r$ so that the total energy is zero. The resulting escape velocity is

$$v_e = \sqrt{\frac{2GM}{r}}$$

If we set the escape velocity equal to the speed of light and solve for the radius, we obtain the critical radius $R_G$, called the *Schwarzschild radius*:

$$R_G = \frac{2GM}{c^2} \qquad 35\text{-}37$$

For an object of mass equal to that of our sun to be a black hole, its radius must be about 3 km. Since no radiation is emitted from a black hole and its radius is expected to be small, the detection of such an object is not easy. The best chance of detection would occur if a black hole were a companion to a normal star in a binary star system. Measurements of the doppler shift of the light from the normal star might then allow a computation of the mass of the unseen companion to determine whether it is great enough to be a black hole. At present the binary x-ray source Cygnus X-1 (in the constellation Cygnus) appears to be an excellent candidate to be a black hole, but the evidence is not conclusive.

# Black Holes

Alan P. Lightman
*Cornell University*

Imagine a region of space where attractive gravitational force is so intense that light rays venturing too near can be bent into circular orbit—a region from which no matter, radiation, or communication of any kind can ever escape. Such is a black hole, one of the most exciting creatures of theoretical physics and possibly the most bizarre object of space.

Although they were implicitly predicted by Einstein's 1915 theory of gravity, general relativity, black holes were first theoretically "discovered" by Oppenheimer and Snyder in 1939. Because of their highly nonintuitive properties, however, black holes were not taken seriously by most physicists and astronomers until the mid-1960s. We are now on the verge of confirming the discovery in space of the first black hole.

The concept of a black hole stretches our ideas of time and space to their limits. The surface of a black hole, called its *horizon,* is a closed boundary within which the escape speed exceeds the speed of light. The prediction of such a surface for sufficiently compact bodies can be made just on the basis of Newton's theory of gravity together with special relativity: the escape speed of a particle launched from the surface of a spherical mass $M$ of radius $R$ is $v_e = \sqrt{2GM/R}$ (see Chapter 7). When $M/R$ satisfies $2GM/R > c^2$, $v_e$ exceeds the speed of light and no particle or photon may escape, as required by special relativity. As a remarkable result, the interior of a black hole is causally disconnected from the rest of the universe; no physical process occurring inside the horizon can communicate its existence or effects to the outside. For a spherical black hole of mass $M$, the horizon is a sphere of circumference equal to $2\pi$ times the *Schwarzschild radius* of the hole $R_G$, where $R_G = 2GM/c^2$ (exact numerical coincidence of this radius with the newtonian analog above is accidental). A black hole with a mass equal to that of our sun would have a Schwarzschild radius of 2.95 km.

According to general relativity, time and space are warped by the gravitational field of massive bodies, with the warping most severe near a black hole. Gravity affects all physical systems in a universal way so that all clocks (whether transitions of an ammonia molecule or heartbeats of a human being) and all rulers would indicate that time is slowed down and space stretched out near a black hole (see Figure 1). Alternatively, one can describe the effects of the hole's gravitational field on a local measurement of time and space intervals as an acceleration of the local Lorentz frame (in which special relativity is valid) with respect to other local Lorentz frames at different locations.

Black holes are formed when massive stars undergo total gravitational collapse. While emitting heat and light into space, stars balance themselves against their own gravity with the outward thermal pressure force generated by heat from nuclear reactions in their deep interiors. But every star must die. When its nuclear fuel is exhausted, a star will contract. If its mass is less than about three times the mass of our sun, the shrinking star will stabilize at a smaller size when the inward pull of gravity cannot force the star's constituent particles any closer together. Such a star will spend eternity as a *white dwarf* or *neutron star.* But if the star's mass exceeds about 3 solar masses, theory indicates that no amount of outward pressure force can stave off the overwhelming crush of gravity and the star will plunge in upon itself, disappearing forever from sight and giving birth to a black hole.

Recent mathematical proofs, using Einstein's general relativity theory, demonstrate that a black hole is one of the simplest objects of nature and can be completely characterized by only three quantities: its mass, angular momentum, and net electrical charge. Other than these three quantities, all information of the progenitor star, e.g., whether it was composed of particles or antiparticles, whether it was pancake-shaped or spherical, is lost via gravita-

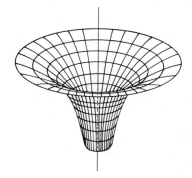

**Figure 1**
How the geometry of space is warped by a spherical (nonrotating) black hole. The radial distance from any point to the center of the black hole can be measured in two ways. One way is to measure the circumference of a circle passing through the point and centered on the hole, and divide by $2\pi$. A second way is to measure the distance from the point to the center along a radial line on the curved surface. These two measures will be nearly equal far from the hole where space is "flat," but highly discrepant near the black-hole horizon where the funnel narrows to a vertical tube.

tional and electromagnetic waves shortly after formation of the hole. Rotating black holes, called *Kerr holes* (nonrotating holes are called *Schwarzschild holes*), surround themselves with a region called the *ergosphere* in which space and time are so strongly twisted that all particles, photons, and even local Lorentz frames are forced to rotate about the hole.

At great distances from a black hole, its gravitational field behaves as if generated by an ordinary star of mass $M$, obeying the newtonian inverse-square law. Close to a black hole, the gravitational field is far stronger than would be predicted by newtonian theory. A human being sucked into a solar-mass black hole would be ripped to shreds by the differential force of gravity across his body long before he reached the horizon.

Once the hole's horizon has formed we cannot receive any information about the ultimate fate of the collapsing star inside. Calculations of the collapse (idealized to be spherical) indicate that the star is crushed to zero volume and infinite density at the center of the black hole, forming there a point of infinite gravitational force called a *singularity*. For a collapsing star of a few solar masses, the final stellar death throes would be over in a few hundred-thousandths of a second (as measured locally). Quantum effects, omitted in the classical theory of general relativity, could possibly halt the stellar collapse at the unimaginable density $\rho \sim c^5/hG^2 \sim 5 \times 10^{93}$ g/cm$^3$, where $h$ is Planck's constant, and thus prevent creation of singularities; but such effects could not prevent the formation of black holes.

Current theories of stellar evolution suggest that there could be as many as 100 million black holes in our galaxy. The search for these black holes is not easy. We could never detect a tiny black speck a few miles across against the night sky. Instead, we must search for signs of black holes interacting with their astrophysical neighborhoods. Most vulnerable to discovery would be a black hole orbiting a normal star in a binary system. Using Kepler's laws, analysis of the magnitude and period of the doppler shift of the visible normal star allows us to compute whether the unseen companion is massive enough to be a black hole. In addition we might observe intense, flickering x-rays produced when gas from the normal star is sucked toward the black hole and heated to a billion degrees in its spiraling path to destruction in the hole. Such are the telltale signs of the binary x-ray source Cygnus X-1, an excellent candidate for a black hole about 8000 light-years from earth in the constellation Cygnus.

Black holes are a fundamental phenomenon of nature. Space may be littered with black holes, tempting us with their secrets of time and space behind a cloak of impenetrable darkness, a challenge and a prize for the persistence of astronomers and physicists.

## Review

A. Define, explain, or otherwise identify:

Einstein postulates, 931
Ether, 931
Time dilation, 938
Proper time, 938
Proper length, 938
Length contraction, 939
Synchronized clocks, 941
Lorentz transformation, 947

Twin paradox, 948
Relativistic momentum, 952
Rest energy, 954
Unified mass unit, 956
Equivalence principle, 958
Gravitational red shift, 960
Black hole, 960
Schwarzschild radius, 960

B. True or false:

1. The speed of light is the same in all reference frames.

2. Proper time is the shortest time interval between two events.

3. Absolute motion can be determined by means of length contraction.

4. The light-year is a unit of distance.

5. A particle cannot be accelerated to a speed greater than the speed of light in vacuum.

6. Simultaneous events must occur at the same place.

7. If two events are not simultaneous in one frame, they cannot be simultaneous in any other frame.

8. $\Sigma \mathbf{F} = m\mathbf{a}$ holds at all speeds if the mass is the relativistic mass.

9. Rest mass can sometimes be converted into energy.

10. If two particles are tightly bound together by strong attractive forces, the mass of the system is less than the sum of the masses of the individual particles when separated.

## Exercises

### Section 35-1, The Michelson-Morley Experiment

1. In one series of measurements of the speed of light, Michelson used a path length $L$ of 35.4 km (22 mi). (a) What is the time needed for light to make the round-trip distance of $2L$? (b) What is the classical correction term in seconds in Equation 35-1 assuming the earth's speed is $v = 10^{-4}c$? (c) From about 1600 measurements, Michelson quoted the result for the speed of light as 299,796 ± 4 km/s. Is this experiment accurate enough to be sensitive to the correction term in Equation 35-1?

2. An airplane flies with speed $c$ relative to still air from point $A$ to point $B$ and returns. Compare the time required for the round trip when the wind blows from $A$ to $B$ with speed $v$ with that when the wind blows perpendicularly to the line $AB$ with speed $v$.

### Section 35-2, Consequences of Einstein's Postulates

*There are no exercises for this section.*

### Section 35-3, Time Dilation and Length Contraction

3. Derive the following results for $v$ much less than $c$ and use when applicable in the exercises and problems.

(a) $\gamma \approx 1 + \dfrac{1}{2}\dfrac{v^2}{c^2}$

(b) $\dfrac{1}{\gamma} \approx 1 - \dfrac{1}{2}\dfrac{v^2}{c^2}$

(c) $\gamma - 1 \approx 1 - \dfrac{1}{\gamma} \approx \dfrac{1}{2}\dfrac{v^2}{c^2}$

4. How great must the relative speed of two observers be for their time-interval measurements to differ by 1 percent (see Exercise 3)?

5. The proper mean life of $\pi$ mesons is $2.6 \times 10^{-8}$ s. If a beam of such particles has speed $0.9c$, (a) what would their mean life be as measured in the laboratory? (b) How far would they travel on the average before they decay? (c) What would your answer be to part (b) if you neglected time dilation?

6. (a) In the reference frame of the $\pi$ meson in Exercise 5, how far does the laboratory travel in a typical lifetime of $2.6 \times 10^{-8}$ s? (b) What is this distance in the laboratory frame?

7. A metrestick moves with speed $v = 0.6c$ relative to you in the direction parallel to the stick. (a) Find the length of the stick measured by you. (b) How long does it take for the stick to pass you?

8. Supersonic jets achieve maximum speeds of about $3 \times 10^{-6}c$. (a) By what percentage would you see such a jet contracted in length? (b) During a time of $1 \text{ y} = 3.15 \times 10^7$ s on your clock, how much time would elapse on the pilot's clock? How many minutes are lost by the pilot's clock in 1 y of your time?

## Section 35-4, Clock Synchronization and Simultaneity

*Exercises 9 to 13 refer to the following situation: an observer in S' lays out a distance L' = 100 c·min between points A' and B' and places a flashbulb at the midpoint C'. He arranges for the bulb to flash and for clocks at A' and B' to be started at 0 when the light from the flash reaches the clocks (see Figure 35-19). Frame S' is moving to the right with speed 0.6c relative to an observer C in S who is at the midpoint between A' and B' when the blub flashes and sets his clock to zero at that time.*

**Figure 35-19**
Exercises 9 to 13.

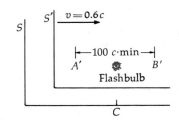

9. What is the separation distance between clocks A' and B' according to the observer in S?

10. As the light pulse from the flashbulb travels toward A' with speed c, A' travels toward C with speed 0.6c. Show that the clock in S reads 25 min when the flash reaches A'.

11. Show that the clock in S reads 100 min when the light flash reaches B', which is traveling away from C with speed 0.6c.

12. The time interval between reception of the flashes is 75 min according to the observer in S. How much time does he expect to have elapsed on the clock at A' during this 75 min?

13. The time interval calculated in Exercise 12 is the amount that the clock at A' leads that at B' according to observers in S. Compare this result with $L_0 v/c^2$.

## Section 35-5, The Doppler Effect

14. (a) Show that Equations 35-14a and b are identical and can both be written

$$f' = f_0 \sqrt{\frac{1+\beta}{1-\beta}} \qquad \text{where } \beta = \frac{v}{c}$$

(b) Show that Equation 35-15 can be written

$$f' = f_0 \sqrt{\frac{1-\beta}{1+\beta}}$$

15. How fast must you move toward a red light ($\lambda = 650$ nm) for it to appear green ($\lambda = 525$ nm)?

16. A distant galaxy is moving away from us at speed $2 \times 10^7$ m/s. Calculate the fractional red shift $(\lambda' - \lambda_0)/\lambda_0$ in the light from this galaxy.

17. A distant galaxy is moving away from the earth with speed such that each wavelength is shifted by a factor of 2; that is, $\lambda' = 2\lambda_0$. What is the speed of the galaxy relative to us (see Exercise 14)?

## Section 35-6, The Lorentz Transformation

18. Two events happen at the same point $x_0'$ in frame $S'$ at times $t_1'$ and $t_2'$. (a) Use Equations 35-18 to show that in frame $S$ the time interval between the events is greater than $t_2' - t_1'$ by the factor $\gamma$. (b) Why are Equations 35-19 less convenient than Equations 35-18 for this problem?

19. The length of an object moving in your frame is measured by recording the positions of each end $x_2$ and $x_1$ at the same time $t_0$. (a) Use Equations 35-19 to show that the result is smaller than the proper length $x_2' - x_1'$ measured in the frame in which the object is at rest. (b) Why are Equations 35-19 more convenient for this problem than Equations 35-18?

20. Two spaceships are approaching each other. (a) If the speed of each is $0.9c$ relative to the earth, what is the speed of one relative to the other? (b) If the speed of each relative to the earth is 30,000 m/s (about 100 times the speed of sound), what is the speed of one relative to the other?

21. A light beam moves along the $y'$ axis with speed $c$ in frame $S'$ which is moving to the right with speed $v$ relative to frame $S$. (a) Find $u_x$ and $u_y$, the $x$ and $y$ components of the velocity of the light beam in frame $S$. (b) Show that the magnitude of the velocity of the light beam in $S$ is $c$.

22. A particle moves with speed $0.9c$ along the $x''$ axis of frame $S''$, which moves with speed $0.9c$ along the $x'$ axis relative to frame $S'$. Frame $S'$ moves with speed $0.9c$ along the $x$ axis relative to frame $S$. (a) Find the speed of the particle relative to frame $S'$. (b) Find the speed of the particle relative to frame $S$.

## Section 35-7, The Twin Paradox

*There are no exercises for this section.*

## Section 35-8, Relativistic Momentum, and Section 35-9, Relativistic Energy

23. Show that $p_{yA} = -p_{yB}$, where $p_{yA}$ and $p_{yB}$ are the relativistic momenta of the balls in Figure 35-15, given by

$$p_{yA} = \frac{mu_0}{\sqrt{1 - u_0^2/c^2}} \qquad p_{yB} = \frac{mu_{yB}}{\sqrt{1 - (u_{xB}^2 + u_{yB}^2)/c^2}}$$

$$u_{yB} = -u_0\sqrt{1 - \frac{v^2}{c^2}} \qquad \text{and} \qquad u_{xB} = v$$

24. Combine Equations 35-24 and 35-28 to derive the equation $E^2 = p^2c^2 + m^2c^4$.

25. Make a sketch of the total energy $E$ of an electron as a function of its momentum $p$. (See Equations 35-30 and 35-31 for the behavior of $E$ at the limits of large and small values of $p$.)

26. Make a sketch of the kinetic energy of an electron $E_k$ versus its speed $u$ (see Equation 35-27).

27. An electron of rest energy 0.511 MeV has a total energy of 5 MeV. (a) Find its momentum in units of MeV/$c$ from Equation 35-29. (b) Find the ratio of its speed $u$ to the speed of light.

28. How much energy would be required to accelerate a particle of mass $m$ from rest to speeds of (a) $0.5c$, (b) $0.9c$, (c) $0.99c$? Express your answers as multiples of the rest energy.

29. The rest energy of a proton is about 938 MeV. If its kinetic energy is also 938 MeV, find (a) its momentum and (b) its speed.

30. If the kinetic energy of a particle equals its rest energy, what error is made by using $p = mu$ for its momentum?

## Section 35-10, Mass and Binding Energy

31. How much energy is required to remove one of the neutrons from $^3$H to yield $^2$H plus the neutron?

32. The rest mass of $^3$He (two protons and one neutron) is 3.01440 u. (a) What is the rest energy of $^3$He in millions of electronvolts? (b) How much energy is needed to remove a proton to make $^2$H plus the proton?

33. The energy released when sodium and chlorine combine to form NaCl is 4.2 eV. (a) What is the increase in mass (in unified mass units) when a molecule of NaCl is dissociated into an atom of Na and an atom of Cl? (b) What percentage error is made in neglecting this mass difference? (The mass of Na is about 23 u, and that of Cl is about 35.5 u.)

34. In nuclear fusion, two $^2$H atoms are combined to produce $^4$He. (a) Calculate the decrease in rest mass in unified mass units. (b) How much energy is released in this reaction? (c) How many such reactions must take place per second to produce 1 W of power?

## Section 35-11, General Relativity

*There are no exercises for this section.*

## Problems

1. If a plane flies at a speed of 2000 km/h, how long must it fly before its clock loses 1 s because of time dilation?

2. (a) Show that the speed $u$ of a particle of mass $m$ and total energy $E$ is given by $u/c = [1 - (mc^2/E)^2]^{1/2}$ and that if $E$ is much greater than $mc^2$, the approximation $u/c \approx 1 - \frac{1}{2}(mc^2/E)^2$ holds. (b) Find the speed of an electron of kinetic energy 0.51 MeV and that of an electron of kinetic energy 10 MeV.

3. What percent error is made in using $\frac{1}{2}mu^2$ for the kinetic energy of a particle if its speed is (a) $u = 0.1c$, (b) $u = 0.9c$?

4. A stick has proper length $L$ and makes an angle $\theta$ with the $x$ axis in frame $S$. Show that the angle made with the $x'$ axis in frame $S'$ moving along the $+x$ axis with speed $v$ is $\theta'$ given by $\tan \theta' = \gamma \tan \theta$ and that the length of the stick in $S'$ is $L' = L[(\cos \theta/\gamma)^2 + \sin^2 \theta]^{1/2}$.

5. Two spaceships each 100 m long when measured at rest travel toward each other with speeds $0.8c$ relative to earth. (a) How long is each ship as measured by someone on earth? (b) How fast is each ship traveling as measured by the other? (c) How long is one ship when measured by the other? (d) At some time $t = 0$ (on earth clocks) the fronts of the ships are together as they begin to pass each other. At what time (on earth clocks) are their backs together? (e) Sketch diagrams in the frame of one of the ships showing the passing of the other ship.

6. Show that if a particle moves at an angle $\theta$ with the $x$ axis with speed $u$ in frame $S$, it moves at an angle $\theta'$ with the $x'$ axis in $S'$ given by

$$\tan \theta' = \frac{\sin \theta}{\gamma(\cos \theta - v/u)}$$

7. For the special case of a particle moving with speed $u$ along the $y$ axis in $S$, show that the momentum and energy in frame $S'$ are related to the momentum and energy in $S$ by the transformation equations

$$p'_x = \gamma\left(p_x - \frac{vE}{c^2}\right) \qquad p'_y = p_y$$

$$p'_z = p_z \qquad \frac{E'}{c} = \gamma\left(\frac{E}{c} - \frac{vp_x}{c}\right)$$

Compare these equations with the Lorentz transformation for $x'$, $y'$, $z'$, and $t'$.

These show that the quantities $p_x$, $p_y$, $p_z$, and $E/c$ transform in the same way as $x$, $y$, and $ct$.

8.  The equation for a spherical wavefront of a light pulse which begins at the origin at time $t = 0$ is $x^2 + y^2 + z^2 - (ct)^2 = 0$. Using the Lorentz transformation equations, show that such a light pulse also has a spherical wavefront in frame $S'$ by showing that $x'^2 + y'^2 + z'^2 - (ct')^2 = 0$ in $S'$.

9.  In Problem 8 you showed that the quantity $x^2 + y^2 + z^2 - (ct)^2$ has the same value (0) in both $S$ and $S'$. Such a quantity is called an *invariant*. From the results of Problem 7 the quantity $p_x^2 + p_y^2 + p_z^2 - (E/c)^2$ must also be invariant. Show that this quantity has the value $-m^2c^2$ in both the $S$ and $S'$ reference frames.

10.  Show that if $v$ is much less than $c$, the doppler frequency shift is approximately given by $\Delta f/f = \pm v/c$, both classically and relativistically. A radar transmitter-receiver bounces a signal off an aircraft and observes a fractional increase in the frequency of $\Delta f/f = 8 \times 10^{-7}$. What is the speed of the aircraft? (Assume the aircraft to be moving directly toward the transmitter.)

11.  Two observers agree to test time dilation. They use identical clocks, and one observer in frame $S'$ moves with speed $v = 0.6c$ relative to the other observer in frame $S$. When their origins coincide, they start their clocks. They agree to send a signal when their clocks read 60 min and to send a confirmation signal when each receives the other's signal. (a) When does the observer in $S$ receive the first signal from the observer in $S'$? (b) When does he receive the confirmation signal? (c) Make a table showing the times in $S$ when the observer sent the first signal, received the first signal, and received the confirmation signal. How does this table compare with one the observer in $S'$ might have constructed?

12.  A double star rotates about its center of mass such that one member is moving toward the earth and the other away from the earth. The center of mass is fixed relative to the earth. The hydrogen line of wavelength 656.3 nm in the laboratory has wavelength 660.0 nm from one of the stars and 650.0 nm from the other. What is the ratio of the masses of the stars? *Hint:* Show that the nonrelativistic doppler-effect equations have sufficient accuracy for this problem.

13.  Two events in $S$ are separated by a distance $D = x_2 - x_1$ and time $T = t_2 - t_1$. (a) Use the time-transformation equations to show that in frame $S'$ moving with speed $v$ relative to $S$ the time separation is $t'_2 - t'_1 = \gamma(T - vD/c^2)$. (b) Show that the events can be simultaneous in frame $S'$ only if $D$ is greater than $cT$. (c) If one of the events is the *cause* of the other, the separation $D$ must be less than $cT$ since $D/c$ is the smallest time a signal can take to travel from $x_1$ to $x_2$ in frame $S$. Show that if $D$ is less than $cT$, $t'_2$ is greater than $t'_1$ in all reference frames. This shows that the cause must precede the effect in all reference frames (assuming that it does in one frame). (d) Suppose that a signal could be sent with speed $c'$ greater than $c$ so that in frame $S$ the cause precedes the effect by the time $T = D/c'$, which is less than $D/c$. Show that there is then a reference frame moving at speed $v$ less than $c$ in which the effect precedes the cause.

14.  Two identical particles each have rest mass $m$. In reference frame $S$ which is the center-of-mass frame, the two particles are moving toward each other with speed $u$. According to the mass-energy relation, the rest mass of this two-particle system is now $M = \gamma 2m$, where $\gamma = (1 - u^2/c^2)^{-1/2}$ because of the internal (kinetic) energy of the system. The two particles collide inelastically with a spring which locks shut (Figure 35-20) and come to rest in $S$ with their initial kinetic energy transformed into potential energy. In this problem you are going to show that $M = \gamma 2m$ is a necessary consequence of momentum conservation in a reference frame $S'$ moving with speed $v = u$ relative to frame $S$ so that one of the particles is initially at rest in $S'$. (a) Show that the speed of the particle not at rest in frame $S'$ is $u'$, where $u' = 2u/(1 + u^2/c^2)$. Use this result to show that

$$\sqrt{1 - \frac{u'^2}{c^2}} = \frac{1 - u^2/c^2}{1 + u^2/c^2}$$

(b) Show that the initial momentum in frame $S'$ can be written $P_i' = 2mu/(1 - u^2/c^2)$. (c) After the collision, the two masses move with speed $u$ in $S'$ (since they are at rest in $S$). Write the total momentum in $S'$ after the collision in terms of the mass of the system $M$ and show that this equals the initial momentum only if $M = \gamma 2m$. (d) Find the total energy before and after the collision in each reference frame and show that the total energy is conserved in each frame.

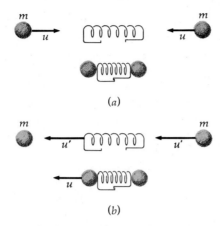

(a)

(b)

**Figure 35-20**
Problem 14. Inelastic collision of two equal masses in (a) center-of-mass frame $S$ and (b) frame $S'$ moving to right with speed $u$ relative to frame $S$ so that one of the particles is initially at rest. The spring is assumed to be massless.

15. A horizontal turntable rotates with angular speed $\omega$. There is a clock at the center of the turntable and one at a distance $r$ from the center. In an inertial reference frame the clock at distance $r$ is moving with speed $v = r\omega$. (a) Show that, from time dilation in special relativity, time intervals $\Delta t_0$ for the clock at rest and $\Delta t_r$ for the moving clock are related by

$$\frac{\Delta t_r - \Delta t_0}{\Delta t_0} \approx -\frac{r^2\omega^2}{2c^2} \quad \text{if} \quad r\omega \ll c$$

(b) In a reference frame attached to the turntable, both clocks are at rest. Show that the clock at distance $r$ experiences a pseudo (centrifugal) force $F_r = mr\omega^2$ in this accelerated frame. Show that this is equivalent to a difference in gravitational potential between $r$ and the origin of $\phi_r - \phi_0 = \frac{1}{2}r^2\omega^2$. Use this potential difference in Equation 35-36 to show that in this frame the difference in time intervals is the same as in the inertial frame.

# CHAPTER 36  Quantization

---

**Objectives**  After studying this chapter you should:

1. Be able to sketch the spectral distribution curve for blackbody radiation and the curve predicted by the Rayleigh-Jeans law.

2. Be able to discuss which features of the photoelectric effect are in accord with the predictions of classical physics and which are not.

3. Be able to state the Einstein equation for the photoelectric effect.

4. Be able to discuss how the photon concept explains all the features of the photoelectric effect and Compton scattering of x rays.

5. Be able to state the Bohr postulates and use them to derive the energy levels of the hydrogen atom and the wavelengths of the spectral lines of hydrogen.

6. Be able to draw an energy-level diagram for hydrogen and indicate on it transitions involving the emission of a photon.

7. Be able to state the de Broglie relations for the frequency and wavelength of electron waves and use them and the standing-wave condition to derive the Bohr condition for the quantization of angular momentum in the hydrogen atom.

8. Be able to discuss the experimental evidence for the existence of electron waves.

9. Be able to discuss wave-particle duality.

10. Be able to discuss the uncertainty principle.

---

In Chapter 31 we showed that a wave equation for electromagnetic waves traveling at the speed of light can be derived from Maxwell's equations, which summarize the experimental laws of electricity and magnetism. This unification of the previously separate subjects of optics and electromagnetism through Maxwell's equations was one of the great triumphs of classical physics in the nineteenth century. Other predictions of electromagnetism, some of which we did not study, are also important for understanding the modifications of classical physics in the twentieth century.

1. Light and other electromagnetic waves carry momentum as well as energy. The momentum $p$ and the energy $E$ are related by $p = E/c$, where $c$ is the speed of the waves.

2. If a charged particle has acceleration $a$, it radiates electromagnetic energy at a rate proportional to $a^2$.

3. If the acceleration of the charged particle is associated with periodic motion, e.g., simple harmonic oscillation or circular motion, the particle emits electromagnetic radiation with a frequency equal to that of the motion.

All these predictions have been verified in experiments involving macroscopic phenomena. Near the end of the nineteenth century it was generally believed that all natural phenomena could be described by Newton's laws, the laws of thermodynamics, and the laws of electromagnetism (as expressed in Maxwell's equations). There was nothing left for scientists to do but apply these laws to various phenomena and measure the next decimal in the fundamental constants (the speed of light, the electron charge, the gas constant, Avogadro's number, etc.). However, this optimism (or pessimism if you were a scientist hoping to

**Table 36-1**
Approximate dates of some important experiments and theories, 1881–1932

| | |
|------|--------------------------------------------------------------------------------------------------------------------------------------|
| 1881 | Michelson obtains null result for absolute velocity of earth |
| 1884 | Balmer finds empirical formula for spectral lines of hydrogen |
| 1887 | Hertz produces electromagnetic waves, verifying Maxwell's theory and accidently discovering photoelectric effect |
| 1887 | Michelson repeats his experiment with Morley, again obtaining null result |
| 1895 | Röntgen discovers x rays |
| 1896 | Becquerel discovers nuclear radioactivity |
| 1897 | J. J. Thomson measures $e/m$ for cathode rays, showing that electrons are fundamental constituents of atoms |
| 1900 | Planck explains blackbody radiation using energy quantization involving new constant $h$ |
| 1900 | Lenard investigates photoelectric effect and finds energy of electrons independent of light intensity |
| 1905 | Einstein proposes special theory of relativity |
| 1905 | Einstein explains photoelectric effect by suggesting quantization of radiation |
| 1907 | Einstein applies energy quantization to explain temperature dependence of heat capacities of solids |
| 1908 | Rydberg and Ritz generalize Balmer's formula to fit spectra of many elements |
| 1909 | Millikan's oil-drop experiment shows quantization of electric charge |
| 1911 | Rutherford proposes nuclear model of atom based on alpha-particle scattering experiments of Geiger and Marsden |
| 1912 | Friedrich and Knipping and von Laue demonstrate diffraction of x rays by crystals showing that x rays are waves and crystals are regular arrays |

make important discoveries) proved premature. We have already described how newtonian mechanics must be replaced with the special theory of relativity when the speed of a particle is not small compared with the speed of light, and we showed that when the particle's speed is small, the special theory of relativity reduces to Newton's laws. Many startling discoveries, both experimental and theoretical, in the last 20 years of the nineteenth century and first 30 years of the twentieth century demonstrated that the laws of classical physics also break down when applied to microscopic systems such as the particles within an atom. This failure is even more drastic than the failure of newtonian mechanics at high speeds. The interior of the atom can be described only in terms of quantum theory, which requires the modification of some of our fundamental ideas about the relationships between physical theory and the physical world. As with special relativity, quantum theory reduces to classical physics when applied to macroscopic systems. Table 36-1 lists the approximate dates of some of the important experiments and theories from 1881 to 1932.

A proper study of quantum theory is beyond the scope of this book. In this chapter we shall concentrate on the fundamental ideas of energy quantization and the wave-particle duality of nature.

**Table 36-1**
(*continued*)

|  |  |
|------|-------------|
| 1913 | Bohr proposes model of hydrogen atom |
| 1914 | Moseley analyzes x-ray spectra using Bohr model to explain periodic table in terms of atomic number |
| 1914 | Franck and Hertz demonstrate atomic energy quantization |
| 1915 | Duane and Hunt show that the short-wavelength limit of x rays is determined from quantum theory |
| 1916 | Wilson and Sommerfeld propose rules for quantization of periodic systems |
| 1916 | Millikan verifies Einstein's photoelectric equation |
| 1923 | Compton explains x-ray scattering by electrons as collision of photon and electron and verifies results experimentally |
| 1924 | De Broglie proposes electron waves of wavelength $h/p$ |
| 1925 | Schrödinger develops mathematics of electron wave mechanics |
| 1925 | Heisenberg invents matrix mechanics |
| 1925 | Pauli states exclusion principle |
| 1927 | Heisenberg formulates uncertainty principle |
| 1927 | Davisson and Germer observe electron wave diffraction by single crystal |
| 1927 | G. P. Thomson observes electron wave diffraction in metal foil |
| 1928 | Gamow and Condon and Gurney apply quantum mechanics to explain alpha-decay lifetimes |
| 1928 | Dirac develops relativistic quantum mechanics and predicts existence of positron |
| 1932 | Chadwick discovers neutron |
| 1932 | Anderson discovers positron |

## 36-1    The Origin of the Quantum Constant: Blackbody Radiation

One of the most puzzling phenomena studied near the end of the nineteenth century was the spectral distribution of blackbody radiation. A blackbody is an ideal system which absorbs all the radiation incident on it; it can be approximated by a cavity with a very small opening (see Section 18-4). The characteristics of the radiation in such a cavity of a body in thermal equilibrium depend only on the temperature of the walls. The fraction of the radiant energy density of wavelength $\lambda$ in the interval $d\lambda$ is called the *spectral distribution* $f(\lambda, T)\, d\lambda$. The function $f(\lambda, T)$ was calculated from classical physics in a straightforward way and the result compared with experiment. The calculation involved finding the number of standing waves in a three-dimensional cavity in the wavelength interval $d\lambda$ and multiplying by the average energy per wave, which the equipartition theorem predicts to be $kT$. The result, known as the *Rayleigh-Jeans law*, is

$$f(\lambda, T) = 8\pi kT\lambda^{-4} \qquad \text{36-1} \qquad \textit{Rayleigh-Jeans law}$$

This result agrees with experiment in the region of long wavelengths but disagrees violently at short wavelengths. As $\lambda$ approaches zero, the experimentally determined $f(\lambda, T)$ also approaches zero but the calculated function becomes infinite as $\lambda^{-4}$. This result was known as the *ultraviolet catastrophe*. In 1900 the German physicist Max Planck announced that by making a somewhat peculiar modification in the classical calculation he could derive a function $f(\lambda, T)$ that agreed with the experimental data at all wavelengths. He first found an empirical function that fitted the data and then searched for a way to modify the usual calculation. His empirical function is

$$f(\lambda, T) = \frac{8\pi hc\lambda^{-5}}{e^{hc/\lambda kT} - 1} \qquad \text{36-2} \qquad \textit{Planck's law}$$

where $h$ is an adjustable constant. Figure 36-1 compares this function with the Rayleigh-Jeans law and with experimental data. At large wavelengths, the exponential in the denominator of Equation 36-2 is small, and we can use the approximation

$$e^{hc/\lambda kT} \approx 1 + \frac{hc}{\lambda kT}$$

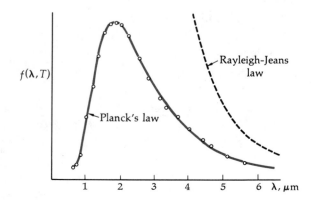

$f(\lambda, T)$

Rayleigh-Jeans law

Planck's law

1    2    3    4    5    6    $\lambda, \mu m$

**Figure 36-1**
Spectral distribution of blackbody radiation versus wavelength at $T = 1600$ K.
(*Adapted from F. K. Richtmyer, E. H. Kennard, and J. N. C. Lauritsen,* Introduction to Modern Physics, *5th ed., McGraw-Hill Book Company, New York, 1955, by permission.*)

Then

$$f(\lambda,\ T) \rightarrow \frac{8\pi hc\lambda^{-5}}{hc/\lambda kT} = 8\pi kT\lambda^{-4}$$

which is the Rayleigh-Jeans formula. For small wavelengths, the exponential is very large, and we can neglect the 1 in the denominator of Equation 36-2, giving

$$f(\lambda,\ T) \rightarrow 8\pi hc\lambda^{-5}e^{-hc/\lambda kT} \qquad 36\text{-}3$$

This approaches zero as $\lambda$ approaches zero. The constant $h$ can be found by fitting this formula to the data points at small $\lambda$. Its presently accepted value is

$$h = 6.626 \times 10^{-34}\ \text{J·s} = 4.136 \times 10^{-15}\ \text{eV·s} \qquad 36\text{-}4$$

*Planck's constant*

Planck found that he could derive Equation 36-2 by modifying the calculation of the average energy per wave in the cavity. Instead of dealing directly with the radiation in the cavity, Planck considered the emission and absorption of radiation by the cavity walls, which he represented by a set of oscillators of all frequencies. He reasoned that, in equilibrium, the average energy of an oscillator of frequency $f$ would be associated with the average energy of the electromagnetic radiation of that frequency in the cavity. In the usual calculation of the average energy of an oscillator in thermal equilibrium with its surroundings the energy $E$ is a continuous variable, and integration is used. The result for a one-dimensional oscillator is $kT$. Planck modified this calculation by assuming that the energy of an oscillator is discrete; i.e., it can take on only certain values $E_n$, given by

$$E_n = nhf \qquad 36\text{-}5$$

where $n$ is an integer, $f$ is the frequency, and $h$ is a constant. With this assumption Planck obtained for the average energy of an oscillator

$$E_{\text{av}} = \frac{hc/\lambda}{e^{hc/\lambda kT} - 1} \qquad 36\text{-}6$$

Replacing $kT$ in Equation 36-1 by this expression gives Equation 36-2. Although Planck tried to fit the constant $h$ into the framework of classical physics, he was unable to do so. The fundamental importance of his assumption of energy quantization, implied by Equation 36-5, was not generally appreciated until Einstein applied similar ideas to explain the photoelectric effect and suggested that quantization is a fundamental property of electromagnetic radiation.

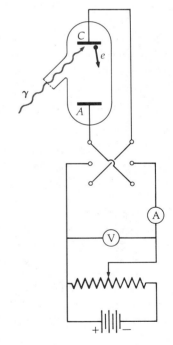

**Figure 36-2**
Schematic drawing of photoelectric-effect apparatus. Light strikes the cathode $C$ and ejects electrons. The number of electrons which reach the anode $A$ is measured by the current in the ammeter. The anode can be made positive or negative with respect to the cathode to attract or repel the electrons.

## 36-2   Quantization of Electromagnetic Radiation: Photons

As mentioned in Chapter 32, the photoelectric effect was discovered by Hertz in 1887 and studied by Lenard in 1900. Figure 36-2 shows a schematic diagram of the basic apparatus. When light is incident on a clean metal surface, the cathode $C$, electrons are emitted. If some of these electrons strike the anode $A$, there is a current in the external circuit. The number of the emitted electrons reaching the anode can be increased or decreased by making the anode positive or negative with respect to the cathode. Let $V$ be the increase in potential from the cathode to anode. Figure 36-3 shows the current versus $V$ for two values of the intensity of light incident on the cathode. When $V$ is positive, the electrons are attracted to the anode. At sufficiently large $V$ all the emitted electrons reach the anode and the current reaches its maximum value. A further increase in $V$ does not affect the current. Lenard observed that the maximum current is proportional to the light intensity, an expected result since doubling the energy per unit time incident on the cathode should double the number of electrons emitted. When $V$ is negative, the electrons are repelled from the anode. Only electrons with initial kinetic energy $\frac{1}{2}mv^2$ greater than $|eV|$ can then reach the anode.

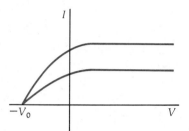

From Figure 36-3 we see that if $V$ is less than $-V_0$, no electrons reach the anode. The potential $V_0$ is called the *stopping potential*. It is related to the maximum kinetic energy of the emitted electrons by

$$(\tfrac{1}{2}mv^2)_{\text{max}} = eV_0$$

*Stopping potential*

The experimental result that $V_0$ is independent of the incident light intensity was surprising. Apparently increasing the rate of energy falling on the cathode does not increase the maximum kinetic energy of the electrons emitted. In 1905 Einstein demonstrated that this result can be understood if light energy is not distributed continuously in space but quantized in small bundles called *photons*. The energy of each photon is $hf$, where $f$ is the frequency and $h$ is Planck's constant. An electron emitted from a metal surface exposed to light receives its energy from a single photon. When the intensity of the light of a given frequency is increased, more photons fall on the surface in a unit time but the energy absorbed by each electron is unchanged. If $\phi$ is the energy necessary to remove an electron from the surface of a metal, the maximum kinetic energy of the electrons emitted will be

$$(\tfrac{1}{2}mv^2)_{\text{max}} = eV_0 = hf - \phi \qquad\qquad 36\text{-}7$$

*Einstein's photoelectric equation*

The quantity $\phi$, called the *work function*, is a characteristic of the metal. Some electrons will have kinetic energy less than $hf - \phi$ because of energy loss traversing the metal. Equation 36-7 is known as *Einstein's photoelectric equation*.

> If the derived formula is correct, then $V_0$ when represented in cartesian coordinates as a function of the frequency of the incident light must be a straight line whose slope is independent of the nature of the emitting substance.[*]

From Equation 36-7 we see that the slope of $V_0$ versus $f$ should equal $h/e$. Einstein's equation was a bold prediction, for at that time there

---

[*] From Einstein's original paper translated by A. B. Arons and M. B. Peppard, *American Journal of Physics*, vol. 33, p. 367, 1965.

was no evidence that Planck's constant had any applicability outside of blackbody radiation, and there were no experimental data on the stopping potential $V_0$ as a function of frequency. The experimental verification of Einstein's theory was quite difficult. Careful experiments by R. C. Millikan reported in 1914 and in more detail in 1916 showed that the Einstein equation was correct, and measurements of $h$ agreed with the value found by Planck. Figure 36-4 shows a plot of Millikan's data.

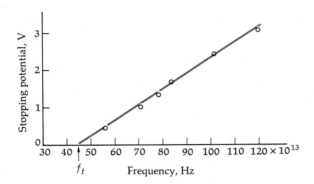

**Figure 36-4**
Millikan's data for the stopping potential versus frequency for the photoelectric effect. The data fall on a straight line which has slope $h/e$ as predicted by Einstein a decade before the experiment. (*From R. C. Millikan,* Physical Review, *vol. 7, p. 362, 1916.*)

The threshold frequency $f_t$ and the corresponding threshold wavelength $\lambda_t$ are related to the work function $\phi$ by setting $V_0$ equal to zero in Equation 36-7. Then

$$\phi = hf_t = \frac{hc}{\lambda_t} \qquad\qquad 36\text{-}8$$

*Threshold frequency and wavelength*

Photons of frequency less than $f_t$ (and therefore wavelengths greater than $\lambda_t$) do not have enough energy to eject an electron from the metal. Work functions for metals are typically a few electronvolts. Since wavelengths are usually given in nanometres and energies in electronvolts, it is useful to have the value of $hc$ in electronvolt-nanometres. We have

$$hc = (4.14 \times 10^{-15} \text{ eV·s})(3 \times 10^8 \text{ m/s})$$
$$= 1.24 \times 10^{-6} \text{ eV·m}$$

or

$$hc = 1240 \text{ eV·nm} \qquad\qquad 36\text{-}9$$

**Example 36-1** The threshold wavelength for potassium is 564 nm. What is the work function for potassium? What is the stopping potential when light of wavelength 400 nm is used?

$$\phi = hf_t = \frac{hc}{\lambda_t} = \frac{1240 \text{ eV·nm}}{564 \text{ nm}} = 2.20 \text{ eV}$$

The energy of a photon of wavelength 400 nm is

$$E = \frac{hc}{\lambda} = \frac{1240 \text{ eV·nm}}{400 \text{ nm}} = 3.10 \text{ eV}$$

The maximum kinetic energy of the emitted electrons is then

$$(\tfrac{1}{2}mv^2)_{\max} = hf - \phi = 3.10 \text{ eV} - 2.20 \text{ eV} = 0.90 \text{ eV}$$

The stopping potential is therefore 0.90 V.

Another interesting feature of the photoelectric effect is the absence of lag between the time the light is turned on and the time the electrons appear. In the classical theory, given the intensity (power per unit area), the time can be calculated for enough energy to fall on the area of an atom to eject an electron. However, even when the intensity is so small that such a calculation gives time lags of hours, essentially no time lag is observed. The explanation of this result is that although the number of photons hitting the metal per unit time is very small when the intensity is low, each photon has enough energy to eject an electron and there is a chance that one photon will be absorbed immediately. The classical calculation gives the correct average number of protons absorbed per unit time.

Further evidence of the correctness of the photon concept was furnished by Arthur H. Compton, who measured the scattering of x rays by free electrons. According to classical theory, when an electromagnetic wave of frequency $f_1$ is incident on material containing charges, the charges will oscillate with this frequency and reradiate electromagnetic waves of the same frequency. Compton pointed out that if the scattering process were considered to be a collision between a photon and electron, the electron would absorb energy due to recoil and the scattered photon would have less energy and therefore a lower frequency than the incident photon. According to classical theory, the energy and momentum of an electromagnetic wave are related by $E = pc$. This result is also consistent with the relativistic expression relating the energy and momentum of a particle (Equation 35-29):

$$E^2 = p^2 c^2 + (mc^2)^2 \qquad \text{36-10}$$

if the mass of the photon is assumed to be zero. Figure 36-5 shows the geometry of a collision between a photon of wavelength $\lambda_1$ and an electron at rest. Compton related the scattering angle $\theta$ to the incident and scattered wavelengths $\lambda_1$ and $\lambda_2$ by treating the scattering as a relativistic-mechanics problem using conservation of energy and momentum. Let $\mathbf{p}_1$ be the momentum of the incident photon, $\mathbf{p}_2$ that of the scattered photon, and $\mathbf{p}_e$ that of the recoiling electron. Conservation of momentum gives

$$\mathbf{p}_1 = \mathbf{p}_2 + \mathbf{p}_e$$

or

$$p_e^2 + (p_1 - p_2)^2 = p_1^2 + p_2^2 - 2\mathbf{p}_1 \cdot \mathbf{p}_2$$
$$= p_1^2 + p_2^2 - 2p_1 p_2 \cos \theta \qquad \text{36-11}$$

The energy before the collision is $p_1 c + mc^2$, where $mc^2$ is the rest energy of the electron. After the collision the electron has energy $\sqrt{(mc^2)^2 + p_e^2 c^2}$. Conservation of energy then gives

$$p_1 c + mc^2 = p_2 c + \sqrt{(mc^2)^2 + p_e^2 c^2} \qquad \text{36-12}$$

Compton eliminated the electron momentum $p_e$ from Equations 36-11 and 36-12 and expressed the photon momenta in terms of the wavelengths, to obtain an equation relating the incident and scattered wavelengths $\lambda_1$ and $\lambda_2$ and the angle $\theta$. The algebraic details are left as a problem (Problem 4). Compton's result is

$$\lambda_2 - \lambda_1 = \frac{h}{mc} (1 - \cos \theta) \qquad \text{36-13}$$

The change in wavelength is independent of the original wavelength.

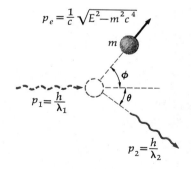

**Figure 36-5**
Compton scattering of an x ray by an electron. The scattered photon has less energy and therefore greater wavelength than the incident photon because of the recoil energy of the electron. The change in wavelength $\lambda_2 - \lambda_1$ is found from conservation of energy and momentum.

The quantity $h/mc$ depends only on the mass of the electron. It has dimensions of length and is called the *Compton wavelength*. Its value is

$$\lambda_c = \frac{h}{mc} = \frac{hc}{mc^2} = \frac{1240 \text{ eV·nm}}{5.11 \times 10^5 \text{ eV}} = 2.43 \text{ pm} \qquad 36\text{-}14$$

*Compton wavelength*

Because $\lambda_2 - \lambda_1$ is small, it is difficult to observe unless $\lambda_1$ is so small that the fractional change $(\lambda_2 - \lambda_1)/\lambda_1$ is appreciable. Compton used x rays of wavelength 71.1 pm. The energy of a photon of this wavelength is $E = hc/\lambda = (1240 \text{ eV·nm})/(0.0711 \text{ nm}) = 17.4 \text{ keV}$. Since this is much greater than the binding energy of the valence electrons in carbon, these electrons can be considered to be essentially free. Compton's experimental results for $\lambda_2 - \lambda_1$ as a function of scattering angle $\theta$ agreed with Equation 36-13, thereby confirming the correctness of the photon concept.

## 36-3  Quantization of Atomic Energies: The Bohr Model

In Planck's treatment of blackbody radiation he represented the walls of the body by a set of oscillators of frequency $f$ whose energy is quantized in units of $hf$. In 1907 Einstein applied similar reasoning to explain the temperature dependence of the heat capacity of a solid. The most famous application of energy quantization to microscopic systems was that of Niels Bohr. In 1913 Bohr proposed a model of the hydrogen atom which had spectacular successes in calculating the wavelengths of lines in the known hydrogen spectrum and in predicting new lines (later found experimentally) in the infrared and ultraviolet spectra.

Many data were collected near the turn of the century on the emission of light by atoms in a gas when excited by an electric discharge. Viewed through a spectroscope with a narrow-slit aperture, this light appears as a discrete set of lines of different color or wavelength; the spacing and intensities of the lines are characteristic of the element. It was possible to determine the wavelengths of these lines accurately, and much effort went into finding regularities in the spectra. In 1884 a Swiss schoolteacher, Johann Balmer, found that the wavelengths of some of the lines in the spectrum of hydrogen can be represented by the formula

$$\lambda = 364.6 \frac{m^2}{m^2 - 4} \text{ nm} \qquad 36\text{-}15$$

*Balmer formula*

where $m$ is a variable integer which takes on the values $m = 3, 4, 5, \ldots$. Figure 36-6 shows the set of spectral lines of hydrogen, now known as the *Balmer series*, whose wavelengths are given by Equation 36-15. Balmer suggested that his formula might be a special case of a more general expression applicable to the spectra of other elements.

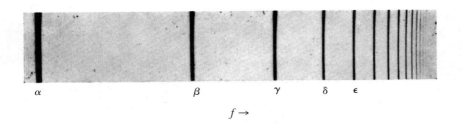

α        β        γ    δ   ε

$f \rightarrow$

**Figure 36-6**
Balmer series in hydrogen. The wavelengths of these lines are given by Equation 36-15 for different values of the integer $m$. [*From G. Herzberg, Annalen der Physik, vol. 84, p. 565 (1927).*]

Such an expression, found by Johannes R. Rydberg and Walter Ritz, gives the reciprocal wavelength as

$$\frac{1}{\lambda} = R \left( \frac{1}{m^2} - \frac{1}{n^2} \right) \qquad n > m \qquad \qquad 36\text{-}16$$

*Rydberg-Ritz formula*

where $R$, called the *Rydberg constant* or the *Rydberg,* is the same for all series of the same element and varies only slightly in a regular way from element to element. For hydrogen, the value of $R$ is $R_H = 10.96776 \ \mu m^{-1}$. For very massive elements, $R$ approaches the value $R_\infty = 10.97373 \ \mu m^{-1}$. (The quantity $1/\lambda$ was used instead of the frequency $f = c/\lambda$ because $\lambda$ can be measured much more accurately than the speed of light $c$.) Such empirical expressions were successful in predicting other spectra; e.g., other hydrogen lines outside the visible spectrum were predicted and found.

Many attempts were made to construct a model of the atom that would yield these formulas for its radiation spectrum. The most popular model, due to J. J. Thomson, considered various arrangements of electrons embedded in some kind of fluid that contained most of the mass of the atom and had enough positive charge to make the atom electrically neutral. Since classical electromagnetic theory predicted that a charge oscillating with frequency $f$ would radiate light of that frequency, Thomson searched for configurations that were stable and had normal modes of vibration of frequencies equal to those of the spectrum of the atom. A difficulty of this model and all others was that electric forces alone cannot produce stable equilibrium. Thomson was unsuccessful in finding a model which predicted the observed frequencies for any atom.

The Thomson model was essentially ruled out by a set of experiments by H. W. Geiger and E. Marsden under the supervision of E. Rutherford about 1911, in which alpha particles from radioactive radium were scattered by atoms in a gold foil. Rutherford showed that the number of alpha particles scattered at large angles could not be accounted for by an atom in which the positive charge was distributed throughout the atomic size (known to be about 0.1 nm in diameter) but required that the positive charge and most of the mass of the atom be concentrated in a very small region, now called the nucleus, of diameter of the order of $10^{-6}$ nm = 1 fm. Niels Bohr, working in the Rutherford laboratory at the time, proposed a model of the hydrogen atom which combined the work of Planck, Einstein, and Rutherford and successfully predicted the observed spectra. Bohr assumed that the electron in the hydrogen atom moved under the influence of the Coulomb attraction to the positive nucleus according to classical mechanics, which predicts circular or elliptical orbits with the force center at one focus, as in the motion of the planets about the sun. For simplicity he chose a circular orbit. Although mechanical stability is achieved because the Coulomb attractive force provides the centripetal force necessary for the electron to remain in orbit, such an atom is unstable electrically according to classical theory because the electron must accelerate when moving in a circle and therefore radiate electromagnetic energy of frequency equal to that of its motion. According to classical electromagnetic theory, such an atom would quickly collapse, the electron spiraling into the nucleus as it radiates away its energy. Bohr "solved" this difficulty, modifying the laws of electromagnetism by *postulating* that the electron could move in certain orbits without radiating. He called these stable orbits *stationary states.*

*Bohr's first postulate: stationary states*

The atom radiates when the electron somehow makes a transition from one stationary state to another. The frequency of radiation is not the frequency of motion in either stable orbit but is related to the energies of the orbits by

$$f = \frac{W_i - W_f}{h} \qquad 36\text{-}17$$

Bohr's second postulate: photon frequency from energy conservation

where $h$ is Planck's constant and $W_i$ and $W_f$ are the total energies in the initial and final orbits. This assumption, which is equivalent to that of energy conservation with the emission of a photon, is a key one in the Bohr theory because it deviates from the classical theory, which requires the frequency of radiation to be that of the motion of the charged particle. Planck made a similar assumption in his blackbody radiation theory, but in that case the energy of the oscillators was $nhf$ and the energy difference between two adjacent states $hf$, so that when the oscillator emitted or absorbed radiation, the frequency of the radiation was the same as that of the oscillator.

If the nuclear charge is $+Ze$ and the electron charge $-e$, the potential energy at a distance $r$ is

$$U = -\frac{kZe^2}{r}$$

where $k = 1/4\pi\epsilon_0$ is the Coulomb constant. (For hydrogen, $Z = 1$, but it is convenient to not specify $Z$ at this time so that the results can be applied to other atoms.) The total energy of the electron moving in a circular orbit with speed $v$ is then

$$W = \tfrac{1}{2}mv^2 + U = \tfrac{1}{2}mv^2 - \frac{kZe^2}{r}$$

The kinetic energy can be obtained as a function of $r$ by using Newton's law $F = ma$. Setting the Coulomb attractive force equal to the mass times the centripetal acceleration, we have

$$\frac{kZe^2}{r^2} = m\frac{v^2}{r}$$

or

$$\tfrac{1}{2}mv^2 = \frac{1}{2}\frac{kZe^2}{r} \qquad 36\text{-}18$$

For circular orbits the kinetic energy equals half the magnitude of the potential energy, a result which holds for circular motion in any inverse-square-law force field. The total energy is then

$$W = -\frac{1}{2}\frac{kZe^2}{r} \qquad 36\text{-}19$$

Using Equation 36-17 for the frequency of radiation when the electron changes from orbit 1 of radius $r_1$ to orbit 2 of radius $r_2$, we obtain

$$f = \frac{W_1 - W_2}{h} = \frac{1}{2}\frac{kZe^2}{h}\left(\frac{1}{r_2} - \frac{1}{r_1}\right) \qquad 36\text{-}20$$

To obtain the Balmer-Ritz formula $f = c/\lambda = cR(1/m^2 - 1/n^2)$ it is evident that the radii of stable orbits must be proportional to the squares of integers. Bohr searched for a quantum condition for the radii of the stable orbits which would yield this result. After much trial and error, he found that he could obtain the correct results if he postulated that in a stable orbit the angular momentum of the electron equals an integer times Planck's constant divided by $2\pi$. Since the angular momentum of a circular orbit is just $mvr$, this postulate is

$$mvr = \frac{nh}{2\pi} = n\hbar \qquad 36\text{-}21$$

*Bohr's third postulate: quantized angular momentum*

where $\hbar = h/2\pi$. (The constant $\hbar = h/2\pi$ is often more convenient than $h$ itself, just as angular frequency $\omega = 2\pi f$ is often more convenient than the frequency $f$.) We can determine $r$ by eliminating $v$ between Equations 36-18 and 36-21:

$$v^2 = n^2 \frac{\hbar^2}{m^2 r^2} = \frac{kZe^2}{mr}$$

or

$$r = n^2 \frac{\hbar^2}{mkZe^2} = n^2 \frac{r_0}{Z} \qquad 36\text{-}22$$

where

$$r_0 = \frac{\hbar^2}{mke^2} \approx 0.0529 \text{ nm} \qquad 36\text{-}23$$

is called the *first Bohr radius*. Combining Equations 36-22 and 36-20, we get

$$f = Z^2 \frac{mk^2 e^4}{4\pi\hbar^3} \left( \frac{1}{n_2^2} - \frac{1}{n_1^2} \right) \qquad 36\text{-}24$$

When we compare this for $Z = 1$ with the empirical Balmer-Ritz formula (Equation 36-16), we have for the Rydberg constant

$$R = \frac{mk^2 e^4}{4\pi ch^3} \qquad 36\text{-}25$$

Using the values of $m$, $e$, and $\hbar$ known in 1913, Bohr calculated $R$ and found his result to agree (within the limits of the uncertainties of the constants) with the value obtained from spectroscopy.

The possible values of the energy of the hydrogen atom predicted by the Bohr model are given by Equation 36-19, with $r$ given by Equation 36-22. They are

$$W_n = -\frac{k^2 e^4 m}{2\hbar^2} \frac{Z^2}{n^2} = -Z^2 \frac{E_0}{n^2} \qquad 36\text{-}26$$

where $E_0 = k^2 e^4 m / 2\hbar^2 \approx 13.6$ eV. It is convenient to represent these energies in an energy-level diagram, as in Figure 36-7. Various series of transitions are indicated in this diagram by vertical arrows between the levels. The frequency of the light emitted in one of these transitions is the energy difference divided by $h$, according to Equation 36-17. At the time of Bohr's paper (1913) the Balmer series, corresponding to $n_2 = 2$, $n_1 = 3, 4, 5, \ldots$ , and the Paschen series, corresponding to $n_2 = 3$, $n_1 = 4, 5, 6 \ldots$ , were known. In 1916 T. Lyman found the series corresponding to $n_2 = 1$, and in 1922 and 1924 F. Brackett and H. A. Pfund, respectively, found series corresponding to $n_2 = 4$ and $n_2 = 5$. As can be determined by computing the wavelengths of these series, only the

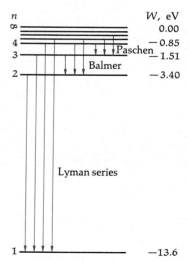

**Figure 36-7**
Energy-level diagram for hydrogen showing a few transitions in each of the Lyman, Balmer, and Paschen series. The energies of the levels are given by Equation 36-26.

Balmer series lies in the visible portion of the electromagnetic spectrum.

In our derivations we have assumed the electron to revolve around a stationary nucleus. This is equivalent to assuming the nucleus to have infinite mass. Since the mass of the hydrogen nucleus is not infinite but only about 2000 times that of the electron, a correction must be made for the motion of the nucleus. In the center-of-mass frame of the atom, the nucleus and electron have equal and opposite momenta, giving zero total momentum to the atom. If $p$ is the magnitude of the momentum of either, the kinetic energy of the nucleus is $p^2/2M$, where $M$ is its mass and the kinetic energy of the electron is $p^2/2m$. The total kinetic energy is

$$E_k = \frac{p^2}{2M} + \frac{p^2}{2m} = \frac{M + m}{mM}\frac{p^2}{2} = \frac{p^2}{2\mu}$$

where $\mu$, called the *reduced mass*, is given by

$$\mu = \frac{mM}{m + M} = \frac{m}{1 + m/M} \qquad\qquad \text{36-27} \qquad \textit{Reduced mass}$$

This is slightly different from the kinetic energy of just the electron because the reduced mass $\mu$ differs slightly from the electron mass $m$. We can correct any of our results for the motion of the nucleus by replacing the electron mass $m$ with the reduced mass $\mu$. (The validity of this procedure is shown in most intermediate and advanced mechanics books.) This correction leads to a very slight dependence of the Rydberg constant as given in Equation 36-25 on the nuclear mass, in precise agreement with the observed variation.

---

**Example 36-2** Find the energy and wavelength of the longest-wavelength line in the Lyman series.

From Figure 36-7 we see that the Lyman series corresponds to transitions ending at the ground state of energy, $W_1 = -13.6$ eV. Since $\lambda$ varies inversely with energy, the longest-wavelength transition is the lowest-energy transition from the first excited state $n = 2$ to the ground state. The first excited state has energy $W_2 = -13.6$ eV$/4 = -3.40$ eV. Since this is 10.2 eV above the ground-state energy, the energy of the photon emitted is 10.2 eV. The wavelength of this photon is

$$\lambda = \frac{hc}{E} = \frac{1240 \text{ eV·nm}}{10.2 \text{ eV}} = 121.6 \text{ nm}$$

This photon is outside the visible spectrum toward the ultraviolet. Since all the other lines in the Lyman series have even greater energies and shorter wavelengths, the Lyman series is completely in the ultraviolet region.

---

**Questions**

1. If an electron moves to a larger orbit, does its total energy increase or decrease? Does its kinetic energy increase or decrease?

2. How does the spacing of adjacent energy levels change as $n$ increases?

3. What is the energy of the shortest-wavelength photon emitted by the hydrogen atom?

## 36-4   Electron Waves

In 1924 a French student, L. de Broglie, suggested in his dissertation that since light is known to have both wave and particle properties, perhaps matter—especially electrons—may also have wave as well as particle characteristics. This suggestion was highly speculative; there was no evidence at that time for any wave aspects of electrons. For the frequency and wavelength of electron waves de Broglie chose the equations

$$f = \frac{E}{h} \qquad\qquad\qquad\qquad 36\text{-}28$$

*De Broglie relations*

$$\lambda = \frac{h}{p} \qquad\qquad\qquad\qquad 36\text{-}29$$

where $p$ is the momentum and $E$ the energy of the electron. These equations also hold for photons, as we have seen. De Broglie pointed out that with these relations, the Bohr quantum condition (Equation 36-21) is equivalent to a standing-wave condition. We have

$$mvr = n \frac{h}{2\pi}$$

Substituting $h/\lambda$ for the momentum $mv$ gives

$$\frac{h}{\lambda} r = n \frac{h}{2\pi}$$

or

$$n\lambda = 2\pi r = C \qquad\qquad\qquad 36\text{-}30$$

where $C$ is the circumference of the circular Bohr orbit. Thus Bohr's quantum condition is equivalent to saying that an integral number of electron waves must fit into the circumference of the circular orbit (Figure 36-8). The idea of explaining the discrete energy states of matter by standing waves seemed promising. In classical wave theory applied to waves on a string, for example, application of the boundary conditions that the string be fixed at both ends leads to a discrete set of allowed frequencies—those for which an integral number of half wavelengths fit into the length of the string. If energy is associated with the frequency of a standing wave, as in Equation 36-28, standing waves imply quantized energies. These ideas were developed into a detailed mathematical theory by Erwin Schrödinger in 1925. Schrödinger found a wave equation for electron waves and solved the standing-wave problem for the hydrogen atom, the simple harmonic oscillator, and other systems of interest. He found that the allowed frequencies, combined with the de Broglie relation $E = hf$, led to the same set of energy levels for the hydrogen atom as found by Bohr (Equation 36-26) and to almost the same set of energies for the simple harmonic oscillator as assumed by Planck and Einstein.* Schrödinger's *wave mechanics* therefore gave a general method of finding the quantization condition for a given system.

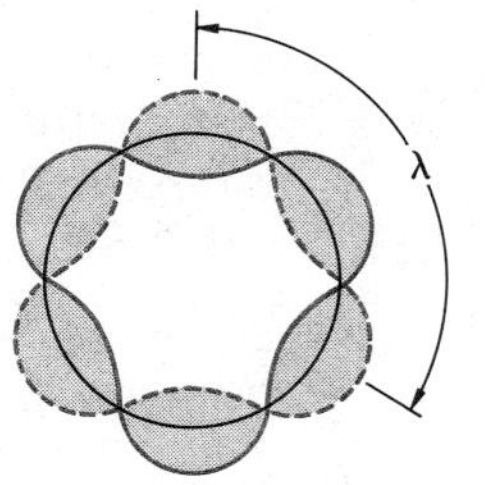

**Figure 36-8**
Standing waves around the circumference of a circle.

---

* Schrödinger found $E_n = (n + \frac{1}{2})hf$ for the energy levels of the simple harmonic oscillator, which differs slightly from $E_n = nhf$ assumed by Planck and Einstein. This difference amounts to a shift of each energy level by $\frac{1}{2}hf$ and has no effect on their results, which depend only on the distance between energy levels.

The first direct measurements of the wavelengths of electrons were made in 1927 by C. J. Davisson and L. H. Germer, who were studying electron scattering from a nickel target at Bell Telephone Laboratories. After heating the target to remove an oxide coating that had accumulated during an accidental break in the vacuum system, they found that the scattered-electron intensity as a function of the scattering angle showed maxima and minima. Their target had crystallized, and by accident they had observed electron diffraction. They then prepared a target consisting of a single crystal of nickel and investigated this phenomenon extensively. From measurements of the location of the observed diffraction maxima and minima, the wavelength of the electrons could be calculated, and agreement with the de Broglie relation (Equation 36-29) was found in all cases. In the same year G. P. Thomson (son of J. J. Thomson) also observed electron diffraction in transmission of electrons through thin metallic foils. Since then, diffraction has been observed for neutrons, protons, and other particles of small mass. Figure 36-9 shows a diffraction pattern produced by electrons incident on two narrow slits. This pattern is identical to that observed with photons of the same wavelength. Figures 36-10 and 36-11 compare the diffraction patterns of x rays, electrons, and neutrons of similar wavelength in transmission through thin foils.

**Figure 36-9**
Two-slit electron diffraction pattern. This pattern is the same as that usually obtained with photons. (*Courtesy of Claus Jönsson.*)

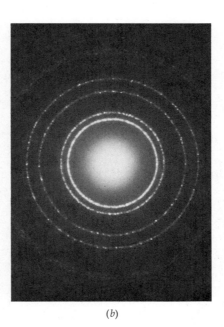

(*a*)                                    (*b*)

**Figure 36-10**
Diffraction pattern produced by (*a*) x rays and (*b*) electrons on an aluminum foil target. (*Courtesy of Film Studio, Education Development Center, Newton, Mass.*)

**Figure 36-11**
Diffraction pattern produced by neutrons on a target of polycrystalline copper. Note the similarity in the patterns produced by x rays, electrons, and neutrons. (*Courtesy of C. G. Shull.*)

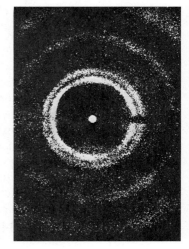

We can understand why the wave properties of matter are not observed in macroscopic experiments by calculating the de Broglie wavelength for a macroscopic object.

---

**Example 36-3** Find the de Broglie wavelength of a particle of mass $10^{-6}$ g moving with speed $10^{-6}$ m/s.

From Equation 36-29 we have

$$\lambda = \frac{h}{p} = \frac{h}{mv} = \frac{6.63 \times 10^{-34}\ \text{J·s}}{(10^{-9}\ \text{kg})(10^{-6}\ \text{m/s})}$$

$$= 6.63 \times 10^{-19}\ \text{m}$$

---

Since this wavelength is much smaller than any possible apertures or obstacles (the diameter of the nucleus of the atom is about $10^{-15}$ m, roughly 10,000 times this wavelength), diffraction or interference of such waves cannot be observed. As we have discussed, the propagation of waves of very small wavelength is indistinguishable from the propagation of particles. Note that we have chosen an extremely small value for the momentum in Example 36-3. Any other macroscopic particle with greater momentum will have an even smaller de Broglie wavelength. However, the situation is different for low-energy electrons. Consider an electron that has been accelerated through a potential difference $V$. Its energy is then (assuming the electron to be nonrelativistic)

$$E = \frac{p^2}{2m} = eV$$

and its wavelength is

$$\lambda = \frac{h}{p} = \frac{h}{\sqrt{2mE}} = \frac{h}{\sqrt{2meV}}$$

Putting in the values for $m$, $e$, and $h$, we obtain an expression for $\lambda$ in nanometres when $V$ is in volts:

$$\lambda = \frac{1.226}{\sqrt{V}} \text{ nm} \qquad\qquad 36\text{-}31$$

---

**Example 36-4** Find the de Broglie wavelength of an electron whose kinetic energy is 13.6 eV.

From Equation 36-31 we have for $V = 13.6$ V,

$$\lambda = \frac{1.226 \text{ nm}}{\sqrt{13.6}} = 0.332 \text{ nm} = 2\pi(0.0529 \text{ nm})$$

which is the circumference of the first Bohr orbit in the hydrogen atom.

---

From Equation 36-31 we see that electrons with energies of the order of tens of electronvolts have de Broglie wavelengths of the order of nanometres, which is the order of magnitude of the size of the atom and of the spacing of atoms in a crystal.

## 36-5   Wave-Particle Duality

We have seen that light, which we ordinarily think of as a wave motion, exhibits particle properties when it interacts with matter, as in the photoelectric effect or in Compton scattering, whereas electrons, which we usually think of as particles, exhibit the wave properties of interference and diffraction. All phenomena—electrons, atoms, light, sound, etc.—have both particle and wave characteristics. It might be tempting to say that an electron, for example, is both a wave and a particle, but the meaning of such a statement is not clear. In classical physics the concepts of waves and particles are mutually exclusive. A *classical particle* behaves like a piece of shot: it can be localized and scattered; it exchanges energy suddenly in lumps; and it obeys the laws of conservation of energy and momentum in collisions. It does *not* exhibit interference or diffraction. A *classical wave*, on the other hand, behaves like a water wave; it exhibits diffraction and interference, and its energy is spread out continuously in space and time. Nothing can be both a classical particle and a classical wave at the same time.

Until the twentieth century, it was thought that light was a classical wave and electrons were classical particles. We now see that these concepts of classical waves and particles do not adequately describe the complete behavior of any phenomenon. Everything propagates like a classical wave and exchanges energy like a classical particle. Of course, there are times when the classical-particle and classical-wave concepts give the same results. When the wavelength is very small, the propagation of a classical wave cannot be distinguished from that of a classical particle. Similarly, the time averages of energy and momentum exchanges are the same for classical waves as for classical particles. For example, the wave theory of light correctly predicts that the total electron current in the photoelectric effect is proportional to the intensity of light.

## 36-6   The Uncertainty Principle

The wave-particle duality of nature has many important consequences. In newtonian mechanics the rate of change of momentum of a particle is related to the resultant force acting on the particle by Newton's second law. To determine the position of the particle at any time we must know the forces acting and the position and velocity of the particle at some time; i.e., we must know the initial conditions. Although there are always experimental uncertainties in any measurement of the initial position and velocity, it is assumed in classical mechanics that such uncertainties can in principle be made as small as desired. Because of the wave-particle duality of both radiation and matter, we now understand that it is impossible in principle to measure both the position and velocity of a particle simultaneously with infinite precision. This result is known as the *uncertainty principle*, first enunciated by Werner Heisenberg in 1927. For one-dimensional motion it is expressed in terms of the momentum $p$ and the position $x$ of a particle. Let $\Delta x$ and $\Delta p$ be the uncertainties in the position and momentum, respectively. According to the uncertainty principle, the product of $\Delta x$ and $\Delta p$ can never be less than $h/4\pi = \frac{1}{2}\hbar$:

$$\Delta x \, \Delta p \geq \tfrac{1}{2}\hbar \qquad\qquad 36\text{-}32$$

*Uncertainty principle*

Usually the uncertainty product is much greater than $\frac{1}{2}\hbar$. The equality holds only if the measurements of both $x$ and $p$ have ideal gaussian distributions and the experiments are ideal.

We can get a qualitative understanding of the uncertainty principle by considering the measurement of the position and momentum of a particle. If we know the mass of the particle, we can determine its momentum by measuring its position at two nearby times and computing its velocity. A common way to measure the position of an object is to look at it with light. When we do this, we scatter light from the object and determine the position by the direction of the scattered light. If we use light of wavelength $\lambda$, we can measure the position only to an uncertainty of the order of $\lambda$ because of diffraction effects. To reduce the uncertainty in position we therefore use light of very short wavelength, perhaps even x rays. In principle, there is no limit to the accuracy of such a position measurement because there is no limit on how short a wavelength of electromagnetic radiation can be used. However, since all electromagnetic radiation carries momentum, the scattering of the radiation by the particle will deflect the radiation and change its original momentum in an uncontrollable way. By momentum conservation, the momentum of the particle also changes in an uncontrollable way. According to classical wave theory, this effect on the momentum of the particle could be reduced by reducing the intensity of the radiation. However, the energy and momentum of the radiation are quantized; each photon has momentum $h/\lambda$. When the wavelength of the radiation is small, the momentum of each photon will be large and the momentum measurement will have a large uncertainty. This uncertainty cannot be eliminated by reducing the intensity of light; such a reduction merely reduces the number of photons in the beam. To "see" the particle we must scatter at least one photon. Therefore, the uncertainty in the momentum measurement of the particle will be large if $\lambda$ is small, and the uncertainty in the position measurement will be large if $\lambda$ is large. A detailed analysis shows that the product of these uncertainties will always be at least of the order of Planck's constant $h$. Of course we could "look at" the particle by scattering electrons instead of photons, but we have the same difficulty. If we use low-momentum electrons to reduce the uncertainty in the momentum measurement, we have a large uncertainty in the position measurement because of diffraction of the electrons. The relation between the wavelength and momentum $\lambda = h/p$ is the same for electrons as for photons.

# The Expanding Universe

Martin Rees
*Institute of Astronomy, Cambridge University*

Over 400 years has passed since Copernicus argued that the earth must be dethroned from the privileged central position accorded to it by Ptolemy's cosmology, and described the general layout of the solar system in the form accepted today. Only within the twentieth century, however, have we fully recognized the sun's status as just one of the 100 billion ($10^{11}$) stars in the Milky Way. Harlow Shapley, Jan Oort, and others showed that the Milky Way is a flat disk-shaped system of stars spinning around its axis about once every $2 \times 10^8$ years. The sun lies near the edge of the disk, roughly 30,000 light-years from its center. Furthermore, the Milky Way, our own galaxy, is merely one of the 100 billion galaxies on the cosmic scene. If our galaxy were viewed from a distance of several million light-years, it would look rather like the spiral galaxies illustrated (Figures 1 and 2).

Photographs of the sky taken with a large telescope may show many thousands of galaxies, some so far away that the light we receive has taken billions of years on its journey toward us. Although individual stars cannot be resolved, it is possible to obtain spectra of the integrated light from all the stars in a remote galaxy. Such spectra imply that the galaxies contain more or less the same mixture of chemical elements as the sun and that the basic physical constants (e.g., the masses and charges of electrons and protons and Planck's constant) whose values determine the properties of the radiation from stellar surfaces do indeed have uniform values throughout the observed part of the universe.

Such spectra do, however, reveal a significant systematic trend: the spectral lines are all shifted toward the red, i.e., to longer wavelengths. This effect was studied by Edwin Hubble, who interpreted the red shift as a doppler effect and inferred that distant galaxies are receding from us. Moreover, Hubble showed that the recession velocity of a given galaxy is proportional to its distance. The velocity-distance relationship is also found to be the same in all parts of the sky; in other words, the apparent "expansion of the universe" is *isotropic* (the same in all directions). At first sight, one might wonder whether this implies that the earth (or at least our galaxy) is in a privileged central position; but it is easy to convince oneself that this is not so and that a hypothetical astronomer on another galaxy could also have discovered Hubble's law: he would find all the other galaxies apparently receding isotropically from *him.*

If the galaxies had always had the same relative velocities, Hubble's law implies that at some definite time in the past (about 15 billion years ago) they would all have been "squashed together" at one point. Georges Lemáitre and others proposed that the universe actually evolved from a primordial state of high density. The primary aim of observational cosmology since the 1930s has therefore been to test whether there was indeed a "big bang" or whether (as postulated by a rival theory) the universe has existed forever in a steady state, new material and new galaxies being continually created so that its average properties remain unaltered despite the Hubble expansion.

By observing distant objects, one can in principle probe the early history of the universe and perhaps learn whether galaxies were more crowded together or looked systematically different in the past. But ordinary galaxies become invisibly faint, even with the largest optical telescopes, before one reaches the distance where such effects might really be expected to show up. Fortunately, however, galaxies occasionally undergo mysterious violent explosions, in the course of which a small central region flares up until it outshines all the rest of the galaxy by a factor of 100. These rare bright sources of light, called *quasars,* permit optical astronomers to probe much deeper into space—and thus much farther into the past—than is possible using normal galaxies. These "galactic explosions" are sometimes strong sources of radio-frequency emission and are one of the important cosmic phenomena that can be investigated by the techniques of radio astronomy. Very little is understood about the physical nature of quasars. The problems they pose are among the most important and challenging that confront modern astrophysicists. However, by comparing the number of quasars observed at different distances one can infer that quasars (whatever they may actually be) were very much commoner in early epochs than they are now. This is quite incompatible with the steady-state model; in

**Figure 1**
The *Andromeda* galaxy, Messier 31. (*Courtesy of the Mount Wilson and Palomar Observatories.*)

an evolutionary cosmology, however, there are reasons for expecting that galaxies might indeed have been more prone to violent explosions when they were young.

In 1965, Arno Penzias and Robert Wilson made a discovery (more or less by accident) which is generally regarded as the most compelling evidence for a big bang. They found that the earth is apparently bathed in microwave radiation which comes from all directions but has no apparent source. Subsequent measurements have established that this microwave background seems to have the spectrum of blackbody radiation at a temperature of 2.7 K (see Section 18-4). Most cosmologists interpret this background as a relic of a hot, dense early phase when the universe was opaque and any radiation would have established the blackbody spectrum characteristic of thermal equilibrium. This possibility had in fact been predicted by George Gamow in 1948.

By synthesizing the disparate strands of evidence bearing on the problem, most cosmologists have reached a tentative consensus on how the universe might have evolved to its present state. About 15 billion years ago, according to this picture, all the material in the universe—all the stuff of which galaxies are now composed—constituted an exceedingly compressed and hot gas (hotter, in fact, than the center of the sun). The intense radiation in this fireball, though cooled and diluted by the expansion, would still be around, pervading the whole universe: this is the interpretation of the microwave background radiation discovered by Penzias and Wilson—an "echo," as it were, of the "explosion" which initiated the universal expansion. The early universe would not have been *completely* smooth and homogeneous (it may even have been turbulent), and the primordial irregularities eventually developed into galaxies. It may, however, have taken a billion years for these fluctuations to condense out into gravitationally bound systems. The most distant quasars are so far away that the radiation now reaching us has been traveling toward us for 80 or even 90 percent of the time elapsed since the initial big bang. That is, it set out toward us when the universe was only 10 or 20 percent of its present age. The universe would have appeared much more violent and active at these early epochs: the cosmic radio background would have been 100 times more intense; and an astronomer then would have found his nearest quasar perhaps 50 times closer to him than in these relatively quiescent times.

The microwave background is a "fossil" from even earlier stages in the expansion. There seems to be another important vestige of the primordial fireball in the present universe: the element helium, which constitutes about a quarter of the mass of most stars, including the sun. The other chemical elements could have been synthesized via the nuclear reactions which provide the power source in the cores of ordinary stars. It is believed that all the carbon, nitrogen, oxygen, and iron on the earth—and indeed in our own bodies—was manufactured in stars which exhausted their energy supply and exploded as *supernovae* before the sun formed. The solar system then condensed from gas contaminated by debris ejected from early generations of stars. One of the major achievements of theoretical astrophysics has been to understand some quantitative details of these processes of *cosmic nucleosynthesis*, e.g., the relative abundances of different elements. But it proved hard to account in this fashion for all the observed helium; it was therefore gratifying both to cosmologists and to experts on nucleosynthesis when the expected composition of material emerging from the big bang was calculated and found to be about 75 percent hydrogen and 25 percent helium. The synthesis of helium would have occurred within only a *few minutes* of the big bang. This epitomizes how the discovery of the microwave background has extended the scope of cosmology by bringing remote eras that were previously entirely speculative within the scope of quantitative scientific discussion. Such discussions do, however, entail extrapolating the locally determined physical laws into domains where one cannot be overwhelmingly sure of their continued validity.

The consistency of most of the available data (limited though it is) with this general picture encourages most cosmologists to adopt the hot big-bang theory as a working hypothesis, which should form the basis for interpreting new observations until some better theory emerges or until some glaring contradiction reveals itself.

Having drawn some tentative conclusions about how the universe has evolved from the primordial fireball to its present state, one is tempted to speculate about its future and its eventual fate. The conventional theories offer two alternative scenarios: either the universe will continue expanding forever, or else the expansion is slowing down to such an extent that it will eventually stop and be followed by a recontraction. This question can in principle be tackled observationally by extending Hubble's work out to very large distances: the velocity-distance relationship obviously refers to the velocity of the galaxies at the time when the light we now receive was emitted, so that any deceleration should be measurable if one observes sufficiently remote objects. In practice, however, this technique has yet to provide reliable results, because the observational difficulties are severe and also because of various uncertain corrections which must be incorporated into the analysis.

But there is another, more indirect, way of trying to determine how much the universal expansion is slowing down. Imagine that a big sphere is shattered by an explosion, the debris flying off in all directions. Each fragment feels the gravitational pull of all the others, and this causes the expansion to decelerate. If the explosion were sufficiently violent, the debris would fly apart forever; but if the fragments were not moving quite so fast, gravity might bind them together strongly enough to bring the expansion to a halt. The material would then collapse again. More or less the same argument probably holds for the universe. One might feel somewhat uneasy about applying a result based on Newton's theory of gravity to the whole universe. But even though one cannot describe the global properties of the universe properly (or the propagation of light) without using a more sophisticated theory such as Einstein's general relativity, the dynamics of the expansion are thought to be the same as in Newton's theory. One can therefore rephrase our earlier question: Does the universe have the escape velocity or not?

In the case of the galaxies (which, for the purposes of this argument, are regarded as fragments of the expanding universe) we know the expansion velocity from Hubble's law. What we do *not* know is the amount of gravitating matter that is causing the deceleration. It is a straightforward procedure to calculate how much material would be needed in order to halt the expansion: it works out at about one atom per cubic metre. If the average concentration of material were *below* this so-called critical density, we would expect the uni-

**Figure 2**
NGC 4594, spiral galaxy in *Virgo,* seen edge on. (*Courtesy of Hale Observatories.*)

verse to continue expanding forever; but if the mean density *exceeded* the critical density, the universe would seem destined eventually to recontract. There
are various ways of estimating the masses of individual galaxies. Also, of
course, one knows roughly how many galaxies there are in a typical volume
of space. These estimates are bedeviled by many uncertainties, but it looks as
though the material in galaxies, if spread uniformly through space, would fall
short of the critical density by a factor of at least 30. At first sight, one might
accept this as evidence that the universe will go on expanding forever; but
this inference would really be unjustified, because there may be a lot *more*
material embodied in some form *other than* ordinary galaxies. Galaxies are, ad-
mittedly, the most prominent features in the sky when we look with an optical
telescope, but there is no reason to believe that everything in the universe
shines. There may be many objects so cool that they radiate predominantly in
the infrared and *absorb* light instead of emitting it—"dead" galaxies, for example, whose stars have all exhausted their nuclear energy, or objects
shrouded by opaque clouds of dust. Alternatively, some objects may be so hot
that their emission is concentrated in the ultraviolet and x-ray bands of the
electromagnetic spectrum. This kind of astronomical work was not feasible
until quite recently because since the air is very opaque in these wavebands,
one must send equipment above the earth's atmosphere to make observations.
Space astronomy is still in its pioneering stage, and one would not be sur-
prised if it were to disclose many unsuspected sorts of objects. So our present
inventory of the contents of the universe may well prove exceedingly biased
and incomplete. For instance, there may be a large amount of diffuse gas
*between* the galaxies. (There is, after all, no reason to expect all, or even most,
of the primordial hydrogen and helium to have condensed into galaxies.) This
intergalactic gas may in fact be responsible for emitting most of the cosmic
x-rays reaching us from beyond our own galaxy.

At present, therefore, there is no definite reason for believing that there is
enough gravitating material in the universe to bring the expansion to a halt.
But it remains quite conceivable that the universe contains a great deal of stuff
even more elusive than intergalactic gas. For instance, the critical density
could be provided by neutrinos, gravitational waves, or black holes without
there being the slightest chance of detection by present techniques.

We therefore cannot rule out the possibility that eventually the expansion
will stop and turn into a contraction. Distant galaxies, displaying *blue* shifts
instead of red shifts, would eventually collide and merge with one another. As
the contraction proceeded further, the sky would become brighter and
brighter; and eventually all the stars would explode (because the sky would be
hotter than the fuel in their interiors!), everything in the universe being finally
engulfed in a fireball like that from which, according to most cosmologists, it
emerged: the ultimate, universal, gravitational collapse. But there is no immediate cause for concern; our breathing space before this cataclysm should be
billions of years, at the very least. If the universe did *not* have the critical density, it would continue to expand forever. Each galaxy would fade to a dull
glow as its constituent stars exhausted their available energy and the supply of
gas from which new bright stars can condense was inexorably depleted.

An important aim of current cosmological research is to decide between
these two contrasting fates for the universe. Also, we hope to eventually fill
in more details of the big-bang picture and to understand how the primordial
material aggregates into galaxies; why galaxies have the sizes and shapes that
are observed; how they evolve; and why they sometimes explode. But one
should remain aware that present data in cosmology are still limited, ambigu-
ous, and fragmentary; and they all depend on complex instruments stretched
right to the limits of their sensitivity and performance. Therefore, even if this
general picture seems consistent with what is known at the moment, it would
be rash to bet *too* heavily on its being correct. Moreover, self-consistency is, of
course, no guarantee of truth in itself. Many more observations are essential
before we can be sure whether we really know, even in outline, the basic
overall structure of the physical universe or whether our current ideas will
eventually be discarded and superseded as surely as Ptolemy's epicycles were.

Review

A. Define, explain, or otherwise identify:

Blackbody radiation, 972
Ultraviolet catastrophe, 972
Photoelectric effect, 973
Work function, 974
Einstein's photoelectric equation, 974
Compton wavelength, 977
Balmer series, 977
The Rydberg, 978
Stationary states, 978

Bohr quantization condition, 980
Energy-level diagram, 980
Reduced mass, 981
De Broglie wavelength, 982
Classical particle, 984
Classical wave, 984
Wave-particle duality, 984
Uncertainty principle, 985

B. True or false:

1. The spectral distribution of radiation in a blackbody depends only on the temperature of the body.

2. In the photoelectric effect, the maximum current is proportional to the intensity of the incident light.

3. The work function of a metal depends on the frequency of the incident light.

4. The maximum kinetic energy of electrons emitted in the photoelectric effect varies linearly with the frequency of the incident light.

5. In the Bohr model of the hydrogen atom the electron moves in circular orbits.

6. One of Bohr's assumptions is that atoms never radiate light.

7. In a stationary state in the Bohr model the total energy is negative and has the same magnitude as the kinetic energy.

8. In the ground state of the hydrogen atom the potential energy is $-27.2$ eV.

9. The de Broglie wavelength of an electron varies inversely with its momentum.

10. Electrons can be diffracted.

11. Neutrons can be diffracted.

12. Classical particles can be diffracted.

13. The position and momentum of an object cannot be measured simultaneously.

14. The uncertainty principle could be violated in principle if the experimental measurements were ideal.

Exercises

## Section 36-1, The Origin of the Quantum Constant: Blackbody Radiation

*There are no exercises for this section.*

## Section 36-2, The Quantization of Electromagnetic Radiation: Photons

1. Find the range of photon energies in the visible spectrum from about 400 to 700 nm wavelength.

2. Find the photon energy if the wavelength is (a) 0.1 nm (about 1 atomic diameter) and (b) 1 fm ($10^{-15}$ m, about 1 nuclear diameter).

3. Find the photon energy for an electromagnetic wave in the FM band of frequency 100 MHz.

4. The work function for tungsten is 4.58 eV. (*a*) Find the threshold frequency and wavelength for the photoelectric effect. Find the stopping potential if the wavelength of the incident light is (*b*) 200 nm; (*c*) 250 nm.

5. When light of wavelength 300 nm is incident on potassium, the emitted electrons have maximum kinetic energy of 2.03 eV. (*a*) What is the energy of the incident photon? (*b*) What is the work function for potassium? (*c*) What would be the stopping potential if the incident light had a wavelength of 400 nm?

6. The threshold wavelength for the photoelectric effect for silver is 262 nm. (*a*) Find the work function for silver. (*b*) Find the stopping potential if the incident radiation has a wavelength of 200 nm.

7. The work function for cesium is 1.9 eV. (*a*) Find the threshold frequency and wavelength for the photoelectric effect. Find the stopping potential if the wavelength of the incident light is (*b*) 300 nm; (*c*) 400 nm.

8. The wavelength of Compton-scattered photons is measured at $\theta = 90°$. If $\Delta\lambda/\lambda$ is to be 1 percent, what should the wavelength of the incident photons be?

9. Compton used photons of wavelength 0.0711 nm. (*a*) What is the energy of these photons? (*b*) What is the wavelength of the photon scattered at $\theta = 180°$? (*c*) What is the energy of the photon scattered at $\theta = 180°$? (*d*) What is the recoil energy of the electron for $\theta = 180°$?

## Section 36-3, Quantization of Atomic Energies: The Bohr Model

10. Use the known values of the constants in Equation 36-23 to show that $r_0$ is approximately 0.0529 nm.

11. The wavelength of the longest wavelength in the Lyman series was calculated in Example 36-2. Find the wavelengths for the transitions (*a*) $n = 3$ to $n = 1$ and (*b*) $n = 4$ to $n = 1$. (*c*) Find the shortest wavelength in the Lyman series.

12. (*a*) Use the Balmer formula (Equation 36-15) to calculate the three longest wavelengths in the Balmer series. (*b*) Find the photon energies corresponding to each of these wavelengths.

13. Calculate the longest three wavelengths and the series limit (shortest wavelength) for the Paschen series ($n_2 = 3$) and indicate their positions on a horizontal linear scale.

14. Repeat Exercise 13 for the Brackett series ($n_2 = 4$).

## Section 36-4, Electron Waves

15. Use Equation 36-31 to calculate the de Broglie wavelength for an electron of kinetic energy (*a*) 1 eV, (*b*) 100 eV, (*c*) 1 keV, and (*d*) 10 keV.

16. When the kinetic energy of an electron is much greater than its rest energy, the relativistic approximation $E \approx pc$ is good. (*a*) Show that in this case photons and electrons of the same energy have the same wavelength. (*b*) What is the de Broglie wavelength of an electron of energy 100 MeV?

17. A thermal neutron in a reactor has kinetic energy of about 0.02 eV. Calculate the de Broglie wavelength of this neutron.

18. Find the de Broglie wavelength of a proton of energy 1 MeV.

## Section 36-5, Wave-Particle Duality

*There are no exercises for this section.*

## Section 36-6, The Uncertainty Principle

19. A particle of mass $10^{-6}$ g moves with speed 1 cm/s. If its speed is uncertain by 0.01 percent, what is the minimum uncertainty in its position?

20. The uncertainty in the position of an electron in an atom cannot be greater than the diameter of the atom. (a) Calculate the minimum uncertainty in momentum associated with an uncertainty in position of 0.1 nm. (b) If an electron has momentum $p$ equal in magnitude to the uncertainty $\Delta p$ found in (a), what is its kinetic energy? Express your answer in electronvolts.

21. An electron has kinetic energy 25 eV. If its momentum is uncertain by 10 percent, what is the minimum uncertainty in its position?

## Problems

1. Data for stopping potential versus wavelength for the photoelectric effect using sodium are

| $\lambda$, nm | 200 | 300 | 400 | 500 | 600 |
|---|---|---|---|---|---|
| $V_0$, V | 4.20 | 2.06 | 1.05 | 0.41 | 0.03 |

Plot these data so as to obtain a straight line and from your plot find (a) the work function, (b) the threshold frequency, and (c) the ratio $h/e$.

2. A light bulb radiates 100 W uniformly in all directions. (a) Find the intensity at a distance of 1 m. (b) Find the number of photons per second that strike a 1-cm² area oriented so that its normal is along the line to the bulb if the wavelength of the light is 600 nm.

3. This problem is one of estimating the time lag (expected classically but not observed) in the photoelectric effect. Let the intensity of the incident radiation be 0.01 W/m². (a) Assuming that the area of an atom is 0.01 nm² find the energy per second falling on an atom. (b) If the work function is 2 eV, how long would it take classically for this much energy to fall on one atom?

4. (a) Solve Equation 36-12 for $p_e^2$ to obtain $p_e^2 = p_1^2 + p_2^2 - 2p_1p_2 + 2mc \times (p_1 - p_2)$. (b) Eliminate $p_e^2$ from your result in part (a) and Equation 36-11 to obtain $mc(p_1 - p_2) = p_1p_2(1 - \cos\theta)$. (c) Multiply both sides of your result in part (b) by $h/mcp_1p_2$ and use the de Broglie relation $h/p = \lambda$ to obtain the Compton formula (Equation 36-13).

5. The photoelectric effect can occur only with electrons that are bound (in atoms); i.e., a photon cannot be absorbed by a single free electron. Prove this by considering the problem of conservation of energy and momentum in the reference frame in which the total momentum of the initial photon and electron is zero.

6. The total energy density of radiation in a blackbody is given by $\eta = \int_0^\infty f(\lambda, T)\, d\lambda$, where $f(\lambda, T)$ is given by the Planck formula (Equation 36-2). Change the variable to $x = hc/\lambda kT$ and show that the total energy density can be written

$$\eta = \left(\frac{kT}{hc}\right)^4 8\pi hc \int_0^\infty \frac{x^3}{e^x - 1}\, dx = \alpha T^4$$

where $\alpha$ is a constant independent of $T$. This result, that the total energy density is proportional to $T^4$, is the *Stefan-Boltzmann law*.

7. Find the ratio of the reduced mass for the hydrogen atom to the electron mass $\mu/m$. Show that this ratio equals the ratio of the Rydberg constant for hydrogen to that for an infinitely massive nucleus $R_H/R_\infty$, and compare your results with the values of $R_H$ and $R_\infty$ given in the text.

8. The frequency of revolution of an electron in a circular orbit of radius $r$ is $f_{rev} = v/2\pi r$, where $v$ is the speed. (a) Show that in the $n$th stationary state

$$f_{rev} = \frac{k^2 Z^2 e^4 m}{2\pi\hbar^3} \frac{1}{n^3}$$

(b) Show that when $n_1 = n$, $n_2 = n - 1$, and $n$ is much greater than 1,

$$\frac{1}{n_2^2} - \frac{1}{n_1^2} \approx \frac{2}{n^3}$$

(c) Use your result of part (b) in Equation 36-24 to show that in this case the frequency of radiation emitted equals the frequency of motion. This result is an example of Bohr's correspondence principle: when $n$ is large, so that the energy difference between adjacent states is a small fraction of the total energy, classical and quantum physics must give the same result.

9. The kinetic energy of rotation of a diatomic molecule can be written $E_k = L^2/2I$, where $L$ is its angular momentum and $I$ the moment of inertia. (a) Assuming that the angular momentum is quantized, as in the Bohr model of the hydrogen atom, show that the energy is given by $E_n = n^2 E_1$, where $E_1 = \hbar^2/2I$, and make an energy-level diagram for such a molecule. (b) Estimate $E_1$ for the hydrogen molecule assuming the separation of the atoms to be 0.1 nm and considering rotation about an axis through the center of mass and perpendicular to the line joining the atoms. Express your estimate in electronvolts. (c) When $E_1$ is greater than $kT$, molecular collisions do not excite rotation and rotational energy does not contribute to the internal energy of the gas. Use your result of part (b) to find the critical temperature $T_c$, defined by $kT_c = E_1$. (d) Estimate $E_1$ for rotation about the line joining the atoms. In this case, the moment of inertia $I$ is of the order $m_e r_a^2$, where $m_e$ is the electron mass and $r_a$ is the radius of the atom. Find $T_c$ for rotation about this axis.

10. A particle moves about in a one-dimensional box of length $L$. Its energy is quantized by the condition $n(\lambda/2) = L$, where $\lambda$ is the de Broglie wavelength of the particle. (a) Show that the allowed energies are given by $E_n = n^2 E_1$, where $E_1 = h^2/8mL^2$ and $m$ is the mass of the particle. Evaluate $E_1$ for (b) an electron in a box of size $L = 0.1$ nm and (c) a proton in a box of size $L = 10^{-15}$ m. (Take the potential energy of the particle in the box to be zero so that its total energy is its kinetic energy $p^2/2m$.)

11. An important consequence of the uncertainty principle is that a particle confined to some region of space has a minimum kinetic energy greater than zero. If $\Delta p$ is taken to be the standard deviation, it is given by $(\Delta p)^2 = [(p - p_{av})^2]_{av}$. (a) Show that this is equivalent to $(\Delta p)^2 = (p^2)_{av} - (p_{av})^2$. (b) Consider a particle of mass $m$ moving in a box of length $L$ such that $p_{av} = 0$. Its average kinetic energy is then $E_{k,av} = (p^2)_{av}/2m = (\Delta p)^2/2m$. Show that its average kinetic energy must be at least as large as $\hbar^2/8mL^2$ by using the fact that $\Delta x$ cannot be larger than $L$.

CHAPTER 37    Nuclear Physics

---

**Objectives**   After studying this chapter you should:

1.   Be able to give the order of magnitude of the radius of an atom and of a nucleus.

2.   Be able to sketch the $N$-versus-$Z$ curve for stable nuclei.

3.   Be able to sketch the binding energy per nucleon versus $A$ and discuss the significance of this curve for fission and fusion.

4.   Know the exponential law of radioactive decay and be able to work problems using it.

5.   Be able to describe the nuclear-fission chain reaction and discuss the advantages and disadvantages of fission reactors.

6.   Be able to state the Lawson criterion for nuclear-fusion reactors.

7.   Be able to discuss the chief mechanisms of energy loss of particles in matter and explain why some particles have well-defined ranges and others do not.

8.   Be able to discuss the radiation dosage units of roentgen, rad, and rem.

---

The first information about the atomic nucleus came from the discovery of radioactivity by A. H. Becquerel in 1896. The rays emitted by radioactive nuclei were studied by many physicists in the early decades of the twentieth century. They were first classified by E. Rutherford as $\alpha$, $\beta$, and $\gamma$ rays, according to their ability to penetrate matter and ionize air: $\alpha$ radiation penetrates the least and produces the most ionization and $\gamma$ penetrates the most with the least ionization. It was later found that $\alpha$ rays are helium nuclei, $\beta$ rays are electrons or positrons, and $\gamma$ rays are high-energy photons, i.e., electromagnetic radiation of very short wavelength. (A positron is an antielectron, a particle identical to the electron except that its charge is positive.) The $\alpha$-particle scattering experiments of H. W. Geiger and E. Marsden in 1911 and the successes of the Bohr model of the atom led to the modern picture of an atom as consisting of a tiny massive nucleus with a radius of the order of 1 to 10 fm (1 fm = $10^{-15}$ m) surrounded by a cloud of electrons at a relatively great distance, of the order of 0.1 nm = 100,000 fm, from the nucleus. In 1919 Rutherford bombarded nitrogen with $\alpha$ particles and observed scintillations on a zinc sulfide screen due to protons, which have a much longer range in air than $\alpha$ particles. This was the first observation of artificial nuclear disintegration. Such experiments were extended to many other elements in the next few years.

In 1932 the neutron was discovered by Chadwick and the positron by Anderson, and the first nuclear reaction using artificially accelerated particles was observed by Cockcroft and Walton. It is quite reasonable to mark this year as the beginning of modern nuclear physics. With the discovery of the neutron, it became possible to understand some of the properties of nuclear structure; and the advent of nuclear accelerators made many experimental studies possible without the severe limitations on particle type and energy imposed by naturally radioactive sources.

In this chapter we first discuss some of the general properties of the atomic nucleus and then consider the important features of radioactivity. We then look at some nuclear reactions, including the important reactions of fission and fusion. Finally, we look at the interactions of nuclear particles with matter, a subject important for understanding the detection of nuclear particles, the shielding of reactors, and the effects of radiation on the human body. Our discussions will be descriptive and phenomenological, with the aim of presenting general information rather than a theoretical understanding of nuclear physics.

## 37-1    Atoms

Slightly more than 100 different elements have been discovered, 92 of which are found in nature. Each is characterized by an atom which contains a number $Z$ of electrons, an equal number of protons, and a number $N$ of neutrons. The number $Z$ of electrons or protons is called the *atomic number*. The lightest atom, hydrogen (H), has $Z = 1$; the next lightest, helium (He), has $Z = 2$; the next lightest, lithium (Li), has $Z = 3$, etc. The structure of an atom is somewhat like that of our solar system. Nearly all the mass of the atom is in a tiny nucleus, which contains the protons and neutrons. The nuclear radius is typically about 1 to 10 fm. The distance between the nucleus and the electrons in the atom is about 0.1 nm = 100,000 fm. This distance determines the "size" of the atom. The chemical and physical properties of an element are determined by the number and arrangement of the electrons in the atom. Because each proton has a positive charge $+e$, the nucleus has a total positive charge $+Ze$. The electrons are negatively charged $(-e)$ and so are attracted to the nucleus and repelled by each other. The electrons are arranged in shells: the first shell has up to 2 electrons; the second shell, about 4 times farther out, can contain up to 8 electrons; etc. This shell structure accounts for the periodic nature of the properties of the elements as shown in the periodic table. Elements with a single electron in an outer shell (hydrogen, lithium, sodium, etc.) or those with a single vacancy in an outer shell (fluorine, chlorine, bromine, etc.) are very active chemically; those with completely filled outer shells (helium, neon, argon, etc.) are chemically inert. The calculation of the electron configurations of the atoms and the resulting chemical properties was one of the great triumphs of quantum mechanics in the 1920s.

Since electrons and protons have equal but opposite charges and there are an equal number of electrons and protons in an atom, atoms are electrically neutral. Atoms that lose or gain one electron (or more) are then electrically charged and are called *ions*. Atoms with just 1 electron in an outer shell (such as sodium which has 11 electrons) tend to lose it readily and become positive ions; those lacking just one electron to have a complete shell tend to gain an electron to become negative ions, e.g., chlorine. Atoms bond together to form molecules, e.g., $H_2O$, or solids. This bonding involves only the outer electrons. In a molecule or solid, the separation of the atomic nuclei is about 1 atomic diameter, i.e., of the order of 0.1 nm.

## 37-2  Properties of Nuclei

The nucleus of an atom contains just two kinds of particles, protons and neutrons (aside from hydrogen, which contains a single proton). These particles are similar except that the proton has a charge $+e$ and the neutron is uncharged. They have approximately equal masses, the neutron being about 0.2 percent more massive; the mass of the electron is smaller by a factor of about 2000. A convenient unit of mass for discussing nuclei is the unified mass unit u (formerly the atomic mass unit amu), which is defined so that Avogadro's number of unified mass units equals 1 g and Avogadro's number of $^{12}C$ atoms has a mass of 12 g. The value of Avogadro's number is

*Unified mass unit*

$$N_A = 6.02204 \times 10^{23} \qquad\qquad 37\text{-}1$$

*Avogadro's number*

Another unit of mass convenient for nuclear physics is the $\text{MeV}/c^2$, which comes from the relativistic expression for rest energy $E = mc^2$. It is related to the unified mass unit by

$$1 \text{ u} = 931.50 \text{ MeV}/c^2 \qquad\qquad 37\text{-}2$$

The masses of the proton, neutron, and electron are

$$m_p = 1.6726 \times 10^{-27} \text{ kg} = 1.007276 \text{ u} = 938.28 \text{ MeV}/c^2$$

$$m_n = 1.6750 \times 10^{-27} \text{ kg} = 1.008665 \text{ u} = 939.573 \text{ MeV}/c^2 \qquad 37\text{-}3$$

$$m_e = 9.109 \times 10^{-31} \text{ kg} = 5.486 \times 10^{-4} \text{ u} = 0.511 \text{ MeV}/c^2$$

The lightest element, hydrogen, has in its most common form a nucleus consisting of just a single proton. Another less common form of hydrogen, called deuterium, has a nucleus with 1 proton and 1 neutron in it. Although the deuterium atom is about twice as massive as the ordinary hydrogen atom, its chemical behavior is the same because they each have 1 electron. A particular nuclear species is called a *nuclide*. Two or more different nuclides corresponding to the same atom are called *isotopes*. Isotopes have the same $Z$ value (because they correspond to the same atom) but different $N$ values and therefore different atomic mass numbers $A$. The two isotopes of hydrogen just mentioned are written $^1H$ and $^2H$ with the $A$ number as a presuperscript. The $Z$ number is sometimes written as a presubscript as in $^1_1H$ or $^2_1H$, but this is not necessary because the chemical symbol H tells you that $Z = 1$. Another even less common isotope of hydrogen is tritium, $^3H$, which has a nucleus consisting of 1 proton and 2 neutrons. On the average, there are about three stable isotopes for each atom (some have only one; others have five or six) so there are about 300 stable nuclei for the 100 different atoms. The most common isotope of the second lightest atom, helium, is $^4He$ (which is also known as an $\alpha$ particle), but $^3He$ also exists.

*Isotopes*

Inside the nucleus, the nucleons exert strong attractive forces on their nearby neighbors. This force, called the *strong nuclear force* or the *hadronic force,* is much stronger than the electric force of repulsion between the protons and very much stronger than the gravitational forces between the nucleons. (Gravity is so weak it can always be neglected in nuclear physics.) The force between two neutrons is roughly the same as that between two protons or between a neutron and proton, except that between two protons there is also the electric force due to their charges, which tends to weaken the attractions somewhat. The hadronic force decreases rapidly with distance, so that when two nucleons are more than a few femtometres apart, the force is negligible.

## Size and Shape

The size and shape of the nucleus can be determined by bombarding it with high-energy particles and observing the scattering, similar to the experiments of Geiger and Marsden; or in some cases from measurements of radioactivity. The results depend somewhat on the kind of experiment. For example, a scattering experiment with electrons measures the charge distribution of the nucleus, whereas an experiment done with neutrons determines the region of influence of the hadronic force. Despite these differences, a wide variety of experiments suggest that most nuclei are approximately spherical, with radii given by

$$R = R_0 A^{1/3} \qquad\qquad 37\text{-}4 \qquad \textit{Nuclear radius}$$

where $R_0$ is about 1.5 fm. (Some nuclei in the rare-earth region of the periodic table are ellipsoidal, the major and minor axes differing by about 20 percent or less.) The fact that the radius of a spherical nucleus is proportional to $A^{1/3}$ implies that the volume of the nucleus is proportional to $A$, which in turn implies that the density is the same for all nuclei since the mass is approximately proportional to the number of nucleons $A$. The fact that a liquid drop has a constant density independent of its size has led to the analogy of the nucleus with a liquid drop, a model that has proved quite successful in understanding nuclear behavior, especially the fission of heavy nuclei.

## $N$ and $Z$ Numbers

One characteristic of the nuclear force is that for light nuclei greatest stability is achieved with approximately equal proton and neutron numbers, $N = Z$. For heavier nuclei, the electric repulsion of the protons leads to greater stability with more neutrons than protons. You can see this by looking at $N$ and $Z$ numbers for the most abundant isotopes of various elements, for example, $^{40}_{20}\text{Ca}$, $^{16}_{8}\text{O}$, $^{56}_{26}\text{Fe}$, $^{207}_{82}\text{Pb}$, and $^{238}_{92}\text{U}$. Figure 37-1 shows a plot of $N$ versus $Z$ for the known stable nuclei. The straight line $N = Z$ is followed for small $N$ and $Z$.

## Mass and Binding Energy

The mass of a nucleus is not equal to the sum of the masses of the nucleons making up the nucleus. When two or more nucleons fuse together to form a nucleus, the total mass decreases and energy is given off. Conversely, to break up a nucleus into its parts, energy must be put into the system to be changed into the increase in rest mass. The energy

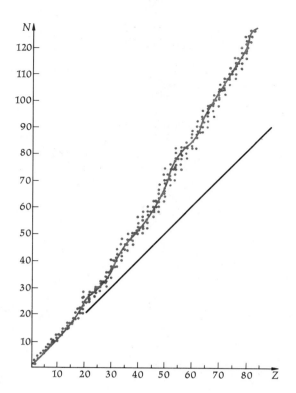

**Figure 37-1**
Plot of neutron number $N$ versus proton number $Z$ for the stable nuclides. The straight line is $N = Z$.

involved is $c^2$ times the change in mass, where $c$ is the speed of light in vacuum. The difference between the rest energy of the parts of a nucleus and the rest energy of the nucleus is called the total *binding energy* of the nucleus. Consider $^4$He, for example, which consists of two protons and two neutrons. The mass of an atom can be accurately measured in a mass spectrometer, which measures $q/M$ for ions by bending them in a magnetic field. The mass of the $^4$He atom is 4.00263 u. This includes the masses of the two electrons in the atom. The mass of the $^1$H atom is 1.007825 u, and that of the neutron is 1.008665 u. The sum of the masses of two $^1$H atoms plus two neutrons is $2(1.007825\ \mathrm{u}) + 2(1.008665\ \mathrm{u}) = 4.03298\ \mathrm{u}$, which is greater than the mass of the $^4$He atom by 0.030377 u. Note that by using the masses of two $^1$H atoms rather than two protons, the masses of the electrons in the atom cancel out. We do this because it is atomic masses that are measured directly and listed in mass tables.*

This mass difference of 0.030377 u can be converted into the binding energy of the He atom by using the mass conversion factor $1\ \mathrm{u}{\cdot}c^2 = 931.50$ MeV from Equation 37-2. Then $0.030377\ \mathrm{u}{\cdot}c^2 = 0.030377(931.5) = 28.20$ MeV, which is the total binding energy of $^4$He. In general, the binding energy of a nucleus of atomic mass $M_A$ containing $Z$ protons and $N$ neutrons is found from

$$E_b = ZM_Hc^2 + Nm_nc^2 - M_Ac^2 \qquad\qquad 37\text{-}5$$

*Binding energy*

where $M_H$ is the mass of the $^1$H atom and $m_n$ that of the neutron. The atomic masses for some selected nuclei are listed in Table 37-1. Once the

---

\* A slight error is made because the mass of the $^1$H atom is slightly less than the sum of the mass of an electron plus a proton because of the binding energy of the atom. However, since atomic binding energies are very small compared with nuclear binding energies, this error is negligible.

**Table 37-1**
Atomic masses of selected isotopes

| Element | Symbol | $Z$ | Atomic Mass, u |
|---------|--------|-----|----------------|
| Hydrogen | $^1$H | 1 | 1.007 825 |
|          | $^2$H | 1 | 2.014 102 |
|          | $^3$H | 1 | 3.016 050 |
| Helium | $^3$He | 2 | 3.016 030 |
|        | $^4$He | 2 | 4.002 603 |
| Lithium | $^6$Li | 3 | 6.015 125 |
| Boron | $^{10}$B | 5 | 10.012 939 |
| Carbon | $^{12}$C | 6 | 12.000 000 |
|        | $^{14}$C | 6 | 14.003 242 |
| Oxygen | $^{16}$O | 8 | 15.994 915 |
| Sodium | $^{23}$Na | 11 | 22.989 771 |
| Potassium | $^{39}$K | 19 | 38.963 710 |
| Iron | $^{56}$Fe | 26 | 55.939 395 |
| Copper | $^{63}$Cu | 29 | 62.929 592 |
| Silver | $^{107}$Ag | 47 | 106.905 094 |
| Gold | $^{197}$Au | 79 | 196.966 541 |
| Lead | $^{208}$Pb | 82 | 207.976 650 |
| Polonium | $^{212}$Po | 84 | 211.989 629 |
| Radon | $^{222}$Rn | 86 | 222.017 531 |
| Radium | $^{226}$Ra | 88 | 226.025 360 |
| Uranium | $^{238}$U | 92 | 238.048 608 |
| Plutonium | $^{242}$Pu | 94 | 242.058 725 |

atomic mass has been determined, the binding energy can be computed from Equation 37-5. It is found that the total binding energy of a nucleus is approximately proportional to the total number of nucleons $A$ in the nucleus. Figure 37-2 shows the binding energy per nucleon $E_b/A$ versus $A$. The mean value is about 8.3 MeV. The fact that this curve is approximately constant (meaning that $E_b$ is approximately proportional to $A$) indicates that there is saturation of nuclear forces in the nucleus; i.e., each nucleon bonds to only a certain number of other nucleons independent of the total number of nucleons in the nucleus. If, for example, there were no saturation and each nucleon bonded to each other nucleon, there would be $A - 1$ bonds for each nucleon and a total of $A(A - 1)$ bonds altogether. The total binding energy, which is a measure of the energy needed to break all these bonds, would then be proportional to $A(A - 1)$ and $E_b/A$ would not be approximately constant. Figure 37-2 indicates that, instead, there is a fixed number of bonds per nucleon, as would be the case if each nucleon were attracted only to its nearest neighbors. Such a situation also leads to a constant nuclear density consistent with the measurements of radius. If the binding energy per nucleon were proportional to the number of nucleons, the radius would be approximately constant, as it is for atoms.

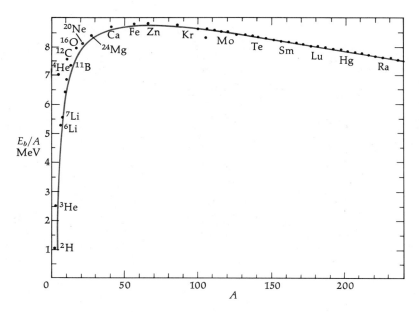

**Figure 37-2**
The binding energy per particle versus mass number $A$. (*From R. Leighton,* Principles of Modern Physics, *McGraw-Hill Book Company, New York, 1959, by permission.*)

## Questions

1. How does the nuclear force differ from the electromagnetic force?

2. What property of the nuclear force is indicated by the fact that all nuclei have about the same density?

3. The mass of $^{12}$C which contains 6 protons and 6 neutrons is exactly 12.000 u, by the definition of the unified mass unit. Why isn't the mass of $^{16}$O which contains 8 protons and 8 neutrons exactly 16.0000 u?

## 37-3  Radioactivity

Nuclei that are not stable are radioactive. The three kinds of radioactivity, alpha ($\alpha$) decay, beta ($\beta$) decay, and gamma ($\gamma$) decay, were named before it was known that $\alpha$ particles are $^4$He nuclei, $\beta$ particles are electrons ($e^-$) or positrons ($e^+$), and $\gamma$ rays are photons ($\gamma$). All very heavy nuclei ($Z > 83$) are theoretically unstable to $\alpha$ decay; i.e., for heavy nuclei the mass of the original radioactive nucleus is greater than the sum of the masses of the decay products consisting of an $\alpha$ particle and what is called the daughter nucleus. An example of $\alpha$ decay is $^{232}$Th ($Z = 90$), which decays into $^{228}$Ra ($Z = 88$) plus an $\alpha$ particle, written

$$^{232}\text{Th} \rightarrow {}^{228}\text{Ra} + {}^4\text{He} \qquad\qquad 37\text{-}6$$

The mass of the $^{232}$Th atom is 232.038124 u. The mass of the daughter atom $^{228}$Ra is 228.031139 u. Adding this to the mass of $^4$He, we get 232.03374 for the total mass on the right side of the reaction. This is less than that of $^{232}$Th by 0.00438 u, which multiplied by 931.50 MeV/$c^2$ gives 4.08 MeV/$c^2$ for the excess rest mass of $^{232}$Th over the decay products. $^{232}$Th is therefore theoretically unstable to $\alpha$ decay, and this decay does in fact occur in nature with the emission of an $\alpha$ particle of energy 4.08 MeV. (The energy is actually somewhat less than 4.08 MeV because some of the decay energy is shared by the recoiling $^{228}$Ra nucleus.) Examination of the possible $\alpha$ decays of other heavy nuclei shows that all nuclei heavier than bismuth ($Z = 83$) are theoretically unstable to $\alpha$ decay whereas all lighter nuclei are stable; i.e., the rest mass of the decay products is greater than that of the original nucleus for $Z$ less than 83.

Beta decay occurs for light or intermediate nuclei that have too many or too few neutrons for stability, as indicated on the $N$-versus-$Z$ curve of Figure 37-1. The energy released in $\beta$ decay can also be determined by computing the rest mass of the original nucleus and that of the decay products. Note that in $\alpha$ decay the daughter nucleus has a mass number $A$ that is less by 4 than that of the original nucleus and an atomic number $Z$ that is less by 2, whereas in $\beta$ decay $A$ remains the same while $Z$ either increases by 1 ($e^-$ decay) or decreases by 1 ($e^+$ decay). $\gamma$ decay usually happens very quickly and is observed only because it follows either $\alpha$ or $\beta$ decay. In $\gamma$ decay, the radioactive nucleus remains the same nucleus as it decays from an excited state to the ground state or an excited state of lower energy than that of the original nucleus. The question of stability of a nucleus to $\alpha$ or $\beta$ decay can always be answered by looking at the rest masses of the nuclei and particles involved.

In 1900 Rutherford discovered that the rate of emission of radioactive particles from a substance is not constant in time but decreases exponentially. This exponential time dependence is characteristic of all radioactivity and indicates that the decay is a statistical process. Because each nucleus is well shielded from others by the atomic electrons, pressure and temperature changes have no effect on nuclear properties. For a statistical decay, in which the decay of any individual nucleus is a random event, the number of nuclei decaying in a time interval $dt$ is proportional to $dt$ and to the number of radioactive nuclei. If $N(t)$ is the number of radioactive nuclei at time $t$ and $-dN$ is the number that decay in time $dt$ (the negative sign is necessary because $N$ decreases), we have

$$-dN = \lambda N \, dt \qquad\qquad 37\text{-}7$$

where the constant of proportionality $\lambda$ is called the *decay constant*. It is the probability per unit time of the decay of any given radioactive nucleus. The solution of this equation is

$$N = N_0 e^{-\lambda t} \qquad\qquad 37\text{-}8$$

where $N_0$ is the number of nuclei at $t = 0$. The decay rate is

$$R = \frac{-dN}{dt} = \lambda N_0 e^{-\lambda t} = R_0 e^{-\lambda t} \qquad\qquad 37\text{-}9 \qquad \textit{Decay rate}$$

where $R_0 = \lambda N_0$ is the initial decay rate. Note that both the number of nuclei and the rate of decay decrease exponentially. It is the decrease in the rate of decay that is determined experimentally. Figure 37-3 shows $N$ versus $t$. If we multiply the numbers on the $N$ axis by $\lambda$, this figure becomes a graph of $R$ versus $t$.

An important parameter of radioactive decay related to the decay constant is the half-life $t_{1/2}$, defined as the average time for half of a given number of radioactive nuclei to decay. It is also the time for which the decay rate decreases to half its original value. If we set $N$ equal to $\frac{1}{2}N_0$ and $t = t_{1/2}$ in Equation 37-8, we obtain

$$\tfrac{1}{2}N_0 = N_0 e^{-\lambda t_{1/2}}$$

$$e^{+\lambda t_{1/2}} = 2$$

$$\lambda t_{1/2} = \ln 2 = 0.693$$

or

$$t_{1/2} = \frac{\ln 2}{\lambda} = \frac{0.693}{\lambda} \qquad\qquad 37\text{-}10 \qquad \textit{Half-life}$$

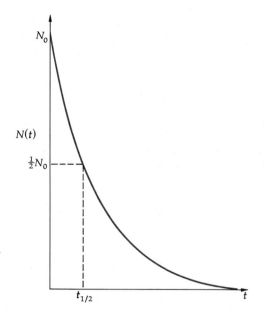

**Figure 37-3**
Exponential radioactive-decay law. The number of nuclei remaining at time $t$ decreases exponentially with time $t$. The half-life $t_{1/2}$ is indicated. The decay rate $R(t) = \lambda N(t)$ has the same time dependence.

The decay of any one nucleus is completely random; all we can say is that, on the average, half of some original number of radioactive nuclei will decay in one half-life, then half of those left will decay in the next half-life, etc. The half-lives of various radioactive nuclei vary from very small times (less than a microsecond) to very large times, for example $10^{16}$ y. Some nuclei that are theoretically unstable to $\alpha$ decay as calculated from the mass-energy relations have a half-life that is long compared with the age of the universe; for all practical purposes such nuclei can be considered to be stable.

**Example 37-1** A radioactive source has a half-life of 1 min. At time $t = 0$ it is placed near a detector, and the counting rate is observed to be 2000 counts/s. Find the decay constant and the counting rate at times $t = 1, 2, 3,$ and 10 min.

From Equation 37-10 we find for the decay constant:

$$\lambda = \frac{0.693}{t_{1/2}} = \frac{0.693}{1 \text{ min}} = 1.16 \times 10^{-2} \text{ s}^{-1}$$

Since the half-life is 1 min, the counting rate will be half as great at $t = 1$ min as at $t = 0$, so $R_1 = 1000$ counts/s at 1 min, $R_2 = 500$ counts/s at 2 min, and $R_3 = 250$ counts/s at 3 min. At $t = n$ min $= nt_{1/2}$ the rate will be

$$R = (\tfrac{1}{2})^n R_0 \qquad\qquad 37\text{-}11$$

so that after 10 min $= 10t_{1/2}$ the rate will be $R_{10} = (\tfrac{1}{2})^{10}(2000) = 1.95$ counts/s.

**Example 37-2** If the detection efficiency in Example 37-1 is 20 percent, how many radioactive nuclei are there at time $t = 0$? At time $t = 1$ min? How many decay in the first minute?

The detection efficiency depends on the distance from the source to the detector and the chance that a radioactive decay particle entering the detector will produce a count. If the source is smaller than the detector and placed very close, about half the emitted particles will enter the detector. If the counting rate at $t = 0$ is 2000 counts/s and the efficiency is 20 percent, the decay rate at $t = 0$ must be 10,000 $s^{-1}$. The number of radioactive nuclei can be found from Equation 37-9 using $\lambda = 0.693$ $min^{-1}$ from Example 37-1:

$$N_0 = \frac{R_0}{\lambda} = \frac{10,000 \ s^{-1}}{0.693 \ min^{-1}} = 8.66 \times 10^5$$

At time $t = 1$ min $= t_{1/2}$ there are half as many nuclei as at $t = 0$; so $N_1 = \frac{1}{2}(8.66 \times 10^5) = 4.33 \times 10^5$. The number that decay in the first minute is therefore $4.33 \times 10^5$. Note that this is not equal to the initial decay rate of 10,000 $s^{-1}$ times 60 s because the decay rate is not constant during the 1-min interval but decreased by half during each minute. Note also that the average decay rate is not the mean of 10,000 and 5000 $s^{-1}$ because the decrease is exponential, not linear in time.

---

The SI unit of radioactive decay is the becquerel (Bq), which is defined as one decay per second:

$$1 \ Bq = 1 \ decay/s \qquad\qquad 37\text{-}12 \qquad \textit{Becquerel}$$

An historical unit that applies to all types of radioactivity is the Curie (Ci), defined as

$$1 \ Ci = 3.7 \times 10^{10} \ decays/s = 3.7 \times 10^{10} \ Bq \qquad 37\text{-}13 \qquad \textit{Curie}$$

The curie is the amount of radiation emitted by 1 g of radium. Since this is a very large unit, the millicurie (mCi) or microcurie ($\mu$Ci) is often used.

### Alpha Decay

When a nucleus emits an $\alpha$ particle, both $N$ and $Z$ decrease by 2 and $A$ decreases by 4. The daughter of a radioactive nucleus is often itself radioactive, decaying either by $\alpha$ or $\beta$ decay or both. If the original decaying nucleus has a mass number $A$ that is 4 times an integer, the daughter nucleus and all those in the chain will also have mass numbers equal to 4 times an integer. Similarly, if the mass number of the original nucleus is $4n + 1$, where $n$ is an integer, all the nuclei in the decay chain will have mass numbers given by $4n + 1$. We see therefore that there are four possible $\alpha$-decay chains, depending on whether $A$ equals $4n$, $4n + 1$, $4n + 2$, or $4n + 3$, where $n$ is an integer. All but one of these are found in nature. The $4n + 1$ series is not, because its longest-lived member (other than the stable end product $^{209}$Bi), $^{237}$Np, has a half-life of only $2 \times 10^6$ y, which is much smaller than the age of the earth.

Figure 37-4 illustrates the thorium series, which has $A = 4n$ and begins with an $\alpha$ decay from $^{232}$Th to $^{228}$Ra. The daughter nuclide of an $\alpha$ decay is on the left or neutron-rich side of the stability curve (dashed line), so that it often decays by $\beta^-$ decay, in which one neutron changes to a proton by emitting an electron. In Figure 37-4, $^{228}$Ra decays by $\beta$ decay to $^{228}$Ac, which in turn decays to $^{228}$Th. There are then four $\alpha$ decays to $^{212}$Pb, which decays by $\beta^-$ to $^{212}$Bi. There is a branch point at $^{212}$Bi which decays either by $\alpha$ decay to $^{208}$Tl or by $\beta$ decay to $^{212}$Po. The branches meet at the stable lead isotope $^{208}$Pb.

The energy released in $\alpha$ decay is determined by the difference in mass of the parent nucleus and the decay products, which include the daughter nucleus and the $\alpha$ particle. If all the $\alpha$ decays proceeded from the ground state of the parent nucleus to the ground state of the daugh-

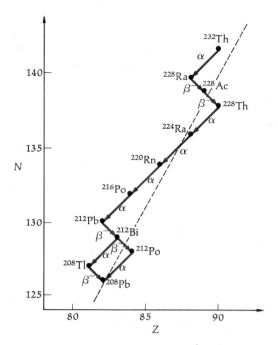

**Figure 37-4**
The thorium (4*n*) α-decay series.

ter nucleus, the emitted α particles would all have the same energy (slightly less than the energy released because of the small recoil energy of the daughter nucleus needed to conserve momentum). When the energies of the emitted α particles are measured with high resolution, a spectrum of energies is observed, as shown in Figure 37-5. The peak in the spectrum labeled $\alpha_0$ corresponds to α decays to the ground state of the daughter nucleus with a total energy of 6.04 MeV, as determined from the masses of the particles. The peak labeled $\alpha_{30}$ indicates α particles with energy 30 keV less than those of maximum energy, indicating that the decay is to an excited state of the daughter nucleus at 30 keV above the ground state. The energy spectrum of α particles then indicates the energy levels in the daughter nucleus, as in the case shown in Figure 37-5. This interpretation of the particle energy spectrum is confirmed by the observation of a 30-keV γ ray emitted as the daughter nucleus decays to its ground state. Because the half-life for γ decay is very short, this decay follows essentially immediately after the α decay. Figure 37-6 shows an energy-level diagram for $^{223}$Ra obtained from the measurement of the α-particle energies in the decay of $^{227}$Th

**Figure 37-5**
The α-particle spectrum from $^{227}$Th. The highest-energy α particles correspond to decay to the ground state of $^{223}$Ra with a transition energy of $Q = 6.04$ MeV. The next-highest-energy particles, $\alpha_{30}$, result from transitions to the first excited state of $^{223}$Ra, 30 keV above the ground state. The energy levels of the daughter nucleus, $^{223}$Ra, can be determined by measurement of the α-particle energies. (*Data taken from R. C. Pilger, Ph.D. thesis, University of California Radiation Laboratory Report UCRL-3877, 1957.*)

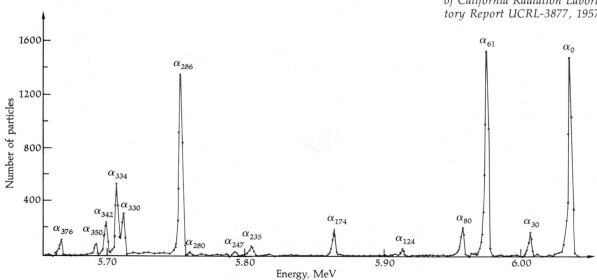

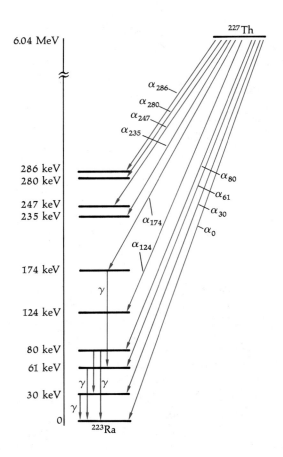

**Figure 37-6**
Energy levels of $^{223}$Ra determined by measurement of $\alpha$-particle energies from $^{227}$Th, as shown in Figure 37-5. Only the lowest-lying levels and some of the $\gamma$-ray transitions are shown.

and from the observed $\gamma$-decay energies. Only the lowest-lying levels and some of the $\gamma$-ray transitions are indicated. This nuclear energy-level diagram is similar to the atomic energy-level diagram for hydrogen in our discussion of the Bohr model (Chapter 36).

### Beta Decay

The simplest example of $\beta$ decay is that of the free neutron, which decays into a proton plus an electron with a half-life of about 10.8 min. The energy of decay is 0.78 MeV, which is the difference between the rest energy of the neutron (939.57 MeV) and that of the proton plus electron (938.28 + 0.511 MeV). More generally, in $\beta^-$ decay, a nucleus of mass number $A$ and atomic number $Z$ decays into one with mass number $A$ and atomic number $Z' = Z + 1$ with the emission of an electron. If the decay energy is shared by the daughter nucleus and the emitted electron, the energy of the electron is uniquely determined by conservation of energy and momentum, just as in $\alpha$ decay. Experimentally, however, the energies of the electrons emitted in decay are observed to vary from zero to the maximum energy available. A typical energy spectrum is shown in Figure 37-7; compare it with the discrete spectrum of particle energies of Figure 37-5.

To explain the apparent nonconservation of energy in $\beta$ decay, W. Pauli in 1930 suggested that a third particle, which he called the *neutrino*, is emitted. The mass of the neutrino was assumed to be very much less than that of the electron because the maximum energy of the electrons emitted is nearly equal to the total available for the decay. The neutrino was observed experimentally in 1957. It is now known that there are at least two kinds of neutrinos, one ($\nu_e$) associated with elec-

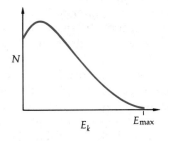

**Figure 37-7**
Energy spectrum of electrons emitted in $\beta$ decay. The number of electrons per unit energy interval is plotted versus energy. The fact that all the electrons do not have the same energy $E_{max}$ suggests that there is another particle emitted which shares the energy available for decay.

*Neutrino*

trons and one ($\nu_\mu$) associated with muons, and possibly one associated with a new particle, the heavy lepton, recently discovered. Moreover, each neutrino has an antiparticle, written $\bar{\nu}_e$ and $\bar{\nu}_u$. It is the antineutrino that is emitted in the decay of the neutron, which is written

$$n \rightarrow p + \beta^- + \bar{\nu}_e \qquad \qquad 37\text{-}14$$

In $\beta^+$ decay a proton changes into a neutron with the emission of a positron (and a neutrino). A free proton cannot decay by positron emission because of conservation of energy (the rest mass of the neutron plus positron is greater than that of the proton), but because of binding-energy effects, a proton inside a nucleus can. A typical decay is

$$^{13}_{7}\text{N} \rightarrow {}^{13}_{6}\text{C} + \beta^+ + \nu_e \qquad \qquad 37\text{-}15$$

The electrons or positrons emitted in decay do not exist inside the nucleus. They are created in the process of decay, just as photons are created when an atom makes a transition from a higher to a lower energy state.

An important example of decay is that of $^{14}\text{C}$, which is used in radioactive dating:

$$^{14}\text{C} \rightarrow {}^{14}\text{N} + \beta^- + \bar{\nu}_e \qquad \qquad 37\text{-}16 \qquad \textit{Carbon dating}$$

The half-life for this decay is 5730 y. The radioactive isotope $^{14}\text{C}$ is produced in the upper atmosphere from nuclear reactions caused by cosmic rays. The chemical behavior of carbon atoms with $^{14}\text{C}$ nuclei is the same as those with ordinary $^{12}\text{C}$ nuclei; e.g., atoms with these nuclei combine with oxygen to form $CO_2$ molecules. Since living organisms continually exchange $CO_2$ with the atmosphere, the ratio of $^{14}\text{C}$ to $^{12}\text{C}$ in a living organism is the same as the equilibrium ratio in the atmosphere, which is about $1.3 \times 10^{-12}$. When an organism dies, it no longer absorbs $^{14}\text{C}$ from the atmosphere, so that the ratio $^{14}\text{C}/^{12}\text{C}$ continually decreases due to the radioactive decay of $^{14}\text{C}$. A measurement of the decay rate per gram of carbon makes it possible to calculate when the organism died.

---

**Example 37-3** Calculate the decay rate of a living organism per gram of carbon, assuming that the ratio $^{14}\text{C}/^{12}\text{C} = 1.3 \times 10^{-12}$.

From Equation 37-9 the decay rate is related to the number of radioactive nuclei as follows:

$$R = -\frac{dN}{dt} = \lambda N = \frac{0.693}{t_{1/2}} N$$

The number of $^{12}\text{C}$ nuclei in 1 g of carbon is

$$N_{^{12}\text{C}} = \frac{6.02 \times 10^{23} \text{ nuclei/mol}}{12 \text{ g/mol}} = 5.02 \times 10^{22} \text{ nuclei/g}$$

The number of radioactive $^{14}\text{C}$ nuclei is $1.3 \times 10^{-12}$ times this number. The decay rate is therefore

$$R = \frac{0.693}{5730 \text{ y}} (1.3 \times 10^{-12}) \frac{5.02 \times 10^{22}}{\text{g}} \frac{1 \text{ y}}{3.16 \times 10^7 \text{ s}} \frac{60 \text{ s}}{\text{min}}$$

$$= 15 \text{ decays/min·g}$$

The decay rate for a living organism is therefore 15 decays per minute per gram of carbon.

**Example 37-4** A bone containing 200 g of carbon has a $\beta$-decay rate of 400 decays/min. How old is the bone?

We first obtain a rough estimate. If the bone were from a living organism, we would expect the decay rate to be (15 decays/min·g) (200 g) = 3000 decays/min. Since 400/3000 is roughly $\frac{1}{8}$ (actually 1/7.5), the sample must be about 3 half-lives old, or about 3(5730) y. To find the age more accurately, we note that after $n$ half-lives, the decay rate decreases by a factor of $(\frac{1}{2})^n$. We therefore find $n$ from

$$(\tfrac{1}{2})^n = \frac{400}{3000}$$

or

$$2^n = \frac{3000}{400} = 7.5$$

$$n \ln 2 = \ln 7.5$$

$$n = \frac{\ln 7.5}{\ln 2} = 2.91$$

The age is therefore $t = n t_{1/2} = 2.91(5730 \text{ y}) = 16,700 \text{ y}$.

### Gamma Decay

In $\gamma$ decay a nucleus in an excited state decays to a lower-energy state by the emission of a photon. This decay is the nuclear analog of the emission of light by atoms. Since the spacing of the nuclear energy levels is of the order of millions of electronvolts (compared with electronvolts in atoms), the wavelengths of the emitted photons are of the order of

$$\lambda = \frac{hc}{E} \sim \frac{1240 \text{ MeV·fm}}{1 \text{ MeV}} = 1240 \text{ fm}$$

Emission of $\gamma$ rays usually follows $\beta$ or $\alpha$ decay. For example, if a radioactive parent nucleus decays by $\beta$ decay to an excited state of the daughter nucleus, the daughter nucleus often decays down to its ground state by $\gamma$ emission. The mean life for $\gamma$ decay is often very short. Direct measurements of mean lives down to about 1 $\mu$s are possible, and sophisticated coincidence techniques can be used to measure mean lives as short as $10^{-11}$ s. Measurements of lifetimes smaller than $10^{-11}$ s are difficult but can sometimes be accomplished by determining the natural line width $\Gamma$ and using the uncertainty relation $\tau = \hbar/\Gamma$.

### Questions

4. Why do extreme changes in temperature and pressure of a radioactive sample have little or no effect on the radioactivity?

5. Why is the decay series $A = 4n + 1$ not found in nature?

6. A decay by $\alpha$ emission is often followed by $\beta$ decay. When this occurs, it is by $\beta^-$ and not $\beta^+$ decay. Why?

7. The half-life of $^{14}$C is much less than the age of the universe, yet $^{14}$C is found in nature. Why?

8. What effect would a long-term variation in the cosmic-ray activity have on the accuracy of $^{14}$C dating?

# 37-4 Nuclear Reactions

Most of the information about nuclei is obtained by bombarding them with various particles and observing the results. Although the first experiments were limited by the need to use radiation from naturally occurring sources, they produced many important discoveries. In 1932 J. D. Cockcroft and E. T. S. Walton succeeded in producing the reaction

$$p + {}^7\text{Li} \rightarrow {}^8\text{Be} \rightarrow {}^4\text{He} + {}^4\text{He}$$

using artificially accelerated protons. At about the same time, the Van de Graaff electrostatic generator was built (by R. Van de Graaff in 1931), as was the first cyclotron (by E. O. Lawrence and M. S. Livingston in 1932). Since then, an enormous technology has been developed for accelerating and detecting particles, and many nuclear reactions have been studied.

When a particle is incident on a nucleus, several different things can happen. The particle may be scattered elastically or inelastically (in which case the nucleus is left in an excited state and decays by emitting photons or other particles), or the original particle may be absorbed and another particle or particles emitted.

Consider a general reaction of particle $x$ on nucleus $X$, resulting in nucleus $Y$ and particle $y$. The reaction is usually written

$$x + X \rightarrow Y + y + Q$$

or, for short, $X(x, y)$ $Y$. The quantity $Q$, defined by

$$Q = (m_x + m_X - m_y - m_Y)c^2 \qquad\qquad \text{37-17} \qquad \textit{Q value}$$

is the energy released in the reaction. The $Q$ value for the reaction $n + {}^1\text{H} \rightarrow {}^2\text{H} + \gamma + Q$ is +2.22 MeV, whereas for the inverse reaction, $\gamma + {}^2\text{H} \rightarrow n + {}^1\text{H} + Q$, it is −2.22 MeV. If $Q$ is positive, the reaction is said to be *exothermic*. It is energetically possible even if the particles are at rest. If $Q$ is negative, the reaction is *endothermic* and cannot occur below a threshold energy. In the reference frame in which the total momentum is zero (the center-of-mass frame) the threshold energy for the two particles is just $|Q|$. However, most reactions occur with one particle $X$ at rest. In this frame, called the *laboratory frame*, the incident particle must have energy greater than $|Q|$ because, by conservation of momentum, the kinetic energy of $y$ and $Y$ cannot be zero. Consider the nonrelativistic case of $x$, of mass $m$, incident on $X$, of mass $M$. We showed in Equation 8-44 that if $|Q|$ is the threshold in the center-of-mass reference frame, the threshold in the laboratory frame is

$$E_{\text{th}} = \frac{m + M}{m} |Q| \qquad\qquad \text{37-18}$$

A measure of the effective size of a nucleus for a particular reaction is the *cross section* $\sigma$, defined as *the number of reactions per unit time per nucleus divided by the incident intensity* (number of incident particles per unit time per unit area). Consider, for example, the bombardment of ${}^{13}\text{C}$ by protons. A number of reactions might occur. Elastic scattering is written ${}^{13}\text{C}(p, p)$ ${}^{13}\text{C}$: the first $p$ indicates an incident proton; the second indicates that the particle that leaves is also a proton. If the scattering is inelastic, the outgoing proton is indicated by $p'$ and the nucleus in an excited state by ${}^{13}\text{C}^*$; one then writes ${}^{13}\text{C}(p, p')$ ${}^{13}\text{C}^*$. Some other pos-

*Cross section*

The Cockcroft-Walton accelerator. Walton is sitting in the foreground. J. D. Cockcroft and E. T. S. Walton produced the first transmutation of nuclei with artificially accelerated particles in 1932, for which they received the Nobel prize (1951). (*Courtesy of Cavendish Laboratory.*)

sible reactions are

$$(p, n): {}^{13}\text{C}(p, n)\ {}^{13}\text{N}$$

$$\text{Capture: } {}^{13}\text{C}(p, \gamma)\ {}^{14}\text{N}$$

$$(p, \alpha): {}^{13}\text{C}(p, \alpha)\ {}^{10}\text{B}$$

The total cross section is the sum of the partial cross sections

$$\sigma = \sigma_{p,p} + \sigma_{p,p'} + \sigma_{p,n} + \sigma_{p,\gamma} + \sigma_{p,\alpha} + \cdots$$

Cross sections have the dimensions of area. Since nuclear cross sections are of the order of the square of the nuclear radius, a convenient unit for them is the *barn,* defined by

$$1 \text{ barn} = 10^{-28} \text{ m}^2 \qquad\qquad 37\text{-}19$$

The cross section for a particular reaction is a function of energy. For an endothermic reaction, it is zero for energies below the threshold.

### Reactions with Neutrons

Nuclear reactions involving neutrons are important for understanding nuclear reactors. The most likely reaction with a nucleus for a neutron of energy of more than about 1 MeV is scattering. However, even if the scattering is elastic, the neutron loses some energy to the nucleus. If a neutron is scattered many times in a material, its energy decreases until it is of the order of $kT$. The neutron is then equally likely to gain or lose energy from a nucleus when it is elastically scattered. A neutron with energy of the order of $kT$ is called a *thermal neutron.* At low energies, a neutron is likely to be captured, with the emission of a $\gamma$ ray from the excited nucleus:

$$n + {}^{A}_{Z}M \rightarrow {}^{A+1}_{Z}M + \gamma \qquad\qquad 37\text{-}20$$

Since the binding energy of a neutron is of the order of 6 to 10 MeV and the kinetic energy of the neutron is negligible in comparison, the excitation energy of the nucleus after capturing the neutron is from 6 to 10 MeV and $\gamma$ rays of this energy are emitted. If there are no resonances, the cross section $\sigma(n, \gamma)$ varies smoothly with energy, decreasing with increasing energy roughly as $E^{-1/2} \propto 1/v$, where $v$ is the speed of the neutron. This energy dependence is easily understood because the time spent by a neutron near a nucleus is inversely proportional to the speed of the neutron and the capture cross section is therefore proportional to this time. Superimposed on the $1/v$ dependence are large fluctuations in the capture cross section due to resonances. Figure 37-8 shows the neutron-capture cross section for silver as a function of energy. The dashed line indicates the $1/v$ dependence. At the maximum of the resonance, the value of the cross section is very large ($> 5000$ barns) compared with the value of about 10 barns just past the resonance. Many elements show similar resonances in the neutron-capture cross section.

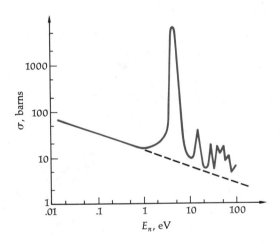

**Figure 37-8**
Neutron-capture cross section for Ag versus energy. The dashed-line extension would be expected if there were no resonances and the cross section were merely proportional to the time spent near the nucleus, i.e., proportional to $1/v$. (*From R. Evans*, The Atomic Nucleus, *McGraw-Hill Book Company, New York, 1955, by permission.*)

Since the maximum cross section for $^{113}$Cd is about 57,000 barns, this material is useful for shielding against low-energy neutrons.

An important nuclear reaction that involves neutrons is fission. Very heavy nuclei (atomic numbers greater than 100) are subject to spontaneous fission; i.e., they break apart into two nuclei even if left to themselves with no outside disturbance. We can understand this by considering the analogy of a charged liquid drop. If the drop is not too large, surface tension can overcome the repulsive forces of the charges and hold the drop together, but there is a certain maximum size beyond which the drop will be unstable and spontaneously break apart. Spontaneous fission puts an upper limit on the size of a nucleus and therefore on the number of elements that are possible. Although the very heavy elements of uranium and polonium are stable as far as spontaneous fission is concerned, they can be induced to fission by the capture of a neutron. We shall study fission and another important nuclear reaction, fusion, in the next section.

George Gerster, Photo Researchers

Three-mile-long linear accelerator at Stanford University, used for accelerating electrons to very high energies.

## Questions

9. What is meant by the cross section for a nuclear reaction?

10. Why is the neutron-capture cross section (excluding resonances) proportional to $E^{-1/2}$?

11. What is meant by the $Q$ value of a reaction? Why is the reaction threshold not equal to $Q$?

12. Why isn't there an element with $Z = 130$?

# 37-5  Fission, Fusion, and Nuclear Reactors

Two nuclear reactions, fission and fusion, are of particular importance. In the fission of $^{235}U$, for example, the uranium nucleus is excited by the capture of a neutron and splits into two nuclei, each with about half the total mass. The Coulomb force of repulsion drives the fission fragments apart, the energy eventually showing up as thermal energy. In fusion, two light nuclei such as those of deuterium and tritium ($^2H$ and $^3H$) fuse together to form a heavy nucleus (in this case $^4He$ plus a neutron). A typical fusion reaction is

$$^2H + {}^3H \rightarrow {}^4He + n + 17.6 \text{ MeV} \qquad\qquad 37\text{-}21$$

A plot of the mass difference per nucleon $(M - Zm_p - Nm_n)/A$ versus $A$, in units of $\text{MeV}/c^2$, is shown in Figure 37-9. This is just the negative of the binding-energy curve of Figure 37-2. From this figure we see that the rest mass per particle for both very heavy ($A \approx 200$) and very light ($A \approx 20$) nuclides is more than that for nuclides of intermediate mass.

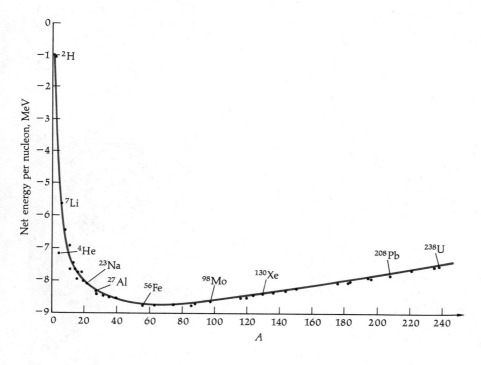

**Figure 37-9**
Plot of the mass difference $(M - Zm_p - Nm_n)/A$ versus $A$ in units of $\text{MeV}/c^2$. The rest mass per particle is less for intermediate-mass nuclei ($A \approx 80$) than for either very light or very heavy nuclei.

Thus in both fission and fusion the total rest mass decreases and energy is released. Since for $A = 200$ the rest energy is about 1 MeV per nucleon greater than for $A = 100$, about 200 MeV is released in the fission of a heavy nucleus. This is much more energy than is released in atomic or chemical reactions such as combustion (about 4 eV). The energy release in fusion depends on the particular reaction. For the $^2H + {}^3H$ reaction above, 17.6 MeV is released. Although this is less than the energy released in fission, it is a greater amount of energy per unit mass.

**Example 37-5** Calculate the total energy released in kilowatt-hours in the fission of 1 g of $^{235}$U, assuming that 200 MeV is released per fission.

The number of $^{235}$U nuclei in 1 g is

$$N = \frac{6.02 \times 10^{23}}{235} = 2.56 \times 10^{21}$$

The energy released per gram is then

$$\frac{200 \text{ MeV}}{\text{Nucleus}} \frac{2.56 \times 10^{21} \text{ nuclei}}{\text{g}} \frac{1.6 \times 10^{-19} \text{ J}}{\text{eV}} \frac{1 \text{ h}}{3600 \text{ s}} \frac{1 \text{ kW}}{1000 \text{ J/s}}$$

$$= 2.28 \times 10^4 \text{ kW·h}$$

**Example 37-6** Compare the energy released per gram in the fusion of deuterium and tritium with that of fission.

We can compare the energy per gram by first comparing the energy per nucleon. In fission this is about 1 MeV per nucleon. In the fusion of $^2$H + $^3$H it is (17.6 MeV)/5 = 3.52, or about $3\frac{1}{2}$ times as great. The energy released per gram in this fusion reaction is therefore about $3\frac{1}{2}$ times that in fission.

The application of both fission and fusion to the development of nuclear weapons has had a profound effect on our lives in the past 40 years. The peaceful application of these reactions to the development of our energy resources may have an even greater effect on the future. In this section we shall look at some of the features of fission and fusion that are important for their application in reactors to generate power.

The fission of uranium was discovered in 1939 by Hahn and Strassmann, who found, by careful chemical analysis, that medium-mass elements such as barium and lanthanum were produced in the bombardment of uranium with neutrons. The discovery that several neutrons are emitted in the fission process led to speculation concerning the possibility of using these neutrons to cause further fissions, thereby producing a chain reaction. Just 3 years later, in 1942, a group led by Fermi at the University of Chicago constructed the first nuclear reactor, using such a chain reaction. When $^{235}$U captures a neutron, the nucleus emits $\gamma$ rays as it deexcites to the ground state about 15 percent of the time and fissions about 85 percent of the time. The fission process is analogous to the oscillation of a liquid drop. If the oscillations are violent enough, the drop splits in two, as shown in Figure 37-10. Using a liquid-drop model, Bohr and Wheeler calculated the critical energy $E_c$ needed by the compound nucleus $^{236}$U to fission. For this nucleus the energy is 5.3 MeV, which is less than the binding energy of the last neutron in $^{236}$U, 6.4 MeV. The addition of a neutron to $^{235}$U therefore

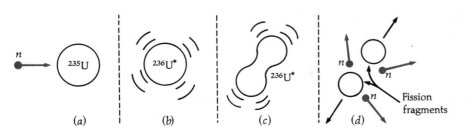

**Figure 37-10**
Schematic description of nuclear fission. (a) Absorption of a neutron by $^{235}$U leads to $^{236}$U* in excited state (b). In (c), the oscillation of $^{236}$U* has become unstable. (d) The nucleus splits apart, emitting several neutrons which can produce fission in other nuclei.

produces an excited state of the $^{236}$U nucleus with more than enough energy to break apart. On the other hand, the critical energy for the $^{239}$U nucleus is 5.9 MeV, while the binding energy of the last neutron is only 5.2 MeV, so that when a neutron is captured by $^{238}$U to form $^{239}$U, the excitation energy is not great enough for fission and the nucleus deexcites by $\gamma$ emission.

A fissioning nucleus can break into two medium-mass fragments in many different ways. A typical breakup is

$$n + {}^{235}U \rightarrow ({}^{236}U^*) \rightarrow {}^{141}_{56}Ba + {}^{92}_{36}Kr + 3n + Q \qquad \text{37-22}$$

For this reaction $Q$ is 175 MeV. Other fission reactions induced by neutron capture of $^{235}$U lead to other end products.

Figure 37-11 shows the distribution of fission products versus mass number. The average number of neutrons emitted in the fission of $^{235}$U is about 2.5. In order to sustain a chain reaction, one of these neutrons (on the average) must be captured by another $^{235}$U nucleus and cause it to fission. The *reproduction constant k* of a reactor is defined as the average number of neutrons from each fission that cause a further fission. The maximum possible value of $k$ is 2.5, but it is less than this for two important reasons: (1) neutrons may escape from the region containing fissionable nuclei, and (2) neutrons may be captured by other nuclei in the reactor not leading to fission. If $k$ is exactly 1, the reaction will be self-sustaining. If $k$ is less than 1 it will die out. If $k$ is significantly greater than 1, the reaction rate will increase rapidly and "run away." In the design of nuclear bombs, such a runaway reaction is desired; *in power reactors, the value of k must be kept very nearly equal to 1.*

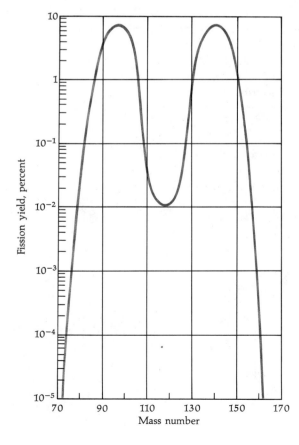

**Figure 37-11**
Distribution of fission fragments of $^{235}$U. Symmetric fission, in which the uranium nucleus splits into two nuclei of equal mass, is less likely than asymmetric fission, in which the fragments have unequal masses.

Since the neutrons emitted in fission have energies of the order of 1 MeV whereas the cross section for capture leading to fission in $^{235}$U is largest at small energies, the chain reaction is most easily sustained if the neutrons are slowed down before they escape from the reactor. At high energies (1 to 2 MeV) neutrons lose energy rapidly by inelastic scattering from $^{238}$U, leaving the nucleus in an excited state that decays by $\gamma$ emission. Once the neutron energy is below the excitation energies of the reactor elements (about 1 MeV), the main process of energy loss is by elastic scattering, in which a fast neutron collides with a nucleus essentially at rest and transfers some of its kinetic energy. Such energy transfers by elastic scattering are efficient only if the masses of the two bodies are comparable, e.g., a neutron will not transfer much energy in an elastic collision with a heavy $^{238}$U nucleus. (Natural uranium contains 99.3 percent $^{238}$U and only 0.7 percent fissionable $^{235}$U.) A *moderator* consisting of material, such as water or carbon, containing light nuclei is therefore placed around the fissionable material in the core of a reactor to slow the neutrons down. Neutrons are slowed down by elastic collisions with the nuclei of the moderator until the neutrons are in thermal equilibrium with the moderator. Because of the relatively large neutron-capture cross section of the hydrogen in water, reactors using ordinary water as a moderator must use enriched uranium, in which the $^{235}$U content is increased from 0.7 to 1 to 4 percent. Natural uranium can be used if heavy water ($D_2O$) replaces ordinary (light) water ($H_2O$) as the moderator. Although heavy water is expensive, most Canadian reactors use it for a moderator to avoid the cost of constructing uranium-enrichment facilities.

The ability to control the reproduction factor $k$ is important if a power reactor is to operate with any degree of safety. There are both natural negative-feedback mechanisms and mechanical methods of control. If $k$ is greater than 1 and the reaction rate increases, the temperature of the reactor increases. If water is used as a moderator, its density decreases with increasing temperature and it becomes a less effective moderator. A second important method is the use of control rods made of a material such as cadmium, which has a very large neutron-capture cross section. When a reactor is started up, the control rods are inserted, so that $k$ is less than 1. As they are gradually withdrawn from the reactor the neutron capture decreases and $k$ increases to 1. If $k$ becomes greater than 1, the rods are inserted again.

Control of the reaction rate of a nuclear reactor with mechanical control rods is possible only because some of the neutrons emitted in the fission process are *delayed*. The time needed to slow down a neutron from 1 or 2 MeV to thermal energy is only of the order of a millisecond. If all the neutrons emitted in fission were prompt neutrons, i.e., emitted immediately in the fission process, mechanical control would not be possible because the reactor would run away before the rods could be inserted. However, about 0.65 percent of the neutrons emitted are delayed by an average time of about 14 s. These neutrons are emitted not in the fission process itself but in the decay of the fission fragments. A typical decay is

$$^{87}\text{Br} \rightarrow {}^{87}\text{Kr}^* + \beta^-$$
$$\downarrow$$
$$^{86}\text{Kr} + n$$

In this decay, which has a 56-s half-life, the $^{87}\text{Kr}^*$ nucleus has enough excitation energy to emit a neutron. This neutron is thus delayed by 56 s on the average. The effect of the delayed neutrons can be seen in the following examples.

**Example 37-7** If the average time between fission generations (i.e., the time for a neutron emitted in one fission to cause another) is 1 ms and the reproduction factor is 1.001, how long would it take for the reaction rate to double?

If $k = 1.001$, the rate after $N$ generations is $1.001^N$. Setting this rate equal to 2 and solving for $N$, we obtain $N = 700$ generations for the rate to double. The time is then $700(1\text{ ms}) = 0.70$ s. This is not enough time for the mechanical control of such an excursion.

**Example 37-8** Assuming that 0.65 percent of the neutrons emitted are delayed by 14 s, find the average generation time and the doubling time if $k = 1.001$.

Since 99.35 percent of the generations are $10^{-3}$ s and 0.65 percent are 14 s, the average time is

$$t_{av} = 0.9935 \times 10^{-3} + (0.0065)(14) = 0.092 \text{ s}$$

The time for 700 generations is then

$$700(0.092 \text{ s}) = 64.4 \text{ s}$$

Because of the limited supply of the small fraction of $^{235}$U in natural uranium and the limited capacity of enrichment facilities, reactors using the fission of $^{235}$U cannot meet our energy needs for very long. Two possibilities hold much promise for the future: breeder reactors and controlled nuclear-fusion reactors.

The breeder reactor makes use of the fact that when the relatively plentiful but nonfissionable $^{238}$U nucleus captures a neutron, it decays by $\beta$ decay (half-life 20 min) to $^{239}$Np, which in turn decays by $\beta$ decay (half-life 2 days) to the fissionable nuclide $^{239}$Pu. Since plutonium fissions with fast neutrons, no moderator is needed. A reactor with a mixture of $^{238}$U and $^{239}$Pu will breed as much fuel as it uses or more if one or more of the neutrons emitted in the fission of $^{239}$Pu is captured by $^{238}$U. Practical studies indicate that a typical breeder reactor can be expected to double its fuel supply in 7 to 10 years.

There are several safety problems inherent with breeder reactors. Since fast neutrons are used, the time between generations is essentially determined by the fraction of delayed neutrons, which is only 0.3 percent for the fission of $^{239}$Pu. Mechanical control is therefore much more difficult. Also, since the operating temperature of a breeder reactor is relatively high and a moderator is not desired, a heat exchanger such as liquid sodium metal is used rather than water (which is the moderator as well as the heat-transfer material in an ordinary reactor). When the temperature of the reactor increases, the decreased density of the heat exchanger now leads to positive feedback since it now absorbs fewer neutrons than before. There is also the general safety problem with the large-scale production of plutonium, which is extremely poisonous; this is the material used in nuclear bombs. In addition, there is the problem of storage of the radioactive waste products produced in any reactor. For example, a single 1-GW ordinary reactor produces in 1 y about 3 million Ci of the radioactive $^{90}$Sr (half-life 28.8 y), enough to contaminate Lake Michigan above the legal limit if it were mixed uniformly in the lake. Despite elaborate storage methods of this and other long-lived radioactive waste products, the long-term safety of large-scale production of these wastes is always open to question.

The production of power from the fusion of light nuclei holds great promise because of the relative abundance of the fuel and the lack of some of the dangers presented in fission reactors. Unfortunately, the technology has not yet been developed sufficiently for use of this plentiful energy source. We shall consider the $^2$H + $^3$H reaction (Equation 37-21); other reactions present similar problems.

Because of the Coulomb repulsion between the $^2$H and $^3$H nuclei, kinetic energies of the order of 10 keV or greater are needed to get the nuclei close enough for the attractive nuclear forces to become effective in causing fusion. Such energies can be obtained in an accelerator, but since the scattering of one nucleus by the other is much more probable than fusion, the bombardment of one nucleus by another from an accelerator requires the input of more energy than is recovered. To obtain energy from fusion, the particles must be heated to a temperature great enough for the fusion reaction to occur in random thermal collisions. The temperature corresponding to $kT = 10$ keV is of the order of $10^8$ K. (This is roughly the temperature in the interiors of stars, where such reactions actually take place.) At such temperatures a gas consists of positive ions and negative electrons; such a gas is called a *plasma*. One of the problems arising in the attempt to produce controlled fusion reactions is that of confining such a plasma long enough for the reactions to take place. The energy required to heat a plasma is proportional to the density of the ions $n$, whereas the collision rate is proportional to the square of the density $n^2$. If $\tau$ is the confinement time, the output energy is proportional to $n^2\tau$. If the output energy is to exceed the input energy, we have

$$C_1 n^2 \tau > C_2 n$$

where $C_1$ and $C_2$ are constants. In 1957 the British physicist J. D. Lawson evaluated these constants from estimates of efficiencies of various hypothetical fusion reactors and derived the following relation between density and confinement time, known as *Lawson's criterion*:

$$n\tau > 10^{14} \text{ s·particles/cm}^3 \qquad \text{37-23}$$

*Lawson's criterion*

Two schemes for achieving the Lawson criterion are currently under investigation. (At present, the product of density and confinement time attainable by either scheme is several orders of magnitude too small.) In one scheme a magnetic field is used to confine a hot plasma. Densities of $3 \times 10^{13}$ particles/cm$^3$ at temperatures corresponding to $kT \approx 1$ keV have been achieved for times up to about 0.03 s, each of which is about a factor of 10 too low. In a second scheme, called *inertial confinement*, a pellet of solid deuterium and tritium is bombarded from all sides by intense pulsed laser beams of energy of the order of $10^4$ J lashing about $10^{-8}$ s. Computer simulation studies indicate the pellet should be compressed to about $10^4$ times its normal density and heated to a temperature greater than $10^8$ K, producing about $10^6$ J of fusion energy in $10^{-11}$ s, which is so brief that confinement is achieved by inertia alone.

### Questions

13. Why is a moderator needed in an ordinary nuclear-fission reactor?

14. What happens to neutrons produced in fission that do not produce another fission?

15. What is the advantage of using $^{238}$U as a fuel to breed $^{239}$Pu rather than using $^{235}$U? What are the disadvantages?

16. Why does fusion occur spontaneously in the sun but not on earth?

## 37-6   The Interaction of Particles with Matter

In this section we shall discuss briefly the main interactions of charged particles, neutrons, and photons with matter. Understanding these interactions is important for the study of nuclear detectors, shielding, and the effects of radiation. We shall not attempt to give a detailed theory, but shall indicate instead the principal factors involved in stopping or attenuating a beam of particles.

### Charged Particles

When a charged particle traverses matter, it loses energy mainly by excitation and ionization of electrons in the matter. If the particle energy is large compared with the ionization energies of the atoms, the energy loss in each encounter with an electron will be only a small fraction of the particle energy. (A heavy particle cannot lose a large fraction of its energy to a free electron because of conservation of momentum.) Since the number of electrons in matter is so large, we can treat the problem as that of a continuous loss of energy. After a fairly well-defined distance, called the *range*, the particle has lost all its kinetic energy and stops. Near the end of the range, the continuous picture of energy loss is not valid because individual encounters are important. The statistical variation of the path length for monoenergetic particles is called *straggling*. For electrons, this can be quite important; however, for heavy particles of several million electronvolts or more, the path lengths vary by only a few percent or less.

*Range*

Figure 37-12 shows the rate of energy loss per unit path length $-dE/dx$ versus the energy of the ionizing particle. We see from this figure that the rate of energy loss is maximum at low energies and that for high energies the rate is approximately independent of energy. Particles with kinetic energy greater than their rest energy are called *minimum ionizing particles*. Their energy loss per unit path length or range is

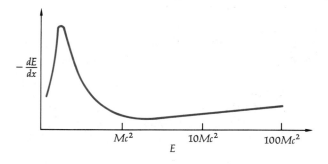

**Figure 37-12**
Energy loss $-dE/dx$ versus energy for a charged particle.

roughly proportional to the energy. Figure 37-13 shows the range-versus-energy curve for protons in air.

Since the energy loss of a charged particle is due to interactions with the electrons in the material, the greater the number of electrons, the greater the rate of energy loss. The energy loss rate $-dE/dx$ is approximately proportional to the density of the material. For example, the range of a 6-MeV proton is about 40 cm in air. In water, which is about 800 times denser, the range is only 0.5 mm.

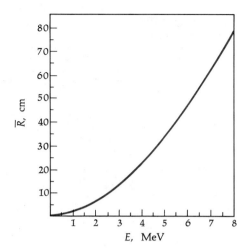

**Figure 37-13**
Mean range versus energy for protons in dry air. Except at low energies, the range is approximately linear with energy.

If the energy of the charged particle is large compared with its rest energy, the energy loss due to radiation as the particle slows down is important. This radiation is called *Bremsstrahlung*. The ratio of the energy lost by radiation and that lost through ionization is proportional to the energy of the particle and to the atomic number $Z$ of the stopping material. This ratio equals 1 for electrons of about 10 MeV in lead.

The fact that the rate of energy loss for heavy charged particles is very great at very low energies (as seen from the low energy peak in Figure 37-12) has important applications in nuclear radiation therapy. Figure 37-14 shows the energy loss versus penetration distance of charged particles in water. Most of the energy is deposited near the end of the range. The peak in this curve is called the *Bragg peak*. A beam of heavy charged particles can be used to destroy cancer cells at a given depth in the body without destroying other, healthy cells if the energy is carefully chosen so that most of the energy loss occurs at the proper depth.

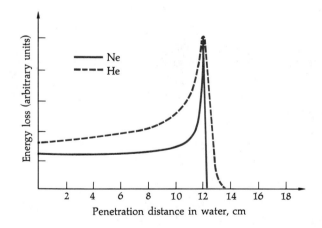

**Figure 37-14**
Energy loss of helium ions and neon ions in water versus depth of penetration. Most of the energy loss occurs near the end of the path in the Bragg peak. In general, the heavier the ion, the narrower the peak.

### Neutrons

Since neutrons are uncharged, they do not interact with electrons in matter. *Neutrons lose energy by nuclear scattering, or they may be captured.* For energies large compared with $kT$, the most important processes are elastic and inelastic scattering. If we have a collimated beam of intensity $I$, any scattering or absorption will remove neutrons from the beam. If the sum of the cross sections for all such processes is $\sigma$, the number removed from the beam is $\sigma I$ per nucleus. If $n$ is the number of nuclei per cubic centimetre, the number of neutrons removed in a distance $dx$ is

$$-dN = \sigma In A\, dx$$

where $A$ is the area of the neutron beam and $N = IA$ is the number of neutrons. Thus

$$-dN = \sigma n N\, dx$$

which gives

$$N = N_0 e^{-\sigma n x} \qquad\qquad\qquad 37\text{-}24$$

or

$$I = I_0 e^{-\sigma n x} \qquad\qquad\qquad 37\text{-}25$$

The number of neutrons in the beam decreases exponentially with distance, and there is no range.

The main source of energy loss for a neutron is usually elastic scattering. (In materials of intermediate weight, such as iron and silicon, inelastic scattering is also important. We shall neglect inelastic scattering here.) The maximum energy loss possible in one elastic collision occurs when the collision is head on. This can be calculated by considering a neutron of mass $m$ with speed $v_L$ making a head-on collision with a nucleus of mass $M$ at rest in the laboratory frame (see Problem 8). The result is that the fractional energy lost by a neutron in one such collision is

$$-\frac{\Delta E}{E} = \frac{4mM}{(M + m)^2} = \frac{4\,m/M}{(1 + m/M)^2} \qquad\qquad 37\text{-}26$$

This fraction has a maximum value of 1 when $M = m$ and approaches $4\,m/M$ for $M \gg m$.

### Photons

The intensity of a photon beam, like that of a neutron beam, decreases exponentially with distance through an absorbing material. The intensity is given by Equation 37-25, where $\sigma$ is the cross section for absorption per atom. *The important processes that remove photons from a beam are the photoelectric effect at low energies, Compton scattering at intermediate energies, and pair production at high energies.* The total cross section for absorption is the sum of the partial cross sections $\sigma_{pe}$, $\sigma_{cs}$, $\sigma_{pp}$ for these three processes. The photonuclear cross sections, such as $\sigma(\gamma, n)$, are very small compared with these atomic cross sections and can usually be neglected. The cross section for the photoelectric effect dominates at very low energy but decreases rapidly with increasing energy. The photoelectric effect cannot occur unless the electron is bound in an atom, which can recoil to conserve momentum.

If the photon energy is large compared with the binding energy of the electrons, the electrons can be considered to be free and Compton scattering is the principal mechanism for the removal of photons from the beam. If the photon energy is greater than $2m_e c^2 = 1.02$ MeV, the photon can disappear, with the creation of an electron-positron pair, a process called *pair production*.

The cross section for pair production increases rapidly with the photon energy and is the dominant term in $\sigma$ at high energies. Like the photoelectric effect, pair production cannot occur in free space. If we consider the reaction $\gamma \rightarrow e^+ + e^-$, there is some reference frame in which the total momentum of the electron-positron pair is zero but there is no reference frame in which the photon's momentum is zero. Thus a nucleus is needed nearby to absorb the momentum by recoil. The cross section for pair production is proportional to $Z^2$ of the absorbing matter. The three partial cross sections $\sigma_{pe}$, $\sigma_{cs}$, and $\sigma_{pp}$ are shown with the total cross section as functions of energy in Figure 37-15.

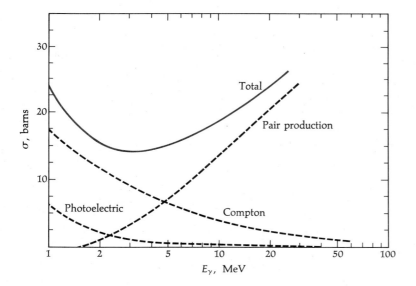

**Figure 37-15**
Photon attenuation cross section versus energy for lead. The cross sections for the photoelectric effect, Compton scattering, and pair production are shown by the dashed curves. The total cross section (solid curve) is the sum of these. (*From C. Davisson and R. Evans,* Reviews of Modern Physics, *vol. 24, p. 79, 1952.*)

## Dosage

The biological effects of radiation are principally due to the ionization produced. Three different units are used to measure these effects. The roentgen (R) is defined as the amount of radiation that produces $\frac{1}{3} \times 10^{-9}$ C of electric charge (either positive ions or electrons) in 1 cm³ of dry air at standard conditions. The roentgen has been largely replaced by the rad (*r*adiation *a*bsorbed *d*ose), which is defined in terms of the energy absorbed. One rad is the amount of radiation that deposits $10^{-2}$ J/kg of energy in any material.* Since 1 R is equivalent to the deposit of about $8.7 \times 10^{-3}$ J/kg, these units are roughly equal. The amount of biological damage depends not only on the energy absorbed, which is equivalent to dependence on the number of ion pairs formed, but also on the spacing of the ion pairs; if the ion pairs are closely spaced, as in ionization by $\alpha$ particles, the biological effect is enhanced. The unit rem (*r*oentgen *e*quivalent in *m*an) is the dose that has the same biological effect as 1 rad of $\beta$ or $\gamma$ radiation:

*Rad*

$$1 \text{ rem} = 1 \text{ rad} \times \text{RBE} \qquad\qquad 37\text{-}27$$

*Rem*

where the factor RBE (*r*elative *b*iological *e*ffectiveness factor) is tabulated for $\alpha$ particles, protons, and neutrons of various energies. For example, RBE is 1 for $\gamma$ and $\beta$ rays, about 4 or 5 for slow neutrons, about 10 for fast neutrons, and between 10 and 20 for $\alpha$ particles of energies from 5 to 10 MeV. Table 37-2 compares the various radiation units we have discussed. Some typical human radiation exposures are listed in Table 37-3, and some dose-limit recommendations are listed in Table 37-4.

---

* The SI unit joules per kilogram is now called a gray (Gy); 1 rad = $10^{-2}$ Gy.

**Table 37-2**
Radiation and dose units*

| Quantity | Customary unit | | SI unit | | Conversion |
|---|---|---|---|---|---|
| | Name | Symbol | Name | Symbol | |
| Energy | Electronvolt | eV | Joule | J | $1 \text{ MeV} = 1.602 \times 10^{-13} \text{ J}$ |
| Exposure | Roentgen | R | Coulomb per kilogram | C/kg | $1 \text{ R} = 2.58 \times 10^{-4} \text{ C/kg}$ |
| Absorbed dose | Rad | rad or rd | Gray | Gy = J/kg | $1 \text{ rad} = 10^{-2} \text{ J/kg} = 10^{-2} \text{ Gy}$ |
| Dose equivalent | Rem | rem | Rem | rem | |
| Activity | Curie | Ci | Becquerel | Bq = 1/s | $1 \text{ Ci} = 3.7 \times 10^{10} \text{ decays/s}$ $= 3.7 \times 10^{10} \text{ Bq}$ |

* From S. C. Bushong, *The Physics Teacher*, vol. 15, no. 3, p. 135, 1977.

**Table 37-3**
Sources and average intensities of human radiation exposure*

| Source | mrad/y | |
|---|---|---|
| **Natural** | | |
| Cosmic rays | 45 | |
| Terrestrial, external exposure | 60 | |
| Terrestrial, internally deposited radionuclides | 25 | |
| | | 130 |
| **Artificial** | | |
| Diagnostic x rays | 70 | |
| Weapons testing | <1 | |
| Power generation | <1 | |
| Occupational | ≤1 | |
| | | 70 |
| Total | | 200 |

* From S. C. Bushong, *The Physics Teacher*, vol. 15, no. 3, p. 135, 1977.

## Review

A. Define, explain, or otherwise identify:

**Table 37-4**
Recommended dose limits*

| | Maximum permissible dose equivalent for occupational exposure |
|---|---|
| Combined whole-body occupational exposure | |
|    Prospective annual limit | 5 rems in any one year |
|    Retrospective annual limit | 10–15 rems in any one year |
|    Long-term accumulation to age $N$ years | $(N - 18) \times 5$ rems |
| Skin | 15 rems in any one year |
| Hands | 75 rems in any one year (25 per quarter) |
| Forearms | 30 rems in any one year (10 per quarter) |
| Other organs, tissue, and organ systems | 15 rems in any one year (5 per quarter) |
| Fertile women (with respect to fetus) | 0.5 rem in gestation period |
| Dose limits for nonoccupationally exposed | |
|    Population average | 0.17 rem in any one year |
|    An individual in the population | 0.5 rem in any one year |
|    Students | 0.1 rem in any one year |

* Adapted from *NCRP Rep.* 39, 1971, as given in S. C. Bushong, *The Physics Teacher,* vol. 15, no. 3, p. 135, 1977.

B. True or false:

1. The atomic nucleus contains protons, neutrons, and electrons.

2. The mass of $^2$H is less than the mass of a proton plus a neutron.

3. All nuclei are radioactive.

4. After two half-lives, all the radioactive nuclei in a given sample have decayed.

5. Exothermic reactions have no threshold energy.

6. In a breeder reactor, fuel can be produced as fast as it is consumed.

7. The number of protons in a beam decreases exponentially with distance in matter.

8. The number of photons in a beam decreases exponentially with distance in matter.

9. The rad is a measure of the energy deposited per unit volume in matter.

Exercises

**Section 37-1, Atoms**

*There are no exercises for this section.*

**Section 37-2, Properties of Nuclei**

1. Give the symbols for two other isotopes of (*a*) $^{14}$N, (*b*) $^{56}$Fe, (*c*) $^{118}$Sn.

2. Use Table 37-1 to calculate the binding energy and the binding energy per nucleon for (a) $^{12}$C, (b) $^{56}$Fe, (c) $^{208}$Pb.

3. Repeat Exercise 2 for (a) $^{6}$Li, (b) $^{39}$K, and (c) $^{238}$U.

4. Given that the mass of a nucleus of mass number $A$ is approximately $m = CA$, where $C$ is a constant, find an expression for the nuclear density in terms of $C$ and the constant $R_0$ in Equation 37-4. Compute the value of this density in grams per centimetre using the fact that $C$ has the approximate value of 1 g per $N_A$ nucleon.

5. Find the energy needed to remove a neutron from $^{4}$He.

6. Use Equation 37-4 to compute the radii of the following nuclei: (a) $^{16}$O, (b) $^{56}$Fe, (c) $^{108}$Ag, (d) $^{197}$Au.

## Section 37-3, Radioactivity

7. The stable isotope of sodium is $^{23}$Na. What kind of radioactivity would you expect of (a) $^{22}$Na, (b) $^{24}$Na?

8. The half-life of radium is 1620 y. Calculate the number of disintegrations per second of 1 g of radium and show that the disintegration rate is approximately 1 Ci.

9. The counting rate from a radioactive source is 8000 counts/s at time $t = 0$, and 10 min later the rate is 1000 counts/s. (a) What is the half-life? (b) What is the decay constant? (c) What is the counting rate after 1 min?

10. A radioactive silver foil ($t_{1/2} = 2.4$ min) is placed near a Geiger counter, and 1000 counts/s are observed at time $t = 0$. (a) What is the counting rate at $t = 2.4$ min and $t = 4.8$ min? (b) If the counting efficiency is 20 percent, how many radioactive nuclei are there at time $t = 0$? At time $t = 2.4$ min? (c) At what time will the counting rate be about 30 counts/s?

11. Use Table 37-1 to calculate the energy of decay for the alpha decay of (a) $^{226}$Ra and (b) $^{242}$Pu.

12. The counting rate from a radioactive source is measured every minute. The resulting counts per second are 1000, 820, 673, 552, 453, 371, 305, 250, . . . Plot the counting rate versus time and use your graph to find the half-life.

13. A sample of wood contains 10 g of carbon and shows a $^{14}$C decay rate of 100 counts/min. What is the age of the sample?

14. A bone claimed to be 10,000 y old contains 15 g of carbon. What should the decay rate of $^{14}$C be from this bone?

## Section 37-4, Nuclear Reactions

15. (a) Find the $Q$ value for the reaction $^{2}$H + $^{1}$H → $^{3}$He + $n$ + $Q$ using Table 37-1. (b) Find the threshold energy for this reaction if stationary $^{1}$H nuclei are bombarded with $^{3}$H nuclei from an accelerator. (c) Find the threshold energy for this reaction if stationary $^{3}$H nuclei are bombarded with $^{1}$H nuclei from an accelerator.

16. Find the $Q$ values for the following fusion reactions:

(a) $^{2}$H + $^{2}$H → $^{3}$He + $n$ + $Q$
(b) $^{2}$H + $^{2}$H → $^{3}$H + $^{1}$H + $Q$
(c) $^{2}$H + $^{3}$He → $^{4}$He + $^{1}$H + $Q$
(d) $^{6}$Li + $n$ → $^{3}$H + $^{4}$He + $Q$

## Section 37-5, Fission, Fusion, and Nuclear Reactors

17. Assuming an average energy of 200 MeV per fission, calculate the number of fissions per second needed for a 500-MW reactor.

18. If the reproduction factor in a reactor is $k = 1.1$, find the number of generations needed for the power level to (a) double, (b) increase by a factor of 10, and (c) increase by a factor of 100. Find the time needed in each case if (d) there were no delayed neutrons so that the time between generations is 1 ms and (e) if the delayed neutrons make the average time between generations 100 ms.

19. Compute the temperature $T$ corresponding to $kT = 10$ keV, where $k$ is Boltzmann's constant.

## Section 37-6, The Interaction of Particles with Matter

20. The range of 4-MeV $\alpha$ particles in air is 2.5 cm ($\rho = 1.29$ mg/cm$^3$). Assuming the range to be inversely proportional to the density of matter, find the range of 4-MeV $\alpha$ particles in (a) water and (b) lead (density 11.2 g/cm$^3$).

21. The range of 6-MeV protons in air is approximately 45 cm. Find the approximate range of 6-MeV protons in (a) water and (b) lead (see Exercise 20).

22. If the absorption cross section for neutrons in iron is 2.5 barns, find the thickness of iron that will reduce the intensity of a neutron beam by a factor of 2 through absorption.

23. The total absorption cross section for $\gamma$ rays in lead is about 20 barns at 15 MeV. (a) What thickness of lead will reduce the intensity of 15-MeV $\gamma$ rays by $1/e$? (b) What thickness will reduce the intensity by a factor of 100?

## Problems

1. Show that

$$ke^2 = \frac{1}{4\pi\epsilon_0} e^2 = 1.44 \text{ MeV·fm}$$

where $k$ is the Coulomb constant.

2. Show that if the decay rate is $R_0$ at $t = 0$ and $R_1$ at some later time $t_1$, the decay constant $\lambda$ and the half-life $t_{1/2}$ are given by

$$\lambda = t_1^{-1} \ln \frac{R_0}{R_1} \quad \text{and} \quad t_{1/2} = \frac{0.693t_1}{\ln (R_0/R_1)}$$

3. (a) Calculate the radius of $^{141}_{56}$Ba and of $^{92}_{36}$Kr from Equation 37-4. (b) Assume that after fission of $^{235}$U into $^{141}$Ba and $^{92}$Kr, the two nuclei are momentarily separated by a distance $d$ equal to the sum of the radii found in (a). Calculate the electrostatic potential energy $U = k(Z_1e)(Z_2e)/d$ for these two nuclei at this separation and compare with the fission energy of 175 MeV.

4. Calculate the radii of $^2$H and $^3$H from Equation 37-4 and find the electrostatic energy when these two nuclei are "just touching"; i.e., when their centers are separated by $d$ equal to the sum of their radii.

5. The intensity of a neutrino beam decreases exponentially with distance according to Equation 37-25, just like that of a beam of neutrons or photons. The absorption cross section for neutrinos is of the order of $10^{-20}$ barn. Find the thickness of iron needed to reduce the intensity of a neutrino beam by $1/e$. Compare this thickness with the distance from the earth to the sun (about 150 Gm).

6. Radioactive nuclei with decay constant $\lambda$ are produced in an accelerator at a constant rate $R_p$. The number of radioactive nuclei $N$ then obeys the equation $dN/dt = R_p - \lambda N$. (a) If the number $N$ is zero at $t = 0$, sketch $N$ versus $t$ for this situation. (b) $^{62}$Cu is produced at a rate of 100 per second by placing ordinary copper ($^{63}$Cu) in a beam of high-energy photons (the reaction is $\gamma + {}^{63}$Cu $\rightarrow$ $^{62}$Cu $+ n$). $^{62}$Cu decays by $\beta$ decay with a half-life of 10 min. After time long enough so that $dN/dt \approx 0$, how many $^{62}$Cu nuclei are there?

7. Use the results of Problem 2 to find the decay constant and half-life if the counting rate is 1000 counts/s at time $t = 0$ and 800 counts/s 1 min later.

8. Consider a neutron of mass $m$ moving with speed $v_L$ and colliding head on with a nucleus of mass $M$. (a) Show that the speed of the center of mass in the lab frame is $V = mv_L/(m + M)$. (b) What is the speed of the nucleus in the center-of-mass frame before the collision? After the collision? (c) What is the speed of the nucleus in the original lab frame after the collision? (d) Show that the energy of the nucleus after the collision is $\frac{1}{2}M(2V)^2 = [4mM/(m + M)^2] \times \frac{1}{2}mv_L^2$ and use it to obtain Equation 37-26.

9. (a) Show that after $N$ head-on collisions of a neutron with a carbon nucleus at rest, the energy of a neutron is approximately $0.72^N E_0$, where $E_0$ is its original energy. (b) How many head-on collisions are required to reduce the energy of a neutron from 2 to 0.02 MeV, assuming stationary carbon nuclei?

10. Energy is generated in the sun and other stars by fusion. One of the fusion cycles, the proton-proton cycle, consists of the following reactions:

$$^1H + {}^1H \rightarrow {}^2H + \beta^+ + \nu$$

$$^1H + {}^2H \rightarrow {}^3He + \gamma$$

followed by either

$$^3He + {}^3He \rightarrow {}^4He + {}^1H + {}^1H$$

or

$$^1H + {}^3He \rightarrow {}^4He + \beta^+ + \nu$$

(a) Show that the net effect of these reactions is

$$4\,{}^1H \rightarrow {}^4He + 2\beta^+ + 2\nu + \gamma$$

(b) Show that the rest-mass energy of 24.7 MeV is released in this cycle, not counting the $2 \times 0.511$ MeV released when each positron meets an electron and is annihilated according to $e^+ + e^- \rightarrow 2\gamma$. (c) The sun radiates energy at the rate of about $4 \times 10^{26}$ W. Assuming that this is due to the conversion of four protons into helium plus $\gamma$ rays and neutrinos, which releases 26.7 MeV, what is the rate of proton consumption in the sun? How long will the sun last if it continues to radiate at its present level? (Assume that protons constitute about half the total mass of the sun, which is about $2 \times 10^{30}$ kg.)

# APPENDIX A    SI Units

**Basic units**

| | |
|---|---|
| Length | The *metre* (m) is the length equal to 1,650,763.73 wavelengths in vacuum of the radiation corresponding to the transition between the $2p_{10}$ and $5d_5$ levels of the $^{86}Kr$ atom |
| Time | The *second* (s) is the duration of 9,192,631,770 periods of the radiation corresponding to the transition between the two hyperfine levels of the ground state of the $^{133}Cs$ atom |
| Mass | The *kilogram* (kg) is the mass of the international standard body preserved at Sèvres, France |
| Current | The *ampere* (A) is that current in two very long parallel wires 1 m apart that gives rise to a magnetic force per unit length of $2 \times 10^{-7}$ N/m |
| Temperature | The *kelvin* (K) is 1/273.16 of the thermodynamic temperature of the triple point of water |
| Luminous intensity | The *candela* (cd) is the luminous intensity, in the perpendicular direction, of a surface of area 1/600,000 m² of a blackbody at the temperature of freezing platinum at pressure of 1 atm |

**Derived units**

| | | |
|---|---|---|
| Force | newton (N) | $1 \text{ N} = 1 \text{ kg} \cdot \text{m/s}^2$ |
| Work, energy | joule (J) | $1 \text{ J} = 1 \text{ N} \cdot \text{m}$ |
| Power | watt (W) | $1 \text{ W} = 1 \text{ J/s}$ |
| Frequency | hertz (Hz) | $1 \text{ Hz} = \text{s}^{-1}$ |
| Charge | coulomb (C) | $1 \text{ C} = 1 \text{ A} \cdot \text{s}$ |
| Potential | volt (V) | $1 \text{ V} = 1 \text{ J/C}$ |
| Resistance | ohm ($\Omega$) | $1 \text{ } \Omega = 1 \text{ V/A}$ |
| Capacitance | farad (F) | $1 \text{ F} = 1 \text{ C/V}$ |
| Magnetic induction | tesla (T) | $1 \text{ T} = 1 \text{ N/A} \cdot \text{m}$ |
| Magnetic flux | weber (Wb) | $1 \text{ Wb} = 1 \text{ T} \cdot \text{m}^2$ |
| Inductance | henry (H) | $1 \text{ H} = 1 \text{ J/A}^2$ |

## APPENDIX B        Numerical Data

For additional data see the front and back endpapers and the following tables in the text:

## Physical constants

| | | |
|---|---|---|
| Gravitational constant | $G$ | $6.672 \times 10^{-11}$ N·m$^2$/kg$^2$ |
| Speed of light | $c$ | $2.997925 \times 10^8$ m/s |
| Electron charge | $e$ | $1.60219 \times 10^{-19}$ C |
| Avogadro's number | $N_A$ | $6.0220 \times 10^{23}$ particles/mol |
| Gas constant | $R$ | $8.314$ J/mol·K |
| | | $1.9872$ cal/mol·K |
| | | $8.206 \times 10^{-2}$ L·atm/mol·K |
| Boltzmann's constant | $k = R/N_A$ | $1.3807 \times 10^{-23}$ J/K |
| | | $8.617 \times 10^{-5}$ eV/K |
| Unified mass unit | $u = 1/N_A$ g | $1.6606 \times 10^{-24}$ g |
| Coulomb constant | $k = 1/4\pi\epsilon_0$ | $8.98755 \times 10^9$ N·m$^2$/C$^2$ |
| Permittivity of free space | $\epsilon_0$ | $8.85419 \times 10^{-12}$ C$^2$/N·m$^2$ |
| Magnetic constant | $k_m = \mu_0/4\pi$ | $10^{-7}$ N/A$^2$ (exact) |
| Permeability of free space | $\mu_0$ | $4\pi \times 10^{-7}$ N/A$^2$ |
| | | $1.256637 \times 10^{-6}$ N/A$^2$ |
| Planck's constant | $h$ | $6.6262 \times 10^{-34}$ J·s |
| | | $4.1357 \times 10^{-15}$ eV·s |
| | $\hbar = h/2\pi$ | $1.05459 \times 10^{-34}$ J·s |
| | | $6.5822 \times 10^{-16}$ eV·s |
| Mass of electron | $m_e$ | $9.1095 \times 10^{-31}$ kg |
| | | $511.0$ keV/$c^2$ |
| Mass of proton | $m_p$ | $1.67265 \times 10^{-27}$ kg |
| | | $938.28$ MeV/$c^2$ |
| Mass of neutron | $m_n$ | $1.67495 \times 10^{-27}$ kg |
| | | $939.57$ MeV/$c^2$ |
| Bohr magneton | $m_B = e\hbar/2m_e$ | $9.2741 \times 10^{-24}$ A-m$^2$ |
| | | $5.7884 \times 10^{-5}$ eV/T |

## Terrestrial data

| | |
|---|---|
| Acceleration of gravity $g$ | 9.80665 m/s² standard value |
| | 32.1740 ft/s² |
| At sea level, at equator* | 9.7804 m/s² |
| At sea level, at poles* | 9.8322 m/s² |
| Mass of earth $M_E$ | $5.98 \times 10^{24}$ kg |
| Radius of earth $R_E$, mean | $6.37 \times 10^6$ m |
| | 3960 mi |
| Escape speed $\sqrt{2R_E g}$ | $1.12 \times 10^4$ m/s |
| | 6.95 mi/s |
| Solar constant** | 1.35 kW/m² |
| Standard temperature and pressure (STP): | |
| $T$ | 273.15 K |
| $P$ | 101.325 kPa |
| | 1.00 atm |
| Molecular mass of air | 28.97 g/mol |
| Density of air (STP), $\rho_{air}$ | 1.293 kg/m³ |
| Speed of sound (STP) | 331 m/s |
| Heat of fusion of $H_2O$ (0°C, 1 atm) | 333.5 kJ/kg |
| Heat of vaporization of $H_2O$ (100°C, 1 atm) | 2.257 MJ/kg |

* Measured relative to the earth's surface.
** Average power incident normally on 1 m² outside the earth's atmosphere at the mean distance from the earth to the sun.

## Astronomical data

| | |
|---|---|
| Earth, distance to moon* | $3.844 \times 10^8$ m |
| | $2.389 \times 10^5$ mi |
| Distance (mean) to sun* | $1.496 \times 10^{11}$ m |
| | $9.30 \times 10^7$ mi |
| | 1.00 AU |
| Orbital speed (mean) | $2.98 \times 10^4$ m/s |
| Moon, mass | $7.35 \times 10^{22}$ kg |
| Radius | $1.738 \times 10^6$ m |
| Period | 27.32 d |
| Acceleration of gravity at surface | 1.62 m/s² |
| Sun, mass | $1.99 \times 10^{30}$ kg |
| Radius | $6.96 \times 10^8$ m |

* Center to center.

# APPENDIX C    Conversion Factors

Conversion factors are written as equations for simplicity; relations marked with an asterisk are exact.

**Length**

1 km = 0.6215 mi

1 mi = 1.609 km

1 m = 1.0936 yd = 3.281 ft = 39.37 in

*1 in = 2.54 cm

*1 ft = 12 in = 30.48 cm

*1 yd = 3 ft = 91.44 cm

1 light-year = 1 $c \cdot y$ = 9.461 × $10^{15}$ m

*1 Å = 0.1 nm

**Area**

*1 $m^2$ = $10^4$ $cm^2$

1 $km^2$ = 0.3861 $mi^2$ = 247.1 acres

*1 $in^2$ = 6.4516 $cm^2$

1 $ft^2$ = 9.29 × $10^{-2}$ $m^2$

1 $m^2$ = 10.76 $ft^2$

*1 acre = 43,560 $ft^2$

1 $mi^2$ = 640 acres = 2.590 $km^2$

**Volume**

*1 $m^3$ = $10^6$ $cm^3$

*1 L = 1000 $cm^3$ = $10^{-3}$ $m^3$

1 gal = 3.786 L

1 gal = 4 qt = 8 pt = 128 oz = 231 $in^3$

1 $in^3$ = 16.39 $cm^3$

1 $ft^3$ = 1728 $in^3$ = 28.32 L = 2.832 × $10^4$ $cm^3$

**Time**

*1 h = 60 min = 3.6 ks

*1 d = 24 h = 1440 min = 86.4 ks

1 y = 365.24 d = 31.56 Ms

**Speed**

1 km/h = 0.2778 m/s = 0.6215 mi/h

1 mi/h = 0.4470 m/s = 1.609 km/h

1 mi/h = 1.467 ft/s

**Angle and Angular Speed**

*$\pi$ rad = 180°

1 rad = 57.30°

1° = 1.745 × $10^{-2}$ rad

1 rev/min = 0.1047 rad/s

1 rad/s = 9.549 rev/min

**Mass**

*1 kg = 1000 g

*1 tonne = 1000 kg = 1 Mg

1 u = 1.6606 × $10^{-27}$ kg

1 kg = 6.022 × $10^{23}$ u

1 slug = 14.59 kg

1 kg = 6.852 × $10^{-2}$ slugs

1 u = 931.50 MeV/$c^2$

**Density**

*1 $g/cm^3$ = 1000 $kg/m^3$ = 1 kg/L

(1 $g/cm^3$)$g$ = 62.4 $lb/ft^3$

**Force**

1 N = 0.2248 lb = $10^5$ dyn

1 lb = 4.4482 N

(1 kg)$g$ = 2.2046 lb

**Pressure**

*1 Pa = 1 $N/m^2$

*1 atm = 101.325 kPa = 1.01325 bars

1 atm = 14.7 $lb/in^2$ = 760 mmHg

= 29.9 inHg = 33.8 ftH$_2$O

1 $lb/in^2$ = 6.895 kPa

1 torr = 1 mmHg = 133.32 Pa

1 bar = 100 kPa

**Energy**

*1 kW·h = 3.6 MJ

*1 cal = 4.1840 J

1 ft·lb = 1.356 J = 1.286 × $10^{-3}$ Btu

*1 L·atm = 101.325 J

1 L·atm = 24.217 cal

1 Btu = 778 ft·lb = 252 cal = 1054.35 J

1 eV = 1.602 × $10^{-19}$ J

1 u·$c^2$ = 931.50 MeV

*1 erg = $10^{-7}$ J

**Power**

1 horsepower = 550 ft·lb/s = 745.7 W

1 Btu/min = 17.58 W

1 W = 1.341 × $10^{-3}$ horsepower

= 0.7376 ft·lb/s

**Magnetic Induction**

*1 G = $10^{-4}$ T

*1 T = $10^4$ G

**Thermal Conductivity**

1 W/m·K = 6.938 Btu·in/h·ft²·°F

1 Btu·in/h·ft²·°F = 0.1441 W/m·K

**APPENDIX D**    Mathematical Symbols and Formulas

## Mathematical Symbols and Abbreviations

| | |
|---|---|
| $=$ | is equal to |
| $\neq$ | is not equal to |
| $\approx$ | is approximately equal to |
| $\sim$ | is of the order of |
| $\propto$ | is proportional to |
| $>$ | is greater than |
| $\gg$ | is much greater than |
| $<$ | is less than |
| $\ll$ | is much less than |
| $\Delta x$ | change in $x$ |
| $n!$ | $n(n-1)(n-2) \cdots 1$ |
| $\Sigma$ | sum |
| $\lim$ | limit |
| $\Delta t \to 0$ | $\Delta t$ approaches zero |
| $\int$ | integral |
| $\dfrac{dx}{dt}$ | derivative of $x$ with respect to $t$ |
| $\dfrac{\partial y}{\partial x}$ | partial derivative of $y$ with respect to $x$ |

## Roots of the Quadratic Equation

If

$$ax^2 + bx + c = 0$$

then

$$x = \frac{-b}{2a} \pm \frac{1}{2a} \sqrt{b^2 - 4ac}$$

## Binomial Expansion

$$(1 + x)^n = 1 + nx + \frac{n(n-1)}{2!} x^2 + \frac{n(n-1)(n-2)}{3!} x^3 + \cdots$$

To compute $(a + b)^n$ write

$$(a + b)^n = a^n(1 + x)^n \qquad \text{with } x = \frac{b}{a}$$

or

$$(a + b)^n = b^n(1 + y)^n \qquad \text{with } y = \frac{a}{b}$$

## Trigonometric Formulas

See also Sec. 1-8.

$$\sin^2 \theta + \cos^2 \theta = 1 \qquad \sec^2 \theta - \tan^2 \theta = 1 \qquad \csc^2 \theta - \text{ctn}^2 \theta = 1$$

$$\sin 2\theta = 2 \sin \theta \cos \theta$$

$$\cos 2\theta = \cos^2 \theta - \sin^2 \theta = 2 \cos^2 \theta - 1 = 1 - 2 \sin^2 \theta$$

$$\tan 2\theta = \frac{2 \tan \theta}{1 - \tan^2 \theta}$$

$$\sin \tfrac{1}{2}\theta = \sqrt{\frac{1 - \cos \theta}{2}} \qquad \cos \tfrac{1}{2}\theta = \sqrt{\frac{1 + \cos \theta}{2}} \qquad \tan \tfrac{1}{2}\theta = \sqrt{\frac{1 - \cos \theta}{1 + \cos \theta}}$$

$$\sin (A \pm B) = \sin A \cos B \pm \cos A \sin B$$

$$\cos (A \pm B) = \cos A \cos B \mp \sin A \sin B$$

$$\tan (A \pm B) = \frac{\tan A \pm \tan B}{1 \mp \tan A \tan B}$$

$$\sin A \pm \sin B = 2 \sin \left[\tfrac{1}{2}(A \pm B)\right] \cos \left[\tfrac{1}{2}(A \mp B)\right]$$

$$\cos A + \cos B = 2 \cos \left[\tfrac{1}{2}(A + B)\right] \cos \left[\tfrac{1}{2}(A - B)\right]$$

$$\cos A - \cos B = 2 \sin \left[\tfrac{1}{2}(A + B)\right] \sin \left[\tfrac{1}{2}(B - A)\right]$$

$$\tan A \pm \tan B = \frac{\sin (A \pm B)}{\cos A \cos B}$$

## Exponential and Logarithmic Functions

$$e = 2.71828 \qquad e^0 = 1$$

If $y = e^x$, then $x = \ln y$.

$$e^{\ln x} = x$$

$$e^x e^y = e^{(x+y)}$$

$$(e^x)^y = e^{xy} = (e^y)^x.$$

$$\ln e = 1 \qquad \ln 1 = 0$$

$$\ln xy = \ln x + \ln y$$

$$\ln \frac{x}{y} = \ln x - \ln y$$

$$\ln e^x = x \qquad \ln a^x = x \ln a$$

$$\ln x = (\ln 10) \log x = 2.3026 \log x$$

$$\log x = \log e \ln x = 0.43429 \ln x$$

$$e^x = 1 + x + \frac{x^2}{2!} + \frac{x^3}{3!} + \cdots$$

$$\ln (1 + x) = x - \frac{x^2}{2} + \frac{x^3}{3} - \frac{x^4}{4} + \cdots$$

**APPENDIX E**         # Derivatives and Integrals

Table E-1 lists some important properties of derivatives and the derivatives of some particular functions which often occur in physics. It is followed by comments aimed at making these properties and rules plausible. More detailed and rigorous discussions can be found in most calculus books. Table E-2 lists some important integration formulas. More extensive lists of differentiation and integration formulas can be found in handbooks, e.g., Herbert Dwight, *Tables of Integrals and Other Mathematical Data*, 4th ed., Macmillan Publishing Co., Inc., New York, 1961.

---

**Table E-1**
Properties of derivatives and derivatives of particular functions

---

Linearity

---

1. The derivative of a constant times a function equals the constant times the derivative of the function:

$$\frac{d}{dt}[Cf(t)] = C\frac{df(t)}{dt}$$

2. The derivative of a sum of functions equals the sum of the derivatives of the functions:

$$\frac{d}{dt}[f(t) + g(t)] = \frac{df(t)}{dt} + \frac{dg(t)}{dt}$$

---

Chain rule

---

3. If $f$ is a function of $x$ and $x$ is in turn a function of $t$, the derivative of $f$ with respect to $t$ equals the product of the derivative of $f$ with respect to $x$ and the derivative of $x$ with respect to $t$:

$$\frac{d}{dt}f(x) = \frac{df}{dx}\frac{dx}{dt}$$

---

Derivative of a product

---

4. The derivative of a product of functions $f(t)g(t)$ equals the first function times the derivative of the second plus the second function times the derivative of the first:

$$\frac{d}{dt}[f(t)g(t)] = f(t)\frac{dg(t)}{dt} + \frac{df(t)}{dt}g(t)$$

---

Reciprocal derivative

---

5. The derivative of $t$ with respect to $x$ is the reciprocal of the derivative of $x$ with respect to $t$, assuming that neither derivative is zero:

$$\frac{dx}{dt} = \left(\frac{dt}{dx}\right)^{-1} \quad \text{if} \quad \frac{dt}{dx} \neq 0$$

## Derivatives of particular functions

6. $\dfrac{dC}{dt} = 0$    where $C$ is a constant

7. $\dfrac{d(t^n)}{dt} = nt^{n-1}$

8. $\dfrac{d}{dt} \sin \omega t = \omega \cos \omega t$

9. $\dfrac{d}{dt} \cos \omega t = -\omega \sin \omega t$

10. $\dfrac{d}{dt} e^{bt} = be^{bt}$

11. $\dfrac{d}{dt} \ln bt = \dfrac{1}{t}$

**Comment on Rules 1 to 5**

Rules 1 and 2 follow from the fact that the limiting process is linear. We can understand rule 3, the chain rule, by multiplying $\Delta f / \Delta t$ by $\Delta x / \Delta x$ and noting that since $x$ is a function of $t$, both $\Delta x$ and $\Delta f$ approach zero as $\Delta t$ approaches zero. Since the limit of a product of two functions equals the product of their limits, we have

$$\lim_{\Delta t \to 0} \frac{\Delta f}{\Delta t} = \lim_{\Delta t \to 0} \frac{\Delta f}{\Delta x} \frac{\Delta x}{\Delta t} = \left( \lim_{\Delta x \to 0} \frac{\Delta f}{\Delta x} \right) \left( \lim_{\Delta t \to 0} \frac{\Delta x}{\Delta t} \right) = \frac{df}{dx} \frac{dx}{dt}$$

Rule 4 is not immediately apparent. The derivative of a product of functions is the limit of the ratio

$$\frac{f(t + \Delta t)g(t + \Delta t) - f(t)g(t)}{\Delta t}$$

If we add and subtract the quantity $f(t + \Delta t)g(t)$ in the numerator, we can write this ratio as

$$\frac{f(t + \Delta t)g(t + \Delta t) - f(t + \Delta t)g(t) + f(t + \Delta t)g(t) - f(t)g(t)}{\Delta t}$$

$$= f(t + \Delta t) \frac{g(t + \Delta t) - g(t)}{\Delta t} + g(t) \frac{f(t + \Delta t) - f(t)}{\Delta t}$$

As $\Delta t$ approaches zero, the terms in brackets become $dg(t)/dt$ and $df(t)/dt$, respectively, and the limit of the expression is

$$f(t) \frac{dg(t)}{dt} + g(t) \frac{df(t)}{dt}$$

Rule 5 follows directly from the definition:

$$\frac{dx}{dt} = \lim_{\Delta t \to 0} \frac{\Delta x}{\Delta t} = \lim_{\Delta t \to 0} \left(\frac{\Delta t}{\Delta x}\right)^{-1} = \lim_{\Delta x \to 0} \left(\frac{\Delta t}{\Delta x}\right)^{-1} = \left(\lim_{\Delta x \to 0} \frac{\Delta t}{\Delta x}\right)^{-1}$$

## Comment on Rule 7

We can obtain this important result using the binomial expansion (Appendix D). We have

$$f(t) = t^n$$

$$f(t + \Delta t) = (t + \Delta t)^n = t^n \left(1 + \frac{\Delta t}{t}\right)^n$$

$$= t^n \left[1 + n\frac{\Delta t}{t} + \frac{n(n-1)}{2!}\left(\frac{\Delta t}{t}\right)^2 + \frac{n(n-1)(n-2)}{3!}\left(\frac{\Delta t}{t}\right)^3 + \cdots\right]$$

Then

$$f(t - \Delta t) - f(t) = t^n \left[n\frac{\Delta t}{t} + \frac{n(n-1)}{2!}\left(\frac{\Delta t}{t}\right)^2 + \cdots\right]$$

and

$$\frac{f(t - \Delta t) - f(t)}{\Delta t} = nt^{n-1} + \frac{n(n-1)}{2!}t^{n-2}\,\Delta t + \cdots$$

The next term omitted from the last sum is proportional to $(\Delta t)^2$, the following to $(\Delta t)^3$, and so on. Each term except the first approaches zero as $\Delta t$ approaches zero. Thus

$$\frac{df}{dt} = \lim_{\Delta x \to 0} \frac{f(t + \Delta t) + f(t)}{\Delta t} = nt^{n-1}$$

## Comment on Rules 8 and 9

We first write $\sin \omega t = \sin \theta$ with $\theta = \omega t$ and use the chain rule,

$$\frac{d \sin \theta}{dt} = \frac{d \sin \theta}{d\theta} \frac{d\theta}{dt} = \omega \frac{d \sin \theta}{d\theta}$$

We then use the trigonometric formula for the sine of the sum of two angles $\theta$ and $\Delta\theta$:

$$\sin (\theta + \Delta\theta) = \sin \Delta\theta \cos \theta + \cos \Delta\theta \sin \theta$$

Since $\Delta\theta$ is to approach zero, we can use the small-angle approximations

$$\sin \Delta\theta \approx \Delta\theta \quad \text{and} \quad \cos \Delta\theta \approx 1$$

Then

$$\sin (\theta + \Delta\theta) \approx \Delta\theta \cos \theta + \sin \theta$$

and

$$\frac{\sin (\theta + \Delta\theta) - \sin \theta}{\Delta\theta} \approx \cos \theta$$

Similar reasoning can be applied to the cosine function to obtain rule 9.

## Comment on Rule 10

Again we use the chain rule,

$$\frac{de^\theta}{dt} = b\frac{de^\theta}{d\theta} \quad \text{with } \theta = bt$$

and use the series expansion for the exponential function (Appendix D):

$$e^{\theta + \Delta\theta} = e^\theta e^{\Delta\theta} = e^\theta \left[1 + \Delta\theta + \frac{(\Delta\theta)^2}{2!} + \frac{(\Delta\theta)^3}{3!} + \cdots\right]$$

Then

$$\frac{e^{\theta + \Delta\theta} - e^{\theta}}{\Delta\theta} = e^{\theta} + e^{\theta} \frac{\Delta\theta}{2!} + e^{\theta} \frac{(\Delta\theta)^2}{3!} + \cdots$$

As $\Delta\theta$ approaches zero, the right side of the equation above approaches $e^{\theta}$.

**Comment on Rule 11**

Let

$$y = \ln bt$$

Then

$$e^y = bt \qquad \text{and} \qquad \frac{dt}{dy} = \frac{1}{b} e^y = t$$

Then using rule 5, we obtain

$$\frac{dy}{dt} = \left(\frac{dt}{dy}\right)^{-1} = \frac{1}{t}$$

---

**Table E-2**
Integration formulas*

1. $\displaystyle\int A \, dt = At$

2. $\displaystyle\int At \, dt = \tfrac{1}{2} At^2$

3. $\displaystyle\int At^n \, dt = A \frac{t^{n+1}}{n+1} \qquad n \neq -1$

4. $\displaystyle\int At^{-1} \, dt = A \ln t$

5. $\displaystyle\int e^{bt} \, dt = \frac{1}{b} e^{bt}$

6. $\displaystyle\int \cos \omega t \, dt = \frac{1}{\omega} \sin \omega t$

7. $\displaystyle\int \sin \omega t \, dt = -\frac{1}{\omega} \cos \omega t$

---

* In these formulas $A$, $b$, and $\omega$ are constants. An arbitrary constant $C$ can be added to the right side of each equation.

# APPENDIX F

# Trigonometric Tables

| Angle $\theta$ | | | | | Angle $\theta$ | | | | |
|---|---|---|---|---|---|---|---|---|---|
| Degree | Radian | $\sin \theta$ | $\cos \theta$ | $\tan \theta$ | Degree | Radian | $\sin \theta$ | $\cos \theta$ | $\tan \theta$ |
| 0 | 0.0000 | 0.0000 | 1.0000 | 0.0000 | | | | | |
| 1 | 0.0175 | 0.0175 | 0.9998 | 0.0175 | 46 | 0.8029 | 0.7193 | 0.6947 | 1.0355 |
| 2 | 0.0349 | 0.0349 | 0.9994 | 0.0349 | 47 | 0.8203 | 0.7314 | 0.6820 | 1.0724 |
| 3 | 0.0524 | 0.0523 | 0.9986 | 0.0524 | 48 | 0.8378 | 0.7431 | 0.6691 | 1.1106 |
| 4 | 0.0698 | 0.0698 | 0.9976 | 0.0699 | 49 | 0.8552 | 0.7547 | 0.6561 | 1.1504 |
| 5 | 0.0873 | 0.0872 | 0.9962 | 0.0875 | 50 | 0.8727 | 0.7660 | 0.6428 | 1.1918 |
| 6 | 0.1047 | 0.1045 | 0.9945 | 0.1051 | 51 | 0.8901 | 0.7771 | 0.6293 | 1.2349 |
| 7 | 0.1222 | 0.1219 | 0.9925 | 0.1228 | 52 | 0.9076 | 0.7880 | 0.6157 | 1.2799 |
| 8 | 0.1396 | 0.1392 | 0.9903 | 0.1405 | 53 | 0.9250 | 0.7986 | 0.6018 | 1.3270 |
| 9 | 0.1571 | 0.1564 | 0.9877 | 0.1584 | 54 | 0.9425 | 0.8090 | 0.5878 | 1.3764 |
| 10 | 0.1745 | 0.1736 | 0.9848 | 0.1763 | 55 | 0.9599 | 0.8192 | 0.5736 | 1.4281 |
| 11 | 0.1920 | 0.1908 | 0.9816 | 0.1944 | 56 | 0.9774 | 0.8290 | 0.5592 | 1.4826 |
| 12 | 0.2094 | 0.2079 | 0.9781 | 0.2126 | 57 | 0.9948 | 0.8387 | 0.5446 | 1.5399 |
| 13 | 0.2269 | 0.2250 | 0.9744 | 0.2309 | 58 | 1.0123 | 0.8480 | 0.5299 | 1.6003 |
| 14 | 0.2443 | 0.2419 | 0.9703 | 0.2493 | 59 | 1.0297 | 0.8572 | 0.5150 | 1.6643 |
| 15 | 0.2618 | 0.2588 | 0.9659 | 0.2679 | 60 | 1.0472 | 0.8660 | 0.5000 | 1.7321 |
| 16 | 0.2793 | 0.2756 | 0.9613 | 0.2867 | 61 | 1.0647 | 0.8746 | 0.4848 | 1.8040 |
| 17 | 0.2967 | 0.2924 | 0.9563 | 0.3057 | 62 | 1.0821 | 0.8829 | 0.4695 | 1.8807 |
| 18 | 0.3142 | 0.3090 | 0.9511 | 0.3249 | 63 | 1.0996 | 0.8910 | 0.4540 | 1.9626 |
| 19 | 0.3316 | 0.3256 | 0.9455 | 0.3443 | 64 | 1.1170 | 0.8988 | 0.4384 | 2.0503 |
| 20 | 0.3491 | 0.3420 | 0.9397 | 0.3640 | 65 | 1.1345 | 0.9063 | 0.4226 | 2.1445 |
| 21 | 0.3665 | 0.3584 | 0.9336 | 0.3839 | 66 | 1.1519 | 0.9135 | 0.4067 | 2.2460 |
| 22 | 0.3840 | 0.3746 | 0.9272 | 0.4040 | 67 | 1.1694 | 0.9205 | 0.3907 | 2.3559 |
| 23 | 0.4014 | 0.3907 | 0.9205 | 0.4245 | 68 | 1.1868 | 0.9272 | 0.3746 | 2.4751 |
| 24 | 0.4189 | 0.4067 | 0.9135 | 0.4452 | 69 | 1.2043 | 0.9336 | 0.3584 | 2.6051 |
| 25 | 0.4363 | 0.4226 | 0.9063 | 0.4663 | 70 | 1.2217 | 0.9397 | 0.3420 | 2.7475 |
| 26 | 0.4538 | 0.4384 | 0.8988 | 0.4877 | 71 | 1.2392 | 0.9455 | 0.3256 | 2.9042 |
| 27 | 0.4712 | 0.4540 | 0.8910 | 0.5095 | 72 | 1.2566 | 0.9511 | 0.3090 | 3.0777 |
| 28 | 0.4887 | 0.4695 | 0.8829 | 0.5317 | 73 | 1.2741 | 0.9563 | 0.2924 | 3.2709 |
| 29 | 0.5061 | 0.4848 | 0.8746 | 0.5543 | 74 | 1.2915 | 0.9613 | 0.2756 | 3.4874 |
| 30 | 0.5236 | 0.5000 | 0.8660 | 0.5774 | 75 | 1.3090 | 0.9659 | 0.2588 | 3.7321 |
| 31 | 0.5411 | 0.5150 | 0.8572 | 0.6009 | 76 | 1.3265 | 0.9703 | 0.2419 | 4.0108 |
| 32 | 0.5585 | 0.5299 | 0.8480 | 0.6249 | 77 | 1.3439 | 0.9744 | 0.2250 | 4.3315 |
| 33 | 0.5760 | 0.5446 | 0.8387 | 0.6494 | 78 | 1.3614 | 0.9781 | 0.2079 | 4.7046 |
| 34 | 0.5934 | 0.5592 | 0.8290 | 0.6745 | 79 | 1.3788 | 0.9816 | 0.1908 | 5.1446 |
| 35 | 0.6109 | 0.5736 | 0.8192 | 0.7002 | 80 | 1.3963 | 0.9848 | 0.1736 | 5.6713 |
| 36 | 0.6283 | 0.5878 | 0.8090 | 0.7265 | 81 | 1.4137 | 0.9877 | 0.1564 | 6.314 |
| 37 | 0.6458 | 0.6018 | 0.7986 | 0.7536 | 82 | 1.4312 | 0.9903 | 0.1392 | 7.115 |
| 38 | 0.6632 | 0.6157 | 0.7880 | 0.7813 | 83 | 1.4486 | 0.9925 | 0.1219 | 8.144 |
| 39 | 0.6807 | 0.6293 | 0.7771 | 0.8098 | 84 | 1.4661 | 0.9945 | 0.1045 | 9.514 |
| 40 | 0.6981 | 0.6428 | 0.7660 | 0.8391 | 85 | 1.4835 | 0.9962 | 0.0872 | 11.430 |
| 41 | 0.7156 | 0.6561 | 0.7547 | 0.8693 | 86 | 1.5010 | 0.9976 | 0.0698 | 14.301 |
| 42 | 0.7330 | 0.6691 | 0.7431 | 0.9004 | 87 | 1.5184 | 0.9986 | 0.0523 | 19.081 |
| 43 | 0.7505 | 0.6820 | 0.7314 | 0.9325 | 88 | 1.5359 | 0.9994 | 0.0349 | 28.636 |
| 44 | 0.7679 | 0.6947 | 0.7193 | 0.9657 | 89 | 1.5533 | 0.9998 | 0.0175 | 57.290 |
| 45 | 0.7854 | 0.7071 | 0.7071 | 1.0000 | 90 | 1.5708 | 1.0000 | 0.0000 | $\infty$ |

# APPENDIX G  Periodic Table of the Elements

The values listed are based on $^{12}_{6}C = 12$ u exactly. For artificially produced elements, the approximate atomic mass of the most stable isotope is given in brackets.

| PERIOD | SERIES | GROUP I | II | III | IV | V | VI | VII | VIII | | | 0 |
|---|---|---|---|---|---|---|---|---|---|---|---|---|
| 1 | 1 | 1 H 1.00797 | | | | | | | | | | 2 He 4.003 |
| 2 | 2 | 3 Li 6.942 | 4 Be 9.012 | 5 B 10.81 | 6 C 12.011 | 7 N 14.007 | 8 O 15.9994 | 9 F 19.00 | | | | 10 Ne 20.183 |
| 3 | 3 | 11 Na 22.990 | 12 Mg 24.31 | 13 Al 26.98 | 14 Si 28.09 | 15 P 30.974 | 16 S 32.064 | 17 Cl 35.453 | | | | 18 Ar 39.948 |
| 4 | 4 | 19 K 39.102 | 20 Ca 40.08 | 21 Sc 44.96 | 22 Ti 47.90 | 23 V 50.94 | 24 Cr 52.00 | 25 Mn 54.94 | 26 Fe 55.85 | 27 Co 58.93 | 28 Ni 58.71 | |
| | 5 | 29 Cu 63.54 | 30 Zn 65.37 | 31 Ga 69.72 | 32 Ge 72.59 | 33 As 74.92 | 34 Se 78.96 | 35 Br 70.909 | | | | 36 Kr 83.80 |
| 5 | 6 | 37 Rb 85.47 | 38 Sr 87.62 | 39 Y 88.905 | 40 Zr 91.22 | 41 Nb 92.91 | 42 Mo 95.94 | 43 Tc [98] | 44 Ru 101.1 | 45 Rh 102.905 | 46 Pd 106.4 | |
| | 7 | 47 Ag 107.870 | 48 Cd 112.40 | 49 In 114.82 | 50 Sn 118.69 | 51 Sb 121.75 | 52 Te 127.60 | 53 I 126.90 | | | | 54 Xe 131.30 |
| 6 | 8 | 55 Cs 132.905 | 56 Ba 137.34 | 57-71 Lanthanide series* | 72 Hf 178.49 | 73 Ta 180.95 | 74 W 183.85 | 75 Re 186.2 | 76 Os 190.2 | 77 Ir 192.2 | 78 Pt 195.09 | |
| | 9 | 79 Au 196.97 | 80 Hg 200.59 | 81 Tl 204.37 | 82 Pb 207.19 | 83 Bi 208.98 | 84 Po [210] | 85 At [210] | | | | 86 Rn [222] |
| 7 | 10 | 87 Fr [223] | 88 Ra [226] | 89-103 Actinide series** | | | | | | | | |

| *Lanthanide Series | 57 La 138.91 | 58 Ce 140.12 | 59 Pr 140.91 | 60 Nd 144.24 | 61 Pm [147] | 62 Sm 150.35 | 63 Eu 152.0 | 64 Gd 157.25 | 65 Tb 158.92 | 66 Dy 162.50 | 67 Ho 164.93 | 68 Er 167.26 | 69 Tm 168.93 | 70 Yb 173.04 | 71 Lu 174.97 |
|---|---|---|---|---|---|---|---|---|---|---|---|---|---|---|---|
| **Actinide Series | 89 Ac [227] | 90 Th 232.04 | 91 Pa [231] | 92 U 238.03 | 93 Np [237] | 94 Pu [242] | 95 Am [243] | 96 Cm [247] | 97 Bk [247] | 98 Cf [251] | 99 Es [254] | 100 Fm [253] | 101 Md [256] | 102 No [254] | 103 Lw [257] |

# Answers

These answers are calculated using $g = 9.81$ m/s² unless otherwise specified in the exercise or problem. The results are usually rounded to three significant figures. Differences in the last figure can easily result from differences in rounding the input data and are not important.

**Chapter 1**

**True or False**

1. True 2. False; e.g., $x = vt$, where the speed $v$ and the time $t$ have different dimensions 3. True 4. True 5. False; $\cos \theta \approx 1$ for small $\theta$

**Exercises**

1. (a) 1 MW (b) 2 mg (c) 3 $\mu$m (d) 30 ks

3. (a) 1 picaboo (b) 1 gigalow (c) 1 microphone (d) 1 attoboy (e) 1 megaphone (f) 1 nanogoat (g) 1 terabull

5. (a) $C_1$, length; $C_2$, length/time (b) length/(time)² (c) length/(time)² (d) $C_1$, length; $C_2$, (time)$^{-1}$ (e) $C_1$, length/time; $C_2$, (time)$^{-1}$

7. (a) $4 \times 10^7$ m (b) $6.37 \times 10^6$ m (c) 24,860 mi; 3957 mi

9. (a) unitless (b) s (c) m

11. (a) $3.16 \times 10^7$ s/y (b) 31.7 y (c) $1.90 \times 10^{16}$ y

13. (a) 1.61 km/mi (b) 2.20 lb

15. (a) $1.00 \times 10^6$ (b) $3.03 \times 10^{-8}$ (c) $6.02 \times 10^{23}$ (d) $1.4 \times 10^{-3}$

17. (a) $1.14 \times 10^5$ (b) $2.25 \times 10^{-8}$ (c) $8.27 \times 10^3$ (d) $6.27 \times 10^2$

19. (a) 10 (b) $-\frac{5}{22}$ (c) $\frac{1}{9}$

21. (a) 2.25 mW/m² (b) 1.00 mW/m² (c) 0.36 mW/m² (d) 9.00 mW/m²

23. (a) 1.00 (b) 0.00 (c) 0.00 (d) 1.00 (e) 0.707 (f) $-0.707$

25. (a) $\pi/3$ rad (b) $\pi/2$ rad (c) $\pi/6$ rad (d) $\pi/4$ rad (e) $\pi$ rad (f) 0.643 rad (g) $4\pi$ rad

27. 3.49 rad = 200°

29. $\sin \theta_1 = 0.6$; $\cos \theta_1 = 0.8$; $\tan \theta_1 = 0.75$; $\sin \theta_2 = 0.8$; $\cos \theta_2 = 0.6$; $\tan \theta_2 = 1.33$; $\theta_1 = 37°$; $\theta_2 = 53°$

31. (a) 0.14 (b) 0.087

33. (a) 9.95 (b) 0.99 (c) 4.99

35. (a) 96.0 (b) 0.995 (c) 0.996

**Problems**

1. 3.51 Mm

5. 62.1°

7. (a) $n = 0.5$; $C = 1.8$ s·kg$^{-1/2}$ (b) $m = 0.2$ kg and $m = 1.0$ kg

9. (a) $\sqrt{L/g}$ (c) $T = 2\pi \sqrt{L/g}$

11. (a) increase with increasing $H$; increase with increasing $v$ (b) $R$ proportional to $v\sqrt{H/g}$

13. 236 m

15. (a) 8.66 km; 5.0 km (b) 17.3 km

17. (a) 0.04%; 0.001%; 0.0012% (b) 0.042%; 0.001%; 0.0012%

**Chapter 2**

**True or False**

1. False; holds only for constant acceleration 2. False; e.g., motion with constant velocity 3. False; holds only for constant acceleration 4. True

**Exercises**

1. (a) 24 km/h (b) $-12$ km/h (c) 0 km/h (d) 16 km/h

3. (a) $-2$ m/s (b) 2.25 m/s (c) $-0.3$ m/s

5. (a) 260 km (b) 65 km/h

7. (a) velocity and speed at $t_2$ less than at $t_1$ (b) both equal (c) velocity greater at $t_2$ but speed less at $t_2$ (d) velocity less at $t_2$ but speed greater at $t_2$

9. about 2 m/s; 2.7 m/s; 3.2 m/s; 4.0 m/s; $v \approx 4.2$ m/s from measurement of slope at 0.75 s

11. (a) 2 m, 2 m/s (b) $\Delta x = 2t\,\Delta t - 5\,\Delta t + (\Delta t)^2$ m (c) $v = (2t - 5)$ m/s

13. (a) 5.0 m/s$^2$ (b) 0.51

15. $-2$ m/s$^2$

17. (a) $a = 0$ (b) $a > 0$ (c) $a < 0$ (d) $a = 0$

19. $v = (16$ m/s$^2)t - 6$ m/s; $a = 16$ m/s$^2$

21. (a) 80 m/s (b) 400 m (c) 40 m/s

23. $15\frac{5}{8}$ m/s$^2$

25. 4.47 s; 44.7 m/s; 22.4 m/s

27. (a) $-7.7$ m/s (b) $+6.3$ m/s (c) $7.0 \times 10^2$ m/s$^2$ up

29. $x(t) = x_0 + (6$ m/s$)t$

31. (a) $x(t) = x_0 + v_0 t + \frac{1}{6}Ct^3$ (b) 62.5 m; 37.5 m/s

33. (a) $-36$ m (b) $\Delta x = -36$ m (c) $-9$ m/s

35. (a) 0.25 m/s (b) about 0.95 m/s; 3.2 m/s; 6.2 m/s (c) about 7 m

37. (a) 60.45 m (b) 60 m

39. 2%

**Problems**

1. (b) 15 s (c) 300 m from the intersection

3. $x = 124$ m

7. (b) 20 s (c) 160 km/h

9. (b) 65 km/h; 5.54 s

11. (a) yes (c) 8 m/s; 4 m/s; 32 m

13. 4.8 m/s

15. 40 cm/s; $-6.88$ cm/s$^2$

17. $\frac{2}{3}$

21. (b) $v_{av} = 0.72$ cm/s; 0.83 cm/s; 0.855 cm/s; 0.868 cm/s; 0.878 cm/s; 0.872 cm/s (c) 0.875 cm/s

23. (b) distance run during time $T$ is $x_1 = \frac{1}{2}aT^2 = \frac{1}{2}v_0 T$; from time $T$ to $t$ distance is $x_2 = v_0(t - T)$; total distance is $x_1 + x_2 = v_0(t - \frac{1}{2}T)$ (c) $T \approx 2.4$ s; $v_0 \approx 11.5$ m/s; $a = 4.8$ m/s$^2$

25. (a) $g/B$

**Chapter 3**

**True or False**

1. True 2. False; e.g., projectile moving up, **a** down 3. False; e.g., circular motion 4. True 5. False 6. False; e.g., magnitude of **A** $+ (-\mathbf{A})$ is zero 7. True 8. True unless speed is zero 9. True

**Exercises**

1. resultant displacement vector is 18.6 m, 22° north of east

3. (a) 2 km at 0°; 1 km at 135°; 3 km at 90° (b) $A_x = 2$ km, $A_y = 0$ km; $A_x =$

$-0.707$ km, $A_y = +0.707$ km; $A_x = 0$ km, $A_y = 3$ km (c) 3.93 km, 70.8° north of east (d) no; possibly but not necessarily

5.  $(-5$ m, 5 m); $\mathbf{A} = -5\mathbf{i} + 5\mathbf{j}$ m

7.  (a) $A = 5.83$; $\phi = 31°$ (b) $B = 12.2$; $\phi = 325°$ or $-35°$ (c) $C = 5.39$; $\phi = 236°$; $\theta = 42°$; the angles $\theta$ and $\phi$ are those shown in Fig. 3-2

9.  $g_x = 4.9$ m/s², $g_y = -8.49$ m/s²

11.  (a) $\mathbf{v} = 5\mathbf{i} + 8.66\mathbf{j}$ m/s (b) $\mathbf{A} = -3.54\mathbf{i} - 3.54\mathbf{j}$ m (c) $\mathbf{r} = 14\mathbf{i} - 6\mathbf{j}$ m

15.  $5\mathbf{A} = 10\mathbf{i} - 30\mathbf{j}$; $-7\mathbf{A} = -14\mathbf{i} + 42\mathbf{j}$

17.  $\mathbf{B} = -1.5\mathbf{A}$

19.  $\Delta\mathbf{A}$ and $\mathbf{A}$ are approximately perpendicular

21.  (a) $2\mathbf{i} + 2\mathbf{j}$ m/s (b) $\frac{11}{5}\mathbf{i} + \frac{11}{5}\mathbf{j}$ m/s

23.  (b) $\mathbf{v} = 5\mathbf{i} + 10\mathbf{j}$ m/s; $v = 11.2$ m/s

25.  (a) 90°, 45°, 0°, $-45°$, and $-90°$ with $x$ axis (b) $AB$ up; $BC$, toward center of arc; $DE$, toward center of arc; $EF$ up (c) $a$ is greater along $DE$

27.  (a) $\mathbf{v}_{\text{av}} = 33.3\mathbf{i} + 26.7\mathbf{j}$ m/s (b) $\mathbf{a}_{\text{av}} = -3\mathbf{i} - 1.77\mathbf{j}$ m/s²

29.  0.553 s

31.  (a) $2.30 \times 10^3$ m (b) 43.3 s (c) $9.18 \times 10^3$ m

33.  $H = (v_0^2 \sin^2\theta_0)/2g$

35.  45 m/s²

37.  $v^2/r = 2.47$ cm/s²

39.  8.33 m/s² up

41.  (a) $v = 10$ m/s; $dv/dt = 0$ (b) $v = 11.4$ m/s; $dv/dt = 15$ m/s² (c) $v = 13.3$ m/s; $dv/dt = -35.4$ m/s²

**Problems**

1.  (a) 31.6 m/s² (b) 26.2 m/s² (c) 23.9 m/s²   Note: It is easiest to express $\mathbf{v}_2$ and $\mathbf{v}_2 - \mathbf{v}_1$ in rectangular components and then find $|\Delta\mathbf{v}/\Delta t|$.

3.  (a) 12.0 m (b) 10 m/s; 14.7 m/s (c) 17.8 m/s (d) 55.8°

5.  3.19 m/s toward the ball

7.  (a) 8.14 m/s (b) 23.2 m/s

9.  51.0 km/h

11.  (a) $x \tan\theta$ (b) $\sqrt{gx^2/[2\cos^2\theta(x\tan\theta - H)]}$

13.  76°

15.  (a) $25\mathbf{i} + 7.07\mathbf{j} + 7.07\mathbf{k}$ m/s (b) 11.3 m from the road at the point $40.0\mathbf{i} + 11.3\mathbf{j}$ m from the car's original position or $0\mathbf{i} + 11.3\mathbf{j}$ m from the car's present position

17.  (b) $\mathbf{v} = 8\pi\cos 2\pi t\,\mathbf{i} - 8\pi\sin 2\pi t\,\mathbf{j}$ m/s (c) $\mathbf{a} = -(4\pi^2/\text{s}^2)\mathbf{R}$

19.  (b) $\mathbf{v} = 6\pi\cos 2\pi t\,\mathbf{i} - 4\pi\sin 2\pi t\,\mathbf{j}$ m/s (c) $\mathbf{a} = -(4\pi^2/\text{s}^2)\mathbf{R}$ (d) $v$ is maximum at $t = 0, \frac{1}{2}$ s, 1 s, $\frac{3}{2}$ s, 2 s, . . . , and minimum at $t = \frac{1}{4}$ s, $\frac{3}{4}$ s, $\frac{5}{4}$ s . . . ; show this by writing $v = 2\pi\sqrt{9\cos^2\theta + 4\sin^2\theta} = 2\pi\sqrt{4 + 5\cos^2\theta}$, where $\theta = 2\pi t$; then $v$ is maximum when $\cos\theta = \pm 1$ and minimum when $\cos\theta = 0$

21.  408 m

23.  34.6 m/s

**Chapter 4**    **True or False**    1. True 2. False; could be balancing forces 3. False; e.g., circular motion 4. True 5. False 6. True 7. False; this is a common misconception 8. True 9. False

**Exercises**

1. (*a*) 8 m/s² (*b*) second object has half the mass of first (*c*) $\frac{8}{3}$ m/s²

3. (*a*) 7.07 m/s² (*b*) 14.0 m/s²

5. *AB*: toward *B*; *BC*: toward center of arc; *CD*: *a* = 0; *DE*: toward center of arc; *EF*: toward *E*

7. from 2 s to about 5 s (−); from about 6 s to about 8 s (+)

9. 3 kg

11. 10 kg

13. (*a*) 300 N (*b*) 500 N

15. (*a*) 5.43 slugs (*b*) 79.3 kg (*c*) 7.93 × 10⁴ g

17. 4.9 × 10⁴ dyn; 0.49 N

19. (*a*) 785 N; 176 lb (*b*) 716 N; 161 lb (*c*) 80 kg

21. (*a*) 12 kg·m/s (*b*) 18 kg·m/s (*c*) 6 kg·m/s; −3 kg·m/s

23. (*a*) 1.67 m/s² (*b*) 1.67 m/s²; 1.67 N; contact force exerted by the 2-kg object (*c*) 3.33 N

25. (*a*) momentum changes are equal in magnitude and opposite in direction (*b*) velocity change of 1-kg body is twice that of 2-kg body and opposite in direction

27. 1.2 kg

29. (*a*) 2.60**i** + 0.500**j** m/s² (*b*) −26.0**i** − 5.00**j** N

31. (*a*) 1.5**i** + 2**j** m/s² (*b*) 4.5**i** + 6**j** m/s (*c*) 6.75**i** + 9**j** m

33. (*a*) 19.6 N in each wire (*b*) $w/(2 \sin \theta)$ (*c*) *T* least for $\theta = 90°$; $T \to \infty$ as $\theta \to 0$

35. (*a*) 4.90 m/s²; 24.5 N

37. (*a*) $a_1 = 0$; $a_2 = 0.75$ m/s² (*b*) 0.50 m/s² (*c*) $F_1 = 1$ N to right; $F_2 = 2$ N to right (*d*) 1 N to left

39. (*a*) 19.6 N (*b*) 19.6 N (*c*) 39.6 N (*d*) 19.6 N from 0 to 2 s; 9.62 N from 2 to 4 s

41. 2.47 s

43. (*a*) $v'_x + 3$ m/s (*b*) 2.5 m/s; 5 m/s; 7.5 m/s; (*c*) 5.5 m/s; 8 m/s; 10.5 m/s (*d*) 2.5 m/s²; 2.5 m/s²

**Problems**

1. (*b*) 7.2 N (*c*) 9.3 cm

3. 3.75 kN

5. 3.82 kN; 5.73 kN

7. $a = 17.0$ m/s²; for greater *a*, block would slide up incline

9. for $a = \frac{1}{2}g \approx 5$ m/s², $t = 5$ s, and $F = 400$ N for an 80-kg driver and 300 N for a 60-kg driver

11. $F = 202$ N; $T = 100$ N at *A*, 101 N at *B*, and 201 N at *C*

13. 80 kg; 785 N; 2.19 m/s²

15. 492 N; 3.71 m/s; 21.0 m/s

17. (*a*) 1.37 m/s²; 61.4 N (*b*) $m_1 = 1.19 m_2$

19. (*b*) apparent weight is less than force of gravity (*c*) in an inertial frame, $g = 981.4$ cm/s² at equator and 982.7 cm/s² at $\theta = 45°$  *Note:* The angle between the force of gravity (radial line) and the apparent free-fall acceleration at $\theta = 45°$ is small and can be neglected here.

21. (*a*) deflection is opposite to acceleration (*c*) 9.3° forward

23. (*a*) 0.5 m/s²; 2 N (*b*) $m_2 F/(m_1 + m_2)$

**Chapter 5**

**True or False**     1. True 2. True 3. False; the strong nuclear force is attractive 4. True 5. True 6. True 7. False 8. True 9. True 10. False

**Exercises**

1.  $5.97 \times 10^{24}$ kg

3.  $8.67 \times 10^{-7}$ N

5.  (a) $6.24 \times 10^{12}$ electrons (b) 0.225 N

7.  $2.69 \times 10^{-2}$ N

9.  (a) $2.31 \times 10^3$ N (b) ratio is $2.81 \times 10^{10}$

11.  (a) $F_g = F_s = 49.1$ N (b) 24.5 cm

13.  (a) 327 N/m (b) 8.18 m/s²

15.  (a) 5.89 m/s² (b) 76.4 m

17.  83.9 m

19.  (b) pushing at 30°, 520 N; pulling at 30°, 252 N; pushing or pulling at 0°, 294 N

21.  22.7 N; 1.51 m/s

23.  0.544

25.  31°

27.  42.2 Mm

29.  1.4 m from pivot

31.  (a) 181 N (b) 457 N up (c) $F = 157$ N; $f_{\text{hinge}} = 502$ N up, 78.6 N right

33.  117 N at left and 333 N at right

35.  (a) $f_x = f_y = 30$ N (b) $f_x = 35$ N; $f_y = 45$ N

37.  (a) mass/time; kg/s (b) mass/length; kg/m

39.  (a) $x = x_0 + 10t - 2.5\ t^2$; $v = 10 - 5t$ relative to car (b) $t = 4$ s

41.  $x = x_0 + 10t - 4\ t^2$; $v = 10 - 8t$ for $0 < t < 1.25$ s; object stops at $t = 1.25$ s and remains at rest relative to car

43.  6.7 cm

**Problems**

1.  0.241 $\mu$C

3.  $\mu_s = 0.577$; $\mu_k = 0.401$

5.  (a) 400 N (b) 625 N

7.  (a) 5.00 m (b) 4.87 m

9.  (a) 0.6g (b) 22.5 kg (c) 0.7g for $m_2$ and $m_3$, 0.4g for $m_1$; 88.2 N

11.  (b) $mv^2/R - mg$ (c) $\sqrt{Rg}$; block leaves track and moves in a parabolic path as a projectile until hitting track

13.  (a) $F = \mu mg/(\mu \sin \theta + \cos \theta)$

15.  44.1 N left at top; 44.1 N right at bottom

17.  $T_2 = m_2(L_1 + L_2)\omega^2$; $T_1 = m_1 L_1 \omega^2 + m_2(L_1 + L_2)\omega^2$

21.  from 20.1 to 56.0 km/h

23.  18.2°

25.  $F = mg \sqrt{2Rh - h^2}/(R - h)$

27.  26.6°

29.  (a) circular motion with angular velocity $\omega$ and linear velocity $r\omega$; resultant

force is $mr\omega^2$ inward (b) centrifugal force is $mr\omega^2$ outward; Coriolis force is $2mv = 2mr\omega^2$ inward

**Chapter 6**    **True or False**    1. False 2. True 3. False 4. True 5. False; unit of energy 6. False 7. False 8. False; only with conservative forces 9. False; the total energy does not change 10. True 11. True

**Exercises**

1.  (a) 490 J (b) −490 J

3.  (a) 296 J (b) −196 J

5.  (a) 5000 J (b) 1250 J

7.  400 J for 2-kg body; 4 J for 200-kg body

9.  (a) 320 J; −196 J (b) 124 J

11.  (a) 74 N (b) 222 J (c) −147 J (d) 75 J

13.  (a) 9 J (b) 10.5 J (c) 4.42 m/s

15.  $20\frac{1}{4}C$

17.  (a) normal force, no work; gravity, 102 J; friction, −11.8 J (b) 90.2 J (c) 5.48 m/s

19.  (a) 327 N; 245.25 N; 196.2 N (b) 981 J; 981 J; 981 J

21.  18 m²

23.  (a) 142° (b) 101° (c) 90°

25.  83°

27.  conservative: (b), (c), and (e); nonconservative: (a), (d), (f), and (g)

29.  (a) 392 J (b) 2.45 m; 4.90 m/s (c) 368 J; 24 J; 392 J (d) 392 J; 19.8 m/s

31.  20.4 m

33.  (a) 6x (b) 6(x − 5) (c) 6x + 50

35.  (a) 1.215 J; −1.215 J (b) 1.215 J; 1.215 J

37.  (a) A, zero; B, zero; C, positive; D, zero; E, negative; F, negative (b) F (c) A, D, stable; B, unstable

39.  $-4Ax^3$ (b) x = 0

43.  (a) 1.67 m/s (b) 15 J

45.  (a) 17.6 N; 0.197 (b) 1.73 W (c) −5.20 J

47.  (a) 6 W (b) −49 W (c) 0

**Problems**

1.  (a) −11 J, −10 J, −7 J, −3 J, 0, +1 J, 0, −2 J, −3 J (b) $F_x$ is given as a function of x only (c) graph should be negative of that in (a)

3.  (a) 2.7 J (b) 4.2 J (c) 1.7 m/s (d) 3.5 J (e) 1.8 m/s

5.  263 W

7.  (a) 1.41 MJ (b) 12.0 kW

9.  (a) 6.74 m (b) 0.15

11.  (a) $-\frac{3}{8}mv_0^2$ (b) $\mu = 3v_0^2/(16\pi rg)$ (c) $\frac{1}{3}$ rev

13.  (a) assuming a weight of 700 N is raised 3 cm for each step, $mgh = 21$ J per step; using 1250 steps per kilometre gives 26 kJ (b) 22 W

15.  (a) from (0, 0) to (2, 0) W = 0; from (2, 0) to (2, 2) W = 10 J; $W_{total}$ = 10 J (b) from (0, 0) to (0, 2) W = 10 J; from (0, 2) to (2, 2) W = 16 J; $W_{total}$ = 26 J

17. (a) along direct path from (4, 1) to (4, 4) on line $x = 4$ m the work is 36 J; along path from (4, 1) to (0, 1) to (0, 4) to (4, 4) the work is zero; other similar paths give different results because work along paths parallel to $x$ axis adds to zero while work along a vertical path depends on value of $x$ (b) second path given in (a).

19. (a) $g \sin \theta \, (m_2 l_2 - m_1 l_1)$ (b) $+90°$ if $m_1 l_1 > m_2 l_2$; $-90°$ if $m_1 l_1 < m_2 l_2$

21. (b) $F_x = -(U_0/x^2)e^{-x/a} (x + a)$ (c) at $x = a$, $F_x = -0.736 \, U_0/a = F_x(a)$; at $x = 2a$, $F_x = 0.14 F_x(a)$; at $x = 5a$, $F_x = 0.002 F_x(a)$

| Chapter 7 | True or False | 1. False 2. True 3. False 4. False |

Exercises

1. (a) 75 J (b) $-29.4$ J (c) 45.6 J (d) 4.27 m/s

3. (a) 4.43 m/s (b) $-19.6$ J (c) 0.167

5. 2.55 m

7. (a) 294 J (b) 147 J (c) decrease by 147 J (d) $-147$ J

9. (a) 62.5 J (b) $\frac{1}{2}mv_x^2 + \frac{1}{2}mv_y^2 = 22.6$ J $+ 39.9$ J; minimum $E_k$ is 22.6 J at top of path (c) 39.9 J (d) 20.3 m

11. 3.61 m/s

15. (a) $-6.27$ GJ (b) $-3.14$ GJ (c) 7.92 km/s

17. 2.38 km/s

19. 51.2 km/s

21. If $r$ increases, $U$ increases, $E_k$ decreases, and $E_{total}$ increases because $\Delta U = -2 \, \Delta E_k$

Problems

1. (a) $5mg$ (b) $3.5mg$ (c) $0.5mg$

3. (a) $60°$ (b) 6.26 m/s

5. $x_0$

7. (a) 0.989 m   Note: The mass falls a distance $h = (4 + x) \sin 30°$, where $x$ is the compression of the spring. If you neglect $x$ in calculating the potential-energy loss, you get 0.885 for $x$. (b) 0.783 m (c) 1.54 m above spring (d) it will slide up and down a distance that decreases each time, and will eventually come to rest with the spring compressed about 8.6 cm

9. $8.83 \times 10^{-2}$ J (b) $y = (0.9)^N H$ (c) 44

11. (a) $\frac{1}{2}mgR_E$ (b) $\sqrt{gR_E} = 7.91$ km/s

13. 6 m

15. 46 cm

17. (a) 5.10 m (b) 10.2 m

19. (a) $v = \sqrt{2g(2x - L)}$ (b) $a \geq g$

21. (b) when $r$ decreases, $E_k$ increases but $U$ and $E_{total}$ both decrease

23. (a) 463 m/s (b) 10.74 km/s (c) about 8%

| Chapter 8 | True or False | 1. True 2. True 3. True 4. True 5. False; true only in the center-of-mass frame 6. False; true only in the center-of-mass frame 7. True 8. True 9. True |

Exercises

1. (a) $-2$ m/s (b) $-12$ kg·m/s (c) $-3$ m (d) 2 kg is at 15 m; 4 kg at $-21$ m; CM at $-9$ m

3. (2 m, 1 m)

5. (a) $3\mathbf{i} - 1.5\mathbf{j}$ m/s (b) $18\mathbf{i} - 9\mathbf{j}$ kg·m/s

7. (a) $23\mathbf{i} - 1.5\mathbf{j}$ m/s (b) (65 m, −7.5 m)

9. (a) $3L\lambda_0/2$ (b) $5L/9$

11. (a) 50 kg·m/s (b) 50 kg·m/s (c) 3.57 m/s

13. 4.55 m/s

15. 3.33 m/s

17. $1.77 \times 10^{-2}$ J

19. (a) 39 J (b) 3 m/s (c) 2 m/s to right and 2 m/s to left (d) 12 J (6 J each) (e) $\frac{1}{2}Mv_{\mathrm{CM}}^2 = 27$ J $= 39$ J $- 12$ J

21. (a) 25 J (b) 2 m/s (c) +3 m/s; −2 m/s (d) 15 J (e) $\frac{1}{2}M_{\mathrm{CM}}^2 = 10$ J $= 25$ J $- 15$ J

23. (a) 2 m/s; 2 m/s (b) −1 m/s; +4 m/s

25. (a) 2 m/s; 2 m/s (b) −1 m/s; +7 m/s

27. 0.459 m

29. 4.00 m

31. (a) 3 m/s; 8 m/s (b) 32.0 J

33. (a) 20% (b) internal energy of ball and surroundings (c) 0.89

35. (a) $180\mathbf{i} + 80\mathbf{j}$ Mg·km/h (b) 49.2 km/h; 24° north of east

37. (a) $v_2 = \sqrt{2}\,v_0$; $\theta_2 = 45°$ (b) $E_f = \frac{3}{2}mv_0^2 = E_i$

39. 37.5 kN·s

41. 600

43. (a) 2 N (b) 4.7 m/s

45. (a) 1.72 (b) 6.39 (c) 147

47. (a) 0.8 m/s (b) 0.2 m/s; −0.3 m/s; 0; 0 (c) $E_1 = 1.2$ kJ, $E_2 = 1.8$ kJ before; $E_1 = E_2 = 0$ after

49. (a) $u_{1i} = 4$ m/s, $u_{2i} = -2$ m/s before; $u_{1f} = -4$ m/s, $u_{2f} = +2$ m/s after (b) $v_{1f} = -2$ m/s; $v_{2f} = +4$ m/s (c) 32 J

51. 3.30 MeV

53. 6.16 eV

**Problems**

1. (a) about 4670 km from center of earth or 1700 km below earth's surface (b) gravitational attraction of sun and other planets; neglecting other planets, $\mathbf{a}_{\mathrm{CM}}$ is toward sun (c) about 9340 km

3. (a) 2 m/s (b) 8 m/s (c) $2\frac{2}{3}$ m/s (d) frictional force exerted by ground on man (e) 1080 J is total increase in kinetic energy

5. center of mass is $r/6$ from center of plate along common diameter and away from hole

7. 162 m from launch; 5.51 kJ

9. (a) 3.5 m/s (b) 10.5 J (c) 0.75

11. −0.238 m/s

13. assuming hand moves 1 m, $a = 40g$ and $F = 40mg = 58.9$ N neglecting weight ($F = 41mg = 60.3$ N including weight), and $t = 0.07$ s

15. 205 m/s; 589 J

17. $v_1 = 5\sqrt{3}$ m/s; $v_2 = 5$ m/s at 90° to $\mathbf{v_1}$

19. $x_{CM} = 0$; $y_{CM} = 4R/3\pi$

21. $x_{CM} = 0$; $y_{CM} = 3R/2\pi$

23. (a) 4.8 N·s (b) 1600 N (c) 2.4 N·s (d) 19.2 N

25. 36.4 m/s

27. $v = 2\sqrt{gL}\,(m_1 + m_2)/m_1$

29. (a) 7 m/s (b) $p_1 = 12$ N·s; $p_2 = -12$ N·s (c) $p_1 = -6$ N·s; $p_2 = +6$ N·s (d) $v_1 = 5.5$ m/s; $v_2 = 8$ m/s (e) 22.5 J

31. 0.85; 6.7%

33. (a) 9.66 km/s (b) 8.68 km/s (c) yes, the rocket travels only a small distance compared with $R_E$ in the 100 s it takes to burn the fuel

35. (a) $0.72E_1$ (b) 42

## Chapter 9

### True or False

1. False 2. True 3. True 4. True 5. False

### Exercises

1. (a) 0.2 rad/s (b) 0.955 rev

3. (a) 3.49 rad/s (b) 52.4 cm/s; 1.83 m/s²

5. (a) $-2.91 \times 10^{-2}$ rad/s² (b) 1.75 rad/s (c) 33.3 rev

7. 0.625 s

9. (a) 56.0 kg·m² (b) 112 J

11. 21.9 kg·m²

13. (a) 2.4 N·m (b) 66.7 rad/s² (c) 200 rad/s (d) 720 J (e) 7.2 kg·m² s (f) 300 rad

15. (a) 1.5 N·m; $1.5 \times 10^{-2}$ kg·m²; 100 rad/s² (b) 400 rad/s; 6 kg·m²/s; 600 W

17. 183 kW

19. (a) 28.0 kg·m² (b) 28.0 kg·m² (c) 56.0 kg·m²

21. $1.5MR^2$

23. $1.4MR^2$

25. $2MR^2$

### Problems

1. (a) $\pi$ rad/s² (b) $3\pi/2$ rad/s

3. (a) 197 kJ (b) 26.2 N·m; 52.5 N (c) 1200 rev

5. 1.95 m

7. (a) 2.73 m/s; 27.3 rad/s (b) $T_L = 233.5$ N; $T_R = 238.2$ N (c) 1.47 s

9. (a) $\sqrt{8mg/(2m + M)R}$ (b) $mg(10m + M)/(2m + M)$

11. (a) $(2m_2g \sin \theta)/(2m_2 + m_1)$ (b) $(m_1m_2g \sin \theta)/(2m_2 + m_1)$ (c) $m_2gh$ (d) $m_2gh$ (e) $\sqrt{2gh/(1 + m_1/2m_2)}$ (f) for $\theta = 0$: $a = 0$, $T = 0$, $E = 0$; for $\theta = 90°$: $a = 2m_2gh/(2m_2 + m_1)$, $T = m_1m_2g/(2m_2 + m_1)$, $E = m_2gh$; for $m_1 = 0$: $a = g \sin \theta$, $T = 0$, $E = m_2gh$

13. (a) 72 kg (b) 1.37 rad/s²; $T_1 = 294$ N; $T_2 = 746$ N

15. (a) $4.00 \times 10^{-2}$ kg·m² (b) $4.145 \times 10^{-2}$ kg·m²; approximation is good to about 3.5%

19. $a = [5m/(5m + 2M)]g$; $T = [2M/(5m + 2M)]mg$

**Chapter 10**    **True or False**    1. False; $\mathbf{A} \times \mathbf{B} = -\mathbf{B} \times \mathbf{A}$ 2. True 3. False 4. False; $\mathbf{L}$ must be constant 5. True 6. True 7. False; $f$ is less than $\mu_s N$ except at limiting angle

**Exercises**    1. $\mathbf{F} = -F\mathbf{i}$, $\mathbf{r} = R\mathbf{j}$, $\boldsymbol{\tau} = RF\mathbf{k}$

3. $\boldsymbol{\tau} = -mgx\mathbf{k}$

5. (a) $25\mathbf{k}$ (b) $-25\mathbf{j}$ (c) $8\mathbf{k}$

9. (a) 60 kg·m²/s (b) 75 kg·m² (c) 0.8 rad/s

11. (a) doubled (b) doubled

13. (a) 4 N·m (b) $\frac{2}{3}t$ rad/s

15. $L_{earth} = 7.15 \times 10^{33}$ kg·m²/s; $L_{sun} = 2.69 \times 10^{40}$ kg·m²/s

17. (a) 132 g·cm²/s (b) 132 g·cm²/s

19. (a) 0.784 N·m (b) $0.26v$ (c) 3.02 m/s²

21. (a) $\frac{1}{3}Ml^2\omega$ (b) $\frac{1}{2}l\omega$ (c) $\frac{1}{12}Ml^2\omega$

23. (a) 600 rev/min (b) since $E = L^2/2I$ and $I_f = 3I_i$, $E_f = E_i/3$

25. (a) 14.3 rad/s = 2.27 rev/s (b) $E_i = 39.9$ J; $E_f = 181$ J (c) 141 J

27. without smaller rotor, any variation in angular speed of main rotor would product an opposite rotation in helicopter body because of conservation of angular momentum

29. 1.26 kJ

31. 81.5 m

33. (a) $\frac{5}{7}g \sin \theta$ (b) $\frac{2}{7}mg \sin \theta$ (c) $\tan \theta \leqslant 7\mu/2$

35. (a) right, to give small angular momentum in downward direction to compensate for upward angular momentum due to propeller (b) down

37. (a) 2.88 rad/s (b) 0.144 m/s (c) 0.417 m/s² toward pivot (d) $F_{vert} = 19.6$ N; $F_{horiz} = 0.834$ N

39. (a) $\mathbf{L} = -2.91\hat{\mathbf{r}} - 1.68\mathbf{k}$ kg·m²/s (b) 9.81 N·m

**Problems**    3. (a) sphere, 5.29 m/s; disk, 5.11 m/s; hoop, 4.43 m/s (b) sphere, 4.20 N; disk, 4.90 N; hoop, 7.36 N (c) sphere, 1.51 s; disk, 1.56 s; hoop, 1.81 s

5. $2.7R$; $2.5R$

7. (e) $E_k = \frac{3}{4}MV^2$

9. (b) $\frac{1}{3}Mg$

11. $s_1 = 12v_0^2/49\mu g$, $t_1 = 2v_0/7\mu g$, $v_f = 5v_0/7$; $s_1 = 3.99$ m, $t_1 = 0.582$s, $v_f = 5.71$ m/s

13. $\frac{2}{7}R\omega_0$

15. $\omega_0 = v_0/3L$; $\frac{2}{3}$

17. (a) $f = \frac{2}{3}(\frac{1}{2} - r/R)T$ assuming $\mathbf{f}$ opposite to $\mathbf{T}$; (b) $a = (2T/3mR)(R + r)$ (c) yes, if $r$ is greater than $\frac{1}{2}R$ (d) $\mathbf{f}$ in same direction as $\mathbf{T}$

19. 0.6 s longer using $\frac{2}{5}M_E R_E^2$ for moment of inertia $I_E$ of earth, which overestimates $I_E$ because earth is denser near its center; estimate is therefore somewhat low

**Chapter 11**    **True or False**    1. False 2. True 3. False; period is independent of amplitude 4. True 5. True 6. False; it is periodic but not simple harmonic for large amplitudes 7. True 8. True 9. True 10. True 11. True 12. False 13. True 14. True 15. True 16. True 17. False 18. True 19. True

**Exercises**

1. (a) 2.5 Hz; 0.4 s (b) 3 m (c) $x(0) = -3$ m; $x(\frac{1}{2}) = 0$

3. $x = (6$ cm$)$ cos $\pi t$; $v = (-6\pi$ cm/s$)$ sin $\pi t$; $a = (-6\pi^2$ cm/s$^2)$ cos $\pi t$

5. (a) 4 m; $t = \pi/4$ s (b) $v = (8$ m/s$)$ cos $2t$; 8 m/s (c) $a = (-16$ m/s$^2)$ sin $2t$; 0; $a_{max} = 16$ m/s$^2$

7. (a) 2 rad/s (b) 0.318 Hz; 3.14 s (c) $x = (40$ cm$)$ cos $(2t + \delta)$; $y = (40$ cm$) \times$ sin $(2t + \delta)$, where $\delta$ depends on initial position

9. (a) 474 N/m (b) 2.37 J (c) $x(t) = (10$ cm$)$ cos $(4\pi t + \delta)$; no

11. (a) 1580 N/m (b) 79.0 m/s$^2$ (c) 0.0197 J

13. $\frac{3}{4}$; $A/\sqrt{2}$

15. (a) 5.61 cm; 0.550 J (b) $\Delta U_{spring} = +0.746$ J; $\Delta U_{gravity} = -0.589$ J; $\Delta U_{net} = 0.157$ J $= \frac{1}{2}(350)(0.03)^2$ J (c) 0.475 s; 2.10 Hz; 3 cm

17. 24.8 cm

19. 9.79 m/s$^2$

23. 2.0 s

25. (a) 0.444 s; 0.180 J (b) 0.045 kg/s; 628

27. (a) 10% (b) 29.2 s (c) 62.8

29. (a) 34.6 (b) 314

31. 25.1

33. (a) 4.98 cm (b) $\omega_0 = 14.1$ rad/s $(f_0 = 2.25$ Hz$)$ (c) 35.4 cm (d) 1.00 rad/s

35. 0.736 W

**Problems**

1. (a) equal (b) equal (c) 24 times as far during first second

3. (a) 1.51 kg (b) 0.82 cm (c) $x(t) = (2.5$ cm$)$ cos $34.5t$; $v(t) = -(86.4$ cm/s$) \times$ sin $34.5t$; $a(t) = -(29.9$ m/s$^2)$ cos $34.5t$

5. about $3 \times 10^5$

7. (a) $kA/(m_1 + m_2)g$ (b) $A$ and $E$ remain unchanged; $\omega$ decreases, $\omega_f = \sqrt{m_1/(m_1 + m_2)}\ \omega_i$; $T$ increases, $T_f = \sqrt{(m_1 + m_2)/m_1}\ T_i$

9. 0.262 s

11. (a) 13 cm (b) 6.5 cm (c) 0.51 s (d) 0.32 J (e) 0.8 m/s; 0.13 s (f) 0.12 s; 1.13 m/s

13. (a) 219 s (b) $\omega = 24.5$ rad/s; $f = 3.90$ Hz

15. 1.28 s

17. (a) $v_{SHM} = \theta_0\sqrt{gL}$ (b) $= \sqrt{2gL(1 - \cos\ \theta_0)}$ (d) $v_{SHM} = 0.5464$ m/s; $v = 0.5457$ m/s; difference is 0.07 cm/s, about 0.13%

19. (a) $23.7T = 119$ s (b) 149 (c) 1/23.7 (d) 1.24 cm; $1.22 \times 10^{-5}$ J

21. 9.1 Hz

23. (a) 29.4 (b) $t = 3.54$ s (c) 6 J

25. (a) 14.1 cm; 0.444 s (b) 23.1 cm; 0.363 s (c) inelastic: 4 N·s; elastic: 8 N·s

29. $A(t + T) = 6.46 \times 10^{-2}A(t)$; $E(t + T) = 4.18 \times 10^{-3}E(t)$

31. (b) 2.04 cm/s$^2$

**Chapter 12**

**True or False**

1. False 2. True 3. False 4. True 5. True 6. True 7. True

**Exercises**

1. $4.37 \times 10^{11}$ m

3. $1.50 \times 10^{15}$ m

5.  84.6 y

7.  (a) 1.88 Gm (b) 1.90 × 10²⁷ kg

9.  (a) 8.18 × 10⁴ s (b) 1.22 Gm

11.  1.99 × 10³⁰ kg

13.  (a) 0 (b) $\mathbf{g} = [4Gmax/(x^2 - a^2)^2]\mathbf{i}$

15.  $-2Gmm_0/a$

17.  (a) $V(x) = -GM/(x^2 + r^2)^{1/2}$ (c) at $x = 0$, $V$ is maximum and $\mathbf{g} = 0$

21.  (a) $\mathbf{g} = -2.0 \times 10^{-10}\hat{\mathbf{r}}$ N/kg (b) $\mathbf{g} = 0$

23.  $-4.00 \times 10^{-10}$ J/kg both inside and outside

**Problems**

3.  (a) 3.31 y (b) 1.49 × 10²⁹ kg (c) $m_2$ has greater speed and greater total energy (d) $v_P = 1.8v_A$

5.  186 m/s = 670 km/h

7.  $g_r = 0$, $r < R_1$;

$$g_r = \begin{cases} -\dfrac{GM}{r^2}\dfrac{r^3 - R_1^3}{R_2^3 - R_1^3} & R_1 < r < R_2 \\[2ex] -\dfrac{GM}{r^2} & r > R_2 \end{cases}$$

9.  $g_x = -GM/(x_0^2 - L^2/4)$

11.  $g_x = (\pi G\rho_0 R^3/6)[1/(x - \tfrac{1}{2}R)^2 - 8/x^2]$

13.  $\omega = \sqrt{4\pi G\rho_0/3}$

15.  (a) 179 (c) 0.46

**Chapter 13**

**True or False**

1. False; only on the volume submerged 2. True 3. False; an object can be supported by surface tension 4. True

**Exercises**

1.  300 g

3.  1.03 kg/L

5.  13,620.5 kg/m³

7.  (a) 41.6 N (b) 0.136%

9.  1.024.4 kg/m³

11.  23.7 cm

13.  (a) 150 kPa (b) 49 kPa

15.  ±0.44 Pa

17.  (a) 8.72 N (b) block moves up relative to jar

19.  (a) 3.0 (b) 2.04 L

21.  5

23.  sphere's radius is 1.23 mm

25.  (a) 65 m/s (b) 2.21 MPa = 21.9 atm

27.  (a) 12 m/s (b) 132.5 kPa (c) they are equal

29.  1.47 kPa

31.  8 × 10⁻⁴ N/m²

**Problems**

1. 3.88 kg

3. upper scale 12.35 N; lower scale 36.7 N

5. (*a*) 58.9 N (*b*) about twice the weight of Al (26.5 N); no; it would take $6mg = 159$ N to accelerate Al upward at $a = 5g$

7. (*a*) 1.82 MPa (*b*) 11.8 mJ

9. (*a*) 25 N (*b*) 0.566 J (*c*) 2.25 J

11. (*a*) 5.13 cm (*b*) 173 J

13. $2.06 \times 10^7$ kg

15. 14.9 cm

19. (*a*) $2\sqrt{h(H - h)}$ (*c*) $H$


**Chapter 14**

**True or False**

1. True 2. False 3. False 4. False 5. True 6. True 7. False 8. True 9. False; it has 1000 times the intensity 10. True 11. False

**Exercises**

3. segments from $x = 2$ cm to $x = 3$ cm are moving up; segments from $x = 1$ cm to $x = 2$ cm are moving down; segment at $x = 2$ cm is at rest

7. 0.252 s

9. 251 m/s

11. $1.25 \times 10^3$ km/h

13. 5.09 km/s

15. $2.70 \times 10^{10}$ N/m$^2$

17. (*a*) 1.30 m (*b*) 0.649 m

19. $4.3 \times 10^{14}$ to $7.5 \times 10^{14}$ Hz

21. (*a*) left; 5.00 m/s (*b*) 0.1 m; 50 Hz; 20 ms (*c*) 1 mm

23. 82.7 mPa

25. (*a*) 0 (*b*) 3.70 $\mu$m

27. (*a*) 15 m; 20 min$^{-1}$ (*b*) 13.5 m; 22.2 min$^{-1}$ (*c*) 15 m; 22.0 min$^{-1}$

29. (*a*) 1.3 m (*b*) 262 Hz

31. (*a*) 2.10 m (*b*) 162 Hz

33. (*a*) 80 m/s toward listener; (*b*) 420 m/s (*c*) 1.7 m (*d*) 247 Hz

35. 529 Hz; 474 Hz

37. 1.53 kN

39. 180 m/s

41. (*a*) 2 m; 25 Hz (*b*) $y = (4 \text{ mm}) \cos 50\pi t \sin \pi x$

43. (*a*) 200 m/s; 0.25 cm (*b*) 1.26 m (*c*) 1.26 m

45. (*a*) 16 m; 5.33 m; 3.2 m (*b*) 6.25 Hz; 18.75 Hz; 31.25 Hz

47. (*a*) 4 m; 1.57 m$^{-1}$ (*b*) $800\pi$ rad/s (*c*) $y = (0.03 \text{ m}) \sin \frac{1}{2}\pi x \cos 800\pi t$

49. 17 Hz; 8.5 Hz

51. (*a*) 2.27 kHz (*b*) eighth or ninth

53. (*b*) 4.74 J (*c*) 30.6 W

55. (*a*) 138 Pa (*b*) 21.6 W/m$^2$ (*c*) 0.216 W

57.  (a) $10^{-11}$ W/m² (b) $2 \times 10^{-12}$ W/m² (c) 926 mPa; 414 mPa

59.  99%

**Problems**

1.  0.34 km/s; in 3 s sound travels about 1.02 km, so estimate is off by about 2%; correction time for light negligible

3.  (a) 78.5 m (b) 69.7 m (c) 70.5 m (using $v = 340$ m/s)

5.  (a) 2.27 ms; 440 Hz (b) 316 m/s (c) 0.719 m; 8.74 m$^{-1}$ (d) $y = (5 \times 10^{-4}$ m) $\times$ sin $2\pi(1.39x - 440t)$ (e) 1.38 m/s; 3.82 km/s² (f) 3.02 W

7.  (a) 1.615 m; 211 Hz (b) 221 Hz

9.  (a) $\frac{1}{9}$ reflected; $\frac{8}{9}$ transmitted; (b) reflected: $\frac{1}{3}A$; transmitted: $\frac{2}{3}A$

11.  3.27 cm; 6.19 cm; 7.56 cm; 10 cm

13.  (a) the powder moves about due to the motion of the gas and collects at the displacement nodes where the gas is at rest (b) $v = 2fx$, where $x$ is distance between piles

15.  (a) eighth and ninth (b) 2.16 m (c) 4.32 m

17.  338 m/s (not accurate because of end correction)

19.  345 m/s; 1.25 cm

21.  (a) $v_y = -(7.54$ m/s) sin 2.36x sin 377t (b) free end at $x = 0.666$ m; maximum speed of this point is 7.54 m/s (c) $a_y = -(2.84$ km/s²) sin 2.36x cos 377t; the free end at $x = 0.666$ m has the greatest acceleration, which has a maximum value of 2.84 km/s²

23.  yes; 1.54%

25.  (a) 17.7 $\mu$W/m² (b) 72.5 dB (c) 1.26 km

27.  39.5 km/h

29.  (a) 55.1 Pa (b) 3.46 W/m² (c) 0.245 W

31.  (a) 100 dB (b) 25.1 W (c) 2.00 m (d) 96.5 dB

33.  87.8 dB

35.  (b) 2.21 s

37.  20 $\mu$W

39.  about 26%; doubling time of about 3.0 y

**Chapter 15**

**True or False**

1. True 2. False 3. False 4. True 5. True 6. False 7. True 8. True 9. True 10. False 11. True 12. False

**Exercises**

1.  0.035 m

3.  $A$

5.  85 Hz, 255 Hz; the amplitudes will be different because (1) the distances to the speakers are different, and (2) indirect sound due to reflections will not cancel

7.  (a) 8.3° (b) 16.5° (c) 3 maxima beyond the $\theta = 0$ maximum

9.  (a) 0.278 m (b) 1.22 kHz (c) 24.7°, 33.8°, 44.1°, 56.6°, 77.0° (d) 4°

11.  (a) 0 (b) $2I_0$ (c) $4I_0$

13.  (a) $(n + \frac{1}{4})\lambda$ (b) $(n - \frac{1}{4})\lambda$, $n = 0, 1, 2, \ldots$

15.  (a) 113 Hz for 30-cm speaker; 567 Hz for 6-cm speaker (b) 11.3 kHz for 30-cm speaker; 56.7 kHz for 6-cm speaker

17.  437 Hz

19.  (a) 10 $\mu$s (b) $\Delta f \sim 10^5$ Hz

21.  (a) 0.4 cm (b) $y(x, t) = (0.4\ \text{cm}) \cos (0.1x - 10t) \cos (5.9x - 590t)$; $v_p = 100$ m/s (c) $v_g = 100$ m/s (d) 62.8 cm (e) nondispersive

23.  (a) $s_1 = s_0 \cos (9.24x - 1000\pi t)$; $s_2 = s_0 \cos (9.33x - 1010\pi t)$; $s = 2s_0 \times \cos (0.046x - 5\pi t) \cos (9.29x - 1005\pi t)$

25.  $v_p = v_g = \sqrt{\gamma RT/M}$; no

**Problems**

1.  (a) 0 (b) 66 dB (c) 63 dB

3.  (a) $I_1 = 19.9\ \mu\text{W/m}^2$; $I_2 = 8.84\ \mu\text{W/m}^2$ (b) 55.3 $\mu\text{W/m}^2$ (c) 2.2 $\mu\text{W/m}^2$ (d) 28.7 $\mu\text{W/m}^2$

5.  Speakers are connected properly if they move in phase when driven by a monophonic source. A stereo source is not used because the speakers would not be coherent (they would be driven by different sources). Bass is used so that the listener's position is not critical for detection of constructive interference from in-phase speakers and destructive interference from 180°-out-of-phase speakers. For example, at 100 Hz, the wavelength is 34 m, and the listener must be equidistant from the speakers only to within a few metres; but at 1 kHz, the wavelength is only 34 cm and the listener's position is critical.

7.  (a) 4.50 m/s (b) 22 (c) 22

11.  $\Delta T/T = 2f_b/f_0$; 0.15%

**Chapter 16**

**True or False**

1. True 2. True 3. False 4. False 5. False 6. True

**Exercises**

1.  The system is not in equilibrium when the temperature and pressure readings are taken.

3.  (a) adiabatic (b) diathermic (c) adiabatic

5.  $t_A \neq t_B$ and $t_B \neq t_C$; no, $t_A$ and $t_C$ may or may not be equal

7.  (a) 8.4 cm (b) 107°C

9.  (a) 54.9 mm (b) 371 K

11.  10.4°F–19.4°F

13.  (a) 6173 K (b) 10,650°F

15.  37.0°C

17.  (a) $10^7$ °C (b) about $1.8 \times 10^7$ °F

19.  (a) 12.3 L (b) 24.6 L (c) 375 K

21.  1.15

23.  $5.40 \times 10^3$ N due to air inside box; if box is in air at pressure of 1 atm, there will be an opposing force of $4.05 \times 10^3$ N due to air outside, giving a net force of $1.35 \times 10^3$ N outward

25.  1.79 mol; $1.08 \times 10^{24}$ molecules

27.  275 m/s for argon and 870 m/s for helium

29.  8.8 K

**Problems**

1.  (a) 6.37 atm (b) 21.4% escapes (c) 3.93 atm

3.  (a) 231 kPa (b) 200 kPa

5.  507 m/s

7.  (a) $1.6 \times 10^5$ K (b) $1.0 \times 10^4$ K (c) yes (see essay in Chapter 7 for complete discussion) (d) about $7.3 \times 10^3$ K for $O_2$ and $4.6 \times 10^2$ K for $H_2$; yes

9.  $3.94 \times 10^{-29}$ m³; 0.2 nm

11.  5.09 m/s

| | | |
|---|---|---|
| **Chapter 17** | **True or False** | 1. False  2. False  3. False  4. False  5. True  6. True  7. True  8. False  9. True |

**Exercises**

1.  837 kJ

3.  1440 Btu

5.  1.70 kJ/kg·K

7.  21.54°C

9.  (a) 0.12°C (b) 1.74°C

11.  (a) 608 J (b) 1064 J

13.  (a) 334 J (b) 790 J

15.  (a) 608 J (b) 1200 K (c) 300 K (d) 608 J

17.  2.04 kJ/kg·K

21.  (a) 1.57 kJ (b) 1.57 kJ

23.  (a) 555 J (b) 555 J

25.  (a) 192°C (b) 114°C

27.  (a) 3.50 mol (b) 43.65 J/mol·K; 72.75 J/mol·K (c) 72.75 J/mol·K; 101.85 J/mol·K

29.  (b) a rigid dumbbell

31.  (a) $\Delta U = 3.74$ kJ; $W = 0$; $Q = 3.74$ kJ (b) $\Delta U = 3.74$ kJ; $W = 2.50$ kJ; $Q = 6.24$ kJ (c) 2.49 kJ (differs only because of round-off differences)

33.  55.8 g/mol; iron

**Problems**

1.  28.5°C; $t_i$ should be about 16°C

3.  171 K

5.  (a) $U_i = 3.40$ kJ; $U_f = 3.70$ kJ; $W = 200$ J (b) $U_i = 3.40$ kJ; $U_f = 3.90$ kJ; $W = 0$

7.  3.41 cm

9.  (b) calculation gives $-65.1$ L·atm $= -6.6$ kJ done by gas; that is, 65.1 L·atm done on gas (c) 6.6 kJ removed from gas during cycle (d) work done by gas for each step is 22.2 L·atm during adiabatic expansion, 8.9 L·atm during expansion at 1 atm, 0 during heating at constant volume, and $-96.2$ L·atm during compression at 5 atm

11.  (a) 92.0 mJ/kg·K (b) 58.4 mJ/kg

| | | |
|---|---|---|
| **Chapter 18** | **True or False** | 1. False; e.g., water contracts when heated if the temperature is between 0 and 4°C. 2. True 3. True 4. True 5. True 6. True 7. True |

**Exercises**

1.  30.026 cm

3.  7.7 cm

7.  3.70 mL

9.  (a) 0°C (b) 125.6 g

11.  310.9 kJ

13.  134 g

15.  99.8 g

17.  argon, helium, hydrogen, neon, nitric oxide, oxygen

19.  $2.6 \times 10^4$ Btu/h (using $R_f = 0.9$ from Table 18-6; the thickness of the glass is unimportant for this calculation)

21.  (a) 15.9 K/W (b) 6.30 W (c) 50 K/m (d) 87.5°C

23.  (a) 1.53 kW (b) $5.22 \times 10^{-2}$ K/W (c) 1.69

25.  380 kJ/m

27.  9.47 $\mu$m

29.  16

**Problems**

1.  145°C

3.  34 km; variation in $g$ can be neglected since $H/R_E$ is only about 0.5%; if most of the energy dissipated because of air resistance were absorbed by the ice, air resistance could be neglected; since much of the eergy is probably absorbed by the surrounding air, $H$ would be larger if air resistance were taken into account

5.  (a) 40.6 kJ (b) 30.6 L; 3.10 kJ (c) 37.5 kJ (d) slightly greater because the volume is slightly reduced

7.  (b) 70 $\mu$J/mol·K

9.  665 m

11.  0.134 K/min

13.  0.42°C

15.  (a) $T_1 = 76.3$°C; $T_2 = 55.4$°C (b) $T_1 = 79.1$°C; $T_2 = 55.4$°C

17.  46 W to 50°C reservoir, 23 W to 100°C reservoir; 150°C

19.  (a) $R = 1.66$ K/W; $C = 3003$ J/K; $RC = 4985$ s $= 83$ min (c) 58 min

21.  134 kN

23.  (b) 65.6°C

**Chapter 19**

**True or False**

1. False 2. False; e.g., isothermal expansion of an ideal gas 3. False 4. False 5. False 6. True 7. False 8. True 9. False 10. True

**Exercises**

1.  (a) 500 J (b) 400 J

3.  (a) 40% (b) 80 W

5.  66.7 J absorbed and 46.7 J rejected per cycle

13.  (a) $33\frac{1}{3}$% (b) 33.3 J; 66.7 J (c) 2

15.  a 5-K decrease in temperature of the cold reservoir

17.  224 K

19.  (a) 11.5 J/K (b) 0

21.  (a) $+50$ J (b) $+0.167$ J/K (c) 0 (d) parts (a) and (b) have the same answers but for part (c) $\Delta S_u > 0$

23.  (a) 11.5 J/K (b) 11.5 J/K

25. 0.417 J/K

27. 6.05 kJ/K

29. (a) 400 J (b) 0.50 J/K (c) 400 J; 200 J less than in (a) (e) 100 J could be used, leaving 100 J completely wasted

31. (a) 5.76 J/K (b) 1.73 kJ

33. still air at 25°C has about 102 J more energy, but only air with center-of-mass motion can drive a windmill

**Problems**

1. (a) 1. $W = 0$; $Q = +3.74$ kJ; $\Delta U = +3.74$ kJ; 2. $W = 4.99$ kJ; $Q = 12.47$ kJ; $\Delta U = +7.48$ kJ; 3. $W = 0$; $Q = -7.48$ kJ; $\Delta U = -7.48$ kJ; 4. $W = -2.495$ kJ; $Q = -6.24$ kJ; $\Delta U = -3.74$ kJ (b) 15.4%

3. (c) 65% (d) the expansion and compression are not adiabatic, and none of the processes is quasi-static

5. (a) best possible efficiency is 41% (b) 1.68 GJ

7. (a) 373 K (b) 1. $Q = +3.12$ kJ; 2. $Q = 0$; 3. $Q = -2.91$ kJ (c) 6.7% (d) 35.5%

9. 100 W

13. (a) $T_1 = 300.7$ K; $T_2 = 601.4$ K; $T_3 = 601.4$ K (b) 3.75 kJ of heat enters during the first part and 3.466 kJ enters during the second part; 6.25 kJ leaves during the last part (compression) (c) 13.4%

15. 313 K

17. (a) 10.1°C (b) 22.0 J/K

19. $+10.7$ J/K

21. $+1.98$ kJ/K

**Chapter 20**

**True or False**

1. False; only if the charge is positive 2. True 3. False; they diverge from a positive point charge. 4. True

**Exercises**

1. (a) $6.24 \times 10^{12}$ (b) $6.24 \times 10^6$

3. 230 N

5. $1.5 \times 10^{-2}\mathbf{i}$ N

7. $0.914 kq^2/L^2$ away from negative charge

9. (a) $8.00\mathbf{i}$ kN/C at $x = 2$ m; $9.36\mathbf{i}$ kN/C at $x = 10$ m (b) $x = 4$ m (c) $\mathbf{E}$ points in the positive $x$ direction just to the right of the origin and in the negative $x$ direction just to the left of the origin.

11. (a) $-25.9\mathbf{j}$ kN/C (b) $-51.8\mathbf{j}$ $\mu$N

13. (a) $|q_{\text{left}}| = 4|q_{\text{right}}|$ (b) left positive; right negative (c) The field is strong close to charges and between them where the lines are most dense. The field is weak away from charges, particularly to the right of the negative charge, where the lines are less dense.

17. (a) $1.42\mathbf{i}$ MN/C (b) $1.92\mathbf{i}$ MN/C (c) $1.77\mathbf{i}$ MN/C (d) $125\mathbf{i}$ N/C

19. (a) $\dfrac{k\lambda}{y}\left(\dfrac{b}{\sqrt{y^2 + b^2}} + \dfrac{a}{\sqrt{y^2 + a^2}}\right)$

21. 1.8 MN/C, assuming an infinite line charge; 30 N/C, assuming a point charge

23. (a) 1.69 MN/C (b) 1.69 MN/C (c) 1.69 MN/C (d) 122 N/C

25. (a) 0 (b) 0 (c) 1.79 MN/C (d) 449 kN/C (e) 112 kN/C

27. (a) 176 GC/kg (b) 17.6 Tm/s² opposite the direction of $\mathbf{E}$ (c) 0.17 $\mu$s (d) 25.6 cm

29. (a) $-70.4\mathbf{j}$ Tm/s$^2$ (b) 50 ns (c) 8.8 cm in the negative $y$ direction

31. The electric force is about $2.7 \times 10^{12}$ times $mg$.

33. $8 \times 10^{-18}\mathbf{i}$ C·m

**Problems**

1. (a) $0.914\ kq^2/L^2$ toward other positive charge

3. (a) 4 and 2 $\mu$C (b) $3 + \sqrt{17}\ \mu$C $= +7.12\ \mu$C and $3 - \sqrt{17}\ \mu$C $= -1.12\ \mu$C

5. (c) For $x \gg a$, the separation of charges is not important, and so you expect the field to be like that of a point charge $+2q$.

7. (a) stable for displacements of $q_0$ along $y$ axis and unstable for displacements along $x$ axis, assuming other charges held fixed (b) stable for displacements of $q_0$ along $x$ axis and unstable for displacements along $y$ axis, assuming other charges held fixed (c) $q_0 = -\frac{1}{4}q$. If none of the charges is held fixed, the equilibrium is unstable for displacement of any of the charges. A stable equilibrium distribution of electric charges is not possible under electric forces only.

9. (a) $dE_x/dx = kQ/a^3$ (b) $E_x = E_x(0) + x(dE_x/dx)$ for small $x$, so $F_x = qE_x = -qx(kQ/a^3) = ma_x$ for small $x$. This is of the form $a_x = -\omega^2 x$ with $\omega = (kqQ/ma^3)^{1/2}$, the angular frequency of simple harmonic motion.

11. $E_x = kQx(x^2 + \frac{1}{4}L^2)^{-1}(x^2 + \frac{1}{4}L^2)^{-1/2}$, where $Q = 4L\lambda$. For a ring of radius $r = \frac{1}{2}L$ but carrying the same charge, $E_x = kQx(x^2 + \frac{1}{4}L^2)^{-3/2}$. *Hint*: Use Equation 20-16 for the magnitude of the field of each segment, replacing $y$ with $r = (x^2 + \frac{1}{4}L^2)^{1/2}$. Then take 4 times the $x$ component of this field.

13. (a) left: $-4\pi k\sigma$; right: $+4\pi k\sigma$; between: zero (the $+$ direction is to the right) (b) left: 0; right: 0; between: $+4\pi k\sigma$

17. (a) 46° replacing $\tan \theta$ with $\sin \theta$, or 36° replacing $\sin \theta$ with $\tan \theta$ (b) 40.7°

**Chapter 21**

**True or False**

1. False 2. False 3. True 4. False; it is zero in electrostatics. 5. False 6. False 7. False 8. True

**Exercises**

1. (a) 20.0 N·m$^2$/C (b) 17.3 N·m$^2$/C

3. (a) 3.14 m$^2$ (b) 71.9 kN/C (c) 226 kN·m$^2$/C (d) No (e) 226 kN·m$^2$/C

5. (a) $N$ (b) $N/6$ (c) $q/\epsilon_0$ (d) $q/6\epsilon_0$ (e) (b) and (d)

7. (a) 1.57 N·m$^2$/C (b) 0 (c) 3.14 N·m$^2$/C (d) 27.8 pC

9. (a) $E_r = kQr/R^3$ (b) $E_r = kQ/r^2$ where $Q = \frac{4}{3}\pi R^3\rho$

11. (a) $2.85 \times 10^{22}$ electrons (b) 4.56 kC

13. 11.3 kN/C

15. (a) 7.07 $\mu$C/m$^2$ (b) 800 kN/C outward (c) 0 (d) 400 kN/C outward both just inside and just outside the hole. If the portion with its charge were replaced, its field would cancel the field just inside the shell and add to the field just outside.

**Problems**

1. (a) $Q = \pi AR^4$ (b) inside: $E_r = Ar^2/4\epsilon_0$; outside: $E_r = AR^4/4\epsilon_0 r^2 = kQ/r^2$

3. (a) $Q = 4\pi CR$ (b) inside: $E_r = C/\epsilon_0 r$; outside: $E_r = CR/\epsilon_0 r^2$

5. (a) for $r < R_1$, $E_r = 0$; for $R_1 < r < R_2$, $E_r = kq_1/r^2$; for $r > R_2$, $E_r = k(q_1 + q_2)/r^2$ (b) $q_2 = -q_1$

7. $Q = \rho\frac{4}{3}\pi(r_2^3 - r_1^3)$; for $r < r_1$, $E_r = 0$; for $r_1 < r < r_2$, $E_r = \rho(r^3 - r_1^3)/3\epsilon_0 r^2$; for $r > r_2$, $E_r = \rho(r_2^3 - r_1^3)/3\epsilon_0 r^2$

9. for $r < R$, $g_r = -GMr/R^3$; for $r > R$, $g_r = -GM/r^2$

11. (a) for $r < a$, $E_r = kq/r^2$; for $a < r < b$, $E_r = 0$; for $r > b$, $E_r = 0$ (c) uniform; the charge density would not change

15. for $r < a$, $E_r = 0$; for $a < r < b$, $E_r = (\rho/2\epsilon_0)(r^2 - a^2)/r$; for $r > b$, $E_r = (\rho/2\epsilon_0)(b^2 - a^2)/r$

**Chapter 22**

**True or False**

1. False; $V$ must be constant in the region but need not be zero. 2. True; but if $V = 0$ at a *point*, **E** need *not* be zero at that point. If $V$ is zero or some other constant in a *region* of space, it cannot be changing from point to point, so **E** must also be zero. 3. False 4. True 5. True

**Exercises**

1. (a) 24 mJ (b) $-24$ mJ (c) $-8$ kV (d) $V(x) = -2000x$ (e) $V(x) = 4000 - 2000x$ (f) $V(x) = -2000(x - 1)$ ($V$ is in volts and $x$ is in metres)

3. (a) positive (b) 25 kV/m

5. (a) V/m² (b) $\frac{1}{2}ax^2q_0$ (c) $V(x) = -\frac{1}{2}ax^2$

7. (a) 25.4 kV (b) 12.7 kV (c) 0

9. (a) 48.7 mJ (b) 0 (c) $-12.7$ mJ if like charges are adjacent, $-23.3$ mJ if like charges are diagonally opposite

11. (a) 12.0 kV (b) 60 mJ (c) 0; 0

13. (a) 0 (b) $E_x = 6000$ V/m (c) $V(3.01) = -60$ V; $\Delta V/\Delta x = 6000$ V/m, the same as $E_x$ to this accuracy

15. (a) $-3$ kV/m (b) $-3$ kV/m (c) $+3$ kV/m (d) 0

17. 1.77 mm

21. $V(x)$ is maximum at $x = 0$, at which point $E_x = 0$

23. (a) 9 kV/m just outside; 0 just inside (b) 900 V both just inside and just outside (c) $V = 900$ V, $E = 0$ at center

25. 200 W

27. 300 kV

**Problems**

1. (a) $E_k = ke^2/2r$ (b) $\frac{1}{2}mv^2 = 2.18 \times 10^{-18}$ J = 13.6 eV; $E = \frac{1}{2}mv^2 + U = -13.6$ eV (c) 13.6 eV

3. (a) $V = \begin{cases} \sigma x/\epsilon_0 & \text{for } x < 0 \\ 0 & \text{for } 0 \leqslant x \leqslant a \\ -\sigma(x - a)/\epsilon_0 & \text{for } x > a \end{cases}$  (b) $V = \begin{cases} 0 & \text{for } x < 0 \\ -\sigma x/\epsilon_0 & \text{for } 0 \leqslant x \leqslant a \\ -\sigma a/\epsilon_0 & \text{for } x > a \end{cases}$

5. 44.6 fm

7. $-4.82kq^2/a$

9. (a) $\sigma d/\epsilon_0$ (b) $\sigma(d - a)/\epsilon_0$

13. $kq(b - a)/ab$

15. for $r < a$, $V = kq/r - kq(b - a)/ab$; for $a < r < b$, $V = kq/b$; for $r > b$, $V = kq/r$

17. (a) $V = \pm\dfrac{ke}{a}\left(1 - \dfrac{1}{2} + \dfrac{1}{3} - \dfrac{1}{4} + \cdots \pm \dfrac{1}{n}\right)$; $+$ if the $n$th charge is $+e$, $-$ if it is $-e$ (b) $(ke^2/a)2 \ln 2 = 20.0$ eV

**Chapter 23**

**True or False**

1. False 2. False 3. False 4. True 5. True 6. False 7. True

**Exercises**

1. $1.13 \times 10^7$ m²; 3.36 km

3. 44.6 $\mu$F

5. (a) 2 pF (b) 200 pF (c) 0.2 $\mu$F

7. (a) 9000 Mm (b) 1410

9. (a) 6.67 $\mu$F (b) 40 $\mu$C (c) 4 V across 10-$\mu$F capacitor and 2 V across 20-$\mu$F capacitor

11. (a) 14 $\mu$F (b) $\frac{8}{7}$ $\mu$F

13. (a) 100 (b) 10 V (c) $q = 10$ $\mu$C; $V = 1$ kV

15. 10 $\mu$F

17. $5 \times 10^{-5}$ J

19. (a) 0.625 J (b) 1.875 J

21. (a) $\epsilon_0 A/2d$ (b) $2V$ (c) $\epsilon_0 AV^2/d$ (d) $\epsilon_0 AV^2/2d$

23. (a) $10^5$ V/m (b) $4.42 \times 10^{-2}$ J/m$^3$ (c) $8.85 \times 10^{-5}$ J (d) $8.85 \times 10^{-5}$ J, the same

25. (a) $E = 2.5 \times 10^4$ V/m; $\sigma = 2.21 \times 10^{-7}$ C/m$^2$; $U = 6.64 \times 10^{-7}$ J (b) 6.25 kV/m; 25 V (c) $1.66 \times 10^{-7}$ J (d) $1.66 \times 10^{-7}$ C/m$^2$

**Problems**

1. (a) 0.242 $\mu$F (b) 2.42 $\mu$C on 0.30-$\mu$F capacitor; 1.94 $\mu$C on 1.0-$\mu$F capacitor; 0.484 $\mu$C on 0.25-$\mu$F capacitor (c) 12.1 $\mu$J

3. (a) 15.2 $\mu$F (b) $Q_{12} = 2400$ $\mu$C; $Q_{15} = Q_4 = 632$ $\mu$C (c) 0.303 J

5. (a) $Q_{50} = 42.9$ nC; $Q_{20} = 17.1$ nC (b) $U_i = 90$ $\mu$J; $U_f = 25.7$ $\mu$J; energy is lost

7. Each capacitor must have $C = 5$ $\mu$F. If all are in series, $C_{eff} = 1.67$ $\mu$F; if one is in series with the other two in parallel, $C_{eff} = 3.33$ $\mu$F; if one is in parallel with the other two in series, $C_{eff} = 7.50$ $\mu$F.

9. (a) 6.0 V (b) 864 $\mu$J

11. (a) 1.2 kV (b) 640 $\mu$J

13. (a) The potential difference between the plates is $Ed$, where $E$ can be the field either in the dielectric or in the free space.

15. (a) 5.0 $\mu$F (b) 133 V

17. (a) 2.28 nF  *Note:* The approximation that this is a parallel-plate capacitor with area $2\pi rL$ and separation 2 mm gives 2.22 nF. (b) about 67 $\mu$C

19. Use 16 capacitors with 4 parallel sets of 4 in series.

21. (a) for $r < R_1$, $E = 0$, $\eta = 0$; for $R_1 < r < R_2$, $E = kQ/r^2$, $\eta = kQ^2/8\pi r^4$; for $r > R_2$, $E = 0$, $\eta = 0$ (b) $dU = (Q^2/8\pi\epsilon_0 r^2)\,dr$ (c) $\dfrac{Q^2}{8\pi\epsilon_0}\left(\dfrac{1}{R_1} - \dfrac{1}{R_2}\right) = \frac{1}{2}QV$

23. (a) for $r < R$, $\eta = kQ^2 r^2/8\pi R^6$; for $r > R$, $\eta = kQ^2/8\pi r^4$ (b) for $r < R$, $dU = (kQ^2 r^4/2R^6)\,dr$; for $r > R$, $dU = (kQ^2/2r^2)\,dr$ (c) $3kQ^2/5R$. For a conducting sphere, $E$ is the same and therefore the energy is the same for $r > R$; but $E = 0$ for $r < R$, so there is no field energy inside the sphere.

**Chapter 24**

**True or False**

1. False 2. True 3. False; this is the *definition* of resistance. 4. False; true only for conductors obeying Ohm's law 5. True 6. True

**Exercises**

1. (a) 600 C (b) $3.75 \times 10^{21}$ electrons

3. (a) 0.40 A (since ions and electrons flow in opposite directions, their currents add) (b) 566 A/m$^2$

5. (a) $f = v/2\pi r$ (b) $I = qv/2\pi r$

7. (a) $3.2 \times 10^{13}$ protons per cubic metre (the proton speed is $6.2 \times 10^7$ m/s, obtained from $E_k = \frac{1}{2}mv^2$) (b) $3.75 \times 10^{17}$ protons (c) $Q = 1.0 \times 10^{-3}t$ C, since current is 1.0 mA

9. (a) 1.0 V (b) 0.10 V/m

11. (a) $V_{Cu}/V_{Fe} = 0.17$ (b) $E$ greater in iron

13. (a) $2.75 \times 10^{-2}$ Ω (b) $3.0 \times 10^{-2}$ Ω

15. 1.2 Ω

17. 46°C

19. 0.031 Ω

21. (a) 2.42 kW (b) 1.21 kW

23. (a) 48.4 Ω; 4.55 A (b) 250 W

25. (a) 0.364 cents (b) 96 cents

27. 0.03 Ω

29. (a) 30 A (b) 4 V

31. (a) 5000 A/m² (b) 370 nm/s (c) $5.0 \times 10^{-14}$ s (d) about 5 nm

**Problems**

1. (a) 3.12 V (b) $J = 4.59 \times 10^6$ A/m²; $E = 7.8 \times 10^{-2}$ V/m (c) 18.7 W

3. (a) $10^{12}$ electrons (b) $1.6 \times 10^{-4}$ A (c) 64 kW (d) 640 MW (e) 1/10,000

5. (a) $2.55 \times 10^6$ A/m² (b) $E_{Cu} = 0.0433$ V/m; $E_{Fe} = 0.255$ V/m (c) $V_{Cu} = 3.46$ V; $V_{Fe} = 12.48$ V (d) 7.97 Ω $= R_{Cu} + R_{Fe}$, where $R_{Cu} = 1.73$ Ω and $R_{Fe} = 6.24$ Ω

7. (c) $l_{Cu}/l_C = 264$

**Chapter 25**

**True or False**

1. True 2. True 3. True 4. True 5. False 6. False

**Exercises**

1. (a) 1.0 Ω (b) $I_{2\Omega} = 6$ A; $I_{3\Omega} = 4$ A; $I_{6\Omega} = 2$ A

3. (a) 3.6 Ω (b) $I_{2\Omega} = 1.33$ A; $I_{7\Omega} = 1.33$ A; $I_{6\Omega} = 2$ A

5. (a) 18 V (b) 2 A

7. (a) 5 Ω (b) upper branch: $I_{4\Omega} = 1.2$ A with 0.72 A in upper subbranch (10-Ω resistor) and 0.48 A in lower subbranch (5- and 10-Ω resistors); lower branch: $I_{5\Omega} = 1.2$ A with 0.6 A in each 10-Ω resistor

9. (a) 2 A through 1.5-Ω resistor; 0.75 A through each 4-Ω resistor; 0.5 A through 6-Ω resistor (b) 12 W

11. (a) 5 Ω (b) 30 V

13. (a) 4 A (b) 2 V (c) 1 Ω

15. (c) 3 Ω; 27.0 W

17. (a) $I_{4\Omega} = 0.667$ A; $I_{6\Omega} = 1.56$ A; $I_{3\Omega} = 0.889$ A (b) 9.33 V (c) 8 W by left battery; 10.7 W by right battery

19. (a) 600 μC (b) 0.2 A (c) 3 ms (d) 81.2 μC

21. 57.7 MΩ

23. (a) 3.79 μC (b) 2.21 μA (c) 2.21 μA (d) 13.2 μW (e) 4.87 μW (f) 8.37 μW

25. (a) 0.050 Ω (b) 4.9 kΩ

27. (a) 0.100 Ω (b) 0.100 Ω (c) 3.00 MΩ

29. (a) 1.40 kΩ (b) 1.50 kΩ (c) 13.5 kΩ

31. $R_1 = 9.90$ kΩ; $R_2 = 90.0$ kΩ; $R_3 = 900$ kΩ

33. (a) 43.9 Ω (b) 300 Ω (c) 3800 Ω

**Problems**

3. 63.0 A in second battery; 57.0 A in sick battery; 5.99 A in load. The emf of the second battery delivers 793.7 W, of which 39.7 W goes into heat in its own

internal resistance, 32.5 W goes into heat in the internal resistance of the sick battery, 71.6 W goes into heat in the load resistance, and 650 W goes into internal energy of the sick battery.

5. (*a*) 2 A to the right through the 1-$\Omega$ and first 2-$\Omega$ resistor; 1 A down through the second 2-$\Omega$ resistor; 1 A down through the 6-$\Omega$ resistor (*b*) 16 W by the 8-V battery; 8 W by the upper 4-V battery and 4 W *to* the lower 4-V battery (*c*) 4 W in the 1-$\Omega$ resistor; 8 W in the horizontal 2-$\Omega$ resistor; 2 W in the vertical 2-$\Omega$ resistor; 6 W in the 6-$\Omega$ resistor

7. $V_a - V_b = 2.4$ V

9. (*a*) 1.974 A (*b*) 1.3% (*c*) 1.48 V (*d*) 0.001%

11. 600 $\Omega$

13. (*b*) method *a* gives $R_c = 0.498$ $\Omega$, 2.91 $\Omega$, 44.4 $\Omega$; method *b* gives 0.6 $\Omega$, 3.1 $\Omega$, 80.1 $\Omega$

17. (*a*) $I = \dfrac{R_1 + R_2}{R_1 R_2}\,\varepsilon$ (*b*) $I = \varepsilon/R_2$ (*c*) $I(t) = \varepsilon\left(\dfrac{1}{R_2} + \dfrac{1}{R_1}\,e^{-t/R_1 C}\right)$ (*d*) $2.3 \times 10^{-2}$ s

**Chapter 26**

**True or False**

1. True 2. True 3. True 4. False; the period is independent of the radius. 5. True

**Exercises**

1. $-1.28 \times 10^{-12}\mathbf{j}$ N

3. (*a*) $-7.2 \times 10^{-13}\mathbf{j}$ N (*b*) $4.8 \times 10^{-13}\mathbf{i}$ N (*c*) 0 (*d*) $9.6 \times 10^{-13}\mathbf{i} - 7.2 \times 10^{-13}\mathbf{j}$ N

5. 1.0 N

7. $20\mathbf{k}$ N/m

9. (*a*) 2 A·m² (*b*) 50 A·m

11. (*a*) 1 dyn·cm/G $= 10^{-3}$ A·m² (*b*) 1 eV/G $= 1.6 \times 10^{-15}$ A·m²

13. (*a*) 0.30 A·m² (*b*) 0.13 N·m

15. (*a*) 37°; $\hat{\mathbf{n}} = 0.8\mathbf{i} - 0.6\mathbf{j}$ (*b*) $0.48\hat{\mathbf{n}}$ A·m² $= 0.38\mathbf{i} - 0.29\mathbf{j}$ A·m² (*c*) $0.58\mathbf{k}$ N·m

17. $3.93 \times 10^{-6}$ N·m

19. (*a*) 0.131 $\mu$s (*b*) $3.83 \times 10^7$ m/s (*c*) 7.66 MeV

21. (*a*) 2.7 mm (*b*) $3.5 \times 10^{10}$ rad/s; 0.18 ns

23. (*a*) 0.13 $\mu$s (*b*) $2.41 \times 10^7$ m/s (*c*) 12.0 MeV

25. (*a*) $-10^4\mathbf{k}$ V/m (*b*) no

27. (*a*) 6.9 mm (*b*) 75 $\mu$T

29. (*a*) $1.44 \times 10^8$ rad/s (*b*) 27.1 MeV (*c*) assuming $m_d = 2m_p$, both answers are halved

31. (*a*) $1.07 \times 10^{-2}$ cm/s (*b*) $5.84 \times 10^{22}$ electrons/cm³

33. (*a*) $3.68 \times 10^{-3}$ cm/s (*b*) 1.47 $\mu$V

**Problems**

1. 0.98 A

3. (*a*) $B = (Mg/IL) \tan \theta$ (*b*) $g \sin \theta$ up the rails

5. (*a*) 190 km/s (*b*) 9.9 mm

**Chapter 27**

**True or False**

1. True 2. False 3. False 4. True 5. True

**Exercises**

1. (*a*) $4.44 \times 10^{-11}\mathbf{j}$ T (*b*) $-1.11 \times 10^{-11}\mathbf{j}$ T (*c*) 0 (*d*) $-4.44 \times 10^{-11}\mathbf{i}$ T

3. (*a*) $2 \times 10^{-11}$ T in direction of the unit vector $\hat{\mathbf{n}} = -(2/\sqrt{5})\mathbf{i} + (1/\sqrt{5})\mathbf{j}$ (*b*) $2/\sqrt{5} \times 10^{-11}\mathbf{j}$ T

5. $(a) -88.9\mathbf{k}\ \mu T$ $(b)$ 0 $(c)$ 88.9$\mathbf{k}\ \mu T$ $(d) -160\mathbf{k}\ \mu T$

7. $(a) -178\mathbf{k}\ \mu T$ $(b) -133\mathbf{k}\ \mu T$ $(c) -178\mathbf{k}\ \mu T$ $(d) +107\mathbf{k}\ \mu T$

9. $(a)$ 64$\mathbf{j}\ \mu T$ $(b) -48\mathbf{k}\ \mu T$

11. 0 due to segments along line through $P$; 56.6 $\mu T$ inward due to each vertical segment; 113 $\mu T$ inward due to horizontal segment above $P$; resultant $\mathbf{B}$ is 226 $\mu T$ inward

13. 23.6 $\mu T$

15. $(a)$ 32$\sqrt{2}\ k_m I/L = 45.3 k_m I/L$ $(b)$ $4\pi^2 k_m I/L = 39.5 k_m I/L$ $(c)$ the square

17. $(a)$ antiparallel $(b)$ 44.7 mA

19. 28 A

21. $(a)$ $4.5 \times 10^{-4}$ N/m to right $(b)$ 30 $\mu T$ down

23. $(a)$ $+10\mu_0$ for $C_1$; 0 for $C_2$; $-10\mu_0$ for $C_3$ $(b)$ None of these paths can be used because of lack of symmetry.

25. $(a)$ 8 G $=$ 800 $\mu T$ $(b)$ 40 G $(c)$ 28.6 G

27. $(a)$ 6.03 mT $(b)$ 0.377 A·m² $(c)$ 1.51 A·m $(d)$ 7.9 nT using the point-pole approximation

29. $(a)$ 0.40 A·m $(b)$ 7.76 nN

31. 4800 A/m

33. $(a)$ 1.54 MA/m $(b)$ 1.93 T

35. $(a)$ 0 $(b)$ $1.37 \times 10^{-5}$ T·m² $= 1.37 \times 10^{-5}$ Wb $(c)$ 0 $(d)$ $1.19 \times 10^{-5}$ Wb

37. $6.74 \times 10^{-3}$ Wb

39. $(a)$ $1.80 \times 10^{15}$ N/C·s $(b)$ $I_d = \epsilon_0 A\ dE/dt = 5$ A $(c)$ $1.60 \times 10^4$ A/m²

**Problems**

1. $(a)$ $B_x = (\mu_0/4\pi L)(I_2 + 2I_3)$; $B_y = -(\mu_0/4\pi L)(2I_1 + I_2)$
$(b)$ $B_x = (\mu_0/4\pi L)(2I_3 - I_2)$; $B_y = (\mu_0/4\pi L)(I_2 - 2I_1)$
$(c)$ $B_x = (\mu_0/4\pi L)(I_2 - 2I_3)$; $B_y = -(\mu_0/4\pi L)(2I_1 + I_2)$

3. At the earth's surface the magnetic field due to the cable is 0.05 G, which is about 7% of the magnitude of the earth's field.

5. $(a)$ $3.2 \times 10^{-16}$ N directed opposite current $(b)$ $3.2 \times 10^{-16}$ N directed radially away from wire $(c)$ 0

7. $v = (2\pi m/\mu_0 q I)v_t^2$

9. where $J = I/(\pi R^2 - \pi a^2)$, $(a)$ $B_x = 0$; $B_y = \dfrac{\mu_0 J}{2}\left(\dfrac{a^2}{2R - b} - \dfrac{R}{2}\right)$

$(b)$ $B_x = \dfrac{\mu_0 J}{2}\left(\dfrac{R}{2} - \dfrac{2Ra^2}{4R^2 + b^2}\right)$; $B_y = -\dfrac{\mu_0 J}{2}\left(\dfrac{a^2 b}{4R^2 + b^2}\right)$

13. $(a)$ $5.40 \times 10^{-2}$ T at $x = 5$ cm; $5.39 \times 10^{-2}$ T at $x = 7$ cm; $5.26 \times 10^{-2}$ T at $x = 9$ cm; $4.86 \times 10^{-2}$ T at $x = 11$ cm

15. $(a)$ $5.01 \times 10^{-7}$ Wb $(b)$ $7.14 \times 10^{-5}$ N away from wire

17. $1.97 \times 10^{-6}$ N·m

19. $k_m I$

**Chapter 28**

**True or False**

1. False; it is proportional to the rate of change of the flux. 2. True 3. False 4. False; it is an independent law. 5. True

**Exercises**

1. $(a)$ $6.4 \times 10^{-20}$ N $(b)$ 0.40 V/m $(c)$ 0.12 V

3. $(a)$ 1.6 V $(b)$ 0.8 A counterclockwise $(c)$ 0.128 N $(d)$ 1.28 W; 1.28 W

5. (a) clockwise; repel (b) counterclockwise; attract

7. (a) counterclockwise (b) clockwise

9. 199 T/s

11. (a) $\varepsilon = 0.4 - 0.2t$ V (b) at $t = 0$, $\phi_m = 0$, $\varepsilon = 0.4$ V; at $t = 2$ s, $\phi_m = -0.4$ T·m², $\varepsilon = 0$; at $t = 4$ s, $\phi_m = 0$, $\varepsilon = -0.4$ V; at $t = 6$ s, $\phi_m = 1.2$ T·m², $\varepsilon = -0.8$ V

13. (a) $1.26 \times 10^{-3}$ C (b) 12.6 mA (c) 0.628 V

15. 0.80 G

17. (a) 24 Wb (b) (−)1.60 kV

19. (a) 60.3 G (b) $7.58 \times 10^{-4}$ Wb (c) 253 $\mu$H (d) 37.9 mV

21. (b) $M = \mu_0 N_1 N_2 A_1 / \ell_2$ (c) The flux through the solenoid depends on $B$ due to the coil, and $B$ is not uniform, especially outside the coil.

23. (a) $I = 0$; $dI/dt = 25$ A/s (b) $I = 2.26$ A; $dI/dt = 20.5$ A/s (c) $I = 7.90$ A; $dI/dt = 9.20$ A/s (d) $I = 10.8$ A; $dI/dt = 3.38$ A/s

25. (a) 13.5 mA (b) $7.44 \times 10^{-44}$ A

27. 4.4 ms

29. (a) 44.06 W (b) 40.44 W (c) 3.62 W

31. (a) 2.0 A (b) 4.0 J

33. (a) $W_m = 3.98 \times 10^5$ J (b) $W_e = 4.43 \times 10^{-4}$ J (c) $3.98 \times 10^5$ J

35. 1.26 ms

37. (b) in the third circuit, since $I_{max} = \omega Q_0 = \omega CV$ and $\omega$ and $V$ are the same for all three circuits

**Problems**

3. (a) for $0 \leq t \leq 5$ s, $\phi_m = 10^{-3}t$ Wb; for $5$ s $\leq t \leq 10$ s, $\phi_m = 5 \times 10^{-3}$ Wb; for $10$ s $\leq t \leq 15$ s, $\phi_m = (15 - t) \times 10^{-3}$ Wb; for $15$ s $\leq t$, $\phi_m = 0$ (b) for $0 \leq t < 5$ s, $\varepsilon = -1$ mV; for $5$ s $< t < 10$ s, $\varepsilon = 0$; for $10$ s $< t < 15$ s, $\varepsilon = +1$ mV; for $15$ s $< t$, $\varepsilon = 0$

5. $\mu_0 n_1 n_2 \pi r_1^2 \ell$

7. $R = 20$ $\Omega$; $L = 4$ H

9. (a) 3.53 J (b) 1.92 J

11. (b) 275 rad/s

13. (a) $I^2 R = (2400$ W$)e^{-6000t}$ (b) 0.40 J (c) 0.40 J

**Chapter 29**

**True or False**

1. True; but it is often masked by paramagnetism. 2. True 3. False 4. True 5. True 6. False

**Exercises**

1. (a) $\mathbf{H} = 8.00 \times 10^3\mathbf{k}$ A/m; $\mathbf{B} = 101\mathbf{k}$ G (b) $\mathbf{H} = 8.00 \times 10^3\mathbf{k}$ A/m; $\mathbf{B} = 1.52\mathbf{k}$ T

3. (a) 75 A·m (b) $\mathbf{H}_p = -1.19 \times 10^3\mathbf{k}$ A/m (c) $\mathbf{H} = +6.81 \times 10^3\mathbf{k}$ A/m (d) $\mathbf{B} = 714\mathbf{k}$ G

5. diamagnetic: $H_2$, $CO_2$, and $N_2$; paramagnetic: $O_2$

7. $\mathbf{H} = 8.00 \times 10^3\mathbf{k}$ A/m; $\mathbf{M} = 0.544\mathbf{k}$ A/m; $\mathbf{B} = 101\mathbf{k}$ G

9. $1.69m_B$

11. (a) $M_s = 5.6 \times 10^5$ A/m; $\mu_0 M_s = 0.70$ T (b) $5.2 \times 10^{-4}$ (c) The measured magnetization includes that due to induced moments (diamagnetism), which reduces the net magnetization.

13. $\mu = 1.48 \times 10^{-5}$ H/m; $K_m = 11.7$

15. 11 A

17. (a) 1000 A/m (b) $1.26 \times 10^6$ A/m (c) $1.26 \times 10^3$

**Problem**

1. (b) $H = NI/2\pi R$, the same as without core since there are no end effects; $B = \mu_0 NI/2\pi R + \mu_0 M$

**Chapter 30**

**True or False**

1. True 2. True 3. True 4. False 5. True 6. True

**Exercises**

1. (a) 250 rad/s (b) $4.8\pi$ V = 15.1 V

3. (b) 170 V

5. (a) $120\pi$ rad/s = 377 rad/s (b) $I_{max} = 4.0$ A; $I_{rms} = I_{max}/\sqrt{2} = 2.83$ A (c) 48 W (d) 0 (e) $P_{av} = I_{rms}^2 R = P_{max}/2 = 24$ W

7. (a) 2.65 M$\Omega$ (b) 26.5 k$\Omega$ (c) 26.5 $\Omega$

11. (a) 15.9 kHz; angular frequency, $10^5$ rad/s (b) 159 Hz; angular frequency, $10^3$ rad/s (c) 1.59 MHz; angular frequency, $10^7$ rad/s

13. (a) 0.377 $\Omega$ (b) 3.77 $\Omega$ (c) 37.7 $\Omega$

15. 1.59 kHz

17. (a) $7.07 \times 10^3$ rad/s (b) 14.1 A (c) $X_C = 62.5$ $\Omega$; $X_L = 80$ $\Omega$ (d) $Z = 18.2$ $\Omega$; $I_{rms} = 3.89$ A (e) $\phi = 74.1°$

19. (a) 1.13 kHz (b) $X_C = 79.6$ $\Omega$; $X_L = 62.8$ $\Omega$ (c) $Z = 17.5$ $\Omega$; $I_{rms} = 4.04$ A (d) $-73.4°$

21. 2000

23. (a) 14.1 (b) 500 rad/s (c) 0.275

25. (a) 0.539 (b) 95.3 mA (c) 0.726 W

27. (a) 0.553 (b) 663 mA (c) 44.0 W

29. (a) step-down (b) 2.4 V (c) $(-)5$ A

31. 10.4 turns for 2.5 V; 31.3 turns for 7.5 V; 37.5 turns for 9 V

**Problems**

1. (a) 0.333 (b) 26.7 $\Omega$ (c) 200 mH (d) lags; 70.5°

3. (a) $L = 0.8$ mH; $C = 12.5$ $\mu$F (b) 1.6 (c) for the given values of $X_L$ and $X_C$, $I_{max} = 2.0$ A; at resonance, $I_{max} = 5.2$ A

5. (a) 3.00 A (b) 8.00 to 8.38 mH or 31.6 to 40.0 mH

7. (a) 933 W (b) 7.71 $\Omega$ (c) 99.9 $\mu$F (d) increase capacitance by 40.8 $\mu$F

11. (a) 12 $\Omega$ (b) $R = 7.2$ $\Omega$; $X = 9.6$ $\Omega$ (c) capacitive

13. $L = 4.00$ mH; $I_{max} = 100$ mA

15. $\cos\phi = Z/R$

**Chapter 31**

**True or False**

1. False 2. True 3. True 4. True 5. False; $B = E/c$ 6. True

**Exercises**

1. (a) Equation 31-4 (b) Equation 31-5 (c) Equation 31-2 (d) Equation 31-3

3. (a) 333 nPa (b) 19.4 V/m (c) 647 nT

5. (a) 283 V/m (b) 943 nT (c) 212 W/m² (d) 708 nPa

7. (a) 40 nN (b) 80 nN

**Problems**

1. $I = 1.91$ kW/m$^2$; $E_{rms} = 849$ V/m; $B_{rms} = 2.83$ $\mu$T; $p_r = 6.37$ $\mu$Pa

3. (a) $I = 553$ mW/m$^2$; $p_r = 1.84$ nPa (b) $E_{rms} = 14.4$ V/m; $B_{rms} = 48.1$ nT

5. (a) $E = \rho I/\pi a^2$ parallel to current (b) $B = \mu_0 I/2\pi a$ encircling current (c) $S = \rho I^2/2\pi^2 a^3$ radially in towards wire (d) $SA = \rho I^2 L/\pi a^2 = I^2 R$

**Chapter 32**  **Exercises**

3. 12.4 keV

5. 10 GHz

7. AM: 300 m for $f = 1000$ kHz; FM: 3.00 m for $f = 100$ MHz

9. (a) $I_1/9$ (b) $I_1/2$ (c) $I_1/16$

11. (a) $9.46 \times 10^{12}$ km (b) $2.11 \times 10^6$ light years

13. 10.8 min

15. 92%

17. image is 5 cm below mirror surface

19. (a) 27.1° (b) 41.7° (c) 70.1° (d) no refracted ray; total internal reflection

21. $2.25 \times 10^8$ m/s in water; $2.00 \times 10^8$ m/s in glass

23. 2.46

25. blue: 40.8°; red: 41.8°

27. (a) $I_0/8$ (b) $3I_0/32$

29. (a) 30° (b) 1.73

31. 56.3°

**Problems**

5. (a) 62.7° (b) No, the critical angle for glass-air refraction is 41.8°. For angles greater than 41.8° but less than 62.7°, the light rays will enter the water, but they cannot get into the air.

7. (a) 37.0°

9. (a) 15 m (b) 15.4 m (c) 13.86 s, which is 0.19 s less than the time for the straight-line path

**Chapter 33**  **True or False**

1. True 2. True 3. False 4. True 5. False 6. True 7. False 8. True 9. False; it is smaller than the object. 10. True

**Exercises**

7. (a) $s' = 33.3$ cm; real; inverted; reduced, $y' = -y/3$
(b) $s' = 50$ cm; real; inverted; same size
(c) $s' = \infty$; no image
(d) $s' = -16.7$ cm; virtual; erect; enlarged, $y' = 5y/3$

9. (a) $s' = -20$ cm; virtual; erect; reduced, $y' = y/5$
(b) $s' = -16.7$ cm; virtual; erect; reduced, $y' = y/3$
(c) $s' = -12.5$ cm; virtual; erect; reduced, $y' = y/2$
(d) $s' = -7.14$ cm; virtual; erect; reduced, $y' = 0.714y$

13. 1.33 cm

15. $s = 10$ cm

17. (a) $s' = -46.15$ cm; virtual (b) $s' = -6.47$ cm; virtual (c) $s' = 44.1$ cm; real

19. (a) 13.6 cm (b) 20 cm (c) $-10$ cm (d) $-40$ cm

21. (a) $-40$ cm (b) $+40$ cm (c) $-40$ cm

23. A negative object distance occurs when a surface interrupts light that was converging to a point beyond the surface. The image point is then a virtual object, and the object distance is negative. (a) $s' = +10$ cm; real; erect (b) $s' = +15$ cm; real; erect

25. $s' = +10$ cm; $y' = -1$ cm

27. There are two solutions: (1) $s = 15$ cm produces a real, inverted image at $s' = 30$ cm; (2) $s = 5$ cm produces a virtual, erect image at $s' = -10$ cm.

31. The image is 30 cm to the right of the second lens. It is erect and twice as large as the object.

33. The computed distances from the vertex of the mirror are (to four significant figures): for 0.5 cm, $s' = 3.000$ cm; for 1.0 cm, $s' = 2.998$ cm; for 2.0 cm, $s' = 2.975$ cm; and for 4.0 cm, $s' = 2.674$ cm.

35. The lens should move 2.78 mm forward.

37. $\theta = 8 \times 10^{-5}$ rad; $\Delta x = 1.6$ mm

39. (a) 5.0 (b) 6.0

41. (a) 20.0 cm (b) $-4$ (c) $-20$ (d) 6.25 cm

43. (a) 0.9 cm (b) 0.180 rad

**Problems**

1. 0.129 m

3. concave; 90 cm

5. (a) $s' = -128$ cm (b) $s' = 14.7$ cm (c) real

7. 19.3 cm

9. The final image is the one formed by the mirror; it is real, erect, and at the center of the lens.

11. 1.6

13. The final image is virtual and 15 cm in front of the second lens.

15. (b) 2.55 m

17. $f_{blue} = 17.4$ cm

**Chapter 34**    **True or False**    1. False 2. True 3. True 4. True 5. True

**Exercises**

1. (a) 376 nm (b) 5.32 waves (c) 295°

3. (a) dark; the ray reflecting from air-glass surface suffers a phase change of 180° (b) $1.75 \times 10^{-4}$ rad

5. 5.46 $\mu$m to 5.75 $\mu$m

7. (a) 84,848 (b) 84,873 (c) 1.0003

9. (a) 0.15 mm (b) 33.3 bands per centimetre

11. 695 nm

13. (a) 9.29 $\mu$m (b) 29

15. If $m = 0, 1, 2, \ldots$, then the maxima is $y_m = 1.2m$ cm, and the minima is $y_m = (m + \frac{1}{3})1.2$ cm or $y_m = (m + \frac{2}{3})1.2$ cm.

17. (a) $\theta_1 = \lambda/d$; $\theta_{min} = \theta_1/6$

19. (a) $6.00 \times 10^{-3}$ rad (b) $6.00 \times 10^{-4}$ rad (c) $6.00 \times 10^{-2}$ rad

21. 1.68 cm

23. 20.0 $\mu$m; 9

25. (a) $8.54 \times 10^{-3}$ rad (b) 6.83 cm

27. (a) 55.6 km (b) 55.6 m

29. 33.6 mm

31. 484 m

33. $\lambda_1 = 486$ nm; $\lambda_2 = 660$ nm

35. (a) $\theta_{579} = 0.1158$ rad; $\theta_{576} = 0.1154$ rad; $\Delta\theta = 4 \times 10^{-4}$ rad (b) 1.44 nm

**Problems**

1. (a) 0.300° (b) 8th order

3. 8

5. about $6.5 \times 10^6$ km using light with $\lambda = 700$ nm

7. (a) minimum thickness: 600 nm (b) 400, 514, and 720 nm (c) 400, 514, and 720 nm

9. (a) $y_1 = 35.3$ cm; $y_2 = 70.7$ cm (b) 88.4 $\mu$m (c) 8000

11. 625 and 417 nm

13. (a) 1.000292 (b) $3.7 \times 10^{-5}$ % ($n - 1$ is accurate to 0.126%)

15. (b) 491 (c) 0.99 mm

17. 12 mm for 2 slits; 6 mm for 4 slits

19. 0.37 mm

**Chapter 35**

**True or False**

1. True  2. True  3. False  4. True  5. True  6. False  7. False  8. False  9. True  10. True

**Exercises**

1. (a) 236 $\mu$s (b) $2.36 \times 10^{-12}$ s (c) no

3. *Hint:* Use the binomial expansion.

5. (a) $5.96 \times 10^{-8}$ s (b) 16.1 m (c) 7.02 m

7. (a) 0.8 m (b) 4.44 ns

9. 80 $c$-min

13. $L_0 v/c^2 = 60$ min, which is the result of Exercise 12

15. about $0.21c = 63,000$ km/s

17. $0.6c$

19. (a) $x_2' - x_1' = \gamma(x_2 - x_1)$ (b) Times $t_1$ and $t_2$ in Equations 35-19 are equal to $t_0$, but times $t_1'$ and $t_2'$ in Equations 35-18 are not equal and are not known.

21. (a) $u_x = v$; $u_y = c/\gamma$

27. (a) 4.974 MeV/c (b) 0.9948

29. (a) $938 \sqrt{3}$ MeV/c = 1625 MeV/c (b) $v/c = \frac{1}{2}\sqrt{3} = 0.866$; $v = 2.60 \times 10^8$ m/s

31. 6.26 MeV

33. (a) $4.5 \times 10^{-9}$ u (b) about $8 \times 10^{-9}$%, a very small error

**Problems**

1. 18,500 y

3. (a) 0.76% [$E_k = (\gamma - 1)mc^2 = 5.038 \times 10^{-3}mc^2$; $\frac{1}{2}mu^2 = 5.00 \times 10^{-3}mc^2$] (b) 68% ($E_k = 1.29mc^2$; $\frac{1}{2}mu^2 = 0.402mc^2$)

5. (a) 60 m (b) $0.9756c$ (c) 21.95 m (d) 0.25 $\mu$s

11.  (a) 120 min (b) 240 min (c) 60 min; 120 min; 240 min; observer in $S'$ makes the same table

**Chapter 36**    **True or False**    1. True 2. True 3. False; it is a property of the metal. 4. True 5. True 6. False 7. True 8. True 9. True 10. True 11. True 12. False 13. True 14. False

**Exercises**    1.  1.77 to 3.10 eV

3.  $4.14 \times 10^{-7}$ eV

5.  (a) 4.13 eV (b) 2.10 eV (c) 1.00 V

7.  (a) $f_t = 4.59 \times 10^{14}$ Hz; $\lambda_t = 653$ nm (b) 2.23 V (c) 1.20 V

9.  (a) 17.44 keV (b) 0.0760 nm (c) 16.32 keV (d) 1.12 keV

11.  (a) 103 nm (b) 97.2 nm (c) 91.2 nm

13.  $\lambda_{4\to3} = 1876$ nm; $\lambda_{5\to3} = 1282$ nm; $\lambda_{6\to3} = 1094$ nm; $\lambda_{\infty\to3} = 820.5$ nm

15.  (a) 1.23 nm (b) 0.123 nm (c) 0.0388 nm (d) 0.0123 nm

17.  0.202 nm

19.  about $5 \times 10^{-20}$ m $= 5 \times 10^{-5}$ fm, where 1 fm $= 10^{-15}$ m is of the order of the diameter of the nucleus

21.  0.195 nm

**Problems**    1.  Plot $c/\lambda$ versus $V_0$. (a) 2.05 eV (b) about $5 \times 10^{14}$ Hz (c) $4.14 \times 10^{-15}$ V·s

3.  (a) $10^{-22}$ W (b) about 53 min

7.  $\mu/m = 0.99946 = R_H/R_\infty$

9.  (b) $E_1 = h^2/m_p r^2 = 6.25 \times 10^{-3}$ eV (c) $T_c \approx 73$ K (d) about 13 eV using $r_a = 0.053$ nm; $T_c \approx 1.5 \times 10^5$ K

**Chapter 37**    **True or False**    1. True 2. True 3. False 4. False 5. True 6. True 7. False 8. True 9. True

**Exercises**    1.  (a) $^{13}$N, $^{15}$N (b) $^{55}$Fe, $^{58}$Fe (or $^{57}$Fe, $^{59}$Fe, etc.) (c) $^{116}$Sn, $^{120}$Sn

3.  (a) 31.99 MeV; 5.33 MeV (b) 333.7 MeV; 8.56 MeV (c) 1803.7 MeV; 7.58 MeV

5.  20.6 MeV

7.  (a) $\beta^+$ (b) $\beta^-$

9.  (a) 3.33 min (b) 0.208/min (c) 6500 counts per second

11.  (a) 4.87 MeV (b) 7.00 MeV

13.  3350 y

15.  (a) $-0.764$ (b) 3.05 MeV (c) 1.02 MeV

17.  $1.56 \times 10^{19}$ fissions per second

19.  $1.2 \times 10^8$ K

21.  (a) 0.58 mm (b) 51.4 $\mu$m

23.  (a) 1.54 cm (b) 7.1 cm

**Problems**    3.  (a) $R_{Ba} = 7.8$ fm; $R_{Kr} = 6.8$ fm (b) 199 MeV, slightly greater than 175 MeV

5.  $1.18 \times 10^{19}$ m, about $7.9 \times 10^7$ times distance from earth to sun

7.  $\lambda = 3.72 \times 10^{-3}$/s; $t_{1/2} = 3.11$ min

9.  (b) 55

# Index

**Some Conversion Factors**

1 m = 39.37 in = 3.281 ft = 1.094 yd

1 m = $10^{15}$ fm = $10^{10}$ Å = $10^9$ nm

1 km = 0.6215 mi

1 mi = 5280 ft = 1.609 km

1 in = 2.540 cm

1 L = $10^3$ cm$^3$ = $10^{-3}$ m$^3$ = 1.057 qt

1 y = 365.24 d = 3.156 × $10^7$ s

1 km/h = 0.278 m/s = 0.6215 mi/h

1 ft/s = 0.3048 m/s = 0.6818 mi/h

1 rev = $2\pi$ rad = 360°

1 rad = 57.30°

1 rev/min = 0.1047 rad/s

1 slug = 14.59 kg

1 tonne = $10^3$ kg = 1 Mg

1 atm = 101.3 kPa = 1.013 bar = 76.00 cmHg = 14.70 lb/in$^2$

1 N = $10^5$ dyn = 0.2248 lb

1 lb = 4.448 N

1 Pa·s = 10 poise

1 J = $10^7$ erg = 0.7373 ft·lb = 9.869 × $10^{-3}$ L·atm

1 kW·h = 3.6 MJ

1 cal = 4.184 J = 4.129 × $10^{-2}$ L·atm

1 L·atm = 101.3 J = 24.22 cal

1 eV = 1.602 × $10^{-19}$ J

1 Btu = 778 ft·lb = 252 cal = 1054 J

1 horsepower = 550 ft·lb/s = 746 W

1 W/m·K = 6.938 Btu·in/h·ft$^2$·°F

1 T = $10^4$ G

1 kg weighs about 2.205 lb